SIDE EFFECTS OF DRUGS ANNUAL

VOLUME 40

SIDE EFFECTS OF DRUGS ANNUAL

VOLUME 40

A Worldwide Yearly Survey of New Data in Adverse Drug Reactions

Editor

SIDHARTHA D. RAY, PHD., FACN

Department of Pharmaceutical and Biomedical Sciences, Touro University College of Pharmacy and School of Osteopathic Medicine, Manhattan, NY, United States

Elsevier
Radarweg 29, PO Box 211, 1000 AE Amsterdam, Netherlands
The Boulevard, Langford Lane, Kidlington, Oxford OX5 1GB, United Kingdom
50 Hampshire Street, 5th Floor, Cambridge, MA 02139, United States

First edition 2018

ISBN: 978-0-444-64119-9
ISSN: 0378-6080

For information on all Elsevier publications visit our website at https://www.elsevier.com/books-and-journals

Publisher: Zoe Kruze
Acquisition Editor: Zoe Kruze
Editorial Project Manager: Shellie Bryant
Production Project Manager: Vignesh Tamil
Cover Designer: Mark Rogers

Typeset by SPi Global, India

Contributors

Adam Abba-Aji Complex Psychosis Program, Department of Psychiatry, University of Alberta, Edmonton, AB, Canada

Olawale D. Afolabi Milken Institute School of Public Health, The George Washington University, Washington, DC, United States

Vincent Agyapong Complex Psychosis Program, Department of Psychiatry, University of Alberta, Edmonton, AB, Canada

Sara Al-Dahir Division of Clinical and Administrative Sciences, College of Pharmacy, Xavier University of Louisiana, New Orleans, LA, United States

Asima N. Ali Campbell University College of Pharmacy and Health Sciences, Buies Creek; Wake Forest Baptist Health, Outpatient Internal Medicine Clinic, Winston-Salem, NC, United States

Natalia Amacher Department of Pharmaceutical Sciences, Manchester University College of Pharmacy, Fort Wayne, IN, United States

Vasilios Athans Department of Pharmacy, Cleveland Clinic, Cleveland, OH, United States

Manyo Ayuk-Tabe University of Maryland Eastern Shore School of Pharmacy, Princess Anne, MD, United States

Jennifer L. Babin University of Utah College of Pharmacy; University of Utah Health, Salt Lake City, UT, United States

Mohamed Bayoumi-Ali Abertawe Bro Morgannwg University Health board (ABMU), Singleton hospital, Swansea, United Kingdom

Robert D. Beckett Manchester University College of Pharmacy, Natural and Health Sciences, Fort Wayne, IN, United States

Nika Bejou University of Pittsburgh Medical Center Pinnacle, Harrisburg, PA, United States

Renee A. Bellanger Department of Pharmacy Practice, University of the Incarnate Word, Feik School of Pharmacy, San Antonio, TX, United States

Nicholas T. Bello Department of Animal Sciences, Nutritional Sciences Graduate Program, School of Environmental and Biological Sciences, Rutgers, The State University of New Jersey, New Brunswick, NJ, United States

Lydia Benitez University of Kentucky HealthCare; University of Kentucky College of Pharmacy, Lexington, KY, United States

Vishakha Bhave Department of Pharmaceutical Sciences, PCOM School of Pharmacy, Suwanee, GA, United States

Brittany D. Bissell Medical Intensive Care Unit/Pulmonary; Department of Pharmacy Practice and Science, University of Kentucky HealthCare, University of Kentucky College of Pharmacy, Lexington, KY, United States

Adrienne T. Black 3E Services, Verisk 3E, Warrenton, VA, United States

Joshua Brockbank Department of Pharmacotherapy, University of Utah College of Pharmacy, Salt Lake City, UT, United States

Rebecca A. Buckler Jefferson Health—Methodist Hospital Division, Philadelphia, PA, United States

Pierre Chue Complex Psychosis Program, Department of Psychiatry, University of Alberta, Edmonton, AB, Canada

Jessica Cox University of Kentucky HealthCare; University of Kentucky College of Pharmacy, Lexington, KY, United States

Katherine A. Curtis St. Vincent Indianapolis Hospital, Indianapolis, IN, United States

Stacey D. Curtis Pharmacotherapy and Translational Research, University of Florida College of Pharmacy, Gainesville, FL, United States

Frank Davis University of Kentucky HealthCare, Lexington, KY, United States

Rahul Deshmukh Department of Pharmaceutical Sciences, Rosalind Franklin University of Medicine and Science, College of Pharmacy, North Chicago, IL, United States

Tyler S. Dougherty Department of Pharmacy Practice, South College School of Pharmacy, Knoxville, TN, United States

Manoranjan S. D'Souza Department of Pharmaceutical and Biomedical Sciences, Raabe College of Pharmacy, Ohio Northern University, Ada, OH, United States

Victoria L. Dzurinko The Pennsylvania College of Optometry, Salus University, Elkins Park, PA, United States

Alex Ebied Pharmacotherapy and Translational Research, University of Florida College of Pharmacy, Gainesville, FL, United States

Kirk E. Evoy College of Pharmacy, University of Texas at Austin, Austin; School of Medicine, University of Texas Health at San Antonio; University Health System, San Antonio, TX, United States

Doaa Farag Abertawe Bro Morgannwg University Health board (ABMU), Neath Port-Talbot hospital, Port Talbot, United Kingdom

Dana R. Fasanella Department of Pharmacy Practice and Administration, University of Maryland Eastern Shore School of Pharmacy, Princess Anne, MD, United States

Roman Fazylov Department of Pharmacy Practice, Touro College of Pharmacy, New York, NY, United States

Christina Ford Department of Pharmaceutical Sciences, Manchester University College of Pharmacy, Fort Wayne, IN, United States

Heath Ford Department of Pharmacy Practice, South College School of Pharmacy, Knoxville, TN, United States

Michelle Friedman-Jakubovics Department of Pharmacy Practice, Touro College of Pharmacy, New York, NY, United States

Hannah R. Fudin University of Utah College of Pharmacy; University of Utah Health, Salt Lake City, UT, United States

Jason C. Gallagher Temple University, Philadelphia, PA, United States

Caitlin M. Gibson System College of Pharmacy, University of North Texas Health Science Center, Fort Worth, TX, United States

Angela L. Goodhart St. Vincent Indianapolis Hospital, Indianapolis, IN, United States

David L. Gordon Department of Microbiology and Infectious Diseases, Flinders Medical Centre, Bedford Park, SA, Australia

Joshua P. Gray Department of Science, United States Coast Guard Academy, New London, CT, United States

Holly Gurgle Department of Pharmacotherapy, University of Utah College of Pharmacy, Salt Lake City, UT, United States

Alison Hall Royal Liverpool Hospital, Liverpool, United Kingdom

F. Scott Hall Department of Pharmacology and Experimental Therapeutics, University of Toledo College of Pharmacy and Pharmaceutical Sciences, Teledo, OH, United States

Lisa T. Hong Loma Linda University School of Pharmacy, Loma Linda, CA, United States

Crystal K. Howell The University of North Texas Health Science Center System College of Pharmacy, Fort Worth; Medical City Dallas, Dallas, TX, United States

Dalton K. Hudgins Campbell University College of Pharmacy and Health Sciences, Buies Creek, NC, United States

Jason Isch College of Pharmacy, Natural & Health Sciences, Manchester University, Fort Wayne, IN, United States

Alyssa Johnson Manchester University College of Pharmacy, Natural and Health Sciences, Fort Wayne, IN, United States

Jackie Johnston Ernest Mario School of Pharmacy, Rutgers University, Piscataway; St. Joseph's University Medical Center, Paterson, NJ, United States

Sipan Keshishyan Department of Pharmaceutical Sciences, Manchester University College of Pharmacy, Natural and Health Sciences, Fort Wayne, IN, United States

Jennifer J. Kim Greensboro Area Health Education Center, Greensboro; University of North Carolina Eshelman School of Pharmacy, Chapel Hill, NC, United States

Dushyant Kshatriya Department of Animal Sciences, Nutritional Sciences Graduate Program, School of Environmental and Biological Sciences, Rutgers, The State University of New Jersey, New Brunswick, NJ, United States

Jennifer Ku University of Utah Health, Salt Lake City, UT, United States

Justin G. Kullgren The Ohio State University Wexner Medical Center, Columbus, OH, United States

Dirk W. Lachenmeier Chemisches und Veterinäruntersuchungsamt (CVUA) Karlsruhe, Karlsruhe, Germany

Shandrika Landry Division of Clinical and Administrative Sciences, College of Pharmacy, Xavier University of Louisiana, New Orleans, LA, United States

Amy Lehnert Department of Pharmacy Practice, KGI School of Pharmacy, Claremont, CA, United States

Sarah E. Luttrell Department of Pharmacy Practice and Administration, University of Maryland Eastern Shore School of Pharmacy, Princess Anne, MD, United States

Alisyn L. May University of Utah College of Pharmacy; ARUP Laboratories Family Health Clinic, Salt Lake City, UT, United States

Dianne May University of Georgia College of Pharmacy on Augusta University Campus, Augusta, Georgia

Dayna S. McManus Department of Pharmacy Services, Yale-New Haven Hospital, New Haven, CT, United States

Meghan T. Mitchell Thomas Jefferson University Hospital, Philadelphia, PA, United States

Philip B. Mitchell School of Psychiatry, University of New South Wales, and Black Dog Institute, Sydney, NSW, Australia

W. Cary Mobley University of Florida College of Pharmacy, Gainesville, FL, United States

Vicky Mody Department of Pharmaceutical Sciences, PCOM School of Pharmacy, Suwanee, GA, United States

Carmen Moore Department of Internal Medicine, Guy's and St. Thomas' Hospital, London, United Kingdom

Toshio Nakaki Department of Pharmacology, Teikyo University School of Medicine, Tokyo, Japan

Heather D. Nelkin Novant Health Oncology Specialists, Winston-Salem, NC, United States

Harris Ngokobi University of Maryland Eastern Shore School of Pharmacy, Princess Anne, MD, United States

Henry L. Nguyen Department of Dermatology, Mayo Clinic, Rochester, MN, United States

Zachary R. Noel Department of Pharmacy Practice and Science, University of Maryland School of Pharmacy, Baltimore, MD, United States

John D. Noti Allergy and Clinical Immunology Branch, Health Effects Laboratory Division, National Institute for Occupational Safety and Health, Centers for Disease Control and Prevention, Morgantown, WV, United States

Anayo Ohiri Manchester University College of Pharmacy, Natural and Health Sciences, Fort Wayne, IN, United States

Mark E. Olah Department of Pharmaceutical and Biomedical Sciences, Raabe College of Pharmacy, Ohio Northern University, Ada, OH, United States

Igho J. Onakpoya Nuffield Department of Primary Care Health Sciences, Oxford, United Kingdom

Michael G. O'Neil Department of Pharmacy Practice, South College School of Pharmacy, Knoxville, TN, United States

Sreekumar Othumpangat Allergy and Clinical Immunology Branch, Health Effects Laboratory Division, National Institute for Occupational Safety and Health, Centers for Disease Control and Prevention, Morgantown, WV, United States

Anuj Patel Department of Pharmaceutical Sciences, PCOM School of Pharmacy, Suwanee, GA, United States

Michelle M. Peahota Thomas Jefferson University Hospital, Philadelphia, PA, United States

Mary Ellen Pisano Novant Health, Winston Salem, NC, United States

Jasmine M. Pittman Parkwest Medical Center, Knoxville, TN, United States

Hanna Raber College of Pharmacy, University of Utah, Salt Lake City, UT, United States

Sara Radparvar RWJ Barnabas Health, Saint Barnabas Medical Center, Livingston; Ernest Mario School of Pharmacy, Rutgers University, Piscataway, NJ, United States

Meenakshi R. Ramanathan The University of North Texas Health Science Center System College of Pharmacy, Fort Worth; Medical City Arlington, Arlington, TX, United States

Kerry Anne Rambaran Department of Pharmacy Practice, Keck Graduate Institute School of Pharmacy and Health Sciences, Claremont, CA, United States

Kelley Ratermann University of Kentucky HealthCare; University of Kentucky College of Pharmacy, Lexington, KY, United States

Sidhartha D. Ray Department of Pharmaceutical and Biomedical Sciences, Touro University College of Pharmacy and School of Osteopathic Medicine, Manhattan, NY, United States

Matthew B. Roberts Department of Microbiology and Infectious Diseases, Flinders Medical Centre, Bedford Park, SA, Australia

Connie Rust Department of Pharmacy Practice, South College School of Pharmacy, Knoxville, TN, United States

James M. Sanders The University of North Texas Health Science Center System College of Pharmacy; JPS Health Network, Fort Worth, TX, United States

Arindam Basu Sarkar Department of Pharmaceutical Sciences, College of Pharmacy, University of Findlay, Findlay, OH, United States

Laura A. Schalliol South College School of Pharmacy, Knoxville, TN, United States

Bridgette K. Schroader University of Kentucky Markey Cancer Center, Lexington, KY, United States

Christina M. Seeger Pharmacy Librarian, University of the Incarnate Word, Feik School of Pharmacy, San Antonio, TX, United States

Ajay Singh Department of Pharmaceutical Sciences, South University School of Pharmacy, Savannah, GA, United States

Brittany Singleton Division of Clinical and Administrative Sciences, College of Pharmacy, Xavier University of Louisiana, New Orleans, LA, United States

Sunil Sirohi Laboratory of Endocrine and Neuropsychiatric Disorders, Division of Basic Pharmaceutical Sciences, College of Pharmacy, Xavier University of Louisiana, New Orleans, LA, United States

Brian W. Skinner Manchester University College of Pharmacy, Natural & Health Sciences, Fort Wayne; St. Vincent Indianapolis Hospital, Indianapolis, IN, United States

Keaton S. Smetana Department of Pharmacy, The Ohio State University Wexner Medical Center, Columbus, OH, United States

Helen E. Smith Department of Pharmaceutical Science, University of the Incarnate Word, Feik School of Pharmacy, San Antonio, TX, United States

Thomas R. Smith Manchester University College of Pharmacy, Natural and Health Sciences, Fort Wayne, IN, United States

Jonathan Smithson School of Psychiatry, University of New South Wales, and Black Dog Institute, Sydney, NSW, Australia

Kristine Sobolewski RWJ Barnabas Health, Saint Barnabas Medical Center, Livingston, NJ, United States

Lisa V. Stottlemyer Wilmington VA Medical Center, Wilmington, DE, United States

Nicola A. Sweeney Department of Microbiology and Infectious Diseases, Flinders Medical Centre, Bedford Park, SA, Australia

Richard John Sweeney Wales Deanery and ABMU, Singleton hospital, Swansea, United Kingdom

Thomas Syratt Royal Liverpool Hospital, Liverpool, United Kingdom

Amulya Tatachar System College of Pharmacy, University of North Texas Health Science Center, Fort Worth, TX, United States

Frederick R. Tejada Department of Pharmaceutical Sciences, University of Maryland Eastern Shore School of Pharmacy, Princess Anne, MD, United States

Favas Thaivalappil Abertawe Bro Morgannwg University Health board (ABMU), Singleton hospital, Swansea, United Kingdom

Amar P. Thakkar Department of Pharmacy Practice, South College School of Pharmacy, Knoxville, TN, United States

Melissa L. Thompson Bastin Medical Intensive Care Unit/ Pulmonary; Department of Pharmacy Practice and Science, University of Kentucky HealthCare, University of Kentucky College of Pharmacy, Lexington, KY, United States

Rebecca Tran Department of Pharmacy Practice, Keck Graduate Institute School of Pharmacy, Claremont, CA, United States

Katie Traylor Department of Pharmacotherapy, University of Utah College of Pharmacy, Salt Lake City, UT, United States

Emily C. Tucker Department of Microbiology and Infectious Diseases, Flinders Medical Centre, Bedford Park, SA, Australia

Michael P. Veve Department of Clinical Pharmacy and Translational Science, University of Tennessee Health Science Center College of Pharmacy; Department of Pharmacy, University of Tennessee Medical Center, Knoxville, TN, United States

Luqman Hussain Wali Wales Deanery and ABMU, Singleton hospital, Swansea, United Kingdom

Marley L. Watson Emory Healthcare, Winship Cancer Institute, Atlanta, GA, United States

Kirby Welston Charlie Norwood VA Medical Center, Augusta, Georgia

Maame Wireku Manchester University College of Pharmacy, Health & Natural Sciences, Fort Wayne, IN, United States

Alexander Wisner Department of Pharmacology and Experimental Therapeutics, University of Toledo College of Pharmacy and Pharmaceutical Sciences, Teledo, OH, United States

Cecilia Wong Ernest Mario School of Pharmacy, Rutgers University, Piscataway; Saint Peter's University Hospital, New Brunswick, NJ, United States

Contents

10. General Anesthetics and Therapeutic Gases

THOMAS SYRATT, AND ALISON HALL

11. Local Anesthetics

HENRY L. NGUYEN

12. Neuromuscular Blocking Agents and Skeletal Muscle Relaxants

ALEX EBIED

13. Drugs That Affect Autonomic Functions or the Extrapyramidal System

TOSHIO NAKAKI

14. Dermatological Agents

STACEY D. CURTIS, AND W. CARY MOBLEY

15. Antihistamines (H1 Receptor Antagonists)

TYLER S. DOUGHERTY

16. Drugs That Act on the Respiratory Tract

MOHAMED BAYOUMI-ALI, RICHARD JOHN SWEENEY, LUQMAN HUSSAIN WALI, DOAA FARAG, AND FAVAS THAIVALAPPIL

17. Positive Inotropic Drugs and Drugs Used in Dysrhythmias

KERRY ANNE RAMBARAN, AND AMY LEHNERT

18. Beta Adrenergic Antagonists and Antianginal Drugs

ASIMA N. ALI, JENNIFER J. KIM, MARY ELLEN PISANO, DALTON K. HUDGINS, MAAME WIREKU, AND SIDHARTHA D. RAY

19. Drugs Acting on the Cerebral and Peripheral Circulations

KEATON S. SMETANA, ZACHARY R. NOEL, AND SIDHARTHA D. RAY

20. Antihypertensive Drugs

KATIE TRAYLOR, HOLLY GURGLE, AND JOSHUA BROCKBANK

21. Diuretics

MICHELLE FRIEDMAN-JAKUBOVICS, AND ROMAN FAZYLOV

22. Metal Antagonists and Metals

JOSHUA P. GRAY, NATALIA AMACHER, CHRISTINA FORD, AND SIDHARTHA D. RAY

23. Antiseptic Drugs and Disinfectants

DIRK W. LACHENMEIER

24. Beta-Lactams and Tetracyclines

REBECCA A. BUCKLER, MEGHAN T. MITCHELL, MICHELLE M. PEAHOTA, AND JASON C. GALLAGHER

25. Miscellaneous Antibacterial Drugs

EMILY C. TUCKER, MATTHEW B. ROBERTS, NICOLA A. SWEENEY, AND DAVID L. GORDON

26. Antifungal Drugs

DAYNA S. MCMANUS, AND NIKA BEJOU

27. Antiprotozoal Drugs

ADRIENNE T. BLACK, DAYNA S. MCMANUS, OLAWALE D. AFOLABI, AND SIDHARTHA D. RAY

28. Antiviral Drugs

SREEKUMAR OTHUMPANGAT, AND JOHN D. NOTI

29. Drugs in Tuberculosis and Leprosy

MEENAKSHI R. RAMANATHAN, CRYSTAL K. HOWELL, AND JAMES M. SANDERS

35. Gastrointestinal Drugs

KIRBY WELSTON, AND DIANNE MAY

36. Drugs That Act on the Immune System: Cytokines and Monoclonal Antibodies

KELLEY RATERMANN, JESSICA COX, LYDIA BENITEZ, AND FRANK DAVIS

37. Drugs That Act on the Immune System: Immunosuppressive and Immunostimulatory Drugs

MARLEY L. WATSON, BRIDGETTE K. SCHROADER, AND HEATHER D. NELKIN

38. Corticotrophins, Corticosteroids, and Prostaglandins

MELISSA L. THOMPSON BASTIN, AND BRITTANY D. BISSELL

39. Sex Hormones and Related Compounds, Including Hormonal Contraceptives

CAITLIN M. GIBSON, AND AMULYA TATACHAR

40. Miscellaneous Hormones

AMULYA TATACHAR, SIDHARTHA D. RAY, AND CAITLIN M. GIBSON

41. Thyroid Hormones, Iodine and Iodides, and Antithyroid Drugs

VISHAKHA BHAVE, ANUJ PATEL, RAHUL DESHMUKH, AJAY SINGH, AND VICKY MODY

42. Insulin and Other Hypoglycemic Drugs

LAURA A. SCHALLIOL, JASMINE M. PITTMAN, AND SIDHARTHA D. RAY

Preface

Side Effects of Drugs: Annual (SEDA) is a yearly publication focused on existing, new and evolving side effects of drugs encountered by a broad range of healthcare professionals including physicians, pharmacists, nurse practitioners and advisors of poison control centres. This 40th edition of SEDA includes analyses of the side effects of drugs using both clinical trials and case-based principles which include encounters identified during bedside clinical practice over the 12–14 months since the previous edition. *SEDA* seeks to summarize the entire body of relevant medical literature into a single volume with dual goals of being comprehensive and of identifying emerging trends and themes in medicine as related to side effects and adverse drug effects (ADEs).

With a broad range of topics authored by practicing clinicians and scientists, *SEDA* is a comprehensive and reliable reference to be used in clinical practice. The majority of the chapters include relevant case studies that are not only peer-reviewed but also have a forward-looking, learning-based focus suitable for practitioners as well as for students in training. The nationally known active practitioners believe that this educational resource can be used to stimulate an active learning environment in a variety of settings. Each chapter in this volume has been reviewed by the editor, experienced clinical educators, actively practicing clinicians and scientists to ensure the accuracy and timeliness of the information. The overall objective is to provide a framework for further understanding the intellectual approaches in analyzing the implications of the case studies and their appropriateness when dispensing medications, as well as interpreting adverse drug reactions (ADRs), toxicity and outcomes resulting from medication errors.

This issue of SEDA has included new perspectives from pharmacogenomics/pharmacogenetics and personalized medicine and USFDA's advisories. Due to the advances in science, the genetic profiles of patients must be considered in the aetiology of side effects, especially for medications provided to very large populations. This marks the first phase of genome-based personalized medicine, in which side effects of common medications are linked to polymorphisms in one or more genes. A focus on personalized medicine should lead to major advances for patient care and awareness among clinicians to deliver the most effective medication for the patient. This modality should considerably improve 'appropriate medication use' and enable the clinicians to predetermine 'good vs the bad responders', and help reduce ADRs. Overall, clinicians will have a better control on 'predictability and preventability' of ADEs induced by certain medications. Over time, it is anticipated that pharmacogenetics and personalized medicine will become an integral part of the practice sciences. *SEDA* will continue to highlight the genetic basis of side effects in future editions.

The collective wisdom, expertise and experience of the editor, authors and reviewers were vital in the creation of a volume of this breadth. Reviewing the appropriateness, timeliness and organization of this edition consumed an enormous amount of energy by the authors, reviewers and the editor, which we hope will facilitate the flow of information both interprofessionally among health practitioners, professionals in training and students, and will ultimately improve patient care. Scanning for accuracy, rebuilding and reorganizing information between each edition is not an easy task; therefore, the editor had the difficult task of accepting or rejecting information. The editor will consider this undertaking worthwhile if this publication helps to provide better patient care; fulfils the needs of the healthcare professionals in sorting out side effects of medications, medication errors or adverse drug reactions; and stimulates interest among those working and studying medicine, pharmacy, nursing, physical therapy, chiropractic, and those working in the basic therapeutic arms of pharmacology, toxicology, medicinal chemistry and pathophysiology.

Editor of this volume gratefully acknowledges the leadership provided by the former editor Prof. J. K. Aronson and will continue to maintain the legacy of this publication by building on their hard work. The editor would also like to extend special thanks for the excellent support and assistance provided by Zoe Kruze (Publisher, Serials and Series) and Shellie Bryant (Senior Editorial Project Manager) during the compilation of this work.

Sidhartha D. Ray
Editor

Special Reviews in SEDA 40

Table of Essays, Annuals 1–39

SEDA	Author	Country	Title
1	M.N.G Dukes	The Netherlands	The moments of truth
2	K.H. Kimbel	Germany	Drug monitoring: why care?
3	L. Lasagna	USA	Wanted and unwanted drug effects: The need for perspective
4	M.N.G. Dukes	The Netherlands	The van der Kroef syndrome
5	J.P. Griffin, P.F. D'Arcy	UK	Adverse reactions to drugs—the information lag
6	I. Bayer	Hungary	Science vs practice and/or practice vs science
7	E. Napke	Canada	Adverse reactions: some pitfalls and postulates
8	M.N.G. Dukes	Denmark	The seven pillars of foolishness
9	W.H.W. Inman	UK	Let's get our act together
10	S. Van Hauen	Denmark	Integrated medicine, safer medicine and "AIDS"
11	M.N.G. Dukes	Denmark	Hark, hark, the fictitious dogs do bark
12	M.C. Cone	Switzerland	Both sides of the fence
13	C. Medawar	UK	On our side of the fence
14	M.N.G. Dukes, E. Helsing	Denmark	The great cholesterol carousel
15	P. Tyrer	UK	The nocebo effect—poorly known but getting stronger
16	M.N.G. Dukes	Denmark	Good enough for Iganga?
17	M.N.G. Dukes	Denmark	The mists of tomorrow
18	R.D. Mann	UK	Databases, privacy, and confidentiality—the effect of proposed legislation on pharmacoepidemiology and drug safety monitoring
19	A. Herxheimer	UK	Side effects: Freedom of information and the communication of doubt
20	E. Ernst	UK	Complementary/alternative medicine: What should we do about it?
21	H. Jick	USA	Thirty years of the Boston Collaborative Drug Surveillance Program in relation to principles and methods of drug safety research
22	J.K. Aronson, R.E. Ferner	UK	Errors in prescribing, preparing, and giving medicines: Definition, classification, and prevention
23	K.Y. Hartigan-Go, J.Q. Wong	Philippines	Inclusion of therapeutic failures as adverse drug reactions
24	IPalmlund	UK	Secrecy hiding harm: case histories from the past that inform the future
25	L. Marks	UK	The pill: untangling the adverse effects of a drug
26	D.J. Finney	UK	From thalidomide to pharmacovigilance: a Personal account
26	L.L.Iversen	UK	How safe is cannabis?
27	J.K. Aronson	UK	Louis Lewin—Meyler's predecessor
27	H. Jick	USA	The General Practice Research Database
28	J.K. Aronson	UK	Classifying adverse drug reactions in the twenty-first century
29	M. Hauben, A. Bate	USA/Sweden	Data mining in drug safety
30	J.K. Aronson	UK	Drug withdrawals because of adverse effects
31	J. Harrison, P. Mozzicato	USA	MedDRA®: The Tale of a Terminology
32	K. Chan	Australia	Regulating complementary and alternative medicines

SEDA	Author	Country	Title
33	Graham Dukes	Norway	Third-generation oral contraceptives: time to look again?
34	Yoon K. Loke	UK	An agenda for research into adverse drug reactions
35	J.K. Aronson	UK	Observational studies in assessing benefits and harms: Double standards?
36	J.K. Aronson	UK	Mechanistic and Clinical Descriptions of Adverse Drug Reactions
			Definitive (Between-the-Eyes) Adverse Drug Reactions
37	Sidhartha D. Ray	USA	ADRs, ADEs and SEDs: A Bird's Eye View
38	Sidhartha D. Ray	USA	ADRs, ADEs and SEDs: A Bird's Eye View
39	Sidhartha D. Ray	USA	ADRs, ADEs and SEDs: A Bird's Eye View

Abbreviations

The following abbreviations are used throughout the SEDA series.

2,4-DMA	2,4-Dimethoxyamfetamine
3,4-DMA	3,4-Dimethoxyamfetamine
3TC	Lamivudine (dideoxythiacytidine)
ADHD	Attention deficit hyperactivity disorder
ADP	Adenosine diphosphate
ANA	Antinuclear antibody
ANCA	Antineutrophil cytoplasmic antibody
aP	Acellular pertussis
APACHE	Acute physiology and chronic health evaluation (score)
aPTT	Activated partial thromboplastin time
ASA	American Society of Anesthesiologists
ASCA	*Anti-Saccharomyces cerevisiae* antibody
AUC	The area under the concentration versus time curve from zero to infinity
$AUC_{0\rightarrow x}$	The area under the concentration versus time curve from zero to time x
$AUC_{0\rightarrow t}$	The area under the concentration versus time curve from zero to the time of the last sample
AUC_{τ}	The area under the concentration versus time curve during a dosage interval
AVA	Anthrax vaccine adsorbed
AZT	Zidovudine (azidothymidine)
BCG	Bacillus Calmette Guérin
bd	Twice a day (bis in die)
BIS	Bispectral index
BMI	Body mass index
CAPD	Continuous ambulatory peritoneal dialysis
CD [4, 8, etc]	Cluster of differentiation (describing various glycoproteins that are expressed on the surfaces of T cells, B cells and other cells, with varying functions)
CI	Confidence interval
C_{max}	Maximum (peak) concentration after a dose
$C_{ss.max}$	Maximum (peak) concentration after a dose at steady state
$C_{ss.min}$	Minimum (trough) concentration after a dose at steady state
COX-1 and COX-2	Cyclo-oxygenase enzyme isoforms 1 and 2
CT	Computed tomography
CYP (e.g. CYP2D6, CYP3A4)	Cytochrome P450 isoenzymes
D4T	Stavudine (didehydrodideoxythmidine)
DDC	Zalcitabine (dideoxycytidine)
DDI	Didanosine (dideoxyinosine)
DMA	Dimethoxyamfetamine; *see also* 2,4-DMA, 3,4-DMA
DMMDA	2,5-Dimethoxy-3,4-methylenedioxyamfetamine
DMMDA-2	2,3-Dimethoxy-4,5-methylenedioxyamfetamine
DTaP	Diphtheria + tetanus toxoids + acellular pertussis
DTaP-Hib-IPV-HB	Diphtheria + tetanus toxoids + acellular pertussis + IPV + Hib + hepatitis B (hexavalent vaccine)
DT-IPV	Diphtheria + tetanus toxoids + inactivated polio vaccine
DTP	Diphtheria + tetanus toxoids + pertussis vaccine
DTwP	Diphtheria + tetanus toxoids + whole cell pertussis
eGFR	Estimated glomerular filtration rate
ESR	Erythrocyte sedimentation rate
FDA	(US) Food and Drug Administration
FEV_1	Forced expiratory volume in 1 s
FTC	Emtricitabine
FVC	Forced vital capacity
G6PD	Glucose-6-phosphate dehydrogenase
GSH	Glutathione
GST	Glutathione S-transferase
HAV	Hepatitis A virus
HbA_{1c}	Hemoglobin A_{1c}
HbOC	Conjugated Hib vaccine (Hib capsular antigen polyribosylphosphate covalently linked to the nontoxic diphtheria toxin variant CRM197)

HBV	Hepatitis B virus
HDL, LDL, VLDL	High-density lipoprotein, low-density lipoprotein, and very low density lipoprotein (cholesterol)
Hib	*Haemophilus influenzae* type b
HIV	Human immunodeficiency virus
hplc	High-performance liquid chromatography
HPV	Human papilloma virus
HR	Hazard ratio
HZV	Herpes zoster virus vaccine
ICER	Incremental cost-effectiveness ratio
Ig (IgA, IgE, IgM)	Immunoglobulin (A, E, M)
IGF	Insulin-like growth factor
INN	International Nonproprietary Name (rINN = recommended; pINN = provisional)
INR	International normalized ratio
IPV	Inactivated polio vaccine
IQ [range], IQR	Interquartile [range]
JE	Japanese encephalitis vaccine
LABA	Long-acting beta-adrenoceptor agonist
MAC	Minimum alveolar concentration
MCV4	4-valent (Serogroups A, C, W, Y) meningococcal Conjugate vaccine
MDA	3,4-Methylenedioxyamfetamine
MDI	Metered-dose inhaler
MDMA	3,4-Methylenedioxymetamfetamine
MenB	Monovalent serogroup B meningococcal vaccine
MenC	Monovalent serogroup C meningococcal conjugate vaccine
MIC	Minimum inhibitory concentration
MIM	Mendelian Inheritance in Man (see http://www.ncbi.nlm.nih.gov/omim/607686)
MMDA	3-Methoxy-4,5-methylenedioxyamfetamine
MMDA-2	2-Methoxy-4,5-methylendioxyamfetamine
MMDA-3a	2-Methoxy-3,4-methylendioxyamfetamine
MMR	Measles + mumps + rubella
MMRV	Measles + mumps + rubella + varicella
MPSV4	4-Valent (serogroups A, C, W, Y) meningococcal polysaccharide vaccine
MR	Measles + rubella vaccine
MRI	Magnetic resonance imaging
NMS	Neuroleptic malignant syndrome
NNRTI	Non-nucleoside analogue reverse transcriptase inhibitor
NNT, NNT_B, NNT_H	Number needed to treat (for benefit, for harm)
NRTI	Nucleoside analogue reverse transcriptase inhibitor
NSAIDs	Nonsteroidal anti-inflammatory drugs
od	Once a day (omne die)
OMIM	Online Mendelian Inheritance in Man (see http://www.ncbi.nlm.nih.gov/omim/607686)
OPV	Oral polio vaccine
OR	Odds ratio
OROS	Osmotic-release oral system
PCR	Polymerase chain reaction
PMA	Paramethoxyamfetamine
PMMA	Paramethoxymetamfetamine
PPAR	Peroxisome proliferator-activated receptor
ppb	Parts per billion
PPD	Purified protein derivative
ppm	Parts per million
PRP-CRM	*See* HbOC
PRP-D-Hib	Conjugated Hib vaccine(Hib capsular antigen polyribosylphosphate covalently Linked to a mutant polypeptide of diphtheria toxin)
PT	Prothrombin time
PTT	Partial thromboplastin time
QALY	Quality-adjusted life year
qds	Four times a day (quater die summendum)
ROC curve	Receiver-operator characteristic curve
RR	Risk ratio or relative risk
RT-PCR	Reverse transcriptase polymerase chain reaction
SABA	Short-acting beta-adrenoceptor agonist
SMR	Standardized mortality rate
SNP	Single nucleotide polymorphism
SNRI	Serotonin and noradrenaline reuptake inhibitor

SSRI	Selective serotonin reuptake inhibitor
SV40	Simian virus 40
Td	Diphtheria + tetanus toxoids (adult formulation)
Tdap:	Tetanus toxoid + reduced diphtheria toxoid + acellular pertussis
tds	Three times a day (ter die summendum)
TeMA	2,3,4,5-Tetramethoxyamfetamine
TMA	3,4,5-Trimethoxyamfetamine
TMA-2	2,4,5-Trimethoxyamfetamine
t_{max}	The time at which C_{max} is reached
TMC125	Etravirine
TMC 278	Rilpivirine
V_{max}	Maximum velocity (of a reaction)
wP	Whole cell pertussis
VZV	*Varicella zoster* vaccine
YF	Yellow fever
YFV	Yellow fever virus

ADRs, ADEs and SEDs: A Bird's Eye View

*Sidhartha D. Ray**[,1], *Adrienne T. Black*[†], *Samit K. Shah*[‡], *Sidney J. Stohs*[§]

*Department of Pharmaceutical and Biomedical Sciences, Touro University College of Pharmacy and School of Osteopathic Medicine, Manhattan, NY, United States

[†]3E Services, Verisk 3E, Warrenton, VA, United States

[‡]Keck Graduate Institute College of Pharmacy, Claremont, CA, United States

[§]Dean Emeritus, Creighton University College of Pharmacy & Health Professions, Omaha, NE, United States

[1]Corresponding author: micrornagenomics@gmail.com

INTRODUCTION

Adverse drug events (ADEs), drug-induced toxicity and side effects are a significant concern. ADEs are known to pose significant morbidity, mortality, and cost burdens to society; however, there is a lack of strong evidence to determine their precise impact. The landmark 1999 Institute of Medicine (IOM) report, "*To Err is Human*" implicated adverse drug events in 7000 annual deaths at an estimated cost of $2 billion [1]. Similarly, a second landmark study from 1995 suggested that approximately 28% of ADEs were preventable through optimization of medication safety and distribution systems, provision and dissemination of timely patient and medication information, and staffing assignments [2]. Subsequent investigations suggest these numbers are most likely conservative estimates of the morbidity and mortality impact of ADEs [3]. This concern, however, has not been resolved as demonstrated in a 2015 report from the National Patient Safety Foundation, "*Free from Harm: Accelerating Patient Safety Improvement Fifteen Years after To Err Is Human*" [4]. The report found adverse drug events play a role in 50% of surgeries and that more than 700000 outpatients annually are treated in emergency departments for a drug-induced adverse event and that 120000 of these cases require hospitalization.

In 2001, the US Department of Health and Human Services estimated 770000 people were injured or died each year in hospitals from ADEs, which cost up to $5.6 million each year per hospital excluding the other accessory costs (e.g., hospital admissions due to ADEs, malpractice and litigation costs, or the costs of injuries) [5]. Nationally, hospitals spend $1.56–$5.6 billion each year, to treat patients who suffer ADEs during hospitalization [5]. In response, the Department of Health and Human Services issued the *National Action Plan for Adverse Drug Event Prevention* (ADE Action Plan) in 2014 that identified the means to measure and prevent ADEs and describes future goals to improve patient safety [6].

Analysis of ADEs, ADRs, Side Effects and Toxicity

A recent report suggested that ADEs and/or side effects of drugs occur in approximately 30% of hospitalized patients [7]. The American Society of Health-System Pharmacists (ASHP) defines medication mishap as unexpected, undesirable, iatrogenic hazards or events where a medication was implicated [8]. These events can be broadly divided into two categories: (i) medication errors (i.e., preventable events that may cause or contain inappropriate use), (ii) adverse drug events (i.e., any injury, whether minor or significant, caused by a medication or lack thereof). Another significant ADE-generating category that can be added to the list is: lack of incorporation of pre-existing condition(s) or pharmacogenetic factors. This work focuses on adverse drug events; however, it should be noted that adverse drug events may or may not occur secondary to a medication error.

The lack of more up-to-date epidemiological data regarding the impact of ADEs is largely due to challenges with low adverse drug event reporting. ASHP recommends that health systems implement adverse drug reaction (ADR) monitoring programs in order to (i) mitigate ADR risks for specific patients and expedite reporting to clinicians involved in care of patients who do experience ADRs and (ii) gather pharmacovigilance information that can be reported to pharmaceutical companies and regulatory bodies [9]. Factors that may increase the risk for ADEs include polypharmacy, multiple concomitant disease states, pediatric or geriatric status, female gender, genetic variance, and drug factors, such as class and route

Following table provides most recent CDC estimates (updated June 7, 2018) on Adverse Drug Events, hospitalizations and/or ED visits. Corresponding references are provided below

ADEs (age groups)	ADEs from specific medicines	No. of ED visits/year (approx.)	No. of hospitalizations/ year	Reference
Total		@1.3 million	450000	[a]https://www.cdc.gov/medicationsafety/adult_adversedrugevents.html @ https://www.ncbi.nlm.nih.gov/pubmed/27893129
Elderly (65 + years)	[b]Blood thinners, Diabetic medications, Seizure medications and Heart medicines	450000		[a]https://www.cdc.gov/medicationsafety/adult_adversedrugevents.html
Children (17 years or younger)		200000[c]		[c]https://www.cdc.gov/medicationsafety/parents_childrenadversedrugevents.html

Note: Meta-analysis of Interventions to Reduce ADRs in Older Adults: https://onlinelibrary.wiley.com/doi/abs/10.1111/jgs.15195.

[a] *Approximately 150000 adults are treated in emergency departments each year because of adverse events from antibiotics.*

[b] *https://www.cdc.gov/medicationsafety/adverse-drug-events-specific-medicines.html.*

[c] *Finding and eating or drinking medicines, without adult supervision, is the main cause of emergency visits for adverse drug events among children less than 5 years old. Approximately 60 000 children less than 5 years old are brought to emergency departments each year because of unsupervised ingestions. Nearly 70% of emergency department visits for unsupervised medication ingestions by young children involve 1- or 2-year-old children.*

of administration. The Institute for Safe Medication Practices (ISMP) defines high-alert medications as those with high risk for harmful events, especially when used in error [10]. Examples of high-alert medications include antithrombotic agents, cancer chemotherapy, insulin, opioids, and neuromuscular blockers. Meta-analysis of intervention studies is also underway to reduce ADRs in certain populations (see below for reference).

Terminology

ADEs may be further classified based on expected severity into adverse drug reactions (ADRs) or adverse effects (also known as side effects). ASHP defines ADRs as an "unexpected, unintended, undesired, or excessive response to a drug" resulting in death, disability, or harm [9]. The World Health Organization (WHO) has traditionally defined an ADR as a "response to a drug which is noxious and unintended, and which occurs at doses normally used"; however, another proposed definition, intended to highlight the seriousness of ADRs, is "an appreciably harmful or unpleasant reaction, resulting from an intervention related to the use of a medicinal product, which predicts hazard from future administration and warrants prevention or specific treatment, or alteration of the dosage regimen, or withdrawal of the product" [11]. Under all definitions, ADRs are distinguished from side effects in that they generally necessitate some type of modification to the patient's therapeutic regimen. Such modifications could include discontinuing treatment, changing medications, significantly altering the dose, elevating or prolonging care received by the patient, or changing diagnosis or prognosis. ADRs include drug allergies, immunologic hypersensitivities, and idiosyncratic reactions. In contrast, side effects, or adverse effects, are defined as "expected, well-known reactions resulting in little or no change in patient management" [9]. Side effects occur at predictable frequency and are often dose-related, whereas ADRs are less foreseeable [11, 12].

Two additional types of adverse drug events are drug-induced diseases and toxicity. Drug-induced diseases are defined as an "unintended effect of a drug that results in mortality or morbidity with symptoms sufficient to prompt a patient to seek medical attention, require hospitalization, or both" [13]. In other words, a drug-induced disease has elements of an ADR (i.e., significant severity, elevated levels of patient care) and adverse effects (i.e., predictability, consistent symptoms). Toxicity is a less precisely defined term referring to the ability of a substance "to cause injury to living organisms as a result of physicochemical interaction" [14]. This term is applied to both medication and non-medication types of substances, while "ADRs", "side effects", and "drug-induced diseases" typically only refer to medications. When applied to medication use, toxicity typically refers to use at higher than normal dosing or accumulated supratherapeutic exposure over time, while ADRs, side effects, and drug-induced diseases are associated with normal therapeutic use.

Although the title of this monograph is "Side Effects of Drugs", this work provides emerging information for all adverse drug events including ADRs, side effects, drug-induced diseases, toxicity, and other situations less clearly classifiable into a particular category, such as effects subsequent to drug interactions with other drugs, foods, and cosmetics. Pharmacogenetic considerations have been incorporated in several chapters as appropriate and subject to availability of literature.

Adverse drug reactions are described in Side Effects of Drugs Annual (SEDA) using two complementary systems, EIDOS and DOTS [15–17]. These two systems are illustrated in Figs 1 and 2 and general templates for describing reactions in this way are shown in Figs 3–5. Examples of their use have been discussed elsewhere [18–22]. As clinicians are becoming more cognizant about different types of ADRs, reports in this arena are growing faster than one can imagine; few recent articles are listed for reference [23–34].

EIDOS

The EIDOS mechanistic description of adverse drug reactions [17] has five elements:

- the Extrinsic species that initiates the reaction (Table 1);
- the Intrinsic species that it affects;
- the Distribution of these species in the body;
- the (physiological or pathological) Outcome (Table 2), which is the adverse effect;
- the Sequela, which is the adverse reaction.

- Extrinsic species

  This can be the parent compound, an excipient, a contaminant or adulterant, a degradation product, or a derivative of any of these (e.g. a metabolite) (for examples see Table 1).
- Intrinsic species

  This is usually the endogenous molecule with which the extrinsic species interacts; this can be a nucleic acid, an enzyme, a receptor, an ion channel or transporter, or some other protein.
- Distribution

  A drug will not produce an adverse effect if it is not distributed to the same site as the target species that mediates the adverse effect. Thus, the pharmacokinetics of the extrinsic species can affect the occurrence of adverse reactions.
- Outcome

  Interactions between extrinsic and intrinsic species in the production of an adverse effect can result in physiological or pathological changes (for examples see Table 2). Physiological changes can involve either increased actions (e.g. clotting due to tranexamic acid) or decreased actions (e.g. bradycardia due to β(beta)-adrenoceptor antagonists). Pathological changes can involve cellular adaptations (atrophy, hypertrophy, hyperplasia, metaplasia and neoplasia), altered cell function (e.g. mast cell degranulation in IgE-mediated anaphylactic reactions) or cell damage (e.g. cell lysis, necrosis or apoptosis).
- Sequela

  The sequela of the changes induced by a drug describes the clinically recognizable adverse drug reaction, of which there may be more than one. Sequelae can be classified using the DoTS system.

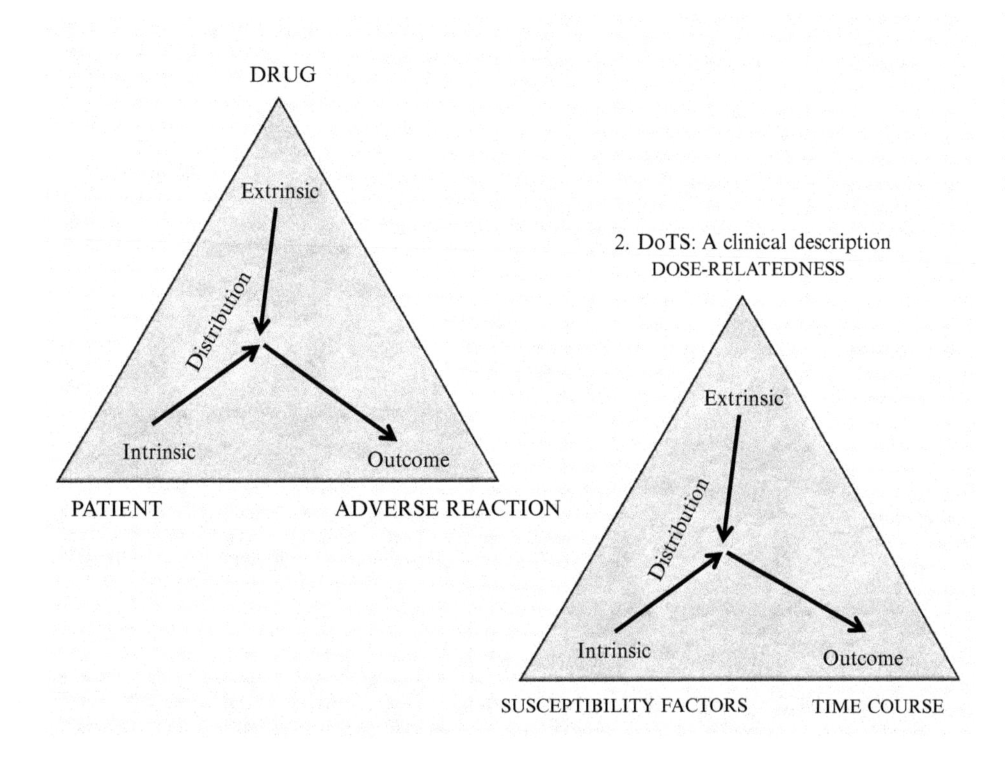

FIG. 1 Describing adverse drug reactions using two complementary systems. Note that the triad of drug–patient–adverse reaction appears outside the triangle in EIDOS and inside the triangle in DoTS, which leads to Fig. 2.

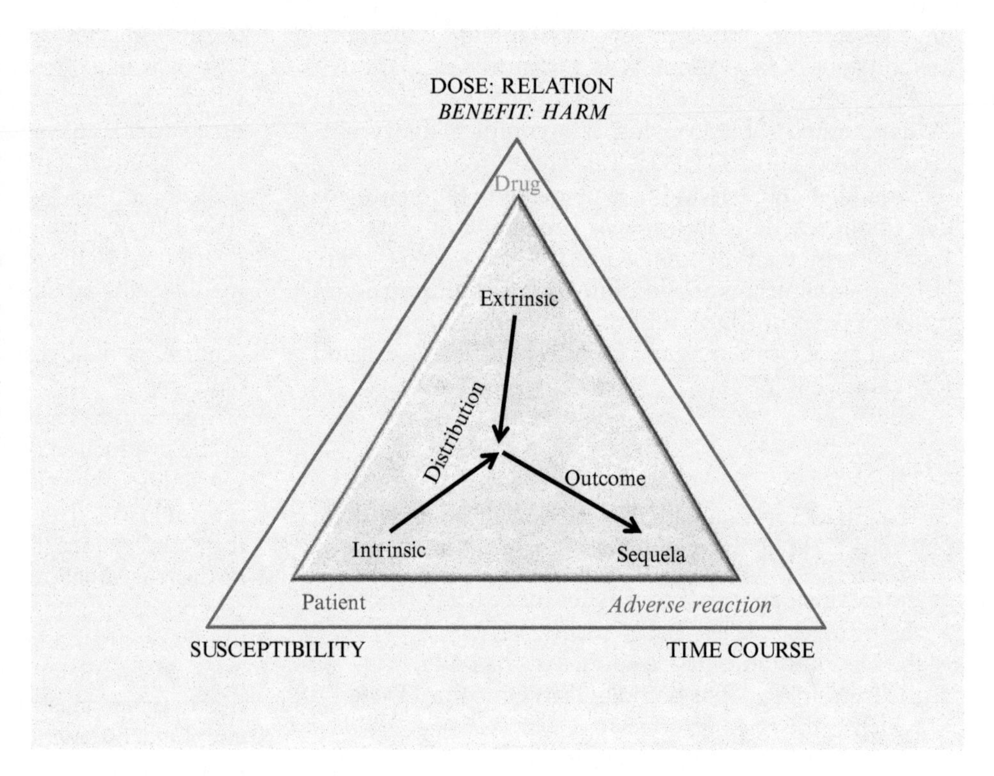

FIG. 2 How the EIDOS and DoTS systems relate to each other. Here the two triangles in Fig. 1 are superimposed, to show the relation between the two systems. An adverse reaction occurs when a drug is given to a patient. Adverse reactions can be classified mechanistically (EIDOS) by noting that when the Extrinsic (drug) species and an Intrinsic (patient) species are co-Distributed, a pharmacological or other effect (the Outcome) results in the adverse reaction (the Sequela). The adverse reaction can be further classified (DoTS) by considering its three main features—its Dose-relatedness, its Time-course, and individual Susceptibility.

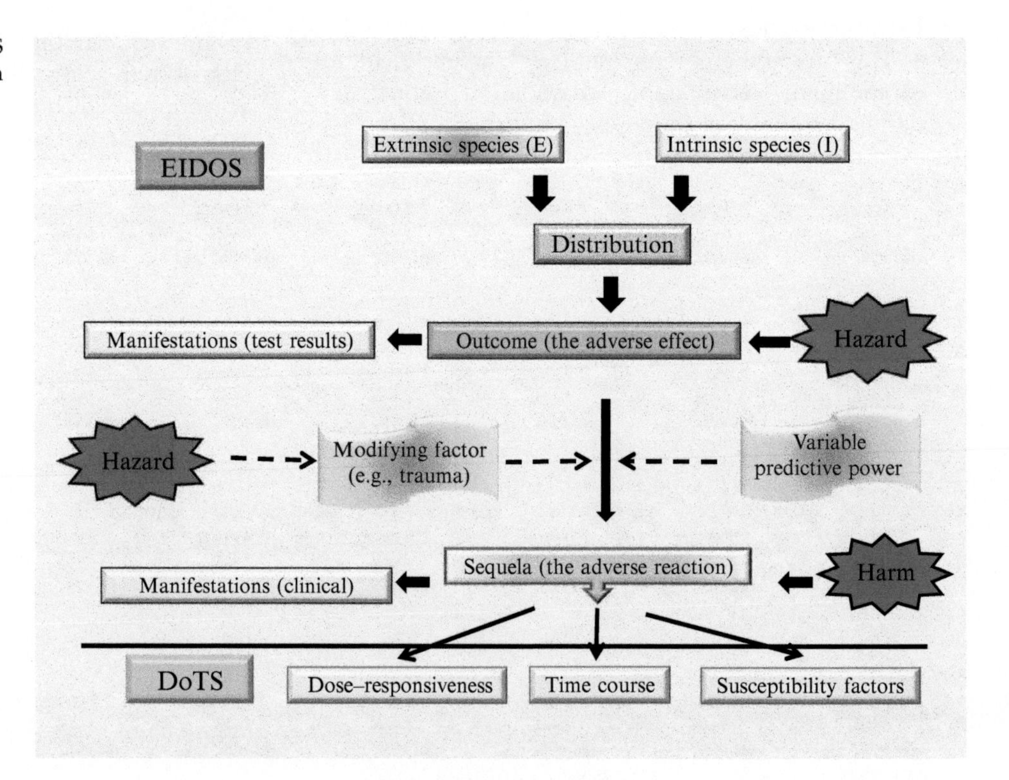

FIG. 3 A general form of the EIDOS and DoTS template for describing an adverse effect or an adverse reaction.

DoTS

In the DoTS system (SEDA-28, xxvii–xxxiii; 1,2) adverse drug reactions are described according to the Dose at which they usually occur, the Time-course over which they occur, and the Susceptibility factors that make them more likely, as follows:

- Relation to dose
  - Toxic reactions (reactions that occur at supratherapeutic doses)
  - Collateral reactions (reactions that occur at standard therapeutic doses)
  - Hypersusceptibility reactions (reactions that occur at subtherapeutic doses in susceptible individuals)

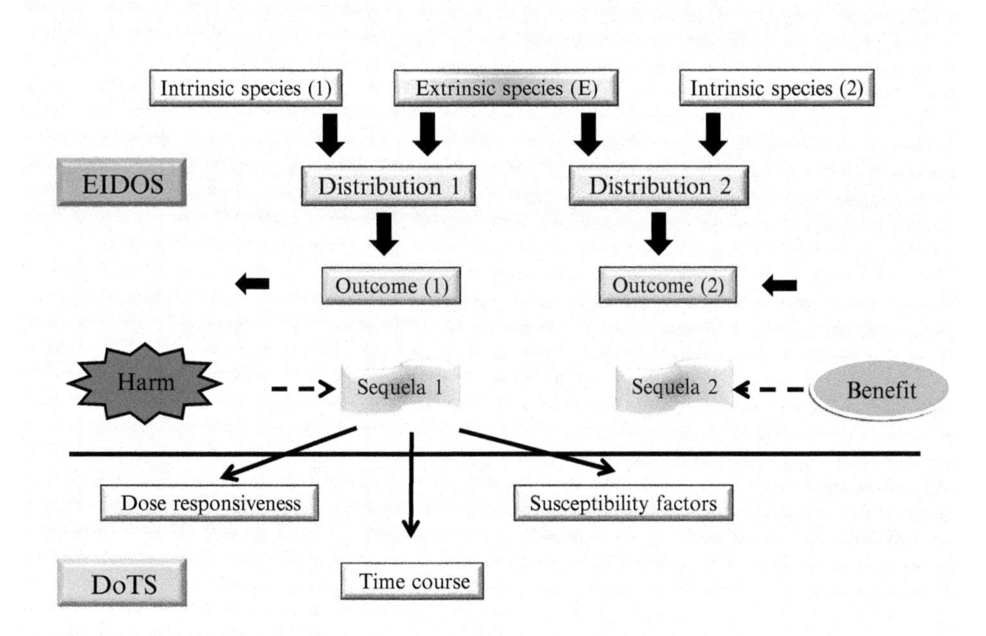

FIG. 4 A general form of the EIDOS and DoTS template for describing two mechanisms of an adverse reaction or (illustrated here) the balance of benefit to harm, each mediated by a different mechanism.

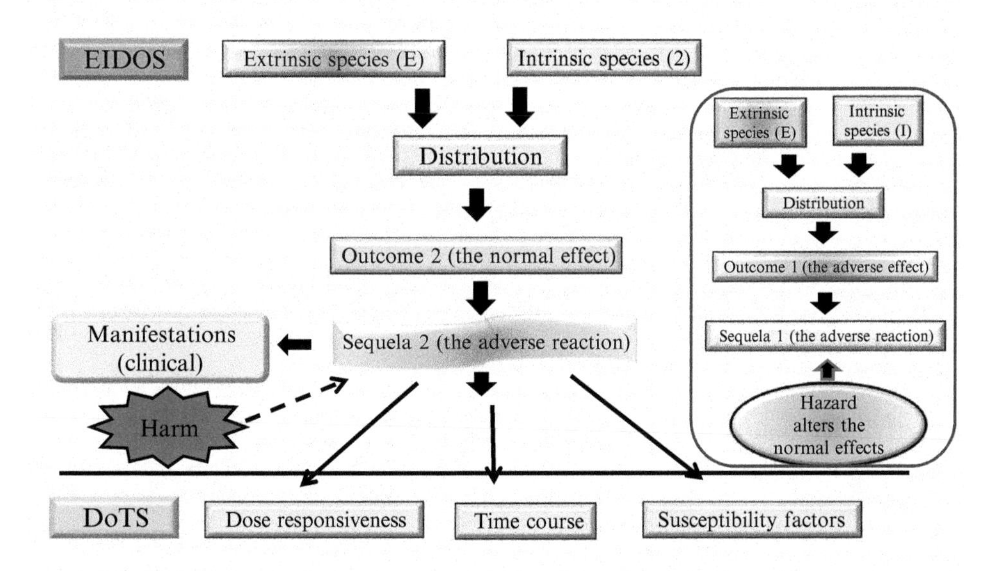

FIG. 5 A general form of the EIDOS and DoTS template for describing an adverse drug interaction.

- Time course
  - Time-independent reactions (reactions that occur at any time during a course of therapy)
- Time-dependent reactions
  - Immediate or rapid reactions (reactions that occur only when drug administration is too rapid)
  - First-dose reactions (reactions that occur after the first dose of a course of treatment and not necessarily thereafter)
  - Early tolerant and early persistent reactions (reactions that occur early in treatment then either abate with continuing treatment, owing to tolerance, or persist)

TABLE 1 The EIDOS Mechanistic Description of Adverse Drug Effects and Reactions

Feature	Varieties	Examples
E. Extrinsic species	**1.** The parent compound	Insulin
	2. An excipient	Polyoxyl 35 castor oil
	3. A contaminant	1,1-Ethylidenebis [1-tryptophan]
	4. An adulterant	Lead in herbal medicines
	5. A degradation product formed before the drug enters the body	Outdated tetracycline
	6. A derivative of any of these (e.g. a metabolite)	Acrolein (from cyclophosphamide)
1. The intrinsic species and the nature of its interaction with the extrinsic species		
(a) Molecular	**1.** Nucleic acids	
	(a) DNA	Melphalan
	(b) RNA	Mitoxantrone
	2. Enzymes	
	(a) Reversible effect	Edrophonium
	(b) Irreversible effect	Malathion
	3. Receptors	
	(a) Reversible effect	Prazosin
	(b) Irreversible effect	Phenoxybenzamine
	4. Ion channels/transporters	Calcium channel blockers; digoxin and Na^+–K^+–ATPase
	5. Other proteins	
	(a) Immunological proteins	Penicilloyl residue hapten
	(b) Tissue proteins	*N*-acetyl-*p*-benzoquinone-imine (paracetamol [acetaminophen])
(b) Extracellular	**1.** Water	Dextrose 5%
	2. Hydrogen ions (pH)	Sodium bicarbonate
	3. Other ions	Sodium ticarcillin
(c) Physical or physicochemical	**1.** Direct tissue damage	Intrathecal vincristine
	2. Altered physicochemical nature of the extrinsic species	Sulindac precipitation
D. Distribution	**1.** Where in the body the extrinsic and intrinsic species occur (affected by pharmacokinetics)	Antihistamines cause drowsiness only if they affect histamine H_1 receptors in the brain
O. Outcome (physiological or pathological change)	The adverse effect (see Table 2)	
S. Sequela	The adverse reaction (use the Dose, Time, Susceptibility [DoTS] descriptive system)	

- Intermediate reactions (reactions that occur after some delay but with less risk during longer term therapy, owing to the 'healthy survivor' effect)
- Late reactions (reactions the risk of which increases with continued or repeated exposure)
- Withdrawal reactions (reactions that occur when, after prolonged treatment, a drug is withdrawn or its effective dose is reduced)
- Delayed reactions (reactions that occur at some time after exposure, even if the drug is withdrawn before the reaction appears)

TABLE 2 Examples of Physiological and Pathological Changes in Adverse Drug Effects (Some Categories Can Be Broken Down Further)

Type of change	Examples
1. Physiological changes	
(a) Increased actions	Hypertension (monoamine oxidase inhibitors); clotting (tranexamic acid)
(b) Decreased actions	Bradycardia (beta-adrenoceptor antagonists); QT interval prolongation (antiarrhythmic drugs)
2. Cellular adaptations	
(a) Atrophy	Lipoatrophy (subcutaneous insulin); glucocorticosteroid-induced myopathy
(b) Hypertrophy	Gynecomastia (spironolactone)
(c) Hyperplasia	Pulmonary fibrosis (busulfan); retroperitoneal fibrosis (methysergide)
(d) Metaplasia	Lacrimal canalicular squamous metaplasia (fluorouracil)
(e) Neoplasia	
– Benign	Hepatoma (anabolic steroids)
– Malignant	
– Hormonal	Vaginal adenocarcinoma (diethylstilbestrol)
– Genotoxic	Transitional cell carcinoma of bladder (cyclophosphamide)
– Immune suppression	Lymphoproliferative tumors (ciclosporin)
3. Altered cell function	IgE-mediated mast cell degranulation (class I immunological reactions)
4. Cell damage	
(a) Acute reversible damage	
– Chemical damage	Periodontitis (local application of methylenedioxymethamfetamine [MDMA, 'ecstasy'])
– Immunological reactions	Class III immunological reactions
(b) Irreversible injury	
– Cell lysis	Class II immunological reactions
– Necrosis	Class IV immunological reactions; hepatotoxicity (paracetamol, after apoptosis)
– Apoptosis	Liver damage (troglitazone)
5. Intracellular accumulations	
(a) Calcification	Milk-alkali syndrome
(b) Drug deposition	Crystal-storing histiocytosis (clofazimine) Skin pigmentation (amiodarone)

- Susceptibility factors
  - Genetic (e.g. Variations in expression of certain drug-metabolizing enzymes)
  - Age (newborn, pediatric, young adult, adult and old age)
  - Sex (gender differences—hormonal variations)
  - Physiological variation (e.g. weight, pregnancy)
  - Exogenous factors (for example the effects of other drugs, devices, surgical procedures, food, phytochemicals & nutraceuticals, alcoholic beverages, smoking, etc.)
  - Diseases (ongoing but latent with no clinical signs, pre-existing and obvious)
  - Environmental factors (drinking water containing trace chemicals; breathing polluted air)

WHO Classification

Although not systematically used in Side Effects of Drugs Annual, the WHO classification, used at the Uppsala Monitoring Center, is a useful schematic to consider in assessing ADRs and adverse effects. Possible classifications include:

- Type A (dose-related, "augmented"): more common events that tend to be related to the pharmacology of the drug, have a mechanistic basis, and result in lower mortality
- Type B (non-dose-related, "bizarre"): less common, unpredictable events that are not related to the pharmacology of the drug

- Type C (dose-related and time-related, "chronic"): events that are related to cumulative dose received over time
- Type D (time-related, "delayed"): events that are usually dose-related but do not become apparent until significant time has elapsed since exposure to the drug
- Type E (withdrawal, "end of use"): events that occur soon after the use of the drug
- Type F (unexpected lack of efficacy, "failure"): common, dose-related events where the drug effectiveness is lacking, often due to drug interactions

REFERENCES ON ADVERSE DRUG REACTIONS

[1] Kohn, LT, Corrigan JM, Donaldson MS, editors. *To Err is Human: Building a Safer Health System.* Washington, DC National Academy Press; 1999: 1–8.

[2] Leape LL, Bates DW, Cullen DJ, et al. Systems analysis of adverse drug events. JAMA. 1995; 274 (1):35–43.

[3] James JT. A new, evidence-based estimate of patient harms associated with hospital care. J Patient Saf. 2013; 9(3):122–128.

[4] National Patient Safety Foundation. *Free from Harm: Accelerating Patient Safety Improvement Fifteen Years after To Err Is Human.* Boston, MA: National Patient Safety Foundation; 2015: 1–59.

[5] US Department of Health & Human Services. Reducing and Preventing Adverse Drug Events To Decrease Hospital Costs: Research in Action. 2001. http://archive.ahrq.gov/research/findings/factsheets/errors-safety/aderia/ade.html.

[6] US Department of Health and Human Services. *National Action Plan for Adverse Drug Event Prevention* (ADE Action Plan). 2014: 1–190.

[7] Wang G, Jung K, Winnenburg R, Shah NH. A method for systematic discovery of adverse drug events from clinical notes. J Am Med Inform Assoc. 2015 Jul 31. pii: ocv102. https://doi.org/10.1093/jamia/ocv102.

[8] American Society of Health-Systems Pharmacists. Positions. Medication Misadventures. http://www.ashp.org/DocLibrary/BestPractices/MedMisPositions.aspx.

[9] American Society of Health-Systems Pharmacists. Guidelines. ASHP Guidelines on adverse drug reaction monitoring and reporting. http://www.ashp.org/DocLibrary/BestPractices/MedMisGdlADR.aspx.

[10] Institute for Safe Medication Practices. ISMP list of high-alert medications in acute care settings. http://www.ismp.org/Tools/institutionalhighAlert.asp.

[11] Edwards IR, Aronson JK. Adverse drug reactions: Definitions, diagnosis, and management. Lancet. 2000; 356:1255–59.

[12] Cochrane ZR, Hein D, Gregory PJ. Medication misadventures I: adverse drug reactions. In: Malone PM, Kier KL, Stanovich JE, Malone MJ, editors. Drug Information: A Guide for Pharmacists, 5th edition. New York, NY: McGraw-Hill; 2013.

[13] Tisdale JE, Miller DA, editors. Drug-Induced Diseases: Prevention, Detection, and Management. 2nd edition. Bethesda, MD: American Society of Health-System Pharmacists; 2010.

[14] Wexler P, Abdollahi M, Peyster AD, et al., editors. Encyclopedia of Toxicology. 3rd edition. Burlington, MA: Academic Press, Elsevier; 2014.

[15] Aronson JK, Ferner RE. Joining the DoTS. New approach to classifying adverse drug reactions. BMJ. 2003; 327:1222–1225.

[16] Aronson JK, Ferner RE. Clarification of terminology in drug safety. Drug Saf. 2005; 28(10):851–870.

[17] Ferner RE, Aronson JK. EIDOS: a mechanistic classification of adverse drug effects. Drug Saf. 2010; 33(1):13–23.

[18] Callréus T. Use of the dose, time, susceptibility (DoTS) classification scheme for adverse drug reactions in pharmacovigilance planning. Drug Saf. 2006; 29(7):557–566.

[19] Aronson JK, Price D, Ferner R.E. A strategy for regulatory action when new adverse effects of a licensed product emerge. Drug Saf. 2009; 32(2):91–98.

[20] Calderón-Ospina C, Bustamante-Rojas C. The DoTS classification is a useful way to classify adverse drug reactions: a preliminary study in hospitalized patients. Int J Pharm Pract. 2010; 18(4):230–235.

[21] Ferner RE, Aronson JK. Preventability of drug-related harms. Part 1: A systematic review. Drug Saf. 2010; 33(11):985–994.

[22] Aronson JK, Ferner RE. Preventability of drug-related harms. Part 2: Proposed criteria, based on frameworks that classify adverse drug reactions. Drug Saf. 2010; 33(11):995–1002.

[23] Saini VK, Sewal RK, Ahmad Y, Medhi B. Prospective Observational Study of Adverse Drug Reactions of Anticancer Drugs Used in Cancer Treatment in a Tertiary Care Hospital. Indian J Pharm Sci., 2015; 77(6):687–93.

[24] White RS; Thomson Reuters Accelus. Pharmaceutical and Medical Devices: FDA Oversight. Issue Brief Health Policy Track Serv. 2015; 28:1–97.

[25] Davies EA, O'Mahony MS. Adverse drug reactions in special populations—the elderly. Br J Clin Pharmacol., 2015; 80(4):796–807.

[26] Mouton JP, Mehta U, Parrish AG, et al. Mortality from adverse drug reactions in adult medical inpatients at four hospitals in South Africa: a cross-sectional survey. Br J Clin Pharmacol., 2015; 80(4):818–26.
[27] Bouvy JC, De Bruin ML, Koopmanschap MA. Epidemiology of adverse drug reactions in Europe: a review of recent observational studies. Drug Saf., 2015; 38(5):437–53.
[28] Bénard-Laribière A, Miremont-Salamé G, Pérault-Pochat MC, et al. Incidence of hospital admissions due to adverse drug reactions in France: the EMIR study. (EMIR Study Group on behalf of the French network of pharmacovigilance centres). Fundam Clin Pharmacol., 2015; 29(1):106–11.
[29] Coleman JJ, Pontefract SK. Adverse drug reactions. Clin Med (Lond). 2016;16(5):481–485. Review. https://www.ncbi.nlm.nih.gov/pubmed/27697815.

FURTHER READING

[30] Castillon G, Salvo F, Moride Y. The Social Impact of Suspected Adverse Drug Reactions: An analysis of the Canada Vigilance Spontaneous Reporting Database. Drug Saf. 2018 Aug 19. doi: 10.1007/s40264-018-0713-8.
[31] Gray SL, Hart LA, Perera S, et al. Meta-analysis of Interventions to Reduce Adverse Drug Reactions in Older Adults. J Am Geriatr Soc. 2018, 66(2):282–288. doi: 10.1111/jgs.15195.
[32] Patton K, Borshoff DC. Adverse drug reactions. Anaesthesia. 2018 Jan;73 Suppl 1:76–84. doi: 10.1111/anae.14143.
[33] Onakpoya IJ, Heneghan CJ, Aronson JK. Post-marketing withdrawal of analgesic medications because of adverse drug reactions: a systematic review. Expert Opin Drug Saf. 2018, 17(1):63–72. doi: 10.1080/14740338.2018.1398232.
[34] Alfirevic A, Pirmohamed M. Genomics of Adverse Drug Reactions. Trends Pharmacol Sci. 2017; 38(1):100–109. doi: 10.1016/j.tips.2016.11.003.

PHARMACOGENOMICS CONSIDERATIONS

It has long been known that individuals respond differently to the same medication regardless of dose and that these differences may result in adverse drug reactions. These differences in drug response may be explained, in part, by pharmacogenomics (previously known as pharmacogenetics), the study of how genes affect the individual responses to drugs. A primary goal of this field is to characterize the relationship between genetic variations and drug responses and the role of genetic variations in the development of adverse drug reactions is now well recognized. Greater use of pharmacogenomics may replace, in part or in full, the "one-size-fits-all" approach to drug prescription in favor of a more tailored pharmacological approach for each individual that may increase efficacy of the medication while reducing the risk of adverse drug reactions. This is the promise of individualized or personalized medicine.

The genetic variations between individuals may be described by polymorphisms in their genetic code, also known as genetic variants. These variations may be caused by the insertion or deletion of a few nucleotides, entire gene deletions, copy number variations or, more typically, are the result of single nucleotide polymorphisms (SNPs). SNPs are variations in a single nucleotide within a gene that may result in the production of a protein with a different amino acid sequence and altered activity. Depending on the kind of protein that is affected by the polymorphisms, drug pharmacokinetics or pharmacodynamics may be altered. Polymorphisms in proteins such as drug-metabolizing enzymes or transporters may affect drug absorption, metabolism, distribution and excretion. Polymorphisms in proteins such as membrane and intracellular receptors may affect drug binding to the receptor. Changes in drug pharmacokinetics and pharmacodynamics affect both the efficacy of the drug and the development of adverse drug reactions. Of particular interest in the field of pharmacogenomics is the role of polymorphisms in drug-metabolizing enzymes, ion transporters and human l-eukocyte antigens (HLA).

Enzymes and Transporters

It has been reported that polymorphisms in one gene superfamily, the cytochrome P450s (CYP) are associated with up to 60% of drug-induced toxicity. Other drug-metabolizing enzymes with variation-inducing toxicity include *N*-acetyl transferase type 2 (NAT2), thiopurine methyltransferases (TPMT), dihydropyrimidine dehydrogenase (DPD), uridine diphosphateglucuronosyl transferases (UGT) and organic anion transporters (OAT). As an illustration of the importance of pharmacogenetics in this area, genomic analysis has recently confirmed the association of long QT syndrome (LQTS) induced by a number of drugs including antibiotics, antipsychotics, chemotherapeutics, antiemetics, opioid analgesics and anti-arrhythmics with *KCNH2* and *SCN5A* mutations. Examples of the genetic variants with the primary gene and the associated drug with adverse effects are shown in Table 3.

TABLE 3 Examples of Genetic Polymorphisms Leading to Increased Risk of Drug-Induced Adverse Reactions

Gene	Drug	Comments
CYP2D6	Tricyclic antidepressants	Increased risk of adverse effects in poor metabolizers
TPMT	Thiopurines	Increased risk of myelosuppression in individuals with deficient or intermediate activity
DPYD	Fluoropyrimidines	Increased risk for toxicity in DPYD intermediate or poor metabolizers
SLCO1B1	Simvastatin	Increased myopathy risk with low or intermediate function SLCO1B1 phenotypes
CYP2D6	Codeine	Potential for toxicity in ultrarapid metabolizers
CYP2C9	Phenytoin	Increased risk of SJS/TEN in CYP2C9 poor or intermediate metabolizers
CYP2C19	SSRIs	Increased risk of adverse effects in CYP2C19 poor metabolizers
UGT1A1	Atazanavir	Increased probability of hyperbilirubinemia in UGT1A1 poor metabolizers

TABLE 4 Examples of HLA Allele Variants and Associated Drug-Induced Adverse Reactions

Reaction	Drug	HLA variant(s)
Hypersensitivity	Abacavir	*HLA-B*5701, HLA-DR7, HLA-DQ3*
Stevens–Johnson syndrome (SJS)/toxic epidermal necrolysis (TEN) and Severe cutaneous adverse reactions	Allopurinol	*HLA-B*5801*
Hypersensitivity and SJS/TEN	Carbamazepine	*HLA-B*1502, HLA-A3101*
Hepatoxicity	Co-Amoxiclav	*HLA-DRB1*15:01, HLA-A*0201, HLA-B*1801*
Hepatotoxicity	Flucloxacillin	*HLA-B*5701*
Systemic lupus erythematosus	Hydralazine	*HLA-DR4*
SJS/TEN	Lamotrigine	*HLA-B*3801*
Hepatoxicity	Lumiracoxib	*HLA-DRB*1501, HLA-DQA*0102*
Systemic lupus erythematosus	Minocycline	*HLA-DQB1*
SJS/TEN	Nevirapine	*HLA-B*3505, HLA-C0401*
SJS/TEN	Phenytoin	*HLA-B*1502*
Hepatoxicity	Ticlopidine	*HLA-A*3303*
Hepatoxicity	Ximelagatran	*HLA-DRB1*0701*

Human Leukocyte Antigens (HLA)

The HLA family comprises over 200 genes, forming the human major histocompatibility complex (MHC). Associated adverse drug reactions may occur through direct or indirect interaction of the drug with a specific HLA allele, thereby initiating an immune response and causing an adverse effect. These HLA-mediated reactions are typically hypersensitivity reactions but they also include drug-induced injury to such organs as the liver, kidney, skin, muscle, and heart. As these reactions are dependent upon the presence of specific HLA alleles, the phenotype or ethnicity of an individual is of particular interest in determining if a medication may cause unwanted effects.

Due to the risk of adverse drug reactions, HLA genotyping is commonly done prior to prescribing many medications. In the case of abacavir, screening for the presence of *HLA-B*5701* is required by American and European regulatory authorities prior to the initiation of treatment. Similarly, genetic screening for *HLA-B*1502* and *HLA-B*5801* before administration of carbamazepine or allopurinol has become commonplace in several Asian countries. A number of drug-induced reactions have been associated with specific variants in the HLA genes and several examples are shown in Table 4. The HLA Adverse Drug Reaction Database also provides an up-to-date listing of HLA alleles and the associated adverse drug reactions (http://www.allelefrequencies.net/hla-adr/default.asp).

Genetic Testing Technology

The ability to conduct genetic testing may be limited by cost and/or computing power. In addition, it is critical to know how the drug is metabolized, transported, etc., and what key enzymes or molecules are involved in these processes. It must also be noted that even strong associations from genomic studies are not the equivalent of clinical relevancy as degree of relevancy is dependent upon the relative frequency of the genetic variation in the affected population, the disease phenotype and the severity of the outcome/reaction.

The following section describes the general categories of genomic testing:

1. Linkage mapping

   This analysis is a tracking method to identify patterns of DNA phenotypes or markers within families (i.e., hemophilia, brown eye color, etc.). Linkage mapping is generally not a useful tool for the detection of a genetic basis for the occurrence of an adverse drug reaction, although identification of an individual with a suspect phenotype may lead to more focused genetic testing.

2. Genome-wide association (GWA) studies
   GWA studies analyze data from large cohorts to detect genetic variations via high-throughput analysis of SNPs in each individual's genome. Although genotyping of SNPs was initially used as a means to better understand human disease, this method has increasingly been used to study the genetic basis for adverse drug reactions. It is likely that as costs and application limitations decrease, whole genomic analysis will become the preferred method.
3. Next-Generation Sequencing Methods
   While whole genome sequencing is currently possible and may be preferential, the processing power for analysis of the data is prohibitive for the majority of users. As a result, a subset of the genome, the exome or all exons (protein coding regions), is typically analyzed in whole exome sequencing (WES). This technique allows identification of any variations within this section of the genome; a major drawback, however, is that any variations in other areas of the genome such as in introns (regulatory coding regions) are not included.

The US FDA has provided a "Table of Pharmacogenomic Biomarkers in Drug Labeling" with current labeling information for numerous drugs that includes pharmacogenetics information (genetic variations, polymorphisms, etc.) for each medication. This table also provides a recommendation on genetic testing and/or screening to be conducted prior to prescription (https://www.fda.gov/downloads/Drugs/ScienceResearch/ResearchAreas/Pharmacogenetics/UCM545881.pdf). An additional resource, "Actionable Pharmacogenetics in FDA Labeling", succinctly describes the recommended actions to be taken for each medication based on the FDA information (http://www.wolterskluwercdi.com/blog/actionable-pharmacogenetics-fda-labeling/).

The FDA has also issued a guidance document for recommended pharmacogenetic and genetic tests intended for drug sponsors and FDA reviewers involved in preparing and reviewing premarket approval applications (PMA) and premarket notification (510(k)) submissions (https://www.fda.gov/RegulatoryInformation/Guidances/ucm077862.htm).

Conclusion

The availability of genetic testing technology and increasing knowledge about genetic variation-associated adverse drug reactions has elevated the role of pharmacogenomics in designing the drug regimen tailored to the individual patient (individualized or personalized medicine). An individual's genetic profile would, therefore, determine what medication would be most appropriate along with the most effective dosing regimen to increase the drug effectiveness and reduce the probability of adverse drug reactions. To this goal, the Clinical Pharmacogenetics Implementation Consortium (CPIC) has issued genetic testing and dosing guidelines for over 200 drugs (https://cpicpgx.org/genes-drugs).

The cost of conducting genetic testing to identify polymorphisms that influence drug response has been steadily declining. There are several companies that now provide pharmacogenetic testing, offering testing that provide results for a single gene or more typically for an entire panel of genes. Turnaround time to receive results from tests ranges from 48 hours to 2 weeks, and typically only require a saliva sample, blood sample or a buccal swab.

Although pharmacogenetic testing is becoming more widespread in application, it is not expected that this approach will completely eliminate standard therapeutic monitoring or measurement of other phenotype variables. It is known that adverse reactions may be affected by other factors such as drug–drug interactions, drug–food and drug–dietary supplement interactions, age, gender, ethnicity and comorbities. In the future, it is likely that there will be a blending of the different methods to provide the most appropriate therapeutic approach.

Additional information on this topic can be found in these reviews [9R, 10R].

REFERENCES

[1] Ray SD, Thomas, K and Kisor DF. ADRs, ADEs and SEDs: A Bird's Eye View. Editorial Article—In: Ray SD, Editor. Side Effects of Drugs: Annual: A worldwide yearly survey of new data in adverse drug reactions, Vol. 38: pp. xxvii–xli, 2016. [R] http://www.sciencedirect.com/science/article/pii/S0378608016300605.

[2] Ray SD, Beckett RD, Kisor DF, Gray JP and Kiersma ME. Editorial article—ADRs, ADEs and SEDs: A Bird's Eye View In: Ray SD, Editor. Side Effects of Drugs: Annual: A worldwide yearly survey of new data in adverse drug reactions, Vol. 37, Pp. XXVII–XXXIX, Elsevier; 2015. [R]. http://www.sciencedirect.com/science/article/pii/S0378608015000616.

[3] Kullak-Ublick GA, Andrade RJ, Merz M, et al. Drug-induced liver injury: Recent advances in diagnosis and risk assessment. 2017. Gut. 66(6): 1154–1164. [3R]

[4] Lauschke VM, Ingelman-Sundberg M. The importance of patient-specific factors for hepatic drug response and toxicity. 2016. Int. J. Mol. Sci. 17(10): 1714–1741. [R]

[5] Maggo SD, Savage RL, Kennedy MA. Impact of new genomic technologies on understanding adverse drug reactions. 2016. Clin. Pharmacokinet. 55(4): 419–436. [R]

[6] Maagdenberg H, Vijverberg SJ, Bierings MB, et al. Pharmacogenomics in pediatric patients: towards personalized medicine. 2016. Paediatr. Drugs. 18(4): 251–260. [R]

[7] O'Donnell PH, Danahey K, Ratain MJ. The outlier in all of us: why implementing pharmacogenomics could matter for everyone. 2016. Clin. Pharmacol. Ther. 99(4): 401–404. [R]

[8] Su SC, Hung SI, Fan WL, et al. Severe cutaneous adverse reactions: the pharmacogenomics from research to clinical implementation. 2016. Int. J. Mol. Sci. 17(11): 1890–1990. [R]

[9] Ray SD, Black A. ADRs, ADEs and SEDs: A Bird's Eye View. Editorial Article—In: Ray SD, Editor. Side Effects of Drugs: Annual: A worldwide yearly survey of new data in adverse drug reactions, Vol. 39: pp. xxvii–xlii, 2017 [R]. https://www.sciencedirect.com/science/article/pii/S0378608017300570.

FURTHER READING

[10] Osanlou O, Pirmohamed M, Daly AK. Pharmacogenetics of Adverse Drug Reactions. Adv Pharmacol. 2018; 83:155–190. doi: 10.1016/bs.apha.2018.03.002.

IMMUNOLOGICAL REACTIONS

The immunological reactions are diverse and varied but considered specific. Nearly five decades ago, Karl Landsteiner's ground-breaking work "The Specificity of Serological Reactions" set the standard in experimental immunology. Several new discoveries in immunology in the 20th century, such as, 'CD' receptors (cluster of differentiation), recognition of 'self' vs 'non-self', a large family of cytokines and antigenic specificity became instrumental in describing immunological reactions. The most widely accepted classification divides immunological reactions (drug allergies or otherwise) into four pathophysiological types:

A. Type I hypersensitivity: (anaphylaxis, immediate type)
B. Type II hypersensitivity: (antibody-mediated cytotoxic reactions, cytotoxic type) or
C. Type III hypersensitivity: (immune complex-mediated reactions, toxic-complex syndrome)
D. Type IV hypersensitivity: (cell-mediated immunity, delayed-type hypersensitivity)

Although this classification was proposed more than 30 years ago, it is still widely used today [1–3].

Type I Reactions (IgE-Mediated Anaphylaxis; Immediate Hypersensitivity)

In type I reactions, the drug or its metabolite interacts with immunoglobulin (IG) IgE molecules bound to specific type of cells (mast cells and basophils). This triggers a process that leads to the release of pharmacological mediators (histamine, 5-hydroxytryptamine, kinins, and arachidonic acid derivatives) which cause the allergic response. The development of such a reaction depends exclusively upon exposure to the same assaulting agent (antigen, allergen or metabolite) for the second time and the severity depends on the level of exposure. The clinical effects [2] are due to smooth muscle contraction, vasodilatation, and increased capillary permeability. The symptoms include faintness, light-headedness, pruritus, nausea, vomiting, abdominal pain, and a feeling of impending doom (angor animi). The signs include urticaria, conjunctivitis, rhinitis, laryngeal edema, bronchial asthma, pulmonary edema, angioedema, and anaphylactic shock. In addition, takotsubo cardiomyopathy can occur as well as Kounis syndrome (an acute coronary episode associated with an allergic reaction). Not all type I reactions are IgE-dependent; however, adverse reactions that are mediated by direct histamine release have conventionally been called anaphylactoid reactions but are better classified as non-IgE-mediated anaphylactic reactions. Cytokines, such as, interleukin (IL)-4, IL-5, IL-6 and IL-13, either mediate or influence this class of hypersensitivity reaction. Representative agents that are known to induce such reactions include: gelatin, gentamicin, kanamycin, neomycin, penicillins, polymyxin B, streptomycin and thiomersal [1–3].

Type II Reactions (Cytotoxic Reactions)

Type II reactions involve circulating immunoglobulins G (IgG) or M (IgM) (or rarely IgA) binding with cell surface antigens (membrane constituent or protein) and interacting with an antigen formed by a hapten (drug or metabolite) and subsequently fixing complement. Complement is then activated leading to cytolysis. Type II reactions often involve antibody-mediated cytotoxicity directed to the membranes of erythrocytes, leukocytes, platelets, and probably hematopoietic precursor cells in the bone marrow. Drugs that are typically involved are methyldopa (hemolytic anemia), aminopyrine (leukopenia), and heparin (thrombocytopenia) with mostly hematological consequences, including thrombocytopenia, neutropenia, and hemolytic anemia [1–3].

Type III Reactions (Immune Complex Reactions)

In type III reactions, formation of an immune complex and its deposition on tissue surface serve as primary initiators. Occasionally, immune complexes bind to endothelial cells and lead to immune complex deposition with subsequent complement activation in the linings of blood vessels. Circumstances that govern immune formation or immune complex disease remain unclear to date, and it usually occurs without symptoms. The clinical symptoms of a type III reaction include serum sickness (β-lactams), drug-induced lupus erythematosus (quinidine), and vasculitis (minocycline). Type III reactions can result in acute interstitial nephritis or serum sickness (fever, arthritis, enlarged lymph nodes, urticaria, and maculopapular rashes) [1–3].

Type IV Reactions (Cell-Mediated or Delayed Hypersensitivity Reactions)

Type IV reactions are initiated when a hapten–protein antigenic complex sensitizes T lymphocytes (T cells). Upon re-exposure to the immunogen, the activity of the sensitized T-cells usually results in severe inflammation in the affected areas. Type IV reactions are exemplified by contact dermatitis while pseudoallergic reactions may resemble allergic reactions clinically but are not immunologically mediated. Examples of type IV reactions include asthma and rashes caused by aspirin and maculopapular erythematous rashes due to ampicillin or amoxicillin in the absence of penicillin hypersensitivity. This reaction may also be caused by sulfonamides and sulfites, anticonvulsants (phenytoin, carbamazepine, and phenobarbital), NSAIDs (aspirin, naproxen, nabumetone, and ketoprofen), antiretroviral agents and cephalosporins [1–4].

Other Types of Reactions

Several types of adverse drug reactions do not easily fit into the general classification scheme. These include most cutaneous hypersensitivity reactions (such as toxic epidermal necrolysis), 'immune-allergic' hepatitis and hypersensitivity pneumonitis. Another difficulty is that allergic drug reactions can occur via more than one mechanism; picryl chloride in mice induces both type I and type IV responses. Several articles are included in this review to serve as a pointer to this field [4–12].

References

[1] Coombs RRA, Gell PGH. Classification of allergic reactions responsible for clinical hypersensitivity and disease. In: Gell PGH, Coombs RRA, Lachmann PJ, editors. Clinical Aspects of Immunology. London: Blackwell Scientific Publications; 1975. pp. 761–81.

[2] Schnyder B, Pichler W. Mechanisms of Drug-Induced Allergy. Mayo Clin Proc. Mar 2009; 84(3): 268–272.

[3] Boyman O, Comte D, Spertini F. Adverse reactions to biologic agents and their medical management. Nat Rev Rheumatol., 2014 Aug 12. https://doi.org/10.1038/nrrheum.2014.123. [Epub ahead of print].

[4] Brown SGA. Clinical features and severity grading of anaphylaxis. J Allergy Clin Immunol 2004; 114(2): 371–6.

[5] Johansson SGO, Hourihane JO, Bousquet J, Bruijnzeel-Koomen C, Dreborg S, Haahtela T, Kowalski ML, Mygind N, Ring J, van Cauwenberge P, van Hage-Hamsten M, Wüthrich B. A revised nomenclature for allergy. An EAACI position statement from the EAACI nomenclature task force. Allergy 2001; 56(9): 813–24.

[6] Uzzaman A, Cho SH. Chapter 28: Classification of hypersensitivity reactions. Allergy Asthma Proc. 2012 May–Jun; 33 Suppl 1:S96–9.

[7] Descotes J, Choquet-Kastylevsky G. Toxicology, 2001; 158(1–2):43–9. Gell and Coombs's classification: is it still valid?

[8] Corominas M, Andrés-López B, Lleonart R. Severe adverse drug reactions induced by hydrochlorothiazide: A persistent old problem. Ann Allergy Asthma Immunol., 2016; 117(3):334–5.

[9] Velicković J, Palibrk I, Miljković B, et al. Self-reported drug allergies in surgical population in Serbia. Acta Clin Croat. 2015; 54(4):492–9.

[10] Yip VL, Alfirevic A, Pirmohamed M. Genetics of immune-mediated adverse drug reactions: a comprehensive and clinical review. Clin Rev Allergy Immunol., 2015; 48(2–3):165–75.

[11] Agúndez JA, Mayorga C, García-Martin E. Drug metabolism and hypersensitivity reactions to drugs. Curr Opin Allergy Clin Immunol. 2015;15 (4):277–84.

[12] Wheatley LM, Plaut M, Schwaninger JM. et al. Report from the National Institute of Allergy and Infectious Diseases workshop on drug allergy. J Allergy Clin Immunol., 2015; 136(2):262–71.e2.

Further Reading

[13] Ramsbottom KA, Carr DF, Jones AR, Rigden DJ. Critical assessment of approaches for molecular docking to elucidate associations of HLA alleles with adverse drug reactions. Mol Immunol. 2018 Aug 17;101:488–499. doi: 10.1016/j.molimm.2018.08.003. [Epub ahead of print].

[14] Leon L, Gomez A, Vadillo C, et al. Severe adverse drug reactions to biological disease-modifying anti-rheumatic drugs in elderly patients with rheumatoid arthritis in clinical practice. Clin Exp Rheumatol. 2018; 36(1):29–35.

ANALYSIS OF TOXICOLOGICAL REACTIONS

Potentiation Reactions

This type of reaction occurs when either one non-toxic chemical interacts with another non-toxic chemical or one non-toxic chemical interacts with another toxic chemical at low doses (subtoxic, acutely toxic) results in a greater level of toxicity. An alternate interpretation could be when two drugs are taken together and one of them intensifies the action of the other. In such scenarios, if the final result is high toxicity, then the final outcome is called potentiation (increasing the toxic effect of 'Y' by 'X'). Results usually lead to unanticipated level of cell death in the form of apoptosis, necrosis, autophagy and apocrosis (or necraptosis, aponecrosis). Theoretically, it can be expressed as: $x+y=M$ $(1+0=4)$.

Examples

(i) When chronic or regular alcohol drinkers consume therapeutic doses of acetaminophen, it can lead to alcohol-potentiated acetaminophen-induced hepatotoxicity (cause: ethanol-induced massive CYP2E1 induction in the liver)
(ii) Administration of iron supplements in patients on doxorubicin therapy may cause potentiation of doxorubicin-induced cardiotoxicity (cause: hydroxyl radical formation and redox cycling of doxorubicin)
(iii) Phenergan®, an antihistamine, when given with a painkilling narcotic such as Demerol® can intensify the narcotic effect; reducing the dose of the narcotic is advised
(iv) Ethanol potentiation of CCl4-induced hepatotoxicity
(v) Use of phenytoin and calcium-channel blockers combination should be used with caution
(vi) Potentiation of warfarin by dietary supplements and foods as garlic, ginger, ginkgo and grapefruit

Synergistic Effect

Synergism is somewhat similar to potentiation. When two drugs are taken together that are similar in action, such as barbiturates and alcohol, which are both depressants, an effect exaggerated out of proportion to that of each drug taken separately at the given dose may occur (mathematically: $1+1=4$). Normally, taken alone, neither substance would cause serious harm, but if taken together, the combination could cause coma or death. Another example is when smokers get exposed to asbestos, resulting in the development of lung cancer.

Additive Effect

Additive effect is defined as a consequence which follows exposure to two or more agents which act jointly but do not interact; the total effect is the simple sum of the effects of separate exposure to the agents under the same conditions. This could be represented by $1+1=2$:

Examples

(i) a barbiturate and a tranquilizer given together before surgery to relax the patient
(ii) the toxic effect on bone marrow resulting after AZT+ganciclovir or AZT+clotrimazole administration.

Antagonistic Effects

Antagonistic effects are when two drugs/chemicals are administered simultaneously or one closely followed by the other with the net effect of the final outcome of the reaction being negligible or zero. This could be expressed by $1+1=0$. An example might be the use of a tranquilizer to stop the action of LSD.

Examples

(i) When ethanol is administered to methanol poisoned patient
(ii) NSAIDs administered to diuretics (hydrochlorothiazide/furosemide): Reduce diuretics effectiveness
(iii) Certain β-blockers (INDERAL®) taken to control high blood pressure and heart disease, counteract β-adrenergic stimulants, such as albuterol®.
(iv) St. John's wort in combination with drugs as digoxin, indinavir, nifedipine and alprazolam.

Cross References

[1] Ray SD, Mehendale HM. Potentiation of CCl4 and CHCl3 hepatotoxicity and lethality by various alcohols. Fundam Appl Toxicol. 1990; 15(3):429–40.
[2] Gammella, E., Maccarinelli, F., Buratti, P., et al. The role of iron in anthracycline cardiotoxicity. Front Pharmacol. 2014; 5:25. doi: https://doi.org/10.3389/fphar.2014.00025. eCollection 2014.
[3] NLM's Toxlearn tutorials: http://toxlearn.nlm.nih.gov/Module1.htm.
[4] NLM's Toxtutor (visit interactions): http://sis.nlm.nih.gov/enviro/toxtutor/Tox1/a42.htm.
[5] Smith MA, Reynolds CP, Kang MH, et al. Synergistic activity of PARP inhibition by talazoparib (BMN 673) with temozolomide in pediatric cancer models in the pediatric preclinical testing program. Clin Cancer Res., 2015; 21(4):819–32.
[6] Niu F, Zhao S, Xu CY, et al. Potentiation of the antitumor activity of adriamycin against osteosarcoma by cannabinoid WIN-55,212-2. Oncol Lett. 2015; 10(4):2415–2421.
[7] Calderon-Aparicio A, Strasberg-Rieber M, Rieber M. Disulfiram anti-cancer efficacy without copper overload is enhanced by extracellular H2O2 generation: antagonism by tetrathiomolybdate. Oncotarget. 2015; 6(30):29771–81.

[8] Zajac J, Kostrhunova H, Novohradsky V, et al. Potentiation of mitochondrial dysfunction in tumor cells by conjugates of metabolic modulator dichloroacetate with a Pt(IV) derivative of oxaliplatin. J Inorg Biochem. 2016; 156:89–97.
[9] Nurcahyanti AD, Wink M. Cytotoxic potentiation of vinblastine and paclitaxel by L-canavanine in human cervical cancer and hepatocellular carcinoma cells. Phytomedicine. 2015; 22(14):1232–7.
[10] Lu CF, Yuan XY, Li LZ, et al. Combined exposure to nano-silica and lead-induced potentiation of oxidative stress and DNA damage in human lung epithelial cells. Ecotoxicol Environ Saf. 2015; 122:537–44.
[11] Kuchárová B, Mikeš J, Jendželovský R, et al. Potentiation of hypericin-mediated photodynamic therapy cytotoxicity by MK-886: focus on ABC transporters, GDF-15 and redox status. Photodiagnosis Photodyn Ther. 2015; 12(3):490–503.
[12] Djillani A, Doignon I, Luyten T, et al. Potentiation of the store-operated calcium entry (SOCE) induces phytohemagglutinin-activated Jurkat T cell apoptosis. Cell Calcium. 2015; 58(2):171–85.
[13] Yoon E, Babar A, Choudhary M. et al. Acetaminophen-Induced Hepatotoxicity: a Comprehensive Update. J Clin Transl Hepatol. 2016; 4(2): 131–142.
[14] Ray SD, Corcoran, G.B. "Apoptosis and Cell Death", 3rd Edition, Vol. I, Chapter-11, pp. 247–312, in Ballantyne, Marrs and Syversen Eds., 'General and Applied Toxicology'; Wylie Publishing, UK, 2009. http://onlinelibrary.wiley.com/doi/10.1002/9780470744307.gat015/abstract.
[15] Betharia S, Farris FF, Corcoran GB, Ray SD. 'Mechanisms of Toxicity' In: Wexler, P. (Ed.), Encyclopedia of Toxicology, 3rd edition vol 3. Elsevier Inc., Academic Press, pp. 165–175, 2014. http://www.sciencedirect.com/science/article/pii/B9780123864543003298.
[16] Ge B, Zhang Z, Zuo Z. Updates on the clinical evidenced herb-warfarin interactions. Evid Based Compl Alt Med. 2014; https://doi.org/10.1155/2014/957362. 18 pp.
[17] Tsai HH, Lin HW, Pickard AS, Tsai HY, Mahady GB. Evaluation of documented drug interactions and contraindications associated with herbs and dietary supplements: a systematic literature review. Int J Clinical Practice 2013; 66: 1056–1078.
[18] Onakpoya IJ, Heneghan CJ, Aronson JK. Postmarketing withdrawal of human medicinal products because of adverse reactions in animals: a systematic review and analysis. Pharmacoepidemiol Drug Saf. 2017, 26(11):1328–1337. https://doi.org/10.1002/pds.4256.

GRADES OF ADVERSE DRUG REACTIONS

Drugs and chemicals may exhibit adverse drug reactions (ADRs or adverse drug effects) that may include unwanted (side effects), uncomfortable (system dysfunction), or dangerous effects (toxic). ADRs are a form of manifestation of toxicity which may occur after overexposure or high-level exposure, or in some circumstances ADRs may also occur after exposure to therapeutic doses but often an underlying cause (pre-existing condition) is present. In contrast, 'Side effect' is an imprecise term often used to refer to a drug's unintended effects that occur within the therapeutic range [1]. Risk–benefit analysis provides a window into the decision-making process prior to prescribing a medication. Patient characteristics such as age, gender, ethnic background, pre-existing conditions, nutritional status, genetic pre-disposition or geographic factors, as well as drug factors (e.g., type of drug, administration route, treatment duration, dosage, and bioavailability) may profoundly influence ADR outcomes. Drug-induced adverse events can be categorized as unexpected, serious or life-threatening.

Adverse drug reactions are graded according to intensity, using a scheme that was originally introduced by the US National Cancer Institute to describe the intensity of reactions to drugs used in cancer chemotherapy [2]. This scheme is now widely used to grade the intensity of other types of adverse reactions, although it does not always apply so clearly to them. The scheme assigns grades as follows:

- Grade 1 ≡ mild;
- Grade 2 ≡ moderate;
- Grade 3 ≡ severe;
- Grade 4 ≡ life-threatening or disabling;
- Grade 5 ≡ death.

Then, instead of providing general definitions of the terms "mild", "moderate", "severe", and "life-threatening or disabling", the system describes what they mean operationally in terms of each adverse reaction, in each case the intensity being described in narrative terms. For example, hemolysis is graded as follows:

- Grade 1: Laboratory evidence of hemolysis only (e.g. direct antiglobulin test; presence of schistocytes).
- Grade 2: Evidence of red cell destruction and ≥2 g/dL decrease in hemoglobin, no transfusion.
- Grade 3: Transfusion or medical intervention (for example, steroids) indicated.
- Grade 4: Catastrophic consequences (for example, renal failure, hypotension, bronchospasm, emergency splenectomy).
- Grade 5: Death.

Not all adverse reactions are assigned all grades. For example, serum sickness is classified as being of grade 3 or grade 5 only; i.e. it is always either severe or fatal.

The system is less good at classifying subjective reactions. For example, fatigue is graded as follows:

- Grade 1: Mild fatigue over baseline.
- Grade 2: Moderate or causing difficulty performing some activities of daily living.
- Grade 3: Severe fatigue interfering with activities of daily living.
- Grade 4: Disabling.

Attribution categories can be defined as follows:

(i) Definite: The adverse event is clearly related to the investigational agent(s).
(ii) Probable: The adverse event is likely related to the investigational agent(s).
(iii) Possible: The adverse event may be related to the investigational agent(s).
(iv) Unlikely: The adverse event is doubtfully related to the investigational agent(s).
(v) Unrelated: The adverse event is clearly NOT related to the investigational agent(s).

References

[1] Merck Manuals: http://www.merckmanuals.com/professional/clinical_pharmacology/adverse_drug_reactions/adverse_drug_reactions.html.

[2] National Cancer Institute. Common Terminology Criteria for Adverse Events v3.0 (CTCAE). 9 August, 2006. http://ctep.cancer.gov/protocolDevelopment/electronic_applications/docs/ctcaev3.pdf.

FDA PREGNANCY CATEGORIES/ CLASSIFICATION OF TERATOGENICITY

The FDA has established five categories to indicate the potential of a drug to cause birth defects if used during pregnancy. The categories are determined by the reliability of documentation and the risk to benefit ratio. They do not take into account any risks from pharmaceutical agents or their metabolites in breast milk. The pregnancy categories are:

Category A

Adequate and well-controlled studies have failed to demonstrate a risk to the fetus in the first trimester of pregnancy and there is no evidence of risk in later trimesters.

Example drugs or substances: levothyroxine, folic acid, magnesium sulfate, liothyronine.

Category B

Animal reproduction studies have failed to demonstrate a risk to the fetus and there are no adequate and well-controlled studies in pregnant women.

Example drugs: metformin, hydrochlorothiazide, cyclobenzaprine, amoxicillin, pantoprazole.

Category C

Animal reproduction studies have shown an adverse effect on the fetus and there are no adequate and well-controlled studies in humans, but potential benefits may warrant use of the drug in pregnant women despite potential risks.

Example drugs: tramadol, gabapentin, amlodipine, trazodone, prednisone.

Category D

There is positive evidence of human fetal risk based on adverse reaction data from investigational or marketing experience or studies in humans, but potential benefits may warrant use of the drug in pregnant women despite potential risks.

Example drugs: lisinopril, alprazolam, losartan, clonazepam, lorazepam.

Category X

Studies in animals or humans have demonstrated fetal abnormalities and/or there is positive evidence of human fetal risk based on adverse reaction data from investigational or marketing experience, and the risks involved in use of the drug in pregnant women clearly outweigh potential benefits.

Example drugs: atorvastatin, simvastatin, warfarin, methotrexate, finasteride.

Category N

FDA has not classified the drug.

Example drugs: aspirin, oxycodone, hydroxyzine, acetaminophen, diazepam.

Examples of drugs approved since June 30th, 2015 showing various new pregnancy and lactation subsections in their labels [3]:

- Addyi (flibanserin)—indicated for generalized hypoactive sexual desire disorder (HSDD) in premenopausal women.
- Descovy (emtricitabine and tenofovir alafenamide fumarate)—indicated for HIV-1 infection.

- Entresto (sacubitril and valsartan)—indicated for heart failure.
- Harvoni (ledipasvir and sofosbuvir)—indicated for chronic viral hepatitis C infection (HCV).
- Praluent (alirocumab)—indicated for heterozygous familial hypercholesterolemia, or patients with atherosclerotic heart disease who require additional lowering of LDL-cholesterol.
- Vosevi (sofosbuvir, velpatasvir and voxilaprevir)—indicated for chronic hepatitis C infection (HCV).
- Nerlynx (neratinib)—indicated for early stage HER2-overexpressed breast cancer, following adjuvant trastuzumab-based therapy.
- Rituxan Hycela (rituximab and hyaluronidase human)—indicated for follicular lymphoma, diffuse large B-cell lymphoma (DLBCL), and chronic lymphocytic leukemia (CLL).
- Mydayis (amphetamine mixed salts)—indicated for attention deficit hyperactivity disorder (ADHD).
- Kevzara (sarilumab)—indicated for rheumatoid arthritis.
- Radicava (edaravone)—indicated for amyotrophic lateral sclerosis (AML).
- Imfinzi (durvalumab)—indicated for urothelial carcinoma.

On December 3, 2014, the FDA issued a final rule for the labeling of drugs during pregnancy and lactation, titled "Content and Format of Labeling for Human Prescription Drug and Biological Products; Requirements for Pregnancy and Lactation Labeling"; this rule is also informally known as the "Pregnancy and Lactation Labeling Rule (PLLR)".

The rule changes the content and format for drug labeling information and the new labeling requirements include:

- *Elimination the pregnancy letter categories (A, B, C, D and X)*
  - *Text provides specific information in each section to assist with making benefit-risk decisions when medication is needed*
- *Labeling sections are changed*
  - *Old: Pregnancy, Labor and Delivery, Nursing Mothers*
  - *New: Pregnancy (includes L&D), Lactation (includes nursing mothers), Females and Males of Reproductive Potential*
- *Requirement that the label is updated as new information becomes available*

The PLLR changes are effective as of June 30, 2015. Prescription medications and biologics approved after this date will use the new format while older material will have a 3-year phase-in for the new labeling. These changes are not applicable to over-the-counter (OTC) products.

References

[1] Doering PL, Boothby LA, Cheok M. Review of pregnancy labeling of prescription drugs: is the current system adequate to inform of risks? Am J Obstet Gynecol 2002; 187(2): 333–9.

[2] Ramoz LL, Patel-Shori NM. Recent changes in pregnancy and lactation labeling: retirement of risk categories. Pharmacotherapy, 2014; 34(4):389–95. doi: 10.1002/phar.

[3] Drugs.com: http://www.drugs.com/pregnancy-categories.html.

[4] FDA. Pregnancy and Lactation Labeling (Drugs) Final Rule. https://www.fda.gov/downloads/drugs/guidancecomplianceregulatoryinformation/guidances/ucm450636.pdf; https://www.fda.gov/Drugs/DevelopmentApprovalProcess/DevelopmentResources/Labeling/ucm093307.htm.

Clinicians are suggested to be aware of the information contained in the following literature originating from regulatory agencies

[5] FDA/CDER SBIA Chronicles. Drugs in Pregnancy and Lactation: Improved Benefit-Risk Information. January 22, 2015. URL: http://www.fda.gov/downloads/Drugs/DevelopmentApprovalProcess/SmallBusinessAssistance/UCM431132.pdf.

[6] FDA Consumer Articles. Pregnant? Breastfeeding? Better Drug Information Is Coming. Updated: December 17, 2014. URL: https://www.drugs.com/fda-consumer/pregnant-breastfeeding-better-drug-information-is-coming-334.html.

[7] FDA News Release. FDA issues final rule on changes to pregnancy and lactation labeling information for prescription drug and biological products. December 3, 2014. URL: http://www.fda.gov/NewsEvents/Newsroom/PressAnnouncements/ucm425317.htm.

[8] Mospan C. New Prescription Labeling Requirements for the Use of Medications in Pregnancy and Lactation. CE for Pharmacists. Alaska Pharmacists Association. April 15, 2016. URL: http://www.alaskapharmacy.org/files/CE_Activities/0416_State_CE_Lesson.pdf.

[9] Australian classification: https://www.tga.gov.au/australian-categorisation-system-prescribing-medicines-pregnancy#.U038WfmSx8E.

[10] Schaefer C. Drug safety in pregnancy-a particular challenge. Bundesgesundheitsblatt Gesundheitsforschung Gesundheitsschutz. 2018 Aug 9. doi: 10.1007/s00103-018-2798-8. https://www.ncbi.nlm.nih.gov/pubmed/30094469 [Epub ahead of print].

Conclusion

Adverse drug events, including ADRs, side effects, drug-induced diseases, toxicity, pharmacogenetics and immunologic reactions, represent a significant burden to patients, health care systems, and society. It is the goal of *Side Effects of Drugs Annual* to summarize and evaluate important new evidence-based information in order to guide clinicians in the prevention, monitoring, and assessment of adverse drug events in their patients. The work provides not only a summary of this essential new data, but suggestions for how it may be interpreted and implications for practice.

CHAPTER

1

Central Nervous System Stimulants and Drugs That Suppress Appetite

Dushyant Kshatriya, Nicholas T. Bello[1]

Department of Animal Sciences, Nutritional Sciences Graduate Program, School of Environmental and Biological Sciences, Rutgers, The State University of New Jersey, New Brunswick, NJ, United States

[1]**Corresponding author: ntbello@rutgers.edu**

Abbreviations

ADHD attention deficit hyperactive disorder
EMA European Medicines Agency
FDA U.S. Food and Drug Administration
LDX lisdexamfetamine dimesylate
MA methamphetamine
MDMA 3,4-methylenedioxymetamphetamine
MAS mixed amphetamine salts
MPH methylphenidate
RCT randomized controlled trial
TEAE treatment emergent adverse events

AMPHETAMINE AND AMPHETAMINE DERIVATES [SEDA-35, 1; SEDA-36, 1; SEDA-37, 1; SEDA-38, 1; SEDA-39, 1]

An open-label trial assessed the duration of effectiveness of lisdexamfetamine dimesylate (LDX) in adults with ADHD ($n=40$; 18–55 years old; 63.6% male). Study duration was 12 weeks with LDX, which included a 4-week dose optimization (initiating dose was 30mg/day) followed by an 8-week dose maintenance phase. The most common TEAE was insomnia (80%) at multiple time points during the 12 weeks. Sleep disturbances were the TEAE that were reported by one subject that discontinued the study and another subject that refused dose-escalation [1c].

In two head-to-head randomized, double-blind placebo-controlled studies, LDX was compared with osmotic-release oral system methylphenidate (OROS-MPH). The multicenter trials in children and adolescents with ADHD (13–17 years old; 61.9%–68.5% male) included an 8-week flexible dosing of LDX (30–70mg/day; $n=186$), OROS-MPH (18–72mg/day; $n=185$) and placebo ($n=93$) and a 6-week forced dosing of LDX (70mg/day; $n=219$), OROS-MPH (72mg/day; $n=220$), and placebo ($n=110$). For the flexible-dose study, severe TEAEs were 5.4% for LDX, 3.8% for OROS-MPH and 2.2% for placebo, whereas, in the forced-dose study, the TEAEs were 1.4% for LDX, 2.7% for OROS-MPH 0.9%. In the flexible- and forced-doses studies, TEAEs that occurred ≥5% were decreased appetite, decreased weight, and insomnia in both LDX and OROS-MPH. In the flexible-dose study, TEAEs that occurred ≥5% were initial insomnia, and nasopharyngitis with both medications. Dizziness only occurred with LDX and increased heart rate occurred only with OROS-MPH in the flexible dosing study. In the forced dose studies, TEAE that occurred ≥5% was dizziness with both medications, whereas dry mouth and upper abdominal pain only occurred with LDX [2C].

A long-term (104 weeks) open-label dose optimization of LDX (30–70mg/day) Phase-IV trial (clinical trial identifier: NTC01328756) was conducted in children and adolescents with ADHD (6–17 years old; $n=314$; 79.6% male). TEAEs were reported in most subjects (89.8%) and most were considered mild or moderate. A total of 36 serious TEAEs were reported in 28 subjects. Serious TEAE reported in more than one subject were syncope (7 events, 6 subjects), appendicitis (3 events, 3 subjects), and pyelonephritis (2 events, 2 subjects). TEAEs that led to study discontinuation were experienced in 39 subjects, 2 were serious (arrhythmia and suicide attempt). The most common TEAEs that led to discontinuation were decreased appetite ($n=7$), self-reported medication

Side Effects of Drugs Annual, Volume 40
ISSN: 0378-6080
https://doi.org/10.1016/bs.seda.2018.08.014

ineffectiveness ($n=6$; 5 additional subjects were discontinued for investigator-determined medication ineffectiveness), depressed mood ($n=4$), irritability ($n=4$), tic ($n=3$), insomnia ($n=3$), aggression ($n=2$), tachycardia ($n=2$) and weight loss ($n=2$). The most common ($\geq$ 5%) drug-related TEAEs were reduced appetite (49.4%), decreased weight (18.2%), and insomnia (13.1%) [3C].

Two studies examined the pharmacokinetics of a novel once-daily extended-release amphetamine oral suspension formulation (AMP XR-OS). The first study was an open-label, randomized, cross-over study in healthy adults ($n=30$; 20–68 years old; 56.7% male) that examined bioavailability of a single oral dose AMP XR-OS (15 mL liquid) in fed and fasted conditions on two separate trials 7-day apart. In addition, the bioavailability was compared with a single commercially available oral formulation of mixed amphetamine salt extended release (MAS ER; 30 mg) in fed conditions administered in a separate trial. One subject experienced emesis during the AMP XR-OS fasted trial and withdrew from the study. A total of 36 TEAEs occurred in 14 subjects, 9 subjects during the AMP XR-OS fasted trial, 5 during the AMP-OS fed trial, and 5 during the MAS ER trial. The most common TEAEs in the AMP XR-OS fasted trial were anxiety (10%), tachycardia (6.7%), and paresthesia (6.7%), in the AMP XR-OS fed trial decreased appetite (6.9%), musculoskeletal (6.9%), and headache (6.9%), and in the MAS ER nausea (6.9%) and feeling hot (3.4%) [4c]. In the second study, AMO XR-OS was formulated into an orally disintegrating tablet (AMP XR-ODT) and examine in children with ADHD ($n=28$; 6–12 years old; 64.3% male). A total of 34 TEAEs (31 mild, 3 moderate) were observed in 11 subjects. The moderate TEAEs were rhinitis, headache, and syncope. The most common TEAEs that occurred were vomiting (3 events, 2 subjects), abdominal pain (2 events, 2 subjects), dry mouth (1 event, 1 subject) and insomnia (1 event, 1 subject) [5c].

Organs and Systems

Cardiovascular

Cardiovascular adverse events are often reported because of use and misuse of amphetamines and its derivates. An analysis of cases from the National Coronial Information System (NCIS) in Australia revealed 893 methamphetamine-related coroner cases with a full quantitative toxicological assessment. A majority of the cases were male (78.5%) with a mean age of 37.9 years old (15–69 years old) and the most frequent cause of death was accidental drug toxicity (52.8%). All cases had detectable levels of MA (99.9%) and/or amphetamine (75.6%). The most common cardiopathological findings were cardiomegaly (26%), replacement fibrosis (19.8%), coronary artery stenosis (19.0%), and left ventricular hypertrophy (18.9%) [6MC].

A 15-year-old female presented with tachycardia and chest pain 12 h after intentional consumption of 6 extended release amphetamine/dextroamphetamine tablets (20 mg), 15 sertraline tablets (25 mg) and unknown amount of loratadine. Examination was significant for S3 gallop, and soft systolic murmur, absent for hyperreflexia and clonus. Urine drug screen was positive for amphetamines (1958 ng/mL). EKG showed sinus tachycardia, prolonged QTc, and lowered left ventricular ejection fraction of 19%. Takotsubo cardiomyopathy was diagnosed. Patient was started on heparin infusion; patient's condition improved by day 4. Normalization of cardiac function was observed at a 2-month follow-up [7A].

A 24-year-old-male (YOM) arrived at the emergency department presenting with chest pain and anxiety. The symptoms emerged 4 h after intranasal ingestion of methamphetamine and cocaine. He had increased blood pressure and elevated heart rate, with an EKG revealing sinus tachycardia. Two intravenous doses of lorazepam (1 mg) were administered which failed to reduce agitation and anxiety. Oral aspirin (325 mg) and 2 doses of intravenous labetalol (5 mg) 30 minutes apart were administered, which resolved the symptoms after the second dose of labetalol. His condition normalized and he was released from the hospital 7 h later. Resolution of MA- and cocaine-induced cardiac impairments with labetalol were attributed to on the mixed actions of labetalol on alpha- and beta-adrenergic receptors [8A].

A 27 YOM presented with vomiting, fever, dyspnea and chest pain. An EKG revealed sinus tachycardia with a prolonged QTc. His condition worsened after IV fluids, and developed bilateral inspiratory crepitations with a gross elevation of troponin level of >10 000 ng/L. His hypotension was unresponsive to noradrenaline, and a transthoracic ECG revealed severe reduction in left ventricular ejection fraction, with mid-left ventricular wall akinesia and sparing of apical function. Patient was diagnosed with reverse Takotsubo cardiomyopathy and cardiogenic shock. Urinalysis confirmed methamphetamines. Patient responded well to ionotropic support and showed recovery of cardiac function and structure at a 6-month follow-up [9A].

Nervous System

Six-month-old male infant presented at the emergency department in Arizona with inconsolable agitation, flailing of head and extremities, dilated pupils, constant motion, and increased tone. Symptoms were suggestive of a bite by locally prevalent bark scorpion (*Centuroides sculpturatus*); however, antivenin therapy was unsuccessful. Intravenous lorazepam and midazolam therapies were initiated. Urinalysis screen showed methamphetamine, confirmed by gas chromatography. Symptoms gradually resolved after 36 hours [10A].

A 34 YOM experienced limb weakness and eating pathology following cessation of methamphetamine use. Patient had a 3-year history of MA use and his forced MA cessation was a consequence of imprisonment. His initial blood chemistry revealed metabolic acidosis and an infection. Brain CT revealed patchy shadows of low density white matter. Patient deteriorated over the following 4 days and died after failed resuscitation attempt. Autopsy revealed extensive atrophy of the cerebral and cerebellar cortex with soft and discolored white matter. Histological observations revealed apoptotic oligodendrocytes and macrophages in the white matter. MA-induced toxic leukoencephalopathy was determined as the cause of death. Authors suggest MA toxicity resulted in quinones and reactive oxidative species production that resulted in bouts of widespread hypoxia and mitochondrial dysfunction resulting in leukoencephalopathy [11A].

A 30 YOM presented with fever, agitation, tachycardia, elevated blood pressure, and elevated pulse was admitted to an intensive care unit due to a change in mental status resulting from sepsis. Patient had a history of methamphetamine use. The patient had been suffering from MA-associated hallucinations of "ants crawling in his mouth and nose". In attempt to kill the ants he ingested an unknown quantity of a commercial pesticide containing boric acid salts. The combined ingestion of MA and boric acid resulted in pseudosepsis and sustained delirium. Patient was transferred to inpatient psychiatric facility [12A].

Gastrointestinal

A 42 YOM presented with increased abdominal pain and distention. Patient was nauseated and had multiple bouts of watery brown vomitus. Patient had a chronic history of methamphetamine use. Abdominal film radiography revealed multiple dilated small bowel loops. Patient was treated for paralytic ileus using a nasogastric tube for decompression. After 48h, patient resumed full diet without abdominal symptoms. Patient was asymptomatic at 2-month follow-up. The authors suggest that the MA-induced vasoconstriction of the mesenteric artery resulted in the paralytic ileus [13A].

A 44-year-old woman suffered a major cardiac arrest in transit to the hospital. The patient self-reported a 1-day history of headache, neck pain, abdominal pain, vomiting and increasing agitation. Abdominal CT revealed bowel dilation with thrombi in the mesenteric and portal veins, with signs of an ischemic bowel. A laparotomy was performed, followed by a subtotal colectomy. Patient continued to bleed despite ionotropic support and suffered cardiac arrest and died in immediate post recovery period. Blood amphetamine levels were at twice the therapeutic dose at 0.52mg/L which was most probable cause of the vasospasm [14A].

Death

A 40-year-old obese woman presented to emergency department complaining of severe chest pain with radiating lumbar pain, nausea and agitation. Cardiac enzymes were normal, and EKG revealed a sinus bradycardia. Within few hours, she displayed signs of vasomotor instability and lethargy. She had a history of appetite suppressant use. Urinary analysis tested positive for methamphetamine, morphine and benzodiazepines. Ultrasonography revealed abdominal free fluid, mild intestinal distension, and intestinal intramural distension. Patient did not consent to an emergent laparotomy and suffered cardiopulmonary arrest soon after. Serosanguinous intraperitoneal fluid and gangrene at distal ileum were found at autopsy. Toxicological analysis of viscera revealed methamphetamine. Intestinal gangrene and perforation caused by methamphetamine toxicity were determined as the cause of death [15A].

ECSTASY (3,4-METHYLENEDIOXY-N-METHYLAMPHETAMINE, MDMA)

Organs and Systems

Pulmonary

A 19 YOM presented to the medical tent during a concert and dance festival for an altered mental state. He confirmed use of MDMA, marijuana and alcohol. Two doses of intravenous midazolam (2.5mg) were administered 15 minutes apart to resolve persistent and increasing agitation. Following this, the agitation worsened, and he developed acute respiratory distress with tachypnea and 50% oxygen saturation. Bilateral rales were observed in both lungs. Patient displayed MDMA-induced clonus, bruxism, and trismus, which prevented the use of supraglottic air way support. He was administered an intravenous dose of rocuronium (100mg) and was able to be intubated. Patient was brought to a local hospital, given Lasix (40mg), midazolam (4mg), and started on a propofol drip along with endotracheal intubation and mechanical ventilation. A chest X-ray revealed interstitial and alveolar edema, with airspace disease, consistent with acute respiratory distress syndrome (ARDS). Patient was discharged after 5 days and his condition was normalized at the 4-month follow-up. The authors attribute the ARDS and pulmonary dysfunction to MDMA-induced multi-organ toxicity [16A].

Liver

A 25 YOM presented with nausea, abdominal distension and pain for 2 days. He had a 3-month history of MDMA use. Abdominal ultrasonography revealed ascites, hepatomegaly and splenomegaly, with an endoscopy revealing esophageal varices. Copper Doppler ultrasonography revealed thrombi in the right and mid-hepatic veins. Patient was diagnosed with venous hepatic flow obstruction, specifically primary Budd–Chiari Syndrome. The patient received transjugular intrahepatic portosystemic shunt procedure to restore hepatic portal function [17A].

Nervous System

A 23 YOM presented with right eye retrobulbar pain, proptosis, and eyelid bruising. Conditions emerged shortly after physical exertion from dancing. Patient confirmed consumption of unknown quantity of MDMA. Upon examination a subacute hemorrhage was observed in the retrobulbar extraconal region. Symptoms resolved within 5 days without intervention. MDMA-induced vasculature alterations and neuritis were discussed as possible mechanisms for acute ocular hemorrhaging [18A].

Death

A 19-year-old female was found dead at her residence. Patient had a history of insulin-dependent diabetes mellitus, which required several hospitalizations resulting from poor management of blood glucose. The autopsy showed brain swelling and micro hemorrhages. Renal pathology revealed prominent Armanni–Ebstein change in kidney tubules around the corticomedullary junction. Tissue analysis confirmed ketoacidosis with a blood glucose level of 46.5 mmol/L and a betahydroxybutyrate of 13.86 mmol/L. Toxicological analysis showed a low level of MDMA (0.01 mg/L) with acetone. Cause of death was diabetic ketoacidosis exacerbated by MDMA and alcohol [19A].

A 25 YOM was unresponsive and asystolic. Spontaneous circulation was restored after 30 minutes of cardiopulmonary resuscitation. Urinalysis showed 1 mg/mL of MDMA. Patient was known to ingest MDMA by inhalation. CT scan revealed brain edema and diffuse alveolar infiltrates in both lungs. After 12 hours in the hospital, patient developed subcutaneous emphysema of the thorax and neck. A bronchoscopy showed extensive tracheal and bronchial inflammation and necrosis. An analysis of fluid from a bronchoalveolar lavage revealed detectable levels of MDMA. Patient was declared brain dead. Although an autopsy was refused, the authors concluded that MDMA-induced potent vasoconstriction of the airway produced widespread bronchial necrosis [20A].

METHYLPHENIDATE [SEDA-35, 1; SEDA-36, 1; SEDA-37, 1; SEDA-38, 1; SEDA-39, 1]

A meta-analysis of RCTs (4 placebo-controlled crossover studies; $n=113$) examined the effectiveness of methylphenidate for the management of ADHD-like symptoms and/or the core symptoms of autism spectrum disorder (ASD) in children (5–13 years old; 83% male) diagnosed with ASD or pervasive developmental disorder. Each RCTs study length was 4–6 weeks and include at least two doses of MPH. Reduced appetite (risk ratio: 8.28 with 95% CI; 2.57, 26.73; $P<0.001$) was only significant TEAE. Based on the sample size, however, this was considered by the authors as very low-quality evidence for the adverse effects of MPH in ASD population [21M].

A placebo-controlled double-blind RCT examined the effectiveness and safety of an evening-dosed, delayed-release, extended-release of methylphenidate (DR/ER-MPH; 40–80 mg/day) in ADHD children ($n=161$; 6–12 years old). The study had a 2-week screening/washout phase, which had subjects stop taking their ADHD medication, prior to the 3-week study phase. Subjects receiving DR/ER-MPH reported more TEAE than placebo (69.1% compared with 48.8%). The most common TEAEs ($\geq$ 10%) associated with DR/ER-MPH were insomnia (33.3%) and decreased appetite (18.5%). These TEAEs were mild or moderate in severity. The DR/ER-MPH subjects had an increase (1.1 mmHg) in systolic blood pressure from baseline [22C].

A retrospective open-label study examined the effectiveness of extended-release methylphenidate in adults diagnosed with social anxiety disorder (SAD) and ADHD ($n=20$; 26.6 years old; 55% male). The mean study duration was 6.6 (1.94 SD) weeks and mean dose was 27.5 mg/day (10.45 SD). A total 3 subjects discontinued the medication (2 subjects complained headaches, loss of appetite, and anxiety, 1 subject was a MPH non-responder). The most commonly reported TEAE was transient loss of appetite (8 out of 17), 1 subject had severe loss of appetite. Other reported TEAE were insomnia ($n=2$), transient dryness of mouth ($n=7$), and transient headaches ($n=2$) [23c].

Organs and Systems

Nervous System

An 8 YOM presented with combative behavior and agitation, along with excessive thirst. He was prescribed lisdexamfetamine dimesylate (20 mg/day) for the management of ADHD, but accidently ingested his older brother's daily dose of long-acting methylphenidate (36 mg). Onset of symptoms began 3 hour after dosing. It was noted by his parents that he likely consumed 3 or 4 gal of water. Patient was administered lorazepam, but later experienced urinary incontinence, multiple episodes of emesis, altered mental state, and hypothermia. He also had brief bouts of tonic–clonic seizure activity. Further clinical testing revealed he suffered from hyponatremia. Antiepileptic dosing regimen (fosphenytoin;

20 mg/kg), antibiotics, and 3% NaCl were administered. Patient stabilized 36 hours after accidental ingestion and discharged 24 hours later. The clinical resemblance to syndrome of inappropriate diuretic hormone (SIADH) secretion is noted [24A].

Hair

Two siblings with ADHD (10 and 12-year-olds) began their second week of treatment with osmotic release oral system methylphenidate (OROS-MPH; 18 and 27 mg/day, respectively). Hair loss without inflammation was noted and they were diagnosed with MPH-induced anagen effluvium type non-scarring alopecia areta. Hair growth gradual returned in both children after cessation of OROS-MPH. Hair loss symptoms resolved when MPH was resumed at a higher dose (27 and 54 mg/day respectively) a month later. The mechanisms for this transient alopecia with OROS-MPH remain unknown [25A].

Skin

A search of the Food and Drug Administration Adverse Event Reporting System (FAERS) revealed 51 cases of chemical leukoderma associated with methylphenidate transdermal system (MTS). Most of the cases chemical leukoderma were localized to the application site (84.3%). The search was from data collected in the FDA repository over an 8-year time span (4/2006–12/2014). None of the cases reported resolution of the depigmentation following cessation of MTS [26C].

Drug–Drug Interaction

A 16 YOM presented with behavioral problems and conduct disorder. He had a 5-year history of extended-release methylphenidate (54 mg/day) for ADHD. He recently developed more defiant behavior and was prescribed aripiprazole (2.5 mg/day). After the initial dose of aripiprazole, 12 hours later he had a 4-hour bout of hiccups. Hiccups were only observed when aripiprazole was taken on same day as MPH. Genotyping for *CYP2D6* revealed this patient to be a normal/extensive metabolizer [27A].

A 15-year-old female diagnosed with adjustment disorder with depressive symptoms and trichotillomania. She had been on osmotic release oral system methylphenidate (OROS-MPH; 54 mg/day) for 2 years for ADHD. She was started on sertraline (uptitrated to 100 mg/day) but haloperidol (0.5 mg/day) was added because of her persistence trichotillomania. This regimen resolved her depression and trichotillomania, but her ADHD symptoms escalated. OROS-MPH was switched to modified release MPH (uptitrated to 50 mg/day). After 3 days of the MPH uptitration, she developed spontaneous galactorrhea. Modified release MPH and haloperidol were discontinued, and galactorrhea ceased after 3 days, and the patient was maintained on sertraline. Prolactin levels normalized after 2 weeks. The formulation switch to MPH with haloperidol was suspected to increase dopamine levels to promote hyperprolactinemia galactorrhea [28A].

METHYLXANTHINES

Caffeine [SEDA-33, 11; SEDA-34, 6; SEDA-36, 1; SEDA-37, 1; SEDA-38, 1; SEDA-39, 1]

Organs and Systems

NERVOUS SYSTEM

A 45-year-old female experienced sudden vision acuity and aniscoria in the right eye. She reported use of a facial jelly containing coffee extract. Use of facial jelly was discontinued, and symptoms improved 24 hours after presentation. Methylxanthine-induced increased sympathetic tone was suspected to cause the aniscoria [29A].

Pure caffeine is readily available commercially and is often used a stimulant. Three cases described toxicity from pure caffeine. A 36 YOM experienced severe vomiting, arterial hypotension, sinus tachycardia, elevated levels of serum nitrogen metabolites, and muscle tremors after consuming approximately 30 g of pure caffeine. His plasma caffeine level was in the range of fatal toxicity (80.16 μg/mL). Hemodialysis was started which reduced plasma caffeine levels to 3.82 μg/mL, and soon resolved symptoms. Another case reported a 27 YOM consumed approximately 45 g pure caffeine. He presented with severe nausea, vomiting, muscle tremors, and sinus tachycardia, and fatally toxic levels of plasma caffeine (140.64 μg/mL). He died from cardiac arrest 10 minutes after arriving at the hospital. A 20-year-old female was found dead with apparent caffeine poisoning. Autopsy revealed signs of acute cardiopulmonary failure and her plasma caffeine level exceeded fatal toxic level (613 μg/mL) [30A].

DRUG–DRUG INTERACTION

A 54 YOM developed symptoms of irritability, lack of need from sleep, overeating, depressed mood, and suicidal ideation. He was diagnosed with bipolar disorder and was receiving lithium for a month (uptitrated to 1400 mg/day). His blood level of lithium was 0.38 mEq/L and his bipolar disorder was difficult to manage. Notably, he consumed 13–20 cups of coffee a day, with an approximate caffeine intake of 1300–2000 mg/day. His coffee was reduced to 10 cups a day which led to his manageable psychiatric symptoms. Elevated blood lithium concentration (1.07 mEq/L) was noted after the coffee intake was substantially reduced. The potential interaction of lithium blood levels and caffeine intake was discussed, but the mechanism is not known [31A].

Selective Norepinephrine Reuptake Inhibitors Atomoxetine [SEDA-36, 1; SEDA-37, 1; SEDA-38, 1; SEDA-39, 1]

Electrocardiogram parameters were measured in subjects diagnosed with ADHD ($n=41$; 63.4%) that initiated atomoxetine. EKG and plasma levels of atomoxetine were measured at baseline (24h after dose initiation) and dose stabilization (≥2 weeks treatment; 20–120mg/day). There was a main finding only in female subjects, such that atomoxetine significantly prolonged QTcB ($P=0.039$). In females there was also a significant positive correlation between atomoxetine dosage and QTcB interval ($r=0.632$, $P=0.012$) [32c].

Organ and Systems

NERVOUS SYSTEM

An 8 YOM with ADHD was prescribed initially atomoxetine (12.5mg/day). He developed tic symptoms such as abdominal contractions and vocal tic-like hiccups when atomoxetine was uptitrated to 25mg/day in the second week. Symptoms resolved within 10 days following the cessation of atomoxetine [33A].

DRUG–DRUG INTERACTIONS

A 55 YOM had initiated atomoxetine treatment to improve his akinetic mutism following recovery from a hemorrhagic stroke. Patient was also on a regimen of methylphenidate (uptitrated to 40mg/day) and levodopa/benserazide (uptitrated to 450/112.5mg/day) post-stroke. After 55 days, the atomoxetine dose was uptitrated from 40 to 60mg/day. Within 5 hours of dose escalation of atomoxetine, the patient developed diaphoresis, fever, tachycardia, hypertension, and recurrent bouts of diarrhea. He also developed inducible clonus and tremors in the upper extremities. Symptoms resolved after 4 days following cessation of all three medications [34A].

VIGILANCE PROMOTING DRUGS

Modafinil and Armodafinil [SEDA-36, 1; SEDA-37, 1; SEDA-38, 1; SEDA-39, 1]

Prader–Willi Syndrome (PWS) is a genetic disorder associated with obesity. PWS patients often report obstructive sleep apnea and consequently day-time sleepiness. Sleep disorders were characterized in a population of PWS patients ($n=60$; 43% male; 16–54 years old). Modafinil was taken by 16 PWS patients ($n=8$ at 100mg/day; $n=8$ at 200mg/day). Panic attacks ($n=1$) and depressive mood ($n=1$) were experienced after initiation of treatment, whereas 4 patients experienced anxiety aggressiveness, and delirium with longer treatment (0.1- to 1.5-year period). In the remaining PWS patients receiving modafinil ($n=10$) the average length of treatment were 4.9±2.8 years [35c].

Organs and Systems

DRUG–DRUG INTERACTIONS

A 58-year-old female was resistant to anesthesia induction with intravenous propofol (incremental dosing to 6mg), for an ophthalmologic surgery. She had a 5-year history of 200mg daily modafinil for narcolepsy, and last dose was taken 4 hours before surgery. General anesthesia was induced by 0.3% sevoflurane by face mask, after which the anesthesia and procedure was uneventful. The mechanism by which modafinil caused propofol resistance is unclear [36A].

PITOLISANT

Pitolisant is an EMA-approved (in 2016) medication for the treatment of narcolepsy. A double-blinded, placebo-controlled RCT for safety and efficacy was conducted in adult (range; 18–66 years old) subjects with narcolepsy with cataplexy (clinical identifier: NTC01800045). Pitolisant (5–40mg/day; $n=54$; 48% male) and placebo ($n=51$; 53% male) were administered over a 7-week period with 1-week withdrawal. The TEAs were mild-to-moderate in severity and were at a 28% frequency for pitolisant and 12% for placebo ($P=0.048$). A severe case ($n=1$) of nausea was observed in the pitolisant group, but the symptoms resolved following cessation of treatment [37C].

DRUGS THAT SUPPRESS APPETITE [SEDA-36, 1; SEDA-37, 1; SEDA-38, 1; SEDA-39, 1]

A 12-week, double-blinded RCT pilot study was performed to assess lorcaserin (10mg BID) alone or in combination with phentermine (15mg QD or 15mg BID) for weight management in overweight or obese adults ($n=235$; 10.3%–21.5% male). Approximately 64% of all subjects experienced TEAE. The TEAEs that had a frequency >5% were headache, fatigue, insomnia, dry mouth, diarrhea, cough, constipation, dizziness, and nausea. Headache led to discontinuation in lorcaserin only group ($n=2$), lorcaserin+phentermine BID group ($n=2$), whereas dizziness led to discontinuation in lorcaserin+phentermine BID group ($n=2$). One severe TEAE occurred due to atrial fibrillation in the lorcaserin+phentermine BID group [38C].

Phentermine [SEDA-36, 1; SEDA-37, 1; SEDA-38, 1; SEDA-39, 1]

Organs and Systems

RENAL

A 43-year-old female reported a 5-week history of lethargy, nausea, left abdominal pain, vomiting, progressive polydipsia and polyuria. Onset of symptoms was coincident with her cessation of 9-month treatment of phentermine for weight loss. Her history included long-term use of citalopram for depression, bilateral mastectomy for Paget's disease, and a BMI of 32.9 kg/m^2. High serum creatinine of 629 μmol/L and elevated protein creatinine ratio were recorded. Renal biopsy revealed lymphohistiocytic inflammatory cells into the tubulointerstitium. Tubulitis was also noted and she was diagnosed with phentermine-induced acute interstitial nephritis. Patient was treated with 3-day course of methylprednisone (500 mg) and discharged with pantoprazole (20 mg/day) and sodium bicarbonate (840 mg BID). At her 1-month follow-up, she had a recovery of renal measurements. The mechanism for phentermine-induced acute interstitial nephritis is unknown [39A].

Naltrexone Sustained-Release/Bupropion Sustained-Release [SEDA-37, 1]

Organs and Systems

SKIN

A 55-year-old female (BMI 47.7 kg/m^2) presented with a widespread rash 3 weeks after initiating naltrexone/bupropion (180 mg in the morning and 90 mg in the evening). She was also taking esomeprazole (40 mg/day), cholecalciferol, and cyanocobalamin. She had a history of obesity and gastroesophageal reflux disease, and 2 weeks prior to starting naltrexone/bupropion, she had started phentermine/topiramate, but discontinued due to difficulty sleeping and body ache. Topical hydrocortisone and antihistamines failed to control the rash, and a skin biopsy ruled out Stevens–Johnson syndrome. The rash included erythema, minimal sloughing and formation of pustules. Patient was hypersensitive to neomycin and was treated with mupirocin. The mechanism by which bupropion caused the generalized psoriasis is unclear [40A].

Drug–Drug Interactions

A 42-year female (BMI 30.2 kg/m^2) with anxiety was receiving spine surgery for a large C5–C6 disk herniation and secondary cervical radiculopathy. In addition to alprazolam, the patient was taking naltrexone/bupropion for 6 months. In order to achieve an optimal level of surgical anesthesia, she received acetaminophen (1000 mg), gabapentin (600 mg) preoperatively and an intravenous ketamine (500 mg), fentanyl (500 μg), and midazolam (2 g) intraoperatively. Anesthesia was maintained with a continuous infusion of ketamine (10 μg/kg/min), propofol (125 μg/kg/min), and remifentanil (0.2 μg/kg/min). Post-operative pain management was maintained hydromorphone patient-controlled pump (0.3 mg/8 minutes), acetaminophen (1000 mg/6 hours), and diazepam (5 mg as needed). The use of an opioid-tolerant protocol is warranted in patients receiving naltrexone/bupropion, since the extended-release opioid antagonist, naltrexone, impairs pain management [41A].

PARASYMPATHOMIMETICS [SEDA-34, 9; SEDA-36, 1; SEDA-37, 1; SEDA-38, 1]

Donepezil [SEDA-36, 1; SEDA-37, 1; SEDA-38, 1; SEDA-39, 1]

Organ and Systems

CARDIOVASCULAR

An 84 YOM with Alzheimer's disease arrived at the ED who was consuming 7 times (35 mg) his daily dose of donepezil (5 mg/day). He arrived 6–7 h after the accidental ingestion and presented with nausea, vomiting, fatigue, excessive sweating. His EKG had showed premature atrial contractions and prolonged QTc at 502 ms. He was observed for 48 hours and intravenously administered atropine (0.5 mg) for bradycardia. Symptoms resolved when donepezil was withheld, and fluid support was initiated. This was a case of donepezil toxicity related cardiac dysrhythmias in elderly patients because of excessive cholinergic activation [42A].

MUSCULOSKELETAL

An 87 YOM with mild Alzheimer's disease complained of recurrent episodes of partial seizures, agitation, sweating, restlessness, and a feeling of leaning to the right. His medications included losartan (50 mg/day), memantine (20 mg/day), donepezil (10 mg/day), citalopram (10 mg/day), omeprazole (40 mg BID) lorazepam (1 mg/day), lamotrigine (25 mg/day) and simvastatin (10 mg/day). Due to the suspected serotonin syndrome citalopram and lamotrigine were discontinued without improvement of symptoms. Further examination revealed his axial posture to abnormally flex to the right. Donepezil was discontinued, omeprazole was reduced to 40 mg, and symptoms resolved after 3 months. This case of Pisa syndrome is thought to be due to pharmacokinetic interactions of common medications with acetylcholine esterase inhibitors such as donepezil [43A].

Memantine [SEDA-34, 1; SEDA-39, 1]

Organ and Systems

SKIN

An 89 YOM was diagnosed with Alzheimer's disease and was treated with memantine and donepezil. Two months after treatment initiation he reported erythematous eruptions on trunk and extremities. A laboratory lymphocyte stimulation test revealed that his mononuclear cells were affected by memantine. Skin symptoms resolved after topical application of betamethasone butyrate propionate ointment and stopping memantine. Donepezil was continued. This was a rare case of a cutaneous adverse reaction attributable to memantine [44A].

MUSCULOSKELETAL

A 76-year-old female accidentally consumed a double dose of memantine extended release (21 mg/day) for 40 days after switching from a twice daily dose of immediate release memantine (10 mg). She developed a movement disorder characterized by movement induced chorea or jerky involuntary movements and dystonic posture. The chorea was thought to be due a rare dopaminergic effect of memantine overdose. Symptoms stopped after memantine was discontinued [45A].

Rivastigmine [SEDA-36, 1; SEDA-37, 1; SEDA-38, 1; SEDA-39, 1]

Organ and Systems

CARDIOVASCULAR

A 91-year-old female with Alzheimer's disease presented with vomiting, and somnolent. She was found to have erroneously applied 10 transdermal rivastigmine patches (18 mg each), and later experienced excessive constriction of pupils, sweating, increased bowel sound, hypertension and sinus tachycardia at 122 beats/min. Acute cholinergic syndrome due to rivastigmine poisoning was concluded. All patches removed; however, blood pressure and heart rate peaked at 206/89 mmHg and 177 beats/min, respectively, a few hours after patch removal. Symptoms resolved spontaneously 17 hours after presentation [46A].

MUSCULOSKELETAL

A 57-year-old female with early onset Alzheimer's disease was taking rivastigmine (4.5 mg BID) for 19 months and experienced right axial deviation or Pisa syndrome. Rivastigmine was downtitrated to a maintenance dose of once daily 4.5 mg to resolve the axial deviation. Underlying mechanism of acetylcholine esterase induced Pisa syndrome is obscure, but a cholinergic–dopaminergic imbalance may play a role [47A].

References

[1] Thornton SL, Pchelnikova JL, Cantrell FL. Characteristics of pediatric exposures to antidementia drugs reported to a poison control system. J Pediatr. 2016;172:147–50 [c].

[2] Newcorn JH, Nagy P, Childress AC, et al. Randomized, double-blind, placebo-controlled acute comparator trials of lisdexamfetamine and extended-release methylphenidate in adolescents with attention-deficit/hyperactivity disorder. CNS Drugs. 2017;31:999–1014 [C].

[3] Coghill DR, Banaschewski T, Nagy P, et al. Long-term safety and efficacy of lisdexamfetamine dimesylate in children and adolescents with adhd: a Phase IV, 2-year, open-label study in Europe. CNS Drugs. 2017;31:625–38 [C].

[4] Sikes C, Stark JG, McMahen R, et al. Pharmacokinetics of a new amphetamine extended-release oral liquid suspension under fasted and fed conditions in healthy adults: a randomized, open-label, single-dose, 3-treatment study. Clin Ther. 2017;39:2389–98 [c].

[5] Stark JG, Engelking D, McMahen R, et al. Pharmacokinetics of a novel amphetamine extended-release orally disintegrating tablet in children with attention-deficit/hyperactivity disorder. J Child Adolesc Psychopharmacol. 2017;27:216–22 [c].

[6] Darke S, Duflou J, Kaye S. Prevalence and nature of cardiovascular disease in methamphetamine-related death: a national study. Drug Alcohol Depend. 2017;179:174–9 [MC].

[7] Toce MS, Farias M, Bruccoleri R, et al. A case report of reversible Takotsubo cardiomyopathy after amphetamine/dextroamphetamine ingestion in a 15-year-old adolescent girl. J Pediatr. 2017;182:385–388e3 [A].

[8] Richards JR, Lange RA, Arnold TC, et al. Dual cocaine and methamphetamine cardiovascular toxicity: rapid resolution with labetalol. Am J Emerg Med. 2017;35:519e1–4 [A].

[9] Chehab O, Ioannou A, Sawhney A, et al. Reverse Takotsubo cardiomyopathy and cardiogenic shock associated with methamphetamine consumption. J Emerg Med. 2017;53:e81–3 [A].

[10] Pariury HE, Steiniger AM, Lowe MC. A stinging suspicion something was just not right: methamphetamine toxicity in infant mimics scorpion envenomation. Pediatr Emerg Care. 2017;33: e124–5 [A].

[11] Mu J, Li M, Guo Y, et al. Methamphetamine-induced toxic leukoencephalopathy: clinical, radiological and autopsy findings. Forensic Sci Med Pathol. 2017;13:362–6 [A].

[12] Johnson K, Stollings JL, Ely EW. Breaking bad delirium: methamphetamine and boric acid toxicity with hallucinations and pseudosepsis. South Med J. 2017;110:138–41 [A].

[13] McKelvie MA, Gercek Y. Paralytic ileus secondary to methamphetamine abuse: a rare case. Case Rep Surg. 2017;2017:9762803 [A].

[14] Green PA, Battersby C, Heath RM, et al. A fatal case of amphetamine induced ischaemic colitis. Ann R Coll Surg Engl. 2017;99:e200–1 [A].

[15] Attaran H. Fatal small intestinal ischemia due to methamphetamine intoxication: report of a case with autopsy results. Acta Med Iran. 2017;55:344–7 [A].

[16] Haaland A, Warman E, Pushkar I, et al. Isolated non-cardiogenic pulmonary edema—a rare complication of MDMA toxicity. Am J Emerg Med. 2017;35:1385e3–6 [A].

[17] Tas A, Kara B, Yilmaz C, et al. Fulminant Budd-Chiari syndrome due to ecstasy. Clin Res Hepatol Gastroenterol. 2017;41: e12–3 [A].

[18] Chervenkoff JV, Rajak SN, Selva D, et al. A case of a spontaneous self-resolving retrobulbar hemorrhage following 3,4-methylenedioxy-methamphetamine use. Ophthalmic Plast Reconstr Surg. 2017;33:e100–1 [A].

[19] Gilbert JD, Byard RW. Fatal diabetic ketoacidosis—a potential complication of MDMA (Ecstasy) use. J Forensic Sci. 2018;63:939–41 [A].

[20] Van den Kerckhove E, Roosens L, Siozopoulou V, et al. Airway necrosis and barotrauma after ecstasy inhalation. Am J Respir Crit Care Med. 2017;196:105–6 [A].
[21] Sturman N, Deckx L, van Driel ML. Methylphenidate for children and adolescents with autism spectrum disorder. Cochrane Database Syst Rev. 2017;11:CD011144 [M].
[22] Pliszka SR, Wilens TE, Bostrom S, et al. Efficacy and safety of HLD200, delayed-release and extended-release methylphenidate, in children with attention-deficit/hyperactivity disorder. J Child Adolesc Psychopharmacol. 2017;27:474–82 [C].
[23] Koyuncu A, Celebi F, Ertekin E, et al. Extended-release methylphenidate monotherapy in patients with comorbid social anxiety disorder and adult attention-deficit/hyperactivity disorder: retrospective case series. Ther Adv Psychopharmacol. 2017;7:241–7 [c].
[24] Patel V, Krishna AS, Lefevre C, et al. Methylphenidate overdose causing secondary polydipsia and severe hyponatremia in an 8-year-old boy. Pediatr Emerg Care. 2017;33:e55–7 [A].
[25] Ardic UA, Ercan ES. Resolution of methylphenidate osmotic release oral system-induced hair loss in two siblings after dose escalation. Pediatr Int. 2017;59:1217–8 [A].
[26] Cheng C, La Grenade L, Diak IL, et al. Chemical leukoderma associated with methylphenidate transdermal system: data from the US food and drug administration adverse event reporting system. J Pediatr. 2017;180:241–6 [C].
[27] Kutuk MO, Guler G, Tufan AE, et al. Hiccup due to aripiprazole plus methylphenidate treatment in an adolescent with attention deficit and hyperactivity disorder and conduct disorder: a case report. Clin Psychopharmacol Neurosci. 2017;15:410–2 [A].
[28] Ekinci O, Gunes S, Ekinci N. Galactorrhea probably related with switching from osmotic-release oral system methylphenidate (MPH) to modified-release MPH: an adolescent case. Clin Psychopharmacol Neurosci. 2017;15:282–4 [A].
[29] Martins T, Costa A, Martins T. Anisocoria after use of dermatological product. Einstein (Sao Paulo). 2017;15:514–5 [A].
[30] Magdalan J, Zawadzki M, Skowronek R, et al. Nonfatal and fatal intoxications with pure caffeine—report of three different cases. Forensic Sci Med Pathol. 2017;13:355–8 [A].
[31] Kunitake Y, Mizoguchi Y, Sogawa R, et al. Effect of excessive coffee consumption on the clinical course of a patient with bipolar disorder: a case report and literature review. Clin Neuropharmacol. 2017;40:160–2 [A].
[32] Suzuki Y, Tajiri M, Sugimoto A, et al. Sex differences in the effect of atomoxetine on the QT interval in adult patients with attention-deficit hyperactivity disorder. J Clin Psychopharmacol. 2017;37:27–31 [c].
[33] Yang R, Li R, Gao W, et al. Tic symptoms induced by atomoxetine in treatment of ADHD: a case report and literature review. J Dev Behav Pediatr. 2017;38:151–4 [A].
[34] Jeon DG, Kim YW, Kim NY, et al. Serotonin syndrome following combined administration of dopaminergic and noradrenergic agents in a patient with akinetic mutism after frontal intracerebral hemorrhage: a case report. Clin Neuropharmacol. 2017;40:180–2 [A].
[35] Ghergan A, Coupaye M, Leu-Semenescu S, et al. Prevalence and phenotype of sleep disorders in 60 adults with Prader-Willi syndrome. Sleep. 2017;40:1–10 [c].
[36] Harwood TN. Resistance to propofol induction in a patient taking modafinil: a case report. A A Case Rep. 2017;9:322–3 [A].
[37] Szakacs Z, Dauvilliers Y, Mikhaylov V, et al. Safety and efficacy of pitolisant on cataplexy in patients with narcolepsy: a randomised, double-blind, placebo-controlled trial. Lancet Neurol. 2017;16:200–7 [C].
[38] Smith SR, Garvey WT, Greenway FL, et al. Coadministration of lorcaserin and phentermine for weight management: a 12-week, randomized, pilot safety study. Obesity (Silver Spring). 2017;25:857–65 [C].
[39] Shao EX, Wilson GJ, Ranganathan D. Phentermine induced acute interstitial nephritis. BMJ Case Rep. 2017;2017:1–3 [A].
[40] Singh PA, Cassel KP, Moscati RM, et al. Acute generalized erythrodermic pustular psoriasis associated with bupropion/naltrexone (Contrave((R))). J Emerg Med. 2017;52:e111–3 [A].
[41] Ninh A, Kim S, Goldberg A. Perioperative pain management of a patient taking naltrexone HCL/bupropion HCL (Contrave): a case report. A A Case Rep. 2017;9:224–6 [A].
[42] Pourmand A, Shay C, Redha W, et al. Cholinergic symptoms and QTc prolongation following donepezil overdose. Am J Emerg Med. 2017;35:1386 e1–e3 [A].
[43] Pollock D, Cunningham E, McGuinness B, et al. Pisa syndrome due to donepezil: pharmacokinetic interactions to blame? Age Ageing. 2017;46:529–30 [A].
[44] Saito R, Sawada Y, Yamaguchi T, et al. Drug eruption caused by memantine. Ann Allergy Asthma Immunol. 2017;119:89–90 [A].
[45] Borges LG, Bonakdarpour B. Memantine-induced chorea and dystonia. Pract Neurol. 2017;17:133–4 [A].
[46] Suzuki Y, Kamijo Y, Yoshizawa T, et al. Acute cholinergic syndrome in a patient with mild Alzheimer's type dementia who had applied a large number of rivastigmine transdermal patches on her body. Clin Toxicol (Phila). 2017;55:1008–10 [A].
[47] Hsu CW, Lee Y, Lee CY, et al. Reversible pisa syndrome induced by rivastigmine in a patient with early-onset Alzheimer disease. Clin Neuropharmacol. 2017;40:147–8 [A].

CHAPTER

2

Antidepressants

Jonathan Smithson, Philip B. Mitchell[1]

School of Psychiatry, University of New South Wales, and Black Dog Institute, Sydney, NSW, Australia

[1]**Corresponding author: phil.mitchell@unsw.edu.au**

GENERAL

Hematological

Upper Gastrointestinal Bleeding

Laporte et al. conducted a meta-analysis of 42 observational studies to quantify the risk of severe bleeding, irrespective of site, associated with SSRIs [1R]. The investigators included both cohort and case–control reports. They found a significant association between SSRI use and the risk of bleeding [OR 1.41 (95% CI 1.27–1.57), random effect model, $P<0.0001$]. The association was significant in both the 31 case–control studies and the 11 cohort studies. The increased risk was mainly related to GI bleeding (OR 1.55 (95% CI 1.32–1.81)). The association remained significant but was weaker when considering only intracerebral hemorrhage (OR 1.16 (95% CI 1.01–1.33)). Of note, the authors emphasized that despite the significantly increased risk, the absolute risk of abnormal bleeding is low, with the crude incidence of upper GI bleeding being one event per ~8000 SSRI prescriptions.

Postpartum Hemorrhage

In a study that sought to clarify the influence of serotonergic medication on the development of postpartum hemorrhage (PPH), Heller et al. conducted a matched observational cohort study which included 578 consecutively identified pregnant women who used serotonergic medications, and 50 who used other psychotropic medications (such as lithium and antipsychotics). The investigators compared the incidence of PPH in each group with that in 641 364 pregnant women not using psychiatric medication [2c]. The serotonergic medication group included those taking any serotonergic antidepressant (SSRIs, TCAs or SNRIs) with or without other psychotropics. Subjects were matched on nine factors: parity, maternal age, ethnicity, socioeconomic status, macrosomia, gestational duration, history of postpartum hemorrhage, labour induction and hypertensive disorder. The investigators found that PPH occurred in 9.7% of the women using serotonergic medication compared with 6.6% of matched controls (adjusted odds ratio (aOR) 1.5 (95% CI 1.1–2.1)). In women using other psychopharmacological medication, the incidence of PPH before matching was 12.1% compared to 4.4% (aOR 3.3 (95% CI 1.1–9.8)). A higher prevalence of PPH among women using serotonergic medication in the third trimester of pregnancy was found. The authors concluded that women taking serotonergic medication have an increased risk of PPH but that this higher risk is also seen in those taking other psychopharmacological medications. They suggested that the higher risk may not be solely explained by serotonergic effects but also by other mechanisms including the underlying psychiatric disorder.

Venous Thromboembolism

The literature examining risk of venous thromboembolism (VTE) with antidepressants has been limited, with few studies and small samples. Parkin et al., using the Million Women Study cohort, sought to re-examine the association using a large population [3C]. In an attempt to isolate the effects of specific factors, the authors examined four groups: (i) women on no treatment, (ii) women being treated for depression without psychotropic drugs, (iii) women treated with antidepressant drugs, and (iv) women treated with other psychotropic drugs. The authors identified 734 092 women with depression and regular antidepressant use, 3922 of whom were later admitted to hospital and/or died as a result of VTE. Women who reported use of antidepressants had neither a significantly higher risk of VTE than those who reported either depression nor use of psychotropic drugs (aHR: 1.39; 95% CI 1.23–1.56; $P<0.0001$). Those who

Side Effects of Drugs Annual, Volume 40
ISSN: 0378-6080
https://doi.org/10.1016/bs.seda.2018.08.009

reported being treated for depression or anxiety without antidepressants or other psychotropic drugs did not have an increased risk of VTE, suggesting that antidepressants may play a role in the development of VTE in older women. The type of antidepressant did not seem to affect the magnitude of the increased risk of VTE, which was similar for users of tricyclic antidepressants (aHR: 1.32), SSRIs (HR: 1.40), and other antidepressants (HR: 1.61). The authors concluded that the main influence on development of VTE is the use of an antidepressant rather than the depression itself, consistent with other studies with prospectively collected data.

Nervous System

Stroke

The literature on the relationship between antidepressants and stroke has been inconsistent. Although SSRIs have been linked to an increased risk of intracranial hemorrhage (ICH), the increased risk is generally seen with ischemic, rather than hemorrhagic stroke. Using a nationwide database, examining patients with newly-diagnosed bipolar disorders, Wu et al. [6c] undertook a nested case–control study to investigating the association of psychotropic agents, including antidepressants with stroke—both hemorrhagic and ischemic. The reference group of non-users was defined by having no exposure to any psychotropic drug within the 6 months prior to the index date of their stroke. A total of 1232 patients with newly diagnosed bipolar disorder suffering a new stroke were matched to 5314 controls. There was a significant association between any psychotropic medication increased risk of stroke (AOR=1.82; 95% CI: 1.56–2.13). With respect to antidepressants alone, there was a significantly elevated risk of any stroke (AOR=1.44; 95% CI=1.16–1.79) and ischemic (AOR=1.60; 95% CI=1.25–2.05) but not hemorrhagic stroke. A significant association was also found for antipsychotics (AOR=1.98; 95% CI=1.53–2.56), and mood stabilizers (AOR=1.89; 95% CI=1.22–2.93). The combined use of antipsychotics, antidepressants, and mood stabilizers was found to be associated with the highest risk of stroke (AOR=2.62; 95% CI=1.98–3.45). With respect of individual agents, only TCAs were associated with stroke risk. No SSRIs were associated with increased risk of stroke. Considering, other antidepressants, only bupropion (AOR=1.76; 95% CI=1.16–2.69) and mirtazapine (AOR=1.41; 95% CI=1.04–1.92) increased risk of stroke. Renoux et al. examined the association of SSRIs—focusing on the relative serotonin transporter affinity—and ICH in a population-based cohort study of 1 363 990 new adult users of antidepressants using a nested case–control design, identifying 3036 incident cases of ICH identified during follow-up [4C]. They found that current SSRI use was associated with a small increased risk for ICH (RR, 1.17; 95% CI, 1.02–1.35) relative to TCAs. The risk was increased by 25% for those SSRIs with the highest serotonin transporter affinity (RR, 1.25; 95% CI, 1.01–1.54) and was highest during the first 30 days of use (RR, 1.68; 95% CI, 0.90–3.12). In keeping with previous findings, the concomitant use of anticoagulants was associated with a substantially increased risk (RR, 1.73; 95% CI, 0.89–3.39). Using a national database, Chan et al. conducted an observational cohort study [5C], finding a greater probability of first onset stroke in those exposed to SSRIs ($P<0.001$) which persisted for 3 years but reducing thereafter. The cumulative incidence ratios of both ischemic and hemorrhagic stroke were higher during the first 3 years for exposed subjects. Analysis showed that younger exposed subjects were more likely to experience stroke. There was a greater increase in risk for ischemic than hemorrhagic stroke.

Seizures

Using a national health insurance database, Wu et al. identified patients visiting the emergency department or hospitalized with a new-onset seizure after receiving antidepressants for depression [6c]. From a total of 10 002 patients included over a 10-year period, the authors found that antidepressant exposure was associated with increased seizure risk (OR=1.48, 95% CI, 1.33–1.64) with the greatest risk being for bupropion (OR=2.23, 95% CI, 1.58–3.16). Additionally, SSRIs (OR=1.76, 95% CI, 1.55–2.00), serotonin and norepinephrine reuptake inhibitors (SNRIs) (OR=1.40, 95% CI, 1.10–1.78), and mirtazapine (OR=1.38, 95% CI, 1.08–1.77) all showed clear dose–response effects. The seizure risk was highest in those aged between 10 and 24 years and in patients with major depression.

Jitteriness

'Jitteriness' is a term used to describe the anxiety, agitation and restlessness associated with antidepressants. Sinha et al. conducted a prospective study of 209 patients with any anxiety or depressive disorder, who were prescribed mirtazapine, sertraline, desvenlafaxine, escitalopram or fluoxetine [7c]. Subjects were assessed for the predefined categories of "Jitteriness Syndrome" (JS), the operational definition of which comprised four categories (i) subjective anxiety, (ii) subjective anxiety with observable motor restlessness, (iii) visible anxiety or depression with clinically significant agitation, and (iv) akathisia. The investigators found an incidence of JS during the 6-week study of 27.7%; however, only 6.7% of subjects experienced this within the first 2 weeks of treatment. Rates were similar in anxiety and depressive disorders. Mirtazapine was associated with the lowest rate of JS—14.3%. Higher doses were significantly associated with JS (OR, 2.68; 95% CI, 1.37–5.25) with around 10% of those on high dose experiencing this effect.

Emotional Blunting

A reduction in the subjective experience of positive and negative emotions, emotional detachment, a sense that the medications prevent "natural sadness", changes to personality, diminished reactivity in relationships, caring less about self and others, fear of addiction, and suicidality has been regularly described with antidepressants [8R, 9R, 10R, 11R, 12c, 13c, 14R]. To assess the phenomena of emotional blunting, Goodwin et al. used qualitative methods to develop an appropriate scale, resulting in the Oxford Questionnaire on the Emotional Side-Effects of Antidepressants (OQESA) [15A]. Using an internet-based survey in subjects previously identified as having depression, the authors examined 669 depressed patients receiving antidepressant treatment and compared them with 150 formerly depressed, recovered adult controls [16c]. Depressed patients with emotional blunting had much higher total blunting scores on OQESA than controls (42.83 ± 14.73 vs 25.73 ± 15.00, $P < 0.0001$). Goodwin et al. found a rate of emotional blunting in treated depressed patients of 46%. They found no difference according to the type of antidepressant, although this was somewhat less frequent with bupropion. The OQESA scores were highly correlated with HAD depression score, suggesting that emotional blunting may not simply be a side effect of antidepressants, but also a symptom of depression.

Cardiovascular

Vascular Outcomes

Biffi et al. aimed to quantify the risk of cardiovascular and cerebrovascular disease in a systematic review and meta-analysis of 22 observational studies with a total of 99 367 patients taking antidepressants [17M]. The investigators found that, compared with non-users of antidepressants, the use of SSRIs was associated with an increased risk of cerebrovascular disease (RRs, 1.24; 95% CI, 1.15–1.34), and the use of TCAs with an increased risk of acute heart disease (RRs, 1.29; 95% CI, 1.09–1.54). The authors themselves noted that their results should be interpreted with caution due to high study heterogeneity and the probability of confounding by indication; the inability to distinguish between the effects of antidepressants and depression itself.

Acute Myocardial Infarction

To date, studies examining the risk of acute myocardial infarction (AMI) associated with SSRIs have been inconclusive. Wu et al. conducted a case-crossover study using a nationwide health insurance database investigating the association between antidepressant use and the risk of hospitalization for acute myocardial infarction (AMI; [18C]). A total of 18 631 patients with incident AMI were included. The authors found that the risk of AMI was not associated with antidepressant use. Similarly, the class, the dose, the binding affinity of serotonin transporter and the norepinephrine transporter were associated with the risk of AMI.

Endocrine

Type 2 Diabetes

Salvi et al. conducted a systematic review and meta-analysis of 20 studies examining the association of incident diabetes and antidepressants, confirming a significant link. This association remained significant even after inclusion of recent negative studies [pooled relative risk = 1.27, 95% CI; 1.19–1.35; $P < 0.001$]. When the analyses were restricted to the (six) high-quality studies, as defined by a high Newcastle-Ottawa Scale score of 8 or 9, relative risk further increased to 1.40. The authors noted that it remains contentious whether the association is causal. Burcu et al. studied the association between use of SSRIs or SNRIs and risk of type 2 diabetes in a large cohort of American youths [19R]. They found that with both SSRIs and SNRIs, the risk of type 2 diabetes was significantly greater during current compared to prior use (absolute risk, 1.29 per 10 000 person-months vs 0.64 per 10 000 person-months; adjusted relative risk [RR], 1.88; 95% CI, 1.34–2.64). The risk was also elevated for tricyclic or other cyclic antidepressants (absolute risk, 0.89 per 10 000 person-months vs 0.48 per 10 000 person-months; RR, 2.15; 95% CI, 1.06–4.36). For youths who were currently using SSRIs or SNRIs, the risk of type 2 diabetes increased with the duration of use (RR, 2.66; 95% CI, 1.45–4.88 for >210 days and RR, 2.56; 95% CI, 1.29–5.08 for 151–210 days compared with 1–90 days) and with cumulative dose (RR, 2.44; 95% CI, 1.35–4.43 for >4500 mg and RR, 2.17; 95% CI, 1.07–4.40 for 3001–4500 mg compared with 1–1500 mg in fluoxetine hydrochloride dose equivalents).

Pancreatitis

The association between treatment with SSRIs and acute pancreatitis remains unclear. Using national registry data Lin et al. performed a population-based case–control study to examine the relationship between current and former use of SSRIs and acute pancreatitis [20c]. They compared 4631 cases with their first attack of acute pancreatitis and 4631 matched controls from a randomly sampled cohort of one million health insurance subjects. Cases and controls were aged 20–84 years and matched according to sex, age, comorbidities, and the index year of diagnosis of acute pancreatitis. Current use of SSRIs was defined as those whose last tablet of SSRIs was recorded ≤7 days before their diagnosis of acute pancreatitis, and late use of SSRIs defined as those whose last tablet of SSRIs was ≥8 days before diagnosis. Using multivariate unconditional logistic regression analysis, after adjusting for covariables, they found that, compared to patients who had never used SSRIs, the adjusted OR of acute pancreatitis associated with current use of SSRIs was 1.7 (95% CI, 1.1–2.5), whereas that for patients with "late use" of SSRIs was not significant.

Reproductive System (Pregnancy, Development and Infancy)

Pregnancy

Although untreated depression appears to be associated with some adverse outcomes for both mother and child [21C], the literature regarding the potential negative effects of antidepressants during pregnancy remains conflicting, and concerns regarding confounding by indication persist. A recent meta-analysis examining 23 studies of 25663 pregnant women whose antenatal depression was untreated found a significantly greater risk of preterm birth and offspring with low birth weight [22R]. Cantarutti et al., in a population-based cohort study of 9825 deliveries from 354735 single births, sought to examine the relationship between antidepressants in pregnancy and neonatal outcomes [23C]. The investigators found that, when compared to infants whose mothers only took antidepressants before pregnancy, infants exposed during pregnancy showed a significant increase in the adjusted prevalence ratio of intrauterine asphyxia and birth asphyxia [PSS1 PR=1.39, 95% CI, 1.08–1.73], neonatal convulsion (PSS1 PR=2.81, 95% CI, 1.07–7.36) and other respiratory conditions (PSS1 PR=1.24, 95% CI, 1.00–1.52). There was no increase in the prevalence ratio of small for gestational age. The subgroup analysis restricted to women exposed to an SSRI during pregnancy found markedly increased adjusted prevalence ratios of intrauterine asphyxia and birth asphyxia (PSS1 PR=1.39, 95% CI, 1.07–1.81), low Apgar score (PSS1 PR=1.69, 95% CI, 1.02–2.79) and other respiratory conditions (PSS1 PR=1.37, 95% CI, 1.08–1.74). Interestingly, the authors found a clear association between antidepressant exposure in the first trimester and an increased prevalence ratio for neonatal convulsion but not for exposure in the second and third trimesters. Conversely, intrauterine asphyxia and birth asphyxia, a low Apgar score, and other respiratory problems were more likely to be associated with exposure in the third trimester.

Fetal Malformations

Zhang et al. conducted a systematic review and meta-analysis to examine the risk between first-trimester exposure to SSRIs and cardiac malformations in the offspring [24M]. Using 18 studies meeting their inclusion criteria they found that pregnant women who received SSRIs during the first trimester had a statistically significantly increased risk of infant cardiovascular-related malformations (RR=1.26, 95% CI=1.13–1.39). The corresponding RRs of atrial septal defects (ASD), ventricular septal defects (VSD), and ASD and/or VSD were 2.06 (95% CI=1.40–3.03), 1.15 (95% CI=0.97–1.36), and 1.27 (95% CI=1.14–1.42), respectively.

Shen et al. performed a meta-analysis of cohort studies examining for an association between sertraline use in the first trimester and congenital anomalies in offspring in studies to the end of 2015 [25M]. Twelve studies involving 6468241 pregnant women were included in their meta-analysis. The authors found sertraline in the first trimester had a statistically significant increased risk of infant cardiovascular-related malformations (OR=1.36; 95% CI=1.06–1.74) as well as atrial and/or ventricular septal defects (OR=1.36, 95% CI=1.06–1.76).

In a similar review and meta-analysis of 16 cohort studies to March 2016, Gao et al. found that offspring of pregnant women exposed to fluoxetine during the first trimester had a small, statistically significant, increased risk of major malformations (RR=1.18, 95% CI=1.08–1.29), cardiovascular malformations (RR=1.36, 95% CI=1.17–1.59), septal defects (RR=1.38, 95% CI=1.19–1.61), and non-septal defects (RR=1.39, 95% CI=1.12–1.73) [26M].

Bérard et al., in a longitudinal prospective cohort study of depressed or anxious pregnant women, sought to examine the association between first-trimester exposure to a range of antidepressants and the subsequent risk of major congenital malformations [27C]. The investigators used a group of depressed/anxious pregnant women who did not use antidepressants as the reference category, thereby adjusting for indication (depression/anxiety) and lifestyle factors associated with the indication (smoking, alcohol use) in the design.

A total of 18487 pregnant women, 3640 of whom were exposed to a single antidepressant in the first trimester, were included in the study. Adjusting for potential confounders, the investigators found that use of SSRI, SNRI, TCA or other antidepressants was not associated with an increased risk of major malformation. Antidepressant users, however, had lower weight newborns than non-users ($P<0.001$).

When examining specific types of antidepressant used during the first trimester, only citalopram, in 88 exposed cases, was found to increase the risk of major congenital malformations (adjusted OR, (aOR) 1.36, 95% CI 1.08–1.73). Antidepressants with a serotonin reuptake inhibition effect (SSRI, SNRI, and amitriptyline) were found to increase the risk of certain organ-specific defects: paroxetine increased the risk of cardiac defects (aOR 1.45, 95% CI 1.12–1.88) and ventricular/atrial septal defects (aOR 1.39, 95% CI 1.00–1.93); citalopram increased the risk of musculoskeletal defects (aOR 1.92, 95% CI

1.40–2.62) and craniosynostosis (aOR 3.95, 95% CI 2.08–7.52); and TCAs were associated with eye, ear, face and neck defects (aOR 2.45, 95% CI 1.05–5.72) and digestive defects (aOR 2.55, 95% CI 1.40–4.66). Venlafaxine was associated with respiratory defects (aOR 2.17, 95% CI 1.07–4.38).

Preeclampsia

It remains unclear whether the increased risk of preeclampsia associated with antidepressant treatment in some studies is independent of any underlying depression [28C]. Lupattelli et al. conducted a prospective population-based study of 5887 depressed women and sought to evaluate the risk of early and late onset preeclampsia according to the stage of pregnancy and duration of exposure to antidepressants [29C]. The authors found no association between SSRI exposure and early- or late-onset preeclampsia. In a detailed systematic review of the literature examining the possible relationship between use of antidepressants and preeclampsia or gestational hypertension, Uguz identified 7 studies that compared users with nonusers [30M]. He noted that at least half of the available studies suggest a positive association between the usage of antidepressants during pregnancy and preeclampsia or gestational hypertension with the adjusted relative risk ranging from 1.28 to 1.53 for any antidepressant, 1.05–3.16 for selective serotonin reuptake inhibitors, 1.49–1.95 for selective serotonin-norepinephrine reuptake inhibitors, and 0.35–3.23 for tricyclic antidepressants. He observed that risk was consistently associated with duration of use and particularly that antidepressant use during the second trimester was the most consistent association with increased risk.

Persistent Pulmonary Hypertension of the Newborn

Persistent pulmonary hypertension of the newborn (PPHN) occurs when the newborn's pulmonary vascular resistance fails to decrease after birth, and has been associated with the use of selective serotonin receptor inhibitors (SSRIs) in late pregnancy [31M, 32C]. Limited data are available with respect to SNRIs. In a further large population-based study, Berard et al. [33C] examined three exposure categories (SSRI, SNRI and other antidepressants)/Non-users formed the reference category, and the authors attempted to minimize confounding by indication by adjusting for history of maternal depression/anxiety prior to pregnancy. PPHN was identified in 0.2% of newborns. SSRI use during the second half of pregnancy was associated with an increased risk of PPHN [adjusted odds ratio (aOR) 4.29, 95% CI 1.34, 13.77] compared with non-use of antidepressants, after adjusting for maternal depression, and other potential confounders. SNRI use over the same period was not statistically associated with the risk of PPHN (aOR 0.59, 95% CI 0.06, 5.62). For both SSRIs and SNRIs, their use in the first half of pregnancy was not associated with the risk of PPHN.

Childhood Psychiatric Disorders

Liu et al., in a study of 900 women identified through national registries who continued antidepressants during pregnancy and 1400 women who discontinued treatment, found an association between prenatal exposure to antidepressants and psychiatric disorders in childhood, attributing the relationship as being possibly due to underlying maternal illness, antidepressants, or a combination of both [34C]. Liu et al.'s study is noteworthy as it included a disease comparison group of women discontinuing antidepressants before pregnancy, thereby facilitating the disentanglement of potential effects of antidepressants from effects of the underlying maternal psychiatric condition for which drugs were prescribed. Unlike the majority of studies, the investigators explored a range of psychiatric outcomes including autism, mental retardation, and mood disorders. They found that the children of women who continued using antidepressants had a 30% higher risk of psychiatric disorders (HR 1.3; 95% CI 1.2–1.4). Interestingly, the largest effect size was found for mood disorders (HR 2.8; 95% CI 1.6–4.8). They calculated that, in absolute terms, if the association between antidepressant treatment and psychiatric disorders in childhood was causal, and mothers in the antidepressant continuation group ceased their antidepressants before pregnancy, 0.5% of psychiatric disorders could be prevented. Liu et al. also reported absolute risks for specific disorders. For example, if prenatal exposure to antidepressants is associated with a 23% increased risk of autism in children, and assuming a baseline prevalence of autism of 1%, then for every 10000 women who continue treatment during pregnancy 23 additional cases of autism would be expected to occur as a result.

Autism

In recent years the risk of autism spectrum disorder (ASD) in exposed offspring has been an intense and controversial area. To date, most case–control studies and their pooled results in meta-analyses have consistently found a small increase in risk of autism in the exposed offspring of mothers treated with antidepressant drugs.

Andalib et al. undertook a meta-analysis of 7 studies [35M], each of which adjusted for maternal depression. After such adjustment, they found a persisting significant association between prenatal SSRI exposure and ASD (OR = 1.82, 95% CI = 1.59–2.10, Z = 8.49, P = 0.00).

Two recent cohort studies have also noted increased risk but came to different conclusions. First, in an observational prospective cohort Scandinavian national register study, Rai et al. examined 254610 individuals aged 4–17, of whom 5378 had autism [36C]. Groups of subjects who were not exposed to antidepressants, and whose mothers did not have psychiatric disorder but mothers who took antidepressants during pregnancy, and mothers with psychiatric disorders who did not take antidepressants during pregnancy, were identified and compared. The investigators found that 4.1% of the 3342 children exposed to antidepressants during pregnancy were diagnosed with ASD, compared with 2.9% of the 12325 children not so exposed and whose mothers had a history of a psychiatric disorder (aOR 1.45, 95% CI 1.13–1.85). The authors concluded the association between antidepressant use during pregnancy and autism might not be entirely due to confounding by maternal depression. They noted, however, the absolute risk of autism was small, and that if no pregnant women took antidepressants, the number of cases that could potentially be prevented would be small.

Second, using a national register cohort study, Viktorin et al. [37C] found no significant overall association between in utero antidepressant exposure and risk of ASD. Interestingly, subgroup analyses of specific antidepressants revealed increased RRs of ASD in offspring exposed to citalopram and escitalopram (RR: 1.47; 95% CI 0.92–2.35) and clomipramine (RR: 2.86; 95% CI 1.04–7.82). Overall, the authors concluded that antidepressants during pregnancy do not appear to be causally associated with the observed increased risk of ASD in the offspring, and that the association is probably explained by factors associated with the vulnerability to psychiatric disorders.

Recent studies have particularly sought to address this potential importance of parental diagnostic confounders. Brown et al. [38C] found that in 2837 pregnancies exposed to antidepressants, 2.0% of offspring were diagnosed with ASD, corresponding to an incidence of 4.51 per 1000 person-years compared with 2.03 per 1000 person-years among those unexposed (adjusted HR, 1.59 [95% CI, 1.17–2.17]). However, after considering a number of potential confounding variables, the association just failed to reach significance (HR, 1.61 [95% CI, 0.997–2.59]) [39C].

One area of focus in this literature has been upon the particular trimester of exposure to antidepressants. In a meta-analysis of 10 studies, Mezzacappa et al. [40M] found a significant association with ASD for all three trimesters which remained significant after controlling for past maternal mental illness (first trimester: OR, 1.79; 95% CI, 1.27–2.52, second: OR, 1.67, 95% CI, 1.14–2.45, and third: OR, 1.54, 95% CI, 0.82–2.90). Kaplan et al. performed a meta-analysis on 6 case–control studies, finding the risk little different for each trimester of exposure, although not significant in the third [41M]. Of note, the increased risk was almost identical for pre-conception exposure, thus strongly suggesting that confounding by indication could be driving the increased risk.

To further investigate this possibility of confounding by indication, Kaplan et al. undertook another meta-analysis of four cohort studies examining the possible link between prenatal SSRIs and ASD [42M]. The pooled point estimates for pregnant women exposed to SSRIs during pregnancy, and pregnant women with maternal psychiatric disorder without SSRI exposure were both significantly associated with ASD. The investigators interpreted these findings as supportive of confounding by indication, in addition to further confirmation of the association between prenatal exposure and ASD.

Using a large retrospective national cohort register, Sujan et al. evaluated alternative hypotheses for previously observed associations between first-trimester antidepressant exposure and the study outcomes of preterm birth, small for gestational age (SGA), ASD and ADHD by addressing confounding variables [43C]. In models that compared siblings, while adjusting for pregnancy, maternal, and paternal traits, first-trimester antidepressant exposure was only associated with preterm birth (OR, 1.34 [95% CI, 1.18–1.52]) but not with SGA, ASD or ADHD.

ADHD

Compared to the ASD literature, research into the association between antidepressant use in pregnancy and ADHD has been limited but has had similarly conflicting results [44C, 45C, 46C]. Recently two large cohort studies using national registry data have examined this association.

Boukhris et al. used large administrative, hospital, prescription and clinical registers in a single province, examining for the effects of different classes of antidepressants, gestational stage and maternal history of psychiatric illness including ADHD [47C]. After adjusting for potential confounders, they found that antidepressant use in pregnancy was significantly associated with an increased risk of ADHD (aHR 1.2, 95% CI; 1.0–1.4). Antidepressant use during the second or third trimester was associated with a significantly increased risk (aHR 1.3, 95% CI; 1.0–0.6; 134 exposed cases), though use in the first trimester was not. TCAs was associated with significantly increased risk (aHR 1.8, 95% CI; 1.0–3.1), unlike use of SSRIs or SNRIs. Man et al. addressed this question in a retrospective population-based cohort study of 190618 births [48C]. After adjustment for potential confounders such as maternal psychiatric disorders or the use of other psychiatric drugs, there was a persisting significantly increased risk of ADHD in mothers using antidepressants [aHR 1.39 (95% CI; 1.07–1.82, $P=0.01$)].

Falls and Fractures

Antidepressants are known to be associated with an increased risk of falls [49C, 50C]. Macri et al., in a matched retrospective cohort study of long-term care, examined the association between the new use of antidepressants and fall-related injuries in older adults in long-term care (LTC) [51C]. They studied any fall resulting in an emergency department (ED) visit or hospitalization within 90 days and secondary outcomes including hip and wrist fractures were examined against exposure to antidepressants. They found that new users of any antidepressant had an increased risk of ED visits or hospitalizations for falls within 90 days compared with those not receiving antidepressants (5.2% vs 2.8%; adjusted OR: 1.9, 95% CI: 1.7–2.2). In addition, they found that antidepressants were associated with an increased risk of all secondary outcomes. They concluded that the new use of antidepressants was associated with a significantly increased risk of falls and fall-related injuries among long-term care residents across different patient subgroups and antidepressant classes.

Hung et al. sought to investigate the association in elderly (≥65 years) patients using a national health insurance database random sample of 1 million insurance enrollees [52C]. The investigators identified 4891 patients with newly diagnosed hip fractures and 4891 controls over an 11-year period. They matched cases and controls were matched by sex, age, comorbidities, year of hip fracture diagnosis. Current SSRI users were defined as having taken medication ≤7 days before the diagnosis of hip fracture diagnosis. Late use was defined as taking the last SSRI tablet ≥8 days before the hip fracture diagnosis. Non-SSRI users had never been prescribed SSRIs. Although the adjusted OR for late users of SSRIs was not significant, a significant association between use of SSRI within 7 days and hip fracture diagnosis was found with an adjusted OR of hip fracture, compared with those who never used SSRIs, of 2.17 for current SSRI users (95% CI: 1.60–2.93).

While tricyclic antidepressants have been linked to the risk of falls through their potential for postural hypotension, and SSRIs via their potential effects on bone metabolism, mirtazapine has been less studied in this respect. Leach et al. in a matched case–control study of 8828 cases, with a median age of 88 years of age, and 35 310 controls, aimed to examine the risk of hip fracture in veteran's affairs patients associated with mirtazapine use as well as those switching between, or concurrently using mirtazapine and other antidepressant agents [53C]. The investigators found that for patients continuously using mirtazapine, the risk of hip fracture was increased (odds ratio [OR] 1.27, 95% confidence interval [CI] 1.12–1.44). Perhaps unsurprisingly, the risk was higher when antidepressants were used in combination; the increased odds of hip fracture with the addition of SSRIs to mirtazapine was 11 (95% CI 2.2–51; RERI 7.7, 95% CI −9.0 to 24). Where tricyclic antidepressants (TCAs) were added to mirtazapine the risk was similarly elevated (OR 14, 95% CI 1.4–132; RERI 12, 95% CI −19 to 43). Continuous use of both SSRIs and mirtazapine was also associated with an increased risk of hip fracture (OR 2.4, 95% CI 1.4–4.2; RERI 0.4, 95% CI −0.9 to 1.7).

In an interesting comparison of new users of antidepressants and antipsychotics in nursing home residents with Alzheimer's disease and related dementias (ADRD) and who had moderate-to-severe behavioral symptoms, Wei et al., in a retrospective, longitudinal cohort study, of subjects with dementia who had spent more than 100 days in a nursing home, used health insurance claims data to estimate the risks of falls and fractures [54M]. A total of 1964 patients were identified who received antipsychotic monotherapy and 4680 who received antidepressant monotherapy. The primary outcomes of interest were incident fractures, accidental falls, and a composite outcome of a fracture or fall event requiring inpatient or outpatient (i.e., physician visit, emergency room visit) care. Among their subjects with moderate-to-severe behavioral symptoms of ADRD, there were 1040 fall and 529 fracture events. The investigators found that, compared to antipsychotic users, antidepressant users had a significantly higher risk for fractures (aHR = 1.35, 95% CI = 1.10–1.66). In addition, the overall risk of falls or fractures remained significant in the antidepressant vs antipsychotic group (aHR = 1.16, 95% CI = 1.02–1.32). They concluded antidepressants are associated with higher fall and fracture risk when compared to antipsychotics in the management of older adults with Alzheimer's disease and related dementias with moderate-to-severe behavioral symptoms.

Sensory Systems

Glaucoma

SSRIs have been associated with glaucoma in multiple case reports and previous studies. In one recent study, SSRI initiation was associated with a 5.8-fold elevation in the risk of acute angle closure glaucoma within 7 days of commencement [55C]. Chen et al. aimed to investigate whether a possible association exists between SSRI exposure and the incidence of glaucoma incidence using a national database [56C]. New primary diagnoses (15 865) of glaucoma were compared against 77 014 sex, age, residence and insurance premium matched controls not diagnosed with glaucoma. Index cases were compared in terms of prescribed duration and dosage of SSRIs up to 365 days before the index date to proxy SSRIs exposure. The investigators found that individuals receiving SSRIs were at greater risk of glaucoma incidence (OR = 1.39; 95% CI = 1.29–1.50), which was still significant after adjusting for confounding variables (aOR = 1.09; 95% CI = 1.00, 1.18). They found that longer duration of SSRI treatment (i.e. >365 days) and higher doses (≥1 defined daily dose) was associated with greater risk of glaucoma incidence (aOR=1.36; 95% CI=1.08–1.71). Interestingly, subgroup analysis found that the effect of SSRIs on glaucoma was restricted to those younger than 65 years of age (aOR=1.37; 95% CI=1.25–1.50), without diabetes (aOR=1.39; 95% CI=1.27–1.52), and without hypertension (aOR=1.46; 95% CI=1.31–1.63) or hypercholesterolemia (aOR=1.35; 95% CI=1.23–1.48). The authors concluded SSRIs were associated with a greater risk of glaucoma, especially where the duration or dosage was longer or higher.

Cataracts

Two studies have been performed examining risk of cataract development with antidepressant treatment. In a nested case–control study using study of 14288 patients, using national health insurance database information, Chou et al. examined the association between the development of cataract in patients taking antidepressant drugs, and sought to investigate whether use of antidepressants, including those other than SSRIs, was associated with an increased risk of cataract development [57C]. The relationship between binding affinities for the serotonin transporter (SERT) and risk of cataracts was also examined. A total of 7651 patients with cataracts were identified and compared against 6637 control subjects according to antidepressant exposure type, duration of use, and their SERT binding affinities. The investigators found that the adjusted odds ratios (AORs) for developing cataracts in continuous users of SSRIs, SNRIs, and other antidepressants overall were 1.26 (95% confidence interval (CI): 1.12–1.41, $P<0.001$), 1.21 (95% CI: 1.02–1.43, $P=0.027$), and 1.18 (95% CI: 1.04–1.34, $P=0.009$), respectively. Sub-analyses by drug revealed that continuous uses of fluoxetine (AOR: 1.21; 95% CI: 1.01–1.46, $P=0.042$), fluvoxamine (AOR: 1.47; 95% CI: 1.01–2.12, $P=0.043$) and venlafaxine (AOR: 1.44; 95% CI: 1.19–1.74, $P<0.001$) were at significantly increased risk of cataract development. Also, continuous users of antidepressants with intermediate SERT binding affinities (AOR: 1.68; 95% CI: 1.10–2.56, $P=0.017$) were found to be at significantly increased risks of cataract development.

Becker et al. examined cataract risk after exposure to SSRI or other antidepressant drugs in a large primary care database in a younger population of 206931 first time cataract patients aged 40 or older, comparing them to an equal number of cataract-free controls matched on age, sex, general practice, date of cataract recording (i.e., index date), and years of history in the database before the index date using conditional logistic regression analyses adjusted for body mass index, smoking, hypertension, diabetes, and systemic steroid use [58A]. The exposure of interest was the number of SSRI prescriptions and prescriptions for other antidepressant drugs. Analyses were restricted to cases and controls without a prior glaucoma diagnosis. The investigators found that current long-term use of SSRI (≥20 prescriptions) was not associated with an increased cataract risk; however, in patients aged 40–64 years, they found a slightly increased risk of cataract for long-term SSRI users (adjusted OR, 1.24; 95% CI, 1.15–1.34) compared with non-users.

SELECTIVE SEROTONIN REUPTAKE INHIBITORS

Fluoxetine

Interstitial Lung Disease

Data related to adverse effects of fluoxetine related to the respiratory tract are scarce. Deidda et al. reviewed the literature in relation to interstitial lung disease, as a pulmonary adverse drug reaction possibly induced by fluoxetine via a systematic review of four domains: (i) published case reports in PubMed, OvidSp, Scopus and Web of Science, (ii) the World Health Organization VigiAccess database, (iii) the European EudraVigilance database, and (iv) a national Pharmacovigilance database [59S]. The authors found seven cases in the medical literature where interstitial lung disease was linked to fluoxetine. The age of the seven patients ranged from 40 to 86 years and six were female. In all but one case, the duration of ranged from 7 weeks to 1 year, and in one case was 20 years. Dyspnoea (with or without a non-productive cough) was the commonest clinical manifestation. Only in one case was asymptomatic ILD an incidental radiological finding (multiple pulmonary nodules resembling metastatic lesions). Bronchoalveolar lavage (BAL) showed lymphocytic (>15% lymphocytes) fluid characteristics in four of seven cases suggesting an inflammatory reaction. All cases showed a resolution on withdrawal, but re-challenge was positive in only two cases, and not feasible or reported in the other five cases.

With respect to the VigiAccess database (to July 2016) and the EudraVigilance database (to June 2016), 36 cases of interstitial lung disease related to fluoxetine reported by health professionals were retrieved. The national Pharmacovigilance database (to August 2016) revealed only one case which was codified as "pulmonary disease". Despite the widespread use of fluoxetine in teenagers and young adults, analysis of the cases of suspected fluoxetine-induced ILD from EudraVigilance found most involve female patients (86%), which was higher than might be expected, even when allowing for a higher prevalence of depression in females. The majority of these cases (61%) involved patients older than 65 years.

The single case reported to the national ADR authority was of a 68-year-old woman who had received fluoxetine for 8 years before developing a severe dry cough, and later, exertional dyspnoea. BAL revealed a lymphocytic infiltrate and thoracic CT showed evidence of bilateral diffuse interstitial disease. Spirometry demonstrated a restrictive ventilatory defect. Application of the Naranjo algorithm [59S] resulted in a determination of "probable" causation, based on previous reports of lung disease associated with fluoxetine, presence of objective data confirming the adverse event, and presence of chronological criteria in her case.

From their review, the authors noted female and elderly patients were at particular risk of fluoxetine associated ILD. They concluded that fluoxetine might be considered as a potential cause of Interstitial Lung Disease in patients receiving fluoxetine treatment who present with dyspnoea, with or without a dry cough. They noted in the reports that prompt withdrawal of fluoxetine was frequently followed by a complete remission.

Additional literature on this topic can be found in these reviews [60R, 61R, 62R].

References

[1] Laporte S, Chapelle C, Caillet P. Bleeding risk under selective serotonin reuptake inhibitor (SSRI) antidepressants: a meta-analysis of observational studies. Pharmacol Res. 2017;118:19–32 [R].

[2] Heller HM, Ravelli AC, Bruning AH, et al. Increased postpartum haemorrhage, the possible relation with serotonergic and other psychopharmacological drugs: a matched cohort study. BMC Pregnancy Childbirth. 2017;17(1):166 [c].

[3] Parkin L, Balkwill A, Sweetland S, et al. Antidepressants, depression, and venous thromboembolism risk: large prospective study of UK women. J Am Heart Assoc. 2017;6(5):e005316 [C].

[4] Renoux C, Vahey S, Dell'Aniello S. Association of selective serotonin reuptake inhibitors with the risk for spontaneous intracranial hemorrhage. JAMA Neurol. 2017;74(2):173–80 [C].

[5] Chan CH, Hung HH, Lin CH, et al. Risk of first onset stroke in ssri-exposed adult subjects: survival analysis and examination of age and time effects. J Clin Psychiatry. 2017;78(8):e1006–12 [C].

[6] Wu CS, Liu HY, Tsai HJ. Seizure risk associated with antidepressant treatment among patients with depressive disorders: a population-based case-crossover study. J Clin Psychiatry. 2017;78(9):e1226–32 [c].

[7] Sinha P, Shetty DJ, Bairy LK. Antidepressant-related jitteriness syndrome in anxiety and depressive disorders: incidence and risk factors. Asian J Psychiatr. 2017;29:148–53 [c].

[8] Gibson K, Cartwright C, Read J. Patient-centred perspectives on antidepressant use: a narrative review. Int J Ment Health Nurs. 2014;43:81–99 [R].

[9] Gibson K, Cartwright C, Read J. Patient-centred perspectives on antidepressant use: a narrative review. Int J Ment Health Nurs. 2014;43:81–99 [R].

[10] Moncrieff J. Antidepressants: misnamed and misrepresented. World Psychiatry. 2015;14:302–3 [R].

[11] Gibson K, Cartwright C, Read J. 'In my life antidepressants have been …': a qualitative analysis of users' diverse experiences of antidepressants. BMC Psychiatry. 2016;16:135 [R].

[12] Read J, Cartwright C, Gibson K. Adverse emotional and interpersonal effects reported by 1,829 New Zealanders while taking antidepressants. Psychiatry Res. 2014;216:67–73 [c].

[13] Price J, Cole V, Goodwin GM. Emotional side-effects of selective serotonin reuptake inhibitors: qualitative study. Br J Psychiatry. 2009;195:211–7 [c].

[14] Sansone RA, Sansone LA. SSRI-induced indifference. Psychiatry (Edgmont). 2010;7:14–8 [R].

[15] Price J, Cole V, Doll H, et al. The Oxford questionnaire on the emotional side-effects of antidepressants (OQuESA): development, validity, reliability and sensitivity to change. J Affect Disord. 2012;140(1):66–74 [A].

[16] Goodwin GM, Price J, De Bodinat C, et al. Emotional blunting with antidepressant treatments: a survey among depressed patients. J Affect Disord. 2017;221:31–5 [c].

[17] Biffi A, Scotti L, Corrao G. Use of antidepressants and the risk of cardiovascular and cerebrovascular disease: a meta-analysis of observational studies. Eur J Clin Pharmacol. 2017;73(4):487–97 [M].

[18] Wu CS, Wu HT, Tsai YT, et al. Use of antidepressants and risk of hospitalization for acute myocardial infarction: a nationwide case-crossover study. J Psychiatr Res. 2017;94:7–14 [C].

[19] Burcu M, Zito JM, Safer DJ, et al. Association of antidepressant medications with incident type 2 diabetes among medicaid-insured youths. JAMA Pediatr. 2017;171(12):1200–7 [R].

[20] Lin HF, Liao KF, Chang CM, et al. Association of use of selective serotonin reuptake inhibitors with risk of acute pancreatitis: a case-control study in Taiwan. Eur J Clin Pharmacol. 2017;73(12):1615–21 [c].

[21] Ross LE, Grigoriadis S, Mamisashvili L, et al. Selected pregnancy and delivery outcomes after exposure to antidepressant medication: a systematic review and meta-analysis. JAMA Psychiat. 2013;70(4):436–43 [C].

[22] Jarde A, et al. Neonatal outcomes in women with untreated antenatal depression compared with women without depression: a systematic review and meta-analysis. JAMA Psychiat. 2016;73:826–37 [R].

[23] Cantarutti A, Merlino L, Giaquinto C, et al. Use of antidepressant medication in pregnancy and adverse neonatal outcomes: a population-based investigation. Pharmacoepidemiol Drug Saf. 2017;26(9):1100–8 [C].

[24] Zhang TN, Gao SY, Shen YQ, et al. Use of selective serotonin-reuptake inhibitors in the first trimester and risk of cardiovascular-related malformations: a meta-analysis of cohort studies. Sci Rep. 2017;7:43085 [M].

[25] Shen ZQ, Gao SY, Li SX, et al. Sertraline use in the first trimester and risk of congenital anomalies: a systemic review and meta-analysis of cohort studies. Br J Clin Pharmacol. 2017;83(4):909–22 [M].

[26] Gao SY, Wu QJ, Zhang TN, et al. Fluoxetine and congenital malformations: a systematic review and meta-analysis of cohort studies. Br J Clin Pharmacol. 2017;83(10):2134–47 [M].

[27] Bérard A, Zhao J, Sheehy O. Antidepressant use during pregnancy and the risk of major congenital malformations in a cohort of depressed pregnant women: an updated analysis of the Quebec pregnancy cohort. BMJ Open. 2017;7(1)e013372 [C].

[28] Qiu C, Williams MA, Calderon-Margalit R, et al. Preeclampsia risk in relation to maternal mood and anxiety disorders diagnosed before or during early pregnancy. Am J Hypertens. 2009;22:397–402 [C].

[29] Lupattelli A, Wood M, Lapane K, et al. Risk of preeclampsia after gestational exposure to selective serotonin reuptake inhibitors and other antidepressants: a study from the Norwegian mother and child cohort study. Pharmacoepidemiol Drug Saf. 2017;26(10):1266–76 [C].

[30] Uguz F. Is there any association between use of antidepressants and preeclampsia or gestational hypertension?: a systematic review of current studies. J Clin Psychopharmacol. 2017 Feb;37(1):72–7 [M].
[31] Grigoriadis S, Vonderporten EH, Mamisashvili L, et al. Prenatal exposure to antidepressants and persistent pulmonary hypertension of the newborn: systematic review and meta-analysis. BMJ. 2014;348:f6932 [M].
[32] Huybrechts KF, Bateman BT, Palmsten K, et al. Antidepressant use late in pregnancy and risk of persistent pulmonary hypertension of the newborn. JAMA. 2015;313:2142–51 [C].
[33] Berard A, Sheehy O, Zhao JP, et al. SSRI and SNRI use during pregnancy and the risk of persistent pulmonary hypertension of the newborn. Br J Clin Pharmacol. 2017;83(5):1126–33 [C].
[34] Liu X, Agerbo E, Ingstrup KG, et al. Antidepressant use during pregnancy and psychiatric disorders in offspring: Danish nationwide register based cohort study. BMJ. 2017;358:j3668. [C].
[35] Andalib S, Emamhadi MR, Yousefzadeh-Chabok S, et al. Maternal SSRI exposure increases the risk of autistic offspring: a meta-analysis and systematic review. Eur Psychiatry. 2017;45:161–6 [M].
[36] Rai D, Lee BK, Dalman C, et al. Antidepressants during pregnancy and autism in offspring: population based cohort study. BMJ. 2017;358:j2811 [C].
[37] Viktorin A, Uher R, Reichenberg A, et al. Autism risk following antidepressant medication during pregnancy. Psychol Med. 2017;47(16):2787–96 [C].
[38] Brown HK, Ray JG, Wilton AS, et al. Association between serotonergic antidepressant use during pregnancy and autism spectrum disorder in children. JAMA. 2017;317(15): 1544–1552 [C].
[39] Vigod SN, Gomes T, Ray JG. Prenatal antidepressant use and autism spectrum disorder-reply. JAMA. 2017;318(7): 665 [C].
[40] Mezzacappa A, Lasica PA, Gianfagna F, et al. Risk for autism spectrum disorders according to period of prenatal antidepressant exposure: a systematic review and meta-analysis. JAMA Pediatr. 2017;171(6):555–63 [M].
[41] Kaplan YC, Keskin-Arslan E, Acar S, et al. Prenatal selective serotonin reuptake inhibitor use and the risk of autism spectrum disorder in children: a systematic review and meta-analysis. Reprod Toxicol. 2016;66:31–43 [M].
[42] Kaplan YC, Keskin-Arslan E, Acar S, et al. Maternal SSRI discontinuation, use, psychiatric disorder and the risk of autism in children: a meta-analysis of cohort studies. Br J Clin Pharmacol. 2017;83(12):2798–806 [M].
[43] Sujan AC, Rickert ME, Öberg AS, et al. Associations of maternal antidepressant use during the first trimester of pregnancy with preterm birth, small for gestational age, autism spectrum disorder, and attention-deficit/hyperactivity disorder in offspring. JAMA. 2017;317(15):1553–62 [C].
[44] Laugesen K, Olsen MS, Telen Andersen AB, et al. In utero exposure to antidepressant drugs and risk of attention deficit hyperactivity disorder: a nationwide Danish cohort study. BMJ Open. 2013;3:e003507 [C].
[45] Clements CC, Castro VM, Blumenthal SR, et al. Prenatal antidepressant exposure is associated with risk for attention deficit hyperactivity disorder but not autism spectrum disorder in a large health system. Mol Psychiatry. 2015;20:727–34 [C].
[46] Castro VM, Kong SW, Clements CC, et al. Absence of evidence for increase in risk for autism or attention-deficit hyperactivity disorder following antidepressant exposure during pregnancy: a replication study. Transl Psychiatry. 2016;6:e708 [C].
[47] Boukhris T, Sheehy O, Berard A. Antidepressant use in pregnancy and the risk of attention deficit with or without hyperactivity disorder in children. Paediatr Perinat Epidemiol. 2017;31(4):363–73 [C].
[48] Man KKC, Chan EW, Ip P, et al. Prenatal antidepressant use and risk of attention-deficit/hyperactivity disorder in offspring: population based cohort study. BMJ. 2017;357:j2350 [C].
[49] Cox CA, van Jaarsveld HJ, Houterman S, et al. Psychotropic drug prescription and the risk of falls in nursing home residents. J Am Med Dir Assoc. 2016;17:1089–93 [C].
[50] Quach L, Yang FM, Berry SD, et al. Depression, antidepressants, and falls among community-dwelling elderly people: the MOBILIZE Boston study. J Gerontol A Biol Sci Med Sci. 2013;68:1575–81 [C].
[51] Macri JC, Iabone A, Kirkham JG, et al. Association between antidepressants and fall-related injuries among long-term care residents. Am J Geriatr Psychiatry. 2017;25(12):1326–36 [C].
[52] Hung SC, Lin CH, Hung HC, et al. Use of selective serotonin reuptake inhibitors and risk of hip fracture in the elderly: a case-control study in Taiwan. J Am Med Dir Assoc. 2017;18(4):350–4 [C].
[53] Leach MJ, Pratt NL, Roughead EE. The risk of hip fracture due to mirtazapine exposure when switching antidepressants or using other antidepressants as add-on therapy. Drugs Real World Outcomes. 2017;4(4):247–55 [C].
[54] Wei YJ, Simoni-Wastilia L, Lucas JA, et al. Fall and fracture risk in nursing home residents with moderate-to-severe behavioral symptoms of Alzheimer's disease and related dementias initiating antidepressants or antipsychotics. J Gerontol A Biol Sci Med Sci. 2017;72(5):695–702 [M].
[55] Chen HY, Lin CL, Lai SW, et al. Association of selective serotonin reuptake inhibitor use and acute angle-closure glaucoma. J Clin Psychiatry. 2016;77(6)e692-6 [C].
[56] Chen VC, Ng MH, Chiu WC, et al. Effects of selective serotonin reuptake inhibitors on glaucoma: a nationwide population-based study. PLoS One. 2017;12(3)e0173005 [C].
[57] Chou PH, Chu CS, Chen YH, et al. Antidepressants and risk of cataract development: a population-based, nested case-control study. J Affect Disord. 2017;215:237–44 [C].
[58] Becker C, Jick SS, Meier CR. Selective serotonin reuptake inhibitors and cataract risk: a case-control analysis. Ophthalmology. 2017;124(11):1635–9 [A].
[59] Deidda A, Pisanu C, Micheletto L, et al. Interstitial lung disease induced by fluoxetine: systematic review of literature and analysis of vigiaccess, eudravigilance and a national pharmacovigilance database. Pharmacol Res. 2017;120:294–301 [S].
[60] Smithson J, Mitchell PB. Antidepressants. In: Ray SD, editor. Side effects of drugs annual: a worldwide yearly survey of new data in adverse drug reactions, vol. 37. Elsevier; 2015. p. 15–31 [chapter 2]. [R].
[61] Smithson J, Mitchell PB. Antidepressant drugs. In: Ray SD, editor. Side effects of drugs annual: a worldwide yearly survey of new data in adverse drug reactions, vol. 38. Elsevier; 2016. p. 11–9 [chapter 2]. [R].
[62] Smithson J, Mitchell PB. Antidepressants. In: Ray SD, editor. Side effects of drugs annual: a worldwide yearly survey of new data in adverse drug reactions, vol. 39. Elsevier; 2017. p. 9–21 [chapter 2]. [R].

CHAPTER

3

Lithium

*Connie Rust**[,1], *Heath Ford**, *Sidhartha D. Ray*[†]

*Department of Pharmacy Practice, South College School of Pharmacy, Knoxville, TN, United States

[†]Department of Pharmaceutical and Biomedical Sciences, Touro University College of Pharmacy and School of Osteopathic Medicine, Manhattan, NY, United States

[1]Corresponding author: crust@southcollegetn.edu

INTRODUCTION

Lithium has been used as a mood-stabilizer for decades and represents a first-line treatment for patients with bipolar disorder [1R]. Twelve-hour plasma lithium concentrations are considered normal at 0.6–1.2 mmol/L for patients treated for acute mania and 0.4–1 mmol/L for chronic prophylaxis [2R]. Adverse effects of lithium use, such as thirst, excessive urination, nausea, diarrhea, and tremor, are relatively common and can be easily managed. Other adverse effects, particularly weight gain and cognitive impairment (or "brain fog"), while common, are also somewhat serious due to their adverse impact on patient medication adherence and thus treatment success [3R, 4R]. Other important adverse effects of therapy include thyroid and parathyroid gland dysfunction and renal impairment [3R]. While hypothyroidism can be expected with lithium use, hyperparathyroidism and renal insufficiency are less common and may, in the case of renal insufficiency, warrant discontinuation [3R]. Cardiovascular and gastrointestinal symptoms have also been observed and include sinus bradycardia, ST-segment elevation, prolonged QT interval, nausea, vomiting, diarrhea, and ileus [2R].

Neurologic manifestations, including lethargy, ataxia, confusion, agitation, and neuromuscular excitability including hyperreflexia, myoclonic jerks, and muscle fasciculations, represent the most important adverse effects of lithium therapy and typically occur at supratherapeutic plasma lithium concentrations [2R]. The time course of lithium dosing is a very important consideration in this context, particularly since chronic overdose has a poor prognosis due to prolonged drug distribution to the brain and other tissues [2R]. Even at normal plasma concentrations, however, lithium can exert paradoxical effects on neurocognition—enhancing some functions while impairing others [4R].

Because lithium is excreted almost exclusively by the kidneys, factors that impact kidney function also affect the disposition of lithium in the body (e.g., diuretic and renin-angiotensin system medication use, dehydration, electrolyte concentrations, age) [5R]. Specifically, it is thought that predisposing factors for lithium toxicity include (1) nephrogenic diabetes insipidus (where activation of the renin-angiotensin system promotes reabsorption of sodium and lithium secondary to volume depletion), (2) age older than 50 years (which is associated with polypharmacy and reduced muscle mass), and (3) renal impairment (since lithium is excreted by the kidneys) [2R]. In addition to diuretic and renin-angiotensin drugs, other medications that may affect plasma lithium concentrations include sodium bicarbonate (which, in large quantities, decreases lithium concentrations by increasing renal clearance) and topiramate (which decreases lithium concentrations by increasing renal clearance due to weak inhibition of carbonic anhydrase) [5R].

Given this background, the following pages present evidence of adverse drug reactions and toxic manifestations of lithium therapy published in English between January 1 and December 31 of 2017.

ORGANS AND SYSTEMS

Nervous System

Berk and colleagues [6c] conducted a study to compare the effects of lithium and quetiapine on brain white and grey matter volume. To this end, a total of 19 first-episode

Side Effects of Drugs Annual, *Volume 40*
ISSN: 0378-6080
https://doi.org/10.1016/bs.seda.2018.06.001

mania (FEM) patients were randomized to the lithium group and 20 to the quetiapine group; a control group was also formed and consisted of 30 healthy subjects matched on age, sex, handedness (i.e., right- vs left-handedness), and intelligent quotient. Statistical significance was set at an alpha level less than 0.01. Results were suggestive of significantly less grey and white matter in FEM patients compared to healthy controls at baseline. Further, lithium appeared to slow progression of white matter volume loss compared to quetiapine over a period of 12 months. While the results appear significant, it must be emphasized that two primary hypotheses were tested: (1) that white and grey matter volume differences would be observed between the FEM group and the control group and (2) that lithium and quetiapine would be equivalent in terms of brain volume effects over 12 months. With two different primary hypotheses tested, two different study designs were therefore required. While titled *randomized, single-blind experimental trial*, the study first employed a mixed design (i.e., prospective cohort and cross-sectional analyses) to establish significant differences between the FEM (i.e., lithium and quetiapine groups) and control groups. In this regard, randomization is no longer an important consideration, and causation, therefore, cannot be inferred. The second study design was an experimental analysis, which randomized FEM patients between lithium and quetiapine groups. Even as the experimental component of the study may allow inferences in causation, it must be emphasized that (1) power was not established (i.e., neither sample size nor effect size were established a priori) and (2) given the small sample size of both the mixed- and experimental analyses, meaningful conclusions regarding the effects of lithium and quetiapine can only be interpreted as suggestive (and not conclusive).

Rossi and colleagues [7A] recently discussed a case of permanent cerebellar syndrome in a patient with therapeutic levels of lithium after urosepsis and hyperthermia. A 33-year-old male diagnosed with bipolar disorder and treated with lithium presented with fever of 104 (Fahrenheit), encephalopathy, urinary obstruction, and acute urosepsis and neurogenic bladder caused by traumatic injury of the cauda equine. The patient also had an extended history of intermittent urinary tract infections (UTI). The patient was treated with gentamicin, a combination of ampicillin/sulbactam, and cefazolin resulting in the abatement of fever 2 days later. The patient's plasma lithium level was 1.18 mEq/L at admission. Two weeks after admission, patient was dysarthric and ataxic. Neurological assessment showed bilateral dysmetria, scanning speech, bilateral dysdiadochokinesis, and cerebellar ataxic gait. Lab tests indicated leukocytosis, and other routine laboratory tests as well as thyroid function tests, cerebrospinal fluid analysis, and Purkinje cell antibodies were normal. A brain MRI showed cerebellar atrophy 2 weeks after hyperthermia, which resulted in persistent cerebellar dysfunction 20 years after the episode of urosepsis and fever. The authors postulate that chronic exposure to lithium may have contributed to an accumulation of lithium in the brain, the effects of which were precipitated by the admission diagnoses. The authors emphasize the importance of avoiding permanent neurotoxicity through early recognition and treatment of lithium-associated neurotoxicity in the presence of fever in patients taking non-toxic levels of lithium. It is worth noting, however, that the dose, dosing frequency, and duration of lithium therapy were not described by the authors, and therefore, it can be difficult, in the case, to draw parallels between length of exposure to lithium and the adverse outcomes explained here.

Albright and Way [8r] describe a similar case involving chronic use of lithium and the development of prolonged delirium. A 63-year-old female in rural Australia with a diagnosis of bipolar disorder was hospitalized for depression for 3 months. Upon admission, lithium toxicity was discovered [plasma concentration not provided], for which she was treated and transferred to the psychiatric unit on low-dose lithium. The depressive episode was accompanied by signs and symptoms of delirium, including cognitive impairment, social withdrawal, and sleep disturbances. Nonpharmacological treatments were denied by patient. Psychotropic medication dosages were changed, including lithium, due to worsening confusion upon dosage increases. The patient's prolonged delirium was resolved after 5 weeks of treatment. The authors attribute the delirium to low resilience to external triggers and brain vulnerability and emphasized the need for a multifaceted approach to prevention and treatment of delirium in such cases.

Smith and Cipriani [9R] conducted a meta-review to determine the effect of lithium on suicide rates, suicidal ideation, and self-harm. Databases including PubMed, PsycINFO, and the Cochrane Library were searched for reviews published in any language between January 1, 1980 and June 1, 2017. The initial search produced 1008 total references; these were evaluated by title and abstract to reduce the overall number of citations to those most relevant to the research question. This yielded a total of 16 reviews that included information published from randomized controlled trials. Significant heterogeneity was noted among these studies in terms of participants, diagnoses, comparators, study duration, and phase of illness. Therefore, of these 16, only four reviews were included in the final analysis—three that evaluated the effect of lithium on suicide rate, and one that evaluated lithium in terms self-harm. The authors concluded that, in addition to mood-stabilizing properties, lithium also exerts significant protective effects against suicidal action.

Cardiovascular

Maddala and colleagues [10A] describe a case report of probable lithium-induced cardiotoxicity in an elderly female. A 77-year-old female with a history of bipolar disorder presented to emergency personnel with ataxia, tremor, and giddiness. She had received chronic lithium therapy at 600mg orally daily (in two divided doses) for 30 years. She also received drug therapy with losartan 25mg, metoprolol 25mg, and levothyroxine 12.5mcg [no dosing route, frequency, or duration were provided]. The patient had a pulse rate of 38 beats per minute, consistent with blood pressure readings indicative of hypotension. Initial ECG showed junctional rhythm. On day 1 of admission, plasma lithium concentration was 2.6mmol/L (normal range 0.5–1mmol/L). All other lab values, including thyroid stimulating hormone (TSH) and serum sodium, were within normal limits. All drugs were withheld and the patient was treated with fluid resuscitation. On day 2 of admission, the patient had a plasma lithium concentration of 2.1mmol/L, which was repeated approximately 12h later and found to be 1.8mmol/L. An ECG on day 2 indicated atrial fibrillation, which was controlled by drug therapy. On admission day 3, the patient developed complete heart block (plasma lithium concentration was 1.4mmol/L). On admission day 4, the plasma lithium concentration dipped to 0.9mmol/L and the patient reverted to normal sinus rhythm with bradycardia. Lithium was never restarted; valproate, instead, was initiated in addition to her previous hypertensive medications. The Naranjo Adverse Drug Reaction Probability Scale rated the causal link between lithium therapy and adverse cardiac events as "probable" (score of 5).

A recent case report by Bains [11A] suggests an association between short-term lithium exposure and the syndrome known as "drug reaction with eosinophilia and systemic symptoms" (DRESS). A 40-year-old female patient with a history of bipolar disorder presented to a dermatology outpatient clinic with whole-body urticated maculopapular rash with scaling after approximately three and one half weeks of lithium and lamotrigine therapies. The patient also received concurrent therapy with mirtazapine, amisulpride, and trihexyphenidyl, which the patient had taken in the past without incident. Clinical examination revealed fever, facial edema (no lymphadenopathy), and maculopapular rash, the extent of which was more severe on the face, trunk, and arms. Laboratory examination was significant for a hemoglobin of 9.8g, total leukocyte count of 15,400 per cubic millimeter, and an absolute eosinophil count of 1100 per microliter. Liver enzymes were slightly elevated (i.e., serum glutamic oxaloacetic transaminase 45units per liter, serum glutamic pyruvic transaminase 57units per liter) and a peripheral smear was positive for microcytic hypochromic anemia with neutrophilic leukocytosis. Other lab values were within normal limits. The European Registry of Severe Cutaneous Adverse Reaction (RegiSCAR) classified the event as a "possible" reaction secondary to lamotrigine use—particularly since lamotrigine and other aromatic drugs are more frequently associated with DRESS. All psychotherapeutic drugs were stopped and the patient was treated with a 6-week steroid taper. The patient recovered and all drugs except lamotrigine were restarted. After 3 days, the patient again reported whole-body rash and facial edema, with a total leukocyte count of 14900 per cubic millimeter and an absolute eosinophil count of 1300 per microliter. It was then determined that DRESS was due to lithium exposure. The patient was treated with a 4-week steroid taper and antihistamines resulting in symptom improvement. No information regarding lithium dose, formulation (i.e., immediate or controlled-release), dosing frequency, or serum level was reported.

Research by Saad and colleagues [12E] is suggestive of renal and cardiac protective effects of diets enriched by cactus extract (*Opuntia ficus indica*) in rat models compared to those with lithium-induced oxidative stress. Differences between groups were assessed by body weight, antioxidative activity, lipid peroxidation, creatinine and urea, histopathological examination.

An experimental study conducted by Bhardwaj and colleagues [13E] in rats suggests that, in places where arsenic can be considered a drinking water contaminant, supplemental lithium therapy may result in enhanced lithium-related adverse drug effects involving red blood cells, including reduced glutathione and lipid peroxidation levels, reduced total red blood cell count, reduced activity of sodium–potassium ATPase, and increased counts of neutrophils and total leukocytes.

Renal

Decloedt and colleagues [14c] conducted a randomized controlled trial in a small sample of patients with HIV who were also being treated with lithium for bipolar disorder. The purpose of the study was to determine if the combination of lithium and tenofovir disoproxil fumarate (TDF) resulted in changes in glomerular filtration rate. Patients receiving drugs known to precipitate lithium toxicity (i.e., nonsteroidal anti-inflammatory drugs, renin-angiotensin system drugs and diuretics), those with a glomerular filtration rate (GFR) less than 60mL/min, and those with dehydration and diarrhea were excluded from the study. A total of 53 patients were recruited (mostly female) and randomly assigned to either a TDF plus placebo or TDF plus lithium (titrated to a plasma concentration between 0.6 and 1 mmol/L) group. Medication adherence between groups was similar; no statistical differences were detected between groups based on age, gender, weight, months on TDF, and renal function. In terms of the proportion of patients who experienced a reduction in GFR at 24 weeks, results were not significant for differences between groups. The authors concluded that concurrent treatment with TDF and lithium does not result in nephrotoxicity at 24 weeks. Several limitations to the study were discussed, including small sample size (i.e., low power to detect differences between groups), short follow-up period (i.e., 24 weeks), and failure to account for subtle changes in renal function, such as urine markers for tubulopathy.

A review by Gupta [15R] suggests that the risk of chronic kidney disease (and end-stage renal disease [ESRD]) associated with lithium therapy may be narrowly applicable to a small subset of patients. The risk of adverse renal effects may be minimized by lowering the therapeutic target lithium concentration, avoiding acute overdoses, and dosing lithium only once daily. While continuing lithium therapy in patients with renal insufficiency can represent a difficult therapeutic management strategy, it can be argued that, for patients where the risk of adverse psychiatric manifestations outweighs the risk of progressive renal insufficiency, a prudent course may be to continue lithium therapy (with a lower therapeutic plasma concentration target) while regularly monitoring renal function. The same risk-to-benefit consideration may also apply to patients with renal insufficiency who may otherwise be candidates for lithium therapy.

Tondo and colleagues [16A] evaluated the long-term effects of lithium carbonate therapy on renal function in an international retrospective cohort study employing multivariate regression modeling. Data from a total of 312 patients were gathered for the study, most of which were derived from women (57.7 percent) with diagnoses of bipolar disorder I (78.2 percent) in 12 western European countries and Canada. Independent variables evaluated in the study included age, sex, years of lithium therapy exposure, mean daily dose of lithium carbonate, mean daily lithium trough concentration, body-mass index, white blood cell count, blood glucose concentration, blood urea nitrogen, and serum creatinine. The dependent variable was estimated glomerular filtration rate (eGFR). Results from multivariate regression modeling suggest that risk factors for eGFR reduction include longer lithium treatment, lower lithium dose, higher serum lithium concentration, older age, and medical comorbidity. Limitations to the study include (i) eGFR calculation (eGFR represents an estimation of kidney function; more accurate assessments of renal function could have been employed), (ii) the strong potential for biases inherent in retrospective observational study designs, and (iii) lack of a comparator group to determine the effect of aging on renal function.

Pattanayak and colleagues [17r] discuss a case of lithium-induced polyuria in a young female patient. A 24-year-old female patient in India presented with a history of bipolar disorder II with symptoms that included thirst and increased frequency of nocturia and micturition. In addition to bipolar disorder, the patient had a past medical history of obesity. The patient's medication profile included lithium 1200 mg daily, on which the patient had been stable for approximately 10 months, aripiprazole 20 mg daily, and escitalopram [no dosing information provided]. At the time of presentation, laboratory values were within normal limits. These included urine protein (24-h) (70 mg), urinary creatinine (24-h) (1000 mg), urinary sodium/potassium (103/33.4 mEq/L), and serum sodium/potassium (135/4.1 mEq/L). Serum osmolality (302 mOsm/kg) was slightly elevated and urine osmolality was low (180 mmol/L). In terms of polyuria, the patient was treated with a combination of amiloride and hydrochlorothiazide (5/50 mg) owing to the fact that, in India, amiloride is not available as a non-combination product. The total daily lithium dosage was lowered by approximately 30percent to 750 mg daily to decrease the risk for thiazide-induced lithium toxicity. A marketed reduction in urine output was noted (5–1.8 L daily) over the course of a week, and the patient was later discharged without further incident.

Masiran and Aziz [18A] present a case report of lithium-induced renal failure aggravated (or perhaps precipitated) by concomitant use of an angiotensin converting enzyme (ACE) inhibitor. A 58-year-old male with a 30-year medical history of bipolar I disorder presented to the emergency department at the insistence of his wife with signs of fidgeting, disorientation to time and place, difficulty focusing, confusion, anxiety. Even with a stable 20-year history on lithium and haloperidol, the patient had begun to experience, within the previous 3 months, auditory and visual hallucinations, insomnia, and tremors. The patient had a past medical history positive for hypertension (15 years) and diabetes (20 years). The patient's medication profile included haloperidol 5 mg daily, lithium carbonate 300 mg twice daily, benzhexol (trihexyphenidyl hydrochloride) 2 mg three times daily, perindopril 4 mg daily, simvastatin 10 mg daily, aspirin 100 mg daily, metformin 500 mg twice daily, Actrapid® (insulin) 18 units three times daily, and Insulatard® (isophane insulin) 34 units daily. (With the exception of lithium and haloperidol, which the patient had been taking for 20 years, no route, frequency, or duration of drug therapies were provided.) At admission, the patient's plasma lithium level was found to be 2.3 mmol/L. (It was noted that 2 months prior, the patient's lithium level was 2.1 mmol/L, and due to difficulties contacting the patient, the patient did not receive further medical evaluation and treatment.) Laboratory values were consistent with elevated levels of urea (18.8 mmol/L), creatinine (426 micromol/L), and potassium (6.2 mmol/L). Blood counts were significant for anemia (hemoglobin 11.4 g/dL), thrombocytopenia (platelets 133 000/L), and leukocytosis (17.4×10^9/L). Venous blood gas suggested metabolic acidosis. Thyroid stimulating hormone and T4 levels were within normal limits. A baseline ECG was insignificant for adverse cardiac effects. An ultrasonography of the ureters, kidney, and bladder was positive for bilateral renal parenchymal disease with left simple renal cyst. The patient was diagnosed with lithium toxicity with acute renal failure. Other conditions ruled-out included neuroleptic malignant syndrome, delirium secondary to uncontrolled diabetes, and delirium due to a cardiovascular accident. Due to the patient's initial aggressiveness, hemodialysis was ruled-out as a treatment option, and instead fluid therapy was initiated. Lithium and perindopril were discontinued as were his other medications. Sodium bicarbonate was added to treat metabolic acidosis. Later, haloperidol was restarted with valproate to prevent the onset of adverse psychiatric manifestations. By day 8 of admission, the patient's lithium level decreased to 0.8 mmol/L; his mental status at that time improved. Laboratory values, excepting hemoglobin, platelets, urea, and serum creatinine, also improved. On day 24 of admission, the patient was discharged in stable condition with elevated urea and creatinine concentrations (16 mmol/L and 296 μmol/L, respectively). The Naranjo Adverse Drug Reaction Probability Scale rated lithium-induced renal failure as "probable" with a score of 6.

Endocrine

Shanks and colleagues [19A] discuss a case of suspected lithium-induced endocrine dysfunction (i.e., nephrogenic diabetes insipidus, hyperparathyroidism, and hyperthyroidism). A 58-year-old female with a 26-year history of bipolar disorder treated by lithium presented to the emergency department with leg weakness, poor appetite, slurred speech, and worsening anxiety and depression. Her medication profile included lithium carbonate modified release 400 mg three times daily, which she had recently restarted after it had been discontinued 3 weeks earlier. Other medications included chlorpromazine and cephalexin, which had been prescribed for a urinary tract infection. [The time course for these medications was not provided.] Laboratory values upon admission were significant for hypernatremia (151 mmol/L), hypercalcemia with hyperparathyroidism (calcium [adjusted] = 3.46 mmol/L, parathyroid hormone = 8.8 pmol/L), hyperthyroidism (free T4 = 80.8 pmol/L, free T3 = 13.4 pmol/L), and lithium toxicity (2.06 mmol/L). The patient was treated with three liters of normal saline over 24 h (to address hypercalcemia and hypernatremia) followed by zoledronic acid 4 mg in 500 mL of normal saline (for hypercalcemia) and 5 percent dextrose for hypernatremia. Propranolol 40 mg and carbimazole 40 mg once daily were added to address hyperthyroidism. During the course of the hospitalization, the patient's condition worsened. Hypernatremia (166 mmol/L around days 8 and 9) was unresponsive to continued rehydration therapy (6 L of 5 percent dextrose per day), and the serum osmolality ratio was noted to be 0.76, which supported a diagnosis of nephrogenic diabetes insipidus. The patient developed flaccid paralysis and experienced varying levels of responsiveness, with wide shifts in Glasgow Coma Scale scores ranging between 11 and 3. Neurologic tests, including a magnetic resonance imaging electroencephalopgraphy, electromyography, and nerve conduction studies, revealed marked cerebral atrophy and myopathy. Cerebrospinal fluid was positive for oligoclonal bands type 2. Consequently, the patient was started on immunoglobulin therapy and showed improvement in muscle paresis and weakness over the following 3 weeks. Over the remaining course of the hospitalization, endocrine values normalized, and the patient was subsequently discharged to a rehabilitation facility for further care. The authors considered chronic lithium therapy to be the cause of the nephrogenic diabetes insipidus, hyperparathyroidism, and though rare (i.e., between 0.1 and 1.7 percent annually), hyperthyroidism observed in this patient case.

Dermatologic

Pinna and colleagues [20A] discuss the presence of an erythematous maculopapular rash after challenge, de-challenge, and subsequent re-challenge with lithium carbonate in a 31-year-old female patient with bipolar disorder I. The patient voluntarily presented to the psychiatric ward of a local hospital with signs suggestive of mania. Her past medical history was positive for a family history of mood disorder and diagnoses of depression (treated with a selective serotonin reuptake inhibitor otherwise unspecified) and bipolar disorder (treated with carbamazepine 400 mg daily, lithium carbonate 450 mg daily, and risperidone 4 mg daily). This case presentation represents the patient's third medical evaluation regarding bipolar disorder. During the first evaluation, lithium, carbamazepine, and risperidone (at the doses listed above) were prescribed during a weeklong hospitalization secondary to psychosis. After remission of psychosis, the patient was discharged and seen by a private clinician 12 days later, where a cutaneous reaction was noted (the second medical evaluation). The practitioner discontinued carbamazepine therapy, increased the lithium carbonate dose to 750mg daily, and continued risperidone therapy at 4mg daily. The cutaneous reaction persisted for approximately 8 more days, after which lithium carbonate was discontinued. Six days after the discontinuation of lithium, the cutaneous reaction subsided. Sixteen days after the discontinuation of lithium and now concurrent with the case presentation, the patient presented to the psychiatric ward with signs suggestive of mania (i.e., marked anxiety, persecutory delusions, insomnia, flight of ideas, pressured speech). Lithium carbonate was again restarted. The following day (admission day 1), the patient developed a pruritic erythematous maculopapular rash. Lithium was stopped and resolution of the rash occurred the following day. The authors rated the association between lithium (challenge, de-challenge, and re-challenge) and the maculopapular rash as "probable" using the Naranjo Adverse Drug Reaction Probability Scale.

Priyadharshini and Ummar [21c] conducted a prospective observational study (otherwise unspecified) to assess the prevalence of skin reactions associated with lithium use in a small convenience sample of 52 patients with bipolar disorder at a single clinic in India. While results suggest a very high prevalence of lithium-induced skin reactions (38 percent), it must be emphasized that (1) causal links between lithium and various skin reactions observed in the study could not be established and (2) the probability of lithium-induced skin reactions was not assessed using validated instruments (e.g., the Naranjo Adverse Drug Reaction Probability Scale).

PREGNANCY

Results from a very large retrospective cohort study by Patorno and colleagues [22c] suggest that the risk for congenital cardiac malformations in infants exposed to lithium in utero increased between 2 and 168 percent when exposure occurs during the first trimester compared to non-exposed infants. The study was conducted based on a subset of women included in the US Medicaid Analytic eXtract (MAX) database between 2000 and 2010. Data from women who gave birth between 12 and 55 years of age whose pregnancies were financed by Medicaid funds were included in the study. Exposure to lithium was defined as having at least one prescription for lithium filled within the first trimester (90days) of pregnancy; this included a total 663 women. The reference group consisted of women not exposed to lithium (or lamotrigine) during the first trimester (approximately 1.3 million women). A total of 1945 data points from women who were exposed to lamotrigine during the first trimester (90days) of pregnancy constituted a second reference group. Compared to those who had not been exposed to either lithium or lamotrigine, women who were exposed to lithium were at a 65 percent higher risk of delivering infants with congenital cardiac malformations (relative risk = 1.65 [95 percent confidence interval 1.02–2.68]). Similarly, relative to women exposed to lamotrigine during the first trimester, those who received lithium were at a 125 percent higher risk of giving birth to infants with cardiac malformations (relative risk = 2.25 [95 percent confidence interval 1.17–4.34]). Stratified analyses suggest that lithium exposure greater than 900mg daily is associated with significantly elevated risk compared to unexposed women (relative risk = 3.22 [95 percent confidence interval 1.47–7.02]); this was not observed with doses between 601 and 900mg daily (relative risk = 1.6 [95 percent confidence interval 0.67–3.8]) or 600mg daily or less (relative risk = 1.11 [95 percent confidence interval 0.46–2.64]).

DRUG–DRUG INTERACTIONS

Kayipmaz and colleagues [23A] describe the presence of a possible drug interaction between lithium and moxifloxacin in an elderly patient in Turkey. A 74-year-old female patient with a history of bipolar disorder presented to the emergency department with severe tremors in both hands, slurred speech, myoclonic jerks in the upper extremities, exhaustion, and drowsiness. Her list of medications included lithium carbonate 300mg orally twice daily, haloperidol 10mg orally daily, risperidone 2mg orally daily, and sertraline 25mg orally daily (for depression). The patient had been stable on this regimen since 2011. Moxifloxacin 400mg orally daily was added to the regimen 1 day prior to the presence of symptoms (described above). On evaluation, body temperature and blood pressure were within normal limits. Ten days prior to the onset of symptoms, the patient's plasma lithium concentration was 0.8mEq/L; the day of presentation, lithium concentration as found to be 1.7mEq/L (normal range 0.5–1mEq/L). There were no signs or lab tests indicative of renal failure or dehydration; all other laboratory values were within normal limits. An ECG was insignificant for cardiac abnormalities. Lithium and moxifloxacin were stopped, and the patient was rehydrated with normal saline. Myoclonic jerks and hand tremors subsided within a day of treatment, and all neurologic signs normalized within a week. The authors postulated that (1) since the patient had been stable on risperidone, haloperidol, and sertraline for a number of years without incident and (2) since signs of toxicity developed within a day of starting moxifloxacin, that, therefore, using the Naranjo Adverse Drug Reaction Probability Scale, the likelihood of a lithium–moxifloxacin drug–drug interaction was "probable" (i.e., a score of 6). The authors concluded that the mechanism of the drug–drug interaction may be similar to that of levofloxacin, which slows renal excretion of lithium by decreasing glomerular filtration or tubular secretion.

Topiramate (TPM) has been shown to improve symptoms of obesity and binge eating disorder. Khan and Shah [24A] reported the case of topiramate-induced lithium toxicity in a 47-year-old African American male with a 20-year diagnosis of schizoaffective disorder and bipolar disorder type I. The patient was admitted from a group home upon increased aggression and violence to residents and staff in addition to nonadherence to medications prior to relapse. Upon admission the patient was treated with lithium and paliperidone for symptoms of psychosis and mania, to which he responded quickly and was stabilized. The patient's lithium level and comprehensive metabolic panel were monitored weekly and thyroid panel was monitored every 6 months. While hospitalized, the patient experienced compulsive urges to eat and drink fluids in excess, resulting in a 40-pound weight gain over 8 months. Water restriction and weight monitoring were implemented. Nonpharmagologic treatments and therapies were tried in an effort to control compulsive food and fluid consumption without success. The patient was started on TPM. [The dose, frequency, and duration of TPM were not reported]. Prior to initiation of TPM therapy, lab values were within reference range (lithium 0.47 mEq/L, Na 141 mEq/L, Cr 1.1 mg/dL, BUN 15 mg/dL, K 3.8 mEq/L). On day 4 of treatment, the patient presented with symptoms including slurred speech, hand tremors, unsteady gait, and rapid, shallow breathing. The patient was also disoriented, confused, and delirious. Both lithium and TPM were stopped due to suspected lithium toxicity. Examination of lab values indicated lithium levels in toxic range (1.46 mEq/L), and creatinine (3.4 mg/dL) and BUN (47 mg/dL) were above reference ranges. The patient was diagnosed and treated for renal insufficiency due to lithium toxicity. The patient was provided supportive, conservative treatment (including electrolyte imbalance correction and rehydration) and showed signs of considerable improvement. In light of lithium's narrow therapeutic index, the authors note that an awareness of the pharmacokinetic and pharmacodynamic properties of TPM as well as its inhibitory action on carbonic anhydrase need to be considered when prescribed concurrently with lithium.

℞ Foulser and colleagues emphasize the importance of close monitoring of patients—including those with normal plasma lithium concentrations—for signs of toxicity [25A]. A 62-year-old male stabilized for 20 years on lithium for bipolar disorder presented with acute mania and course tremor suggestive of lithium toxicity. Upon admission, lithium levels were normal. [The exact plasma lithium concentration was not provided.] Medical history included low-risk prostate cancer managed with regular prostate-specific antigen monitoring. Current medications were lithium 800 mg at night, trazodone 50 mg at night, and tadalafil 20 mg as needed. Blood tests revealed that plasma lithium levels were at the high end of normal range (0.95 mmol/L) with elevated creatine (123 μmol/L), urea 10.1 mmol/L, and eGFR of 52 mL/min. The patient's PSA lab was 6.82 μmol/L. The patient was rehydrated with fluids, and urea dropped to within normal range. Lithium was discontinued and olanzapine 10 mg was initiated. As that patient's mania was uncontrolled, lithium was restarted resulting in the return of tremor, indicative of lithium toxicity. Lithium was stopped and sodium valproate was initiated. Marked improvement in tremor was observed over a 2-week period. The authors suggest that the patient developed renal impairment during episodes of lithium toxicity, and that the recommend normal range of plasma lithium, particularly in the elderly, should be revisited. Furthermore, the authors recommend that consideration of what constitutes a "normal" lithium concentration should be approached in terms of the wellbeing of whole patient.

℞ A review by Pickard [26R] discusses single nucleotide polymorphism (SNP) variants and associated chromosomes and genes that may be predictive of lithium response. Several genes were identified at a statistical significance (P-value) level of less than 10^{-8} in four genome-wide studies of lithium response factors. These genes include the GADL1 gene on chromosome 3 (P-value of 10^{-37}), AL157359 gene on chromosome 21 (P-value of 10^{-9}), and the PTS, PLET1 gene on chromosome 11 (P-value of 10^{-9}).

Additional case studies on this topic can be found in these reviews [27R, 28R].

References

[1] Finley PR. Drug interactions with lithium: an update. Clin Pharm. 2016;55:925–41 [R].

[2] Baird-Gunning J, Lea-Henry T, Hoegberg LCG, et al. Lithium poisoning. J Intensive Care Med. 2017;32(4):249–63 [R].

[3] Gitlin M. Lithium side effects and toxicity: prevalence and management strategies. Int J Bipolar Disord. 2016;4:27 [R].

[4] Malhi GS, Outhred T. Lithium therapy, bipolar disorder—and neurocognition, Psychiatric Times. 2017;34(1):1–5. http://www.psychiatrictimes.com/special-reports/lithium-therapy-bipolar-disorder-and-neurocognition [R].

[5] Adis Medical Writers. Take care when using lithium with other drugs, as clinically relevant interactions may occur. Drugs Ther Perspect. 2017;33:66–9.

[6] Berk M, Dandash O, Daglas R, et al. Neuroprotection after a first episode of mania: a randomized controlled maintenance trial comparing the effects of lithium and quetiapine on grey and white matter volume. Transl Psychiatry. 2017;7:e1011 [c].

[7] Rossi FH, Rossi EM, Hoffmann M, et al. Permanent cerebellar degeneration after acute hyperthermia with non-toxic Lithium

levels: a case report and review of literature. Cerebellum. 2017;16(5–6):973–8 [A].
[8] Albright A, Way RT. Lithium toxicity and prolonged delirium. Aust N Z J Psychiatry. 2017;51(9):945. https://doi.org/10.1177/0004867417698230 [r].
[9] Smith KA, Cipriani A. Lithium and suicide in mood disorders: updated meta-review of the scientific literature. Bipolar Disord. 2017;19:575–86. https://doi.org/10.1111/bdi.12543 [R].
[10] Maddala R, Ashwal A, Rao M, et al. Chronic lithium intoxication: varying electrocardiogram manifestations. Indian J Pharmacol. 2017;49(1):127 [A].
[11] Bains A. Lithium-induced drug reaction with eosinophilia and systemic symptom syndrome. Indian J Dermatol. 2017;62(5):532 [A].
[12] Saad AB, Rjeibi I, Ncib S, et al. Ameliorative effect of cactus (Opuntia ficus indica) extract on lithium-induced nephrocardiotoxicity: a biochemical and histopathological study. Biomed Res Int. 2017;1–8 [E].
[13] Bhardwaj P, Jain K, Dhawan DK. Lithium treatment aggregates the adverse effects on erythrocytes subjected to arsenic exposure. Biol Trace Elem Res. 2017;184(1):206–13. https://doi.org/10.1007/s12011-017-1168-y [E].
[14] Decloedt EH, Lesoky M, Maartens G, et al. Renal safety of lithium in HIV-infected patients established on tenofovir disoproxil fumarate containing antiretroviral therapy: analysis from a randomized placebo-controlled trial. AIDS Res Ther. 2017;14:6 [c].
[15] Gupta S, Khastgir U. Drug information update. Lithium and chronic kidney disease: debates and dilemmas. BJPsych Bulletin. 2017;41:216–20 [R].
[16] Tondo L, Abramowicz M, Alda M, et al. Long-term lithium treatment in bipolar disorder: effects on glomerular filtration rate and other metabolic parameters. Int J Bipolar Disord. 2017; 5:27 [A].
[17] Pattanayak RD, Rajhans P, Shakya P, et al. Lithium-induced polyuria and amiloride: key issues and considerations. Indian J Psychiatry. 2017;59(3):391–2. 2p. (Case Study—case study, letter), ISSN: 0019-5545 [r].
[18] Masiran R, Aziz MFA. Hypertensive bipolar: chronic lithium toxicity in patients taking ACE inhibitor. BMJ Case Rep. 2017. https://doi.org/10.1136/bcr-2017-220631 [A].
[19] Shanks G, Mishra V, Nikolova S. Endocrine abnormalities in lithium toxicity. Clin Med. 2017;17(5):434–6 [A].
[20] Pinna M, Manchia M, Pudda S, et al. Cutaneous adverse reaction during lithium treatment: a case report and updated systematic review with meta-analysis. Int J Bipolar Disord. 2017;5:20 [A].
[21] Priyadharshini BSS, Ummar IS. Prevalence and sociodemographic profile of lithium-induced cutaneous side effects in bipolar affective disorder patients: a 1-year prospective observational study in South India. Indian J Psychol Med. 2017;39(5):648–52 [c].
[22] Patorno E, Huybrechts KF, Bateman BT, et al. Lithium use in pregnancy and the risk of cardiac malformations. N Engl J Med. 2017;376:2245–54 [c].
[23] Kayipmaz S, Altinoz AE, Ok NEG. Lithium intoxication: a possible interaction with moxifloxacin. Clin Psychopharmacol Neurosci. 2017;15(4):407–9 [A].
[24] Khan AH, Shah SQ. Topiramate-induced lithium toxicity. Cureus. 2017;9(10): e1804. https://doi.org/10.7759/cureus.1804 [A].
[25] Foulser P, Abbasi Y, Mathilakath A, et al. Do not treat the numbers: lithium toxity. BMJ Case Rep. 2017. https://doi.org/10.1136/bcr-2017-220079 [A].
[26] Pickard BS. Genomics of lithium action and response. Neurotherapeutics. 2017;14:582–7 [R].
[27] Maroney M, Liu MT, Smith T, et al. Chapter 3—Lithium. In: Ray SD, editor. Side effects of drugs: Annual: A worldwide yearly survey of new data in adverse drug reactions. Elsevier, 2016;38:21–28 [R].
[28] Thomas K, Darougdar S. Chapter 3—Lithium. In: Ray SD, editor. Side effects of drugs: Annual: A worldwide yearly survey of new data in adverse drug reactions. Elsevier, 2017;39:23–30 [R].

CHAPTER

4

Drugs of Abuse

Hannah R. Fudin*,†,1, Jennifer L. Babin*,†, Lisa T. Hong‡, Jennifer Ku†, Alisyn L. May*,§, Alexander Wisner¶, F. Scott Hall¶, Sidhartha D. Ray||

*University of Utah College of Pharmacy, Salt Lake City, UT, United States

†University of Utah Health, Salt Lake City, UT, United States

‡Loma Linda University School of Pharmacy, Loma Linda, CA, United States

§ARUP Laboratories Family Health Clinic, Salt Lake City, UT, United States

¶Department of Pharmacology and Experimental Therapeutics, University of Toledo College of Pharmacy and Pharmaceutical Sciences, Toledo, OH, United States

||Department of Pharmaceutical and Biomedical Sciences, Touro University College of Pharmacy and School of Osteopathic Medicine, Manhattan, NY, United States

[1]Corresponding author: hannah.fudin@pharm.utah.edu

INTRODUCTION

Drug use and abuse continues to be a significant problem affecting the population worldwide, represented by ~5% (250 million people), using at least one drug in 2015, of which 29.5 million (0.6% global adult population) suffer from drug use disorders (DUD). 'Health life' estimated by disability-adjusted life years (DALYs) resulted in an estimated 28 million years lost in 2015 due to harm from drug use, of which 17 million were solely attributed to DUD (all drug types). Overdose deaths in the United States (US) have increased more than threefold (1999–2015) from 16849 to 52404 annually, accounting for an 11.4% increase over the past year, reaching the highest level ever recorded. Access and availability to science-based services for treating DUD and related conditions remain limited, represented by one-in-six people provided drug treatment each year. In 2015, 12 million DALYs or 70% of the global disease burden attributed to opioids. The US accounts for the largest (~25%) estimated drug-related deaths worldwide, significantly resulting from opioid-use. Amphetamine-related DUDs are second to opioids for global health burden, while methamphetamine represents the greatest global health threat. Novel Psychoactive Substances (NPS) belong to diverse chemical groups. 106 countries and territories (2009–2016) reported the emergence of 739 different NPS to the United Nations, Office on Drugs and Crime (UNODC). Fentanyl analogues are associated with rising overdoses among opioid users due to increasing serious adverse events (AE) and deaths. Effects are compounded by the quantity and potency variation in active components and unawareness about the content and dosage in substances, coupled by little to no information available to determine their effects and how to counteract them. People who inject drugs (PWID) are faced with severe consequences represented by the number living with Human Immunodeficiency Virus (HIV) (1.6 million) and Hepatitis C Virus (HCV) (6.1 million). Globally, deaths caused by HCV for PWID are greater than any other cause of death related to drug use. Although DUDs in men compared to women are 2:1, once initiated women tend to increase their consumption rate, thereby progressing more rapidly to DUDs than men. The increased DALY rate attributed to DUD in 2015 (opioid and cocaine use) was greater for women (25% vs 40%) compared to men (17% vs 26%) respectfully. Drug use and abuse continues to cause an overall increase in morbidity and mortality in the changing culture worldwide. The community is faced with new obstacles including constant NPS variability, increased availability through online access channels, increased death rates due to spreading HIV and HCV, and limited access to treatment for DUD. Global standards developed by UNODC and the World Health Organization (WHO) for DUD treatment need to be better integrated within health-care systems across all countries ensuring evidence-based policies and interventions are available and effective to prevent unnecessary AE, toxicity, and mortality [1S].

Side Effects of Drugs Annual, Volume 40
ISSN: 0378-6080
https://doi.org/10.1016/bs.seda.2018.08.015

CANNABINOIDS: SYNTHETIC (HONG)

Synthetic cannabinoids (SCs) were initially developed as research tools for studying the endocannabinoid system. Recreational use was first reported in 2004 and SC became drugs of abuse (DOA) in 2008 [2c]. SCs are now the fastest growing psychoactive drug class with three generations and over 177 compounds identified by the United Nations (UN) Office on Drugs and Crime [2c,3c]. Many SCs are Drug Enforcement Administration (DEA) schedule 1 substances, but with continuous chemical composition evolution, some are not yet listed as such [4A,5R]. Currently available SC products have various names including, but not limited to: "Spice", "K2", "Black Mamba", "Blaze", "Skunk" and "Kronic" [6A,5R]. SCs are usually dissolved in solvent, applied to an inactive herbal substrate, and smoked, similar to cannabis (Δ9-tetrahydrocannabinol), but have much greater potency [2c]. Given the higher affinity for cannabinoid receptors, SCs lead to higher toxicity and hospitalization rates [7A]. Since July 2016, the US northeast region has seen more than 1200 emergency department (ED) visits related to SCs every month [8A]. The most common side effects (SE) are nausea, vomiting, anxiety, agitation, drowsiness, lightheadedness, tachycardia and mydriasis [2c,8A]. More serious intoxication symptoms include psychosis, delirium, auditory and/or visual hallucinations, hyperemesis, cardiotoxicity, seizures, acute kidney injury (AKI), severe lung injury, stroke, hyperthermia, and even death [2c,8A,9R]. Despite these risks, users may select SCs because they are able to bypass drug screening and are lower in cost compared to other drugs of abuse [2c,8A,5R,10r].

Cardiovascular

Andonian and colleagues authored a case series including three male patients 41, 35, and 53-years-old (YO) who presented with hypotension and bradycardia, seemingly on the verge of cardiovascular collapse after using "K2". Vital signs (VS) upon admission included heart rates (HR) in the 40s and blood pressure (BP) readings 66/35, 74/33 and 96/51 mmHg. Respiratory and neurologic function was maintained and patients could converse appropriately. All three patients had improved HR and BP within 7 h of intravenous (IV) hydration and walked out of the ED [6A].

A 50-YO African American male was brought to the ED with altered mental status (AMS), somnolence, agitation, and chest pain after smoking "Scooby Snax Limited Edition Blueberry Potpourri", later identified by gas chromatography-mass spectrometry (GC-MS) as AMB-FUBINACA and PB-22. BP and HR were 87/52 mmHg and 52 bpm, respectively, and he was given 1 L of normal saline. An inferior wall myocardial infarction (MI) was found on electrocardiogram (ECG). He was intubated and brought to the cardiac catheterization laboratory for drug-eluting stent placement in an occluded obtuse marginal second branch. His admission troponin = 0.048 ng/mL, peaked at 32.95 ng/mL and trended back down over 72 h after percutaneous coronary intervention (PCI). Patient had a normal echocardiogram. He reported a 17-pack-year smoking history and his urine drug screen (UDS) was positive for benzodiazepines (BZD), buprenorphine, and opiates, but negative for THC. The authors believe this is the first SC consumption report, resulting in MI requiring PCI [4A].

A 23-YO male (YOM) with no past medical history (PMH) presented to ED with shortness of breath (SOB), confusion, and palpitations after reportedly smoking "Bonzai", a SC, for the first time. No alcohol or other DOA were used. The patient was tachycardic (HR = 137 bpm) with atrial fibrillation (AF) identified by ECG and echocardiography showing normal cardiac function and structure. Serum Creatinine Kinase-Myocardial Band (72 U/L) and troponin I (1.49 ng/mL) were elevated, but all other laboratory findings were normal. Sinus rhythm restored spontaneously [11A].

A 24-YOM without any PMH arrived in the ED confused and agitated with visual hallucinations and palpitations 30 min after reportedly ingesting two drops of e-cigarette fluid containing a "VaporFi" mixture (propylene glycol, glycerin, and natural and artificial flavorings) and "liquid cannabis", which high-performance liquid chromatography and MS (HPLC-MS) identified as SCs: AB-FUBINACA (*N*-[(2S)-1-Amino-3-methyl-1-oxobutan-2-yl]-1-[(4-fluorophenyl)methyl]indazole-3-carboxamide) and ADB-FUBINACA (*N*-(1-Amino-3,3-dimethyl-1-oxobutan-2-yl)-1-[(4-fluorophenyl) methyl]indazole-3-carboxamide). The patient had an elevated BP (163/93 mmHg) and supraventricular tachycardia (SVT) (172 bpm), which resolved spontaneously. Serum potassium was slightly low (3.2 mmol/L). About 19 h later, repeat ECG showed normal sinus rhythm (NSR) at 82 bpm. UDS was negative, but urine and serum analyses confirmed ADB-FUBINACA in the urine and both AB-FUBINACA and ADB-FUBINACA concentrations in the serum (5.6 and 15.6 ng/mL, respectively). This case illustrates the oral AB-FUBINACA and ADB-FUBINACA bioavailability and rapid onset causing both cardiac toxicity and neurotoxic effects. The authors state that delayed drug exposure identification from a toxicology laboratory emphasizes the challenge posed to clinicians to promptly recognize similar clinical presentations and consider the differential diagnosis of SC toxicity [12A].

Respiratory

A case series described acute respiratory failure developing in four patients requiring intubation after reported SC use a few hours prior to presentation. None of the patients had pulmonary conditions. On admission, all patients (three males 27-, 28-, and 30-YO and one female 55-YO) had respiratory acidosis (arterial blood gas pH [6.98–7.198]) which resolved after 24 h. Chest radiographs were unremarkable. The authors concluded SC consumption may result in debilitating consequences such as respiratory depression [8A].

Another case series with three patients 21-, 23- and 43-YOMs, hospitalized from prison after MDMB-CHMICA intoxication report finding hypercapnia ($PaCO_2 = 48$, 55, and 95.27 for each patient, respectively) with reduced consciousness and seizures in the latter two patients. The first patient was bradycardic with a transient improvement to HR = 47 bpm after atropine 600 mcg treatment. He took propranolol before admission and his bradycardia resolved after 24 h. The other two patients were tachycardic (HR = 105 and 114 bpm). Drug screening detected their prescribed antidepressants or antipsychotics (mirtazapine, quetiapine, olanzapine for each patient, respectively). All symptoms resolved between 6 and 24 h [13A].

Nervous System

An incident in Brooklyn, NY was described as "Zombieland". Serum testing in eight patients who reported taking "AK-47 24 Karat Gold" identified methyl 2-(1-(4-fluorobenzyl)-1H-indazole-3-carboxamido)-3-methylbutanoate (AMB-FUBINACA) as causing the "Zombie" outbreak. Each aliquot of product contained 14.5–15.2 mg per gram. AMB-FUBINACA is a SC with a potency 85 times Δ9-tetrahydrocannabinol (Δ-9 THC) and 50 times JWH-018 (found in "K2"). The index patient had normal vitals, labs, and ECG, but was lethargic with a blank stare, occasional groaning and slowed arm and leg movements, which normalized about 9 h after being hospitalized. The authors note that central nervous system (CNS) depression from AMB-FUBINACA intoxication is unusual since it is not associated with seizures, cardiotoxicity, nor AKI, unlike other potent SCs [2c].

A 25-YOM woke up with left-sided hemiparesis, hypesthesia, dysarthria and visual neglect after smoking 3 g of "Freeze" (later identified as ADB-FUBINACA) the evening prior. He denied any other drug abuse for the past 2 years (previous amphetamine, cocaine, heroin, opioid, and lorazepam use), but reported alcohol consumption up to twice weekly and a 10-pack-year smoking history. An acute ischemic infarct in the proximal right middle cerebral artery was found on magnetic resonance imaging (MRI) with right cerebral brain edema and hemorrhagic transformation of the lentiform nucleus without signs of herniation seen on a cerebral computerized tomography (CT) scan. Transesophageal echocardiography, spinal fluid analysis and diagnostic testing for thrombophilia and vasculitis were all negative. Serum screening was negative, including ethanol, but ADB-FUBINACA and MDMB-CHMICA were detected in the patient's UDS. After 9 days, he was discharged to a rehabilitative center with unresolved paresis, hypesthesia and dysarthria. The authors believe this case is similar to previous reports and that it suggests a cardioembolic etiology from transient tachyarrhythmias associated with SC use, given the lack of other vascular risk factors. However, unlike previous reports, this case provides laboratory confirmation for the product ingested and a temporal relationship with ischemic stroke [7A].

Wolff and colleagues reviewed 9 articles regarding neurovascular complications during or following SC use in 13 patients. On average, patients were 29.3 ± 10.4-YO and the male to female ratio was 1.6:1. Ischemic stroke was described in eight cases with five patients using tobacco concurrently and five cases describing hemorrhagic stroke with only one patient using tobacco. Persistent neurologic symptoms were seen in two patients who had an ischemic stroke and one patient who had a hemorrhagic stroke [10r].

In Europe, AB-CHMINACA and MDMB-CHMICA were first reported in 2014. These SC products represent a third generation with higher potency or half maximal effective concentration ($EC_{50} = 0.278$ nM and $EC_{50} = 0.142$ nM, respectively) compared with THC ($EC_{50} = 15.19$) and even in comparison with JWH-018 ($EC_{50} = 1.132$ nM). In a multicenter, prospective case series, Hermanns-Clausen et al. evaluated 44 patients with confirmed AB-CHMINACA and/or MDMB-CHMICA consumption without high concentrations of other psychoactive substances or DOA. The patient majority was male ($n = 39$; 89%) with a median age = 20.5-YO (IQR 16–25) who smoked ($n = 41$; 93%), ate, or sniffed SC. Symptoms from these SCs were similar with the exception that MDMB-CHMICA was associated with a greater hallucination incidence ($n = 3$[15%] vs $n = 12$[50%], $P < 0.025$). Other symptoms that occurred in ≥20% of patients were mostly neurologic, including CNS depression ($n = 27$[61.4%]), disorientation ($n = 20$[45.5%]), agitation ($n = 15$[34.1%]), seizures ($n = 12$[27.3%]), mydriasis ($n = 13$ [29.6%]), tachycardia ($n = 21$[47.7%]), vomiting ($n = 16$ [36.4%]), and hypokalemia ($n = 10$[22.7%]). The poisoning severity was minor, moderate and severe in 4(9%), 31(70.5%), and 9(20.5%) patients, respectively, without any correlation to SC serum levels. Most symptoms were short-lived, lasting 2–16 h; however, 15(34.1%) patients had prolonged symptoms lasting 1 day to 3 weeks. One patient had short-term memory impairment that lasted 3 weeks and another developed psychosis and self-mutilating behavior leading to a 5-day hospitalization for pneumomediastinum. This study also compared results to another publication evaluating 29 similar patients who consumed SCs (aminoalkylindole compounds such as JWH-081). As expected given the higher potency, the newer SC generation had a greater incidence of several neuropsychiatric clinical effects including somnolence ($n = 27/44$[61%] vs $n = 10/29$[34%], $P = 0.0323$), disorientation ($n = 20$[45%] vs $n = 4$[14%], $P = 0.0054$), seizures ($n = 12$[27%] vs $n = 1$[3%], $P = 0.0112$), and combativeness ($n = 9$[21%] vs $n = 1$[3%], $P < 0.0001$) with more patients experiencing symptoms for ≥24h ($n = 15$[34%] vs $n = 1$[3%], $P = 0.0015$). Tachycardia ($n = 21$[48%] vs $n = 22$[76%], $P = 0.0279$) and hypertension ($n = 6$; 14 vs $n = 10$[34%], $P = 0.0457$) actually occurred more among those consuming a first generation SC in the aminoalkylindole group [3c].

A 40-YOM, previously healthy with a PMH including uncomplicated febrile seizure early in life without any recurrence presented with stupor, unresponsiveness, left gaze deviation, lip smacking and was found to be in status epilepticus. He was intubated, given IV lorazepam, valproic acid, levetiracetam, lacosamide, and clobazam. Electroencephalography identified right temporal seizures, generalized periodic discharges, and diffuse slowing. Lumbar puncture and workup for viral encephalitis, prion disease and paraneoplastic meningoencephalitis were unremarkable, as were labs, serum and UDS, as well as imagining including chest, abdomen and pelvis CT. Hyperintensities within the limbic structures were seen bilaterally on brain MRI, which may have been related to epileptiform activity. The patient was also given methylprednisolone 250mg IV every 6h and two doses of IV immunoglobulin. After extubation, the patient exhibited aggression and severe psychosis, including violent and homicidal behavior leading to use of physical restraints in addition to antipsychotic, sedative, and hypnotic medications. During follow-up 1-month post-discharge, the patient reported experimentation with Spice days prior to his hospitalization. He did not have any permanent neurologic sequelae. The authors believe this report is the first case identifying new-onset refractory status epilepticus associated with SC use [14A].

Social web data analysis, supported by the National Institute on Drug Abuse, was conducted to better understand the trends in SC-related posts. Of 19052 posts from international users regarding SCs only and no other drugs, 5160 contained mention of an effect, including feeling "high" ($n = 2155$[43.2%]), hallucinations ($n = 560$[11.2%]), anxiety ($n = 470$[9.4%]), overdose ($n = 307$[6.2%]), euphoria ($n = 286$[5.7%]), seizures ($n = 267$[5.4%]), and nausea ($n = 246$, [4.9%]). When evaluating trends (2008–2015), the proportion of positive posts mentioning highs and euphoria decreased significantly ($P = 0.003$ and $P < 0.001$, respectively). On the other hand, a significant increase was seen in trends of negative posts regarding seizures ($P < 0.001$), overdose ($P < 0.001$), nausea ($P < 0.001$) and anxiety ($P = 0.0013$). A non-statistically significant increase was seen in posts regarding hallucinations ($P = 0.1497$). The authors concluded that these posts are consistent with the literature on effects seen with SCs and that the trends suggest increased harmful risk with newer generation SC products [15MC].

Urinary Tract

Park et al. summarized several AKI cases associated SC use. Most kidney biopsies were consistent with acute tubular necrosis ($n = 18$), but there were also acute interstitial nephritis ($n = 3$) cases. Cannabis use does not seem to impair renal function; however, additional research is needed to fully explain the effects of cannabinoids on the development of kidney injury or disease [16R].

Hepatic

In a review by Solimini and colleagues, multiple hepatotoxicity reports associated with SC use are summarized. A multicenter (60 EDs in Japan), retrospective study ($n = 518$) identified 25 patients with complications concerning liver dysfunction. Most patients were 20–30-YOM users. Toxicology screening identified various products including: 5F-PB-22, PB-22 (quinolinyl-ester-indole cannabinoid); UR-144, XLR-11 (tetramethylcyclopropylketone indole); MAM-2201, JWH122N-(4-pentenyl) analog, JWH-081, JWH-122, JWH-210, AM-2232 (naphthylindole); AKB-48, AB-PINACA (indazolecarboxamide); and AM-694 (benzoylindole) [5R].

Drug–Drug Interactions

The hepatic cytochrome P450 (CYP) enzyme system is involved in SC metabolism. JWH-018 and AM-2201 (its fluorinated analogue) are metabolized in the liver by CYP2C9 and 1A2. Thus, there is the potential to cause AE when given concurrently with medications such as valproic acid, warfarin, phenytoin, and ciprofloxacin. CYP2C9 is also present in the intestines and is involved in the metabolism of orally ingested SCs whereas CYP1A2 is present in the lungs and involved in the metabolism of smoked SCs. CYP2D6 is thought to be involved in metabolism of these products in the cortex, hippocampus and cerebellum. Glucuronic acid conjugation facilitates elimination of these products in the urine and glucuronide metabolite formation involves UGT isoforms. Given the variability in SCs including products combined with other drugs, the possibility of within product interactions and interactions with prescribed medications exist and should be considered. However, additional studies are needed to better understand the clinical significance of these interactions [17R].

CANNABINOIDS: NON-SYNTHETIC (HONG)

While Colorado and Washington were the first states to legalize recreational marijuana in 2012, there are 28 states that have since followed suit [16R]. With use becoming more widespread, we must recognize the potentially harmful effects associated with cannabis use.

Cardiovascular

Desai et al. analyzed almost 500000 hospitalizations in the US (2010–2014) with the National Inpatient Sample database, which includes a 20% inpatient sample from 1000 hospitals in >40 states, to evaluate the relationship between marijuana use and acute MI (AMI). Researchers evaluated hospitalized AMI patients and compared those with and without a marijuana use history. In comparison to the non-marijuana group ($n=487317$), the marijuana group ($n=7202$) consisted of younger patients (49.34 ± 10.8 vs 57.79 ± 8.98 years), more males (76.9% vs 66%), Medicaid enrollees (20.6% vs 36.6%), weekend hospitalizations (28.4% vs 25.8%) and disposition against medical advice (4% vs 1.2%). Marijuana use was associated with an increased AMI risk (3%) (OR=1.031 [95% CI: 1.018–1.045], $P<0.001$). The study also found a yearly increase in the prevalence of AMI, dysrhythmias, respiratory failure, cardiogenic shock, congestive heart failure (CHF), and mortality in marijuana users during the 5-year analysis period. The authors believe these findings may be due to the generation of carboxyhemoglobin and interference of cellular oxygenation secondary to marijuana inhalation as well as increased sympathetic activity by THC [18MC].

According to a recent systematic review, ischemic stroke is the most commonly reported cardiovascular effect of cannabis use and the evidence linking cannabis use to ischemic stroke is strong. In a population survey in Australia ($n=7455$), a 2.3-fold increase in the risk for cerebrovascular ischemic events was seen in those who used cannabis during the past year and a 4.7-fold increase in those who used cannabis weekly or more frequently. To further support the association between cannabis consumption and ischemic stroke, some patients have demonstrated recurring cerebrovascular ischemic symptoms following cannabis use resumption. The responsible mechanism for stroke is thought to be reversible cerebrovascular spasm (RCVS) and 32% of 67 patients with angiographically confirmed RCVS were positive for cannabis on toxicology screening. Of interest, cerebral artery narrowing in the posterior cerebral circulation is correlated with cannabis use [19R].

In a review, Wolff and colleagues aimed to describe cannabis-related strokes and found that most cases involving chronic cannabis smoking were associated with ischemic strokes, while only four were hemorrhagic strokes. The patients were 32.7 ± 12-YO with a 3.7:1 male to female ratio. One-quarter of cases reported a recent cannabis consumption increase. Other risk factors were tobacco ($n=59$[69%]) and alcohol ($n=25$ [29%]) use as well as concurrent vasoactive or illicit drugs in 10 cases. The stroke location was posterior ($n=27$[37%]) and anterior ($n=39$[53.4%]); however, when looking only at the patients with RCVS ($n=23$[31.5%], $n=11$[48%]) were posterior and $n=7$[30%] were anterior. Persistent neurologic deficits occurred in 69 of 85 patients and 9 had a recurrent stroke, all of whom continued cannabis use [10r].

A case report of a 51-YO female (YOF) with a PMH including hypertension and asthma, presented to the ED 2h after the onset of left-sided weakness. She was tachycardic (102bpm) and hypertensive (256/112mmHg), had a right cerebral infarct for which hemorrhage was ruled out via head CT, a UDS positive for cannabis, and normal labs. Her friend reported long-term and heavy cannabis consumption prior to symptom onset. IV labetalol was given for BP control and recombinant tissue plasminogen activator (rtPA) was given within 1h of presentation after which she developed confusion and slurred speech within 30min. A new hemorrhage in the left pons was found on repeat head CT and the patient was soon declared brain dead. The authors concluded that heavy cannabis use might augment the effects of rtPA [20A]. However, the patient's National Institute of Health Stroke Scale score and BP during the hemorrhagic transformation were not reported.

Gastrointestinal Tract

A case series involving three adults with chronic inhaled cannabis use describes several abdominal pain episodes resulting from intussusception in various locations within the small intestine without evidence of an underlying lesion. The abdominal pain resolved following the cannabis smoking discontinuation. A 27-YOF had 9 months without another episode when cannabis consumption was stopped, but returned to the ED with a new sub-occlusive episode when she resumed cannabis smoking. A 23-YOM who continued smoking cannabis daily was hospitalized 7 times and presented to the ED around 30 times during a 10-year follow-up period. While there were two previous case reports of intussusception with mention of cannabis consumption, a relationship was not endorsed. The authors proposed that cannabis plays a potential role involving intussusception onset [21A].

Dezieck and colleagues reviewed 13 cannabinioid hyperemesis syndrome (CHS) cases treated successfully with topical capsaicin. Nine men and four women reported chronic cannabis use (mean age=34-YO). Most ($n=10$) patients did not have relief with IV antiemetics and were discharged after using capsaicin. Capsaicin activates the transient receptor potential vanilloid 1 (TRPV1) receptor, which is thought to play a role in the cyclic abdominal pain with nausea and vomiting. This is seen with CHS as receptor overstimulation leads to their desensitization [22c].

Mineral Balance

A case series including six men (27–68-YO) admitted to a Veteran Affairs Medical Center (VAMC) for CHS, with symptom onset ranging from 7 h to 3 days, had positive hypophosphatemia. Serum phosphate levels were <1.3 mg/dL in all patients despite normal calcium levels, no alcoholism or malnutrition history, no concern for refeeding syndrome, and no phosphate binder use. In the previous year, four patients had ≥1 admission for CHS with serum phosphate <2.5 mg/dL. Additionally, five patients had normal serum phosphate levels prior to admission indicating an acute process. This is the first report of hypophosphatemia in patients presenting with CHS. Half of the cases required phosphate supplementation while the other half self-corrected within hours, which the author believes is reflective of phosphate redistribution secondary to respiratory alkalosis given the patient's underlying anxiety disorders or possible anxiety associated with acute cannabis discontinuation since most reported daily use. Notably, two patients who did not receive phosphate supplementation did received lorazepam, which may have assisted in the resolution of the underlying anxiety and respiratory alkalosis [23c].

Nervous System

A retrospective cohort ($n=478\,650$ hospitalized patients with acute ischemic stroke) found that $n=11\,320$[2.36%] patients reported cannabis use, which was identified as anindependent predictor of hospitalization for acute ischemic stroke (OR=1.17, [95% CI: 1.15–1.20]; $P<0.0001$) [24MC]. In another retrospective analysis, 25.9% of 213 patients with an aneurysmal subarachnoid hemorrhage were cannabis users detected by UDS. Patients taking cannabinoids (~50%), and non-cannabinoid users (~25%) had delayed cerebral ischemia (DCI) defined as a new neurological sign, and/or worsening level of consciousness ≥48 h after presentation. Cannabinoid use was independently associated with an increased risk for development of DCI (OR=2.68, [95% CI: 1.03–6.99]; $P=0.01$) [24MC, 25C].

While we already have data to support cannabis use increasing the developing risk for psychotic disorders (twofold) without increasing the risk of transitioning to psychosis, McHugh et al. prospectively evaluated adults ($n=190$) at ultra-high risk (patients with a genetic vulnerability or with attenuated or intermittent psychotic symptoms in the past year) with a 5-year mean follow-up period. The study was unable to demonstrate an association between first cannabis use age or cannabis use severity with conversion to psychosis. However, 28 patients developed psychosis and an increased risk for this transition was seen among cannabis users ($n=18/28$; 64.3%) who experienced attenuated psychotic symptoms associated with cannabis use ($n=10/18$; 55.6%) with a hazard ratio (HR)=4.9 (95% CI: 1.93–12.44) [26C].

In a study involving 48 heavy chronic and systematic cannabis users, 45 were male with a 5 g median daily cannabis dose and mean age=30.2±10-YO. Upon clinical psychiatric evaluation, patients reported hallucinations (39.6%; 57.9% auditory, 21.1% visual, and 21.1% with both), delusions (54.2%; 57.7% had both reference and persecution symptoms), and organic brain dysfunction (85.4%; 5% decision making, 22.5% memory, 7.5% concentration, and 7.5% abstract thinking). In patients <35-YO (72.9%), a correlation was seen between cannabinoid amount identified in hair samples and those with hallucinations and delusions compared with those who did not experience these effects. The cannabinoid levels were ≥0.2 ng/mg greater in those with paranoid schizophrenia-like symptoms. Of note, the symptom onset occurred ≥1 after starting heavy cannabis use. Additionally, in eight patients who developed psychiatric symptoms and discontinued use, all of them continued to have symptoms during a follow-up evaluation 3 months later [27c].

A literature review including 13 studies evaluating the effect of Δ9-THC and cannabidiol on cognition suggested that preparations with greater Δ9-THC concentrations are associated with increased memory disruption. However, this effect is absent when Δ9-THC is combined with cannabidiol [28r].

Drug–Drug Interactions

In a review including clinical and animal studies by Iffland and colleagues, two clinical trials were summarized demonstrating the cytochrome P450 3A4 and 2C19 isozyme-mediated interaction between cannabidiol and clobazam in pediatric patients with epilepsy. Cannabidiol was initially dosed at 5 mg/kg and up-titrated to 25 mg/kg, leading to an increase in clobazam bioavailability, reduction in clobazam dosing and decreased clobazam-related SEs. While inhibition of several other cytochrome P450 isozymes by cannabidiol has been seen in vitro, the clinical impact in humans has yet to be fully elucidated [29r].

Pregnancy

In a retrospective review including 191 maternal–infant cases where opioid-dependent pregnant women were receiving buprenorphine, marijuana was found in the UDS of 76 women during the third trimester. No differences were seen in gestational age, birth weight (BW),

Apgar scores, peak neonatal abstinence syndrome (NAS), and time to NAS symptom onset between infants exposed to marijuana during the third trimester and infants not exposed to marijuana. There was a trend toward increased length of hospital stay and pharmacotherapy for NAS; however, additional research is needed to further evaluate findings [30C].

Similarly, another review including marijuana use during pregnancy found a 100 g lower BW, which is not clinically significant. No association was identified between prenatal marijuana use and preterm birth rates, major congenital anomalies, newborn behavior, or marijuana use at 14-YO. However, some found lower memory and verbal scores in children ≤3-YO and ≤10% greater impulsivity and hyperactivity scores in 6-year-old children as well as 5%–10% difference in Stanford–Binet IQ test scores in the verbal and/or quantitative components when comparing marijuana use ≥once daily vs < once daily among 10-year-old children [31r].

Age

A retrospective study was conducted to assess accidental or unintentional cannabis exposure (confirmed by UDS in children (n = 29) <3-YO in Toulouse, France. The average age and weight were 16.5 ± 5.2 months and 11.2 ± 2.1 kg, respectively. The most common exposure was ingested hashish (resin). Children presented with tachycardia (n = 17[58.6%]), mydriasis (n = 14[48.3%]), drowsiness (n = 7[24%]), hypoventilation (n = 6[20.6%]) (of which half required assisted ventilation), agitation (n = 3[10.3%]), seizures (n = 4[13.8%]) and decreased level of consciousness with Glasgow Coma Score (GCS) < 12 (n = 10[34%]) [32r].

A review of cannabis influence on adolescent brains found that heavy cannabis use along with alcohol use is associated with deficits in attention, memory, processing speed and visuospatial performance. While some evidence suggests verbal learning and working memory improvement after 3 weeks of abstinence, others have seen neuropsychological changes, including attention deficits, persisting even >1 month after abstinence when compared to nonusers. Neuroanatomical changes (e.g., reduced grey matter volume in the hippocampus, prefrontal cortex and amygdala) support these findings [33r].

CANNABINOIDS (KU)

Pharmacogenetics

ABCB1 is a gene that encodes for P-glycoprotein 1 (PgP, also known as multidrug resistant protein 1, MDR1), which is known to modify drug pharmacokinetics. Kebir et al. analyzed serum THC concentrations in chronic heavy cannabis users (individuals using >7 joints per week) to determine if the *ABCB1* polymorphism C3435T (rs1045642) modulates cannabinoid pharmacokinetics. In a sample of 39 French Caucasians, the authors found that T allele carriers had significantly lower plasma THC concentrations than non-T carriers (P = 0.043). However, the C variant allele is associated with higher intestinal PgP expression and thus, more rapid drug efflux. The authors hypothesize that decreased THC concentrations seen with chronic users carrying the T allele may be due to higher THC body stores that are slowly released back into circulation overtime, which suggests that T allele carriers may have milder cannabis withdrawal symptoms than non-T carriers [34c].

THC, the primary psychoactive ingredient in cannabis, produces its psychoactive effects by binding to cannabinoid 1 (CB1) receptors in CNS. The rewarding cannabis effects may be associated with genetic variations in *CNR1* rs2023239, a known SNP polymorphism that regulates CB1 functioning and CB1 concentrations. To assess if the rs2023239 polymorphism is associated with CB1 levels, Ketcherside et al. quantified the CB1 levels in 58 daily cannabis users and 47 nonusers. The authors found that cannabis users carrying the G allele had greater normalized CB1 density than non-G carrier users (P = 0.038), which suggests that G allele carriers of rs2023239 may be more susceptible to the rewarding cannabis effects and exhibit greater cannabis withdrawal symptoms during abstinence [35C].

ALCOHOL (BABIN)

Cardiovascular

Moderate alcohol consumption has been associated with both protective and harmful cardiovascular effects in the literature, but less is known about the population-level effects of alcohol abuse. In order to determine if alcohol abuse increases cardiovascular risk, Whitman et al. utilized a large statewide healthcare database to identify patients with alcohol abuse and a cardiovascular diagnosis across a 5-year period. Of the >14 million patients in the database with AF, myocardial MI, or CHF, 268 084 patients (1.8%) were coded with an alcohol abuse diagnosis. After multivariable adjustment, the risks of AF (HR = 1.93, 95% CI 1.88–1.98), MI (HR = 1.45, 95% CI 1.4–1.51), and CHF (HR = 2.34, 95% CI 2.29–2.39) in patients with alcohol abuse were all increased to a similar extent compared to other established modifiable risk factors for these conditions. The authors hypothesized that the mechanism of alcohol-related AF is electrical rather than structural, and the mechanism for alcohol-related CHF is distinct from age-related or hypertensive cardiomyopathy [36MC].

In light of the increased AF risk with current alcohol use, Dixit et al. sought to determine if risk decreases with alcohol cessation. The authors conducted a population-based prospective cohort study ($n=15222$; aged 45–64-YO) including patients who were former drinkers ($n=2886$[19%]), current drinkers ($n=8546$[56%]), and never drinkers ($n=3790$ [25%]). The median follow-up period was 19.7 years, during which there were new AF cases ($n=1631$), with 370 occurring in former alcohol drinkers. While former drinkers had an increased AF risk compared to never drinkers (HR = 1.46, [95% CI 1.26–1.68]) in an unadjusted analysis, this association became statistically insignificant after adjustment for multiple variables. Former drinkers, who were abstinent for longer periods, drank for a shorter period of time, and consumed smaller alcohol quantities were less likely to develop AF. Interestingly, the authors did not find a relationship between current alcohol use and AF and suggested this finding could be due to unrecognized AF diagnoses using the dataset [37MC].

Huang et al. also conducted a retrospective, population-based cohort study in patients diagnosed with alcohol intoxication to elucidate peripheral arterial disease (PAD) risk. Previous studies have shown a lower PAD risk in light to moderate drinkers than non-drinkers, and a higher PAD risk in heavy drinkers. This study identified 56544 patients with healthcare visits for alcohol intoxication over a 10-year period in Taiwan. In comparison to patients with no alcohol intoxication history, the incidence of PAD development in patients with alcohol intoxication was 12.8- vs 3.74-per 100000 person-years. After adjustment, the HR = 3.80 (95% CI 3.35–4.32). Study limitations were a lack of available details regarding long-term alcohol use patterns and tobacco use [38MC].

HEENT

Previous studies have failed to demonstrate a strong association between alcohol use and voice disorders. Bainbridge et al. analyzed data from 14794 US adults 24–34-YO. The authors estimated a 6% voice disorder prevalence of ≥3 days' duration over a 12-month period in this group. Compared to abstinence, occasional or moderate alcohol use increased the likelihood of having a voice disorder (OR = 1.57, 95% CI 1.14–2.15 for drinking 0–3 days/month). The mechanism, quantity, and duration of alcohol consumption relating to voice disorder development are still unclear [39MC].

Nervous System

Studies in young adult animals and humans have raised concerns about alcohol's effect on the brain. Carbia et al. specifically examined the relationship between binge drinking and verbal episodic memory in 155 university students 18–23-YO over a 6-year period. The authors found that participants who continued binge drinking from adolescence through early adulthood experienced deficits related to poor verbal memory performance [40C].

Meda et al. studied gray matter volume change in young adults (mean age 18.5-YO at baseline), comparing non- and light-drinkers ($n=45$) to sustained heavy drinkers ($n=84$). Light drinking was defined as binge drinking <13 of the past 26 weeks, and heavy drinking was defined as meeting criteria for alcohol abuse or consuming ≥4 drinks/occasion for females and ≥5 drinks/occasion for males during ≥13 of the previous 26 weeks. Over a 2-year period, both groups exhibited a significant decline in gray matter volume via MRI with a more widespread loss in participants in the heavy drinking group. Gray matter decline mainly occurred in brain areas responsible for memory, emotion, decision-making, and mental flexibility [41C].

Another observational cohort study examined the effects of alcohol consumption on brain structure and function in 550 participants (mean age = 43-YO). Based on MRI results, participants with higher alcohol use over a 30-year follow-up period had increased odds of hippocampal atrophy. Risk increased according to alcohol amount consumed (OR = 5.8, [95% CI 1.8–18.6] for alcohol consumption >30 units/week (240 g/week) and OR = 3.4, [95% CI 1.4–8.1] for 14–21 units/week (112–168 g/week)) [42C].

Sensory Systems

Haldar et al. reported the first published case of microcystic macular edema related to excess alcohol consumption and tobacco use. A 48-YOM with a 50 pack-year smoking history and unquantified heavy alcohol use presented with reduced visual acuity and color vision. When the patient stopped drinking alcohol and smoking, remission was achieved and visual function improved [43A].

Endocrine

The effect of alcohol use on blood glucose in patients with type 1 diabetes (T1DM) has been inconsistent. A recent study used a registry with data from German and Austrian diabetes centers to study 29630 patients with T1DM (12–29-YO; median age = 17). Overall, $n=3213$ [10.8%] reported regular alcohol use, and $n=143$[0.5%] were considered to be at-risk drinkers (≥12 g/day consumed for females and ≥24 g/day for males). Adjusted HbA1c was highest in patients with at-risk drinking behavior (9.3%) and lowest in patients who abstained from alcohol (8.6%). The adjusted rates for severe hypoglycemia and diabetic ketoacidosis were also lower in patients who reported no alcohol consumption [44MC].

Gastrointestinal

Alcohol use has been associated with esophageal cancer, but data regarding the relationship between alcohol and Barrett's esophagus and reflux esophagitis are conflicting. Filiberti et al. conducted a multicenter case–control study in Italy that compared wine-, beer-, and liquor-use among patients with endoscopically confirmed Barrett's esophagus ($n=339$) or erosive esophagitis ($n=462$) with patients without disease ($n=619$). The results suggested a reduced Barrett's esophagus and erosive esophagitis risk with beer intake and low wine consumption. With higher alcohol intake, there was a non-statistically significant increased disease risk. Therefore, the relationship between alcohol and these esophageal disorders remains uncertain based on the study results [45C].

Skin

Case–control studies have reported different results regarding the relationship between alcohol use and rosacea. Li et al. conducted an epidemiologic study to determine the association between alcohol intake and rosacea in US women using a large dataset, including 82737 women with >14 years of follow-up. Women with increased alcohol intake, especially white wine and liquor, had a significantly increased rosacea risk compared to women who never drank alcohol ($P<0.0001$). The multivariate-adjusted HR=1.12 (95% CI 1.05–1.20) for 1–4g/day alcohol intake which increased to 1.53 (95% CI 1.26–1.84) when using ≥30g/day [46MC].

A case involving ethanol-induced systemic allergic contact dermatitis has also been published. An Asian 30-YOF reported a previous history of a widespread, painful papular rash 8–12h after drinking liqueur. On separate occasions, she developed rashes after accidental exposure to a spilled alcoholic beverage, alcohol swabs, and an ethanol-based perfume. Skin testing confirmed ethanol-induced hypersensitivity, and the patient has been managed with alcohol avoidance [47A].

Musculoskeletal

Chronic alcohol abuse is associated with osteopenia and fractures, but the mechanism is not clear. Ulhoi et al. recently conducted a case–control autopsy study in 84 individuals. Fifty-two were males (mean age=44) and 32 females (mean age=46). Patients with chronic alcohol abuse had a mean daily alcohol intake ≥72g (males) or 48g (females) for ≥5 years immediately preceding death. Patients with other potential osteopenia causes were excluded from the study. The authors found a significantly lower bone mineral density (BMD) in the femoral neck and iliac crest in patients with alcohol abuse history compared to controls. No difference in BMD was seen in the lumbar vertebrae. A bone sample analysis suggested that decreased BMD is due to decreased bone formation, and not trabecular bone structure destruction [48c].

A 63-YOM presented with pain and limited right hip movement for 4 months, with a palpable soft tissue mass deep in his groin. The patient had a history of daily alcohol intake (300mL of clear spirits) for 40 years. He was diagnosed with femoral head necrosis due to heavy alcohol use and an associated iliopsoas cyst. Total hip arthroplasty was performed, and the patient was free from complications after 1 year [49A].

Immunologic

Barbhaiya et al. examined the association between alcohol use and systemic lupus erythematosus (SLE) development in a pooled cohort ($n=204055$ women). Within the two groups, 244 SLE cases developed. The mean age at diagnosis in the first cohort was 55.8 years and 43.4 in the second cohort. The majority of women included were Caucasian. The authors found an inverse association between cumulative average alcohol intake and SLE risk in women who drank alcohol ≥5g/day compared to non-drinkers (multivariable-adjusted HR=0.61 [95% CI 0.41–0.89]). The authors suggested possible alcohol-induced mechanisms to explain these results are diminished cellular response to immunogens, suppressed proinflammatory cytokine synthesis, antioxidant effects, reduced serum IgG, and epigenetic changes [50MC].

Tumorigenicity

It is generally accepted that alcohol use is a risk factor for oral cavity, pharynx, larynx, esophagus, liver, colon/rectum, and breast cancers [51S]. Evidence for lung and prostate cancer, however, is less robust. Betts et al. analyzed a cohort (9003 adults) in the United Kingdom (UK) with 25-years of follow-up and found that alcohol consumption (15–28units/week (120–224g/week)) was associated with an increased risk for cancer, specifically lung (HR=2.23 [95% CI 1.18–4.24]) and colorectal (HR=2.28 [95% CI 1.13–4.57]) in men. The study did not find a significantly increased risk based on weekly alcohol quantity consumed for lung, breast, or colorectal cancers in women and prostate cancer in men. This finding may have been due to the low numbers of women who reported high levels of drinking. Additional study limitations include the possibility that the association between alcohol use and lung cancer was influenced by residual confounding by cigarette smoking, and the alcohol use assessment at only one time point in the beginning of the follow-up period [52MC].

Fehringer et al. also examined the association between alcohol and lung cancer risk using 22 case–control and cohort studies for patients (n=2548) with lung cancer and controls (n=9362) without lung cancer who had never smoked. Without the smoking confounder, authors found that low to moderate wine and liquor consumption (0–4.9 g/day) was inversely associated with lung cancer risk. However, the authors acknowledge that other confounders may have been present that affected the results [53MC].

While alcohol consumption is an established risk factor for breast cancer, most studies have not examined women younger than 40-YO. A large US cohort study involving 93835 women aged 27–44-YO concluded that alcohol consumption ≥10 g/day alone in younger women is not associated with breast cancer. However, the breast cancer risk is increased in women with both a breast cancer family history and low folate intake (<400 μg/day) [54MC].

White et al. examined breast cancer risk in relation to binge drinking in 50884 US women aged 35–74-YO who had a sister diagnosed with breast cancer. Over a mean 6.4-year follow-up period, higher lifetime alcohol intake (≥230 drinks/year vs <60 drinks/year) was associated with increased breast cancer risk (HR=1.35 [95% CI 1.15–1.58]). Compared to women who consumed <60 drinks/year, women with a binge drinking history also had an increased breast cancer risk (HR=1.29 [95% CI 1.15–1.45]). However, this association was not significant in low-level drinkers who reported ever binging [55MC].

A prospective population-based cohort study that examined adults (n=23323730) in South Korea found that light/moderate (<30 g/day) and heavy (≥30 g/day) alcohol consumption both significantly increased esophageal, stomach, and colorectal cancer risk after a median 5.4 years of follow-up. This study was unique because most data regarding alcohol use and cancer risk has been obtained from Western populations [56MC].

A case–control study that included inpatients in Japan (1569 cases and 506797 controls) concluded that alcohol consumption is also a risk factor for upper-tract urothelial cancer (UTUC), which includes renal, pelvic and ureter cancer. Compared to individuals with no alcohol consumption, alcohol drinkers with consumption >15 g/day had an OR=1.23 (95% CI 1.08–1.40) for developing UTUC [57MC].

Bladder cancer risk with alcohol has also been recently studied. Botteri et al. followed a cohort (n=476160; 35–70-YO) for an average of 13.9 years. Although no dose–response relationship was seen, the authors found that men with high baseline alcohol intake (>96 g/day) had an increased urothelial cell carcinoma (UCC) risk (HR=1.57 [95% CI 1.03–2.40]) compared to men with moderate alcohol intake (<6 g/day). This association was also seen for smokers with high alcohol intake (HR=1.82 [95% CI 1.19–2.79]). No association between alcohol and UCC was observed in women. Overall, the association between alcohol consumption and UCC remains unclear [58MC].

Age

Studies have shown that moderate alcohol consumption is associated with a decreased risk of functional limitations in older adults. León-Muñoz et al. sought to assess this association, specifically in adults ≥60-YO in Spain (n=6622 for the pooled cohort). Former drinkers showed a higher instrumental activities of daily living (IADL) limitation risk compared to non-drinkers (OR=1.63 [95% CI 1.04–2.21]). However, moderate drinkers (<40 g/day in males and <24 g/day in females) were less likely to develop IADL limitations vs non-drinkers (OR=0.54 [95% CI 0.39–0.69]). Heavy drinkers also had a decreased mobility limitation (OR=0.67 [95% CI 0.40–0.94]) and agility (OR=0.71 [95% CI 0.44–0.98]) risk. Despite the beneficial effects seen with alcohol consumption, the authors' caution against promoting alcohol use in older adults due to concerns for other harmful effects [59MC].

ALCOHOL (KU)

Pharmacogenetics

Alcohol dehydrogenase and aldehyde dehydrogenase are involved in alcohol metabolism. Polymorphisms in the genes encoding these enzymes, *ADH1B* and *ALDH2*, have been shown to affect leukocyte counts and development of macrocytosis and macrocytic anemia in male Japanese drinkers. These polymorphisms may contribute to changes observed in platelet count in alcohol drinkers. Thrombocytopenia during intoxication and rebound thrombocytosis during the first few abstinence weeks are common in alcoholics. To determine genetic polymorphisms of alcohol dehydrogenase-1B (*ADH1B*; rs1229984) and aldehyde dehydrogenase-2 (*ALDH2*; rs671) effects on platelet counts, Yokoyama et al. evaluated 989 Japanese alcoholic men during an 8-week in-hospital abstinence period. Reversible suppression and rebound increase in platelet counts was more significant in slow-metabolizing *ADH1B**1/*1* than the fast-metabolizing *ADH1B**2* group. During the abstinence period, mean platelet count was lower in *ALDH2**1/*1* vs *ALDH2**1/*2* group (OR=1.73 [95% CI 1.06–2.82], P=0.027). The authors found that

*ADH1B*1/*1* was associated with increased thrombocytopenia risk on admission (OR=1.61 [95% CI 1.14–2.27], $P=0.007$) and rebound platelet increase 2 weeks into abstinence period (OR=3.86 [95% CI 2.79–5.34], $P<0.001$) compared to *ADH1B*2* allele carriers [60C].

Otto et al. further investigated *OPRM1* polymorphisms effect on alcohol response. The rs3778150 and rs1799971 polymorphisms of *OPRM1* were studied in 152 regular drinkers recruited from Columbia, MO. C allele carriers for rs3778150 exhibited lower retrospective self-report to EtOH sensitivity ($P=0.002$) and BAES alcohol-related stimulation ($P=0.032$) during an alcohol challenge. G allele carriers for rs1799971 reported lower alcohol-related sedation levels compared to individuals with the A/A genotype at that locus ($P=0.015$). The authors conclude that rs2778150 polymorphisms may affect alcohol use phenotypes and may also account for the association previously seen between rs1799971 polymorphisms and substance use phenotypes [61C].

Polymorphisms in the *OPRM1* gene are also related to increased sensitivity to alcohol's rewarding effects. Korucuoglu et al. analyzed young drinkers 17–20-YO and found that the A/G genotype of rs1799971 had decreased activation in prefrontal and parietal regions than AA genotype at baseline. Increased connectivity from the ventral-striatum to frontal regions during alcohol exposure was seen in the AG group vs the AA group. The authors concluded that these results suggest that adolescents carrying the 118G allele may be more genetically susceptible to drinking [62c].

BENZODIAZEPINES (MAY)

BZD medications are classified as sedative-hypnotic medications. These medications have been heavily used for treating common conditions such as insomnia, anxiety, and epilepsy [63R]. While it is clear that numerous detrimental AEs have been shown following BZD use, we will further detail new and understudied AEs associated with BZD use.

Cardiovascular

An 87-YOF diagnosed with severe dementia was admitted due to cerebral infarction and symptomatic epilepsy. Upon admission, HR was low (31 beats/min). This severe bradycardia was linked to clonazepam use and resolved 3 days after clonazepam withdrawal. Although it is unknown exactly why this bradyarrhythmia occurred, the proposed mechanism is due to T-type calcium channel blockage [64c].

Oncology

BZD use was linked to increased cancer risk as published in a meta-analysis by Kim and colleagues. The pooled OR/RR of 22 studies was 1.19 (95% CI 1.15–1.20) and increases with high dose BZDuse (OR/RR=2.93 [95% CI 2.45–3.52]). The associated risk was linked to an increase in cancers of the breast, brain, esophagus, renal cell, prostate, liver, stomach, pancreas, and lungs. No association was noted to an increased ovarian cancer, melanoma, or colon cancer risk [65M].

CNS

Migraine

BZD use was linked to increased migraine risk in a study completed by Harnod et al. This study found an OR=1.73 (95% CI 1.63–1.84) of migraine with BZD use. The migraine incidence is higher in patients with underlying sleep and anxiety disorders, possibly the reason for BZD use in these patient populations [66C].

Cognitive Decline

A meta-analysis published in 2015 suggests that BZD use is linked to dementia risk [67M]. A retrospective case–control study, evaluating Chinese adults confirmed findings that high-dose BZD use may be associated with dementia. This study calls for further investigations into the topic [68C].

Orthopedic

While falls have been commonly associated with BZD use [62R], a study completed by Donnelly et al. reviewed BZD use and its associated risk with hip fracture. The pooled risk found a significantly increased hip fracture incidence (RR=1.52, 95% CI 1.37–1.68) with BZD use. Furthermore, this study concluded that patients newly prescribed BZDs are at the highest risk [69M].

Hepatic

Drug induced liver injury is a serious, sometimes fatal outcome of medication use. A case report detailing a 63-YOM presenting with acute liver failure following 3 months of trazodone and diazepam administration was linked as "probable" regarding trazodone use and "possible" in regards to hepatotoxicity with diazepam. The group assessed causality using the Council for International Organizations of Medical Sciences/Roussell Uclaf Causality Assessment Method. The patient presented 3 months following diazepam and trazodone therapy initiation for his depression. The patient presented with jaundice and marked elevation in hepatic function.

Workup for infectious disease induced hepatitis was negative. Within 2 weeks after medication discontinuation, the patient underwent liver transplantation and fully recovered. It is estimated that hepatotoxicity due to BZD use is due to a rarely produced intermediate metabolite [70c].

HEENT

A retrospective review of US and Japanese AE-reporting databases found gingival hyperplasia associated with clobazam use. The study published the ratio of odds of reporting (ROR) compared to other AE for the given drug. The ROR for gingival hyperplasia with clobazam was 65.4 (95% CI 33.8–126.7) [71R].

SYMPATHOMIMETICS: COCAINE (MAY)

Cocaine remains a common illicit substance abused in the US [72S]. Also known as blow, coke, crack, rock, and snow, cocaine is abused due to its effect on dopamine, leading to increased energy and mental alertness, which can progress to irritability and paranoia [73S]. This chapter describes case reports detailing rare, recently published, AE associated with cocaine use.

HEENT

Cocaine is known to cause mucous membrane destruction, typically in the nasal cavity [74c]. A case report and review article by Siemerink et al. described orbital complications associated with intranasal cocaine abuse. Two cases were described as extensive unilateral chronic orbital inflammatory disease and optic neuropathy. Though rare, this complication was linked to long-term intranasal cocaine abuse and in both cases, destruction of the nasal cavity bone was described [75c].

A detailed report of a pre-diabetic 60-YOF with a PMH of cocaine use discussed the link between cocaine use and abnormally high tortuosity levels in the corneal sub-basal nerve plexus, a change seen in diabetes-induced nerve damage. This report suggests that past history of cocaine use may have longstanding effects on corneal nerves. Thus, it is imperative to review past drug abuse when evaluating diabetes-induced corneal changes [76c].

Cocaine inhalation is linked to burn injuries in airways. Singh et al. described a 73-YOM in a case report, who developed laryngeal injury following cocaine and cannabinoid use. The patient presented with respiratory distress, elevated BP (213/102 mmHg) and physical examination showed diffuse bilateral rhonchi. Following bronchoscopy, supraglottic, glottis and arytenoid burns with edema were confirmed [77c].

Splenic

A detailed report by Khan et al. described a rare, yet serious, splenic rupture complication tied to cocaine abuse. The case report details a 39-YOM presenting with atraumatic splenic rupture. Symptom presentation included diaphoresis, abdomen distention and tenderness but there was no palpable lymphadenopathy or hepatosplenomegaly. Laboratory abnormalities included leukocytosis and decreased hemoglobin and hematocrit. The report details that cocaine's effects on vasoconstriction within the spleen are likely the cause for splenic hemorrhage and arteriolar rupture [78c].

CNS

A 17-YOM presented with acute onset of weakness following nasal inhalation of cocaine. The described weakness was primarily noted in the right upper and lower limbs. A detailed work up confirmed sensory and motor neurological deficits. The authors conclude that the described neuropathy is believed to be due to cocaine-induced vasculitis leading to acute multiple mononeuropathy. The patient was treated with intravenous methylprednisolone followed by oral prednisone and intensive physiotherapy. Substantial improvement was noted after 15 days of treatment [79c].

Thrombosis

Upon presentation with peritonitis and sepsis, a 41-YOM with history of cocaine use was diagnosed with a giant aortic thrombus leading to bowel perforation. In the absence of any coagulation disorder, the authors conclude that cocaine abuse is a potential cause of aortic thrombus and bowel perforation [80c].

SYMPATHOMIMETICS: METHAMPHETAMINE

Methamphetamine is often abused due to its ability to increase mood, alertness, and concentration. However, long-term use has been linked to many CNS deficits such as psychosis, depression, and violent behavior [81R]. This section will detail three recently published case reports detailing the effects of methamphetamine in the lung, thyroid and bowel.

Pulmonary

Albanese et al. described a rare complication of methamphetamine use—spontaneous pneumomediastinum. This case report details a 27-YOM with a PMH of marijuana and methamphetamine use. His presentation included diaphoresis, tachycardia, muscle tenderness, and the following lab abnormalities: creatinine = 2.2, BUN = 34, CPK <40000. The pneumomediastinum was confirmed via chest X-ray [82c].

Endocrine

Viswanath et al. described thyroid storm induced by methamphetamine use in a detailed case report discussing a 47-YOF presenting with various complaints including palpitations, emesis, and SOB, all of which happened in the 24h following methamphetamine use. Laboratory data and VS lead to a Burch–Wartofsky Score = 35, suggestive of thyroid storm. The authors concluded that methamphetamine use was a likely trigger for this exacerbation of thyrotoxicosis [83c].

Gastrointestinal

Green et al. described a fatal case of ischaemic colitis. A 44-YOF presented after cardiac arrest and amphetamine use. Her presentation detailed headache, neck pain, abdominal pain, vomiting and increased agitation [84c].

CATHINONES: NON-SYNTHETIC

General Information

Cathinones were first used by humans by extraction from the leaves and young shoots of Khat (*Catha edulis*), an evergreen shrub cultivated in regions extending from southern Africa to the Arabian Peninsula, commonly referred to as qat, q'at, kat, kath, gat, chat, tschat, miraa, or murungu. Immigration and improved methods of khat distribution from these regions have led to an increase in reported habitual khat chewing in Europe, the US, and Australia [85R,86S]. "Graba", a dried form of khat, may also be smoked. The major psychoactive compounds in khat are the phenylalkylamines cathinone [S-(−)-cathinone], cathine [1S,2S-(+)-norpseudoephedrine], and norephedrine [1R,2S-(−)norephedrine]. As these compounds are structurally related to amphetamine, they are known to have similar subjective effects, which is the reason for their licit and illicit use. Khat use in the US is largely limited to immigrant populations from regions where its use is more widespread, while synthetic cathinone use (discussed in a subsequent section) has increased in the general population as an alternative to using methamphetamine (METH) and methylenedioxymethamphetamine (MDMA) [85R,86S].

Adverse consequences of khat use may be more limited than adverse consequences of synthetic cathinone use, in part based upon the typical mode of administration. The AEs of khat have been described in some detail [87R], and primarily include orodental and sympathomimetic effects associated with chronic and acute usage, respectively, although clinical reports also include other effects. The orodental effects reported include keratotic white lesions, leukoedema, frictional keratosis, mucosal pigmentation, periodontal disease, tooth loss, plasma cell stomatitis, and xerostomia [88C,89M,90R,91C].

Khat Overdose

In 2017, a khat overdose was reported, describing symptoms similar to synthetic cathinones. A 45-YO Somalian male with a PMH of hypertension and chronic kidney disease (CKD) was experiencing elevated BP (220/130mmHg), agitation, confusion, and severe headaches upon presentation to a hospital in Dublin, Ireland [92A]. Shortly after arrival, the patient developed left hemiparesis, followed by seizures. Brain tomography showed a large right temporo-parietal intraparenchymal hematoma with intraventricular extension, a small left parietal hematoma, and hydrocephalus. Angiography could not identify an aneurysm or arteriovenous malformation, but did suggest vasoconstriction of the pericallosal artery. Serum and urine toxicology were negative for amphetamines or cocaine metabolites. Family members and neighbours confirmed the patient had been consuming large amounts of khat prior to hospital admission.

CATHINONES: SYNTHETIC (WISNER, HALL)

General Information

Cathinones are the β-keto-derivatives of amphetamines, and have similar properties, acting primarily as monoamine releasers or monoamine reuptake inhibitors, with subjective properties similar to METH, 3,4-MDMA, or cocaine [93R,94E,95R]. Similar to amphetamine neurotoxicity, cathinone neurotoxicity is also observed, although this varies substantially depending on the particular cathinone examined [96E,97E,98R]. A much wider array of cathinone derivatives are now being used compared to amphetamines, and have been developed for a variety of reasons, including novelty, evading existing legal restrictions and toxicological tests, ambiguous legal status, global internet marketing, and reduced public knowledge of potential adverse effects [99R]. While the standard administration methods include oral consumption, insufflation, and intravenous injection, use of pyrovalerones in "herbal" preparations in Poland (2013–2015) that are intended to be smoked, has been increasing [100c]. Discarded injecting paraphernalia were analyzed (2015–2016) in Hungary, which were found to commonly contain multiple cathinones, including pentadrone, mephedrone, α-pyrrolidinovalerophenone (α-PVP), α-pyrrolidinohexanophenone (α-PHP), and pyrrolidinoenanthophenone (α-PEP) [101C]. Most syringes contained several compounds, indicating that they were either reused or the substances were administered in mixtures. An HIV outbreak (2012–2013) was also associated with increased intravenous cathinone use, due to increased heroin prices, and often occurring in combination with buprenorphine [102c].

Various cathinones are discussed below, for which clinical overdoses were reported in the last year. The symptoms in such cases include varying neurological and cardiac effects and, at higher doses, multiple organ failure; a syndrome of effects previously reported for other stimulant drugs termed, "Excited Delirium". Some reports involve ingesting specific confirmed substances, while others do not. Several cases identify the likelihood that individuals may have ingested cathinones unknowingly, or believing them to be other drugs. A study of electronic dance music scene patrons (summer 2016) found that 27.8% tested positive for ≥1 cathinone (including butylone, ethylone, pentylone, or α-PVP) of which half tested positive for a non-reported drug (i.e., ingested unknowingly), and of which 72% were positive for cathinones [103c]. Reports [104MC] were examined for AE resulting in admission to the ED and intensive care unit (ICU) in Sweden (2011–2016) involving methylenedioxypyrovalerone (MDPV) derivatives. These cases were confirmed by urine and blood toxicology, identifying 11 pyrovalerone derivatives in addition to MDPV and α-PVP. Most cases (n = 114) (16–66-YO, average age = 30, 84% male) involved ingesting multiple pyrovalerones (and other drugs, including other classical stimulants, opiates, BZD and alcohol), with only eight cases involving a single pyrovalerone derivative (five unique derivatives). Most patients showed stimulant-like intoxications/delirium symptoms on admission, including agitation, delirium, hallucinations, excessive motor activity, seizures, tachycardia, and hypertension. Hyperthermia was observed in most cases (50%). Some patients presented with CNS depression, perhaps associated with co-ingesting other drugs, such as opiates. Prior to 2014, most cases involved MDPV and α-PVP, while after 2013 a more diverse range of pyrovalerone derivatives were involved. A summary involving 13 α-PVP overdose cases [105A] identified similar symptoms: tachycardia, hypertension, delusions/paranoia, hallucinations, and mydriasis. For all patients, urine concentration was >200 μg/L, and >2500 μg/L for 6/13 cases. In all cases other drugs were present, but in two cases the only other drug present was caffeine.

Cardiovascular

A 41-YOF with no significant PMH presented to the ED with anterior ST-elevation myocardial infarction (STEMI) and multiple intracardiac thrombi [106A]. Cardiac catheterization showed that the anterior descending artery was completely occluded and a left ventricular apical thrombus had formed. The patient reported ingesting a cathinone derivative, α-PVP. Pyrovalerones have pharmacological effects different from other cathinones, primarily affecting monoamine uptake rather than acting as monoamine releasers [107R,108E]. Death associated with acute heart failure and pulmonary edema has been previously reported for α-PVP [109A,110A,111A]. Cardiac events have also been reported for other cathinones that act primarily as monoamine releasers. Overdose deaths involving two men (37- and 42-YO) with heart blood methylone concentrations 0.41 and 0.6 mg/L, respectively, were reportedly attributed to myocardial infarction due to complicating atherosclerotic cardiovascular disease from coronary artery thrombosis [112M].

This data would seem to suggest that cardiac events may occur in cases of cathinone overdose, particularly perhaps for pyrovalerones. Moreover, these events derive not just from the sympathomimetic effects of these drugs, but also from thrombotic conditions.

Nervous System/Psychological

Drug-induced psychosis has been reported for many cathinones [113A], sometimes with death subsequently resulting from consequent accidental injuries, but sometimes resulting from the end course of excited delirium, i.e., cardiorespiratory collapse [114R]. A case series (n = 50 cases) from Germany of individuals arrested for various crimes found to be positive for MDPV [115c]. Most cases in which plasma MDPV levels were ≥30 μg/L involved psychotic and aggressive behavior, resulting in violent crimes. In a clinical case report, a 24-YOM suffered fatal blunt force injuries after jumping from a height during a multi-drug-induced psychotic episode (heart blood methylone = 0.49 mg/L, MDPV = 0.14 mg/L, ketamine = 0.71 mg/L) [112M]. A great difficulty in evaluating risks associated with particular cathinones is that they are often taken in combination with other cathinones, or other licit and illicit drugs.

A 39-YOM patient presented with inarticulate speech, bizarre behavior, fear, anxiety, restlessness and psychotic symptoms, including hallucinations and delusions [116A]. Temperature was normal but HR and BP were slightly elevated, and blood oxygenation was low. Patient responded to treatment, with mild hepatic dysfunction. His screening panel was negative for alcohol and standard DOA; however, the subject tested positive for α-PHP. Psychiatric symptoms normalized by hospital day-3, in parallel with decreasing plasma α-PHP levels, that were initially 175ng/L, but took nearly 10 days to drop below detectable levels.

A 40-YO patient, with no previous psychiatric history, but a positive marijuana use history, was admitted for psychological observation by police after exhibiting paranoia, persecutorial thinking, homicidal thoughts, reduced sleep, increased energy, increased sex drive, weight loss, and self-mutilation (hair-pulling). Police were involved over concerns that he would harm others, and the patient had recently purchased a gun for protection from "trespassers" (wearing camouflage and clown makeup). Patient tested positive for marijuana and MDMA. Wife and patient admitted that the patient had recently used "bath salts", although these were of undetermined type [116A].

Patients were screened and admitted ($n=305$) for drug overdose, 11 patients were found to be positive for mexedrone (3-methoxy-2-(methylamino)-1-(4-methylphenyl) propan-1-1; the alpha methoxy-derivative of mephedrone (4-methyl-*N*-methyl cathinone) in UDS [117C]. Agitation was present in all but one patient (which was comatose), requiring physical restraint or sedation in six of these patients. This state was accompanied by delusions, hallucinations, paranoia, and confusion. No patients had a PMH for psychosis. Other symptoms present included tachycardia (except for one patient that had bradycardia), mydriasis, acidosis, increased creatine kinase activity, urinary incontinence/retention, pyrexia (although hypothermia was also observed and no patients had severe hyperpyrexia), and coma (two cases). Most patients were polydrug abusers, and all patients had multiple drugs detected in urine, including other stimulants, but mexedrone was found in two cases with much higher concentrations than any other drug. All patients were released after symptoms abated, (~24–48h). A similar symptom pattern was reported for hospitalizations after acute cathinone overdose in Southern Germany for 81 patients (2010–2016) [118c]. The primary cathinones found in patients were methylone, and subsequently MDPV, and 3-methylmethcathinone (3-MMC). A high polydrug abuse incidence was also observed.

Two hospital admission cases for severe psychological disturbance (agitation and aggression) were described for patients with positive UDS, 2-(ethylamino-1-(4-methylphenyl)-1-pentanone (4-MEAP; 4–7mg/L), in high concentrations with several other drugs. Neither case demonstrated extreme hyperpyrexia, but was tachycardic, showing evidence for acute renal dysfunction, and acidosis in one patient. Both patients recovered and were released [119A].

It is clear from these reports that cathinones can induce psychotic disturbances, and even Excited Delirium, similar to other psychostimulant drugs. It remains to be seen whether cathinones overall, or particular cathinones, have a greater likelihood to induce these states.

Death

21% of hair samples from cases ($n=112$), primarily from the UK, submitted for forensic analysis to a French laboratory, were positive ($n=24$) for mephedrone, the β-ketone analogue of 4-methyl-methamphetamine [120c]. A large proportion also contained MDMA (and other stimulants). Although most samples were tested primarily to verify drug use, three cases resulted in death: one from apparent overdose (plasma mephedrone concentration [PMC]=5432ng/mL), one from a motor vehicle accident (PMC=103ng/mL), and one from drowning (PMC=88ng/mL) (urine concentration=2430ng/mL). These numbers are in line with a literature study that identified 97 fatal mephedrone and 57 non-fatal intoxications reported since 2010 [121M]. The mean plasma concentration was 2663 and 166ng/mL for fatal vs non-fatal cases, respectively, while reported cases with no acute toxicity had a mean concentration of 135ng/mL. As in most other analyses, myriad other drugs were detected simultaneously in these subjects, including ethanol, tetrahydrocannabinol, MDMA, stimulant drugs, opiates, gamma-hydroxybutyric acid, BZDs, and other cathinones.

An acute fatality after ingesting *N*-ethylpentylone was reported [122A]. The individual was transported to hospital by police after exhibiting agitation, aggression, and unusual behavior after being found naked, wandering on the side of the road. Temperature on arrival was 105.5°F. The patient was hypotensive and tachycardic, blood pH 6.7, with elevated troponins, rhabdomyolysis, hypoglycemia, indications of hepatic and renal impairment, respiratory failure, and disseminated intravascular coagulation. Patient went into cardiac arrest three times and died 36h after admission. Although initial toxicology found polydrug use, the only drug present at time of death in high concentrations was *N*-ethylpentylone, and a "rock" of *N*-ethylpentylone was subsequently found in the patient's car.

Two similar deaths attributed to methylone and ethylone ingestion were reported in a retrospective review: a 20-YOM was unresponsive at a rave party, hyperthermic (107°F), and died after 10h with a blood methylone concentration=0.71mg/L. A 25-YOF complained of a headache after returning from a rave party, and was later found unresponsive, with a postmortem blood ethylone concentration=1.7mg/L [112M].

Pharmacogenetics

Genetic polymorphisms have been shown to affect the metabolism and cardiovascular SEs of MDMA. MDMA is primarily metabolized by CYP2D6 followed by *COMT* to its main inactive metabolite 4-hydroxy-3-methoxymethamphetamine (HMMA). MDMA is also metabolized by CYP3A4, CYP1A2, CYP2B6, and CYP2C19 to its minor active metabolite 3,4-methylenedioxyamphetamine (MDA). Genetic polymorphisms in CYP2D6 have been shown to influence MDMA metabolism in humans but limited data exist regarding the polymorphism effect in CYP1A2, CYP2B6, and CYP2C19 on MDMA SEs. Vizeli et al. sought to assess if genetic variations in these CYP enzymes may alter MDMA pharmacodynamics. In a pooled population ($n=142$ healthy European Caucasian adults), the authors determined that CYP2C19 poor-metabolizers had increased cardiovascular SEs compared to intermediate- or ultra-metabolizers ($P<0.05$). CYP2C19 genotype did not subjectively affect MDMA significantly in association with body temperature. Individuals with the T/T genotype for CYP2B6 had higher plasma MDMA concentrations than those with the G/G genotype ($P<0.05$). Individuals with the T/T genotype for CYP2B6 also had higher MDMA/MDA AUC ratios than CYP2B6 G allele carriers ($P<0.05$ for both G/T and G/G genotypes). Polymorphism rs762551 for the CYP1A2 gene increased MDA peak concentrations ($P<0.001$) and AUC ($P<0.01$) in those with the inducible expression A/A genotype. The authors suggest that polymorphisms in CYP2C19, CYP2B6, and CYP1A2 may affect the metabolism and the cardiovascular toxicity of MDMA [123C].

MDMA may interact with presynaptic monoamine transporters and affect norepinephrine efflux, which may contribute to MDMA's cardiovascular effects. The *SLC6A2* gene encodes for one of the neurotransmitter transporters involved in MDMA's mechanism of action. Vizeli et al. examined 124 healthy Swiss adults pooled from eight double-blinded, placebo-controlled crossover studies with similar methodologies to determine if *SLC6A2* polymorphisms may explain inter-individual differences in acute stimulant-type responses to MDMA. Specific polymorphisms included in this analysis are rs168924, rs47958, rs1861647, rs2242446, and rs36029. The authors found that the individuals with the G/G genotype at rs1861647 had higher HRe and rate-pressure product after MDMA administration compared to individuals with the A/A ($P<0.02$) and A/G ($P<0.04$) genotypes. Individuals with the T/T genotype for rs2242446 though had lower HR elevation and rate-pressure product than individuals with the C/T and C/C genotypes ($P<0.04$ for both). Individuals with the A/A genotype for rs36029 had increased elevations in mean arterial pressure (MAP) ($P<0.01$), HR ($P<0.05$), and rate-pressure product ($P<0.01$) than G allele carriers. The authors suggest that polymorphisms in the *SLC6A2* gene may moderate acute cardiovascular SEs of MDMA [124C].

HEROIN (BABIN)

Cardiovascular

A 28-YOF was admitted to the hospital after intentional repeated radial artery injections of 4–5 bags of heroin/day. Presenting symptoms were acute right hand pain and color change in the thumb 1 day after heroin injection into her right wrist. The patient was diagnosed with a radial artery occlusion and underwent tissue plasminogen activator (tPA) catheter thrombolysis. The following day a repeat arteriogram showed continued artery occlusion. One week after discharge, the patient was readmitted with an occluded right brachial artery, likely due to tPA catheterization injury. After intervention, the patient continued to have blue thumb discoloration. The authors note it is unusual that the patient did not experience any complications from intra-arterial injection prior to this hospitalization, since this practice can be associated with severe complications [125A].

Respiratory

A 22-YOF with a history of smoking heroin daily for 2 months presented with fever, cough, pleuritic chest pain, and severe dyspnea. A chest CT scan showed bilateral interstitial infiltration with ground glass opacity and consolidation. Piperacillin–tazobactam, clarithromycin, and methylprednisolone were initiated. On hospital day 3, bronchoalveolar lavage revealed 6% eosinophils and the patient was diagnosed with acute eosinophilic pneumonia. The patient was discharged with normalized laboratory and clinical findings after 4 days [126A].

CNS

Fareed et al. conducted a systematic review to investigate heroin use impact on brain function using resting state functional MRI. The review included 21 studies and concluded that heroin use may have a direct damaging effect on certain brain functions, which may contribute to impulsive and unhealthy decision making. A longer duration of heroin use appears to be associated with greater damage [127M].

Gerra et al. compared oxytocin levels in 18 abstinent heroin dependent males with levels in 18 healthy subjects and no history of psychotropic drug use or alcohol abuse. Former heroin users had increased oxytocin serum levels, which were positively correlated with psychiatric symptoms, aggressiveness, and mother neglect scores on a questionnaire. The amount and duration of heroin exposure were not correlated with oxytocin levels in this small sample [128c].

A 29-YOM with a history of cannabis, alcohol, and heroin abuse for many years was admitted with coma due to sniffing heroin. He quickly improved with naloxone and did not continue heroin use after discharge. Three-weeks later, he was readmitted with severe cognitive impairment and was diagnosed with heroin-induced leukoencephalopathy based on brain MRI. The patient was hospitalized for 1 month with no cognitive improvement. During a follow-up appointment 4 months after discharge, the patient's cognitive function was dramatically improved [129A].

Another case report describes a 33-YOM who was found unconscious with a history of injecting heroin 0.5–2 g/day. The patient had almost complete paraplegia, and cerebrospinal fluid analysis showed extremely elevated nerve injury markers, including glial fibrillary acidic protein, which indicates a toxic effect on astrocytes. After methylprednisolone treatment for 5 days, the patient was markedly improved [130A].

A 23-YOM presented with a 1 month history of headache and visual disturbances. He was initially empirically treated for meningitis with ceftriaxone, vancomycin, and prednisone until an eye exam suggested possible candida endophthalmitis and amphotericin B was started. After discharge, the patient completed 2 weeks of voriconazole. Several months later his vision had returned to baseline and his headaches were improved, although not entirely resolved. All collected cultures were negative, but based on clinical findings and the patient's recovery the authors diagnosed him with candida meningitis, ventriculitis, uveitis, and endophthalmitis due to brown heroin injection [131A].

Hematologic

Chouk et al. reported a puffy hand syndrome case, which is characterized by non-pitting, painless edema of the hands that becomes permanent and symmetrical over months. One key mechanism is likely chronic destruction of the lymphatic network due to heroin or additives. The patient, a 40-YOM, used intravenous heroin from age 17 to 23. Two years after stopping heroin, edema developed in his hands that were treated with corticosteroids during a 15-year period with no improvement. After trying compression sleeves for 3 months, the patient noted improvement in the edema [132A].

Musculoskeletal

A single center, retrospective study examined patients undergoing operative decompression for compartment syndrome over a 6-year period. Of the 213 patients included, 22 were determined to have heroin-related compartment syndrome (HRCS). Overall, heroin overdose was the second most common compartment syndrome (10.3%) etiology after trauma (64.8%). Patients with HRCS presented with AKI ($n=12$[54.5%]), required dialysis ($n=8$[36.4%]), and developed significant complications from compartment syndrome ($n=6$[27%]) [133c].

HEROIN (KU)

Pharmacogenetics

The mu-opioid receptor gene *OPRM1* has been found to influence abstinence maintenance from heroin. Levran et al. found that the 118G allele may be associated with reduced vulnerability for heroin relapse in European/Middle Eastern former heroin users not currently treated with agonist therapy. The proportion of 118G allele carriers in the untreated abstinence group was higher compared to the abstinence group treated with methadone (44% vs 26%, OR=2.2 [95% CI 1.5–3.3]). The authors hypothesize that the *OPRM1* SNP may blunt the endogenous stress response once abstinence is achieved to contribute to a reduced vulnerability for heroin relapse [134C].

PRESCRIPTION/OVER-THE-COUNTER OPIOIDS AND NPS (FUDIN)

Population

Polysubstance use is a consistent predictor for problematic opioid analgesic use but limited data exist in countries outside the US. Morley et al. chose to investigate the relationship with participants from the Global Drug Survey 2015 resident in the US ($N=1334$), UK ($n=1199$), France ($n=1258$), Germany ($n=866$), and Australia ($n=1013$) who had used at least one prescription opioid analgesic medication in the past year. Adjusting for country of residence and sociodemographic factors, illicit drug and BZDs use were associated with a sixfold greater odds for opioid abuse compared with not using either drug (OR=6.49, 95%; CI 4.0–10.48), but strength of association varied by country [135MC].

Zawilska et al. presented a study based on survey results to provide an overview on NPS for pharmacological properties, pattern of use, and desired and unwanted effects. The total NPS ($n=644$) were reported to the UNODC (102 countries; 2008–2015) [1S], but by the end of 2015, the European Monitoring Centre for Drugs and Drug Addiction (EMCDDA) were notified that more NPS ($n=561$) cases had been reported [136S]. Due to the narrow therapeutic index, fentanyl used as a recreational drug is especially deadly to opioid intolerant users. High doses could result in death due to respiratory arrest and pulmonary edema. Fentanyl exposures reported to the American Association of Poison Control Centers increased from 300 (2010) to 1724 (2011), but have since remained steadily high (1632 in 2012; 1486 in 2013; 1418 in 2014) [137S]. In Canada (2009–2014) at least 1019 drug-poisoning deaths were reported to have postmortem toxicological screening indicating fentanyl presence of which more than half occurred in the latter two years (2013 and 2014) [138S]. In Europe, the first fatal intoxication cases with fentanyl were reported in Sweden [139c]. Later, illicit fentanyl use became a serious health problem in Estonia, with an estimated 1100 deaths (2005–2013) [140H]. According to epidemiological data, the increased fentanyl use in Europe and the US was largely associated with the low availability, low purity, and/or high price of heroin, features, at least partially linked to Taliban control on opium production [141H,140H].

More recently reports, from the US and Canada, show that carfentanil has been increasingly laced with or disguised as heroin. Carfentanil has already been connected to hundreds of overdose cases, many of which are fatal [142S,143H,144H,145H,146H]. The estimated carfentanil potency is much greater than morphine (10000 times), heroin (4000 times), and fentanyl (100 times), making it one of the most potent known and commercially used opioids. Several fentanyl analogs are synthesized for recreational use [141H,147S,140H,148A,149S,150H]. Modifying or replacing fentanyl's propionyl chain or its ethylphenyl moiety completes the compound development process. The obtained analogs have been further modified by substitution with fluoro-, chloro-, or methoxy-groups at the *N*-phenyl ring. Table 1 provides examples of fentanyls that have not been approved for medical use [151M].

More specific case data and toxicological findings are found in Table 2.

Concerns about misusing codeine-containing combination analgesics have led many countries to review codeine availability in over-the-counter (OTC) preparations due to increasing presentation rates to primary care and addiction treatment service for dependence. In Australia, New Zealand, Canada, South Africa, France, Ireland and the UK codeine is widely available OTC without a prescription. France also has codeine available as a single-ingredient OTC product, removing the concerns associated with consuming high doses of combination products. Common harms seen with codeine abuse are available in Table 3 [152H].

Wang et al. used a systematic review and meta-analysis to identify 118904 patients with opioid use disorder (OUD) in China (1990–2015). Researchers aimed to identify poly-drug use prevalence within this population. The pooled prevalence of prescription opioids was 23.2%. From most to least common use in OUD were, tramadol (27.3%), methadone (16.8%), buprenorphine (12.6%), pethidine (8.9%), morphine (6.5%), dihydroetorphine (3.9%), and codeine-containing cough syrup (3.7%). Prescription opioids were primarily co-administered for nonmedical purposes (69.4%) in China [153M]. In South-East Hungary (2008–2015), medical opiates including

TABLE 1 Fentanyls Not Approved for Medical Use (NPF) [151M]

	Opioids: Synthetic Prescription/OTC opioids and NPS
	Chemical name
Acetylfentanyl	*N*-phenyl-*N*-[1-(2-phenylethyl)piperidin-4-yl] acetamide; acetyl fentanyl, desmethyl fentanyl, MCV 4848, NIH 10485
Acryloylfentanyl	*N*-phenyl-*N*-[1-(2-phenylethyl)piperidin-4-yl] prop-2-enamide, acrylfentanyl, acryloyl-F, Acr-F, ACF
α-Methylfentanyl	*N*-phenyl-*N*-[1-(1-phenyl-2-propanyl) piperidin-4-yl]propanamide
3-Methylfentanyl	*N*-[3-methyl-1-(2-phenylethyl)piperidin-4-yl]-*N*-phenylpropanamide; mefentanyl, 3-MF
Butyrylfentanyl	*N*-phenyl-*N*-[1-(2-phenylethyl)piperidin-4-yl]butanamide; butyl fentanyl; BF
4-Methoxybutyrylfentanyl	*N*-(4-methoxyphenyl)-*N*-[1-(2-phenylethyl)piperidin-4-yl]butanamide; 4-MeO-BF
4-Fluorobutyrylfentanyl	*N*-(4-fluorophenyl)-*N*-[1-(2-pheny-lethyl)piperidin-4-yl]butanamide; 4-FBF
4-Fluoroisobutyrylfentanyl	*N*-(4-fluorophenyl)-2-methyl-*N*- [1-(2-phenylethyl)piperidin-4-yl]propanamide; 4F-iBF
4-Chloroisobutyrylfentanyl	*N*-(4-chlororophenyl)-2-methyl-*N*-[1-(2-phenylethyl)piperidin-4-yl]propanamide; 4F-iBF
Furanylfentanyl	*N*-phenyl-*N*-[1-(2-phenylethyl)piperidin-4-yl]-2-furancarboxamide; furafentanyl
Cyclopentylfentanyl	*N*-(1-phenylethylpiperidin-4-yl)-*N*-phenylcyclopentanecarboxamide; CP-F
Tetrahydrofuranylfentanyl	*N*-(1-phenethylpiperidin-4-yl)-*N*-phenyltetrahydrofuran-2-carboxamide; tetrahydrofuran-fentanyl, THF-F
Ocfentanil	*N*-(2-fluorophenyl)-2-methoxy-*N*-[1-(2-pheny-lethyl)piperidin-4-yl]acetamide; ocfentanil, A-3217

TABLE 2 Case Reports of Fatalities Involving Novel Synthetic Opioids [151M]

Gender/ age	Case data	Toxicological findings
Acetylfentanyl		
M/32	A deceased was found dead in the bed in a supine position. Snorting at least 12h before death. Insufflation straws were found in his bag and in the drawers of a chest. At autopsy: pulmonary edema with mild to severe intraalveolar hemorrhage	Acetylfentanyl was detected in heart blood, urine and gastric contents
M/early 30s	A deceased was found at home, not breathing. A small plastic bag with a pale brown white powder and a syringe with a small amount of liquid were found at the scene. Acetylfentanyl and 4-methoxy-PV8 were detected in both the powder and the liquid. At autopsy: congested lungs, petechiae on eyelid conjunctiva, capsula cordis and pleura, fluidity of the heart blood, and two very recent forearm needle marks. Habitual "bath salt" use history	Acetylfentanyl: femoral blood, 153ng/mL; urine, 240ng/mL; gastric contents, 880ng/mL. 4-MethoxyPV8: femoral blood, 389ng/mL; urine, 245ng/mL; gastric contents, 500ng/mL. Additionally in femoral blood: 7-aminonitrazepam (200ng/mL), phenobarbital (7700ng/mL), methylphenidate (30ng/mL), chlorpromazine, and risperidone
M/24	A deceased was found unresponsive with uncapped syringe and rubber tourniquet. At autopsy: pulmonary congestion and edema, three recent punctures in left forearm. History of heroin abuse, with two previous overdoses	Acetylfentanyl: peripheral blood, 260ng/mL; heart blood, 250ng/mL; vitreous humor, 240ng/mL; urine, 2600ng/mL
M/28	A deceased was found in the bathroom with a tourniquet secured around his arm and a syringe nearby. At autopsy: marked pulmonary and cerebral edema and needle track marks. Illicit drug abuse history	Acetylfentanyl: subclavian blood, 235ng/mL; vitreous humor, 131ng/mL; urine, 234ng/mL; liver, 2400ng/g
M/20	A deceased was found dead at home. Illicit drug abuse history	Acetylfentanyl: heart blood, 285ng/mL; femoral blood, 192ng/mL; urine, 3420ng/mL; liver, 1100ng/g; brain, 620ng/g. Additionally in heart blood: methoxetamine and fluoxetine
F/50	A deceased was found unresponsive in bed. History of bilateral knee replacement, chronic pain, depression and seizures, prescription drug abuse, and ethanol abuse	Acetylfentanyl: heart blood, 219ng/mL; femoral blood, 255ng/mL; urine, 2720ng/mL. Additionally in heart blood: venfalaxine, nordiazepam, and chlordiazepoxide
Butyrylfentanyl		
M/23	A deceased was found unresponsive in the bathroom. A tray with traces of white powder and a tube were found in the bedroom. At autopsy: cerebral edema and small amounts of residual white powder in the nose. Drug abuse history	Butyrylfentanyl: peripheral blood, 66ng/mL; heart blood, 39ng/mL; liver, 57ng/g; kidney, 160ng/g, muscle, 100ng/g
F53	A deceased was found unresponsive in the bathroom. At autopsy: edematous and congested lungs. History of smoking, prescription drug abuse, and psychiatric disorder hospitalization	Butyrylfentanyl: peripheral blood, 99ng/mL; heart blood, 220ng/mL; vitreous humor, 32ng/mL; bile, 260ng/mL; urine, 64ng/mL; gastric contents, 590ng/mL; brain, 93ng/g; liver, 41ng/g
Butyrylfentanyl and acetylfentanyl		
F/49	A deceased was found unresponsive and not breathing on the bed. At autopsy: edematous and congested lungs. History of anxiety, bipolar disorder, and two previous suicide attempts	Acetylfentanyl: peripheral blood, 21ng/mL; heart blood, 95ng/mL; vitreous humor, 68ng/mL; bile, 330ng/mL; urine, 8ng/mL; gastric contents, 28000ng/mL; brain, 200ng/g; liver, 160ng/g. Butyrylfentanyl: peripheral blood, 3.7ng/mL; heart blood, 9.2ng/mL; vitreous humor, 9.8ng/mL; bile, 49ng/mL; urine, 2ng/mL; gastric contents, 4000ng/mL; brain, 63ng/g; liver, 39ng/g Additionally in peripheral blood: alprazolam, 40ng/mL and ethanol, 0.11g/dL
M/44	A deceased was found unresponsive on the bathroom floor. A box with drug paraphernalia (used syringes, aluminum foil with black residue, scissors, and alcohol wipes) was found elsewhere. At autopsy: pulmonary edema and congestion, evidence of subacute and chronic intravenous drug use in the antecubital fosse, forearms, left wrist, and ankles. Heroin use history	Butyrylfentanyl: peripheral blood, 58ng/mL; heart blood, 97ng/mL; vitreous humor, 40ng/mL; urine 670ng/mL; gastric contents, 170mg/mL; liver, 320ng/g. Acetylfentanyl: peripheral blood, 38ng/mL; heart blood, 32ng/mL; vitreous humor, 38ng/mL; urine, 690ng/mL; gastric contents, <170mg/mL; liver, 110ng/g

Continued

TABLE 2 Case Reports of Fatalities Involving Novel Synthetic Opioids [151M]—cont'd

Gender/ age	Case data	Toxicological findings
4-Fluorobutyrylfentanyl		
M/26	A deceased was found dead at home. History of drug abuse	4-Fluorobutyrylfentanyl: blood, 91 ng/mL; urine, 200 ng/mL; liver, 902 ng/g; kidney, 136 ng/g
F/25	A deceased was found dead at home. History of occasional drug and novel psychoactive substances use	4-Fluorobutyrylfentanyl: blood, 112 ng/mL; urine, 414 ng/mL; liver, 411 ng/g; kidney, 197 ng/g
Furanylfentanyl		
M/26	A deceased was found dead in the bathroom. A tourniquet was found around his arm and a used needle next to the body. At autopsy: brain edema and pulmonary edema. Drug abuse history	Blood (ng/mL): furanylfentanyl = 1.05; Δ9-tetrahydrocannabinol (THC) = 0.63; mirtazapine = 74.1; desmethylnitrazapine = 31.7; pregabalin = 6032; buprenorphine = 2.01; norbuprenorphine = 2.86; clonazepam = 21.1; 7-aminoclonazepam = 624. Urine (ng/mL): buprenorphine = 30; norbuprenorphine = 180
M/36	A deceased was found lying on the floor of the bathroom. At autopsy: pulmonary edema and froth in the airways. Drug abuse history	Blood (ng/mL): furanylfentanyl = 7.66; pregabalin = 14 815
M/37	A deceased was found lying in the ditch, with a body temperature of 25°C. An empty strip of zopiclone was found nearby. Resuscitation for 35 min was unsuccessful. At autopsy: generalized visceral congestion	Blood (ng/mL): furanylfentanyl = 0.95; carbamazepine = 9524; venlafaxine = 9480; alimemazine = 317; promethazine = 63.5; desmethylpromethazine = 106; methyphenidate = 28.6; ritalinic acid = 762; acetaminophen = 8466; pregabalin = 33 862; amphetamine = 116; 7-aminoclonazepam = 95
M/26	A deceased was found dead on the couch. A used needle, a spoon, and a suspected drug were found at the scene. At autopsy: brain edema and pulmonary edema. Drug abuse history	Blood (ng/mL): furanylfentanyl = 0.43
M/26	A deceased was found dead in his apartment. Three nasal sprays suspected to contain fentanyl were found at the scene. At autopsy: pulmonary edema and froth in the airways. Drug abuse history	Blood (ng/mL): furanylfentanyl = 0.78; carbamazepine = 14 815; pregabalin = 28 481; gabapentin, 94 937; norbuprenorphine, 1.37; fentanyl = 0.4; alprazolam = 42.2; alimemazine = 211; desmethylalimemazine = 211; diazepam = 31.6; methylphenidate = 4.2; ritalinic acid = 232. Urine (ng/mL): buprenorphine = 6; norbuprenorphine = 30
M/27	A deceased was found dead in an apartment shared by drug abusers. History of suicide attempts	Blood (ng/mL): furanylfentanyl, 1.16
M/24	A deceased was found dead on the couch. Drug paraphernalia were found nearby. At autopsy: congested and edematous lungs. History of drug abuse and recent treatment in an addiction center	Blood (ng/mL): furanylfentanyl, 0.4; fentanyl, 1.27. Urine (ng/mL): fentanyl, 150
Octfentanil		
M/16	A deceased was found dead at home, seated and leaning forward on the toilet. Drug paraphernalia, brown powder in a small zip-locked plastic bag lying on a card with a straw were found at the scene. Illicit drug abuse and depression history	Ocfentanil: femoral blood, 15.3 ng/mL; heart blood, 23.3 ng/mL; vitreous humor, 12.5 ng/mL; urine 6.0 ng/mL; bile, 13.7 ng/mL; liver, 31.2 ng/g; kidney, 51.2 ng/g; brain, 37.9 ng/g; nose mucus membrane, 2999 ng/swab. Additionally in peripheral blood: acetaminophen, 45 μg/mL; caffeine, 230 ng/mL
M/24	A deceased was found dead in his apartment. Drug paraphernalia, plastic zipper bag with brown powder, identified as ocfentanil, were found at the scene. At autopsy: lung congestion and edema, brain congestion and edema. Illicit drug use history	Ocfentanil: peripheral blood, 9.1 ng/mL; heart blood, 27.9 ng/mL; urine, 480 ng/mL. Additionally in peripheral blood: citalopram (130 ng/mL); quetiapine (<10 ng/mL), THC (2.8 ng/mL), and carboxy-THC (<5 ng/mL)
AH-7921		
M/early 20s	A deceased, victim of a minor traffic accident, was discharged from hospital the following day with a prescription for 30 mg codeine/400 mg acetaminophen. He ingested six tablets and some powder from zip-lock bags marked 3-methylmetcathinon (3-MMC) and 4-fluoromethamphetamine (4-FMA) bought on the internet. Soon after ingestion, when lying on the floor, he began to snore and was unresponsive. At autopsy: pulmonary edema	Peripheral blood: AH-7921, 430 ng/mL; 2-FMA, 6.9 ng/mL; 3-MMC, 2.1 ng/mL; codeine, 420 ng/mL; acetaminophen, 18 700 ng/mL

TABLE 2 Case Reports of Fatalities Involving Novel Synthetic Opioids [151M]—cont'd

Gender/ age	Case data	Toxicological findings
F/young	A deceased was found dead at home. Used needles and small plastic bags labeled "AH-7921" and "etizolam" were found in waste bins. At autopsy: needle marks in various stages of healing on the right cubital fossa	Peripheral blood: AH-7921, 330 ng/mL; methoxetamine, 64 ng/mL; etizolam, 270 ng/mL; phenazepam, 1330 ng/mL; 7-aminonitrazepam, 43 ng/mL; diazepam, 46 ng/mL; oxazepam, 18 ng/mL; nordiazepam 73 ng/mL
M/19	A deceased was found dead on the bed. Frosty substance around the mouth. At autopsy: pulmonary congestion and edema	AH-7921: peripheral blood, 6600 ng/mL; heart blood, 3900 ng/mL; urine, 6000 ng/mL; bile, 17 000 ng/mL, liver, 26 000 ng/g; kidney, 7200 ng/g; brain, 7700 ng/g
F/22	A deceased was found dead in the bedroom of her apartment. A plastic bag labeled "AH-7921" was found in the apartment. At autopsy: cerebral edema with increased intracranial pressure, the internal organs full of blood. Drug abuse history and AH-7921 use	AH-7921: femoral blood, 450 ng/mL; heart blood, 480 ng/mL; urine, 760 ng/mL; vitreous humor, 190 ng/mL; stomach content, 40 µg/mL; liver, 530 ng/g
U-47700		
M/20	A deceased was found dead with a syringe clutched in his hand. Drug paraphernalia were located to his proximity. Drug abuse history	Blood: U-47700, 382 ng/mL; amphetamine, 12 ng/mL
M/39	A deceased was found unresponsive lying on the sofa; a syringe was found on the floor. History of ordering designer drugs on the internet	Blood: U-47700, 217 ng/mL; mephedrone, 22 ng/mL
M/25	A deceased was found unresponsive with symptoms of pulmonary edema. A white powder, determined to be U-47700, was found at the scene. Polydrug abuse history	Blood: U-47700, 334 ng/mL
M/23	A deceased was found on the bathroom floor with a ligature around his arm. Syringe and a pocket containing a powdery substance labeled "U-47700" were found at the scene. Drug abuse history	Blood: U-47700, 252 ng/mL; citalopram, 43 ng/mL
M/29	A deceased was complaining of a headache the day of his death and suddenly collapsed. At the autopsy: pulmonary edema and brain edema	Blood: U-47700, 453 ng/mL
M/29	A deceased was found unresponsive with the evidence of pulmonary edema. A rolled-up 10 dollar bill with a residue of white powder and series of packets with white powder were found at the scene. Drug abuse history	Blood: U-47700, 242 ng/mL; carboxy-THC, 5.3 ng/mL
M/26	A deceased was found dead at home. Five syringes, benadryl and etizolam pills, diphenhydramine tables, and three glass dropper bottles were found at the scene. Drug abuse history	Blood: U-47700, 103 ng/mL; diphenhydramine, 694 ng/mL
M/21	A deceased was found dead at home with an injection site in the right arm containing a needle. Drug abuse history	Blood: U-47700, 299 ng/mL; tramadol <250 ng/mL; alprazolam, 47 ng/mL; lorazepam, 11 ng/mL; 3-methoxyphencyclidine, 180 ng/mL
M/24	A deceased was found unconscious and unresponsive at home. History of "U-47700" abuse	Blood: U-47700, 487 ng/mL; etizolam, 86 ng/mL; chlorpheniramine <250 ng/mL; diphenhydramine, 250 ng/mL
M/23	A deceased was found dead sitting up in a chair	Blood: U-47700, 311 ng/mL; oxycodone, 11 ng/mL; venlafaxine, 2600 ng/mL; *O*-desmethylvenlafaxine, 380 ng/mL
M/24	A deceased was found unresponsive with a syringe in his arm. Drug abuse history	Blood: U-47700, 59 ng/mL
M/46	A deceased was snorting a compound from an envelope labeled "U-47700." At autopsy: pulmonary congestion and edema	Peripheral blood: U-4770, 190 ng/mL; alprazolam, 120 ng/mL; doxylamine, 300 ng/mL; diphenhydramine, 140 ng/mL; carboxy-THC, 2.4 ng/mL. Urine: U-47700, 360 ng/mL. Liver: U-47700, 1700 ng/g

Continued

TABLE 2 Case Reports of Fatalities Involving Novel Synthetic Opioids [151M]—cont'd

Gender/ age	Case data	Toxicological findings
Not provided	A deceased was found on the bed. At autopsy: pulmonary congestion	U-47700: femoral blood, 525 ng/mL; heart blood, 1347 ng/mL; urine, 1393 ng/mL; kidney, 270 ng/g; liver, 430 ng/g; lung, 320 ng/g; brain, 97 ng/g. Additionally in blood: diphenidine (ca. 1.7 ng/mL); methoxphenidine (ca. 26 ng/mL); ibuprofen (ca. 1.8 μg/mL); naloxone (1.9 ng/mL)
Not provided	A deceased was found on the bed. At autopsy: pulmonary congestion	U-47700: femoral blood, 819 ng/mL; heart blood, 1043 ng/mL; urine, 1848 ng/mL; kidney, 140 ng/g; liver, 3100 ng/g; lung, 240 ng/g; brain, 110 ng/g. Additionally in blood: diphenhydramine (ca. 45 ng/mL) and methylphenidate (ca. 2.5 ng/mL)
U-47700 and furanylfentanyl or fentanyl or butyrlfentanyl		
M/36	A deceased was found unresponsive in the bathroom with a syringe cup in his mouth. History of drug abuse	Blood: U-47700, 135 ng/mL; furanylfentanyl, 26 ng/mL
M/33	Heroin and cocaine abuse history	Blood: U-47700, 167 ng/mL; furanylfentanyl, 56 ng/mL; morphine, 48 ng/mL
M/29	A deceased was found unresponsive	Blood: U-47700, 490 ng/mL; furanylfentanyl, 76 ng/mL
M/40	History of heroin/opioid abuse	Blood: U-47700, 105 ng/mL; furanylfentanyl, 2.5 ng/mL
M36	At autopsy: pulmonary edema. Drug abuse history and experimentation with substances purchased over the internet	Blood: U-47700, 13.8 ng/mL; fentanyl, 10.9 ng/mL. Urine: U-47700, 71 ng/mL
M/18	A deceased was found unresponsive in the bed. Syringes and two white powders, determined to be butyrylfentanyl and U-47700, were found at the scene. Drug abuse history	Blood: U-47700, 17 ng/mL; butyrylfentanyl, 26 ng/mL; ethanol, 0.03 g/dL
MT-45		
M/24	A deceased was found dead sitting on the chair in front of the desk. An e-cigarette with unknown fluid, drug paraphernalia, and several bags of white powder labeled "Methoxphenidine," "Methoxmetamine," and "MT-45" were found at the scene. At autopsy: brain edema, hemorrhagic pulmonary edema, and hyperemia of the internal organs. History of amphetamine abuse	MT-45: femoral blood, 660 ng/mL; heart blood, 1300 ng/mL; urine, 370 ng/mL; vitreous humor, 260 ng/mL; gastric content, 49 μg/mL; liver, 24 μg/g. Also in femoral blood: lidocaine and two synthetic cannabinoids—PB-22 and 5F-APINACA
M/35	A deceased was found dead at home. Drug paraphernalia (scale, spoon, pipe, and lighter) and two packets of white powder, one testing positive for MT-45 and the other for etizolam, were found at home. At autopsy: pulmonary congestion and edema. History of substance abuse	Peripheral blood: MT-45, 520 ng/mL; etizolam, 35 ng/mL; diphenhydramine, 220 ng/mL

TABLE 3 Common Harms Associated With Over-the-Counter Codeine Use [152H]

Harm	Description	Main products	Published cases
Opioid dependence	Features of dependence include escalating doses, withdrawal symptoms on discontinuation, and continued use of codeine despite experiencing harms. A number of cases of dependence to OTC codeine products have been successfully treated with Opioid Agonist Treatments such as methadone or buprenorphine	OTC codeine combination products	Frei et al. (2010), McDonough (2011), Robinson et al. (2010), and Evans et al. (2010)
Gastrointestinal harm	Gastric and peptic ulcers, perforation, haemorrhage, pyloric stenosis, gastrectomy and other bowel surgery	OTC codeine–ibuprofen	Dutch (2008), Frei et al. (2010), McDonough (2011), Evans et al. (2010), and Robinson et al. (2010)
Anemia	Resulting from blood loss, often secondary to gastric ulcers	OTC codeine–ibuprofen	Frei et al. (2010), Evans et al. (2010), Robinson et al. (2010)
Renal effects	Renal tubular acidosis due to prolonged high doses of ibuprofen, renal failure	OTC codeine–ibuprofen	Frei et al. (2010)
Hypokalaemia	Low potassium leading to muscle weakness and respiratory difficulties, one case of rhabdomyolysis and quadriparesis. Some cases requiring ICU admission for life support	OTC codeine–ibuprofen	Ernest et al. (2010), Frei (2010), and Ng et al. (2011)

codeine without morphine, methadone, tramadol (4.03%), and morphine with or without 6-acethylmorphine (1.98%), tested positive for urine and/or blood samples in 1777 confirmed drug users [154MC].

On average, 16 Canadians are hospitalized for opioid poisoning every day (2016–2017). About half were considered accidental and one-third self-harm. Older adults and patients 45–64-YO have the highest hospitalization rates, but rates are increasing for ages 15–24-YOs [155S]. Vogel et al. reported that adults (aged ≥65) have the highest opioid poisoning rates despite only representing 16% of Canadian population. For any poisoning in this population, one-third (33%) was intentional and accounted for 14% of total hospitalizations. Psychological therapies including cognitive behavioral therapy (CBT) has proven effective for pain and depression but in many instances, public insurance doesn't cover this service, and patients aren't able to pay for the service out-of-pocket. Authors also report that improving mental health among youth aged 15–24 may also be beneficial since they have the fastest growing rate of hospitalizations (increasing by 62% for 10/100 000 people in 2014–2015) related to intentional poisonings [156c].

Male Afghanistan/Iraq-era US military veterans' who have overdosed on opioids participated in a qualitative research study to better understand factors surrounding overdose in this population. Participants' accounts indicated that background experiences, such as self-medication for social and psychological pain, trauma, social alienation and isolation, and illicit drug use history, precondition the more immediate factors and behaviors that precipitate overdose (including binging and mixing drugs, naiveté about dosage, and ambivalence about life/death). Overdose experiences ($n=22/36$[61%]) reported by participants' involved heroin. Reported overdose involved oral (PO) opioids ($n=20/36$[56%]), cocaine plus PO opioids and/or heroin ($n=6/36$[17%]) and <15% had knowledge about naloxone during their interview [157c].

Although previous research has examined the association of childhood abuse with opioid misuse and dependence in adulthood, insufficient research focuses specifically on prescription opioids, and no studies have examined associations with prescription opioid use, considered a potential pathway to later opioid misuse and dependence. Austin et al. explored childhood emotional, physical, and sexual abuse with prescription opioid use in early adulthood. Researchers used data from the National Longitudinal Study of Adolescent to Adult Health Waves I (12–18-YO) and IV (24–3-YO). Respondents in Wave IV reported childhood abuse experiences occurring <18-YO and prescription opioid use in the last four weeks. Through multivariable logistic regression, childhood abuse was examined in association with recent prescription opioid use. Multivariable models adjusted for respondent sex, race/ethnicity, age, and socioeconomic status demonstrated significant association with childhood emotional abuse (OR=1.57, 95% CI 1.29, 1.90), physical abuse (OR=1.46, 95% CI 1.14, 1.87), and any childhood abuse (OR=1.51, 95% CI 1.24, 1.82) with recent prescription opioid use. Prescription opioid use with related morbidity and mortality continues to increase in the US, requiring a need to strengthen and expand current prevention and intervention strategies related to social and environment factors. Through future research, factors mediating childhood abuse and prescription opioid use could provide the information needed to provide such strategies [158MC].

Several case–control studies have shown associations between self-reported opium use and cancer risk but inquiring into these sensitive issues (drug use) is prone to information bias. In order to justify the findings, authors planned to quantify the negative bias. Rashidian et al. aimed to evaluate self-reported opioid use sensitivity and suggest suitable control groups for case–control studies for opioid use and cancer risk. To maximize the response validity, trained interviewers questioned 176 hospitalized patients and 186 healthy individuals, with self-reported opioid use. Most participants denied (43.3%) or reported mild (27%) pain. Self-reported opioid used during the past 72 h, compared to UDS showed sensitivity, 77% (CI 65.8%–89.2%) and 69% (CI 52.8%–84.9%) for hospitalized vs healthy individuals for self-reports (P value = 0.4). Based on corrected sensitivity, regular opioid use frequency was 47% [CI 39.7–54.8] and 28% [CI 21.6–35] for hospitalized vs healthy participants, respectively. Regular opioid use among hospitalized patients was significantly higher than in healthy individuals ($P<0.001$). Authors reported that although opioid use is underreported, it is comparable in both hospitalized and healthy individuals. Corrections are needed in studies evaluating these sensitive issues to prevent distortion of results [159C].

Law (Regulations)

The aggregate production quotas for 2018 were released by the US DEA mandating opioid (oxycodone, hydrocodone, oxymorphone, hydromorphone, morphine, codeine, meperidine, fentanyl) manufacturing for 2018 to be reduced by 20% [160S].

US Federal Drug Administration (FDA) granted fast track approval to the first once-monthly injectable buprenorphine product, Sublocade, for OUD treatment in adult patients who have initiated treatment with a transmucosal buprenorphine-containing product. This product has a delivery system that forms a solid deposit (depot), which is broken down to release buprenorphine. The most common AEs were constipation, nausea, vomiting, headache, drowsiness, injection site pain, itching at the injection site, and abnormal liver function tests. Safety and efficacy have not been established for patients <17-YO or >65-YO. Sublocade boxed warning has an IV self-administration risk and must be prescribed and dispensed as a risk evaluation and mitigation strategy, provided to health professionals through a restricted program. Sublocade should only be injected by a health professional and should be used as part of a complete treatment program that includes counseling and psychosocial support [161S].

All 50 US states have established prescription drug monitoring programs to collect information about individuals' prescription drug history in an electronic database. Eleven states have laws regulating pain-management clinics [162S] and several states have enacted laws limiting opioid dosage or duration. Six states declared their opioid overdose situation an emergency so the state can act temporarily to mitigate emergency resources that are not otherwise available.

The opioid-overdose crisis has led to emergency declarations for a non-communicable health condition, but the impact remains unclear. July 2017, the US President's Commission on Combating Drug Addiction and the Opioid Crisis called for a national emergency declaration [162S].

Within the preliminary report, the commission stated, issuing this declaration was the, "first and most urgent recommendation," since doing so could allow the federal government's executive and legislative branches to respond to the crisis with additional resources and policies. On October 26, 2017, the opioid crisis was declared a national public health emergency under the federal Public Health Services Act. Federal emergency powers have the potential to cover different actions, such as deploying providers from the Public Health Service or steps to reduce key medication cost, including naloxone. The emergency declaration in conjunction with a standing order allowed pharmacists to dispense naloxone to people who previously needed a prescription to obtain it. States could also allocate funds for community-based naloxone education and training for laypeople or to purchase naloxone for distribution to schools or other state facilities. The federal emergency declaration will supplement but not replace state declarations [163S].

The global commission on drug policy reviewed the global response to the opioid epidemic. Switzerland has offered access to opioid substitution therapy (OST), mainly with methadone and buprenorphine but also medical heroin, and substance analysis (drug checking). They have since created supervised injection facilities by introducing a balanced four-pillar strategy focusing on prevention, treatment, repression, and harm reduction. Since the strategy has been introduced, no one has died from overdose to date. The results from changes show that drug-related deaths are reduced by 50% and HIV rates have decreased. Beyond health-related outcomes, property crime by people using drugs reduced significantly (90%) for patients enrolled in OST. Although there was opposition at first, the policies are now well-supported [164S]. Portugal has also embraced the harm reduction approach by decriminalizing users. Drug consumption and small quantity possession is treated as a public health issue, not as a criminal offense, so people needing health and social services can access them freely and without fear of persecution. Thousands of lives have been saved by needle and syringe programs, supervised injection facilities, drug checking and heroin-assisted treatment for patients not responding methadone maintenance therapy (MMT) [164S]. In Norway, the Storting (Norwegian parliament) agreed to decriminalize illicit drugs in order to 'stop punishing people who struggle but instead give them help and treatment.' Labor Party (Ap), Conservatives (Høyre), Socialist Left (SV) and Liberals (Venstre) supported the proposal in order to shift the focus away from the courtroom and instead to focus on health services. This change was driven by positive influence and results from Portugal's policies [165S].

In contrast, although training prescribers is essential, problematic relationships with doctors in the US health system continue to hinder progression towards the public health initiative. Additionally, advertising medications to consumers in the US has contributed to the crisis by exponentially expanding opioid prescription supply. The US has reduced the prescription opioid supply without first putting supporting measures in place leading to opioid-dependent individuals to "turn to street heroin, which is increasingly cut with fentanyl". It is cheaper to make and ~50 times more potent than morphine, easier to hide and transport [164S].

Cardiovascular

Wentlandt et al. presented a case report of a patient experiencing a seizure whose EKG led to diagnosing intermittent monomorphic and polymorphic ventricular tachycardia (torsades de pointes [TdP]). A 41-YOF presented to the ED with concerning seizure activity. Family witnessed at least five unresponsive episodes with the entire body tensing and shaking or upper extremity tremor, without gaze deviation. Medical history included alcohol-related cirrhosis, ongoing daily alcohol-use, and alcohol withdrawal. Chronic medications included methadone 82 mg daily, furosemide, and spironolactone. Monitor leads during subsequent events demonstrated polymorphic ventricular tachycardia and TdP. The patient continued to have several brief monomorphic and polymorphic ventricular tachycardia and TdP episodes associated with clinical seizure-like activity. Repeat

EKG showed normal sinus rhythm (NSR) with markedly prolonged QT-interval (789 ms). Providers should carefully consider other alternative causes to seizures even in a patient with known epilepsy to avoid inappropriate diagnosis and appropriately treat the underlying cause [166A].

Tramadol is classically associated with neurological and respiratory SEs but cardiac effects are poorly documented in the literature. Belin et al. reported a case involving severe tramadol intoxication, with plasma concentration 20 times the toxic threshold, complicated by refractory cardiogenic shock. A 50-YOF found in the woods 24 h after leaving home, was conscious without confusion and had boxes of tramadol 100 mg (supposed ingested dose [SID] = 18 g), alprazolam 0.25 mg (SID = 15 mg) and 1 L (15%) alcohol. In the ED, she presented with clonus epileptic seizure, non-reactive bilateral mydriasis, and respiratory failure requiring orotracheal intubation and mechanical ventilation. The transthoracic echocardiography showed increased cardiac output with hyperkinetic profile and preserved left ventricular ejection fraction (LVEF). Plasma tramadol concentration = 46.5 mg/L (toxic if >2 mg/L) measured 12 h post hospital admission and ECG identified first-degree atrioventricular block. Patient developed new onset seizure and rapidly evolved towards cardiogenic shock (EF reduction to 20%). Acute renal and liver failures required a venoarterial femorofemoral extra corporeal life support (ECLS), which was successfully implanted by surgeons. After 72 h of circulatory incompetence, with the electrical cardiac response alternating between sustained ventricular arrhythmias and asystole, the patient began recovering to effective cardiac activity. On ICU day-8 LV systolic function improved, and ECLS was removed day-10 without hemorrhagic or ischemic complications. Although, in this case the patient co-administerd high-dose alprazolam and alcohol, poisoning caused by high-dose tramadol could lead to refractory cardiogenic shock for which ECLS is strongly recommended as effective rescue therapy associated with drug intoxication [167A].

Multiple studies have reported increased events involving prolonged QTc-interval when BZDs are used concomitantly with methadone. Mijatović et al. studied the role of diazepam concentrations with low-concentration-methadone in patients with OUD during MMT induction and QTc-prolongation, which has not been previously discussed. Individuals ($n = 30$) with addiction disorder on MMT were studied before initiating MMT and after 1- and 6-months treatment. Mean methadone dose at 1- and 6-month time points was 50 ± 17.55 and 78.63 ± 18.14 mg, while the mean diazepam dose was 28.33 ± 11.55 and 28.12 ± 11.67 mg. QTc-intervals before MMT initiation, 1- and 6-months after were 412 ± 27, 425 ± 18 and 424 ± 15 ms, respectively, showing statistically significant increased QTc-interval after MMT for 1- and 6-months. Statistically significant correlation between the methadone concentration and QTc-interval length at observed time points ($R^2 = 0.239$ [$P = 0.018$]; $R^2 = 0.513$ [$P = 0.006$]) was shown, and it remained so if the diazepam concentration was included ($R^2 = 0.347$ [$P = 0.026$], $R^2 = 0.513$ [$P = 0.009$]). The QTc-prolongation was below the risk threshold in low methadone therapeutic doses and concomitant diazepam use, which could be an associated co-factor risk [168c].

Literature has recently described opiate withdrawal as a possible trigger for the high catecholamine state needed to induce Takotsubo Cardiomyopathy (TC) suggesting that TC is an opportunity for improving patient safety and clinical outcomes in early substance-withdrawal treatment. Olsen et al. described a case involving TC precipitated by opiate withdrawal [169H,170H]. A 66-YOF with coronary artery disease (CAD), hypertension, type-2 diabetes mellitus (T2DM), and rheumatoid arthritis presented to the ED with a 3 h history of severe substernal chest and epigastric pain waking her from sleep. She was chronically taking oxycodone/acetaminophen 5 mg/325 mg. Additional cardiovascular risk factors included sedentary lifestyle and obesity. On admission, the patient was pale, diaphoretic, and in severe distress. VS included BP = 142/77 mmHg, HR = 110 beats/min, RR = 22/min and admission troponin levels <0.02. Recent cardiac catheterization revealed no obstructive CAD. ECG identified AF with rapid ventricular response and LVEF = 55%–65% with a large pericardial effusion suggesting early tamponade. Repeat ECG showed effusion increased in size. The patient received pericardiocentesis, removing yellowish fluid (500-cc). Pericardial effusion was deemed to be secondary to suspected viral inflammation. Arterial blood gas, post-procedure, showed respiratory acidosis with increased lactic acid levels leading to subsequent intubation [171A].

Day-2 after opioid cessation, the patient was extubated. Patient presented with behaviors including reorienting conversations frequently about her medications, challenging staff members, anxiety, agitation, and tachycardia. Day-3, patient had mild respiratory distress. Repeat ECG revealed LVEF = 20%–25% with hypokinesis of mid-apical anterior, mid-anteroseptal, mid-inferoseptal, mid-apical inferior, mid-inferolateral, apical septal, apical lateral, and apical walls; LV hypertrophy; moderate mitral and tricuspid regurgitation; and dilated left and right atriums. Findings were consistent with TC. At that time troponin levels (4.42) peaked. Patient was then re-started on opiates, similar to home doses. Withdrawal symptoms and associated stress resolved and medication was switched to MMT. Seven-days after diagnosing TC and replacing equivalent opioid therapy, a repeat ECG showed LVEF = 40% with improved wall motion abnormality and mild apical hypokinesis.

Thirty-three days after, ECG revealed LVEF = 55%–65% with no wall motion abnormalities. At that time, the patient's acute heart failure symptoms had resolved. With increased opiate use prevalence, suspicion for opiate-withdrawal induced cardiomyopathy should be high in the presence of cardiac symptomatology [171A].

Khaja et al. described a patient who was found in her hospital room, unresponsive with a tapentadol bottle containing crushed pills suspended in a liquid with a syringe. The patient received naloxone during resuscitation and EKG revealed QRS = 74 ms, corrected QT-interval = 430 ms. Patient went into asystole, leading to cardiac arrest and autopsy confirmed tapentadol overdose as cause of death. Despite tapentadol's popularity as an alternative to opioids, IV tapentadol drug abuse can still cause respiratory depression leading to cardiac arrest [172A].

Shattuck et al. described a 23-YOF, 10–12 weeks pregnant with IV drug use history. She was recently diagnosed with pneumonia and found deceased in her bed. Four-days prior she was diagnosed with pneumonia and empirically treated with azithromycin. Two-days prior to death, alprazolam tablets (#90) were refilled and the bottle was found empty in the home. 0.2×0.2 cm tear in the ascending thoracic aorta was located 0.3 cm superior to the aortic valve, non-coronary cusp. Masson's trichrome staining revealed severely disrupted collagenous fibers surrounded by neutrophils in the abscess area. Gram staining identified both gram-positive and -negative bacteria but results may have been altered due to antibiotic therapy. Postmortem serum toxicology tested positive for alprazolam, tetrahydrocannabinol, and morphine and the cause of death was cardiac tamponade secondary to an isolated abscess in the ascending aorta. Risk factors for aortic rupture and cardiac tamponade included IVDU, pneumonia, and pregnancy. Autopsy revealed an isolated abscess in the ascending aorta without any signs for infective endocarditis, coronary artery rupture, aortic aneurysm, or aortic dissection, making this an unusual cardiac tamponade case. The responsible organism was not identified since tissue cultures were not obtained due to fatal drug overdose suspicion. Authors advise providers to obtain tissue cultures in aortic abscesses to identify the organisms involved, helping to guide antimicrobial treatment in future presentations [173A].

Hunt et al. described the first cardiac involvement and atypical clinical features of thrombotic microangiopathy (TMA) (e.g., pulmonary involvement). Dyspnea patients had serum compliment decline which did not occur concomitantly with the disease onset and therefore complement activation/consumption was likely the secondary consequence of TMA [174E,175H,176c]. Since 2012, case reports have described TMA occurrence following IV abuse with extended-release oxymorphone hydrochloride (Opana ER®). Authors present three patients with unique clinical features and investigated IV exposure to the tablet's inert ingredients as a possible causal mechanism. Patient's had microangiopathic hemolytic anemia, thrombocytopenia, and AKI following recent adulterated injection of Opana® tablets. Kidney biopsy revealed varying cardiac involvement and retinal ischemia with TMA. Authors advised all physicians should be aware of IV drug abuse when presented with cases of TMA [177A].

Patients inadvertently inject buprenorphine/naloxone via the intra-arterial route, which can produce acute ischemic insult to the hand due to gelatin embolism. Wilson et al. reviewed a patient series to better describe the clinical entity, review the outcomes, and propose a rational treatment algorithm. Patient records from hand surgery (2011–2015) were reviewed for hand ischemia after buprenorphine/naloxone injection. Ten patients presented during the review period, averaging 13-week follow-up period. Eight had radial side ischemia, one on the ulnar side, and one with bilateral ischemia. Three patients were treated by IV heparin, five with oral agents (aspirin or clopidogrel) while two presented with dry gangrene and did not receive anticoagulation. All patients experienced tissue loss and no outcome difference was found, despite selected treatment. Authors concluded that further studies are needed to determine the most effective treatment for these injuries [178c].

Loperamide: Through post-marketing reports, Swank et al. set out to identify and characterize cardiotoxicity, including TdP, associated with loperamide use (Table 4). Researchers used US FDA Event Reporting System (FAERS) database from December 28, 1976 (US drug approval date) to December 14, 2015. Forty-eight serious cardiac AE cases were associated with loperamide use but greater than one-half were reported after 2010. Most frequently reported cardiac AE were syncope ($n = 24$), cardiac arrest ($n = 13$), QT-interval prolongation ($n = 13$), ventricular tachycardia ($n = 10$), TdP ($n = 7$) and death ($n = 10$). The most common reasons for use were drug abuse ($n = 22$) and diarrhea treatment ($n = 17$). In drug abuse cases, the median daily dose was 250 mg (70–1600 mg) and events occurred as early as 6 h after a dose and as long as 18 months after loperamide initiation. In 13/22 drug abuse cases, patients used loperamide for euphoric or analgesic effects, and to prevent opioid withdrawal symptoms ($n = 9$) [179c].

Sheng et al. defined loperamide-hERG channel interaction characteristics and used a ventricular myocyte action potential model to compare loperamide's proarrhythmia propensity to 12 drugs with a defined clinical risk. Electrophysiological data were integrated into a myocyte model, stimulating dynamic drug-hERG channel interaction to estimate TdP risk through comparisons with reference drugs and their known defined clinical risk. In the overdose setting, loperamide is placed in the high-risk category, alongside quinidine, bepridil, dofetilide, and sotalol for proarrhythmia risk [180E].

Wu et al. reported new treatment options for loperamide toxicity. Thirteen patients with TdP appeared to benefit from overdrive pacing using transvenous electrical pacing or isoproterenol to suppress ventricular ectopy and prevent recurrent dysrhythmias. The second therapy reported was IV lipid emulsion. Since loperamide is highly protein-bound (97%) [181C,182H] it is not markedly cleared by hemodialysis but IV lipid emulsion could be considered for patients with severe cardiotoxicity. Wu et al. reported that treatment has been used in three cases. Lipid doses were not specified and had limited clinical details. Authors concluded that IV lipid emulsion is justified for trial in patients who remain severely unstable despite otherwise optimal care. The third therapy discussed was venoarterial extracorpeal membrane oxygenation (ECMO), which has been reported in one published case for a patient presenting with cardiac arrest. It is unknown whether care was withdrawn or if ECMO was ineffective (Table 4). Authors suggest using an ECMO trial, if available for patients

TABLE 4 Reported Cases of Loperamide-Associated Cardiac Toxicity [183c]

Case, year	Age, sex	Dose	Clinical presentation	ECG changes	Treatment	Loperamide plasma conc. (μg/L)	Outcome
2017	28, M	400 mg/day × 5 month	Presyncope, palpitations	QTc = 601 ms, ventricular tachycardia	Magnesium, isoproterenol	N/A	Survived
	37, F	400 mg	Loss of consciousness	QTc > 600 mg	Naloxone 0.4 mg, isoproterenol	Undetectable	Survived
	41, M	>100 mg/day	Recurrent syncope	QTc = 702 ms, TdP	Magnesium, amiodarone, cardioversion, electrical pacing	N/A	Survived
	28, M	N/A	Dyspnea, presyncope	QRS = 168 ms, QTc = 693 ms, TdP	Magnesium, sodium bicarbonate, lidocaine, isoproterenol, electrical pacing	120	Survived
	39, F	N/A	Seizure-like activity	QRS = 142 ms, QTc = 687 ms, TdP	Magnesium, sodium bicarbonate, isoproterenol	Undetectable	Survived
	24, M	N/A	Loss of consciousness, then cardiac arrest	N/A	Naloxone, CPR	77	Died
	39, M	N/A	Loss of consciousness	Asystole	CPR	140	Died
2016	28, F	400–600 mg/day + cimetidine	Recurrent syncope, palpitations	QRS = 192 ms, QTc = 642 ms, ventricular tachycardia, TdP	Lidocaine, amiodarone, isoproterenol	83.2	Survived
	24, M	400 mg/day + cimetidine	Generalized weakness	QRC = 146 ms, QTc = 544 ms, ventricular tachycardia, TdP	Amiodarone, sodium bicarbonate, isoproterenol	N/A	Survived
	46, M	100 mg/day × 5 years, 200 mg/day × 2 days	Syncope, cardiac arrest	QTc > 600 ms, ventricular fibrillations, TdP	Defibrillation, magnesium, amiodarone, lidocaine, metoprolol	N/A	Survived
	20, M	288 mg/day	Recurrent syncope, palpiations, dyspnea	QRS = 200 ms, QTc = 600 ms, TdP	Defibrillation, electrical pacing, isoproterenol	N/A	Survived
	26, M	192 mg/day	Syncpe meiosis	Ventricular tachycardia, TdP	CPR, defibrillation, magnesium, isoproterenol	2	Survived
	48, F	40–80 mg/day	Decreased level of consciousness	QRS = 164 ms, QT = 582–614 ms, ventricular tachycardia	Magnesium, self-limited	210	Survived

Continued

TABLE 4 Reported Cases of Loperamide-Associated Cardiac Toxicity [183c]—cont'd

Case, year	Age, sex	Dose	Clinical presentation	ECG changes	Treatment	Loperamide plasma conc. (μg/L)	Outcome
2015	54, F	144 mg/day × 2 years	Recurrent syncope	Junctional escape, ventricular tachycardia, TdP	Lidocaine, amiodarone cardioversion, electrical pacing	N/A	Survived
	30, M	400 mg/day × 7days	Syncope, then cardiac arrest	QRS = 192 ms, QT = 704 ms, TdP	Defibrillation, isoproterenol	120	Survived
	26, M	100–250 mg/day + cimetidine	Recurrent syncope	QTc > 700 ms, TdP	Cardioversion, isoproterenol	N/A	Survived
	26, M	800 mg/day × 18 months	Cardiac arrest	PEA	N/a	N/A	Died
	25, F	N/A	Syncope, nausea, vomiting, then cardiac arrest	QRS 140–170 ms, QTc 490–527 ms, TdP	Sodium bicarbonate, magnesium, atropine, lipid emulsion, electrical pacing, ECMO	32	Died
2014	30, M	400 mg/day × weeks	Recurrent syncope, then cardiac arrest	QRS "wide," QTc > 500 ms, TdP	N/A	22	Survived
	43, F	288 mg/day	Syncope	QRC = 130 ms, QTc = 684, TdP	Sodium bicarbonate, magnesium, lidocaine, amiodarone, lipid emulsion, cardioversion, electrical pacing	N/A	Survived
	28, M	792 mg/day	Syncope	RS 162 ms, QT/QTc 670647 ms, ventricular tachycardia, TdP	Magnesium, potassium, sodium bicarbonate, lidocaine, lipid emulsion, defibrillation, electrical pacing, isoproterenol	N/A	Survived
	33, M	120–200 mg/day	Dyspnea	QRS = 128 ms, QT/QTc = 566/636 mg	Sodium bicarbonate	77	Left against medical advice
	33, M	70–140 mg/day × month	Anxiety, chest tightness	QRS = 118 ms, QTc = 490 ms	Magnesium, potassium	33	Survived

PEA = Pulseless electrical activity.

with severe loperamide cardiotoxicity who are refractory to other measures [183c].

Zabeer et al. discussed a 37-YOF with hypothyroidism and drug abuse (marijuana, cocaine, heroin) presenting to the hospital after being found unconscious at home. She reported attempted suicide two days earlier with ~200 loperamide (2 mg) tablets. EKG on arrival identified prolonged QTc-interval (>600 ms) and normal HR (61 bpm). On admission serum loperamide was not detectable but loperamide metabolite levels, desmethylloperamide = 32 ng/mL (limit quantification = 5 ng/mL) were detectable. Laboratory results demonstrate delayed toxicity response resulting from desmethylloperamide, suggesting it plays a major role in arrhythmia pathogenesis [184A].

Endocrine

Glucose Metabolism (Methadone)

Although uncommon, hypoglycemia resulting from methadone use has been reported but the mechanism is still unknown [185A,186A,187c,188c,189E,190A,191A,192A]. Masharani et al. presented a 39-YOF complaining of fatigue and confusion when fasting longer than 4–6 h. She constantly wakes with blood glucose (BG) ranging 40–50 mg/dL and reports recovering from confusion after consuming rapidly absorbed carbohydrates. The patient does not take insulin and oral hypoglycemic agents nor has a history of multiple endocrine neoplasia-1 syndrome. Current medications include baclofen, lidocaine patches, duloxetine and methadone (160 mg every

6h as needed) for chronic pain with a medical history including stage II CKD of unknown etiology. Labs include SCr = 1.77 mg/dL, eGFR = 32 mL/min/1.73 m^2, no proteinuria and serum albumin range = 2.5–3 g/dL. She was evaluated for protein losing enteropathy. She had normal upper/lower endoscopies and liver function. She was monitored during a 6h fasting period which resulted in abnormal labs including BG = 44 mg/dL with elevated insulin 8.5 mU/L, C-peptide (2.7 ng/mL), proinsulin (49 pmol/L), and negative insulin autoantibodies, cosyntropin stimulation test, endoscopic ultrasound and gallium-labeled [1, 4, 7, 10-tetraazacyclododecane-1, 4, 7, 10-tetraacetic acid]-Phe1-Tyr3-octreotide PET/CT.

A few months after the hypoglycemia evaluation, chronic pain therapy was changed to buprenorphine and methadone was tapered. Hypoglycemic episodes resolved one day after discontinuing methadone. While taking methadone, mean glucose level = 68 mg/dL (±13 mg/dL), which increased (114 mg/dL (±18 mg/dL; $P < 0.0001$, Student *t*-test) when methadone was discontinued. While taking buprenorphine, she fasted for 10h resulting in normal labs including, BG = 97 mg/dL and insulin = 45 pmol/L. Renal failure may exacerbate the hypoglycemia risk. Providers should be aware that patients taking high methadone doses could result in hypoglycemia and the dose should be reduced or replaced with another opioid before an extensive evaluation for inappropriate hyperinsulinism is considered [191A].

Additionally, Toce et al. reported an accidental methadone ingestion resulting in hypoglycemia and for the first time, propose a mechanism for this reaction. A previously healthy 11-month-old male was intubated after presenting with respiratory failure with notable miosis, BG = 17 mg/dL, an elevated insulin level and suppressed serum beta-hydroxybutyrate. Patient was given a dextrose bolus and IV naloxone, which improved BG, miosis and mental status. The quantitative methadone level from arrival was 123 ng/mL (lower limit of detection = 10 ng/mL). Testing was negative for ethanol, salicylates, sulfonylureas, and metabolic causes of hypoglycemia. A fasting study showed euglycemia with insulin suppression and appropriate ketosis. Based on this case, Toce et al. reported that the patient's hypoglycemia resulted from methadone-induced insulin secretion and proposed that hyperinsulinism is the mechanism responsible for methadone-associated hypoglycemia. Methadone exposure should be included in the differential diagnosis for new onset hypoglycemia [192A].

Glucose Metabolism (Buprenorphine)

Tilbrook et al. sought to measure the effect of buprenorphine–naloxone as opioid substitution therapy on glycemic control in patients with T2DM and OUD in Northwestern Ontario ($n = 573$[1982–2015]). Over a 2-year continuum, mean glycated hemoglobin A1c values revealed an absolute decrease (1.20%) for T2DM patients receiving buprenorphine–naloxone compared to T2DM patients not being treated for opioid dependence, whose values rose by 0.02% ($P = 0.011$). Treating OUD with buprenorphine–naloxone substitution therapy could have an unintended positive effect on diabetes management [193C].

Testosterone

Effects from OST on skeletal health in men have limited data. Gotthardt et al. developed a cross-sectional study aiming to determine the low bone mass prevalence in male drug users to evaluate the relationship between endogenous testosterone and bone mass. 144 males (opioid dependent >10 years) on long-term OST followed at the Center of Addiction Medicine in Basel, Switzerland were recruited. 35 healthy age- and BMI-matched men served as the control group. The study participants received OST with methadone (69%), morphine (25%) or buprenorphine (6%). 74.3% had low bone mass, with comparable bone mass irrespective of OST type. Older men (≥40 years) $n = 106$[29.2%]) were osteoporotic (mean *T*-score = −3.0 ± 0.4 SD) and 48.1% were diagnosed with osteopenia (mean *T*-score = −1.7 ± 0.4 SD). Younger men ($n = 38$), 65.8% had low bone mass. In all age groups, BMD was significantly lower than in age- and BMI-matched controls. In multivariate analyses, serum free testosterone (fT) was significantly associated with low BMD at the lumbar spine ($P = 0.02$), but not at the hip. When analyzed by fT quartiles, lumbar spine BMD decreased progressively with decreasing testosterone levels. Authors concluded that low bone mass is highly prevalent in middle-aged men taking opioids long-term which may be explained partially by androgen deficiency [194C].

To the best of the authors' knowledge, Stange et al. were the first to attempt investigating the associations between empathic abilities with current testosterone levels as a potential biomarker for socio-cognitive deficits in opiate addicted patients. Using a cross-sectional approach, 27 opiate-addicted, diacetylmorphine-maintained patients (21 males, age mean = 41.67 years, SD = 8.814) and 31 healthy controls (23 males, age mean = 40.77 years, SD = 8.401) matched in age, sex, and educational level were examined. Results identified higher personal distress ($P < 0.01$, $d = 0.817$) and testosterone ($P < 0.001$, $d = 1.093$) scores in the patient group compared to controls. Testosterone and personal distress among the patient group resulted in a positive correlation ($r = 0.399$, $P < 0.05$). Opiate-addicted patients show specific impairments in emotional empathy, specifically with elevated personal distress, which has clinical implications regarding social cognition rehabilitation and relapse prevention. Authors conclude that this data could suggest high personal distress and testosterone during withdrawal as potential biomarkers for severe opiate addiction [195c].

Gastrointestinal

Stercoral colon perforation is a rare pathology believed to be caused by increased intraluminal pressure created by a fecaloma. Opioid induced constipation could cause colonic perforation albeit rare and often unsuspected [196A,197r]. Poitras et al. reported a 58-YOF presenting to the ED, with severe abdominal pain, discomfort, and a syncopal episode (thought to be related to patient's severe hypotension). She had GI-bleeding with diffusely tender and distended abdomen. She had no drug abuse history but had long-term suboxone use and chronic constipation. Abdominopelvic CT scan revealed a bowel perforation, ascites and fecal impaction. Emergency laparotomy revealed hemoperitoneum and extensive stool in the peritoneal cavity. Fecal bolus with perforation was located in the sigmoid colon. Post-op day-6, a second CT scan revealed multiple loculated abscesses within the small bowel and other areas, which were opened and washed out. A large fecal mass was located in the sigmoid colon causing pressure necrosis on surrounding tissues, progressing to bowel perforation. Fecal material was protruding through the perforation site into the peritoneal cavity and affected bowel was resected. Asplenic rupture was noted resulting in a splenectomy. The postoperative diagnosis was intra-abdominal abscesses and sepsis. After re-operation patient had a steady recovery with conservative treatment. Authors note that recognizing this pathology has become increasingly important with increased opioid drug use since any patients taking opioids are at risk for constipation progressing to stercoral perforation and should be monitored closely. Stercoral perforation, as a differential diagnosis should be considered for any patient taking this therapy, presenting with chronic constipation, fecal impaction on imaging and clinical signs of peritonitis or sepsis to reduce mortality rates in these cases [198A].

HEENT

Dental

Prescription opioid abuse may have irreversible adverse dental effects resulting in substantially decayed dentitions in the younger population. Fraser et al. explored the damaging effects of a 3-year prescription opioid abuse history in a 26-YO's dentition and oral health. A Caucasian male presented to dental medicine with a chief complaint of "I need a better smile. I think my teeth are preventing me from getting job interviews." PMH included oxycodone/acetaminophen abuse, consuming ~20 tablets per day (~6.5 g acetaminophen, 100 mg oxycodone daily) for three years without seeing dentistry for five years. Patient had a confirmed history of MI at 22-YO. Social history included smoking cigarettes two packs-per-day (ppd) and marijuana frequently, but denied current alcohol use. Patient typically consumed meals high in carbohydrate content and drank sweet tea at least one quart per day. Patient claimed to have stopped prescription opioid abuse for >2 years. Past dental history included regular visits to a private dentist as a child and teenager. Patient presented with gross generalized decay and plaque accumulation. He was diagnosed with severe plaque induced gingivitis with localized chronic periodontitis and xerostomia. Treatment included periodontal maintenance and control, caries excavation, root canal therapy, extractions of non-restorable teeth and continuous dental education. Opioid abusers have diminished salivary flow, often compounded by poor oral hygiene. Typically, opioid abusers present with poor overall health and oral hygiene habits [199C,200MC]. An even greater risk for caries is presented by xerostomia when patients consume a "high-sugar diet," which allows sugar to serve as a reservoir for cariogenic bacteria, thereby inducing dental decay [201M]. There is a well-established link between chronic opioid abuse and increased sugar consumption [202H,203H,204c,205A].

Infants born with NAS have an elevated cavity development risk. The infant's caregiver should be encouraged to establish a dental home ≤1-YO which enables the dental provider to assess the child's behavior over time and provide education regarding early childhood caries, good oral hygiene, diet, and non-nutritive habits. A caries risk assessment should be completed to develop an individualized prevention plan to avoid caries, including using the appropriate quantity of age-appropriate toothpaste with fluoride [206H].

Eye

The infectious ocular complication, endogenous endophthalmitis, is rare but feared by IVDUs. Researchers reported the first endogenous endophthalmitis case in the US caused by the emerging fungal pathogen, Rhodotorula, in a patient with IVDU history leading to no light perception vision. Worldwide experience with Rhodotorula endogenous endophthalmitis is limited, but existing cases suggest infection by this particular fungal genus has an ominous prognosis (Table 5).

Increasing opioid use in the last decade predicts increased risk for endogenous endophthalmitis, emerging rare infectious agents and poor visual outcomes. Researchers suggest that Rhodotorula should be considered in the differential diagnosis for IVDU patients with sudden profound visual loss, especially if response to initial therapy is suboptimal. Given the association with endocarditis, blood cultures should be drawn in all

TABLE 5 Rhodotorula Patient Case Presentation [207A]

	Symptoms	Assessment	Treatment
Initial presentation (21-YOM)	Left-sided vision loss, eye pain, and conjunctival injection (4 days) Floaters (2 weeks)	Previously healthy with current IVDU *Best corrected visual acuity*: Counting fingers at one foot *Slit lamp examination:* 4/4 conjunctival injection 4/4 anterior chamber cells with hypopyon Posterior synechiae *Ophthalmoscopic exam:* Identified two white, hazy collections in the posterior vitreous No issues with right eye	Pars plana vitrectomy *Intravitreal injection*: ○ Amphotericin B ○ Ceftazidime ○ Vancomycin *Discharged home with:* ○ Atropine 1% ○ Moxifloxacin 0.5% ○ Prednisolone 1% ○ Oral azithromycin
1-month follow-up	Pain completely subsided Visual acuity unchanged at hand motion Non-clearing vitritis Posterior pole white mass	Symptoms indicative of abscess	Pars plana vitrectomy with abscess debridement *Intravitreal injections*: ○ Ceftazidime ○ Vancomycin *Discharge home with*: ○ Moxifloxacin 0.5% ○ Prednisolone 1%
		Abscess culture: *Rhodotorula* species colonies susceptible to amphotericin B and posaconazole	Posaconazole × 14 days
		Species identified (morphology): *Rhyacophila mroczkowskii* or *Rhynchospio glutaea*	
10-days post-surgery	Funnel retinal detachment	Funnel retinal detachment deemed inoperable by two retinal specialists because of extensive retinal ischemia	None
3-year follow-up	No light perception in left pthisis bulbi		

patients with endogenous endophthalmitis prior to initiating broad-spectrum antibiotics. Hepatitis C and HIV coinfection is common in these cases and should also be evaluated [208A,209A]. The prognosis in Rhodotorula endogenous endophthalmitis is poor, so increased prevention in patients with IVDU is critical.

Nasal

Effectiveness for intranasal naloxone ranges from 70% to 80% [210C,211C,212A,213H]; however, features within the nasal cavity have been known to affect the medication absorption through the nasal mucosa [214c,215c,216E]. Septal abnormalities, nasal trauma, epistaxis, excessive nasal mucous, and intranasal damage which are caused by cocaine use, are contraindicated for intranasal naloxone use [217AH]. Weiner et al. aimed to determine obstructive nasal pathology incidence in patients who experienced serious opioid-induced respiratory depression (OIRD). Using a retrospective, multivariable analysis through IMS LifeLink: Health Plan Claims Database, researchers aimed to detect serious OIRD (2009–2013). Patients ($n=7234$) who experienced a serious OIRD event had obstructive nasal pathology ($n=840$[11.6%]) including deviated nasal septum ($n=20$[2.4%]), nasal cavity polyp ($n=246$[29.3%]), nasal turbinate hypertrophy ($n=130$[15.5%]), or other nasal cavity disease ($n=659$[78.5%]). Patients with concurrent obstructive nasal pathology who experienced serious OIRD had an adjusted odds ratio (aOR)=1.28 (95% CI 1.13–1.46). Obstructive nasal pathology is common in patients who experience OIRD and in itself is associated with a higher OIRD risk [218MC].

Immunologic

Methicillin Resistant Streptococcus Aureus (MRSA) is increasing globally but Sweden is a low prevalence country, with *Staphylococcus aureus* isolates resulting in 1% of

MRSA positive results [219MC]. Since 2000, MRSA notification rates in Sweden increased continuously from 3.7/100000 to 30/100000 inhabitants (years 2000–2014), respectively [219MC,220MC]. A high-risk factor for MRSA carriage and infection in other countries results from substance dependence and IVDU [221c,222c,223c]. A retrospective, longitudinal register study, investigated MRSA epidemiology (1997–2013) in opioid and amphetamine-dependent individuals, in comparison with alcohol-dependent subjects. Data from the national Swedish in- and outpatient registers included 73201 individuals (1997, 1999, 2004, 2009, 2013) which were analyzed for SUD and demographic MRSA predictors using generalized estimating equations. Both opioid (aOR=2.82; 95% [CI]=2.16, 3.67) and amphetamine dependence (aOR=2.71; 95% CI=1.70, 4.16) were significantly associated with MRSA diagnosis' compared with alcohol dependence, when adjusting for age, sex and year [224MC].

Opioid-use effects on T-cell abnormalities are consistent with the immune risk phenotype among HIV-infected individuals which is understudied [225H,226H]. Edelman et al. aimed to assess the associations between illicit opioid use and T-cell characteristics (CD4/CD8 ratio, memory profiles) based on CD45RO and CD28 expression, and senescence based on CD57 expression. Individuals were recruited from an exploratory cross-sectional analysis of Russian ARCH [cohort of antiretroviral therapy (ART)-naïve HIV-infected) (11/2012–10/2014)] in St. Petersburg, Russia to evaluate past 30-day illicit opioid use. 186 participants reported any illicit opioid use (38%), intermittent (18%) and persistent (20%). Among ART-naïve HIV-infected Russians, any illicit opioid use was not significantly associated with T-cell abnormalities although intermittent illicit opioid use may be associated with CD8+ T-cell abnormalities ($P=0.02$) and borderline significantly increased senescent T-cells ($P=0.05$). Given increased infection and comorbidity risks seen among HIV-infected individuals with illicit opioid use, longitudinal studies are warranted to confirm results [227C].

Schuppener et al. aimed to describe the investigation and control of a Serratia marcescens cluster bacteremia in a 505-bed tertiary-care center. All patients with S. marcescens bacteremia (March 2–April 7, 2014) were identified (usually <10 infections per year). In total, six patients developed S. marcescens bacteremia <48h after admission within the 5-week period. 5/6 had identical Serratia isolates determined by molecular typing, and were included in a case–control study. Exposure to the post-anesthesia care unit was a risk factor identified in bivariate analysis. Tampered opioid-containing syringes on several hospital units were discovered directly proceeding the initial cluster case and a full narcotic diversion investigation was conducted. A nurse was identified as responsible for the drug diversion and was epidemiologically linked to all five patients in the cluster. Once the implicated employee was terminated, no further cases were identified. Illicit drug use by healthcare employees is a mechanism that can result in transmitting bloodstream infections in hospitalized patients [228c].

Individual and structural factors have restricted HIV prevention from expanding for PWID using opioid agonist therapies (OAT) in Ukraine but extended-release naltrexone (XRNTX) provides new opportunities for treating OUDs in this region, where there is continued increases in HIV incidence and mortality. The HIV incidence and mortality only continue to increase in Eastern European and Central Asian regions (Joint United Nations Programme on HIV/AIDS) [229C,230C]. The Ukrainian HIV epidemic remains concentrated with PWID, specifically with opioids, and their sexual partners [231MC,232MC]. Marcus et al. randomly selected and surveyed PWID ($n=1613$) from five Ukraine regions, who were currently, previously or never on OAT and analyzed their pharmacological therapy preference for treating OUDs. For those preferring XRNTX, correlating independent willingness was examined to initiate XRNTX. Among the PWID, $n=449$[27.8%] were interested in initiating XRNTX. Independent correlates associated with interest in XRNTX included being from Mykolaiv (aOR=3.7, [95% CI=2.3–6.1]) or Dnipro (aOR=1.8, [95% CI=1.1–2.9]); never taking OAT (AOR=3.4, [95% CI =2.1–5.4]); shorter-term injectors (AOR=0.9, [95% CI 0.9–0.98]); and inversely for both positive (aOR=0.8, [CI=0.8–0.9]), and negative attitudes toward OAT (AOR=1.3, [CI=1.2–1.4]), respectively. Due to the concentrated OUD in the HIV population for PWID, new options for treating OUD are urgently needed [233MC].

Hematologic

Vivaldelli et al. reviewed a 45-YOM who presented to rheumatology with bilateral swelling in his hands from the proximal segments in his fingers to the wrist. Edema developed gradually over three years (non-pitting and not affected by raising hand) but his hand function and fingers were not impaired. Patient reported drinking wine 1–1.5L/day but otherwise his PMH was unremarkable. Patient had confirmed hepatitis with ultrasound, elevated liver enzymes (ALT=175IU/L, AST=170IU/L, alb=4.68mg/dL) and a positive HCV antibody test. Rheumatoid factor=29IU/mL (normal <14), antinuclear antibodies, anti-citrullinated peptide antibodies, complement, protein electrophoresis and total bilirubin were normal. Patient admitted to injecting high-dose sublingual buprenorphine for >5 years, preferable in the wrist 3–4 times per day. "Puffy hand syndrome" is a complication associated with IVDU. Although the risk is generally known among users, most physicians including addiction specialists and general practitioners rarely encounter the presentation and are typically unfamiliar with it. Pathology associated with the edema includes venous and lymphatic insufficiency and the direct toxicity from injected drugs. Biopsy samples from the skin, subcutaneous tissue and lymphangiograms identify destruction to the lymphatics leading subcutaneous tissue fibrosis most likely due to the drug's soluble compounds damaging the lymphatic vessels [234A].

Nervous System

Serotonin Syndrome

To their knowledge, Mastroianni et al. reported the first serotonin syndrome (SS) case associated with linezolid and methadone [235A]. SS has been reported in patients receiving linezolid concomitantly with serotonergic drugs [236R]. Methadone is a serotonin reuptake inhibitor which may be associated with serotonin toxicity reactions [237A]. A 39-YO HIV-positive male presented with drug-addiction, sepsis, osteomyelitis, multiple muscle and metastatic skin abscesses caused by MRSA, under combined antibiotic treatment. Current medications included MMT and antiretroviral therapy (lamivudine, tenofovir, darunavir/ritonavir). Upon admission, clinical examination revealed patient was dehydrated with high fever (VS: BP=90/50mmHg, HR=125/min, RR=35/min), multiple erythematous and tender swellings over the right elbow and arm, left side of the chest wall, and abdomen, bilateral coarse crepitations. Based on blood and pus cultures (Hospitalization day-5), antibiotic treatment was changed to linezolid plus clindamycin intravenously, previous antibiotic combination was suspended, and repeated surgical drainage was performed on muscle abscess. After three days, patient became increasingly confused, disoriented and agitated. His vitals fluctuated (HR=95–120beats/min; BP=110/70–100/60mmHg). Patient was unable to sustain a conversation (GCS=10). He had generalized abdominal tenderness, dilated reactive pupils, an increased tone in both legs, with brisk reflexes and clonus at both ankles but no meningism. White cell count (21.1×109/L) was predominantly neutrophils, and C-reactive protein was elevated (15mg/dL). SS was suspected and after neurological consultation linezolid was immediately suspended, patient was initiated on rehydration therapy, steroids and BZDs, with progressive improvement over the next 48h. Researchers ruled out other CNS pathologies and other possible drug interactions. Due to the quick recovery after linezolid discontinuation and the knowledge that linezolid and methadone are weak serotonin reuptake inhibitors, the neurological team associated symptoms to serotonin toxicity. Researchers advise healthcare professionals to take caution in patients on MMT and linezolid [235A].

CNS

Since limited information is available about opioid prenatal exposure and its effect on long-term neurocognitive function in children, Skoylund et al. aimed to determine the association between prenatal exposure to analgesic opioids and it relation to language competence and communication skills at 3-YO. Norwegian Mother and Child Cohort Study prospectively included pregnant women (1999–2008). Women ($n=45211$) with singleton pregnancies ($n=51679$) were included who reported opioid analgesic use during pregnancy ($n=892$[1.7%]). Using adjusted analyses, no association was found between opioid use and reduced language competence or communication skills (OR=1.04 [95% CI: 0.89–1.22] and OR=1.10 [95% CI: 0.95–1.27], respectively) but pain and paracetamol use were associated with a small reduction in communication skills [238MC].

Animal and human studies have suggested a link between prenatal opioid exposure and negative impact to the developing fetal brain, but results are conflicting. Sirnes et al. aimed to investigate the association between prenatal opioid exposure and brain morphology in school-aged children by using a cross-sectional MRI study with prenatally opioid-exposed children and matched controls. Children ($n=16$) 10 to 14-YO with prenatal opioid exposure were evaluated for automated brain volume measures (T1-weighted MRI scans). Basal ganglia, thalamus, and cerebellar white matter regional volumes were significantly reduced in the opioid-exposed group, whereas global brain measures (total brain, cerebral cortex, and cerebral white matter volumes) did not demonstrate significance. In the opioid-exposure group compared to control, participants demonstrated higher ADHD prevalence. Eleven children were exposed to multiple drugs that cause NAS including BZDs ($n=8$[50%]), cannabis ($n=4$[25%]), amphetamines ($n=4$[25%]), alcohol ($n=1$[6%]). AEs caused by opioids on the developing fetal brain may explain the association but further research is needed [239c].

Sayulich et al. sought to better understand the relationship between the endogenous opioid system in emotion regulation since the mu-opioid receptor blockade effect on brain systems and underlying negative emotional processing are not clear in addiction. Researchers found that naltrexone modulated task-related activation in the medial prefrontal cortex and functional connectivity between the anterior cingulate cortex and the hippocampus as a function of childhood adversity (for aversive vs neutral images) in all groups. Authors concluded that pharmacological response to naltrexone is one environmental factor that is influenced by early childhood adversity [240c].

Cognition

Schmidt et al. evaluated cognitive functioning in 45 formerly opioid-dependent adults after abstaining ≥1 year compared to 45 matched healthy controls in Germany. Compared to controls, abstainers had significantly more clinical symptoms, physical diseases, psychiatric history, worse outcomes in both verbal capability tests, and worse outcomes on learning performance and recognition. Abstainers' attention did not diminish over time compared with controls in tasks using low attention requirements (stimulus response) but abstainers performed worse in tasks with higher attention requirements. Abstainers performed better in the complex Planning Test (executive function), which may be explained by a lack of objectivity application. Overall, abstainers demonstrated a good level of cognitive functioning [241c].

To their knowledge, Shrestha et al. explored the first study about neuropsychological factors operating together and their effect on health-related quality of life (HRQoL). Baseline data (collected January 2010–December 2014) for a two-by-two multi-factorial, randomized controlled trial (Project Harapan) were used to examine the relationships between neurocognitive impairment (NCI), depression, alcohol use disorders (AUDs), and HRQoL in HIV-infected male prisoners ($N=301$) in Malaysia. A pre-release MMT program was evaluated and compared to an adapted, evidence-based behavior, based on relative effectiveness. Results demonstrated increased NCI ($P<0.001$) and depression ($P<0.001$) were negatively associated with HRQoL. NCI effect on HRQoL was significantly mediated via depression ($P<0.001$). Conditional indirect NCI effect on HRQoL via depression for individuals with AUDs was significant ($P=0.0087$), suggesting a moderated mediation effect. The findings disentangled the complex relationship using a moderated mediation model, demonstrating that increasing NCI levels, which can be reduced with HIV treatment, negatively influenced HRQoL via depression for individuals with AUDs and opioid dependence [242C].

Behavioral

Stein et al. examined the relationship between adverse experiences and three opioid use landmarks: opioid initiation age, injection drug use, and lifetime overdose. Researchers interviewed consecutive persons seeking inpatient opioid detoxification (May–December 2015). Participants responded to questions regarding opioid initiation age, last months injection drug use and lifetime overdose history. Participants ($n=457$) completed the 10-item Adverse Childhood Experience (ACE) questionnaire. They averaged 32.2 (±8.64)-YO, 71.3% male, and 82.5% non-Hispanic White. Their mean ACE scale score was 3.64 (±2.75), mean age when initiating opioid use was 21.7 (±7.1)-YO, 68.7% had injected drugs within the past month, and 39.0% had overdosed. When adjusting for age, gender, and ethnicity, the ACE score was associated inversely with opioid initiation age ($b=-0.5$ [95% CI −0.70; −0.29] [$P<0.001$]) but positive for recent IVDU (OR = 1.11, [95% CI 1.02; 1.20] [$P=0.014$]) and likeliness to experience an overdose (OR = 1.10 [95% CI 1.02; 1.20] [$P=0.015$]) in a graded dose–response manner. Authors concluded that increased adverse childhood experiences are associated with three opioid use landmark risks. ACE screening could be useful to identify high-risk opioid-using population subsets [243C].

Mental Illness Associated Overdose

Suicide risk associated with substance use disorders (SUDs) may differ for men and women. Crude suicide rates are greater in male compared to female Veterans Health Association (VHA) patients. Sex-specific, age-adjusted standardized mortality ratios indicate that suicide rates among female VHA patients are considerably higher compared to women in the general US population [244MC,245MC]. Bohnert et al. aimed to estimate the association between SUDs and suicide for men and women receiving VHA care by conducting a cohort study using national administrative health records from all VHA users (FY2005) alive beginning FY2006 ($n=4\,863\,086$) [244MC,245MC]. Although the unadjusted diagnosis analysis with any SUD (alcohol, cocaine, cannabis, opioid, amphetamine, sedative use disorders) had a significantly increased risk for both males and females, HRs ranged from 1.35 (cocaine use disorder) to 4.74 (sedative use disorder) for men and 3.89 (cannabis use disorder) to 11.36 (sedative use disorder) for women. The estimated HR relationship between any SUD (alcohol, cocaine, opioid use disorders) and suicide was significantly elevated for women compared to men ($P<0.05$). Further adjustment, most notably for psychiatric diagnosis, associations linking SUDs with suicide were markedly attenuated and greater suicide risk among females was observed only for any SUD and OUD ($P<0.05$). SUD should signal increased suicide risk, especially among VHA women and may be important markers to consider in suicide risk assessment but co-occurring psychiatric disorders could explain the association between SUD and suicide especially among SUDs in women [246MC].

Trist et al. reported an attempted suicide with fentanyl patches in a 70-YOF suffering from chronic back pain. Authors state that no other cases report a patient surviving such a lethal opioid dose without medical intervention. Patient gradually accumulated 100 mcg fentanyl patches from repeat prescriptions, applying 14-patches with fatal intent, in addition to mirtazapine (90 mg), tramadol and morphine sulfate oral solution at therapeutic doses. 24 h later she awoke from a deep sleep, amazed that her efforts had failed. There was no evidence the patient had insensitivity to opioids outside chronic opioid use. The patient could have a polymorphism, favoring rapid metabolism from active fentanyl to inactive norfentanyl metabolites [247R]. Authors concluded that patient's mental well-being and education about opioid lethality should be regularly assessed and opioids should be used with further caution in patients who have a mental illness history evaluating early exploration into suicidal intent [248A].

Respiratory

There is limited literature about the concomitant use with opioid- and gabapentinoid-use and the administration requirements needed for naloxone to reverse respiratory depression. Rahmati et al. conducted a retrospective study (March 2014–September 2016) to compare respiratory depression frequency among patients who received naloxone and opioids (non-gabapentinoid group) to those who received naloxone, opioids, and gabapentinoids. The secondary objective was to compare oversedation association using the Pasero Opioid-induced sedation scale and various risk factors with the gabapentinoid group. 153 patient episodes with naloxone administration (non-gabapentinoid ($n=102$); gabapentinoid ($n=51$)) in 125 patients. Respiratory depression was associated more with the non-gabapentinoid group ($n=33/102$ [32.4%]) vs the gabapentinoid group ($n=17/51$[33.3%]) ($P=0.128$). There was significant association between respiratory depression and surgery in the previous 24h ($P=0.036$) and respiratory depression plus age >65 years ($P=0.031$) for patients in the non-gabapentinoid group compared to the gabapentinoid group. Naloxone was given 38.2% vs 45.1% of the time in non-gabapentinoid vs gabapentinoid group and the average MME daily dose was 63.9mg (±70.5) vs 89.9mg (±115), respectively, for non-gabapentinoid vs gabapentinoid participants. Although there wasn't significant respiratory depression associated in the gabapentinoid vs the non-gabapentinoid group, there was an increased respiratory depression risk for the gabapentinoid group, specifically in patients who had surgery within the previous 24h [249C].

Strang et al. investigated buprenorphine and norbuprenorphine plasma levels and their relationship to respiratory depression. Elevated active metabolite (norbuprenorphine) levels have been reported in postmortem analyses and could be the main association to observed deaths [250A,251E,252E,253H,254E]. Animal studies have also found substantial respiratory depression with norbuprenorphine unlike buprenorphine itself. Buprenorphine may diminish norbuprenorphine's respiratory depression effect by competing with it at receptors in the lung and antagonizing depressant effects in a dose-dependent manner [253H]. Opioid-dependent subjects were randomized 2:1 to novel lyophilized rapid-disintegrating tablet ("bup-lyo") Espranor® or standard sublingual buprenorphine tablet ("bup-SL") Subutex®. Respiratory depression (cumulative duration of SpO_2 <90% over 30min periods) increased with corresponding buprenorphine exposure levels (AUC_{30min}), but more specifically with norbuprenorphine. Lower buprenorphine/norbuprenorphine ratio predicted respiratory depression and the observed ratio (mean[SD]) was significantly higher for "bup-lyo" (3.4[2.8]) compared to "bup-SL" (1.7[0.77]), $P<0.0001$, indicating that "bup-lyo" is less likely to cause respiratory depression than "bup-SL." The linear regression lines identified a relationship between buprenorphine bioavailability and $SpO_2<90\%$ duration for "bup-lyo" and "bup-SL" but the slope was steeper for "bup-SL" than for "bup-lyo", (an increased bioavailability, greater respiratory depression). The relationship between plasma norbuprenorphine and $SpO_2<90\%$ duration was similar for both treatments. Buprenorphine plasma exposure levels within a given period was determined by "bup-SL" more greatly impacting respiratory depression than for "bup-lyo" ($P=0.0935$), although not apparent for norbuprenorphine ($P=0.663$). Researchers report that respiratory depression is more strongly associated with norbuprenorphine than with buprenorphine [255E].

Due to lacking evidence for detailed data about buprenorphine toxicity in children related to clinical presentation and management, Toce et al. sought to examine a single center cohort who were exposed to buprenorphine and provide descriptive analysis about respiratory depression rates, time to respiratory depression, interventions, disposition, and outcomes. Children were 6-months to 7-YO and hospitalized (January 2006–September 2014) with reported buprenorphine or buprenorphine/naloxone exposure. Eighty-eight patients were treated with activated charcoal ($n=20$[23%]) and/or naloxone ($n=48$[55%]), were admitted to the ICU ($n=36$ [41%]), and presented with clinical respiratory depression symptoms (83%), oxygen saturation, Sp02 < 93% (28%), depressed mental status (80%), miosis (77%), and emesis (45%). The median time from exposure to respiratory depression was 263min [IQR 105–486] and median hospital stay was 22h [IQR 20–26] which was positively associated with estimated exposure dose ($P=0.002$). Compared to adults, pediatric patients exposed to buprenorphine are less likely to exhibit opioid toxicity signs including respiratory depression, altered mental status and miosis. Although the patient majority developed clinical toxicity signs within 8h from reported exposure, the optimum monitoring duration remains unclear [256c].

Heppell et al. investigated whether paracetamol–codeine combination overdoses caused respiratory depression more than paracetamol alone by reviewing deliberate self-poisoning admissions with paracetamol (>2g) and paracetamol–codeine presenting to a tertiary toxicology unit (1987–2013). Researchers evaluated naloxone requirement or ventilation for respiratory depression. 1376 admissions were included with paracetamol alone ($n=929$), paracetamol–codeine ($n=346$) or paracetamol–codeine–doxylamine ($n=101$) without co-ingestants. Patients ($n=7/1376[0.5\%]$) were intubated ($n=3$) or received naloxone ($n=3$), were both intubated and recieved naloxone ($n=1$) (95% CI 0.2%–1.1%). Patients ingesting paracetamol alone ($n=3/929[0.3\%]$ (95% CI 0.1%–1%) requiring intubation or naloxone was similar to ingesting paracetamol–codeine combinations ($n=2/346[0.6\%]$ (95% CI: 0.1%–2.3%; absolute difference, 0.26%; 95% CI: 0.7%–1.2%; $P=0.62$). Similarly, few patients ($n=2/101[2\%]$) ingesting paracetamol–codeine–doxylamine combinations (95% CI: 0.3%–8%) required intubation or naloxone. Paracetamol–codeine combination overdoses are rarely associated with severe respiratory depression [257MC].

Renal

Babak et al. aimed to investigate hospitalized patients with acute opioid toxicity and rhabdomyolysis through clinical and demographic characteristics and laboratory findings. They conducted a cross-sectional study with hospitalized patients ($n=354$; males$=291$) with acute illicit drug toxicity in Tehran (2014). Seventy-six (21.5%) patients developed rhabdomyolysis (male, $n=69$ [90.8%]) and most cases were associated with methadone followed by opium abuse. Rhabdomyolysis was most common in 20–29 and 30–39-YO patients. Mean blood urea was low (3.8 ± 1.0 mg/dL) but mean potassium and sodium were 3.8 ± 0.3 mg/dL and 140.4 ± 4.0 mg/dL, respectively. Five (6.5%) male patients passed away due to severe renal failure. Opioid toxicity can lead to abnormal electrolyte disorders, like rhabodomyolisis caused by severe renal failure, which could result in death or life-threatening complications. Abnormal laboratories should be identified in patients with opioid toxicity to efficiently initiate treatment [258C].

NPS

Fentanyl Analogs

On August 15, 2016, the Mayor's Office of Drug Control Policy in Huntington, WV notified the Cabell-Huntington Health Department (CHHD) that they received multiple calls that day from EMS between 3 p.m. and 8 p.m. CHHD identified 20 opioid overdose cases within 53 h in Cabell County. First responders administered naloxone to $n=16[80\%]$ patients. 6/16 were administered multiple doses, which suggested they were exposed to a highly potent and/or long-acting opioid. The clinical specimens were identified as furanylfentanyl and carfentanil (100 times more potent than fentanyl). Authors suggest that developing public health partnerships to identify substance identification strategies could help link patient's overdosing to recovery at the point of resuscitation [259A].

In April 2015, the DEA and CDC reported an increase in fentanyl seizures by law enforcement in Massachusetts, mostly believed to be illicitly manufactured fentanyl (IMF) [260S]. To guide overdose prevention and response activities, the Massachusetts Department of Public Health and the Office of the Chief Medical Examiner collaborated with the CDC to investigate fentanyl overdose characteristics in three Massachusetts counties with high opioid overdose death rates (April 2016). Medical examiner charts were reviewed for opioid overdose decedents who passed away October 1, 2014–March 31, 2015. April 2016, semi-structured, in-person interviews were conducted with persons who used illicit opioids and witnessed or experienced an opioid overdose within six months. A nonrandom sample ($n=64$) was recruited for 20 knowledgeable respondents per county by using a harm-reduction program to identify persons during the previous 12 months. Six-months prior to the interview respondents either witnessed an overdose (95%) or overdosed themselves (42%). Respondents (88%) attributed increased opioid overdose deaths to suspected fentanyl and reported (69%) that suspected fentanyl was now available for purchase in powdered form obtained alone or mixed with heroin, (consistent with IMF preparation) and not as diverted prescription medications (e.g., Duragesic) but persons using heroin often did not know whether fentanyl was mixed into the heroin they purchase. Although some persons sought out fentanyl, others attempted to avoid it but the majority reported that opioid-seeking behaviors were not altered in response to rapidly emerging fentanyl analogs. Respondents witnessing a suspected fentanyl overdose (75%), described symptoms as occurring rapidly, within seconds to minutes. Respondents (25%) witnessed or experienced an overdose when fentanyl was insufflated (snorted), and the remainder reported the overdose always involved injecting fentanyl. Atypical overdose characteristics described were immediate blue discoloration of the lips (20%), gurgling sounds with breathing (16%), body stiffening or seizure-like activity (13%), foaming at the mouth (6%) and confusion or strange affect before unresponsiveness (6%). Respondents (65%) witnessed naloxone administration, self administering naloxone, or receiving naloxone to successfully reverse an opioid or fentanyl overdose. Respondents (83%) reported that ≥ 2 naloxone doses per suspected fentanyl overdose were used before the person responded and heroin or fentanyl (30%) was used with others present to help protect them from a fatal overdose [261c].

Among 196-opioid overdose decedent records reviewed, 64% had a positive fentanyl test, increasing from 44% to 76% (October 2014–March 2015). Fentanyl deaths were suspected to involve IMF (82%), prescription fentanyl (4%), and unknown fentanyl sources (14%). Most deaths occurred in a hotel, motel or private residence (91%) and few (6%) had evidence that a lay bystander-administered naloxone. The likelihood that a bystander administered naloxone was decreased due to lack of bystanders (18%), spatial separation (58%), awareness of decedent's drug use (24%), intoxicated bystanders (12%), failure to recognize overdose symptoms (11%), assumption that decedent had gone to sleep (15%). Clear evidence that a bystander was unimpaired, witnessed the drug consumption and was present during an overdose was rarely reported (1%) [261c].

Although illicit fentanyl synthesis does not require any special equipment nor a high technical knowledge level, when amateurs mix it with heroin, the results are unpredictable and potentially deadly [262R]. Soloman et al. reported a 20-YOF who was dropped off at the ED where acquaintances informed the valet that patient had been using heroin. Patient presented cyanotic, apneic, and unresponsive to painful stimuli and 1 mg intranasal naloxone was then administered in each nostril. Bag-mask ventilation was initiated (Vitals: RR = 10 breaths/min; 02-sat = 100%; HR = 110 beats/min). An additional 2 mg naloxone was given and IV access was established. Over 10 min periods, naloxone (14 mg) was given in divided doses until the patient exhibited spontaneous respirations. Although UDS was not taken authors theorize that this woman was exposed to fentanyl-tainted heroin since the 2–4 mg naloxone dose resulted in an absent response. Researchers advise EMS and ED staff to be prepared to adjust naloxone doses with the increased fentanyl-tainted heroin overdoses. Since this phenomenon is so new no research data is available to support best practice [263A].

MT-45

Helander et al. attempted to find a common underlying cause for the complex skin–hair–eye symptoms observed in three Swedish men aged 23–34 (late 2013–mid-2014). The common clinical signs included hair depigmentation, hair loss, widespread folliculitis and dermatitis, painful intertriginous dermatitis, dry eyes, and elevated liver enzymes. Case-1 was a 23-YOM who developed folliculitis-like rash on his legs, hair depigmentation and loss of scalp hair. He had dry skin and dermatitis on his groin and was referred to dermatology clinic. Clinical exam revealed dry and scaly skin, angular cheilitis, cracks on fingers and under the feet, groin and armpits exhibiting redness and moist maceration, and total alopecia. As the dermatitis deteriorated, the patient reported severe generalized skin pain requiring morphine treatment. Patient was hospitalized a few months later and reported taste and smell loss, and constant chills. His hair (eyebrows and eyelashes) started to grow back completely white. He experienced irritated and dry eyes and an elevated liver panel (AST = 493.98 U/L; ALT = 1210.84 U/L). Liver biopsy revealed active portal inflammation and blood sample identifying elevated MT-45 [1-cyclohexyl-4-(1,2-diphenylethyl)piperazine] serum concentration (280 ng/mL). One-year later the patient experienced rapidly progressive vision loss and became almost blind. He was treated with cataract surgery on both eyes. Symptoms originally alerted providers to Thallium poisoning but it was not detected in biological samples. Two cases showed transverse white Mees' lines (leukonychia striata) on fingernails and toenails suggesting temporary, drug induced, disorganized keratinization. Clinical signs gradually resolved over time but later on, both cases developed severe bilateral secondary cataracts, requiring surgery. Since drug tests within the Swedish STRIDA project had demonstrated MT-45 intake in all patients, MT-45 should be suspected, as the common causative agent. Researchers recommend that health professionals consider the increasing untested recreational drugs as a potential cause for unusual and otherwise unrecognized clinical signs and symptoms [264A].

U-47700

More recently, novel opioids have become available including U-47700 (3,4-dichloro-*N*-[2-(dimethylamino) cyclohexyl]-*N*-methylbenzamide) which is a synthetic, selective l-opioid agonist developed in the 1970s [265E]. Domanski et al. reported two cases, 26-YOM and 24-YOF who insufflated a substance they believed to be "synthetic cocaine." They presented with cyanosis, agonal respirations, anxiety, tremors and drowsiness. Urine samples from both patients were analyzed using GC/MS/MS and LC/QToF, and U-47700 was isolated in both cases. No other opioids were detected [266A].

3-Fluorophenmetrazine (3-FPM) is a stimulant-like NPS and fluorinated phenmetrazine analog recently appearing on the recreational drug market, with limited published information while U-47700 continues to gain popularity. Ellefsen et al. presented a case with 3-FPM and U-47700 use for the first time in the literature. A 34-YOM was found unresponsive and pronounced dead after EMS performed CPR. Patient had PMH including depression, bipolar disorder and expressed suicidal ideation weeks prior to death. He had an enlarged heart, which is more susceptible to cardiac arrhythmias, especially when using stimulant type drugs, found in his blood. 3-FPM was detected with U-47700 and other drugs including amitriptyline, nortriptyline, methamphetamine, amphetamine, diazepam, nordiazepam, temazepam, and the designer BZDs, flubromazolam and delorazepam. 3-FPM was quantified in the decedent's femoral (2.4 mg/L) and aortic (2.6 mg/L) blood, respectively. These concentrations are similar to reported concentrations in nonfatal intoxications. U-47700 was present in peripheral blood at a semi-quantitative concentration (0.36mg/L), consistent with reported U-47700 postmortem concentrations. The cause of death was considered multiple drug-toxicity and death was ruled an accident. This case highlights the risk of polysubstance use while discussing overlapping recreational and fatal concentrations for some NPS [267A].

To the knowledge of Rambaran et al., there are no clinical trial data assessing U-47700 safety. Authors aimed to describe signs and symptoms of ingestion, laboratory testing, and treatment modalities for U-47700 intoxication. Researchers identified and extracted data from relevant articles, which included 16 patients. Patients dying from U-47700 overdose typically presented to the hospital with pulmonary edema but patients who survived presented with tachycardia, decreased mental status and respiratory rate suggesting an opioid toxidrome (Table 6a). Since immunoassays failed to identify U-47700, chromatographic and spectral techniques were required [268c] (Table 6b).

To their knowledge, Seither et al. reported the first DUI case where carfentanil, U-47700 and other synthetic opioids were confirmed and identified in a human performance blood sample. A 28-YOM involved in a single-vehicle incident was found in the driver's seat unconscious with a needle and two clear plastic bags containing suspected heroin. The subject admitted that another male was buying heroin for him. While the other male left to purchase more drugs, the subject took a "bump of heroin" and then passed out behind the wheel. Initial routine blood testing identified alprazolam = 55ng/mL and fentanyl <0.5ng/mL. Researchers then used a validated LC–MS–MS assay, identifying carfentanil, furanylfentanyl, parafluoroisobutyryl fentanyl, U-47700 and its metabolite. The four NPS opioids would not have been detected without additional targeted methodologies identifying fentanyl as the NPS compound accounting for the opioid-like condition. The case demonstrates further need to supplement routine toxicological analyses with increased sensitivity and specificity [269A].

Opioid Withdrawal

Symptoms and AE

Prilutskaya et al. aimed to prospectively assess whether regular SC use affects opioid-related withdrawal duration and craving symptoms in patients undergoing drug detoxification treatments. Patients ($n = 193$) with OUD underwent drug detoxification therapies, one-quarter ($n = 47$) of which regularly used SC. The Clinical Opiate Withdrawal Scale (COWS) and visual analogue scale (VAS) were used to assess opioid withdrawal and craving symptoms over time. Subjects using SCs had significantly longer withdrawal duration and craving symptoms ($P < 0.001$). Greater SC intake within the past 30 days ($P = 0.045$), less time since the last SC intake ($P = 0.033$), greater SC use duration ($P < 0.001$), and greater SC dosage ($P < 0.001$) were associated with greater withdrawal symptom duration. This study highlights the negative effects with concomitant SC use while withdrawing from opioids [270C].

Ovietdo-Joekes et al. aimed to review injectable hydromorphone and diacetylmorphine safety profiles and explore dose and attendance pattern association with AEs or serious adverse events (SAEs). Researchers conducted a non-inferiority randomized double-blind controlled trial (Vancouver, Canada) testing hydromorphone ($n = 100$) and diacetylmorphine ($n = 102$) for severe OUD treatment. Most common related AE included immediate post-injection or injection site pruritus reactions, somnolence, and opioid overdose. Hydromorphone group participants were less likely to have any related AE or SAE compared to the diacetylmorphine group. Diacetylmorphine group participants (5/11) had SAE including opioid overdoses (requiring naloxone) occurring in the first 30 days since most recent treatment initiation. A non-linear relationship was identified for somnolence and opioid overdose (AEs and SAEs) event rates by received dose. Overall when injectable opioids are medically prescribed and administered with supervision, opioid-related AE including potentially fatal overdoses, are safely mitigated and treated by health care providers [271C].

OUD Withdrawal treatment

Since withdrawal management from abrupt loperamide discontinuation has not been discussed in the literature, Leo et al. presented a patient with longstanding loperamide associated OUD requiring methadone managed withdrawal. Specific risks for using extra-medical loperamide at 40–100 times the recommended antidiarrheal dosing results in sodium and potassium channel blockade precipitating QRS-widening, QT-prolongation, and ventricular arrhythmia including TdP [272c,273A, 274E,275A]. Patient was a 38-YOF presenting to the ED with nausea, vomiting, restlessness, diaphoresis, tremulousness, generalized body aches, restless legs, and anxiety. She reported experiencing anxiety, dysphoria, restlessness, irritability, abdominal cramping, and lower-extremity cramping with obvious mydriasis, but no "gooseflesh," nor diaphoresis; mild tachycardia (HR ~ 100beats/min), but no significant hypertension. Patient admitted to symptoms associated with drug use including dysphoria, decreased appetite, decreased sleep, and guilty feelings but denied suicidal thoughts, anhedonia or hopelessness (COWS = 12). She used loperamide to induce euphoria over 10 months, starting with 48 (2mg) and increasing to 144 (2mg) tablets daily but managed constipation with periodic enemas. She admitted to "problems" related to paroxysmal

TABLE 6A Case Reports Involving the Use of U-47700 and Outcome [268c]

Country	Age, year	Sex	Drug(s)	Route	Presentation	Toxicity Duration	Treatment/Treatment outcomes
USA	41	F	U-47700 (3-tablets) Home medications: baclofen, gabapentin, sertraline Alcohol Smokes Marijuana Methamphetamine abuse history	PO	*Arrival to ED*: depressed level of consciousness. Pinpoint pupils, mildly responsive to sternal rub. 10 min after naloxone: BP = 141/98 mmHg, HR = 134 beats/min, RR = 19, SPO_2 = 100 on nonrebreather mask. Serum chemistry 48 min into ED arrival, significant for acetaminophen <10 mg/mL and glucose 64 mg/dL. No drug abuse screen was performed	1.5 h	Two minutes into ED arrival received 0.4 mg naloxone IV, which resulted in arousal and regain of consciousness. Pruritus and anxiety occurred shortly after naloxone and was resolved after receiving lorazepam 1 mg IV and diphenhydramine 50 mg IV. She remained somnolent for 2 h but then recovered to normal coherent state. She attempted to DAMA and was discharged home 4 h after arrival with a complete recovery.
USA	26	M	Alcohol Alprazolam U-47700	Insufflation	*EMS*: Face down with agonal breathing, cyanotic with O_2 sat = 50% on ambient air. *ED*: GCS = 3, pinpoint pupils. HR = 125 beats/min, BP = 150/63 mmHg, RR = 14 breaths/min, T = 97.4°F or 36.3°C, ABG pH 6.97, PCO_2 > 90, PO_2 = 80, HCO_3 = 21.4. CXR = bilateral consolidation; CT = patchy consolidation on right upper lobe and bilateral lower lobes. SCr = 1.5 mg/dL, lactate = 4.4 mmol/L, WBC = 11.2 K/mL, Glu = 94 mg/dL, CK = 130 U/L. ECG = sinus tachycardia, 125 beats/min with normal intervals and nonspecific ST changes. Ethanol level = 55 mg/dL	3 h	Intubation by EMS: ketamine, lorazepam, rocuronium. From ED, he was transferred to ICU where he was sedated on propofol and given antibiotics for presumed pneumonia. He self-extubated in the ICU and was discharged 3 days after presentation with a normal exam
USA	24	F	Alcohol, alprazolam, U-47700	Insufflation	*ED*: She reported anxiety, nausea, and abdominal pain. T = 97.8°F or 36.5°C. Drowsy upon physical exam but no dyspnea; HR = 97 beats/min, BP = 111/77 mmHg, RR = 18 breaths/min, O_2 sat = 100 on room air (RA). Ethanol level = 11 mg/dL. CXR = normal. ECG = NSR at 87 beats/min with normal intervals, no ST changes	3 h	Kept for 24 h observation then discharged
USA	23	F	U4 or U-47700	IV. The previous two times via insufflation but it caused epistaxis	*EMS*: She was cyanotic, RR = 4 breaths/min and O_2 sat: 60% *ED*: HR = 104 beats/min, O_2 sat = 77% on 5 L O_2 via nasal cannula (NC), RR = 30 breaths/min. *PE*: normal with dried blood at nares and mouth, increased respiratory effort and crackles in both lung bases. CXR = patchy infiltrates and mild congestion consistent with pulmonary edema. BMP, CBC normal, WBC = 18.3, VBG with venous O_2 sat = 65.4%. UDS negative for recreational drugs	Minutes	*EMS*: Given naloxone 2 mg intranasal without response followed by 2 mg naloxone IV, whereby she became more responsive. *ED*: Placed on BiPAP, which improved O_2 sat to 100 and decreased RR to normal limits. She came off BiPAP later that night and was discharged the following day after her pulmonary function improved
USA	22	M	U-47700 (250 mg) divided into five separate doses. Only utilized one, utilized two in the past without adverse effect	Mixed with water and applied into nostril	EMS: Found him cyanotic with agonal respirations, 4 breaths/min and pulse oxygenation = 60%. BP = 138/88 mmHg, HR = 134 beats/min and GCS = 3. *ED*: BP = 112/71 mmHg, HR = 92 beats/min, *T* = 97.6°F (36.4°C), pulse oxygenation = 97% on RA. CXR = normal, ECG = unremarkable with normal QRS width, QTc = 442 ms. Leukocytosis (16 000/mm^3) with differential, neutrophils = 88%, lymphocytes = 8%, monocytes = 2%, eosinophils = 1%. UDS positive for benzodiazepines	Time prior to being found was not reported	Naloxone (2 mg) was given by EMS, which reversed the coma and bradypnea. He was discharged after staying 5 h in the ED. One week follow-up revealed he was asymptomatic
USA	29	M	U-47700 on day of presentation and phenazepam a few days prior	IV	Initially was unresponsive but regained consciousness prior to ED transport. *ED*: BP = 157/105 mmHg, HR = 112 beats/min. PE unremarkable, CBC = normal except elevated neutrophils (79.6%). Elevated SCR = 1.4 mg/dL, normal eGFR	3 h	Monitored in ED and provided with oral hydration, resulting in VS improvement. He was discharged 3 h post observation

ABG = arterial blood gases; *BiPAP* = bilevel positive airway pressure; *BMP* = basic metabolic panel; *BP* = blood pressure; *CBC* = complete blood count; *CK* = creatine kinase; *CT* = computed tomography; *CXR* = chest X-ray; *DAMA* = discharge against medical advice; *ECG* = electrocardiogram; *ED* = emergency department; *EMS* = emergency medical services;*GCS* = Glasgow Coma Scale score; *GFR* = glomerular filtration rate; *Glu* = glucose; HCO_3 = bicarbonate; *h* = hour; *HR* = heart rate; *ICU* = intensive care unit; *i.v.* = intravenous; O_2 = oxygen; PCO_2 = partial pressure of carbon dioxide; *PE* = physical examination; PO_2 = partial pressure of oxygen; *RR* = respiratory rate; *SCr* = serum creatinine; SPO_2 = peripheral capillary oxygen saturation; *T* = temperature; *USA* = United States of America; *VBG* = venous blood gas; *WBC* = white blood cells.

TABLE 6B Case Reports Identified in Which Use of U-47700 Resulted in Death [268c]

Country	Age, years	Sex	Drugs	Route	Cumulative dose	Presentation
Belgium	30	M	Fentanyl, U-47700	Vaporization on foil	~36 g	Found deceased, pulmonary edema reported no apparent injection sites
UK	27	M	Possible mirtazapine History using: cannabis, ketamine, methcathinone "legal highs"	Insufflation	NP	Found deceased
USA	46	M	U-47700	Insufflation	NP	*Postmortem findings*: Lungs showed edema, congestion, and expanded alveoli, with terminal septa clubbing. Liver and spleen were enlarged. Microscopically the liver showed macrovesicular steatosis, but no significant fibrosis, and the spleen showed sinusoidal congestion
USA	25	M	U-47700	NS	NP	Found deceased. Pulmonary edema was present. U-47700 peripheral blood concentration (PBC) was 334 ng/mL
USA	29	M	U-47700, Chiesel 60% sativa, 1G322	NS	NP	Subject complained of a headache and then collapsed. EMS attempted resuscitation. *Postmortem findings*: cerebral and pulmonary edema. U-47700 PBC = 453 ng/mL
USA	29	M	U-47700	NS	NP	Found deceased. Pulmonary edema was present. Postmortem U-47700, PBC = 242 ng/mL
USA	30	M	U-47700	NS	NP	Found deceased. Postmortem findings: pulmonary edema, gastritis and chronic hepatitis
USA	26	M	U-47700, hydrocodone/acetaminophen, alprazolam, venlafaxine	NS	NP	Found deceased. *Postmortem findings*: cerebral and pulmonary edema
USA	27	M	U-47700	NS	NP	Found deceased. *Postmortem findings*: pulmonary edema and cardiomegaly
USA	20	M	U-47700	NS	NP	Found deceased. *Postmortem findings*: pulmonary edema, cardiomegaly and hepatosplenomegaly

NP = not provided; *NS* = not specified.

tachyarrhythmia necessitating medical attention and in recent weeks reported feeling dysphoria with vomiting, constipation, dizziness and malaise associated with loperamide use. UDS failed to reveal opioids, BZDs, cocaine, or cannabinoids and upon admission EKG revealed an elevated QT-interval (497 milliseconds). Methadone was initially administered which successfully mitigated withdrawal symptoms including nausea, rhinorrhea, abdominal cramping, sweating, restlessness, joint aches, yawning, and anxiety. Methadone was weaned by 5 mg daily over the next four days. Patient tolerated the taper well without need for clonidine due to lack of significant autonomic arousal [276A].

Sullivan et al. agreed that presently there is no established optimal approach for transitioning opioid-dependent adults to XRNTX while preventing relapse. Researchers conducted a trial, examining efficacy with two outpatient methods initiating XRNTX for opioid detoxification. Opioid-dependent adults ($n = 150$) were assigned (2:1) to naltrexone-assisted or buprenorphine-assisted detoxification, followed by XRNTX injection. Naltrexone-assisted detoxification (seven days) included buprenorphine (one day), oral naltrexone up-titration plus clonidine and other adjunctive medications. Buprenorphine-assisted detoxification included buprenorphine taper (seven days), then one-week delay before initiating XRNTX, in accordance with official prescribing information for XRNTX. Both groups received behavioral therapy focusing on medication adherence and a second XRNTX dose. Compared to participants in the buprenorphine-assisted group, participants assigned to naltrexone-assisted detoxification were significantly more likely to be successfully induced on XRNTX (56.1% vs 32.7%) and to receive the second injection at week-five (50.0% vs 26.9%). Results demonstrate safety, efficacy and tolerability with low-dose naltrexone in conjunction with single-day buprenorphine dosing and adjunctive non-opioid medication. These results demonstrate that initiating low-dose naltrexone in conjunction with single-day buprenorphine dosing plus adjunctive non-opioid medications to transition patients to XRNTX is safe, efficacious and tolerable. The strategy identified offers promise in decreasing attrition and relapse rates with an alternative agent compared to agonist tapers in inpatient and outpatient settings [277C].

XRNTX is FDA-approved to prevent relapse in patients with OUD. However, since knowledge is limited regarding long-term use among community-based outpatients, Williams et al. conducted a retrospective chart review and long-term follow-up survey among individuals ($N=57$) who entered an outpatient XRNTX trial (2011 and 2015). Participants were offered three XRNTX monthly injections at no cost. The survey evaluated four domains including substance use, treatment continuation, barriers and attitudes. Participants who completed the study had superior outcomes and a lower likelihood to relapse (defined as daily use), with lengthened time to relapse, despite higher concurrent non-opioid substance use rates. At the study completion most participants discontinued XRNTX due to internal barriers including feeling 'cured' or 'wanted to do it on my own' as opposed to external reasons including cost or AE. Since many patients experience internal barriers, they may benefit from anticipatory guidance and motivation techniques to encourage long-term adherence. Study results reiterate that few patients experienced AE from XRNTX long-term use, ensuring positive efficacy and safety [278c].

Abuse-Deterrent

Miyazaki et al. presented in vitro and in vivo studies, which demonstrate that NKTR-181 is the first selective mu-opioid agonist to combine analgesic efficacy and reduced abuse liability through the alteration of brain-entry kinetics. NKTR-181 had a reduced brain entry rate compared to oxycodone and hydrocodone, measuring 71-fold less than oxycodone and 13-fold less than hydrocodone. At the mu-opioid receptor (MOR), NKTR-181 had a lower affinity compared to traditional opioids (morphine and oxycodone), between 15- and 28-fold less potent. The traditional MOR-agonists exhibit full [3H] naloxone displacement and authors observed the same result for NKTR-181. The polyethylene glycol functional group added to a small molecule, not only changed its oral availability, but also its distribution across the blood–brain barrier [279E].

Webster et al. evaluated NKTR-181, a novel mu-opioid agonist molecule for the human abuse potential pharmacokinetics, pharmacodynamics, and safety, relative to oxycodone. They conducted a randomized, single-center, double blind, active- and placebo-controlled five-period crossover study with healthy, adult, non-physically dependent recreational opioid users. Forty-two randomized subjects (male = 73.8%, White = 81%, mean age = 25-YO) were evaluated using the Drug Liking VAS for single orally administered 100, 200, and 400 mg NKTR-181 doses in solution compared with 40 mg oxycodone and placebo solutions. Secondarily, researchers evaluated participants with the Drug Effects Questionnaire, Addiction Research Center Inventory/Morphine Benzedrine Group Subscale, Price Value Assessment Questionnaire, Global Assessment of Overall Drug Liking, and Take Drug Again. Pupillometry was used to assess CNS mu-opioid effects. Subjects received five treatments using a crossover design. At all dose levels NKTR-181 had significantly lower Drug Liking E_{max} than oxycodone ($P<0.0001$). Drug Liking scores for oxycodone increased rapidly within 15 min with a 1 h post-dose peak, whereas Drug Liking (and most secondary abuse potential measures) for all NKTR-181 doses were comparable with placebo for the first hour. The Drug Liking scores were differentiated compared to placebo (between 1.5 and 4 h) but remained significantly lower than oxycodone ($P<0.003$). NKTR-181 treatment-related AEs were mild and occurred less often compared with oxycodone (sedation and respiratory depression). Authors concluded that for recreational opioid users, NKTR-181 demonstrated delayed onset for CNS effects and significantly lower abuse potential compared to oxycodone [280c].

Through retrospective review Wilson et al. sought to analyze specific complications and sequelae requiring ICU resources for patients intravenously abusing extended-release oral oxymorphone (January 2012–December 2015). Medical charts were reviewed for patients abusing intravenously extended-release oral oxymorphone, to identify associated sequelae and patients requiring intensive care. Fifty-three patients required treatment in an ICU setting resulting from intravenously abusing extended-release oral oxymorphone. Patients ($n=28$[52.8%]) required endotracheal intubation with mechanical ventilation for either acute hypoxic respiratory failure or airway protection. Patients ($n=48$[90.6%]) developed AKI, and some (28.3%) failed to regain renal function, requiring renal replacement therapy. Patients ($n=36$[67.9%]) were diagnosed with bacteremia, some ($n=30$[56.6%]) of which developed acute infective bacterial endocarditis. Clinicians should be aware that clinical decompensation often results when patients report extended-release oral oxymorphone intravenous abuse [281c].

Setnick et al. aimed to assess the intranasal human abuse potential of ELI-200, a novel immediate-release (IR) oxycodone formulation containing sequestered naltrexone. Authors performed a randomized, double-blind, double-dummy, active and placebo-controlled, five-way crossover study. Pharmacodynamics, safety, and pharmacokinetics were evaluated for up to 36 h post-dose. Healthy, non-dependent recreational opioid users underwent a naloxone challenge and drug discrimination qualification test where a single intranasal dose of ground ELI-200 (30 mg oxycodone hydrochloride [HCl]/3 mg naltrexone HCl), crushed oxycodone HCl IR (30 mg) (Roxicodone®), placebo, fixed placebo, or single oral dose of intact ELI-200 (30 mg/3 mg) were given.

Subjects ($n = 36$) who completed all five-treatment periods and all active treatments showed significantly higher ($P < 0.001$) median Drug Liking E_{max} relative to placebo. Significant reductions ($P < 0.001$) in median Drug Liking (E_{max}) were observed for intranasal ELI-200 [56.0] compared to intranasal oxycodone IR [100.0] and secondary positive or objective measures (High, Good Drug Effects, Overall Drug Liking, Take Drug Again, and maximum pupil constriction) showed significantly lower E_{max} for intranasal ELI-200 ($P < 0.001$) compared to intranasal oxycodone IR. Significantly decreased effects on subjective and physiological measures, and greater nasal irritation were demonstrated by intranasal ELI-200 compared to intranasal oxycodone IR. The pharmacokinetic properties for the naltrexone component demonstrated rapid release and meaningful abuse-deterrent properties [282c].

Setnick et al. aimed to evaluate the abuse potential for ALO-02, an abuse-deterrent formulation comprised of extended-release oxycodone hydrochloride pellets, surrounding sequestered naltrexone hydrochloride. They conducted a randomized, double-blind, placebo-active controlled, six-way crossover study, with a naloxone challenge, drug discrimination, and treatment phases. Drug Liking and High maximum effect (E_{max}) for crushed oxycodone IR 40 mg was significantly higher compared to placebo, confirming study validity ($P < 0.0001$). Drug Liking and High (area under effect curve [AUE]) (E_{max}, AUE_{0-2h}) for crushed ALO-02 (40/4.8 mg and 60/7.2 mg) were significantly lower compared to corresponding crushed oxycodone IR doses (40 and 60 mg; $P < 0.0001$). Drug Liking and High (E_{max} and AUE_{0-2h}) for intact ALO-02 60 mg/7.2 mg were significantly lower compared with crushed oxycodone IR 60 mg ($P < 0.0001$). Secondary pharmacodynamic endpoints for oxycodone and naltrexone plasma concentrations were consistent with these results. Fewer participants experienced AEs after ALO-02 (crushed or intact: 71.1%–91.9%) compared with crushed oxycodone IR (100%). Most common AEs following crushed ALO-02 and oxycodone IR were euphoric mood, pruritus, somnolence, and dizziness. Authors concluded that ALO-02 (crushed or intact) demonstrated lower abuse potential than crushed oxycodone IR when administered orally in nondependent, recreational opioid users [283c].

Tapentadol

Vosburg et al. conducted a retrospective cohort study to further identify tapentadol abuse liability as an Active Pharmaceutical Ingredient (API) when both IR and ER formulations were on the market together (4Q2011–2Q2016). Tapentadol API was compared to tramadol, hydrocodone, morphine, oxycodone, hydromorphone, and oxymorphone across Poison Center, Drug Diversion and Treatment Center Programs Combined data streams from the RADARS System. Average quarterly event rate for tapentadol intentional abuse per 100000 dosing units dispensed was 0.028 (95% CI 0.023–0.035), which was neither the largest nor smallest intentional abuse event rate. Event rates were greatest for oxymorphone (0.166, 95% CI 0.153–0.179), and lowest for hydrocodone (0.020, 95% CI 0.019–0.020). Intentional abuse rate ratios per 100000 prescriptions dispensed comparing tapentadol to other opioids, revealed that hydrocodone and tramadol were intentionally abused less than tapentadol (0.703 and 0.766 times the rate of tapentadol abuse), whereas the remainder of comparators were abused from 1.275 (oxycodone) to 5.886 (oxymorphone) times the intentional abuse rate of tapentadol. Overall the findings suggest that public health burden related to tapentadol is low, but present [284MC].

Buprenorphine

Coplan et al. sought to compare poison center call rates categorized as intentional abuse, suspected suicidal intent or fatality for the seven-day buprenorphine transdermal system/patch (BTDS) vs other extended-release and long-acting (ER/LA) opioids indicated for chronic pain. Researchers aimed to assess intentional abuse, suspected suicidal intent or fatality for the seven-day BTDS reported to National Data Poison System, compared to other Schedule II ER opioid analgesics and methadone. Through a 24-month cohort, researchers identified numbers and rates of calls for intentional abuse, suspected suicidal intent and fatalities to evaluate BTDS, ER morphine, ER oxycodone, fentanyl patch, ER oxymorphone and methadone tablets/capsules, using prescription adjustment to account for community availability. Absolute numbers and prescription-adjusted rates for intentional abuse and suspected suicidal intent with BTDS were significantly lower ($P < 0.0001$) than for all other ER/LA opioid analgesics examined and there were no reported fatalities associated with BTDS exposure. Postmarketing BTDS evaluation resulted in infrequent poison center calls for intentional abuse and suspected suicidal intent events, thereby concluding that BTDS has lower risk rates compared to other ER/LA opioids [285MC] (Table 7).

Strang et al. aimed to test the new buprenorphine oral lyophilisate wafer ("bup-lyo") vs standard sublingual buprenorphine ("bup-SL") for safety. They conducted a randomized (2:1) open label study with opioid-dependent subjects and subsequent partial cross-over. Opioid-dependent subjects ($n = 36$) initiated buprenorphine maintenance including patients concomitantly

TABLE 7 Absolute Counts and Prescription-Adjusted Rates of Abuse, Cases of Suspected Suicidal Intent, and Fatalities by Opioid Type (National Poison Data System, 3Q2012–2Q2014) [285MC]

Opioid	Cases (N)	Prescriptions (N)	Cases/1 000 000 prescriptions	Rate ratio	95% CI	*P* value
Intentional abuse						
Methadone	1255	7653314	164	35.2	14.63–84.76	<0.0001
ER oxycodone	379	11278171	34	7.2	2.99–17.4	<0.0001
ER oxymorphone	203	1882585	108	23.2	9.53–56.2	<0.0001
ER morphine	182	14226318	13	2.7	1.13–6.68	0.0258
Fentanyl Patch	663	13237727	50	10.8	4.46–25.9	<0.0001
BDTS	5	1073812	5	Ref	Ref	Ref
Suspected suicidal intent						
Methadone	1891	7653314	247	22.1	12.53–39	<0.0001
ER oxycodone	782	11278171	69	6.2	3.51–11	<0.0001
ER oxymorphone	164	1882585	87	7.8	4.34–14	<0.0001
ER morphine	596	14226318	42	3.7	2.12–6.64	<0.0001
Fentanyl Patch	409	13237727	31	2.8	1.56–4.91	0.0005
BDTS	12	1073812	11	Ref	Ref	Ref
Fatality						
Methadone	51	7653314	7	5.1	2.99–8.98	<0.0001
ER oxycodone	13	11278171	1	0.9	0.44–1.85	0.7693
ER oxymorphone	3	1882585	2	1.2	0.36–4.23	0.7304
ER morphine	11	14226318	1	0.6	0.28–1.29	0.1898
Fentanyl Patch	17	13237727	1	Ref	Ref	Ref
BDTS	0	1073812	0	N/A	N/A	N/A

All *P* value comparisons are to the reference (Ref) opioid.
Q=quarter within the fiscal year.

using alcohol, cocaine and BZDs (below thresholds). Oral lyophilised buprenorphine ("bup-lyo") completely dissolved within 2 min for 58% vs 5% for "bup-SL." Dose titration resulted in similar maintenance dosing (10.8 mg vs 9.6 mg) without significant between-group differences in opiate-withdrawal phenomena, craving, adequacy of "hold," respiratory function. No SAE nor "severe" AEs were reported but more AEs and Treatment-Emergent AEs (TEAE) were seen with "bup-lyo" (mostly "mild"). The most frequently reported TEAE was headache for "bup-lyo" participants ($n=4$[17.4%]) and less reported self-limiting oral hypoesthesia (three events) from two "bup-lyo" subjects at five and 10 min post-administration. Pharmacokinetics resulted in substantially increased buprenorphine (but not norbuprenorphine) bioavailability with "buplyo" relative to "bup-SL." For supervised dosing programs, rapidly disintegrating formulations may enable broader buprenorphine prescribing [286c].

Buprenorphine is a widely used for OUD treatment but the transmucosal formulation can be associated with misuse, diversion, and non-adherence, which may be obviated by the sustained-release formulation. Fluid-Crystal injection depot technology was developed to address these concerns with a sustained-release subcutaneous (SQ) buprenorphine formulation (CAM2038; weekly and monthly depots) [287H,288E]. Walsh et al. aimed to evaluate novel CAM2038 to block euphorigenic opioid effects and suppress withdrawal in non-treatment-seeking individuals with OUD. In a multisite, double-blind, randomized, within-patient study, 46 adults with DSM-V moderate to severe OUD participated in the study (October 12, 2015 to April 21, 2016). The primary endpoint was maximum rating on the VAS for drug liking while secondary endpoints included other VAS (e.g., high and desire to use), opioid withdrawal scales, and physiological and pharmacokinetic outcomes. Both weekly CAM2038 doses significantly produced immediate and sustained blockade against hydromorphone effects (liking maximum effect for 24mg [effect size] 0.813; $P<0.001$, and 32mg [effect size] 0.753; $P<0.001$) and withdrawal suppression (COWS: CAM2038, 24mg [effect size], 0.617; $P<0.001$, vs CAM2038, 32mg [effect size] 0.751; $P<0.001$). CAM2038 produced a rapid initial plasma increase, reached maximum concentration at 24h, with a prolonged half-life (4–5 days) and 50% trough concentration accumulation from first to second dose (trough concentration 24mg=0.822ng/mL (week 1) and 1.23ng/mL (week 2) compared to 32mg trough concentration=0.993ng/mL (week 1) and 1.47ng/mL (week 2). CAM2038 was safely tolerated and produced immediate and sustain opioid blockage and withdrawal suppression. Buprenorphine depot formulation for OUD may further obviate misuse and diversion risk compared to daily buprenorphine while retaining therapeutic benefits [289c].

Common treatment for patients who are opioid dependent presenting with respiratory depression is naloxone but this treatment has the potential to cause severe, life-threatening complications in this population including acute respiratory distress syndrome and MI [290H,291C]. Buprenorphine administration for an opioid-intoxicated patient may reverse respiratory depression with less severe withdrawal SEs. Buprenorphine's longer half-life compared to naloxone may reduce the need for repeated antidote administration. Zamani et al. evaluated a 20-YOM addicted to morphine who presented with methadone-induced respiratory depression (Table 8). He was safely and effectively treated with IV buprenorphine [292A].

Patient ingested methadone syrup (30mL[150mg]), 10 (20mg) fluoxetine capsules, and 10 (2mg) clonazepam tablets approximately 2h prior to ED admission. Upon EMS arrival patient was administered naloxone 0.4mg for apnea. 2.5h later, vitals declined, patient became unresponsive and cyanotic with COWS score=14. Conversely when patient was administered buprenorphine bolus (0.6mg) slowly via IV, patient then became conscious, but remained calm. Although he continued to have diaphoresis, he didn't display any other withdrawal signs or symptoms (COWS=3). Cyanosis resolved and vitals normalized. The patient remained clinically stable, with normal VS and two normal venous blood gas results until approximately 24h later. Respiratory depression reoccurred and patient was switched back to naloxone therapy (0.4mg bolus, 0.2mg/h infusion). Patient started experiencing moderate–severe withdrawal symptoms (COWS=26), mandating naloxone drip discontinuation. One-hour later patient had respiratory acidosis, prompting naloxone drip re-initiation. This cycle continued for three days while the patient was minimally conscious with mild respiratory acidosis and hypoxemia, after which time, naloxone infusion was discontinued and patient had near-normal vitals and venous blood gas. Patient had positive UDS results for methadone and positive serum for methadone and BZDs. Buprenorphine may be a useful alternative for opioid reversal in opioid-dependent patients.

NAS

Loudin et al. reported a retrospective case series ($n=19$) investigating infant exposure to opioids and gabapentin prenatally. A positive drug screen at delivery, during pregnancy, or a drug abuse history resulted in testing an umbilical cord tissue sample ($n=420$) for non-prescribed drug (methadone, buprenorphine) analysis. Currently, gabapentin levels are not offered for cord tissue by the United States Drug Testing Laboratories so researchers incorporated the gabapentin homogeneous enzyme immunoassay into the UDS (ARKDiagnostics, Inc., Sunnyvale, California). About three-quarters ($n=188$[73.4%]) of cord tissue samples ($n=288$) were positive for opioids. All 19 infants who had an opioid positive cord analysis and maternal admission to using non-prescription gabapentin during pregnancy required treatment. Nine were managed with methadone alone and 10 required additional therapy. Patient 1 was delivered (37 weeks) to a mother (28-YOF) with a heroin use history, positive HCV infection, and no prenatal care. Cord drug screen was positive (heroin and hydromorphone), infant developed NAS and was

TABLE 8 Laboratory Values for Buprenorphine Treated Opioid Toxicity [292A]

Time of venous blood gas measurement	pH	PCO_2 (mmHg)	HCO_3 (mEq/L)	Base excess	Oxygen saturation
On admission	7.33	51.4	25.5	−0.1	91
Time of administration of buprenorphine (2.5h after admission)	7.29	66.4	22.3	1.7	76
12h after buprenorphine administration	7.36	46.7	24.8	1.3	95
28h after buprenorphine administration (second episode of apnea)	7.20	103	40.1	9.2	87
During naloxone drip (with mask)	7.33	81	42	6.3	90
On discharge	7.46	56.2	39.5	4.5	99

treated with methadone on day of life (DOL)-3. Clonidine was added on DOL-7 due to persistent Finnegan scores >12. The infant had an unusual behavioral pattern including tongue thrusting, wandering eye movements, back arching, and continuous extremity movement, even while being held. Electroencephalography and MRI were normal. Mother admitted to using gabapentin (600–3000 mg/day) during her first pregnancy month. Due to persistent, continuous extremity movements and Finnegan scores >12, gabapentin was initiated (5 mg/kg/day in two-doses), within 48 h Finnegan scores decreased to <8 and abnormal neurologic movements ceased. Clonidine was discontinued followed by gabapentin down-titration by 25% every 4-days coordinated with methadone weaning every 48 h. Infant completed therapy by DOL-45.

Overall, 19 infants were born to mothers taking opioids and gabapentin (dose range: 600–12 500 mg/day) during pregnancy. 15/19 infants demonstrated the same behaviors (indicated by case above) and normal electroencephalograms. Ten infants were unable to wean from methadone, and gabapentin was added to their treatment (DOL: 7–26), all of which manifested tongue thrusting, wandering eye movements, back arching, and continuous extremity movements. Gabapentin initiation depended on admitted prenatal use, a failure to wean or methadone escalation with sustained behavioral changes. All infants ($n=10$) were weaned to methadone (0.02 mg/8 h) every 48 h coordinated with gabapentin weaning (25%/4 days). The remaining nine infants were managed with methadone weaning alone (Table 9). The average medication therapy duration for infants treated with gabapentin was 47 days (±8.8 days) [293c].

Hensley et al. aimed to evaluate whether depression in pregnant women with opioid dependency negatively impacted adherence with prenatal care using a retrospective chart analysis and including opioid-dependent pregnant women (six-years). Researchers evaluated prenatal care adherence based on concurrent depression diagnosis, which was assessed for observed vs expected prenatal visits. Secondary outcomes included NAS incidence and severity and neonatal ICU (NICU) duration (days). 45/74 (60.8%) opioid-dependent pregnant patients were diagnosed with depression ($n=41$), anxiety ($n=2$), or scored >10 on the Edinburgh Prenatal Depression Scale ($n=1$). Patients with depression vs anxiety were significantly less adherent to prenatal care; 80% adherent (73% vs 93%; $P=0.03$) vs 90% adherent (62% vs 93%; $P=0.003$). More patients in the depression group had an infant treated for withdrawal (62% vs 38%; $P=0.041$), and lengthier NICU stays (27% vs 21%; $P=0.018$). For the entire cohort analysis, opioid-dependent gravidas revealed that buprenorphine compared to MMT and no maintenance, had the lowest mean NAS score (6.5[±4.4] compared to 10.6[±3.6]) and 9.4 [±4.0] [$P=0.008$]). Authors concluded that depression negatively impacts adherence with prenatal care indicated by the significant association with greater neonatal withdrawal incidence and lengthier NICU stays. Buprenorphine therapy had the lowest NAS incidence compared to MMT and no maintenance therapy [294c].

Buprenorphine is a highly effective treatment for OUDs, but its continuation in the perioperative setting remains controversial, unlike the accepted preoperative methadone treatment continuation. In the US, methadone dispensed from a licensed opioid treatment program was considered the gold standard for OAT in pregnant women but more recently buprenorphine has become more appealing due to office-based availability and the potential for an attenuated NAS symptoms compared to methadone [295c,296C, 297c,298M]. Vilkins et al. conducted a retrospective cohort study (2006–2014) comparing post-cesarean section opioid analgesic requirements for women with OUD treated

TABLE 9 NAS and Therapy Course With Gabapentin [293c]

Patient	Maternal exposure	Mother's gabapentin dose (mg/day)	DOL gabapentin started	Days(#) of gabapentin treatment	EEG results	Unique symptoms
1	Heroin, hydromorphone	600–3000	15	30	Normal	Yes
2	Methadone	Unknown	15	37	—	Yes
3	Subutex	4800	11	45	—	Yes
4	Subutex	1800	11	37	—	Yes
5	Hydrocodone, heroin, vistaril, flexeril	Unknown	10	40	Normal	Yes
6	Heroin, hydromorphone	Unknown	23	60	Normal	Yes
7	Subutex, marijuana	Unknown; used daily	19	30	—	Yes
8	Heroin, oxycodone	400–1200	7	63	Normal	Yes
9	Subutex	12500	15	52	—	Yes
10	Subutex, marijuan, heroin use (<26 weeks gestation)	Unknown, used daily	26	73	Normal	Yes

with methadone or buprenorphine. During the nine-year study period, women ($n=185$) were treated with methadone (mean dose= 93.7 mg[SD 2.6]) and buprenorphine ($n=88$) (mean dose=16.1 mg[SD 7.8]). There were no statistically significant differences in maximum effective dose (MED) requirements for methadone vs buprenorphine treatment groups, for postoperative complications LOS. Authors concluded that buprenorphine treatment will not interfere more than methadone with pain management after a cesarean section but future studies are needed to investigate the generalizability to other surgeries [299C].

Brain-derived neurotrophic factor (BDNF) is a growth factor that regulates and promotes neurite outgrowth through maintaining synaptic connectivity in the nervous system and promotes neuronal growth and survival [300H,301H,302H]. Several studies demonstrated altered serum or plasma BDNF levels in drug abusers [303c,304c,305H], and implicated BDNF for addiction development [305H]. According to the authors, there is no available data about opiate exposure effects on BDNF levels in infants who are exposed to opiates in utero and whether BDNF levels may correlate with NAS. Subedi et al. aimed to compare plasma BDNF levels among NAS and non-NAS infants to determine correlation between BDNF levels and NAS severity. Through a prospective cohort infants ($n=67$) ≥35 weeks' gestation were enrolled (NAS [$n=34$]; non-NAS [$n=33$]). BDNF levels were measured and NAS severity was determined with hospital LOS and medications required to treat NAS. For NAS compared to non-NAS infants mean BW was significantly lower (3070 g [±523 g] vs 3340 g [±459 g] [$P=0.028$]) and mean BDNF level was significantly higher (252.2 [±91.6 ng/mL] vs 211.3 [±66.3 ng/mL] [$P=0.04$]), respectively. There were no differences for BDNF levels between NAS infants requiring one vs >1 medication (254 ng/mL [±91] vs 218 ng/mL [±106] [$P=0.47$]), correlation between the BDNF levels and hospital LOS ($P=0.68$) nor between BDNF levels and NAS scores except at around 15 h post-admission (correlation 0.35[$P=0.045$]). Overall the BDNF level during the first 48 h was significantly increased in NAS vs non-NAS infants but correlation between plasma BDNF levels and NAS severity warrant further study. Authors suggest that BDNF may play a neuromodulatory role during withdrawal after in utero opiate exposure [306MC].

Zuzarte et al. aimed to examine the therapeutic potential of stochastic vibrotactile stimulation (SVS) as a complementary non-pharmacological intervention for withdrawal in opioid-exposed newborns. Researchers prospectively evaluated opioid-exposed newborns ($n=26$) (>37 weeks; 16 male), who were hospitalized since birth and treated pharmacologically for NAS. SVS therapy demonstrated significant reduction (35%) in movement activity ($P<0.001$), and fewer movement periods >30 s duration for ON than OFF ($P=0.003$). Tachypneic breaths and tachycardic incidence were each significantly reduced, whereas eupneic breaths and eucardic HR incidents each significantly increased with SVS ($P<0.03$). No AEs were associated with SVS in the current study, though more studies to establish the safety and efficacy are needed. Nurses and parents reported anecdotally that infants seemed "less irritable", "calmer" and "slept better" during periods with SVS therapy. Mean and max Finnegan scores were not significantly different between the day preceding, the day of, or the day following the study session. Overall, SVS reduced hyperirritability and pathophysiological instabilities commonly observed in pharmacologically-managed opioid-exposed newborns. SVS may provide an effective complementary therapeutic intervention for improving autonomic function in newborns with NAS [307c].

Death

Fatal poisonings in children are only represented by a small proportion investigated by Australian coroners, but the cases present a major opportunity to prevent death in this population. Pilgrim et al. aimed to examine fatal child poisonings in Australia to estimate acute poisoning death rates in children, describe key cohort characteristics and describe coronial recommendations made as a death prevention measure. Researchers reviewed the Australian National Coronial Information System (NCIS) for poisoning deaths ($n=90$) reported to an Australian coroner (January 2003–December 2013) involving children (≤16-YO). Data were analyzed using logistic regression, Pearson's correlation coefficient and descriptive statistics examining significance of associations. Most cases occurred in children 13–16-YO (58.9%) (CI 7.5–12.5) and were typically unintentional (61.1% [CI 17.9–27.1]) and occurring at home (68.9% [CI 6.8–15.7]). The most common poisoning resulted from opioids (24.4%), followed by carbon monoxide (20%) and volatile substances (18.9% [CI 18.5–19.6]). Males were slightly more likely to die from prescription opioids compared with females (OR=1.9 [CI 0.7–5.1]). The coroner made recommendations for 12 cases, 10/12 were due to poisoning (including drugs), of which eight recommendations were implemented. On average (2003 – 2013) eight acute poisoning deaths occurred in children each year, most commonly involving prescription opioids and adolescents. Mortality has followed a downward trend since 2003, but cases generated more than twice as many recommendations for public safety compared with other Australian coroners' cases [308c].

Fels et al. reported two fatalities associated with the synthetic opioids AH-7921 and MT-45 which have emerged over the past few years on the recreational drug market and are sold as "research chemicals" on the internet. Case-1 was a 22-YOF found dead in the bedroom with a plastic bag labeled "AH-7921." Two friends stated that the deceased had consumed AH-7921 prior to her death. AH-7921 femoral blood concentration was 450 mg/L. Autopsy revealed cerebral edema with moderate to clearly increased intracranial pressure and signs for an incipient pneumonia in the central lung sections. The internal organs were filled with blood. Case-2 was a 24-YOM who was a known amphetamine abuser, found dead in his room slumped over. An e-cigarette with unknown liquid was found nearby with a spoon, a glass, ascorbic acid, and several bags containing white powder, labeled "MT-45", "methoxmetamine" and "methoxphenidine." MT-45 concentration in the femoral blood was 2900 mg/L. Autopsy revealed brain and hemorrhagic pulmonary edema, hyperemia of internal organs, and urine (450 mL) contained in the urinary bladder. No methoxphenidine was found in the decedent's femoral blood, urine or stomach contents [309A].

Guerrieri et al. investigated the first lethal report involving furanylfentanyl. They evaluated seven fatal intoxications involving the opioid-analogue furanylfentanyl occurring in Sweden over four months (2015–2016). Toxicology verified furanylfentanyl either alone or in combination with other illicit substances. Decedents were 24–37-YOMs with varying femoral furanyl-fentanyl blood concentration (0.41–2.47 ng/g). Decedents ($n=5$) had a well-documented drug abuse history, presenting with drug panels involving complex drug of abuse and prescription drugs ($n=5$) and concurrent pregabalin ($n=5$), which corroborates previous observations indicating pregabalin as a possible contributing factor in polydrug intoxications. In one case pregabalin was the only drug detected in addition to furanylfentanyl, but there are no present pharmacological data showing that pregabalin aggravates furanylfentanyl toxicity or other MORs. In January 2016, furanylfentanyl was scheduled as a narcotic in Sweden and only one fatality associated to furanylfentanyl toxicity has since been reported. Since concomitant drugs were found in the blood at low concentrations, they were unlikely the primary cause of death. Lethal furanylfentanyl concentrations are difficult to establish due to lacking evidence about pharmacokinetic and pharmacodynamic interactions with other drugs and unknown opioid tolerance. Researchers advise health-care providers to perform a full toxicological screen with hair analysis to better estimate interactions and tolerance [310c].

As new fentanyl analogues and other synthetic opioids emerge, laboratories continue to identify them. To their knowledge, Krotulski et al. documented the first fatality from Tetrahydrofuranylfentanyl (THFF) ingestion in the US and the first fatality reported with THFF and U-49900, worldwide. THFF has been identified in Europe and the US as a NPS (tetrahydrofuran fentanyl or *N*-phenyl-*N*-[1-(2-phenylethyl)piperidin-4-yl]tetrahydrofuran-2-carboxamide) and has been linked to 15-deaths to date [149S,148A]. THFF is structurally similar to furanylfentanyl, responsible for numerous AEs, including death.

A 39-YOM was found vomiting and convulsing in his residence. He had PMH positive for schizophrenia, bipolar and recreational drug use. EMS administered Narcan® (naloxone) but revival was unsuccessful and decedent was pronounced dead at the scene without pulse or respiration. At autopsy, decedent had pulmonary congestion and edema and mild cerebral edema. THFF, U-49900 and methoxyphencyclidine were identified in postmortem blood and urine. An unknown substance collected from the scene was analytically confirmed as THFF and U-49900. Metabolic profiling was then conducted using pooled human liver microsomes, to assist other laboratories in identifying THFF. Since THF-norfetanyl is a unique THFF metabolite, it serves as a great biomarker for identification [311A].

Furanylfentanyl prevalence in postmortem casework contributes to opioid-related deaths (~50%) for drug-induced fatalities in the US. Opioids (2015) caused the majority of drug-induced deaths (73%) in the US (with heroin [24.8%]; natural and semisynthetic opioids [24.3%]; methadone [6.3%]; and synthetic opioids excluding methadone [18.3%]) [312MC]. The number of individuals testing positive for fentanyl more than doubled (5343–13 882 encounters from 2014 to 2016) in the US [313MC]. Martucci et al. presented a case of a 23-YOM found dead in San Francisco after ingesting blue pills imitating oxycodone. Vomitus near subject's mouth had a blue color. Roommates revealed that the subject admitted to obtaining oxycodone and using it recently with alcohol and marijuana. Pathological findings included pulmonary and cerebral edema, focal (60%) atherosclerotic narrowing and gastric material aspiration through the lungs. The initial blood toxicology screening did not detect oxycodone. The immunoassay used to identify the blue pills, revealed furanylfentanyl presence. Postmortem blood samples identified furanylfentanyl concentrations (peripheral blood [1.9 ng/mL], cardiac blood [2.8 ng/mL], and gastric contents [55 000 ng]). The metabolite 4-anilino-*N*-phenethyl-piperidine (4-ANPP) was confirmed (peripheral blood [4.3 ng/mL], cardiac blood [5.8 ng/mL], liver [conc. > 40 ng/g]). Both analytes were found in trace amounts in urine and vitreous humor. Drug users are subject to unknown drugs so further analysis is typically warranted to identify specific drugs causing toxicity [314A].

Austin et al. aimed to examine characteristics for self-inflicted drug overdose deaths, overall and in comparison to unintentional drug overdose deaths. Among patients using commonly prescribed opioids (oxycodone, hydrocodone) or BZDs (alprazolam, clonazepam) self-inflicted drug overdose decedents with a prescription in the past 30 days, had a significantly higher incidence of death compared to unintentional drug overdose decedents. North Carolina (2012) reported 1250 drug overdose deaths, the majority resulting from unintentional overdose ($n=1014$[81.1%]) compared to self-inflicted ($n=207$[16.6%]). Substances contributing to the majority of self-inflicted and unintentional drug overdose deaths were prescription opioids (50.0% vs 72.3%), respectively. Decedents with ≤1 controlled substance prescription prior to death were dispensed hydrocodone (59.6%) and oxycodone (47.0%). Authors identify that self-inflicted compared to unintentional drug overdose decedents were more likely to have contact with the health care system in the weeks preceding death, offering an opportunity for professionals to identify and intervene on risk factors or distress signals and potential for self-harm [315MC].

To the best of their knowledge, Rojkiewicz et al. aimed to characterize 4-flurorbutryfentanyl (4-FBF) with the first two analytically confirmed cases resulting in fatal intoxication. The first case was reported by the Swedish Poison Information Centre {(January 2015) [316H]} but the pharmacological, pharmacokinetic, and toxicological data are lacking for this fentanyl derivative. Limited physicochemical data can be accessed from an analytical report by the National Forensic Laboratory in Ljubljana, Slovenia [317S], which provides identification and analytical characterization for 4-FBF from the detection in two lethal intoxication cases in Poland (2015). Two cases presented in this study identify and analytically characterize a new fentanyl derivative, 4-fluorobutyrfentanyl (1-((4-fluorophenyl)(1-phenethylpiperidin-4-yl)amino) butan-1-one, 4-FBF). Apart from the seized powder, 4-FBF was also identified in the e-cigarette liquid (compound concentration = 35 mg/mL) secured in Case-1. The liquid mostly contained glycerol, and nicotine, which was detected in seized material originating from the illegal drug market in Poland. In case-1, a 26-YOM drug user was found dead at home with an e-cigarette and e-liquid secured nearby. In case-2, a 25-YOF, with occasional NPS use was found dead at home with a plastic bag containing light-yellow powder secured nearby. 4-FBF blood concentrations were identified for the two decedents (blood = 91 ng/mL; 112 ng/mL, urine = 200 ng/mL; 414 ng/mL, liver = 902 ng/g; 411 ng/g, kidney = 136 ng/g; 197 ng/g, respectively) [318A].

Abrahamsson et al. aimed to assess whether sedative prescriptions may be associated with mortality in patients taking opioid maintenance treatment (OMT) through a retrospective register-based open cohort study including a nation-wide Swedish register data and all individuals dispensed methadone or buprenorphine as OMT for opioid dependence (July 2005–December, 2012 [$n = 4501$]). Extended Cox regression analyses examined associations between sedative type and death, controlling for sex, age, previous overdoses, suicide attempts, psychiatric in-patient treatment and OMT status. OMT was assumed to last 90 days (30 days in a sensitivity analysis) after the last methadone or buprenorphine prescription. BZD prescriptions were associated with non-overdose death (HR = 2.02 [95% CI 1.29–3.18]) but not significantly associated with overdose death (1.49 [0.97–2.29]). Z-drug (1.60 [1.07–2.39]) and pregabalin prescriptions (2.82 [1.79–4.43]) were associated with overdose death. In the sensitivity analysis, all sedative categories, including BZDs, were significantly associated with overdose death. Authors suggest using caution when prescribing sedative drugs, including BZDs, z-drugs and pregabalin, to patients on OMT [319MC].

Haukka et al. aimed to estimate non-medical substance use prevalence and predictors to assess the association between non-medical substance use and fatal poisoning or drug abuse in Finland. Researchers used a retrospective cohort to study all medico-legally investigated death cases in Finland through the postmortem toxicology database. All postmortem cases (2011–2013) positive for ≥1 drug including: oxycodone, fentanyl, tramadol, clonazepam, gabapentin, pregabalin, tizanidine, olanzapine, quetiapine, risperidone, alprazolam, zolpidem, mirtazapine and bupropion ($n = 2974$) were analyzed. Studied cases (50.4%) had ≥1 drug detected without a prescription. Clonazepam, alprazolam and tramadol were the most prevalent non-medical findings in these cases (6.6%, 6.1% and 5.6%, respectively). Using prescription drugs for a non-medical purpose was significantly more elevated for decedents with a drug abuse history (88.5%) and in fatal poisonings (71.0%). Non-medical use for substances studied varied (5.9% [95%; CI 3.1%–10.1%] risperidone; 55.7% [95%; CI 44.1%–66.9%] fentanyl). Any psychoactive substance with ≥1 valid prescription was associated with lower odds for non-medical use. With psychoactive drugs, the larger proportion prescribed by a psychiatrist, resulted in a lower probability for non-medical use. Overall, non-prescribed psychoactive drugs are commonly found postmortem in drug poisoning deaths in Finland, in decedents with drug abuse history as a major contributing factor [320MC].

Hopkins et al. aimed to identify unintentional deaths in Australia associated with using combination analgesic products containing paracetamol, codeine and doxylamine; describe case characteristics, including demographics and additional medication use; and identify common factors with using these medications, associated with misuse and mortality in Australia. Researchers conducted a retrospective case series by analyzing the National Coronial Information System data to identify unintentional death cases attributable to paracetamol, codeine and doxylamine products (2002 and 2012) in three Eastern Australian states (New South Wales, Queensland, Victoria), with 18.6-million people. Unintentional deaths ($n = 441$) attributed to paracetamol/codeine products and doxylamine were detected in 102-cases (23%). Unintentional death rates rose from 0.9-per-million (2002) to 3.6-per-million (2009), declining to 1.9-per-million (2012). At death, median age = 48-YO with >50% 35–54-YO, majority female (57%). Concomitant medication use was detected (79%), including BZDs, other opioids, psychiatric medications, alcohol and illicit drugs. Behaviors consistent with drug misuse included doctor/pharmacy shopping, excessive dosages and extended use (24%). Researchers identified deaths associated with codeine-combination analgesic products across three Australian states averaging 40 deaths/year, most commonly involving multiple substance use and abuse behaviors, consistent misuse and dependence [321C].

Overdose mortality varying both by time since release from prison and time of release has not been investigated sufficiently, so Bukten et al. aimed to estimate and compare overdose death rates at time intervals after prison release and effect on overdose death rates over calendar time. Researchers conducted a 15-year cohort study including all individuals ($n=91090$) released from prison (January 2000–December 2014) obtained from the Norwegian prison registry, linked to the Norwegian Cause of Death Registry. Overdose deaths accounted for most deaths ($n=123$[85%]) one-week following release peaking two-days immediately following release ($n=145$). Overdose risk during week-two was one-half compared to week-one [IRR=0.43; 95% CI=0.31–0.59] and reduced to one-fifth in weeks three to four (IRR 0.22; 95% CI 0.16–0.31). Overdose mortality risk during the first six-months post-release was almost twofold higher in 2000–04 compared with 2005–09 (IRR 0.53; 95% CI 0.43–0.65) and 2010–14 (IRR 0.47; 95% CI 0.37–0.59) and was highest for those incarcerated for 3–12 months compared to those incarcerated for shorter or longer periods [322MC].

Forsyth et al. aimed to estimate the incidence and identify risk factors for mortality in adults released from prisons in Queensland, Australia by designing a prospective cohort study, linking baseline survey data with a national death register. Adults ($n=1320$) recruited in Queensland prisons within 6 weeks of expected release, (August 2008–July 2010), were followed in the community for up to 4.7 years. The mortality rate for this cohort was higher than in the age- and sex-matched general Queensland population for all causes (Standardized Mortality Ratio [SMR]=4.0, 95% CI 2.9–5.4) and drug-related causes (SMR=32, 95% CI 19–55). The multivariable model, adjusting for age, sex and indigenous status was associated with increased mortality risk included expecting to have average or better funds available on release ([aHR]=2.9, 99% CI 1.2–7.1), poor mental health (aHR=2.6, 99% CI 1.1–6.1) and self-reported lifetime overdose history (aHR=2.5, 99% CI 1.04–6.2). Prisoners released in Queensland, Australia are at increased mortality risk, particularly due to drug-related causes, characterized by "risky" substance use and poor physical/mental health [323MC].

Nationwide data regarding drug abuse (DA)-associated mortality are lacking and studies are needed to further understand whether individual DA characteristics or DA itself results in mortality. Kendler et al. assessed DA from medical, criminal, and prescribed drug registries to obtain relative pairs discordant for DA from the Multi-Generation and Twin Registers. Researchers examined all individuals born in Sweden (1955–1980) ($n=2696253$), whom developed ($n=75061$) DA. The mortality hazard ratio (mHR) (95% CIs) for DA (11.36 [95% CIs, 11.07–11.66]) was substantially higher in non-medical (18.15, 17.51–18.82) than medical causes (8.05, 7.77–8.35) and stronger for women (12.13, 11.52–12.77) than men (11.14, 10.82–11.47). Co-relative analyses demonstrated substantial familial confounding in the DA–mortality associated with the strongest direct effects seen in middle and late-middle ages. The mHR was highest for opiate abusers (24.57, 23.46–25.73), followed by sedatives (14.19, 13.11–15.36), cocaine/stimulants (12.01, 11.36–12.69), and cannabis (10.93, 9.94–12.03). Authors concluded that excess mortality originates both indirectly from drug-abusing characteristics and directly from DA effects. Mortality from opiate abuse was significantly higher than any the drug class. Findings could result in conducting interventions to reduce DA-associated premature mortality [324MC].

Shipton et al. aimed to investigate the opioid-related death rate in New Zealand (January 2008–December 2012) with a population-based cohort study. Researcher's secondarily aimed to compare the opioid-related death rate per population in 2001–2002 with that between 2011 and 2012; investigate the opioid prescription amount (2001–2012); and compare the opioid-related death rate per population with the opioid prescription amount in New Zealand (2001–2012). Opioid-related deaths rates in New Zealand increased by 33% (2001–2012). Unintentional opioid-related deaths rates ($n=179$) were >50% (2008–2012) which could have been avoided. Opioid analgesic deaths most likely resulted from methadone, morphine and codeine prescribed by healthcare professionals. Steady annual increases in opioid prescriptions in New Zealand (2001 – 2012) are associated with the risk of death associated with opioid analgesics. A multifaceted, targeted national public health initiative is needed to bring together stakeholders involved in pain management, opioid dependence, availability and diversion. Current and future medical practitioners need to be educated regarding opioid use for pain management and provide patients with primary, secondary and tertiary resources to support patients who suffer from pain [325MC].

PRESCRIPTION/OVER-THE-COUNTER OPIOIDS AND FENTANYL ANALOGS (KU)

Pharmacogenetics

Genetic polymorphisms identified in pharmacogenetic studies that affect metabolism and response to synthetic opioids involve genes encoding for drug transporters, enzymes involved in opioid metabolism, opioid receptors, and proteins modulating opioid-induced neuroadaptation and drug addiction. Table 10 outlines the

TABLE 10 Synthetic Opioid Pharmacogenetics [326C,327c,328C,329c,330C,331C,332C,333c,334C,335C]

Opioids: Pharmacogenetics				
Objectives	**Gene/allele**	**Population**	**Clinical effect**	**Practical implication**
To identify genetic factors associated with intraoperative and postoperative opioid requirements in a genome-wide pharmacogenetic study [333c]	*TMEM8A* rs199670311 *SLC9A9* rs4839603	Japanese adults undergoing laparoscopic-assisted colectomy (*n*=350) Patients received continuous remifentanil infusion and fentanyl boluses during surgery. Fentanyl PCA was used for postoperative pain	G/A genotypes of rs199670311 compared to G/G genotypes: – Required more intraoperative remifentanil ($Q<0.05$) – Used more postoperative fentanyl over 24h ($P=0.04$) T allele carriers of rs4839603 compared to non-carriers: – Required more intraoperative remifentanil ($Q<0.05$) – Experienced less analgesia from preoperative fentanyl ($P=0.013$) – Used more total perioperative analgesia ($P=0.045$)	A allele carriers of rs199670311 and T allele of rs4839603 may have reduced opioid sensitivity and require higher perioperative opioid doses
1. To evaluate the association of genetic factors with morphine requirements in end-of-life cancer patients **2.** To evaluate the relationship between genetic factors with pain intensity and quality of life measures [327c]	*OPRM1* A118G *COMT* Val158Met *ABCB1* C3435T	Adults receiving pumped IV morphine for palliative cancer pain management (*n*=89)	– *OPRM1* A/G genotype required higher morphine dose at 24h than A/A genotype ($P<0.001$) – *OPRM1* A/A genotype had lower cognitive function than A/G genotype (61.59 ± 32.38 vs 80 ± 28.41; $P=0.014$) – *COMT* and *ABCB1* polymorphisms were not associated with differences in morphine dose at 24h and quality of life measures – Changes in pain intensity were not associated with any of the polymorphisms studied	Patients carrying a *OPRM1* G allele require higher morphine doses for cancer pain relief. Patients with the *OPRM1* A/A genotype have better activation of the mu-receptor and may be more sensitive to morphine's cognitive side effects
To evaluate the association of *OPRM1* polymorphism with intraoperative remifentanil consumption under general anesthesia [328C]	*OPRM1* A118G	Adults who underwent septoplasty surgery under general anesthesia with IV remifentanil (*n*=121)	*OPRM1* A/A genotype used less remifentanil than A/G genotype (0.173 ± 0.063 mcg/kg/min vs 0.316 ± 0.1 mcg/kg/min; $P<0.0001$)	Patients with *OPRM1* A/G genotype may require higher remifentanil doses than A/A genotype patients
To determine associations between fentanyl requirements and *ABCB1* polymorphisms [329c]	*ABCB1* rs1045642, rs1128503, and rs2229109	Children who received a fentanyl infusion while in a pediatric ICU (*n*=61)	Rs1045642 A/A genotype received 18.6 mcg/kg/day less fentanyl than G allele carriers (95% CI, −33.4 to −3/8 mcg/kg/day; $P=0.014$)	*ABCB1* rs1045642 A/A genotype may have decreased PgP expression and activity, and may have require less fentanyl
To evaluate the association of genetic polymorphisms in *ABCB1*, *ARRB2*, *DRD1* and *OPRD1* with methadone requirements among opioid-dependent Han Chinese patients (Luo, 2017 [C])	*ABCB1* (rs1128503, rs2032582, and rs1045642) *ARRB2* (rs1045280) *DRD1* (rs686) *OPRD1* (rs529520, rs2234918, and rs581111)	Opioid-dependent Han Chinese patients on methadone maintenance treatment (*n*=257)	– OPRD1 T/G genotype rs529520 required higher methadone dosages (adjusted OR 2.24; 95% CI, 1.06–5.1) – Methadone dosage requirements were associated with a 3-locus SNP-SNP interaction between *ABCB1* rs1128503, *OPRD1* rs529520, and *ARRB2* rs1045280 ($P=0.029$)	*ABCB1, ARRB2,* and *OPRD1* polymorphisms may affect methadone dosage requirements. Identified genetic interactions may not be clinically significant

To identify the association of *OCT1* polymorphisms affecting morphine PKs with postoperative morphine-related adverse outcomes [331C]	*OCT1* rs12208357, rs34130495, rs72552763, and rs34059508	311 Caucasian ($n=262$) and African American ($n=49$) children 6–15 YO who underwent tonsillectomy ($n=311$)	– OCT1 rs12208357 T allele carriers experienced higher incidences of PONV ($P<0.05$) and PONV leading to prolonged PACU stay ($P<0.05$) – GAT deletion of OCT1 rs72552763 was associated with a higher incidence of respiratory depression ($P=0.007$) – *Note*: Total mg/kg morphine requirement ($P=0.028$) and incidence of OSA ($P<0.001$) differed between Caucasian and African American patients	OCT1 rs12208357 and rs72552763 polymorphisms may increase risk of morphine-related AE
To determine if CYP2B6 SNPs are associated with a poor metabolizer phenotype in fatal methadone cases [332C]	CYP2B6 SNPs rs2279344, rs3211371, rs3745274, rs4803419, rs8192709, rs8192719, rs12721655, and rs35979566	125 Caucasian methadone-only fatalities obtained from West Virginia ($n=78$) and Kentucky ($n=47$) and 25.5 control cases obtained from West Virginia	– SNP rs3211371 homozygous genotype had higher methadone serum concentrations ($P<0.05$) – SNPs rs3745274 ($P<0.005$) and rs8192719 ($P<0.001$) were more frequent in the methadone-only group ($P<0.005$)	CYP2B6 polymorphisms at SNPs rs3211371, rs3745274, and rs8192719 may result in a poor methadone-metabolizer phenotype. These individuals may have increased risk of fatal methadone intoxication
To evaluate the association between CYP2D6 genotype and oxycodone metabolism [333c]	CYP2D6 polymorphism	2–17-YO children administered oral oxycodone postoperatively ($n=30$) Participants: 5 CYP2D6 EM, 5 CYP2D6 IM, and 1 CYP2D6 PM phenotypes Oxycodone was dosed based on weight and age: – 2–6-YO: 0.1 mg/kg – 7–12-YO: 0.08 mg/kg – 13–17-YO: 0.07 mg/kg	CYP2D6 EM compared to CYP2D6 IM and PM phenotypes: – Elevated oxymorphone peak concentration ($P=0.028$) – Greater AUC over 6 h ($P=0.0162$) – Greater AUC over 24 h ($P=0.026$) – Higher oxymorphone/oxycodone peak concentration ratio ($P=0.0007$) – Higher oxymorphone/oxycodone AUC over 6 h ratio ($P=0.001$) – Higher oxymorphone/oxycodone AUC over 24 h ratio ($P=0.004$)	Extensive CYP2D6 metabolizers have higher exposure to oxymorphone. Oxycodone dosing based on CYP2D6 may improve pain control and decrease side effects
To examine the association between CYP2D6 phenotype and poor outcomes related to opioid use in primary care patients [334C]	CYP2D6 polymorphism	Patients from the Mayo Clinic RIGHT Protocol cohort who received at least one prescription of codeine, tramadol, oxycodone, or hydrocodone ($n=257$) Participants: 40 CYP2D6 PM, 146 CYP2D6 IM and EM, and 71 CYP2D6 UM	– *Note*: 38 patients used a CYP2D6 inhibitor (includes 5 CYP2D6 PM, 22 CYP2D6 IM and EM, and 11 CYP2D6 UM) – Analysis including CYP2D6 inhibitor patients: CYPD26 PM and UM were more likely to experience poor pain control (OR 2.63; 95% CI, 1.19–5.83; $P=0.02$) – Analysis excluding CYP2D6 inhibitor patients: CYP2D6 PM and UM were more likely to experience opioid-related adverse reaction (OR 2.53; 95% CI, 1.12–5.7; $P=0.03$) or any problem (OR 2.68; 95% CI, 1.39–5.17; $P=0.003$)	CYP2D6 PM or UM phenotypes may have an increased risk of adverse outcomes when using opioids
To evaluate the association between CYP3A4 polymorphism and sufentanil doses used in general anesthesia during lung resection [335C]	CYP3A4 polymorphism	Han Chinese adults who underwent lung resection ($n=191$) General anesthesia was induced and supplemented by IV sufentanil boluses	CYP3A4*1G allele carriers used less sufentanil compared to non-carriers (30.43 ± 3.69 mcg vs 41.07 ± 5.49 mcg; $P<0.001$)	CYP3A4*1G variants may have impaired CYP3A4 activity that may results in oversedation with sufentanil. CYP3A4 genotyping may help guide sufentanil use

Abbreviations: *EM* = extensive metabolizer; *IM* = intermediate metabolizer; *OSA* = obstructive sleep apnea; *PCA* = patient-controlled analgesia; *PM* = poor metabolizer; *PONV* = postoperative nausea and vomiting; *UM* = ultra-rapid metabolizer.

details and practical implications of different studies analyzing synthetic opioid pharmacogenetics.

CONCLUSION (PRECAUTIONARY NOTE TO CLINICIANS)

Cannabinoids: Large studies support the association between non-synthetic cannabinoid use and MI as well as ischemic stroke. However, only case reports and case series describe these AE rates among synthetic cannabinoid users. Additional reports have described tachyarrhythmias, respiratory or CNS depression, status epilepticus and psychiatric AEs. Given the challenge of timely identification of synthetic cannabinoid use, awareness of these possible clinical presentations is key. Clinicians should also be aware that potential CYP interactions might exist.

Alcohol: Alcohol consumption has been associated with an increased risk of certain cardiovascular diseases, neurocognitive changes, and other disorders in both large population studies and case reports. Much of the emerging data regarding alcohol AEs is focused on malignancy risk.

Benzodiazepines: BZD use and misuse has been associated with many undesirable AEs. Emerging data examine the correlation between BZD use with bradyarrhythmia, increased cancer risk, cognitive decline, migraines, fracture risk, hepatotoxicity, and gingival hyperplasia.

Psychostimulants: Psychostimulants, specifically cocaine and methamphetamine, have recently been reported in small case reports to be linked to AEs such as orbital injury, spleen rupture, acute multiple mononeuropathy, thrombosis, thyroid storm, and ischaemic colitis.

Synthetic Psychoactive Cathinones: SPC overdose is associated with similar symptoms as other types of psychomotor stimulants, including cardiac and neurological symptoms, potentially culminating in psychotic delirium and multi-organ failure at very high doses. Based on current evidence, it is uncertain whether some SPCs may present a greater likelihood of inducing psychotic delirium and other adverse effects than other psychomotor stimulants.

Heroin: While a few small studies have been conducted, most reports of heroin-induced SEs are described in case reports and include pneumonia, cardiovascular complications, compartment syndrome, and neurologic diseases.

Opioids: Strategies are needed to decrease the morbidity and mortality related to opioid misuse and abuse through affording accessibility to those in need of prevention and intervention strategies related to social and environmental factors. Providers should be aware of the continued chemical changes to NPS in order to better identify and treat opioid-related complications and overdoses. Through multiple case reports described, providers are recommended to consider traditionally uncommon differential diagnoses when presented with specific symptoms for patients with OUD.

Pharmacogenetics: Although there are limited data, genetic variation may affect pharmacodynamics and pharmacokinetics of different drugs of abuse.

Additional literature in this topic can be found in these reviews [336R,337R].

Acknowledgements

Authors gratefully acknowledge the assistance provided by Ms. Tisha Mentnech, MSLIS, Camryn Froerer, PharmD and Jina Kim, PharmD.

References

Introduction

[1] United Nations Office on Drugs and Crime. World Drug Report 2017. United Nations publication; 2017. Sales No. E.17.XI.6. [S].

Cannabinoids

[2] Adams AJ, Banister SD, Irizarry L, et al. "Zombie" outbreak caused by the synthetic cannabinoid AMB-FUBINACA in New York. N Engl J Med. 2017;376(3):235–42 [c].

[3] Hermanns-Clausen M, Müller D, Kithinji J, et al. Acute side effects after consumption of the new synthetic cannabinoids AB-CHMINACA and MDMB-CHMICA. Clin Toxicol (Phila). 2017;56(6):404–11 [c].

[4] Hamilton RJ, Keyfes V, Banka SS. Synthetic cannabinoid abuse resulting in ST-segment elevation myocardial infarction requiring percutaneous coronary intervention. J Emerg Med. 2017;52(4):496–8 [A].

[5] Solimini R, Busardò FP, Rotolo MC, et al. Hepatotoxicity associated to synthetic cannabinoids use. Eur Rev Med Pharmacol Sci. 2017;21(1 Suppl):1–6 [R].

[6] Andonian DO, Seaman SR, Josephson EB. Profound hypotension and bradycardia in the setting of synthetic cannabinoid intoxication—a case series. Am J Emerg Med. 2017;35(6). 940. e5-6. [A].

[7] Moeller S, Lücke C, Struffert T, et al. Ischemic stroke associated with the use of a synthetic cannabinoid (spice). Asian J Psychiatr. 2017;25:127–30 [A].

[8] Alon MH, Saint-Fleurb MO. Synthetic cannabinoid induced acute respiratory depression: case series and literature review. Respir Med Case Rep. 2017;22:137–41 [A].

[9] Pintori N, Loi B, Mereu M. Synthetic cannabinoids: the hidden side of Spice drugs. Behav Pharmacol. 2017;28(6):409–19 [R].

[10] Wolff V, Jouanjus E. Strokes are possible complications of cannabinoids use. Epilepsy Behav. 2017;70(Pt B):355–63 [r].

[11] Efe TH, Felekoglu MA, Çimen T, et al. Atrial fibrillation following synthetic cannabinoid abuse. Turk Kardiyol Dern Ars. 2017;45(4):362–4 [A].

[12] Lam RPK, Tang MHY, Leung SC, et al. Supraventricular tachycardia and acute confusion following ingestion of e-cigarette fluid containing AB-FUBINACA and ADB-FUBINACA: a case report with quantitative analysis of serum drug concentrations. Clin Toxicol (Phila). 2017;55(7):662–7 [A].

[13] Meyyappan C, Ford L, Vale A. Poisoning due to MDMB-CHMICA, a synthetic cannabinoid receptor agonist. Clin Toxicol (Phila). 2017;55(2):151–2 [A].
[14] Babi MA, Robinson CP, Maciel CB. A spicy status: synthetic cannabinoid (spice) use and new-onset refractory status epilepticus—a case report and review of the literature. SAGE Open Med Case Rep. 2017;5:2050313X17745206 [A].
[15] Lamy FR, Daniulaityte R, Nahhas RW, et al. Increases in synthetic cannabinoids-related harms: results from a longitudinal web-based content analysis. Int J Drug Policy. 2017;44:121–9 [MC].
[16] Park F, Potukuchi PK, Moradi H, et al. Cannabinoids and the kidney: effects in health and disease. Am J Physiol Renal Physiol. 2017;313(5):F1124–32 [R].
[17] Tai S, Fantegrossi WE. Pharmacological and toxicological effects of synthetic cannabinoids and their metabolites. Curr Top Behav Neurosci. 2017;32:249–62 [R].
[18] Desai R, Patel U, Sharma S, et al. Recreational marijuana use and acute myocardial infarction: insights from nationwide inpatient sample in the United States. Cureus. 2017;9(11):e1816 [MC].
[19] Singh A, Saluja S, Kumar A, et al. Cardiovascular complications of marijuana and related substances: a review. Cardiol Ther. 2017;7(1):45–59 [R].
[20] Shere A, Goyal H. Cannabis can augment thrombolytic properties of rtPA: intracranial hemorrhage in a heavy cannabis user. Am J Emerg Med. 2017;35(12). 1988.e1-2. [A].
[21] Fernández-Atutxa A, de Castro L, Arévalo-Senra JA, et al. Cannabis intake and intussusception: an accidental association? Rev Esp Enferm Dig. 2017;109(2):157–9 [A].
[22] Dezieck L, Hafez Z, Conicella A, et al. Resolution of cannabis hyperemesis syndrome with topical capsaicin in the emergency department: a case series. Clin Toxicol (Phila). 2017;55(8):908–13 [c].
[23] Cadman PE. Hypophosphatemia in users of cannabis. Am J Kidney Dis. 2017;69(1):152–5 [c].
[24] Rumalla K, Reddy AY, Mittal MK. Recreational marijuana use and acute ischemic stroke: a population-based analysis of hospitalized patients in the United States. J Neurol Sci. 2016;364:191–6, Epub 2016 Feb 4. https://doi.org/10.1016/j.jns.2016.01.066 [MC].
[25] Behrouz R, Birnbaum L, Grandhi R, et al. Cannabis use and outcomes in patients with aneurysmal subarachnoid hemorrhage. Stroke. 2016;47:1371–3. https://doi.org/10.1161/strokeaha.116.013099 [C].
[26] McHugh MJ, McGorry PD, Yung AR, et al. Cannabis-induced attenuated psychotic symptoms: implications for prognosis in young people at ultra-high risk for psychosis. Psychol Med. 2017;47(4):616–26 [C].
[27] Nestoros JN, Vakonaki E, Tzatzarakis MN, et al. Long lasting effects of chronic heavy cannabis abuse. Am J Addict. 2017;26(4):335–42 [c].
[28] Colizzi M, Bhattacharyya S. Does cannabis composition matter? Differential effects of delta-9-tetrahydrocannabinol and cannabidiol on human cognition. Curr Addict Rep. 2017;4(2):62–74 [r].
[29] Iffland K, Grotenhermen F. An update on safety and side effects of cannabidiol: a review of clinical data and relevant animal studies. Cannabis Cannabinoid Res. 2017;2(1):139–54 [r].
[30] O'Connor AB, Kelly BK, O'Brien LM. Maternal and infant outcomes following third trimester exposure to marijuana in opioid dependent pregnant women maintained on buprenorphine. Drug Alcohol Depend. 2017;180:200–3 [C].
[31] Zhang A, Marshall R, Kelsberg G. Clinical inquiry: what effects—if any—does marijuana use during pregnancy have on the fetus or child? J Fam Pract. 2017;66(7):462–6 [r].
[32] Claudet I, Le Breton M, Bréhin C, et al. A 10-year review of cannabis exposure in children under 3-years of age: do we need a more global approach? Eur J Pediatr. 2017;176(4):553–6 [r].
[33] Meruelo AD, Castro N, Cota CI, et al. Cannabis and alcohol use, and the developing brain. Behav Brain Res. 2017;325(Pt A):44–50 [r].
[34] Kebir O, Lafaye G, Blecha L, et al. ABCB1 C3435T polymorphism is associated with tetrahydrocannabinol blood levels in heavy cannabis users. Psychiatry Res. 2018;262:357–8 [c].
[35] Ketcherside A, Noble LJ, McIntyre CK, et al. Cannabinoid receptor 1 gene by cannabis use interaction on CB1 receptor density. Cannabis Cannabinoid Res. 2017;2(1):202–9 [C].

Alcohol

[36] Whitman IR, Agarwal V, Nah G, et al. Alcohol abuse and cardiac disease. J Am Coll Cardiol. 2017;69(1):13–24 [MC].
[37] Dixit S, Alonso A, Vittinghoff E, et al. Past alcohol consumption and incident atrial fibrillation: The Atherosclerosis Risk in Communities (ARIC) study. PLoS One. 2017;12(10): e0185228 [MC].
[38] Huang JY, Chen WK, Lin CL, et al. Increased risk of peripheral arterial disease in patients with alcohol intoxication: A population-based retrospective cohort study. Alcohol. 2017;65:25–30 [MC].
[39] Bainbridge KE, Roy N, Losonczy KG, et al. Voice disorders and associated risk markers among young adults in the United States. Laryngoscope. 2017;127(9):2093–9 [MC].
[40] Carbia C, Cadaveira F, Caamano-Isorna F, et al. Binge drinking during adolescence and young adulthood is associated with deficits in verbal episodic memory. PLoS One. 2017;12(2):e00171393 [C].
[41] Meda SA, Dager AD, Hawkins KA, et al. Heavy drinking in college students is associated with accelerated gray matter volumetric decline over a 2 year period. Front Behav Neurosci. 2017;11:176 [C].
[42] Topiwala A, Allan CL, Valkanova V, et al. Moderate alcohol consumption as risk factor for adverse brain outcomes and cognitive decline: longitudinal cohort study. BMJ. 2017;357: j2353 [C].
[43] Haldar S, Mukherjee R, Elston J. Microcystic macular edema in a case of tobacco-alcohol optic neuropathy. Can J Ophthalmol. 2017;52(1):e19–22 [A].
[44] Hermann JM, Meusers M, Bachran R, et al. Self-reported regular alcohol consumption in adolescents and emerging adults with type 1 diabetes: a neglected risk factor for diabetic ketoacidosis? Multicenter analysis of 29 630 patients from the DPV registry. Pediatr Diabetes. 2017;18(8):817–23 [MC].
[45] Filiberti RA, Fontana V, De Ceglie A, et al. Alcohol consumption pattern and risk of Barrett's oesophagus and erosive oesophagitis: an Italian case-control study. Br J Nutr. 2017;117(8):1151–61 [C].
[46] Li S, Cho E, Drucker AM, et al. Alcohol intake and risk of rosacea in US women. J Am Acad Dermatol. 2017;76(6):1061–7 [MC].
[47] Chu GJ, Murad A. A case of ethanol-induced systemic allergic dermatitis. Contact Dermatitis. 2017;76(3):182–4 [A].
[48] Ulhøi MP, Meldgaard K, Steiniche T, et al. Chronic alcohol abuse leads to low bone mass with no general loss of bone structure or bone mechanical strength. J Forensic Sci. 2017;62(1):131–6 [c].
[49] Li C, Liu H, Wang C, et al. A rare case report: enlarged iliopsoas cystic solid mass associated with femoral head necrosis induced by heavy alcohol consumption. Medicine (Baltimore). 2017;96(26): e7341 [A].
[50] Barbhaiya M, Lu B, Sparks JA, et al. Influence of alcohol consumption on the risk of systemic lupus erythematosus among women in the Nurses' Health Study cohorts. Arthritis Care Res (Hoboken). 2017;69(3):384–92 [MC].
[51] World Health Organization. Cancer prevention [Internet], 2018, Geneva: The World Health Organization; 2018. c2018 [cited 2018 May 10]. Available from, http://www.who.int/cancer/prevention/en [S].

[52] Betts G, Ratschen E, Breton MO, et al. Alcohol consumption and risk of common cancers: evidence from a cohort of adults from the UK. J Public Health (Oxf). 2017;11:1–9 [MC].

[53] Fehringer G, Brenner DR, Zhang ZF, et al. Alcohol and lung cancer risk among never smokers: a pooled analysis from the international lung cancer consortium and the SYNERGY study. Int J Cancer. 2017;140(9):1976–84 [MC].

[54] Kim HJ, Jung S, Eliassen AH, et al. Alcohol consumption and breast cancer risk in younger women according to family history of breast cancer and folate intake. Am J Epidemiol. 2017;186(5):524–31 [MC].

[55] White AJ, DeRoo LA, Weinberg CR, et al. Lifetime alcohol intake, binge drinking behaviors, and breast cancer risk. Am J Epidemiol. 2017;185(5):541–9 [MC].

[56] Choi YJ, Lee DH, Han KD, et al. The relationship between drinking alcohol and esophageal, gastric, or colorectal cancer: a nationwide population-based cohort study of South Korea. PLoS One. 2017;12(10):e0185778 [MC].

[57] Zaitsu M, Kawachi I, Takeuchi T, et al. Alcohol consumption and risk of upper-tract urothelial cancer. Cancer Epidemiol. 2017;48:36–40 [MC].

[58] Botteri E, Ferrari P, Roswall N, et al. Alcohol consumption and risk of urothelial cell bladder cancer in the European prospective investigation into cancer and nutrition cohort. Int J Cancer. 2017;141(10):1963–70 [MC].

[59] León-Muñoz LM, Guallar-Castillón P, García-Esquinas E, et al. Alcohol drinking patterns and risk of functional limitations in two cohorts of older adults. Clin Nutr. 2017;36(3):831–8 [MC].

[60] Yokoyama A, Yokoyama T, Mizukami T, et al. Platelet Counts and Genetic Polymorphisms of alcohol dehydrogenase-1B and aldehyde dehydrogenase-2 in Japanese alcoholic men. Alcohol Clin Exp Res. 2017;41(1):171–8 [C].

[61] Otto JM, Gizer IR, Deak JD, et al. A cis-eQTL in OPRM1 is associated with subjective response to alcohol and alcohol use. Alcohol Clin Exp Res. 2017;41(5):929–38 [C].

[62] Korucuoglu O, Gladwin TE, Baas F, et al. Neural response to alcohol taste cues in youth: effects of the OPRM1 gene. Addict Biol. 2017;22(6):1562–75 [c].

Benzodiazepine

[63] Brandt J, Leong C. Benzodiazepines and z-drugs: an updated review of major adverse outcomes reported on in epidemiologic research. Drugs R D. 2017;17(4):493–507 [R].

[64] Maruyoshi H, Maruyoshi N, Hirosue M, et al. Clonazepam-associated bradycardia in a disabled elderly woman with multiple complications. Intern Med. 2017;56(17):2301–5 [c].

[65] Kim HB, Myung SK, Park YC, et al. Use of benzodiazepine and risk of cancer: a meta-analysis of observational studies. Int J Cancer. 2017;140(3):513–25 [M].

[66] Harnod T, Wang YC, Lin CL, et al. Association between use of short-acting benzodiazepines and migraine occurrence: a nationwide population-based case-control study. Curr Med Res Opin. 2017;33(3):511–7 [C].

[67] Zhong G, Wang Y, Zhang Y, et al. Association between benzodiazepine use and dementia: a meta-analysis. PLoS One. 2015;10:e0127836 [M].

[68] Chan TT, Leung WC, Li V, et al. Association between high cumulative dose of benzodiazepine in Chinese patients and risk of dementia: a preliminary retrospective case-control study. Psychogeriatrics. 2017;17(5):310–6 [C].

[69] Donnelly K, Bracchi R, Hewitt J, et al. Benzodiazepines, z-drugs and the risk of hip fracture: a systematic review and meta-analysis. PLoS ONE. 2017;12(4):e0174730 [M].

[70] Carvalhana S, Oliveira A, Ferreira P, et al. Acute liver failure due to trazodone and diazepam. GE Port J Gastroenterol. 2017;24(1):40–2 [c].

[71] Hatahira H, Abe J, Hane Y, et al. Drug-induced gingival hyperplasia: a retrospective study using spontaneous reporting system databases. J Pharm Health Care Sci. 2017;3:19 [R].

Cocaine

[72] Center for Behavioral Health Statistics and Quality. Behavioral health trends in the United States: results from the 2014 national survey on drug use and health. HHS publication; 2015. no. SMA 15–4927, NSDUH Series H-50. Retrieved from http://www.samhsa.gov/data/ [S].

[73] National Institute on Drug Abuse. DrugFacts: Cocaine [Internet]. Bethesda, MD: National Institute on Drug Abuse; 2016. c2016 [cited 2018 May 10]. Available from, https://www.drugabuse.gov/publications/drugfacts/cocaine [S].

[74] Trimarchi M, Bussi M, Sinico RA, et al. Cocaine-induced midline destructive lesions—an autoimmune disease? Autoimmun Rev. 2013;12(4):496–500 [c].

[75] Siemerink MJ, Freling NJM, Saeed P. Chronic orbital inflammatory disease and optic neuropathy associated with long-term intranasal cocaine abuse: 2 cases and literature review. Orbit. 2017;36(5):350–5 [c].

[76] Stuard WL, Gallerson BK, Robertson DM. Alterations in corneal nerves following crack cocaine use mimic diabetes-induced nerve damage. Endocrinol Diabetes Metab Case Rep. 2017; [c].

[77] Singh A, Thawani R, Thankur K. Crack cocaine-induced laryngeal injury. Am J Emerg Med. 2017;35(2):381.e385-7 [c].

[78] Khan AN, Casaubon JT, Paul Regan J, et al. Cocaine-induced splenic rupture. J Surg Case Rep. 2017;2017(3):rjx054 [c].

[79] de Souza A, Desai PK, de Souza RJ. Acute multifocal neuropathy following cocaine inhalation. J Clin Neurosci. 2017;36:134–6 [c].

[80] Poon SS, Nawaytou O, Hing A, et al. Giant aortic thrombus in the ascending aorta and perforation of bowel associated with cocaine use. Ann Thorac Surg. 2017;104(3):e219–20 [c].

[81] Prakash MD, Tangalakis K, Antonipillai J, et al. Methamphetamine: effects on the brain, gut, and immune system. Pharmacol Res. 2017;120:60–7 [R].

[82] Albanese J, Gross C, Azab M, et al. Spontaneous pneumomediastinum: a rare complication of methamphetamine use. Respir Med Case Rep. 2017;21:25–6 [c].

[83] Viswanath O, Menapace DC, Headley DB. Methamphetamine use with subsequent thyrotoxicosis/thyroid Storm, agranulocytosis, and modified total thyroidectomy: a case report. Clin Med Insights Ear Nose Throat. 2017;10: 1179550617741293 [c].

[84] Green PA, Battersby C, Heath RM, et al. A fatal case of amphetamine induced ischaemic colitis. Ann R Coll Surg Engl. 2017;99(7):e200–1 [c].

Cathinones

[85] Engidawork E. Pharmacological and toxicological effects of Catha edulis F. (Khat). Phytother Res. 2017;31(7):1019–28 [R].

[86] WHO Expert Committee on Drug Dependence. WHO expert committee on drug dependence. World Health Organ Tech Rep Ser. 2006;942(1–21):23–4 [S].

[87] Abebe W. Khat and synthetic cathinones: emerging drugs of abuse with dental implications. Oral Surg. 2018;125(2):140–6 [R].

[88] Al-Maweri SA, Al-Jamaei A, Saini R, et al. White oral mucosal lesions among the Yemeni population and their relation to local oral habits. J Investig Clin Dent. 2017;9(2):e12305 [C].

[89] Kalakonda B, Al-Maweri SA, Al-Shamiri HM, et al. Is Khat (Catha edulis) chewing a risk factor for periodontal diseases? A systematic review. J Clin Exp Dent. 2017;9(10):e1264–70 [M].

[90] Al-Maweri SA, Warnakulasuriya S, Samran A. Khat (Catha edulis) and its oral health effects: an updated review. J Investig Clin Dent. 2018;9(1):e12288 [R].

[91] Al-Maweri SA, AlAkhali M. Oral hygiene and periodontal health status among khat chewers. A case-control study. J Clin Exp Dent. 2017;9(5):e629–34 [C].
[92] Bede P, El-Kininy N, O'Hara F, et al. 'Khatatonia'—cathinone-induced hypertensive encephalopathy. Neth J Med. 2017;75(10):448–50 [A].
[93] Baumann MH, Partilla JS, Lehner KR. Psychoactive "bath salts": not so soothing. Eur J Pharmacol. 2013;698(1–3): 1–5 [R].
[94] Zwartsen A, Verboven AHA, van Kleef R, et al. Measuring inhibition of monoamine reuptake transporters by new psychoactive substances (NPS) in real-time using a high-throughput, fluorescence-based assay. Toxicol In Vitro. 2017;45(Pt 1):60–71 [E].
[95] Watterson LR, Olive MF. Reinforcing effects of cathinone NPS in the intravenous drug self-administration paradigm. Curr Top Behav Neurosci. 2017;32:133–43 [R].
[96] Anneken JH, Angoa-Perez M, Sati GC, et al. Assessing the role of dopamine in the differential neurotoxicity patterns of methamphetamine, mephedrone, methcathinone and 4-methylmethamphetamine. Neuropharmacology. 2017;134(Pt A):46–56 [E].
[97] Anneken JH, Angoa-Perez M, Sati GC, et al. Dissecting the influence of two structural substituents on the differential neurotoxic effects of acute methamphetamine and mephedrone treatment on dopamine nerve endings with the use of 4-methylmethamphetamine and methcathinone. J Pharmacol Exp Ther. 2017;360(3):417–23 [E].
[98] Angoa-Perez M, Anneken JH, Kuhn DM. Neurotoxicology of synthetic cathinone analogs. Curr Top Behav Neurosci. 2017;32:209–30 [R].
[99] Madras BK. The growing problem of new psychoactive substances (NPS). Curr Top Behav Neurosci. 2017;32:1–18 [R].
[100] Byrska B, Stanaszek R, Zuba D. Alpha-PVP as an active component of herbal highs in Poland between 2013 and 2015. Drug Test Anal. 2017;9(8):1267–74 [c].
[101] Gyarmathy VA, Peterfi A, Figeczki T, et al. Diverted medications and new psychoactive substances—a chemical network analysis of discarded injecting paraphernalia in Hungary. Int J Drug Policy. 2017;46:61–5 [C].
[102] Katchman E, Ben-Ami R, Savyon M, et al. Successful control of a large outbreak of HIV infection associated with injection of cathinone derivatives in Tel Aviv, Israel. Clin Microbiol Infect. 2017;23(5) 336 e5-8. [c].
[103] Palamar JJ, Salomone A, Gerace E, et al. Hair testing to assess both known and unknown use of drugs amongst ecstasy users in the electronic dance music scene. Int J Drug Policy. 2017;48:91–8 [c].
[104] Beck O, Backberg M, Signell P, et al. Intoxications in the STRIDA project involving a panorama of psychostimulant pyrovalerone derivatives, MDPV copycats. Clin Toxicol (Phila). 2018;56(4):256–63 [MC].
[105] Patel N, Ford L, Jones R, et al. Poisoning to alpha-pyrrolidinovalerophenone (alpha-PVP), a synthetic cathinone. Clin Toxicol (Phila). 2017;55(2):159–60 [A].
[106] Cherry SV, Rodriguez YF. Synthetic stimulant reaching epidemic proportions: flakka-induced ST-elevation myocardial infarction with intracardiac thrombi. J Cardiothorac Vasc Anesth. 2017;31(1):e13–4 [A].
[107] Baumann MH, Bukhari MO, Lehner KR, et al. Neuropharmacology of 3,4-methylenedioxypyrovalerone (MDPV), its metabolites, and related analogs. Curr Top Behav Neurosci. 2017;32:93–117 [R].
[108] Marusich JA, Antonazzo KR, Wiley JL, et al. Pharmacology of novel synthetic stimulants structurally related to the "bath salts" constituent 3,4-methylenedioxypyrovalerone (MDPV). Neuropharmacology. 2014;87:206–13 [E].
[109] Eiden C, Mathieu O, Cathala P, et al. Toxicity and death following recreational use of 2-pyrrolidino valerophenone. Clin Toxicol (Phila). 2013;51(9):899–903 [A].
[110] Nagai H, Saka K, Nakajima M, et al. Sudden death after sustained restraint following self-administration of the designer drug alpha-pyrrolidinovalerophenone. Int J Cardiol. 2014;172(1):263–5 [A].
[111] Sykutera M, Cychowska M, Bloch-Boguslawska E. A fatal case of pentedrone and alpha-pyrrolidinovalerophenone poisoning. J Anal Toxicol. 2015;39(4):324–9 [A].
[112] de Roux SJ, Dunn WA. "Bath Salts" the New York City medical examiner experience: a 3-year retrospective review. J Forensic Sci. 2017;62(3):695–9 [M].
[113] Penders TM, Gestring R. Hallucinatory delirium following use of MDPV: "bath salts" Gen Hosp Psychiatry. 2011;33(5):525–6 [A].
[114] Mash DC. Excited delirium and sudden death: a syndromal disorder at the extreme end of the neuropsychiatric continuum. Front Physiol. 2016;7:435 [R].
[115] Diestelmann M, Zangl A, Herrle I, et al. MDPV in forensic routine cases: psychotic and aggressive behavior in relation to plasma concentrations. Forensic Sci Int. 2018;283:72–84 [c].
[116] Fujita Y, Mita T, Usui K, et al. Toxicokinetics of the synthetic cathinone alpha-pyrrolidinohexanophenone. J Anal Toxicol. 2018;42(1):e1–5 [A].
[117] Roberts L, Ford L, Patel N, et al. 11 analytically confirmed cases of mexedrone use among polydrug users. Clin Toxicol (Phila). 2017;55(3):181–6 [C].
[118] Romanek K, Stenzel J, Schmoll S, et al. Synthetic cathinones in southern Germany—characteristics of users, substance-patterns, co-ingestions, and complications. Clin Toxicol (Phila). 2017;55(6):573–8 [c].
[119] Varma A, Patel N, Ford L, et al. Misuse of 2-(ethylamino)-1-(4-methylphenyl)-1-pentanone (4-MEAP), a synthetic cathinone. Clin Toxicol (Phila). 2017;55(3):231–2 [A].
[120] Kintz P. Evidence of 2 populations of mephedrone abusers by hair testing. Application to 4 forensic expertises. Curr Neuropharmacol. 2017;15(5):658–62 [c].
[121] Papaseit E, Olesti E, de la Torre R, et al. Mephedrone concentrations in cases of clinical intoxication. Curr Pharm Des. 2017;23(36):5511–22 [M].
[122] Thirakul P, Hair LS, Bergen KL, et al. Clinical presentation, autopsy results and toxicology findings in an acute N-ethylpentylone fatality. J Anal Toxicol. 2017;41(4):342–6 [A].
[123] Vizeli P, Schmid Y, Prestin K, et al. Pharmacogenetics of ecstasy: CYP1A2, CYP2C19, and CYP2B6 polymorphisms moderate pharmacokinetics of MDMA in healthy subjects. Eur Neuropsychopharmacol. 2017;27(3):232–8 [C].
[124] Vizeli P, Meyer Zu Schwabedissen HE, Liechti ME. No major role of norepinephrine transporter gene variations in the cardiostimulant effects of MDMA. Eur J Clin Pharmacol. 2017;74(3):275–83 [C].

Heroin

[125] Suzuki J, Valenti ES. Intentional intra-arterial injection of heroin: a case report. J Addict Med. 2017;11(1):77–9 [A].
[126] Eyüpoğlu D, Ortaç Ersoy E, Rollas K, et al. Acute eosinophilic pneumonia secondary to heroin inhalation. Tuberk Toraks. 2017;65(2):154–6 [A].
[127] Fareed A, Kim J, Ketchen B, et al. Effect of heroin use on changes of brain functions as measured by functional magnetic resonance imaging, a systematic review. J Addict Dis. 2017;36(2):105–16 [M].
[128] Gerra LM, Gerra G, Mercolini L, et al. Increased oxytocin levels among abstinent heroin addicts: association with aggressiveness, psychiatric symptoms and perceived childhood neglect. Prog Neuropsychopharmacol Biol Psychiatry. 2017;75:70–6 [c].

[129] Lefaucheur R, Lebas A, Gérardin E, et al. Leukoencephalopathy following abuse of sniffed heroin. J Clin Neurosci. 2017; 35:70–2 [A].
[130] Sveinsson O, Herrman L, Hietala MA. Heroin-induced acute myelopathy with extreme high levels of CSF glial fibrillar acidic protein indicating a toxic effect on astrocytes. BMJ Case Rep. 2017; pii: bcr-2017-219903 [A].
[131] Melnychuk EM, Sole DP. A rare central nervous system fungal infection resulting from brown heroin use. J Emerg Med. 2017;52(3):314–7 [A].
[132] Chouk M, Vidon C, Deveza D, et al. Puffy hand syndrome. Joint Bone Spine. 2017;84(1):83–5 [A].
[133] Benns M, Miller K, Harbrecht B, et al. Heroin-related compartment syndrome: an increasing problem for acute care surgeons. Am Surg. 2017;83(9):962–5 [c].
[134] Levran O, Peles E, Randesi M, et al. The mu-opioid receptor nonsynonymous variant 118A > G is associated with prolonged abstinence from heroin without agonist treatment. Pharmacogenomics. 2017;18(15):1387–91 [C].

Opioid (synthetic)

[135] Morley KI, Ferris JA, Winstock AR, et al. Polysubstance use and misuse or abuse of prescription opioid analgesics: a multi-level analysis of international data. Pain. 2017;158(6):1138–44 [MC].
[136] European Monitoring Centre for Drugs and Drug Addiction. European Drug Report 2016: Trends and Developments. Lisbon (Portugal): European Monitoring Centre for Drugs and Drug Addiction. Retrieved from, http://www.emcdda.europa.eu/edr2016, 2016 [S].
[137] Drug Enforcement Administration. 2016 National Drug Threat Assessment Summary (DEA-DCT-DIR-001–17), Retrieved from https://www.dea.gov/resource-center/2016%20NDTA%20Summary.pdf; 2017 [S].
[138] Canadian Centre on Substance Abuse and Addiction. CCENDU Bulletin: Deaths involving fentanyl in Canada, 2009–2014. Ottawa (Canada): Canadian Centre on Substance Abuse. Retrieved from, http://www.ccsa.ca/Resource%20Library/CCSA-CCENDU-Fentanyl-Deaths-Canada-Bulletin-2015-en.pdf; 2015 [S].
[139] Kronstrand R, Druid H, Holmgren P, et al. A cluster of fentanyl-related deaths among drug addicts in Sweden. Forensic Sci Int. 1997;88(3):185–93 [c].
[140] Mounteney J, Giraudon I, Denissov G, et al. Fentanyl: are we missing the signs? Highly potent and on the rise in Europe. Int J Drug Policy. 2015;26(7):626–31 [H].
[141] Suzuki J, El-Haddad S. A review: fentanyl and non-pharmaceutical fentanyls. Drug Alcohol Depend. 2016;171:107–16 [H].
[142] Drug Enforcement Administration. DEA issues carfentanil warning to police and public, https://www.dea.gov/divisions/hq/2016/hq092216.shtml; 2016. DEA. Released 9/22/2016. [S].
[143] Kreeger TJ, Arnemo J, Raath JP. Handbook of wildlife chemical immobilization. International ed. Fort Collins: Wildlife Pharmaceuticals; 1999. p. 36–9 125–32. [H].
[144] Rotuno-Johnson M. Ohio has most carfentanil seizures in United States, 2016, Nexstar Broadcasting Channel 4 Columbus; 2016. Nov 3. Retrieved from, http://nbc4i.com/2016/11/03/ohio-has-most-carfentanil-seziures-in-united-states [H].
[145] Ougler J, Star S. Town hall meeting hears dangers of fentanyl, which is in algoma, as well as threat of carfentanil, 100 times more powerful and spotted in southern Ontario, 2017, Sault Star; 2017. Apr 4. Retrieved from, http://www.saultstar.com/2017/04/03/town-hall-meeting-hears-dangers-of-fentanyl-which-is-in-algoma-as-well-as-threat-of-carfentanil-100-times-more-powerful-and-spotted-in-southern-ontario [H].
[146] Olson J. Minnesota deaths rise despite attention, intervention. StarTribune, May 28. Retrieved from, http://www.startribune.com/minnesota-opioid-deaths-rise-despite-attention-intervention/424836053/; 2017 [H].
[147] United Nations Office on Drugs and Crime. Fentanyl and its analogues—50 years on. Glob Smart Update. 2017;17:3–7 [S].
[148] Helander A, Bäckberg M, Singell P, et al. Intoxications involving acrylfentanyl and other novel designer fentanyls—results from the Swedish STRIDA project. Clin Toxicol (Phila). 2017;55(6):589–99 [A].
[149] European Monitoring Centre for Drugs and Drug Addiction. EMCDDA–Europol joint report on a new psychoactive substance: N-(1-phenylethylpiperidin-4-yl)-N-phenylacrylamide (acryloylfentanyl). Lisbon (Portugal): European Monitoring Centre for Drugs and Drug Addiction. Retrieved from, http://www.emcdda.europa.eu/publications/joint-reports/acryloylfentanyl; 2017 [S].
[150] Lucyk SN, Nelson LS. Novel synthetic opioids: an opioid epidemic within an opioid epidemic. Ann Emerg Med. 2017;69(1):91–3 [H].
[151] Zawilska JB. An expanding world of novel psychoactive substances: opioids. Front Psychiatry. 2017;8:110 [M].
[152] Nielsen S, Van Hout MC. Over-the-counter codeine—from therapeutic use to dependence, and the grey areas in between. Curr Top Behav Neurosci. 2017;34:59–75 [H].
[153] Wang T, Ma J, Wangg R, et al. Poly-drug use of prescription medicine among people with opioid use disorder in China: a systematic review and meta-analysis. Subst Use Misuse. 2018;53(7):1117–27 [M].
[154] Árok Z, Csesztregi T, Sija E, et al. Changes in illicit, licit and stimulant designer drug use patterns in south-east Hungary between 2008 and 2015. Leg Med (Tokyo). 2017;28:37–44 [MC].
[155] Collier R. 16 hospitalizations daily for opioid poisoning. CMAJ. 2017;189:E1248 [S].
[156] Vogel L. Seniors and self-harm factor in the opioid crisis. CMAJ. 2017;189:E42–3 [c].
[157] Bennett AS, Elliott L, Golub A, et al. Opioid-involved overdose among male Afghanistan/Iraq-era U.S. military veterans: a multidimensional perspective. Subst Use Misuse. 2017;52(13):1701–11 [c].
[158] Austin AE, Shanahan ME, Zvara BJ. Association of childhood abuse and prescription opioid use in early adulthood. Addict Behav. 2018;76:265–9 [MC].
[159] Rashidian H, Hadji M, Marzban M, et al. Sensitivity of self-reported opioid use in case-control studies: healthy individuals versus hospitalized patients. PLoS One. 2017;13(2):e0192814 [C].
[160] Drug Enforcement Administration. DEA proposes reduction to amount of controlled substances to be manufactured in 2018, https://www.dea.gov/divisions/hq/2017/hq080417.shtml; 2017. DEA. Released 8/4/2017. [S].
[161] Food Drug Administration. FDA approves first once-monthly buprenorphine injection, a medication-assisted treatment option for opioid use disorder, https://www.fda.gov/NewsEvents/Newsroom/PressAnnouncements/ucm587312.htm; 2017. Released 11/30/2017. [S].
[162] President's Commission on Combating Drug Addiction and the Opioid Crisis. Interim report draft, Retrieved from, https://www.whitehouse.gov/sites/whitehouse.gov/files/ondcp/commission-interim-report.pdf; 2017 [S].
[163] Rutkow L, Vernick JS. Emergency legal authority and the opioid crisis. N Engl J Med. 2017;377:2412–4 [S].
[164] Dreifuss R, Sampaio J. Our international perspective on America's response to the opioid epidemic, 2017, Global commission on drug policy; 2017. Nov 3. Retrieved from, http://www.globalcommissionondrugs.org/our-international-perspective-on-americas-response-to-the-opioid-epidemic/ [S].
[165] Norway decriminalizes drug use. The Nordic Page. 2017, Norway: The Nordic Page; 2017. Dec 13. Retrieved from, https://www.tnp.no/norway/panorama/norway-decriminalize-drug [S].

[166] Wentlandt M, Morris SC, Mitchell SH. Ventricular tachycardia and prolonged QT interval presenting as seizure-like activity. Am J Emerg Med. 2017;35(5) 804.e5-6. [A].
[167] Belin N, Clairet A, Chocron S, et al. Refractory cardiogenic shock during tramadol poisoning: a case report. Cardiovasc Toxicol. 2017;17:219–22 [A].
[168] Mijatović V, Samojlik I, Petković S, et al. Cardiovascular effects of methadone and concomitant use of diazepam during methadone maintenance treatment induction: low concentration risk. Expert Opin Drug Saf. 2017;16(12):1323–8 [c].
[169] Prasad A, Lerman A, Rihal CS. Apical ballooning syndrome (Tako-Tsubo or stress cardiomyopathy): a mimic of acute myocardial infarction. Am Heart J. 2008;155(3):408–17 [H].
[170] Koulouris S, Pastromas S, Sakellariou D, et al. Takotsubo cardiomyopathy: the "broken heart" syndrome. Hellenic J Cardiol. 2010;51(5):451–7 [H].
[171] Olsen PC, Agarwal V, Lafferty JC, et al. Takotsubo cardiomyopathy precipitated by opiate withdrawal. Heart Lung. 2018;47(1):73–5 [A].
[172] Khaja M, Lominadze G, Millerman K. Cardiac arrest following drug abuse with intravenous tapentadol: case report and literature review. Am J Case Rep. 2017;18:817–21 [A].
[173] Shattuck S, Livingstone J. A case of cardiac tamponade due to an isolated abscess in the ascending aorta of a pregnant woman with a history of intravenous substance abuse. Forensic Sci Med Pathol. 2017;13:226–9 [A].
[174] Frimat M, Tabarin F, Dimitrov JD, et al. Complement activation by heme as a secondary hit for atypical hemolytic uremic syndrome. Blood. 2013;122(2):282–92 [E].
[175] Gordon RJ, Lowy FD. Bacterial infections in drug users. N Engl J Med. 2005;353(18):1945–54 [H].
[176] Thorlacius S, Mollnes TE, Garred P, et al. Plasma exchange in myasthenia gravis: changes in serum complement and immunoglobulins. Acta Neurol Scand. 1988;78(3):221–7 [c].
[177] Hunt R, Yalamanoglu A, Tumlin J, et al. A mechanistic investigation of thrombotic microangiopathy associated with IV abuse of Opana ER. Blood. 2017;129(7):896–905 [A].
[178] Wilson RM, Elmaraghi S, Rinker BD. Ischemic hand complications from intra-arterial injection of sublingual buprenorphine/naloxone among patients with opioid dependency. Hand (N Y). 2017;12(5):507–11 [c].
[179] Swank KA, Wu E, Kortepeter C, et al. Adverse event detection using the FDA post-marketing drug safety surveillance system: cardiotoxicity associated with loperamide abuse and issue. J Am Pharm Assoc. 2017;57(2S):S63–7 [c].
[180] Sheng J, Tran PN, Li Z. Characterization of loperamide-mediated block of hERG channels at physiological temperature and its proarrhythmia propensity. J Pharmacol Toxicol Methods. 2017;88:109–22 [E].
[181] Litovitz T, Clancy C, Korberly B, et al. Surveillance of loperamide ingestions: an analysis of 216 poison center reports. J Toxicol Clin Toxicol. 1997;35:11–9 [C].
[182] Regnard C, Twycross R, Mihalyo M, et al. Loperamide. J Pain Symptom Manage. 2011;42:319–23 [H].
[183] Wu PE, Juurlink DN. Clinical review: loperamide toxicity. Ann Emerg Med. 2017;70(2):245–52 [c].
[184] Bhatti Z, Norsworthy J, Szombathy T. Loperamide metabolite-induced cardiomyopathy and QTc prolongation. Clin Toxicol (Phila). 2017;55(7):659–61 [A].
[185] Fung HT, Cheung KH, Lam SK, et al. A case of unintentional methadone overdose followed by hypoglycaemia. Hong Kong J Emerg Med. 2011;18:239–42 [A].
[186] Gjedsted J, Dall R. Severe hypoglycemia during methadone escalation in an 8-year-old child. Acta Anaesthesiol Scand. 2015;59:1394–6 [A].
[187] Moryl N, Pope J, Obbens E. Hypoglycemia during rapid methadone dose escalation. J Opioid Manag. 2013;9:29–34 [c].
[188] Maingi S, Moryl N, Faskowitz A. Symptomatic hypoglycemia due to escalating doses of intravenous methadone. J Pain. 2008;9(2):37 [c].
[189] Faskowitz AJ, Kramskiy VN, Pasternak GW. Methadone-induced hypoglycemia. Cell Mol Neurobiol. 2013;33:537–42 [E].
[190] Li ATY, Chu FKC. A case of massive methadone overdose presented with refractory hypoglycemia. Clin Toxicol (Phila). 2017;55(3):233 [A].
[191] Masharani U, Alba D. Methadone-associated hypoglycemia in chronic renal failure masquerading as an insulinoma. Pain Med. 2018;19:1876–8 [A].
[192] Toce MS, Stefater MA, Breault DT, et al. A case report of methadone-associated hypoglycemia in an 11-month-old male. Clin Toxicol (Phila). 2018;56(1):74–6 [A].
[193] Tilbrook D, Jacob J, Parsons R, et al. Opioid use disorder and type 2 diabetes mellitus: effect of participation in buprenorphine-naloxone substitution programs on glycemic control. Can Fam Physician. 2017;63:e350–4 [C].
[194] Gotthardt F, Huber C, Thierfelder C, et al. Bone mineral density and its determinants in men with opioid dependence. J Bone Miner Metab. 2017;35:99–107 [C].
[195] Stange K, Krüger M, Janke E, et al. Positive association of personal distress with testosterone in opiate-addicted patients. J Addict Dis. 2017;36(3):167–74 [c].
[196] Kang J, Chung M. A stercoral perforation of the descending colon. J Korean Surg Soc. 2012;82(2):125 [A].
[197] Alexander-Williams J, Hollingworth J. Stercoral perforation of the colon. Br J Surg. 1991;68(6):763 [r].
[198] Poitras R, Warren D, Oyogoa S. Opioid drugs and stercoral perforation of the colon: case report and review of literature. Int J Surg Case Rep. 2017;42:94–7 [A].
[199] Havens JR, Oser CB, Leukefeld CG. Injection risk behaviors among rural drug users: implications for HIV prevention. AIDS Care. 2011;23(5):638–45 [C].
[200] Dickson-Spillmann M, Haug S, Uchtenhagen A, et al. Rates of HIV and hepatitis infections in clients entering heroin-assisted treatment between 2003 and 2013 and risk factors for hepatitis C infection. Eur Addict Res. 2015;22(4):181–91 [MC].
[201] Burt BA, Pai S. Sugar consumption and caries risk: a systematic review. J Dent Educ. 2001;65:1017–23 [M].
[202] Mysels DJ, Sullivan MA. The relationship between opioid and sugar intake: review of evidence and clinical applications. J Opi Manag. 2010;6(6):445–52 [H].
[203] Worobey J, Tepper BJ, Kanarek RB, editors. Dietary sugar and behavior. In: Nutrition and behavior: a multidisciplinary approach. Wallingford: CABI; 2006. p. 172–3 [H].
[204] Bogucka-Bonikowska A, Baran-Furga H, Chmielewska K, et al. Taste function in methadone maintained opioid dependent men. Drug Alcoh Depend. 2002;68(1):113–7 [c].
[205] Fraser AD, Zhang B, Khan H, et al. Prescription opioid abuse and its potential role in gross dental decay. Curr Drug Saf. 2017;12(1):22–6 [A].
[206] Cully JL. Born into addiction: neonatal abstinence syndrome and pediatric dentistry. Pediatr Dent. 2017;39(5):358–60 [H].
[207] Luong PM, Kalpakian B, Jaeger LJ, et al. Rhodoturula endogenous endophthalmitis: a novel harbinger of the injection drug epidemic in the United States. Case Rep Infect Dis. 2017;2017: Epub ahead of print. [A].
[208] Pinna A, Carta F, Zanetti S, et al. Endogenous rhodotorula minuta and Candida albicans endophthalmitis in an injecting drug user. Br J Ophthalmol. 2001;85(6):754 [A].
[209] Merkur AB, Hodge WG. Rhodotorula rubra endophthalmitis in an HIV positive patient. Br J Ophthalmol. 2002;86(12):1444–5 [A].
[210] Kelly AM, Kerr D, Dietze P, et al. Randomised trial of intranasal versus intramuscular naloxone in prehospital treatment for suspected opioid overdose. Med J Aust. 2005;182(1):24–7 [C].

[211] Kerr D, Kelly AM, Dietze P, et al. Randomized controlled trial comparing the effectiveness and safety of intranasal and intramuscular naloxone for the treatment of suspected heroin overdose. Addiction. 2009;104(12):2067–74 [C].
[212] Zuckerman M, Weisberg SN, Boyer EW. Pitfalls of intranasal naloxone. Prehosp Emerg Care. 2014;18(4):550–4 [A].
[213] Arora P, Sharma S, Garg S. Permeability issues in nasal drug delivery. Drug Discov Today. 2002;7(18):967–75 [H].
[214] Weber R, Keerl R, Radziwill R, et al. Videoendoscopic analysis of nasal steroid distribution. Rhinology. 1999;37(2):69–73 [c].
[215] Dowley AC, Homer JJ. The effect of inferior turbinate hypertrophy on nasal spray distribution to the middle meatus. Clin Otolaryngol Allied Sci. 2001;26(6):488–90 [c].
[216] Chen XB, Lee HP, Chong VF, et al. Impact of inferior turbinate hypertrophy on the aerodynamic pattern and physiological functions of the turbulent airflow—a CFD simulation model. Rhinology. 2010;48(2):163–8 [E].
[217] Robinson A, Wermeling DP. Intranasal naloxone administration for treatment of opioid overdose. Am Health Syst Pharm. 2014;71:2129–35 [AH].
[218] Weiner SG, Joyce AR, Thomson HN. The prevalence of nasal obstruction as a consideration in the treatment of opioid overdose. J Opioid Manag. 2017;13(2):69–76 [MC].
[219] Public Health Agency of Sweden and National Veterinary Institute. Swedres-Svarm. Consumption of antibiotics and occurrence of antibiotic resistance in Sweden, 2014, Folkhalsomyndigheten publication; 2014. 14027. Retrieved from, http://www.sva.se/globalassets/redesign2011/pdf/om_sva/publikationer/swedres_svarm2014.pdf [MC].
[220] Stenhem M, Ortqvist A, Ringberg H, et al. Epidemiology of methicillin-resistant Staphylococcus aureus (MRSA) in Sweden 2000–2003, increasing incidence and regional differences. BMC Infect Dis. 2006;6:30 [MC].
[221] El-Sharif A, Ashour HM. Community-acquired methicillin-resistant Staphylococcus aureus (CA-MRSA) colonization and infection in intravenous and inhalational opiate drug abusers. Exp Biol Med (Maywood). 2008;233:874–80 [c].
[222] Atkinson SR, Paul J, Sloan E, et al. The emergence of meticillin-resistant Staphylococcus aureus among injecting drug users. J Infect. 2009;58:339–45 [c].
[223] Saravolatz LD, Markowitz N, Arking L, et al. Methicillin-resistant Staphylococcus aureus. Epidemiologic observations during a community acquired outbreak. Ann Intern Med. 1982;96:11–6 [c].
[224] Dahlman D, Berge J, Nilsson AC, et al. Opioid and amphetamine dependence is associated with methicillin-resistant Staphylococcus aureus (MRSA): an epidemiological register study with 73,201 Swedish in- and outpatients 1997–2013. Infect Dis (Lond). 2017;49(2):120–7 [MC].
[225] Strioga M, Pasukoniene V, Characiejus D. CD8+ CD28- and CD8+ CD57+ T cells and their role in health and disease. Immunology. 2011;134(1):17–32 [H].
[226] Tsoukas C. Immunosenescence and aging in HIV. Curr Opin HIV AIDS. 2014;9(4):398–404 [H].
[227] Edelman EJ, So-Armah K, Cheng DM, et al. Impact of illicit opioid use on T cell subsets among HIV-infected adults. PLoS One. 2017;12(5):e0176617 [C].
[228] Schuppener LM, Pop-Vicas AE, Brooks EG, et al. Serratia marcencens bactermia: nosocomial cluster following narcotic diversion. Infect Control Hosp Epidemiol. 2017;38(9):1027–31 [c].
[229] Joint United Nations Programme on HIV/AIDS (UNAIDS) (2016), Global AIDS Update 2016. UNAIDS, Geneva, Switzerland. Accessed on May 28, 2016 at: http://www.unaids.org/sites/default/files/media_asset/global-AIDS-update-2016_en.pdf [C].
[230] Polonsky M, Azbel L, Wegman MP, et al. Pre-incarceration police harassment, drug addiction and HIV risk behaviours among prisoners in Kyrgyzstan and Azerbaijan: results from a nationally representative cross-sectional study. J Int AIDS Soc. 2016;19(4 Suppl 3):20880 [C].
[231] Kiriazova TK, Postnov OV, Perehinets IB, et al. Association of injecting drug use and late enrolment in HIV medical care in Odessa Region. Ukraine HIV Med. 2013;14(3):38–41 [MC].
[232] Mazhnaya A, Andreeva TI, Samuels S, et al. The potential for bridging: HIV status awareness and risky sexual behaviour of injection drug users who have non-injecting permanent partners in Ukraine. J Int AIDS Soc. 2014;17:18825 [MC].
[233] Marcus R, Makarenko I, Mazhnaya A, et al. Patient preferences and extended-release naltrexone: a new opportunity to treat opioid use disorders in Ukraine. Drug Alcohol Depend. 2017;179:213–9 [MC].
[234] Vivaldellli E, Sansone S, Maier A, et al. Puffy hand syndrome. Lancet. 2017;389:298 [A].
[235] Mastroianni A, Ravaglia G. Serotonin syndrome due to co-administration of linezolida and methadone. Le Infezioni in Medicina. 2017;3:263–6 [A].
[236] Ramsey TD, Lau TT, Ensom MH. Serotonergic and adrenergic drug interactions associated with linezolid: a critical review and practical management approach. Ann Pharmacother. 2013;47(4):543–60 [R].
[237] Rastogi R, Swarm R, Patel TA. Case scenario: opioid association with serotonin syndrome: implications to the practitioners. Anesthesiology. 2011;115(6):1291–8 [A].
[238] Skovlund E, Handal M, Selmer R, et al. Language competence and communication skills in 3-year-old children after prenatal exposure to analgesic opioids. Pharmacoepidemiol Drug Saf. 2017;26(6):625–34 [MC].
[239] Sirnes E, Oldedal L, Bartsch H, et al. Brain morphology in school-aged children with prenatal opioid exposure: a structural MRI study. Early Hum Dev. 2017;106–107:33–9 [c].
[240] Savulich G, Riccelli R, Passamonti L, et al. Effects of naltrexone are influenced by childhood adversity during negative emotional processing in addiction recovery. Transl Psychiatry. 2017;7(3): E1054 [c].
[241] Schmidt P, Haberthür A, Soyka M. Cognitive functioning in formerly opioid-dependent adults after at least 1 year of abstinence: a naturalistic study. Eur Addict Res. 2017;23:269–75 [c].
[242] Shrestha R, Weikum D, Copenhaver M, et al. The influence of neurocognitive impairment, depression, and alcohol use disorders on health-related quality of life among incarcerated, HIV-infected, opioid dependent Malaysian men: a moderated mediation analysis. AIDS Behav. 2017;21:1070–81 [C].
[243] Stein MD, Conti MT, Kenney S. Adverse childhood experience effects on opioid use initiation, injection drug use and overdose among persons with opioid use disorder. Drug Alcohol Depend. 2017;179:325–9 [C].
[244] Mccarthy JF, Valenstein M, Kim HM, et al. Suicide mortality among patients receiving care in the veterans health administration health system. Am J Epidemiol. 2009;169:1033–8 [MC].
[245] Hoffmire CA, Kemp JE, Bossarte RM. Changes in suicide mortality for veterans and nonveterans by gender and history of VHA service use, 2000–2010. Psychiatr Serv. 2015;66:959–65 [MC].
[246] Bohnert KM, Ilgen MA, Louzon S, et al. Substance use disorders and the risk of suicide mortality among men and women in the US Veterans Health Administration. Addiction. 2017;112(7):1193–201 [MC].
[247] Kuip EJ, Zandvliet ML, Koolen SL, et al. A review of factors explaining variability in fentanyl pharmacokinetics; focus on implications for cancer patients. Br J Clin Pharmacol. 2016;83(2):294–313 [R].

[248] Trist AJ, Sahota H, Williams L. Not so patchy story of attempted suicide…leading to 24 hours of deep sleep and survival! BMJ Case Rep. 2017;1–3 [A].
[249] Rahmati A, Shakeri R, Khademi H, et al. Mortality from respiratory diseases associated with opium use: a population-based cohort study. Thorax. 2017;72:1028–34 [C].
[250] Kintz P. A new series of 13 buprenorphine-related deaths. Clin Biochem. 2002;35:513–6 [A].
[251] Pirnay S, Borron SW, Giudicelli CP, et al. A critical review of the causes of death among post-mortem toxicological investigations: analysis of 34 buprenorphine-associated and 35 methadone-associated deaths. Addiction. 2004;99:978–88 [c].
[252] Ohtani M, Kotaki H, Nishitateno K, et al. Kinetics of respiratory depression in rats induced by buprenorphine and its metabolite, norbuprenorphine. J Pharmacol Exp Ther. 1997;281:428–33 [E].
[253] Mégarbane B, Hreiche R, Pirnay S, et al. Does high-dose buprenorphine cause respiratory depression? Possible mechanisms and therapeutic consequences. Toxicol Rev. 2006;25:79–85 [H].
[254] Alhaddad H, Cisternino S, Declèves X, et al. Respiratory toxicity of buprenorphine results from the blockage of p-glycoprotein-mediated efflux of norbuprenorphine at the blood-brain barrier in mice. Crit Care. 2012;40:3215–23 [E].
[255] Strang J, Knight A, Baillie S, et al. Norbuprenorphine and respiratory depression: exploratory analyses with new lyophilized buprenorphine and sublingual buprenorphine. Int J Clin Pharmacol Ther. 2018;56(2):81–5 [E].
[256] Toce MS, Burns MM, O'Donnell KA. Clinical effects of unintentional pediatric buprenorphine exposure at a single tertiary care center. Clin Toxicol. 2017;55(1):12–7 [c].
[257] Heppell SPE, Isbister GK. Lack of respiratory depression in paracetamol-codeine combination overdoses. Br J Clin Pharmacol. 2017;83:1273–8 [MC].
[258] Babak K, Mohammad A, Mazaher G, et al. Clinical and laboratory findings of rhabdomyolysis in opioid overdose patients in the intensive care unit of a poisoning center in 2014 in Iran. Epidemiol Health. 2017;39:e2017050 [C].
[259] Massey J, Kilkenny M, Batdorf S, et al. Opioid overdose outbreak—West Virginia, August 2016. MMWr Morb Mortal Wkly Rep. 2017;66(37):975–80 [A].
[260] Centers for Disease Control and Prevention. Increases in fentanyl drug confiscations and fentanyl-related overdose fatalities, 2015, Centers for Disease Control and Prevention; 2015. https://emergency.cdc.gov/han/han00384.asp. Released 10/26/2015. [S].
[261] Somerville NJ, O'Donnell J, Gladden RM, et al. Characteristics of fentanyl overdose—Massachusetts, 2014–2016. MMWR Morb Mortal Wkly Rep. 2017;66(14):382–6 [c].
[262] Hempstead K, Yildirim EO. Supply-side response to declining heroin purity: fentanyl overdose episode in New Jersey. Health Econ. 2013;23(6):688–705 [R].
[263] Soloman R. 20-year-old woman with severe opioid toxicity. J Emerg Nurs. 2018;44(1):77–8 [A].
[264] Helander A, Bradley M, Hasselblad A, et al. Acute skin and hair symptoms followed by severe, delayed complication in subjects using the synthetic opioid MT-45. Br J Pharmacol. 2017;176:1021–7 [A].
[265] Szmuszkovicz J. U-50,488 and the receptor: a personalized account covering the period 1973 to 1990. Prog Drug Res. 1999;52:167–95 [E].
[266] Domanski K, Kleinschmidt KC, Schulte JM, et al. Two cases of intoxication with new synthetic opioid, U47700. Clin Toxicol. 2017;55(1):46–50 [A].
[267] Ellefsen KN, Taylor EA, Simmons P, et al. Multiple drug-toxicity involving novel psychoactive substances, 3-fluorophenmetrazine and U-47700. J Anal Toxicol. 2017;41(9):765–70 [A].
[268] Rambaran KA, Fleming SW, An J, et al. U-47700: a clinical review of the literature. J Emerg Med. 2017;53(4):509–19 [c].
[269] Seither J, Reidy L. Confirmation of carfentanil, U-47700 and other synthetic opioids in a human performance case by LC-MS-MS. J Anal Toxicol. 2017;41(6):493–7 [A].
[270] Prilutskaya M, Bersani FS, Coazza O, et al. Impact of synthetic cannabinoids on the duration of opioidrelated withdrawal and craving among patients of addiction clinics in Kazakhstan—a prospective case-control study. Hum Psychopharmacol. 2017;32(3):1–9 [C].
[271] Oviedo-Joekes E, Brissett S, MacDonald S, et al. Safety profile of injectable hydropmorphine and diacetylmorphine for long term severe opioid use disorder. Drug Alcohol Depend. 2017;176:55–62 [C].
[272] Marraffa JM, Holland MG, Sullivan RW, et al. Cardiac conduction disturbance after loperamide abuse. Clin Toxicol. 2014;52(9):952–7 [c].
[273] Enakpene EO, Riaz IB, Shirazi FM, et al. The long QT teaser: loperamide abuse. Am J Med. 2015;128(10):1083–6 [A].
[274] Klein MG, Haigney MCP, Mehler PS, et al. Potent inhibition of hERG channels by the over-the-counter antidiarrheal agent loperamide. JACC Clin Electrophysiol. 2016;2(7):784–9 [E].
[275] Eggleston W, Clark KH, Marraffa JM. Loperamide abuse associated with cardiac dysrhythmia and death. Ann Emerg Med. 2017;69(1):83–6 [A].
[276] Leo RJ, Ghazi MA, Jaziri KS. Methadone management of withdrawal associated with loperamide-related opioid use disorder. J Addict Med. 2017;11(5):402–4 [A].
[277] Sullivan M, Bisaga A, Pavlicova M. Long-acting injectable naltrexone induction: a randomized trial of outpatient opioid detoxification with naltrexone versus buprenorphine. Am J Psychiatry. 2017;174(5):459–67 [C].
[278] Williams AR, Barbieri V, Mishlen K. Long-term follow-up study of community-based patients receiving XR-NTX for opioid use disorders. Am J Addict. 2017;26(4):319–25 [c].
[279] Miyazaki T, Choi IY, Rubas W, et al. NKTR-181: a novel mu-opioid analgesic with inherently low abuse potential. J Pharmacol Exp Ther. 2017;363(1):104–13 [E].
[280] Webster L, Henningfield J, Buchhalter AR, et al. Human abuse potential of the new opioid analgesic molecule NKTR-181 compared with oxycodone. Pain Med. 2018;19(2):307–18 [c].
[281] Wilson MW, Bonnecaze AK, Dharod A, et al. Analysis of intensive care unit admission and sequelae in patients intravenously abusing extended-release oral oxymorphone. South Med J. 2017;11(3):217–22 [c].
[282] Setnick B, Schoedel K, Bartlett C, et al. Intranasal abuse potential of an abuse-deterrent oxycodone formulation compared to oxycodone immediate release and placebo in nondependent, recreational opioid users. J Opioid Manag. 2017;13(6):449–64 [c].
[283] Setnick B, Bass A, Bramson C, et al. Abuse potential study of ALO-02 (extended-release oxycodone surrounding sequestered naltrexone) compared with immediate-release oxycodone administered orally to nondependent recreational opioid users. Pain Med. 2017;18(6):1077–88 [c].
[284] Vosburg SK, Severtson SG, Dart RC, et al. Assessment of tapentadol API abuse liability with the researched abuse, diversion and addiction-related surveillance (RADARS) system. J Pain. 2018;19(4):439–53 [MC].
[285] Coplan MP, Sessler NE, Harikrishnan V, et al. Comparison of abuse, suspected suicidal intent, and fatalities related to the 7-day buprenorphine transdermal patch versus other opioid analgesics in the National Poison Data System. Postgrad Med. 2017;129(1):55–61 [MC].

[286] Strang J, Reed K, Bogdanowicz K, et al. Randomised comparison of a novel buprenorphine oral lyophilisate versus existing buprenorphine sublingual tablets in opioid-dependent patients—a first-in-patient phase II randomised open label safety study. Eur Addict Res. 2017;23(2):61–70 [c].
[287] Tiberg F, Johnsson M, Nistor C, et al. Self-assembling liquid formulations. In: Wright JC, Burgess DJ, editors. Long-acting injections and implants. New York: Springer; 2012. p. 315–34 [H].
[288] Tiberg F, Johnsson M, Jankunec M, et al. Phase behavior, functions, and medical applications of soy phosphatidylcholine and diglyceride lipid compositions. Chem Lett. 2012;41(10):1090–2 [E].
[289] Walsh SL, Comer SD, Lofwall MR, et al. Effect of buprenorphine weekly depot (CAM20138) and hydropmorphone blockade in individuals with opioid use disorder: a randomized clinical trial. JAMA Psychiatry. 2017;74(9):894–902 [c].
[290] Howland MA, Neslon L. Opioid antagonists. In: Nelson LS, Lewin NA, Howland MA, et al., editors. Goldfrank's toxicologic emergencies. 9th ed. New York: McGraw-Hill; 2011. p. 579–85 [H].
[291] Hassanian-Moghaddam H, Afzali S, Pooya A. Withdrawal syndrome caused by naltrexone in opioid abusers. Hum Exp Toxicol. 2013;33(6):561–7 [C].
[292] Zamani N, Hassanjan-Moghaddam H. Intravenous buprenorphine: a substitute for naloxone in the methadone-overdosed patients? Ann Emerg Med. 2017;69(6):737–9 [A].
[293] Loudin S, Murray S, Prunty L, et al. Atypical withdrawal syndrome in neonates prenatally exposed to gabapentin and opioids. J Pediatr. 2017;181:286–8 [c].
[294] Hensley L, Sulo S, Kozmic S, et al. Opioid addiction in pregnancy: does depression negatively impact adherence with prenatal care? J Addict Med. 2018;12(1):61–4 [c].
[295] Jones H, Johnson R, Jasinski D, et al. Buprenorphine versus methadone in the treatment of pregnant opioid-dependent patients: effects on the neonatal abstinence syndrome. Drug Alcohol Depend. 2005;79(1):1–10 [c].
[296] Jones H, Kaltenbach K, Heil S, et al. Neonatal abstinence syndrome after methadone or buprenorphine exposure. N Engl J Med. 2010;363(24):2320–31 [C].
[297] Jones H, Heil S, Baewert A, et al. Buprenorphine treatment of opioid-dependent pregnant women: a comprehensive review. Addiction. 2012;107(1):5–27 [c].
[298] Zedler BK, Mann AL, Kim MM, et al. Buprenorphine compared with methadone to treat pregnant women with opioid use disorder: a systematic review and meta-analysis of safety in the mother, fetus and child. Addiction. 2016;111(12):2115–28 [M].
[299] Vilkins AL, Baley SM, Hahn KA, et al. Comparison of post-cesarean section opioid analgesic requirements in women with opioid use disorder treated with methadone or buprenorphine. J Addict Med. 2017;11(5):397–401 [C].
[300] Leibrock J, Lottspeich F, Hohn A, et al. Molecular cloning and expression of brain-derived neurotrophic factor. Nature. 1989;341(6238):149–52 [H].
[301] Hohn A, Leibrock J, Bailey K, et al. Identification and characterization of a novel member of the nerve growth factor/brain-derived neurotrophic factor family. Nature. 1990;344(6264):339–41 [H].
[302] Twiss JL, Chang JH, Schanen NC. Pathophysiological mechanisms for actions of the neurotrophins. Brain Pathol. 2006;16(4):320–32 [H].
[303] Corominas-Roso M, Roncero C, Eiroa-Orosa FJ, et al. Brain-derived neurotrophic factor serum levels in cocaine-dependent patients during early abstinence. Eur Neuropsychopharmacol. 2013;23(9):1078–84 [c].
[304] Kim DJ, Roh S, Kim Y, et al. High concentrations of plasma brain-derived neurotrophic factor in methamphetamine users. Neurosci Lett. 2005;388(2):112–5 [c].
[305] Ghitza UE, Zhai H, Wu P, et al. Role of BDNF and GDNF in drug reward and relapse: a review. Neurosci Biobehav Rev. 2010;35(2):157–71 [H].
[306] Subedi L, Huang H, Pant A, et al. Plasma brain-derived neurotrophic factor levels in newborn infants with neonatal abstinence syndrome. Front Pediatr. 2017;5:238 [MC].
[307] Zuzarte I, Indic P, Barton B, et al. Vibrotactil stimulation: a non-pharmacological intervention for opioid-exposed newborns. PLoS One. 2017;12(4):e0175981 [c].
[308] Pilgrim JL, Jenkins EL, Baber Y, et al. Fatal acute poisonings in Australian children (2003 – 13). Addiction. 2017;112(4):627–39 [c].
[309] Fels H, Krueger J, Sachs H, et al. Two fatalities associated with synthetic opioids: AH-7921 and MT-45. Forensic Sci. 2017;277: e30–5 [A].
[310] Guerrieri D, Rapp E, Roman M, et al. Postmortem and toxicological findings in a series of furanylfentanyl-related deaths. J Anal Toxicol. 2017;41(3):242–9 [c].
[311] Krotulski AJ, Papsun DM, Friscia M, et al. Fatality following ingestion of tetrahydrofuranylfentanyl, U-49900 and methoxy-phencyclidine. J Anal Toxicol. 2018;42(3): e27–e32 [A].
[312] Centers for Disease Control and Prevention, National Center for Health Statistics. Drug Overdose Deaths in the United States, 1999–2015. NCHS; 2017 Data Brief No. 273, February. [MC].
[313] CDC WONDER [Internet]. Multiple causes of death 1999–2014. Atlanta (GA): Centers for Disease Control and Prevention; 2018. [cited 2018 May 10]. Available from, https://wonder.cdc.gov/wonder/help/mcd.html [MC].
[314] Martucci HFH, Ingle EA, Hunter MD, et al. Distribution of furanyl fentanyl and 4-ANPP in an accidental acute death: a case report. Forensic Sci Int. 2018;283:e13–7 [A].
[315] Austin AE, Proescholdbell SK, Creppage KE, et al. Characteristics of self-inflicted drug overdose deaths in North Carolina. Drug Alcohol Depend. 2017;181:44–9 [MC].
[316] Vardanyan RS, Hruby VJ. Fentanyl-related compounds and derivatives: current status and future prospects for pharmaceutical applications. Future Med Chem. 2014;6(4):385–412 [H].
[317] European Monitoring Centre for Drugs and Drug Addiction. EMCDDA-Europol 2014 annual report on the implementation of council decision 2005/387/JHA. Lisbon (Portugal): European Monitoring Centre for Drugs and Drug Addiction. Retrieved from, http://www.emcdda.europa.eu/system/files/publications/1018/TDAN15001ENN.pdf; 2015 [S].
[318] Rojkiewicz M, Majchrzak M, Celiński R, et al. Identification and physicochemical characterization of 4-fluorobutyrfentanyl (1-((4-fluorophenyl)(1-phenethylpiperidin-4-yl)amino)butan-1-one, 4-FBF) in seized materials and post-mortembiological samples. Drug Test Anal. 2017;9(3):405–14 [A].
[319] Abrahamsson T, Berge J, Öjehagen A, et al. Benzodiazepine, z-drug and pregabalin prescriptions and mortality among patients in opioid maintenance treatment—a nation-wide register-based open cohort study. Drug Alcohol Depend. 2017;174:58–64 [MC].
[320] Haukka J, Kriikku P, Mariottini C, et al. Non-medical use of psychoactive prescription drugs is associated with fatal poisoning. Addiction. 2017;113(3):464–72 [MC].
[321] Hopkins RE, Dobbin M, Pilgrim JL. Unintentional mortality associated with paracetamol and codeine preparations, with and without doxylamine, in Australia. Forensic Sci Int. 2017;282:122–6 [C].
[322] Bukten A, Stavseth MR, Skurtveit S, et al. High risk of overdose death following release from prision: variations in mortality during a 15-year observational period. Addiction. 2017;112(8):1432–9 [MC].

[323] Forsyth SJ, Carroll M, Lennox N, et al. Incidence and risk factors for morality after release from prison in Australia: a prospective cohort study. Addiction. 2017;113(5):937–45 [MC].

[324] Kendler KS, Ohlsson H, Sudquist K, et al. Drug abuse-associated mortality across the lifespan: a population-base longitudinal cohort and co-relative analysis. Soc Psychiatry Psychiatr Epidemiol. 2017;52(7):877–86 [MC].

[325] Shipton EE, Shipton AJ, Williman JA, et al. Deaths from opioid overdosing: implications of coroners' inquest reports 2008–2012 and annual rise in opioid prescription rates: a population-based cohort study. Pain Ther. 2017;6:203–15 [MC].

[326] Nishizawa D, Mieda T, Tsujita M, et al. Genome-wide scan identifies candidate loci related to remifentanil requirements during laparoscopic-assisted colectomy. Pharmacogenomics. 2018;19(2):113–27 [C].

[327] Hajj A, Halepian L, Osta NE, et al. OPRM1 c.118A>G polymorphism and duration of morphine treatment associated with morphine doses and quality-of-life in palliative cancer pain settings. Int J Mol Sci. 2017;18(4):27 [c].

[328] Al-Mustafa MM, Al Oweidi AS, Al-Zaben KR, et al. Remifentanil consumption in septoplasty surgery under general anesthesia. Association with humane mu-opioid receptor gene variants. Saudi Med J. 2017;38(2):170–5 [C].

[329] Horvat CM, Au AK, Conley YP, et al. ABCB1 genotype is associated with fentanyl requirements in critically ill children. Pediatr Res. 2017;82(1):29–35 [c].

[330] Luo R, Li X, Qin S, et al. Impact of SNP-SNP interaction among ABCB1, ARRB2, DRD1 and OPRD1 on methadone dosage requirement in Han Chinese patients. Pharmacogenomics. 2017;18(18):1659–70 [C].

[331] Balyan R, Zhang X, Chidambaran V, et al. OCT1 genetic variants are associated with postoperative morphine-related adverse effects in children. Pharmacogenomics. 2017;18(7):621–9 [C].

[332] Ahmad T, Sabet S, Primerano DA, et al. Tell-tale SNPs: the role of CYP2B6 in methadone fatalities. J Anal Toxicol. 2017;41(4):325–33 [C].

[333] Balyan R, Mecoli M, Venkatasubramanian R, et al. CYP2D6 pharmacogenetic and oxycodone pharmacokinetic association study in pediatric surgical patients. Pharmacogenomics. 2017;18(4):337–48 [c].

[334] St Sauver JL, Olson JE, Roger VL, et al. CYP2D6 phenotypes are associated with adverse outcomes related to opioid medications. Pharmacogenomics Pers Med. 2017;10:217–27 [C].

[335] Zhang H, Chen M, Wang X, et al. Patients with CYP3A4 1G genetic polymorphism consumed significantly lower amount of sufentanil in general anesthesia during lung resection. Medicine (Baltimore). 2017;96(4):e6013 [C].

[336] Patel J. Drugs of abuse. In: Ray SD, editor. Side effects of drugs annual: a worldwide yearly survey of new data in adverse drug reactions. vol. 36. Elsevier; 2014. p. 37–52. Chapter 4. [R].

[337] Fudin HR, Babin JL, Hansen AL, et al. Drugs of abuse. In: Ray SD, editor. Side effects of drugs annual: a worldwide yearly survey of new data in adverse drug reactions. vol. 39. Elsevier; 2017. p. 31–55. Chapter 4. [R].

CHAPTER

5

Hypnotics and Sedatives

*Brian W. Skinner**,†,1, *Katherine A. Curtis*†, *Angela L. Goodhart*†

*Manchester University College of Pharmacy, Natural & Health Sciences, Fort Wayne, IN, United States

†St. Vincent Indianapolis Hospital, Indianapolis, IN, United States

[1]Corresponding author: bwskinner@manchester.edu

BENZODIAZEPINES AND NON-BENZODIAZEPINE GAMMA-AMINOBUTYRIC ACID (GABA) AGONISTS [SEDA-37, 5; SEDA-38, 4; SEDA-39, 5]

Review

Benzodiazepines and non-benzodiazepine GABA agonists, commonly referred to as z-hypnotics due to their generic names (i.e. zolpidem, zaleplon, zopiclone, etc.), are utilized as potential treatment options for insomnia. Additionally, benzodiazepines are commonly used for the treatment of anxiety. Benzodiazepines have been associated with increased risk of cardiac arrest or transfer to an ICU during a hospitalization, and both benzodiazepines and z-hypnotics have been associated with more serious dispositions for emergency department patients compared to visits due to other agents classified as "anxiolytics, sedatives, and hypnotics" by the Drug Abuse Warning Network *Multum* Lexicon Drug Reference Vocabulary [1C, 2MC].

A review article published in September 2017 evaluated the evidence for adverse effects commonly reported with benzodiazepines and z-hypnotics [3R]. Based on the evidence presented in 143 articles, the authors concluded that psychomotor impairment due to benzodiazepine or z-hypnotic use has a strong causal relationship to motor vehicle accidents, falls, and fractures. Other adverse effects reported have mixed evidence regarding their association with benzodiazepine and z-hypnotic use and have no widely accepted proposed mechanism for why they occur; these include dementia, infections and immune dysregulation, acute pancreatitis, exacerbation of respiratory diseases, and cancer.

Musculoskeletal System

A systematic review and meta-analysis of 18 studies reported increased risk of hip fractures in elderly patients taking benzodiazepines or z-hypnotics [4M]. The relative risk of hip fracture was significantly higher with use of both benzodiazepines and z-hypnotics (RR 1.52, 95% CI 1.37–1.68; RR 1.90, 95% CI 1.68–2.13) in populations over the age of 50, although no studies directly compared this outcome between the two drug classes. A subgroup analysis showed increased risk of fractures with short-term use of either drug class compared to long-term use. The authors attribute the increase in risk of fractures to the increased fall risk associate with the drug classes.

Infection Risk

A case-control study used information from the Taiwan National Health Insurance Database to look at the possible association between benzodiazepine or z-hypnotic use and development of pneumonia in patients with chronic kidney disease [5C]. The study classified use of benzodiazepines or z-hypnotics as current, recent, past, or remote use. Increased risk of pneumonia was associated with current use of benzodiazepines, new initiation of benzodiazepines, new use of z-hypnotics, and new use of benzodiazepines and z-hypnotics combined with significant odds ratios of 1.31, 2.47, 2.94, and 2.47, respectively, when compared to nonusers.

Another case-control study in Finland used data from the Medication Use and Alzheimer Disease Cohort to assess benzodiazepine or z-hypnotic use and risk of hospitalization or death due to pneumonia in community-dwelling adults with recent diagnosis of Alzheimer

Side Effects of Drugs Annual, Volume 40
ISSN: 0378-6080
https://doi.org/10.1016/bs.seda.2018.06.002

disease [6C]. Use of benzodiazepines and z-hypnotics was associated with an increased risk of pneumonia when analyzed collectively. However, when the drugs were separated, only benzodiazepines were significantly associated with pneumonia (HR 1.28, CI 1.07–1.54), while z-hypnotics did not have a significant difference (HR 1.10, CI 0.84–1.44). It was also found that the first 30 days of benzodiazepine use carried the greatest risk of pneumonia (HR 2.09, CI 1.26–3.48).

BENZODIAZEPINES [SEDA-37, 5; SEDA-38, 4; SEDA-39, 5]

Special Review

Drug addiction and abuse has been on the rise in the United States over the past few decades. From 1999 to 2010, there has been a fourfold increase in the number of deaths attributed to benzodiazepine overdose in the United States (0.58 deaths per 100,000 adults to 3.07 deaths per 100,000 adults). This risk is compounded substantially when benzodiazepines are used in combination with other substances, especially opioids. This prompted the USFDA to release a drug-safety communication in 2016 regarding the risks of combining opioids with benzodiazepines. The abrupt discontinuation of benzodiazepines can cause the development of a potentially fatal withdrawal syndrome, similar in clinical presentation to alcohol withdrawal. Onset of withdrawal develops within 2–3 days after discontinuation of short-acting benzodiazepines and within 5–10 days after discontinuation of long-acting benzodiazepines. Both somatic symptoms (i.e. weakness, flu-like symptoms, tachycardia, nausea/vomiting) and psychological symptoms (i.e. anxiety, hallucinations, depression) may be present. The most serious symptoms of withdrawal, however, include seizures which can potentially be fatal. Treatment of benzodiazepine withdrawal requires gradual discontinuation of benzodiazepines over 4–8 weeks to minimize or prevent withdrawal syndromes. Various strategies have been employed with reductions ranging from 10% to 50% per week, though it is recommended to avoid prolonged reductions over several months. Additionally, while it is known that patients withdrawing from short-acting benzodiazepines have a greater risk of relapse, little evidence suggests utilization of a benzodiazepine with a longer half-life is associated with increased rates of successful withdrawal or a reduction in relapse. In patients taking multiple benzodiazepines, a moderate amount of evidence suggests it is best to consolidate to one benzodiazepine, preferably diazepam [7R].

For those who successfully discontinue use of benzodiazepines, long-term cognitive effects persist well after benzodiazepine discontinuation. A meta-analysis published in 2017 compared patients without a history of substance abuse with current benzodiazepine users and users that were experiencing short-term withdrawal (<3 months), medium-term withdrawal (3–12 months), or long-term withdrawal (>12 months). Compared to benzodiazepine non-users, current benzodiazepine use was associated with decreased working memory, processing speed, divided attention, visuoconstruction, recent memory, and expressive language ($P<0.05$). After sustained cessation of benzodiazepine use (>12 months), subjects were found to continue to have deficits in recent memory, processing speed, visuoconstruction, divided attention, and working memory ($P<0.05$). This analysis underscores the potential long-term cognitive effects that can persist after benzodiazepine cessation [8M].

Central Nervous System

A retrospective case-control study examined the risk of falls in 132 hospitalized patients over the age of 45 receiving benzodiazepines [9C]. The ratio of patients included who did not fall to those who experienced a fall was 3:1. Patients who experienced falls were found to have a higher likelihood of receiving either a new benzodiazepine or an increased dose of a benzodiazepine compared to the patients who did not fall (63.6% vs 41.4%, $P=0.043$). The half-life of the specific benzodiazepine received, total daily dose in lorazepam equivalents, and age of the patient had no significant effects on fall risk. While benzodiazepines are commonly attributed to increasing the risk of fall, the results of this study suggest the risk is greatest upon initiation or upon dose titration.

Cardiovascular System

Maruyoshi and colleagues detailed a case report of an 87-year-old woman who experienced bradycardia secondary to clonazepam [10A]. The woman received clonazepam 0.5 mg by mouth three times daily for the treatment of epilepsy. Upon presentation to the hospital, her pulse was noted to be 32 beats per minute. An electrocardiogram demonstrated first-degree atrioventricular block, p wave abnormalities, and ST elevations in leads II, III, aVF, and V6. Additionally, the patient's serum clonazepam level was noted to be within the therapeutic target of 20–70 ng/mL (34.8 ng/mL). The only medication change that occurred during the initial episode was the discontinuation of clonazepam. Her pulse gradually improved to 47, 53, 64, and 78 beats per minute over the course of 12, 24, 48, and 72 h, respectively. It is theorized that clonazepam may block T-type calcium channels in the sinoatrial and atrioventricular nodes which could lead to bradycardia.

Respiratory System

Procedural sedation and analgesia is common practice in the emergency department (ED) for many painful procedures that require a short duration of sedation.

A retrospective cohort study of procedural sedation in the ED examined the safety of midazolam for procedural sedation compared to propofol [11c]. Patients receiving midazolam were less likely to achieve a deep level of sedation (25% vs 45%, $P<0.001$), experienced less successful procedures (81% vs 92%, $P<0.001$), and were sedated for a longer duration (17min vs 1min, $P<0.001$). Additionally, more patients receiving midazolam experienced oxygen desaturation (defined as a drop to <90%) when compared with propofol (8% vs 1%, $P=0.001$).

A case non-case study examined the incidence of sleep apnea syndrome (SAS) adverse drug reactions to identify drugs and classes of drugs associated with SAS using the World Health Organization pharmacovigilance database [12c]. Of the 3325 reported cases of SAS included in the study, 65 (2%) were related to benzodiazepine derivatives. The authors attribute this incidence of SAS adverse drug reaction reports to the depressive effects of benzodiazepines on the respiratory system.

Immunologic System

Perioperative hypersensitivity reactions are often attributed to antibiotics. Thirty four patients experiencing periprocedural hypersensitivity reactions underwent a comprehensive evaluation that included IgE testing of common periprocedural medications including neuromuscular blockers, anesthetic agents (including propofol, etomidate, midazolam, and ketamine), opiates, beta-lactam antibiotics, local anesthetics, and ondansetron [13c]. IgE-mediated causes were noted in 22 patients (64.7%) with a total of 25 positive tests. Induction agents were the most frequently implicated medications (36%) with midazolam being implicated in 6 positive tests (24%).

Prenatal Exposure

Brandlistuen and colleagues used data from the Norwegian Mother and Child Cohort Study to perform a cohort and sibling control study on the effects of prenatal exposure to benzodiazepines or z-hypnotics [14C]. The study used The Child Behavior Checklist to measure internalizing and externalizing behavior problems at 0.5, 1.5, and 3years of age. Results were adjusted using scores of siblings without exposure to either drug class in order to control potential bias from unobservable family variables such as environment and genetic predisposition. It was found that children with long-term prenatal exposure to benzodiazepines had increased internalizing behaviour problems at 1.5 and 3years old. The same increase was not seen with z-hypnotics.

NON-BENZODIAZEPINE GABA AGONISTS [SEDA-37, 5; SEDA-38, 4; SEDA-39, 5]

Central Nervous System

Researchers in Taiwan sought to identify if zolpidem use was a risk factor for the development of Alzheimer's disease [15C]. A retrospective cohort study matched 3461 patients who received zolpidem with those who did not receive zolpidem in a 1:1 fashion based on age, sex, and similar propensity scores. In addition to zolpidem exposure, the authors categorized patients based on the cumulative defined daily dose (cDDD) received per year of zolpidem therapy. Each DDD was defined as the average maintenance dose of zolpidem (i.e. 10mg). Zolpidem exposure was not found to increase the risk of Alzheimer's disease development (HR=1.35, 95% CI=0.85–2.13); however, patients who received greater than 180 cDDD were at a significant risk of Alzheimer's disease development when compared to patients with less than 28 cDDD (HR=4.18; 95% CI=1.61–5.49). While the mechanism of this is not well known, the authors note that zolpidem may inhibit hippocampus activity, which is necessary for forming new memories. While zolpidem does not serve as an independent risk factor, this study suggests that higher cumulative doses of zolpidem may increase the risk of Alzheimer's disease, and future studies are necessary to further demonstrate this finding.

A systematic review published in 2017 examined the prevalence of medication induced sleepwalking, or somnambulism [16R]. While multiple drugs and drug classes were reviewed, the majority of the evidence (43.5%) indicates a significant concern with z-hypnotics, including 24 studies or case reports with zolpidem use, 2 with zaleplon use, and 1 with zopiclone. Additionally, several clinical trials support this association with z-hypnotic usage; however, the evidence for other medications listed in the review primarily relies on case reports as no other clinical trials were noted in the study.

Infection Risk

A retrospective case-control study examined the association of zolpidem use and acute pyelonephritis (APN) [17c]. Females with APN were matched with randomly selected control patients without APN based on age, comorbidity index, and date of encounter in a 1:2 fashion. Patients with current zolpidem use (defined as zolpidem use within 7days of APN onset) had significantly higher odds of developing APN (4.1% vs 1.4%; adjusted OR=2.2, 95% CI 1.7–2.8). The authors postulate that modulation of the benzodiazepine receptor is associated with impaired immunity which may explain the increased risk of APN.

KETAMINE [SEDA-39, 5]

Review

A clinical summary on the use of ketamine for depression discusses a review of adverse effects associated with ketamine in depression clinical trials [18R]. The review found that doses of ketamine used for depression can cause temporary dissociative symptoms, cognitive impairment, and psychomimetic symptoms. These most commonly manifest in the form of drowsiness, dizziness, poor coordination, blurred vision, and feelings of unreality, although these effects are often transient and do not often cause ketamine to be stopped in these patients. However, the review also reports almost one-third of patients experienced transient increases in systolic blood pressure as great as 20 mmHg and in diastolic blood pressure as great as 13 mmHg with ketamine use. The incidence and severity of all side effects reported with ketamine when used for depression have been found to be dose-dependent.

Cardiovascular System

A retrospective chart review of nine patients who received ketamine for refractory status epilepticus (RSE) identified a case of one patient who experienced repetitive arrhythmias with recurrent asystole [19A]. The patient, a 72-year-old female, had no prior history of seizures or cardiovascular disease and had undergone frontosphenoidal temporal craniotomy due to a right middle cerebral artery aneurysm and received phenytoin and levetiracetam for seizure prophylaxis. Her routine EEG showed non-convulsive status epilepticus (NCSE) 48 h postadmission, so she was treated with lacosamide in addition to the phenytoin and levetiracetam. Midazolam and propofol were initiated when the NCSE persisted, with a failed attempt to wean the sedatives 48 h later, at which point ketamine was started at a rate of 0.07 mg/kg/h. Several hours after the ketamine infusion began, the patient was found in atrial fibrillation with a rapid ventricular response and treated with a bolus of metoprolol followed by diltiazem 3 h later. She then experienced sinus bradycardia and three episodes of asystole lasting 1 min each. Ketamine was discontinued, she was resuscitated with chest compressions, received magnesium and calcium gluconate, was transcutaneously paced, started on amiodarone, and then therapeutically cooled. The authors of the case report acknowledge that both RSE and subarachnoid hemorrhages can be associated with arrhythmias, but it is possible that ketamine may augment these risk factors.

Endocrine System

An 18-year-old female with a history of Moyamoya disease and diabetes mellitus developed temporary central diabetes insipidus (DI) while being treated for cerebral ischemia with a left superficial temporal artery to middle cerebral artery bypass [20A]. During the course of her surgery she received a low dose ketamine infusion at 10 mcg/kg/min as adjunctive anesthesia (in addition to sevoflurane, propofol, and remifentanil infusions). About 2 h into the procedure, she had a 2 min episode of hypertension and bradycardia that was treated with atropine, nicardipine, and esmolol. After the hypertensive episode resolved, she had excessive urine output, and her urine osmolality was found to be only 117 mOsm/kg, her serum sodium 152 mEq/L, and her arterial blood gases showed metabolic acidosis. She was treated with 4 mcg of desmopressin and her urine output decreased. It is speculated that as an *N*-methyl-D-aspartate receptor antagonist, ketamine may block the release of vasopressin, causing symptoms consistent with central DI. As there have previously been two case reports of ketamine associated with central DI, the authors conclude that urine output and osmolality should be monitored in patients receiving ketamine for anesthesia.

Infant Exposure

A previously healthy 10-month-old female was found unresponsive and originally suspected to have carbon monoxide poisoning [21A]. She was cyanotic with poor respiratory effort and was intubated immediately. Her blood glucose was 400 mg/dL, sodium bicarbonate was 8 mmol/L, and PCO_2 was over 52 mmHg, however, her carbon monoxide percentage was normal. Urine drug studies were negative for amphetamines, barbiturates, cannabinoids, cocaine, opiates, phencyclidine, and oxycodone, and positive for benzodiazepines (midazolam administered after presentation), levetiracetam (administered after presentation), and ketamine. It was later discovered that ketamine was being produced in the child's home and she had been served eggs using the same dishes that had been used to handle the ketamine. She had eaten the eggs, vomited, and fallen asleep. Imaging studies showed bilateral cerebellar edema causing obstructive hydrocephalus. She had an emergency external ventricular drain placed, which was changed to a cerebral spinal fluid shunt 15 days later. The patient went on to exhibit normal movement of all extremities at 1-year follow-up and met all cognitive developmental milestones with the exception of a mild speech delay. The authors conclude that with increasing potential for ketamine abuse, it is important to be able to recognize signs of toxicity such as bilateral cerebellar edema.

Drug–Drug Interaction

Cyclosporine was observed to exacerbate the toxic effects of ketamine in zebrafish embryos [22E]. Zebrafish embryos were exposed to placebo; 2mM ketamine; 10mcM cyclosporine; 2mM ketamine and 10mcM cyclosporine; 2mM ketamine and 1.0mM acetyl-L-carnitine (ALCAR); or 2mM ketamine, 10mcM cyclosporine and 1mM ALCAR for 72h starting at 24h post fertilization. The specific concentrations were chosen based on the equivalent serum concentrations attained in humans. Compared to the embryos receiving ketamine alone or cyclosporine alone, mortality was significantly higher in those who received the combination of ketamine and cyclosporine. The ketamine and the combination groups both had significantly shorter body lengths than the groups that did not receive ketamine. These toxic effects were prevented in embryos concomitantly treated with ALCAR, as both of the groups that received ALCAR had no significant differences in body length or mortality compared to the control group. The authors surmise that the increased toxicities could be due to decreased adenosine triphosphate (ATP) availability, as both ketamine and cyclosporine can reduce the production of ATP. Based on these findings, it may be prudent to use caution when ketamine is used in conjunction to cyclosporine, although this effect has not been studied in humans.

MELATONIN RECEPTOR AGONISTS

Walker and colleagues conducted a systematic review of melatonin receptor agonists utilized to prevent dementia [23M]. Of the six included studies, four examined the use of melatonin, whereas one examined the use of ramelteon and one examined the use of L-tryptophan. Only two of the studies, one involving melatonin and the other involving L-tryptophan, reported adverse effects.

L-Tryptophan

The first study evaluated L-tryptophan 1g given three times daily compared with placebo beginning after surgery for at least 3 days or until out of the intensive care unit [23M]. No significant differences in adverse events were noted, however, more patients receiving LL-tryptophan experienced nausea compared to placebo (12% vs 9%, $P=0.45$). Additionally, patients receiving L-tryptophan were more likely to report a bad taste (3% vs 0%, $P=0.06$).

Melatonin [SEDA-37, 5]

The second study utilized melatonin 0.5mg vs placebo for 14 days for the prevention of delirium [23M]. Notable adverse effects occurred in two patients, both receiving melatonin. The first patient developed nightmares that required melatonin discontinuation, and the second patient developed hallucinations yet continued melatonin therapy. No adverse effects were reported in the placebo group.

OREXIN RECEPTOR ANTAGONISTS

Almorexant

Almorexant, along with the other orexin receptor antagonists, is utilized in the treatment of insomnia [24C]. A randomized, placebo-controlled trial evaluated the safety and efficacy of almorexant in adults aged 18–64 with chronic insomnia. Participants were randomized to receive almorexant 200mg, almorexant 100mg, placebo, or zolpidem 10mg in a 1:1:1:1 fashion. A total of 707 patients were included in the safety analysis ($N=176, 186, 177$, and 168, respectively). Adverse effects were most common in the zolpidem group (42.9%), whereas the incidence of adverse effects was lower in patients receiving 100mg almorexant (35.5%), 200mg almorexant (35.2%), and placebo (36.2%). Common side effects reported were headache, fatigue, dizziness, somnolence, nausea, and diarrhea.

Elderly

A multi-center, double-blind, randomized, placebo-controlled trial was utilized to determine the safety and efficacy of almorexant in patients 65 years of age or older [25C]. Patients with a BMI less than 18.5 or greater than or equal to 34 were excluded from the trial. The study utilized a five-way crossover design to examine the effects of almorexant at various doses (25, 50, 100, and 200mg, as well as placebo). Total adverse effects were comparable between placebo (12.1%) and almorexant across all doses (9.4%–11.3%). Somnolence was the most common adverse effect, occurring in 3.8%, 0.9%, 1.9%, and 1.9%, respectively, in patients receiving almorexant, whereas no incidence of somnolence was reported in the placebo period. One female patient discontinued the trial due to urinary frequency, urinary incontinence, gait disturbance, and nausea that began 2.5h after taking her first dose of almorexant 200mg; however, given the timing, these adverse events were not attributed to almorexant by the authors. Another patient experienced transient muscular weakness in the jaw and knees after almorexant 200mg, but this resolved without treatment, and the patient remained enrolled in the trial. Lastly, decreased neutrophil counts were reported in three patients taking almorexant 25mg and two patients taking placebo. Of these, four patients had repeat neutrophil labs that demonstrated a return to baseline after an unspecified period of time.

Filorexant

A phase II trial was conducted to explore the potential benefit of filorexant as an adjunctive medication in patients with major depressive disorder [26C]. A total of 128 patients, aged 18–64, were enrolled and randomized 1:1 to filorexant 10mg or placebo. Severity of depression symptoms were measured using the Montgomery Asberg Depression Rating Scale (MADRS) and the Hamilton Depression Rating Scale 17-item assessment (HAMD-17). Filorexant failed to provide a significant benefit compared to placebo in reducing MADRS (mean difference: −0.7, $P=0.679$) or HAMD-17 (−0.5, $P=0.697$) after 6weeks. Additionally, adverse events occurred in 42.2% of patients taking filorexant, compared with only 26.6% taking placebo. As with the other orexin antagonists, somnolence was commonly reported in the filorexant group (7.8% vs 0%). In this population, however, filorexant users were more likely to report suicidal ideation (10.9% vs 4.7%). The authors note that the investigators did not believe this was related to the study drug at all but one patient had a history of suicidal ideation prior to enrollment in the study.

Suvorexant

A systematic review and meta-analysis examined the safety and efficacy of suvorexant across four clinical trials [27M]. Suvorexant was associated with a clinically significant increase in the risk of any adverse events when compared to placebo (Relative Risk [95% CI]: 1.07 [1–1.15]), somnolence (3.53 [2.19–5.69]), excessive daytime sleepiness (3.05 [1.1–8.48]), fatigue (2.09 [1.08–8.48]), abnormal dreams (2.08 [1.13–3.84]), and dry mouth (1.99 [1.10–3.61]). Additionally, suicidal ideation, complex sleep-related behaviors, hypnagogic hallucinations, hypnopompic hallucinations, and sleep paralysis were all reported with suvorexant use, but were not reported in patients on placebo (0.4%, 0.1%, 0.3%, 0.1%, and 0.3%, respectively).

Elderly

A pooled analysis of two phase III clinical trials reported the safety and efficacy of suvorexant use in the elderly population [28C]. In both studies, patients were discontinued from maintenance pharmacological sleep therapy. Doses of 30 and 15mg were utilized in the elderly (age >65) compared with 40 and 20mg in non-elderly patients. A total of 1298 elderly patients were enrolled in the trials, with 627 patients receiving 30mg, 202 patients receiving 15mg, and 469 patients receiving placebo. The most common side effect reported was somnolence, occurring in 8.8%, 5.4%, and 3.2% of patients, respectively. Other common adverse events included headache, fatigue, and dry mouth. Excessive daytime sleepiness was not reported in the placebo arm, but was reported infrequently with suvorexant use (0.6% in the 30mg arm and 0.5% in the 15mg arm). Additionally, sleep-related hallucinations (0.4%) and sleep paralysis (0.1%) were reported with suvorexant use whereas no reports were seen with placebo. Lastly, self-reported motor vehicle accidents were more common in the suvorexant group when compared with placebo (1.6% vs 1.0%).

Adolescents

In a small study in Japan, 30 adolescent patients (8 male and 22 female) were given suvorexant for the treatment of insomnia [29c]. Patients were all between the ages of 10–20 (Mean ± SD: 15.7 ± 2.4years). A total of 8 patients discontinued the medication (1 male, 7 female) with 4 discontinuing the medication for unspecified reasons, 2 discontinuing due to lack of efficacy, and 2 discontinuing due to abnormal dreams. The patients were 16- and 17-year-old female patients who utilized suvorexant for 207 and 215days, respectively. Of the patients who were continued on suvorexant, another 17-year-old female reported excessive sleep and daytime somnolence that persisted initially; however, this subsided after the first month.

PROPOFOL [SEDA-38, 4; SEDA-39, 5]

Cardiovascular System

A meta-analysis of studies from many areas of the world compared cardiopulmonary adverse events during sedation with propofol to those seen with traditional agents during gastrointestinal endoscopy [30M]. The analysis included hypoxia, hypotension, and arrhythmias as cardiopulmonary events and compared propofol to midazolam, meperidine, pethidine, remifentanil, and/or fentanyl. The odds ratios of developing hypoxia in propofol vs traditional agents was reported as 0.82 (95% CI, 0.63–1.07), hypotension was 0.92 (95% CI, 0.64–1.32), arrhythmia was 1.07 (95% CI, 0.68–1.68), and any of the three complications was 0.77 (95% CI 0.56–1.07). The meta-analysis concluded that there is a similar risk of cardiopulmonary adverse events in both groups. However, two letters to the editor question the validity of the results from this meta-analysis [31r, 32r]. The letters express concern for the heterogeneity between trials included, statistical method used to obtain pooled odds ratios, inclusion and exclusion criteria and implementation, omission of data, use of nonsignificant results to make conclusions, and more.

In contrast to the meta-analysis, a retrospective analysis of data from 5years of gastrointestinal endoscopies at a tertiary center found increased rates of many adverse events when use of propofol was compared to intravenous conscious sedation (IVCS), which is usually accomplished with the use of a short-acting benzodiazepine and short-acting opioid [33c]. Reactions that had statistically higher rates in propofol compared to IVCS include cardiac arrest (6.069 vs 0.666), pancreatitis (5.698 vs 0.222), post procedure pain (3.561 vs 1.322), febrile reaction

(1.780 vs 0.222), and iatrogenic reaction (2.137 vs 0.222). This study shows concern for increased adverse events with use of propofol for sedation during gastrointestinal endoscopy.

Infection Risk

A study done in mice found that brief periods of sedation with propofol increased the severity of methicillin-resistant *Staphylococcus aureus* (MRSA) bloodstream infections [34E]. The kidneys of mice were examined 7, 10, 14, or 32 days after being infected with MRSA. Mice in the treatment group were given doses of propofol adjusted for body weight and metabolism to be similar to human anesthetic induction. Mice treated with propofol had decreased clearance of MRSA, increased abscess formation, and more MRSA and neutrophil dissemination in the kidney compared to the kidneys of mice without propofol treatment. The study also examined mice that were given vancomycin as prophylaxis before infection with MRSA. While vancomycin prophylaxis did not seem to change bacterial burden, mice given both vancomycin and propofol showed a significant increase in inflammation, abscesses, and necrosis earlier than other treatment groups. Although more studies are needed, this study suggests that using propofol may increase kidney damage and dissemination with MRSA bloodstream infections.

Additional case study reviews can be found in previous SEDA publication [35R].

References

[1] Lyons PG, Snyder A, Sokol S, et al. Association between opioid and benzodiazepine use and clinical deterioration in ward patients. J Hosp Med. 2017;12(6):428–34 [C].

[2] Kaufmann CN, Spira AP, Alexander C, et al. Emergency department visits involving benzodiazepines and non-benzodiazepine receptor agonists. Am J Emerg Med. 2017;35(10):1414–9 [MC].

[3] Brandt J, Leong C. Benzodiazepines and z-drugs: an updated review of major adverse outcomes reported on in epidemiologic research. Drugs R D. 2017;17:493–507 [R].

[4] Donnelly K, Bracchi R, Hewitt J, et al. Benzodiazepines, z-drugs and the risk of hip fracture: a systematic review and meta-analysis. PLoS One. 2017;12(4):e0174730 [M].

[5] Wang MT, Wang YH, Chang HA, et al. Benzodiazepine and z-drug use and risk of pneumonia in patients with chronic kidney disease: a population-based nested case control study. PLoS One. 2017;12(7):e0179472 [C].

[6] Taipale H, Tolppanen AM, Koponen M, et al. Risk of pneumonia associated with incident benzodiazepine use among community-dwelling adults with Alzheimer's disease. CMAJ. 2017;189(14):E519–29 [C].

[7] Soyka M. Treatment of benzodiazepine dependence. N Engl J Med. 2017;376(12):1147–57 [R].

[8] Crowe SF, Stranks EK. The residual medium and long-term cognitive effects of benzodiazepine use: an updated meta-analysis. Arch Clin Neuropsychol. 2017. https://doi.org/10.1093/arclin/acx120 [M].

[9] Skinner BW, Johnston EV, Saum LM. Benzodiazepine initiation and dose escalation. Ann Pharmacother. 2017;51(4):281–5 [C].

[10] Maruyoshi H, Maruyoshi N, Hirosue M, et al. Clonazepam-associated bradycardia in a disabled elderly woman with multiple complications. Intern Med. 2017;56(17):2301–5 [A].

[11] Lameijer H, Sikkema YT, Pol A, et al. Propofol versus midazolam for procedural sedation in the emergency department: a study on efficacy and safety. Am J Emerg Med. 2017;35(5):692–6 [c].

[12] Linselle M, Sommet A, Bondon-Guitton E, et al. Can drugs induce or aggravate sleep apneas? a case-noncase study in VigiBase®, the WHO pharmacovigilance database. Fundam Clin Pharmacol. 2017;31(3):359–66 [c].

[13] Iammatteo M, Keskin T, Jerschow E. Evaluation of periprocedural hypersensitivity reactions. Ann Allergy Asthma Immunol. 2017;119(4):349–55 [c].

[14] Brandlistuen RE, Ystrom E, Hernandez-Diaz S, et al. Association of prenatal exposure to benzodiazepines and child internalizing problems: a sibling-controlled cohort study. PLoS One. 2017;12(7):e0181042 [C].

[15] Cheng HT, Lin FJ, Erickson SR, et al. The association between the use of zolpidem and the risk of Alzheimer's disease among older people. J Am Geriatr Soc. 2017;65(11):2488–95 [C].

[16] Stallman HM, Kohler M, White J. Medication induced sleepwalking: a systematic review. Sleep Med Rev. 2017;37:105–13. pii: S1087-0792(17)30020-5, https://doi.org/10.1016/j.smrv.2017.01.005 [R].

[17] Hsu FG, Sheu MJ, Lin CL, et al. Use of zolpidem and risk of acute pyelonephritis in women: a population-based case-control study in Taiwan. J Clin Pharmacol. 2017;57(3):376–81 [c].

[18] Andrade C. Ketamine for depression, 1: clinical summary of issues related to efficacy, adverse effects, and mechanism of action. J Clin Psychiatry. 2017;78(4):e415–9 [R].

[19] Koffman L, Yan Yiu H, Forrokh S, et al. Ketamine infusion for refractory status epilepticus: a case report of cardiac arrest, J Clin Neurosci. 2017;47:149–51. pii: S0967-5868(17)31064-0, https://doi.org/10.1016/j.jocn.2017.10.044 [A].

[20] Gaffar S, Eskander JP, Beakley BD, et al. A case of central diabetes insipidus after ketamine infusion during an external to internal carotid artery bypass. J Clin Anesth. 2017;36:72–5 [A].

[21] Villelli N, Hauser N, Gianaris T, et al. Severe bilateral cerebellar edema from ingestion of ketamine: case report. J Neurosurg Pediatr. 2017;20(4):393–6 [A].

[22] Robinson BL, Dumas M, Ali SF, et al. Cyclosporine exacerbates ketamine toxicity in zebrafish: mechanistic studies on drug-drug interaction. J Appl Toxicol. 2017;37(12):1438–47 [E].

[23] Walker CK, Gales MA. Melatonin receptor agonists for delirium prevention. Ann Pharmacother. 2017;51(1):72–8 [M].

[24] Black J, Pillar G, Hedner J, et al. Efficacy and safety of almorexant in adult chronic insomnia: a randomized placebo-controlled trial with an active reference. Sleep Med. 2017;36:86–94 [C].

[25] Roth T, Black J, Cluydts R, et al. Dual orexin receptor antagonist, almorexant, in elderly patients with primary insomnia: a randomized, controlled study. Sleep. 2017;40(2):1–8 [C].

[26] Connor KM, Ceesay P, Hutzelmann J, et al. Phase II proof-of-concept trial of the orexin receptor antagonist filorexant (MK-6096) in patients with major depressive disorder. Int J Neuropsychopharmacol. 2017;20(8):613–8 [C].

[27] Kuriyama A, Tabata H. Suvorexant for the treatment of primary insomnia: a systematic review and meta-analysis. Sleep Med Rev. 2017;35:1–7 [M].

[28] Herring WJ, Connor KM, Snyder E, et al. Suvorexant in elderly patients with insomnia: pooled analyses of data from phase III

randomized controlled clinical trials. Am J Geriatr Psychiatry. 2017;25(7):791–802 [C].

[29] Kawabe K, Horiuchi F, Ochi M, et al. Suvorexant for the treatment of insomnia in adolescents. J Child Adolesc Psychopharmacol. 2017;27(9):792–5 [c].

[30] Wadhwa V, Issa D, Garg S, et al. Similar risk of cardiopulmonary adverse events between propofol and traditional anesthesia for gastrointestinal endoscopy: a systematic review and meta-analysis. Clin Gastroenterol Hepatol. 2017;15(2):194–206 [M].

[31] Goudra B, Singh PM. More questions than answers: comparison of the risk of cardiopulmonary adverse events between propofol and traditional anesthesia for gastrointestinal endoscopy. Clin Gastroenterol Hepatol. 2017;15(3):468 [r].

[32] Hoaglin DC. A flawed meta-analysis? Similar risk of cardiopulmonary adverse events between propofol and traditional anesthesia for gastrointestinal endoscopy. Clin Gastroenterol Hepatol. 2017;15(10):1640 [r].

[33] Goudra B, Nuzat A, Singh PM, et al. Association between type of sedation and the adverse events associated with gastrointestinal endoscopy: an analysis of 5 years' data from a tertiary center in the USA. Clin Endosc. 2017;50(2):161–9 [c].

[34] Visvabharathy L, Freitag NE, et al. Infect Immun. 2017;85(7): e00097–e000117 [E].

[35] Beck T, Garcia-Trevino A, Ramirez BA. Hypnotics and sedatives. In: Side effects of drugs annual. Elsevier; 2017, Chapter 5, vol. 39: p. 57–63 [R].

CHAPTER

6

Antipsychotic Drugs

Pierre Chue[1], *Vincent Agyapong*, *Adam Abba-Aji*
Complex Psychosis Program, Department of Psychiatry, University of Alberta, Edmonton, AB, Canada
[1]Corresponding author: pchue@ualberta.ca

GENERAL [SEDA-35, 85; SEDA-36, 59; SEDA-37, 63; SEDA-38, 35; SEDA-39, 65]

Comparative Studies

A randomized, controlled trial (RCT) of second-generation antipsychotics (SGAs) in patients with first episode psychosis (FEP) ($n=200$) found less *extrapyramidal symptoms (EPS)* with quetiapine and olanzapine [1C].

A 26-week RCT ($n=461$) in patients with predominant negative symptoms of schizophrenia found that *treatment-emergent adverse events (TEAEs)* were reported in 54% of cariprazine-treated patients and in 57% of risperidone-treated patients; and included *insomnia, akathisia, schizophrenia worsening, headache*, and *anxiety* [2C].

A RCT ($n=113$) in children and adolescents (12–17 years) with psychosis found quetiapine XR was associated with more *metabolic AEs*, while aripiprazole had more initial *akathisia* and more *sedation* [3C]. Other *AEs* included *tremor* (79% for quetiapine XR vs 91% for aripiprazole), increased *sleep duration* (92% vs 71%), *orthostatic dizziness* (78% vs 81%), *depression* (80% vs 77%), *tension/inner unrest* (69% vs 88%), failing *memory* (76% vs 77%), and *weight gain (WG)* (87% vs 68%).

A RCT ($n=247$) of delirium in palliative care found that both haloperidol and risperidone caused more *EPS* than placebo [4C]. Another RCT ($n=100$) in delirium reported *parkinsonism* ($n=3$; haloperidol), *akathisia* ($n=1$; olanzapine) and *sedation* ($n=1$; olanzapine) [5C].

Observational Studies

A study ($n=53$) in treatment-resistant schizophrenia (TRS) found that clozapine had significantly more *AEs* as measured on the Glasgow Antipsychotic Side-effect Scale (GASS) than quetiapine [6c].

A study of the Number Needed to Harm (NNH) reported that the predominantly *activating* SGAs were lurasidone (NNH=11 for *akathisia* vs 20 for *somnolence*) and cariprazine (NNH=15 for *akathisia* vs 65); the equally *activating* and *sedating* SGAs were risperidone (NNH=15 for *akathisia* vs 13 for *sedation*) and aripiprazole (NNH=31 for *akathisia* vs 34 for *somnolence*); and the predominantly *sedating* were olanzapine, quetiapine IR and XR, ziprasidone, asenapine, and iloperidone [7C].

A retrospective study ($n=16935$) of U.S. National Poison Data System in children (<6 years) of unintentional general single substance exposure found the most frequent *AEs* were *drowsiness/lethargy* (35.6%), *tachycardia* (6.9%), *agitation* (4.0%), and *ataxia* (3.3%). *Drowsiness/lethargy* occurred most with aripiprazole (47.6%), ziprasidone (46.5%) and olanzapine (45.1%), and least with quetiapine (20.5%) and risperidone (28.6%); *tachycardia* and *agitation* occurred most often with olanzapine (11.4% and 12.7%, respectively) [8C].

A study ($n=87$) found higher *prolactin* levels with blonanserin, olanzapine, risperidone, quetiapine, olanzapine compared with aripiprazole, and a correlation between *prolactin elevation* and *sexual dysfunction* [9c].

A study comparing adult (18–65 years), pediatric (<18 years), and geriatric (>65 years) populations found statistically significant differences between the number of *AEs* in the pediatric vs adult populations for aripiprazole, clozapine, fluphenazine, haloperidol, olanzapine, quetiapine, risperidone, and thiothixene, and between the geriatric and adult populations for aripiprazole, chlorpromazine, clozapine, fluphenazine, haloperidol, paliperidone, promazine, risperidone, thiothixene, and ziprasidone; *diabetes (DM)* was the most common *AE* in the adult population, compared to *behavioral problems* in the pediatric population and *neurologic symptoms* in the geriatric population [10C].

A study ($n=57$) of children with Tourette's initiated on aripiprazole or risperidone found there were smaller

Side Effects of Drugs Annual, Volume 40
ISSN: 0378-6080
https://doi.org/10.1016/bs.seda.2018.08.010

changes in *BMI* initially with aripiprazole, followed by a faster rate of increase than with risperidone. There were significant increases in *EPS, insulin* and *prolactin* after initiation [11c].

An 8-week study comparing risperidone to aripiprazole in schizophrenia found *AEs* occurred more frequently with risperidone and specifically *EPS* and *WG* were also greater with risperidone [12c].

A pharmacovigilance study ($n=3325$) found an association of *sleep apnea AEs* with quetiapine, and clozapine for individual drugs, and antipsychotics (APs) for drug class [13C].

Systematic Reviews

A Cochrane review ($n=130$) of chlorpromazine and penfluridol in schizophrenia found *antiparkinsonian* medication use was less with chlorpromazine, but the incidence of *akathisia* was similar [14M]. A Cochrane review ($n=276$) of chlorpromazine and clotiapine in schizophrenia did not find any difference in *akathisia* [15M].

A systematic review and meta-analysis (19 studies; $n=2669$) in FEP found quetiapine was associated with less *akathisia* than haloperidol, aripiprazole, risperidone, and olanzapine; molindone with less *WG* than risperidone, haloperidol, and olanzapine; and lower *prolactin* than risperidone [16M].

A network meta-analysis (60 RCTs; $n=6418$) found the best and worst drugs in terms of *WG, EPS* and *somnolence* were aripiprazole and olanzapine, clozapine and amisulpride, aripiprazole and clozapine, respectively [17M].

A network meta-analysis ($n=3446$) found the lowest incidence of individual *AEs* for olanzapine for *EPS*, quetiapine for *hyperprolactinemia*, paliperidone for *sedation*, and blonanserin for *WG* [18M].

A systematic review in children found *WG* was primarily associated with olanzapine; *EPS* and *akathisia* were associated with molindone; and *prolactin* increase with risperidone, paliperidone, and olanzapine. [19M].

A systematic review (95 RCTs; 40 observational studies) of children and young adults (<24 years) found olanzapine was more associated with *WG* and *BMI* increase than other SGAs except for clozapine, and SGAs, compared to placebo, were associated with increased short-term risk for elevated *triglyceride* levels, *EPS, sedation*, and *somnolence* [20M].

A systematic review and meta-analysis of AP augmentation found that that addition of a D_2 antagonist AP increased *prolactin* while aripiprazole decreased *prolactin* and *weight* [21M].

A review of *TEAEs* of FGAs and SGAs and concluded that FGAs were associated with more *neuromotor* and *cognitive* disorders and SGAs with *cardiometabolic* effects [22R].

A review of APs in dementia reported that SGAs were associated with lower risk of *all-cause mortality* and but higher risk of *stroke* when compared with FGAs. Aripiprazole, risperidone, quetiapine, and olanzapine were associated with increased odds of acute *myocardial infarction*, and risperidone and olanzapine were associated with increased odds of *hip fracture* [23R].

A review of SGAs in bipolar disorder found that olanzapine, asenapine, quetiapine, risperidone had significant *WG* compared to placebo in both short- and long-term trials, while aripiprazole caused significant *WG* in long-term studies [24R].

A review of LAIs reported the most frequent *AEs* with aripiprazole LAI were *psychotic symptoms, EPS* and *WG*; olanzapine LAI were *psychosis, metabolic disturbances, hyperprolactinemia* and *post-injection delirium/sedation syndrome* (PDSS); paliperidone LAI were *WG* and *hyperprolactinemia*, and risperidone LAI and *suicide* [25R].

CARDIOVASCULAR

A study of the SafEty of NeurolepTics in Infancy and Adolescence (SENTIA) registry ($n=101$; mean age= 11.5 years) reported that 6.9% of patients had abnormal *QT* changes [26C].

A study of inpatients ($n=31$) found an inverted-U relationship between α_2-adrenergic receptor antagonism and *resting heart rate* attributed to increased norepinephrine release affecting β_1-adrenergic receptors [27c].

A study of male patients ($n=184$) found *QTc* values >430 ms with paliperidone ($n=2$), risperidone ($n=1$), and olanzapine ($n=1$) [28C].

A study ($n=1059$) of psychiatric inpatients found that clozapine increased *T-peak to T-end interval* and piperazine phenothiazines increased *QT dispersion* [29C]. Another study ($n=225$) of psychiatric inpatients found 4% had *QT prolongation;* olanzapine and risperidone were the most implicated APs [30C].

EAR, NOSE, THROAT

Two cases of a 5-year-old male and a 7-year-old male who developed *epistaxis* with risperidone and subsequently with aripiprazole are reported [31A].

NERVOUS SYSTEM

A study ($n=157$) of *Parkinsonism* found the most commonly associated APs were haloperidol, levomepromazine, and chlorpromazine for the SGAs; and risperidone and olanzapine for the SGAs [32C].

A study ($n=23888$) of movement disorders in children and youth (≤19 years) found an increased risk of *EPS* with risperidone [33C].

A pooled analysis of three 8-week trials (n=497) reported an overall incidence of *akathisia* of 19.5% with significant incidence differences between haloperidol (57%), risperidone (20%), aripiprazole (18.2%), ziprasidone (17.2%), olanzapine (3.6%), and quetiapine (3.5%) [34C].

A case series reported *oculogyric crises* occurring with SGAs (quetiapine, olanzapine, amisulpride and lurasidone, amisulpride, aripiprazole) [35A].

A case of *tardive dystonia* a 20-year-old male with intellectual disability exposed to risperidone and olanzapine is reported [36A].

SENSORY SYSTEMS

Two cases of *central retinal vein occlusion* are reported; one in a 62-year-old female on quetiapine and risperidone, and the second in a 43-year-old male on sulpiride [37A].

ENDOCRINE

A prospective, randomized, open-label study (n=141) found a reduced risk of *hyperprolactinemia* for aripiprazole (19.6%), compared to quetiapine (44.4%) and ziprasidone (32.7%) [38C].

A prospective study (n=174) in FEP found higher *prolactin* levels with risperidone, amisulpride and FGAs, compared with other APs [39C].

An observational study (n=845) found that the prevalence of *metabolic syndrome* (*MetS*) was 18.8% for quetiapine, 22.0% for aripiprazole, 33.3% for amisulpride and paliperidone, 34.0% for olanzapine, 35% for risperidone, 39.4% for haloperidol, and 44.7% for clozapine [40C].

A systematic review of population-based studies (n=15) found clozapine and olanzapine were most strongly associated with *DM* [41M]. A literature review found that the majority of patients who developed *DKA* were treated with olanzapine and clozapine [42R].

METABOLISM

A 1-year study of adolescents on APs (risperidone, olanzapine, quetiapine) found *leptin* and *TSH* levels significantly increased; all APs were associated with *WG* and *BMI/z-BMI* and significantly higher for quetiapine [43c].

A study (n=465) found the risk of *DM* was higher with SGAs compared with FGAs, and higher with quetiapine compared with haloperidol [44C].

A 2-phase, retrospective chart study (n=284) of geriatric psychiatric inpatients found that olanzapine was associated with greater *WG* than risperidone or aripiprazole, and more clinically significant *WG* ($\geq$7%) [45C].

A systematic review and meta-analysis (47 studies) found olanzapine was associated with significantly increased *glucose* levels compared to ziprasidone, lurasidone, risperidone and placebo [46M].

HEMATOLOGIC

A study (n=83) found higher doses of APs (risperidone, quetiapine, olanzapine, aripiprazole) correlated with *leukocytopenia* and significantly with high-dose combination aripiprazole and quetiapine [47c].

SKIN

A case of a 21-year-old female with persistent severe *psoriasis* lesions following treatment with quetiapine, and switch to aripiprazole because of *WG* [48C].

MUSCULOSKELETAL

A systematic review and meta-analysis (n=544811) found both FGAs and SGAs were associated with an increased risk of *fractures*, especially among the elderly; FGA users were at a higher risk of *hip fracture* than SGA users with greatest risk for chlorpromazine and least for risperidone [49M].

URINARY TRACT

A study found that the risk of *acute kidney injury* (*AKI*) was significantly increased in patients taking olanzapine, quetiapine, and ziprasidone relative to haloperidol [50c].

SEXUAL FUNCTION

A post hoc analysis of a RCT of aripiprazole and paliperidone LAI found increased *prolactin* and *prolactin-related AEs* and *sexual dysfunction* with paliperidone LAI [51c].

REPRODUCTIVE FUNCTION

A study of Japanese Adverse Drug Event Report database found a potential signal for *miscarriage* for aripiprazole, in contrast to other SGAs [52C].

BREASTS

A study of female schizophrenia patients on APs ($n = 29641$) found the risk of *breast cancer* was slightly higher for those patients on a combination of FGAs and SGAs; patients on risperidone, paliperidone, and amisulpride had a 1.96-fold risk of *breast cancer* compared to the non-schizophrenic cohort [53C].

IMMUNOLOGIC

A study of *C-reactive protein* HS-CRP in schizophrenia found quetiapine and cyamemazine associated with increased peripheral low-grade *inflammation* and aripiprazole was associated with decreased peripheral low-grade *inflammation* [54C].

DEATH

A cohort study ($n = 29823$) of patients with schizophrenia found the highest *mortality* in patients not receiving APs; the lowest *mortality* was associated with SGA-LAIs and oral aripiprazole [55C].

A study ($n = 70718$) in Alzheimer's found AP use was associated with an increased risk of *mortality*; polypharmacy was associated with an increased risk of *mortality* than monotherapy compared to non-use [56C]. Haloperidol was associated with higher risk and quetiapine with lower risk of *mortality*, compared with risperidone.

In a study ($n = 6930$) of dementia patients (>60 years) AP users had an increased risk of requiring more *care* (twofold for long-term *care*), with the exception of quetiapine, compared with non-users; among those living in private homes ($n = 9950$) the risk of moving into a nursing home was increased by 50% compared with non-users [57C]. Risk of *death* was significantly higher with haloperidol, melperone, and risperidone, but not for quetiapine.

DRUG–DRUG INTERACTIONS

An inpatient study ($n = 600$) of *drug–drug interactions (DDIs)* causing *QT interval prolongation* found that olanzapine ($n = 142$), haloperidol ($n = 138$), risperidone ($n = 91$), zuclopenthixol ($n = 87$), quetiapine ($n = 80$) were commonly involved [58C].

DIAGNOSIS OF ADVERSE DRUG REACTIONS

A 3-month prospective study ($n = 224$) identified 38 *AEs* including *sexual dysfunction, decreased libido, WG* and *menstrual problems*; while the number of *AEs* was greatest with risperidone and olanzapine, the incidence rate was the lowest as these were the most commonly prescribed APs [59C].

INDIVIDUAL DRUGS

Amisulpride *[SEDA-36, 59; SEDA-37, 67; SEDA-38, 38; SEDA-39, 69]*

Controlled Studies

A RCT ($n = 40$) of intravenous (iv) amisulpride compared to placebo found a plasma concentration-dependent effect on *QTc* [60c].

Observational Studies

An 8-week, Chinese switch study ($n = 109$) in schizophrenia found that most common *AEs* included *EPS, prolactin increase* and *WG* [61C].

NERVOUS SYSTEM

A case of *oropharyngeal dyskinesia* is reported in a 39-year-old male [62A]. A case of *neuroleptic malignant syndrome* (*NMS*) in a 70-year-old female with amisulpride and sertraline is reported [63A]. A case of *withdrawal dyskinesia* in a 63-year-old male after 1 year on amisulpride is reported [64A].

PSYCHIATRIC

A case of *mania* in an 18-year-old male developing after 8 months on amisulpride is reported [65A].

URINARY TRACT

A case of *urinary incontinence* in a 41-year-old female after 2 weeks on amisulpride is reported [66A].

Aripiprazole *[SEDA-35, 96; SEDA-36, 59; SEDA-37, 68; SEDA-38, 38; SEDA-39, 69]*

Controlled Studies

A 52-week RCT of aripiprazole LAI reported *TEAEs* occurring at higher rates than placebo were *weight increase, akathisia, insomnia,* and *anxiety* [67C].

In an 8-week RCT in children and youth with Tourette's randomized to low-dose aripiprazole ($n = 44$), high-dose aripiprazole ($n = 45$), or placebo ($n = 44$) the most common *AEs* were *sedation, somnolence,* and *fatigue* [68C].

Observational Studies

A 52-week study ($n=478$) of aripiprazole lauroxil reported 18% gained ≥7% *body weight*, and 12% lost ≥7% *body weight* [69C]. A post-hoc analysis of a 12-week study with aripiprazole lauroxil found that the most common *AEs* (>5%) were *schizophrenia, akathisia, headache, insomnia*, and *anxiety* [70C].

Systematic Reviews

A systematic review and meta-analysis ($n=20$ studies) in bipolar disorder reported aripiprazole was associated with higher levels of *high density lipoprotein* (HDL) but similar *EPS* to placebo and haloperidol [71M].

A systematic review ($n=1305$) of tic disorders in children and adolescents reported the most common *AEs* of aripiprazole were *drowsiness, nausea/vomiting* and increased *appetite* [72M].

CARDIOVASCULAR

A case of asymptomatic *hypertension* in a 56-year-old male after 9 days on aripiprazole is reported [73A].

RESPIRATORY

A case of *respiratory distress* and *hypersensitivity pneumonitis* in a 36-year-old female is reported [74A].

NERVOUS SYSTEM

A case series of 14 patients with *parkinsonism, tardive dyskinesia* (*TD*) and *akathisia* is reported [75A]. A case of *acute dystonia* in an 8-year-old female is reported [76A]. Two cases of acute *dystonia* in a 4-year-old female and a 3.5-year-old male are reported [77A]. A case of acute *dystonia* associated with aripiprazole overdose in an adolescent boy is reported [78A]. A case of *tardive dystonia* in a 28-year-old male treated for 5 years is reported [79A]. A case of *tardive dystonia* in a 13-year-old female treated for 5 years is reported [80A]. A case report of *TD* in twins is reported [81A]. A case on *NMS* in a 26-year-old male after 4 days on aripiprazole is reported [82A]. A case of a 64-year-old male who developed brief *NMS* after surgery is reported [83A]. A case of severe *parkinsonism* and *creatine kinase* increase after low-dose aripiprazole is reported [84A].

PSYCHIATRIC

A nested case–control study ($n=4696$) found that the use of aripiprazole was associated with an increased risk of *pathologic gambling* and *impulse control disorder* [85C]. A case of a 35-year-old male who developed *pathological gambling* and *compulsive shopping* after 1 month on aripiprazole is reported [86A]. A case of *problem gambling* and *impulse control deficit* is reported [87A]. A case of *pathological gambling* in a 50-year-old male is reported [88A]. Two cases of *hypersexuality* are reported [89A]. A case of *OCS* is reported [90A].

ELECTROLYTE BALANCE

A case of a 35-year-old female who developed *hyponatremia* after 1 month on aripiprazole is reported [91A].

HEMATOLOGIC

A case of *neutropenia* in a 7-year-old male is reported [92A].

URINARY TRACT

Two cases of *diurnal enuresis* in a 9-year-old male and a 16-year-old male are reported [93A]. A case of *urinary retention* is reported [94A].

SKIN

A case of a 17-year-old female who developed *psoriatic lesions* with quetiapine and aripiprazole is reported [95A].

SWEAT GLANDS

Two cases of *hyperhidrosis* in a 36-year-old female and 40-year-old female are reported [96A].

MUSCULOSKELETAL

A case of dose-dependent *Achilles tendinopathy* is reported [97A].

SUSCEPTIBILITY FACTORS

A study found associations of ABCB1 gene polymorphisms with *autonomic nervous system dysfunction* with aripiprazole in schizophrenia [98c].

DRUG–DRUG INTERACTIONS

A case of a 10-year-old male with low serum aripiprazole *concentrations* due to oxcarbazepine CYP3A4 induction is reported [99A].

Asenapine [*SEDA-37, 69; SEDA-38, 39; SEDA-39, 70*]

Controlled Studies

A 6-week RCT ($n=311$) in acute schizophrenia found higher incidences of *oral hypoesthesia* and *dysgeusia* with asenapine, and greater *WG* with olanzapine [100C]. A 26-week extension study found the most common *TEAE* ($\geq$5% in every group) was *worsening of schizophrenia*, and *WG* greatest with olanzapine [101C].

Observational Studies

A 26-week extension study ($n=164$) in bipolar disorder found the most common *TEAEs* ($\geq$1 incidence) were *sedation, headache, somnolence, akathisia,* and *dizziness*; WG ($\geq$7% increase) was more frequent for the highest dose of asenapine [102C].

A 12-week study in borderline personality disorder found that asenapine was associated with *oral hypoesthesia* and *akathisia/restlessness*, and olanzapine was associated with *WG, somnolence* and *fatigue* [103c].

Systematic Reviews

A systematic review and meta-analysis (8 studies; $n=3765$) found that asenapine caused less clinically significant *WG* or *triglyceride increase* than olanzapine, but was more likely to cause *EPS* than olanzapine [104M].

NERVOUS SYSTEM

A case of *acute laryngeal dystonia* in a 56-year-old female after 5 years on asenapine is reported [105A].

A case of a 32-year-old male who developed *NMS* after an asenapine overdose of 50 mg is reported [106A].

Brexpiprazole [*SEDA-38, 40; SEDA-39, 71*]

Observational Studies

A post hoc analysis of 2 pivotal trials in MDD found the most common *TEAEs* (incidence $\geq$5%) were *akathisia* (7.8%), *WG* (7.2%), and *headache* (7.0%); there were dose-dependent increases in *akathisia* incidence [107c].

Systematic Reviews

A systematic review and meta-analysis of in MDD found that the incidences of discontinuation due to *akathisia*, and *WG* were higher than with placebo; the incidence of *akathisia* was dose-related [108M].

A review of three short-term and one long-term study in schizophrenia found a low prevalence of *activating* and *sedating AEs* and *WG* of ~1 kg with brexpiprazole greater than placebo in the short-term studies [109R].

A review of RCTs in schizophrenia found rates of *akathisia* were lower with brexpiprazole compared to aripiprazole and cariprazine; *WG* was more prominent compared to aripiprazole, cariprazine, or ziprasidone; and *sedation* was less than with aripiprazole, but more than with cariprazine [110R].

Cariprazine (*SEDA-38, 40; SEDA-39, 40*)

Observational Studies

A post-hoc analysis of 4 acute trials in schizophrenia found a dose–response relationship for *akathisia, EPS,* and *diastolic BP (DBP)*; there were small *WG* increases (~1–2 kg) vs placebo [111c]. In a post-hoc analysis ($n=679$) of two 48-week trials in schizophrenia *TEAEs* ($\geq$10%) included *akathisia, insomnia, WG,* and *headache*; and mean *WG* was 1.58 kg; *body weight increase* and *decrease* $\geq$7% occurred in 27% and 11% of patients, respectively. *EPS-related TEAEs* that occurred in $\geq$5% of patients were *akathisia, tremor, restlessness,* and *EPS* [112C]. In a 48-week study ($n=93$) in schizophrenia the most common *TEAEs* were *akathisia* (14%), *insomnia* (14%), and *WG* (12%); mean *WG* was *body weight* increased by 1.9 kg [113c].

In a post hoc analysis of 3 short-term trials in bipolar disorder *TEAEs* ($\geq$5%; 2×) increases in *fasting glucose* were greater than with placebo and mean *WG* was 0.54 and 0.17 kg for cariprazine and placebo, respectively; *WG* $\geq$7% was <3% for all treatment groups [114C].

Chlorpromazine [*SEDA-35, 85; SEDA-36, 59; SEDA-37, 69; SEDA-38, 40; SEDA-39, 71*]

Sensory Systems

A case series (44-year-old male, 34-year-old female, 55-year-old female) of *kerato-lenticular ocular deposits* and *visual impairment* with prolonged chlorpromazine use is reported [115A].

Clozapine [*SEDA-35, 99; SEDA-36, 59; SEDA-37, 69; SEDA-38, 40; SEDA-39, 72*]

Observational Studies

A 21-year naturalistic study ($n=96$) found that the Kaplan–Meier estimates for 21-year *cardiovascular events* and *mortality* were 29% and 10%, prospectively; the prevalence of *DM* was 42.7% [116c].

Systematic Reviews

A Cochrane review (5 studies) found limited evidence that the incidence of some *AEs* (*WG, glucose levels, lethargy, hypersalivation, dizziness, tachycardia*) was lower at lower doses [117M].

CARDIOVASCULAR

A chart review ($n=18$) found that *DBP* and *HR* were significantly increased after 24 weeks of clozapine treatment; the proportion of patients meeting criteria for *hypertension* increased from 22% to 67% [118c].

A study found that 33% of patients on clozapine ($n=174$) had *tachycardia* compared with 16% on LAIs ($n=87$) [119C].

A study ($n=503$) found the incidence of sudden *death* and *myocarditis* were 2% and 3%, respectively; mean time to *myocarditis* post clozapine commencement was 15 ± 7 days, and the reduction in *left ventricular ejection fraction* in those with *myocarditis* was $11\pm2\%$ [120C].

A case of successful rechallenge after *myocarditis* is reported [121A]. A case of myocarditis presenting with *fever*, increased *CRP* and high *troponin* after 13 days of clozapine is reported [122A]. A case of *atrial flutter* and successful rechallenge without recurrence is reported [123A]. A case of a 25-year-old male who developed dilated *cardiomyopathy* is reported [124A]. A case of a 63-year-old female who developed dilated *cardiomyopathy* and was successful rechallenge without recurrence is reported [125A]. A case of a 44-year-old man who developed *Brugada syndrome* after 13 days is reported [126A]. A case of a 66-year-old male who developed a *thromboembolic* event after an overdose of clozapine is reported [127A]. A case of 29-year-old male who developed *myocarditis* after 13 days but was successfully rechallenged 2 years later is reported [128A]. A case of a 42-year-old female who developed a *pericardial effusion* is reported [129A]. A case of a 16-year-old female who developed *tachycardia* is reported [130A]. A case of *hypertension* in a 32-year-old male is reported [131A].

RESPIRATORY

A study ($n=465$) of general medicine inpatients found an increased risk of *pneumonia* with clozapine, but not risperidone compared with the untreated general population [132C].

NERVOUS SYSTEM

A case of a 44-year-old male who developed withdrawal *catatonia* and *NMS* is reported [133A]. A case of recurrent withdrawal *catatonia* in a 38-year-old male is reported [134A]. A case of NMS leading to abrupt cessation and consequent withdrawal is reported [135A]. A case of atypical *NMS* in a 55-year-old male is reported [136A]. A case of acute buccal *dystonia* in a 44-year-old male is reported [137A]. A case of a 51-year-old female who developed *TD* after 1 year is reported [138A]. A case of a 52-year-old female who developed *TD* (*Pisa syndrome*) is reported [139A]. A case of a 40-year-old male who developed *TD* after 7 years is reported [140A]. A case report of *EPS* developing after 32 months is reported [141C]. A case of a 29-year-old male who experienced a *seizure* with low dose clozapine (200 mg/day) is reported [142A]. A case of *microseizures*, orofacial *dyskinesia*, and *stuttering* in a 17-year-old male is reported [143A].

NEUROMUSCULAR FUNCTION

A case of a 41-year-old female who developed elevated *CPK* after 4 months on clozapine is reported [144A].

EYES

Two cases of *maculopathy* are reported [145A,146A].

ENDOCRINE

A case of a 48-year-old male who experienced rapid-onset loss of *glycemic control* with clozapine twice is reported [147A].

A study ($n=188$) in schizophrenia found a prevalence of *DM* of 14.3% with clozapine treatment; males were $2.3\times$ more likely and females were $4.4\times$ more likely to have developed *DM* than controls [148C].

METABOLIC

A study ($n=251$) of community patients on clozapine found 45% met criteria for *MetS* [149C].

A case of a 34-year-old female who developed *hypertriglyceridemia* after 2 weeks on clozapine is reported [150A].

HEMATOLOGIC

A retrospective chart study ($n=285$) after 18 weeks of AP treatment found a higher risk of developing both transient and persistent *anemia, neutrophilia* and *eosinophilia* with clozapine compared to controls, while with other SGAs only transient *neutrophilia* and *eosinophilia* were observed [151C]. A case of a 54-year-old male who developed *eosinophilia* which resolved after 3 weeks is reported [152A]. A case series of 8 patients with *neutropenia* and *agranulocytosis* treated with GSF is reported [153A]. A case of a 24-year-old male who developed severe systemic *eosinophilia* resulting in *eosinophilic GI tract* infiltration, *myocarditis, pericardial* and *pleural effusions* is reported [154A]. A case of a 38-year-old female who developed *neutropenia* after 4 years who was successfully rechallenged is reported [155A]. A case of a 38-year-old female successfully rechallenged after 3 episodes of *agranulocytosis* with clozapine is reported [156A].

GASTROINTESTINAL

A pharmacovigilance study ($n=43\,132$) found a *gastrointestinal hypomotility* (*GIH*) rate of 37/10000 clozapine users, and a *mortality* rate of 18% [157C]. A literature review ($n=104$) of GIH reported a *mortality* rate of 38% [158R]. A study ($n=14$) of patients, not on laxative, found a median colonic transit time was 110; the prevalence of gastrointestinal hypomotility was 86% [159c].

A case of a 66-year-old male who died from *septic shock* due to acute *gastrointestinal necrosis* is reported [160A].

SEXUAL FUNCTION

A case of a 32-year-old male who developed *delayed ejaculation* after 7 months is reported [161A].

IMMUNOLOGIC

A case of *Drug Reaction with Eosinophilia and Systemic Symptom (DRESS)* syndrome is reported [162A].

BODY TEMPERATURE

A study ($n=43$) of new users vs chronic use of clozapine found an increased rate of fever in the first group (47.1%) and higher levels of *tumor necrosis factor-α (TNF-α), interferon-γ (INF-γ), interleukin-2 (IL-2),* and *interleukin-6 (IL-6)* [163c]. A case of a 46-year-old male who developed a *pyrexia* with elevation of *CRP* and *procalcitonin* is reported [164A].

SECOND-GENERATION EFFECTS

A review of clozapine in pregnancy found the rates of gestational *DM* were twice as high; and increased rates of *floppy infant syndrome* at delivery, decreased *foetal heart rate* variability, *seizures* in infancy and a case of *agranulocytosis* in a breast-fed infant are reported following maternal exposure to clozapine [165R].

SUSCEPTIBILITY FACTORS

A Dutch study ($n=310$) of patients on clozapine found *NRH quinone oxidoreductase 2 (NQO2 1541AA)* and *ABC-transporter-B1 (ABCB1 3435TT)* were present at a higher frequency in *agranulocytosis* patients ($n=38$) compared with controls [166C]. In patients developing *neutropenia,* ABCB1 3435TT and homozygosity for glutathione-*S*-transferase ($GSTT1^{null}$) were more frequent, and $GSTM1^{null}$ less frequent compared with controls.

A systematic review and meta-analysis of associations with *BMI* increase and *MetS* for variants in genes such as LEP and HTR2C revealed that the presence of the T allele of rs381328 in HTR2C led to a decrease in *BMI* compared to the C allele [167M].

A study ($n=187$) found that the CYP2C 19*17 polymorphism was associated with significantly lower *fasting glucose* levels, while CYP2C19*17/*17 polymorphism was associated with a significantly lower prevalence of *DM* [168C].

A study of *ABCB1* rs1045642 and *ABCC1* rs212090 single-nucleotide polymorphisms (SNPs) found an influence on baseline and 3-month and 12-month *BMI* in males, and *diastolic blood pressure* (DBP) [169C].

DRUG–DRUG INTERACTIONS

A case of *psychotic exacerbation* in a 29-year-old male following the addition of oxcarbazepine and subsequent decrease in clozapine levels is reported [170A].

Fluphenazine *[SEDA-37, 71; SEDA-39, 74]*

Musculoskeletal

A case of late-onset *rhabdomyolysis* with fluphenazine decanoate in a 24-year-old female is reported [171A].

Haloperidol *[SEDA-35, 107; SEDA-36, 59; SEDA-37, 72; SEDA-38, 42; SEDA-39, 74]*

Systematic Reviews

A Cochrane review (41 studies) of rapid tranquilization found that haloperidol was associated with more *dystonia* than aripiprazole, adjunctive promethazine or placebo [172M].

CARDIOVASCULAR SYSTEM

A study found greater *QTc prolongation* with haloperidol than benperidol [173c].

NERVOUS SYSTEM

A case of cervical and limb *dystonia, uterine contractions* and increased *fetal movements* in a 27-year-old female is reported [174A]. A case of persistent oromandibular *dystonia* and angioedema in a male is reported [175A].

A case of *urticaria* and *angioedema* in an 11-year-old female is reported [176A].

OVERDOSE

A case of overdose resulting in *drowsiness, tremor;* increased *muscle tone* with *contractures* and *hyperreflexia* in an 11-year-old male is reported [177A]. A case of an oral overdose leading to *tremor* and *hypothermia* in a female with haloperidol LAI is reported [178A].

DRUG–DRUG INTERACTIONS

A study ($n=64$) of DRD2, SLC6A3 (DAT) and COMT genetic polymorphisms found a statistically significant difference in AEs with genotypes 9/10 and 10/10 of polymorphic marker SLC6A3 rs28363170. There were statistically significant increases in UKU Side-Effect Rating Scale (UKU) and Simpson-Angus Scale (SAS) scores [179c].

Iloperidone *[SEDA-35, 109; SEDA-36, 59; SEDA-37, 72; SEDA-38, 43; SEDA-39, 75]*

Observational Studies

A 5-month study ($n=35$) in bipolar disorder mixed episodes found the most common AEs were increased *heart rate/palpitations* and *urinary incontinence/intense urge to urinate;* 39% discontinued due to AEs [180c].

Levosulpiride *[SEDA-37, 72; SEDA-38, 43; SEDA-39, 75]*

Nervous System

A case series including *parkinsonism* ($n=1$) and *dyskinesia* ($n=10$) is reported [181A].

Lurasidone *[SEDA-37, 72; SEDA-38, 43; SEDA-39, 75]*

Controlled Studies

A 6-week RCT ($n=347$) in young patients (10–17 years) with bipolar depression found the most common *AEs* (≥5% of patients) were *nausea* and *somnolence* [182C].

A 6-week RCT ($n=347$) of two doses of lurasidone vs placebo in young patients (13–17 years) with schizophrenia found the most common *AEs* (≥5%; 2 × placebo) were *nausea, somnolence, akathisia, vomiting*, and *sedation* [183C].

HEMATOLOGIC

Two cases of *neutropenia are reported,* including one in 41-year-old male with previous neutropenia on risperidone [184A].

NERVOUS SYSTEM

A case of *oculogyric crisis* in 32-year-old male is reported [185A]. A case of *NMS* is reported [186A]. A case of *dystonia* in a 28-year-old female is reported [187A]. A case of *rabbit syndrome* in a 31-year-old female is reported [188A].

PSYCHIATRIC

A case of *mania* in a 23-year-old male is reported [189A].

Olanzapine *[SEDA-32, 99; SEDA-33, 104; SEDA-34, 66; SEDA-35, 108; SEDA-36, 59; SEDA-37, 73; SEDA-38, 43; SEDA-39, 75]*

Observational Studies

A 10-year study ($n=285$) in a pediatric trauma center emergency department reported respiratory AEs including *hypoxia* (2.4%), use of *supplemental oxygen* (3.2%), and *intubation* (0.7%), and 1 case of *dystonia* [190C].

CARDIOVASCULAR

A case of acute massive *pulmonary embolism* in a 74-year-old female is reported [191A].

A case of spontaneous *intracranial hemorrhage* and *thrombocytopenia* in a 77-year-old male is reported [192A].

NERVOUS SYSTEM

A case of *NMS* in a 49-year-old male after treatment of catatonia is reported [193A]. A case of *NMS* in a 19-year-old male is reported [194A]. A case of *NMS* in a 27-year-old male is reported [195A]. A case of NMS in a 30-year-old male after 7 years is reported [196A]. Two other cases of *NMS* are reported [197A]. A case of secondary *tics* is reported [198A]. A case of *restless legs* in a 20-year-old male is reported [199A]. A case of *blepharospasm* in a 30-year-old female after 6 months is reported [200A]. A case of *TD* in a 66-year-old male after 12 years is reported [201A].

PSYCHIATRIC

Two cases of *delirium*, one in 69-year-old male and another in a 75-year-old male, are reported [202A].

A case of *somnambulism* (*sleepwalking*) in a 20-year-old male after 2 weeks is reported [203A]. A case of *somnambulism* associated with coadministration of olanzapine and propranolol is reported [204A].

ENDOCRINE

A 7-day study ($n=15$) found that olanzapine resulted in an increase in acute *insulin* response to glucose, which was not seen with placebo, and was blocked by atropine [205C].

A study ($n=35$) in schizophrenia diagnosed 8 patients with *DM; insulin sensitivity* correlated with BMI and TGs, while *insulin resistance* and FPG correlated with TGs [206C].

A case of *BMI* increase from 16.3 to 23.5 kg/m^2 in a 41-year-old male after 3 months on olanzapine is reported [207A].

A case of insulin-dependent *DM* developing after 5 months with olanzapine is reported [208A]. A case of insulin-dependent *DM* developing after 4 months with olanzapine in a 16-year-old boy is reported [209A].

ELECTROLYTES

A case of *hyponatremia* occurring within 24 hours in a 45-year-old female is reported [210A]. A case of recurrent *hyperammonemia* is reported [211A].

HEMATOLOGIC

A case of *neutropenia* and *thrombocytopenia* in a 50-year-old male is reported [212A].

GASTROINTESTINAL

A case of *hepatotoxicity* with elevated *liver enzymes* is reported [213A].

MOUTH AND TEETH

A case of *black hairy tongue* in a 45-year-old male is reported [214A].

ENDOCRINE

A case of painful *gynecomastia* in a 20-year-old male after 6 months is reported [215A].

SKIN

A case of *psoriasis* in a 42-year-old female after 1 month of treatment is reported [216A]. A case of *fixed drug eruption* in a 42-year-old male is reported [217A].

SEXUAL

A case of recurrent *priapism* in a 9-year-old male with Asperger's is reported [218A].

IMMUNOLOGIC

A case of a *DRESS* syndrome with olanzapine and sodium valproate is reported [219A].

BODY TEMPERATURE

A case of *fever* in a 40-year-old female is reported [220A].

SECOND-GENERATION EFFECT

A case of *tracheo-esophageal fistula* in a newborn following maternal antenatal exposure to olanzapine in a 29-year-old female [221A].

DRUG ADMINISTRATION

An international study ($n=3858$) in schizophrenia with olanzapine LAI reported 46 confirmed *postinjection delirium/sedation syndrome (PDSS)* events (0.044% of 103,505 injections) in 45 patients; 91.1% ($n=41$) occurred within 1 hour of injection and with a time-to-recovery of 72 hours for 95.6% of patients (range 6 hours–11 days) [222C]. Risk factors (per-injection) included high dose and male gender.

A 12-month study ($n=30$) reported 1 case of reversible *PDSS* [223A]. A case of *PDSS* in a 55-year-old male is reported [224A]. A case of *PDSS* occurring after the fifth injection of olanzapine LAI in a 29-year-old female is reported [225A]. A case of severe *PDSS* is reported [226A].

OVERDOSE

A case of overdose presenting as acute *muscle toxicity* with elevated *CPK* in a 20-year-old male is reported [227A].

Paliperidone *[SED-33, 108; SEDA-35, 85; SEDA-36, 59; SEDA-37, 74; SEDA-38, 44; SEDA-39, 77]*

Observational Studies

A 25-week, Chinese study ($n=353$) with paliperidone LAI in schizophrenia reported *TEAEs* (≥5%) including *EPS* (15.3%), *akathisia* (10.5%), *prolactin* increase (8.8%), insomnia (5.4%); there were 8 *deaths* including 4 completed *suicides* [228C].

CARDIOVASCULAR

A case of *circulatory failure* resulting in *multiple organ failure* and *death* in a 44-year-old male with paliperidone LAI is reported [229A].

NERVOUS SYSTEM

A case of *EPS* in a 23-year-old female on paliperidone 6 mg/day with improvement on dose reduction is reported [230A]. A case of a 15-year-old male with *head tremors* after 3 weeks of treatment, responding to gabapentin, is reported [231A]. A case of *EPS* persisting 5 months after a single injection of paliperidone LAI in a 53-year-old female is reported [232A]. Two cases of *EPS* attributed to increased exercise and blood flow are reported [233A]. A case of *orofacial dystonia* (*Meige* or *Brueghel Syndrome*) occurring after 1 month with paliperidone LAI is reported [234A].

ENDOCRINE

A 24-month study ($n=400$) of youth and adolescents (12–17 years) treated with paliperidone ER found that females (18.5%) had a higher incidence of *prolactin-related TEAEs* than males (3.3%); mean ± SD maximum changes in prolactin levels were higher in those with *prolactin-related TEAEs* [235C]. Risk factors included female sex, age at diagnosis (13–14 years), girls of Hispanic ethnicity, and region (EU and North America) and higher baseline Tanner stages.

ELECTROLYTES

A case of *hyponatremia* with paliperidone LAI is reported [236A]. A case of *hyponatremia* after 3 days treatment with paliperidone ER is reported [237A].

IMMUNOLOGIC

A case of *angioedema* with paliperidone LAI is reported [238A]. A case of bilateral *pretibial edema* occurring 2 weeks after initiation of paliperidone LAI, but not with paliperidone ER, is reported [239A].

SECOND GENERATION

A case of placental transfer of paliperidone LAI is reported [240A].

Quetiapine *[SEDA-36, 59; SEDA-37, 74; SEDA-38, 45; SEDA-39, 78]*

Controlled Studies

A 12-week placebo-controlled RCT ($n=32$) with quetiapine XR in bipolar disorder found the most common AEs were somnolence (9.1%), increased appetite, dry mouth and dizziness (6.8%) [241c].

CARDIOVASCULAR

A case of *ventricular fibrillation* and *cardiac arrest* due to *torsades de pointes* from acquired *QTc prolongation* secondary to quetiapine is reported [242A]. A case of *cardiomyopathy* in a 34-year-old female with high-dose quetiapine is reported [243A]. A case of *myopericarditis* is reported [244A]. A case of *LBBB* following an overdose of quetiapine is reported [245A]. A case of *hypotension* and *convulsive syncope* in a 16-year-old male is reported [246A].

PSYCHIATRIC

A case series of 3 patients experiencing *hypomania* after 7–10 days of quetiapine XR monotherapy (all subjects had a quetiapine/norquetiapine plasma concentration ratio < 1) is reported [247A]. A case of reversible *global aphasia* occurring 3 days in an 83-year-old female is reported [248A].

METABOLIC

A case of recurrent *anorexia* and *weight loss* in a 37-year-old male with 22q11.2DS is reported [249A].

HEMATOLOGIC

A case of fatal *agranulocytosis* is reported [250A]. A case of *thrombotic thrombocytopenic purpura* is reported [251A]. A case of *neutropenia* associated with quetiapine and sertraline is reported [252A].

GASTROINTESTINAL

A case of bowel obstruction resulting in death after 20 days on quetiapine and tropatepine in a 34-year-old male is reported [253A].

LIVER

A case of *liver injury* after dose increase in a 45-year-old female is reported [254A].

MUSCULOSKELETAL

A case of *fall* and *skull fracture* in a 43-year-old female after 1 day of quetiapine is reported [255A].

Risperidone *[SEDA-36, 59; SEDA-37, 75; SEDA-38, 46; SEDA-39, 80]*

Observational Studies

A study ($n=42$) in children and youth (12–17 years) with risperidone LAI found the most common AEs included *WG, daytime somnolence, muscle stiffness/spasms*, impaired *concentration*, and *fatigue*; *menstrual problems* were common in females; and treatment was terminated in 26.2% of the patients, due to *WG, dystonia*, and *galactorrhea* [256c].

A 12-month study ($n=168$) in ASD found dose-dependent increases in *glucose, insulin, prolactin, leptin* and *HOMA-IR* [257C].

CARDIOVASCULAR

A retrospective chart review ($n=1115$) of ED patients found that risperidone use in elderly patients ($\geq$65 years; $n=48$) was significantly associated with *hypotension*, even with lower mean doses [258C].

Two cases of life-threatening *arrhythmia*, one in a 20-year-old female following an overdose, and one in a 54-year-old female after surgery, are reported [259A].

NERVOUS SYSTEM

A 2-year, Indian pharmacovigilance study ($n=1830$) detected 10 cases of *EPS* with high doses resulting in early *EPS* and lower doses resulting in later *EPS* [260A]. A case of *NMS* in a 66-year-old male with risperidone LAI is reported [261A]. A case of *Pisa syndrome* in a 51-year-old female with history of benzodiazepine and opioid use with risperidone LAI and clotiapine is reported [262A].

SENSORY SYSTEMS

A case of *hyperacusis* in an 11-year-old male with autism is reported [263A].

ENDOCRINE

A retrospective chart study of prolactin laboratory testing found that *prolactin* levels were increased twofold in those experiencing *menstrual disorders*, of which 70% of patients were receiving risperidone [264A].

A case of *hyperprolactinemia* and *amenorrhea* in a 44-year-old female is reported [265A]. A case of *hyperprolactinemia, amenorrhea* and *akathisia* in a 22-year-old female with a history of schizophrenia, methamphetamine use, and fetal alcohol syndrome is reported [266A]. A case of *granulomatous mastitis* secondary to *hyperprolactinemia* in a non-pregnant 30-year-old female is reported [267A].

METABOLISM

A study ($n=22$) in children and adolescents (mean follow-up=7.6 months) found significant increases in mean values of *waist circumference, BMI, BMI percentile, BMI z score, total cholesterol*, and *prolactin*; pubertal/post-pubertal stage and female patients were more susceptible to developing *cardiometabolic* changes, and there were changes observed in *liver parenchyma* and *abdominal fat* distribution mass [268c].

A study ($n=203$) in children and adolescents with ASD found *fasting uric acid* levels were significantly higher in those treated with risperidone and increased with changes in other metabolic parameters; higher risperidone dose and longer duration of treatment were associated with higher *uric acid* levels [269C].

HEMATOLOGIC

A case of *neutropenia* in a 44-year-old male is reported [270A]. A case of *leukopenia* in a 66-year-old male is reported [271A].

SKIN

A case of 23-year-old male who developed *erythema multiforme minor* after 2 days on risperidone is reported [272A].

MUSCULOSKELETAL

A case of *rhabdomyolysis* leading to acute *kidney injury* with risperidone LAI added to oral risperidone, asenapine and aripirazole in a 57-year-old male is reported [273A].

SEXUAL FUNCTION

A case of *priapism* in an 11-year-old male following the addition of risperidone to methylphenidate is reported [274A].

BODY TEMPERATURE

A case of *hypothermia* after 6 days in an 11-year-old male is reported [275A].

SUSCEPTIBILITY FACTORS

An Indian case–control study ($n=52$) of 2 functional polymorphisms in the 5-HTR2C gene (rs3813929 and rs518147) found an association with *insulin resistance* [276c].

A study ($n=289$) of risperidone-related *AEs* (*hyperprolactinemia* [87.2%], *EPS* [36.7%] *tremor* [31.8%], *WG* [53.6%], *sedation* [9.7%], *gastrointestinal effects* [5.1%], and *amenorrhea* [6.1%]) and polymorphisms in patients with schizophrenia found an association with dopamine receptor (*DRD2* -141 C Ins/Del/rs1799732) and increased *prolactin* [277C].

Ziprasidone *[SEDA-36, 59; SEDA-37, 75; SEDA-38, 47; SEDA-39, 81]*

Controlled Studies

A 12-week study ($n=38$) in the elderly with late-onset FEP found there were significant differences in *FBG*, *TG*, and *low-density lipoprotein (LDL)* between olanzapine and ziprasidone [278c].

Zuclopenthixol

Systematic Reviews

A Cochrane review ($n=1850$) found zuclopenthixol caused more *EPS* than clozapine, risperidone or perphenazine, but no difference compared to placebo or chlorpromazine; use of *hypnotics/sedatives* was similar to sulpiride [279M].

Additional case studies can be found in these reviews [280R,281R,282R].

Acknowledgements

The authors thank Jonathan Chue BSc for his editorial assistance.

References

[1] Wang C, Shi W, Huang C, et al. The efficacy, acceptability, and safety of five atypical antipsychotics in patients with first-episode drug-naïve schizophrenia: a randomized comparative trial. Ann Gen Psychiatry. 2017;16:47 [C].
[2] Németh G, Laszlovszky I, Czobor P, et al. Cariprazine versus risperidone monotherapy for treatment of predominant negative symptoms in patients with schizophrenia: a randomised, double-blind, controlled trial. Lancet. 2017;389(10074):1103–13 [C].
[3] Pagsberg AK, Jeppesen P, Klauber DG, et al. Quetiapine extended release versus aripiprazole in children and adolescents with first-episode psychosis: the multicentre, double-blind, randomised tolerability and efficacy of antipsychotics (TEA) trial. Lancet Psychiatry. 2017;4(8):605–18 [C].
[4] Agar MR, Lawlor PG, Quinn S, et al. Efficacy of oral risperidone, haloperidol, or placebo for symptoms of delirium among patients in palliative care: a randomized clinical trial. JAMA Intern Med. 2017;177(1):34–42 [C].
[5] Jain R, Arun P, Sidana A, et al. Comparison of efficacy of haloperidol and olanzapine in the treatment of delirium. Indian J Psychiatry. 2017;59(4):451–6 [C].
[6] Kumar M, Chavan BS, Sidana A, et al. Efficacy and tolerability of clozapine versus quetiapine in treatment-resistant schizophrenia. Indian J Psychol Med. 2017;39(6):770–6 [c].
[7] Citrome L. Activating and sedating adverse effects of second-generation antipsychotics in the treatment of schizophrenia and major depressive disorder: absolute risk increase and number needed to harm. J Clin Psychopharmacol. 2017;37(2):138–47 [C].
[8] Stassinos G, Klein-Schwartz W. Comparison of pediatric atypical antipsychotic exposures reported to U.S. poison centers. Clin Toxicol (Phila). 2017;55(1):40–5 [C].
[9] Kirino E. Serum prolactin levels and sexual dysfunction in patients with schizophrenia treated with antipsychotics: comparison between aripiprazole and other atypical antipsychotics. Ann Gen Psychiatry. 2017;16:43 [c].
[10] Sagreiya H, Chen YR, Kumarasamy NA, et al. Differences in antipsychotic-related adverse events in adult, pediatric, and geriatric populations. Cureus. 2017;9(2):e1059. https://doi.org/10.7759/cureus.1059 [C].
[11] Pringsheim T, Ho J, Sarna JR, et al. Feasibility and relevance of antipsychotic safety monitoring in children with Tourette syndrome: a prospective longitudinal study. J Clin Psychopharmacol. 2017;37(5):498–504 [c].

[12] Kumar PBS, Pandey RS, Thirthalli J, et al. A comparative study of short term efficacy of aripiprazole and risperidone in schizophrenia. Curr Neuropharmacol. 2017;15(8):1073–84 [c].
[13] Linselle M, Sommet A, Bondon-Guitton E, et al. Can drugs induce or aggravate sleep apneas? A case-noncase study in VigiBase(®), the WHO pharmacovigilance database. Fundam Clin Pharmacol. 2017;31(3):359–66 [C].
[14] Nikvarz N, Vahedian M, Khalili N. Chlorpromazine versus penfluridol for schizophrenia. Cochrane Database Syst Rev. 2017;9:CD011831 [M].
[15] Mazhari S, Esmailian S, Shah-Esmaeili A, et al. Chlorpromazine versus clotiapine for schizophrenia. Cochrane Database Syst Rev. 2017;4:CD011810. https://doi.org/10.1002/14651858.CD011810.pub2 [M].
[16] Zhu Y, Krause M, Huhn M, et al. Antipsychotic drugs for the acute treatment of patients with a first episode of schizophrenia: a systematic review with pairwise and network meta-analyses. Lancet Psychiatry. 2017;4(9):694–705 [M].
[17] Bai Z, Wang G, Cai S, et al. Efficacy, acceptability and tolerability of 8 atypical antipsychotics in Chinese patients with acute schizophrenia: a network meta-analysis. Schizophr Res. 2017;185:73–9 [M].
[18] Kishi T, Ikuta T, Matsunaga S, et al. Comparative efficacy and safety of antipsychotics in the treatment of schizophrenia: a network meta-analysis in a Japanese population. Neuropsychiatr Dis Treat. 2017;13:1281–302. https://doi.org/10.2147/NDT.S134340. eCollection 2017. [M].
[19] Pagsberg AK, Tarp S, Glintborg D, et al. Acute antipsychotic treatment of children and adolescents with schizophrenia-spectrum disorders: a systematic review and network meta-analysis. J Am Acad Child Adolesc Psychiatry. 2017;56(3):191–202 [M].
[20] Pillay J, Boylan K, Carrey N, et al. First- and second-generation antipsychotics in children and young adults: systematic review updates, 2017, Rockville, MD: Agency for Healthcare Research and Quality (US); 2017. Available from http://www.ncbi.nlm.nih.gov/books/NBK442352/ [M].
[21] Galling B, Roldán A, Hagi K, et al. Antipsychotic augmentation vs. monotherapy in schizophrenia: systematic review, meta-analysis and meta-regression analysis. World Psychiatry. 2017;16(1):77–89 [M].
[22] Solmi M, Murru A, Pacchiarotti I, et al. Safety, tolerability, and risks associated with first- and second-generation antipsychotics: a state-of-the-art clinical review. Ther Clin Risk Manag. 2017;13:757–77 [R].
[23] Farlow MR, Shamliyan TA. Benefits and harms of atypical antipsychotics for agitation in adults with dementia. Eur Neuropsychopharmacol. 2017;27(3):217–31 [R].
[24] Fang F, Wang Z, Wu R, et al. Is there a 'weight neutral' second-generation antipsychotic for bipolar disorder? Expert Rev Neurother. 2017;17(4):407–18 [R].
[25] Gentile S. Safety concerns associated with second-generation antipsychotic long-acting injection treatment. A systematic update. Horm Mol Biol Clin Investig. 2017. https://doi.org/10.1515/hmbci-2017-0004. pii:/j/hmbci.ahead-of-print/hmbci-2017-0004/hmbci-2017-0004.xml, Epub ahead of print. [R].
[26] Palanca-Maresca I, Ruiz-Antorán B, Centeno-Soto GA, et al. Prevalence and risk factors of prolonged corrected QT interval among children and adolescents treated with antipsychotic medications: a long-term follow-up in a real-world population. J Clin Psychopharmacol. 2017;37(1):78–83 [C].
[27] Kim DD, Lang DJ, Warburton DER, et al. Heart-rate response to alpha(2)-adrenergic receptor antagonism by antipsychotics. Clin Auton Res. 2017;27(6):407–10 [c].
[28] Miniati M, Simoncini M, Vanelli F, et al. QT and QTc in male patients with psychotic disorders treated with atypical neuroleptics. Scientific World Journal. 2017;2017:1951628. https://doi.org/10.1155/2017/1951628 Epub 2017 Jul 12. [C].
[29] Acciavatti T, Martinotti G, Corbo M, et al. Psychotropic drugs and ventricular repolarisation: the effects on QT interval, T-peak to T-end interval and QT dispersion. J Psychopharmacol. 2017;31(4):453–60 [C].
[30] Rodríguez-Leal CM, López-Lunar E, Carrascosa-Bernáldez JM, et al. Electrocardiographic surveillance in a psychiatric institution: avoiding iatrogenic cardiovascular death. Int J Psychiatry Clin Pract. 2017;21(1):64–6 [C].
[31] Binici NC, Güney SA. Epistaxis as an unexpected side effect of aripiprazole and risperidone treatment in two children with two different psychiatric diagnosis. J Child Adolesc Psychopharmacol. 2017;27(8):759–60 [A].
[32] Munhoz RP, Bertucci Filho D, Teive HA. Not all drug-induced parkinsonism are the same: the effect of drug class on motor phenotype. Neurol Sci. 2017;38(2):319–24 [C].
[33] Biscontri RG, Jha S, Collins DM, et al. Movement disorders in children and adolescents receiving antipsychotic pharmacotherapy: a population-based study. J Child Adolesc Psychopharmacol. 2017;27(10):892–6 [C].
[34] Juncal-Ruiz M, Ramirez-Bonilla M, Gomez-Arnau J, et al. Incidence and risk factors of acute akathisia in 493 individuals with first episode non-affective psychosis: a 6-week randomised study of antipsychotic treatment. Psychopharmacology (Berl). 2017;234(17):2563–70 [C].
[35] Nebhinani N, Suthar N. Oculogyric crisis with atypical antipsychotics: a case series. Indian J Psychiatry. 2017;59(4):499–501 [A].
[36] Chandra NC, Sheth SA, Mehta RY, et al. Severe tardive dystonia on low dose short duration exposure to atypical antipsychotics: factors explored. Indian J Psychol Med. 2017;39(1):96–8 [A].
[37] Taki K, Kida T, Fukumoto M, et al. Central retinal vein occlusion in 2 patients using antipsychotic drugs. Case Rep Ophthalmol. 2017;8(2):410–5 [A].
[38] Crespo-Facorro B, Ortiz-Garcia de la Foz V, Suarez-Pinilla P, et al. Effects of aripiprazole, quetiapine and ziprasidone on plasma prolactin levels in individuals with first episode nonaffective psychosis: analysis of a randomized open-label 1 year study. Schizophr Res. 2017;189:134–41 [C].
[39] Lally J, Ajnakina O, Stubbs B, et al. Hyperprolactinaemia in first episode psychosis—a longitudinal assessment. Schizophr Res. 2017;189:117–25 [C].
[40] Lee JS, Kwon JS, Kim D, et al. Prevalence of metabolic syndrome in patients with schizophrenia in Korea: a multicenter nationwide cross-sectional study. Psychiatry Investig. 2017;14(1):44–50 [C].
[41] Hirsch L, Yang J, Bresee L, et al. Second-generation antipsychotics and metabolic side effects: a systematic review of population-based studies. Drug Saf. 2017;40(9):771–81 [M].
[42] Vuk A, Kuzman MR, Baretic M, et al. Diabetic ketoacidosis associated with antipsychotic drugs: case reports and a review of literature. Psychiatr Danub. 2017;29(2):121–35 [R].
[43] Baeza I, Vigo L, de la Serna E, et al. The effects of antipsychotics on weight gain, weight-related hormones and homocysteine in children and adolescents: a 1-year follow-up study. Eur Child Adolesc Psychiatry. 2017;26(1):35–46 [c].
[44] Nishtala PS, Chyou TY. Real-world risk of diabetes with antipsychotic use in older New Zealanders: a case-crossover study. Eur J Clin Pharmacol. 2017;73(2):233–9 [C].
[45] Yeung EY, Chun S, Douglass A, et al. Effect of atypical antipsychotics on body weight in geriatric psychiatric inpatients. SAGE Open Med. 2017;5. https://doi.org/10.1177/2050312117708711. 2050312117708711. eCollection 2017. [C].

[46] Zhang Y, Liu Y, Su Y, et al. The metabolic side effects of 12 antipsychotic drugs used for the treatment of schizophrenia on glucose: a network meta-analysis. BMC Psychiatry. 2017;17(1):373 [M].

[47] Tomita T, Goto H, Chiba T, et al. Leukocytopenia in patients treated with multiple antipsychotics, including aripiprazole and quetiapine. Psychiatry Clin Neurosci. 2017;71(1):71–2 [c].

[48] Bujor CE, Vang T, Nielsen J, et al. Antipsychotic-associated psoriatic rash—a case report. BMC Psychiatry. 2017;17(1):242 [C].

[49] Lee SH, Hsu WT, Lai CC, et al. Use of antipsychotics increases the risk of fracture: a systematic review and meta-analysis. Osteoporos Int. 2017;28(4):1167–78 [M].

[50] Jiang Y, McCombs JS, Park SH. A retrospective cohort study of acute kidney injury risk associated with antipsychotics. CNS Drugs. 2017;31(4):319–26 [c].

[51] Potkin SG, Loze JY, Forray C, et al. Reduced sexual dysfunction with aripiprazole once-monthly versus paliperidone palmitate: results from QUALIFY. Int Clin Psychopharmacol. 2017;32(3):147–54 [c].

[52] Sakai T, Ohtsu F, Mori C, et al. Signal of miscarriage with aripiprazole: a disproportionality analysis of the Japanese adverse drug event report database. Drug Saf. 2017;40(11):1141–6 [C].

[53] Wu Chou AI, Wang YC, Lin CL, et al. Female schizophrenia patients and risk of breast cancer: a population-based cohort study. Schizophr Res. 2017;188:165–71 [C].

[54] Fond G, Resseguier N, Schürhoff F, et al. FACE-SZ (FondaMental Academic Centers of Expertise for Schizophrenia) group. Relationships between low-grade peripheral inflammation and psychotropic drugs in schizophrenia: results from the national FACE-SZ cohort. Eur Arch Psychiatry Clin Neurosci. 2018;268:541–53 [C].

[55] Taipale H, Mittendorfer-Rutz E, Alexanderson K, et al. Antipsychotics and mortality in a nationwide cohort of 29,823 patients with schizophrenia. Schizophr Res. 2018;197:274–80. pii: S0920-9964(17)30762-4. [C].

[56] Koponen M, Taipale H, Lavikainen P, et al. Risk of mortality associated with antipsychotic monotherapy and polypharmacy among community-dwelling persons with Alzheimer's disease. J Alzheimers Dis. 2017;56(1):107–18 [C].

[57] Nerius M, Johnell K, Garcia-Ptacek S, et al. The impact of antipsychotic drugs on long-term care, nursing home admission and death among dementia patients. J Gerontol A Biol Sci Med Sci. 2018;73(10):1396–402. https://doi.org/10.1093/gerona/glx239 [C].

[58] Khan Q, Ismail M, Haider I, et al. Prevalence of QT interval prolonging drug-drug interactions (QT-DDIs) in psychiatry wards of tertiary care hospitals in Pakistan: a multicenter cross-sectional study. Int J Clin Pharmacol. 2017;39(6):1256–64 [C].

[59] Chawla S, Kumar S. Adverse drug reactions and their impact on quality of life in patients on antipsychotic therapy at a tertiary care center in Delhi. Indian J Psychol Med. 2017;39(3):293–8 [C].

[60] Täubel J, Ferber G, Fox G, et al. Thorough QT study of the effect of intravenous amisulpride on QTc interval in Caucasian and Japanese healthy subjects. Br J Clin Pharmacol. 2017;83(2):339–48 [c].

[61] Liang Y, Yu X. The effectiveness and safety of amisulpride in Chinese patients with schizophrenia who switch from risperidone or olanzapine: a subgroup analysis of the ESCAPE study. Neuropsychiatr Dis Treat. 2017;13:1163–73 [C].

[62] Gowda GS, Sastry Nagavarapu LS, Reddy Mukka SS, et al. Amisulpride induced oropharyngeal dyskinesia in a patient with schizophrenia: a case report and review of literature, Asian J Psychiatr. 2017; pii: S1876-2018(17)30640-8. https://doi.org/10.1016/j.ajp.2017.10.010, Epub ahead of print. [A].

[63] Uvais NA, Gangadhar P, Sreeraj VS, et al. Neuroleptic malignant syndrome (NMS) associated with amisulpride and sertraline use: a case report and discussion. Asian J Psychiatr. 2017. https://doi.org/10.1016/j.ajp.2017.10.021. pii: S1876-2018(17)30627-5. Epub ahead of print. [A].

[64] Lo YC, Peng YC. Amisulpride withdrawal dyskinesia: a case report. Ann Gen Psychiatry. 2017;16:25 [A].

[65] Thapa P, Sharma R. A case of probable amisulpride induced mania after eight months of therapy. Case Rep Psychiatry. 2017;2017:6976917 [A].

[66] Niranjan V, Bagul KR, Razdan RG. Urinary incontinence secondary to amisulpride use: a report. Asian J Psychiatr. 2017;29:190–1 [A].

[67] Calabrese JR, Sanchez R, Jin N, et al. Efficacy and safety of aripiprazole once-monthly in the maintenance treatment of bipolar I disorder: a double-blind, placebo-controlled, 52-week randomized withdrawal study. J Clin Psychiatry. 2017;78(3):324–31 [C].

[68] Sallee F, Kohegyi E, Zhao J, et al. Randomized, double-blind, placebo-controlled trial demonstrates the efficacy and safety of oral aripiprazole for the treatment of Tourette's disorder in children and adolescents. J Child Adolesc Psychopharmacol. 2017;27(9):771–81 [C].

[69] Nasrallah HA, Aquila R, Stanford AD, et al. Metabolic and endocrine profiles during 1-year treatment of outpatients with schizophrenia with aripiprazole lauroxil. Psychopharmacol Bull. 2017;47(3):35–43 [C].

[70] Potkin SG, Risinger R, Du Y, et al. Efficacy and safety of aripiprazole lauroxil in schizophrenic patients presenting with severe psychotic symptoms during an acute exacerbation. Schizophr Res. 2017;190:115–20 [C].

[71] Li DJ, Tseng PT, Stubbs B, et al. Efficacy, safety and tolerability of aripiprazole in bipolar disorder: an updated systematic review and meta-analysis of randomized controlled trials. Prog Neuropsychopharmacol Biol Psychiatry. 2017;79(Pt B):289–301 [M].

[72] Wang S, Wei YZ, Yang JH, et al. The efficacy and safety of aripiprazole for tic disorders in children and adolescents: a systematic review and meta-analysis. Psychiatry Res. 2017;254:24–32 [M].

[73] Seven H, Ayhan MG, Kürkcü A, et al. Aripiprazole-induced asymptomatic hypertension: a case report. Psychopharmacol Bull. 2017;47(2):53–6 [A].

[74] Gunasekaran K, Murthi S, Jennings J, et al. Aripiprazole-induced hypersensitivity pneumonitis. BMJ Case Rep. 2017. https://doi.org/10.1136/bcr-2017-219929. pii: bcr-2017-219929. [A].

[75] Selfani K, Soland VL, Chouinard S, et al. Movement disorders induced by the "atypical" antipsychotic aripiprazole. Neurologist. 2017;22(1):24–8 [A].

[76] Sridaran R, Nesbit CE. Acute dystonia versus neuroleptic malignant syndrome without fever in an eight-year-old child. Pediatr Emerg Care. 2017;33(1):38–40 [A].

[77] Küçükköse M, Kabukçu BB. Acute dystonia due to aripiprazole use in two children with autism spectrum disorder in the first five years of life. Turk Psikiyatri Derg. 2017;28(1):71–3 [A].

[78] Kubota K, Yamamoto T, Orii K, et al. Acute dystonia associated with aripiprazole overdose in an adolescent boy. Asian J Psychiatr. 2017;29:183–4 [A].

[79] Kim S, Lee SY, Kim M, et al. Tardive dystonia related with aripiprazole. Psychiatry Investig. 2017;14(3):380–2 [A].

[80] Pitzer M, Engelmann G, Stammschulte T. Tardive movement disorders with antipsychotics—a case of aripirazole-induced tardive dystonia and review of the literature. Z Kinder Jugendpsychiatr Psychother. 2017;45(4):325–34 [A].

[81] She S, Kuang Q, Zheng Y. Aripiprazole-induced tardive dyskinesia in patients with schizophrenia: a case report of twins. Schizophr Res. 2017. https://doi.org/10.1016/j.schres.2017.10.036. pii: S0920-9964(17)30664-3, Epub ahead of print. [A].
[82] Tibrewal P, Bastiampillai T, Kannampuzha C, et al. Aripiprazole-induced neuroleptic malignant syndrome. Asian J Psychiatr. 2017;27:5–6 [A].
[83] Mizumura N, Uematsu M, Ito A, et al. "Brief" aripiprazole-induced neuroleptic malignant syndrome with symptoms that only lasted a few hours. Intern Med. 2017;56(22):3089–92 [A].
[84] Jørgensen A, Thorleifsson A, Jimenez-Solem E, et al. Severe parkinsonism and creatine kinase increase after low-dose aripiprazole treatment in a patient of African descent. J Clin Psychopharmacol. 2017;37(5):630–1 [A].
[85] Etminan M, Sodhi M, Samii A, et al. Risk of gambling disorder and impulse control disorder with aripiprazole, pramipexole, and ropinirole: a pharmacoepidemiologic study. J Clin Psychopharmacol. 2017;37(1):102–4 [C].
[86] Dhillon R, Bastiampillai T, Cao CZ, et al. Aripiprazole and impulse-control disorders: a recent FDA warning and a case report. Prim Care Companion CNS Disord. 2017;19(1). https://doi.org/10.4088/PCC.16l02003 [A].
[87] Peterson E, Forlano R. Partial dopamine agonist-induced pathological gambling and impulse-control deficit on low-dose aripiprazole. Australas Psychiatry. 2017;25(6):614–6 [A].
[88] Mohan T, Dolan S, Mohan R, et al. Aripiprazole and impulse-control disorders in high-risk patients. Asian J Psychiatr. 2017;27:67–8 [A].
[89] Reddy B, Ali M, Guruprasad S, et al. Hypersexuality induced by aripiprazole: two case reports and review of the literature, Asian J Psychiatr. 2017; pii: S1876-2018(17)30692-5. https://doi.org/10.1016/j.ajp.2017.10.008, Epub ahead of print. [A].
[90] Gupta P, Gupta R, Balhara YP. Obsessive-compulsive symptoms associated with aripiprazole treatment in bipolar disorder: a case report. J Clin Psychopharmacol. 2017;37(1):108–9 [A].
[91] Lin MW, Chang C, Yeh CB, et al. Aripiprazole-related hyponatremia and consequent valproic acid-related hyperammonemia in one patient. Aust N Z J Psychiatry. 2017;51(3):296–7 [A].
[92] Gunes S. Aripiprazole-related diurnal enuresis in children: 2 cases (aripiprazole-related enuresis). Clin Neuropharmacol. 2017;40(4):175–6 [A].
[93] Majeed MH, Ali AA. Aripiprazole-induced neutropenia in a seven year-old male: a case report. Cureus. 2017;9(8)e1561. https://doi.org/10.7759/cureus.1561 [A].
[94] Boyer MG, Kheloufi F, Denis J, et al. Urinary retention associated with aripiprazole: report of a new case and review of the literature, Therapie. 2018;73(3):287–9. pii: S0040-5957(17)30157-9. https://doi.org/10.1016/j.therap.2017.09.001, Epub 2017 Oct 23. [A].
[95] Bujor CE, Vang T, Nielsen J, et al. Antipsychotic-associated psoriatic rash—a case report. BMC Psychiatry. 2017;17(1):242 [A].
[96] Vohra A. Aripiprazole-induced hyperhidrosis: two case reports. Turk Psikiyatri Derg. 2017;28(2):132–4 [A].
[97] Javanbakht A, Ajluni V. Dose-dependent Achilles tendinopathy in a patient on aripiprazole. J Clin Psychopharmacol. 2017;37(6):747–8 [A].
[98] Hattori S, Suda A, Kishida I, et al. Associations of ABCB1 gene polymorphisms with aripiprazole-induced autonomic nervous system dysfunction in schizophrenia, Schizophr Res. 2017; pii: S0920-9964(17)30716-8. https://doi.org/10.1016/j.schres.2017.11.020, Epub ahead of print. [c].
[99] McGrane IR, Loveland JG, de Leon J. Possible oxcarbazepine inductive effects on aripiprazole metabolism: a case report, J Pharm Pract. 2018;31(3):361–3. 897190017710523. https://doi.org/10.1177/0897190017710523, Epub 2017 May 24 [A].
[100] Landbloom R, Mackle M, Wu X, et al. Asenapine for the treatment of adults with an acute exacerbation of schizophrenia: results from a randomized, double-blind, fixed-dose, placebo-controlled trial with olanzapine as an active control. CNS Spectr. 2017;22(4):333–41 [C].
[101] Durgam S, Landbloom RP, Mackle M, et al. Exploring the long-term safety of asenapine in adults with schizophrenia in a double-blind, fixed-dose, extension study. Neuropsychiatr Dis Treat. 2017;13:2021–35 [C].
[102] Ketter TA, Durgam S, Landbloom R, et al. Long-term safety and tolerability of asenapine: a double-blind, uncontrolled, long-term extension trial in adults with an acute manic or mixed episode associated with bipolar I disorder. J Affect Disord. 2017;207:384–92 [C].
[103] Bozzatello P, Rocca P, Uscinska M, et al. Efficacy and tolerability of asenapine compared with olanzapine in borderline personality disorder: an open-label randomized controlled trial. CNS Drugs. 2017;31(9):809–19 [c].
[104] Orr C, Deshpande S, Sawh S, et al. Asenapine for the treatment of psychotic disorders. Can J Psychiatry. 2017;62(2):123–37 [M].
[105] Collins N, Sager J. Acute laryngeal dystonia associated with asenapine use: a case report. J Clin Psychopharmacol. 2017;37(6):738–9 [A].
[106] Das S, Purushothaman ST, Bc M, et al. Oral asenapine overdose leading to neuroleptic malignant syndrome (NMS). Asian J Psychiatr. 2017;30:31–2. https://doi.org/10.1016/j.ajp.2017.07.006 Epub 2017 Jul 8. [A].
[107] Fava M, Weiller E, Zhang P, et al. Efficacy of brexpiprazole as adjunctive treatment in major depressive disorder with irritability: post hoc analysis of 2 pivotal clinical studies. J Clin Psychopharmacol. 2017;37(2):276–8 [c].
[108] Yoon S, Jeon SW, Ko YH, et al. Adjunctive brexpiprazole as a novel effective strategy for treating major depressive disorder: a systematic review and meta-analysis. J Clin Psychopharmacol. 2017;37(1):46–53 [M].
[109] Marder SR, Hakala MJ, Josiassen MK, et al. Brexpiprazole in patients with schizophrenia: overview of short- and long-term phase 3 controlled studies. Acta Neuropsychiatr. 2017;29(5):278–90 [R].
[110] Parikh NB, Robinson DM, Clayton AH. Clinical role of brexpiprazole in depression and schizophrenia. Ther Clin Risk Manag. 2017;13:299–306 [R].
[111] Earley W, Durgam S, Lu K, et al. Safety and tolerability of cariprazine in patients with acute exacerbation of schizophrenia: a pooled analysis of four phase II/III randomized, double-blind, placebo-controlled studies. Int Clin Psychopharmacol. 2017;32(6):319–28 [c].
[112] Nasrallah HA, Earley W, Cutler AJ, et al. The safety and tolerability of cariprazine in long-term treatment of schizophrenia: a post hoc pooled analysis. BMC Psychiatry. 2017;17(1):305 [C].
[113] Durgam S, Greenberg WM, Li D, et al. Safety and tolerability of cariprazine in the long-term treatment of schizophrenia: results from a 48-week, single-arm, open-label extension study. Psychopharmacology (Berl). 2017;234(2):199–209 [c].
[114] Earley W, Durgam S, Lu K, et al. Tolerability of cariprazine in the treatment of acute bipolar I mania: a pooled post hoc analysis of 3 phase II/III studies. J Affect Disord. 2017;215:205–12 [C].
[115] Gowda GS, Hegde A, Shanbhag V, et al. Kerato-lenticular ocular deposits and visual impairment with prolonged chlorpromazine use: a case series. Asian J Psychiatr. 2017;25:188–90 [A].
[116] Nemani KL, Greene MC, Ulloa M, et al. Clozapine, diabetes mellitus, cardiovascular risk and mortality: results of a 21-year

naturalistic study in patients with schizophrenia and schizoaffective disorder. Clin Schizophr Relat Psychoses. 2017. https://doi.org/10.3371/CSRP.KNMG.111717, Epub ahead of print [c].
[117] Subramanian S, Völlm BA, Huband N. Clozapine dose for schizophrenia. Cochrane Database Syst Rev. 2017;6:CD009555 [M].
[118] Norman SM, Sullivan KM, Liu F, et al. Blood pressure and heart rate changes during clozapine treatment. Psychiatr Q. 2017;88(3):545–52 [c].
[119] Nilsson BM, Edström O, Lindström L, et al. Tachycardia in patients treated with clozapine versus antipsychotic long-acting injections. Int Clin Psychopharmacol. 2017;32(4):219–24 [C].
[120] Khan AA, Ashraf A, Baker D, et al. Clozapine and incidence of myocarditis and sudden death—long term Australian experience. Int J Cardiol. 2017;238:136–9 [C].
[121] Nguyen B, Du C, Bastiampillai T, et al. Successful clozapine re-challenge following myocarditis. Australas Psychiatry. 2017;25(4):385–6 [A].
[122] Gerasimou C, Vitali GP, Vavougios GD, et al. Clozapine associated with autoimmune reaction, fever and low level cardiotoxicity—a case report. In Vivo. 2017;31(1):141–3 [A].
[123] Perdue M, Botello T. Development of atrial flutter after initiation of clozapine and successful rechallenge without recurrence. J Clin Psychopharmacol. 2017;37(4):475–7 [A].
[124] Longhi S, Heres S. Clozapine-induced, dilated cardiomyopathy: a case report. BMC Res Notes. 2017;10(1):338 [A].
[125] Nederlof M, Benschop TW, de Vries Feyens CA, et al. Clozapine re exposure after dilated cardiomyopathy, BMJ Case Rep. 2017; pii: bcr-2017-219652. https://doi.org/10.1136/bcr-2017-219652 [A].
[126] Sawyer M, Goodison G, Smith L, et al. Brugada pattern associated with clozapine initiation in a man with schizophrenia. Intern Med J. 2017;47(7):831–3 [A].
[127] Sackey B, Miller LJ, Davis MC. Possible clozapine overdose-associated thromboembolic event. J Clin Psychopharmacol. 2017;37(3):364–6 [A].
[128] Sarathy K, Alexopoulos C. A successful re-trial after clozapine myopericarditis. J R Coll Physicians Edinb. 2017;47(2):146–7 [A].
[129] Trincas G, Lampugnani D, Beretta S, et al. Late onset clozapine-induced sierositis: the case of ms C. Riv Psichiatr. 2017;52(3):126–8 [A].
[130] Kolli V, Bourke D, Ngo J, et al. Clozapine-related tachycardia in an adolescent with treatment-resistant early onset schizophrenia. J Child Adolesc Psychopharmacol. 2017;27(2):206–8 [A].
[131] Grover S, Sahoo S, Mahajan S. Clozapine-induced hypertension: a case report and review of literature. Ind Psychiatry J. 2017;26(1):103–5 [A].
[132] Stoecker ZR, George WT, O'Brien JB, et al. Clozapine usage increases the incidence of pneumonia compared with risperidone and the general population: a retrospective comparison of clozapine, risperidone, and the general population in a single hospital over 25 months. Int Clin Psychopharmacol. 2017;32(3):155–60 [C].
[133] Ingole A, Bastiampillai T, Tibrewal P. Clozapine withdrawal catatonia, psychosis and associated neuroleptic malignant syndrome. Asian J Psychiatr. 2017;30:96–7 [A].
[134] Bilbily J, McCollum B, de Leon J. Catatonia secondary to sudden clozapine withdrawal: a case with three repeated episodes and a literature review. Case Rep Psychiatry. 2017;2017:2402731 [A].
[135] Modak A, Åhlin A. The treatment of clozapine-withdrawal delirium with electroconvulsive therapy. Case Rep Psychiatry. 2017;2017:1783545 [A].
[136] Leonardo QF, Juliana GR, Fernando CJ. Atypical neuroleptic malignant syndrome associated with use of clozapine. Case Rep Emerg Med. 2017;2017:2174379 [A].
[137] Kaplan AM, Pitts WB, Ahmed I. An unexpected circumstance: acute dystonic reaction in the setting of clozapine administration, J Pharm Pract. 2017; 897190017737696. https://doi.org/10.1177/0897190017737696, Epub ahead of print. [A].
[138] Huh L, Lee BJ. Efficacy of low-dose aripiprazole to treat clozapine-associated tardive dystonia in a patient with schizophrenia. Turk Psikiyatri Derg. 2017;28(3):208–11 [A].
[139] Suresh Kumar PN, Gopalakrishnan A. Clozapine-associated Pisa syndrome: a rare type of tardive dystonia. Indian J Psychiatry. 2017;59(3):390–1 [A].
[140] Das S, Purushothaman ST, Rajan V, et al. Clozapine-induced tardive dyskinesia. Indian J Psychol Med. 2017;39(4):551–2 [A].
[141] Gallo M, Squarcione C, Bersani FS, et al. Clozapine-related extrapyramidal side effects: a case report. Riv Psichiatr. 2017;52(4):172–3 [C].
[142] Bolu A, Akarsu S, Pan E, et al. Low-dose clozapine-induced seizure: a case report. Clin Psychopharmacol Neurosci. 2017;15(2):190–3 [A].
[143] Rachamallu V, Haq A, Song MM, et al. Clozapine-induced microseizures, orofacial dyskinesia, and speech dysfluency in an adolescent with treatment resistant early onset schizophrenia on concurrent lithium therapy. Case Rep Psychiatry. 2017;2017:7359095 [A].
[144] Takahashi Y, Ogihara T, Sasayama D, et al. Successful readministration of clozapine in a patient with a history of clozapine-induced elevation of creatine phosphokinase, Prim Care Companion CNS Disord. 2017;19(4) pii: 16l02084. https://doi.org/10.4088/PCC.16l02084 [A].
[145] Tong JY, Pai A, Young SH. Clozapine-induced maculopathy. Med J Aust. 2017;207(7):316 [A].
[146] Mack HG, Symons RCA. Clozapine-induced maculopathy. Med J Aust. 2017;207(7):316 [A].
[147] Ingimarsson O, MacCabe JH, Haraldsson M, et al. Risk of diabetes and dyslipidemia during clozapine and other antipsychotic drug treatment of schizophrenia in Iceland. Nord J Psychiatry. 2017;71(7):496–502 [A].
[148] Porras-Segovia A, Krivoy A, Horowitz M, et al. Rapid-onset clozapine-induced loss of glycaemic control: case report. BJPsych Open. 2017;3(3):138–40 [C].
[149] Tso G, Kumar P, Jayasooriya T, et al. Metabolic monitoring and management among clozapine users. Australas Psychiatry. 2017;25(1):48–52 [C].
[150] Kumar M, Sidana A. Clozapine-induced acute hypertriglyceridemia. Indian J Psychol Med. 2017;39(5):682–4 [A].
[151] Ho YC, Lin HL. Continuation with clozapine after eosinophilia: a case report. Ann Gen Psychiatry. 2017;16:46 [C].
[152] Fabrazzo M, Prisco V, Sampogna G, et al. Clozapine versus other antipsychotics during the first 18 weeks of treatment: a retrospective study on risk factor increase of blood dyscrasias. Psychiatry Res. 2017;256:275–82 [A].
[153] Andres E, Mourot-Cottet R. Clozapine-associated neutropenia and agranulocytosis. J Clin Psychopharmacol. 2017;37(6):749–50 [A].
[154] Marchel D, Hart AL, Keefer P, et al. Multiorgan eosinophilic infiltration after initiation of clozapine therapy: a case report. BMC Res Notes. 2017;10(1):316 [A].
[155] Yamaki N, Hishimoto A, Otsuka I, et al. Optimizing outcomes in clozapine rechallenge following neutropenia using human leukocyte antigen typing: a case report. Psychiatry Clin Neurosci. 2017;71(4):289–90 [A].
[156] Foster J, Lally J, Bell V, et al. Successful clozapine re-challenge in a patient with three previous episodes of clozapine-associated blood dyscrasia. BJPsych Open. 2017;3(1):22–5 [A].

[157] Osterman MT, Foley C, Matthias I. Clozapine-induced acute gastrointestinal necrosis: a case report. J Med Case Reports. 2017;11(1):270 [C].
[158] Every-Palmer S, Ellis PM. Clozapine-induced gastrointestinal hypomotility: a 22-year bi-national pharmacovigilance study of serious or fatal 'slow gut' reactions, and comparison with international drug safety advice. CNS Drugs. 2017;31(8):699–709 [R].
[159] Every-Palmer S, Ellis PM, Nowitz M, et al. The Porirua protocol in the treatment of clozapine-induced gastrointestinal hypomotility and constipation: a pre- and post-treatment study. CNS Drugs. 2017;31(1):75–85 [c].
[160] West S, Rowbotham D, Xiong G, et al. Clozapine induced gastrointestinal hypomotility: a potentially life threatening adverse event: a review of the literature. Gen Hosp Psychiatry. 2017;46:32–7 [A].
[161] Das S, Agrawal AK. Clozapine-induced delayed ejaculation. Indian J Psychol Med. 2017;39(6):828 [A].
[162] Hassine H, Ouali U, Ouertani A, et al. Clozapine-induced DRESS syndrome with multiple and rare organ involvement. Asian J Psychiatr. 2017;28:146–7 [A].
[163] Duarte TA, Godinho FDG, Ferreira ALB. Clozapine-induced procalcitonin elevation, Prim Care Companion CNS Disord. 2017;19(3) pii: 16l02054. https://doi.org/10.4088/PCC.16l02054 [c].
[164] Hung YP, Wang CS, Yen CN, et al. Role of cytokine changes in clozapine-induced fever: a cohort prospective study. Psychiatry Clin Neurosci. 2017;71(6):395–402 [A].
[165] Mehta TM, Van Lieshout RJ. A review of the safety of clozapine during pregnancy and lactation. Arch Womens Ment Health. 2017;20(1):1–9 [R].
[166] van der Weide K, Loovers H, Pondman K, et al. Genetic risk factors for clozapine-induced neutropenia and agranulocytosis in a Dutch psychiatric population. Pharmacogenomics J. 2017;17(5):471–8 [C].
[167] Suetani RJ, Siskind D, Reichhold H, et al. Genetic variants impacting metabolic outcomes among people on clozapine: a systematic review and meta-analysis. Psychopharmacology (Berl). 2017;234(20):2989–3008 [M].
[168] Piatkov I, Caetano D, Assur Y, et al. CYP2C19*17 protects against metabolic complications of clozapine treatment. World J Biol Psychiatry. 2017;18(7):521–7 [C].
[169] Piatkov I, Caetano D, Assur Y, et al. BCB1 and ABCC1 single-nucleotide polymorphisms in patients treated with clozapine. Pharmgenomics Pers Med. 2017;10:235–42 [C].
[170] Yousra H, Pierrick L, Laurent L, et al. Interaction between clozapine and oxcarbazepine: a case report. Ther Adv Psychopharmacol. 2017;7(2):95–9 [A].
[171] Umino M, Kobayashi R, Nisijima K, et al. Late-onset rhabdomyolysis associated with an intramuscular injection of fluphenazine decanoate, Prim Care Companion CNS Disord. 2017;19(5) pii: 16l02078. https://doi.org/10.4088/PCC.16l02078 [A].
[172] Ostinelli EG, Brooke-Powney MJ, Li X, et al. Haloperidol for psychosis-induced aggression or agitation (rapid tranquillisation). Cochrane Database Syst Rev. 2017;7:CD009377 [M].
[173] Schmidt A, Fischer P, Wally B, et al. Influence of intravenous administration of the antipsychotic drug benperidol on the QT interval. Neuropsychiatr. 2017;31(4):172–5 [c].
[174] Ridout KK, Ridout SJ. Haloperidol-associated uterine dystonia. Am J Psychiatry. 2017;174(3):296 [A].
[175] Masiran R. Persistent oromandibular dystonia and angioedema secondary to haloperidol, BMJ Case Rep. 2017;2017: pii: bcr-2017-220817. https://doi.org/10.1136/bcr-2017-220817 [A].
[176] Balai M, Ansari F, Gupta LK, et al. Urticaria and angioedema associated with haloperidol. Indian J Dermatol. 2017;62(5):539–40 [A].
[177] López-Valdés JC. Haloperidol poisoning in pediatric patients. Gac Med Mex. 2017;153(1):125–8 [A].
[178] Dekkers BGJ, Eck RJ, Ter Maaten JC, et al. An acute oral intoxication with haloperidol decanoate. Am J Emerg Med. 2017;35(9) 1387.e1-1387.e2. [A].
[179] Zastrozhin MS, Brodyansky VM, Skryabin VY, et al. Pharmacodynamic genetic polymorphisms affect adverse drug reactions of haloperidol in patients with alcohol-use disorder. Pharmgenomics Pers Med. 2017;10:209–15 [c].
[180] Singh V, Arnold JG, Prihoda TJ, et al. An open trial of iloperidone for mixed episodes in bipolar disorder. J Clin Psychopharmacol. 2017;37(5):615–9 [c].
[181] Choudhury S, Chatterjee K, Singh R, et al. Levosulpiride-induced movement disorders. J Pharmacol Pharmacother. 2017;8(4):177–81 [A].
[182] DelBello MP, Goldman R, Phillips D, et al. Efficacy and safety of lurasidone in children and adolescents with bipolar I depression: a double-blind, placebo-controlled study. J Am Acad Child Adolesc Psychiatry. 2017;56(12):1015–25 [C].
[183] Goldman R, Loebel A, Cucchiaro J, et al. Efficacy and safety of lurasidone in adolescents with schizophrenia: a 6-week, randomized placebo-controlled study. J Child Adolesc Psychopharmacol. 2017;27(6):516–25 [C].
[184] Sood S. Neutropenia with multiple antipsychotics including dose dependent neutropenia with lurasidone. Clin Psychopharmacol Neurosci. 2017;15(4):413–5 [A].
[185] Das S, Agrawal A. Lurasidone-induced oculogyric crisis. Indian J Psychol Med. 2017;39(5):719–20 [A].
[186] Lee M, Marshall D, Saddichha S. Lurasidone-associated neuroleptic malignant syndrome. J Clin Psychopharmacol. 2017;37(5):639–40 [A].
[187] Tibrewal P, Cheng R, Bastiampillai T, et al. Lurasidone-induced dystonia, Prim Care Companion CNS Disord. 2017;19(3) pii: 16l02053. https://doi.org/10.4088/PCC.16l02053 [A].
[188] Reichenberg JL, Ridout KK, Rickler KC. Lurasidone-induced rabbit syndrome: a case report. J Clin Psychiatry. 2017;78(5)e553 [A].
[189] Kanzawa M, Hadden O. Case report of a switch to mania induced by lurasidone. Ther Adv Psychopharmacol. 2017;7(2):91–3 [A].
[190] Cole JB, Klein LR, Strobel AM, et al. The use, safety, and efficacy of olanzapine in a level I pediatric trauma center emergency department over a 10-year period. Pediatr Emerg Care. 2017. https://doi.org/10.1097/PEC.0000000000001231, Epub ahead of print. [C].
[191] Keles E, Ulasli SS, Basaran NC, et al. Acute massive pulmonary embolism associated with olanzapine. Am J Emerg Med. 2017;35(10) 1582.e5–1582.e7. [A].
[192] Cruz MD, Danoff R. Thrombocytopenia and spontaneous intracranial hemorrhage after olanzapine therapy. J Am Osteopath Assoc. 2017;117(7):473–5 [A].
[193] Reilly TJ, Cross S, Taylor DM, et al. Neuroleptic malignant syndrome following catatonia: vigilance is the price of antipsychotic prescription. SAGE Open Med Case Rep. 2017;5: 2050313X17695999. [A].
[194] Saha PK, Chakraborty A, Layek AK, et al. Olanzapine-induced neuroleptic malignant syndrome. Indian J Psychol Med. 2017;39(3):364–5 [A].
[195] Hosseini S, Elyasi F. Olanzapine-induced neuroleptic malignant syndrome. Iran J Med Sci. 2017;42(3):306–9 [A].
[196] Sahoo MK, Kamath S, Sharan A. Neuroleptic malignant syndrome in a patient with stable dose of olanzapine. J Family Med Prim Care. 2017;6(1):158–60 [A].
[197] Ali M, Das S, Thirthalli J, et al. Olanzapine induced neuroleptic malignant syndrome, treated with electroconvulsive therapy (ECT): a case report. Asian J Psychiatr. 2017;30:230–1 [A].

[198] Mills EW, Shaffer LS, Goes FS, et al. Case of secondary tics associated with olanzapine in an adult. Front Psychiatry. 2017;8:150. https://doi.org/10.3389/fpsyt.2017.00150. eCollection 2017. [A].
[199] Kar SK, Singh A. Management dilemma in olanzapine induced restlessness and cramps in legs. Clin Psychopharmacol Neurosci. 2017;15(1):87–8 [A].
[200] Arora T, Maharshi V, Rehan HS, et al. Blepharospasm: an uncommon adverse effect caused by long-term administration of olanzapine. J Basic Clin Physiol Pharmacol. 2017;28(1):85–7 [A].
[201] Sangroula D, Virk I, Mohammad W, et al. Clozapine treatment of olanzapine-induced tardive dyskinesia: a case report. J Psychiatr Pract. 2017;23(1):53–9 [A].
[202] Park JI. Delirium associated with olanzapine use in the elderly. Psychogeriatrics. 2017;17(2):142–3 [A].
[203] Das S, Gupta R, Dhyani M, et al. Role of 5HT1A and 5HT2A receptors and default mode network in olanzapine-induced somnambulism. Neurol India. 2017;65(2):373–4 [A].
[204] Deng S, Hu J, Zhu X, et al. Sleepwalking is associated with coadministration of olanzapine and propranolol: a case report. J Clin Psychopharmacol. 2017;37(5):622–3 [A].
[205] Rickels MR, Perez EM, Peleckis AJ, et al. Contribution of parasympathetic muscarinic augmentation of insulin secretion to olanzapine-induced hyperinsulinemia. Am J Physiol Endocrinol Metab. 2018;315(2):E250–7. https://doi.org/10.1152/ajpendo.00315.2017, Epub 2017 Dec 19. [C].
[206] Guina J, Roy S, Gupta A. Oral glucose tolerance test performance in olanzapine-treated schizophrenia-spectrum patients is predicted by BMI and triglycerides but not olanzapine dose or duration. Hum Psychopharmacol. 2017;32(4). https://doi.org/10.1002/hup.2604 Epub 2017 Jun 1. [C].
[207] Okazaki K, Yamamuro K, Kishimoto T. Reversal of olanzapine-induced weight gain in a patient with schizophrenia by switching to asenapine: a case report. Neuropsychiatr Dis Treat. 2017;13:2837–40 [A].
[208] Iwaku K, Otuka F, Taniyama M. Acute-onset Type 1 diabetes that developed during the administration of olanzapine. Intern Med. 2017;56(3):335–9 [A].
[209] Ripoli C, Pinna AP, Podda F, et al. Second-generation antipsychotic and diabetes mellitus in children and adolescents. Pediatr Med Chir. 2017;39(4):149 [A].
[210] Anil SS, Ratnakaran B, Suresh N. A case report of rapid-onset hyponatremia induced by low-dose olanzapine. J Family Med Prim Care. 2017;6(4):878–80 [A].
[211] Wu YF. Recurrent hyperammonemia associated with olanzapine. J Clin Psychopharmacol. 2017;37(3):366–7 [A].
[212] Pang N, Thrichelvam N, Naing KO. Olanzapine-induced pancytopenia: a rare but worrying complication. East Asian Arch Psychiatry. 2017;27(1):35–7 [A].
[213] Dönmez YE, Özcan Ö, Soylu N, et al. Management of hepatotoxicity induced by the use of olanzapine. J Child Adolesc Psychopharmacol. 2017;27(3):293–4 [A].
[214] Kanodia S, Giri VP, Veerabhadrappa RS, et al. Black hairy tongue with olanzapine: a rare case report. Indian J Psychiatry. 2017;59(2):249–50 [A].
[215] Shahi MK, Kar SK, Singh A. Asymmetric, tender gynecomastia induced by olanzapine in a young male. Indian J Psychol Med. 2017;39(2):215–6 [A].
[216] Chepure AH, Ungratwar AK. Olanzapine-induced psoriasis. Indian J Psychol Med. 2017;39(6):811–2 [A].
[217] Chawla N, Kumar S, Balhara YPS. Olanzapine-induced skin eruptions. Indian J Psychol Med. 2017;39(4):537–8 [A].
[218] Bozkurt H, Şahin S. Olanzapine-induced priapism in a child with Asperger's syndrome. Balkan Med J. 2017;34(1):85–7 [A].
[219] Penchilaiya V, Kuppili PP, Preeti K, et al. DRESS syndrome: addressing the drug hypersensitivity syndrome on combination of sodium valproate and olanzapine. Asian J Psychiatr. 2017;28:175–6 [A].
[220] Yang CH, Chen YY. A case of olanzapine-induced fever. Psychopharmacol Bull. 2017;47(1):45–7 [A].
[221] Maharshi V, Banerjee I, Nagar P, et al. Tracheo-esophageal fistula (TEF) in a newborn following maternal antenatal exposure to olanzapine. Drug Saf Case Rep. 2017;4(1):2 [A].
[222] Meyers KJ, Upadhyaya HP, Landry JL, et al. Postinjection delirium/sedation syndrome in patients with schizophrenia receiving olanzapine long-acting injection: results from a large observational study. BJPsych Open. 2017;3(4):186–92 [C].
[223] Uzun S, Kozumplik O, Ćelić I, et al. Occurrence of post-injection delirium/sedation syndrome after application of olanzapine long-acting injection during one year period. Psychiatr Danub. 2017;29(4):497–9 [A].
[224] Uglešić L, Glavina T, Lasić D, et al. Postinjection delirium/sedation syndrome (PDSS) following olanzapine long-acting injection: a case report. Psychiatr Danub. 2017;29(1):90–1 [A].
[225] Upadhyay A, Bhandari SS, Sharma V, et al. Postinjection delirium/sedation syndrome with long-acting olanzapine pamoate in a middle-aged female. Indian J Psychiatry. 2017;59(4):517–9 [A].
[226] Descusse A, Chebili S, Artiges E. Post-injection syndrome and olanzapine pamoate: a severe case report. Encephale. 2017;43(4):405 [A].
[227] Keyal N, Shrestha GS, Pradhan S, et al. Olanzapine overdose presenting with acute muscle toxicity. Int J Crit Illn Inj Sci. 2017;7(1):69–71 [A].
[228] Zhao J, Li L, Shi J, et al. Safety and efficacy of paliperidone palmitate 1-month formulation in Chinese patients with schizophrenia: a 25-week, open-label, multicenter, phase IV study. Neuropsychiatr Dis Treat. 2017;13:2045–56 [C].
[229] Omi T, Kanai K, Kiguchi T, et al. The possibility of the treatment for long-acting injectable antipsychotics induced severe side effects. Am J Emerg Med. 2017;35(8) 1211.e1–1211.e2. [A].
[230] Suzuki H, Hibino H, Inoue Y, et al. One patient with schizophrenia showed reduced drug-induced extrapyramidal symptoms as a result of an alternative regimen of treatment with paliperidone 3 and 6 mg every other day, SAGE Open Med Case Rep. 2017;5: 2050313X17742836. https://doi.org/10.1177/2050313X17742836, eCollection 2017. [A].
[231] Ahmed N. Add-on gabapentin alleviates paliperidone-induced head tremors and boosts antipsychotic response in early-onset schizophrenia. Isr J Psychiatry Relat Sci. 2017;54(2):59–60 [A].
[232] Jang S, Woo J. Five month-persistent extrapyramidal symptoms following a single injection of paliperidone palmitate: a case report. Clin Psychopharmacol Neurosci. 2017;15(3):288–91 [A].
[233] Kim DD, Lang DJ, Warburton DER, et al. Exercise-associated extrapyramidal symptoms during treatment with long-acting injectable antipsychotic medications: a case report. Clin Schizophr Relat Psychoses. 2017. https://doi.org/10.3371/CSRP.DKDL.071317, Epub ahead of print. [A].
[234] Marques JG. Oral facial dystonia (Meige or Brueghel Syndrome) induced by paliperidone palmitate, Prim Care Companion CNS Disord. 2017;19(1). https://doi.org/10.4088/PCC.16l01997 [A].
[235] Isaacs AN, Eaves SM, Ott CA. Hyponatremia associated with once-monthly paliperidone palmitate use. Ann Pharmacother. 2017;51(9):817–8 [C].
[236] Tibrewal P, Dhillon R, Sharma J, et al. Paliperidone-induced hyponatremia, Prim Care Companion CNS Disord. 2017;19(4). pii: 16l02088. https://doi.org/10.4088/PCC.16l02088 [A].
[237] Gopal S, Lane R, Nuamah I, et al. Evaluation of potentially prolactin-related adverse events and sexual maturation in

adolescents with schizophrenia treated with paliperidone extended-release (ER) for 2 years: a post hoc analysis of an open-label multicenter study. CNS Drugs. 2017;31(9):797–808 [A].
[238] Papadopoulou A, Gkikas K, Efstathiou V, et al. Angioedema associated with long-acting injectable paliperidone palmitate: a case report. J Clin Psychopharmacol. 2017;37(6):730–2 [A].
[239] Cicek E, Cicek IE, Uguz F. Bilateral pretibial edema associated with paliperidone palmitate long-acting injectable: a case report. Clin Psychopharmacol Neurosci. 2017;15(2):184–6 [A].
[240] Binns R, O'Halloran SJ, Teoh S, et al. Placental transfer of paliperidone during treatment with a depot formulation. J Clin Psychopharmacol. 2017;37(4):474–5 [A].
[241] Garriga M, Solé E, González-Pinto A, et al. Efficacy of quetiapine XR vs. placebo as concomitant treatment to mood stabilizers in the control of subthreshold symptoms of bipolar disorder: results from a pilot, randomized controlled trial. Eur Neuropsychopharmacol. 2017;27(10):959–69 [c].
[242] Giancaterino S, Solimine S. Probable acquired QTc prolongation and subsequent torsades de pointes attributable to quetiapine, Prim Care Companion CNS Disord. 2017;19(6) pii: 17l02106. https://doi.org/10.4088/PCC.17l02106 [A].
[243] Smolders DME, Smolders WAP. Case report and review of the literature: cardiomyopathy in a young woman on high-dose quetiapine. Cardiovasc Toxicol. 2017;17(4):478–81 [A].
[244] Bhogal S, Ladia V, Paul TK. Quetiapine-associated myopericarditis. Am J Ther. 2018;25(5):e578–9. https://doi.org/10.1097/MJT.0000000000000671 [A].
[245] Khalid M, Bakhit A, Dufresne A, et al. LBBB induced by quetiapine overdose: a case report and literature review. Am J Ther. 2017;24(5):e618–20 [A].
[246] Lai J, Lu Q, Huang T, et al. Convulsive syncope related to a small dose of quetiapine in an adolescent with bipolar disorder. Neuropsychiatr Dis Treat. 2017;13:1905–8 [A].
[247] Rovera C, Esposito CM, Ciappolino V, et al. Quetiapine-induced hypomania and its association with quetiapine/norquetiapine plasma concentrations: a case series of bipolar type 2 patients. Drug Saf Case Rep. 2017;4(1):13 [A].
[248] Chien CF, Huang P, Hsieh SW. Reversible global aphasia as a side effect of quetiapine: a case report and literature review. Neuropsychiatr Dis Treat. 2017;13:2257–60 [A].
[249] Demily C, Poisson A, Thibaut F, et al. Weight loss induced by quetiapine in a 22q11.2DS patient. Mol Genet Metab Rep. 2017;13:95–6 [A].
[250] Glocker C, Grohmann R, Schulz H. Fatal agranulocytosis associated with quetiapine in monotherapy: a case report. J Clin Psychopharmacol. 2017;37(5):625–7 [A].
[251] Husnain M, Gondal F, Raina AI, et al. Quetiapine associated thrombotic thrombocytopenic purpura: a case report and literature review. Am J Ther. 2017;24(5):e615–6. https://doi.org/10.1097/MJT.0000000000000456 [A].
[252] Somani A, Sharma M, Singh SM. Neutropenia associated with quetiapine and sertraline: a case report and review of literature. Asian J Psychiatr. 2017;26:129–30 [A].
[253] Cuny P, Houot M, Ginisty S, et al. Quetiapine and anticholinergic drugs induced ischaemic colitis: a case study. Encephale. 2017;43(1):81–4 [A].
[254] Das A, Guarda LA, Allen LG. Liver injury associated with quetiapine: an illustrative case report. J Clin Psychopharmacol. 2017;37(5):623–5 [A].
[255] Tay JL. Quetiapine causing severe fall and skull fracture: a case report. Asian J Psychiatr. 2017;28:150–1 [A].
[256] Ceylan MF, Erdogan B, Tural Hesapcioglu S, et al. Effectiveness, adverse effects and drug compliance of long-acting injectable risperidone in children and adolescents. Clin Drug Investig. 2017;37(10):947–56 [c].
[257] Srisawasdi P, Vanwong N, Hongkaew Y, et al. Impact of risperidone on leptin and insulin in children and adolescents with autistic spectrum disorders. Clin Biochem. 2017;50(12):678–85 [C].
[258] Wilson MP, Nordstrom K, Hopper A, et al. Risperidone in the emergency setting is associated with more hypotension in elderly patients. J Emerg Med. 2017;53(5):735–9 [C].
[259] Ito A, Enokiya T, Kawamoto E, et al. Two cases of life-threatening arrhythmia induced by risperidone: evaluation of risperidone and 9-hydroxy-risperidone concentrations. Acute Med Surg. 2017;4(3):341–3 [A].
[260] Thomson SR, Chogtu B, Bhattacharjee D, et al. Extrapyramidal symptoms probably related to risperidone treatment: a case series. Ann Neurosci. 2017;24(3):155–63 [A].
[261] Mitchell BG, McMahon BC, Mitchell CW. Neuroleptic malignant syndrome after early administration of risperidone long-acting injection, Prim Care Companion CNS Disord. 2017;19(6). pii: 17l02125. https://doi.org/10.4088/PCC.17l02125 [A].
[262] Sutter M, Walter M, Dürsteler KM, et al. Psychosis after switch in opioid maintenance agonist and risperidone-induced Pisa syndrome: two critical incidents in dual diagnosis treatment. J Dual Diagn. 2017;13(2):157–65 [A].
[263] Sürer Adanir A, Gizli Çoban Ö, Özatalay E. Increased hyperacusis with risperidone in an autistic child. Noro Psikiyatr Ars. 2017;54(2):187–8 [A].
[264] Takechi K, Yoshioka Y, Kawazoe H, et al. Psychiatric patients with antipsychotic drug-induced hyperprolactinemia and menstruation disorders. Biol Pharm Bull. 2017;40(10):1775–8 [A].
[265] Shagufta S, Farooq F, Khan AM, et al. Risperidone-induced amenorrhea in floridly psychotic female. Cureus. 2017;9(9)e1683. https://doi.org/10.7759/cureus.1683 [A].
[266] Boothby A, Shad MU. Hyperprolactinemia with low dose of risperidone in a young female. Asian J Psychiatr. 2017;27:69–70 [A].
[267] Holla S, Amberkar MB, Kamath A, et al. Risperidone-induced granulomatous mastitis secondary to hyperprolactinemia in a non-pregnant woman-a rare case report in a bipolar disorder. J Clin Diagn Res. 2017;11(1):FD01–3. https://doi.org/10.7860/JCDR/2017/20733.9278 [A].
[268] Matera E, Margari L, Palmieri VO, et al. Risperidone and cardiometabolic risk in children and adolescents: clinical and instrumental issues. J Clin Psychopharmacol. 2017;37(3):302–9 [c].
[269] Vanwong N, Srisawasdi P, Ngamsamut N, et al. Hyperuricemia in children and adolescents with autism spectrum disorder treated with risperidone: the risk factors for metabolic adverse effects. Front Pharmacol. 2017;7:527 [C].
[270] Kattalai Kailasam V, Chima V, Nnamdi U, et al. Risperidone-induced reversible neutropenia. Neuropsychiatr Dis Treat. 2017;13:1975–7 [A].
[271] Morrison M, Schultz A, Sanchez DL, et al. Leukopenia associated with risperidone treatment. Curr Drug Saf. 2017. https://doi.org/10.2174/1574886312666170531072837, Epub ahead of print. [A].
[272] Pendharkar SS, Telgote SA, Jadhav A, et al. Risperidone-induced erythema multiforme minor. Indian J Psychol Med. 2017;39(6):808–10 [A].
[273] Look ML, Boo YL, Chin PW, et al. Risperidone-associated rhabdomyolysis without neuroleptic malignant syndrome: a case report. J Clin Psychopharmacol. 2017;37(1):105–6 [A].

[274] Unver H, Memik NC, Simsek E. Priapism associated with the addition of risperidone to methylphenidate monotherapy: a case report. North Clin Istanb. 2017;4(1):85–8 [A].

[275] Grau K, Plener PL, Gahr M, et al. Mild hypothermia in a child with low-dose risperidone. Z Kinder Jugendpsychiatr Psychother. 2017;45(4):335–7 [A].

[276] Das S, Dey JK, Prabhu Ss N, et al. Association between 5-HTR2C -759C/T (rs3813929) and -697G/C (rs518147) gene polymorphisms and risperidone-induced insulin resistance syndrome in an Indian population. J Clin Pharmacol. 2017. https://doi.org/10.1002/jcph.1012, Epub ahead of print. [c].

[277] Alladi CG, Mohan A, Shewade DG, et al. Risperidone-induced adverse drug reactions and role of DRD2 (-141 C Ins/Del) and 5HTR2C (-759 C>T) genetic polymorphisms in patients with schizophrenia. J Pharmacol Pharmacother. 2017;8(1):28–32 [C].

[278] Chen J, Pan X, Qian M, et al. Efficacy and metabolic influence on blood-glucose and serum lipid of ziprasidone in the treatment of elderly patients with first-episode schizophrenia. Shanghai Arch Psychiatry. 2017;29(2):104–10 [c].

[279] Bryan EJ, Purcell MA, Kumar A. Zuclopenthixol dihydrochloride for schizophrenia. Cochrane Database Syst Rev. 2017;11:CD005474. https://doi.org/10.1002/14651858.CD005474.pub2 [M].

[280] Chue P, Baker G. Antipsychotic drugs. In: Ray SD, editor. Side effects of drugs annual: a worldwide yearly survey of new data in adverse drug reactions, vol. 37. Elsevier; 2015. p. 63–83 [chapter 6]. [R].

[281] Chue P, Chue J. Antipsychotic drugs. In: Ray SD, editor. Side effects of drugs annual: a worldwide yearly survey of new data in adverse drug reactions, vol. 38. Elsevier; 2016. p. 35–54 [chapter 5]. [R].

[282] Chue PS, Siraki AG. Antipsychotic drugs. In: Ray SD, editor. Side effects of drugs annual: a worldwide yearly survey of new data in adverse drug reactions, vol. 39. Elsevier; 2017. p. 65–90 [chapter 6]. [R].

CHAPTER

7

Antiepileptics

Robert D. Beckett[1], *Anayo Ohiri*, *Alyssa Johnson*, *Thomas R. Smith*

Manchester University College of Pharmacy, Natural and Health Sciences, Fort Wayne, IN, United States

[1]Corresponding author: rdbeckett@manchester.edu

GABA RECEPTOR AGONISTS [1R, 2R]

Clobazam

No relevant studies identified from the search period.

Clonazepam

A 63-year-old female began treatment with clonazepam for facial myoclonus; after 9 months, the patient presented with pruritic, scaly erythema on the back and upper extremities (symptoms were reported to have occurred concurrently with initiation of therapy and the patient was on no other medications) [3A]. The clonazepam was discontinued, with symptoms resolving within 2 months. The patient was re-challenged on clonazepam due to re-exacerbation of myoclonus. The symptoms reoccurred, and the patient was diagnosed with lichenoid drug eruption following a negative drug-induced lymphocyte stimulation test and lack of prodromal or infectious symptoms, as well as considering epidermis histopathology consistent with the diagnosis. The patient was switched to treatment with diazepam, which resulted in complete remission of symptoms. The authors reported this was the first reported case of lichenoid drug eruption associated with clonazepam, although it has been associated with other dermatological reactions including alopecia, bullous dermatosis, erythema multiforme, exfoliating eruptions, and pseudomycosis fungoides.

Diazepam

A randomized, controlled trial of 273 pediatric patients with status epilepticus, published in 2014, found similar results for IV lorazepam vs IV diazepam (rates of status epilepticus cessation with one or two doses (0.1 mg/kg lorazepam up to 4 mg, 0.2 mg/kg diazepam up to 8 mg) were 72.9% and 72.1%, and rates of respiratory depression were 16% and 17%, respectively) [4C]. A secondary Bayesian analysis focused on evaluating risk for respiratory depression was conducted to determine the probabilities that lorazepam is superior, non-inferior, and equivalent to diazepam. The authors calculated a median likelihood of significant respiratory depression, defined as need for ventilation, of 0.174 (95% high density interval [HDI] 0.112–0.240) for lorazepam vs 0.173 (95% HDI 0.116–0.234) for diazepam. The calculated difference in median likelihood of respiratory depression was zero (95% HDI −0.087 to 0.089), with similar results obtained in a sensitivity analysis considering different assumption of prior risk. The authors calculated 98.9% probability that lorazepam is non-inferior and 97.6% probability that lorazepam is equivalent to diazepam in terms of risk for respiratory depression using a delta value of ±0.10.

Lorazepam

In a single-center, double-blind, randomized controlled trial, 90 patients with agitated delirium refractory to scheduled haloperidol were randomized to receive lorazepam 3 mg IV or placebo with haloperidol 2 mg IV upon the next episode of agitation ($n = 58$) [5c]. Patients who received combination therapy had a higher mean increase in the Edmonton Symptom Assessment System (ESAS) score for drowsiness (1.9, 95% confidence interval [CI] 0.2–3.7) compared to monotherapy with haloperidol (0.2, 95% CI −4.9 to 0.9) ($P = 0.03$), but there were no differences in terms of overall respiratory score. There were also unexpectedly higher rates of akathisia (although not statistically significant) with combination therapy (18.8% vs 6.7%, $P = 0.60$), considering benzodiazepines are commonly used to treat this effect. The study was likely underpowered to detect statistically significant differences in adverse effects. There were no other adverse effects associated with combination therapy.

Side Effects of Drugs Annual, *Volume 40*
ISSN: 0378-6080
https://doi.org/10.1016/bs.seda.2018.07.004

Phenobarbital

No relevant studies identified from the search period.

Primidone

No relevant studies identified from the search period.

GABA REUPTAKE INHIBITORS [1R, 2R]

Tiagabine

No relevant studies identified from the search period.

GABA TRANSAMINASE INHIBITORS [1R, 2R]

Vigabatrin

A prospective cross-sectional study included 24 children previously treated with vigabatrin to determine whether vigabatrin-attributed retinal toxicity is associated with visual system defect in adolescents after vigabatrin withdrawal [6c]. Among the 24 children monitored and assessed during infancy, 10 had been diagnosed with vigabatrin-attributed retinal defect in one or both eyes (Group I) and 14 had no vigabatrin-attributed retinal defect (Group II). The mean of the extent of the peripheral visual field (target size IV 4e) including all four meridians was smaller in participants in Group I when compared with participants in Group II ($P<0.01$). Attenuation of the retinal nerve fiber (RNFL) was demonstrated in all six affected eyes of Group I participants and in only one eye in Group II participants ($P<0.0001$). The study concluded vigabatrin-attributed retinal toxicity in infancy was associated with both visual field loss and RNFL attenuation of the retinal nerve.

A multicenter, open-label, randomized trial assessed whether combination therapy of hormonal therapy (i.e., prednisolone or tetracosactide depot) and vigabatrin would be more effective for infantile spasms than hormonal therapy alone [7C]. A total of 377 infants were randomized to hormonal therapy with vigabatrin ($n=186$) or hormonal therapy alone ($n=161$). Adverse reactions were reported in 228 infants (111, 49%, in the hormonal therapy group vs 117, 51%, in the combination therapy group). Out of 33 infants who experienced serious adverse reactions, 16 occurred in the hormonal therapy group and 17 in the combination therapy group. Infection was the most common serious adverse reaction which occurred more frequently in the hormonal therapy alone than the combination therapy (10% vs 8%). Due to adverse reactions, a lower dose was given to 17 infants (three in the hormonal therapy group and 14 in the combination therapy group). Movement disorders were reported more frequently in the combination group (8% vs 1%) than the hormonal therapy group.

DRUGS WITH POTENTIAL GABA MECHANISM OF ACTION [1R, 2R]

Gabapentin

A systematic review including 37 studies with 5914 subjects assessed the analgesic efficacy and adverse events of gabapentin in adults with chronic neuropathic pain [8M]. Adverse event withdrawals were more common with gabapentin 1200 mg (11%) than with placebo (8.2%) (relative risk [RR] 1.4, 95% CI 1.1–1.7; number needed to harm [NNH] 30, 95% CI 20–65; 22 studies, 4346 participants). There was no difference in serious adverse events between gabapentin (3.2%) and placebo (2.8%) (RR 1.2, 95% CI 0.8–1.7; 19 studies, 3948 participants). Participants were more likely to experience at least one adverse event with gabapentin (63%) than with placebo (49%) (RR 1.3, 95% CI 1.2–1.4; NNH 7.5, 95% CI 6.1–9.6; 18 studies, 4279 participants). Somnolence, drowsiness, or sedation occurred in 14% of participants with gabapentin at doses of 1200 mg daily or more, compared to 5.2% with placebo. Other common adverse events reported by patients taking gabapentin were dizziness (19%), peripheral edema (7%), and gait disturbance (14%).

A systematic review of two double-blind, randomized, controlled trials evaluated the use of GABA modulator, gabapentin, for the treatment of amyotrophic lateral sclerosis (ALS). The trials enrolled a total of 355 participants with ALS: 80 in the gabapentin group and 72 in the placebo group in the phase II trial and 101 patients in the gabapentin group and 102 in the placebo group in the phase III trial [9M]. In the phase II trial, the intervention consisted of oral gabapentin 800 mg or placebo three times daily for 6 months. In the phase III trial, participants received oral gabapentin 1200 mg three times daily or placebo for 9 months. Adverse events reported more often in those taking gabapentin were lightheadedness (RR 2.80, 95% CI 1.79–4.40), drowsiness (RR 2.64, 95% CI 1.61–4.33), and limb swelling (RR 2.70, 95% CI 1.45–5.02). Fatigue and falls occurred more frequently with gabapentin than with placebo in one trial, but combined data showed no clear difference between the groups (RR from the combined analysis 1.65, 95% CI 1.06–2.57). Falls occurred more frequently with gabapentin than placebo in the phase III trial, but not in the phase II trial. In the phase II trial, 149 patients reported a greater frequency of fatigue with gabapentin than with placebo (RR 3.84, 95% CI 1.14–12.92) and a lower frequency of headache with gabapentin than with placebo (RR 0.51, 95% CI 0.26–1.00).

Other reported adverse events showed no difference between the groups.

An observational cohort study evaluated the impact of in utero exposure to psychotropic agents (i.e., antidepressants, atypical antipsychotics, benzodiazepines, gabapentin, and non-benzodiazepine hypnotics, "Z drugs") with concomitant opioid prescriptions filled during the same period on the incidence and severity of neonatal drug withdrawal in 201275 pregnant women [10MC]. The absolute risk for neonatal drug withdrawal was higher in infants exposed to prescription opioids and gabapentin (11.4%, 95% CI 8.60–14.16%) than to those exposed to opioids alone (1.25%, 95% CI 1.20–1.30%). Among neonates exposed to prescription opioids, the relative risk adjusted for propensity score was 1.34 (95% CI 1.22–1.47) with concomitant exposure to antidepressants, 1.49 (1.35–1.63) with benzodiazepines, 1.61 (1.26–2.06) with gabapentin, 1.20 (0.95–1.51) with antipsychotics, and 1.01 (0.88–1.15) with "Z drugs". Exposure to two or more psychotropic medications in combination with opioids was associated with a twofold increased risk of withdrawal (2.05, 95% CI 1.77–2.37). Overall, frequency and severity of withdrawal were increased in neonates exposed to both opioids and psychotropic medications compared with opioids alone.

A case report described a 29-year-old male presenting with whitish discoloration and wrinkling of the palms after sweating or immersion in water for 5–10 minutes [11A]. He also had tingling and pruritus. After drying, the skin normalized within 10 minutes. He was subsequently diagnosed with aquagenic wrinkling of the palms/soles (APW) where etiology is unknown. Onset usually includes whitish papules, edema, thickening, and hyperwrinkling with or without desquamation of the palms and/or soles after immersion in water. The patient had begun treatment with gabapentin 3 weeks before his cutaneous symptoms first appeared with no history of other risk factors for the adverse effect. Once treatment with gabapentin was discontinued, the patient's symptoms improved. The case report concludes gabapentin induced APW may be due to the sodium lauryl sulfate included in this patient's brand of gabapentin. The exact mechanism of action which may have contributed to the induced APW is unknown but could be connected to increased sodium retention of epidermal cells.

A case report described a 32-year-old female who visited the emergency room with complaints of muscle weakness, fatigue, somnolence, gait instability, and numbness in both legs which developed 3 hours prior to her presentation to the emergency room and was associated with recent gabapentin use [12A]. The patient had been taking gabapentin 600 mg three times daily (as well as hydromorphone 4 mg twice daily) for the past month. Electromyography showed a decreased action potential of the compound muscle of the bilateral peroneal nerves, revealing possible rhabdomyolysis. Sensory nerve action potential in the bilateral sural nerve was also decreased in addition to a positive sharp wave in left tibialis anterior and biceps femoris long head, consistent with herniated nucleus pulposus. Neuropathic pain was controlled with oral morphine due to the suspected association of gabapentin and rhabdomyolysis. The patient was discharged 25 days after the initial visit and was asymptomatic at her 5-month follow-up visit.

Pregabalin

A systematic review and meta-analysis of eight randomized, controlled trials (RCTs) was conducted to assess the effectiveness and safety of gabapentinoids (i.e., gabapentin and pregabalin) in adult patients diagnosed with chronic lower back pain (CLBP) [13M]. Compared with placebo, the common adverse events reported with gabapentin were dizziness (RR 1.99, 95% CI 1.17–3.37, I^2 49%), fatigue (RR 1.85, 95% CI 1.12–3.05, I^2 0%), difficulties with mentation (RR 3.34, 95% CI 1.54–7.25, I^2 0%), and visual disturbances (RR 5.72, 95% CI 1.94–16.91, I^2 0%). There was an absolute increase in rates of dizziness (14%, NNH 7, 95% CI 4–30), fatigue (13%, NNH 8, 95% CI 4–44), mental difficulties (16%, NNH 6, 95% CI 4–15), and visual disturbances (15%, NNH 6, 95% CI 4–13). With pregabalin, dizziness was more common compared to the active comparator (i.e., pooled results for amitriptyline, celecoxib, and tramadol/acetaminophen) (RR 2.70, 95% CI 1.25–5.83, I^2 0%), with an absolute increased risk of 9% (NNH 11, 95% CI 6–30). This review found there are very few RCTs that have attempted to assess the benefit of using gabapentin or pregabalin in adult patients with CLBP, despite widespread use. Use of gabapentin and pregabalin, compared to placebo and active analgesic comparators, respectively, was associated with significant increases in adverse events.

A randomized, double-blind, placebo-controlled trial examined whether pregabalin reduces the intensity of sciatica [14C]. A total of 209 patients were randomly assigned to receive either pregabalin ($n = 108$) at a dose of 150 mg per day titrated to a maximum dose of 600 mg per day or matching placebo ($n = 101$) for up to 8 weeks. The incidence of serious adverse events (i.e., deemed medically important, requiring hospitalization, life-threatening, resulting in death) in the pregabalin group (1.9%) was similar to the placebo group (5.9%, $P = 0.16$). The study found a significantly higher incidence of adverse events in the pregabalin group (64.2%) than in the placebo group (42.6%, $P = 0.002$). Dizziness was the most commonly reported adverse event in each group and was more common in the pregabalin group than in the placebo group (39.6% vs 12.9%, *P*-value not reported).

A prospective, randomized, placebo-controlled, double-blind study evaluated preemptive analgesia efficacy of pregabalin at 75, 150, and 300mg doses for postoperative pain management after laparoscopic hysterectomy [15c]. A total of 96 women with American Association of Anesthesiologist (ASA) physical status I and II underwent an elective laparoscopic hysterectomy and were randomly assigned to four groups. Groups 1–3 (treatment groups; $n=20$) received each dose the night before surgery, 30 minutes before surgery and 6 hours after surgery, and group 4 (control group; $n=22$) received matching placebo following the same schedule. There was no difference between the groups in terms of incidence of the most common adverse effects of nausea/vomiting, dizziness, headache, visual disturbance, or itching, suggesting lack of significant dose-related adverse effects when pregabalin is prescribed in this manner.

A double-blind, randomized, placebo-controlled cross-over study evaluated the efficacy and safety of pregabalin combined with morphine in 40 patients with severe neuropathic cancer pain [16c]. Patients were randomized into two groups: the pregabalin-placebo group ($n=20$) received pregabalin plus oral morphine in phase I and placebo plus oral morphine in phase II and the placebo-pregabalin group ($n=20$) received placebo plus oral morphine in phase I and pregabalin plus oral morphine in phase II. Two patients from the pregabalin-placebo group withdrew during treatment periods due to adverse effects. Compared to the placebo-control group, pregabalin treatment was associated with a higher frequency of dry mouth (20% vs 2.5%, $P=0.029$) and somnolence (40% vs 10%, $P=0.004$), and lower scores on the constipation assessment scale (CAS) (6.4 ± 4.4 vs 8.6 ± 3.7, $P=0.017$). The study concluded rates of dry mouth and somnolence are usually higher at the titration period and are generally tolerable, suggesting adequate safety of combination therapy of pregabalin and morphine.

A 4-week single-blind, multicenter, randomized trial compared the efficacy and side effects of pregabalin and doxepin [17c]. A total of 72 patients with uremic pruritus were randomized to receive pregabalin 50mg every other day ($n=37$) or doxepin 10mg per day ($n=35$) for 4 weeks. Three patients in the pregabalin group and one patient in the doxepin group discontinued the medication due to adverse effects of somnolence and drowsiness. Somnolence was the most reported adverse effect in both groups (16.2% vs 14.2%). Patients also reported edema ($n=3$), drowsiness ($n=3$), imbalance during walking ($n=1$) and numbness ($n=1$) in the pregabalin group and nervousness ($n=1$) in the doxepin group.

A case report described a 55-year-old male with a 1-year history of smoking marijuana who reported taking five or six pregabalin 300mg capsules with alcohol in an attempt to commit suicide [18A]. He experienced vision loss in the right eye the next morning, which persisted for about 1 week, at which time the patient presented for treatment. Fluorescein angiography of his eye revealed a large area of capillary non-perfusion. The authors concluded the high dose of pregabalin, which may have caused hypotension, and marijuana-associated arteritis with impaired vascular autoregulation may have caused a macular infarction. This case report is the first to associate marijuana and pregabalin misuse with a hemorrhagic macular infarction. Although macular infarction associated with marijuana has not been previously reported, its occurrence may have been potentiated by the high dosages of pregabalin and alcohol.

Valproic Acid/Divalproex Sodium

A cross-sectional study compared the effect of psychotropic medications on T-peak to T-end interval, QT dispersion, and QT interval in psychiatry inpatients ($n=1059$) [19C]. Treatment with valproic acid was associated with T-peak to T-end interval reduction (OR 0.6, 95% CI 0.37–0.98, $P=0.04$), which could imply increased risk for sudden cardiac death. There were no significant changes to QT dispersion or the QT interval.

In an open-label, randomized controlled trial, newly diagnosed or untreated female patients (age 12–40 years) with epilepsy were assigned to valproic acid ($n=34$) or lamotrigine ($n=32$) to evaluate metabolic changes [20C]. The most common diagnoses were generalized motor epilepsy (32% vs 44%) and focal aware seizures (32% vs 28%). Following 1 year of study, there were small but significant ($P<0.05$) differences between groups in patients with a baseline body mass index (BMI) of 18–22.9kg/m^2 in several measures (valproic acid vs lamotrigine): weight (51.95 ± 5.49kg vs 48.45 ± 5.58kg), BMI (23.37 ± 1.99kg/m^2 vs 20.34 ± 0.02kg/m^2), fasting insulin (9.07mIU/L vs 6.89mIU/L), homeostasis model assessment of insulin resistance (HOMA-IR) (2.17 vs 1.66), adiponectin (4.57 ± 1.39mcg/mL vs 6.56 ± 1.82mcg/mL), and triglycerides (97.8mg/dL vs 96.0mg/dL). There was a similar pattern for patients with baseline BMI 23–24.9kg/m^2. For patients with baseline BMI greater than 25kg/m^2, differences in these parameters were more substantial: weight (72.01 ± 5.6kg vs 63.44 ± 4.63kg, $P<0.001$), BMI (31.94 ± 2.37kg/m^2 vs 25.69 ± 1.42kg/m^2, $P<0.001$), fasting insulin (10.6mIU/L vs 8.17mIU/L), HOMA-IR (3.24 vs 1.73), adiponectin (2.71 ± 1.16mcg/mL vs 6.25 ± 1.63mcg/mL), high density lipoprotein (HDL) (41 ± 2.4mg/dL vs 45 ± 2.7mg/dL), triglycerides (132.6mg/dL vs 101.0mg/dL). There were no differences between groups in terms of fasting blood glucose, leptin, total cholesterol, or low-density lipoprotein.

Neurological

A case report described a 61-year-old Chinese female who presented with dizziness for 3 weeks after a bifrontal mass was found via magnetic resonance imaging

(MRI) [21A]. The patient was initiated on valproic acid 500 mg twice daily 3 days before a scheduled craniotomic meningioma resection. The patient received valproic acid 800 mg IV 30 minutes before the end of the procedure. After the operation, the patient's blood ammonia level was elevated and fluctuated (144.8–207.7 mmol/L). She remained unconscious for 3 days. Although the serum valproate level was within range, a multidisciplinary team diagnosed her with valproate-induced hyperammonemic encephalopathy. The patient's mental state began to improve from the first 24 hours after valproic acid was discontinued. The patient was discharged 8 days after discontinuation of valproic acid with no complications; the patient reported full recovery at 5-month follow-up.

A case report described a 75-year-old female presenting with complaints of freezing gait, stiffness of extremities and tremor [22A]. She was previously admitted to the psychiatric ward due to depression and complaints of tremor. She was given valproate 1000 mg once daily, which improved her symptoms. Four months after discharge parkinsonism worsened. She was readmitted 6 months after original discharge with a 25.27 mcg/mL plasma concentration of valproate suggesting nonadherence. Her dose was reduced to 800 mg per day and resulted in a plasma concentration of 60 mcg/mL. Her Unified Parkinson's Disease Rating Scale score was 98. The dose of valproate was reduced to 400 mg per day and eventually withdrawn due to suspected drug-induced parkinsonism. Three days after dose reduction, the patient's range of motion of upper extremities improved. After discontinuation of valproate, her rigidity and intention tremor improved.

Dermatological

A case report described a 73-year-old female with a case of acute generalized exanthematous pustulosis (AGEP) secondary to dual antiepileptic therapy with valproic acid and levetiracetam [23A]. The patient's past medical history included a recent meningioma resection and she presented to the emergency department with leukocytosis and a diffuse pustular rash on her face, torso, back and scalp. The onset of AGEP was after discharge from a previous admission, in which the patient was discharged with levetiracetam 1500 mg twice daily and valproic acid 800 mg three times daily for postoperative seizure activity. The patient noticed an erythematous pustular rash which started on her back and spread to her scalp, face, chest, and abdomen 5 days after antiepileptic treatment was initiated. The patient's symptoms quickly resolved with the application of triamcinolone and discontinuation of both antiepileptic medications.

Endocrine/Reproductive

A series of four cases of human fetal valproate syndrome with minor skeletal abnormalities in the offspring of mothers who took valproic acid during pregnancy for treatment of epilepsy was described [24A]. A 16-month-old female, whose mother took valproic acid 1500 mg once daily up to the 10th gestational week and at a dosage of 1000 mg per day throughout the remainder of her pregnancy, presented with facial dysmorphism and finger abnormalities. The case series also described a 5-year-old male with speech disability, bilateral cryptorchidism, facial dysmorphism, and finger abnormalities whose mother also took valproic acid 1000 mg once daily throughout her pregnancy. The third patient described was a 19-month-old patient who was the brother of the 5-year-old patient previously described who had facial dysmorphism, bilateral cryptorchidism, and finger abnormalities. This patient's mother also took valproic acid 1000 mg once daily throughout her pregnancy. The fourth case described was a 3-year and 6-month-old boy with minor facial dysmorphism and sternum deformity who were exposed to valproic acid in utero from their mother's 500 mg once daily dose. The case series found a correlation between in utero exposure to valproic acid and facial dysmorphic features with minor skeletal abnormalities.

Toxicological

A case report described a 29-year-old male suffering from bipolar disorder and substance use disorder who presented with sudden altered mental status [25A]. The patient was started on valproic acid 1000 mg per day (later increased to 1800 mg) which did not improve symptoms of agitation. Plasma valproic acid levels were found to be elevated, with ammonia levels highly elevated at 594 mcg/dL (baseline not described). Valproic acid was withdrawn and the patient received lactulose and IV L-carnitine supplementation of 4.5 g per day which reduced ammonia levels to 99 mcg/dL. Mental status was restored after 24 hours of L-carnitine treatment.

A case series described five patients with symptoms of valproic acid overdose (i.e., coma, drowsiness, serum ammonia greater than 80 mcg/dL) treated with L-arginine, which can theoretically help correct valproate-associated hyperammonemia by increasing *N*-acetylglutamate synthetase activity [26A]. Serum ammonia levels substantially decreased in each patient to below the threshold for hyperammonemia (i.e., 381 to 39 mcg/dL, 281 to 50 mcg/dL, 669 to 74 mcg/dL, 447 to 56 mcg/dL, 202 to 60 mcg/dL). Other interventions administered prior to L-arginine (i.e., hemodialysis, L-carnitine) were ineffective, although it should be noted the study lacks the control necessary to determine the true effect of L-arginine in this setting.

Miscellaneous

A case series evaluated the range of valproate plasma protein binding in 15 critically ill patients [27A]. An index case report described a 72-year-old male who remained comatose after cardiac arrest. He was given 30 mg/kg/day of valproate on day four of his hospital stay. On day seven, he was given a 2000 mg dose of valproate with an increased maintenance dose of 40 mg/kg/day after his serum showed a valproate level of 71 mg/L of. After worsening of transaminases and an elevated free fraction of 86%, valproate was discontinued, and his family requested withdrawal from life-support. The median valproate dose was 3.0 g per day (interquartile range [IQR] 2.0–4.0 g per day) or 35 mg/kg/day (IQR 27–43 mg/kg/day). Adverse drug events occurred in 10 patients (68%). Out of 12 patients, 7 (58%) had hyperammonemia. Out of 15 patients, 2 (13%) developed elevated transaminases and 5 (33%) had thrombocytopenia.

℞

Special Review

A systematic review and network meta-analysis of experimental and observational studies indexed on Cochrane CENTRAL, Embase, or MEDLINE through 2015 assessed the association of first-generation (i.e., carbamazepine, clobazam, clonazepam, ethosuximide, phenobarbital, phenytoin, primidone, valproate) or newer-generation (i.e., gabapentin, lamotrigine, levetiracetam, oxcarbazepine, topiramate, vigabatrin) in pregnancy with incidence of overall and specific types of major congenital malformations (CM), combined fetal losses, prenatal growth retardation, preterm birth, and minor CM [28M]. The meta-analysis pooled results from 96 studies and 58461 patients. Increased incidence of major CMs was associated with ethosuximide (OR 3.04, 95% credible interval [CrI] 1.23–7.07), valproate (OR 2.93, 95% CrI 2.36–3.69), topiramate (OR 1.90, 95% CrI 1.17–2.97), phenobarbital (OR 1.83, 95% CrI 1.35–2.47), phenytoin (OR 1.67, 95% CrI 1.30–2.17), and carbamazepine (OR 1.37, 95% CrI 1.10–1.71). These results are contrasted with a smaller study evaluating a pregnancy registry of 1688 fetuses exposed to antiepileptics in South India that only found increased risk for CM with clobazam compared to unexposed healthy patients (RR 6.44, 95% CI 1.67–24.94) and unexposed patients with epilepsy (RR 4.0, 95% CI 1.06–15.03) and valproic acid compared to unexposed healthy patients (RR 2.60, 95% CI 1.30–5.20) [29MC]. Results for valproate suggested a dose relationship, with 33.3% of patients on a total daily dose of 801 mg or higher experiencing a CM compared to 3.2% of patients on a dose up to 400 mg (P=0.001). These results are likely limited and underpowered due to the relatively smaller number of evaluated pregnancies.

A further analysis of the specific types of major CMs was performed in the meta-analysis. There were increased cardiac malformations with gabapentin (OR 5.98, 95% CrI 1.37–19.73), carbamazepine plus phenytoin (OR 6.58, 95% CrI 2.25–18.97), phenobarbital plus valproate (OR 8.01, 95% CrI 1.17–35.40), phenytoin plus valproate (OR 8.88, 95% CrI 2.62–30.65), and carbamazepine plus clonazepam (OR 10.08, 95% CrI 1.40–51.22) [28M]. Hypospadias were associated with gabapentin (OR 16.54, 95% CrI 2.50–121.70), clonazepam (OR 6.17 95% CrI 1.17–24.80), primidone (OR 5.92, 95% CrI 1.01–23.77), and valproate (OR 2.58, 95% CrI 1.24–5.76). Cleft lip/palate was associated ethosuximide (OR 22.22, 95% CrI 4.56–87.64), primidone (OR 7.68, 95% CrI 1.41–29.27), topiramate (OR 6.12, 95% CrI 1.89–19.05), phenobarbital (OR 5.75, 95% CrI 2.41–14.08), phenytoin (OR 3.11, 95% CrI 1.31–7.72), valproate (OR 3.26, 95% CrI 1.38–5.58), phenobarbital plus phenytoin plus primidone (OR 11.50, 95% CrI 1.70–63.48), phenytoin plus primidone (OR 16.75, 95% CrI 3.02–77.19), carbamazepine plus phenobarbital (OR 18.51, 95% CrI 3.34–94.21), and carbamazepine plus valproate (OR 19.12, 95% CrI 3.74–88.68). Club foot was associated with phenytoin (OR 2.73, 95% CrI 1.13–6.18), valproate (OR 3.26, 95% CrI 1.43–8.25), primidone (OR 4.71, 95% CrI 1.11–17.24), ethosuximide (OR 12.99, 95% CrI 1.66–76.39), carbamazepine plus phenobarbital (OR 7.30, 95% CrI 1.29–32.31), and phenobarbital plus phenytoin plus primidone (OR 13.46, 95% CrI 1.45–132.80). Inguinal hernia was associated with phenobarbital plus phenytoin (OR 5.51, 95% CrI 1.25–34.61) and phenobarbital plus primidone (OR 534.20, 95% CrI 14.39–1.31 × 10^5).

Topiramate (OR, 23.58, 95% CrI 1.18–549.60), primidone (OR 2.81, 95% CrI 1.21–6.28), valproate (OR 1.83, 95% CrI 1.04–3.45), carbamazepine plus valproate (OR 5.09, 95% CrI 1.35–16.79), and phenytoin plus valproate (OR 8.96, 95% CrI 1.77–37.95) were associated with increased odds for total fetal loss [28M]. Therapies associated with risk for prenatal growth retardation include clobazam (OR 4.47, 95% CrI 1.60–11.18), topiramate (OR 2.64, 95% CrI 1.41–4.63), and phenobarbital (OR 1.88, 95% CrI 1.07–3.32). Increased preterm birth rate was associated with clobazam (OR 3.42, 95% CrI 1.41–7.92) and primidone (OR 2.12, 95% CrI 1.01–4.27). Finally, carbamazepine (OR 10.81, 95% CrI 1.40–373.90), carbamazepine plus phenytoin (OR 12.46, 95% CrI 1.17–438.90), valproate (OR 17.76, 95% CrI 1.60–633.30), phenobarbital plus phenytoin (OR 20.14, 95% CrI 1.96–764.20), and carbamazepine plus phenobarbital plus valproate (OR 122.20, 95% CrI 2.09–9539.00) were associated with increased odds for minor CMs.

Overall, monotherapies associated with a significant risk of CMs and prenatal harms compared to control included carbamazepine, clobazam, ethosuximide, gabapentin, phenobarbital, phenytoin, topiramate, and valproate, with clobazam and valproate also suggesting risk in a retrospective cohort study of a pregnancy registry [28M, 29MC]. Newer generation antiepileptics, including lamotrigine, levetiracetam, oxcarbazepine, and vigabatrin were not associated with statistically significant risks to physical development when compared to control; however, fetal risk of these agents has not been ruled out [28M]. Other agents (e.g., felbamate, lacosamide, pregabalin, rufinamide, tiagabine, zonisamide) were minimally addressed. Lamotrigine and levetiracetam were associated with lower risk of overall major and minor CMs. Application of these results in practice may be limited by the lack of presented heterogeneity results, imprecise odds estimates particularly with combination therapies, and lack of absolute risk information.

SODIUM CHANNEL BLOCKERS [1R, 2R]

Carbamazepine

A cross-sectional surveillance study evaluated reports submitted to the National Poison Data System from regional poison centers across the US from 2000 to 2014 involving patients being treated for depression [30C]. There were 962222 reports involving 48 medications. Mean morbidity index per 1000 exposures was considered high for carbamazepine (223.7, 95% CI 218.3–229.2). Mean mortality index per 1000 exposures was considered high for carbamazepine (mean 5.6, 95% CI 3.2–9.0), gabapentin (mean 5.9, 95% CI 2.5–11.4), and valproic acid (mean 8.1, 95% CI 5.7–11.2).

A multi-center, case–control study examined human leukocyte antigen (HLA) loci involved in aromatic antiepileptic-induced Stevens–Johnson syndrome (SJS) or toxic epidermal necrolysis (TEN) in southern Han Chinese patients [31C]. Overall, 91 cases occurred in the aromatic antiepileptics which included carbamazepine (56 SJS/TEN), lamotrigine (22 SJS), and phenytoin (13 SJS). The control group included 322 patients exposed to carbamazepine (180), lamotrigine (102), and phenytoin (40). HLA-A, HLA-B, HLA-C, and HLA-DRB1 genotype were analyzed in the 91 cases of SJS/TEN. Significant alleles in carbamazepine-induced SJS/TEN included HLA-B*15:02, HLA-C*08:01, HLA-DRB1*12:02, HLA-A*24:02, HLA-B*15:11, and HLA-DRB1*01:01. In HLA-B*15:02 negative patients, HLA-A*24:02 was significantly associated with SJS/TEN (7 of 17 vs 24 of 149 in carbamazepine controls, odds ratio 3.65, 95% CI 1.26–10.52, $P = 0.029$) which suggests the HLA-A*24:02 allele is independently and significantly associated with carbamazepine-induced SJS/TEN. For lamotrigine and phenytoin-induced SJS, significant alleles include HLA-A*24:02, and HLA-*24:02 and HLA*02:01, respectively. There were more HLA-B*15:02 genotypes in the carbamazepine-induced SJS/TEN group compared to control ($P = 5.63 \times 10^{-15}$). When compared to control, HLA-B*15:02 was present in more lamotrigine-induced SJS (22.7% vs 18.6%, $P = 0.89$) and more phenytoin-induced SJS (46.2% vs 22.5%, $P = 0.22$), which suggests HLA-B*15:02, is a high-risk but relatively specific factor for carbamazepine induced SJS/TEN. When compared to control, HLA-A*24:02 was present in more carbamazepine-induced SJS (30.4% vs 15.7%, $P = 0.015$), more phenytoin-induced SJS (46.2% vs 12.5%, $P = 0.027$), and more lamotrigine-induced SJS (45.4% vs 15.7%, $P = 0.005$), which suggests presence of HLA-A*24:02 is a potential shared risk factor for carbamazepine, lamotrigine, and phenytoin-induced SJS/TEN. Carbamazepine-induced SJS/TEN is strongly associated with HLA-B*15:02 in southern Han Chinese. HLA-A*24:02 is a shared risk factor for carbamazepine, lamotrigine, and phenytoin-induced SJS/TEN. HLA-B*15:02 is a high-risk factor for carbamazepine induced SJS/TEN.

An observational, cross-sectional study investigated 213 subjects with schizophrenia and the factors associated with the occurrence of extrapyramidal symptoms (EPS) in users of second-generation antipsychotics (SGA) with the administration of adjunctive drugs [32C]. Administered SGA agents were olanzapine (35.7%), risperidone (34.3%), quetiapine (12.2%), ziprasidone (10.8%), and clozapine (6.1%). Benzodiazepines and carbamazepine comprised 31.5% and 24.4% of adjunctive therapy, respectively. Extrapyramidal manifestations, indicated by a score greater than 0.3 on the Simpson-Angus Scale, were observed in 81 patients (38.0%, 95% CI 31.5 44.9%). Carbamazepine was associated with EPS (adjusted OR 3.677, 95% CI 1.627–8.310, $P = 0.002$), but benzodiazepines were not (OR 1.510, 95% CI 0.838–2.724, $P = 0.170$). Adjunctive use of carbamazepine may predispose the user of atypical antipsychotics to the occurrence of EPS.

A cross-sectional, observational study collected blood samples from 112 North Indian children (ages 2–12 years) receiving either monotherapy phenytoin ($n = 62$) or carbamazepine ($n = 50$) for greater than 6 months to determine the prevalence of hyperhomocysteinaemia compared to 50 control children [33C]. Hyperhomocysteinaemia occurred in 101 (90%) children in the study group and 17 (34%) in the control group ($P = 0.001$). The mean serum homocysteine concentration in children receiving monotherapy phenytoin (20.3 ± 11.5 mcmol/L) or carbamazepine (17.2 ± 7.9 mcmol/L) was greater than that of the control group (9.1 ± 3 mcmol/L, $P = 0.001$). Those with raised homocysteine concentrations received higher doses of carbamazepine (15.6 ± 3.8 ng/mL vs 9.1 ± 0.9 ng/mL, $P < 0.0004$). When compared to children with normal levels, longer mean duration of antiepileptics was associated with hyperhomocysteinaemia (21.4 ± 13.6 months vs 10.8 ± 3.3 months, $P = 0.01$) and low folic acid (24 ± 11.1 months vs 18.2 ± 12.5 months, $P = 0.001$). Long-term phenytoin or carbamazepine therapy may be associated with hyperhomocysteinaemia and reduced serum folic acid concentrations.

Dermatological

A case report described a 27-year-old female receiving levetiracetam (up to 3 g per day) and brivaracetam (up to 200 mg per day) for treatment of focal epilepsy [34A]. Due to the refractory course, carbamazepine 400 mg per day was added which resulted in cessation of seizures. Three weeks later, the patient presented with generalized maculopapular exanthema. Blood tests revealed a normal carbamazepine level (6.4mg/L), elevated carbamazepine-10,11-epoxide (5.5mg/L), and slightly elevated liver enzymes. Human leukocyte antigen testing revealed an HLA-A*31:01 haplotype. Carbamazepine was discontinued, and the seizures recurred. Four weeks later, eslicarbazepine 200mg per day was added and titrated to 800mg per day. Seizure frequency decreased, and no adverse events developed. Authors concluded eslicarbazepine may be considered in patients with the HLA-A*31:01 haplotype and a hypersensitivity reaction to carbamazepine if the benefits exceed the risks.

Rheumatological

A case report described a 9-year-old female with Rolandic epilepsy managed with carbamazepine 200mg per day for 3 years who presented with a 3-month history of fever, asthenia, loss of appetite, weight loss, skin lesions, oral ulcers, arthralgias, and myalgias [35A]. Laboratory data revealed anemia (hemoglobin 11g/dL), leukocytopenia (3.5×10^9/L), positive antinuclear antibody, positive anti-histone antibodies, and reduced serum C3 levels (45.4mg/dL). Liver function values were within normal ranges. Chest radiography, electrocardiogram, and abdominal ultrasound showed no abnormal findings. A diagnosis of carbamazepine-induced systemic lupus erythematosus was made. Cutaneous and systemic symptoms improved after discontinuation of carbamazepine and initiation of prednisone 1mg/kg, tapered over 3 months. During a 7-year follow-up, the patient had no evidence of disease recurrence. Although uncommon, drug-induced lupus erythematosus may occur in patients receiving carbamazepine.

Hematological

A case report described a 55-year-old-male who presented with a 2-day history of pain and swelling in the right leg [36A]. For the past 11 years, the patient has managed his structural epilepsy with carbamazepine 900mg per day. For the past 4 months, the patient has managed an unprovoked venous thrombosis in the right leg with rivaroxaban 20mg daily. Duplex sonography revealed thrombosis of the right popliteal and femoral vein. Serum carbamazepine level was normal and anti-Xa activity was less than 20ng/mL. Rivaroxaban therapy was stopped and low-molecular-weight heparin, followed by phenprocoumon, was initiated. Leg swelling and pain decreased and the patient was discharged after 5 days. Co-administration of rivaroxaban with permeability glycoprotein or cytochrome P 450 3A4-inducing agents, like carbamazepine, may lead to a decrease in serum levels of rivaroxaban and decreased anticoagulant activity.

Eslicarbazepine acetate

No relevant studies identified from the search period.

Lacosamide

In a prospective, open-label trial, lacosamide was added to levetiracetam after down titration of a concomitant sodium channel blocker among 120 patients with focal epilepsy not adequately controlled on levetiracetam and a sodium channel blocker, including carbamazepine, lamotrigine, oxcarbazepine, phenytoin, or eslicarbazepine [37C]. Patients were started on lacosamide 50mg twice daily and increased to 200mg per day after 1 week. The dose was increased by 100mg per day/week as needed (maximum 600mg per day) over the remaining 8 weeks. When patients reached lacosamide 200mg per day, sodium channel blocker down-titration began and was fully discontinued by the end of cross-titration. During the maintenance period, lacosamide and levetiracetam doses remained stable. Treatment emergent adverse events (TEAEs) occurred in 90 of 120 patients (75%) during the 21-week treatment period of which 64.2% occurred during the 9-week cross titration and 40.8% during the 12-week maintenance period. The most common TEAEs during cross-titration were dizziness (20.8%), headache (11.7%), fatigue (7.5%), and nausea (5.8%) which decreased during the maintenance period to 4.9%, 5.8%, 1.9%, and 0%, respectively. TEAEs during the treatment period included dizziness (23.3%), headache (15.0%), and fatigue (8.3%). Two patients discontinued during cross-titration due to convulsion. Eight patients discontinued therapy due to TEAEs during the treatment period with the most common reason being convulsion (2.5%) and suicidal ideation (1.7%). The most common TEAEs, including dizziness, headache, and fatigue, were consistent with the lacosamide safety profile and were most reported during cross-titration. The combination of lacosamide and levetiracetam was well tolerated, demonstrated by the low discontinuation rate due to TEAEs (6.7% overall population vs 3.6% those who discontinued a sodium channel blocker).

Lamotrigine

A retrospective surveillance study compared rates of events submitted to the US Food and Drug Administration Adverse Event Reporting System between brand, generic, and authorized generic formulations of lamotrigine ($n = 27150$), carbamazepine ($n = 13950$), and oxcarbazepine (5077) from January 2004 to March 2015 [38C]. Reports were more common for brand formulations than generic formulations, and more common for generic formulations than authorized generic formulations for lamotrigine (71.32% vs 27.04% vs 1.64%), carbamazepine (57.01% vs 40.82% vs 2.17%), and oxcarbazepine (66.36% vs 32.46% vs 1.18%). Generic formulations of lamotrigine (reporting OR 4.6, 95% CI 4.4–4.9) and carbamazepine (reporting OR 2.8, 95% CI 2.5–3.1) showed significant results compared to both brand and authorized generic formulations (adjusted $P < 0.001$, adjusted $P = 0.001$, respectively) suggesting a potential safety signal.

Hematological

A case report described a 17-year-old female with a history of seizure disorder controlled with oxcarbazepine and lamotrigine who presented with bilateral cervical neck lymphadenopathy [39A]. A biopsy found reactive lymphoid hyperplasia with atypical paracortical hyperplasia, which could be seen in anticonvulsant lymphadenopathy. Five months prior, lamotrigine had been started and 11 months after discontinuation, the lymphadenopathy fully resolved. Lamotrigine induced lymphadenopathy should be considered in patients with lymphadenopathy due to an unknown cause.

Neurological

A case report described a 19-year-old female who presented with decreasing visual acuity, central scotomas, and photosensitivity in both eyes [40A]. The patient was hospitalized for a concomitant acute kidney injury, transaminitis (AST 107 U/L, ALT 249 U/L, alkaline phosphatase 249 U/L), leukocytosis (12.5×10^9) with eosinophilia (16.6%), and elevated serum β2 microglobulin (15.2 mg/L). Three weeks prior, a maculopapular rash and arthralgias developed after starting lamotrigine for bipolar disorder. An infection workup was negative. Based on an abnormal serum creatinine (3.1 mg/dL), elevated β2 microglobulin, and systemic symptoms, the patient was diagnosed with tubulointerstitial nephritis and uveitis syndrome (TINU) with findings of acute posterior multifocal placoid pigment epitheliopathy (APMPPE). TINU was likely precipitated by lamotrigine, which was discontinued, and the patient was treated with topical prednisolone acetate and oral prednisone 60 mg daily; the patient had a full recovery.

A 34-year-old male with history of complex partial seizures with gestural/verbal automatisms and sporadic secondary generalization was treated with valproic acid up to 1000 mg/day starting at age 11 years followed by addition of lamotrigine up to 400 mg/day starting at age 17 years [41A]. At age 32, he presented with a complex partial seizure with verbal automatisms on EEG and asynchronous myoclonic jerks involving upper and lower limbs, with no correlate between myoclonus and EEG activity. Valproic acid serum levels were normal, but lamotrigine was slightly elevated (12.6 mcg/mL compared to a normal range of 4–12 mcg/mL). It was discovered myoclonic jerks had been common since lamotrigine was initiated (most often during drowsiness and post-ictal state), but never reported. Lamotrigine was tapered to 300 mg/day with resolution of symptoms. Symptoms returned upon an increase in lamotrigine serum levels (15.8 mcg/mL) related to recent weight loss.

Toxicological

A 3-year-old previously healthy male whose father was taking lamotrigine and valproic acid was found unconscious with generalized tonic–clonic jerks [42A]. He was refractory to initial treatment with diazepam and controlled in the emergency room with midazolam (IV push 0.25 mg/kg followed by 0.15 mg/kg/h continuous infusion). Other symptoms in the emergency room included sinus tachycardia and right bundle branch block on electrocardiogram and post-ictal pattern on EEG. The patient was able to be weaned to midazolam 0.03 mg/kg/h, at which point he became arousable with restlessness, malaise, dystonic movements (trunk, limbs, neck), severe oculogyric crisis, ataxia, and excessive thirst and appetite (that continued for 3 days). Serum levels (following 10 hours of IV fluids) revealed an elevated lamotrigine level (28.4 mg/L, therapeutic range 1–14 mg/L) and normal valproic acid. It was determined one blister containing eight 200 mg lamotrigine tablets was missing from the father's supply. The patient's ataxia and dysmetria persisted for 72 hours, and he was discharged with persistent increased appetite, but normal electrocardiogram and EEG.

Oxcarbazepine

No relevant studies identified from the search period.

Phenytoin

A case report described a 70-year-old Canadian male presenting with weakness and fatigue with a history of seizure disorder, atrial fibrillation, heterozygous factor V Leiden, recurrent deep vein thrombosis, and a pulmonary embolus [43A]. Transesophageal echocardiology showed a large thrombus in the left atrium. Four weeks prior to admission, the patient experienced a seizure and phenytoin 300 mg daily was reinitiated (the patient had been seizure-free and off medication for an extended period). More than 1 year prior to admission, dabigatran 150 mg orally twice daily was initiated. Other preadmission medications included atenolol, betahistine, diltiazem, and valsartan. Phenytoin was changed to levetiracetam. At 12 weeks there was full clot resolution. Authors concluded coadministration of dabigatran and phenytoin may result in decreased efficacy of dabigatran through phenytoin p-glycoprotein induction.

Zonisamide

No relevant studies identified from the search period.

GLUTAMATE BLOCKERS [1R, 2R]

Felbamate

No relevant studies identified from the search period.

Perampanel

An observational, retrospective review evaluated the efficacy and tolerability of perampanel in 26 female patients [44A]. Psychiatric adverse events, including irritability ($n = 6$), aggression ($n = 4$), increased sensitivity ($n = 3$), mood instability ($n = 2$), and suicidal ideation/behavior ($n = 2$) occurred in 50% of the patients. Out of 13 cases, adverse events in 11 cases were reasons to discontinue perampanel treatment, occurring usually in the first 6 months of treatment. Slower titration schemes were paradoxically associated with more psychiatric adverse events (8 out of 13 patients; 2 mg over more than 4.4 weeks) compared to rapid titration schemes (5 out of 13 patients; 2 mg over less than 4.4 weeks). Following psychiatric events, grouped by the investigators as a whole, the second most common adverse event was dizziness. However, no dizziness or related adverse events required withdrawal. The range of dosages at which adverse events occurred was from 4 to 12 mg per day.

Topiramate

A network meta-analysis included 32 studies with 6052 subjects to compare medications for migraine prophylaxis [45M]. Overall adverse events were most strongly associated with topiramate compared to placebo (OR 2.44, 95% CrI 1.55–3.88) and amitriptyline (OR 4.66, 95% CrI 1.74–12.93), but not gabapentin or divalproex. However, divalproex was associated with increased odds for nausea (OR 3.20, 95% CrI 1.70–6.30) and withdrawal (OR 1.10, 95% CrI 1.14–3.67). Both topiramate (OR 2.50, 95% CrI 1.50–3.50) and divalproex (OR 2.30, 95% CrI 1.00–5.60) were associated with increased withdrawal due to adverse events. Gabapentin was associated with dizziness (OR 3.70, 95% CI 1.20–9.80).

A systematic review and meta-analysis included three studies with 83 participants to assess topiramate for treatment of juvenile myoclonic epilepsy [46M]. There was no difference compared to placebo in terms of abnormal vision, diarrhea, nausea, or respiratory illness. However, there was increased risk for paresthesia (17.1% vs 0%, $P = 0.0024$, $I^2 = 0\%$), and decreased risk for weight gain (0% vs 30.8%, $P = 0.0025$, $I^2 = 0\%$) and tremor (0% vs 23.5%, $P = 0.0032$, $I^2 = \text{N/A}$) compared to valproate. There was no difference between topiramate and valproate in terms of abnormal vision, alopecia, anorexia, appetite increase, concentration diarrhea, difficulty, dizziness, fatigue, hallucination, headache, insomnia, nausea, somnolence, psychomotor slowing, rash, or weight loss. These results help clarify the comparative tolerability profile in a less studied diagnosis.

A randomized, double-blind, placebo-controlled trial of 361 patients evaluated the safety of amitriptyline ($n = 144$), topiramate ($n = 145$), and placebo ($n = 72$) for pediatric migraine prophylaxis [47C]. There were a total of 852 adverse events in 272 patients: 301 with amitriptyline, 419 with topiramate, and 132 with placebo. Adverse events that occurred significantly more often in the topiramate group than in the placebo group were paresthesia (31% vs 8%, $P < 0.001$) and decreased weight (8% vs 0%, $P = 0.02$). Other common adverse events reported in the topiramate group included fatigue (25%, which also occurred in 30% of patients who received amitriptyline), dry mouth (18%, which also occurred in 25% of patients who received amitriptyline), memory impairment (17%), aphasia (16%), cognitive disorder (16%), and upper respiratory tract infection (12%). Although not formally compared, aphasia (16% vs 9%), cognitive disorder (16% vs 10%), memory impairment (17% vs 8%), and paresthesia (31% vs 7%) were more common with topiramate than amitriptyline.

A prospective randomized open-label, blinded-endpoint trial assessed the efficacy of flunarizine compared to topiramate for chronic migraine prophylaxis [48c]. Patients were randomized to flunarizine 10 mg once daily ($n = 29$) or topiramate 50 mg once daily ($n = 27$). There was not a statistically significant difference in the rate of TEAEs between the groups (37.9% vs 51.9%, $P = 0.295$); however, the study was likely underpowered to detect relevant differences in this endpoint. Patients receiving a lower dose of topiramate 50 mg per day (51.9%) had a lower TEAE rate than reported for a dosage of 100 mg per day (66%–82.5%). Patients receiving flunarizine had a significantly lower rate of paresthesia development than the topiramate group (0% vs 25.9%, $P = 0.004$). The safety profile of this study suggests patients receiving topiramate were significantly more likely to have paresthesia.

A case–control clinical study of 38 patients evaluated the accommodation function in patients receiving topiramate [49c]. Patients were divided to receive topiramate 100 mg per day ($n = 22$) or placebo ($n = 16$). In most of the accommodation stimuli ranges (0D, 2.5D, 3D, and 5D), topiramate users had a significantly higher accommodative lag compared with controls ($P = 0.028$, $P = 0.014$, $P = 0.011$, and $P = 0.011$, respectively). In a univariate analysis, the most important causes of accommodative lag were accommodation stimulus and topiramate use (R^2 0.32, 95% CI 0.22–0.37 and 0.42–0.91, respectively, $P < 0.001$). In a multivariate linear regression accounting for covariates, accommodation stimulus and age (r 0.51, 95% CI 0.31–0.32 and 0.67–0.69, respectively, $P < 0.001$) predicted accommodative lag.

A case report described a 44-year-old female presenting with lower back pain in May 2012 [50A]. The patient had been taking topiramate 200 mg once daily since 2007 for migraine prophylaxis. In April 2012, lab findings demonstrated presence of kidney stones and urolithiasis (3 mm) which was confirmed by computed tomography. Upon admission, arterial blood gas analysis revealed metabolic acidosis and urine analysis showed alkaline urine (pH 6.39) with low urine citrate concentration (0.3 mmol/24 h). The patient was diagnosed with topiramate induced metabolic acidosis and kidney stones. The patient refused discontinuation of topiramate due to fear of migraine recurrence. Potassium citrate was given to correct metabolic acidosis. Patient eventually stopped topiramate in December 2014 after her stones enlarged. Her bicarbonate concentration and urine citrates normalized after this withdrawal. Extracorporeal shock wave lithotripsy was performed on the kidney to treat the kidney stones.

NEURONAL POTASSIUM CHANNEL OPENERS

Ezogabine (Retigabine)

A case report described a 48-year-old female presenting with new macular pigmentary abnormalities [51A]. She was initiated on ezogabine 50 mg orally three times daily approximately 10 months prior to presentation, with a dose increase to 250 mg three times daily about 7 months prior to presentation. At the time of presentation, she had consumed a cumulative dose of 142.8 g of ezogabine over 238 days. Her eye examination (indicating 20/20 vision) follow-up showed macular pigmentary changes, to which she was referred to a specialist. Ezogabine was discontinued 4 months later once the patient successfully transitioned to another antiepileptic medication. Two months after ezogabine discontinuation, her vision was 20/25 in both eyes. Eight months after stopping ezogabine, her vision reverted to 20/20 in both eyes. Retinal pigment abnormalities visualized through infrared imaging, ocular coherence tomography, and multifocal electroretinography showed improvement after ezogabine discontinuation.

OTHER MECHANISMS OF ACTION [1R, 2R]

Brivaracetam

A meta-analysis included eight randomized, controlled trials, totaling 2505 patients, to evaluate serious adverse events significantly associated with brivaracetam [52M]. A total of 1178 patients were randomized to brivaracetam. There was no significant difference in adverse event-related withdrawal rate between brivaracetam and placebo groups (RR 1.36, 95% CI 0.91–2.04, $P = 0.14$). Among the 17 reported adverse events, dizziness (RR 1.57, 95% CI 1.13–2.18, $P = 0.008$) and fatigue (RR 1.98, 95% CI 1.32–2.97, $P = 0.0010$) were significantly associated with brivaracetam treatment. However, there was no association between the risk of adverse events and dose increase. The meta-analysis concluded brivaracetam treatment was reasonably tolerated by patients and rarely caused serious adverse events.

Levetiracetam

A multicenter, retrospective, observational study was conducted to assess the risk for angioedema in 276 665 levetiracetam users compared to 74 682 phenytoin users [53MC]. The per-protocol analysis included 59 367 levetiracetam users and 74 550 phenytoin users in which 54 and 71 events of angioedema occurred, respectively, (HR 0.72, 95% CI 0.39–1.31). However, the intent-to-treat analysis included 75 056 levetiracetam patients and 95 598 phenytoin patients in which 248 and 435 events of angioedema occurred, respectively (HR 0.64, 95% CI 0.52–0.79). These findings show no evidence of increased angioedema risk with levetiracetam use compared with phenytoin use; however, an increased risk was associated with phenytoin use compared to levetiracetam in the intent-to-treat population.

A multicenter prospective controlled study of 200 patients evaluated the efficacy and safety of prophylactic levetiracetam administration in adults undergoing cranioplasty [54C]. Patients were randomized into one of two groups: levetiracetam (at doses 500, 750, and 1000 mg per day for 24 weeks) or control (placebo). Among patients without postoperative seizures, 20 (20.8%) patients reported 33 side effects in the levetiracetam group and 17 (18.7%) patients reported 35 side effects in the control group at the 2-week follow-up. At the 24-week follow-up, 13 (13.7%) patients reported 22 side effects in the levetiracetam group and 12 (14.5%) patients reported 19 side effects in the control group. At the 2-week and 24-week follow-ups, anemia was the most common side effect in the levetiracetam and control groups (10 and 7 vs 10 and 5 patients). Levetiracetam was not associated with increased risk for side effects in this setting.

A retrospective cohort study evaluated outpatient new-generation antiepileptic use and the related adverse event rate compared to old-generation antiepileptics in 115 epilepsy patients at least 60 years of age [55C]. Old-generation antiepileptics included phenytoin, valproic acid, carbamazepine, phenobarbital, clobazam, and clonazepam while new-generation antiepileptics included gabapentin, levetiracetam, lamotrigine, topiramate, oxcarbazepine, lacosamide, and perampanel. Overall, adverse events occurred in 40 (34.8%) patients and included fatigue ($n=21$), dizziness ($n=7$), tremor ($n=6$), memory deterioration ($n=2$), nervousness ($n=1$), headache ($n=1$), weight loss ($n=1$), and dyspepsia ($n=1$). Adverse events were more frequent with new-generation antiepileptics compared to old-generation antiepileptics or combinations of new and old-generation antiepileptics (49.0% vs 24.4%, $P=0.044$). Adverse events were significantly more frequent in patients treated with levetiracetam compared to other agents (62.1% vs 27.2%, $P=0.001$) but did not significantly differ for lamotrigine (41.7% vs 34.9%, $P=0.633$) or gabapentin (33.3% vs 37.2%, $P=0.813$). Adverse events were associated with moderate levetiracetam titration (i.e., up to 250mg/day/week) (OR 16.35, 95% CI 2.94–90.98, $P=0.001$), low dose (i.e., up to 1500mg per day) (OR 5.68, 95% CI 1.40–22.95, $P=0.015$), and high dose (i.e., greater than 1500mg per day) (OR 4.24, 95% CI 1.28–14.02, $P=0.018$) compared to lamotrigine or gabapentin. A slower titration rate (i.e., up to 125mg/day/week) and moderate maximal dose (i.e., 1500mg per day) was recommended to decrease risk for adverse events.

A case report described a 59-year-old non-smoking female presenting with several weeks of night sweats and exertional breathlessness [56A]. She was initiated on levetiracetam 8 weeks prior as treatment for meningioma-related absence seizures. After analysis of her blood, new eosinophilia (8.38×10^9/L) was detected. Chest radiography revealed asymmetrical "reverse batwing" infiltrates of the lung. The patient was diagnosed with eosinophilic pneumonia secondary to levetiracetam. Levetiracetam was discontinued and replaced with phenytoin. The patient showed clinical improvement 1 month later.

A case report described a 10-year-old female presenting with an osteochondritis dissecans lesion after undergoing a prolonged 7-year treatment with levetiracetam for sporadic convulsions secondary to enterovirus encephalitis [57A]. The patient had a negative anterior drawer test and limited ankle dorsiflexion due to evidence of swelling and tenderness of tarsal navicular joint. Radiography showed an increased focal uptake in the talar head, which is a rare condition. She was diagnosed with juvenile osteochondritis dissecans (OCD) and treated with cast immobilization for 3 weeks. She was treated conservatively with analgesics, rest, nonweight-bearing walking, and a custom ankle brace for 2 months. The patient was asymptomatic, her foot was without swelling, and she had a full range of motion. The patient remained asymptomatic at her 3-year follow-up. The authors associated this event with levetiracetam prior association with impaired skeletal growth and metabolism in animal studies.

Ethosuximide

No relevant studies identified from the search period.

Rufinamide

A multicenter, retrospective chart review of 58 patients assessed the long-term effectiveness of rufinamide in managing Lennox–Gastaut Syndrome (LGS), other epileptic encephalopathies, and intractable focal epilepsies in adults and children [58M]. The mean daily rufinamide dose was 32mg/kg (range 12.5–66.7mg/kg) in children and 24.7mg/kg (range 5–47mg/kg) in adults. Most of the adverse events observed were considered mild. Out of 58 patients, rufinamide was discontinued due to side effects in 8.6% of patients. A total of 21 (36.2%) patients reported adverse events, with the most common being nausea, vomiting and weight loss. Adverse event related withdrawal occurred in five patients (8.6% of total). Adverse events resolved with dose reduction or slower rufinamide titration in seven patients (12.1% of total; 33.3% of patients with adverse events) and with no changes to rufinamide dose in nine patients (15.5% of total; 42.9% of patients with adverse events).

Additional case studies can be found in these reviews [59R, 60R, 61R].

References

[1] Beckett RD, Klemke N, Bessesen M, et al. Antiepileptics. In: Ray SD, editor. Side effects of drugs annual. 39th ed. London, UK: Elsevier Inc; 2017 [R].

[2] Spoelhof B, Frendak L, Kiehle N, et al. Antiepileptics. In: Ray SD, editor. Side effects of drugs Annual. 38th ed. London, UK: Elsevier Inc.; 2016 [R].

[3] Muramatsu K, Ujiie H, Natsuga K, et al. Lichenoid drug eruption caused by clonazepam. J Eur Acad Dermatol Venereol. 2017;31(2):e117–8 [A].

[4] Chamberlain DB, Chamberlain JM. Making sense of a negative clinical trial result: a Bayesian analysis of a clinical trial of lorazepam and diazepam for pediatric status epilepticus. Ann Emerg Med. 2017;69(1):117–24 [C].

[5] Hui D, Frisbee-Hume S, Wilson A, et al. Effect of lorazepam with haloperidol vs haloperidol alone on agitated delirium in patients with advanced cancer receiving palliative care: a randomized clinical trial. JAMA. 2017;318(11):1047–56 [C].

[6] Wright T, Kumarappah A, Stavropoulos A, et al. Vigabatrin toxicity in infancy is associated with retinal defect in adolescence: a prospective observational study. Retina. 2017;37(5):858–66 [C].

[7] O'Callaghan FJ, Edwards SW, Alber FD, et al. Safety and effectiveness of hormonal treatment versus hormonal treatment

with vigabatrin for infantile spasms (ICISS): a randomised, multicentre, open-label trial. Lancet Neurol. 2017;16(1):33–42 [C].

[8] Wiffen PJ, Derry S, Bell RF. Gabapentin for chronic neuropathic pain in adults. Cochrane Database Syst Rev. 2017;6: CD007938 [M].

[9] Diana A, Pillai R, Bongioanni P, et al. Gamma aminobutyric acid (GABA) modulators for amyotrophic lateral sclerosis/ motor neuron disease. Cochrane Database Syst Rev. 2017;1: CD006049 [M].

[10] Huybrechts KF, Bateman BT, Desai RJ. Risk of neonatal drug withdrawal after intrauterine co-exposure to opioids and psychotropic medications: cohort study. BMJ. 2017;358:j3326 [MC].

[11] Emiroglu N, Cengiz FP, Su O, et al. Gabapentin-induced aquagenic wrinkling of the palms. Dermatol Online J. 2017;23(1) pii: 13030/ qt64k739q5. [A].

[12] Choi MS, Jeon H, Kim HS, et al. A case of gabapentin-induced rhabdomyolysis requiring renal replacement therapy. Hemodial Int. 2017;21(1):E4–8 [A].

[13] Shanthanna H, Gilron I, Rajarathinam M, et al. Benefits and safety of gabapentinoids in chronic low back pain: a systematic review and meta-analysis of randomized controlled trials. PLoS Med. 2017;14(8):e1002369 [M].

[14] Mathieson S, Maher CG, McLachlan AJ, et al. Trial of pregabalin for acute and chronic sciatica. N Engl J Med. 2017;376(12):1111–20 [C].

[15] Asgari Z, Rouholamin S, Nataj M, et al. Dose ranging effects of pregabalin on pain in patients undergoing laparoscopic hysterectomy: a randomized, double blinded, placebo controlled, clinical trial. J Clin Anesth. 2017;38(May):13–7 [C].

[16] Dou Z, Jiang Z, Zhong J, et al. Efficacy and safety of pregabalin in patients with neuropathic cancer pain undergoing morphine therapy. Asia Pac J Clin Oncol. 2017;13(2):e57–64 [C].

[17] Foroutan N, Etminan A, Nikvarz N, et al. Comparison of pregabalin with doxepin in the management of uremic pruritus: a randomized single blind clinical trial. Hemodial Int. 2017;21(1):63–71 [C].

[18] Aktaş S, Tetiko⊠lu M, ⊠nan S, et al. Unilateral hemorrhagic macular infarction associated with marijuana, alcohol and antiepileptic drug intake. Cutan Ocul Toxicol. 2017;36(1):88–95 [A].

[19] Acciavatti T, Martinotti G, Corbo M, et al. Psychotropic drugs and ventricular repolarization: the effects on QT interval, T-peak to T-end interval and QT dispersion. J Psychopharmacol. 2017;31(4):453–60 [C].

[20] Sidhu HS, Srinivas R, Sadhotra A. Evaluate the effects of long-term valproic acid treatment on metabolic profiles in newly diagnosed or untreated female epileptic patients: a prospective study. Seizure. 2017;48(May):15–21 [C].

[21] Guo X, Wei J, Gao L, et al. Hyperammonemic coma after craniotomy: hepatic encephalopathy from upper gastrointestinal hemorrhage or valproate side effect?: case report and literature review. Medicine (Baltimore). 2017;96(15):e6588 [A].

[22] Tada H, Ogihara T, Nakamura T, et al. A case of severe parkinsonism in an elderly person induced by valproic acid. Psychogeriatrics. 2017;17(1):76–7 [A].

[23] Levy ZD, Slowey M, Schulder M. Acute generalized exanthematous pustulosis secondary to levetiracetam and valproic acid use. Am J Emerg Med. 2017;35(7):1036.e1–2 [A].

[24] Mutlu-Albayrak H, Bulut C, Çaksen H. Fetal valproate syndrome. Pediatr Neonatol. 2017;58(2):158–64 [A].

[25] Cattaneo CI, Ressico F, Valsesia R, et al. Sudden valproate-induced hyperammonemia managed with L-carnitine in a medically healthy bipolar patient: essential review of the literature and case report. Medicine (Baltimore). 2017;96(39):e8117 [A].

[26] Schrettl V, Felgenhauer N, Rabe C, et al. L-Arginine in the treatment of valproate overdose—five clinical cases. Clin Toxicol (Phila). 2017;55(4):260–6 [A].

[27] Riker RR, Gagnon DJ, Hatton C, et al. Valproate protein binding is highly variable in ICU patients and not predicted by total serum concentrations: a case series and literature review. Pharmacotherapy. 2017;37(4):500–8 [A].

[28] Veroniki AA, Cogo E, Rios P, et al. Comparative safety of anti-epileptic drugs during pregnancy: a systematic review and network meta-analysis of congenital malformations and prenatal outcomes. BMC Med. 2017;15(1):95 [M].

[29] Thomas SV, Jose M, Divakaran S, et al. Malformation risk of antiepileptic drug exposure during pregnancy in women with epilepsy: results from a pregnancy registry in South India. Epilepsia. 2017;58(2):274–81 [MC].

[30] Nelson JC, Spyker DA. Morbidity and mortality associated with medications used in the treatment of depression: an analysis of cases reported to U.S. poison control centers, 2000–2014. Am J Psychiatry. 2017;174(5):438–50 [C].

[31] Shi YW, Min FL, Zhou D, et al. HLA-A*24:02 as a common risk factor for antiepileptic drug-induced cutaneous adverse reactions. Neurology. 2017;88(23):2183–91 [C].

[32] Barbosa Ribeiro S, Antunes de Araújo A, Addison Xavier Medeiros C, et al. Factors associated with expression of extrapyramidal symptoms in users of atypical antipsychotics. Eur J Clin Pharmacol. 2017;73(3):351–5 [C].

[33] Chandrasekaran S, Patil S, Suther R, et al. Hyperhomocysteinaemia in children receiving phenytoin and carbamazepine monotherapy: a cross-sectional observational study. Arch Dis Child. 2017;102(4):346–51 [C].

[34] Kay L, Willems LM, Zöllner JP, et al. Eslicarbazepine acetate as a therapeutic option in a patient with carbamazepine-induced rash and HLA-A*31:01. Seizure. 2017;47(Apr):81–2 [A].

[35] Molina-Ruiz AM, Lasanta B, Barcia A, et al. Drug-induced systemic lupus erythematosus in a child after 3 years of treatment with carbamazepine. Australas J Dermatol. 2017;58(1): e20–2 [A].

[36] Stöllberger C, Finsterer J. Recurrent venous thrombosis under rivaroxaban and carbamazepine for symptomatic epilepsy. Neurol Neurochir Pol. 2017;51(2):194–6 [A].

[37] Baulac M, Byrnes W, WIlliams P, et al. Lacosamide and sodium channel-blocking antiepileptic drug cross-titration against levetiracetam background therapy. Acta Neurol Scand. 2017;135(4):434–41 [C].

[38] Rahman MM, Alatawi Y, Cheng N, et al. Comparison of brand versus generic antiepileptic drug adverse event reporting rates in the U.S. Food and Drug Administration adverse event reporting system (FAERS). Epilepsy Res. 2017;135(Sep):71–8 [C].

[39] Pomeroy SJ, Ndikumana R, Cavanagh JP. Lamotrigine induced lymphadenopathy: case report and literature review. Int J Pediatr Otorhinolaryngol. 2017;98(Jul):82–4 [A].

[40] Lee AR, Sharma Sm Mahmoud TH. Tubulointerstitial nephritis and uveitis syndrome with a primary presentation of acute posterior multifocal placid pigment epitheliopathy. Retinal Cases Brief Rep. 2017;11(2):100–3 [A].

[41] Tombini M, Pellegrino G, Assenza G, et al. De novo multifocal myoclonus induced by lamotrigine in a temporal lobe epilepsy case. J Neurol Sci. 2017;373(Feb):31–2 [A].

[42] Grosso S, Ferranti S, Gaggiano C, et al. Massive lamotrigine poisoning: a case report. Brain Dev. 2017;39(4): 349–51 [A].

[43] Hager N, Bolt J, Albers L, et al. Development of left atrial thrombus after coadministration of dabigatran etexilate and phenytoin. Can J Cardiol. 2017;33(4):554.e13–4 [A].

[44] Huber B, Schmid G. A two-year retrospective evaluation of perampanel in patients with highly drug-resistant epilepsy and cognitive impairment. Epilepsy Behav. 2017;66(Jan): 74–9 [A].

[45] He A, Song D, Zhang L, et al. Unveiling the relative efficacy, safety and tolerability of prophylactic medications for migraine: pairwise and network-meta analysis. J Headache Pain. 2017;18(1):26 [M].

[46] Liu J, Wang LN, Wang YP. Topiramate monotherapy for juvenile myoclonic epilepsy. Cochrane Database Syst Rev. 2017;4: https://doi.org/10.1002/14651858.CD010008.pub3 [M].

[47] Powers SW, Coffey CS, Chamberlin LA, et al. Trial of amitriptyline, topiramate, and placebo for pediatric migraine. N Engl J Med. 2017;376(2):115–24 [C].

[48] Lai K, Niddam DM, Fuh JL, et al. Flunarizine versus topiramate for chronic migraine prophylaxis: a randomized trial. Acta Neurol Scand. 2017;135(4):476–83 [C].

[49] Çerman E, Akkaya Turhan S, Eraslan M, et al. Topiramate and accommodation: does topiramate cause accommodative dysfunction? Can J Ophthalmol. 2017;52(1):20–5 [C].

[50] Salek T, Andel I, Kurfurstova I. Topiramate induced metabolic acidosis and kidney stones—a case study. Biochem Med (Zagreb). 2017;27(2):404–10 [A].

[51] Zaugg BE, Bell JE, Taylor KY, et al. Ezogabine (potiga) maculopathy. Retin Cases Brief Rep. 2017;11(1):38–43 [A].

[52] Zhu LN, Chen D, Chen T, et al. The adverse event profile of brivaracetam: a meta-analysis of randomized controlled trials. Seizure. 2017;45(Feb):7–16 [M].

[53] Duke JD, Ryan PB, Suchard MA, et al. Risk of angioedema associated with levetiracetam compared with phenytoin: findings of the observational health data sciences and informatics research network. Epilepsia. 2017;58(8):101–6 [MC].

[54] Liang S, Ding P, Zhang S, et al. Prophylactic levetiracetam for seizure control after cranioplasty: a multicenter prospective controlled study. World Neurosurg. 2017;102(Jun):284–92 [C].

[55] Theitler J, Brik A, Shaniv D, et al. Antiepileptic drug treatment in community-dwelling older patients with epilepsy: a retrospective observational study of old- versus new-generation antiepileptic drugs. Drugs Aging. 2017;34(6):479–87 [C].

[56] Fagan A, Fuld J, Soon E. Levetiracetam-induced eosinophilic pneumonia. BMJ Case Rep. 2017;2017(Mar). pii bcr2016219121. [A].

[57] Turati M, Glard Y, Afonso D, et al. Osteochondral alteration in a child treated with levetiracetam: a rare case of juvenile osteochondritis dissecans of the talar head. J Pediatr Orthop B. 2017;26(2):189–92 [A].

[58] Jaraba S, Santamarina E, Miró J, et al. Rufinamide in children and adults in routine clinical practice. Acta Neurol Scand. 2017;135(1):122–8 [M].

[59] Spoelhof B, Frendak L, Rivera Lara L. Chapter 7 - Antiepileptics. In: Ray SD, editor. Side Effects of Drugs- Annual: A worldwide yearly survey of new data in adverse drug reactions. Elsevier. 2015;37:85–106 [R].

[60] Spoelhof B, Frendak L, Kiehle N, et al. Chapter 6 - Antiepileptics. In: Ray SD, editor. Side Effects of Drugs- Annual: A worldwide yearly survey of new data in adverse drug reactions. Elsevier. 2016;38:55–70 [R].

[61] Beckett RD, Klemke N, Bessesen M, et al. Chapter 7 - Antiepileptics. In: Ray SD, editor. Side Effects of Drugs: Annual: A worldwide yearly survey of new data in adverse drug reactions. Elsevier. 2017;39:91–106 [R].

CHAPTER

8

Opioid Analgesics and Narcotic Antagonists

*Justin G. Kullgren**, *Amar P. Thakkar*[†], *Michael G. O'Neil*[†,1]

*The Ohio State University Wexner Medical Center, Columbus, OH, United States
[†]Department of Pharmacy Practice, South College School of Pharmacy, Knoxville, TN, United States
[1]Corresponding author: moneil@southcollegetn.edu

INTRODUCTION

Opioid Abuse Deterrent Formulations: A Response to the Opioid Epidemic

The opioid epidemic has devastated the United States with hundreds of Americans dying daily from opioid drug overdoses [1S]. Over the last several years, steps have been taken to reduce deaths associated with misuse and abuse of opioid analgesics resulting in the Centers for Disease Control to develop guidelines for prescribing opioids in chronic pain. Several states have implemented legislation to define opioid prescribing limits and improved drug technology to deter manipulation of opioid products has been developed [2R]. The latter of these steps has resulted in several new abuse deterrent technologies and the approval of several new formulations of opioid products. The introduction of abuse deterrent formulations (ADFs) has appeared to result in a decreased ability to misuse, abuse, or extract large amounts of active ingredients compared to older formulations of opioid analgesics [3c, 4R].

ADFs employ varying types of technology to make manipulation of opioid products more difficult to insufflate (snort), inhale, inject, or misuse by any means other than the intended route. Unfortunately, these technologies do not seem to deter patients from taking more than prescribed by the oral route at a higher dose or an increased frequency [4R]. Some of the technologies commonly utilized in ADFs include physical barriers, chemical barriers, and agonist/antagonist combinations. Less commonly used technologies or technologies in development include use of aversive agents, products that require transformation after oral absorption, or a combination of multiple technologies [5R].

Physical barriers aim to reduce crushing, chewing, and splitting in addition to other commonly used abuse methods. Chemical barriers can help to minimize extraction of opioids using solvents. One of the first commonly prescribed opioid analgesic ADFs utilized a combination of a physical and chemical barriers [3c]. Both physical and chemical barriers pose little adverse risk to compliant patients or to patients who accidently manipulate the product. Opioid agonist/antagonist combination products contain a pure opioid agonist with an opioid antagonist such as naloxone or naltrexone. The opioid antagonist is intended to only be released if the product is manipulated, at which time blunting the euphoric effect of the pure opioid agonist. When utilized as intended, the opioid antagonist does not interfere with the analgesic effects of the pure opioid. The goal of an aversive agent is to produce, or cause, a negative feeling/sensation if the opioid product is misused or abused. A potential disadvantage of aversive agents is the chance of adverse effects with legitimate dose increases [5R, 6R, 7S]. These are just a few of the technologies utilized in ADFs to help minimize inappropriate use of opioid analgesics. Pharmaceutical manufacturers continue to develop new and innovative opioid ADFs to decrease misuse and abuse of opioids, representing one tool to help combat the opioid crisis in the United States.

OPIOID RECEPTOR AGONISTS

Hydromorphone (SEDA-39, 107–110)

Overdose/Intoxication

This is a case report of a 3-year-old male that was found unresponsive and was unable to be successfully

Side Effects of Drugs Annual, Volume 40
ISSN: 0378-6080
https://doi.org/10.1016/bs.seda.2018.07.018

revived following an accidental hydromorphone overdose. Investigations revealed that two hydromorphone tablets (2mg each) were missing. The postmortem hydromorphone concentrations were reported at 0.03mg/L in the peripheral blood, 0.06mg/L in the central blood, and 0.10mg/kg in the liver. The authors concluded that there was no other cause of death found on the autopsy besides the hydromorphone concentrations. Pediatric hydromorphone postmortem data is limited [8A].

Loperamide

Misuse/Abuse/Toxicity

Loperamide has been traditionally used as an antidiarrheal agent. This review article outlines the toxicities of loperamide seen in patients for euphoric effect or to treat opioid withdrawal symptoms. Traditionally, the extremely low bioavailability of loperamide has limited systemic adverse effects. The maximum recommended dose is 16mg/day. Summary data includes reports of loperamide overdoses with up to 800mg/day for 18 months with lethal consequences. Additionally, authors note intentional use of cimetidine and grapefruit juice to inhibit metabolism of loperamide to enhance the effects. Cardiotoxicity is the hallmark adverse effect of loperamide toxicity best characterized by syncope and lethal dysrhythmias. The authors conclude that the misuse of loperamide is increasing and should be considered in patients presenting with a history of abuse and unexplained syncope or dysrhythmias [9R].

Cardiovascular

Katz et al. discuss two case reports associated with the toxicities of loperamide. The first case involves a 28-year-old male who presented to the emergency department with shortness of breath and lightheadedness who had ingested a large amount of loperamide. The patient's ECG confirmed sinus rhythm, right axis deviation, undetectable PR interval, QRS 168ms, and QTc 693ms that further progressed to Torsades de Pointes. The patient received treatment with intravenous (IV) sodium bicarbonate, magnesium sulfate, lidocaine, isoproterenol and a pacemaker was placed. Treatment was successful, and the patient was discharged in a stable condition on day 16 of his hospital stay. The second case involves a 39-year-old female who was taken to the emergency department after having seizure-like activity from ingesting a large amount of loperamide. She experienced Torsades de Pointes in the emergency department with an ECG that confirmed sinus rhythm, right axis deviation, PR interval 208ms, QRS interval 142ms, and QTc 687ms. The patient also received treatment with IV sodium bicarbonate, magnesium sulfate, lidocaine, and isoproterenol. The patient was discharged on day 6 of her hospital stay. In both case reports, the authors concluded that loperamide led to Torsades de Pointes [10A].

Immunologic

This is a case report of a 48-year-old female that was presenting with a possible immune-mediated allergic reaction due to various medications. The patient reported two previous incidences of hypersensitivity responses to medications in the previous 3 years. Hypersensitivity was characterized by generalized urticaria, angioedema of the lips and dysphagia. The medications included amoxicillin, diclofenac, acetaminophen pantoprazole and loperamide. Skin prick tests, intradermal skin tests or oral rechallenge of these medications resulted in negative results. Loperamide was not retested at this time. The patient returned 4 years later and reported 4 additional hypersensitivity effects that appeared 8–12 hours after taking loperamide requiring urgent treatment. Oral rechallenge with loperamide after a negative skin test produced urticaria, angioedema of the lips and tongue, dysphagia and respiratory distress that were successfully treated. Authors concluded this was not a true type 1 mediated-hypersensitivity reaction. The authors hypothesize that a form of acute gastroenteritis may have altered absorption or metabolism of loperamide leading to the delayed allergic reaction [11A].

Methadone (SEDA-37, 107–114; SEDA-38, 71–76; SEDA-39, 107–110)

Thermoregulation

This is a case report of a 35-year-old male that was initiated on oral methadone for opioid maintenance therapy. Over a 2-week period, doses were titrated to 100mg daily to control opioid withdrawal symptoms. The patient reported excessive sweating after initiating the methadone. The patient denied ever having previous symptoms. Urine drug screens excluded other substances of abuse. His past medical history was significant for long-term substance abuse and hepatitis C. The patient denied use of other medications. Oxybutynin, an agent known to be effective in the treatment of hyperhidrosis, was prescribed with successful resolution of the hyperhidrosis. The authors concluded that hyperhidrosis is a main side effect of methadone therapy and that oxybutynin can control this side effect [12A].

Respiratory

This is a case report of a 38-year-old female that presented to the pulmonary clinic with complaints significant only for dyspnea and dry cough for more than 3 months in duration. Physical examination revealed hypoxemia, oxygen saturation of 92%, and a respiratory rate of 24 breaths/min. Her computed tomography (CT) of the chest showed bilateral centrilobular nodules. Bronchoscopy and culture results were reported negative. Transbronchial biopsy revealed multinucleated giant cell granulomas with birefringent foreign material consistent with talc. The patient stated that she had self-administered IV a crushed methadone tablet for pain. Based on her findings and history she was diagnosed with pulmonary talcosis secondary to IV drug injection [13A].

Endocrine/Metabolic

This is a case report of an 11-month-old male with no known significant past medical history presented with respiratory failure requiring intubation and transport to the emergency room. Physical exam was significant for respiratory depression and miosis. Laboratory values were significant for a blood glucose of 17mg/dL, an elevated insulin level, and suppressed serum beta-hydroxybutyrate. Other toxicology screens and assessments for known causes for hypoglycemia were excluded. The patient was treated with dextrose and naloxone resulting in improvements in symptoms. A serum methadone level of 123ng/mL was reported. The authors concluded that the patient's hypoglycemia was due to methadone intoxication [14A].

Morphine (SEDA-37, 107–114; SEDA-38, 71–76; SEDA-39, 107–110)

Pharmacogenetics

This is a prospective observational study from August 1, 2007 to July 31, 2008 that observed 129 Taiwanese females undergoing hysterectomy and studied the effect of 14 single nucleotide polymorphism (SNPs) on nausea and vomiting induced by intravenous morphine administered via patient controlled analgesia pump. Patients were evaluated by a trained investigator blinded to the patient's genetic profiles. Patient complaints, severity of complaints, and pain control were assessed daily for 72 hours. The authors concluded that there was no association between SNPs and morphine induced nausea or vomiting [15C].

This is a prospective, genotype blinded, clinical observation study that evaluated children undergoing tonsillectomy (ages 6–15 years old) and adolescents undergoing idiopathic scoliosis spine surgery (ages 10–18 years old). The authors evaluated the association of respiratory depression and the ATP binding cassette gene *ABCC3*, which is a gene that helps hepatic morphine efflux. Each patient received a standardized intra-operative intravenous dose of morphine per a predetermined protocol. After studying 316 children undergoing tonsillectomy and 67 adolescents undergoing spine surgery, the authors concluded that the *ABCC3* gene is associated with postoperative respiratory depression [16C].

Oxycodone

Immunologic

This is a prospective study that evaluated the effects of oxycodone injection on the immune function of patients following radical resection of rectal cancer. Eighty patients were randomly divided into two groups: one group of patients received 5mg of intravenous oxycodone vs the other group that received 5mg of intravenous morphine. The authors assessed the changes in the number of T lymphocytes and natural killer cells as determined by flow cytometry at defined time intervals. The authors concluded that both oxycodone and morphine have inhibitory effects on the immune function with patients undergoing radical resection of rectal cancer; however, oxycodone had a lesser effect on the immune function than morphine [17c].

Neurological

This is a case report of an 81-year-old male who had recently undergone a surgical repair of his right distal femoral fracture. Past medical history was significant for schizophrenia, pituitary adenoma, osteopenia, atrial fibrillation and celiac disease. The patient was on multiple medications including levothyroxine, hydrocortisone, testosterone, warfarin, ranitidine, tamsulosin, sertraline, and oxycodone 15mg/day. The patient became worried about becoming addicted to the oxycodone and abruptly stopped the medication. Subsequently, he reported having severe, uncomfortable restlessness in his legs at night that lasted about 2 hours. These symptoms continued for 3 weeks. The patient reinitiated oxycodone 2.5mg TID with significant resolution of the symptoms. The authors conclude that opioid withdrawal from oxycodone may be associated with restless legs syndrome in elderly patients [18A].

Neuromuscular Function

This is a case report of a 52-year-old female that has an extensive medical history. She was taking an SSRI and immediate-release oxycodone for an unknown period. The patient's oxycodone was switched to a fixed-ratio combination of oxycodone/naloxone to help alleviate her constipation and hepatic encephalopathy. After 1 hour of receiving the oxycodone/naloxone (20mg/10mg) she developed psychomotor agitation with rigors, myoclonic jerks, tachycardia, diaphoresis, malar flushing, hyperesthesia, hyperosmia, and frequent yawning. The authors concluded that this was a case of opioid withdrawal [19A].

Tapentadol (SEDA-38, 71–76; SEDA-39, 107–110)

Cardiovascular

This is a case report of a 32-year-old female who presented to the emergency department with complaints of severe intractable headaches for 2 days. Other symptoms included nausea and vomiting. Past medical history was significant for anxiety disorder, migraines, and asthma. Current medications included chlorazepate, albuterol, and tapentadol. She was prescribed tapentadol 75mg every 4 hours for chronic pain a month prior. Upon examination, the patient seemed alert and oriented and had no other clinical problems. Head CT, MRI and all laboratory values were also reported within normal limits. She was admitted to the medicine service for assessment of refractory headaches. She was administered ketorolac, verapamil, and divalproex sodium. The following day the patient was found on the floor unresponsive. The nurse reported finding a tapentadol labeled bottle containing crushed pills suspended in liquid and a syringe next to her bed. Efforts were made to resuscitate the patient. The patient's ECG showed a QRS duration of 74ms and a corrected QT interval of 430ms. The patient was pronounced dead shortly after arrival to the intensive care unit. Autopsy was consistent with tapentadol overdose [20A].

Tramadol

Neuropsychological

A matched-control study evaluated 100 patients that abused tramadol. Participants were subsequently divided into 2 groups: participants who used tramadol alone and participants that used tramadol and other substances. Participants were interviewed to collect socioeconomical and clinical information. Cognitive assessment was preformed utilizing the Montreal cognitive Assessment test. Control subjects were first degree relatives of the participants. Researchers concluded that participants abusing tramadol were nearly three times more likely to demonstrate significant cognitive impairment when compared to controls. Tramadol abuse was more likely in married males that were skilled and living in rural areas [21C].

Cardiovascular

This is a case report of a 50-year-old female who was found conscious and hemodynamically stable after a self-reported ingestion of 18g of tramadol, 15mg alprazolam and 1L 15% alcohol. The patient's clinical course rapidly deteriorated during transport. Tramadol concentration reported 12 hours after admission was 46.5μg/L. Further deterioration was characterized by seizures, mydriasis, hemodynamic instability, lactic acidosis, renal failure, liver failure and respiratory failure. Treatment involved aggressive fluid and pressor support for cardiogenic shock, aggressive seizure management, extracorporeal life support and mechanical ventilation. The patient required a prolonged ICU stay, tracheostomy and hemodialysis and was reported alive 1 year later. The authors concluded that the patient had a severe tramadol intoxication with a plasma concentration 20 times the known toxic dose [22A].

MIXED AGONIST–ANTAGONIST/ PARTIAL OPIOID AGONIST

Buprenorphine (SEDA-37, 107–114; SEDA-39, 107–110)

Fetotoxicity

Forty-nine pregnant women on buprenorphine were evaluated for effects of maternal buprenorphine on fetal neurobehavioral development at predetermined at 24, 28, 32, and 36 weeks gestation during the pregnancy. Fetal neurobehavioral parameters monitored included heart rate, fetal movement and movement-heart rate coupling. Peak and trough levels of buprenorphine were assessed. Fetal heart rate, heart rate variability and heart rate accelerations correlated with buprenorphine peaks and troughs and progression of the pregnancy. The authors concluded that maternal buprenorphine administration has acute suppressive effects on the fetal heart rate and movement as the pregnancy progresses [23c].

Vascular

This is a case report of a 45-year-old male that presented to the rheumatology unit with swelling of his hands that had gradually worsened over the last 3 years. Past medical and social history was significant for hepatitis and heavy alcohol consumption. Examination was significant for bilateral nonpitting edema in the proximal segments of the fingers and wrist that was not altered by elevation of the hands. Liver chemistry values were reported as ALT 175IU/L, AST 170IU/L, and albumin 4.68mg/dL. The patient also had a rheumatoid factor or 29IU/mL. The patient reported injecting high-dose sublingual buprenorphine for more than 5 years 3–4 times a day into the veins around his hands. The authors speculated that venous and lymphatic occlusion from particulate matter in the buprenorphine was likely responsible for the swelling [24A].

In spite of many years of opioid prescribing, a variety of new and unexpected adverse consequences continue to surface. A variety of new abuse patterns have developed with opioids leading to lethal consequences [8A, 9R, 10A]. Clinicians and researchers alike should remain vigilant to adverse consequences to common medications that may be abused. Additional case studies in the same or related topics can be found in other reviews [25R].

References

[1] U.S. Department of Health and Human Services. n.d. About the U.S. Opioid Epidemic. 2017. https://www.hhs.gov/opioids/about-the-epidemic/. Accessed Feb 22, 2018 [S].

[2] Dowell D, Haegerich TM, Chou R. CDC guideline for prescribing opioids for chronic pain—United States, 2016. MMWR Recomm Rep. 2016;65:1–49 [No. RR-1], https://doi.org/10.15585/mmwr.rr6501e1 [R].

[3] Cicero T, Ellis M. Abuse-deterrent formulations and the prescription opioid abuse epidemic in the United States: lessons learned from OxyContin. JAMA Psychiat. 2015;72(5):424–30. https://doi.org/10.1001/jamapsychiatry.2014.3043 [c].

[4] Gasior M, Bond M, Malamut R. Routes of abuse of prescription opioid analgesics: a review and assessment of the potential impact of abuse-deterrent formulations. Postgrad Med. 2016;128(1):85–96. https://doi.org/10.1080/00325481.2016.1120642. Epub 2015 Dec 18. [R].

[5] Hale ME, Moe D, Bond M, et al. Abuse-deterrent formulations of prescription opioid analgesics in the management of chronic noncancer pain. Pain Manag. 2016;6(5):497–508. https://doi.org/10.2217/pmt-2015-0005. Epub 2016 Apr 6. [R].

[6] Pergolizzi J, Taylor Jr. R, LeQuang JA, et al. The evolution of abuse-deterrent extended-release opioid formulations: is there a place in the armamentarium for more products? EC Anaesthesia. 2017;3(2):64–77 [R].

[7] US Food and Drug Administration. Abuse-deterrent opioids—evaluation and labeling guidance for industry, www.fda.gov; 2015 [S].

[8] Cantrell FL, Sherrard J, Andrade M, et al. A pediatric fatality due to accidental hydromorphone ingestion. Clin Toxicol. 2017;55(1):60–2 [A].

[9] Wu PE, Juurlink DN. Loperamide toxicity. Ann Emerg Med. 2017;70(2):245–52 [R].

[10] Katz KD, Cannon RD, Cook MD, et al. Loperamide-induced Torsades de pointes: a case series. J Emerg Med. 2017;53(3):339–44 [A].

[11] Nahas O, Alary V, Chiriac AM, et al. Delayed hypersensitivity reaction to loperamide: an intriguing case report with positive challenge test. Allergol Int. 2017;66(1):139–40 [A].

[12] Hong J, Lee J, Totouom-Tangho H, et al. Methadone-induced hyperhidrosis treated with oxybutynin. J Addict Med. 2017;11(3):237–8 [A].

[13] Escuissato DL, Ferreira RG, de Barros JA, et al. Pulmonary talcosis caused by intravenous methadone injection. J Bras Pneumol. 2017;43(2):154–5 [A].

[14] Toce MS, Stefater MA, Breault DT, et al. A case report of methadone-associated hypoglycemia in an 11-month-old male. Clin Toxicol. 2017;56(1):74–6 [A].

[15] Chen LK, Wang MH, Yang HJ, et al. Prospective observational pharmacogenetic study of side effects induced by intravenous morphine for postoperative analgesia. Medicine (Baltimore). 2017;96(25):e7009 [C].

[16] Chidambaran V, Venkatasubramanian R, Zhang X, et al. *ABCC3* genetic variants are associated with postoperative morphine-induced respiratory depression and morphine pharmacokinetics in children. Pharmacogenomics J. 2017;17(2):162–9 [C].

[17] Cui JH, Jiang WW, Liao YJ, et al. Effects of oxycodone on immune function in patients undergoing radical resection of rectal cancer under general anesthesia. Medicine (Baltimore). 2017;96(31):e7519 [c].

[18] Lee GH, Jaen-Vinuales AV, Stewart JT. A case of restless legs syndrome related to opioid discontinuation. Clin Neuropharmacol. 2017;40(1):50 [A].

[19] Lau F, Gardiner M. Oxycodone/naloxone: an unusual adverse drug reaction. Aust Fam Physician. 2017;46(1):42–3 [A].

[20] Khaja M, Lominadze G, Millerman K. Cardiac arrest following drug abuse with intravenous tapentadol: case report and literature review. Am J Case Rep. 2017;18:817–21 [A].

[21] Bassiony MM, Youssef UM, Hassan MS, et al. Cognitive impairment and tramadol dependence. J Clin Psychopharmacol. 2017;37(1):61–6 [C].

[22] Belin N, Clairet AL, Chocron S, et al. Refractory cardiogenic shock during tramadol poisoning: a case report. Cardiovasc Toxicol. 2017;17(2):219–22 [A].

[23] Jansson LM, Velez M, McConnell K, et al. Maternal buprenorphine treatment and fetal neurobehavioral development. Am J Obstet Gynecol. 2017;216(5):529 [c].

[24] Vivaldelli E, Sansone S, Maier A, et al. Puffy hand syndrome. Lancet. 2017;389:298 [A].

[25] O'Neil MG, Kullgren JG. Opioid analgesics and narcotic antagonists. In: Ray SD, editor. Side effects of drugs. Annual: a worldwide yearly survey of new data in adverse drug reactions, vol. 39. Elsevier; 2017. p. 107–10 [chapter 8]. [R].

CHAPTER

9

Anti-Inflammatory and Antipyretic Analgesics and Drugs Used in Gout

Mark E. Olah[1]

Department of Pharmaceutical and Biomedical Sciences, Raabe College of Pharmacy, Ohio Northern University, Ada, OH, United States

[1]Corresponding author: m-olah@onu.edu

ANILINE DERIVATIVES [SEDA-37, 115; SEDA-38, 77]

Paracetamol (Acetaminophen) [SEDA-37, 115; SEDA-38, 77]

Cardiac

A case of acetaminophen-induced fatal toxic myocarditis was reported [1a]. An 18-year-old female was admitted to the hospital following a suicide attempt that included ingestion of ~40 pills of unspecified dose acetaminophen, quetiapine, aspirin and ethanol. Standard treatment including oral activated charcoal, *N*-acetylcysteine and supportive care was initiated. Hepatic function was severely impaired and the patient was wait-listed for liver transplantation. At 20 hours post-admission, electrocardiogram showed ST segment elevation and atrial fibrillation. Creatinine kinase and troponin I were markedly elevated. At 40 hours, the patient died of cardiac arrest after unsuccessful resuscitation. Histologic examination of the heart revealed interstitial inflammatory infiltrate of the myocardium with the presence of neutrophils, some lymphocytes, eosinophils and macrophages. Contraction band necrosis was present and some hypereosinophilic cardiomyocytes were present. It was concluded that death was due to acute toxic myocarditis due to acetaminophen intoxication as the authors ruled out other ingested substances as a likely cause.

Dermatologic

Sweet syndrome, also referred to as acute neutrophilic dermatosis, is a cutaneous reaction that is typically characterized by painful papules and nodules that may coalesce into plaques. It may be accompanied by fever, arthralgia and other nondermal symptoms. Etiology is often unknown but may be associated with recent infection, cancer or drug administration. A case was reported in which a 32-year-old male received an oral acetaminophen–codeine suspension as an analgesic following surgery for a facial fracture [2a]. After 1 week, he presented to the emergency room with painful dermal eruptions on the forehead and was prescribed 300 mg/30 mg acetaminophen–codeine tablets and topical 1% hydrocortisone cream. After 2 days, the patient returned with worsened cutaneous eruptions and development of an intermittent fever. Painful plaques were present on the forehead, palms, arms and knee. Laboratory values showed a mild anemia and elevated blood glucose and A_{1C} indicating the patient's known diabetes was uncontrolled. Punch biopsy of the forehead revealed minimal to mild papillary dermal edema and a diffuse neutrophilic infiltrate spanning the upper, middle and lower dermis with evidence of mild leukocytoclasia and no evidence of leukocytoclastic vasculitis. The diagnosis of Sweet syndrome due to acetaminophen–codeine was made and intravenous methylprednisolone was initiated in addition to intravenous diphenhydramine and topical desonide ointment. Substantial improvement of dermal eruptions was noted in 48 hours and the patient discharged with topical triamcinolone after 3 days. No recurrence had developed at 2 months of follow-up.

Hematologic

A case report presented the development of hemolytic anemia in a 30-year-old female following ingestion of 150 mg/kg acetaminophen and 15 mg/kg amoxicillin [3a]. The individual had glucose-6-phosphate

Side Effects of Drugs Annual, Volume 40
ISSN: 0378-6080
https://doi.org/10.1016/bs.seda.2018.08.007

dehydrogenase (G6PD) deficiency but there was no description of the precise variant. Following *N*-acetylcysteine administration, the patient was admitted and developed hepatotoxicity and worsening anemia that was consistent with a hemolytic anemia. Over 4 days, the hepatoxicity resolved while hemoglobin values continued to decline. Details of the correction of the anemia were not provided. An increased risk of drug-induced hemolytic anemia in individuals with G6PD deficiency is well-recognized and likely relates to loss of oxidative stress defense mechanisms in erythrocytes. The authors propose multiple mechanisms for acetaminophen-induced hemolytic anemia that primarily focus on the metabolite *N*-acetyl-*p*-benzoquinone imine (NAPQI) that may reduce cellular glutathione levels. Further deficiency of this protective anti-oxidant molecule in G6PD patients may thus predispose these individuals to toxic effects of acetaminophen.

A case report described acetaminophen-induced methemoglobinemia in a 78-year-old woman with a history of autosomal dominant polycystic kidney disease that required hemodialysis for the previous 30 months [4a]. The patient reported to the hospital after developing brownish skin tone earlier in the day with no other complaints. She reported use of acetaminophen (3 g/day) for the previous 7 days for joint pain. Arterial blood gas analysis revealed a 3.3% methemoglobin level (normal = 0%–1.0%). Additionally, there was an increase in total bilirubin and alanine aminotransferase. Acetaminophen was discontinued and the patient remained under observation and discharged after 2 days with a reduction in methemoglobin levels. Nine days after onset of symptoms, the patient had an improvement of skin tone and methemoglobin levels had normalized to 0.6%.

Sensory

A prospective study of 55 850 women in the Nurses' Health Study used self-reporting of analgesic use and hearing loss to investigate the association between use of acetaminophen, NSAIDs and aspirin and hearing impairment [5MC]. At baseline, the mean participant age was 53.9 years and the study was conducted over 22 years. A precise description of the specific NSAIDs identified for study was not provided but analysis included ibuprofen and naproxen. Regular use of study drugs was defined as ≥2 days/week. For longer duration (>6 years compared to <1 year) of regular use of acetaminophen, the multivariable-adjusted relative risk (RR) for hearing loss was 1.09 (95% CI: 1.04–1.14; *P* for trend <0.001). With those same duration parameters, for regular NSAID use the multivariable-adjusted RR was 1.10 (95% CI: 1.06–1.15; *P* for trend <0.001). Increasing duration of aspirin use was not associated with hearing loss. There was a higher risk of hearing loss among women who reported regular acetaminophen use (multivariable-adjusted RR = 1.07, 95% CI: 1.01–1.13) or regular NSAID use (multivariable-adjusted RR = 1.07, 95% CI: 1.01–1.13) compared with individuals who received the medication <2 days/week. Regular use of multiple analgesics was also associated with risk of hearing loss. Doses or ingested cumulative amounts of the analgesics were not reported. This study was limited in that both analgesic use and hearing loss were self-reported by study participants. The authors note that the identified magnitude of the risk of hearing impairment associated with analgesic use was modest but suggest that the impact of this degree of risk could be significant considering the widespread use of the study drugs.

Overdose

Randesi and coworkers [6C] examined a genetic basis that may be associated with acetaminophen overdosage. Specific interest was directed at genes involved in high-risk behavior and in particular substance abuse. From the Acute Liver Failure Study Group registry, 229 patients presenting with acetaminophen overdosage of intentional, unintentional or unknown intent were identified. The control subjects ($n = 208$) were healthy volunteers. Nine genes previously associated with impulsivity and/or stress response were targets for identification of the presence of single nucleotide polymorphisms (SNPs). Two SNPs, rs2282018 in *AVP* (odds ratio 1.64) and rs11174811 in *AVPR1A* (odds ratio 1.89), showed significant association of genotype with all acetaminophen overdose patients as compared to control. Other associations were not identified and there was no correlation when the unintentional overdose group was compared with the intentional overdose or control groups. The *AVP* gene codes for arginine vasopressin, and *AVPR1A* codes for the arginine vasopressin receptor 1A that is expressed in the brain. Both of the gene products are involved in the pituitary axis stress response and SNPs in these genes have been previously associated with substance abuse disorders.

Acetaminophen Use During Pregnancy

Stergiakouli and coworkers [7C] analyzed data from a British prospective birth cohort registry that recruited pregnant women from 1991 to 1992. Focus was on prenatal acetaminophen use and development of behavioral problems in offspring with elimination of distinct potential confounders. Mothers were asked at 18 and 32 weeks of pregnancy if acetaminophen had been used in the previous 12 weeks. To account for postnatal exposure, maternal and paternal recent use of acetaminophen was ascertained when the offspring was 61 months old. The children's behavioral status was determined by using the Strengths and Difficulties Questionnaire, a screening tool that assesses 5 domains: emotional symptoms, conduct problems, hyperactivity symptoms, peer relationship problems and prosocial behaviors. The questionnaire was completed by the mother when the child was 7 years old. Data were also collected to account for potential confounding factors such as maternal age, socioeconomic status, self-reported psychiatric illness, tobacco and alcohol consumption and assessment of probable indications for acetaminophen use such as myalgia, headache, infection and others. Acetaminophen use was reported by 4415 mothers at 18 weeks of pregnancy and by 3381 mothers at 32 weeks. Postnatal use of acetaminophen was reported by 6916 mothers and 3454 of the partners. Maternal prenatal acetaminophen exposure at 18 weeks of pregnancy was associated with higher odds of having conduct problems (RR = 1.20; 95% CI: 1.06–1.37) and hyperactivity symptoms (RR = 1.23; 95% CI: 1.08–1.39). Maternal use at 32 weeks associated with higher odds for emotional symptoms (RR = 1.29; 95% CI: 1.09–1.53), conduct problems (RR = 1.42; 95% CI: 1.25–1.62), hyperactivity symptoms (RR = 1.31; 95% CI: 1.16–1.49) and total difficulties (RR = 1.46; 95% CI: 1.21–1.77). Maternal postnatal acetaminophen use and partner's acetaminophen use were not associated with any of the assessed behavioral outcomes. When the model adjusted for maternal postnatal use or partner use of acetaminophen, the associations between maternal prenatal use and behavioral outcomes were not changed. With the lack of identification of confounding factors, the authors suggest that the association between maternal prenatal use of acetaminophen and behavioral problems in offspring results from the possibility that maternally-ingested acetaminophen has an intrauterine effect to disrupt fetal neurodevelopment. The authors note the requirement not only for future mechanistic studies but also those that would strengthen the causality suggested by this study. Limitations of this study include information regarding doses or duration of acetaminophen treatment or other drug use was not available. The assessment of child behavior was performed solely by the mother and relied on a single evaluation tool. Based on the potential significance and marked concerns of behavioral deficits in children possibly linked to prenatal use of a very commonly employed drug, this publication generated several commentaries including those focusing on data analysis and interpretation [8r,9r,10r,11r].

Ystrom and coworkers [12C] analyzed data from the Norwegian Mother and Child Cohort Study that contains over 110000 offspring to examine the association between prenatal exposure to acetaminophen and the diagnosis of attention-deficit hyperactivity disorder (ADHD) in offspring. In this group, 2246 offspring were diagnosed with ADHD. Adjustments were made for maternal use of acetaminophen prior to pregnancy, familial history of ADHD and indications for acetaminophen use such as pain, fever, and infection. With these adjustments, a modest association between any prenatal maternal use of acetaminophen in 1 (HR = 1.07; 95% CI: 0.96–1.19), 2 (HR = 1.22; 95% CI: 1.07–1.38) and 3 (HR = 1.27; 95% CI: 0.99–1.63) trimesters was observed. There was a stronger correlation with long-term use of acetaminophen as defined by 29 days or more of use (HR = 2.20; 95% CI: 1.50–3.24) while administration for 8 days was negatively associated with ADHD (HR 0.90; 95% CI: 0.81–1.00). For ≥29 days of prenatal acetaminophen use, the risk association was similar across indications though there was not an adjustment for severity of the specific indication. For 22–28 days of acetaminophen use for the indications of fever and infection, there was a strong correlation with ADHD (HR = 6.15; 95% CI: 1.71–22.05). The association between ADHD in offspring and paternal preconceptional use of acetaminophen was similar to that observed for maternal prenatal use of the drug. With the elimination of potential confounders such as indications for acetaminophen use and putative familial risk for ADHD, the authors hypothesize that multiple acetaminophen-induced perturbations in fetal neurodevelopment could contribute to ADHD occurring in children following long-term prenatal acetaminophen exposure.

Gervin and coworkers [13C] conducted an epigenome-wide association study using cord blood samples (n = 384) collected from the Norwegian Mother and Child Cohort. Significant differences in DNA methylation were noted in samples representing children diagnosed with ADHD and whose mothers had prenatally used acetaminophen for > 20 days as compared to control. No differences were evident when prenatal acetaminophen alone or diagnosis of ADHD alone was compared to control. Several of the genes identified in the ADHD–prenatal acetaminophen cohort and as selected by statistical significance and effect size have been previously associated with ADHD and neural development.

ARYLALKANOIC ACID DERIVATIVES [SEDA-37, 116; SEDA-38, 79]

Diclofenac [SEDA-37, 116; SEDA-38, 79]

Gastrointestinal

A case of diclofenac-associated colonic stricture was reported [14a]. A 57-year-old male presented to the hospital with weakness and abdominal pain. Medications included 75 mg diclofenac twice a day, 81 mg aspirin daily and 75 mg clopidogrel daily. Hemoglobin was 8.8 g/dL and fetal occult blood was positive. A computed tomography scan detected luminal narrowing of the bowel at the hepatic flexure and a circumferential stricture in the right colon was observed with colonoscopy. Biopsy of the stricture showed lamina propria fibrosis and infiltration of eosinophils. Following endoscopic dilation of the stricture, two smaller proximal concentric colonic strictures were found but did not require balloon dilation. The patient was diagnosed with NSAID-induced colonic diaphragm disease and diclofenac was discontinued. After 3 months, no further gastrointestinal symptoms were noted.

Immunology

A 67-year-old man reported to the emergency room with an apparent anaphylactic reaction to oral diclofenac potassium [15a]. Medical history included coronary artery disease, hypertension, chronic renal failure and a reported anaphylactic reaction to intramuscular diclofenac. The patient presented with diffuse urticaria, hypotension and complained of chest pain. Electrocardiogram showed progressive ST segment elevation and the cardiac ejection fraction was 25%. Coronary angiography showed total occlusion of the right coronary artery with prominent thrombotic material. Balloon dilatation was performed, however, a stent could not be placed. No cardiovascular complaints were noted 2 weeks after discharge. The authors made a diagnosis of Kounis Syndrome Type-II in which a patient with preexisting coronary atherosclerosis experiences plaque rupture in response to an intense acute allergic reaction. In this individual, the offending stimulus appeared to be diclofenac.

Skeletal Muscle

A study of a single patient was performed in which a 65-year-old male received a magnetic resonance (MR)-guided intragluteal injection of diclofenac (75mg/2mL) for painful enthesopathy of the musculus gluteus minimus [16a]. MR imaging revealed injured muscle with the highest volume of muscle damage detected at 24 hours after injection. At 8 hours postinjection, plasma creatinine kinase was elevated sixfold above baseline and returned to normal levels after 1 week. The authors attribute the indices of muscle damage to direct cytotoxicity of diclofenac and not the hypertonicity of the solution. Interestingly, the patient reported no pain at the injection site which the authors acknowledge may be due to local anesthetic effects of diclofenac.

Ibuprofen [SEDA-37, 117; SEDA-38, 79]

Gastrointestinal

In a case report, short-term use of ibuprofen was associated with early and delayed onset of jejunal perforations that required surgical intervention [17a]. A 48-year-old man underwent laparotomy for peritonitis after a 14-day course of 800mg ibuprofen every 6 hours. He also received 20mg omeprazole daily. Three perforations of the proximal jejunum were detected and resection performed. The patient avoided NSAIDs and 5 months later the individual reported abdominal pain and a 5mm perforated ulcer in the proximal jejunum was detected and repaired. Based on histological evidence of ischemia and infarction at the ulcer margin, the authors hypothesize that despite the significant ibuprofen-free period, the ulcer developed due to previous ibuprofen-induced microvascular injury to the villous circulation of the mucosa in this region.

Overdose

A case was reported of the successful use of therapeutic plasma exchange (TPE) in a 48-year-old male who was admitted following ingestion of ~72g of ibuprofen in a suicide attempt [18a]. Other than ibuprofen, no other substances were known to be ingested. Activated charcoal was administered. Over the next several hours, the patient developed significant hypotension that was refractory to fluid resuscitation and necessitated norepinephrine administration. Based on the worsening hemodynamic status of the patient as well as the high protein binding and low volume of distribution of ibuprofen, TPE was initiated 9 hours after admission. Blood pressure rapidly improved and the dosage of vasopressors was reduced over ~10 hours. During this course, the patient's ibuprofen plasma concentration (IPC) was monitored. Prior to initiation of TPE, the IPC decreased from 550 to 275mcg/mL. Interestingly, changes in IPC during TPE did not correlate with the marked improvement of the patient's hemodynamic status in response to TPE. Rather, IPE increased 10 hours after TPE and then slowly decreased over time. Based on modeling, the authors calculated an overall ibuprofen half-life of 17.2 hours for the observation period of 5 days.

Pulmonary

A retrospective cohort study was performed to determine the frequency of the development of pulmonary arterial hypertension (PAH) in neonates treated with ibuprofen for the closure of patent ductus arteriosus (PDA) as well as to identify risk factors that may be associated with this occurrence [19c]. The study followed 144 neonates: 40 received oral ibuprofen, 100 received intravenous ibuprofen trishydroxyaminomethane (THAM) and 4 were administered intravenous ibuprofen lysine. All regimens were an initial dose of 10mg/kg followed by 5mg/kg at 24 and 48 hours. No other treatment or nontreatment groups were included. PAH was diagnosed by electrocardiography and was considered to be associated with ibuprofen administration if it developed within 24 hour after the last dose. A total of 6.9% of patients developed PAH representing 2 patients, 7 patients and 1 patient receiving oral ibuprofen, I.V. ibuprofen THAM and I.V. ibuprofen lysine, respectively. Severe bronchopulmonary dysplasia or death was significantly more prevalent in the PAH group (2 of the 10 patients; 90.0%) than in those who did not develop PAH (21 of the 134 patients; 15.7%). After controlling for potential confounders, significant risk factors for developing PAH with ibuprofen administration were identified and included a younger gestational age, birth weight of less than the third percentile for gestational age, maternal hypertensive disorders of pregnancy and oligohydramnios.

Ketorolac [SEDA-37, 118]

Pulmonary

A case report described a 65-year-old female who was administered intravenous ketorolac for relief of pain due to biliary colic [20a]. The patient's medical history included gallstone disease, hypertension and asthma. Asthma was partially controlled by medications but there was a hospital admission for asthma exacerbation 6 weeks earlier. Approximately 30 minutes after administration of ketorolac, the patient became tachycardic and complained of worsening dyspnea. The patient was oxemic and oxygen was administered. Chest X-ray was unremarkable; however, computed tomography revealed diffuse bilateral mucoid impaction of lower lobar bronchi and subsegmental atelectasis in upper and right middle lobes. A diagnosis of ketorolac-exacerbated asthma was made and the patient was administered ipratropium bromide, salbutamol and budenoside and intravenous methylprednisolone, hydroxyzine and aminophylline. Over the next few hours the patient improved and after 9 days the patient was discharged with a return to baseline pulmonary function. The patient was advised to avoid all NSAIDs.

Precision Trial

The Prospective Randomized Evaluation of Celecoxib Integrated Safety vs *Ibuprofen or Naproxen (PRECISION) trial [21MC] was implemented following the voluntary withdrawal of the selective cyclooxygenase-2 (COX-2) inhibitor rofecoxib in 2004 due to concerns regarding cardiovascular safety and particularly risk of myocardial infarction with chronic administration of COX-2 inhibitors. PRECISION used a randomized double-blind noninferiority design that included over 24000 participants to examine the cardiovascular safety of the selective COX-2 inhibitor celecoxib compared to the nonselective cyclooxygenase-1 (COX-1) and COX-2 inhibitors, naproxen and ibuprofen. The study population was primarily (90%) osteoarthritis patients with a smaller population (10%) of rheumatoid arthritis patients with both groups requiring NSAIDs for analgesia. For both groups, key inclusion criteria were stabilized cardiovascular disease or an increased risk for the development of cardiovascular disease. Daily doses (mean ± S.D) were: celecoxib 209 ± 37 mg, naproxen 852 ± 103 mg and ibuprofen 2045 ± 46 mg. For all patients, mean durations of treatment and follow-up were 20.3 ± 16.0 and 34.1 ± 13.4 months, respectively, with very minor variation among groups. Primary endpoints were cardiovascular death, nonfatal myocardial infarction or nonfatal stroke. In the intention-to-treat analysis, the primary outcome event was observed in 188 patients (2.3%) in the celecoxib group, 201 patients (2.5%) in the naproxen group and 218 patients (2.7%) of the ibuprofen group. For celecoxib compared to naproxen and celecoxib compared to ibuprofen the hazard ratios were 0.93 (95% CI: 0.76–1.13; P < 0.001 for noninferiority) and 0.85 (95% CI: 0.70–1.04; P < 0.001 for noninferiority), respectively. In the on-treatment analysis, a primary outcome event occurred in 1.7% of celecoxib group, 1.8% of the naproxen group and 1.9% of the ibuprofen group. Thus, this trial indicated that relative to naproxen and ibuprofen, celecoxib is not associated with greater cardiovascular risk.*

PRECISION also examined additional safety parameters of concern with NSAID use. The occurrence of gastrointestinal events with celecoxib was significantly lower than with naproxen (P = 0.01) or ibuprofen (P = 0.002). The proton pump inhibitor esomeprazole was provided to all patients though actual adherence to this gastrointestinal-protective therapy was not known. The development of renal events was lower with celecoxib than with ibuprofen (P = 0.004) but did not significantly differ between celecoxib and naproxen.

It is recognized that the PRECISION trial had several limitations. Perhaps most significant, this study had a drug discontinuation rate of 68.8% and 27.4% of patients were lost in follow-up. Additionally, while mean doses of naproxen and ibuprofen were toward the upper limit of the protocol specified parameters, dosing of celecoxib was toward the lower range as there were restrictions to dosage increases of celecoxib in the osteoarthritis participants that made up the majority of the study population. Reviews of the PRECISION trial provide additional detailed analysis of study design and interpretation of results [22R,23R,24R].

Solomon and coworkers [25MC] conducted a secondary posthoc analysis of composite safety outcomes in the PRECISION trial. The primary outcome was NSAID-associated toxicities including major cardiovascular events, serious gastrointestinal events including hemorrhage and ulceration, renal events including acute renal failure and all-cause mortality. The major toxicities occurred in 4.1% of celecoxib users, 4.8% of naproxen users (P = 0.02) and 5.3% of ibuprofen users (P < 0.001). The adjusted HRs for the primary outcome demonstrated significantly higher risks for naproxen users compared with the celecoxib group (HR = 1.20; 95% CI: 1.04–1.39, P = 0.02) and for ibuprofen users (HR = 1.38; 95% CI: 1.19–1.59, P < 0.001). This risk translated into numbers needed to harm for the primary major toxicity of 135 (95% CI: 72–971) for naproxen compared with celecoxib and 82 (95% CI: 53–173) for ibuprofen compared with celecoxib.

Finally, PRECISION-ABPM was a prespecified substudy of PRECISON that was designed to assess blood pressure effects of the three NSAIDs [26C]. After screening, 444 patients completed this study that had a primary endpoint of change from baseline of 24-hour ambulatory blood pressure at 4 months of treatment. Ibuprofen had a significant effect with a 3.7 mmHg (95% CI: 1.72–5.58) increase in mean 24-hour systolic blood pressure with no statistically significant change for naproxen or celecoxib. The percentage of patients who were normotensive and then subsequently defined as hypertensive (24-hour systolic blood pressure > 130 and/or diastolic blood pressure of > 80) at 4 months was significantly greater for ibuprofen (23.2%) and naproxen (19.0%) compared to celecoxib (10.3%). These findings concur with the primary outcome results in the parent PRECISION trial reflecting cardiovascular and renal events occurring with a higher incidence in ibuprofen-treated patients compared to those receiving celecoxib or naproxen.

With recognition of the limitations of the described studies, it appears that in regard to cardiovascular toxicity, celecoxib has a safety profile not inferior to nonselective COX inhibitors or at least naproxen and ibuprofen. Celecoxib may be preferred when there are confounding concerns of gastrointestinal and renal safety in specific patient groups. Thus, individual risk factors should be considered in the identification of a specific NSAID that may be preferred in that patient requiring analgesic or antiinflammatory medication.

INDOLEACETIC ACIDS [SEDA-37, 121; SEDA-38, 79]

Indomethacin [SEDA-37, 121; SEDA-38, 79]

Pharmacogenetics

Certain NSAIDs including indomethacin are used to promote closure of a patent ductus arteriosus (PDA) in neonates but there is treatment failure with this approach and PDA may then require surgical intervention. In that responsiveness to indomethacin may be heritable, a single-center case–control analysis was conducted to assess the association between success of indomethacin-induced closure of PDA and specific polymorphisms in CYP2C9 as this enzyme is involved in the metabolism of indomethacin [27C]. This study identified 96 responders (no subsequent surgery required) and 52 nonresponders (subsequent surgery required) to indomethacin treatment for PDA. In the single nucleotide polymorphism-outcome association analysis, the G allele of rs2153628 was associated with increased odds of response to indomethacin (OR = 1.918; 95% CI: 1.056–3.483, $P = 0.031$). The transmission disequilibrium test of parent-triad trios of neonates who did respond to indomethacin demonstrated that the G allele of rs2153638 was overtransmitted from parents to child in this group (OR = 2.667; 95% CI: 1.374–5.177, $P = 0.003$). These findings suggest that a polymorphism located in CYP2C9 may influence indomethacin metabolism and thus response to the drug for closure of PDA. As this was a retrospective study, plasma samples were not collected and no direct association could be made between indomethacin levels, CYP2C9 gene status and therapeutic success. Additionally, the rs2183638 variant of CYP2C9 has not been shown to be linked to altered indomethacin metabolism.

OXICAMS [SEDA-37, 121; SEDA-38, 79]

Meloxicam

Cardiovascular

A nested matched case–control study of NSAID users in an electronic medical data base in the United Kingdom was performed to assess the cardiovascular safety of meloxicam which is selective for inhibition of COX-2 over COX-1 [28MC]. The study used data over an ~13-year period from individuals of 35–89 years of age with at least one prescription for NSAIDs and focus was on occurrence of myocardial infarction (MI). Individuals were required to be present in the database for 1 year and those with a history of MI were excluded. Information regarding date and quantity of the NSAID prescription was used to classify patient groups. Categories were "current users", "recent users" (prescription of any NSAID terminated 1–60 days prior to index date) and "remote users" (prescription of any NSAID terminated >60 days but within 1 year prior to the index date). Current users were further divided according to specific NSAID: meloxicam, diclofenac, naproxen and "other" NSAID. The study identified 9291 cases of MI and 30676 matched controls with average age and gender similar between these groups. Multivariable logistic regression analysis demonstrated that compared with remote use of any NSAID, the adjusted odds ratio of MI was 1.38 (95% CI: 1.17–1.63) for current meloxicam use, 1.37 (95% CI: 1.25–1.50) for current diclofenac use and 1.12 (95% CI: 0.96–1.30) for current naproxen use. Recent use of any NSAID was associated with an adjusted odds ratio of MI of 1.25 (95% CI: 1.17–1.33). In an analysis in which patients with a history of ischemic heart disease were excluded, results were very similar. Risk for MI was not observed in current meloxicam users who also received aspirin. A limitation of this study included no assessment of additional nonprescription NSAIDs during the study period. Additionally, all of the analyzed data and classification of patients into study groups relied on information primarily derived from prescription information and actual adherence and durations of use were not known. There was no information regarding NSAID dosage.

Comparative Studies of NSAIDs

Cardiovascular

Gunter and coworkers [29M] performed a meta-analysis involving several NSAIDs with a focus on cardiovascular toxicity and possible association with the relative selectivity of the agents for COX-1 vs *COX-2. Included in the final analysis were 26 randomized controlled trials and prospective cohort studies (representing 228 391 patients) that made comparisons between individual NSAIDs or* vs *placebo. Nonselective NSAIDs included naproxen, ibuprofen and diclofenac while selective COX-2 inhibitors were etoricoxib, celecoxib, lumiracoxib and rofecoxib. Rofecoxib was selected as a comparator due to its documented cardiovascular toxicity. Primary endpoints were any MI, any stroke, cardiovascular death and a composite of all three events. Several comparisons were made for each individual NSAID including that NSAID* vs *placebo, all NSAIDs, nonselective NSAIDs, selective COX-2 inhibitors and selective COX-2 inhibitors with the omission of rofecoxib. As might be expected, rofecoxib showed increased risk of MI and stroke when compared to all NSAIDs and was the only analyzed agent to demonstrate such an association. When compared to selective COX-2 inhibitors, the incidence of MI was decreased by celecoxib (OR=0.583; 95% CI: 0.396–0.857, P=0.006) and naproxen (OR=0.609; 95% CI: 0.375–0.989, P=0.045). However, this difference was lost when compared to selective COX-2 inhibitors with the exclusion of rofecoxib. Celecoxib was associated with a reduced incidence of stroke when compared to all NSAIDs (OR=0.603; 95% CI: 0.410–0.887, P=0.010), nonselective NSAIDs (OR=0.517; 95% CI: 0.287–0.929, P=0.027) and selective COX-2 inhibitors (OR=0.509; 95% CI: 0.280–0.925, P=0.027) but this difference did not exist when rofecoxib was removed from the COX-2 inhibitor category. Celecoxib had no difference compared to placebo in occurrence of stroke. No drug exhibited a significant difference in cardiovascular death. Limitations noted by the authors include the lack of inclusion of drug dose, duration of use or concurrent use of aspirin in the analysis. Additionally, a relatively low overall incidence (0.1% of the study patients) of stroke in this meta-analysis limited the strength of conclusions regarding this event. It is suggested that the cardiovascular toxicity of NSAIDs may not completely depend on the selectivity* vs *nonselectivity of COX isoform inhibition, but rather may be partially determined by other relatively unexplored properties of individual NSAIDs.*

A case–time–control study utilizing data from Danish patient registries was performed to examine the association between multiple selective and nonselective NSAIDs and out-of-hospital cardiac arrest (OHCA) [30MC]. Exposure to NSAIDs 30 days prior to cardiac arrest was defined as the "case period" and a preceding 30-day period for that same patient in which there was not an event was defined as the "control period". For each case of OHCA during 2001–2010, 4 control individuals were matched on age and sex. These controls were used to adjust for changes in NSAID prescribing patterns over time. The study identified 28 947 cases of OHCA and within the 30-day case period, 11.7% of these individuals were treated with an NSAID. Diclofenac (OR=1.50; 95% CI: 1.23–1.82) and ibuprofen [OR=1.31; 95% CI: 1.14–1.51) were associated with a significantly increased risk of OHCA. These were the two most commonly used NSAIDs in the study. The use of naproxen had no significant association with OHCA. Additionally, there was no significant risk with the selective COX-2 inhibitors, celecoxib and rofecoxib though the authors acknowledge a low statistical power for these two agents in this study.

Gastrointestinal

Chan and coworkers [31C] conducted a double-blind, double-dummy, randomized study to assess the gastrointestinal safety of celecoxib and naproxen in arthritis patients requiring NSAIDs for analgesia and who were also receiving aspirin for cardiovascular protection and had recent history of gastrointestinal bleeding. In a single-setting medical center in Hong Kong, patients were identified who presented with upper gastrointestinal bleeding and after endoscopically-verified ulcer healing were randomized to receive either 100 mg celecoxib twice per day or 500 mg naproxen twice per day. Both groups received the proton pump inhibitor esomeprazole (20 mg) daily in addition to 80 mg aspirin daily for cardiothrombotic disease or multiple coronary risk factors. The primary endpoint was recurrence of upper gastrointestinal bleeding due to ulcer or bleeding erosions within 18 months of initiation of celecoxib or naproxen. Over approximately 10 years, a total of 514 patients were enrolled in the study with 257 assigned to either the celecoxib or naproxen group with approximately equal numbers discontinuing treatment prior to 18 months. The incidence of recurrent gastrointestinal bleeding was 5.6% (95% CI: 3.3–9.2) in the celecoxib group and 12.3% (95% CI: 9.8–17.1) in the naproxen group (P=0.008). The study also examined severe cardiovascular events and found that cardiovascular safety did not significantly differ between the two groups. The authors conclude that in this multi-risk patient group requiring both aspirin and an NSAID, celecoxib in combination with a proton pump inhibitor is preferred over naproxen in order to reduce the risk of upper gastrointestinal bleeding.

Immunologic

A retrospective cohort study of an electronic health record system was performed to assess the incidence and identify risk factors for hypersensitivity reactions (HSR) to prescription NSAIDs [32MC]. Analysis was directed at prescription-only nonselective NSAIDs that were used orally. This resulted in inclusion of diclofenac, indomethacin, nabumetone and piroxicam. Subsequently, a separate analysis included the relatively selective COX-2 inhibitors celecoxib and meloxicam. The primary outcome was an adverse drug reaction or HSR to the NSAID within 1 year. HSR included anaphylaxis, hypotension, angioedema, nasal and pulmonary symptoms and dermal manifestations such as hives or rash. Of 62719 patients prescribed diclofenac, indomethacin, nabumetone or piroxicam, 1035 (1.7%) experienced an adverse drug reaction and of this group, 189 (18.3%) had an HSR. Of 32 215 patients receiving celecoxib or meloxicam, 429 (1.3%) had an adverse drug reaction of which 121 (28.2%) were an HSR. Multivariable logistic regression analysis revealed that increased odds of NSAID-associated HSR was associated with history of drug HSR (OR = 1.8; 95% CI: 1.3–2.5), female sex (OR = 1.8; 95% CI: 1.3–2.4), autoimmune disease (OR = 1.7; 95% CI: 1.1–2.7) and prescription of maximum dose of NSAID (OR = 1.5; 95% CI: 1.1–2.0).

SALICYLATES [SEDA-37, 122; SEDA-38, 80]

Acetylsalicylic Acid (Aspirin) [SEDA-37, 122; SEDA-38, 80]

Drug Resistance

Aspirin resistance or nonresponsiveness refers to the apparent lack of pharmacologic effect of aspirin to inhibit platelet cyclooxygenase activity that results in therapeutic failure to provide protection from cardiovascular events. A single-center, randomized, active-control triple-crossover study was performed to examine an association between aspirin nonresponsiveness and drug formulation in diabetic individuals [33C]. Evaluated agents were immediate release aspirin tablets, PL2200 aspirin capsules that are a lipid-based formulation and enteric-coated (EC) aspirin caplets with each administered at a dose of 325 mg once daily for 3 days. The study was conducted in 40 obese Type 2 diabetic patients (average 53 years of age; 65% males) who had no history of cardiovascular disease. Serum thromboxane B2 levels and in vitro platelet aggregation were used to assess aspirin activity and plasma levels of acetylsalicylic acid (ASA) and salicylic acid were determined at defined time points after dosing. Aspirin nonresponsiveness as defined by residual thromboxane B2 levels was 15.8%, 8.1% and 52.8% for immediate release aspirin, PL2200 and EC aspirin, respectively ($P < 0.001$ for both comparisons vs EC aspirin; $P = 0.30$ for comparison between immediate release aspirin and PL2200). This parameter of COX-1 inhibition correlated with variability in platelet aggregation observed with EC aspirin. Pharmacokinetic analysis of plasma ASA levels demonstrated that C_{max} and AUC_{0-t} were significantly lower and the time to reach C_{max} was significantly higher with EC aspirin compared to the other two formulations. The authors suggest that the nonresponsiveness observed with EC aspirin occurs due to reduced bioavailability of this formulation. Limitations of this study include the relatively short treatment period of 3 days, the dose of aspirin (325 mg vs the more typically employed cardioprotective dose of 81 mg) and for safety concerns the study was restricted to individuals without concurrent cardiovascular disease.

Overdose

A secondary analysis of data from a retrospective cohort study of acute overdose patients presenting to the emergency department (ED) was conducted to identify early clinical predictors of severe outcomes following salicylate overdose [34c]. Forty-eight patients met the inclusion criteria of experiencing salicylate overdose within 24 hours of ED presentation and 20.8% of these were classified as severe outcome as defined by occurrence of acidemia, requirement for hemodialysis or death. Univariate analysis revealed that increasing age ($P = 0.04$), respiratory rate ($P = 0.04$), lactate levels ($P = 0.002$), coma ($P = 0.05$) and ingestion of an additional substance ($P = 0.04$) were significantly associated with severe outcome. Salicylate concentration alone, hyperthermia and hypoxemia were not associated with severe outcome. A multivariable logistic regression model that employed literature-derived factors and univariate factors with significance indicated that only age and respiratory rate remained statistically significant. While acknowledging the low sample size and requirement for validation of this study, the authors suggest that older age and increased respiratory rate are early clinical predictors of severe outcomes following salicylate ingestion and may indicate more aggressive therapeutic management in the ED.

DRUGS USED IN THE TREATMENT OF GOUT [SEDA-37, 123; SEDA-38, 80]

Allopurinol [SEDA-37, 123; SEDA-38, 80]

Immunologic

Drug reaction with eosinophilia and systemic symptoms (DRESS) syndrome is a drug hypersensitivity reaction that is frequently characterized by fever, cutaneous events, hematologic abnormalities and internal organ involvement. A 59-year-old man presented with scaly papules and plaques with areas of desquamation on his face, neck, trunk and upper extremities [35a]. He had a history of chronic kidney disease and asymptomatic hyperuricemia for which he received allopurinol 8 weeks earlier. Laboratory results indicated a leukocytosis with marked eosinophilia, elevated hepatic enzymes and acute kidney injury. A diagnosis of DRESS syndrome was made and allopurinol discontinued with initiation of hemodialysis. Kidney biopsy revealed marked tubulointerstitial nephritis with presence of eosinophils and necrotizing vasculitis in at least one interlobular artery. Complex deposition was not detected. Extensive evaluation revealed no evidence of a primary vasculitis. Based on exclusion, it was determined that the renal necrotizing vasculitis was secondary to DRESS syndrome. Cyclophosphamide and pulse corticosteroids were initiated and this resulted in resolution of cutaneous symptoms and hepatic and renal function returned to baseline. The authors note that to their knowledge, this is the first case of biopsy-proven, kidney-limited necrotizing vasculitis occurring as a feature of drug-induced DRESS syndrome.

A case of a 50-year-old woman who experienced allopurinol-induced DRESS syndrome that was exacerbated with initiation of teicoplanin, a glycopeptide antibiotic in many respects similar to vancomycin, was reported [36a]. The hospitalized patient with relapsed acute lymphoblastic leukemia was scheduled to receive fludarabine, cytarabine, and granulocyte colony-stimulating factor. Two days prior, 300mg allopurinol daily was initiated as a preventative for tumor lysis syndrome. Seven days after allopurinol initiation, the patient developed a mild fever, dyspnea, facial edema, generalized pruritus, erythema and macular rash over much of the face, trunk and limbs. Hepatic enzymes were elevated. Blood cell counts decreased presumably as a result of the chemotherapy regimen. DRESS syndrome was suspected and allopurinol was discontinued and intravenous hydrocortisone administered. Within 72 hours, dermal symptoms resolved and liver function tests improved. Due to a possible infectious cause of fever, teicoplanin was started and both cutaneous symptoms and abnormalities in liver enzymes quickly returned. Eosinophilia count was 12%. Teicoplanin was discontinued and the patient remained on glucocorticoids with resolution of DRESS syndrome features within 7 days. The authors suggest discontinuation or avoidance of initiation of drugs associated with hypersensitivity reactions during the acute phase of DRESS syndrome. Due to the structural similarity of teicoplanin to vancomycin, a drug that is one of the more prevalent inducers of DRESS syndrome, it is likely that teicoplanin exacerbated the condition in this patient.

A case suggesting cross-reactivity between allopurinol and febuxostat in the induction of DRESS syndrome was reported [37a]. A 76-year-old woman received allopurinol and colchicine for an episode of gout. After 5 weeks, she presented with a whole body rash and severe pruritus. Serum creatinine was elevated, liver function tests were normal and eosinophil count was 52%. Allopurinol was discontinued and glucocorticoid treatment initiated with resolution of symptoms over 6 weeks. Five months later, the patient was administered febuxostat and 2 days later presented with rash and pruritus and reduced urine output. Laboratory tests showed renal impairment and 7.6% eosinophils. Again, glucocorticoid treatment was initiated with symptoms resolving at 6 weeks. A possible cross-reaction between allopurinol and febuxostat in DRESS syndrome induction is suggested by the temporal features. Specifically, symptoms developed 5 weeks after initial allopurinol administration while only 2 days elapsed between subsequent febuxostat treatment and symptom return. Interestingly, as noted by the authors, febuxostat and allopurinol both act through inhibition of xanthine oxidase yet are chemically dissimilar thus not supportive of a structure-based hypersensitivity mechanism.

Colchicine [SEDA-37, 124; SEDA-38, 82]

Musculoskeletal

Based on case reports of myopathy in patients receiving colchicine and a statin, Kwon and coworkers [38C] performed a retrospective study of a Korean electronic medical record database to determine if risk of myopathy is increased with coadministration of these drugs. The incidence of myopathy was not significantly different between those receiving colchicine with a statin and patients receiving colchicine without a statin (5/188, 2.7% vs 7/486, 1.4%, $P=0.330$). Multivariate analysis identified risk factors in the colchicine without a statin group for developing myopathy including chronic kidney disease, hepatic cirrhosis and higher colchicine doses. However, concomitant statin use was not associated (HR=1.123; 95% CI: 0.262–4.814; $P=0.875$). This study included relatively few patients that developed myopathy and analyzed statins as a group without distinction among individual statins that may vary in hepatic metabolism or other parameters.

A case report was provided that suggested colchicine triggered rhabdomyolysis in a patient who had been on long-term low-dose simvastatin therapy [39a]. A 70-year-old male presented with a 1-year history of progressive muscle weakness and myalgia. Past medical history included chronic kidney disease, chronic heart failure, coronary artery disease with MI and bypass, Type 2 diabetes, gout and hyperlipoproteinemia. He had been receiving 40mg simvastatin over the past 6 years in addition to several other medications for the listed medical conditions including colchicine (0.5–1.0mg) that had been prescribed as required. The patient did not report a previous history of statin-associated muscular symptoms. Creatinine kinase, creatinine, and serum myoglobin were markedly elevated. Histologic examination of a muscle biopsy was consistent with statin-induced myopathy. Simvastatin was discontinued and replaced with ezetimibe. Continuous veno-venous hemofiltration was initiated and after 3 days renal function returned to baseline. After 2 weeks, creatinine kinase and myoglobin returned to normal and the patient discharged without subsequent complaints. The authors report that communication with the patient's physician indicated that colchicine had been prescribed for the myalgia that was diagnosed initially as an acute gout attack. Additionally, colchicine had been initiated "in close temporal vicinity of clinically manifest rhabdomyolysis". The authors conclude that colchicine was likely a trigger to precipitate rhabdomyolysis as its administration was temporally associated with the muscular pathology in this patient who had been receiving simvastatin for several years. Other current medications had negligibly known potential for myopathy or interactions with statins. It was proposed that the interaction may be of a pharmacokinetic basis as both simvastatin and colchicine are metabolized via CYP3A4.

Overdose

A case of accidental colchicine overdose was reported [40a]. A 45-year-old woman presented with severe abdominal pain, diarrhea and vomiting after ingesting homemade vegetable juice. The juice included extract from *Colchicum autumnale* leaves that were identified incorrectly as wild garlic. Three days later, the patient suffered multiple organ failure and hemiparesis. Hematologic workup revealed pancytopenia. Neutrophils and eosinophils had reduced size, loss of segmentation, condensed chromatin and nuclear fragmentation. Cell cytoplasm contained large inclusions. Serum colchicine levels were 12.15 g/L. The patient recovered and 2 months later most of the hematologic abnormalities were no longer apparent. The authors suggest the consideration of colchicine toxicity in patients with reduction in neutrophil size and presence of condensed chromatin.

Febuxostat [SEDA-37, 125; SEDA-38, 82]

Hematologic

A 67-year-old woman presented to the emergency department with fever and sore throat [41a]. She was in end stage renal disease and received hemodialysis 3 times per week for the past 15 years. Additional conditions included gout, hypertension, diabetes and dyslipidemia. White blood cell (WBC) count was 700/μL with 2% (14/μL) neutrophils, 94% lymphocytes and 2% monocytes with a platelet count of 131 000/μL and hemoglobin at 11.1 g/dL. The broad-spectrum antibiotics, piperacillin and tazobactam, were started for febrile neutropenia and acute pharyngitis. The patient had an extensive medication list with all drugs having been used for >1 year with the exception of 40 mg febuxostat daily which had been initiated within the last 2.5 months. Prior to the start of febuxostat, WBC count was 6000/μL. Based on this temporal association, febuxostat was discontinued as it was hypothesized to have induced the agranulocytosis. Subsequent bone marrow examination revealed hypocellularity with a marked decrease in the myeloid component. There was no indication of hematologic malignancy. Seventeen days after febuxostat discontinuation and initiation of antibiotics, blood counts improved with WBC count at 2100/μL with 66% neutrophils. Neutropenia completely resolved at follow-up.

Musculoskeletal

To examine the incidence of febuxostat-induced myopathy in chronic kidney disease patients, a retrospective cohort study of outpatients at a Taiwanese medical center was conducted [42C]. The indication for febuxostat administration was hyperuricemia in gouty arthritis patients who had a glomerular filtration rate (GFR) of <45 mL/min per 1.73 m^2 or those unable to obtain target uric acid levels with uricosuric drugs. For this study, myositis was defined as an increase of 3- to 10-fold over normal of the serum creatinine kinase (CK) level. Rhabdomyolysis was indicated by CK levels of >10-fold over normal. Myopathy represented myositis and rhabdomyolysis collectively. Of 1332 patients receiving febuxostat for greater than 1 month, 41 developed myopathy of which 37 cases were myositis and 4 patients with rhabdomyolysis. All of these patients had CKD (eGFR <60 mL/min per 1.73 m^2). Univariate analysis revealed a significantly increased risk of myopathy with eGFR <12.4 mL/min per 1.73 m^2 compared with eGFR of 29.1 mL/min per 1.73 m^2 ($P < 0.001$). The dose of febuxostat was not associated with myopathy. In subgroup analysis including sex, age, statin or fibrate use and the presence of diabetes, results were similar in that in all subgroups patients with eGFR <12.4 mL/min per 1.73 m^2 demonstrated increased risk of myopathy.

A case was reported of a 50-year-old man with marginal zone lymphoma who was admitted to the hospital for low-grade fever, severe generalized weakness, myalgia and night sweats [43a]. Three weeks prior, chemotherapy with bendamustine and rituximab had been initiated along with 80 mg febuxostat daily for tumor lysis syndrome prophylaxis. Bactrim and valacyclovir had been administered prophylactically for infection. The serum CK level was ~25-fold above normal and blood cell counts showed hypereosinophilia. Transaminases and lactate dehydrogenase were elevated. Biopsy of right quadriceps muscle revealed myonecrosis and eosinophilic inflammatory infiltration of the endomysium and perimysium. The patient did not have renal failure. With the diagnosis of eosinophilic myositis, febuxostat was discontinued and symptoms resolved and laboratory values normalized within 8 days.

Lesinurad [SEDA-38, 83]

Urinary Tract

Lesinurad is a novel first-in-class agent approved by the FDA in December, 2015 for use in gout. Lesinurad at 200 mg daily is approved for use in combination with a xanthine oxidase inhibitor in patients who have not obtained target serum uric acid levels with a first-line xanthine oxidase inhibitor alone. By selectively inhibiting URAT1 and OAT4 transporters responsible for the renal tubular reabsorption of uric acid, lesinurad promotes renal excretion of uric acid and thus lowers serum levels [44R]. The lesinurad dosage limitation of 200 mg daily is attributable to results of the Lesinurad Monotherapy in

Gout Subjects Intolerant to Xanthine Oxidase Inhibitors (LIGHT) study in which 400 mg lesinurad daily was compared to placebo [45C]. LIGHT was a phase 3, randomized, double-blind study that entailed a 6-month treatment with lesinurad (107 patients) or placebo (107 patients) and a 14-day safety follow-up with the majority of patients entering an extension study in which all patients received 400 mg lesinurad daily for ~24 months. In the initial core study, adverse effects occurred in 83 patients (77.6%) in the lesinurad group and 70 patients (65.4%) in the placebo group. Most of these events were classified as grade 1 or grade 2 in severity. Serious adverse effects occurred in 9 patients (8.4%) and 4 patients (3.7%) in the lesinurad and placebo groups, respectively. Renal related adverse effects were reported in 19 patients (17.8%) in the lesinurad group and none in the placebo group. Lesinurad administration was associated with severe renal adverse effects in 5 patients. Serum creatinine levels >1.5 times above baseline occurred in 26 patients (24.3%) receiving lesinurad and were not observed in the placebo group. Based on the high incidence of adverse renal effects, approval of 400 mg lesinurad daily was not sought.

Combining Lesinurad with Allopurinol Standard of Care in Inadequate Responders (CLEAR 2) was an international phase 3 randomized, double-blind, placebo-controlled study to examine 200 or 400 mg lesinurad daily in combination with allopurinol vs allopurinol with placebo in patients in which serum uric acid levels were not controlled by allopurinol alone [46C]. Allopurinol was administered at prestudy dose and for ~84% of patients this was 300 mg daily. Approximately 200 patients were assigned to each of the 3 treatment groups. Adverse effects occurred in 146 patients (70.9%), 152 patients (74.5%) and 161 patients (80.5%) in the allopurinol-alone, 200 mg lesinurad + allopurinol and 400 mg lesinurad + allopurinol groups, respectively. Most of these adverse effects were grade 1 or grade 2 and included upper respiratory tract infections, hypertension, arthralgia and diarrhea. Serious adverse effects were observed in 8 patients (3.9%) receiving allopurinol alone, 9 patients (4.4%) receiving 200 mg lesinurad + allopurinol and 19 patients (9.5%) receiving 400 mg lesinurad + allopurinol. Serum creatinine elevations of >1.5-fold/>2.0-fold of baseline developed in 3.4%/0% (allopurinol-alone), 5.9%/2.0% (200 mg lesinurad + allopurinol) and 15.0%/8.0% (400 mg lesinurad + allopurinol) of patients. These elevations were typically transient and the majority resolved without discontinuation of the study drug. The very similarly designed CLEAR 1 study [47C] that was conducted in the US yielded safety results consistent with those described for CLEAR 2. The possibility of renal side effects with lesinurad dictates the required combined use of this drug with a xanthine oxidase inhibitor to decrease uric acid production. In the absence of this inhibition, lesinurad monotherapy may promote tubular microcrystallization of uric acid that can underlie development of renal pathology [45C].

The Combination Treatment Study in Subjects with Subcutaneous Tophaceous Gout with Lesinurad and Febuxostat (CRYSTAL) trial evaluated safety and efficacy of 200 or 400 mg lesinurad in combination with the xanthine oxidase inhibitor febuxostat compared to febuxostat alone [48C]. The safety findings were very similar to those described above for combination therapy of lesinurad and allopurinol in CLEAR 2. Incidence of adverse effects occurring with 200 mg lesinurad + febuxostat was similar to those with febuxostat alone. The occurrence of elevation in serum creatinine levels was higher in the 200 mg lesinurad + febuxostat group compared to febuxostat monotherapy; however, these increases were transient and resolved even when combination therapy was maintained.

Rasburicase [SEDA-37, 123; SEDA-38, 80]

Hematologic [SEDA-38, 83]

Methemoglobinemia is a rare but recognized adverse event associated with rasburicase administration and use of the drug is contraindicated in G6PD deficient patients. A 50-year-old African-American male with T lymphoblastic leukemia developed spontaneous tumor lysis syndrome and was administered allopurinol and 7.5 mg I.V. rasburicase [49a]. After 6 hours, the patient developed dyspnea and became hypoxic. Blood methemoglobin was 10.3% and I.V. methylene blue was administered to correct methemoglobin levels. The patient's condition worsened and the methemoglobin level continued to increase. With the identification of G6PD deficiency in the patient, methylene blue was discontinued and ascorbic acid therapy initiated yet methemoglobinemia was not corrected. A 1600 mL automated red blood cell exchange was performed and methemoglobin level was reduced to 8.0% and hematocrit was 30%. The patient showed limited improvement in oxygen saturation but this may have reflected the concurrent presence of pleural effusions. After completing chemotherapy, the patient was discharged on room air 30 days later.

Additional literature on this topic can be found in these reviews [50R,51R].

References

[1] Gosselin M, Dazé Y, Mireault P, et al. Toxic myocarditis caused by acetaminophen in a multidrug overdose. Am J Forensic Med Pathol. 2017;38(4):349–52 [a].

[2] Bradley LM, Higgins SP, Thomas MM, et al. Sweet syndrome induced by oral acetaminophen-codeine following repair of a facial fracture. Cutis. 2017;100:E20–3 [a].

[3] Rickner SS, Cao D, Simpson S-E. Hemolytic crisis following acetaminophen overdose in a patient with G6PD deficiency. Clin Toxicol. 2017;55(1):74–5 [a].
[4] Queirós C, Salvador P, Ventura A, et al. Methemoglobinemia after paracetamol ingestion: a case report. Acta Medica Port. 2017;30(10):753–6 [a].
[5] Lin BM, Curhan SG, Wang M, et al. Duration of analgesic use and risk of hearing loss in women. Am J Epidemiol. 2017;185(1):40–7 [MC].
[6] Randesi M, Levran O, Correa da Rosa J, et al. Association of variants of arginine vasopressin and arginine vasopressin receptor 1A with severe acetaminophen liver injury. Cell Mol Gastroenterol Hepatol. 2017;3(3):500–5 [C].
[7] Stergiakouli E, Thapar A, Davey Smith G. Association of acetaminophen use during pregnancy with behavioral problems in childhood: evidence against confounding. JAMA Pediatr. 2017;170(10):964–70 [C].
[8] Beale DJ. Acetaminophen in pregnancy and adverse childhood neurodevelopment. JAMA Pediatr. 2017;171(4):394–5 [r].
[9] Saunders NR, Habgood MD. Acetaminophen in pregnancy and adverse childhood neurodevelopment. JAMA Pediatr. 2017;171(4):394 [r].
[10] Little MO, Wickremsinhe MN, Lyerly AD. Acetaminophen in pregnancy and adverse childhood neurodevelopment. JAMA Pediatr. 2017;171(4):395–6 [r].
[11] Damkier P, Scialli AR, Lusskin SI. Acetaminophen in pregnancy and adverse childhood neurodevelopment. JAMA Pediatr. 2017;171(4):396 [r].
[12] Ystrom E, Gustavson K, Brandlistuen RE, et al. Prenatal exposure to acetaminophen and risk of ADHD. Pediatrics. 2017;140(5):1–9 [C].
[13] Gervin K, Nordeng H, Ystrom E, et al. Long-term prenatal exposure to paracetamol is associated with DNA methylation differences in children diagnosed with ADHD. Clin Epigenetics. 2017;9:77–86 [C].
[14] Nguyen VQ, Grider DJ, Yeaton P. A rare cause of colonic stricture. Gastroenterology. 2017;152:490–1 [a].
[15] Gunes H, Sonmez FT, Saritas A, et al. Kounis syndrome induced by oral intake of diclofenac potassium. Iran J Allergy Asthma Immunol. 2017;16(6):565–8 [a].
[16] Probst M, Kühn J-P, Modeβ C, et al. Muscle injury after intramuscular administration of diclofenac: a case report supported by magnetic resonance imaging. Drug Saf Case Rep. 2017;4:2–7 [a].
[17] Yehiyan A, Barman S, Varia H, et al. Short-course high-dose ibuprofen causing both early and delayed jejunal perforations in a non-smoking man. BMJ Case Rep. 2017. https://doi.org/10.1136/bcr-2017-223644 [a].
[18] Geith S, Renner B, Rabe C, et al. Ibuprofen plasma concentration profile in deliberate ibuprofen overdose with circulatory depression treated with therapeutic plasma exchange: a case report. BMC Pharmacol Toxicol. 2017;18:81–5 [a].
[19] Kim SY, Shin SH, Kim H-S, et al. Pulmonary arterial hypertension after ibuprofen treatment for patent ductus arteriosus in very low birth weight infants. J Pediatr. 2017;179:49–53 [c].
[20] Fernandes V, Alfaro TM, Baptista JP, et al. Severe ketorolac-induced asthma diagnosed by chest computed tomography. J Thorac Dis. 2017;9(Suppl 16):S1567–9 [a].
[21] Nissen SE, Yeomans ND, Solomon DH, et al. Cardiovascular safety of celecoxib, naproxen, or ibuprofen for arthritis. New Eng J Med. 2017;375(26):2519–29 [MC].
[22] Pepine CJ, Gurbel PA. Cardiovascular safety of NSAIDs: additional insights after PRECISION and point of view. Clin Cardiol. 2017;40:1352–6 [R].
[23] Patrono C, Baigent C. Coxibs, traditional NSAIDs, and cardiovascular safety post-PRECISION: what we thought we knew then and what we think we know now. Clin Pharmacol Ther. 2017;102(2):238–45 [R].
[24] Grosser T, Ricciotti E, FitzGerald GA. The cardiovascular pharmacology of nonsteroidal anti-inflammatory drugs. Trends Pharmacol Sci. 2017;38(8):733–48 [R].
[25] Solomon DH, Husni ME, Libby PA, et al. The risk of major NSAID toxicity with celecoxib, ibuprofen, or naproxen: a secondary analysis of the PRECISION trial. Am J Med. 2017;130(12):1415–22 [MC].
[26] Ruschitzka F, Borer JS, Krum H, et al. Differential blood pressure effects of ibuprofen, naproxen, and celecoxib in patients with arthritis: the PRECISION-ABPM (prospective randomized evaluation of celecoxib integrated safety versus ibuprofen or naproxen ambulatory blood pressure measurement) trial. Eur Heart J. 2017;38:3282–92 [C].
[27] Smith CJ, Ryckman KK, Bahr TM, et al. Polymorphisms in CYP2C9 are associated with response to indomethacin among neonates with patent ductus arteriosus. Pediatr Res. 2017;82(5):776–80 [C].
[28] Dalal D, Dubreuil M, Peloquin C, et al. Meloxicam and risk of myocardial infarction: a population-based nested case-control study. Rheumatol Int. 2017;37:2071–8 [MC].
[29] Gunter BR, Butler KA, Wallace RL, et al. Non-steroidal anti-inflammatory drug-induced cardiovascular adverse events: a meta-analysis. J Clin Pharm Ther. 2017;42:27–38 [M].
[30] Sondergaard KB, Weeke P, Wissenberg M, et al. Non-steroidal anti-inflammatory drug use is associated with increased risk of out-of-hospital cardiac arrest: a nationwide case-time-control study. Eur Heart J. 2017;3:100–7 [MC].
[31] Chan FKL, Ching JYL, Tse YK, et al. Gastrointestinal safety of celecoxib versus naproxen in patients with cardiothrombotic diseases and arthritis after upper gastrointestinal bleeding (CONCERN): an industry-independent, double-blind, double-dummy, randomised trial. Lancet. 2017;389:2375–82 [C].
[32] Blumenthal KG, Lai KH, Huang M, et al. Adverse and hypersensitivity reactions to prescription nonsteroidal anti-inflammatory agents in a large healthcare system. J Allergy Clin Immunol Pract. 2017;5(3):737–43 [MC].
[33] Bhatt DL, Grosser T, Dong J-f, et al. Enteric coating and aspirin nonresponsiveness in patients with type 2 diabetes mellitus. J Am Coll Cardiol. 2017;69(6):603–12 [C].
[34] Shively RM, Hoffman RS, Manini AF. Acute salicylate poisoning: risk factors for severe outcome. Clin Toxicol (Phila). 2017;55(3):175–80 [c].
[35] Esposito AJ, Murphy RC, Toukatly MN, et al. Acute kidney injury in allopurinol-induced DRESS syndrome: a case report of concurrent tubulointerstitial nephritis and kidney-limited necrotizing vasculitis. Clin Nephrol. 2017;87(6):316–9 [a].
[36] Masoumi HT, Hadjibabaie M, Zarif-Yeganeh M, et al. Exacerbation of allopurinol-induced drug reaction with eosinophilia and systemic symptoms by teicoplanin: a case report. J Clin Pharm Ther. 2017;42:642–5 [a].
[37] Lien Y-HH, Logan JL. Cross-reactions between allopurinol and febuxostat. Am J Med. 2017;130(2):e67–8 [a].
[38] Kwon OC, Hong S, Ghang B, et al. Risk of colchicine-associated myopathy in gout: influence of concomitant use of statin. Am J Med. 2017;130(5):583–7 [C].
[39] Frydrychowicz C, Pasieka B, Pierer M, et al. Colchicine triggered severe rhabdomyolysis after long-term low-dose simvastatin therapy: a case report. J Med Case Rep. 2017;11(1):8–12 [a].
[40] Kritikos A, Spertini O. Reversible granulocyte abnormalities after accidental ingestion of *Colchicum autumnale*. Blood. 2017;130(1):95 [a].
[41] Poh XE, Lee C-T, Pei S-N. Febuxostat-induced agranulocytosis in an end-stage renal disease patient. Medicine (Baltimore). 2017;96(2):1–3 [a].

[42] Liu C-t, Chen C-Y, Hsu C-Y, et al. Risk of febuxostat-associated myopathy in patients with CKD. Clin J Am Soc Nephrol. 2017;12(5):744–50 [C].
[43] Chahine G, Saleh K, Ghorra C, et al. Febuxostat-associated eosinophilic polymyositis in marginal zone lymphoma. Joint Bone Spine. 2017;84:221–3 [a].
[44] Deeks ED. Lesinurad: a review in hyperuricaemia of gout. Drugs Aging. 2017;34:401–10 [R].
[45] Tausche A-K, Alten R, Dalbeth N, et al. Lesinurad monotherapy in gout patients intolerant to a xanthine oxidase inhibitor: a 6 month phase 3 clinical trial and extension study. Rheumatology. 2017;56(12):2170–8 [C].
[46] Bardin T, Keenan RT, Khanna PP, et al. Lesinurad in combination with allopurinol: a randomized, double-blind, placebo-controlled study in patients with gout with inadequate response to standard of care (the multinational CLEAR 2 study). Ann Rheum Dis. 2017;76:811–20 [C].
[47] Saag KG, Fitz-Patrick D, Kopicko J, et al. Lesinurad combined with allopurinol: a randomized, double-blind, placebo-controlled study in gout patients with an inadequate response to standard-of-care allopurinol (a US-based study). Arthritis Rheumatol. 2017;69(1): 203–12 [C].
[48] Dalbeth N, Jones G, Terkeltaub R, et al. Lesinurad, a selective uric acid reabsorption inhibitor, in combination with febuxostat in patients with tophaceous gout. Arthritis Rheumatol. 2017;69(9):1903–13 [C].
[49] Montgomery KW, Booth GS. A perfect storm: tumor lysis syndrome with rasburicase-induced methemoglobinemia in a G6PD deficient adult. J Clin Apher. 2017;32:62–3 [a].
[50] Raber H, Ali A, Dethloff A, et al. Anti-inflammatory and antipyretic analgesics and drugs used in gout. In: Ray SD, editor. Side effects of drugs annual: a worldwide yearly survey of new data in adverse drug reactions, vol. 37. Elsevier; 2015. p. 115–28 [chapter 9]. [R].
[51] Raber H, Evoy K, Lim L. Anti-inflammatory and antipyretic analgesics and drugs used in gout. In: Ray SD, editor. Side effects of drugs annual: a worldwide yearly survey of new data in adverse drug reactions, vol. 38. Elsevier; 2016. p. 77–85 [chapter 8]. [R].

CHAPTER

10

General Anesthetics and Therapeutic Gases

Thomas Syratt, Alison Hall[1]

Royal Liverpool Hospital, Liverpool, United Kingdom

[1]Corresponding author: alison.hall@rlbuht.nhs.uk

INHALED ANESTHETICS

Metabolism

It has been recognized that volatile anesthesia can cause mitochondrial dysfunction by decreasing oxidative phosphorylation. A retrospective chart review of 26 patients (mean age of 47 months) with genetically proven mitochondrial disorders investigated type, duration and adverse events in these children undergoing general anesthesia (GA) [1M]. Patients received a total of 65 GA episodes either with intravenous (55%) or inhalational induction and maintenance with volatile (51%) or propofol total intravenous anesthesia (TIVA). There were 5 adverse events which were varied and therefore difficult to comment on. They were largely either hypotensive or hypoxic episodes but none led to Intensive Care Unit (ICU) admission or increased length of stay (LOS). There were no cases of malignant hyperthermia despite previous literature expressing a theoretical concern. Whilst the authors comment that there is a higher risk of GA complications in this population, this study is limited by the variety of mitochondrial disorders and the non-standard GA technique.

Nervous System

A prospective, single-centered cohort study was carried out to investigate the incidence of post-operative headache in an adult population undergoing elective surgery ($n=446$) [2E]. Patients were interviewed pre-operatively and 5 days post-operatively using a non-validated questionnaire investigating patients' demographics, lifestyle and habits (e.g. caffeine, alcohol, tobacco, previous post-operative headache, migraine history). The choice of anesthetic technique was determined by the consulting anesthetist. A post-operative questionnaire was completed for up to 5 days (or less if the patient was discharged). This recorded type of GA, drugs used, surgical position and adverse events including headache and other intra-operative events. A significant number of patients had a family history of headache (32%) and 28.3% developed post-operative headache. This was higher in patients with a history of headache and migraine (41% and 21%, respectively) than those without (16%). Multivariate analysis demonstrated increased postoperative headache (all patients) was associated with female gender (Odds ratio (OR) 1.85, $P=0.024$), sevoflurane administration (OR 3.23, $P<0.0001$), intra-operative hypotension (OR 2.11, $P=0.008$) and smoking (OR 1.85, $P=0.006$). In addition, in patients without a previous history of headache, coffee consumption also increased the risk (OR 3.86, $P=0.041$). There were 9 patients with a documented dural puncture, none of whom developed a post dural puncture headache. Due to small numbers, data from those receiving spinal anesthesia were omitted from analysis which may have skewed the results, although other central neuraxial blockade was not associated with an increased risk of headache. A further limitation is the lack of data on other factors such as perioperative and post analgesia and fluids administered. This may have confounded the results.

A small, prospective double-blind, randomized controlled trial investigated the effects of sevoflurane, isoflurane and propofol on the incidence of post-operative cognitive dysfunction (POCD) in an elderly (>65 years) group of patients undergoing laparoscopic cholecystectomy [3c]. All patients ($n=150$) received a standard induction (midazolam, rocuronium and propofol) followed by either target controlled infusion (TCI) propofol 2.5–3.0 mcg/mL, Isoflurane or sevoflurane (1–1.5 Minimum alveolar concentration (MAC)). Depth of anesthesia was monitored using Bispectral (BIS) monitoring

Side Effects of Drugs Annual, Volume 40
ISSN: 0378-6080
https://doi.org/10.1016/bs.seda.2018.08.002

(target 40–50). The primary outcome of POCD as measured using psychomimetric and neurocognitive testing performed at baseline (pre-operative) and on days 1 and 3 post-operatively. The incidence of POCD was significantly lower in the propofol group compared to both the volatile anesthesia groups on both D1 and D3 ($P<0.001$). The incidence of POCD was also significantly lower in the sevoflurane group when compared to the isoflurane group on D1 but not on D3. The authors comment that the incidence of POCD was negatively correlated with level of education but there is no statistical confirmation of this. Inflammatory markers (IL-1β, IL-6) and brain injury markers ($A\beta_{1-40}$, s100β) were also measured and showed that there was a significant decrease in these in the propofol group compared to the volatile groups (↓S-100β at all time points, ↓$A\beta_{1-40}$ at all the time points vs sevoflurane, T1 and T2 vs isoflurane, ↓IL1β vs isoflurane at T1 and T2, ↓IL-6 vs isoflurane at all time points, ↓TNFα vs isoflurane at T1 and T2). There is no long-term follow-up data from these patients which would have been clinically useful.

Tumorigenicity

Natural Killer cells (NKC) are a specific sub group of lymphocytes that lyse tumor cells. Therefore NKC function is critical to minimize post-resection tumor recurrence. Leukocyte function-associated antigen-1 (LFA-1) is an adhesion molecule on NKC causing it to bind to adhesion molecules on tumor cells, inducing lysis. Previously, sevoflurane and isoflurane have been shown to inhibit LFA-1 and this was further investigated in an experimental study using cell lines [4E]. The appropriate cell lines were grown in culture according to standard protocol. Volatile anesthetics (sevoflurane and isoflurane) attenuated tumor cytotoxicity by inhibiting NKC function. This was not observed with any of the intravenous anesthetic agents used (ketamine, etomidate, midazolam, dexmedetomidine, fentanyl, propofol). The observed mechanism was likely to be secondary to reduced conjugation; a cellular event required for NKC mediated toxicity. The inhalational agents did not concurrently increase the proliferation of NKC or tumour cells used or increase the degranulation process observed after lysis has commenced. Whilst an important observation, the authors do comment on the many unmeasured variables, e.g. type and duration of anesthesia and different types of tumour cells as this study was limited to single human and tumour cell lines.

DESFLURANE

Gastrointestinal

A single-center, retrospective cohort study was carried out on 1042 adult patients looking at pain and post-operative nausea and vomiting (PONV) following thyroidectomy for thyroid cancer, particularly comparing desflurane and sevoflurane [5c]. All patients received post-operative fentanyl and prospective pain, operative and post-operative data were collected by an independent practitioner. Due to the retrospective nature of the study, propensity scoring was undertaken to reduce bias which matched the two, previously unmatched, groups. Complete remission (defined as no PONV on day 0) was significantly greater in the sevoflurane group compared to desflurane (81% vs 70%, $P=0.016$). In addition, PONV on day 0, as measured on a numerical rating scale, was significantly better in the sevoflurane than the desflurane group (0.35 vs 0.52, $P=0.041$). There were no differences in pain scores, number of patients with PONV, or rescue anti-emetics required. This study is limited its single-center, retrospective nature and also that data were only collected once daily, thus missing data may be absent.

Ischemic preconditioning (IP) has been postulated to improve ischaemia to abdominal organs known to occur with pneumoperitoneum and anesthetic drugs. A prospective, single-center study ($n=91$) was carried out in ASA 1–3 patients undergoing elective laparoscopic cholecystectomy. Patients were randomized into 4 groups (group I: inhalational GA only, group IIP: inhalation plus IP, group T: TIVA only and group TIP: TIVA plus IP) [6c]. Following a standard induction (propofol, fentanyl and rocuronium), groups I and IIP received desflurane maintenance at 4%–6% titrated to physiological parameters and groups T and TIP received propofol at 12 mg/kg/h reducing to 6 mg/kg/h after 40 minutes with a remifentanil infusion. IP was carried out by deflating the pneumoperitoneum and reinflating again 10 minutes later to 15 mmHg. Physiological parameters were recorded every 10 minutes and blood was sampled 2 minutes post-induction and post-extubation for determination of markers of oxidative and anti-oxidative stress. There were no differences among the group's demographics; however, surgical time in the two IP groups was longer due to the IP process (IIP: 60.2 and TIP: 68.0 minutes vs I: 47 and T: 58 minutes, $P<0.001$, respectively). Blood pressure was higher in the inhalational groups compared to TIVA groups at most time intervals but there were no differences in oxygen saturations (SpO_2) or end tidal CO_2 concentrations ($ETCO_2$). There were no changes in levels of oxidative and anti-oxidative stress markers between different GA protocols. Only paraoxonase levels were decreased post-operatively in the TIP group when compared with the I group, the relevance of which was not commented on by the authors.

Respiratory

A randomized controlled trial investigated the use of desflurane in 200 children undergoing strabismus surgery using a laryngeal mask airway (LMA) [7C]. Using a LMA in this situation is not advised, under manufacturer's guidance, in children under 6 years old due to the increased risk of respiratory adverse events. Following a standardized IV/gas induction using thiopentone and sevoflurane, children were randomized to receive either 2%–3% sevoflurane (S group) or 6%–7% desflurane (D group) following insertion of the LMA to achieve a MAC of 0.8–1.2. A blinded investigator assessed the children post-operatively looking for respiratory adverse events (breath holding, coughing, secretions, laryngospasm, bronchospasm and desaturations (SpO_2 90%–97%)). The incidence of respiratory adverse events was high in both groups (S group: 39% and D group: 41%) but there were no differences between the two groups. There were differences in the incidence of desaturations where D group had a higher incidence (0 vs 7, $P=0.007$). The duration of the desaturation did not exceed 30 seconds and all recovered with supplemental oxygen. The emergence time was significantly longer in S group (8 vs 6.6 minutes, $P=0.003$) but this did not lead to an increased recovery time or increased incidence of delirium, the incidence of which was high (>50%) in both groups. This small study lacked adequate power to draw conclusions about one drug over the other and also used a muscle relaxant to facilitate anesthesia which may have contributed to the high overall incidence of adverse events.

ISOFLURANE

Urinary Tract

A small, prospective randomized single-blinded study was carried out to investigate the effect of xenon and isoflurane on renal function during partial nephrectomy [8c]. Patients ($n=46$) were randomized to receive gaseous maintenance of anesthesia with either 60% xenon or 1.2% isoflurane in 40% oxygen following a standardized IV induction. Primary end point of daily glomerular filtration rate (GFR) was calculated along with secondary endpoints (routine observations, intraoperative data, routine lab data and tumour related data). The isoflurane group had a greater reduction in GFR than the xenon group (19.7 vs 10.9 mL/min/1.73 m^2) but this did not reach statistical significance. There were no differences in other markers of renal function (serum urea and creatinine, NGAL (an early marker of AKI)) between the two groups. The adverse events were higher in the isoflurane group (intraoperative hypotension requiring catecholamine support and anemia) but there were no differences in PONV or the incidence of acute kidney injury or long-term kidney dysfunction. This study excluded any patients with pre-operatively abnormal renal function which therefore limits its generalizability.

Nervous System

An experimental study has looked at the role of isoflurane in the process of neurodegeneration, clinically relevant to the risk of cognitive impairment following GA [9E]. Human Embryonic stem cells (hESCs) were grown in standard culture and then exposed to 5 vol% isoflurane. Isoflurane had no effect on cell survival but did limit the proliferation of the progenitor cells up to day 5. The differences were not evident after this time period. In addition, the differentiation of neural progenitor cells was reduced in the cells exposed to isoflurane (decreased number, density and differentiation of new neurons). However, again, this was a short lived phenomenon with no difference observed by day 7.

A prospective, single-blinded study investigated 25 patients aged 6 months to 11 years undergoing elective MRI scanning under isoflurane anesthesia [10c]. They looked at the impact of isoflurane anesthesia on the stress response clinically relevant as some of these mediators have been implicated in neurocognitive decline. The children all underwent an inhalational induction with 8% sevoflurane in 100% oxygen and then a LMA was inserted. Thereafter, isoflurane was used as the maintenance anesthesia with a mean GA time of 54 minutes. Blood samples were drawn pre- and post isoflurane to investigate concentrations of cytokine mediators of the stress response (pro-inflammatory: IL-1β, TNF-α, IL-6, VEGF. Anti-inflammatory: IL-10). There was a significant increase in the concentration of IL-1β (25.97 vs 38.53 pg/mL, $P=0.0002$) in the post-isoflurane sample but there was no significant difference in the concentration of TNF-α or IL-6. VEGF and IL-10 were not detected in either sample. This study is limited by a short anesthetic time but benefits from the lack of surgical stimulus confounding the results.

Gastrointestinal

A prospective trial looked at 80 ASA 1–2 patients undergoing gynaecological laparoscopy at high altitude (>3000 m above sea level) [11c]. Participants were randomized to receive either IV induction followed by TIVA using propofol or combined IV induction with isoflurane maintenance. Physiological parameters (Mean arterial pressure (MAP), Heart rate (HR), SpO_2, Partial pressure of expired CO_2 (P_ECO_2)) were measured at 30 minutes pre-induction, immediately and 15 minutes after insufflation and 5 minutes post-deflation of pneumoperitoneum. A blinded observer reviewed the patients at 24 hours post-operatively to record awareness, PONV and any other complications. There were no significant differences in any of the physiological parameters measured at any time point throughout the study. All patients had an uneventful emergence from anesthesia but in the TIVA group this was significantly shorter than the group receiving volatile anesthesia at all time points (Eye opening, return to consciousness, extubation and time in operating room). There were no episodes of awareness in either group but the isoflurane group had a significantly higher incidence of PONV (55% vs 35%, $P<0.05$).

NITROUS OXIDE

Nervous System

A prospective, single-center observational study assessed 60 adults (mean age 70.6 years) undergoing painful photodynamic therapy for removal of multiple actinic facial keratoses using nitrous oxide as analgesia [12c]. Following treatment on the first cheek, patients were asked to rate scale on a 0–10 visual analogue scale (VAS) and were allocated a group depending on the result (group A: severe VAS>6, $n=30$, group B: VAS<6, $n=32$, group C: VAS>6 but declined inhaled therapy, $n=6$). Prior to treatment of the second cheek, Group A were given N_2O/O_2 mixture, Group B and C received nothing. Pain scores were reassessed following the second treatment. In group A, there was a significant reduction in pain scores between the two treatments (VAS reduced by 4.2 points (7.8 vs 3.5, $P<0.001$)). In group B, there was an increase of 0.5 VAS points between treatments (3.9 vs 4.4). In group C, there was a smaller but statistically significant reduction in pain scores between the two treatments (7.3 vs 6.6). In addition there was a significant reduction in interruptions to treatment between the two treatment sessions after the use of nitrous oxide (82%). There was very high patient satisfaction and only mild and self-limiting side effects from the N_2O (three had rotational vertigo, two had loss of control feelings and one complained of amplified background noise).

Gastrointestinal

A prospective observational study at 2 tertiary paediatric centers recruited 90 patients between the ages of 4 and 18 years for procedural sedation of orthopaedic injuries in the emergency department (ED) [13c]. Patients received 1.5mcg/kg intranasal fentanyl with additional doses of 0.5mcg/kg at the discretion of the physician. N_2O/O_2 (50/50 or 70/30 mix) was then delivered in addition using an on demand system. Manipulation of the fracture was commenced 3 minutes after N_2O administration. A validated facial pain scale revised (FPS-R) was used to evaluate the efficacy of the procedural sedation immediately before and 10 minutes after the procedure. Patients were also asked about recall. The FLACC score (Face, Leg, activity, cry, consolability) was also used to evaluate the sedation episode, before during and 10 minutes after. 17 patients were excluded due to lack of recall but FLACC score was 3/10 for these patients. 10% patients had a pain score >6 and all of these patients had fentanyl administered <15 minutes prior to N_2O administration. There were no serious adverse events but the incidence of nausea, vomiting and vertigo was 19%, 13% and 23%, respectively. Vomiting was higher in the center where a higher percentage of N_2O was used (27.5% using 70/30 mix vs 10% when 50/50 mix was used). In addition, there were also reported individual cases of bradycardia, paradoxical anxiety, diaphoresis, urticaria and double vision. In total, 62% had side effects during their hospital stay but patient and staff satisfaction was high with over 88% saying they would use the same sedation regime in the future. Unfortunately, recall and the amnesic effect of N_2O limit the effectiveness of self-evaluation.

SEVOFLURANE

Respiratory

A single-center trial looked at optimal exhaled sevoflurane concentration that produces adequate conditions for endotracheal intubation without neuromuscular blockade [14c]. They investigated adult males (aged 18–30), all ASA 1 undergoing elective otolaryngological surgery ($n=68$). The patients were randomized to 3 groups to receive 1.0, 1.5 or 2.0mcg/kg remifentanil in addition to an inhalational induction using 8% sevoflurane. The administering anesthetist was blinded to the group. Following a standardized inhalational induction technique (using the 'up and down' method), remifentanil was administered and endotracheal intubation was attempted 90 seconds later. If unsuccessful, anesthesia was deepened and rocuronium 0.6mg/kg was administered. The effective concentration of sevoflurane for 50% and 95% successful intubation (EC50 and EC95, respectively) was measured in each group. The EC50 and EC95 for group 2.0 were significantly lower than the other two groups (3.00, 2.00, 1.29vol% and 3.45, 2.91 and 1.89vol% for EC50 and EC95, respectively). There were 1, 1, and 3 failed intubations in the 1.0, 1.5 and 2.0mcg/kg remifentanil groups, respectively. This may be due to the muscle rigidity known to occur following remifentanil administration. There was a significant increase in heart rate post induction in the 1.0 group than the other groups ($P=0.014$) but no differences in the other cardiovascular variables measured. There were 3 cardiac adverse events in the 2.0 group (2 bradycardias and 1 arrhythmia requiring atropine injection).

DEXMEDETOMIDINE

Gastrointestinal

A meta-analysis has studied the effects of dexmedetomidine, used in addition to GA, on PONV and cardiovascular side effects [15M]. They identified 24 RCTs published during 2015 and 2016 that were included in the analysis totaling 2046 patients. In each group, heterogeneity was not significant and thus a fixed-effect model was adopted. There was a significant reduction in PONV when using dexmedetomidine (RR 0.53 (95% CI 0.47–0.59, $P<0.00001$)). Subgroup analysis confirmed that this was consistent whether using a loading dose (0.5–1 µg/kg) or continuous infusion (0.1–0.7 µg/kg/h) only or combined. There were no significant differences in the incidence of bradycardia; however, hypotension was increased in studies where a loading dose and a continuous infusion were administered. There were no differences in hypotension in the other groups.

DEXMEDETOMIDINE AND ITS USE AS AN ADJUNCT IN REGIONAL ANESTHESIA

Nervous System

Shivering is major troublesome complication following regional anesthesia. A single-center, prospective, double-blind, parallel controlled trial looked at 50 parturients due to undergo elective caesarean section under spinal anesthesia [16c]. Recruits (aged 18–45 years) were randomized to receive 12.5 mg bupivacaine with or without 5 µg dexmedetomidine intrathecally. All other anesthetic steps were standardized between the two groups. Shivering incidence and intensity were measured using a previously validated tool (0–4, no shivering to shivering involving the whole body) and core and peripheral temperature were measured pre-spinal and at 15-minute intervals thereafter. Standard physiological parameters and complications (bradycardia, nausea, vomiting, sedation) were also measured with any rescue medication required. There was a significantly greater incidence in the incidence of shivering in the control group (52% vs 24%, $P=0.04$). The control group also had a greater intensity of shivering. There was a significant improvement in the incidence of shivering in the dexmedetomidine group (52% vs 24%, $P=0.04$) with a significantly greater intensity also in the control group. There were no differences in the core or peripheral temperature of the two groups. There were no differences in the core or peripheral temperatures of the two groups. There were no differences in the incidence of cardiovascular side effects or nausea and vomiting between the two groups; however, there was a significant increase in the degree of sedation experienced by the patients receiving dexmedetomidine (dex: Grade 4/5 sedation 24%, control: grade 4/5 sedation: 0%, $P=0.004$).

Another study has examined the use of dexmedetomidine as an adjuvant to epidural anesthesia in children [17c]. They included 60 children undergoing elective lower limb orthopaedic surgery. Each child received a standardized GA followed by a lumbar epidural. They received 0.2 mL/kg ropivacaine and either fentanyl 1 µg/kg or dexmedetomidine 1 µg/kg with a patient controlled function set to deliver a bolus of 0.05 mL/kg every 15 minutes. Patients were observed for 1 hour in the recovery room and then for 48 hours post operatively by an independent observer. There was a trend towards faster emergence in the fentanyl group, although this was not statistically significant (9.1 vs 11.9 minutes, $P=0.055$). Pain control was good amongst both groups, as assessed by the r-FLACC score. Although the median pain score at 6 hours was lower in the dexmedetomidine group, these differences were not present after 6 hours and the amount of rescue medication required was not different. There were no differences in the adverse effects between the two groups, although both groups had a moderate incidence of emergence agitation (Fentanyl: 8/29 and dexmedetomidine: 5/28) and nausea and vomiting (Fentanyl:7/29 and dexmedetomidine 10/29).

These studies have been backed up by a systematic review examining the safety and efficacy of dexmedetomidine as an adjuvant to epidural anesthesia [18M]. The review examined 12 RCTs and due to high heterogeneity scores, standard mean difference (SMD) was used to demonstrate differences between the two groups. This study demonstrated that compared to the control groups, dexmedetomidine prolonged duration of anesthesia (8 studies, $n=410$, I^2 97%, SMD 3.5, $P<0.0001$) reduced the time to sensory block (8 studies, $n=510$, I^2 92%, SMD –1.13 $P=0.02$) and decreased the requirement for rescue analgesia (6 studies, $n=382$, I^2 90%, SMD –2.00, P (00001). This was achieved however at the expense of a higher sedation score (9 studies, $n=480$, I^2 90%, SMD 1.41, $P<0.0001$). Whilst dexmedetomidine did not alter blood pressure, it statistically significantly lowered heart rate ($n=360$, I^2 10%, −3.74 bpm, $P=0.0009$). This will likely have marginal clinical significance; however, symptoms of hypotension and bradycardia were more common in the dexmedetomidine group than controls. The incidence of other adverse effects was not different between the dexmedetomidine groups and their controls.

A further meta-analysis looking at the use of clonidine and dexmedetomidine as an adjunct to local anesthesia in supra-clavicular brachial plexus blockade has been carried out [19M]. Data from 868 patients were analyzed (clonidine: $n=419$, dexmedetomidine: $n=419$). The studies, looking at block duration, onset and analgesia, had heterogeneity scores ($I^2>80\%$). Despite this, dexmedetomidine was shown to have a prolonged sensory and motor blockade by a factor of 1.2 ($P<0.00001$) and shortened sensory and motor onset by a factor of 0.9 ($P<0.002$). In addition, dexmedetomidine prolonged the duration of analgesia by a factor of 1.2 (<0.00001). Unfortunately, reporting of side effects was sporadic and largely showed no difference. 3 trials did show an increase in bradycardic episodes in the dexmedetomidine group by a factor of 7.4 (I^2 0%, $P=0.003$). Although the authors made every attempt to exclude publication bias, the heterogeneity amongst the studies included in this report was high and study size was small. This will therefore limit the meaningful conclusions to be drawn by this study. The study was also limited to supraclavicular blocks and not to other brachial plexus blockade techniques.

Nervous System

Two case reports have demonstrated the use of dexmedetomidine in the control of methamphetamine induced agitation [20A]. Both report transient falls in blood pressure (BP) and heart rate (HR) which resolved spontaneously or on cessation of dexmedetomidine.

Case 1

42-year-old male with violent behavior following smoking methamphetamine. On arrival he was agitated and confused, hypertensive and tachycardic but other physiological parameters were normal. There was no improvement with benzodiazepines (10mg diazepam and 332mg midazolam in the first 48 hours). On stopping the midazolam infusion, the patient became agitated once again and therefore dexmedetomidine was commenced (loading dose 1mcg/kg/h followed by an infusion rate of 0.2mcg/kg/h). Midazolam infusion was able to be stopped. Following successful weaning from the dexmedetomidine, the patient experienced rebound agitation after 2 hours which was again successfully controlled by restarting the dexmedetomidine. The patient was physically well during his stay but experienced a reduction in BP (156/111–92/58) and HR (121–65bpm) after the loading dose. The patient was discharged after 79 hours.

Case 2

A 38-year-old male with a history of mental health disorder and substance abuse presented to the ED agitated, confused and with irrelevant speech. He had a urinary positive test for methamphetamine. On admission in addition to the altered mental state, he had normal cardiovascular parameters (BP 130/80, HR 63bpm). Oral benzodiazepines were administered for the first 12 hours; however, at 12 hours post-admission, his agitation worsened and he became hypertensive and tachycardic. He was commenced on dexmedetomidine (no loading dose, infusion dose 0.3mcg/kg/h increased to 0.8mcg/kg/h). The patient became drowsy but rousable and it was noted that his HR and BP returned to normal. The drug was continued for 21 hours before successful weaning and patient discharge at 39 hours.

Respiratory

A small, retrospective cohort study has investigated the utility of dexmedetomidine compared to midazolam for sedation for flexible fibreoptic bronchoscopy (FFB) in sedated patients on the ICU [21c]. Prior to FFB, all drugs were stopped until the Richmond agitation and sedation scale (RASS) was 0 or the patients regained consciousness. There is limited information on the diagnosis and clinical condition of the groups or the rationale for stopping sedation prior to FFB. Patients then received either dexmedetomidine ($n=72$, 0.5mcg/kg loading dose followed by 0.2–0.7mcg/kg/h infusion) or midazolam ($n=76$, 0.05mg/kg loading dose followed by 0.02–0.2mg/kg/h infusion). Propofol boluses were used for remedial sedation if RASS was greater than −3/−4. The patients who received dexmedetomidine had a longer sedation time (27.85 vs 23.04 minutes, $P<0.01$) for a longer procedure time (20.07 vs 16.97 minutes, $P<0.01$) and required significantly more remedial sedation (72.2% vs 22%, $P<0.01$). However, despite this the bronchoscopist's satisfaction did not differ between the two groups. The dexmedetomidine group had more respiratory adverse events (cough 56.9% vs 36.6%, $P=0.014$, bronchospasm (68.1 vs 46.1%, $P=0.007$) and desaturation (44.4% vs 23.7%, $P=0.008$)). There were no differences in the cardiovascular side effects or nausea.

Cardiovascular

A meta-analysis has reviewed the incidence of bradycardia in paediatric patients receiving dexmedetomidine as a sole agent for GA/sedation [22M]. In total, 21 studies ($n=2835$) were identified and included (8 RCTs, 2 prospective non-RCTs and 11 retrospective analyses). IV/IM or intranasal dexmedetomidine was used for different procedures including radiology, endoscopy, dentistry and burns sedation. Mean initial and maintenance doses used were 1.63±0.33mcg/kg and 0.86±0.68mcg/kg/h. 13% received rescue anesthesia after failure of dexmedetomidine to produce adequate anesthesia. There is no comment as to whether this is related to route of administration. Due to the high heterogeneity of the pooled studies, a random effects model was used to analyze the data. Of the 21 studies, 9 reported no incidence of bradycardia. The remainder revealed an incidence of bradycardia (>20% decreased in HR from baseline) of 3% (5% with IV, 2% with IM and 0% with IN) with a drop of HR by 17.26 (12.92–21.6, $P<0.00001$) bpm. Meta-regression analysis did not reveal a relationship between initial or maintenance dose of dexmedetomidine and incidence of bradycardia.

A prospective, randomized study has analyzed the effect of dexmedetomidine on cardiovascular stability post maxillo-facial surgery with nasal intubation [23c]. They randomized 93 adult patients to receive either 0.2

or 0.4 mcg/kg/h dexmedetomidine (or control) immediately before a standard GA induction and sevoflurane/remifentanil maintenance to maintain a BIS of 40–60. Cardiovascular and neurological variables were recorded at five-time intervals (T0: baseline, T1: immediately after drug cessation, T2: 5 minutes later, T3: on eye opening and T4: immediately post-extubation). Primary outcome variable of MAP at T4 was significantly higher in the control group ($P < 0.003$). HR was also higher in the control group than in either dexmedetomidine group (Dex 0.2: $P = 0.014$, Dex 0.4: $P = 0.022$). In the post-operative care unit (PACU), MAP and HR were significantly higher in the control group than either of the two dexmedetomidine groups. Other side effects (shivering, sedation scores, rescue analgesia and antiemetic use) and time to extubation were similar in all groups. Whilst other studies have commented on the incidence of bradycardia being a side effect of dexmedetomidine, this study interpreted these findings as a more controlled, cardiovascularly stable emergence from anesthesia without delaying extubation. There was no longer term follow-up of these patients post-PACU.

Nervous System

A prospective, randomized study investigated the role of dexmedetomidine in the reduction of post-operative analgesic requirements in surgical patients undergoing elective abdominal surgery ($n = 84$, ASA 1–2) [24c]. Control group patients received a standard GA (propofol/remifentanil/ringer's solution PRR, $n = 22$) and treatment groups received additional dexmedetomidine infusion post induction (0.4 mcg/kg/h) with (group PRD_w, $n = 23$) or without (group PRD_0, $n = 25$) a loading dose pre-induction (no dose available). Patients receiving the loading dose had a significant drop in HR of approximately 20 bpm post-dexmedetomidine loading dose. The study claims that more patients experienced bradycardia in the PRD_w group but this is not complete with a definition of bradycardia or P-values. All patients were back at baseline values for both HR and BP, 24 h after surgery. The dexmedetomidine groups both had higher sedation scores in PACU and a greater time to first post-operative analgesia than the control group. There were no differences in other side effects reported (PONV, pruritis, respiratory depression, dizziness and post-operative bradycardia).

Respiratory

2 retrospective studies have investigated the role of dexmedetomidine as procedural sedation for flexible bronchoscopy.

A retrospective non-randomized dose finding trial investigated the safety and efficacy of dexmedetomidine (dex)/remifentanil (RF) infusion for flexible bronchoscopy in children [25c]. They divided 135 children aged 5–10 years into 3 groups (DR1: $n = 47$, dex infusion 0.5 mcg/kg for 10 minutes then 0.5–0.7 mcg/kg/h with RF 0.5 mcg/kg for 2 minutes then 0.05–0.2 mcg/kg/min, DR2: dex infusion 1 mcg/kg for 10 minutes then 0.5–0.7 mcg/kg/h with RF 1 mcg/kg for 2 minutes then 0.05–0.2 mcg/kg/min or DR3: dex infusion 1.5 mcg/kg for 10 minutes then 0.5–0.7 mcg/kg/h with RF 1 mcg/kg for 2 minutes then 0.05–0.2 mcg/kg/min). Standard physiological parameters were measured at multiple time points throughout the procedure. Compared to DR1, patients in DR2 and DR3 had significantly lower HR and MAP at all time points post drug administration. Compared with DR2, group DR3 had significantly lower heart rate and blood pressure during the procedure. Onset of anesthesia was significantly shorter in group DR3 (14.23 vs 14.45 vs 11.13 minutes, respectively, $P = 0.003$) and this group required significantly fewer rescue doses of analgesia. 46% children moved during the procedure in group DR1 compared to 17.7% in group DR3. Whilst most of these movements could be controlled using rescue medication, 12 children needed an alternative sedation regime to complete the procedure. The children in group DR1 had a significantly shorter time to discharge from PACU but this was at the expense of experienced a significantly higher incidence of adverse effects when compared to group DR3 (tachycardia: 46% vs 22%, $P = 0.045$, hypertension 51% vs 20%, $P = 0.007$, Cough 53% vs 26%, $P = 0.033$ and hypoxaemia 38% vs 13%, $P = 0.008$). Bronchoscopist's satisfaction was significantly higher in groups DR2 and DR3.

This retrospective non-randomized trial investigated children aged 5–10 years ASA 1–2 undergoing diagnostic bronchoscopy comparing remifentanil or propofol combined with dexmedetomidine infusions [26c]. 123 patients were divided into two groups (Group DR: $n = 63$, dex infusion at 1 mcg/kg for 10 minutes followed by 0.5–0.7 mcg/kg/h, RF infusion at 1 mcg/kg/h for 5 minutes and then 0.05–0.2 mcg/kg/min and group DP: $n = 60$, dex infusion at 1 mcg/kg/h and propofol for 10 minutes followed by 0.5–0.7 mcg/kg/h and propofol infusion 10 mcg/kg for 5 minutes and 0.05–0.1 mcg/kg/min). Standard monitoring was used and intra-operative haemodynamic variables were recorded at multiple time intervals from arrival in the operating room to arrival in PACU. Both HR and MAP were significantly reduced in the propofol group compared to the remifentanil group at all time points post-administration of sedation. Onset of anesthesia was significantly shorter in the propofol group (8.22 ± 2.48 vs 12.25 ± 6.43 minutes, $P = 0.015$). Total amount of rescue doses of alternative anesthesia were higher in the remifentanil group. More children moved during the procedure in the propofol group, and whilst this was largely controlled with rescue anesthesia, 10/63 children required an alternative sedation regime to complete the procedure. The time to discharge from PACU was significantly shorter in the propofol group but this was at the expense a greater incidence of other side effects: tachycardia (46% vs 28%, $P=0.038$), hypertension (53% vs 31%, $P=0.015$) and hypoxaemia (38% vs 17%, $P=0.01$).

A small, randomized controlled study has compared the use of dexmedetomidine and propofol sedation in patients undergoing elective hand surgery under brachial plexus blockade [27c]. They randomized 87 adult ASA 1–3 patients to receive propofol (P group: effect site concentration (C_e) 1.6μg/mL), dexmedetomidine (D group: 0.4μg/kg/h following a loading dose of 1μg/kg over 10 minutes) or a combination (M group: C_e 0.8μg/mL propofol and Dexmedetomidine 0.5μg/kg loading dose followed by 0.2μg/kg/h). Primary endpoint measured were changes in BP and airway obstruction. Secondary endpoints of time to sedation and time to BIS of 70 and vital signs were also measured. D group had a significantly higher MAP than the other two groups (D group: 96, P group 86.9, M group: 85.6, $P=0.004$), whereas the reduction in HR was least in the P group (P group: 67.3bpm, D group: 57.8, M group: 59.2, $P<0.001$). Measured adverse events were apparent throughout the groups but the M group had a significantly reduced incidence of airway obstruction (3.3 vs 46.4 (P group) and 27.6% (D group), $P=0.001$). The propofol group also had the highest incidence of hypoxia ($SpO_2<90\%$), spontaneous movement and agitation (P all ≤ 0.001). Bradycardic episodes (<45bpm requiring atropine) were significantly higher in the dexmedetomidine group (27.6% vs 3.6% (P group) and 0 (M group)). There were no incidence of awareness and recall in any patient and patient and surgeon satisfaction was high amongst all groups, but significantly higher in M group. These results suggest that a combination of half doses of the two sedatives may show an improved cardiovascular and respiratory side effect profile and improved patient and satisfaction.

Nervous System

A meta-analysis has been carried out comparing propofol and dexmedetomidine in the incidence of delirium following cardiac surgery [28M]. Other secondary endpoints (hypotension, bradycardia, AF, length of intubation and LOS ICU) were also examined. Unfortunately, only 8 studies were eligible reporting on 989 patients all receiving variable doses of either sedation regime. 4 RCTs reported on the incidence of delirium identifying that patients receiving dexmedetomidine had a reduced incidence of delirium (23.5% vs 9.3%, RR 0.4, $P=0.0002$). Pooled data from small studies identified no increased in the incidence of hypotension or atrial fibrillation; however, the relative risk of bradycardia was 3.17 (incidence 9.3% vs 2.9%, $P=0.005$). Analysis of 7 studies ($n=400$) showed a significant decrease in number of intubated days in those receiving dexmedetomidine; however, this did not equate to a reduction in ICU LOS. Unfortunately, many of the included studies are small with high heterogeneity.

Cardiovascular

A small, double-blinded, single-center RCT has investigated the use of dexmedetomidine in patients undergoing elective coronary artery bypass grafting ($n=88$) [29c]. Patients (all ASA 1–3) were randomized to receive dexmedetomidine infusion (0.5μg/kg/h) following induction of a standard anesthesia protocol (which continued until extubation) or placebo (same volume of 0.9% saline). Haemodynamic parameters were measured at 0, 1 and 2 hours post-cardiac bypass and 1, 2, 4, 6 and 12 hours post-operatively. MAP was lower in the dexmedetomidine group in the early post-operative period. HR was significantly reduced in the treatment group at all points up to 6 hours post-operatively. This reduction was small and may not be clinically relevant, although the authors claim that this represents a reduced response to the surgical stress. There were no differences in the amount of inotropes, fluids or beta blockers required between the two groups. In addition, post-operative pain scores were significantly improved in the dexmedetomidine group resulting in a significantly reduced post-operative morphine consumption.

INTRAVENOUS ANESTHETIC AGENTS

Drug Dosing Regimens

A prospective, multicenter, observational cohort study examined the outcomes of paediatric sedation in the ED, looking specifically at serious adverse events (SAE) (e.g. apnea, laryngospasm and hypotension), significant interventions (e.g. intubation, requirement of vasoactive medication), oxygen desaturation or vomiting [30c]. 6295 episodes of sedation were studied. 62.2% received ketamine alone for the procedure and the majority underwent orthopaedic reductions (65.6%). Ketamine alone had the lowest incidence of SAE (0.4%) and of significant interventions (0.9%). Compared with ketamine alone, all categories of medication used were associated with significantly increased odds of an SAE (propofol OR 5.6; 95% CI, 2.3–13.1, ketamine+fentanyl OR 6.5; 95% CI, 2.5–15.2, ketamine+propofol OR 4.4; 95% CI, 2.2–8.7). With regards to significant interventions ketamine+ fentanyl (OR 4.0; 95% CI 1.8–8.1) and ketamine+propofol (OR 2.2; 95% CI, 1.2–3.8) were both at increased odd when compared to ketamine alone. This increased risk was also present for occurrence of desaturation (ketamine+fentanyl OR 2.5; 95% CI, 1.5–3.8, ketamine+propofol OR 2.2; 95% CI, 1.6–3.0). Only ketamine+fentanyl (OR, 1.9; 95% CI, 1.2–2.8) were associated with more emesis than ketamine alone, and ketamine was found to have a significant association with desaturation (OR, 1.3; 95% CI, 1.1–1.6) and vomiting (OR, 1.3; 95% CI, 1.1–1.5) at higher doses. This study was observational and therefore direct causation cannot be drawn. There is also the risk for potential confounders.

A prospective observational study examined adult patients ($n=1711$) undergoing procedural sedation over a 7-year period [31c]. Propofol used alone was associated with a significantly shorter mean duration of sedation (propofol 15 (11–19), midazolam 25 (12.5–37.5), esketamine 25 (16–34), $P<0.001$). Of the 129 patients receiving esketamine, 77.5% achieved dissociation. The most frequent adverse events were hypoxia (4%) and apnea (2.9%), with no significant difference between those receiving propofol, midazolam or esketamine ($P=0.88$), though propofol did have a higher adverse event rate in those patients >65 compared to those below (17% vs 7%, $P<0.001$). Propofol did have a higher procedural success rate (propofol 93% vs midazolam 84% vs esketamine 88%, $P<0.001$). The authors state a significantly greater amnesic effect with esketamine than midazolam (difference 11.7%; 95% CI 0.6%–20.0%) although the figures do not support statistical significance ($P=0.38$); there was no difference between esketamine and propofol. This study is limited as some patients had incomplete data sets.

A randomized, single-blinded prospective study compared the efficacy and safety of ketamine/midazolam vs fentanyl/propofol for children undergoing elective upper gastrointestinal endoscopy [32c]. ASA 1–2 children aged 4–17 (median=12) years were randomized into 2 groups. Group A ($n=119$) midazolam (0.1 mg/kg up to 4 mg) followed by ketamine (1 mg/kg) and then further incremental doses of ketamine (0.5 up to 2 mg/kg) to achieve a Ramsey sedation score (RSS) of 5 (scoring system used to assess depth of sedation 1–6) and group B ($n=119$), fentanyl (1 mcg/kg) followed by propofol (1 mg/kg) with further doses (0.5 mg/kg) until adequate sedation was achieved. The primary aim was to assess the most appropriate sedation protocol, with secondary aims to assess comfort and complications (major: cardiac arrest, apnea or laryngospasm, minor: hypoxia, tachycardia, bradycardia, increased secretions, flushing, vomiting or coughing). The recovery period was also monitored, looking specifically at duration and associated complications. No difference was recorded in endoscopy duration and only one patient had to have the procedure discontinued due to agitation (group B). Significantly more patients in group A were more deeply sedated (sedation score of 6, 34.5% vs 4.2%, no P value stated) and sedation score of 1 (5% vs 43%, no P value stated). No patients in either group developed major complications, though significantly more patients in group B (fentanyl//propofol group) developed peri-procedural complications (27% vs 41%, $P=0.04$) and group A had a much longer recovery time (78 vs 34 minutes, $P=0.001$). Most concerning is that 92% of patients in group A developed complications in recovery (vs 40% group B, $P=0.001$), with six patients experiencing emergence reactions, whilst others experienced headaches, dizziness, hallucinations, anxiety, double vision and nausea and vomiting.

Nervous System

A case series reported three patients, each experiencing transient and repetitive violent motor activity with associated impaired consciousness, following propofol use [33A]. All recovered without adverse sequelae and cause and pathology are unknown.

Case 1

An 18-year-old woman presented for wisdom teeth extraction. She was sedated with fentanyl (100 mcg)/ midazolam 3 mg/propofol 200 mg over a 30-minute period. On emergence from sedation, she developed loud vocalizations and violent thrashing lasting 60 seconds and regaining consciousness between episodes. No resolution with further propofol, naloxone, flumazenil or physostigmine and eventually required intubation and sedation with fentanyl and dexmedetomidine. CT brain was normal and EEG was uninterruptible due to agitation induced artifact. The patient was extubated the following day with return to baseline and no recollection of events.

Case 2

A 27-year-old woman on sertraline and duloxetine for depression presented for wrist arthrodesis. Sedation was performed using fentanyl, midazolam followed by a propofol infusion. Ondansetron was given as prophylaxis for nausea. Following the procedure, she developed whole body jerking, side to side head movements and associated eye movements. Spells lasted 1minute and recurred after 10 minutes. No response was seen to diphenhydramine or benztropine, propofol or fentanyl. The movements did subside with midazolam and spells subsided with minimizing stimulation. The patient had no recollection.

Case 3

A 19-year-old man on fluoxetine and trazodone for depression presented for wisdom tooth extraction. He received fentanyl, midazolam and propofol sedation. Following the procedure he became unresponsive, associated eye fluttering and increased tone.

Symptoms self resolved but reoccurred 10 minutes later which resolved with midazolam. A further episode resolved with lorazepam. Spot video EEG captured a spell but showed no abnormal patterns. After 6 hours he returned to baseline.

Endocrine

Hypothalamic–Pituitary–Adrenal (HPA) axis dissociation in ICU patients has been investigated [34c]. This was a retrospective analysis of samples taken from a previous trial looking at 156 patients on the ICU, and the effects parenteral nutrition had on ACTH–cortisol dissociation. No patient had risk of HPA axis-dysfunction at baseline (e.g. chronic or acute steroid use). Any drugs given in the previous 24 hours were scrutinised and divided into groups known to have an impact on the HPA (anaesthetics, analgesics, sedatives, vasopressors and anticoagulation). There was no comment on any drugs given prior to this time despite 72% patients having had surgery within 24 hours of admission. Samples were taken for free and total cortisol, CBG and ACTH and compared with values from 20 healthy controls. ICU patients had significantly lower mean plasma ACTH (2.7 vs 9.0 pmol/L, $P < 0.0001$) and they remained low for 72 hours compared to controls. On admission, total cortisol levels were the same in the treatment and control arm; however, free cortisol was significantly higher in the ICU group (41.4 vs 5.5 nmol/L). The authors claim that this can be explained by a significant reduction in both plasma CBG and albumin concentrations. From day 2 onwards, both free (58 vs 41.4 nmol/L, $P < 0.0001$) and total cortisol (502.3 vs 336.7 nmol/L. $P < 0.0001$) were significantly raised in the ICU group compared to controls. In multivariate linear regression analysis (corrected for baseline risk factors), propofol (per 100 mg), etomidate (any dose), opioids (per 10 mg morphine) were all independent risk factors for a reduction in plasma cortisol whereas dobutamine (per 4200 mcg given) had a positive effect. This study is limited by its retrospective, ad hoc nature and thus the clinical application of the results may be limited.

ETOMIDATE

Nervous System

This double-blind prospective randomized control trial aimed to identify the effect of pre-treatment propofol on the myoclonus caused by etomidate [35C]. The study enrolled 375 ASA 1 and 2 adults having gastroscopy procedures lasting <30 minutes. 5 sub-groups received propofol alone throughout, etomidate alone throughout, or low (0.25 mg/kg), medium (0.5 mg/kg) or high (0.75 mg/kg) dose propofol prior to administration of etomidate to maintain anaesthesia. All patients had fentanyl and lidocaine IV prior to recruitment. The primary focus was the incidence of myoclonus occurring within 5 minutes of the etomidate administration (graded from 0 to 3 in severity), though other adverse events and vital signs were recorded. The incidence of myoclonus was highest in the etomidate only group at 48.6% reducing as the dose of propofol increased, reaching 0% in the propofol alone group (48.6% vs 26.8% vs 16.4% vs 14.9% vs 0%, $P < 0.05$). Patients in the propofol alone group had the greatest reduction in MAP with HR dropping in all groups. Hypoxia following anaesthesia increased as propofol dose increased ($P < 0.05$). The incidence of adverse events was lowest in the high dose propofol and propofol only groups ($P < 0.05$). The authors report events such as nausea, dizziness, vomiting, bucking and psychiatric symptoms in the etomidate only group, and predominantly nausea and vomiting as the adverse event in the remaining groups. Whilst 0.5 mg/kg propofol pretreatment reduces the incidence of myoclonus, depth of anaesthesia was not recorded; therefore, it may have played a role.

KETAMINE

Respiratory

A case of ketamine use as part of a "delayed-sequence intubation" resulting in apnea during pre-oxygenation of the patient has been described [36A].

Case report

A 60-year-old patient, with asthma and chronic obstructive pulmonary disease, was suffering with acute type 1 respiratory failure requiring intubation. The patient was administered 25 mg of IV ketamine over 20 seconds, becoming apneic 1 minute later, requiring a period of hand ventilation prior to intubation. The authors acknowledge that ketamine is known to cause subclinical respiratory depression, but rarely reported to result in overt apnea. The authors highlight that sedation with ketamine at low doses for pre-oxygenation is not without risk.

Sensory Systems

A secondary analysis of a prospective observational study (Ketamine and intra-ocular pressure in children) aimed at describing the effects of IV ketamine use on systolic blood pressure (BP), diastolic BP and HR in children requiring procedural sedation for orthopaedic reductions in the ED [37c]. The study enrolled 60 children between the ages of 8–18 using convenience sampling, 10 of which were later excluded as no time of the orthopaedic manipulation was recorded. The BP, HR and validated sedation score (University of Michigan sedation score) were recorded prior to sedation, time of administration, after 2 minutes, then at 5-minute intervals until sedation score returned to 0/1 or 30 minutes had passed. Of note, 37/50 children had an elevated BP at baseline. They found a statistically significant increase in HR, systolic and diastolic BP, with a further significant increase in systolic and diastolic BP in those patients that underwent manipulation. Limitations highlighted by the investigators include that this is a convenience sample with a dose of ketamine left to the discretion of the ED physician (median dose 1.47 mg/kg, range 1.25–1.67 mg/kg) and 74% of patients had received opioids for pain management prior to the intervention. The investigators conclude that their study supports the known effects of ketamine increasing HR and BP; however, these effects appear to be less pronounced in children and of statistical rather than clinical significance.

Nervous System

This consensus statement aimed to complement a recent American Psychiatric Association meta-analysis on the use of ketamine treatment for mood disorders [38M]. Currently there is no clearly established indication, with the authors identifying evidence supporting the use of ketamine in those with major depressive episodes without psychotic features, but acknowledging that most studies have previously used a single dose and evaluated over a 1-week period. The most frequently used dose is 0.5 mg/kg administered over a 40-minute period, which results in a peak plasma ketamine concentration well below that required for surgical anaesthesia (70–200 ng/mL vs 2000–3000 ng/mL). At this dose, the authors report a study that found three patients that became nonresponsive to verbal stimuli but remained "medically stable". A second study reported no desaturations or persistent medical complications during 205 infusions; however, they report thirty percent of patients became tachycardic (>110 beats per minute) or BP >180/100 mmHg in approximately 30% of patients with one serious cardiovascular related event. They conclude that ketamine, at the dose they suggest, has little significant effect on respiratory status but does have meaningful effects on BP and HR, suggesting that clinicians delivering the treatment should be prepared to manage potential cardiovascular events, should they occur.

An open-label study aimed to assess the safety and efficacy of repeated dose ketamine infusion in the treatment of major depressive disorder [39c]. 14 patients received 6 ketamine infusions over a 3-week period, 3 lower dose infusions (0.5 mg/kg) followed by three high dose (0.75 mg/kg). The primary outcome measure was reduction in the Hamilton Depression Rating Scale (HDRS—28-item depression rating tool), with several secondary outcome measures including tolerability and self-rating of symptoms. Of the participants, 12 received all 6 doses, with 1 patient stopping after infusion 2 due to unpleasant feelings and dissociative symptoms and 1 patient following infusion 4 as they were unable to attend study visits. Following the sixth infusion, 41.7% were deemed responders (≥50% improvement in HDRS) and 16.7% met criteria for remission (HDRS score ≤ 7). The average HDRS score was significantly improved from baseline to completion of the sixth infusion ($P < 0.001$), due to a significant difference between infusion 3 and 6 ($P = 0.002$). This suggests that the higher dose of ketamine gave a greater improvement. Only a single responder experienced sustained response for 6 weeks following completion of the infusions. There were no serious adverse effects reported, though across both doses patients reported visual and auditory disturbances with mild dissociative symptoms, nausea, headaches and drowsiness, though none persisted for >60 minutes following completion of the infusion. All patients were found to have at least a 10 mmHg rise in systolic BP during infusion. The higher dose infusion, there was an increase of reporting of drowsiness, dizziness, shortness of breath or de-realization, with at least one additional patient reporting headache, cough, congestion, panic, bursitis, easy bruising, amnesia or depersonalization whilst receiving the lower dose ketamine. Over all depersonalization was higher in the lower dose infusion period with no significance reported. This study is limited by a lack of blinding, small numbers and heterogeneity in concomitant anti-depressant regimens used by the patients.

This small study aimed to characterize the dose dependent effects of ketamine at sub-anesthetic and anesthetic doses on cortical oscillations and functional connectivity [40c]. Ten healthy ASA 1 volunteers were administered 0.5 mg/kg of ketamine over 40 minutes (depression treatment dose) followed by a break and then a bolus dose of 1.5 mg/kg of ketamine (anesthetic dose) followed by a recovery period. During this process the patient had EEG monitoring at baseline, following each dose and in recovery. Theta bandwidth power increased most dramatically during ketamine at anesthetic doses when compared to sub-anesthetic doses (4.25 ± 1.9 vs 0.60 ± 0.3, $P < 0.001$). In clinical practice, this could be significant as it allows identification of moving away from sub-anesthetic dose ketamine, which is desired in the treatment of depression. Though all patients received both a dose of ondansetron and a scopolamine patch in attempts to minimize nausea and vomiting, 60% experienced nausea and vomiting and 5 of these required further treatment. A single patient required a jaw thrust following loss of consciousness.

A retrospective review of 77 patients with chronic migraine or new daily persistent headache examined the efficacy of ketamine infusion [41c]. All patients had failed previous trials of preventative and acute therapies. Ketamine was initiated at 0.1 mg/kg/h, increased at 0.05 mg/kg/h with a maximum of 1 mg/kg/h (mean 0.53 mg/kg/h) to a maximum of 5 days (mean 4.8 days). Patients were routinely questioned about adverse events and were administered medications (clonidine, benzodiazepines and antiemetics) in attempts to minimize these. Events reported were suicidality ($n=1$), unsteady gait ($n=6$), falls ($n=2$) and deranged liver enzymes ($n=1$). The overall incidence of adverse event was 86%, with diplopia (36%), confusion (25%) and hallucinations (21%) the most common. Pain scores related to the headache (0 – 10) fell from 7.1 to 3.8 ($P<0.0001$) and 71% were classed as acute responders (defined as at least a 2-point improvement in headache severity score, $P<0.0001$), and 27% were sustained responders (2 point improvement sustained to first follow-up visit within 1 month of admission date). Though the mean difference was 2.63 (95% CI 1.92–3.34), this did not achieve significance based on their pre-defined criteria of a 2-point reduction. It is important to highlight that some patients did receive adjuncts to ketamine during the study (neuroleptics 52%, NSAIDs 36% and dihydroergotamine 16%).

This small single-center prospective observational study examined the use of chemical sedation in the ED in agitated patients [42c]. Five different groups in 98 patients were identified: ketamine, haloperidol, midazolam, lorazepam or benzodiazepine plus haloperidol; however, medication doses were not uniform. The primary outcome was time taken to achieve the desired level of sedation and adverse events and changes to vital signs were also recorded. The ketamine group were significantly younger (median 29 years, $P=0.033$) and although fewer patients were agitated at 5, 10 and 15 minutes post-dose in the ketamine group ($P=0.001$, $P<0.001$, $P=0.032$), there was no significant difference in reported time until agitation was subjectively controlled. There was also no significant difference in requirement for re-dosing of medication between the groups ($P=0.199$). Ketamine caused no significant changes in HR or BP; however, 2/23 required intubation. This study is limited by the lack of standardization of dosing and also by its sample size.

Ketamine has been implicated in a case induced central diabetes insipidus [43A].

Case report

An 18-year-old woman with a history of moyamoya disease, type 2 diabetes and stroke, presented with symptoms of a TIA. She was found to have marked stenosis of her left middle cerebral artery (MCA) and evidence of moyamoya progression, requiring a left superficial temporal artery to MCA bypass. The patient was intubated and ventilated and anaesthesia maintained with sevoflurane, propofol, ketamine and remifentanil. One hour into the procedure, the patient became hypertensive (270/100) and bradycardic with an increased urine output and fall in urine osmolality with hypernatraemia (152 mEq/L). The hypertension was controlled with esmolol and nicardipine and a dose of desmopressin was given with good response. The procedure was abandoned and subsequent CT and MRI showed no changes from baseline. The procedure was repeated the following day without ketamine and was completed without any further problems. There have been two previous reported similar cases, suggesting ketamine may inhibit vasopressin release from the neurohypophysis.

A prospective, randomized, double-blind trial ($n=48$) aimed to compare ketamine administered as an IV bolus (0.3 mg/kg over 5 minutes) vs infusion (0.3 mg/kg over 15 minutes) for pain management in the ED, and the associated side effects and depth of sedation [44c]. The feeling of unreality was significantly higher in the push group (92% vs 54%, $P=0.008$) with 46% describing this as very bothersome (17% in infusion group). The IV push group was found to have a significantly greater depth of sedation at 5 minutes, with median RASS of −2 (0, $P=0.01$ in infusion group). Both groups showed a significant decrease in pain scores ($P=0.026$) but no difference between the two groups at any individual time point during the study ($P=0.14$), and no difference in the use of rescue analgesia. There was also no significant difference in vital signs between the two groups. This study is limited as it was a small convenience sample.

Drug Tolerance

A small observational, retrospective, single study in a 14-bed paediatric intensive care (PICU) aimed to use ketamine to reduce opioid tolerance in children ventilated for a prolonged period [45c]. The study examined 32 patients sedated on PICU who exhibited signs of opioid tolerance (defined as the need to double the dose of fentanyl after exposure for >4 days to achieve same effect). Subjects underwent a change in sedation from fentanyl to ketamine for at least 2 days and continuing other sedation at its existing dose. Dosing and outcome measures were based on COMFORT-B (a validated tool for pain assessment and sedation in PICU), nurse interpretation of sedation score (NISS—reflects nursing opinion on level of analgesia and sedation) and the dose of fentanyl required following the switching of ketamine back to fentanyl. With the rotation to ketamine the authors note

COMFORT-B score was significantly lower (16 vs 14.5, $P<0.001$), indicating a more comfortable patient but no patient was considered over-sedated based on their NISS score (0 patients with a NISS of 3). Following the rotation period, the number of under sedated patients fell (9 vs 1). The authors recognize that this is a small study with a diverse patient group (medical, surgical and cardiac PICU) and they were not able to assess the confounding effects of adjuvant drugs on the doses of fentanyl. They also acknowledge that there is concern about the effects of ketamine infusions and the potential neurotoxicity in the developing brain, especially in the high doses used in this study (1.8–6mg/kg/h), but that this study was not designed to address this concern.

Neuromuscular

Muscle rigidity when using ketamine as procedural sedation has been reported [46A].

Case report

A 70-year-old female presenting to the ED with a right ankle fracture following a mechanical fall requiring manipulation. Medical history included renal transplant, hypothyroidism and diabetes. For the procedure, the patient received a dose of IV morphine followed by IV ketamine (1.5mg/kg) and an intra-articular injection of lidocaine (20mL of 1%). After 2 minutes the patient exhibited bilateral lower limb rigidity from the hip down, preventing reduction of the ankle. There was no rigidity above the waist and no respiratory compromise. The rigidity resolved rapidly with 2mg IV midazolam, allowing completion of the procedure. Whilst causality to ketamine was not determined, there have been previous animal studies showing increased muscular tone when using ketamine, especially when co-administered with morphine.

Genotoxicity

This cross-sectional gene pilot study aimed to identify explore the relationship between *CYP2B6*6* and *GRIN2B* (both thought to increase Emergence phenomena (EP)) and the occurrence of emergence phenomena (EP) following ketamine use for procedural sedation [47c]. EP was classified using the Clinician Administered Dissociative State Scale score ≥4 (CADSS—tool to measure dissociation with higher score representing indicating increased severity). ASA 1 or 2 patients undergoing minor orthopaedic or anorectal surgery were recruited ($n=75$), with all decisions regarding anesthetic left to the anesthetic provider. Of note this led to 98% of all patients involved in the study receiving preoperative midazolam, which may have affected the incidence of EP in the ketamine group. 47 patients received ketamine and there was a statistically significant relationship between ketamine administration and episodes of EP ($P=0.003$). There was no significant link between EP and the named genotypes. The odds of EP occurring experiencing EP increased by 1.14 for every 1mg increase ketamine dose administered. There was no significant difference in patient satisfaction scoring between those patients that received ketamine and those that did and the occurrence of EP did not affect patient satisfaction score.

Cardiovascular

A prospective randomized study performed on patients post single vessel coronary-artery bypass grafting aimed to compare ketamine/propofol sedation vs ketamine/dexmedetomidine sedation in this patient group [48c]. 70 patients were randomized to receive ketamine (1mg/kg bolus followed by 0.25mg/kg/h) plus either propofol (1mg/kg bolus+25–50mcg/kg/h) or dexmedetomidine (1mcg/kg bolus+0.2–0.7mcg/kg/h). Both groups received fentanyl for analgesia, titrated to a non-verbal pain score. There was a significant increase in the total dose of fentanyl (41 ± 20 vs 152.8 ± 51, $P<0.0001$), time of weaning (374 ± 20 vs 445.2 ± 21, $P<0.001$) and time to extubation (432 ± 19 vs 504 ± 28, $P<0.0001$) in the propofol group; however there was no difference in duration of ICU stay (44.97 ± 3 vs 43 ± 3, $P=0.13$). Only two side effects were reported: One patient in the dexmedetomidine group developed nausea, and one in the propofol group developed a skin allergy.

A case report of a hypertensive crisis and subsequent cardiac arrest in a 58-year-old female following ketamine use has been reported [49A].

Case report

A 58-year-old woman with a history of type 2 diabetes, hypertension, coronary artery disease, aortic and pulmonary valve replacements and gout presented with a suspected septic joint. The pain was initially managed with opiates followed by a ketamine infusion (0.15mg/kg/h) to a total dose of 30mg. A further dose of ketamine was administered to achieve sedation in order to allow aspiration of the joint (50mg administered). Between 12 and 15 minutes after this dose, the patient became extremely hypertensive, tachycardic, hypoxic and subsequently had a cardiac arrest following intubation. Following a 7-minute arrest, return of spontaneous circulation was achieved. Following doses of furosemide, the patient was extubated

1 day later and serial echocardiograms in the subsequent weeks showed an ejection fraction improve from 20% to >60%. The authors believe this to be a ketamine driven hypertensive emergency resulting in pulmonary edema.

Urinary Tract

A small randomized double-blind study comparing intra-nasal (IN) ketamine to IV morphine in the management of renal colic [50c]. A total of 40 patients were randomized to receive 1mg/kg IN ketamine (plus IV placebo) or 0.1mg/kg IV morphine (plus IN placebo). The primary endpoints were pain scores before and then at 5, 15 and 30 minutes post-administration. Secondary endpoints were adverse reactions or administration of rescue analgesia. ANCOVA analysis was performed as there was significant difference between the baseline pain scores and following this, there was a significant difference only at the 5-minute pain score as morphine showed a lower score (6.1±0.47 vs 6.9±0.47, $P=0.02$). There was no significant difference ($P=0.37$) in the number of patients requiring rescue analgesia in the two groups. All patients experienced at least one side effect, with only patients in the ketamine group experiencing EP ($n=6$ of 20). Other reported side effects in the ketamine group were nausea (10 vs 6) and dizziness (4 vs 6). The authors state that the sample size was not large enough to detect exact drug effects and adverse events.

PROPOFOL

Immunologic

A study aimed to identify the effects of propofol on patients with different ABO blood groups ($n=72$) [51c]. 18 patients with each blood type group, undergoing cholecystectomy, tonsillectomy or spinal surgery, each received propofol target controlled infusions (TCI) as part of their anesthetic induction. Physiological parameters including BIS were recorded at propofol TCI levels of 0, 1, 2, 3 and 4mcg/mL. Once these were achieved, patients were administered fentanyl and rocuronium and intubated prior to surgery. The authors identified those with blood group B had the largest change in HR and MAP at each time point ($P<0.05$) and those with type A blood had the highest BIS values at 3 and 4 mcg/mL ($P<0.05$). The clinical significance of this remains unclear but the authors suggest that blood group may have an impact of the pharmacokinetics and pharmacodynamics of propofol.

Sensory Systems

In ophthalmic surgery, increased secretions are thought to lead to increased endophthalmitis, whilst increased PONV can lead to wound dehiscence and glaucoma [52c]. This study aimed to identify a difference between propofol TIVA alone and propofol plus sevoflurane in the secretions produced during ophthalmic surgery and the amount of PONV in the 24 hours post-surgery. 50 patients were randomized to each group. Both groups were given 5mg of dexamethasone as an anti-emetic, but no patient in either group suffered with PONV (the authors accept that the study was underpowered to detect changes in PONV). Although there were no episodes of endophthalmitis within the 2 weeks following surgery in either group, the volume of secretions was significantly higher in the propofol only group (31±18mL vs 13±12mL, $P<0.001$). Other precautions were taken in both groups to reduce secretions reaching the eye, including monitoring neck flexion, sealed drapes around the eye and testing to see if fluid from the nose could reach the eye during the procedure.

Nervous System

Pain during administration of propofol is well recognized. A double-blind study randomized patients into 3 intervention groups (with a control receiving normal saline) aiming to identify if esmolol given prior to propofol could reduce the incidence of pain on injection [53c]. 120 ASA 1–2 patients undergoing dental extractions under GA were randomized to receive either remifentanil (0.35mcg/kg) or esmolol (either 0.5mg/kg or 1mg/kg). Each group then had a 30-second pause prior to administration of propofol. Pain was assessed on a 4-point scale (0=no pain to 3=severe pain) by a blinded anaesthetist. Pain was significantly reduced overall by all interventions ($P=<0.05$) when compared to control, as was the incidence of severe pain. There was no significant difference between the three intervention groups in terms of reduction in pain, but also in side effects such as emergence reactions, hypotension and bradycardia.

This meta-analysis investigated the effect of adjunctive lidocaine vs other medications in reducing the pain experienced in the paediatric population during propofol administration [54M]. In total, 11 trials were identified, ASA 1–2 children aged between 2 months and 18 years. The drugs were administered prior to propofol in an attempt to limit pain. They found that in the case of lidocaine vs saline ($n=448$), the overall incidence was reduced (22% vs 67%, RR with 95% 0.34) and the incidence of severe pain also reduced in the lidocaine groups (1.6% vs 16%, RR with 95% 0.12). When lidocaine was compared with ketamine or alfentanil, there were no differences in the incidence or the severity of pain. The final studies were lidocaine vs propofol lipuro (propofol with medium and long chain triglycerides), where lidocaine was found to be more efficient in reducing pain (31% vs 47%, RR with 95% 0.68). Three patients out of all patients studied developed a cutaneous rash (2 received saline/propofol and 1 received propofol lipuro). Reported side effects in the lidocaine groups included, one episode of laryngospasm and frequent bouts of coughing. In the alfentanil study, 7 children developed a junctional rhythm and 2 children had transient severe bradycardias. The authors do state that reporting of side effects was sporadic and diverse, making it difficult to describe them in detail.

A case report has described the development of extra-pyramidal symptoms developing after GA [55A].

Case report

A 17-year-old boy with ulcerative colitis required a GA to facilitate colonoscopy. Current medication regime was omeprazole, mesalamine and mercaptopurine, had no personal or family history of reactions under anaesthesia or extra-pyramidal symptoms. He received 200 mg propofol, 60 mg lidocaine and 100 mg suxamethonium as part of a rapid sequence induction and 2% sevoflurane used to maintain anaesthesia. No anti-emetics were administered during the procedure. On waking the patient developed an oculogyric crisis, torticollis, ballismus and tongue thrusting, though he remained awake with no cardiovascular compromise. There was no response to multiple doses of benzodiazepines and after a neurology review, the patient was given diphenhydramine, which resolved most symptoms and any remaining resolved with benztropine. The patient was treated for 2 days with a diphenhydramine infusion followed by a tapering dose and showed no further extra-pyramidal symptoms.

A prospective randomized single-blinded study investigated the effects of volatile anaesthesia (VIMA) vs propofol TIVA on regional cerebral oxygen saturation ($rcSO_2$) during laparoscopic cholecystectomy ($n=124$, 62 per group) [56c]. The study uses near-infrared spectroscopy (NIRS) to allow continuous noninvasive monitoring of $rcSO_2$. Patients were randomized to induction and maintenance with TIVA or sevoflurane, with both groups given rocuronium and sufentanil. Both groups received midazolam (2.5 mg) 30 minutes prior to the procedure and their baseline $rcS0_2$ recording. The $rcSO_2$ was then recorded at designated time points during and after the surgery, with critical falls in $rcSO_2$ (defined as a decrease in 20% from baseline, or 15% decrease if basal value was lower than 50%) noted. Haemodynamic and respiratory parameters were maintained within a set of target values, and there was no statistically significant difference in the use of ephedrine ($P=0.9$) or urapidil ($P=0.06$) between the groups in order to achieve these values. There were no serious post-operative complications identified in either group. There was a significant increase in the incidence of critical declines in $rcSO_2$ in the reverse Trendelenburg position in the TIVA group (19.4% vs 4.8%, $P=0.013$). In addition, a statistically significant increase in $rcSO_2$ was seen following induction in the VIMA group ($P<0.05$) and was thought to be due to the increase in cerebral blood flow caused by volatile anesthetics. Overall the study identified that the VIMA group had a higher $rcSO_2$ at every stage of induction and maintenance of anaesthesia (all $P<0.05$). This study is limited as NIRS only monitors blood flow to the frontal cortex and not all areas of the brain. Also patients were not monitored for post-operative neuro-cognitive changes, leaving an uncertain clinical impact of the study.

A study aiming to identify rates of awareness in patients undergoing surgery using propofol TIVA compared to a combination of inhalational and intravenous anaesthesia (CIIA) was carried out [57C]. 1244 patients were randomized to receive either TIVA at 3–4 mcg/mL or sevoflurane at 2.5%–4% plus 2 mg/kg/h of propofol. The method of induction was left to the discretion of the anaesthetist but is described as "similar". The patients were then screened 48 hours post-operatively for awareness using the modified Brice Interview (screening tool for awareness). In addition, a Mini-Mental State Examination (MMSE) and a psychiatric evaluation (looking at anxiety and depression scores) were carried out at 2 weeks post-operatively. Gender, age, ASA and type of surgery were not significantly different between the two groups. Awareness was significantly higher in the TIVA group (11 vs 3, $P=0.007$) and a much larger number of possible awareness (15 vs 8). Patients with intraoperative awareness were found to have lower post-operative MMSE ($P<0.001$) scores but higher scores in depression and anxiety assessment ($P<0.001$). Through logistic regression analysis, CIIA was found to be protective against awareness ($P=0.026$) as was older age ($P=0.038$) and midazolam ($P=0.019$), whilst previous awareness ($P=0.003$) and duration of surgery ($P=0.039$) were risk factors. Given the low incidence in awareness, the sample size was relatively small and also the time of performing the Brice Interview may lead to an underestimation.

A randomized control trial comparing the use of sedation (propofol and fentanyl) vs interscalene block for reduction of anterior shoulder dislocations in the ED was carried out ($n=60$) [58c]. There were no significant differences in number of attempts at reduction or number of techniques attempted. There was however a significant reduction in LOS in the interscalene block group (80 vs 108 minutes, $P=0.005$), though this group showed higher pain scores (3.4 vs 0.3,

$P<0.001$) and lower patient satisfaction (3.0 vs 3.6, $P<0.001$). 3 patients in the sedation group developed transient hypoventilation following reduction and one patient in the interscalene block group showed symptoms of local anesthetic toxicity.

Drug Dosing Regimens

A retrospective analysis of prospectively randomized patients ($n=173$) compared the use of bolus dose propofol with TCI propofol to allow drug induced sedation endoscopy (DISE—allows for observation of upper airways sedative induced sleep) [59c]. Primary outcome was to achieve and reproduce "snoring-obstructed hypo/apnea-oxygen desaturation breathing", with secondary outcomes based on stability and safety of TCI-DISE (based on changes in BIS and saturation changes >3). Other outcomes looked more specifically at how the DISE performed against awake procedures and on surgical outcomes after diagnosis. All patients had an apnea–hypopnea index (AHI—a system of diagnosing sleep apnea and grading its severity) of 15–30 and a body mass index (BMI) <30. The bolus group received 1 mg/kg propofol followed by 20 mg increments, whereas the TCI group aimed for a plasma level of 1.5 mcg/mL, then increased in increments of 0.2 mcg/mL aiming for a BIS of 40–60. TCI was superior in allowing reproduction of obstructed breathing (80% vs 30%, $P<0.001$) and also a more stable sedation (5% vs 65% variation in saturations >3%, $P=0.0001$). Only patients in the bolus group required supplemental oxygen (8/50 patients) due to desaturation following the bolus down to 61%–65%. None of the patients experienced nausea, vomiting or severe hypotension. The average BIS score during the procedure was similar between groups (73 in bolus vs 78 in TCI).

This multi center retrospective cohort study investigated the efficacy and safety of propofol vs midazolam sedation when used in the ED ($n=592$) [60c]. The authors measured efficacy (procedure success rate) and safety using sedation adverse events (aspiration, laryngospasm, hospitalization, airway obstruction not relieved by simple maneuvers, intubation or death) and sedation events (agitation, vomiting, apnea >20 seconds, hypotension or desaturation <90% for 60 seconds). The propofol group required more supplemental oxygen during the procedure (87% vs 41%, $P<0.001$) and were less likely to receive supplemental opiates (78% vs 91%, $P<0.001$). Procedure completion was lower in the midazolam group (82% vs 92%, $P<0.001$) and was longer (17 vs 10 minutes). No sedation adverse events were observed, though sedation events were more frequent in the propofol group (23% vs 11%, $P<0.001$), almost entirely attributable to the transient apneas; however, clinically significant desaturations were seen more often in the midazolam group.

A systematic review and meta-analysis of trials investigating the efficacy and safety of etomidate and propofol for patients undergoing gastrointestinal endoscopy were performed [61M]. The primary outcome was anaesthesia duration and recovery time, but secondary outcomes included clinical parameters such as HR and MAP as well as side effects such as injection pain, myoclonus and apnea. 6 studies were included, but only one study compared propofol vs etomidate alone, with the others using adjuncts. There was no significant difference in duration of anaesthesia ($P=0.66$) or recovery time ($P=0.47$), although there was significant heterogeneity in the studies comparing recovery times. In terms of secondary outcomes, there was no difference identified with HR, MAP, or SpO_2 at time of intubation or patient satisfaction. In the etomidate group, there was a significant reduction in the incidence of apnea/hypoxia (OR=0.39, 95% CI, 0.24–0.64 $P=0.0002$) and injection pain ($P<0.00001$); however, there was an increase in episodes of myoclonus ($P<0.00001$). This meta-analysis is limited by the number of studies, 3 of which contained less than 100 participants.

A retrospective chart review of 304 infants aged <6 months receiving propofol sedation for radiology imaging reviewing adverse events. No patients received any other sedation prior to propofol (institution guidelines suggested 3 mg/kg induction dose followed by an infusion at 5–6 mg/kg/h) [62C] aiming for deep sedation (depressed consciousness, during which patients cannot be easily aroused but respond purposely to repeated or painful stimuli). Sedationists were able to adjust this dose and administer bolus doses at their discretion, and ultimately the average induction dose was 4.7 mg/kg and the initial infusion rate was 8 mg/kg. A total of 71 adverse events were recorded in 47 patients, with 57 deemed minor (apnea, hypoxia, desaturation, cough, agitation, hypotension, etc.) and 14 serious adverse events (13 airway obstruction and 1 laryngospasm). A total of 80 interventions were required to manage minor events or prevent them occurring. These included CPAP (8%), jaw-thrust (5%) and bag-mask ventilation (3%). No significant risk factors were identified for occurrence of adverse events. In three cases the images were not obtained, but in all other cases the procedure was completed and images deemed satisfactory.

Respiratory

This randomized trial aimed to determine if moderate sedation achieved with alfentanil compared to propofol would result in a decrease in airway or respiratory adverse events leading to intervention [63c]. Patients were given either alfentanil ($n=52$, 10 mcg/kg bolus plus 5 mcg/kg boluses at 3- to 5-minute intervals as required) or propofol ($n=56$, 1 mg/kg plus 0.5 mg/kg at 3- to 5-minute intervals) to achieve moderate sedation. Patients in both groups may have received opiates prior to this as analgesia, though none were given in the 20 minutes prior to administration of sedation. There was no difference in the primary outcome of airway or respiratory adverse event (23% alfentanil vs 20% propofol, $P=0.657$). One patient in the alfentanil group vomited but this did not lead to an adverse event. Although the alfentanil group had a higher incidence of pain (48% vs 13%) and recall of events (75% vs 23%), the statistical significance was not recorded and overall satisfaction is reported as similar (87% vs 84%).

A retrospective case review of 256 children undergoing sedation for MRI investigated dexmedetomidine in addition to propofol to reduce adverse airway events compared to propofol alone [64c]. Both groups received midazolam and EMLA for cannula insertion prior to sedation. In the propofol group (Pro $n=87$), patients received 1–2 mg/kg/min propofol followed by an infusion of 125–150 mcg/kg/min. In the dexmedetomidine group (D+P), patients received dexmedetomidine dose of 1–2 mcg/kg with 0.05 mg/kg midazolam followed by a propofol infusion. Both groups aimed to achieve a sedation score of 3–4 on the Children's Hospital of Wisconsin Sedation Score (ranges from 0 = anxious/agitated, to 6 = Anaesthesia). The D+P group had significantly fewer adverse events (10 vs 23, $P<0.001$), including reduced episodes of airway obstruction (OR 0.23, 95% CI: 0.09–0.58, $P=0.01$) and no episodes of significant desaturation (0 vs 3, $P=0.04$). There were no aborted procedure in the D+P group (vs 4 in the Pro group, $P=0.01$); however, length of stay was longer (87 vs 70 minutes, $P<0.001$).

METHOHEXITAL

Drug Dosing Regimens

A retrospective review investigated of 240 cases of methohexital for procedural sedation in a paediatric population [65c]. The main reasons for this choice of drug were documented allergy to egg+soy/peanut (38.8%), mitochondrial or metabolic disorder (9%) and egg allergy alone in (30%). The majority of these procedures were carried out in the radiology department (79%), with a median initial bolus of 2.1 mg/kg (IQR 1.9–2.8 mg/kg) methohexital and 77.5% of patients receiving an infusion. Adjunctive medication was given at the clinician's discretion (fentanyl 10%, ketamine 5.8%, lidocaine 14.2%) and a combination of adjuncts was used in 5.4%. 20% had a sedation related adverse event, with hiccups (6.3%), secretions needing suctioning (5.8%) and coughing (5.0%) the most common. 12% had a serious adverse event, with 11.7% experiencing airway obstruction and a single case of anaphylaxis requiring admission to critical care. The most common intervention to manage the minor and serious adverse events was bag-mask ventilation (11.6%), placement of a nasopharyngeal airway (7.1%) and oro-pharyngeal suctioning (7.1%). The overall success rate of the procedures performed was 94%. The authors reference previous studies showing that the incidence of minor and serious adverse events is higher in this study than in previous studies using propofol, with a lower procedural success rate.

Additional case studies can be found in these reviews [66R,67R].

References

[1] Smith A, Dunne E, Mannion M, et al. A review of anaesthetic outcomes in patients with genetically confirmed mitochondrial disorders. Eur J Pediatr. 2017;176:83–8 [M].

[2] Matsota PK, Christodoulopoulou TC, Batistaki CZ, et al. Factors associated with the presence of postoperative headache in elective surgery patients: a prospective single center cohort study. J Anesth. 2017;31:225–36 [E].

[3] Geng YJ, Wu QH, Zhang RQ. Effect of propofol, sevoflurane, and isoflurane on postoperative cognitive dysfunction following laparoscopic cholecystectomy in elderly patients: a randomized controlled trial. J Clin Anesth. 2017;38:165–71 [c].

[4] Tazawa K, Koutsogiannaki S, Chamberlain M, et al. The effect of different anesthetics on tumor cytotoxicity by natural killer cells. Toxicol Lett. 2017;266:23–31 [E].

[5] Yoon IJ, Kang H, Baek CW, et al. Comparison of effects of desflurane and sevoflurane on postoperative nausea, vomiting, and pain in patients receiving opioid-based intravenous patient-controlled analgesia after thyroidectomy: propensity score matching analysis. Medicine (Baltimore). 2017;96:e6681 [c].

[6] Karabayirli S, Surgit O, Kasikara H, et al. The effects of adding ischemic preconditioning during desflurane inhalation anesthesia or propofol total intravenous anesthesia on pneumoperitoneum-induced oxidative stress. Acta Chir Belg. 2017;117:36–44 [c].

[7] Kim EH, Song IK, Lee JH, et al. Desflurane versus sevoflurane in pediatric anesthesia with a laryngeal mask airway: a randomized controlled trial. Medicine (Baltimore). 2017;96:e7977 [C].

[8] Stevanovic A, Schaefer P, Coburn M, et al. Renal function following xenon anesthesia for partial nephrectomy—an explorative analysis of a randomized controlled study. PLoS One. 2017;12:e0181022 [c].

[9] Sohn HM, Kim HY, Park S, et al. Isoflurane decreases proliferation and differentiation, but none of the effects persist in human embryonic stem cell-derived neural progenitor cells. J Anesth. 2017;31:36–43 [E].

[10] Whitaker EE, Christofi FL, Quinn KM, et al. Selective induction of IL-1beta after a brief isoflurane anesthetic in children undergoing MRI examination. J Anesth. 2017;31:219–24 [c].

[11] Xu R, Zhou S, Yang J, et al. Total intravenous anesthesia produces outcomes superior to those with combined intravenous-inhalation anesthesia for laparoscopic gynecological surgery at high altitude. J Int Med Res. 2017;45:246–53 [c].

[12] Fink C, Uhlmann L, Enk A, et al. Pain management in photodynamic therapy using a nitrous oxide/oxygen mixture: a prospective, within-patient, controlled clinical trial. J Eur Acad Dermatol Venereol. 2017;31:70–4 [c].

[13] Hoeffe J, Doyon Trottier E, Bailey B, et al. Intranasal fentanyl and inhaled nitrous oxide for fracture reduction: the FAN observational study. Am J Emerg Med. 2017;35:710–5 [c].

[14] Goo EK, Lee JS, Koh JC. The optimal exhaled concentration of sevoflurane for intubation without neuromuscular blockade using clinical bolus doses of remifentanil: a randomized controlled trial. Medicine (Baltimore). 2017;96:e6235 [c].

[15] Jin S, Liang DD, Chen C, et al. Dexmedetomidine prevent postoperative nausea and vomiting on patients during general anesthesia: a PRISMA-compliant meta analysis of randomized controlled trials. Medicine (Baltimore). 2017;96:e5770 [M].

[16] Nasseri K, Ghadami N, Nouri B. Effects of intrathecal dexmedetomidine on shivering after spinal anesthesia for cesarean section: a double-blind randomized clinical trial. Drug Des Devel Ther. 2017;11:1107–13 [c].

[17] Park SJ, Shin S, Kim SH, et al. Comparison of dexmedetomidine and fentanyl as an adjuvant to ropivacaine for postoperative epidural analgesia in pediatric orthopedic surgery. Yonsei Med J. 2017;58:650–7 [c].

[18] Zhang X, Wang D, Shi M, et al. Efficacy and safety of dexmedetomidine as an adjuvant in epidural analgesia and anesthesia: a systematic review and meta-analysis of randomized controlled trials. Clin Drug Investig. 2017;37:343–54 [M].

[19] El-Boghdadly K, Brull R, Sehmbi H, et al. Perineural dexmedetomidine is more effective than clonidine when added to local anesthetic for supraclavicular brachial plexus block: a systematic review and meta-analysis. Anesth Analg. 2017;124:2008–20 [M].

[20] Lam RP, Yip WL, Wan CK, et al. Dexmedetomidine use in the ED for control of methamphetamine-induced agitation. Am J Emerg Med. 2017;35:665.e1–4 [A].

[21] Gao Y, Kang K, Liu H, et al. Effect of dexmedetomidine and midazolam for flexible fiberoptic bronchoscopy in intensive care unit patients: a retrospective study. Medicine (Baltimore). 2017;96: e7090 [c].

[22] Gong M, Man Y, Fu Q. Incidence of bradycardia in pediatric patients receiving dexmedetomidine anesthesia: a meta-analysis. Int J Clin Pharm. 2017;39:139–47 [M].

[23] Jo YY, Kim HS, Lee KC, et al. CONSORT the effect of intraoperative dexmedetomidine on hemodynamic responses during emergence from nasotracheal intubation after oral surgery. Medicine (Baltimore). 2017;96:e6661 [c].

[24] Fan W, Yang H, Sun Y, et al. Comparison of the pro-postoperative analgesia of intraoperative dexmedetomidine with and without loading dose following general anesthesia: a prospective, randomized, controlled clinical trial. Medicine (Baltimore). 2017;96: e6106 [c].

[25] Li X, Wang X, Jin S, et al. The safety and efficacy of dexmedetomidine-remifentanil in children undergoing flexible bronchoscopy: a retrospective dose-finding trial. Medicine (Baltimore). 2017;96:e6383 [c].

[26] Zhang H, Fang B, Zhou W. The efficacy of dexmedetomidine-remifentanil versus dexmedetomidine-propofol in children undergoing flexible bronchoscopy: a retrospective trial. Medicine (Baltimore). 2017;96:e5815 [c].

[27] Kim KN, Lee HJ, Kim SY, et al. Combined use of dexmedetomidine and propofol in monitored anesthesia care: a randomized controlled study. BMC Anesthesiol. 2017;17:34 [c].

[28] Liu X, Xie G, Zhang K, et al. Dexmedetomidine vs propofol sedation reduces delirium in patients after cardiac surgery: a meta-analysis with trial sequential analysis of randomized controlled trials. J Crit Care. 2017;38:190–6 [M].

[29] Hashemian M, Ahmadinejad M, Mohajerani SA, et al. Impact of dexmedetomidine on hemodynamic changes during and after coronary artery bypass grafting. Ann Card Anaesth. 2017;20:152–7 [c].

[30] Bhatt M, Johnson DW, Chan J, et al. Risk factors for adverse events in emergency department procedural sedation for children. JAMA Pediatr. 2017;171:957–64 [c].

[31] Smits GJ, Kuypers MI, Mignot LA, et al. Procedural sedation in the emergency department by Dutch emergency physicians: a prospective multicentre observational study of 1711 adults. Emerg Med J. 2017;34:237–42 [c].

[32] Akbulut UE, Saylan S, Sengu B, et al. A comparison of sedation with midazolam-ketamine versus propofol-fentanyl during endoscopy in children: a randomized trial. Eur J Gastroenterol Hepatol. 2017;29:112–8 [c].

[33] Carvalho DZ, Townley RA, Burkle CM, et al. Propofol frenzy: clinical spectrum in 3 patients. Mayo Clin Proc. 2017;92:1682–7 [A].

[34] Peeters B, Guiza F, Boonen E, et al. Drug-induced HPA axis alterations during acute critical illness: a multivariable association study. Clin Endocrinol. 2017;86:26–36 [c].

[35] Liu J, Liu R, Meng C, et al. Propofol decreases etomidate-related myoclonus in gastroscopy. Medicine (Baltimore). 2017;96:e7212 [C].

[36] Driver BE, Reardon RF. Apnea after low-dose ketamine sedation during attempted delayed sequence intubation. Ann Emerg Med. 2017;69:34–5 [A].

[37] Patterson AC, Wadia SA, Lorenz DJ, et al. Changes in blood pressure and heart rate during sedation with ketamine in the pediatric ED. Am J Emerg Med. 2017;35:322–5 [c].

[38] Sanacora G, Frye MA, McDonald W, et al. A consensus statement on the use of ketamine in the treatment of mood disorders. JAMA Psychiat. 2017;74:399–405 [M].

[39] Cusin C, Ionescu DF, Pavone KJ, et al. Ketamine augmentation for outpatients with treatment-resistant depression: preliminary evidence for two-step intravenous dose escalation. Aust N Z J Psychiatry. 2017;51:55–64 [c].

[40] Vlisides PE, Bel-Bahar T, Lee U, et al. Neurophysiologic correlates of ketamine sedation and anesthesia: a high-density electroencephalography study in healthy volunteers. Anesthesiology. 2017;127:58–69 [c].

[41] Pomeroy JL, Marmura MJ, Nahas SJ, et al. Ketamine infusions for treatment refractory headache. Headache. 2017;57:276–82 [c].

[42] Riddell J, Tran A, Bengiamin R, et al. Ketamine as a first-line treatment for severely agitated emergency department patients. Am J Emerg Med. 2017;35:1000–4 [c].

[43] Gaffar S, Eskander JP, Beakley BD, et al. A case of central diabetes insipidus after ketamine infusion during an external to internal carotid artery bypass. J Clin Anesth. 2017;36:72–5 [A].

[44] Motov S, Mai M, Pushkar I, et al. A prospective randomized, double-dummy trial comparing IV push low dose ketamine to short infusion of low dose ketamine for treatment of pain in the ED. Am J Emerg Med. 2017;35:1095–100 [c].

[45] Neunhoeffer F, Hanser A, Esslinger M, et al. Ketamine infusion as a counter measure for opioid tolerance in mechanically ventilated children: a pilot study. Paediatr Drugs. 2017;19:259–65 [c].

[46] Vien A, Chhabra N. Ketamine-induced muscle rigidity during procedural sedation mitigated by intravenous midazolam. Am J Emerg Med. 2017;35:200.e3–4 [A].

[47] Aroke EN, Crawford SL, Dungan JR. Pharmacogenetics of ketamine-induced emergence phenomena: a pilot study. Nurs Res. 2017;66:105–14 [c].

[48] Mogahd MM, Mahran MS, Elbaradi GF. Safety and efficacy of ketamine-dexmedetomidine versus ketamine-propofol combinations for sedation in patients after coronary artery bypass graft surgery. Ann Card Anaesth. 2017;20:182–7 [c].

[49] Burmon C, Adamakos F, Filardo M, et al. Acute pulmonary edema associated with ketamine-induced hypertension during procedural sedation in the ED. Am J Emerg Med. 2017;35:522.e1–4 [A].

[50] Farnia MR, Jalali A, Vahidi E, et al. Comparison of intranasal ketamine versus IV morphine in reducing pain in patients with renal colic. Am J Emerg Med. 2017;35:434–7 [c].

[51] Du Y, Shi H, Yu J. Comparison in anesthetic effects of propofol among patients with different ABO blood groups. Medicine (Baltimore). 2017;96:e5616 [c].

[52] Lai HC, Chang YH, Huang RC, et al. Efficacy of sevoflurane as an adjuvant to propofol-based total intravenous anesthesia for attenuating secretions in ocular surgery. Medicine (Baltimore). 2017;96:e6729 [c].
[53] Lee M, Kwon T, Kim S, et al. Comparative evaluation of the effect of remifentanil and 2 different doses of esmolol on pain during propofol injection: a double-blind, randomized clinical consort study. Medicine (Baltimore). 2017;96:e6288 [c].
[54] Lang BC, Yang CS, Zhang LL, et al. Efficacy of lidocaine on preventing incidence and severity of pain associated with propofol using in pediatric patients: a PRISMA-compliant meta-analysis of randomized controlled trials. Medicine (Baltimore). 2017;96:e6320 [M].
[55] Sherer J, Salazar T, Schesing KB, et al. Diphenhydramine for acute extrapyramidal symptoms after propofol administration. Pediatrics. 2017;139:e20161135 [A].
[56] Ruzman T, Simurina T, Gulam D, et al. Sevoflurane preserves regional cerebral oxygen saturation better than propofol: randomized controlled trial. J Clin Anesth. 2017;36:110–7 [c].
[57] Yu H, Wu D. Effects of different methods of general anesthesia on intraoperative awareness in surgical patients. Medicine (Baltimore). 2017;96:e6428 [C].
[58] Raeyat Doost E, Heiran MM, Movahedi M, et al. Ultrasound-guided interscalene nerve block vs procedural sedation by propofol and fentanyl for anterior shoulder dislocations. Am J Emerg Med. 2017;35:1435–9 [c].
[59] De Vito A, Agnoletti V, Zani G, et al. The importance of drug-induced sedation endoscopy (D.I.S.E.) techniques in surgical decision making: conventional versus target controlled infusion techniques—a prospective randomized controlled study and a retrospective surgical outcomes analysis. Eur Arch Otorhinolaryngol. 2017;274:2307–17 [c].
[60] Lameijer H, Sikkema YT, Pol A, et al. Propofol versus midazolam for procedural sedation in the emergency department: a study on efficacy and safety. Am J Emerg Med. 2017;35:692–6 [c].
[61] Ye L, Xiao X, Zhu L. The comparison of etomidate and propofol anesthesia in patients undergoing gastrointestinal endoscopy: a systematic review and meta-analysis. Surg Laparosc Endosc Percutan Tech. 2017;27:1–7 [M].
[62] Jenkins E, Hebbar KB, Karaga KK, et al. Experience with the use of propofol for radiologic imaging in infants younger than 6 months of age. Pediatr Radiol. 2017;47:974–83 [C].
[63] Miner JR, Driver BE, Moore JC, et al. Randomized clinical trial of propofol versus alfentanil for moderate procedural sedation in the emergency department. Am J Emerg Med. 2017;35:1451–6 [c].
[64] Boriosi JP, Eickhoff JC, Klein KB, et al. A retrospective comparison of propofol alone to propofol in combination with dexmedetomidine for pediatric 3T MRI sedation. Paediatr Anaesth. 2017;27:52–9 [c].
[65] Jones NE, Kelleman MS, Simon HK, et al. Evaluation of methohexital as an alternative to propofol in a high volume outpatient pediatric sedation service. Am J Emerg Med. 2017;35:1101–5 [c].
[66] Jha A, Flockton E. General anaesthetics and therapeutic gases. In: Ray SD, editor. Side effects of drugs annual: a worldwide yearly survey of new data in adverse drug reactions, vol. 38. Elsevier; 2016. p. 87–95 [chapter 9]. [R].
[67] Fawkner J, Hall A. General anesthetics and therapeutic gases. In: Ray SD, editor. Side effects of drugs annual: a worldwide yearly survey of new data in adverse drug reactions, vol. 39. Elsevier; 2017. p. 111–21 [chapter 9]. [R].

CHAPTER

11

Local Anesthetics

Henry L. Nguyen[1]

Department of Dermatology, Mayo Clinic, Rochester, MN, United States
[1]Corresponding author: Nguyen.Henry@mayo.edu

INTRODUCTION

Local anesthetics (LA) today are widely used by many clinicians, including dentists, surgeons, anesthesiologists, specialists such as dermatologists, nurses, and others. Therefore, it is important to understand the potential side effects these local anesthetics can cause, which is known as local anesthetic systemic toxicity (LAST) [1R]. The clinical presentation of LAST usually begins right after the administration of an LA. The patient will start experiencing neurological side effects as the LA spreads through the nerves. Symptoms include tongue numbness, metallic taste, lightheadedness, dizziness, visual/auditory disturbances, disorientation, and sleepiness. This is followed by cardiopulmonary symptoms such as hypotension, arrhythmia, and bradycardia, ultimately ending in cardiopulmonary arrest [2R].

The origin of LAST can be traced back to when cocaine was first used as a local anesthetic. Doctors were aware of its potential for addiction and systemic toxicity and so a safer anesthetic was subsequently produced. A further discussion of the anesthetic procaine and LAST is described in the cited paper [3R].

A study on the knowledge base of 104 ophthalmologists regarding LAST and intravenous lipid emulsion (ILE) therapy was conducted in Turkey at various hospitals [4c]. The study showed that bupivacaine was the most commonly used local anesthesia (97.1%) and that the most commonly seen symptoms of LAST included allergy (76%) and hypotension (68.3%) with cardiac arrest during the initial response (57.4%) and hepatotoxicity (56.4%) during the late response. However, 65 of the participating ophthalmologists have never encountered LAST in their medical career, indicating that this adverse effect is not very common. Furthermore, only one person has ever used ILE to treat a patient with LAST.

As mentioned previously, the incidence and prevalence of LAST are rare and recent retrospective review studies of data from big registries or institutions provide a more clear view of the current trend. A retrospective review of 710 327 patients from the US National Inpatient Sample database who underwent hip, knee, or shoulder arthroplasty during 1998–2013 and were administered peripheral nerve block found the incidence to be 1.04 per 1000 peripheral nerve blocks [5C]. Shoulder arthroplasty has a LAST incidence odds ratio of 4.35 over that of hip or knee arthroplasty. The knee arthroplasty has a LAST incidence odds ratio of 0.75 over those of the hip arthroplasty, but this result is not statistically significant ($P=0.452$). Interestingly, large and medium-size hospitals have a LAST incidence odds ratio of 2.40 and 3.35 over that of small hospitals, respectively. Over the 15 years study period, the odds of LAST development decreased by 10% when compared to the beginning of the study period.

Another study of 238 473 eligible patients who received peripheral nerve blocks for total joint arthroplasty from 2006 to 2014 found the incidence of LAST onset to be 1.8 per 1000 peripheral nerve blocks [6C]. Similar to the previous study, there was a decrease in the incidence of LAST from 8.2 per 1000 in 2006 to 2.5 per 1000 in 2014. This change in incidence rate was accompanied by an increased in lipid emulsion therapy incidence from 0.2 per 1000 to 2.6 per 1000 over this period.

A third study examined the incidence of LAST in 80 661 patients who received peripheral nerve blocks at a single institution from 2009 to 2014 [7C]. The primary outcome for LAST was defined as cardiac arrest within 60 min of the peripheral nerve block and the use of lipid emulsion rescue. The secondary outcome was defined as a seizure that lasts 60 s or more within 60 min of the peripheral nerve block. No cases of cardiac arrest occurred

Side Effects of Drugs Annual, Volume 40
ISSN: 0378-6080
https://doi.org/10.1016/bs.seda.2018.07.017

and three cases of seizures were reported, making the total incidence of LAST to be 0.04 per 1000 peripheral nerve blocks.

A Danish study looked at the incidence of local anesthetic (LA) allergy at The Danish Anesthesia Allergy Centre (DAAC) from 2004 to 2013 and found zero cases of LA allergy [8C]. A total of 409 patients were suspected to have LA allergy and 162 patients were tested for LA. All the test results were negative: subcutaneous provocation test to lidocaine ($n=80$), bupivacaine ($n=82$), ropivacaine ($n=31$), and mepivacaine ($n=10$). It was concluded that the occurrence of LA allergy is very rare in the Danish population.

Similarly, the incidence of an immediate-type allergy to local anesthetics in the Danish population is also very rare as shown in a retrospective review [9C]. A total of 168 patients were tested with 195 subcutaneous provocations to LA including lidocaine, mepivacaine, articaine, prilocaine, bupivacaine, and ropivacaine. 164 patients all tested negative for subcutaneous provocations to LA. As a result, immediate-type allergy in the Danish population appears to be very rare.

Chloroprocaine

This literature review looked at the use of 1.5% 2-chloroprocaine infusing at 0.25–1.5mL/kg/h with a mean duration of 50h for regional anesthesia in infants [10R]. The author argued that using local anesthetic is safer than systemic medications because it allowed for earlier tracheal extubation, decreased perioperative stress response, shorter time for the bowel to return to normal and decreased the long-term neurocognitive effects due to local anaesthetics cytotoxicity in this population because children up to 1-year-old have decreased hepatic P450 enzyme function. The author concluded that chloroprocaine is safe to use in epidural infusions for intraoperative anesthesia and postoperative analgesia in infants and children.

This case report is of a 36-year-old woman who was 37 weeks pregnant who presented to the hospital requesting epidural analgesia for her delivery [11A]. She had a history of lidocaine allergy that presented with swelling, pruritus, and hives at the injection site and the surrounding tissues. After a literature search, it was determined to give her an ester anesthetic 2-chloroprocaine. A test dose of 12mL of 1.5% 2-chloroprocaine and 100μg of fentanyl was given and there were no reactions. Therefore, an infusion of 10mL/h of 1.5% 2-chloroprocaine was started for her epidural analgesia for a total of 200min. The authors concluded that, despite not being widely used, 2-chloroprocaine is a safe and effective medication in continuous epidural analgesia for patients with allergy to amide local anesthetics.

Dibucaine

A 7-year-old boy presented to the emergency room with giddiness, altered sensorium, right bundle branch block, arrhythmia, and diplopia after having accidentally been fed 10% ear drop solution containing 0.1% (w/v) of dibucaine [12A]. The amount consumed was equivalent to 0.5mg/kg. ILE was not administered but kept close at hand since the arrhythmia did not lead to a hemodynamic instability. After 72h of close observation in the pediatric intensive care unit, the child was transferred to a normal ward and discharged home on the 5th day at the hospital. This report shows that despite being rarely used as a current treatment choice, dibucaine is a potent anesthetic that can cause LAST.

Lidocaine

A case report described a 52-year-old man who suffered from cervical radiculopathy had a C6-C7 Isovue contrast agent assisted (10mg of dexamethasone and 2mL of preservative-free 1% lidocaine) injection [13A]. Within 5min of entering the post-anesthesia care unit (PACU), the man was unable to swallow, had dizziness, and horizontal nystagmus and it was suggested that the local anesthesia may have affected the glossopharyngeal, abducen, and vestibulocochlear cranial nerves, respectively. After 30min of recovery within the PACU, the symptoms resolved.

A 33-year-old woman with no significant past medical history checked into the surgical unit for an elective nasal septoplasty due to a deviated nasal septum [14A]. She was intubated with propofol and succinylcholine and then injected with 60mL of 2% lidocaine/1% epinephrine subcutaneously in the nasal mucosa. Afterward, bradycardia occurred followed by a pulseless electrical activity. Cardiopulmonary resuscitation was performed for 20min until LAST was suspected. She was then given 100mL of 20% ILE and her sinus rhythm returned to normal within 3min of the ILE infusion.

A 66-year-old man with metastatic melanoma and no knowledge of lidocaine allergy underwent further diagnostic evaluation for a thyroid nodule with fine needle aspiration [15A]. A dose of 1% lidocaine was injected into the anterior neck. Soon afterward, tachycardia and hypotension developed that were attributed to LAST. The biopsy procedure was promptly aborted and the patient was transferred to the intensive care unit for monitoring where he received intravenous fluid to stable his tachycardia and hypotension; no ILE was given. The patient fully recovered and was discharged home the next day.

A 30-year-old woman presented to the hospital in a spontaneous labor for delivery [16A]. She was given a test dose of 3mL of lidocaine 1.5% with epinephrine for a lumbar epidural which was uneventful. The patient was then given bupivacaine 0.0625% with fentanyl 2μg/mL for epidural analgesia. After 10h of labor, deep variable deceleration was encountered and the patient was switched to an urgent non-emergent cesarean delivery for arrest of dilation. A dose of 15mL lidocaine 2% with epinephrine was administered for continued epidural analgesia, but this was not sufficient so 15mL of 3% chloroprocaine was administered epidurally. The patient was then put under general anesthesia for the cesarean section with propofol, succinylcholine, and cricoid pressure. After the procedure, angioedema was noted in the lip and tongue area, but the tryptase level was within normal limits. An allergy consult for hereditary angioedema was considered but all tests were within normal limits. The patient returned to the hospital for allergy testing 6 weeks post-delivery for allergy testing consisting of a simultaneous subcutaneous challenge with chloroprocaine and preservative-free lidocaine. This testing resulted in nasal congestion, a hoarse voice, relative hypotension (90s/60s), and angioedema of the lips and oropharynx consistent with an IgE-mediated anaphylactic reaction. A subcutaneous challenge with bupivacaine resulted in no symptoms. This case demonstrated a potential for anaphylaxis reaction due to local anesthetics used.

Mepivacaine

A case report of a 16-year-old boy who had a history of reddening of the face, pruritus of neck, elbows, and ankles after the administration of Scandonest (mepivacaine) on two separate occasions was referred for further evaluation [17A]. Patch tests were performed using intradermal tests with 1:10 dilution with Scandonest 3% which resulted in a patch of 18mm in the test site. Lignocaine 1.0% and bupivacaine 0.5% were also tested to determine if there were any other dermal allergic reactions and the results were negative. Based on the test results, it was decided that lignocaine and bupivacaine were safe alternative anesthetics to use.

Another case report described an 11-year-old girl who experienced intense muscle twitching after receiving a mandibular nerve block with 1.8mL of 3% mepivacaine [18A]. ILE was immediately administered over a 7h period with a total dose of 3670mL (66mL/kg) of ILE. Her serum triglyceride concentration 2h after stopping ILE was estimated to be 16583mg/dL (a very high value). Her symptoms included hypersomnolence, tachypnea, tachycardia, emesis, headaches, metabolic acidosis, and hyperlactatemia and were resolved within 3 days with discharge the next day. This case demonstrates that while ILE is an effective treatment for LAST, too great an ILE infusion can lead to adverse side effects as well.

Efficacy and Safety of Lidocaine in the Older Population

A narrative review of 12 studies from January 1956 to May 2016 assessed the safety and efficacy of lidocaine in the population who are over 60 years old with a total of 308 cases of systemic lidocaine use [19R]. The limitations of this review include both the small sample size and the lack of data focused solely on patients of 60 years or older; although the average age in the studies was greater than 60 years old, the data included patients who were under 60 years old. The conclusion of the review was that lidocaine was safe to use in short duration (1.33–3mg/kg/h until a plasma concentrations of 2–5μg/mL is reached) for the older population undergoing abdominal, urological surgery, or for opioid refractory pain in malignancy, neuropathic pain, and critical limb ischemia. Three studies included the toxic blood levels of lidocaine, but no adverse side effects were reported at these drug concentrations.

Lignocaine

A 19-year-old man was admitted to the dental and oral surgery department at a hospital in India for bony socket refilling [20A]. Prior to the procedure, lignocaine skin sensitivity testing was negative and 7.5mL of 2% lignocaine with adrenaline was, therefore, injected into his lower lip. The patient was awake and alert throughout the socket refilling; however, he was found unconscious and unresponsive to any verbal or pain stimulus after the procedure. He was then transferred to the intensive care unit (ICU) for observation and treatment where the vital signs were normal. During a discussion with the family, it was discovered that he had a similar incident 2 months earlier under local anesthesia at another dental surgery center, but this event was not revealed prior to this admission. After 4h in the ICU, he was given 200mL of 20% ILE and regained consciousness after 30min but was still drowsy. A second bolus of ILE was given after another 2h and he regained full consciousness. This report highlights a LAST case where the patient lost full consciousness with lignocaine, an adverse effect that is not typical of other LAST reactions.

Bupivacaine Liposome Formulation

The local anesthetic bupivacaine has been on the market since 1963 and is considered a very good anesthetic agent. The FDA approved a new bupivacaine injectable suspension for use called Exparel (bupivacaine liposome) on October 28, 2011. This study used data from the FDA Adverse Event Reporting System database from January 01, 2012 to June 30, 2016, to determine the association between LAST and the new bupivacaine formulation Exparel [21C]. The study found that LAST is statistically significantly associated with Exparel (Chi-squared = 596.66 and Proportional Reporting Ratio = 6.23 [95% CI: 5.41–7.18]). In comparison, the old formulation of bupivacaine HCl also has a statistically significant association with LAST (Chi-squared = 599.98 and Proportional Reporting Ratio = 3.36 [95% CI: 3.05–3.69). The increased in association of Exparel and LAST could be dose related because the recommended dose of 266 mg of Exparel is at least 20% higher than the dosage of bupivacaine HCl. However, this dose-related relationship was never tested. Physicians who use this new bupivacaine formulation Exparel should be aware that it has the increased potential to cause LAST due to its longer duration of action and higher dosage and be prepared to treat it if an incident should arise.

Anesthetics Liposome Delivery System

A review article discussed the benefit of using nanoliposomes as a drug delivery system for local anesthetics [22R]. Such benefits include efficient pain control, rapid patient recovery, increased patient comfort, treatment costs reduction, reduced systemic pain, neurotoxicity, and shortened length of hospitalization.

This paper discussed a new delivery method of using a phospholipid-based phase transition gel (PPTG) injection to administer ropivacaine for anesthesia in order to prolong its effect and to decrease LAST [23E]. Administering the ropivacaine-PPTG in rats led to a smaller release of ropivacaine in vivo than administering ropivacaine by itself. The injection of ropivacaine-PPTG in the nerve of guinea pigs leads to nerve blockade that lasted three times longer than using ropivacaine alone. This delivery system could be a new method to have a sustained delivery of local anesthetic while reducing its systemic toxicity.

Another study used a combined liposomal system composed of two types of liposomal vesicles with transmembrane ionic gradients to trap the ropivacaine inside these vesicles for drug delivery [24E]. An in vitro dialysis study showed that this combined liposomal system was able to prolong the release of ropivacaine over 72 h vs the release times for conventional liposomal system of 4 h. This combined liposomal system was also found to be less toxic to 3T3 cells (mouse fibroblast cell line). This new drug delivery system shows promise due its reduced in vitro toxicity and extended anesthesia time.

Lipid Emulsion Rescue Therapy

A systematic review of 160 cases examined the efficacy of lipid emulsion therapy for oral poisoning and LAST and suggested that there may be a benefit in using lipid rescue in cases of LAST in the clinical settings but not enough evidence is available to support its uses for oral poisoning [25M]. In 130 cases (94 were for oral poisoning and 36 were cases with LAST), lipid emulsion treatment showed a positive effect. However, the evidence of using lipid emulsion to treat oral poisoning was inconclusive since 91% of these cases had a confounding effect due to the simultaneous use of other resuscitative measures. Similarly, there was weak evidence to support its use in LAST as well since other resuscitative measures were also used; albeit stronger justification than for oral poisoning. These findings are consistent with various other reviews [26R, 27R, 28M]. As a result, using lipid rescue for LAST should be limited to the clinical setting.

The mechanism of how lipid emulsion therapy works has only been theorized. One theory suggests that the local anesthetic is bound by lipids in the bloodstream and prevents it from affecting local tissues [29R]. Another theory suggests that the lipid emulsion carries the local anesthetic away from the central nervous system and the heart to be metabolized by the liver.

Another review raised similar concerns about the benefit of using ILE in the treatment of LAST [30R]. As the mechanism of how ILE works is still theoretical, more high quality human clinical data are needed to make a clear medical recommendation. Despite this limitation, lipid emulsion therapy is still the first line treatment for LAST, especially in the case of bupivacaine toxicity.

A cohort study examined the ability for ILE to hemodynamically stabilize the cardiovascular system in medication overdose poisoning [31C]. The study included 36 patients with toxicity due to administration of calcium channel blockers, beta-blockers, tricyclic antidepressants, bupropion, and antiepileptic agents. The primary outcome was the change in mean arterial pressure (MAP) 60 min after ILE administration. A total of 25 patients from the initial cohort survived their overdose incidents. The average MAP increased by 13.79 mmHg which was not clinically significant. It is still unclear if ILE is an effective treatment for the cardiovascular system toxicity in drug overdose.

A case report of a 25-year-old woman who had pulseless electrical activities of the heart due to amitriptyline and propranolol overdose was able to recover following ILE administration [32A]. Despite not being LAST related, it showed the efficacy of ILE in reviving the myocytes following amitriptyline and propranolol toxicity.

A meta-analysis of 16 studies using lipid emulsion therapy in animal models (pigs, dogs, rabbits, rats) showed that this treatment improves the survival rates of animals with LAST [33M]. The analysis indicated that lipid emulsion reduced the odds of death in resuscitative models by 76% but had no effect on survival benefit in cases of asphyxial arrest, lack of CPR, or a delay in treatment.

A case report of an 8-month-old foal (young horse) had an accidental overdose of lidocaine injection [34A]. During the anesthesia procedure, the foal had a drop in mean arterial pressure of 20 mmHg, end-tidal carbon dioxide tension and the ECG wave tracings disappeared, denoting cardiovascular collapse. Lipid emulsion rescue therapy was immediately administered and the blood pressure and heart rate slowly normalized. The foal recovered after 24 h and was able to stand up unassisted. This report indicates that intravenous lipid emulsion therapy may be helpful in the treatment of lidocaine overdose toxicity in a foal.

A rat comparative study investigated ILE treatment in reversing cardiac arrests in the systemic toxicity of levobupivacaine (high lipophilicity) and ropivacaine (low lipophilicity) [35E]. Levobupivacaine 0.2% and ropivacaine 0.2% were injected into rats at a rate of 2 mg/kg/min to induce cardiac arrests. ILE was then administered with chest compression to revive the rats. The heart rate and mean arterial pressure (MAP) of rats who were administered levobupivacaine recovered faster and had higher heart rates and MAPs than the control group. The rats in the ropivacaine also recovered but their heart rates and MAPs were the same as the control group. It is likely that the lipophilicity of the different local anesthetics influences the efficacy of lipid infusion when treating cardiac arrest caused by these two drugs; specifically, ILE is more effective in levobupivacaine-induced cardiac arrests due to its high lipophilicity than that of ropivacaine.

Protective Effect of Intravenous Lipid Emulsion on Mitochondria and Myocytes

This study examined the effect of lipid emulsion administration on myocytes and mitochondria of rat H9C2 myocytes with bupivacaine-induced cardiac arrest [36E]. Lipid emulsion treatment reversed the bupivacaine-induced inhibition of the mitochondrial function and, thereby, regulating calcium ion concentrations in the mitochondria. These actions by ILE, therefore, protected the myocardial cells from toxic effects caused by bupivacaine treatment.

N-Acetyl Cysteine and Chondrocytes

This study examined whether the antioxidant, *N*-acetyl cysteine (NAC), has a cytoprotective effect on human chondrocytes from toxicity caused by local anesthetics (LA) [37E]. There were four study arms: a control group, a NAC group (10 mM NAC was exposed to chondrocytes for 1 h), an LA group, and a NAC-LA group. Cell death was induced in the LA group and the NAC-LA group with exposure to ropivacaine (0.075%), bupivacaine (0.05%), or lidocaine (0.2%) for 24 h. Each group was evaluated for rates of cell viability, apoptosis, necrosis, intracellular reactive oxygen species (ROS) production, and caspase-3/7 activity. The results showed that cell viability in the NAC-LA group was significantly higher than in the LA group ($P<0.001$). The mean percentages of apoptotic/necrosis cells in NAC-LA group were lower than those in the LA groups ($P\leq 0.002$). Intracellular ROS levels were higher in the LA groups than the control or NAC-LA groups (Ropivacaine, $P=0.023$; Bupivacaine, $P=0.002$; Lidocaine, $P=0.008$). Caspase-3/7 activity in the control (ropivacaine and bupivacaine: $P\leq 0.001$) and NAC-LA ($P\leq 0.004$) groups was much lower than the LA groups. These results indicate that just exposing human chondrocytes to 10 mM of NAC for 1 h prior being exposed to local anesthetic agents have a protective effect for these cells in vitro.

L-Carnitine and Bupivacaine Cardiotoxicity

This rat study examined the role of L-carnitine in reversing bupivacaine-induced cardiotoxicity [38E]. Pre-administration of 100 mg/kg L-carnitine intravenously prior to immediate bupivacaine-induced cardiotoxicity increased the chance of survival in the rats by lowering the time to asystole by 33% ($P<0.001$), first dysrhythmia by 65% ($P<0.001$), 50% reduction in heart rate by 71% ($P<0.001$), and 50% reduction in mean arterial pressure by 63% ($P<0.001$). In an additional study phase, the rats were divided into four groups ($n=10$ per group) for four distinct treatment regimens: 30% lipid emulsion; L-carnitine; 30% lipid emulsion plus L-carnitine; and saline. No animals in the saline or L-carnitine groups were successfully resuscitated. All animals in the lipid emulsion and 6 of 10 animals in the lipid + L-carnitine were successfully resuscitated. The probabilities that rats were successfully resuscitated were much higher in the lipid receiving groups vs the non-lipid receiving groups, regardless of whether L-carnitine were added or not ($P<0.001$). In conclusion, this study showed that the pre-administration of L-carnitine decreases the susceptibility to bupivacaine cardiotoxicity but it had no effect in resuscitation of cardiac arrest in rats.

Intravenous Lipid Emulsion Therapy in Term Pregnancy

A 29-year-old primigravid woman who was 39 weeks pregnant presented to the hospital for a planned induction [39A]. She was given 6 mL of bupivacaine 0.1% and fentanyl 2.5 μg/mL combined spinal–epidural analgesia. After 50 min of this injection, she experienced tinnitus, metallic taste in the tongue, tachycardia, and palpitations and it was determined that she had experienced LAST. A 20% intravenous lipid emulsion was given and her symptoms resolved within 10 min. Throughout this incident, the fetal heart tones were normal so the induction continued on as normal and 2 h later, she had a cesarean section due to the arrest of active descent.

Dermatology Allergy

A case report describes a 5-month-old infant presented to the dermatology clinic with a diffuse papulosquamous eruption [40A]. A biopsy was obtained with a suspicion of epidermolysis bullosa simplex. This biopsy indicated the presence of epidermal bulla although the patient had neither clinical bulla anywhere else nor a family history of a blistering disease. A second biopsy obtained without topical anesthetic was negative. It was, therefore, determined that the topical anesthetic, EMLA (2.5% lidocaine and 2.5% prilocaine), that was applied at the initial biopsy site resulted in the dermal pathohistological changes that would have caused a misdiagnosis.

References

[1] Wadlund DL. Local anesthetic systemic toxicity. AORN J. 2017;106:367–77 [R].
[2] Sekimoto K, Tobe M, Saito S. Local anesthetic toxicity: acute and chronic management. Acute Med Surg. 2017;4:152–60 [R].
[3] Jacob JS, Kovac AL. Procaine and local anesthetic toxicity: a collaboration between the clinical and basic sciences. Reg Anesth Pain Med. 2017;42:760–3 [R].
[4] Urfalıoğlu A, Urfalıoğlu S, Öksüz G. The knowledge of eye physicians on local anesthetic toxicity and intravenous lipid treatment: questionnaire study. Turk J Ophthalmol. 2017;47:320–5 [c].
[5] Rubin DS, Matsumoto MM, Weinberg G, et al. Local anesthetic systemic toxicity in total joint arthroplasty: incidence and risk factors in the United States from the National Inpatient Sample 1998–2013. Reg Anesth Pain Med. 2018;43:131–7 [C].
[6] Mörwald EE, Zubizarreta N, Cozowicz C, et al. Incidence of local anesthetic systemic toxicity in orthopedic patients receiving peripheral nerve blocks. Reg Anesth Pain Med. 2017;42:442–5 [C].
[7] Liu SS, et al. Cardiac arrest and seizures caused by local anesthetic systemic toxicity after peripheral nerve blocks: should we still fear the reaper? Reg Anesth Pain Med. 2016;41:5–21 [C].
[8] Kvisselgaard AD, Krøigaard M, Mosbech HF, et al. No cases of perioperative allergy to local anaesthetics in the Danish Anaesthesia Allergy Centre. Acta Anaesthesiol Scand. 2017;61:149–55 [C].
[9] Kvisselgaard AD, Mosbech HF, Fransson S, et al. Risk of immediate-type allergy to local anesthetics is overestimated-results from 5 years of provocation testing in a Danish Allergy Clinic. J Allergy Clin Immunol Pract. 2017. https://doi.org/10.1016/j.jaip.2017.08.010 [C].
[10] Veneziano G, Tobias JD. Chloroprocaine for epidural anesthesia in infants and children. Paediatr Anaesth. 2017;27:581–90 [R].
[11] Lee SC, Moll V. Continuous epidural analgesia using an ester-linked local anesthetic agent, 2-chloroprocaine, during labor: a case report. A A Case Rep. 2017;8:297–9 [A].
[12] Dias R, Dave N, Tullu MS, et al. Local anaesthetic systemic toxicity following oral ingestion in a child: revisiting dibucaine. Indian J Anaesth. 2017;61:587–9 [A].
[13] Viswanath O, Suthar R, Kannan M, et al. Post procedural complication following cervical epidural local anesthetic injection: a case report. Anesth Pain Med. 2017;7:e44636 [A].
[14] Hasan B, Asif T, Hasan M. Lidocaine-induced systemic toxicity: a case report and review of literature. Cureus. 2017;9:e1275 [A].
[15] Liew J, Lundblad J, Obley A. Local anesthetic systemic toxicity complicating thyroid biopsy. Cureus. 2017;9:e1955 [A].
[16] Maxey-Jones CL, Palmerton A, Farmer JR, et al. Difficult airway management caused by local anesthetic allergy during emergent cesarean delivery: a case report. A A Case Rep. 2017;9:84–6 [A].
[17] Allen G, Chan D, Gue S. Investigation and diagnosis of an immediate allergy to amide local anaesthetic in a paediatric dental patient. Aust Dent J. 2017;62:241–5 [A].
[18] Corwin DJ, Topjian A, Banwell BL, et al. Adverse events associated with a large dose of intravenous lipid emulsion for suspected local anesthetic toxicity. Clin Toxicol. 2017;55:603–7 [A].
[19] Daykin H. The efficacy and safety of intravenous lidocaine for analgesia in the older adult: a literature review. Br J Pain. 2017;11:23–31 [R].
[20] Hayaran N, Sardana R, Nandinie H, et al. Unusual presentation of local anesthetic toxicity. J Clin Anesth. 2017;36:36–8 [A].
[21] Aggarwal N. Local anesthetics systemic toxicity association with exparel (bupivacaine liposome)—a pharmacovigilance evaluation. Expert Opin Drug Saf. 2018;17:581–7 [C].
[22] Vahabi S, Eatemadi A. Nanoliposome encapsulated anesthetics for local anesthesia application. Biomed Pharmacother. 2017;86:1–7 [R].
[23] Li H, et al. An in situ-forming phospholipid-based phase transition gel prolongs the duration of local anesthesia for ropivacaine with minimal toxicity. Acta Biomater. 2017;58:136–45 [E].
[24] da Silva CMG, et al. Encapsulation of ropivacaine in a combined (donor-acceptor, ionic-gradient) liposomal system promotes extended anesthesia time. PLoS One. 2017;12:e0185828 [E].
[25] Forsberg M, Forsberg S, Edman G, et al. No support for lipid rescue in oral poisoning: a systematic review and analysis of 160 published cases. Hum Exp Toxicol. 2017;36:461–6 [M].
[26] Karcioglu O. Use of lipid emulsion therapy in local anesthetic overdose. Saudi Med J. 2017;38:985–93 [R].
[27] Hoegberg LCG, Gosselin S. Lipid resuscitation in acute poisoning: after a decade of publications, what have we really learned? Curr Opin Anaesthesiol. 2017;30:474–9 [R].
[28] Rosenberg PH. Current evidence is not in support of lipid rescue therapy in local anaesthetic systemic toxicity. Acta Anaesthesiol Scand. 2016;60:1029–32 [M].
[29] Levine M, et al. Systematic review of the effect of intravenous lipid emulsion therapy for non-local anesthetics toxicity. Clin Toxicol. 2016;54:194–221 [R].
[30] Harvey M, Cave G. Lipid emulsion in local anesthetic toxicity. Curr Opin Anaesthesiol. 2017;30:632–8 [R].
[31] Mithani S, et al. A cohort study of unstable overdose patients treated with intravenous lipid emulsion therapy. CJEM. 2017;19:256–64 [C].
[32] Le Fevre P, Gosling M, Acharya K, et al. A Dramatic resuscitation with Intralipid in an epinephrine unresponsive cardiac arrest following overdose of amitriptyline and propranolol. BMJ Case Rep. 2017;2017 [A].
[33] Fettiplace MR, McCabe DJ. Lipid emulsion improves survival in animal models of local anesthetic toxicity: a meta-analysis. Clin Toxicol. 2017;55:617–23 [M].
[34] Vieitez V, Gómez de Segura IÁ, Martin-Cuervo M, et al. Successful use of lipid emulsion to resuscitate a foal after intravenous lidocaine induced cardiovascular collapse. Equine Vet J. 2017;49:767–9 [A].

[35] Yoshimoto M, Horiguchi T, Kimura T, et al. Recovery from ropivacaine-induced or levobupivacaine-induced cardiac arrest in rats: comparison of lipid emulsion effects. Anesth Analg. 2017;125:1496–502 [E].
[36] Chen Z, et al. The protective effect of lipid emulsion in preventing bupivacaine-induced mitochondrial injury and apoptosis of H9C2 cardiomyocytes. Drug Deliv. 2017;24:430–6 [E].
[37] Kim RJ, Kang J-R, Hah Y-S, et al. N-acetyl cysteine protects cells from chondrocyte death induced by local anesthetics. J Orthop Res. 2017;35:297–303 [E].
[38] Wong GK, Pehora C, Crawford MW. L-carnitine reduces susceptibility to bupivacaine-induced cardiotoxicity: an experimental study in rats. Can J Anaesth. 2017;64:270–9 [E].
[39] Dun-Chi Lin J, Sivanesan E, Horlocker TT, et al. Two for one: a case report of intravenous lipid emulsion to treat local anesthetic systemic toxicity in term pregnancy. A A Case Rep. 2017;8:235–7 [A].
[40] Kieliszak CR, Griffin JR, Pollinger TH, et al. Pseudo-bullous dermatosis induced by topical anesthetic agent-clues to this localized toxic reaction. Am J Dermatopathol. 2017;39:e19–22 [A].

CHAPTER

12

Neuromuscular Blocking Agents and Skeletal Muscle Relaxants

Alex Ebied[1]

Pharmacotherapy and Translational Research, University of Florida College of Pharmacy, Gainesville, FL, United States
[1]Corresponding author: ebiedrx55@ufl.edu

DEPOLARIZING NEUROMUSCULAR BLOCKING AGENTS

General

Neuromuscular blocking agents (NMDA) are used in clinical practice for conditions requiring paralysis such as rapid sequence intubation. A systematic review of 34 trials with 3656 participants evaluated the effects of avoiding NMBA vs using NMBA on difficult tracheal intubation for adults and adolescents allocated to tracheal intubation with direct laryngoscopy. This review supports that use of an NMBA may create the best conditions for tracheal intubation and may reduce the risk of upper airway discomfort or injury following tracheal intubation [1M].

SUCCINYLCHOLINE (SUXAMETHONIUM) (SEDA-36, 173; SEDA-37, 155; SEDA-38, 105)

General

A systematic review was conducted to determine whether rocuronium creates intubating conditions comparable to those of succinylcholine during rapid sequence intubation. Of the 50 trials including 4151 participants evaluated, it was determined succinylcholine was superior to rocuronium for achieving excellent intubating conditions: RR 0.86 (95% confidence interval (CI) 0.81–0.92; $n=4151$) and clinically acceptable intubation conditions (RR 0.97, 95% CI 0.95–0.99; $n=3992$, 48 trials) [2M].

A web-based survey was sent to anesthesiologists working in Turkey regarding succinylcholine frequency of use. 433 physicians participated acknowledging special situations using succinylcholine because of their awareness of side effects. 321 (74.1%) observed side effects with the most common being resistant bradycardia, severe muscle pain, and prolong block [3r].

Neuromuscular

Fasciculation is an adverse effect of succinylcholine. One study aimed to compare the use of atracurium and methocarbamol to decrease the occurrence and severity of succinylcholine-induced muscle fasciculation. The atracurium group of 27 patients (93.1%) suffered from mild fasciculation and 2 (6.9%) from moderate fasciculation. The methocarbamol group of 20 patients (68.9%) suffered from mild fasciculation, 5 (17.2%) from moderate fasciculation, and 4 (13.9%) from severe fasciculation. Atracurium is more effective than methocarbamol in decreasing the occurrence and severity of succinylcholine-induced fasciculations [4c].

Sensory System: Eye

Succinylcholine has a known adverse effect of increased intraocular pressure (IOP). One study evaluated the efficacy of using dexmedetomide (0.5 μg/kg) as premedication in attenuating the rise of IOP and adverse effect if any caused by succinylcholine in patients undergoing rapid sequence intubation for general anesthesia. This double-blind, randomized trial of 60 patients found a better attenuating effect in the dexmedetomide group compared to placebo on increasing IOP [5c].

Susceptibility Factors

Genetic Factors

A case of prolonged paralysis following emergent cesarean section with succinylcholine was reported secondary to a deficiency of pseudocholinesterase activity [6A].

Side Effects of Drugs Annual, Volume 40
ISSN: 0378-6080
https://doi.org/10.1016/bs.seda.2018.07.001

A case of severe masseter spasm related to succinylcholine in a child suffering from Rett syndrome was reported [7A].

Drug Administration: Drug Dosage Regimen

A systematic review was conducted to evaluate the optimal dose of succinylcholine, ≤0.3 or 0.3–1.0mg/kg, for laryngeal mask airway (LMA) insertion and related adverse events. Data from 10 randomized controlled trials comprising 625 participants showed that succinylcholine reduced the first-attempt LMA insertion failure rate (RR, 0.22; 95% CI 0.12–0.43) with a further analysis showing the 0.3–1.0mg/kg dosing having reduced failure rates. The adverse events noted were coughing and gagging (RR, 0.26; 95% CI 0.15–0.45) and laryngospasm (RR, 0.14; 95% CI 0.05–0.39). The use of succinylcholine did not result in a significant increase of postoperative myalgia (RR, 2.58; 95% CI 0.79–8.44) and did not reduce the risk of postoperative sore throat (RR, 0.76; 95% CI 0.55–1.03). Subgroup analysis further showed the 0.3–1.0mg/kg dose has a lower rate of coughing and gagging when compared with ≤0.3mg/kg dose [8M].

Interactions: Drug–Drug Interaction

A case report regarding an interaction between succinylcholine and propranol leading to fatal hyperkalemia in a child was reported [9A].

NON-DEPOLARIZING NEUROMUSCULAR BLOCKING AGENTS

General

Many medications can affect muscle contractility including local anesthetics, antibiotics, magnesium, calcium, anticonvulsants, intravenous anesthetics, inhaled anesthetics, phosphodiesterase inhibitors, statins, toxins, and antiemetics [10R].

A prospective, multicenter survey was conducted of medical intensive care unit intensivists, fellows, nurse practitioners (NPs), physician's assistants (PAs), and pharmacists at 5 tertiary care centers. A total of 335 surveys were sent to providers, with a 47% response rate. Ninety-eight percent of providers correctly identified that NMBAs lack anxiolytic and analgesic properties. The effect of end-organ damage on NMBA clearance was less commonly identified by NPs/PAs for both hepatic ($P=0.0077$) and renal ($P=0.0272$) dysfunction compared with physicians. More NP/PAs identified the association of consciousness with the use of NMBAs than physicians ($P=0.047$). Forty-two percent of prescribers reported always or frequently using continuous-infusion NMBAs in patients with severe ARDS, with 89% initiating NMBAs because of ventilator dyssynchrony. Prescribers perceived continuous NMBAs to be more effective than inhaled prostaglandins (74% vs 56%) in severe ARDS but less safe (45% vs 84%). Train of 4 was identified by 54% of prescribers as their primary method for titration. Perceptions about the efficacy and safety of NMBAs varied among prescribers, and inconsistencies existed in the prioritization of management strategies for ARDS [11C].

Organs and Systems

Respiratory

This study investigated if intermediate-acting NMBA use during surgery is associated with postoperative pneumonia and to investigate if non-reversal of NMBAs is associated with postoperative pneumonia. The authors compared 1455 surgical cases who received an intermediate-acting non-depolarizing NMBA to 1455 propensity score-matched cases who did not and 1320 surgical cases who received an NMBA and reversal with neostigmine to 1320 propensity score-matched cases who did not receive reversal. Patients receiving an NMBA had a higher absolute incidence rate of postoperative pneumonia (9.00 vs 5.22 per 10000 person-days at risk), and the IRR was statistically significant (1.79; 95% bootstrapped CI, 1.08–3.07). Among surgical cases who received an NMBA, cases who were not reversed were 2.26 times as likely to develop pneumonia after surgery compared to cases who received reversal with neostigmine (IRR, 2.26; 95% bootstrapped CI, 1.65–3.03). Intraoperative use of intermediate non-depolarizing NMBAs is associated with developing pneumonia after surgery. Among patients who receive these agents, non-reversal is associated with an increased risk of postoperative pneumonia [12MC].

Long-Term Effects: Death

This retrospective observational study of 11355 adult patients undergoing general anesthesia for non-cardiac surgery at 5 Veterans Health Administration hospitals was designed to assess the association between non-depolarizing NMBA use with or without neostigmine reversal and postoperative morbidity and mortality. Administration of a non-depolarizing NMBA without neostigmine reversal compared with neostigmine reversal was associated with increased odds of respiratory complications (PM odds ratio [OR], 1.75 [95% confidence interval [CI], 1.23–2.50]; MLR OR, 1.71 [CI, 1.24–2.37]) and a marginal increase in 30-day mortality (PM OR, 1.83 [CI, 0.99–3.37]; MLR OR, 1.78 [CI, 1.02–3.13]). However, there were no statistically significant associations with non-respiratory complications or long-term mortality.

The study concludes that reversal of non-depolarizing NMBA should become a standard practice if extubation is planned [13MC].

CISATRACURIUM

Susceptibility Factors

Genetic Factors

A pharmacokinetic/pharmcodynamic study of 43 patients with congenital heart defects found a significantly lowered rate of cisatracurium distribution from the central to peripheral compartment compared to control patients (35%–60%, $P<0.05$). Septal defects caused a marked increase (160%–175%, $P<0.001$) in the distribution half-life. The onset time was prolonged from 2.2 to 5.0 minutes. This study suggests poor distribution of cisatracurium led to a delay in the pharmacodynamics response. Providers should be aware of the use of cisatracurium in patients with septal defects [14c].

CW002

Drug Administration: Drug Dosage Regimen

CW002 is a non-depolarizing neuromuscular-blocking drug found to be inactivated by cysteine in preclinical studies. A 10 patient, human study represents a dose escalation clinical trial designed to describe CW002 potency, duration, cardiopulmonary side effects, and histamine release. Clinical recovery (25% of maximum twitch) occurred in 34 ± 3.4 minutes, with a 5%–95% recovery interval of 35.0 ± 2.7 minutes. The time to a train-of-four ratio greater than 0.9 ranged from 59 to 86 minutes. CW002 did not elicit histamine release or significant (greater than 10%) changes in blood pressure, heart rate, or dynamic airway compliance [15c].

CW002 was studied in dose–response relationship in 8 Rhesus monkeys and 6 cats. A notable lack of autonomic or circulatory effects provided added proof of safety and efficacy [16E].

MIVACURIUM

Susceptibility Factors

Age

The efficacy and safety of mivacurium were studied in 640 pediatric patients. Patients were divided in to age class (2–12m, 13–35m, 3–6 years and 7–14 years), induction dose (0.15, 0.2mg/kg in 2–12m age class; 0.2, 0.25mg/kg in other three age classes), and injection time (20 or 40 seconds). The induction dose and injection time of mivacurium had mostly insignificant effects on onset and recovery times. The main exception to this was that in 2–12m aged patients, increasing the dose of mivacurium from 0.15 to 0.2mg/kg accelerated the onset time by about 30 seconds. Mivacurium produced no significant release of histamine in any age group at the doses studied [17C].

ROCURONIUM (SEDA-38, 107)

Organs and Systems

Musculoskeletal

Rocuronium is associated with intense pain on injection in awake patients and can have withdrawal movements of limbs in anaesthetized patients. A Cochrane review of 66 studies with 7840 participants assessed the ability of both pharmacological and non-pharmacological interventions to reduce or eliminate the pain that accompanies rocuronium administration. The primary outcome of pain was reported in all of the 66 included studies, but few studies reported adverse events and no study reported heart rate and blood pressure changes after administration of rocuronium. The adverse effects reported were associated with the interventions and not rocuronium itself. Those that did so were mainly those investigating opioid drugs. There is low or very low-quality evidence for adverse events, due to risk of bias, inconsistency and imprecision of effect. One adverse event reported was breath holding favoring the control group compared to remifentanil (odds ratio = 1.19). The second adverse event reported was chest tightness favoring the control group compared to remifentanil (odds ratio = 2.61). The third adverse event reported was cough favoring the control group compared to remifentanil (odds ratio = 9.00). The fourth adverse event reported was pain on injection favoring the acetaminophen group compared to the control group (odds ratio = 0.47) [18M].

Drug Administration: Drug Dosage Regimen

The aim of this study was to investigate the relationship between the dose of rocuronium needed to re-establish neuromuscular block and the time interval between sugammadex administration and re-administration of rocuronium. The protocol to re-establish neuromuscular block was an initial rocuronium dose of 0.6mg/kg followed by additional 0.3mg/kg doses every 2minutes until train-of-four responses were abolished. A total of 11 patients were enrolled in this study. Intervals between sugammadex and second rocuronium were 12–465minutes. Total dose of rocuronium needed to re-establish neuromuscular block was 0.6–1.2mg/kg.

0.6 mg/kg rocuronium re-established neuromuscular block in all patients who received initial sugammadex more than 3 h previously. However, when the interval between sugammadex and second rocuronium was less than 2 h, more than 0.6 mg/kg rocuronium was necessary to re-establish neuromuscular block [19c].

Interactions: Drug–Drug Interaction

Statins can cause skeletal muscle myopathy. The neuromuscular effects of rocuronium and muscle injury were observed in 18 patients on long-term statin therapy (at least 3 months) compared to 18 non-statin users. Twelve patients used atorvastatin (10–20 mg/day), four used simvastatin (10–20 mg/day), one used fluvastatin (80 mg/day), and one used rosuvastatin (5 mg/day). The myoglobin and CK concentrations increased after rocuronium administration as compared to baseline in both groups. CK concentration in the statin group was significantly higher than in the non-statin group just at 24 h ($P=0.000003$). However, myoglobin showed no significant difference between the two groups. The onset time of rocuronium decreases and its duration time increases in patients in long-term statin therapy. This study does address the adverse effect profile of both drug classes. However, the various statin doses studied are low potency and may not reflect the current, 2017 dyslipidemia guideline recommendations [20c].

Magnesium sulfate 30 mg/kg accelerated the onset, and improved operating conditions of low-dose rocuronium during laryngeal microsurgery without prolongation of action. This randomized, prospective, double-blinded study of 84 patients compared rocuronium: 0.6 mg/kg (group C, $n=28$), 0.45 mg/kg (group LR, $n=28$), or 0.45 mg/kg plus magnesium sulfate 30 mg/kg (group LM, $n=28$). Group LR showed significantly delayed onset time (group C: 87 ± 22 seconds, group LR: 127 ± 47 seconds, and group LM: 89 ± 32 seconds; $P=0.001$) and maximal suppression than did other groups (group C: 102 ± 30 seconds, group LR: 155 ± 66 seconds, and group LM: 105 ± 36 seconds; $P=0.002$). Duration of action of rocuronium was significantly longer in group C than in other groups (group C: 39 ± 7 minutes, group LR: 28 ± 8 minutes, group LRM: 31 ± 8 minutes; $P<0.001$). Laryngoscope placement score ($P=0.002$), surgeon's satisfaction ($P=0.005$), and sore throat ($P=0.035$) were significantly worse in group LR [21C].

Rapid sequence induction (RSI) should be prompt and have adequate muscle relaxation for tracheal intubation and hemodynamic stability during and after intubation. The purpose of the present study was to investigate the effects of nicardipine and esmolol on the action of rocuronium and intubation conditions during RSI. 82 adult patients were randomly allocated to one of three groups: nicardipine 20 μg/kg, esmolol 0.5 mg/kg, or saline 5 mL. The intubation conditions and score were significantly better in the saline and nicardipine than esmolol ($P<0.001$). The onset time of rocuronium was shortened in the nicardipine group and prolonged in the esmolol group when compared to the saline group ($P<0.001$). A significant attenuation in the increase of MAP immediately after intubation was observed in the nicardipine group as compared with the saline and esmolol groups ($P<0.008$). HR was significantly lower in the esmolol group than in the nicardipine and placebo groups ($P<0.01$). Pretreatment with nicardipine for RSI improved intubation conditions and shortened the onset time of rocuronium and attenuated changes in MAP after intubation. Esmolol may disturb intubation conditions and the onset of action of rocuronium, despite being effective in alleviating responses of HR after RSI [22c].

A case report observed an interaction between magnesium sulfate for severe preeclampsia and rocuronium lead to the development of prolonged and deep neuromuscular blockade requiring the use of neostigmine [23A].

VECURONIUM

Interactions: Drug–Drug Interaction

A prospective, randomized study of 60 patients compared the effects of priming and ephedrine pretreatment on the onset time of intubating dose of vecuronium. One group received 70 μg/kg ephedrine while the priming group received 0.01 mg/kg of vecuronium 3 minutes before intubating dose of vecuronium. The mean time for intubation in ephedrine group (E) was 104 ± 23.282 seconds, and in the priming group (P), it was 142 ± 55.671 seconds ($P=0.001$). Parameters were comparable between groups at all time frames ($P>0.05$). Pretreatment with ephedrine 70 μg/kg shortens the onset time of vecuronium for intubation and is superior to the priming technique [24c].

NEUROMUSCULAR BLOCKERS: REVERSAL AGENTS

Sugammadex (SEDA-37, 157; SEDA-38, 108)

SPECIALTY REVIEW

A Cochrane review of 41 studies with 4206 participants compared the efficacy and safety of sugammadex vs neostigmine in reversing neuromuscular blockade caused by non-depolarizing neuromuscular agents in adults. Sugammadex

2 mg/kg was 10.22 minutes (6.6 times) faster than neostigmine 0.05 mg/kg (1.96 vs 12.87 minutes) in reversing NMB from the second twitch (T2) to TOFR >0.9 (MD 10.22 minutes, 95% CI 8.48–11.96; I^2=84%; 10 studies, n=835; GRADE: moderate quality). Sugammadex 4 mg/kg was 45.78 minutes (16.8 times) faster than neostigmine 0.07 mg/kg (2.9 vs 48.8 minutes) in reversing NMB from post-tetanic count (PTC) 1–5 to TOFR >0.9 (MD 45.78 minutes, 95% CI 39.41–52.15; I^2=0%; two studies, n=114; GRADE: low quality). For our secondary outcomes, we compared sugammadex, any dose, and neostigmine, any dose, looking at risk of adverse and serious adverse events. We found significantly fewer composite adverse events in the sugammadex group compared with the neostigmine group (RR 0.60, 95% CI 0.49–0.74; I^2=40%; 28 studies, n=2298; GRADE: moderate quality). Risk of adverse events was 28% in the neostigmine group and 16% in the sugammadex group, resulting in a number needed to treat for an additional beneficial outcome (NNTB) of 8. When looking at specific adverse events, we noted significantly less risk of bradycardia (RR 0.16, 95% CI 0.07–0.34; I^2=0%; 11 studies, n=1218; NNTB 14; GRADE: moderate quality), postoperative nausea and vomiting (RR 0.52, 95% CI 0.28–0.97; I^2=0%; six studies, n=389; NNTB 16; GRADE: low quality) and overall signs of postoperative residual paralysis (RR 0.40, 95% CI 0.28–0.57; I^2=0%; 15 studies, n=1474; NNTB 13; GRADE: moderate quality) in the sugammadex group when compared with the neostigmine group. Finally, there were no significant differences between sugammadex and neostigmine regarding risk of serious adverse events (RR 0.54, 95% CI 0.13–2.25; I^2=0%; 10 studies, n=959; GRADE: low quality). Application of trial sequential analysis (TSA) indicates superiority of sugammadex for outcomes such as recovery time from T2 to TOFR >0.9, adverse events, and overall signs of postoperative residual paralysis. Overall, patients receiving sugammadex had 40% fewer adverse events compared with those given neostigmine [25M].

Organs and Systems

Endocrine

One study examined the effect of sugammadex on endogenous steroid hormone levels in patients with the assumption that sugammadex could be associated with encapsulation of other steroidal molecules. Fifty male patients between 18 and 45 years of age undergoing elective lower extremity surgery were included in this study. At the termination of surgery, neuromuscular blockade was antagonized using 0.05 mg/kg of neostigmine and 0.01 mg/kg of atropine or 4 mg/kg sugammadex. While there were no differences in serum progesterone levels, patients in neostigmine group had significantly higher cortisol levels at 15 minutes as compared to baseline. Also, patients in sugammadex group had significantly higher serum aldosterone and testosterone levels 15 minutes after antagonism as compared to those in the neostigmine group. The findings suggest that sugammadex is not associated with adverse effects on steroid hormones progesterone and cortisol, while it may lead to a temporary increase in aldosterone and testosterone [26c].

Immunologic

A 50-year-old man developed a severe anaphylactic reaction shortly after the administration of sugammadex at the end of an uneventful laparoscopic appendectomy. Upon allergic skin testing after the event, the patient did not demonstrate a skin reaction to rocuronium or sugammadex individually. However, when the agents were combined together to form a complex, the patient demonstrated a strongly positive skin test reaction [27A].

A second case report of anaphylactic shock after cesarean section that is suggested to be induced by the rocuronium-sugammadex complex. Follow-up skin testing 4 weeks later with sugammadex was negative, as was undiluted rocuronium testing 2 weeks later. However, testing with premixed rocuronium–sugammadex complex resulted in a strong positive reaction suggesting rocuronium–sugammadex complex as the causative agent of anaphylaxis [28A].

A third case report showed positive results on intradermal tests for sugammadex and rocuronium, supporting a diagnosis of allergic reactions to both drugs [29A].

Susceptibility Factors: Physiological Factors

Sugammadex (2 mg/kg) was used successfully in a child with a history of cardiac transplantation. Neostigmine was avoided due to concerns of bradycardia and reports of asystole in heart transplant recipients [30A].

SKELETAL MUSCLE RELAXANTS

General

A Cochrane review assessed the pharmacological interventions used other than botulinum toxin for spasticity after stroke. Seven trials with 403 participants were included. Adverse effects were evident in all interventions but not clinically significant between tizanidine, baclofen, and diazepam [31M].

A Cochrane review of 9 studies and 134 participants evaluated muscle spasticity following traumatic brain injury. The quality of evidence for adverse events was very low yielding a high likelihood of publication bias for baclofen, tizanidine, and botulinum toxin [32M].

BACLOFEN (SEDA-36, 173; SEDA-37, 158; SEDA-38, 110)

Organs and Systems

Nervous System

A randomized, prospective follow-up study was done in spastic cerebral palsy children to assess and compare outcomes of oral diazepam and baclofen for 1 year. Both medications were titrated to maximum dose to 60 children for 3 months. Patients showed significant improvement in spasticity as measured by Mean Modified Ashworth's Scale score and range of motion in both groups. In the baclofen group, 53% reported no side effects by the first month with drowsiness being the most common. By third month, 80% reported no side effects [33c].

Intrathecal baclofen was trialed in 37 patients with cerebral palsy and acquired brain injury for 12 months. Of these 32 (86.5%) showed a positive response of reduced spasticity. However, 8 patients showed adverse effects: 4 patients showed catheter or wound problems, the other 4 patients showed headache, drowsiness, and decreased sitting or standing balance. After the pump implantation, spasticity was significantly reduced within 1 month and the effect maintained for 12 months. Seventeen patients or their caregivers (73.9%) were very satisfied, whereas 5 patients (21.7%) suffered from adverse events showed no subjective satisfaction [34c].

A case report of baclofen toxicity observed a 4-month-old girl who developed encephalopathy, seizures, and respiratory compromise after accidental ingestion of baclofen. An error in compounding the patient's prescribed omeprazole with baclofen rather than sodium bicarbonate at a retail pharmacy. The calculated dose of baclofen the patient received (480mg) was approximately 160 times the dose recommended (3–6mg) for spasticity in a 4-month-old. Baclofen toxicity explained the constellation of signs and symptoms that occurred in the patient, including flaccid paralysis, seizure, respiratory changes, minimally reactive pupils, metabolic acidosis, and slow-wave electroencephalography findings. The timeline of acute presentation and recovery with supportive care was also consistent with oral baclofen toxicity [35A].

Psychological

A case of memory impairment and epileptic amnesia was reported with therapeutic infusion of baclofen [36A].

A case report observed intrathecal baclofen to treat spasticity causes a psychotic episode in a young man with a history of traumatic brain injury [37A].

Long-Term Effects

Drug Dependence

A retrospective study of 113 patients evaluated treatment with baclofen for alcohol dependence in a tertiary hospital of North India. This persistent craving was reported by only 15% of the sample by the end of 4 weeks treatment with baclofen (20–40mg/day). Thirty-four percent of patients reported continued problematic use of alcohol by the end of 4 weeks. No side effects were reported [38C].

The use of high-dose baclofen emerged in 2008 in the treatment of alcohol-use disorders. A case report describes a 32-year-old man treated with a high dose of baclofen 160mg/day for his alcohol-use disorders. Consequently, the patient developed a gambling disorder and sought treatment. According to the Naranjo algorithm, the score was +7, concluding the patient's gambling was probably due to baclofen. The patient was weaned off baclofen and the patient's reported gambling craving was 4–5/10, less than half the rating he assigned during the first presentation [39A].

Drug Administration: Drug Formulation

Compounded topical formulations were studied for the treatment of chronic pain in diabetic neuropathy or neuropathic pain. The objective of this retrospective study was to evaluate the efficacy of a compounded topical cream. Two versions of cream were evaluated: cream 6B ($N=78$) and cream 7B ($N=205$). Both creams contain ketamine (10%), baclofen (2%), gabapentin (6%), amitriptyline (4%), bupivacaine (2%), and clonidine (0.2%). Additionally, one cream (7B) contains nifedipine (2%). The pain score decreased by 2.4 ± 2.4 (35%) with the 6B cream (from 7.8 ± 1.6 to 5.4 ± 2.0, $P<0.001$) and by 3.0 ± 2.4 (40%) with the 7B cream (from 7.5 ± 1.7 to 4.5 ± 2.2, $P<0.001$). Excellent or good effects were reported in 82% of the patients in the 6B and in 70% in the 7B groups. Reduction in oral pain medication was seen in 35% of the patients in the 7B and in 20% in the 6B groups. In the opinion of the treating physicians, the cream therapy caused the avoidance of a pain specialist referral in 53% of the patients in the 6B and in 39% in the 7B groups. This study demonstrates this multi-modal cream therapy may be an effective option for diabetic neuropathy, neuropathic pain, or other chronic pain states [40C].

A case series of 3 patients with radicular pain were successfully treated with a topical formulation of diclofenac, ibuprofen, baclofen, cyclobenzaprine, bupivacaine, gabapentin, and pentoxifylline (T7). One to two grams of T7 was applied to the affected area 3–4 times daily in addition to the patient's baseline pharmacologic management. Three patients with median age of 50 (range, 39–65) and diagnosis of cervical and/or lumbosacral radicular pain participated. Two of the three had chronic radicular pain despite use of analgesic agents, spinal injections and failed spinal surgery syndrome. Each reported subjective improvement in radicular pain, function and sleep. There was an average decrease in pain score consistent with 30%–40% global improvement in symptoms, clinically significant based on the minimal clinically important difference for radicular pain. T7 was well tolerated without adverse reactions. Surgery was prevented or delayed in all cases [41c].

Drug Administration: Drug Dosage Regimen

Twelve patients with severe spasticity received four different doses of intrathecal baclofen (0, 25, 50, 75 μg and an optional dose of 100 μg). All patients achieved an adequate spasmolytic effect with intrathecal baclofen from 50 to 100 μg. No serious side effects were observed. CSF baclofen concentrations, as well as the clinical effects, correlated significantly with intrathecal baclofen [42c].

One study examined the efficacy and safety of high doses of baclofen for treatment of alcohol dependence in a multicenter, double-blind, placebo-controlled trial. 151 patients were randomly assigned to either 6 weeks titration and 10 weeks high-dose baclofen ($N=58$; up to 150 mg), low-dose baclofen ($N=31$; 30 mg), or placebo ($N=62$). Neither low nor high doses of baclofen were effective in the treatment of alcohol dependence. There were frequent dose-related adverse events in terms of fatigue, sleepiness, and dry mouth. One medication related serious adverse event occurred in the high-dose baclofen group. Adverse events were frequent, although generally mild and transient [43C].

Drug Administration: Drug Administration Route

A Cochrane review was conducted to determine whether intrathecal baclofen is an effective treatment for spasticity in children with cerebral palsy. Adverse effects identified were device-related complications requiring further surgery and four were drug-related (dysphagia, dysarthria, excessive hypotonia). One participant had new onset of seizures, although it is not clear that this was related to the use of intrathecal baclofen [44M].

A survey was sent to pediatric physicians regarding intrathecal baclofen use in children with cerebral palsy in France. Children received oral baclofen before the team decided to implant the intrathecal baclofen in 19/24 responses. Reported complications occurred in 16/24 with intrathecal baclofen with no adverse events with the pump itself. The most common complications were local infection (12/24 cases). The authors proposed sub fascial implantation to try to decrease this complication, along with a preoperative course of prophylactic antibiotic therapy [45r].

A retrospective review of 294 children with cerebral palsy was treated with intrathecal baclofen. Infection occurred in 28/294 patients (9.5%) with a 4.9% rate per procedure. The implantation site and patient's weight did note increase risk of infection. The study defined acute infection of less than 90 days and late infection greater than 90 days. MSSA was the most frequent microorganism identified in both acute and late infections; however, MRSA was found in 20% of the acute infections [46C].

A randomized, open-label, dose-escalation, crossover study evaluated the safety profile and pharmacokinetics of oral (PO) and intravenous (IV) baclofen formulations to prevent or treat baclofen withdrawal. Three cohorts of 12 healthy adults received single doses of PO baclofen (10, 15 or 20 mg) and 10-minute infusions of IV baclofen (7.5, 11.5, or 15 mg) with a minimum 48-h wash-out period. The third cohort also received a 60-minute infusion of 15 mg IV baclofen after an additional 48-h wash-out period. None of the PO or IV doses resulted in significant sedation compared to baseline. All subjects could perform tandem gait after each baclofen dose. The most common side effect, transient mild nystagmus, was noted in 4 of 36 and in 13 of 36 subjects after PO and IV administration, respectively. After the 20 mg PO and 15 mg IV doses, mean C_{max} levels were 255 and 722 ng/mL and half-lives were 5.24 and 5.79 h for PO and IV baclofen, respectively. The 80% bioavailability suggests that a 20% reduction in IV dose will produce comparable total drug exposures to that of the PO dose. When PO therapy is interrupted, bridging with IV baclofen may be feasible [47c].

BOTULINUM TOXIN (SEDA-36, 174; SEDA-37, 159; SEDA-38, 110)

Organs and Systems

Neuromuscular

Safety and efficacy of repeated administrations of botulinum toxin A in patients with neuropathic pain were evaluated in a randomized, double-blind, placebo-controlled trial in 152 patients. Botulinum toxin A reduced pain intensity over 24 weeks compared with placebo (adjusted effect estimate −0.77, 95% CI −0.95 to −0.59; $P<0.0001$). Pain on injection was the only adverse effect reported and occurred in 19 (56%) participants in the botulinum toxin A group and 17 (53%) of those in the placebo group ($P=1.0$). Severe pain was experienced by 10 (29%) participants in the botulinum toxin A group and 11 (34%) in the placebo group ($P=0.8$) [48C].

Sensory System

A systematic review of 16 trials with 42405 participants evaluated botulinum toxin A in facial rejuvenation. Adverse effects were significantly observed in the botulinum toxin A group (RR=1.47; 95% CI 1.23–1.77; $P<0.0001$), particularly headaches (RR=1.53; 95% CI 1.15–2.03; $P=0.003$), eyelid ptosis (RR=5.56; 95% CI 1.68–18.38; $P=0.005$), and heavy eyelids (RR=6.94; 95% CI 1.27–37.93; $P=0.03$) [49R].

Gastrointestinal

A Cochrane review of 8 trials with 1010 participants compared the efficacy, safety, and tolerability of botulinum toxin type A (BtA) vs placebo in people with cervical dystonia. Overall, both participants and clinicians reported an improvement of subjective clinical status. There were no differences between groups regarding withdrawals due to adverse events. However, BtA treatment was associated with an increased risk of experiencing an adverse event (risk ratio (RR) 1.19; 95% CI 1.03–1.36; $I^2=16\%$). Dysphagia (9%) and diffuse weakness/tiredness (10%) were the most common treatment-related adverse events (dysphagia: RR 3.04; 95% CI 1.68–5.50; $I^2=0\%$; diffuse weakness/tiredness: RR 1.78; 95% CI 1.08–2.94; $I^2=0\%$) [50M].

Skin

A case report observed OnabotulinumtoxinA used for prophylactic treatment of chronic migraine results in two symmetrical bumps on the upper part of the forehead similar to the horns of a ram [51A].

Organs and Systems

Urinary Tract

Long-term safety of OnabotulinumtoxinA in 227 patients with neurogenic detrusor over activity who completed 4 years of treatment found the most common adverse event was urinary tract infection with no increase incidence with time [52C].

CARISOPRODOL (SEDA-37, 160; SEDA-38, 111)

Long-Term Effects

Drug Withdrawal

A case report observed carisoprodol withdrawal syndrome resembling neuroleptic malignant syndrome. Symptoms included altered sensorium with irrelevant speech progressing to visual hallucinations, restlessness, insomnia, and profuse sweating. Treatment was performed with oral baclofen in the absence of sodium dantrolene [53A].

CYCLOBENZAPRINE (SEDA-36, 175; SEDA-38, 111)

Organs and Systems

Sexual Function

Sexual dysfunction is a well-known side effect of antidepressants. Here, a 55-year-old man was referred to a pain medicine clinic for evaluation and treatment of pain with ejaculation after starting cyclobenzaprine. After discontinuation of cyclobenzaprine, the patient's sexual dysfunction resolved [54A].

Additional case studies can be found in these reviews [55R, 56R].

References

[1] Lundstrom LH, Duez CH, Norskov AK, et al. Avoidance versus use of neuromuscular blocking agents for improving conditions during tracheal intubation or direct laryngoscopy in adults and adolescents. Cochrane Database Syst Rev. 2017;5:CD009237 [M].

[2] Tran DT, Newton EK, Mount VA, et al. Rocuronium versus succinylcholine for rapid sequence induction intubation. Cochrane Database Syst Rev. 2015;3(10):CD002788 [M].

[3] Omur D, Kiraz HA, Sahin H, et al. Use of succinylcholine by anaesthetists in Turkey: a national survey. Turk J Anaesthesiol Reanim. 2015;43(5):323–31 [r].

[4] Shabanian G, Shabanian M, Shabanian A, et al. Comparison of atracurium and methocarbamol for preventing succinylcholine-induced muscle fasciculation: a randomized controlled trial. J Adv Pharm Technol Res. 2017;8(2):59–62 [c].

[5] Singh RB, Choubey S, Mishra S. To evaluate the efficacy of intravenous infusion of Dexmedetomidine as premedication in attenuating the rise of intraocular pressure caused by succinylcholine in patients undergoing rapid sequence induction for general anesthesia: a randomized study. Anesth Essays Res. 2017;11(4):834–41 [c].

[6] Ellison M, Grose B, Howell S, et al. Prolonged paralysis following emergent cesarean section with succinylcholine despite normal dibucaine number. W V Med J. 2016;112(2):44–6 [A].

[7] Jerome R, Sylvain R. Severe masseter spasms in a Rett syndrome during rapid sequence intubation: a succinylcholine severe side effect. Indian J Crit Care Med. 2015;19(9):563–4 [A].

[8] Liao AH, Lin YC, Bai CH, et al. Optimal dose of succinylcholine for laryngeal mask airway insertion: systematic review, meta-analysis and metaregression of randomised control trials. BMJ Open. 2017;7(8)e014274 [M].

[9] Ganigara A, Ravishankar C, Ramavakoda C, et al. Fatal hyperkalemia following succinylcholine administration in a child on oral propranolol. Drug Metab Pers Ther. 2015;30(1):69–71 [A].

[10] Kim YB, Sung TY, Yang HS. Factors that affect the onset of action of non-depolarizing neuromuscular blocking agents. Korean J Anesthesiol. 2017;70(5):500–10 [R].

[11] Torbic H, Bauer SR, Personett HA, et al. Perceived safety and efficacy of neuromuscular blockers for acute respiratory distress syndrome among medical intensive care unit practitioners: a multicenter survey. J Crit Care. 2017;38:278–83 [C].

[12] Bulka CM, Terekhov MA, Martin BJ, et al. Nondepolarizing neuromuscular blocking agents, reversal, and risk of postoperative pneumonia. Anesthesiology. 2016;125:647–55 [MC].

[13] Bronsert MR, Henderson WG, Monk TG, et al. Intermediate-acting nondepolarizing neuromuscular blocking agents and risk of postoperative 30-day morbidity and mortality, and long-term survival. Anesth Analg. 2017;124(5):1476–83 [MC].

[14] Wu Z, Wang S, Peng X, et al. Altered cisatracurium pharmacokinetics and pharmacodynamics in patients with congenital heart defects. Drug Metab Dispos. 2016;44(1):75–82 [c].

[15] Heerdt PM, Sunaga H, Owen JS, et al. Dose–response and cardiopulmonary side effects of the novel neuromuscular-blocking drug CW002 in man. Anesthesiology. 2016;125(6):1136–43 [c].

[16] Sunaga H, Savarese JJ, McGilvra JD, et al. A nondepolarizing neuromuscular blocking drug of intermediate duration, degraded and antagonized by l-cysteine—additional studies of safety and efficacy in the anesthetized rhesus monkey and cat. Anesthesiology. 2016;125:732–43 [E].
[17] Zeng R, Liu X, Zhang J, et al. The efficacy and safety of mivacurium in pediatric patients. BMC Anesthesiol. 2017;17(1):58 [C].
[18] Prabhakar H, Singh GP, Ali Z, et al. Pharmacological and non-pharmacological interventions for reducing rocuronium bromide induced pain on injection in children and adults. Cochrane Database Syst Rev. 2016;2:CD009346 [M].
[19] Iwasaki H, Sasakawa T, Takahoko K, et al. A case series of re-establishment of neuromuscular block with rocuronium after sugammadex reversal. J Anesth. 2016;30(3):534–7 [c].
[20] Ren H, Lv H. Effects of rocuronium bromide in patients in statin therapy for at least three months. Basic Clin Pharmacol Toxicol. 2016;119(6):582–7 [c].
[21] Choi ES, Jeong WJ, Ahn SJ, et al. Magnesium sulfate accelerates the onset of low-dose rocuronium in patients undergoing laryngeal microsurgery. J Clin Anesth. 2017;36:102–6 [C].
[22] Lee JH, Kim Y, Lee KH, et al. The effects of nicardipine or esmolol on the onset time of rocuronium and intubation conditions during rapid sequence induction: a randomized double blind trial. J Anesth. 2015;29(3):403–8 [c].
[23] Berdai MA, Labib S, Harandou M. Prolonged neuromuscular block in a preeclamptic patient induced by magnesium sulfate. Pan Afr Med J. 2016;25:5 [A].
[24] Anandan K, Suseela I, Purayil HV. Comparison of effect of ephedrine and priming on the onset time of vecuronium. Anesth Essays Res. 2017;11(2):421–5 [c].
[25] Hristovska AM, Duch P, Allingstrup M, et al. Efficacy and safety of sugammadex versus neostigmine in reversing neuromuscular blockade in adults. Cochrane Database Syst Rev. 2017;8: CD012763 [M].
[26] Gunduz Gul G, Ozer AB, Demirel I, et al. The effect of sugammadex on steroid hormones: a randomized clinical study. J Clin Anesth. 2016;34:62–7 [c].
[27] Ho G, Clarke RC, Sadleir PH, et al. The first case report of anaphylaxis caused by the inclusion complex of rocuronium and sugammadex. A A Case Rep. 2016;7(9):190–2 [A].
[28] Yamaoka M, Deguchi M, Ninomiya K, et al. A suspected case of rocuronium-sugammadex complex-induced anaphylactic shock after cesarean section. J Anesth. 2017;31(1):148–51 [A].
[29] Yamada Y, Yamamoto T, Tanabe K, et al. A case of anaphylaxis apparently induced by sugammadex and rocuronium in successive surgeries. J Clin Anesth. 2016;32:30–2 [A].
[30] Miller K, Hall B, Tobias JD. Sugammadex to reverse neuromuscular blockade in a child with a past history of cardiac transplantation. Ann Card Anaesth. 2017;20(3):376–8 [A].
[31] Lindsay C, Kouzouna A, Simcox C, et al. Pharmacological interventions other than botulinum toxin for spasticity after stroke. Cochrane Database Syst Rev. 2016;10:CD010362 [M].
[32] Synnot A, Chau M, Pitt V, et al. Interventions for managing skeletal muscle spasticity following traumatic brain injury. Cochrane Database Syst Rev. 2017;11:CD008929 [M].
[33] Goyal V, Laisram N, Wadhwa RK, et al. Prospective randomized study of oral diazepam and baclofen on spasticity in cerebral palsy. J Clin Diagn Res. 2016;10(6) RC01-5. [c].
[34] Yoon YK, Lee KC, Cho HE, et al. Outcomes of intrathecal baclofen therapy in patients with cerebral palsy and acquired brain injury. Medicine (Baltimore). 2017;96(34)e7472 [c].
[35] Lau B, Khazanie U, Rowe E, et al. How a drug shortage contributed to a medication error leading to baclofen toxicity in an infant. J Pediatr Pharmacol Ther. 2016;21(6):527–9 [A].
[36] Zeman A, Hoefeijzers S, Milton F, et al. The GABAB receptor agonist, baclofen, contributes to three distinct varieties of amnesia in the human brain—a detailed case report. Cortex. 2016;74:9–199 [A].
[37] Maneyapanda MB, Driver SP, Ripley DL, et al. Psychosis following an increase in intrathecal baclofen. PM R. 2016;8(12):1222–4 [A].
[38] Rozatkar AR, Kapoor A, Sidana A, et al. Clinical experience of baclofen in alcohol dependence: a chart review. Ind Psychiatry J. 2016;25(1):11–6 [C].
[39] Guillou-Landreat M, Victorri Vigneau C, Gerardin M. Gambling disorder: a side effect of an off-label prescription of baclofen-literature review. BMJ Case Rep. 2017;2017:1–4 [A].
[40] Somberg JC, Molnar J. Retrospective study on the analgesic activity of a topical (TT-CTAC) cream in patients with diabetic neuropathy and other chronic pain conditions. Am J Ther. 2015;22(3):214–21 [C].
[41] Safaeian P, Mattie R, Hahn M, et al. Novel treatment of radicular pain with a multi-mechanistic combination topical agent: a case series and literature review. Anesth Pain Med. 2016;6(2)e33322 [c].
[42] Heetla HW, Proost JH, Molmans BH, et al. A pharmacokinetic-pharmacodynamic model for intrathecal baclofen in patients with severe spasticity. Br J Clin Pharmacol. 2016;81(1):101–12 [c].
[43] Beraha EM, Salemink E, Goudriaan AE, et al. Efficacy and safety of high-dose baclofen for the treatment of alcohol dependence: a multicentre, randomised, double-blind controlled trial. Neuropsychopharmacology. 2016;26(12):1950–9 [C].
[44] Hasnat MJ, Rice JE. Intrathecal baclofen for treating spasticity in children with cerebral palsy. Cochrane Database Syst Rev. 2015;1(11):CD004552 [M].
[45] Mietton C, Nuti C, Dohin B, et al. Clinical practices in intrathecal baclofen pump implantation in children with cerebral palsy in France. Ann Phys Rehabil Med. 2016;59(4):282–4 [r].
[46] Bayhan IA, Sees JP, Nishniandze T, et al. Infection as a complication of intrathecal baclofen treatment in children with cerebral palsy. J Pediatr Orthop. 2016;36(3):305–9 [C].
[47] Schmitz NS, Krach LE, Coles LD, et al. A randomized dose escalation study of intravenous baclofenin healthy volunteers: clinical tolerance and pharmacokinetics. PM R. 2017;9(8):743–50 [c].
[48] Attal N, de Andrade DC, Adam F, et al. Safety and efficacy of repeated injections of botulinum toxin A in peripheral neuropathic pain (BOTNEP): a randomised, double-blind, placebo-controlled trial. Lancet Neurol. 2016;15(6):555–65 [C].
[49] Jia Z, Lu H, Yang X, et al. Adverse events of botulinum toxin type A in facial rejuvenation: a systematic review and meta-analysis. Aesthetic Plast Surg. 2016;40(5):769–77 [R].
[50] Castelao M, Margues RE, Duarte GS, et al. Botulinum toxin type a therapy for cervical dystonia. Cochrane Database Syst Rev. 2017;12(12)CD003633 [M].
[51] Russo A, Silvestro M, Tessitore A, et al. The "Ram's horns sign": a case report of an unusual side effect of OnabotulinumtoxinA in a chronic migraine patient. Headache. 2016;56(10):1656–8 [A].
[52] Rovner E, Kohan A, Chartier-Kastler E, et al. Long-term efficacy and safety of OnabotulinumtoxinA in patients with neurogenic detrusor overactivity who completed 4 years of treatment. J Urol. 2016;196(3):801–8 [C].
[53] Paul G, Parshotam GL, Garg R. Carisoprodol withdrawal syndrome resembling neuroleptic malignant syndrome: diagnostic dilemma. J Anaesthesiol Clin Pharmacol. 2016;32(3):387–8 [A].
[54] Kraus MB, Wie CS, Korlin AW, et al. Painful ejaculation with cyclobenzaprine: a case report and literature review. Sex Med. 2015;3(4):343–5 [A].
[55] Shah MU. Chapter 12 - Neuromuscular blocking agents and skeletal muscle relaxants. In: Ray SD, editor. Side Effects of Drugs-Annual: A worldwide yearly survey of new data in adverse drug reactions. Elsevier. 2015;37:155–61[R].
[56] Shah MU. Chapter 11 - Neuromuscular blocking agents and skeletal muscle relaxants. In: Ray SD, editor. Side Effects of Drugs-Annual: A worldwide yearly survey of new data in adverse drug reactions. Elsevier. 2016;38:105–13 [R].

CHAPTER

13

Drugs That Affect Autonomic Functions or the Extrapyramidal System

Toshio Nakaki[1]

Department of Pharmacology, Teikyo University School of Medicine, Tokyo, Japan
[1]Corresponding author: nakaki@med.teikyo-u.ac.jp

DRUGS THAT STIMULATE BOTH ALPHA- AND β2-ADRENOEPTORS *[SEDA-33, 313; SEDA-34, 233; SEDA-35, 255; SEDA-36, 179; SEDA-37, 163; SEDA-38, 115; SEDA-39, 133; SEDA-40, 000]*

Adrenaline (Adrenaline) and Noradrenaline (Noradrenaline) *[SEDA-32, 281; SEDA-33, 259; SEDA-34, 233; SEDA-35, 255; SEDA-36, 179; SEDA-37, 163; SEDA-38, 115; SEDA-39, 133; SEDA-40, 000]*

Ritodrine [SEDA-34, 318; SEDA-35, 258; SEDA-36, 188; SEDA-40, 000]

MUSCULOSKELETAL

A literature suggested that ritodrine-induced rhabdomyolysis may be likely to occur more acutely after ritodrine administration in myotonic dystrophy compared with non-myotonic dystrophy mothers [1A].

Case Report

A primiparous pregnant woman in remission of myositis suffered very acute-onset ritodrine-induced rhabdomyolysis. At 29 gestational weeks, ritodrine was administered for threatened preterm labor. Just 3h later, she complained of severe limb muscle pain, with serum creatinine phosphokinase elevated to 32019U/L and myoglobinuria. The muscle pain disappeared immediately after ceasing administration of ritodrine. At 31 weeks, premature rupture of the membranes occurred; necessitating cesarean section, yielding a newborn with weak tonus, and the presence of infantile muscle diseases was suspected. Myotonic dystrophy was confirmed in the infant via genetic analysis. Ritodrine can induce rhabdomyolysis even in the prodromal phase with a mild phenotype of myotonic dystrophy.

Pseudoephedrine [SEDA-33, 318; SEDA-34, 236; SEDA-35, 255; SEDA-36, 185; SEDA-39, 133; SEDA-40, 000]

CARDIOVASCULAR

Pseudoephedrine has a property of sympathomimetic α-adrenergic receptor agonist that causes vasoconstriction and reduction in edema throughout the nasal passages. A case of a patient with a new-onset of atrial fibrillation receiving metoprolol for rate control on a background of pseudoephedrine use for allergic rhinitis lead to acute myocardial infarction from multivessel coronary vasospasm. This case illustrates the importance of understanding the pharmacology of potential drug–drug interactions when managing patients with acute cardiovascular syndromes. As a reader of the report, this reviewer feels the diagnosis of acute myocardial infarction could be somewhat peculiar, because nitroglycerin was effective to recover from the symptoms [2A].

Case Report

A 63-year-old woman presented to her physician with dyspnea and palpitations. Her medical history was notable for asthma and seasonal allergies for which she took fexofenadinepseudoephedrine daily as a decongestant. She was referred to the emergency department because of atrial fibrillation. In transit, she received intravenous metoprolol with modest improvement in heart rate. On arrival, she denied

Side Effects of Drugs Annual, Volume 40
ISSN: 0378-6080
https://doi.org/10.1016/bs.seda.2018.07.002

chest pain and the first troponin measurement was normal. Her physical examination was notable for irregular tachycardia. The initial ECG showed atrial fibrillation with nonspecific inferolateral ST-T abnormalities. One hour after receiving additional intravenous and oral metoprolol, she experienced chest pain, bradycardia, and hypotension. A second ECG demonstrated atrial fibrillation, now with a slow ventricular response and inferior ST-segment elevation with reciprocal anterolateral ST depressions that were concerning for acute inferior myocardial infarction. The coronary angiography revealed a long area of narrowing in the left circumflex coronary artery with subtotal occlusion of an obtuse marginal branch, and sequential high-grade stenoses in the right coronary artery. Given the dynamic electrocardiographic changes and the appearance of the lesions, intracoronary nitroglycerin was administered, resulting in complete resolution of the stenoses and establishing a diagnosis of multivessel coronary vasospasm. A subsequent transthoracic echocardiogram showed a left ventricular ejection fraction of 35%–40%, with inferior and inferolateral hypokinesis. The troponin-T level peaked at 2.36ng/mL. She underwent electrical cardioversion under transesophageal echocardiographic guidance before discharge home.

DRUGS THAT PREDOMINATLY STIMULATE α1-ADRENOCEPTORS [SEDA-33, 318; SEDA-34, 236; SEDA-35, 257; SEDA-36, 186; SEDA-37, 163; SEDA-38, 117; SEDA-39, 134; SEDA-40, 000]

Phenylephrine [SEDA-15, 2808; SEDA-32, 283; SEDA-33, 318; SEDA-34, 236; SEDA-38, 117; *SEDA-39, 134; SEDA-40, 000*]

CLINICAL STUDY

Phenylephrine infusions are considered as standard management for obstetric spinal hypotension, but there remains reluctance to implement them in resource-limited contexts. Alternating intervention study of patients undergoing elective or urgent cesarean delivery under spinal anesthesia compared a vasopressor bolus strategy to fixed-rate, low-dose prophylactic phenylephrine infusion with supplemental boluses. The primary outcome was the incidence of severe hypotension (mean arterial pressure <70% baseline or systolic blood pressure <80mmHg). Both groups showed statisticaly no significance in blood pressure changes. It may be recommended to reconsider guidelines for resource-constrained settings to adopt a fixed, low-dose phenylephrine infusion in combination with rescue vasopressor bolus therapy [3c].

RANDOMIZED CONTROLLED TRIAL

A randomized double-blind study aimed to evaluate cardiac output changes with phenylephrine or ephedrine infusions titrated to maintain baseline systolic blood pressure during spinal anesthesia. Women ($n=40$) scheduled for elective cesarean delivery received either phenylephrine 100μg/min or ephedrine 5mg/min infusions. Despite good systolic blood pressure control and increased cardiac output with ephedrine, administration of ephedrine was associated with significantly more fetal acidosis [Median (Interquartile range) umbilical artery pH—phenylephrine=7.33 (7.31–7.34) and ephedrine=7.22 (7.16–7.27), $P<0.05$] [4c].

RANDOMIZED CONTROLLED TRIAL

Eighty-five parturients having spinal anesthesia for elective cesarean delivery were randomized to two groups: phenylephrine 0.1μg/kg/min and noradrenaline 0.05μg/kg/min fixed-rate infusions. Rescue bolus interventions of phenylephrine 100μg for hypotension, or ephedrine 5mg for bradycardia with hypotension, were given as required to maintain systolic blood pressure. Hemodynamic parameters including heart rate, the incidence of bradycardia, blood pressure, cardiac output, cardiac index, stroke volume, and systemic vascular resistance and neonatal outcome were similar between groups (all $P<0.05$) [5c].

DRUGS THAT STIMULATE α1-ADRENOCEPTORS *[SEDA-33, 265; SEDA-34, 285; SEDA-35, 257; SEDA-36, 187; SEDA-37; SEDA-38, 117; SEDA-39, 134; SEDA-40, 000]*

Methoxamine *[SEDA-39, 134; SEDA-40, 000]*

PSYCHIATRIC

Methoxamine potentially promotes TNF-alpha expression, leading to an increased risk of postoperative cognitive dysfunction. This study aimed to investigate the dose-dependent effect of methoxamine on the incidence of early postoperative cognitive dysfunction and blood TNF-alpha level. The intraoperative hemodynamic parameters showed improved stability in the M1 and M2 groups compared with the control group. However, in the M3 group, abnormally increased intraoperative blood pressure, cardiac output, and systolic stroke volume were observed [6c].

DRUGS THAT STIMULATE DOPAMINE RECEPTORS *[SEDA-33, 266; SEDA-34, 283; SEDA-35, 262; SEDA-36, 190; SEDA-37, 167; SEDA-38, 118; SEDA-39, 134; SEDA-40, 000]*

Safinamide (New Listing)

Safinamide is a selective and reversible monoamine oxidase B inhibitor. It also inhibits glutamate release and dopamine and serotonin reuptake. It binds to the σ1 receptor with high affinity. In addition, it blocks sodium and calcium channels. Safinamide tablets as an add-on treatment for patients with Parkinson's disease who are currently taking levodopa/carbidopa and experiencing "off" episodes. Patients who should not take safinamide include those who have severe liver problems, or who take a medicine used to treat a cough or cold called dextromethorphan. It also should not be taken by patients who take a monoamine oxidase inhibitor because it may cause a severe sudden increase in blood pressure. It is also not advisable to patients who take an opioid drug, certain antidepressants (such as serotonin–noradrenaline reuptake inhibitors, tricyclics, tetracyclics, and triazolopyridines), or cyclobenzaprine, because it may cause a life-threatening reaction called serotonin syndrome. The most common adverse reactions observed were uncontrolled involuntary movement, falls, nausea, and insomnia. Serious, but less common, risks include the following: hallucinations and psychotic behavior; problems with impulse control/compulsive behaviors; withdrawal-emergent hyperpyrexia and confusion; and retinal pathology.

SYSTEMIC REVIEWS

Levodopa remains the most effective oral pharmacotherapy for Parkinson's disease; its use is often limited by wearing off effect and dyskinesias. Management of such complications continues to be a significant challenge. Adverse events caused the premature study discontinuation of 12 individuals (4.4%) in the safinamide group and 10 individuals (3.6%) in the placebo group. At 119 centers, 549 patients were randomized: 274 to safinamide and 275 to placebo. Among them, 245 (89.4%) receiving safinamide and 241 (87.6%) receiving placebo completed the study. Mean (SD) change in daily on time without troublesome dyskinesia was +1.42 (2.80) hours for safinamide, from a baseline of 9.30 (2.41) hours, vs +0.57 (2.47) hours for placebo, from a baseline of 9.06 (2.50) hours (least-squares mean difference, 0.96 h; 95% CI, 0.56–1.37 h). The most frequently reported adverse event was dyskinesia (in 40 [14.6%] vs 15 [5.5%] and as a severe event in 5 [1.8%] vs 1 [0.4%]). Overall, safinamide can be used for the treatment of patients with Parkinson's disease without troublesome dyskinesia and reduce wearing off [7C].

Rotigotine [SEDA-35, 264; SEDA-37, 167; SEDA-38, 119; SEDA-39, 136; SEDA-39, 136; SEDA-40, 000].

CLINICAL STUDY

SP1058 was a non-interventional study conducted in routine clinical practice in Germany and Austria in patients experiencing Parkinson disease-associated pain (per the physician's assessment). Data were collected at baseline (i.e., before rotigotine initiation) and at a routine visit after ≥25 days (−3 days allowed) of treatment on a maintenance dose of rotigotine. Adverse drug reactions were consistent with dopaminergic stimulation and transdermal administration [8c].

Dopamine Receptor Agonists *[SEDA-34, 242; SEDA-35, 261; SEDA-36, 191; SEDA-37, 169; SEDA-38, 119; SEDA-39, 000]*

Pramipexole [SEDA-36, 193; SEDA-37, 168; SEDA-38, 124; SEDA-39, 137; SEDA-40, 000]

NERVOUS SYSTEM

The following case may be due to basal ganglia dysfunction, which was probably caused by abnormal activation of dopamine 1-like receptor boosted by pramipexole binding on dopamine 3-like receptor in a situation where dopamine 3-like receptor was overexpressed by the chronic treatment of levodopa. The degree of overexpression of D3 receptor is increased in a high dose of pramipexole, for patients with Parkinson disease who are treated with levodopa chronically, a new use of pramipexole and an increase in dose to alleviate the symptoms of Parkinson disease should be implemented with caution while closely observing the occurrence of drug-induced complications such as dystonia and complex regional pain syndrome [9A].

Case Report

A 71-year-old woman had suffered idiopathic Parkinson's disease since 66 years old with hand resting tremor, cogwheel type rigidity, and bradykinesia. Before transferring to the hospital, she had been treated with 300 mg/day/30 mg/day of levodopa/carbidopa; however, bradykinesia and cogwheel type rigidity were sustained. To improve the remaining bradykinesia and cogwheel type rigidity despite levodopa/carbidopa treatment, pramipexole was initiated at a dose of 0.125 mg twice a day for 1 week and then the dose was increased gradually and carefully over a 3-week period to 1.0 mg 2 times a day (total 2.0 mg). The dose of levodopa/carbidopa was not changed throughout the clinical course. Within 2 weeks of taking pramipexole at 2.0mg/day (5 weeks from initiation of pramipexole), she complained of both hand pain (left more involved) with stiffness and difficulty of using the hands. The dose of pramipexole was increased to 3.0mg/day (1.0mg 3 times a day). However, even 1 week after the increased dose of pramipexole, the pain in both hands had not improved at all, and she began to notice striatal deformities in her hands and feet. After 1 day, she noticed swelling and redness in her left hand and wrist. Clinical examination demonstrated severe tenderness in both wrists and hands, with left-side predominance. Three-phase bone scintigraphy showed increased uptake in the left metacarpal phalangeal and wrist joint in the delayed phase. Then, steroid treatment using prednisolone was started for 7 days. The initial dose was 60mg (1mg per kg of body weight), and the dose was tapered over 1 week. After steroid treatment, the left hand and wrist joint swelling and redness were mildly improved, but the striatal hand and feet deformities were not improved. Considering the possibility of pramipexole-induced dystonia, and tapered the dose of pramipexole to 1.5mg (0.5mg 3 times a day) gradually. After 1 week, however, the left hands and wrist joint swelling and tenderness recurred. This condition led to resumption of steroid treatment for the next 7 days.

With the initiation of the second steroid treatment, the pramipexole was discontinued. One week after the initiation of the second steroid treatment, the pain and swelling in her left hand and wrist joints were markedly improved. The brain PET scan revealed decreased uptake of rostrocaudal gradient in the bilateral posterior putamen, which was consistent with idiopathic Parkinson's disease. However, the electrophysiological study could not reveal any abnormality of the peripheral nerve including the brachial plexus in both upper extremities, and the brain MRI study also could not reveal any evidence of acute cerebral infarction. After discontinuation of pramipexole without changing the dose of levodopa/carbidopa, the striatal deformity of the hands and feet showed mild improvement. Trihexyphenidyl 1.5mg (0.5mg tablet, 3 times a day) was added for the improvement of residual dystonia. Five days after discontinuation of pramipexole, the striatal deformity of both hands and feet disappeared, and she was able to use both hands without any assistance. In this case the use of pramiexole seemed to be related to the deformity of hands and feet.

PSYCHIATRIC

While pathological gambling, excessive shopping and hypersexuality have often been described as adverse effects of dopamine agonists, there are only few reports about impulsive stealing in this context.

Case Report

A case of a 48-year-old female was examined for legal culpability. The proband was taking the dopamine agonist pramipexole as a treatment for Parkinson's disease. In temporal association, she had committed numerous shoplifting offences. The synergy of the pharmacological effects with familial, biographic and social factors, suggesting a bio-psycho-social etiology was demonstrated [10A].

Levodopa *[SEDA-32, 285; SEDA-33, 320; SEDA-34, 286–8; SEDA-35, 259; SEDA-36, 192; SEDA-37, 169; SEDA-38, 122; SEDA-39, 137; SEDA-40, 000]*

Placebo-Controlled Studies

To evaluate the efficacy and safety of 25- and 50-mg/day dosages of opicapone compared with placebo as adjunct to levodopa therapy in patients with Parkinson disease experiencing end-of-dose motor fluctuations. The most common adverse events in the opicapone vs placebo groups were dyskinesia, constipation, and dry mouth. Fifty-one patients (11.9%) discontinued from the study during the double-blind phase. The author concludes opicapone was safe and well tolerated [11C].

Nervous System

Adverse events in carbidopa and levodopa extended-release capsules were consistent with those reported in prior studies, including dyskinesia and psychiatric problems [12c].

Mechanism

Levodopa-induced dyskinesias are motor complications following long-term dopaminergic therapy in Parkinson's disease (PD). Impaired brain plasticity resulting in the creation of aberrant motor maps intended to encode normal voluntary movement is proposed to result in the development of dyskinesias. Traditionally, the various nodes in the motor network like the striato-cortical and the cerebello-thalamic loops were thought to function independent of each other with little communication among them. Anatomical evidence from primates revealed the existence of reciprocal loops between the basal ganglia and the cerebellum providing an anatomical basis for communication between the motor network loops. Dyskinetic patients with Parkinson disease reveal impaired brain plasticity within the motor cortex which may be modulated by cortico-cortical, cerebello-cortical or striato-cortical connections. This review article summarizes the evidence for altered plasticity in the multicomponent motor network in the context of levodopa-induced dyskinesias in Parkinson disease. Current evidence suggests a pivotal role for the cerebellum in the larger motor network with the ability to integrate sensorimotor information and independently influence multiple nodes in this network. Targeting the cerebellum seems to be a justified approach for future interventions aimed at attenuating levodopa-induced dyskinesias [13r].

MISCELLANEOUS OTHER DRUGS THAT INCREASE DOPAMINE ACTIVITY

Amantadine *[SEDA-37, 344; SEDA-39, 140; SEDA-40, 000]*

Placebo-Controlled Studies

This meta-analysis examined the efficacy and safety of amantadine as an adjunctive treatment of weight gain in schizophrenia by systematically searching and analyzing randomized controlled trials, which compared adjunctive amantadine with placebo in adult patients with schizophrenia and were included in the meta-analysis. Except for insomnia ($P=0.007$; number needed to harm, 6; 95% confidence interval, 4–16), all-cause discontinuation (risk ratio, 1.12; $P=0.54$, $I=0\%$) and other adverse events were similar between the amantadine and placebo groups [14M].

Placebo-Controlled Trials

To evaluate the efficacy and safety of ADS-5102 (amantadine) extended-release 274-mg capsules for treatment of levodopa-induced dyskinesia in patients with Parkinson disease. A randomized, double-blind, placebo-controlled clinical trial was conducted, at 44 North American sites among patients with Parkinson disease treated with levodopa who experienced at least 1 h of troublesome dyskinesia per day with at least mild functional impact. The most common adverse events (≥5% in the active arm) included visual hallucinations, peripheral edema, dizziness, dry mouth, and constipation. Other adverse events occurring in less than 5% of patients in the ADS-5102 group included nausea (3 [4.8%]), confusion (2 [3.2%]), and orthostatic hypotension (1 [1.6%]). There were no reports of impulse control disorder in the ADS-5102 group. Thirteen patients (20.6%) in the ADS-5102 group and 4 patients (6.7%) in the placebo group discontinued the study drug because of adverse events. The most common adverse events leading to treatment discontinuation in the ADS-5102 group were visual hallucinations (5 [7.9%]), peripheral edema (3 [4.8%]), and dry mouth (3 [4.8%]). Within the ADS-5102 group, 10 of the 13 patients who discontinued treatment because of adverse events did so during the first month of treatment. For three patients, a dose reduction to 137 mg at bedtime was allowed because of adverse events. In general, vital signs and laboratory test results remained consistent with baseline values and were similar between treatment groups throughout the study [15C].

DRUGS THAT AFFECT THE CHOLIERGIC SYSTEM *[SEDA-31, 272; SEDA-32, 290; SEDA-33, 324; SEDA-34, 290–1318, SEDA-35, 266; SEDA-36, 199; SEDA-37, 172; SEDA-38, 124; SEDA-39, 141; SEDA-40, 000]*

Anticholinergic Drugs *[SEDA-31, 273; SEDA-32, 290; SEDA-33, 324; SEDA-34, 290–1318; SEDA-35, 266; SEDA-36, 199; SEDA-37, 172; SEDA-38, 124; SEDA-39, 141; SEDA-40, 000]*

R *Controversy still exists as to whether initial combination treatment is superior to serial addition of anticholinergics after maintenance or induction of alpha blockers in benign prostatic hyperplasia. A study was undertaken to determine the benefits and safety of initial combination treatment of an alpha blocker with anticholinergic medication in benign prostatic hyperplasia/lower urinary tract symptoms through a systematic review and meta-analysis. A meta-analysis was conducted to assess improvement in lower urinary tract symptoms using International Prostate Symptom Score, maximal urinary flow rate, post-voided residual volume, and quality of life. In total, 16 studies were included in our analysis, with a total sample size of 3548 subjects (2195 experimental subjects and 1353 controls). There was no significant difference in the number of acute urinary retention events or post-voided residual volume [16M].*

To conduct an up-to-date systematic search and meta-analysis on the effectiveness of combined inhaled therapy (short acting anticholinergics + short acting beta2 agonists) vs short acting beta2 agonists alone to reduce hospitalisations in adult patients presenting to the erectile dysfunction with an exacerbation of asthma. Combination inhaled therapy was more effective in preventing hospitalisation in adults with severe asthma exacerbations who are at increased risk of hospitalisation, compared to those with mild–moderate exacerbations, who were at a lower risk to be hospitalised. A single dose of combination therapy and multiple doses both showed reductions in the risk of hospitalisation among adults with acute asthma. Participants receiving combination inhaled therapy were more likely to experience adverse events than those treated with short acting beta2 agonists agents alone (OR 2.03, 95% CI 1.28–3.20; participants=1392; studies=11; I(2)=14%; moderate quality of evidence). Among patients receiving combination therapy, 103 per 1000 were likely to report adverse events (95% 31–195 more) compared to 131 per 1000 patients receiving short acting beta2 agonists alone. The author concluded combination inhaled therapy with short acting anticholinergic agent and short acting beta2 agonists reduced hospitalisation and improved pulmonary function in adults presenting to the erectile dysfunction with acute asthma [17M].

Gastrointestinal

A patient had an occlusive syndrome due to neuroleptics and complications, including mesenteric ischemia with necrotizing colitis. Systematic monitoring of patients treated with quetiapine in combination with other drugs with anticholinergic effects may be necessary [18A].

Case Report

A 34-year-old male adult had been suffering from bipolar disorder since the previous year. It was decided to administrate 600 mg per day of quetiapine in combination with tropatepine consequent to an episode of agitation and aggressiveness. About a month later, while the patient was receiving diazepam and valproic acid, loxapine oral solution was introduced. Nine days later, the patient started mentioning digestive disorders, such as diffuse abdominal pain with constipation but continued to pass gaz. Two days later, at 6:30 am, he declared abdominal pain, followed by malaise and onset of vomiting. His laboratory tests showed leukocytosis 11 000/μL with neutrophils 7700/μL. The abdomen's radiograph without preparation showed small bowel and colonic air-fluid levels. The result of the CT scan confirmed an occlusive syndrome affecting the whole small gut and colon. He received an intramuscular injection of 100 mg of loxapine and an opioid treatment, including tramadol and morphine. At 2:30 pm, the clinical condition further deteriorated with an onset of generalized abdominal contracture, the absence of abdominal breathing, sweating, tachycardia at 104 bpm, and hypothermia. Laboratory tests showed metabolic acidosis, elevated liver enzymes and acute renal failure. He received volume expansion and was treated by renal replacement therapy and antibiotics. He was intubated and transferred to the operating room. At laparotomy, both colonic necrosis with perforation and necrosis of the small bowel were seen. The patient underwent total colectomy with small bowel resection, distal ileostomy and closure of the rectal stump. The onset of septic and hemorrhagic state required further surgery the next day. The evolution was characterized by multi-organ failure with acute anuric renal failure, multiple cardiac arrests, and systemic bacterial and fungal infection. On July 24th, this unfavorable outcome led to death. It is important quetiapine, like all antipsychotics, has anticholinergic effects, including cardiac, psychiatric and digestive. The combination of anticholinergic drugs decreases intestinal peristalsis. Without any prompt management, this decrease can result in a colonic ischemia or necrosis. In patients treated with neuroleptics, the onset of constipation must alert medical staff. The author warned systematic monitoring of bowel movements should be performed in any patient receiving anticholinergic drugs.

Imidafenacin [SEDA-35, 266; SEDA-38, 125; SEDA-40, 000]

CLINICAL STUDY

The safety of mirabegron, a beta3-adrenoceptor agonist, and imidafenacin, an anticholinergic agent, in overactive bladder patients was examined. A multicenter, prospective randomized cross-over study at five hospitals in Japan from December 2012 to June 2015 was conducted. The patients were assigned to Group A or B. Group A patients were administered mirabegron (50 mg per day) for 8 weeks, followed by a 2-week washout period, and then imidafenacin (0.2 mg per day) for 8 weeks. This order of drug administration was reversed in Group B. A total of 33 and 18 patients in Group A and 37 and 26 patients in Group B continued to receive treatment at weeks 8 and 18, respectively. No significant difference was observed in the drug effects between mirabegron and imidafenacin. Although imidafenacin administration significantly increased the scores for dry mouth, blurred vision, and constipation, mirabegron administration did not. The author concluded imidafenacin administration is associated with a higher rate of dry mouth, blurred vision, and constipation as compared to mirabegron administration [19c] (Fig. 1).

Tiotropium [SEDA-35, 319; SEDA-36, 247; SEDA-37, 199; SEDA-38, 124; SEDA-39, 141; SEDA-40, 000]

PLACEBO-CONTROLLED TRIALS

In a multicenter, randomized, double-blind, placebo-controlled trial that was conducted in China, randomly assigned 841 patients with chronic obstructive pulmonary disease of Global Initiative for Chronic Obstructive Lung Disease stage 1 (mild) or 2 (moderate) severity to receive a once-daily inhaled dose (18 mug) of tiotropium (419 patients) or matching placebo (422) for 2 years. The incidence of adverse events was generally similar in the two groups [20C].

CLINICAL STUDY

Long-term cardiovascular safety is of special interest in maintenance treatment of chronic obstructive pulmonary disease with long-acting beta2-agonists and long-acting muscarinic antagonists, given potential cardiovascular effects. Patients (494/3100) contributed to an adjudicated analysis of serious adverse

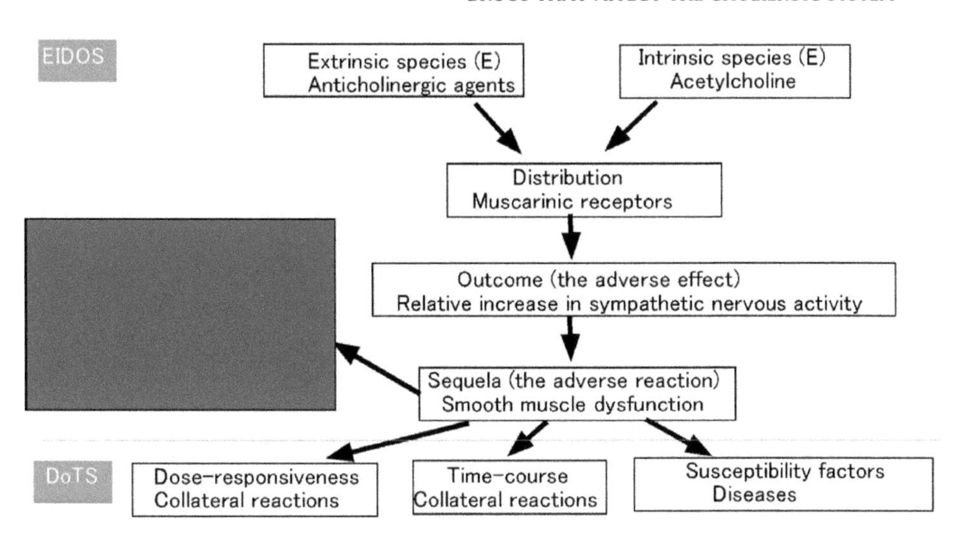

FIG. 1 The EIDOS and DoTS descriptions of anticholinergic agents.

effects: 260 had respiratory-related, 53 had cardiovascular-related and 16 had cerebrovascular-related serious adverse effects. Incidences of these serious adverse effects were comparable between treatments. There was no evidence of any increased risk for the combination compared to the monotherapy groups. The author concluded these data provide confidence for clinicians that tiotropium/olodaterol 5/5 μg can be safely administered once-daily to patients with moderate to very severe chronic obstructive pulmonary disease long-term [21C].

CLINICAL STUDY

The efficacy and safety of once-daily tiotropium + olodaterol (2.5/5 μg or 5/5 μg) for treating chronic obstructive pulmonary disease have been demonstrated in the large, multinational, randomized, Phase III studies TONADO(R) 1 and 2, which included 413 Japanese patients (~80 in each group). The incidence of adverse events was comparable in the tiotropium + olodaterol 2.5/5 μg (75.0%), tiotropium + olodaterol 5/5 μg (85.4%), and olodaterol 5 μg (80.5%) groups, with drug-related adverse events being reported in 5.0%, 7.3%, and 4.9% of patients, respectively. No deaths were reported during the study period. Serious adverse events were reported in 14 out of 122 patients (11.5%). The most common was COPD, reported in seven patients (three in the tiotropiumþolodaterol 2.5/5 mg group, two in the tiotropiumþolodaterol 5/5 mg group, and two in the olodaterol monotherapy group), with all other serious adverse events being single occurrences [22C].

Solifenacin [SEDA-35, 266; SEDA-38, 125; SEDA-39, 000]

Systematic Review

Pooled data from seven randomized placebo-controlled trials were analysed to evaluate relationships between baseline body mass index, gender or age and the efficacy/tolerability of solifenacin (5–10 mg daily) in patients with overactive bladder. The overall incidence of treatment-emergent adverse events was higher in patients receiving solifenacin than in those receiving placebo; the most commonly reported treatment-emergent adverse events were dry mouth and constipation [23M].

Placebo-Controlled Trials

The emergence of urinary retention, specifically acute urinary retention, has been a concern when treating men with lower urinary tract symptoms with antimuscarinic drugs. In NEPTUNE (12-week, double-blind), men (≥45 years) with LUTS were randomized to receive tamsulosin oral-controlled absorption system 0.4 mg, fixed-dose combination of solifenacin 6 mg + tamsulosin oral-controlled absorption system 0.4 mg, fixed-dose combination Soli 9 mg + tamsulosin oral-controlled absorption system 0.4 mg, or placebo. In NEPTUNE II (40-week, open-label extension of NEPTUNE), continuing patients received 4-week fixed-dose combination solifenacin 6 mg + tamsulosin oral-controlled absorption system, then fixed-dose combination solifenacin 6 mg or 9 mg + tamsulosin oral-controlled absorption system for the remainder of the study, switchable every 3 months. Across both studies, 1208 men received ≥1 dose of fixed-dose combination Soli 6 mg or 9 mg + tamsulosin oral-controlled absorption system for up to 52 weeks; 1199 men completed NEPTUNE and 1066 received ≥1 dose in NEPTUNE II. In total, 13 men (1.1%; 95% CI, 0.6%–1.8%) reported a urinary retention event while receiving fixed-dose combination, eight of which were acute urinary retention (0.7%; 95% CI, 0.3%–1.3%, incidence 7/1000 man-years). Six men reported urinary retention events while taking solifenacin 6 mg + tamsulosin oral-controlled absorption system (three acute urinary retention), and seven men reported a urinary retention event while taking solifenacin 9 mg + tamsulosin oral-controlled absorption system (five acute urinary retention). One man developed acute urinary retention while taking tamsulosin oral-controlled absorption system alone and four reported urinary retention (three were acute) during placebo run-in. Most acute urinary retention/urinary retention events occurred within 4 months of treatment initiation. The author concluded fixed-dose combination solifenacin and tamsulosin oral-controlled absorption system was associated with a low rate of urinary retention and acute urinary retention in men with lower urinary tract symptoms [24C].

Clinical Study

The efficacy and tolerability of solifenacin 5 mg fixed dose in children with newly diagnosed idiopathic overactive bladder were investigated. Drug-induced adverse effects were reported in seven patients (20.6%) including constipation, dryness of mouth, fatigue and sleepiness. These results indicate that solifenacin 5 mg fixed dose is effective against overactive bladder symptoms, and its tolerability is acceptable without significant adverse effects in children with overactive bladder [25c].

References

[1] Ogoyama M, Takahashi H, Kobayashi Y, et al. Ritodrine-induced rhabdomyolysis, infantile myotonic dystrophy, and maternal myotonic dystrophy unveiled. J Obstet Gynaecol Res. 2017;43(2):403–7 [A].

[2] Meoli EM, Goldsweig AM, Malm BJ. Acute myocardial infarction from coronary vasospasm precipitated by pseudoephedrine and metoprolol use. Can J Cardiol. 2017;33(5). 688.e1-.e3. [A].

[3] Bishop DG, Cairns C, Grobbelaar M, et al. Prophylactic phenylephrine infusions to reduce severe spinal anesthesia hypotension during cesarean delivery in a resource-constrained environment. Anesth Analg. 2017;125(3):904–6 [c].

[4] Mon W, Stewart A, Fernando R, et al. Cardiac output changes with phenylephrine and ephedrine infusions during spinal anesthesia for cesarean section: a randomized, double-blind trial. J Clin Anesth. 2017;37:43–8 [c].

[5] Vallejo MC, Attaallah AF, Elzamzamy OM, et al. An open-label randomized controlled clinical trial for comparison of continuous phenylephrine versus norepinephrine infusion in prevention of spinal hypotension during cesarean delivery. Int J Obstet Anesth. 2017;29:18–25 [c].

[6] Sun S, Sun D, Yang L, et al. Dose-dependent effects of intravenous methoxamine infusion during hip-joint replacement surgery on postoperative cognitive dysfunction and blood TNF-alpha level in elderly patients: a randomized controlled trial. BMC Anesthesiol. 2017;17(1):75 [c].

[7] Schapira AH, Fox SH, Hauser RA, et al. Assessment of safety and efficacy of safinamide as a levodopa adjunct in patients with parkinson disease and motor fluctuations: a randomized clinical trial. JAMA Neurol. 2017;74(2):216–24 [C].

[8] Timmermann L, Oehlwein C, Ransmayr G, et al. Patients' perception of parkinson's disease-associated pain following initiation of rotigotine: a multicenter non-interventional study. Postgrad Med. 2017;129(1):46–54 [c].

[9] Park D. Pramipexole-induced limb dystonia and its associated complex regional pain syndrome in idiopathic parkinson's disease: a case report. Medicine (Baltimore). 2017;96(28):e7530 [A].

[10] von Hohenberg C. C and H Dressing. [stealing as an impulse control disorder associated with pramipexole—a case report from forensic psychiatric practice]. Psychiatr Prax. 2017;44(3):172–4 [A].

[11] Lees AJ, Ferreira J, Rascol O, et al. Opicapone as adjunct to levodopa therapy in patients with parkinson disease and motor fluctuations: a randomized clinical trial. JAMA Neurol. 2017;74(2):197–206 [C].

[12] Tetrud J, Nausieda P, Kreitzman D, et al. Conversion to carbidopa and levodopa extended-release (ipx066) followed by its extended use in patients previously taking controlled-release carbidopa-levodopa for advanced parkinson's disease. J Neurol Sci. 2017;373:116–23 [c].

[13] Rajan R, Popa T, Quartarone A, et al. Cortical plasticity and levodopa-induced dyskinesias in parkinson's disease: connecting the dots in a multicomponent network. Clin Neurophysiol. 2017;128(6):992–9 [r].

[14] Zheng W, Wang S, Ungvari GS, et al. Amantadine for antipsychotic-related weight gain: meta-analysis of randomized placebo-controlled trials. J Clin Psychopharmacol. 2017;37(3):341–6 [M].

[15] Pahwa R, Tanner CM, Hauser RA, et al. Ads-5102 (amantadine) extended-release capsules for levodopa-induced dyskinesia in parkinson disease (ease lid study): a randomized clinical trial. JAMA Neurol. 2017;74(8):941–9 [C].

[16] Kim HJ, Sun HY, Choi H, et al. Efficacy and safety of initial combination treatment of an alpha blocker with an anticholinergic medication in benign prostatic hyperplasia patients with lower urinary tract symptoms: updated meta-analysis. PLoS One. 2017;12(1)e0169248 [M].

[17] Kirkland SW, Vandenberghe C, Voaklander B, et al. Combined inhaled beta-agonist and anticholinergic agents for emergency management in adults with asthma. Cochrane Database Syst Rev. 2017;1:Cd001284 [M].

[18] Cuny P, Houot M, Ginisty S, et al. Quetiapine and anticholinergic drugs induced ischaemic colitis: a case study. Encéphale. 2017;43(1):81–4 [A].

[19] Torimoto K, Matsushita C, Yamada A, et al. Clinical efficacy and safety of mirabegron and imidafenacin in women with overactive bladder: a randomized crossover study (the micro study). NeurourolUrodyn. 2017;36(4):1097–103 [c].

[20] Zhou Y, Zhong NS, Li X, et al. Tiotropium in early-stage chronic obstructive pulmonary disease. N Engl J Med. 2017;377(10):923–35 [C].

[21] Buhl R, Magder S, Bothner U, et al. Long-term general and cardiovascular safety of tiotropium/olodaterol in patients with moderate to very severe chronic obstructive pulmonary disease. Respir Med. 2017;122:58–66 [C].

[22] Ichinose M, Kato M, Takizawa A, et al. Long-term safety and efficacy of combined tiotropium and olodaterol in japanese patients with chronic obstructive pulmonary disease. Respir Investig. 2017;55(2):121–9 [C].

[23] Cardozo L, Herschorn S, Snijder R, et al. Does bmi, gender or age affect efficacy/tolerability of solifenacin in the management of overactive bladder? Int Urogynecol J. 2017;28(3):477–88 [M].

[24] Drake MJ, Oelke M, Snijder R, et al. Incidence of urinary retention during treatment with single tablet combinations of solifenacin+tamsulosin ocas for up to 1 year in adult men with both storage and voiding luts: a subanalysis of the neptune/ neptune ii randomized controlled studies. PLoS One. 2017;12(2): e0170726 [C].

[25] Lee SD, Chung JM, Kang DI, et al. Efficacy and tolerability of solifenacin 5 mg fixed dose in korean children with newly diagnosed idiopathic overactive bladder: a multicenter prospective study. J Korean Med Sci. 2017;32(2):329–34 [c].

CHAPTER

14

Dermatological Agents

Stacey D. Curtis,[1], W. Cary Mobley*[†]

*Pharmacotherapy and Translational Research, University of Florida College of Pharmacy, Gainesville, FL, United States

[†]University of Florida College of Pharmacy, Gainesville, FL, United States

[1]Corresponding authors: scurtis@cop.ufl.edu

ANTI-INFECTIVE AGENTS

Benzoyl Peroxide

A placebo-controlled, randomized, double-blind, parallel-group, comparative, multicenter study was conducted to investigate the efficacy and safety of benzoyl peroxide (BPO) gel, administered once daily for 12 weeks to 609 Japanese patients with acne vulgaris. Safety was evaluated based on adverse events, local skin tolerability scores and laboratory test values. Patients were randomly assigned to 2.5% BPO, 5% BPO, or placebo. The incidences of adverse events in the 2.5% and 5% BPO groups were 56.4% and 58.8%, respectively. All adverse events were mild or moderate in severity, and most occurred at the site of application. Adverse events with >2% incidence and a possible causal relationship with 2.5% BPO were skin exfoliation (19.1%), application site erythema (13.7%), application site irritation (8.3%), application site pruritus (3.4%), and contact dermatitis (2.5%). Adverse events with >2% incidence and a possible causal relation with 5% BPO were skin exfoliation (23.5%), application site irritation (12.3%), application site erythema (10.8%), and application site pruritus (2.5%). Most patients completed the study without discontinuation and for those who discontinued the study, adverse events resolved quickly. The authors concluded that the tolerability of BPO at these concentrations would be high in actual use settings [1C].

Clindamycin

A case report details a 22-year-old male patient who developed acute generalized exanthematous pustulosis (AGEP) with vancomycin and clindamycin use. Eight days prior to his hospitalization, the patient extracted a pimple on his right hand, which later became infected. He presented and was admitted for cellulitis. After treatment with IV clindamycin for 3 days (dose unknown), IV vancomycin (dose unknown) was added when the patient developed a fever of 103°F with a burning rash of pustules spreading from his groin to his torso. He was given diphenhydramine to treat the rash. Both the clindamycin and vancomycin were continued for 2 additional days for a total of 5 days of IV clindamycin and 2 days of IV vancomycin. The patient was discharged with oral clindamycin for continued treatment of cellulitis. He presented back to the emergency department with a fever of 99.9°F and worsening rash, which had spread up his back and upper extremities. The patient was given a dose of diphenhydramine and ceftriaxone and later discharged. The rash continued to spread to the mucous membranes of his tongue prompting the patient to seek help at a different institution where, upon physical examination, he was admitted. The erythematous rash continued to spread to his proximal lower extremities covering most of his body for management of the acute skin eruption and persistent cellulitis. Based on the rash presentation, time course, lack of eosinophilia, elevated liver enzymes and skin biopsy, the patient's rash was classified as AGEP secondary to clindamycin therapy [2c].

BIOLOGICS

Adalimumab

Investigators analyzed data from the first 7 years of A 10-Year, Post-marketing, Observational Study to Assess Long Term Safety of HUMIRA® (Adalimumab) in Adult Patients With Chronic Plaque Psoriasis (PS) (ESPRIT): randomized controlled trial. This analysis included the assessment of all treatment-emergent

Side Effects of Drugs Annual, Volume 40
ISSN: 0378-6080
https://doi.org/10.1016/bs.seda.2018.08.001

(All-TE) adverse events (AE) since the initial (first ever) dose of adalimumab in 6051 patients, representing 23660.1 patient years (PY) of overall exposure to the drug. During the registry's first 7 years, most of the patients remained free of All-TE cardiovascular events, serious infections, and malignancy: 99.6%, 99.6%, 99.8%, 96.7%, and 97.1% of patients remained free of All-TE myocardial infarction, cerebrovascular accident, congestive heart failure, serious infections, and malignancies, respectively. The incidence rate of All-TE serious AEs was 4.4 events (E)/100 PY of overall exposure to adalimumab. The most common All-TE serious AE was serious infection (1.0 E/100 PY). The All-TE serious AEs with incidence of ≥20 events overall were cellulitis (31 events, 0.1 E/100 PY), pneumonia (29 events, 0.1 E/100 PY), and myocardial infarction (22 events, <0.1 E/100 PY). The standardized mortality ratio for TE deaths in the registry was 0.27 (95% CI 0.18–0.38) while the All-TE cardiovascular events, serious infections, and malignancies were the primary causes of 8, 1, and 12 deaths (<0.1 E/100PY, each), respectively. The investigators reported that the number of TE deaths was below the expected rate. No new safety signals were identified during the first 7 years of the registry, and investigators concluded that the overall safety was consistent with the known safety profile of adalimumab [3MC].

A randomized, double-blind, multiperiod, phase III trial was conducted in 114 patients across 38 clinics in 13 countries. Patients (aged ≥4 to <18 years) with severe plaque psoriasis who had not responded to topical therapy were randomly assigned to receive adalimumab 0.8 mg/kg (n=38) or 0.4 mg/kg (n=39) subcutaneously at week 0, then every other week starting at week 1, or oral methotrexate once weekly (0.1–0.4 mg/kg) (n=37) for 16 weeks. Responders to adalimumab were then withdrawn from treatment for up to 36 weeks and re-treated with adalimumab for 16 weeks if the disease became uncontrolled. During the initial treatment period, 84 (74%) of the patients reported adverse events: 26 (68%) of 38 in the adalimumab 0.8 mg/kg group, 30 (77%) of 39 in the adalimumab 0.4 mg/kg group, and 28 (76%) of 37 in the methotrexate group. The most frequent adverse events were infections [17 (45%) in the adalimumab 0.8 mg/kg group during initial treatment; 22 (56%) in the adalimumab 0.4 mg/kg group; 21 (57%) in the methotrexate group]. Three serious adverse events were reported, all of which were in the adalimumab 0.4 mg/kg group, and were judged as not related to the study drug. During the retreatment period, 25 (66%) of 38 patients reported adverse events. Four (11%) of these patients received adalimumab 0.8 mg/kg and reported severe adverse events. One of these patients who was assigned to methotrexate in the initial treatment period and received re-treatment with adalimumab 0.8 mg/kg reported an adverse event of urticaria that was considered likely related to adalimumab and led to discontinuation from the study. The most frequently reported adverse events during retreatment were infections. No new safety risks were identified [4C].

Adalimumab, Etanercept and Ustekinumab

Researchers applied a propensity-score weighted Cox proportional hazards model with data from the British Association of Dermatologists Biologic Interventions Register (BADBIR), to evaluate the risk of serious infections of biologics used to treat psoriasis in comparison to patients on non-biologic systemic therapies. Data from 9038 patients were analyzed with the following breakdown of cohorts: etanercept (1352 patients), adalimumab (3271 patients), ustekinumab (994 patients), and non-biologics (3421 patients). Analysis of data revealed that 283 patients had a serious infection. Of these patients, the incidence rates with 95% confidence interval (CI)/1000 person-years were: non-biologic 14.2 (11.5–17.4); etanercept 15.3 (11.6–20.1); adalimumab 13.9 (11.4–16.6); ustekinumab 15.1 (10.8–21.1). No significant increases in the risk of serious infection were observed for etanercept, adalimumab, or ustekinumab, compared against non-biologic systemic therapies or against methotrexate only. The researchers concluded that etanercept, adalimumab, and ustekinumab have no significant differential risk of serious infection and recommended that the risk of serious infection should not be a key discriminator for patients and clinicians when choosing between non-biologic systemic therapies, etanercept, adalimumab and ustekinumab for the treatment of psoriasis [5R].

Adalimumab, Etanercept, Infliximab and Ustekinumab

Researchers conducted a multicenter, longitudinal, disease-based analysis of risk of overall infections, serious infections, and infection recurrence among moderate-to-severe psoriasis patients taking both classic and biological systemic drugs. An examination of data from the records of 2153 patients in the BIOBADADERM registry (Spanish Registry of Adverse Events for Biological Therapy in Dermatological Diseases) was used to generate crude rates of infection during therapy with systemic drugs, including biological drugs (infliximab, etanercept, adalimumab, and ustekinumab) and non-biological drugs (acitretin, cyclosporine, and methotrexate). Unadjusted and adjusted risk ratios (RRs) of infection, serious infections, and recurrent infections of systemic therapies compared with methotrexate were also calculated. The researchers found a slight, but significant adjusted increase in the risk of overall infections between the

TNF-antagonist drugs and methotrexate. The adjusted RR for overall infection rate was significantly increased in the groups treated with adalimumab and methotrexate (adjusted RR=2.13, 95% confidence interval [CI]=1.2–3.7), infliximab (adjusted RR=1.71, 95% CI=1.1–2.65), cyclosporine (adjusted RR=1.58, 95% CI=1.17–2.15), ustekinumab and methotrexate (adjusted RR=1.56, 95% CI=1.08–2.25), and etanercept (adjusted RR=1.34, 95% CI: 1.02–1.76) compared with methotrexate alone. For recurrent infections, the combined adalimumab and methotrexate group had the highest risk (adjusted RR=4.33, 95% CI=2.27–8.24), followed by infliximab (adjusted RR=1.98, 95% CI=1–3.94). For serious infections, the number was low among all the treatments, with the highest rate in the combined adalimumab and methotrexate group (adjusted incidence rate=23.3, 95% CI=12.9–42), followed by the cyclosporine group (adjusted incidence rate=20, 95% CI=8.3–47.9) and the infliximab group (adjusted incidence rate=18.9, 95% CI=7.9–45.5). The researchers noted that their study found that psoriasis patients treated with TNF-antagonist drugs (etanercept, infliximab, and adalimumab) had a lower rate of serious infections compared with rheumatoid arthritis patients treated with TNF-antagonist drugs [6MC].

Etanercept

A phase IV, open-label, single-arm estimation study was conducted on 64 patients with moderate-to-severe plaque psoriasis who were treated with etanercept following the loss of satisfactory response to adalimumab. Patients received etanercept 50 mg twice weekly for 12 weeks, followed by 50 mg weekly. A total of 40 patients (63%) reported treatment-emergent adverse events. Of these reported adverse events, those reported by 16 patients (25%) were considered by the investigator to be related to etanercept. Two serious adverse events were reported by two patients, ischemic colitis (considered treatment-related) and diverticulitis (considered non-treatment-related). One patient discontinued the study because of cough. The following adverse events were reported in greater than 5% of the patients: cough, 6 (9%); headache, 6 (9%); nasopharyngitis, 5 (8%); and Vitamin D deficiency, 5 (8%). No patients died during the study and no new safety signals were observed [7c].

A case report describes a 38-year-old male patient with psoriasis who was treated with etanercept for 1 year following failure of initial therapy. The patient discontinued etanercept because he developed lymphadenopathy with an ultimate diagnosis of stage IVB nodular sclerosing Hodgkin's lymphoma. The author noted the rarity of this diagnosis because psoriasis and etanercept seldom cause lymphoma and, in those rare cases, it is most likely Epstein–Barr virus-positive non-Hodgkin's lymphoma. Although the authors concluded that etanercept-induced immunosuppression likely was a key factor in the development of this patient's lymphoma, they cautioned that the exact mechanism is not completely understood, and the potential roles of psoriasis, etanercept or some other underlying trigger are unknown [8c].

Infliximab

A phase III, multicenter, single-arm, 40-week trial study of 51 patients, the Study on Psoriasis Treatment with Remicade Escalating Dosage (SPREAD), was conducted in Japan. Patients with plaque psoriasis, psoriatic arthritis, pustular psoriasis (except for localized pustular psoriasis) or psoriatic erythroderma, aged 16–75 years and showing loss of efficacy to standard-dose infliximab (5 mg/kg every 8 weeks) were included in the study. Patients with guttate psoriasis or drug-induced psoriasis were excluded. The aim of the study was to evaluate efficacy and safety of infliximab dose escalation in Japanese psoriasis patients with loss of efficacy to standard-dose therapy. Patients received infliximab dose escalation (10 mg/kg every 8 weeks) from weeks 0 to 32. The safety of the drug treatment was evaluated up to week 40 for patients who completed the study and up to 8 weeks after the last dose of infliximab for patients who dropped out. Incidences of adverse events for patients who completed the study, serious adverse events, serious infections and serious infusion reactions were 92%, 10%, 4% and 0%, respectively. Common adverse events were abnormal laboratory findings (59%, 30/51), infections and infestations (45%, 23/51) and gastrointestinal disorders (27%, 14/51). Additional adverse events that occurred in at least 5% of patients were double-stranded DNA antibody increase (49%, 25/51), nasopharyngitis (27%, 14/51), headache (8%, 4/51), urticaria (8%, 4/51), and cough, dyspnea and vomiting (6% each, 3/51). The authors concluded that infliximab dose escalation was effective and well-tolerated in psoriasis patients with loss of efficacy to standard-dose therapy [9c].

A case report describes the safety and efficacy of infliximab administered for 11 years to a 55-year-old male patient with a 30-year history of moderate-to-severe plaque psoriasis. Upon initial diagnosis, his PASI (Psoriasis Area and Severity Index) was above 10 and he progressed with joint involvement about 2 months after the appearance of the lesions. For the first 18 years of his therapy, he was treated with topical medications and oral corticosteroids. His joint involvement worsened even during periods of remission and, ultimately, he presented to the author's medical service with psoriatic exfoliative erythroderma and intense arthralgia. After 3 months of treatment with methotrexate 25 mg every week, the patient's erythroderma and joint pain improved, but his lesions remained widespread. He subsequently was placed on infliximab and methotrexate was eventually discontinued. Infliximab administration was maintained at 5mg/kg/day every 8 weeks for the subsequent 11 years. The author reports that there were no complications or side effects throughout 11 years of treatment with infliximab [10c].

Ixekizumab

Efficacy and safety of ixekizumab for the treatment of moderate-to-severe plaque psoriasis: Results through 108 weeks of a randomized, controlled phase III clinical trial (UNCOVER-3) A randomized, double-blind, multicenter, phase III clinical trial (UNCOVER-3) was undertaken to evaluate the long-term efficacy and safety of ixekizumab for the treatment of moderate-to-severe plaque psoriasis. In the study, 1346 patients were randomized at a ratio of 2:2:2:1 in four study arms: 80mg ixekizumab every 2 or 4 weeks, 50mg etanercept twice weekly, or placebo. At week 12, patients switched to ixekizumab every 4 weeks during a long-term extension (LTE) period. During the LTE, 1077 (84.5%) patients reported ≥1 treatment-emergent adverse events (TEAEs), and of these, 85% were mild or moderate in severity. The frequency, severity, and distribution of LTE TEAEs were similar between the treatment arms with the most frequently observed TEAEs (≥5%) being nasopharyngitis, upper respiratory tract infections, injection-site reactions, arthralgia, bronchitis, and headache. Reported TEAEs of special interest included Candida infections (3.8%). Grade 3 or 4 neutropenia occurred at a cumulative rate of 0.6%, with one grade 4 event occurring at week 108. Other TEAEs included malignancies (1.7%), cerebrocardiovascular events (2.4%), Crohn's disease (0.2%) and ulcerative colitis (0.2%) [11C].

An open-label, phase III study was conducted to evaluate the pharmacokinetics, safety, and efficacy of ixekizumab when administered via prefilled syringe (PFS) in patients with moderate-to-severe psoriasis. Patients were randomized to ixekizumab delivery via PFS or autoinjector device. Randomization was stratified by weight (<80kg, 80–100kg, >100kg), injection assistance (yes/no) and injection site (arm, thigh or abdomen). Following a 160-mg initial dose at week 0, patients received subcutaneous 80-mg ixekizumab as a single injection every 2 weeks for 12 weeks. Adverse event reporting, vital signs and clinical laboratory data were used to evaluate safety in 200 patients. Both devices had safety results that were consistent with the known safety profile of ixekizumab. Forty-eight percent of the patients experienced treatment emergent adverse events (TEAEs), the majority of which were mild (n=48, 24%) or moderate (n=42, 21%) in severity. Injection-site reaction was the most frequently reported TEAE (≥5% of patients in either device group). Most TEAEs were judged by the investigators as possibly related to the study drug (n=40, 20%) and few were possibly related to the device (n=11, 5%). Each of the 11 possibly device-related TEAEs involved administration, site conditions, 10 of which occurred in the autoinjector group and one of which occurred in the PFS group. Three patients (1.5%) who used the autoinjector experienced an adverse event that led to study drug discontinuation. Treatment-emergent abnormal laboratory values occurring in ≥5% of patients included low bicarbonate (41.9%), high very-low-density lipoprotein cholesterol (20.7%) and high creatinine clearance (14.0%). The proportion of patients with treatment-emergent high systolic blood pressure was 2.8% and for high diastolic pressure was 4.7% overall. No patient experienced a treatment-emergent low systolic or diastolic blood pressure value or low or high pulse rate [12C].

An international, placebo-controlled, phase III study of ixekizumab in patients with moderate-to-severe psoriasis was conducted (UNCOVER-1). From this study, a subgroup analysis of 33 Japanese patients was undertaken. These patients were randomized at week 0 to a placebo (n=13), ixekizumab 80mg every 4 weeks (IXEQ4W, n=8) or every 2 weeks ((IXEQ2W, n=12). At week 12, the 16 responders (IXEQ4W, n=8, and IXEQ2W, n=8) were re-randomized to a placebo (n=6), IXEQ12 (n=5) or to IXEQ4W, (n=5). The incidence of treatment-emergent adverse events (TEAEs) from weeks 0 to 12 were 76.9% (placebo), 75.0% (IXEQ4W) and 87.5% (IXEQ2W), and from weeks 12 to 60 were 66.7% (placebo) and 100% (IXEQ12W, IXEQ4W). Ixekizumab-treated patients had no severe treatment-emergent adverse events. Nine TEAEs were considered possibly drug related in weeks 0–12, and two in weeks 12–60. Infection was the most frequently reported TEAE of special interest, with all being mild except for moderate cases of bronchopneumonia (n=1), nasopharyngitis (n=2), tinea pedis (n=1), and influenza (n=1). Three IXEG4W-treated patients discontinued the medication due to mild generalized pruritis (n=1), moderate allergic edema (n=1), and moderate bronchopneumonia (considered a serious adverse event, n=1). The investigators concluded that ixekizumab administered for 60 weeks was safe for Japanese patients with moderate-to-severe psoriasis, with no Japanese-specific safety findings, and the results were in line with the overall findings from UNCOVER-1 [13c].

To determine the safety of ixekizumab, an interleukin-17 (IL-17) inhibitor, in psoriasis patients, investigators analyzed the integrated safety data from a 12-week induction period, a 12- to 60-week maintenance period, and from all ixekizumab-treated patients from seven clinical trials. Data analyzed were from a total of 4209 patients and exposure-adjusted incidence rates (IRs) per

100 patient-years were reported. During the induction period, the IRs of patients experiencing one or more treatment-emergent adverse events (TEAEs) were 251 and 236 among ixekizumab- and etanercept-treated patients, respectively, and for serious AEs was 8.3 in both groups. During maintenance, for ixekizumab, the IRs of TEAEs and serious AEs were 100.4 and 7.8, respectively. The IRs of TEAEs of special interest (including serious infections, malignancies and major adverse cardiovascular events) were comparable for ixekizumab and etanercept during the induction period, and the IRs of TEAEs were similar in the ixekizumab groups and the etanercept group. Most TEAEs for both drugs were mild or moderate and the most common TEAE type was infection. The IRs of infections were significantly higher in the total ixekizumab group than in the etanercept group, but this was not true for individual infection types. Cellulitis was the most frequently reported serious infection (0.4%), which was not unexpected, based on known IL-17 inhibition by ixekizumab. The IRs of injection-site reactions (ISRs) were similar in ixekizumab- and etanercept-treated patients. However, among individual types of ISRs, the IR of injection-site pain was higher in patients treated with ixekizumab every 2 weeks compared with etanercept-treated patients, and the IR of injection-site erythema was lower in patients treated with ixekizumab every 4 weeks compared with etanercept-treated patients. The rate of malignancies (excluding non-melanoma skin cancer) among ixekizumab-treated patients was comparable with etanercept-treated patients during the induction period. However, conclusions on malignancy risks were limited by the relatively short-term duration of these trials. New onset and exacerbations of Crohn's disease and ulcerative colitis occurred in ixekizumab-treated patients. Because IL-17 inhibitors appear to be ineffective in the treatment of inflammatory bowel disease and may worsen the condition, additional evaluation is needed to understand the association between IBD and IL-17 inhibitors. No unexpected safety signals were observed during maintenance with ixekizumab [14MC].

Secukinumab

A pooled analysis of four phase III trials; Efficacy of Response and Safety of Two Fixed Secukinumab Regimens in Psoriasis (ERASURE), Full Year Investigative Examination of Secukinumab vs Etanercept Using Two Dosing Regimens to Determine Efficacy in Psoriasis (FIXTURE), Secukinumab administration by pre-filled syringe: efficacy, safety, and usability results from a randomized controlled trial in psoriasis (FEATURE), and Efficacy, safety and usability of secukinumab administration by autoinjector/pen in psoriasis: a randomized, controlled trial (JUNCTURE) was undertaken to evaluate the safety and efficacy of secukinumab in Hispanic and non-Hispanic patients with moderate-to-severe plaque psoriasis. Data up to week 52 of the studies were pooled. Patients who self-identified as being ethnically Hispanic were included in the Hispanic subgroup of this analysis and Hispanic patients accounted for 13.9% (237/1709) of the pooled study population. Administration of secukinumab (150 or 300mg) or etanercept (50mg) through week 52 resulted in similar rates of adverse events (AEs), serious adverse events (SAEs), and discontinuations due to AEs across treatment groups and between the Hispanic and non-Hispanic subgroups. Common adverse events in the Hispanic patients included headache [15 patients (17.0%), 11 (11.3%), 10 (19.6%)], nasopharyngitis [14 (15.9%), 11 (11.3%), 4 (7.8%)], diarrhea [12 (13.6%), 10 (10.3%), 3 (5.9%)], and influenza [11 (12.5%), 15 (15.5%), 4 (7.8%)]. In the Hispanic patient subgroup, SAEs were reported in two patients receiving secukinumab 300mg (overdose and uterine leiomyoma), four patients receiving secukinumab 150mg (cholelithiasis, ischemic stroke, thyroid cancer, and sixth nerve paralysis), and two patients receiving etanercept (acute cholecystitis and clavicle fracture). The investigators concluded that secukinumab is safe in both Hispanic and non-Hispanic patients [15C].

Ustekinumab

Extension of ustekinumab maintenance dosing interval in moderate-to-severe psoriasis: results of a phase IIIb, randomized, double-blinded, active-controlled, multicentre study (PSTELLAR) were undertaken to assess clinical responses of patients with moderate-to-severe psoriasis given extended ustekinumab maintenance dosing intervals at 0, 4 and 16 weeks. Dosing was weight based according to package labeling. A total of 378 patients achieving a week-28 Physician's Global Assessment (PGA) score of cleared/minimal (PGA=0/1) were randomized to group 1 ($n=76$) or group 2 ($n=302$). A higher proportion of patients in group 1 (55%) than group 2 (39%) had PGA=0/1 at all seven visits from weeks 88 to 112. From week 28 to week 124 (core safety analysis), similar proportions of patients in group 1 (72%) and group 2 (73%) had at least one adverse event (AE). Seven patients in group 1 (9%) and 21 patients in group 2 (7%) experienced serious AEs, of which rib fracture ($n=2$, 0.7%) in group 2 was the only AE experienced by more than one patient. Three serious infections were reported in group 2: bacterial infection (20 weeks), cystitis (12 weeks) and urinary tract infection (12 weeks). Two deaths occurred during the double-blinded treatment period (week 28–124). One patient (group 1) died of natural causes (possible cardiovascular AE) approximately 10 weeks after the last dose of ustekinumab 90mg given at week 28. One patient (12 weeks. subgroup of group 2) was diagnosed with acute myeloid leukemia approximately 1 month after the last dose of ustekinumab 90 mg was administered at week 80 and died approximately 5 months later. AEs led to discontinuation of the study agent in 7% of patients in group 1 and 6% of patients in group 2. Overall, 63 of 455 patients (13·8%) treated with ustekinumab, who had evaluable samples, tested positive for antibodies to ustekinumab through week 124. The incidence of antibody development was similar between patients receiving ustekinumab 45 mg ($n=41$, 13.9%) and 90 mg ($n=22$, 13.7%), and between patients in group 1 ($n=7$, 9%) and patients in group 2 ($n=32$, 11%). The investigators concluded that extending the dosing interval did not affect antibody development or safety [16C].

CORTICOSTEROIDS

Betamethasone

A phase IIa, randomized, single-center, investigator-blinded, study compared the efficacy of the fixed combination of calcipotriol as hydrate (Cal) 50 μg/g plus betamethasone as dipropionate (BD) 0.5 mg/g aerosol foam to an alcohol-free treatment vs betamethasone 17-valerate 2.25 mg (BV)-medicated plasters in patients with psoriasis. Both treatments are recommended for treating psoriasis plaques localized in difficult-to-treat areas (elbow, knee, anterior face of the tibia). The study included 35 patients. A total of 15 patients (42.9%) experienced adverse events (AEs) during the trial; none were severe. The most frequent AEs were headache ($n=8$; 22.9%), influenza ($n=3$; 8.6%), and oropharyngeal pain ($n=3$; 8.6%). No cutaneous AEs were observed and no other AEs occurred in more than one patient. No deaths, serious AEs or other significant AEs were observed [17c].

A randomized, parallel-group, investigator-blinded Phase III, PSO-ABLE[a] study compared the efficacy and safety of a fixed combination of calcipotriol 50 μg/g (Cal) plus betamethasone 0.5 mg/g (BD) foam vs gel in 463 patients with psoriasis vulgaris. Adverse events were reported in a similar proportion between the formulations; Cal/BD aerosol foam ($n=77$, 41.6%) and the Cal/BD gel ($n=85$, 45.2%) over the 12-week treatment period. The most common adverse events (AEs) overall were upper respiratory tract infection, nasopharyngitis and vitamin D deficiency. Itch was reported as an adverse event by five patients receiving Cal/BD aerosol foam (2.7%) and two receiving Cal/BD gel (1.1%). Most adverse events were mild or moderate. Serious adverse events were reported in four patients (2.2%) receiving Cal/BD aerosol foam (congestive heart failure, gastroesophageal reflux, prostate cancer, exacerbation of psoriasis) and three (1.6%) receiving Cal/BD gel (post-procedural hemorrhage, type 2 diabetes mellitus, ischemic stroke). One serious adverse event was considered related to Cal/BD aerosol foam treatment (exacerbation of psoriasis after 69 days of treatment). There were no deaths [18C].

An evidence-based review reported that the combination of calcipotriene/betamethasone dipropionate for the treatment of psoriasis vulgaris is well tolerated in patients and has resulted in adverse effects that are rare and not severe. The most common reported adverse effects include erythema, pruritus, and burning near the site of application. Some studies reported the combined medication has less adverse events than the use of calcipotriol as monotherapy. Subjects using betamethasone dipropionate as monotherapy reported the least adverse effects [19R].

Clobetasol

A randomized, single-blind trial for the treatment of keloid scars in 17 patients was conducted using either clobetasol propionate (Dermovate) 0.05% cream occluded with Mepiform silicone occlusion dressing or intralesional (IL) triamcinolone. A significant improvement in the Patient and Observer Scar Assessment Scale (POSAS) at 12 weeks was reported; however, there was no statistically significant difference observed between the two treatments. All patients in the IL triamcinolone group reported pain and 70.6% observed a necrotic skin reaction. There was a significantly higher rate of adverse effects such as erythema (41.2% vs 17.6%), hypopigmentation (35.3% vs 23.5%), telangiectasia (41.2% vs 17.6%) and skin atrophy (23.5% vs 5.9%) documented in the IL triamcinolone group when compared to clobetasol propionate 0.05% cream under occlusion with silicone dressing [20c].

PHOSPHODIESTERASE-4 ENZYME INHIBITORS

Apremilast

Multiple phase III, multicenter, randomized, double-blind, placebo-controlled studies were conducted to demonstrate efficacy, safety and tolerability of apremilast for moderate-to-severe plaque psoriasis and psoriatic arthritis. The studies enrolled 1184 patients who were randomized to apremilast 30 mg twice daily or placebo. The most common adverse effects (AEs) were diarrhea, nausea,

[a] "Calcipotriol plus betamethasone dipropionate aerosol foam provides superior efficacy vs gel in patients with psoriasis vulgaris: randomized, controlled PSO-ABLE study."

upper respiratory tract infection, nasopharyngitis, tension headache, and headache. Rates for overall AEs and most common AEs (particularly diarrhea and nausea) decreased over time. Most cases of diarrhea and nausea were mild-to-moderate in severity, occurred most frequently during the first 2 weeks of apremilast treatment, and generally resolved within 4 weeks. Rates of serious AEs were comparable across exposure periods and did not increase with long-term apremilast exposure. Serious AEs occurring in ≥2 patients in any exposure period (0 to ≥156 weeks) included coronary artery disease ($n=6$), acute myocardial infarction ($n=4$), osteoarthritis ($n=4$), and nephrolithiasis ($n=4$). Discontinuations for AEs were few (<2% for any AE) and occurred primarily during the first few weeks of treatment. A total of three deaths occurred in patients receiving apremilast (one during each year of exposure) due to cardiac failure, cerebrovascular accident, and severe mitral stenosis with congestive heart failure. Most AEs were mild or moderate in severity and did not lead to treatment discontinuation [21MC].

A phase IIIB, randomized, double-blind, placebo-controlled study was conducted to evaluate the efficacy and safety of apremilast vs placebo in biologic-naïve patients with moderate-to-severe plaque psoriasis and to evaluate the safety of switching from etanercept to apremilast. The study enrolled 250 patients who were randomized to placebo, apremilast 30 mg twice daily, or etanercept 50 mg every week. After week 16, all patients continued or switched to apremilast through week 104. Among patients with reported adverse events (AE), ≥95% of AEs were mild or moderate in severity. Serious AEs leading to discontinuation were infrequent and comparable across groups. The most common AEs (in ≥5% of patients in any treatment group) were nausea, diarrhea, upper respiratory tract infection, nasopharyngitis, tension headache and headache. In apremilast-treated patients, more than half of the reported diarrhea and nausea cases occurred within the first 4 weeks of dosing. These were predominantly mild in severity and generally resolved within 1 month. No patient in any group reported severe nausea or severe diarrhea [22C].

A case report details a 71-year-old male with a medical history of hypertension, hyperlipidemia, coronary artery disease status post-percutaneous coronary intervention in 2000, psoriatic arthritis, anemia of chronic disease, and chronic thrombocytopenia. The patient's psoriatic arthritis was well-controlled on etanercept injections for 7 years; however, it was discontinued due to recurrent infections requiring hospitalization. Ten weeks after the discontinuation of etanercept, the patient was started on apremilast. Two weeks after the initiation of the apremilast, the patient was admitted to the hospital with a chief complaint of weakness of a few days duration, worsening to the point that he was unable to stand or feed himself. The patient had a 27-pound weight loss over 2 months. On arrival, laboratory test results were significant for hypokalemia, hyperchloremic metabolic acidosis, low uric acid concentration, positive urine anion gap, and proteinuria, which were resolved on discontinuation of the drug. The patient was subsequently diagnosed with Fanconi syndrome (proximal renal tubular acidosis). Two months after discharge from the hospital, the patient was restarted on apremilast therapy. Seventeen days later, the patient was admitted with similar complaints and laboratory values as before but improved once apremilast treatment was discontinued. No other causes of Fanconi syndrome were evident [23c].

Crisaborole

A long-term multicenter, open-label, 48-week safety study was conducted with 517 patients ≥2 years of age with mild-to-moderate atopic dermatitis, who continued treatment with crisaborole 2% ointment, a topical phosphodiesterase-4 inhibitor, after completing a 28-day phase III pivotal study. In the pivotal and long-term extension studies, treatment-related adverse events (AEs) occurred in 10.3% of patients and most (85.9%) were mild or moderate. The most frequently reported treatment-related AEs were atopic dermatitis (worsening, exacerbation, flare, or flare-up) ($n=16$; 3.1%), application-site pain (burning or stinging) (2.3%), and application-site infection (1.2%). The median duration for the most frequently reported treatment-related AEs was 17.5 (11 − 33) days for atopic dermatitis, 5 (1 − 30) days for application-site pain, and 12 (9–197) days for application-site infection. Nine patients (1.7%) discontinued the long-term study because of treatment-related AEs. The authors concluded that crisaborole ointment had a low frequency of treatment-related AEs over 48 weeks of treatment of patients with atopic dermatitis [24C].

RETINOIC ACID DERIVATIVES

Isotretinoin

A retrospective, comparative study was conducted in 3525 patients to estimate the adverse effects after oral isotretinoin treatment of acne vulgaris over a 5-year observation period. The most common adverse effect observed was dry lip, and was seen in 3525 participants (100%), followed by xerosis in 3348 participants (94.97%), facial erythema in 2334 participants (66.21%), nose bleeds (epistaxis) in 1666 participants (47.26%), cheilitis in 1473 participants (41.78%), and muscle aches (myalgias) in 1367 participants (38.78%). Itching of the skin was observed in 1344 participants (38.12%) and exfoliation of the skin was observed in 1092 participants (30.97%). Other adverse effects were observed in less often: tiredness in 731 participants (20.73%), headaches in 595 participants (16.87%), joint aches (arthralgias) in 433 participants (12.28%), retinoid dermatitis in 412 participants (11.68%), trachyonychia in 368 participants (10.43%), mood change in 335 participants (9.50%), eye lesions in 316 participants(8.96%), dry eyes in 201 participants (5.70%), hair loss only in 153 females (4.34%), abdominal pain in 131 participants (3.71%), vision changes in 99 participants (2.80%), insomnia in 98 participants (2.78%) and sun sensitivity in 92 participants, all patients with retinoid dermatitis (2.60%). Added together, the adverse effects with psychiatric symptoms (mood change, tiredness, insomnia, suicidal ideation, and gastrointestinal upset) accounted for 25.16% of the total AEs (887 participants). In lab investigations, a rise above normal values of total cholesterol and serum triglycerides was seen in 101 participants (3.11%). There was an increase in the level of liver enzymes after treatment and a rise above normal values was seen in 68 participants (2.09%). The vast majority of patients who experienced adverse effects continued therapy to achieve the recommended dose while only five patients (0.14%) discontinued therapy because of side effects: suicidal ideation, intensification of fears, flare of psoriasis, alopecia, herpes zoster [25MC].

A randomized, open therapeutic trial was conducted to compare the efficacy and safety of oral isotretinoin vs topical isotretinoin in the treatment of plane warts. The study enrolled 40 patients who were randomized into two groups. Group A received an oral dose of isotretinoin (0.5 mg/kg/day) and Group B received a topical dose of isotretinoin (0.05% gel). Both groups received their dose once daily at night for 3 months or until the lesion completely cleared. The oral isotretinoin showed better and earlier response than topical isotretinoin. The most common side effect in Group A (oral isotretinoin) was cheilitis, which was manageable in most patients. The most common side effect in Group B (topical isotretinoin) was erythema and burning sensation [26c].

Tazarotene vs Adapalene

A randomized, open-label, parallel design clinical trial was conducted to compare the efficacy and safety of tazarotene 0.1% plus clindamycin 1% gel against adapalene 0.1% plus clindamycin 1% gel for treatment of facial acne vulgaris. The study enrolled 60 patients (30 per treatment group) with facial acne at an outpatient dermatology department of a tertiary healthcare center. Ten patients from the tazarotene plus clindamycin group and four patients from the adapalene plus clindamycin group complained of a burning sensation at the site of application. The severity of burning sensation in 13 patients was mild and tolerable and thus the treatment was continued in those patients. However, one patient on tazarotene plus clindamycin could not tolerate it, so the treatment was withdrawn. Additional adverse effects were mild itching and drying of skin in eight patients. The adverse event profile was statistically similar in both groups [27c].

Tretinoin

A clinical trial, non-blind, comparative study aimed to evaluate the efficacy, tolerability, and safety of salicylic acid peeling in comparison with topical tretinoin in the treatment of post-inflammatory hyperpigmentation (PIH). The study enrolled 45 patients who were classified into three groups according to a therapeutic regimen: Group I (treated with salicylic acid peeling 20%–30%), Group II (treated with topical tretinoin 0.1%), Group III (treated with a combination of salicylic acid peel and topical tretinoin). All patients in Group I complained of stinging and burning sensation during peeling session and postpeel scaliness; four patients (26.7%) experienced itching after the end of session, erythema in four patients (26.7%), dryness in eight patients (53.4%). In Group II, eight patients (53.3%) experienced itching, erythema in five patients (33.3%), dryness in two patients (13.3%), exfoliation in all patients and burning sensation in eight patients (53.3). In Group III, four patients (26.7%) experienced itching, erythema in six patients (40.0%), dryness in two patients (13.3%), exfoliation in all patients and burning sensation in eight patients (53.3%). Patients of Group II and Group III showed more statistically significant dryness, exfoliation and burning than those in Group I. Although most patients complained of exfoliation, itching, burning sensation, erythema, and dryness following topical tretinoin application, these symptoms were alleviated by frequent application of an emollient [28c].

Additional literature on this topic can be found in a review published in SEDA-39 [29R].

References

[1] Kawashima M, Sato S, Furukawa F, et al. Twelve-week, multicenter, placebo-controlled, randomized, double-blind, parallel-group, comparative phase II/III study of benzoyl peroxide gel in patients with acne vulgaris: a secondary publication. J Dermatol. 2017;44(7):774–82 [C].

[2] Croy C, Buehrle K, Austin SJ. Clindamycin-associated acute generalized exanthematous pustulosis. J Clin Pharm Ther. 2017;42(4):499–501 [c].

[3] Menter A, Thaçi D, Wu JJ, et al. Long-term safety and effectiveness of adalimumab for moderate to severe psoriasis: results from 7-year interim analysis of the ESPRIT registry. Dermatol Ther (Heidelb). 2017;7(3)):365–81 [MC].

[4] Papp K, Thaçi D, Marcoux D, et al. Efficacy and safety of adalimumab every other week versus methotrexate once weekly in children and adolescents with severe chronic plaque psoriasis: a randomised, double-blind, phase 3 trial. Lancet. 2017;390(10089):40–9 [C].

[5] Yiu ZZN, Smith CH, Ashcroft DM, et al. Risk of serious infection in patients with psoriasis on biologic therapies: a prospective cohort study from the british association of dermatologists biologic interventions register (BADBIR). J Invest Dermatol. 2018;138(3):534–41 [R].

[6] Dávila-Seijo P, Dauden E, Descalzo MA, et al. Infections in moderate to severe psoriasis patients treated with biological drugs compared to classic systemic drugs: findings from the BIOBADADERM registry. J Invest Dermatol. 2017;137(2):313–21 [MC].

[7] Bagel J, Tyring S, Rice KC, et al. Open-label study of etanercept treatment in patients with moderate-to-severe plaque psoriasis who lost a satisfactory response to adalimumab. Br J Dermatol. 2017;177(2):411–8 [c].

[8] Malik F, Ali N, Jafri SIM, et al. Casual or causal? two unique cases of hodgkin's lymphoma: a case report and literature review. Am J Case Rep. 2017;18:553–7 [c].

[9] Torii H, Nakano M, Yano T, et al. Efficacy and safety of dose escalation of infliximab therapy in Japanese patients with psoriasis: results of the SPREAD study. J Dermatol. 2017;44(5):552–9 [c].

[10] Trídico LA, Antonio JR, Mathias CE, et al. Effectiveness and safety of infliximab for 11 years in a patient with erythrodermic psoriasis and psoriatic arthritis. An Bras Dermatol. 2017;92(5):743–5 [c].

[11] Blauvelt A, Gooderham M, Iversen L, et al. Efficacy and safety of ixekizumab for the treatment of moderate-to-severe plaque psoriasis: results through 108 weeks of a randomized, controlled phase 3 clinical trial (UNCOVER-3). J Am Acad Dermatol. 2017;77(5):855–62 [C].

[12] Callis Duffin K, Bagel J, Bukhalo M, et al. Phase 3, open-label, randomized study of the pharmacokinetics, efficacy and safety of ixekizumab following subcutaneous administration using a prefilled syringe or an autoinjector in patients with moderate-to-severe plaque psoriasis (UNCOVER-A). J Eur Acad Dermatol Venereol. 2017;31(1):107–13 [C].

[13] Imafuku S, Torisu-Itakura H, Nishikawa A, et al. Efficacy and safety of ixekizumab treatment in Japanese patients with moderate-to-severe plaque psoriasis: Subgroup analysis of a placebo-controlled, phase 3 study (UNCOVER-1). J Dermatol. 2017;44(11):1285–90 [c].

[14] Strober B, Leonardi C, Papp KA, et al. Short- and long-term safety outcomes with ixekizumab from 7 clinical trials in psoriasis: etanercept comparisons and integrated data. J Am Acad Dermatol. 2017;76(3):432–440.e17. [MC].

[15] Adsit S, Zaldivar ER, Sofen H, et al. Secukinumab is efficacious and safe in hispanic patients with moderate-to-severe plaque psoriasis: pooled analysis of four phase 3 trials. Adv Ther. 2017;34(6):1327–39 [C].

[16] Blauvelt A, Ferris LK, Yamauchi PS, et al. Extension of ustekinumab maintenance dosing interval in moderate-to-severe psoriasis: results of a phase IIIb, randomized, double-blinded, active-controlled, multicentre study (PSTELLAR). Br J Dermatol. 2017;177(6):1552–61 [C].

[17] Queille-Roussel C, Rosen M, Clonier F, et al. Efficacy and safety of calcipotriol plus betamethasone dipropionate aerosol foam compared with betamethasone 17-valerate-medicated plaster for the treatment of psoriasis. Clin Drug Investig. 2017;37(4):355–61 [c].

[18] Paul C, Leonardi C, Menter A, et al. Calcipotriol plus betamethasone dipropionate aerosol foam in patients with moderate-to-severe psoriasis: sub-group analysis of the PSO-ABLE study. Am J Clin Dermatol. 2017;18(3):405–11 [C].

[19] Patel NU, Felix K, Reimer D, et al. Calcipotriene/betamethasone dipropionate for the treatment of psoriasis vulgaris: an evidence-based review. Clin Cosmet Investig Dermatol. 2017;10:385–91 [R].

[20] Nor NM, Ismail R, Jamil A, et al. A randomized, single-blind trial of clobetasol propionate 0.05% cream under silicone dressing occlusion versus intra-lesional triamcinolone for treatment of keloid. Clin Drug Investig. 2017;37(3):295–301 [c].

[21] Crowley J, Thaçi D, Joly P, et al. Long-term safety and tolerability of apremilast in patients with psoriasis: pooled safety analysis for ≥156 weeks from 2 phase 3, randomized, controlled trials (ESTEEM 1 and 2). J Am Acad Dermatol. 2017;77(2):310–7.e1. [MC].

[22] Reich K, Gooderham M, Green L, et al. The efficacy and safety of apremilast, etanercept and placebo in patients with moderate-to-severe plaque psoriasis: 52-week results from a phase IIIb, randomized, placebo-controlled trial (LIBERATE). J Eur Acad Dermatol Venereol. 2017;31(3):507–17 [C].

[23] Perrone D, Afridi F, King-Morris K, et al. Proximal Renal Tubular Acidosis (Fanconi Syndrome) Induced by Apremilast: A Case Report. Am J Kidney Dis. 2017;70(5):729–31 [c].

[24] Eichenfield LF, Call RS, Forsha DW, et al. Long-term safety of crisaborole ointment 2% in children and adults with mild to moderate atopic dermatitis. J Am Acad Dermatol. 2017;77(4): 641–9.e5. [C].

[25] Brzezinski P, Borowska K, Chiriac A, et al. Adverse effects of isotretinoin: a large, retrospective review. Dermatol Ther. 2017;30(4):e12483. https://doi.org/10.1111/dth.12483 [MC].

[26] Kaur GJ, Brar BK, Kumar S, et al. Evaluation of the efficacy and safety of oral isotretinoin versus topical isotretinoin in the treatment of plane warts: a randomized open trial. Int J Dermatol. 2017;56(12):1352–8 [c].

[27] Maiti R, Sirka CS, Ashique Rahman MA, et al. Efficacy and safety of tazarotene 0.1% plus clindamycin 1% gel versus adapalene 0.1% plus clindamycin 1% gel in facial acne vulgaris: a randomized, controlled clinical trial. Clin Drug Investig. 2017;37(11):1083–91 [c].

[28] Mohamed Ali BM, Gheida SF, El Mahdy NA, et al. Evaluation of salicylic acid peeling in comparison with topical tretinoin in the treatment of postinflammatory hyperpigmentation. J Cosmet Dermatol. 2017;16(1):52–60 [c].

[29] Black AT. Chapter 12—Dermatological drugs, topical agents, and cosmetics. In: Ray SD, editor. Side Effects of Drugs: Annual: A Worldwide Yearly Survey of New Data in Adverse Drug Reactions. 39. Elsevier; 2017. p. 145–55 [R].

CHAPTER

15

Antihistamines (H1 Receptor Antagonists)

Tyler S. Dougherty[1]

Department of Pharmacy Practice, South College School of Pharmacy, Knoxville, TN, United States
[1]Corresponding author: tdougherty@southcollegetn.edu

BILASTINE (SEDA-37, 185; SEDA-38, 143; SEDA-39, 157)

Nervous System

In this randomized, double-blind, multi-center trial, researches prospectively evaluated the efficacy and safety of bilastine for the use of chronic spontaneous urticaria in Japanese patients. Three hundred and four patients were randomly assigned to receive either bilastine 10mg, bilastine 20mg, or placebo for a total of 2 weeks. After 2 weeks, the investigators evaluated the total symptom score of itch and rash as compared to baseline. Adverse events reported in the placebo arm (4/103, 3.9%) were similar to the 10mg (5/100, 5.0%) and 20mg (4/101, 4.0%) bilastine. Of note, the nervous system adverse effects included somnolence with one patient (1.0%) in the placebo group and two patients (2.0%) in the bilastine 10mg group. Somnolence is a well-documented adverse event in histamine antagonists. The authors also noted there were no serious adverse drug events, deaths, or trial withdrawal associated with bilastine [1C].

Safety for Long-Term Use

Researchers performed an open-label, multi-center, phase III study of bilastine for the treatment of skin diseases with pruritus or chronic spontaneous urticaria with the goal of assessing safety and efficacy. One hundred and ninety-seven patients between the ages of 18 and 74 received at least one dose of bilastine 20mg in this two-part trial. In part one of the trial, patients were initially enrolled for a 12-week period. Patients were allowed to continue the trial (part 2) for 40 more weeks if the patients' symptom scores improved from baseline without bilastine associated side effects. Patient demographics included 50.3% male, mean age of 40 years old, and adherence rate of 98%. Study investigators divided adverse events into two categories: adverse events and bilastine-related adverse events. In general, the most common adverse events included nasopharyngitis (56/197, 28.4%), contact dermatitis (8/197, 4.1%), and eczema (8/197, 4.1%). Bilastine-related adverse events included an increase in somnolence (2/197, 1.0%), an elevation of aspartate aminotransferase levels (1/197, 0.5%), an increase in γ-glutamyltransferase levels (1/197, 0.5%), nocturia (1/197, 0.5%), headaches and asteatosis (7/197, 3.6% for both). It was noted that all bilastine-related adverse events occurred during the first 12 weeks of the study period. No patients discontinued therapy due to bilastine. The authors concluded treatment with bilastine 20mg once daily for chronic spontaneous urticaria or skin disease with pruritus was well tolerated and safe [2C].

Nervous System

In this double-blind, randomized, placebo-controlled active comparator trial, patients were given either bilastine, fexofenadine, or a matched placebo for the treatment of perennial allergic rhinitis. The investigators evaluated a total nasal symptom score for 2 weeks, with the primary analysis evaluating the change from baseline. In total, 765 patients were included in the study, with 256 receiving bilastine 20mg once daily, 254 receiving fexofenadine 60mg twice daily, and 255 patients receiving placebo. Overall, the adverse events reported in all three groups were similar. Somnolence was reported more often in the bilastine (2/249, 0.8%) and fexofenadine (1/246, 0.4%) than compared to placebo (0 patients). In addition, postural dizziness was reported in bilastine (1/249, 0.4%), while dizziness was reported in the fexofenadine group (2/246, 0.8%), while zero patients reported dizziness in the placebo group. Finally, headache was also reported in by both bilastine and fexofenadine groups (1/246, 0.4% vs 2/246, 0.8%, respectively) [3C].

Side Effects of Drugs Annual, Volume 40
ISSN: 0378-6080
https://doi.org/10.1016/bs.seda.2018.07.011

DIPHENHYDRAMINE (SEDA-36, 233; SEDA-37, 185; SEDA-38, 143)

Cardiovascular

This is a case report of a 28-year-old pregnant female patient carrying a baby 41 weeks gestation. During labor, the mother requested an epidural in which 100mcg of fentanyl was administered. After having increased pain, the patient requested a second dose of fentanyl. The patient then began complaining of itching without shortness of breath or hives. The doctor then decided to administer 25mg of diphenhydramine to treat the opioid-induced pruritus. Diphenhydramine administration did not change the patient's blood pressure (101/68mmHg) or heart rate (83bpm) from baseline. However, within 14 minutes of diphenhydramine administration, fetal tachycardia was seen with a heart rate increasing from 155 to 205bpm. The fetal heart rate remained at or above 200bpm for 16 minutes, and then remained above 170bpm for another 35 minutes. Fetal tachycardia was present for a total of 51 minutes. The fetus was eventually delivered transverse cesarean section [4A].

Histamine antagonists are known to cause anticholinergic effects, which can include elevated heart rate. The author's noted this is the first report of diphenhydramine-induced fetal tachycardia. The authors did discuss a previous trial of 100mg IV dimenhydrinate, a similar histaminergic antagonist, which resulted in 10% of fetuses experiencing fetal tachycardia lasting on average 36 minutes [5C]. The authors noted the importance of obstetricians being aware of diphenhydramine-induced fetal tachycardia, especially in the presence of fetal insecurity. Finally, the authors suggested using low doses of either naloxone, propofol, or naltrexone in place of diphenhydramine [4A].

Nervous System, Gastrointestinal

This is a case report of a 13-year-old female patient presenting to the emergency department for an intentional overdose. The child's caretaker reported the patient had ingested 4–6 diphenhydramine 25mg tablets. Upon presentation, the child had an elevated heart rate, dilated pupils, and dry mucous membranes. Also, the patient began having active seizures along with hallucinations after an episode of vomiting. The patient was treated with benzodiazepines and physostigmine with minor improvement. After 38 hours of little improvement, the caretakers of the child stated at least 200 tablets of diphenhydramine 25mg were missing from the bottle. An endoscopy was then performed with removal of a pharmacobezoar. The patient improved to normal mental status within 12 hours. The authors emphasized the importance of pharmacobezoar consideration upon intentional overdose of any medication, now including diphenhydramine [6A].

Psychological/Psychiatric

This is a case report of an 82-year-old male patient presenting to the emergency room with hallucinations described as hieroglyphic writings on his walls. The patient's past medical history was negative for dementia, schizophrenia, and depression; however, he was being treated for anxiety with alprazolam. In the previous 2 months, the patient was started on oral 25mg diphenhydramine to help with sleep. All lab values and physical assessments were normal. After discontinuing the diphenhydramine, the patient's hallucinations stopped within 3 days. The author's noted this could potentially be the first case report of diphenhydramine-induced hallucinations in a geriatric patient [7A].

Neuromuscular Function

In this case report, researchers described a case of acute dystonia in a patient with Parkinson's disease followed by the withdrawal of diphenhydramine. The patient was a 58-year-old female currently taking ropinirole 4mg and levodopa 450mg for her 5-year history of Parkinson's and restless leg syndrome. The patient was also taking metformin, atorvastatin, and amitriptyline at home. Upon presenting to the emergency room, the patient had been experiencing bilateral lower extremity muscle spasms, which extended to the upper body and face. These muscle contractions were noted to be involuntary. The patient had also been experiencing excessive tearing, difficulty breathing and perspiring. Other notable findings on physical evaluation included raised and sideways direction of the eyes (oculogyric crisis). It was noted the patient had to forcefully bring her eyes back to normal position, but with maximum effort. Pertinent lab values included a serum creatine kinase of 7964units/L, elevated lactate, and hypomagnesemia. Upon medication reconciliation, it was reported by the husband the patient had been taking diphenhydramine for over 6 years to help with insomnia and tremor but had discontinued the medication 3 days before developing the recent symptoms. The patient was diagnosed with diphenhydramine withdrawal acute dystonic reaction and was treated with benztropine 1mg/day, diphenhydramine 50mg/day, and supportive treatment. Her abnormal movements and lab values quickly normalized within 3 days, and patient was discharged for physical therapy and slowly tapering of diphenhydramine and benztropine. The author's noted the importance of this case report being twofold. First, this is the initial case report of a Parkinson's patient developing acute dystonic reactions from diphenhydramine withdrawal. Secondly, the author's emphasized that clinicians should be aware of sudden medication changes or withdrawals, and their implications for side effects, especially in the Parkinson's disease population [8A].

Skin

This is a retrospective review of 109 individual patients who were administered subcutaneous diphenhydramine in a hospice setting from 2012 to 2015. Previous use of subcutaneous diphenhydramine 5% have resulted in tissue necrosis (reference). The authors noted that subcutaneous diphenhydramine is commonly used in hospice settings due to the difficulty of other routes of drug administration in this patient population. Upon record review, 109 patients were administered at least one diphenhydramine subcutaneous injection, with 50.5% (55/109) being male, with 93% of doses being less than 1 mL (50 mg/mL). Also, the average number of doses per patient was 5.9, with one patient receiving a maximum of 43 doses. Thirty-seven percent (40/109) of patients contained a hospice diagnosis of cancer, while the most common indication for diphenhydramine use was pruritus (61.6%, 69/109) and extrapyramidal symptoms (13.4%, 15/109). Of note, the subcutaneous injections were administered in fatty areas of the body, including upper arms, abdomen, and thighs [9A]. The results of this retrospective review demonstrated zero reports of necrosis resulting from subcutaneous administration of diphenhydramine. The authors noted previous literature demonstrating diphenhydramine skin necrosis resulted from injection sites of fingers, which is much different from fatty tissue areas [10S]. The conclusion of this review was that subcutaneous administration of diphenhydramine is a safe alternative to other routes of diphenhydramine in hospice patients [9A].

FEXOFENADINE (SEDA-39, 158)

Skin, Immunologic

In this case report, researchers described a case of Stevens–Johnson Syndrome/Toxic Epidermal Necrolysis (SJS/TEN) caused by fexofenadine. The patient was a woman with a past medical history positive for allergic rhinitis, but negative for drug allergies. Initially, the patient was being treated for nasal congestion and a sore throat with 625 mg amoxicillin–clavulanic acid three times a day. After worsening of symptoms, the patient began self-medicating with acetaminophen and ibuprofen. On the 6th day of her symptoms, the treating physician prescribed the patient to take fexofenadine–pseudoephedrine twice a day. The patient developed lip swelling after doses of the fexofenadine–pseudoephedrine combination. Two days later, the patient experienced increased lip swelling and blisters on her lips but did not seek treatment until day 8 in an emergency department in which she was treated with oral promethazine, IV hydrocortisone, and cetirizine–pseudoephedrine. However, the patient was admitted to the hospital the following day with red, non-blanching rash affecting approximately 20% of her body and high fever. The patient was treated for Stevens–Johnson Syndrome/Toxic Epidermal Necrolysis with topical steroids, cetirizine, and cyclosporine with full recovery of her skin in 50 days [11A].

The authors noted the patient had never taken the fexofenadine–pseudoephedrine combination medication previously. In addition, possible causes of the SJS/TEN were developed based on the Naranjo and ALDEN models [12H]. These models suspected "penicillin-G, amoxicillin-clavulanic acid, cefazolin, ceftriaxone, clarithromycin, fexofenadine-pseudoephedrine, and ibuprofen" as possible causes. Allergy determination using prick and skin tests was negative for all drugs except for fexofenadine–pseudoephedrine, which resulted in a delayed-hypersensitivity reaction. The patient was then re-tested using pseudoephedrine, fexofenadine–pseudoephedrine, loratadine–pseudoephedrine, clarithromycin, and ibuprofen, resulting in the patient only demonstrating a reaction to the fexofenadine–pseudoephedrine product. The patient was advised to not take fexofenadine containing medications in the future. The authors emphasized this being the first reported case of SJS/TEN caused by fexofenadine [11A].

KETOTIFEN (SEDA-38, 143; SEDA-39, 158)

Immunologic

Ketotifen is a topical second-generation antihistamine that is commonly used for the treatment of allergic conjunctivitis. In this case report, a 32-year-old female patient with a past medical history of atopy, hay fever, and more recent 3-year history of allergic conjunctivitis was being treated with ketotifen fumarate eye drops. Her current symptoms included red, watery conjunctivae, without eyelid involvement. Upon using allergy patch testing, the patient demonstrated an allergy to ketotifen and was advised to no longer use the ketotifen product. This resulted in resolution of her conjunctivitis. The authors noted the rareness of ophthalmic products resulting in conjunctivitis without eyelid involvement. They also emphasized ketotifen being a rare agent that can lead to conjunctivitis, since it is also used for the treatment of conjunctivitis [13A].

LEVOCETIRIZINE (SEDA-36, 233; SEDA-37, 185; SEDA-38, 143; SEDA-39, 158)

Nervous System

The minor sedative properties of cetirizine have been established in previous studies, while the sedation effects of levocetirizine are still undecided [14R]. As stated by the authors, this is the first reported meta-analysis reviewing the sedative effects of levocetirizine. Forty-eight studies were included in the analysis, encompassing 18014 participants. The primary outcome reported was the risk ratio of sedative effects compared to placebo, second-generation antihistamines, and first-generation antihistamines. When compared to placebo, levocetirizine caused modest sedation (RR 1.67; 1.17–2.38, 95% CI), while demonstrating no difference in sedation properties when compared to other second-generation histamine antagonists (fexofenadine, desloratadine, loratadine, bilastine, olopatadine, azelastine, and rupatadine) (RR 1.23; 0.96–1.58, 95% CI). As compared to first-generation antihistamines, levocetirizine had less sedation. Overall, a number needed to harm was calculated for sedation in patients taking levocetirizine as 111, which is comparable to other second-generation antihistamines. The authors concluded that levocetirizine can be used in patients who require an antihistamine with limited sedation effect, while also demonstrating that levocetirizine has less sedation as compared to first-generation antihistamines [15M].

PROMETHAZINE (SEDA-36, 233; SEDA-37, 185; SEDA-38, 143)

Drug Abuse

In 2014, Denmark regulated promethazine, a first-generation histamine antagonist, as a prescription-only-product due to abuse and misuse. In this study, researches retrospectively reviewed poisoning patterns of antihistamines from two independent databases between 2007 and 2013. The four antihistamines reported included promethazine and cyclizine (first-generation), along with cetirizine and loratadine (second-generation). During this time period, first-generation histamine antagonists were reported 61% and 73% of the time in each of the two databases. In one of the databases, 82% of the intentional exposures were due to suicide attempts, of which 78% used a first-generation histamine blocker. However, first-generation histamine antagonists were only used on 29% of overdoses caused by accidents. This implies that patients were choosing to use first-generation antihistamines when attempting to overdose. Also, between the two databases, adults were more likely to use promethazine and cyclizine (66% and 81%) when compared to children (12% and 14%). The authors concluded that limiting promethazine to a prescription-only product was supported based off the findings that a higher percentage of overdoses occurred with first-generation antihistamine, specifically promethazine. Finally, the authors hypothesized that patients could begin replacing promethazine with other available over-the-counter first-generation antihistamines, such as diphenhydramine, due to the comparable side effect profile. These results suggest increasing the regulation of first-generation histamine antagonists [16C].

Additional case studies on this topic can be found in reviews published in previous SEDA editions [17R].

References

[1] Hide M, Yagami A, Togawa M, et al. Efficacy and safety of bilastine in Japanese patients with chronic spontaneous urticaria: a multicenter, randomized, double-blind, placebo-controlled, parallel-group phase II/III study. Allergol Int. 2017;66:317–25 [C].
[2] Yagami A, Furue M, Togawa M, et al. One-year safety and efficacy study of bilastine treatment in Japanese patients with chronic spontaneous urticaria or pruritus associated with skin diseases. J Dermatol. 2017;44:375–85 [C].
[3] Okubo K, Gotoh M, Asako M, et al. Efficacy and safety of bilastine in Japanese patients with perennial allergic rhinitis: a multicenter, randomized, double-blind, placebo-controlled, parallel-group phase II study. Allergol Int. 2017;66:97–105 [C].
[4] Abernathy A, Alsina L, Greer J, et al. Transient fetal tachycardia after intravenous diphenhydramine administration. Obstet Gynecol. 2017;130(2):374–6 [A].
[5] Sosa A, Faneite P. Effects of dimenhydrinate on the fetal heart rate and uterine contractions. Ginecol Obstet Mex. 1977;41:453–8 [C].
[6] Johnson J, Williams K, Banner Jr. W. Adolescent with prolonged toxidrome. Clin Toxicol. 2017;55(5):364–5 [A].
[7] Wongrakpanich S, et al. Diphenhydramine as a cause of visual hallucinations in older adults. Geriatr Gerontol Int. 2017;17(1):172–3 [A].
[8] Jacob S, Ly H, Villenueva R, et al. Acute dystonic reaction in a patient with Parkinson's disease after diphenhydramine withdrawal. Acta Neurol Belg. 2017;117:345–6 [A].
[9] Chen A, Loquias E, Roshan R, et al. Safe use of subcutaneous diphenhydramine in the patient hospice unit. Am J Hosp Palliat Med. 2017;34(10):954–7 [A].
[10] U.S. Food and Drug Administration (FDA). Diphenhydramine hydrochloride injection. (Reference ID: 3243048). Web site http://www.accessdata.fda.gov/drugsatfda_docs/label/2013/091526lbl.pdf; 2013. Published March 26, 2013 [S].
[11] Teo S, Santosa A, Bigliardi P. Stevens-Johnson syndrome/toxic epidermal necrolysis overlap induced by fexofenadine. J Investig Allergol Clin Immunol. 2017;27(3):191–3 [A].
[12] Naranjo CA, Busto U, Sellers EM, et al. A method for estimating the probability of adverse drug reactions. Clin Pharmacol Ther. 1981;30(2):23–45 [H].
[13] Smets K, Werbrouck J, Goossens A, et al. Sensitization from ketotifen fumarate in eye drops presenting as chronic conjunctivitis. Contact Dermatitis. 2017;76:124–6 [A].
[14] Yanai K, Rogala B, Chugh K, et al. Safety considerations in the management of allergic diseases: focus on antihistamines. Curr Med Res Opin. 2012;28(4):623–42 [R].

[15] Snidvongs K, Seresirikachorn K, Khattiyawittayakun L, et al. Sedative effects of levocetirizine: a systematic review and met-analysis of randomized controlled studies. Drugs. 2017;77:175–86 [M].

[16] Jensen L, Romsing J, Dalhoff K. A Danish survey of antihistamine use and poisoning patterns. Basic Clin Pharmacol Toxicol. 2017;120:64–70 [C].

[17] Dougherty TS. Antihistamines (H1 receptor antagonists). In: Ray SD, editor. Side effects of drugs: annual. A worldwide yearly survey of new data in adverse drug reactions, vol. 39; 2017. p. 157–9. chapter 13. [R].

CHAPTER

16

Drugs That Act on the Respiratory Tract

Mohamed Bayoumi-Ali, Richard John Sweeney†, Luqman Hussain Wali†, Doaa Farag‡, Favas Thaivalappil*,[1]*

*Abertawe Bro Morgannwg University Health board (ABMU), Singleton hospital, Swansea, United Kingdom
†Wales Deanery and ABMU, Singleton hospital, Swansea, United Kingdom
‡Abertawe Bro Morgannwg University Health board (ABMU), Neath Port-Talbot hospital, Port Talbot, United Kingdom
[1]Corresponding author: favas.thaivalappil@wales.nhs.uk

INHALED GLUCOCORTICOIDS [SEDA-36, 241; SEDA-37, 195; SEDA-38, 153; SEDA-39, 161]

Inhaled glucocorticoids (ICS) are commonly used in the management of different respiratory diseases including asthma and chronic obstructive pulmonary disease (COPD). It is known from the literature that their adverse effects (AE) can be local or systemic. ICS are often combined with other drugs for the management of different respiratory diseases.

A recent nationwide prospective cohort study by Cho et al. (Korea) sought to highlight the differences in the adverse event profile between, inhaled corticosteroids alone and those delivered as part of a combination inhaler with a long acting beta 2 agonist (LABA). The incidence of pneumonia and fractures was measured in 1995 patients with mild, moderate or severe newly diagnosed COPD. The study demonstrated a reduced all-cause mortality among new ICS/LABA users compared with new ICS users, adjusted hazard ratio (HR) 0.77 (95% CI: 0.62–0.95) for the total population. However, when divided according to COPD severity, there was no statistical significance in all-cause mortality in the severe COPD group, HR 1.07 (95% CI: 0.65–1.76), whereas patients with mild to moderate COPD, there was a statistically significant drop in all-cause mortality, HR 0.70 (95% CI: 0.55–0.89). The study demonstrated a reduced all-cause mortality in newly diagnosed mild to moderate COPD patients using ICS/LABA inhalers compared to those using ICS inhalers alone [1c].

The risk of pneumonia in patients with COPD using inhaled corticosteroids had not been rigorously studied. The SUMMIT study, is a double-blinded randomised control trial on 16568 patients, which investigated the risk of pneumonia with inhaled fluticasone fumarate and vilanterol in COPD patients with moderate airflow limitation. Results of the trial showed a reduction in severe COPD exacerbations when using a combination fluticasone/vilanterol inhaler compared with placebo, HR 0.804 (95% CI: 0.742–0.872, $P<0.001$). However, when compared to inhaled fluticasone alone, the results were not statistically significant, HR 0.977 (95% CI: 0.904–1.057, $P=0.563$). One may deduce that this trial does not demonstrate any evidence to suggest an increased pneumonia risk in moderate/severe COPD exacerbations when using inhaled Corticosteroids vs combined fluticasone/vilanterol inhaler [2C].

This is further demonstrated among the paediatric population in a meta-analysis carried out by Cazeiro et al. of 39 randomised control trials comparing inhaled corticosteroids with placebo in children with asthma. This meta-analysis concluded that there was no statistically significant increase in the risk of pneumonia and the use of ICS, Risk Difference (RD) −0.1% (95% CI: −0.3% to 0.2%, $P=0.72$) [3M].

However, this is in contrast to the findings from the UPLIFT study, which is a double-blinded parallel-group study comparing the incidence of pneumonia in three population groups; those patients not receiving ICS; secondly, patients taking fluticasone propionate and lastly patients on all other ICS. The study found that patients on any ICS other than fluticasone had a higher

Side Effects of Drugs Annual, Volume 40
ISSN: 0378-6080
https://doi.org/10.1016/bs.seda.2018.07.012

incidence of pneumonia than those not using ICS (0.068 vs 0.056, respectively; $P=0.012$). The incidence of pneumonia was even higher in the patient group taking fluticasone propionate compared to those on another ICS (0.077 vs 0.058, respectively; $P<0.001$). Furthermore, COPD exacerbations were more frequent in patients taking ICS, and even more so, those on fluticasone (0.93 vs 0.84, respectively; $P=0.013$) [4C].

A meta-analysis of 10 randomized control trials and observational real-life studies was carried out investigating the effects of withdrawing inhaled corticosteroids on COPD. Overall, no statistically significant increased risk of COPD exacerbations was seen with ICS withdrawal, ($P<0.05$). However, the withdrawal of ICS significantly reduced the FEV1 ($P<0.001$). Furthermore, it was demonstrated that withdrawing ICS significantly shortened the time to first exacerbation, ($P<0.05$) [5M].

A historical cohort study of 23013 patients with obstructive lung disease was carried out in the United Kingdom on the risk of pneumonia when using fine-particle corticosteroids compared with extra-fine particle corticosteroids to treat obstructive lung disease. The extra-fine particle corticosteroids studied were extra-fine beclometasone dipropionate (efBDP) and extra-fine ciclesonide (efCIC). These were compared with fine-particle beclometasone dipropionate and fine-particle ciclesonide. The study reported that patients receiving extra-fine particle inhaled corticosteroids were significantly less likely to be coded for pneumonia, (OR 0.5; 95% CI: 0.27–0.93). Furthermore, they were less likely to experience acute exacerbations, (RR 0.90; 95% CI: 0.84–0.97), as well as less likely to experience acute respiratory events, (RR 0.90; 95% CI: 0.86–0.95). It was also noted that patients receiving daily doses of inhaled corticosteroids more than 700mcg were significantly more likely to develop pneumonia irrespective of particle size (OR 2.38; 95% CI: 1.17–4.83; $P<0.01$) [6C].

A systematic review of randomised control trials was carried out comparing the efficacy and safety of inhaled corticosteroids with fluticasone propionate in patients with asthma. In total, 54 randomised controlled trials were evaluated. Other inhaled corticosteroids compared with fluticasone were beclomethasone dipropionate, budesonide and ciclesonide. Endpoints considered were asthma symptom control, lung function, necessity of reliever, quality of life, frequency of exacerbations and side-effects related to steroid-use. It was noted that across all studies, fluticasone was associated with more favourable, or similar efficacy and safety to budesonide or beclomethasone. Efficacy compared to ciclesonide was comparable, however, in terms of safety, ciclesonide had a smaller impact on cortisol levels [7M].

A systematic review was undertaken to highlight the adverse drug events related to the full range of asthma medications in children. A total of 46 studies were included in the systematic review, including randomised control trials, cohort studies, case–control and quasi-experimental studies. The primary outcome was frequency of adverse drug events. Drug classes assessed included ICS, LABA, short-acting beta-agonists (SABA), combined ICS & LABA, leukotriene receptor antagonists and other. The majority of studies evaluated focus on adverse drug events related to inhaled corticosteroids, there were 174 adverse drug events affecting 13 organ systems related to ICS. Notably, growth suppression was reported with fluticasone and beclomethasone. Adrenal suppression was noted with budesonide, fluticasone and ciclesonide [8M].

Glaucoma Risk

The effect of inhaled corticosteroids on intraocular pressure in patients with open-angle glaucoma and ocular hypertension was investigated in a double-masked, placebo-controlled RCT which involved 22 patients with well controlled open-angle glaucoma or ocular hypertension. The intervention group received a 6-week course of twice-daily inhaled fluticasone propionate 250mcg, and the control group received a saline placebo. The results showed there was no statistically significant increase in mean intraocular pressure [9c].

Low Trauma Fractures

A study was carried out in Taiwan investigating whether the use of inhaled corticosteroids increased the risk of low-energy fractures in patients with chronic airway disease.1182 ICS users were recruited and 1182 controls. The study found an overall increased risk of low-energy fracture in all sites associated with ICS users, the adjusted HR 1.20 (95% CI: 1.10–1.31, $P<0.001$). The risk was significantly higher for fractures of the ulna & radius and hip fractures in ICS users compared to controls, (adjusted HR: 1.24, CI: 1.03–1.48; adjusted HR: 1.63, CI: 1.40–1.90, respectively) [10c].

Case Reports

Idiopathic chronic eosinophilic pneumonia (ICEP) is a rarely encountered form of parenchymal lung disease characterised by constitutional and respiratory symptoms, along with chest X-ray changes and elevated serum and/or bronchoalveolar eosinophilia. A 6 month to 1-year course of systemic corticosteroids dramatically improves symptoms, however, over half the patients relapse upon cessation of treatment or during tapering. Inhaled corticosteroids as monotherapy have been trialed and found not to be successful.

A case report on a 60-year-old female patient was written up detailing a treatment regime.

The patient received 14 days of prednisolone, 40mg daily, reduced to 20mg for 3–4 months with pneumocystis prophylaxis, then gradually weaned off the remaining glucocorticoids over the following 12 months. A CT thorax performed at 8 months into treatment demonstrated resolution of the opacities. The patient was transitioned to mometasone 440mcg MDI, tapered over 12 months to one puff daily for maintenance. At follow-up, the patient felt well and remained symptom free, however, she was found to have eosinophilic infiltration on a nasal polyp biopsy. The case report demonstrates that inhaled corticosteroids may be useful adjuncts in ICEP, in order to reduce risk of relapse [11A].

A case report was published highlighting an episode of iatrogenic Cushing's Syndrome following inhaled budesonide in a 47-year-old HIV positive female on Ritonavir. Inhaled budesonide is used instead of inhaled fluticasone in patients with HIV on Ritonavir due to a lower incidence of iatrogenic Cushing's Syndrome. Fluticasone is metabolised via the cytochrome P450 3A4 (CYP3A4) enzyme system, however, Ritonavir is a known CYP3A4 inhibitor thus leading to impaired metabolism and systemic accumulation, causing iatrogenic Cushing's Syndrome. This case report, however, details the fifth known episode of this adverse drug event and the importance of distinguishing it from HIV lipodystrophy [12A].

LEUKOTRIENE MODIFIERS [SEDA-36, 245; SEDA-37, 197; SEDA-38, 153; SEDA-39, 166]

Leukotriene modifiers (LTRAs) are used for the long-term management of asthma and allergic rhinitis, or exercise-induced asthma.

Montelukast

Although Montelukast is widely prescribed and generally well tolerated, some studies have reported serious neuropsychiatric adverse events among others.

In a recent case report about a 56-year-old male patient who recently started on Montelukast for his mild chronic asthma by his GP. Two to three weeks after initiating Montelukast at 10mg/day he noticed gradually worsening anxiety and waking suddenly from sleep, with increased heart rate, flushing and a loosening sensation in his viscera. After stopping Montelukast, the nocturnal panic symptoms subsided within a few days, while the awakenings themselves took a few weeks to resolve. Several months after the discontinuation of Montelukast, the patient reported that he experienced mild maintenance insomnia less than once a week with no panic symptoms [13A].

A Canadian retrospective cohort study analysed children aged 1–17 years initiated on Montelukast. Out of the 106 participants, most were male (58%), Caucasian (62%) with a median (interquartile range) age of 5 (3–8) years. The incidence (95% CI) of Montelukast discontinuation due to neuropsychiatric adverse events was 16% (10–26), mostly within 2 weeks of starting the drug. The most frequent adverse events were irritability ($n=9$), aggressiveness ($n=8$) and sleep disturbances ($n=7$). The median delay from drug cessation to symptom resolution was 2 (0–3.5) days. Most (70%) of the side-effects were classified as mild, requiring only drug cessation; the remainder (30%) were deemed to be very mild [14c].

Another unique case was described recently for Montelukast-induced angioedema. A 52-year-old man who was commenced on Montelukast for allergies, a week prior to his presentation at the emergency department with dysphagia and a "drowning" sensation while lying supine at night. He was also complaining of shortness of breath and wheezing with no pruritus, rash, nausea or vomiting. There were no other precipitating factors, previous history of angioedema or immunological disorders. If this was indeed an adverse event to his LTRA, it will introduce a potentially life threatening association to the medication [15A].

Pranlukast

A systematic review of 37 randomised control trials reviewed the evidence for the addition of leukotriene-antagonists to standard inhaled corticosteroids in the treatment of asthma in adults and adolescents. Trial participants were divided into three groups, all patients received a LTRA. The first group continued a maintenance dose of ICS, the second had an increased dose of ICS and the third a tapering dose of ICS. The primary outcome was number of exacerbations requiring oral corticosteroids, or percentage reduction in ICS dose from baseline. LTRA included were Montelukast, zafirlukast and pranlukast. Findings showed that addition of LTRA while tapering ICS dose lead to a significant increase in serious adverse events as compared to in patients taking ICS alone, (RR 2.44, 95% CI: 1.52–3.92), although there was no clear definition by authors into what constituted serious adverse events. Data regarding adverse effects were statistically insignificant (RR 1.06, 95% CI: 0.92–1.22). Studies investigating addition of LTRA to the group receiving same dose ICS showed a reduction by half in number of participants requiring oral corticosteroids (RR 0.5, 95% CI: 0.29–0.86). Data regarding adverse effects were statistically insignificant (RR 1.06, 95% CI: 0.92–1.22) in the group of patients on same dose ICS [16M].

A systematic review investigating the occurrence of neuro-psychiatric events with leukotriene-modifying agents was carried out in view of conflicting evidence. Thirty-three studies were included; however, none were RCTs, and only a few observational, therefore high-level evidence is limited. The observational studies did not find a significant association between LTRA and neuro-psychiatric events, although many pharmacovigilance studies have suggested a link. More epidemiological studies are required to evaluate association [17M].

PHOSPHODIESTERASE INHIBOTORS [SEDA-35, 321; SEDA-36, 252; SEDA-37, 201; SEDA-38, 152; SEDA-39, 167]

Roflumilast

Roflumilast is a selective phosphodiesterase-4 inhibitor, used in the treatment of COPD in patients with chronic bronchitis and frequent exacerbations. A letter to the editor was published in the European Journal of Internal Medicine, detailing a retrospective observational analysis of a database of patients on roflumilast for COPD. This was undertaken to document an intolerance to roflumilast in clinical practice. The high adverse event profile of roflumilast was detailed by several RCTs. The most frequently reported adverse events reported by this study were GI disorders (diarrhoea, followed by nausea), neuropsychiatric disorders (nervousness, anxiety, insomnia and depressed mood in order of decreasing frequency), and weight loss. 59% patients discontinued roflumilast in view of the adverse events [18r].

β2-ADRENOCEPTOR AGONISTS [SEDA-36, 245; SEDA-37, 197; SEDA-38, 153; SEDA-39, 163]

Beta$_2$ adrenoreceptor agonists have a significant role in the management of asthma and COPD but are unfortunately associated with a number of AEs, such as cardiovascular risks as well as headache and tremor. They are often used in conjunction with ICS or more recently in the form of LABA/LAMA therapies.

Salbutamol

A double-blind study in 184 young children aged 4–11 years old evaluated the safety of a novel salbutamol inhaler vs placebo. After an initial run-in period of 14 days, participants continued their regular asthma therapy and also received a placebo MDPI; patients were then randomised into placebo and or treatment groups. During the 3-week trial participants either used a placebo or salbutamol in addition to their regular asthma therapy. There was an overall low incidence of AEs (<4%) with no serious AEs, withdrawals due to AEs or deaths occurring during the study period. Most common AEs included headache, cough and pyrexia which occurred in similar numbers across both groups. In addition, the placebo group reported abdominal pain, upper respiratory tract infections and ligament sprain. The incidence of asthma exacerbations was similar in both cohorts [19c].

A prospective study by Ozer and colleagues investigated the incidence of salbutamol induced hypoxia in asthmatic children aged between 1 and 18 presenting to the paediatric emergency department or allergy clinic. In total 304 patients with a confirmed diagnosis asthma or features of probable asthma were included. Pulse oximetry was used to measure Sp028 with hypoxia defined as Sp02 up to 92% and salbutamol induced hypoxia defined as a two unit drop in Sp02. Salbutamol induced hypoxia was found in 14.7%, 3.9% and 1.3% ($P < 0.001$) of patients after the first, second and third dose of nebulised salbutamol. As a control, salbutamol was also administrated in the same cohort 24h and 10 days after presentation; the incidence of hypoxia was found to be 0.7% and 3.5%, respectively. In addition, children with confirmed aeroallergen sensitisation were at more risk of salbutamol induced hypoxia at day 10 [20c].

Formoterol

Reisner and colleagues investigated the safety of glycopyrrolate/formoterol fumarate (GFF), glycopyrrolate (GP), formoterol fumarate (FF), placebo and tiotropium. This multicentre trial consisted of two different parts, phase A and phase B. During phase A, each patient received four different treatments with washout periods in-between. Patients could receive any four out of a possible eight treatments; GFF MDI 72/9.6mcg, GFF MDI 36/9.6mcg, GP MDI 36mcg, FF MDI 9.6mcg, FF MDI 7.2mcg, placebo MDI, FF DPI 12mcg and open label tiotropium DPI 18mcg. Phase B was similar but focussed on four treatments in particular; FF MDI 9.6mcg, FF MDI 7.2mcg, placebo MDI and FF DPI 12mcg. The incidence of drug related AEs ranged from 3.8% in the placebo group to 31.7% in the high dose GFF group (72/9.6mcg). In general, common AEs included dry mouth and headaches, with some reports of tremor, cough and dysphonia. The incidence of dry mouth was highest in the GFF 72/9.6mcg arm (19.5%), followed by GP 36mcg (12.2%) and GFF 36/9.6mcg (7%). The incidence of sore throat in the FF, placebo and tiotropium arms was low. There was a similar trend in terms of headaches which were more common in the GFF, GP and tiotropium groups. Of the six serious AEs reported, none were deemed to be related to the study drug. No changes were identified in laboratory values, ECG characteristics or physical examinations [21c].

A multicentre RCT involving 1765 patients evaluated the efficacy of fluticasone propionate/formoterol (FP/FORM) in COPD. Patients were randomised to receive 52 weeks of either formoterol (12 mcg) or FP/FORM (250/10 mcg) or FP/FORM (500/20 mcg) after completing a run-in period of tiotropium. The incidence of AEs was similar across all three study arms but were least common in the formoterol only group (40.7%) as compared to the high and low dose FP/FORM groups (42.2% and 44.2%, respectively). This was also true of severe AEs and serious AEs, defined as fatal or life-threatening AEs. The most common AEs in the formoterol group were nasopharyngitis (5.1%) followed by dyspnoea (1.7%) which was similar to the other two arms. In terms of cardiovascular AEs, the incidence of atrial fibrillation was highest in patients receiving formoterol alone (1.7%) compared to 0.9% and 0.7%. In contrast, pneumonia was least common in patients receiving formoterol only compared to those receiving ICS therapy. Of the 53 patients that died, none were attributed to the study medication [22C].

Calverley and colleagues carried out a post hoc analysis of 3-month data of three studies in which COPD patients were treated with budesonide/formoterol or placebo. A total of 1571 patient were identified. Pooled analysis of AEs and serious AEs demonstrated no significant difference between treatment and placebo arms across all three studies, after both 3 and 12 months of intervention. The incidence of discontinuation due to AEs was significantly lower in the budesonide/formoterol arm compared to placebo at both 3 and 12 months. Commonly reported AEs include headache, COPD and nasopharyngitis. At 3 months, the incidence of COPD as an AE was 7.1% in the treatment group compared to 13% in placebo. The incidence of headaches and nasopharyngitis was similar in both arms [23c].

Singh et al. investigated the cardiovascular safety profile of pressurised MDI (pMDI) vs DPI containing beclometasone dipropionate/formoterol fumarate (BDP/FF) in 49 COPD patients. Patients received therapeutic BFD/FF (200/12 mcg) and supratherapeutic BFD/FF (800/48 mcg) via MDI and BFD/FF (200/12 mcg), supratherapeutic BFD/FF (800/48 mcg) and placebo via DPI. In terms of AEs, the most common AEs were headache and tremor which were more frequent when receiving supratherapeutic doses especially in the BFD/FF (800/48 mcg) DPI group. Overall, at therapeutic doses, both pMDI and DPI inhalers had a good cardiovascular safety profile when used in COPD [24c].

A multicentre RCT evaluated the use of different ICS/LABA combinations, beclomethasone/formoterol (BDP/F) 100/6 mcg and fluticasone/salmeterol (FP/S) 125/25 mcg, in Taiwanese asthmatic patients aged between 20 and 65 years old. After a 2-week run-in period, a total of 253 patients were randomised to receive BDP/F or FP/S for 12 weeks. Of the BDP/F group 45.9% of patients experienced at least one AE compared to 40% in the FP/S arm. The vast majority of these AEs and all of the seven serious AEs that occurred were not related to the study medication. The most common AE was upper respiratory tract infection occurring in 7.2% of the FP/S group and 5.7% of the BDP/F. No safety concerns were raised in terms of laboratory tests, vital signs and ECG characteristics [25C].

A multicentre RCT, CHASE 3, evaluated the use of budesonide/formoterol MDI, 160/9 or 160/4.5 mcg, against budesonide MDI 160 mcg in asthmatic children aged between 6 and 12 with persistent symptoms when on low dose ICS. Following a run-in period, 279 patients were randomised to one of the three treatment arms for a period of 12 weeks. The incidence of AEs was broadly similar between all three groups with overall incidence being 45%. The most common AEs in the budesonide/formoterol groups, from most frequent to least, include asthma, upper respiratory tract infection, pyrexia, nasopharyngitis, allergic rhinitis, cough, pharyngitis and headache. Asthma was least frequent in children treated with high dose budesonide/formoterol (7.8%). In addition, cough was associated with budesonide/formoterol groups in particular, 12.9% and 10% compared to 4.4% in the budesonide only group. There were no serious AEs in the budesonide/formoterol group and no deaths in the study as a whole. The authors identified headache as a possible AE directly related to formoterol, effecting four patients but none in the budesonide arm. No significant changes in vital signs, laboratory tests or ECG characteristics were detected [26C].

Olodaterol

Koch et al. explored the safety profile of olodaterol 5 mcg in comparison to formoterol and placebo in 1379 patients. During the 48-week study period patients continued their regular COPD management and were randomised to have in addition placebo, formoterol or olodaterol. Patient characteristics were similar in each group.

Approximately 70% of patients, regardless of treatment group, experienced AEs. The incidence of serious AEs, AEs leading to discontinuation and drug related AEs (6.1%) was lowest in those patients who received olodaterol. In addition, the incidence of cardiovascular AEs including arrhythmias, ischaemic heart disease and hypertension was lowest in the olodaterol group. In terms of rate ratios, formoterol was associated with higher rate ratios for cough, gastrointestinal and musculoskeletal disorders. Olodaterol was associated with higher rate ratios for nasopharyngitis, throat irritation metabolic disorders. It should be noted that the majority of rate ratios were not statistically significant [27c].

The TONADO trials were two large phase 3 studies investigating the efficacy of a combined tiotropium/olodaterol inhaler (in either 2.5/5 or 5/5mcg) against each individual component in those with severe COPD. Buhl and colleagues evaluated the safety profile of tiotropium/olodaterol by using data extracted from the TONADO trial especially with regards to cardiovascular health. A total of 3100 patients were deemed suitable for analysis. During the 52-week study, the incidence of overall and treatment related AEs were similar across all groups, approximately 75% and 6.5%, respectively. COPD followed by nasopharyngitis were the most common AEs. In terms of cardiovascular AEs, the incidence of arrhythmias (~4%) and ischaemic heart disease (~2%) was similar across all groups with rarer AEs including cardiac failure, QT prolongation and cerebrovascular disorders. The incidence of major adverse cardiovascular events ranged from 2.4% in olodaterol cohort to 1.8% in the tiotropium cohort. However, when evaluating only those patients with cardiac history, the incidence of major adverse cardiovascular events ranged rose to 3.6%, 4.5%, and 2.7% in the olodaterol, tiotropium and tiotropium/olodaterol groups, respectively [28C].

In a similar study using data from the TONODA trials, Bai et al. focussed on the efficacy of combined tiotropium/olodaterol in COPD patients from East Asia. A total of 700 patients of East Asian origin that met the inclusion criteria were analysed. The incidence of AEs in general and treatment related AEs was similar in the Asian population compared to the general population; most were moderate in nature with similar rates of discontinuation across treatment arms. The incidence of AEs including serious AEs, fatal AEs and those requiring hospitalisation were similar in the olodaterol, tiotropium and tiotropium/olodaterol groups in the Asian population. Common AEs included infections, respiratory disorders and gastrointestinal effects. The East Asian population compared to the overall study population had a lower incidence of respiratory disorders such as COPD exacerbations but a higher incidence of upper respiratory infections. The incidence of cardiovascular AEs was low in both groups and all treatment arms. Generally, the Asian population had lower incidence of tachyarrhythmias and ischaemic heart disease [29c].

Vilanterol

A multicentre double-blind trial by Bhatt and colleagues investigated the role of fluticasone furoate/vilanterol (FF/VI) against placebo in modifying arterial stiffness in patients with COPD. A total of 430 patients were randomised to treatment with either FF/VI 100/25mcg or VI 25mcg or placebo for 24 weeks. Overall, the incidence of AEs was higher in the FF/VI arm (57%) compared to VI (51%) and placebo (41%). Common AEs were nasopharyngitis, headache, back pain and candidiasis. Nasopharyngitis was particular common in the VI only group (8%) compared to FF/VI (6%) and placebo (3%). Candidiasis almost exclusively occurred in the FF/VI group. Two fatal serious AEs were deemed not be treatment related. No changes in physiological parameters was found [30C].

A multicentre RCT involving 1620 patients aimed to evaluate the efficacy and safety of FF/VI 100/25mcg vs VI 25mcg during a 84-day trial period. The overall incidence of AEs in general were similar between both arms. The incidence of serious AEs was 1% in the FF/VI group and 2% in VI only group. The most common AE was nasopharyngitis which occurred in 6% of the study population regardless of intervention. The overall occurrence of pneumonia was very low (<1%) and the incidence of cardiovascular events was 2%–3% [31C].

Theophylline

A single centre retrospective study Rusinowicz et al. analysed 2753 Holter recordings with view to identify cardiac arrhythmias in patients with COPD and associated risk factors. Most patients demonstrated evidence of cardiovascular disease. Approximately a third of all patients were being treated with theophylline. The authors found that there was statistically significant association between theophylline use and the presence of both paroxysmal atrial fibrillation and ventricular premature beats. Of the patients on theophylline, 20% demonstrated paroxysmal atrial fibrillation compared to 8% of non-users ($P=0.04$). Similarly, 57% of users showed ventricular premature beats compared to 49% otherwise ($P=0.02$) [32c].

ANTICHOLINERGIC DRUGS [SEDA-36, 245; SEDA-37, 197; SEDA-38, 153; SEDA-39, 165]

Anticholinergic therapies are a mainstay of treatment of COPD and have recently been used in the management of asthma. Acting on the parasympathetic nervous system they induce relaxation of smooth muscle within the respiratory system.

Glycopyrronium

Hanania et al. investigated the long-term safety of a glycopyrrolate 18mcg (GP) and formoterol fumarate

(FF) combined in a single metred dose inhaler (GFF MDI) vs its individual components and tiotropium in patients with moderate to severe COPD. 778 participants completed the 52 weeks of treatment. Patients were randomised into four different groups, GFF MDI, FF MDI, GP MDI and open label tiotropium. The incidence of AEs was similar across all four groups with the majority being mild in nature. The most common AE was nasopharyngitis which ranged in incidence from 4.3% in the GP MDI group to 6.8% in GFF MDI group. Other common AEs include cough, upper respiratory tract infection and urinary tract infection. The incidence of serious AEs was similar across all groups (approximately 10% of all patients) as was the incidence of AEs leading to early discontinuation (approximately 7% of all patients). Major cardiovascular events were rare, patients who received tiotropium had the highest incidence at 0.9%. No changes were found between the different groups in terms of physical examination data, ECG parameters, vital signs or laboratory tests [33C].

Wedzicha et al. investigated efficacy and safety of the indacaterol/glycopyrronium (IND/GLY) vs salmeterol/fluticasone (SFC) in patients from Asian centres within the FLAME study. A total of 450 patients completed the 52-week study period. Overall, the incidence of AEs was similar between the two groups with the proportion of patients experiencing one AE was 82.4% in the IND/GLY group compared to 85.4% in the SFC group. The most common AE was COPD which was reported in 68% of IND/GLY group which was slight lower than the SFC group at 74.6%. Other common AEs were nasopharyngitis and upper respiratory tract infection. The incidence of pneumonia was significantly higher in the SFC treatment group vs IND/GLY (7.7% vs 3.6%, $P=0.046$). The incidence of major adverse cardiovascular events and cardiovascular deaths was low and similar in both groups [34C].

Tiotropium

Zhou and colleagues investigated the role of tiotropium in early COPD in a multicentre RCT conducted in China. A total of 841 patients with GOLD 1 or 2 disease severity were randomised to receive either once daily tiotropium (18 mcg) or control for a total of 2 years. The largest statistically significant difference in AEs between the two groups was oropharyngeal discomfort, including dry mouth and sore throat. This occurred in 61 (15%) patients assigned to the tioptropium arm as compared to only 28 (6.6%) patients who received placebo, $P<0.001$. Otherwise there was no significant difference between each arm in terms of deaths, serious AEs, cardiovascular events or incidence of other AEs excluding oropharyngeal discomfort [35C].

A RCT conducted by Kerwin et al. aimed to investigate the benefit of umeclidinium/vilanterol (UMEC/VI) inhaler in patients with COPD who are not well controlled on tiotropium alone. A total of 494 patients were prescribed tiotropium for at least 3 months at screening and were then later randomised to tioptropium (18 mcg) or UMEC/VI (62.5/25 mcg) arms for 12 weeks. The incidence of on treatment AEs was near identical between the two study groups, 30% for UMEC/VI compared to 31% for tiotropium. However, the incidence of on treatment drug related AEs was higher in the tiotropium arm (3%) vs UMEC/VI (1%). The most common AEs were nasopharyngitis (7%) followed by headache (~7%) and finally COPD exacerbation. The incidence of these AEs was similar in both groups except for COPD exacerbation which occurred in 3% of the tiotropium group and less than 1% in the UMEC/VI group. Cardiovascular AEs were similar in both arms with an incidence of 1%–2% [36C].

Umeclidinium

A Cochrane review investigated the efficacy umeclidinium vs placebo in patients with stable COPD. Four high quality RCTs involving a total of 3798 patients with moderate to severe COPD were included. All four trials were multicentre randomised, double blind studies ranging in duration from 12 to 52 weeks. Umeclidinium bromide (125 or 62.5 mcg) was administrated once daily using a dry powder inhaler with placebo groups receiving a matching inhaler. Overall, 22%–50% of all participants used an adjuvant ICS during the study period with an equal proportion in treatment and placebo groups throughout all studies. There was no statistically significant difference in terms of all-cause mortality between umeclidinium and placebo in three of the RCTs with no deaths reported in the fourth. In terms of non-fatal serious AEs, pooled analysis of 1922 patients demonstrated no statistically significant difference between placebo and intervention groups (OR 1.33, 95% 0.89–2.00). Similarly, there was no significant difference in AEs (not including serious AEs) between umeclidinium and placebo (OR 1.06, 95% CI 0.85–1.31) [37R].

Lee and colleagues investigated the role a combined ICS/LAMA inhaler consisting of fluticasone furoate (FF) and umeclidinium (UMEC) in patients with features of both asthma and COPD; that is patients FEV1/FVC ratio of <0.7 but also demonstrating reversibility to salbutamol (FEV1 increase of ≥12%). This study was comprised of three separate phases. Phase A of compared the use of FF/UMEC vs a ICS/LABA combined inhaler, FF and vilanterol (VI), for a total of 4 weeks. In addition, various doses of UMEC, ranging from 15.6 to 250 mcg were evaluated with a fixed dose of FF 100 mcg.

Phase B investigated the effect of adding in VI to FF/UMEC therapy for 1 week. Finally, phase C involved withdrawing UMEC entirely to investigate its duration of effect up to 1 week. There were no recorded changes in physiological parameters including vital signs, ECG changes and laboratory measurements in response to treatment throughout this study. During phase A, the incidence of AEs ranged from approximately 15%–30% with the greatest incidence of AEs observed in the FF only and low dose FF/UMEC 15.6mcg groups. Nasopharyngitis was the most common AE in general, occurring most frequently in the FF only group (12%) and less frequently as the dose of UMEC increased (only 2% in the UMEC 250mcg group). The most common drug related AE was coughing which effected 5%–9% of patients depending on FF/UMEC dosage. Other less common AEs included dysphonia, viral respiratory infections and toothache. There were no serious AEs during this phase [38C].

One episode of hypertensive crisis was reported in phase B in patient allocated to FF/UMEC 250mcg. No serious AEs were reported in phase C.

ANTI-FIBROTIC THERAPIES

Anti-fibrotics, in the context of respiratory medicine, are principally used in the treatment of idiopathic pulmonary fibrosis (IPF). Radiological evidence of usual interstitial pneumonitis, restrictive spirometry and an FVC between 50% and 80% may qualify a patient for Pirfenidone or Nintedanib via a respiratory MDT. These are principally targeted at slowing disease progression. Patients are traditionally counselled on the common adverse effects of these drugs, which often dictate compliance with therapy.

Pirfenidone

The RECAP study, an open label extension, published by respiration, in 2017, evaluated the long-term safety profile of pirfenidone, alongside it's efficacy. Treatment emergent adverse events (expressed as a rate per 100 patient exposure years—100 PEY) included; IPF (7.2), pneumonia (0.5), respiratory failure (0.5), acute respiratory failure (0.5), rash (0.5) and nausea (0.4). The total event rate was 17.9. The most frequent reactions that lead to discontinuation of treatment were rashes (1.1% of patients) and nausea (1% of patients). Overall, the adverse event profile is well established with pirfenidone, with gastrointestinal upset and photosensitivity as stated above [39c].

Pirfenidone Case Studies

In July of 2017, a case study was published of a pirfenidone induced eosinophilic pleurisy. This was the first published description of any type of pleural reaction to this anti-fibrotic. The 69-year-old gentleman had a background of reflux oesophagitis but had no preceding respiratory complaints (besides IPF), nor any hazardous inhaled exposures. He was admitted 7 weeks into his pirfenidone treatment with left sided chest pain and a peripheral blood eosinophilia (380/μL) and CRP of 5mg/dL. A non-malignant unilateral pleural effusion was identified alongside a positive QuantiFERON assay. The effusion was both lymphocytic and eosinophilic. The gentleman underwent a medical thoracoscopy which demonstrated pleural thickening and translucent nodules and adhesions. The nodules were demonstrated to be lymphoid with a massive infiltration of eosinophils, plasma cells and lymphocytes, without evidence of malignancy or granuloma. The culture was negative for bacteria and mycobacteria. After discontinuation of Pirfenidone there was a gradual improvement of his chest pain, eosinophilia; with subsequent complete resolution of the effusion at 8 weeks, with no recurrence [40A].

Two thousand and seventeen also saw the "British medical journal" publish a case report of three patients that developed Pirfenidone induced hyponatraemia. The article also offered notes on how it may be managed. The first patient was 72 years of age with a background of hypertension, type 2 diabetes mellitus, hyperlipidaemia, gastro-oesophageal reflux disease and well controlled coeliac disease. His medication history did include omeprazole but he was well established on this for just under a decade without event. He also took triamterene hydrochlorthiazide. No other medication had been started. When he had been up-titrated to a "thrice daily" dosing regime, it was noted that his sodium was dropping. Serum and urine biochemistry identified that this was part of a syndrome of inappropriate ADH secretion. 2 weeks after pirfenidone discontinuation, the sodium was reported to be normal at 137 and remained stable at 4 weeks post-cessation [41A].

The second patient in the report was a lady of 80 years, that had pre-existing mild hyponatraemia (typically 132–138 in the preceding 3 months before pirfenidone initiation). The adverse drug reaction in this case also became apparent at 2 weeks into treatment where the sodium dropped below 130 for the first time (127). This level continued to lower with ongoing pirfenidone use and persisted despite attempts to decrease the dosing. Both this patient and patient one had a normal TSH but unfortunately the latter did not have the appropriate investigations into the possibility of SIADH. The sodium level returned to her baseline when re-checked at 2 weeks from Pirfenidone discontinuation [41A].

The final patient also experienced a similar phenomenon with a hyponatraemia that correlated to the initiation of pirfenidone. Once more, the urine and serum biochemistry illustrated an SIADH picture. Patient three also had rheumatoid arthritis and the pirfenidone was discontinued due to a non-IPF diagnosis (initiated prior to referral to author) and the hyponatraemia. The sodium abnormality improved to acceptable limits at 2 weeks (132) and then furthermore at 4 weeks after stopping the offending drug [41A].

"The British journal of dermatology" published a report in December of 2017 describing the first documented case of drug induced cutaneous lupus erythematosus secondary to pirfenidone. The 69-year-old lady in the case study had suffered with idiopathic pulmonary fibrosis for 6 months, she went on to pirfenidone commencement and subsequently developed (8 weeks after initiation) an intensely pruritic rash, that was painful and distributed in a manner suggestive of photosensitivity. The rash consisted of erythematous and oedematous plaques. The initial suspicion was that of a well described adverse drug reaction to pirfenidone, i.e., Photosensitivity. The patient also had a 30-year background of chronic plaque psoriasis that was very much in remission after UVB phototherapy. Histology of the skin demonstrated "dermatitis with basal vacuolar change and intraepidermal and junctional civatte bodies. Mild superficial perivascular dermal inflammation including lymphocytes and histiocytes." Significant mucin deposition was also noted as well as a peripheral blood lymphopenia. Her other blood markers were grossly normal apart from a strongly positive anti ds-DNA. The diagnosis of cutaneous lupus (pirfenidone induced) was then apparent. The rash responded to potent topical corticosteroids and cessation of pirfenidone [42A].

The "International journal of mycobacteriology" published a case report in 2017 describing incidences of pulmonary mycobacterium tuberculosis re-activation with Pirfenidone therapy. The authors postulated from mouse models, that Pirfenidone may "affect host control of TB infection through marked increase in matrix metalloproteinase." The first patient, a 64-year-old lady, had a radiological and histological picture, suggesting idiopathic pulmonary fibrosis. She was commenced on 1800 mg of pirfenidone in three divided doses. Unfortunately, 1 year later she presented to the emergency department with a cough, fever and weight loss, her chest radiograph demonstrating a large right upper lobe cavitation, which was also seen on CT. The case proceeded to mycobacterial analysis of a brochoalveolar lavage sample, with a smear and PCR both confirming mycobacterium tuberculosis infection. They made a full recovery with anti-tuberculous treatment for a total of 9 months. The authors did not state if the pirfenidone was discontinued. The authors recognise that further prospective study is required in order to explore if pirefenidone is causative in TB re-activation but they postulate that it "certainly may contribute" [43A].

Nintedanib

Nintedanib is also recommended by the national institute of clinical excellence for patients with idiopathic pulmonary fibrosis and a FVC between 50% and 80%.

"Clinical and translational medicine" published the findings of a multi-ILD-centre trial in 2017, highlighting the early experiences of Nintedanib (before it was recommended by NICE in 2016). 187 patients were recruited into the trial, with approximately half of the patients (52%) suffering between 1 and 3 adverse events (AE). The most commonly experienced AE was diarrhoea (50%), with 36% of patients suffering with nausea. The incidence of reduced appetite, tiredness and gastro-oesophageal reflux were all under 25%. Most of the adverse events (64%) did not lead to a change in that patient's management. The trial did not identify any significant risk of cardiovascular morbidity or increased bleeding risk [44c].

"Internal medicine (Tokyo, Japan)" published a case report of a 76-year-old gentleman whom had a radiological diagnosis of idiopathic pulmonary fibrosis 3 years prior. 84 days after the commencement of Nintedanib, he presented with abdominal pain, haematemesis and chronic diarrhoea. A contrasted CT of his abdomen demonstrated bowel thickening, prompting colonoscopy (nintedanib had been stopped 7 days prior) which showed erythema without ulceration and chronic colitis on histology. His symptoms improved 10 days after stopping the offending medication and also relapsed on an attempt to re-start it. One may be able to postulate that this could be related to one of Nintedanib's common adverse events, diarrhoea, but the author's did state that colitis associated with Nintedanib is "relatively rare" [45A].

IMMUNOLOGICAL THERAPIES [SEDA-38, SEDA-39]

Omalizumab

As previously described (SEDA-39), Omalizumab is a monocloncal anti-IgE antibody that directly reduces the levels of circulating IgE, hence its main application in respiratory medicine is severe atopic asthma, refractory to standard inhaled therapies (corticosteroids and long acting beta 2 antagonists).

In September of 2017, Di Bona et al. published the results, in "Respiratory Medicine," of a 9-year retrospective study providing a long-term safety profile on Omalizumab. The study followed 91 patients, with a total of 10472 injections administered. 39.1% of patients dropped out of the study in the 1st year "mainly" due to non-adverse event related issues. The most common cause of adverse event related dropout was myalgia and arthralgia, with fewer patients discontinuing due to urticaria, angioedema, metrorrhagia and relapsing herpes labialis. Mild adverse events that did not lead to discontinuation of therapy included rhinitis/conjunctivitis, injection site reaction, fatigue and thrombosis. Anaphylaxis was not reported [46C].

A review published by Nazir et al. in "Annals of allergy, asthma and immunology" in March of 2017, highlighted a potential association of Omalizumab with eosinophilic granulomatosis and polyangitis (EGPA—formerly known as Churg–Strauss syndrome). Using the American journal of Rheumatologists' criteria for EGPA diagnosis, they conducted an electronic search to identify eight patients with the association. There was a strong male preponderance (six of the eight). The mean number of doses to the diagnosis of EGPA was 21.5 (range 3–81) with a mean time from the first dose being 73 weeks (range 4–352 weeks). Only 5 of the 8 patients had an ANCA status mentioned, 100% of which were negative [47r].

INTERLEUKIN-5 MONOCLOCAL ANTIBODIED [SEDA-39, 168]

Humanised IL-5 monoclonal antibodies (MABs) are selective inhibitors of the main cytokine that activates eosinophils, interleukin 5. In severe eosinophilic asthma, this particular sub-set of the white blood count differential, plays a role in airway hyper-responsiveness. The target of minimising this response is to lessen the patient's asthma exacerbation burden.

Mepolizumab

A "Cochrane systematic review," published by Farne et al. in September 2017 looked at 13 studies measuring various outcomes in IL-5 MABs, 4 of which that were concerned with Mepolizumab. Two of these (Chupp 2017 and Ortega 2014) were compared with regard to adverse events. The secondary outcomes noted that there were no clinically significant adverse events—those that would lead to discontinuation of therapy (risk ratio 0.45, 95% CI 0.11–1.80; participants = 936; studies = 2; $I^2 = 0\%$). There was a comment that there were fewer serious adverse events (risk ratio 0.63, 95% CI 0.41–0.97; participants = 936; studies = 2). It was thought that this was due to a decrease in asthma related serious adverse events although "neither study reached statistical significance alone and therefore this was not commented on by the investigators" [48R].

References

[1] Cho K, Kim Y, Linton J, et al. Effects of inhaled corticosteroids/long-acting agonists in a single inhaler versus inhaled corticosteroids alone on all-cause mortality, pneumonia, and fracture in chronic obstructive pulmonary disease: a nationwide cohort study 2002–2013. Respir Med. 2017;130:75–84 [c].
[2] Crim C, Calverley P, Anderson J, et al. Pneumonia risk with inhaled fluticasone furoate and vilanterol in COPD patients with moderate airflow limitation: the SUMMIT trial. Respir Med. 2017;131:27–34 [C].
[3] Cazeiro C, Silva C, Mayer S, et al. Inhaled corticosteroids and respiratory infections in children with asthma: a meta-analysis. Pediatrics. 2017;139(3):e20163271 [M].
[4] Morjaria J, Rigby A, Morice A. Inhaled corticosteroid use and the risk of pneumonia and COPD exacerbations in the UPLIFT study. Lung. 2017;195(3):281–8 [C].
[5] Calzetta L, Matera M, Braido F, et al. Withdrawal of inhaled corticosteroids in COPD: a meta-analysis. Pulm Pharmacol Ther. 2017;45:148–58 [M].
[6] Sonnappa S, Martin R, Israel E, et al. Risk of pneumonia in obstructive lung disease: a real-life study comparing extra-fine and fine-particle inhaled corticosteroids. PLoS One. 2017;12(6): e0178112 [C].
[7] Yeo S, Aggarwal B, Shantakumar S, et al. Efficacy and safety of inhaled corticosteroids relative to fluticasone propionate: a systematic review of randomized controlled trials in asthma. Expert Rev Respir Med. 2017;11(10):763–78 [M].
[8] McCarter G, Blanchard L. Similar adverse events from two disparate agents implicate lipid inflammatory mediators for a role in anxiety states. Oxf Med Case Rep. 2017;2017(11). https://doi.org/10.1093/omcr/omx060 [M].
[9] Moss E, Buys Y, Low S, et al. A randomized controlled trial to determine the effect of inhaled corticosteroid on intraocular pressure in open-angle Glaucoma and ocular hypertension. J Glaucoma. 2017;26(2):1 [c].
[10] Tsai C, Liao L, Lin C, et al. Inhaled corticosteroids and the risks of low-energy fractures in patients with chronic airway diseases: a propensity score matched study. Clin Respir J. 2017;12:1830–7 [c].
[11] Chan C, DeLapp D, Nystrom P. Chronic eosinophilic pneumonia: adjunctive therapy with inhaled steroids. Respir Med Case Rep. 2017;22:11–4 [A].
[12] Colpitts L, Murray T, Tahhan S, et al. Iatrogenic cushing syndrome in a 47-year-old HIV-positive woman on ritonavir and inhaled budesonide. J Int Assoc Provid AIDS Care. 2017;16(6):531–4 [A].
[13] McCarter G, Blanchford L. Similar adverse events from two disparate agents implicate lipid inflammatory mediators for a role in anxiety states. Oxf Med Case Reports. 2017;2017(11). https://doi.org/10.1093/omcr/omx060 [A].
[14] Benard B, Bastien V, Vinet B, et al. Neuropsychiatric adverse drug reactions in children initiated on montelukast in real-life practice. Eur Respir J. 2017;50(2):1700148 [c].
[15] Gill D, Mann K, Wani L. Montelukast-induced angioedema. Am J Ther. 2017;24(5):e624–5 [A].
[16] Chauhan B, Jeyaraman M, Singh Mann A, et al. Addition of anti-leukotriene agents to inhaled corticosteroids for adults and adolescents with persistent asthma. Cochrane Database Syst Rev. 2017;3. https://doi.org/10.1002/14651858.CD010347.pub2 [M].
[17] Law S, Wong A, Anand S, et al. Neuropsychiatric events associated with leukotriene-modifying agents: a systematic review. Drug Saf. 2017;41:253–65 [M].
[18] Gómez-Rodríguez M, Golpe R. Intolerance to roflumilast in real-life clinical practice. Eur J Intern Med. 2017;43:e28–9 [r].

[19] LaForce C, et al. Albuterol multi-dose dry powder inhaler efficacy and safety versus placebo in children with asthma. Allergy Asthma Proc. 2017;38(1):28–37 [c].
[20] Ozer M, et al. Repeated doses of salbutamol and aeroallergen sensitisation both increased salbutamol-induced hypoxia in children and adolescents with acute asthma. Acta Paediatr. 2017;107:647–52. https://doi.org/10.1111/apa.14202 [c].
[21] Reisner C, et al. A randomized, seven-day study to assess the efficacy and safety of a glycopyrrolate/formoterol fumarate fixed-dose combination metered dose inhaler using novel Co-suspension™ delivery technology in patients with moderate-to-very severe chronic obstructive pulmonary disease. Respir Res. 2017;18(1):8 [c].
[22] Papi A, et al. Fluticasone propionate/formoterol for COPD management: a randomized controlled trial. Int J Chron Obstruct Pulmon Dis. 2017;12:1961–71 [C].
[23] Calverly PM, et al. Early efficacy of budesonide/formoterol in patients with moderate-to-very-severe COPD. Int J Chron Obstruct Pulmon Dis. 2017;12:13–25 [c].
[24] Singh D, et al. Acute cardiovascular safety of two formulations of beclometasone dipropionate/formoterol fumarate in COPD patients: a single-dose, randomised, placebo-controlled crossover study. Pulm Pharmacol Ther. 2017;42:43–51 [c].
[25] Hsieh MJ, et al. Comparative efficacy and tolerability of beclomethasone/formoterol and fluticasone/ salmeterol fixed combination in Taiwanese asthmatic patients. J Formos Med Assoc. 2017. https://doi.org/10.1016/j.jfma.2017.12.005. Dec 29; pii: S0929-6646(17)30549-1, [Epub ahead of print]. [C].
[26] Pearlman DS, et al. Efficacy and safety of budesonide/formoterol pMDI vs budesonide pMDI in asthmatic children (6–<12 years). Ann Allergy Asthma Immunol. 2017;118(4):489–99 [C].
[27] Koch A, et al. Comprehensive assessment of the safety of olodaterol 5 μg in the Respimat® device for maintenance treatment of COPD: comparison with the long-acting β2-agonist formoterol. NPJ Prim Care Respir Med. 2017;27(1):60 [c].
[28] Buhl R, et al. Long-term general and cardiovascular safety of tiotropium/olodaterol in patients with moderate to very severe chronic obstructive pulmonary disease. Respir Med. 2017;122:58–66 [C].
[29] Bai C, et al. Lung function and long-term safety of tiotropium/ olodaterol in east asian patients with chronic obstructive pulmonary disease. Int J Chron Obstruct Pulmon Dis. 2017;12:3329–39 [c].
[30] Bhatt SP, et al. A randomized trial of once-daily fluticasone furoate/vilanterol or vilanterol versus placebo to determine effects on arterial stiffness in COPD. Int J Chron Obstruct Pulmon Dis. 2017;12:351–65 [C].
[31] Siler TM, et al. A randomised, phase III trial of once-daily fluticasone furoate/vilanterol 100/25 μg versus once-daily vilanterol 25 μg to evaluate the contribution on lung function of fluticasone furoate in the combination in patients with COPD. Respir Med. 2017;123:8–17 [C].
[32] Rusinowicz T, et al. Cardiac arrhythmias in patients with exacerbation of COPD. Adv Exp Med Biol. 2017;1022:53–62 [c].
[33] Hanania NA, et al. Long-term safety and efficacy of glycopyrrolate/formoterol metered dose inhaler using novel Co-suspension™ delivery technology in patients with chronic obstructive pulmonary disease. Respir Med. 2017;126:105–15 [C].
[34] Wedzicha JA, et al. Indacaterol/glycopyrronium versus salmeterol/fluticasone in Asian patients with COPD at a high risk of exacerbations: results from the FLAME study. Int J Chron Obstruct Pulmon Dis. 2017;12:339–49 [C].
[35] Zhou Y, et al. Tiotropium in early-stage chronic obstructive pulmonary disease. N Engl J Med. 2017;377(10):923–35 [C].
[36] Kerwin EM, et al. Umeclidinium/vilanterol as step-up therapy from tiotropium in patients with moderate COPD: a randomized, parallel-group, 12-week study. Int J Chron Obstruct Pulmon Dis. 2017;12:745–55 [C].
[37] Ni H, et al. Umeclidinium bromide versus placebo for people with chronic obstructive pulmonary disease (COPD). Cochrane Database Syst Rev. 2017;6:CD011897 [R].
[38] Lee L, et al. The effect of umeclidinium on lung function and symptoms in patients with fixed airflow obstruction and reversibility to salbutamol: a randomised, 3-phase study. Respir Med. 2017;131:148–57 [C].
[39] Costabel U, et al. An open label study of the long-term safety of pirfenidone in patients with idiopathic pulmonary fibrosis (RECAP). Respiration. 2017;94(5):408–15 [c].
[40] Hase I, et al. Pirfenidone-induced eosinophilic pleurisy. Intern Med. 2017;56(14):1863–6 [A].
[41] Silhan L, et al. Pirfenidone-induced hyponatraemia: insight into mechanism, risk factor and management. BMJ Case Rep. 2017;2017. https://doi.org/10.1136/bcr-2017-222734. pii: bcr2017-222734 [A].
[42] Kelly AS, et al. Drug induced lupus erythematosus secondary to pirfenidone. Br J Dermatol. 2017;178:1437–8. https://doi.org/ 10.1111/bjd.16246 [A].
[43] Khan M, et al. Reactivation pulmonary tuberculosis in two patients treated with pirfenidone. Int J Mycobacteriol. 2017;6(2):193–5 [A].
[44] Toellner H, et al. Early clinical experiences with Nintedanib in three UK tertiary interstitial lung disease centres. Clin Transl Med. 2017;6(1):41 [c].
[45] Oda K, et al. Colitis associated with nintedanib therapy for idiopathic pulmonary fibrosis (IPF). Intern Med. 2017;56(10):1267–8 [A].
[46] Bona D, et al. Long-term "real-life" safety of omalizumab in patients with severe uncontrolled asthma: a nine year study. Respir Med. 2017;130:55–60 [C].
[47] Salik N, et al. Omalizumab-associated eosinophilic granulomatosis with polyangitis (Churg-Strauss syndrome). Ann Allergy Asthma Immunol. 2017;118(3) 372-374.e1. [r].
[48] Farne HA, et al. Anti-IL5 therapies for asthma. Cochrane Database Syst Rev. 2017. https://doi.org/10.1002/14651858. CD010834.pub3 [R].

C H A P T E R

17

Positive Inotropic Drugs and Drugs Used in Dysrhythmias

Kerry Anne Rambaran,[1], Amy Lehnert*[†]

*Department of Pharmacy Practice, Keck Graduate Institute School of Pharmacy and Health Sciences, Claremont, CA, United States

[†]Department of Pharmacy Practice, KGI School of Pharmacy, Claremont, CA, United States

[1]Corresponding author: krambaran@kgi.edu

CARDIAC GLYCOSIDES [SEDA-36, 257; SEDA-37, 205; SEDA-15, 173]

Digoxin

Cardiovascular

- Mina and colleagues performed a retrospective review of patients with an implantable cardioverter defibrillator (ICD) between January 1 2012 and January 1 2015. There were two study arms—one with patients who were currently on digoxin or digoxin (55 patients) at any point and the control were those who were not on digoxin (147 patients) who presented to the hospital for device (ICD) interrogation. Their main outcome was ICD shock events, whereby an appropriate shock was one in response to ventricular tachycardia (VT) or ventricular fibrillation (VF) and an inappropriate shock was one in response to a supraventricular arrhythmia or sinus tachycardia. Moreover an "electrical storm was defined as three or more arrhythmias required ICD shocks within 24 hours". They reported that there were more shocks (appropriate and inappropriate) in digoxin users, though not statistically significant. Furthermore, it was found that patients on digoxin had more electrical storms (9.1% vs 1.4%, $P=0.02$) and more hospitalizations (69.1% vs 53.1%, $P=0.04$) which was statistically significant. They further conducted a subgroup analysis of patients with heart failure (HF) without atrial fibrillation (Afib) and patients with HF with Afib, in which they found that patients who had HF and concurrently taking digoxin had statistically significant more shocks than those with only HF ($P=0.03$), and there were more electrical storms in the patients on digoxin ($P=0.008$). Thus, they concluded that digoxin was "independently associated with an increased incidence of ICD shock" [1C].

Whilst this retrospective review contains a small sample size, the findings are corroborated in the MADIT-CRT trial, where the use of digoxin increased the incidence of appropriate shocks [2MC]. This retrospective review highlights the importance of knowing one of the adverse effects of digoxin is VT or VF. Thus, when digoxin is used in patients with an ICD, the device may shock multiple times, but it doesn't imply there is device malfunction.

Electrolyte Balance

- A study was conducted in two groups of patients with rapid atrial fibrillation who received rate control with either intravenous digoxin ($n=9$) or verapamil/control ($n=8$). Patients who received digoxin were administered a loading dose of 1 mg in 24 hours and started with a first dose of 0.25 mg. "Blood and urine samples were obtained at time 0, 60 and 240 minutes after the first dose of digoxin was administered". The patients were comparable between groups with regards to age, gender, pre-existing comorbidities (hypertension, diabetes, hypothyroidism, heart failure) and drugs taken (calcium channel blocker, beta blockers, ACE inhibitors, amiodarone). The laboratory findings between the two groups were similar with the exception of the fractional excretion of magnesium (Fex Mg) being lower in the digoxin group (3.07 ± 1.21 vs 6.21 ± 2.33, $P=0.003$). During the 4 hours post the first dose, there were no laboratory changes in the

Side Effects of Drugs Annual, Volume 40
ISSN: 0378-6080
https://doi.org/10.1016/bs.seda.2018.06.003

verapamil group, compared to the digoxin group. The digoxin concentration was 0 at time 0, 2.87 ± 1.05 ng/mL at 60 minutes and 0.93 ± 0.46 ng/mL at 240 minutes. The Fex Mg was 3.07 ± 1.21 at 0 minutes, 7.58 ± 2.51 at 60 minutes and 6.05 ± 2.30 at 240 minutes. The authors reported correlation between Fex Mg and digoxin concentration ($r = 0.678$, $P < 0.0001$) [3c].

Digoxin induced hypermagnesemia is rare and its mechanism is not fully understood. Prescribers should be mindful of this when monitoring patients on digoxin therapy, especially if they are not on diuretics.

Mouth

- A 28-year-old female with a known history of coronary artery disease (CAD) was initiated on digoxin 0.25 mg by mouth once daily along with furosemide 40 mg by mouth once weekly and acitrom 2 mg by mouth once daily, after undergoing cardiac surgery. A few months into starting the medications, she reported mild swelling of her gingiva which bled upon brushing. She was diagnosed with aggressive periodontitis and drug induced gingival overgrowth. The patient underwent surgical intervention and multiple follow-up visits ensued thereafter. Digoxin was not discontinued and the patient had no relapse when seen a year post-surgical procedure [4A].

Gingival overgrowth is a rare adverse effect of digoxin. Moreover, it should be noted 6/36894 (0.0163%) individuals reported gingival hyperplasia as a side effect of digoxin to the FDA between January 2004 and October 2012. Close monitoring for early signs of adverse effects is necessary to avoid any untoward complications.

Drug–Drug Interaction

- A 44-year-old Caucasian male presented with chest pain at a North American hospital in 2009 and was subsequently diagnosed with diabetes and a myocardial infarction, for which he underwent coronary artery stenting and was initiated on digoxin 0.25 mg of digoxin daily (route not specified). Six years later (August 2015) the patient tested positive for HIV in South Asia after being hospitalized for renal complications secondary to diabetes. In September 2015 he was seen by his general practitioner who initiated him on darunavir 800 mg, ritonavir 100 mg, dolutegravir 50 mg, rilpivirine 25 mg and co-trimoxazole 960 mg. At this time his cardiovascular system and lab work were normal. HIV RNA value was one million IU/mL and his CD lymphocyte count was 05 cells/mm^3. His chest X-ray was normal and the ECG showed sinus rhythm prior to starting antiretroviral therapy. Three weeks later (October 2015) he returned to the HIV clinic with severe anorexia, nausea and vomiting. Atrial flutter was visible on the ECG, digoxin serum level 5 hours from the last dose was >6.4 nmol/mL, Na 124 mmol/L, K 6 mmol/L, creatinine 216 mmol/L, BUN 9.4 mmol/L, eGFR 28, HIV RNA 714 IU/mL. Subsequently, digoxin was stopped and his antiretroviral therapy was changed to efavirenz, abacavir and dolutegravir daily. The next day the digoxin level was 4.2 nmol/L and 1.5 nmol/mL 4 days later. At this time the ECG showed normal sinus rhythm, and he was initiated on aspirin, bisoprolol and atorvastatin. Eight months later, after digoxin was discontinued, the ECG showed sinus rhythm without any evidence of atrial fibrillation or heart failure [5A].

This case showed that digoxin toxicity can manifest as an arrhythmia. Moreover, given that digoxin is renally eliminated and it has a narrow therapeutic index (0.5–2.0 ng/mL), the patients' existing renal insufficiency may have contributed to the toxicity in this case; however, clinical findings of digoxin toxicity did not surface until antiretroviral therapy was initiated. A retrospective review revealed that patients with atrial fibrillation and an eGFR <90 mL/min/1.73 m^2 are at an increased risk for adverse outcomes [6MC]. Thus, caution should be exercised when initiating such medications in patients with compromised renal function.

ANTIDSYRHYTHMIC DRUGS

Adenosine [SEDA-17, 330; SEDA-17, 379; SEDA-15, 36; SEDA-30, 212]

Cardiovascular

- A 26-year-old female presented to the emergency department with sudden onset of palpitations and chest discomfort. The patient was healthy, with no previous medical history or regular medications. An ECG demonstrated rapidly conducted atrial fibrillation with a rate of 194 beats per minute, with no evidence of ventricular pre-excitation. After an unsuccessful attempt at vagal maneuvers, a single dose of adenosine 6 mg was given intravenously, followed by immediate development of coarse ventricular fibrillation (VF) and circulatory collapse. The patient was successfully defibrillated with a single 150 J direct current shock, with immediate return of spontaneous circulation and hemodynamic stability. She was admitted, and no further arrhythmias were identified. Subsequent inpatient workup revealed a normal resting ECG with no evidence of accessory pathway and unremarkable labs. An invasive electrophysiology study was performed and was negative. Cardiac magnetic resonance imaging demonstrated a structurally normal heart with normal coronary artery anatomy. The patient was discharged on no medications and had experienced no further episodes at routine follow-up 3 months later [7A].

Adenosine is considered a safe and effective medication, widely and routinely used in treatment of supraventricular tachycardia. However, proarrhythmic effects of adenosine have been previously recognized, largely in the presence of pre-excited atrial fibrillation. In structurally normal hearts, ventricular ectopy and non-sustained monomorphic and polymorphic ventricular tachycardia (VT) have been reported following adenosine administration [7A]. This is the first documented case of VF with hemodynamic collapse, in structurally normal heart without no demonstrable accessory pathway or use of other anti-arrhythmic drugs. Given the widespread use of adenosine in clinical practice, this unusual case highlights the ability of adenosine to induce circulatory collapse and emphasizes the importance of prompt access to resuscitation, defibrillation, and transcutaneous pacing equipment with drug administration.

Amiodarone [SEDA-36, 259; SEDA-37, 208; SEDA-15, 173]

Skin and Endocrine

- A 65-year-old male presented with a complaint of a grey to violet discoloration on his face and arms that began 6 months ago. He denied any chemical exposure. Past medical history includes of hypertension, type 2 diabetes, atherosclerotic heart disease and bypass surgery that was done 4 years ago. Since his surgery, he was on clopidogrel, metoprolol, amiodarone and nateglinide. His chemistry panel revealed the following: high free triiodothyronine (fT3) 6.14 pg/dL, high free thyroxine (fT4) 4.4.3 ng/dL, and low thyroid stimulating hormone (TSH) level 0.01 mIU/L. All other laboratory markers were normal. Thyroiditis was discovered after a thyroid ultrasonograph was performed. The 24-hour radioioidine uptake was 0.8% and the Technetium-99m pertechnetate thyroid scan disclosed a suppressed thyroid gland. The patient was diagnosed with type 2 amiodarone induced thyrotoxicosis and treatment with glucocorticoid was planned; however, the patient refused. Histopathological examination of facial skin showed "hyperpigmentation in the basal layer of the epidermis, yellow-brownish granules stained with PAS, and Masson Fontana in superficial dermis around vascular structures into histiocytes". Cardiology discontinued the amiodarone and his thyrotoxicosis resolved over 4 weeks. It was further reported the patient is euthyroid with the following laboratory parameters: fT3 2.8 pg/dL, fT4 1.20 ng/dL, TSH 1.58 mIU/L and is currently being followed outpatient for the skin discoloration [8A].

Another case of type 2 amiodarone induced thyrotoxicosis was reported whereby the patient underwent a thyroidectomy [9A]. In another case report, a newborn developed hypothyroidism following treatment with amiodarone [10A]. Hyperpigmentation and thyrotoxicosis are two known adverse effects of amiodarone. Moreover, amiodarone is known to cause both hypo- and hyper thyroidism. These cases highlight the importance of monitoring the thyroid function in patients on amiodarone in order to provide quick and appropriate treatment. Additionally, knowing how to differentiate between amiodarone induced thyrotoxicosis type 1 and 2 is particularly important, as their treatments are vastly different.

Liver

- A 79-year-old female presenting with atrial flutter with rapid ventricular rhythm was admitted to the coronary care unit. Upon admission her blood pressure was 134/91 mmHg and heart rate was 134 beats per minute. An amiodarone 150 mg intravenous (IV) push was initially administered and then an amiodarone infusion was started (infusion rate not indicated). Within the first 24 hours of admission the following total doses were also administered: adenosine 18 mg IV, furosemide 40 mg IV, metoprolol 5 mg IV, diltiazem 20 mg IV, heparin (dose and route not provided), aspirin 325 mg orally and atorvastatin 40 mg orally. A total of 930 mg of amiodarone was administered via the infusion in 24 hours which resulted in her heart rate decreasing to 60 beats per minute with normal sinus rhythm. Upon completion of the infusion, amiodarone was then switched to 200 mg twice daily by mouth. She received a total of 1480 mg of amiodarone by the second day of admission and presented with lethargy, although her vitals remained stable and in sinus rhythm. Her physical examination was significant for a yellow sclera and slowed responses. Her laboratory results were significant for an elevated aspartate aminotransferase (AST) of 2721 U/L, elevated alanine aminotransferase (ALT) of 2188 U/L, elevated total bilirubin 2.8 mg/dL, elevated ammonia level of 134 μmol/L, and prolonged prothrombin time of 27.8 seconds. She was subsequently diagnosed with acute liver failure. Her medication history was insignificant for hepatotoxic agents and included alendronate 70 mg weekly, furosemide 20 mg daily, enalapril 5 mg twice daily, aspirin 81 mg daily, amlodipine 5 mg daily, atorvastatin 40 mg daily and 5% lidocaine patch. Amiodarone was assumed the culprit for the acute liver failure which was then discontinued. She was then treated with *N*-acetylcysteine (NAC). Her AST levels peaked at 6712 U/L 48 hours after amiodarone was discontinued and then trended downwards. She was discharged on the 8th day after admission and her follow-up visit 2 weeks later showed normalized AST, ALT and prothrombin time with no changes in mentation [11A].

Amiodarone induced acute liver failure is very rare and usually reversible once amiodarone is discontinued. Some IV preparations of amiodarone contain polysorbate-80 or benzy-alcohol based vehicles. Literature suggests that polysorbate-80 can cause direct hepatotoxicity and impaired liver blood flow. However, this patient received an IV formulation that had a benzyl-alcohol based vehicle, thus the acute liver failure was not due to the formulation vehicle. *N*-acetylcysteine was utilized in this patient, although the mechanism of action in nonacetaminophen related liver failure is yet to be determined. It is suggested that NAC may improve hemodynamics and liver perfusion. This stresses the importance of monitoring liver enzymes while on amiodarone, particularly if the vehicle in the formulation is polysorbate-80. Moreover, NAC may be a viable option in treating amiodarone induced acute liver failure.

Sensory Systems (Eyes)

- A 66-year-old male presented with increasing physical fatigue and dyspnea for the past 3–4 weeks. His medical history is significant for well-controlled hypertension and persistent atrial fibrillation. His echocardiogram upon admission showed a left ventricular ejection fraction of 40%, "moderate mitral regurgitation and mild tricuspid valve regurgitation". He was cardioverted and started on amiodarone 600mg orally. The patient's eye sight gradually deteriorated 3 weeks post discharge, where he first complained of bilateral "blurred vision and cloudy areas scattered throughout". Subsequently, his primary practitioner discontinued amiodarone. These episodes of bilateral blurred vision and cloudy areas increased in frequency with new events every 2–4 days for approximately 3.5 months post discontinuation of amiodarone. His vision significantly decreased over 4 weeks, where it was "permanently blurry and foggy". "At this point, 4 months after the initial symptoms, his visual acuity was 0.32 (0.5 LogMAR) on the right eye and 0.4 (0.4 LogMAR) on the left eye". He was referred to ophthalmology 3 months later (7 months since initial presentation). He was then diagnosed with "an irreversible toxic optic neuropathy, induced by amiodarone" [12A].

Optic neuropathy is not uncommon with amiodarone. However, once suspected and withdrawn visual deficits may return to normal, though in some cases, visual impairment may be permanent. Current literature states that the time of optic deficits usually occur 9 months post initiation; however, this patient experienced vision changes just 3 weeks after initiation of amiodarone. Thus, it is important that patients are informed about the visual side effects of amiodarone so as to allow for rapid discontinuation if optic neuropathy is suspected. Moreover, it is important that patients get yearly eye exams when receiving amiodarone and perhaps semi-annual exams if there are pre-existing medical conditions such as diabetes.

Pancreas

- A 67-year-old male was admitted to the hospital and underwent radiofrequency ablation for atrial flutter without complications. One-day post procedure he had atrial fibrillation with rapid ventricular response which was unsuccessfully treated with metoprolol intravenously and oral diltiazem. He was transferred to an intensive care unit and an esmolol drip was started. Within the first few hours of the drip, the patient became hypotensive, diaphoretic and there were signs of hypoperfusion. A bolus of amiodarone was administered and phenylephrine was initiated. Once his blood pressure stabilized, he was cardioverted and kept on an amiodarone drip. On the third day, the patient complained of epigastric pain and a computed tomography (CT) scan of the abdomen revealed inflammation of the pancreas which, after examination by a gastroenterologist, led to the diagnosis of acute pancreatitis. Amiodarone was suspected as the causative agent and was discontinued. His pancreatitis was treated and his pain had markedly improved 1 week after the amiodarone drip was stopped [13A].

There is little differentiation in presentation between drug induced pancreatitis and pancreatitis. While amiodarone induced pancreatitis is rare, it can occur within as few as 3 days or may present after years of therapy. Once the offending agent is identified and discontinued, symptoms may resolve in as little as 7 days. It is important that prescribers withdraw the suspected agent if there is clinical suspicion of pancreatitis to allow for rapid treatment of pancreatitis and to avoid any untoward complications.

Respiratory

- A 43-year-old male presented to the pulmonary clinic with complaints of coughing up blood and worsening shortness of breath for 2 weeks. His past medical history is significant for non-ischemic cardiomyopathy (ejection fraction of 20%) and recurrent ventricular tachycardia post biventricular pacemaker implant and was on amiodarone 200mg/day which was started 1 year ago. The patient reported 20 lbs unintentional weight loss over 2 months, fatigue and anorexia. Cardiopulmonary auscultation was positive for "bibasilar inspiratory crackles and a grade 4/6 blowing systolic murmur radiating to the axila". The computed tomography (CT) scan of the chest displayed "bilateral multi-lobar infiltrates and dense consolidation primarily in the mid- to lower lung fields". His transthoracic echocardiogram revealed a significant increase in pulmonary artery pressure, mild mitral regurgitation, and an ejection fraction of 15%–20%. The pulmonary function tests "were consistent with restrictive pattern of lung disease". His free T4 was high and TSH was low. Amiodarone induced pulmonary toxicity was suspected and amiodarone was then discontinued. Six weeks later, the follow-up CT showed significant improvement [14A].

Pulmonary toxicity is fairly common in patients on amiodarone and typically occurs as pulmonary fibrosis. Presentation varies and radiological and histopathological findings may be non-specific. There are other case reports in the literature published in 2017 that report on pulmonary findings with the use of amiodarone [15A–18A]. In practice, physicians may opt to obtain a chest X ray at baseline to ascertain lung function and determine if there are any underlying masses or problems before initiating therapy, particularly in individuals with asthma and chronic obstructive pulmonary disease. Therefore, it is important to screen these individuals if there is clinical suspicion of pulmonary involvement early on to avoid any delay in management and to avoid possible irreversible long-term adverse effects.

Electrolyte Balance

- A 43-year-old male of weight 58.2kg and height 173.4cm was admitted to the hospital in June 2014 for new heart failure (HF). His family history is positive for dilated cardiomyopathy (DCM). Upon presentation his ejection fraction was 17%, blood pressure 96/60mmHg, heart rate 132 beats per minute, temperature of 36.8°C and 96% oxygen saturation on room air. Third and fourth sounds with a gallop rhythm was heard upon cardiac auscultation. His laboratory findings were as follows: hemoglobin 14.8g/dL, serum sodium 139mEq/L, serum potassium 4.4mEq/L, serum chloride 107mEq/L, creatinine of 1.03mg/dL and total bilirubin of 0.8mg/dL. His B-type natriuretic peptide (BNP) was significantly elevated at 2113.1. His ECG showed atrial tachycardia "with 130 beats per minute with left ventricular high voltage" and the echocardiogram showed "dilation of the left ventricle with diffuse hypokensis". A continuous infusion of dobutamine and milrinone, oral enalapril 2.5mg and carvedilol 1.25mg was given. There was an attempt to increase the carvedilol to 2.5mg; however, the blood pressure dropped and the patient became light-headed. As a result the carvedilol was reduced to the initial 1.25mg and he was started on an intravenous amiodarone infusion which corrected the atrial tachycardia into sinus rhythm. Following this, the patient underwent cardiac catheterization and endomyocardial biopsy after which he was diagnosed with DCM. The authors reported that they were unable to wean him off catecholamine infusion and was subsequently placed on the heart transplantation list. The patient remained admitted and received an LVAD implant 9 months later. His carvedilol was increase to 30mg daily, and tolvaptan and furosemide were discontinued post implant. The ECG revealed non-sustained ventricular tachycardia and thus the patient was kept on amiodarone 100mg daily. An echocardiogram one and a half months later showed and ejection fraction of 38%. His laboratory results showed hyponatremia (sodium of 116mEq/L) and an elevated antidiuretic hormone level of 3.6pg/mL (normal ≤2.8pg/mL). The authors suspected syndrome of inappropriate diuretic hormone secretion (SIADH) and restricted his water intake resulting in an improvement of his serum sodium level to 131mEq/L, 3 months post LVAD implant. Upon discharge his amiodarone level was 821ng/mL and desethylamiodarone level was 674ng/mL. An echocardiogram performed 6 months post LVAD implantation reported an ejection fraction of 42% and his chemistry panel was significant for hyponatremia. Ten months post implantation his sodium level was 123mEq/L and was diagnosed with drug induced SIADH. The dose of amiodarone was subsequently reduced to 50mg daily; however, 3 months later the patient was re-admitted with complaints of malaise and lethargy. On this admission his laboratory findings were significant for: low sodium level 108mEq/L, low serum osmolality 220mOsm/kg, high plasma aldosterone concentration 639pg/mL, high cortisol 19.6μg/dL and slightly high BNP level 21.3pg/mL. He was then diagnosed with SIADH secondary to amiodarone, and as a result the amiodarone was discontinued. He was then started on fludrocortisone. His serum sodium improved to 113mEq/L and he was discharged. However, he became delirious, "which resulted in a severe injury of the LVAD drive-line" and was hospitalized. He underwent a pump exchange on the second day of admission. His sodium level increased to 126mEq/L, and 1 month later, he was discharged with a sodium level of 131mEq/L [19A].

This case report brings to light the rare adverse effect of amiodarone induced SIADH. Because the patient was using both amiodarone and tolvaptan concurrently,

SIADH may have been masked. Thus, it is important that practitioners keep in mind that other drugs may mask the symptoms of adverse effects and monitor these individuals closely for possibly delayed presentation of adverse effects.

Diltiazem [SEDA-18, 144; SEDA-11, 154]

Nervous System

- An 85-year-old male receiving electroconvulsive therapy (ECT) for treatment-resistant depression experienced an unwanted reduction in seizure duration thought to be associated with diltiazem use. Initially, his home non-psychiatric medication regimen was continued with diltiazem extended-release (180mg daily), lisinopril, lorazepam, pravastatin, levothyroxine, and terazosin. After his first three inpatient ECT episodes lacked appropriate electroencephalographic (EEG) seizure duration despite appropriate hemodynamic changes, diltiazem was discontinued and stimulus electrode placement was changed to bitemporal electrode placement. For his three subsequent ECT episodes, treatment-induced seizures began to reach more predictable durations [20A].

Although both drug discontinuation and electrode adjustment occurred simultaneously in this patient clouding a true cause-and-effect analysis, diltiazem has previously been reported to shorten EEG seizure duration when compared against placebo in a small sample of patients ($n=18$, $P<0.05$) [21c]. Interestingly, verapamil, another nondihydropyridine calcium channel blocker, has been compared similarly and found no significant reduction in seizure duration. This suggests that this effect may not be class specific [22c].

Skin

- A 53-year-old male with newly diagnosed anal fissures was prescribed topical diltiazem hydrochloride 2% gel. The patient developed local pain around the anus after day 1 of treatment, which progressed to a pruritic maculopapular rash involving most of the body, concerning for a systemic allergic dermatitis reaction. When the gel was discontinued and switched to a zinc cream, the patient's rash and other symptoms cleared within 3 days. Patch testing confirmed positive results for diltiazem gel, in addition to neomycin sulfate 20% pet. All other testing yielded negative results and the patient was diagnosed with contact allergy to diltiazem and neomycin [23A].

Topically applied diltiazem has been previously associated with contact allergy. The authors note this to be the first documented case in which exposure resulted in systemic allergic dermatitis, where the extent of this reaction may have been mitigated by the high degree of absorption in the anal area.

Dofetilide [SEDA-37, 209; SEDA-38, 170; SEDA-15, 175]

Cardiovascular

Dofetilide is well known for causing QT prolongation and torsades de pointes (TdP). It was approved by the US Food & Drug Administration with a strict protocol for initiation or re-initiation, requiring hospitalization with continuous electrocardiogram (ECG) monitoring under the care of personnel specifically trained in dofetilide use. The incidence of TdP with initial dofetilide loading has been documented in multiple randomized controlled trials [24A]. In contrast, there has been a paucity of data describing the risks of TdP in dofetilide reloading after discontinuation of a previously tolerated dose.

- A single-center retrospective cohort study evaluated the incidence of TdP and need for dose adjustment in patients admitted for dofetilide reloading for atrial arrhythmias. Of the 138 patients included, 102 were reloaded with a previously tolerated dose, 30 with a higher dose from a previously tolerated dose, and 2 with a lower dose. Dofetilide dosage adjustment or discontinuation was required in 30 of 102 patients (29.4%) reloaded with a previously tolerated dose, and in 11 of 30 patients (36.7%) admitted for an increase in dose. No TdP occurred in the same-dose reloading group, but TdP did occur in 2 patients admitted to increase dofetilide dosage (0% vs 6.7%, $P=0.05$) [24A].

The sample size of this study is relatively small, and sufficient power to detect significant differences between reloading and initiation patient cohorts is lacking. However, the overall frequency of dose adjustment and possible increased risk of TdP with dose escalation is compelling, and strengthens the evidence supporting hospitalization for dofetilide reloading.

- A 64-year-old female presented to the emergency department with profound dyspnea and intermittent loss of consciousness. The patient had a history of bioprosthetic mitral and tricuspid valve replacement, complete atrioventricular (AV) block requiring dual-chamber pacemaker placement with a conventional right atrial lead and an unfixed left ventricular lead in the coronary sinus, along with atrial fibrillation managed with dofetilide 250mcg twice daily and warfarin at home. On arrival, her ECG demonstrated loss of ventricular capture and non-sustained episodes of TdP. Loss of ventricular capture yielded bradycardia, and treatment with dofetilide resulted in a prolonged QT interval, both of which were critical in generating TdP episodes. After multiple cardioversions with 150J and 4g of magnesium sulfate intravenously, the patient regained consciousness. Overdrive pacing at 90 beats per minute was achieved by temporary reprogramming of the left ventricular lead, resulting in cessation of TdP and clinical stabilization. Other than dofetilide treatment, the patient received no other QT-prolonging medications, and both serum electrolytes and renal function were within normal limits. Removing the offending agent, dofetilide, and addressing the lead dysfunction with eventual right epicardial ventricular lead placement were sufficient to avoid further episodes of TdP [25A].

Strategies for TdP prevention involve strict telemetry monitoring in at-risk patients, routine monitoring of serum potassium and magnesium, and minimizing concomitant use of other QT-prolonging medication. Two important mechanistic concepts of TdP are highlighted in this case—bradycardia and QT prolongation. It is important that clinicians understand the role of these mechanisms in the pathogenesis of TdP to initiate the most appropriate treatment.

Dronedarone [SEDA-17, 205; SEDA-35, 338; SEDA-36, 262]

Hematologic

Dronedarone reduced the rate of stroke and transient ischemic attack among patients with paroxysmal atrial fibrillation (AF) in the ATHENA trial, an affect which appears to distinguish dronedarone from other antiarrhythmic agents studied to date [26MC]. Given these findings, a mechanism independent of rate or rhythm effect may be involved with dronedarone therapy.

- A study using blood samples from 30 patients was performed to investigate any direct effect of dronedarone on blood thrombogenicity, independent of its antiarrhythmic effects. All patients had a history of cardiovascular disease (CVD) and were taking no anticoagulant or antiplatelet medications except aspirin. Blood samples were incubated with dronedarone's active metabolite (SR35021A) at 66ng/mL (am-L) and 119ng/mL (am-H), corresponding to the minimum and maximum mean C_{max} achieved after repeated oral doses of dronedarone 400mg BID. This was then compared against a control sample. Antithrombotic effects were assessed using ThromboElastoMetry with ROTEM® Gamma instrumentation. Compared to control, mean coagulation time was prolonged by am-L and am-H (164 ± 25 seconds vs 180 ± 22 and 182 ± 32 seconds, respectively, $P < 0.01$ for both). Significant reductions were also observed in CFR (am-L and am-H) and MCF (am-H). Platelet aggregation induced by ADP and TRAP was also reduced ($P < 0.05$ for both) by am-H [27c].

The authors concluded that dronedarone exerts direct anticoagulant and antiplatelet effects on human blood in vitro that are independent of its antiarrhythmic actions, suggesting a previously unknown pleotropic effect. Albeit an interesting finding, it is challenging to translate these results into clinical practice. These quantitative in vitro effects may not parallel the magnitude of dronedarone's hemostatic effect in vivo. This study also utilized blood samples from CVD patients, not AF, which may involve different pathophysiology. Further studies are required to better understand the mechanisms involved in dronedarone's antiplatelet and anticoagulant effects.

Flecainide [SEDA-35, 339; SEDA-36, 262; SEDA-15, 175]

Drug Overdose

- A 52-year-old female with a history of paroxysmal atrial fibrillation and major depressive disorder presented with altered mental status, hypotension, and bradycardia after overdosing on her home medications (flecainide and metoprolol). Her ECG showed an irregular slow-wide complex rhythm. Hypertonic sodium bicarbonate, temporary transvenous pacing support and intravenous fat emulsion therapy was administered and the patient was later discharged [28A].

Some medications used for psychiatric conditions are known to cause QTc prolongation. In this case, the medications used for depression were not provided and as such the complex rhythm could have been potentiated or worsened if a cocktail of antidepressants and flecainide was involved. Thus, it is important to closely monitor patients with high risk comorbid conditions. Moreover, this case demonstrated that use of flecainide in patients with depression should be carefully managed with the aid of the psychologist and other health professionals.

- A 38-day-old female infant displayed irritability, poor feeding and increased respiratory effort. At day 40, her respiratory distress worsened, had altered sensorium, and shock. She was intubated and her ECG showed supraventricular tachycardia (SVT) with Atrioventricular reentrant tract with right bundle branch block, which was reverted with adenosine. Thereafter, she was given intravenous metoprolol and amiodarone for 3 days. On day 44, she was started on oral flecainide started at 10.2mg/kg/day in two divided doses (1.7 times the recommended upper limit, i.e. 6 mg/kg/day). Her mother was provided instructions on how to mix and crush the tablet with sterile water and the infant was discharged. At day 47 after the seventh dose of oral flecainide, the infant developed generalized tonic clonic seizures which lasted 5 minutes with post-ictal drowsiness and she was then admitted to the hospital where bloodwork was drawn and no changes to the flecainide dose was made. At day 48, the infant was transferred to another institution (ninth dose of flecainide), where the infant was hemodynamically stable with normal sensorium and widened QRS complex on the ECG and flecainide was discontinued. On day 51 the infant had normalization of the ECG (3 days after flecainide was discontinued) and on day 54 (sixth day after flecainide was withdrawn) amiodarone was initiated [29A].

While digoxin and propranolol are the first line agents, flecainide is suggested in the literature as a second line agent for refractory SVT albeit off label and first line for newborns with SVT. It should be noted, flecainide toxicity is rare, and to date, there are no commercially available reversal agents. Currently, sodium bicarbonate is used as one of the agents to overcome sodium blockade, though results vary among studies. Another case report is provided in the literature on flecainide toxicity secondary to a dosing error [30A]. This case demonstrates the importance of 'double checking' dosing in infants especially as it pertains to narrow therapeutic index drugs to avoid complications that may require emergent attention. Moreover, it stresses on the importance of having a flecainide toxicity treatment protocol in place should toxicities occur.

Hemodynamic Instability

- A 60-year-old male who has a history of borderline hypertension without left ventricular hypertrophy (LVH), obstructive bronchitis, and had paroxysmal atrial fibrillation (PAF) since 2011, presented with an acute attack of PAF in August 2016. His home medications consisted of metoprolol succinate 47.5 mg daily and rivaroxaban 20 mg daily. Upon presentation, his echocardiogram indicated an ejection fraction of 63% and grade 1 diastolic dysfunction without hemodynamic or structural abnormalities. The authors report that the patient was offered cardioversion with flecainide since the PAF was less than 12 hours and he was on a beta-blocker. An IV infusion of flecainide 1 mg/kg over 5 minutes and after 25 minutes another dose of 0.3 mg/kg of flecainide was given to the patient. The patient developed a short episode of atrial flutter then converted to normal sinus rhythm 30 minutes after receiving flecainide. Within 3 minutes of being in sinus rhythm, the patient quickly developed respiratory depression and state he was feeling unwell and lightheaded. The patient became presyncopal, cyanotic, hypotensive (blood pressure of 60/40 mmHg) and bradycardic (heart rate $\leq$30 beats per minute). The patient was immediately started on volume supplementation and changed to a Trendelenburg position; however, his hemodynamic instability was only corrected with an IV injection of 1 mg of atropine and a fractionated injection of 400 μg adrenaline. The patient was hemodynamically stable 15 minutes after and was then observed for 12 hours following which he was transferred to another until and then discharged, without further problems [31A].

This case demonstrates the importance of knowing that hemodynamic instability is a known side effect of flecainide. Moreover, in this case the patient received a standard dose of flecainide in a patient without LVH, whereas in other reports in literature pertain to accidental or suicidal overdoses in patients with LVH. Thus, it is important that prescribers monitor these individuals closely.

Cardiovascular

- A 79-year-old female with a CHA_2DS_2VASc score of 4 was experiencing recurrent episodes of paroxysmal atrial fibrillation (PAF) for 15 months and was subsequently started on flecainide 50 mg twice daily and warfarin 2.5 mg. Her other medications included amlodipine 5 mg, losartan 50 mg, and thiazide 12.5 mg for her hypertension. The initial electrocardiography (ECG) showed a prolonged QT/QTc interval (476/495 ms) while she was in sinus rhythm, and an intermittent sinus pause of up to 1.8 seconds. The echocardiogram reported a left ventricular ejection fraction of 74%. Her PAF episodes continued, despite treatment with flecainide for more than 3 months. Catheter ablation was then performed with uninterrupted strategy of anticoagulation without complications. She was discharged on her home medications including flecainide 50 mg and warfarin. Ten days post ablation, the patient presented to the hospital complaining of palpitation, dizziness and presyncope. At this time her ECG indicated a prolonged QT/QTc (580/590 ms) with T wave inversion at precordial leads, and holter monitoring indicated repeated episodes of torsades de pointes (TdP). Consequently, flecainide was stopped and magnesium administered which provided no resolution. As a result, isoproterenol 1 μg/min was infused with a target heart rate of >80 beats per minute for 11 days and then stopped as there was no further evidence of TdP. On the 25th day postablation, follow-up ECG, QT/QTc interval were at baseline. Significant stenosis (>80%) at the proximal anterior descending artery was seen following a coronary angiogram. She was then discharged with isosorbide dinitrate and aspirin and clopidogrel. A low dose of flecainide (50 mg twice daily) was resumed to prevent PAF recurrence 1 month later as the patient was stable. The patient then presented to the emergency room with non-sustained TdP 2 days later after re-starting flecainide. Additionally, the ECG showed QT/QTc prolongation with U wave. Isoproterenol was given and the arrhythmic presentations resolved. The patient underwent a wash-out period of flecainide. Thereafter the U wave disappeared and the QT/QTc interval returned to baseline after isoproterenol was discontinued. Flecainide was never re-started and the patient remained unproblematic during follow up for more than 3 years [32A].

This case brings to light a rare incidence of flecainide induced TdP following catheter ablation. Severe or life threatening cardiac events occur in less than 1% of patients receiving flecainide. It should be noted that flecainide along with its metabolites is excreted renally, thus patients with compromised renal function would require a dose adjustment and close monitoring. There are numerous case reports on flecainide induced ventricular tachycardia, ECG changes, amongst others and which can also be an indicator of flecainide toxicity [33A–39A].

Mexiletine [SEDA-34, 298; SEDA-35, 341; SEDA-17, 211]

Skin

- A 48-year-old Asian male was transported to the hospital with a history of intermittent fever and rash for 1 month. Two months ago he started therapy with Mexiletine for his ventricular tachycardia. The rash presented as erythematous edematous papules which was present on his "chest, back and proximal extremities". Mexiletine was discontinued and he was treated with "prednisolone 100 mg daily for 9 days", this was tapered to 70 mg daily for 6 days and the patient then subsequently "weaned himself off". Eleven days after the discontinuation of prednisolone the patient had "widespread exfoliative dermatitis" and was readmitted. At this visit he was febrile with a heart rate of 104 beats per minute. His laboratory findings were significant for hypereosinophilia ($3.5 \times 10^3/mm^3$), elevated lactate dehydrogenase level (505 U/L) and elevated aminotransferase level (97 U/L). A skin biopsy was performed which showed findings indicative of a drug hypersensitivity reaction which was confirmed using the RegiSCAR criteria for drug-induced hypersensitivity syndrome (DRESS syndrome). He was then treated with oral cyclosporine dosed at 150 mg twice daily for 7 days. He was given cyclosporine since he had developed diabetes after his recent exposure to prednisolone. The rash responded to therapy within 3 days, and by the completion of the course, the rash had resolved entirely. "The cyclosporine dose was tapered to 100 mg twice daily for 14 days and 150 mg daily for 20 days" for a total duration of therapy of 41 days. Thirty seven days after the onset of the skin rash, the eosinophil counts and alanine aminotransferase level normalized, and his skin remained unproblematic [40A].

This case brings to light a rare, yet important adverse effect of DRESS syndrome with mexiletine. Drug related hypersensitivity has a relatively high mortality rate of approximately 10%–20% and thus delays in treatment can result in untoward effects. Typically the treatment involves discontinuation of the causative agent and using corticosteroids. It should be noted that the majority of previous case reports that speak to mexiletine induced hypersensitivity were in patients of Asian descent. Thus, it is important to monitor patients utilizing this therapy, so rapid treatment can be administered if any hypersensitivity-type reaction is suspected. Moreover, caution should be heeded when initiating in patients of Asian descent, as there may be a genetic predisposition.

Quinidine [SEDA-17, 390; SEDA-15, 2997; SEDA-30, 219]

Quinidine is a cincha alkaloid and a stereoisomer of quinine, an antimalarial agent. The structural similarities between the two compounds yield many similarities in their adverse effect profiles.

Skin

- A 9-month-old boy was prescribed quinidine (40 mg/kg/day; level 3.4 mcg/mL) for treatment of migrating partial seizures of infancy. During the 9-month trial of quinidine, he developed bluish discoloration of the hands, feet, and lips. There was no exposure to other medications that cause pigmentary changes, and quinidine was stopped due to minimal improvement in seizures and development. Skin discoloration persisted at 3 months, but markedly improved by 6 months [41A].

While antimalarial agents have long been known to cause a variety of pigmentation abnormalities, quinidine has only rarely been reported to cause skin discoloration, and all have occurred in adults [42A]. The authors warn epidemiologists to be aware of this uncommon but potential complication of quinidine therapy.

- A 41-year-old male presented with eczematous plaques on the backs of the first and second fingers of both hands, the neck, the lips, and both earlobes developing over the course of 6 months. At home, he administered quinidine sulfate to his child who suffered from Brugada syndrome. The preparation of the medication involved fragmentation of the quinidine capsule with the teeth, and pouring the capsule contents into a syringe with orange juice to facilitate ingestion. Photopatch testing using quinidine sulfate 30% pet confirmed a positive photoallergic reaction, and the patient improved when he stopped handling the medication [43A].

Photoallergic dermatitis secondary to quinidine use is caused by a type IV hypersensitivity reaction and has been described after both oral and systemic administration of quinidine [43A]. Diagnosis of photoallergic contact dermatitis caused by drug handling may often be overlooked, as lesions appear on sun-exposed areas rather than the palms or the palmar sides of fingers commonly used to apply medication.

Hematologic

- In one study, 10 patients with Brugada syndrome and implantable cardioverter defibrillators (ICDs) were treated with low-dose quinidine (≤200 mg/day) to reduce the incidence of ventricular arrhythmias and ICD shocks. One patient developed leukopenia with quinidine 200 mg/day, and the authors note that this resolved with a dose reduction to quinidine 200 mg every other day [44A].

Quinidine has been associated with many hematological side effects. In this patient, the leukopenic effect appeared to be dose-dependent and occurred even at lower-than-normal doses. This case emphasizes the importance of appropriate monitoring while on quinidine therapy, regardless of dose.

Sotalol [SEDA-15, 3170; SEDA-17, 211]

Respiratory

- A 68-year-old female with a history of atrial fibrillation presented with complaints of worsening shortness of breath and cough. In November 2012, she presented with similar symptoms and her computed tomography (CT) scan of the chest showed minimal basilar airspace disease. Since then her dyspnea worsened which required the use of home oxygen and in August 2016 her repeat CT showed worsening of her diffuse parenchymal lung disease (DPLD) with bronchiectasis, which was indicative of usual interstitial pneumonia (UIP). At her current visit, a repeat CT was done which revealed "bilateral diffuse ground glass opacities on the background of DPLD". Minimal diuresis was provided and broad spectrum antibiotics were administered; however, she worsened which necessitated intubation. Following this, a bronchoscopy was performed which revealed "sequentially progressive hemorrhagic bronchoalveolar lavage return". From her home medications, sotalol was thought to be the culprit and was discontinued. The patient was then treated with a pulse dose of methylprednisolone 1 g/day, with which her clinical and radiologic findings improved [45A].

Sotalol is a class II/III antidysrhythmic drug which is classified as a beta blocker. Respiratory problems are not uncommon with beta blockers and may present as drug induced lupus and exacerbations of obstructive lung disease. This case not only highlights the importance of drug induced DLPD but also diffuse alveolar hemorrhage, both of which are rare. Thus, it is important to perform a baseline chest X-ray. If there is suspicion or pulmonary complications, another chest X-ray should be performed.

Cardiovascular

- A 66-year-old female with a history of left atrial myxoma and atrial tachycardia presented to the emergency department with complaints of palpitations, dizziness and near fainting a few hours prior to presentation. She was on Sotalol 120 mg twice daily at home (route not provided). She was admitted to the coronary care unit, and her ECG revealed QT prolongation which was presumed to be secondary to Sotalol. Thus, Sotalol was discontinued and her QT normalized 5 days thereafter. Additionally, she underwent temporary transvenous ventricular pacing for 3 days and she showed high grade antrioventricular blockade even after the Sotalol was discontinued, and she then received a pacemaker implant [46A].

Torsades de Pointes with Sotalol has been reported to be dose related and has an incidence of 1%–4%. Albeit rare, patients receiving therapy with Sotalol should be monitored closely.

INOTROPES

Norepinephrine

Skin

- A 65-year-old male, with a history of smoking and no significant medical history of diabetes, peripheral vascular disease, hypercoagulability, connective tissue disease or vasculitis, presented to the emergency department with acute respiratory distress and significant sweating. The patient denied illicit drug use and wasn't on any medications. Shortly after presenting to the emergency department, he suffered a cardiac arrest and recovered secondary to cardiopulmonary resuscitation. He was hemodynamically unstable and was transferred to the intensive care unit (ICU) and experienced "post-intubation bilateral pneumonia, with multi-organ failure secondary to cardiogenic and septic shock". Upon admission and transfer to the ICU his atrial rhythm was normal on the ECG. An echocardiogram showed "anterioinferior akinesia and low left ventricular ejection fraction". The patient was hemodynamically unstable while on intravenous fluids, antibiotics, dobutamine and norepinephrine (dobutamine ≤10 μg/kg/min, norepinephrine 0.2 μg/kg/min) and the norepinephrine dose was increased to 4.15 μg/kg/min 2 days after admission to maintain a systolic arterial pressure of ≥90 mmHg. His blood cultures grew *Enterococcus faecium* and the bronchial aspirate cultures were positive for *Escherichia coli* and *Serratia marcescens.* "While on norepinephrine, progressive bilateral digital necrosis on his left second third, fourth and fifth fingers and right toe was seen from day 4 following admission". The norepinephrine was discontinued on day 10 with a gradual taper and a partial amputation of his left fingers aforementioned was executed. One-vessel coronary artery disease was seen via coronary angiography and he was later discharged to a rehabilitation institution 74 days following admission [47A].

Norepinephrine induced necrosis often presents bilaterally and symmetrically. Factors that increases the risk of necrosis include concomitant use of dopamine and peripheral vascular disease. Whilst a maximum dose for norepinephrine is yet to be ascertained in adults, a high dose of norepinephrine was utilized in this case. Since norepinephrine and other agents are primarily used in the critical care settings, prescribers should be cognizant of the possibility of necrosis and weigh the risk vs benefit when initiating high doses.

Levosimendan [SEDA-34, 291; SEDA-17, 258]

Levosimendan is an inodilator which was developed for patients with acute decompensated heart failure. It exerts its effects by increasing cardiac contractility, facilitating vasodilation via potassium channel opening and cardioprotection via the opening of mitochondrial potassium channels [48A]. Of note, this drug is not currently approved in the United States. Tenax therapeutics report that it is currently under development for reduction in morbidity and mortality of cardiac surgery patients at risk of low cardiac output syndrome (LCOS). Moreover, it was granted Fast Track status by the Food and Drug Administration (FDA) for its Phase 3 clinical trial under a special protocol assessment (SPA).

Published in 2017, the Effect of Levosimendan on Low Cardiac Output Syndrome in Patients With Low Ejection Fraction Undergoing Coronary Artery Bypass Grafting with Cardiopulmonary Bypass (LICORN) was a doubled blind, multicenter, placebo-controlled randomized trial designed to determine the efficacy of a pre-operative 24-hour Levosimendan infusion in reducing post-operative low cardiac output syndrome in patients with left ventricular ejection fraction ≤40% who were undergoing CABG. This study included 336 patients across 13 French cardiac surgical centers and included patients who were scheduled for a coronary artery bypass graft (CABG) with cardiopulmonary bypass alone or combined with valve surgery, provided they had a left ventricular ejection fraction of 40% or less. The study consisted of two arms—167 patients randomized to receive Levosimendan and 168 randomized to receive placebo. Levosimendan was reconstituted with 500 mL of 5% dextrose immediately before administration to the patient and the placebo was made of riboflavin to reproduce the yellow color of Levosimendan. After anesthetic induction, a continuous prophylactic infusion of 0.1 μg/kg/min was started and maintained for 24 hours provided there were no complications. Both groups were comparable as it pertained to mortality at days 28 and 180. The adverse events were comparable between groups with 57% of the patients in the Levosimendan group being severely hypotensive and 48% in the placebo group. There was no statistical difference between groups as it pertained to needing inotropic support at 48 hours: 49% of Levosimendan group and 59% in placebo group. They concluded that the findings do not support the use of Levosimendan in patients undergoing coronary artery bypass grafting with cardiopulmonary bypass. In this trial while there was no statistically significant difference between groups. There was no bolus dose given to the patients, which could account for the little effect seen [49C].

The Levosimendan in Patients with Left Ventricular Systolic Dysfunction Undergoing Cardiac Surgery Requiring Cardiopulmonary Bypass (LEVO-CTS) trial aimed "to evaluate the efficacy and safety of prophylactic Levosimendan started before and continued after surgery for the prevention of the low cardiac output syndrome and other adverse outcomes in high-risk patients undergoing cardiac surgery with the use of cardiopulmonary bypass". In this multicenter, randomized placebo-controlled phase 3 trial, 882 patients were randomized to receive either placebo or Levosimendan. Patients receiving Levosimendan were initiated at a dose of 0.2 μg/kg/min for 1 hour and then the dose was reduced to 0.1 μg/kg/min for 23 hours. It was reported at 90 days, 4.7% of the patients in the Levosimendan group had died, when compared to 7.1% in the placebo group (unadjusted hazard ratio, 0.64; 95% CI, 0.37–1.13; $P=0.12$). There were no statistically significant differences between the Levosimendan and placebo groups as it pertained to adverse effects (77 vs 70, respectively). Though not statistically significant, the Levosimendan group had the highest incidence of cardiogenic shock (10 patients). Pleural effusion was the most common in respiratory disorders (four patients). Other adverse effects include acute kidney injury (five patients), constipation and upper gastrointestinal hemorrhage of (two patients each), and multiple organ dysfunction syndrome (five patients). Interestingly, heparin induced thrombocytopenia, coagulopathy, thrombocytopenia, anemia, disseminated intravascular coagulation, and increased or decreased international normalized ratio occurred in the patients receiving Levosimendan compared to zero occurrences in the placebo group [50C].

It should be noted some trials utilize higher bolus doses which results in Levosimendan being more effective, but it can also lead to more hypotensive adverse effects. The LEVO-CTS trial was powered to evaluate clinical outcomes, whilst the LICORN trial may not have been sufficiently powered. Moreover, these trials consisted of predominantly white males (at least 80%). Thus, bigger trials with patient's representative of various backgrounds need to be conducted to allow for generalizability.

References

[1] Mina GS, Acharya M, Shepherd T, et al. Digoxin is associated with increased shock events and electrical storms in patients with implantable cardioverter defibrillators. J Cardiovasc Pharmacol Ther. 2018;23:142–8 [C].

[2] Lee AY, Kutyifa V, Ruwald MH, et al. Digoxin therapy and associated clinical outcomes in the MADIT-CRT trial. Heart Rhythm. 2015;12(9):2010–7 [MC].

[3] Abu-Amer N, Priel E, Karlish SJD, et al. Hypermagnesuria in humans following acute intravenous administration of digoxin. Nephron. 2018;138:113–8 [c].

[4] Guru SR, Suresh A, Padmanabhan S, et al. A rare case of digoxin associated gingival overgrowth. J Clin Diagn Res. 2017;11(4): ZD30–2 [A].

[5] Yoganathan K, Roberts B, Heatley MK. Life-threatening digoxin toxicity due to drug–drug interactions in an HIV-positive man. Int J STD AIDS. 2017;28(3):297–301 [A].

[6] Massicotte-Azarniouch D, Kuwornu JP, Carrero J-J, et al. Incident atrial fibrillation and the risk of congestive heart failure, myocardial infarction, end-stage kidney disease, and mortality among patients with a decreased estimated GFR. Am J Kidney Dis. 2018;71:191–9 [MC].

[7] Rajkumar CA, Qureshi N, Ng FS, et al. Adenosine induced ventricular fibrillation in a structurally normal heart: a case report. J Med Case Rep. 2017;11(1):21 [A].

[8] Ünal E, Perkin P, Konca De⊠ertekin C, et al. Hyperpigmentation on face and arms and thyrotoxicosis induced by amiodarone treatment. Cutan Ocul Toxicol. 2017;36(1):98–100 [A].

[9] Tonnelier A, de Filette J, De Becker A, et al. Successful pretreatment using plasma exchange before thyroidectomy in a patient with amiodarone-induced thyrotoxicosis. Eur Thyroid J. 2017;6(2):108–12 [A].

[10] García-García E, De Ceballos RD-G. Acquired hypothyroidism in a newborn treated with amiodarone in the first week of life. Indian Pediatr. 2017;54(5):420–1 [A].

[11] Jaiswal P, Attar BM, Yap JE, et al. Acute liver failure with amiodarone infusion: a case report and systematic review. J Clin Pharm Ther. 2018;43(1):129–33 [A].

[12] Knudsen A. Short-term treatment with oral amiodarone resulting in bilateral optic neuropathy and permanent blindness. BMJ Case Rep. 2017 [A].

[13] Mercogliano C J, Khan M, Ahmad M. A case of probable Amiodarone-induced pancreatitis in the treatment of atrial fibrillation: a literature review and case report. J Community Hosp Intern Med Perspect. 2017;7(6):369–71 [A].

[14] Tariq RZ, Cheema HA, Parikh R. Hemoptysis, a rare presenting symptom of amiodarone induced pulmonary toxicity. Am J Respir Crit Care Med. 2017;195:A5533 [A].

[15] Batnyam U, Rajasekaran A, Mellone J. Short course of amiodarone-induced severe lung toxicity in post-CABG patient. Am J Respir Crit Care Med. 2017;195:A5532 [A].

[16] Jaumally BA, Salem A. Amiodarone-induced pulmonary toxicity. Am J Respir Crit Care Med. 2017;195:C43 [A].

[17] Baumann H, Fichtenkamm P, Schneider T, et al. Rapid onset of amiodarone induced pulmonary toxicity after lung lobe resection—a case report and review of recent literature. Ann Med Surg. 2017;21:53–7 [A].

[18] Kaya SB, Deger S, Hacievliyagil SS, et al. Acute amiodarone toxicity causing respiratory failure. Rev Assoc Med Bras. 2017;63(3):210–2 [A].

[19] Nakamura M, Sunagawa O, Kugai T, et al. Amiodarone-induced hyponatremia masked by tolvaptan in a patient with an implantable left ventricular assist device. Int Heart J. 2017;58(6):1004–7 [A].

[20] Morrisette T, Rice J, Vickery PB. An unwanted reduction of seizure duration during electroconvulsive therapy with diltiazem. J ECT. 2017;33(1):e6–7 [A].

[21] Wajima Z, Yoshikawa T, Ogura A, et al. The effects of diltiazem on hemodynamics and seizure duration during electroconvulsive therapy. Anesth Anal. 2001;92(5):1327–30 [c].

[22] Wajima Z, Yoshikawa T, Ogura A, et al. Intravenous verapamil blunts hyperdynamic responses during electroconvulsive therapy without altering seizure activity. Anesth Anal. 2002;95(2):400–2 [c].

[23] Opstrup MS, Guldager S, Zachariae C, et al. Systemic allergic dermatitis caused by diltiazem. Contact Dermatitis. 2017;76(6):364–5 [A].

[24] Cho JH, Youn SJ, Moore JC, et al. Safety of oral dofetilide reloading for treatment of atrial arrhythmias. Circ Arrhythm Electrophysiol. 2017;10(10):e005333 [A].

[25] Alsaad AA, Silvers SM, Kusumoto F, et al. Dofetilide-induced torsade de pointes in high-grade atrioventricular node dysfunction. Postgrad Med J. 2017;93(1104):635–6 [A].

[26] Connolly SJ, Crijns HJGM, Torp-Pedersen C, et al. Analysis of stroke in ATHENA: a placebo-controlled, double-blind, parallel-arm trial to assess the efficacy of dronedarone 400 mg BID for the prevention of cardiovascular hospitalization or death from any cause in patients with atrial fibrillation/atrial flutter. Circulation. 2009;120(13):1174–80 [MC].

[27] Zafar MU, Santos-Gallego CG, Smith DA, et al. Dronedarone exerts anticoagulant and antiplatelet effects independently of its antiarrhythmic actions. Atherosclerosis. 2017;266:81–6 [c].
[28] Sardana M, McManus D. Depression and atrial fibrillation: a case report of intentional flecainide over dose. J Am Coll Cardiol. 2017;69(11):2338 [A].
[29] Bajaj S, Tullu MS, Khan Z, et al. When potion becomes poison! A case report of flecainide toxicity. J Postgrad Med. 2017;63(4):265–7 [A].
[30] Karmegaraj B, Menon D, Prabhu MA, et al. Flecainide toxicity in a preterm neonate with permanent junctional reciprocating tachycardia. Ann Pediatr Cardiol. 2017;10(3):288–92 [A].
[31] Albakri A, Klingenheben T. Unusual severe hemodynamic failure associated with standard dose of intravenous flecainide for pharmacological cardioversion of atrial fibrillation. HeartRhythm Case Rep. 2017;3(9):440–2 [A].
[32] Park YM, Cha MS, Kang WC, et al. Torsades de pointes associated with QT prolongation after catheter ablation of paroxysmal atrial fibrillation. Indian Pacing Electrophysiol J. 2017;17(5):146–9 [A].
[33] Watts TE, McElderry HT, Kay GN. An irregular wide complex tachycardia. Circulation. 2017;136(8):773–5 [A].
[34] Agrawal Y, Kalavakunta JK, Gupta V. Antiarrhythmic agent induced ventricular tachycardia. Am J Ther. 2017;24(4): e487 [A].
[35] Valentino MA, Panakos A, Ragupathi L, et al. Flecainide toxicity: a case report and systematic review of its electrocardiographic patterns and management. Cardiovasc Toxicol. 2017;17(3):260–6 [A].
[36] Awuor S, Lankford E. Felcainide toxicity. J Am Coll Cardiol. 2017;69(11):2307 [A].
[37] Tessitore E, Ramlawi M, Tobler O, et al. Brugada pattern caused by a flecainide overdose. J Emerg Med. 2017;52(4):e95–7 [A].
[38] Le Conte P, Malliet N, Chapelet G, et al. Flecainide-induced wide complex QRS tachycardia: a case report and review of the literature. Eur Geriatr Med. 2017;8(1):8–9 [A].
[39] Tsai TF, Garcia RT, Bui JQ, et al. Confused and too long: neurotoxicity and cardiac toxicity of flecainide. Crit Pathw Cardiol. 2017;16(1):42–5 [A].
[40] Zhang Z-X, Yang B-Q, Yang Q, et al. Treatment of drug-induced hypersensitivity syndrome with cyclosporine. Indian J Dermatol Venereol Leprol. 2017;83(6):713 [A].
[41] Baumer FM, Sheehan M. Quinidine-associated skin discoloration in KCNT1-associated pediatric epilepsy. Neurology. 2017;89(21):2212 [A].
[42] Conroy EA, Liranzo MO, McMahon J, et al. Quinidine-induced pigmentation. Cutis. 1996;57(6):425–7 [A].
[43] Agudo-Mena JL, Romero-Pérez D, Encabo-Durán B, et al. Photoallergic contact dermatitis caused by quinidine sulfate in a caregiver. Contact Dermatitis. 2017;77(2):131–2 [A].
[44] Shen T, Yuan B, Geng J, et al. Low-dose quinidine effectively reduced shocks in brugada syndrome patients with an implantable cardioverter defibrillator: a Chinese case series report. Ann Noninvasive Electrocardiol. 2017;22(1):e12375 [A].
[45] Pelleg T, Kadakia RS, Siddiqui F, et al. Sotalol induced diffuse alveolar hemorrhage. Am J Respir Crit Care Med. 2017;195:A5569 [A].
[46] Chatterjee D. A case of palpitation and pre-syncope. BMJ. 2017;357: j2343 [A].
[47] Daroca-Pérez R, Carrascosa MF. Digital necrosis: a potential risk of high-dose norepinephrine. Ther Adv Drug Saf. 2017;8(8):259–61 [A].
[48] Nieminen MS, Fruhwald S, Heunks LMA, et al. Levosimendan: current data, clinical use and future development. Heart Lung Vessel. 2013;5(4):227–45 [A].
[49] Cholley B, Caruba T, Grosjean S, et al. Effect of levosimendan on low cardiac output syndrome in patients with low ejection fraction undergoing coronary artery bypass grafting with cardiopulmonary bypass. JAMA. 2017;318(6):548 [C].
[50] Mehta RH, Leimberger JD, van Diepen S, et al. Levosimendan in patients with left ventricular dysfunction undergoing cardiac surgery. N Engl J Med. 2017;376(21):2032–42 [C].

CHAPTER

18

Beta Adrenergic Antagonists and Antianginal Drugs

*Asima N. Ali**,†,1, Jennifer J. Kim‡,§, Mary Ellen Pisano¶, Dalton K. Hudgins*, Maame Wireku‖, Sidhartha D. Ray#*

*Campbell University College of Pharmacy and Health Sciences, Buies Creek, NC, United States
†Wake Forest Baptist Health, Outpatient Internal Medicine Clinic, Winston-Salem, NC, United States
‡Greensboro Area Health Education Center, Greensboro, NC, United States
§University of North Carolina Eshelman School of Pharmacy, Chapel Hill, NC, United States
¶Novant Health, Winston Salem, NC, United States
‖Manchester University College of Pharmacy, Health & Natural Sciences, Fort Wayne, IN, United States
#Department of Pharmaceutical and Biomedical Sciences, Touro University College of Pharmacy and School of Osteopathic Medicine, Manhattan, NY, United States
[1]Corresponding author: ali@campbell.edu

BETA-ADRENORECEPTOR ANTAGONISTS [SEDA-35, 351; SEDA-36, 267; SEDA-37, 215; SEDA-38, 173]

Drug Class

Cardiovascular

INCREASED RISK FOR CARDIOVASCULAR EVENTS AND SEVERE HYPOGLYCEMIA WITH BETA ADRENERGIC ANTAGONIST USE IN DIABETES PATIENTS

In 2008, the Action to Control Cardiovascular Risk in Diabetes (ACCORD) trial revealed an increase in all-cause and cardiovascular mortality in patients receiving intensive therapy with a glycated hemoglobin (A1C) target of <6.0%, compared to standard therapy with an A1C target of 7.0%–7.9%. A recent post hoc analysis of the ACCORD trial revealed new findings. The original ACCORD study included 10 251 patients with a mean age of 62.2 years and a median A1C of 8.1% who had either established cardiovascular (CV) disease or CV risk factors. The medications used were the same in the two treatment groups (metformin, short- and long-acting insulins, sulfonylureas, acarbose, meglitinides, and thiazolidinediones). The recent analysis used propensity matching to divide study patients into two groups, those on beta adrenergic antagonists ($n = 2527$) and those not on beta adrenergic antagonists ($n = 2527$). Cox proportional hazards analyses were performed to determine an association between beta adrenergic antagonist therapy and outcomes [myocardial infarction (MI), unstable angina, nonfatal stroke, and CV death]. For those on beta adrenergic antagonists and those not on beta adrenergic antagonists, the mean follow-up periods were 4.6 ± 1.6 and 4.7 ± 1.6 years, respectively. The CV event rate was significantly higher for beta adrenergic antagonist users compared to non-users [hazard ratio (HR), 1.46; 95% confidence interval (CI), 1.24–1.72; $P < 0.001$]. In patients who received standard diabetes therapy, the CV event rate was higher in beta adrenergic antagonist patients (HR, 1.69; 95% CI, 1.36–2.13, $P < 0.001$). However, there was no significant difference in CV event rates in the intensive therapy patients. In patients with coronary heart disease or heart failure (HF), the cumulative CV event rate was higher in beta adrenergic antagonist users than non-users (HR, 1.27; 95% CI, 1.02–1.60; $P = 0.03$). Severe hypoglycemia (blood glucose <50 mg/dL) was significantly more common in beta adrenergic antagonist users compared to non-users (HR, 1.30; 95% CI, 1.03–1.64; $P = 0.02$). In the intensive diabetes therapy patients, the incidence

Side Effects of Drugs Annual, Volume 40
ISSN: 0378-6080
https://doi.org/10.1016/bs.seda.2018.07.015

of severe hypoglycemia was significantly higher for patients taking beta adrenergic antagonists compared to those not taking beta adrenergic antagonists (HR, 1.36; 95% CI, 1.03–1.81; $P=0.03$). Although it is well-known that beta adrenergic antagonists increase the risk for severe hypoglycemia and hypoglycemia unawareness, there was no significant difference in severe hypoglycemia between the beta adrenergic antagonist users and non-users in standard therapy patients. Thus, the overall increased CV event rate found in this analysis may not be related to the increased hypoglycemia risk. Further studies are needed to explain the increased CV event rates from beta adrenergic antagonist use in diabetes patients with CV risk, with or without established CV disease, that were found in this analysis [1C].

Metoprolol

Psychiatric

WORSENING OF DEPRESSIVE AND HIGH BURNOUT SYMPTOMS ASSOCIATED WITH METOPROLOL IN HEART FAILURE PATIENTS WITH MENTAL HEALTH DISORDERS

A prospective, single-center study investigated the impact of metoprolol on change in mental status of congestive HF patients with psychological disorders. This study included 154 patients with CHF. All patients were screened at baseline for newly diagnosed neuropsychiatric disorder using the Hospital Anxiety and Depression Scale (HADS) questionnaire and the Copenhagen Burnout Inventory (CBI) which were repeated at 1, 3, 6, and 12 months. Patients were divided into eight groups at baseline based on the presence of identified mental health disorder (i.e., depression, anxiety, high burnout, or a combination). The primary outcomes were the changes in HADS depression and HADS anxiety scores and change in CBI average score. Metoprolol therapy was associated with a statistically significant increase in HADS depression score from baseline (10.00 ± 0.20) to 1 month (10.76 ± 1.78; $P \leq 0.05$) in those with depression, but then remained stable through 12 months. A similar result was seen in the group with depression and high burnout with a change from baseline (9.00 ± 0.00) to 1 month (12.00 ± 2.61). At 12 months the HADS depression score was most elevated in patient with both depression and anxiety compared to patients with only CHF (13.7 ± 1.40 vs 7.75 ± 0.96). Interestingly, the four groups with anxiety saw significant decreases from baseline in the HADs anxiety score at 3 months vs patients with only CHF ($P \leq 0.0001$ for all four groups). This change remained stable through 12 months. Regarding CBI scores, patients with high burnout had higher CBI score at baseline vs those with no mental health condition ($P \leq 0.0001$ for all four groups). CBI score increased significantly at month 3 for all groups with a mental health condition and remained consistent through 12 months ($P \leq 0.0001$ for all four groups). Due to methodology, a causal relationship between metoprolol and mental health changes cannot be determined. In addition, there was no evaluation of anti-depressant use. This study supports screening for anxiety, depression, and burnout in CHF patients when considering metoprolol and close monitoring of mental health status [2C].

Drug–Drug Interaction

ACUTE MYOCARDIAL INFARCTION FROM CORONARY VASOSPASM PRECIPITATED BY PSEUDOEPHEDRINE AND METOPROLOL

A 63-year-old woman with dyspnea and palpitations was found to be in atrial fibrillation (AF) with rapid ventricular response and was sent to the emergency department. She had a history of asthma and seasonal allergies for which she treated with fexofenadine–pseudoephedrine daily. In route to the hospital she received intravenous (IV) metoprolol which resulted in modest improvement in her heart rate (HR). On arrival at the hospital she had no clinical symptoms of an acute MI, and her first troponin measure was normal, though she did exhibit irregular tachycardia. An electrocardiogram (ECG) revealed ST-T abnormalities. She then received additional IV and oral metoprolol. One hour later, the patient experienced chest pain, bradycardia, and hypotension. A second ECG revealed continuation of AF but now with slow ventricular response rate and ST elevation indicating acute inferior MI. An urgent coronary angiography established the diagnosis of multivessel coronary vasospasm. Authors concluded that metoprolol may have potentiated the rare but known risk of pseudoephedrine-induced coronary vasospasm by leading to unopposed α-receptor stimulation and vasoconstriction of smooth muscle. The patient underwent electrical cardioversion under the guidance of transesophageal echocardiogram and was discharged home [3A].

Nebivolol

Dermatological

LICHENOID TYPE CUTANEOUS HYPERPIGMENTATION INDUCED BY NEBIVOLOL

A case of lichenoid type cutaneous hyperpigmentation induced by nebivolol has been reported in a 46-year-old female who had been taking nebivolol 5mg daily for a total of 15 months for arterial hypertension. Three months after starting nebivolol, she developed erythematous eruptions on sun-exposed areas of the body including face, hands, and neck. A clinical diagnosis suggestive of hypersensitivity reaction to sunlight was made by dermatology. Hyperpigmentation was present for 1 year and persisted despite patient protecting herself from the sun for the previous 6 months. Other causes of hyperpigmentation, including lupus erythematosus, were fully investigated. Skin biopsy was consistent with drug-induced eruption, revealing lichenoid hyperpigmentation rather than allergic contact dermatitis or generalized pruritis, both of which have been associated with nebivolol therapy. Nebivolol was stopped and ramipril was started. Skin lesions started to resolve about 1 month after cessation of nebivolol and had cleared in 3 month. Lesions did not recur [4A].

Carvedilol

Worsening Heart Failure

WORSENING HEART FAILURE NYHA CLASS I WITH USE OF CARVEDILOL FOR HEART RATE CONTROL IN ATRIAL FIBRILLATION

Adding to the known potential for worsening heart failure (HF) with the use of carvedilol, caution is advised with the administration of carvedilol for rate control in patients with AF and concomitant chronic HF. A 6-week multicenter, randomized double blind study of 127 Japanese patients with persistent or permanent AF investigated the use of three dosing regimens of carvedilol: 5 mg fixed dose, 10 mg dose-escalation, and 30 mg dose-escalation. Primary endpoints were changes in 24-h mean HR on Holter electrocardiograms from baseline to 2, 4, and 6 weeks. Secondary endpoints included proportions of patients achieving target HR, clinical symptoms, and adverse events (AEs). Of the 127 patients enrolled, AEs occurred in 60 patients with no difference among the three dosing groups. No death or serious AEs occurred; however, two patients with NYHA class I HF in the 20-mg dose escalation group withdrew from the study after experiencing worsening chronic HF during the 5 mg treatment phase. These patients were noted to be among 18 patients to have AF with concomitant HF at baseline. Upon discontinuation of carvedilol and with appropriate pharmacological treatment, symptoms improved in these two patients [5C].

Labetalol–Nicardipine

Infection Risk

IMPACT ON OCCURRENCE OF IN-HOSPITAL INFECTION OF LABETALOL VS NICARDIPINE IN HYPERTENSIVE PATIENTS WITH INTRACRANIAL HEMORRHAGE

A single-center retrospective review of patients receiving labetalol, nicardipine, or both agents for management of hypertension (HTN) in context of intracranial hemorrhage (ICH) was conducted to determine which therapy resulted in more favorable mortality and infection rates. Of 1066 admissions identified, 525 were treated with labetalol or nicardipine within the first 3 days of hospitalization; 229 (43.6%) received labetalol only, 107 (20.4%) received nicardipine only, and 189 (36%) received both. Patients who received labetalol only were more likely at baseline to have AF, while patients who received labetalol and nicardipine were more likely to need surgical intervention for elevated intracranial pressure, pointing to a need for more aggressive management of the condition in those cases. Overall mortality rate was 40.2%, and 15.8% of patients were diagnosed with infection after admission. After controlling for covariates, propensity weighted generalized linear models were used for analysis of each outcome. Increased mortality was not associated with use of labetalol alone (OR, 0.93; 95% CI, 0.55–1.57; $P=0.793$) or with use of labetalol and nicardipine (OR 0.97; 95% CI, 0.58–1.62; $P=0.913$) in comparison to nicardipine alone. Labetalol alone was associated with increased infections (OR, 3.12; CI 1.27–7.64; $P=0.013$) but not when used in combination with nicardipine (OR, 2.44; CI, 0.98–6.07; $P=0.055$). Disability was defined as a modified Rankin Scale (mRS) score >2. At discharge, 77.2% of patients overall had an mRS >2. An increase in disability at discharge (mRS >2) was not associated with use of labetalol alone (OR, 1.17; CI, 0.62–2.20; $P=0.634$) or the use of labetalol with nicardipine (OR, 1.17; CI, 0.60–2.29; $P=0.639$). Authors noted that although use of labetalol and nicardipine did not show statistical significance for increased risk of infection, there was a trend toward elevated infection risk, indicating the likelihood that the labetalol component of the two-agent treatment is the one associated with increased risk of infection. Generalizability of this data is limited due to inability to control for all confounders such as concurrent antihypertensive use and other infections that may have occurred during hospitalization [6C].

Propranolol

Body Temperature

TEMPERATRE INSTABILITY ASSOCIATED WITH PROPANOLOL IN INFANTILE HEMAGIOMA

A case of temperature instability has been reported. A 25 and 1/7 week premature female presented with respiratory distress syndrome, possible sepsis, and hyperbilirubinemia, and was noted to have hemangiomas during week 7 of her life. She had an average axilla temperature (AxT) of 36.7°C and an average environmental temperature (EvT) of 23.5°C and was in an open bassinet for 12 days prior to treatment initiation. At 36 5/7 weeks propranolol was started at 0.7 mg orally every 8 h (1 mg/kg/day). After 3 days the AxT decreased to an average of 36.4°C, and the patient was placed in an incubator. Propranolol was increased on day 4 to 1.4 mg orally every 8 h (2 mg/kg/day), and EvT was increased from 27.5°C to 29.0°C in order to maintain appropriate temperatures. Propranolol was thought to have been associated with this temperature instability which persisted despite propranolol dose decreases (0.7 mg orally every 8 h). Propranolol was discontinued on day 19 and the patient transitioned to atenolol 0.7 mg (0.5 mg/kg/day) with titration to 1.4 mg orally twice daily (1 mg/kg/day). After the switch, the patient demonstrated temperature stability at AxT of 36.5°C to 37.2°C. The baby was discharged home with continued improvement seen in outpatient follow-up visits [7A].

Drug Overdose

CARDIAC ARREST FOLLOWING OVERDOSE OF AMITRIPTYLINE AND PROPANOLOL AND PREGABALIN

A case of cardiac arrest following overdose of amitriptyline and propranolol has been reported. A 25-year-old female was found unresponsive surrounded by three empty bottles of amitriptyline, propranolol, and pregabalin. She was intubated in the emergency department and became increasingly bradycardic, hypotensive and, finally, pulseless electrical activity (PEA). She was given calcium chloride, actrapid, sodium bicarbonate, and epinephrine boluses without success. The patient was subsequently infused with Intralipid (1.5 mL/kg, total dose 15 g) over 1 min. Within 10 min her clinical picture improved significantly, and she was thought to have fully recovered 16 h later. Forty-five minutes later she became apneic, bradycardic (heart rate fell from 110 to 64 bpm), and hypotensive (blood pressure of 85/54) with dilated pupils. She was given 100 mL of 8.4% sodium bicarbonate without effect and then administered an infusion of 75 mL of 20% intravenous Intralipid (1.5 mL/kg). Her pupils, blood pressure, and pulse normalized within 2 min. A 75 mL/h Intralipid infusion was started for 2 h and then decreased to 25 mL/h and continued for 14 h. Patient was discharged after 4 days with full recovery. The patient demonstrated toxicological features of all drugs in question. Propranolol toxicity was seen with presentation of 1st degree heart block, bradycardia, hypotension refractory to epinephrine, PEA cardiac arrest, central nervous system (CNS) depression, hypoglycemia, and low peak lactate levels. Authors proposed that despite the different mechanisms of actions of amitriptyline and propranolol, their similar lipophilic physiochemical properties and relationship with sodium channels at high doses may be the reason for intravenous lipid emulsion (ILE) treatment success. This case report supports the thought of ILE use as last line treatment in life threatening lipid soluble beta blocker overdose and provides a mechanistic explanation for therapy success that has been seen with amitriptyline but is not well established with propanolol [8A].

Drug–Drug Interaction

VENTRICULAR EXTRASYSTOLES ASSOCIATED WITH CONCOMITANT SOFOSBUVIR

A case report of ventricular extrasystoles of sofosbuvir and propanolol has been reported. Extrasystoles occurred within 3 h of concomitant medication use and resolved within 24 h after discontinuation. Based on this data it is important to monitor patients for arrhythmias especially during early treatment hours of concurrent sofosbuvir and propranolol use [9A].

Timolol

Psychiatric

HALLUCINATIONS ASSOCIATED WITH OCULAR TIMOLOL USE IN GLAUCOMA

A case series of timolol associated hallucination has been reported. Four cases of 66- to 89-year-old females with glaucoma or suspicion of glaucoma are presented. In all cases, patients developed visual hallucination after use of ophthalmic 0.5% timolol with varying degrees of onset (2 months to 10 years). Hallucinations were reported as "colored dots" or "yellow webs" or "colored flashes" with a consistent theme of alteration in color perception. Symptom resolution occurred after timolol discontinuation and recurred upon re-challenge. In addition to inherent limitation of case series, the presence of cognitive impairment was a potential confounder. Although one patient did not have cognitive impairment she did experience confusion with ocular timolol. Interestingly, the confusion resolved with timolol discontinuation and recurred with re-challenge. While data are limited it does suggest that older women with underlying CNS pathology may be at risk of experiencing hallucinations and possibly confusion with ocular timolol and should be closely monitored [10A].

CALCIUM CHANNEL BLOCKERS [SEDA-35, 354; SEDA-36, 270; SEDA-37, 219; SEDA-38, 175]

Amlodipine

Fluid Balance

PEDIATRIC CASES OF AMLODIPINE-INDUCE EDEMA

The first case was a 12-year-old female with systemic lupus erythematosus and class IV lupus nephritis. Sixteen months after starting amlodipine 10 mg daily, she developed severe generalized edema with pleural effusion and ascites, which resolved 2 weeks after stopping amlodipine. The second case was a 6-month-old girl with familial combined immunodeficiency and nephritic syndrome treated with hydralazine and then amlodipine (5mg daily) for hypertension (HTN). Three days after starting amlodipine, she developed generalized edema and ascites, which resolved 2 weeks after stopping amlodipine. This patient was re-challenged with amlodipine, but the edema returned. The third patient was an 11-year-old female with Fanconi anemia. After her second bone marrow transplant, she developed HTN subsequent to high-dose steroids, which was treated with amlodipine 5mg daily. Four months later, she presented with generalized edema with pleural effusion and ascites. Ten days after amlodipine discontinuation, the edema resolved. The fourth case was a male patient 4 years and 8 months of age. He developed end-stage renal disease secondary to renal dysplasia and required peritoneal dialysis. Two days after starting amlodipine 5mg daily, he presented with generalized edema, which then resolved 3 days after amlodipine cessation [11A].

Mouth and Teeth

PATIENT CASES OF AMLODIPINE-INDUCED GINGIVAL HYPERPLASIA

The first patient, a 60-year-old female, had been taking amlodipine 10mg daily for 9 years, along with other treatments for HTN, HF, and hyperlipidemia. Amlodipine was discontinued and enalapril was increased from 2.5mg twice daily (BID) to 5mg BID. The patient was also instructed to use chlorhexidine 0.2% mouthwash BID. Her gingival enlargement completely resolved 4 months later. The second patient was a 52-year-old male complaining of gum swelling who was on amlodipine 10mg daily for 10 years. The amlodipine was replaced with enalapril 5mg BID, and chlorhexidine 0.2% mouthwash BID was initiated. At 3 month follow-up, there was complete reduction in gingival swelling. The third case was a 42-year-old female who had been taking amlodipine 10mg daily for 1 year. Generalized gingival enlargement was present, and the same regimen of enalapril with chlorhexidine was implemented. Three months later, complete resolution of gingival swelling occurred. All three patients were found to have single nucleotide polymorphisms, indicating the potential role of polymorphism testing for predicting risk of adverse drug reactions [12A].

Infection Risk

NASOPHARYNGITIS ASSOICATED WITH AMLODIPINE USE

Sacubitril/valsartan (LCZ696) was added onto amlodipine in an RCT of 266 Asians with uncontrolled systolic HTN. The adverse effect (AE) incidence was similar between groups (20% with LCZ696/amlodipine 200mg/5mg daily, and 21% with amlodipine 5mg daily), the most common being nasopharyngitis (3.8% and 3.7%, respectively). There were no deaths, edema, hypotension, or SAEs reported [13C]. An open-label, multi-center evaluation of long-term (52-week) therapy with azilsartan/amlodipine/hydrochlorothiazide 20/5/12.5mg was conducted in 341 patients. The overall incidence of drug-related AEs was 38.4%, with the most common being nasopharyngitis (31.1%). No deaths were reported. Thirty-two AEs lead to study drug discontinuation was (9.4%), a variety of which were related to the study drug (hypotension, photosensitivity, anemia, sinus bradycardia, hyponatremia, postural dizziness, chronic kidney disease, drug eruption, eczema, purpura, and orthostatic hypotension) [14C].

Overdose

PATIENT CASE OF SUICIDE ATTEMPT WITH AMLODIPINE

A 17-year-old male was treated for an overdose suicide attempt, in which toxicology results eventually revealed amlodipine, caffeine, and metoprolol in the patient's serum. The patient presented with hypotension, bradycardia, and altered mental status. After receiving aggressive supportive care, consisting of numerous treatments such as glucose, calcium, insulin, and epinephrine, the patient progressed to cardiac arrest. The patient was then treated with extracorporeal membrane oxygenation (ECMO) and survived neurologically intact [15A].

Drug Interactions

RHABDOMYOLYSIS WITH CONCOMITANT AMLODIPINE, ATORVASTATIN AND TICAGRELOR

A 74-year-old female developed severe rhabdomyolysis attributed to an interaction involving atorvastatin (80 mg daily), ticagrelor [90mg BID], and amlodipine (5mg daily). Amlodipine was implicated due to weak cytochrome P450 3A4 (CYP3A4) inhibition, which can enhance rhabdomyolysis risk of atorvastatin and potentiate an interaction with atorvastatin and ticagrelor, another weak CYP3A4 inhibitor. The patient recovered after prolonged intensive care and was discharged for rehabilitation in stable condition [16A].

NEW INSIGHT ON ENANTIOMER ATTRIBUTED TO AMLODIPINE SIDE EFFECTS

A molecular modeling study was conducted to examine amlodipine's enantiomers. *R*-amlodipine (*R*-AML) exhibited more potent time-dependent inhibition of CYP3A than *S*-AML (K_I 8.22μM, K_{inact} 0.065min^{-1} vs K_I 14.06μM, K_{inact} 0.041min^{-1}, respectively) as well as CYP2C9 (K_i 12.11μM vs 21.45μM) and CYP2C19

(K_i 5.97 μM vs 7.22 μM). These results indicate that *R*-AML is responsible for amlodipine side effects and P450 drug interactions [17E].

Verapamil

Drug Interactions

VERAPAMIL–RIVAROXABAN PHARMACOKINETIC MODELLING AND BLEED RISK POTENTIAL

Verapamil may increase bleed risk with rivaroxaban, and this risk may be augmented in renal impairment. Twenty-seven volunteers (mean age 59 years) were split into two renal function cohorts: normal [mean creatinine clearance (CrCl) 105 mL/min] and mild insufficiency (mean CrCl 71 mL/min). Participants received two doses of rivaroxaban 20 mg, one on day 1 and one on day 15. Study subjects also received single morning doses of oral extended-release verapamil 120 mg on day 8, 240 mg on day 9, and 360 mg on days 10–17. Rivaroxaban C_{max} did not differ between the normal and mild insufficiency groups, with or without verapamil. However, within each group, verapamil significantly increased rivaroxaban AUC (by 39% and 42%, respectively), reduced CrCl (by 28% and 32%, respectively), and prolonged $T_{1/2}$ (by 18% and 43%, respectively). Verapamil also enhanced rivaroxaban antithrombotic effects [18c].

Another study of 27 subjects evaluated a pharmacokinetic (PK) model system with verapamil and rivaroxaban in various renal function categories (CrCl 80–130 mL/min, CrCl 50–79 mL/min, CrCl 30–49 mL/min, and CrCl 15–29 mL/min). Addition of verapamil 360 mg daily to rivaroxaban 20 mg daily increased steady-state rivaroxaban AUC by 48%, 49%, 50%, and 51%, respectively. Reducing rivaroxaban to 15 mg daily increased AUC to a lesser extent (11%, 12%, 12%, and 13%, respectively). On the other hand, rivaroxaban 10 mg daily reduced AUC by 26%, 26%, 25%, and 25%, respectively. These models predicted an increased major bleeding risk when combining verapamil 360 mg daily and rivaroxaban 20 mg daily, suggesting that, when given with verapamil 360 mg daily, rivaroxaban be reduced to 15 mg daily in normal renal function and mild renal impairment, or 10 mg daily in moderate to severe renal impairment [19c].

CENTRAL NERVOUS SYSTEM EFFECTS OF VERAPAMIL WITH AN EFFLUX TRANSPORTER INHIBITOR

The impact of drug–drug interactions on efflux transporter P-glycoprotein (ABCB1) activity at the blood–brain barrier (BBB) was examined in five young (mean age 26 years) and five elderly (mean age 68 years) patients. Verapamil was used as the model ABCB1 substrate and was measured through positron emission tomography scans before and after intravenous (IV) administration of a low-dose ABCB1 inhibitor, tariquidar. There was a significant difference between elderly and young subjects in the increase in total verapamil distribution time (40% vs 2%, $P=0.032$). In conclusion, verapamil CNS effects (e.g., dizziness) may be enhanced in elderly patients when co-administered with ABCB1 inhibitors (e.g., cyclosporine) [20c].

Diltiazem

Overdose

CASES OF HYPOTENSION AND SHOCK FROM DILTIAZEM OVERDOSE

A 46-year-old female was treated by intensive care for hypotension and vasoplegic shock after ingesting 8.4 g of extended-release diltiazem. The patient was initially treated with insulin, glucose, calcium, and norepinephrine, but her hemodynamic status worsened, and she went into renal failure with lactic acidosis. Terlipressin 0.1 mg boluses were given intermittently (0.3 mg total dose). The patient's hemodynamic status and acidosis improved dramatically within hours after receipt of the terlipressin [21A]. A 26-year-old female developed hypotension and bradydysrhythmia after ingesting 80 tablets of diltiazem in a suicide attempt. She was given a variety of treatments including calcium, glucagon, vasopressors, insulin, and lipid emulsion. After refractory bradydysrhythmia that responded only to transvenous pacing, ECMO was given and she successfully recovered [22A]. A 53-year-old female presented with cardiogenic shock with complete AV block after allegedly taking 4800 mg of diltiazem. She did not respond to atropine, isoprenaline, and an external pacer. Magnesium sulphate and activated charcoal were given for decontamination, and the patient was later successfully treated with isotone fluids, inotropes, vasopressors, and continuous infusion of high-dose calcium and glucagon. The patient did develop acute, necrotizing pancreatitis thought to be due to iatrogenic high calcium levels but was successfully treated with IV fluid rehydration and analgesics [23A].

Nifedipine

Overdose

TODDLER FATALITY FROM NIFEDIPINE OVERDOSE

A 2-year-old male accidentally ingested an unknown amount of nifedipine. The boy complained of dizziness 1 h after ingestion. Upon hospital arrival, blood pressure (BP) was immeasurable and heart rate was 169 beats per minute, and he received gastric lavage. He had convulsions followed by cardiac arrest which responded to resuscitation, but died the next day [24A].

Drug Interactions

NIFEDIPINE–CANDESARTAN COMBINATION PHARMACOKINETICS AND TOLERABILITY INVESTIGATED

Nifedipine was recently studied in combination with candesartan in two trials [23A, 24A]. The first was a phase I, single-center, non-randomized, non-controlled, non-blinded, observational study in 32 patients with hepatic impairment. Treatment was a single oral dose of nifedipine gastrointestinal therapeutic system (GITS) 30mg with candesartan cilexetil 8mg (N30/C8) fixed-dose combination. Although nifedipine AUC and C_{max} increased in both mild (Child-Pugh A, group 1) and moderate (Child-Pugh B, group 2) hepatic impairment by 93% and 253%, respectively, the AEs were similar to known profiles for nifedipine [25c]. The second nifedipine/candesartan (N/C) study was a subanalysis of the DISTINCT randomized trial, which evaluated the combination at various doses compared to separate agents as monotherapy and placebo in patients with HTN. Specifically, high-risk individuals were examined [422 patients with estimated glomerular filtration rate <90mL/min, 202 with type 2 diabetes mellitus (T2DM), 206 with hypercholesterolemia, and 971 with CV risk factors]. Rates of treatment-related AE were similar (33.1% in the renal impairment group, 29.8% in the T2DM group, 30.5% in the hypercholesterolemia group, and 38.7% in the CV-risk group). Notably, there were less vasodilatory AEs with N/C than N monotherapy in all four groups (12.2% vs 23.0%, 14.4% vs 18.6%, 13.9% vs 18.6%, and 20.3% vs 25.1%, respectively) [26C].

Felodipine

Drug Interaction

FELODIPINE–ENALAPRIL PHARMACOKINETICS AND SIDE EFFECTS INVESTIGATED

A randomized, open-label cross-over study in 12 healthy subjects (mean age 24 years) investigated the possible drug interaction between felodipine and enalapril. Subjects were given felodipine 5mg and enalapril 5mg, alone or in combination, BID for 6 days, and then one dose of each in the morning on day 7. Felodipine PK was significantly altered by enalapril, [18% decrease in C_{max}, point estimate 0.79, 95% CI 0.65–0.95], but not vice versa. However, no adverse events were observed during the study [27c].

Barnidipine

Fluid Balance

RATES OF EDEMA WITH BARNIDIPINE

Barnidipine is a calcium channel blocker (CCB) available in various countries outside the United States. Barnidipine was studied in a post-marketing observational, prospective trial, in which 1710 patients who did not tolerate prior therapy with amlodipine or lercanidipine were examined more closely. In a previous study, edema rates with amlodipine and lercanidipine were 19% and 9.3%, respectively. In the current study, of patients who switched from amlodipine or lercanidipine to barnidipine, 4.8% developed edema with barnidipine. Lower rates of edema with barnidipine were attributed to slow onset of action and lipophilic depot. However, conclusions comparing CCB edema rates are confounded because many patients receive concomitant renin-angiotensin system inhibitors, which can lower these rates by 38%–55% [28MC].

℞

NIFEDIPINE USE IN PREGNANCY

Pregnant women (12^{+0}–27^{+6} weeks' gestation, n=112) with chronic HTN at four United Kingdom centers were enrolled in an open-label RCT. First-line anti-hypertensive therapy was nifedipine-modified release 20–80mg/day, or labetalol 200–1800mg/day. Out of four SAEs reported (epistaxis, gastroenteritis, deep vein thrombosis, influenza), none were considered related to treatment. Fifteen (26%) of nifedipine patients reported an AE compared to 21 (38%) of labetalol patients. Ninety percent of nifedipine patients stated they would definitely take the same treatment again in another pregnancy, compared to 72% of labetalol patients. Five percent of nifedipine patients stated they probably would not take it again compared to 11% of labetalol [29C]. Another RCT compared nifedipine to labetalol, but in postpartum HTN. Fifty women who delivered ≥32 weeks' gestation were given either oral nifedipine extended-release (30–90mg daily) or oral labetalol (200–800mg BID). No SAEs were observed. Minor AEs were significantly more common with nifedipine than labetalol (48% vs 20%, P=0.04), including headache in 24% and constipation in 16% taking nifedipine [30c].

Nifedipine was compared to hydralazine in acute hypertensive emergency during pregnancy in a double-blind RCT. Sixty pregnant women (≥24 weeks' gestation) with sustained increased BP (≥160mmHg systolic or ≥110mmHg diastolic) received oral nifedipine (10mg tablet up to four doses) with IV saline every 20min, or IV hydralazine (in doses of 5, 10, 10, and 10mg) with a placebo tablet, to a target BP of 150/100mmHg. Patients were crossed over if the initial treatment failed. The only AE that was significantly different between the groups was vomiting (7% vs 30% for nifedipine and hydralazine, respectively, P=0.042). A significant increase in maternal heart rate with time in the first 80min was seen with nifedipine (P<0.001) but not for hydralazine. No SAEs were observed [31c]. Nifedipine was utilized for treatment of preterm symptomatic placenta previa in a multicenter, double-blind study. A total of 109 patients were randomized to either nifedipine 20mg slow-release, or placebo, three times *daily. Cesarean due to hemorrhage prior 37 weeks' gestation was more common with nifedipine than placebo (RR, 1.66; 95% CI 1.05–2.72). Similar perinatal AEs were seen between groups (3.8% for nifedipine vs 5.5% for placebo (RR, 0.52; 95% CI 0.10–2.61)]. No maternal mortality or perinatal deaths occurred [32c].*

Adjuvants to Nifedipine for Preeclampsia

Four RCTs evaluated adjuvants [vitamin D, epigallocatechin gallate (EGCG), celastrol, and resveratrol] to nifedipine for preeclampsia. In the first trial, patients received either nifedipine 10mg/vitamin D 200 international units (n=298), or nifedipine 10mg/glucose placebo 20mg (n=304), up to four doses. In the second study, 350 patients were randomized to either nifedipine 10mg/EGCG 100mg, or nifedipine 10mg/placebo 100mg, up to five doses. In the third trial, 552 patients received either nifedipine 10mg/celastrol 10mg, or nifedipine 10mg/glucose placebo 10mg, up to five doses). The fourth trial randomized 349 patients to either nifedipine 10mg/resveratrol 50mg, or nifedipine 10mg/glucose placebo 50mg, up to five doses. In all four studies, patients received study medications every 15min to a target BP of 150/100mmHg. There were no observed significant increases in maternal or neonatal AEs compared to placebo when these adjuvants were given with nifedipine [33C, 34C, 35C, 36C].

Mouth and Teeth

The first case of nifedipine-induced gingival hyperplasia in pregnancy was reported. A 27-year-old female with a history of T2DM presented at 27 weeks' gestation with gingival hyperplasia 9 weeks after initiating nifedipine for HTN (started on extended-release 20mg daily and titrated up to 60mg BID). The patient had also been taking aspirin, metformin, and multivitamins. Nifedipine was replaced with methyldopa, and chlorhexidine mouthwash was started. Within 48h, the gingival hyperplasia improved. Two weeks later, she delivered the baby. The gingival hyperplasia resolved and methyldopa was stopped due to HTN resolution [37A].

POTASSIUM CHANNEL ACTIVATORS [SEDA-35, 353; SEDA-36, 270; SEDA-37, 219; SEDA-38, 176]

Nicorandil

Skin

ULCERATION OF PENILE GLANS ASSOCIATED WITH NICORANDIL USE

Ulceration associated with nicorandil use has been documented in the literature. Additional data have since been reported. A case report of a 78-year-old male with a past medical history significant for T2DM, HTN, symptomatic chest pain, and renal failure presented with penile ulceration after nicorandil use. Ulceration resolved after nicorandil discontinuation. The patient was on 20 more medications that were not identified in report and may potentially confound results. The mechanism of this AE is unclear but thought to be related to either an effect of the drug itself or its metabolites. It is unclear whether frequency or severity of ulceration is dose related [38A].

Nervous System

ACUTE INTRACTABLE HEADACHE AND OCULOMOTOR NERVE PALSY ASSOCIATED WITH NICORANDIL

Headache and abducens nerve palsy with nicroandil use has been documented in the literature. Additional data have since been reported supporting these findings but also sheds light on a new side effect of oculomotor nerve palsy with pupil involvement. A 59-year-old female with a past medical history of AF presented to the emergency department with a headache and right pitosis after prior use of nicorandil 5mg twice daily. She was noted to have a slowed response to light stimuli in her right eye, anisocoria, and after testing was diagnosed with ocular nerve palsy. Both headache and ocular nerve palsy resolved after discontinuation of nicorandil for 3 days [39A].

NITRATES [SEDA-34, 305; SEDA-35, 354; SEDA-38, 176]

Nitroglycerin

Cardiovascular

HYPOTENSION FOLLOWING PREHOSPITAL ADMINISTRATION OF NITROGLYCERIN IN PATIENTS WITH CHEST PAIN AND TACHYCARDIA

A retrospective cohort aimed to determine whether sublingual (SL) nitroglycerin (NTG) spray administration prior to hospitalization was a predictor for hypotension in tachycardic patients compared to those without tachycardia. The study included 10308 subjects of which 2057 (20%) were tachycardic prior to prehospital nitroglycerin administration. Hypotension was defined as systolic blood pressure less than 90mmHg while tachycardia was defined as a heart rate between 50 and 100 beats/min. A total of 320 patients experienced hypotension. More patients within the pre-NTG tachycardia cohort experienced statistically significant hypotension compared to those without tachycardia (3.9% vs 2.9%, $P=0.02$, respectively). A significantly higher odd of post-NTG hypotension was seen in the tachycardic group (AOR: 1.60; 95% CI: 1.23–2.08). Due to the retrospective nature, utilization of only one site, presence of confounders, and imbalance between baseline cohorts (i.e., differences in disease states and baseline line pressures) external validity is limited and causality cannot be proven. However, this study does generate the question regarding the use of continuous blood pressure monitoring in these patients and requires further study may be an area of future study [40C].

Pregnancy

NITROGLYCERIN DOES NOT PROTECT AGAINST EXTENSION, AND SUBLINGUAL NITROGLYCERIN ASSOCIATED WITH ADVERSE NEONATAL OUTCOMES

A retrospective cohort study sought to determine whether NTG was associated with differences in rates of unintended extension of the uterine incision (hysterotomy extension). A total of 391 females who underwent cesarean delivery (CD) during the second stage of labor were included. The average age was 33 years and while body mass indexes remained relatively well balanced between 22.8 and 25.2. Forty nine patients received NTG and of those 24 received intravenous (IV) and 25 received the SL spray. Neither formulation was found to be associated with an increased risk of hemorrhage or extension. However, SL-spray NTG was associated with 4.68-fold increased odds of 5-min Apgar <7 (95% CI 1.42–15.41) and 3.36-fold greater odds of neonatal intensive care unit admission (95% CI 1.20–9.41) both of which were statistically significant. Based on these findings caution is advised regarding use of NTG as a uterine relaxant in these patients, weighing risks vs benefits of treatment considering potential harms to baby. More studies are needed before definitive conclusions can be made [41C].

LATE SODIUM CHANNEL (INA) INHIBITORS [SEDA-38, 176]

Ranolazine

Cardiovascular

Multiple meta-analysis have been published supporting the current body of literature regarding the use of ranolazine's effects on heart rhythm in the treatment of AF with adverse event rates similar to that of controls [42M]. Data have been reported in using ranolazine to prevent AF following cardiac surgery. Results were similar, however, a small number of studies were included with multiple confounding factors (i.e., dose variations), thus, limiting external validity [43M]. Paralleling these studies, multiple meta-analysis demonstrated that ranolazine used both alone and for prevention of AF as an add on to amiodarone therapy was found to be well tolerated with overall all adverse event risks to be similar to controls [44M].

Gastrointestinal

Data supporting the current body of literature have been published indicating ranolazine monotherapy for the treatment of stable angina has been associated with increased risk of mild AEs such as dizziness, nausea and constipation. Authors noted that majority of evidence being of very low, low, and moderate quality [45M].

References

[1] Tsujimoto T, Sugiyama T, Shapiro ME, et al. Risk of cardiovascular events in patients with diabetes mellitus on β-blockers. Hypertension. 2017;70:103–10 [C].
[2] Liu X, Lou X, Cheng X, et al. Impact of metoprolol treatment on mental status of chronic heart failure patients with neuropsychiatric disorders. Drug Des Devel Ther. 2017;11:305–12 [C].
[3] Meoli EM, Goldsweig AA, Malm BJ. Acute myocardial infarction from coronary vasospasm precipitated by pseudoephedrine and metoprolol use. Can J Cardiol. 2017;33: 688e1-e3. [A].
[4] Aslan AN, Guney MC, Akcay M, et al. Lichenoid type cutaneous hyperpigmentation induced by nebivolol. Turk Kardiyol Dern Ars. 2017;45(3):268–70 [A].
[5] Inoue H, Atarashi H, Okumura K, et al. Heart rate control by carvedilol in Japanese patients with chronic atrial fibrillation: the AF carvedilol study. J Cardiol. 2017;69:293–301 [C].
[6] Starr JB, Tirschwell DL, Becker KJ. Labetalol use is associated with increased in-hospital infection compared with nicardipine use in intracerebral hemorrhage. Stroke. 2017;48:2693–8 [C].
[7] Burkey BW, Jacobs JA, Aziz H. Temperature instability in an infant treated with propranolol for infantile hemangioma. J Pediatr Pharmacol Ther. 2017;22(2):124–7 [A].
[8] Fevre PL, Gosling M, Acharya K, et al. Dramatic resuscitation with Intralipid in an epinephrine unresponsive cardiac arrest following overdose of amitriptyline and propranolol. BMJ Case Rep. 2017:1–5. pii: bcr2016218281. https://doi.org/10.1136/bcr-2016-218281 [A].
[9] Rouabhia S, Baghazza S, Sadouki H, et al. Ventricular extrasystoles after first dose of sofosbuvir in a patient treated with propranolol but not with amiodarone: a case report. Rev Recent Clin Trials. 2017;12(3):159–61 [A].
[10] Nanda T, Rasool N, Calahan A, et al. Opthalmic hallucinations: a case series and review of literature. J Glaucoma. 2017;26(9): e214–6 [A].
[11] Rabah F, El-Naggari M, Al-Nabhani D. Amlodipine: the double-edged sword. J Paediatr Child Health. 2017;53(6):540–2 [A].
[12] Kala N, Babu SPKK, Manjeu J, et al. Allele-specific polymerase chain reaction for the detection of single nucleotide polymorphism in amlodipine-induced gingival enlargement. J Clin Pharm Ther. 2018;43:110–3. https://doi.org/10.1111/jcpt.12587 [A].
[13] Wang JG, Yukisada K, Sibulo A, et al. Efficacy and safety of sacubitril/valsartan (LCZ696) add-on to amlodipine in Asian patients with systolic hypertension uncontrolled with amlodipine monotherapy. J Hypertens. 2017;35(4):877–85 [C].
[14] Rakugi H, Shimizu K, Nishiyama Y, et al. A phase III, open-label, multicenter study to evaluate the safety and efficacy of long-term triple combination therapy with azilsartan, amlodipine, and hydrochlorothiazide in patients with essential hypertension. Blood Press. 2017;13:1–9 [C].
[15] Maskell KF, Ferguson NM, Bain J, et al. Survival after cardiac arrest: ECMO rescue therapy after amlodipine and metoprolol overdose. Cardiovasc Toxicol. 2017;17(2):223–5 [A].

[16] Banakh I, Haji K, Kung R, et al. Severe rhabdomyolysis due to presumed drug interactions between atorvastatin with amlodipine and ticagrelor. Case Rep Crit Care. 2017;2017:3801819 [A].
[17] Krasulova K, Holas O, Anzenbacher P. Influence of amlodipine enantiomers on human microsomal cytochromes P450: stereoselective time-dependent inhibition of CYP3A4 enzyme activity. Molecules. 2017;22(11):1879 [E].
[18] Greenblatt DJ, Patel M, Harmatz JS, et al. Impaired rivaroxaban clearance in mild renal insufficiency with verapamil coadministration: potential implications for bleeding risk and dose selection. J Clin Pharmacol. 2018;58:533–40. https://doi.org/10.1002/jcph.1040 [c].
[19] Ismail M, Lee VH, Chow CR, et al. Minimal physiologically based pharmacokinetic and drug-drug-disease interaction model of rivaroxaban and verapamil in healthy and renally impaired subjects. J Clin Pharmacol. 2018;58:541–8. https://doi.org/10.1002/jcph.1044 [c].
[20] Bauer M, Wulkersdorfer B, Karch R, et al. Effect of P-glycoprotein inhibition at the blood-brain barrier on brain distribution of (R)-[11 C] verapamil in elderly vs. young subjects. Br J Clin Pharmacol. 2017;83(9):1991–9 [c].
[21] Ragot C, Gerbaud E, Boyer A. Terlipressin in refractory shock induced by diltiazem poisoning. Am J Emerg Med. 2017;35(7):1032 [A].
[22] Chenoweth JA, Colby DK, Sutter ME, et al. Massive diltiazem and metoprolol overdose rescued with extracorporeal life support. Am J Emerg Med. 2017;35(10). 1581.e3-1581.e5 [A].
[23] Van Veggel M, Van der Veen G, Jansen T, et al. A critical note on treatment of a severe diltiazem intoxication: high-dose calcium and glucagon infusions. Basic Clin Pharmacol Toxicol. 2017;121(5):447–9 [A].
[24] Yamamoto H, Takayasu T, Nosaka M, et al. Fatal acute intoxication of accidentally ingested nifedipine in an infant—a case report. Leg Med (Tokyo). 2017;24:12–8 [A].
[25] Liu Y, Boettcher MF, Schmidt A, et al. Pharmacokinetics and safety of nifedipine GITS/candesartan fixed-dose combination in subjects with hepatic impairment. Int J Clin Pharmacol Ther. 2017;55(3):246–55 [c].
[26] Mancia G, Cha G, Gil-Extremera B, et al. Blood pressure-lowering effects of nifedipine/candesartan combinations in high-risk individuals: subgroup analysis of the DISTINCT randomized trial. J Hum Hypertens. 2017;31(3):178–88 [C].
[27] Li D, Xu S, Wang Y, et al. Pharmacokinetics and drug-drug interaction between enalapril, enalaprilat and felodipine extended release (ER) in healthy subjects. Oncotarget. 2017;8(41):70752–60 [c].
[28] Lins R, Haerden Y, de Vries C. Replacement of amlodipine and lercanidipine by barnidipine: tolerability and effectiveness in a real-life study. High Blood Press Cardiovasc Prev. 2017;24(1):29–36 [MC].
[29] Webster LM, Myers JE, Nelson-Piercy C, et al. Labetalol versus nifedipine as antihypertensive treatment for chronic hypertension in pregnancy: a randomized controlled trial. Hypertension. 2017;70(5):915–22 [C].
[30] Sharma KJ, Greene N, Kilpatrick SJ. Oral labetalol compared to oral nifedipine for postpartum hypertension: a randomized controlled trial. Hypertens Pregnancy. 2017;36(1):44–7 [c].
[31] Sharma C, Soni A, Gupta A, et al. Hydralazine vs nifedipine for acute hypertensive emergency in pregnancy: a randomized controlled trial. Am J Obstet Gynecol. 2017;217(6):687.e1-687.e6 [c].
[32] Verspyck E, de Vienne C, Muszynski C, et al. Maintenance nifedipine therapy for preterm symptomatic placenta previa: a randomized, multicenter, double-blind, placebo-controlled trial. PLoS One. 2017;12(3):e0173717 [c].
[33] Shi DD, Wang Y, Guo JJ, et al. Vitamin D enhances efficacy of oral nifedipine in treating preeclampsia with severe features: a double blinded, placebo-controlled and randomized clinical trial. Front Pharmacol. 2017;8:865 [C].
[34] Shi DD, Guo JJ, Zhou L, et al. Epigallocatechin gallate enhances treatment efficacy of oral nifedipine against pregnancy-induced severe pre-eclampsia: a double-blind, randomized and placebo-controlled clinical study. J Clin Pharm Ther. 2018;43:21–5. https://doi.org/10.1111/jcpt.12597 [C].
[35] Xiao S, Zhang M, Liang Y, et al. Celastrol synergizes with oral nifedipine to attenuate hypertension in preeclampsia: a randomized, placebo-controlled, and double blinded trial. J Am Soc Hypertens. 2017;11(9):598–603 [C].
[36] Ding J, Kang Y, Fan Y, et al. Efficacy of resveratrol to supplement oral nifedipine treatment in pregnancy-induced preeclampsia. Endocr Connect. 2017;6(8):595–600 [C].
[37] Brochet MS, Harry M, Morin F. Nifedipine induced gingival hyperplasia in pregnancy: a case report. Curr Drug Saf. 2017;12(1):3–6 [A].
[38] O'Neil P, Brown R. Penile ulceration secondary to nicorandil use. Urol Case Rep. 2016;10:57–79 [A].
[39] Jeong SH, Kim H. Acute intractable headache and oculomotor nerve palsy associated with nicroandil: a case report. Am J Emerg Med. 2017;35(12). 1988.e3-1988.e5 [A].
[40] Proulx MH, Ross D, Vacon D, et al. Prehospital nitroglycerin in tachycardic chest pain patients: a risk for hypotension or not? Prehosp Emerg Care. 2017;21:68–73 [C].
[41] Isquick S, Henry D, Nakagawa S, et al. The association between nitroglycerin use and adverse outcomes in women undergoing cesarean delivery in the second stage of labor. J Matern Fetal Neonatal Med. 2017;30(11):1297–301 [C].
[42] Guerras F, Romandini A, Barbarossa A, et al. Ranolazine for rhythm control in atrial fibrillation: a systematic review and meta-analysis. Int J Cardiol. 2017;227:284–91 [M].
[43] Trivedi C, Upadhyay A, Solani K, et al. Efficacy of ranolazine in preventing atrial fibrillation following cardiac surgery: results from a meta-analysis. J Arrhythm. 2017;33(3):161–6 [M].
[44] DE Vecchis R, Ariano C, Giasi A, et al. Antiarrhythmic effects of ranolazine used both alone for prevention of atrial fibrillation and as add on to intravenpus amiodarone for its pharmacological cardioversion: a meta-analysis. Minerva Cardioangiol. 2018;66:349–59. https://doi.org/10.23736/S0026-4725.17.04349-3 [M].
[45] Salazar CA, Basillo Flores JE, Veramendi Espinoza LE, et al. Ranolazine for stable angina pectoris. Cochrane Database Syst Rev. 2017;8(2):1–179 [M].

CHAPTER

19

Drugs Acting on the Cerebral and Peripheral Circulations

Keaton S. Smetana[*,1], *Zachary R. Noel*[†], *Sidhartha D. Ray*[‡]

[*]Department of Pharmacy, The Ohio State University Wexner Medical Center, Columbus, OH, United States
[†]Department of Pharmacy Practice and Science, University of Maryland School of Pharmacy, Baltimore, MD, United States
[‡]Department of Pharmaceutical and Biomedical Sciences, Touro University College of Pharmacy and School of Osteopathic Medicine, Manhattan, NY, United States
[1]Corresponding author: keaton.smetana@osumc.edu

DRUGS USED IN THE TREATMENT OF MIGRAINES

Triptans [SEDA-36, 277; SEDA-37, 223; SEDA-39, 180]

Medication overuse headaches (*MOH*) are thought to occur in 1%–2% of society and is the third most common headache following migraine and tension type headaches. While the risk for development is evident in the name, other risk factors associated with MOH are lower socioeconomic status, comorbid psychiatric diseases, genetic predisposition, and previous history of migraines [1R]. Triptan induced MOH are often described with the following symptomatology: unilateral, throbbing, moderate to severe, increase with physical activity, nausea and/or emesis, and photo- or phonophobia. Diagnostic criteria include headaches that occur ≥15 days/month, overuse of one or more symptomatic treatment medications regularly for longer than 3 months, worsening or newly developed headache over course of treatment, and the headaches disappear or return to previous status 2 months after discontinuation of the medication [1R]. The usual approach to treatment of MOH is discontinuation of the offending agent, but few therapies have shown to be helpful in controlling the ongoing headaches.

The greater occipital nerve (GON) arises from between the first and second cervical vertebrae, passes beneath the larger of the two oblique muscles of the neck, and ultimately innervates the scalp at the top of the head. Stimulation of the GON has been shown to affect the ipsilateral trigeminal nerve, which carries cranium-based pain sensations triggering the trigeminocervical complex [2E]. Therefore, a blockade of cervical spine nerves such as the GON would result in a peripheral conduction block and lead to the reduction of the afferent nociceptive inflow [3R]. To evaluate if this intervention would relieve triptan overuse headache (TOH) a prospective randomized trial was performed that included 105 patients between the ages of 18–60 years old [4C]. Subjects were randomized into three groups: Group 1-patients had their triptans abruptly withdrawn without GON blockade, Group 2-patients whose triptans were abruptly withdrawn and who received single GON block, and Group 3-patients whose triptans were abruptly withdrawn and who received a three-stage GON block (once a week for 3 weeks). Greater occipital nerve blockade was achieved through administering 2.5 mL of 1% lidocaine injection 2 cm lateral and 2 cm inferior of the external occipital protuberance [4C]. The number of headache days, severity of pain using the visual analog scale, and number of triptans were evaluated for the duration of the study. The number of headache days in group one were 18.1 ± 3.8 pretreatment, 16.9 ± 3.7 2 months after cessation of the triptan, and 16.9 ± 4.3 at 4 months. The second group, which received one lidocaine injection, reported the number of headache days as 19.1 ± 3.8, 15.9 ± 3.9, and 14.8 ± 4.7. The third group, which received multiple injections, reported the number of headache days as 19.7 ± 3.3, 14.1 ± 4.6, and 9.7 ± 4.5. Comparing the groups independently no differences were seen in group one in reduction of headache days. However, in group two and group three, the number of attacks decreased significantly in the post-treatment period when

Side Effects of Drugs Annual, Volume 40
ISSN: 0378-6080
https://doi.org/10.1016/bs.seda.2018.06.004

compared to the pre-treatment period ($P < 0.001$). A significant reduction in headache days in the second and fourth month was only observed in group three which received multiple lidocaine injections. No serious adverse reactions were observed [4C].

Third cranial nerve paresis was observed in three patients who were treated for repetitive migraines with the oral formulation of sumatriptan [5A]. None of these patients had a previous third nerve paresis. The authors aim to bring awareness that the vasoconstrictive properties of triptans may be a contributing factor for microvascular damage of the cranial nerves [5A].

DRUGS USED IN THE TREATMENT OF ERECTILE DYSFUNCTION

Phosphodiesterase Type 5 Inhibitors [SEDA-37, 275–276; SEDA-39, 180–181]

Numerous genetic polymorphisms have been linked to poor response to PDE-5 inhibitors: NOS3-786T>C [T: poor response], NOS1-84G>A [G: poor response], NOS1 rs2682826 [C: poor response], ACE IN/DEL [DEL: poor response], NOS3 VNTR [4b: poor response], VEGFA g.-2578C>A [A: poor response], VEGFA g.-1154G>A [A: poor response], GNB3 825C>T [C: poor response], NOS3 Glu298Asp [T: poor response] [6R]. There is a dearth of literature assessing the pharmacologic response to polymorphisms specifically for PDE5 (PDE5A-1142T>G). Further research is necessary of these polymorphisms so clinicians can utilize a more individualized approach to patient care. In doing so it may be possible to predict an individuals adverse effect profiles and optimal dose to achieve the expected therapeutic response.

Sildenafil

New Formulation

Sildenafil oro-dispersable (ODF) film is a new formulation and was compared directly to film-coated sildenafil tablets (FCT) in a prospective, cross-over trial of 139 patients with erectile dysfunction (ED) [7c].

Overall sexual satisfaction was improved in those taking the ODF formulation, and no additional adverse effects were observed. A systematic review including 29 studies of sildenafil in ED identified that overall sildenafil was well-tolerated [8R]. Mild reactions occurring more than 2% of time included headache, flushing, visual alterations, congestion, back pain, myalgia, myalgia, nausea, dizziness, and rash. The ODT formulation was associated with similar overall drug exposure as ODF, irrespective of whether the ODF was consumed with water.

Sexual Dysfunction

Herpes simplex encephalitis, albeit rare in immunocompetent individuals, is associated with significant mortality and morbidity. Sildenafil has potential to compromise the effectiveness of the blood–brain barrier secondary to its vasodilatory properties. As such, sildenafil has been studied to help increase penetration of certain drugs (e.g. chemotherapy) across the blood–brain barrier. In one case report, the authors describe herpes simplex encephalitis in a 62-year-old male taking sildenafil during a primary genital herpes simplex virus infection [9H].

Treatment of premature ejaculation was evaluated in a prospective randomized, placebo-controlled trial comparing paroxetine, dapoxetine, sildenafil, or combination dapoxetine with sildenafil [10C]. A total of 150 patients were included during 2015–2016. Those who received combination dapoxetine 30 mg with sildenafil 50 mg experienced improved premature ejaculation diagnostic tool, intravaginal ejaculatory latency time, and patient satisfaction scores. The most adverse effects in different groups were constipation (13.3% with placebo), sleep disturbances (30% with paroxetine), nausea (26.6% with dapoxetine), headache (26.6% with sildenafil), and headache and nausea (33.3% each with dapoxetine and sildenafil combination) [10C].

Cerebral Venous Sinus Thrombosis

Cerebral venous sinus thrombosis (CVST) was associated with sildenafil use in two patient cases. The first was a 29-year-old man who presented with a 2 week history of bilateral visual deterioration, and severe headache [11c]. He reported taking sildenafil for nearly 2 years two to three times daily without medical supervision and was not on any other medications. A fundus exam revealed stage two papilledema, the hypercoagulable work up was negative, and magnetic resonance imaging (MRI) revealed a filling defect in the right transverse sinus, sigmoid sinus, and jugular vein without infarcts. Cerebrospinal fluid revealed normal biochemistry and no abnormal cells, but had an opening pressure of 43 cm of H_2O. The patient was treated with acetazolamide and bridged to warfarin with enoxaparin.

The second case was a 38-year-old man who reported to the emergency department (ED) with meningeal signs (headache, neck stiffness, photophobia) [12c]. Six days prior he had undergone an elective inguinal hernia repair and received spinal anesthesia. His post-op course was complicated by a progressively worsening headache that had minimal response to caffeine. Initially a post-dural puncture headache was suspected, but the patient declined an epidural blood patch and was discharged. Three days after discharge the patient presented to the ED and was found to have a non-aneurysmal subarachnoid hemorrhage in the right sylvian fissure.

A subsequent angiogram revealed thrombosis of the superior sagittal sinus, suggestive of CVST. The patient had initially failed to report taking sildenafil twice a week for the past 4 years for erectile dysfunction prior to the inguinal repair.

While the mechanism underlying PDE-5 induced CVST is unclear there are two proposed mechanisms that potentially overlap: (1) venous dilation results in venous stasis and insufficiency and (2) a biphasic platelet response to cyclic guanosine monophosphate (cGMP) activation of protein kinases exist (initially promoting aggregation then limiting thrombus formation over time) [12c].

Miscellaneous

Hepatotoxicity with sildenafil is considered to be an extremely rare event. A recently published review highlighted five case reports of possible sildenafil-induced hepatotoxicity [13R]. While the underlying mechanism of hepatotoxicity is unclear, some of the reported events involved consumption of herbal products used for sexual enhancements which may have been a contributing factor. The range of toxicity varied from asymptomatic elevations in liver enzymes to fulminant liver failure.

Visual disturbance is a known side effect of sildenafil occurring in about 3% of patients [14E]. This adverse effect results from inhibition of retina-specific phosphodiesterase type 6. The implications of long-term sildenafil usage on the retina have not been well described. An experiment found that rats exposed to 10mg/kg/day of sildenafil experienced toxic effects to the retina and optic nerve. In spite of the observed effects with sildenafil treatment, these effects diminished after sildenafil was discontinued. It is unclear what the effects are in humans at typical doses (0.5–1mg/kg). Another case report described a 32-year-old African American woman who experienced bilateral, asymmetrical outer macular atrophy while receiving sildenafil chronically for pulmonary arterial hypertension (PAH) [15A]. Her partial vision loss began 1 month after initiating sildenafil and was reported to continue for 5 years until discontinuing the medication. This was the first case report of such a finding associated with long-term sildenafil use for the treatment of PAH.

Sildenafil is being studied for the treatment of lymphatic malformations in children. A long-term follow-up study of 12 children with lymphatic malformations was undertaken to evaluate long-term clinical outcomes. The median duration of sildenafil treatment was 9 months and the average follow-up was 4 years. Ten patients reported positive therapeutic responses, and six required further interventions. The most common adverse effects experienced during the treatment period included flushing (20%) and prolonged ejections (50% of males) [16c].

The synergistic anticancer activity of sildenafil was studied in conjunction with doxorubicin in a mouse model of breast cancer [17E]. When added to 1μM of doxorubicin in 4T1 breast cancer cells in vitro a 1.8-fold, 6.2-fold, and 21-fold statistically significant increase in cytotoxic effects were seen with increasing sildenafil doses (1, 30, and 100μM, respectively). Compared to doxorubicin alone sildenafil plus doxorubicin resulted in a 4.7-fold statistically significant reduction in tumor size. The added anticancer effects are possibly due to smooth muscle relaxation in addition to alteration in vascular endothelial permeability [17E].

A systematic review of sildenafil use during pregnancy was conducted to assess the incidence of adverse effects and impact on fetal development and obstetric complications [18R]. A total of 16 studies including 165 pregnant women were included. Stillbirths and neonatal deaths were similar to sildenafil naïve individuals. No congenital defects were observed.

Tadalafil

Genitourinary Disorders

Tadalafil was compared to silodosin and placebo in a prospective, randomized trial evaluating the ease of ureteroscope navigation in management of ureteral stones [19c]. Those receiving tadalafil had greater ureteral dilation, better ureteroscope negotiation, and operating time compared to placebo. Headache and backache were reported more frequently in the tadalafil group compared to the silodosin and placebo groups. Incidence of dizziness, abnormal ejaculation, and orthostatic hypotension was not significantly different among the three groups.

Tadalafil 5mg was studied as a combination product with tamsulosin 0.4mg or tamsulosin 0.2mg compared to only tadalafil 5mg for benign prostatic hyperplasia (BPH) associated lower urinary tract symptoms (LUTS) and erectile dysfunction [20C]. The study demonstrated enhanced efficacy on BPH-associated LUTS, comorbid erectile dysfunction complaints, lower incidence of side effects, and simplification and convenience of therapy in patients taking the combination product. However, the lack of a tamsulosin monotherapy control group was a limitation of this study. Adverse effects were similar across treatment groups. A separate trial evaluating combination tadalafil and tamsulosin in 171 Japanese patients found improved voiding scores and clinically insignificant changes in blood pressure [21C]. In addition, a pooled analysis tadalafil users were stratified according to age ($\geq$75 vs $<$75years-old). The safety profile among older adults was similar to younger adults and there was no increase in cardiovascular adverse effects [22R].

Cerebral Venous Sinus Thrombosis

Cerebral venous sinus thrombosis was attributed to the use of imported tadalafil in Japanese patient. A 45-year-old man with a pertinent past medical history of antiphospholipid antibody syndrome (APS) presented with a 3 day history of posterior headache [23c]. This was initially diagnosed as a tension headache and was discharged the same day. Three days later, the patient presented to an outside hospital with seizures, elevated D-dimer (2.7 mcg/mL), and a brain MRI that indicated edema and thrombosis of the cortical vein. He did not report using the medication the day prior to the headache. The authors commented that the low flow state in the venous system, potentially induced by tadalafil, may have result in increased risk of venous thrombosis given the patients past medical history of APS [23c].

Miscellaneous

Fetal growth restriction is one of the most significant causes of perinatal morbidity and mortality in developed countries. In rat models, the increase in secondary intracellular messengers improves uterine arterial and fetoplacental perfusion, ameliorating fetal growth restriction. A phase-1 study of 12 pregnant women taking tadalafil 10–40mg daily were monitored for adverse effects [24E]. There were no major adverse effects that could be attributed to tadalafil. Mild adverse effects, such as headache, flushing, and nausea, occurred in all patients. Diastolic blood pressure was significantly lower in those receiving tadalafil. Fetal growth rate increased from 9.8 to 18.5g/day 2 weeks after treatment ($P=0.011$).

Udenfail

The Fontan operation is a procedure to separate systemic and pulmonary circulation in patients born with functional single ventricle heart disease. The success of this operation is contingent upon lower pulmonary vascular resistance so that blood can passively flow from the vena cava to the pulmonary circulation. Udenafil is a novel PDE-5 inhibitor recently studied in 30 patients undergoing Fontan operation [25c]. Udenafil has comparable selectivity for PDE5 as sildenafil, but unlike tadalafil does not inhibit PDE11. It has unique pharmacokinetic properties with a Tmax of 1–1.5h and long half-life of 11–13h [26R]. The doses ranged from 37.5 to 85.5mg twice daily and were well-tolerated across all dosing groups [25c]. No series adverse effects were observed. Headache, flushing, nasal congestion, and spontaneous penile erection were the most common mild adverse effects that occurred.

DRUGS USED IN THE TREATMENT OF PULMONARY ARTERIAL HYPERTENSION

Phosphodiesterase Type 5 Inhibitors [SEDA-37, 275–276; SEDA-39, 180–181]

Sildenafil

Pulmonary hypertension is common in children with congenital heart disease and large systemic-to-pulmonary shunts. Sildenafil is traditionally the phosphodiesterase of choice in these children; however, tadalafil offers more convenient once daily dosing and is approved for use in adults with PAH. A comparison of 42 children (age 3–24 months) with large septal defects and PAH were randomly selected for treatment with sildenafil or tadalafil for 7 days prior to and 3–4 weeks following surgery [27A]. There were no difference in pulmonary pressure measurements between the groups, and each treatment group had only mild side effects in a small number of patients. Another study evaluated 40 pediatric patients receiving either intravenous (IV) or enteral sildenafil, with the primary goal of evaluating adverse effects [28c]. Hypotension requiring intervention occurred in 30% of the patients receiving the IV formulation and 10% receiving the enteral formulation ($P=0.24$). Mean arterial pressure was significantly lower in those receiving the IV formulation (44 ± 6.3 vs 65 ± 13.4mmHg, $P=0.0024$).

Inhaled pulmonary vasodilators for pulmonary hypertension offer the benefits of pulmonary vasodilation without as many systemic adverse effects. While prostanoids, such as iloprost or epoprostenol, can be given as inhaled formulations, phosphodiesterase inhibitors are limited to oral or IV routes. In a rat study, an inhalable poly(lactic-*co*-glycolic acid) (PLGA) particle formulation of sildenafil was used to deliver a long acting inhalation to relieve pulmonary hypertension [29E]. Porus PLGA polymer based inhalable microparticles of sildenafil citrate were prepared by a water-in-oil-in-water double emulsion solvent evaporation method. Polyethyleneimine (PEI) was used in the internal aqueous phase as a porosigen. Given the high desnity of cations PEI is thought to damage the lung lactate dehydrogenase, alkaline phosphatase, and total protein were measured as injurious markers in bronchial alveolar lavage fluid collected from the rats. The authors measured these markers in two of the formulations comparing them to saline and found the particles to be safe for inhalational delivery. The caveat being that the safety of the formulation in prolonged use was not evaluated as the delivery of inhalation was given over a short period of time [29E]. This formulation may be eventually a feasible alternative to systemic phosphodiesterase inhibitor administration in patients hemodynamically unstable or with intolerable adverse effects.

Endothelin Receptor Antagonist

Bosentan

Bosentan was recently studied in patients with systemic sclerosis and secondary digital ulcers [30c]. A total of 28 patients were enrolled and followed for 52 weeks. Two major adverse drug effects were observed, and included abnormal liver function tests and pericardial effusion. Elevations in liver enzymes occurred in 43% of patients. Other adverse effects occurring more than 10% of the time included anemia, edema, and diarrhea.

Macitentan

Macitentan is an endothelin receptor antagonist which blocks ET-1 from binding ET_A and ET_B receptors. Endothelin is responsible for pulmonary artery vasoconstriction and hypertrophy, contributing to the pathology of PAH. Other endothelin receptor antagonists, particularly bosentan, have been associated with hepatoxocity. Until recently, macitentan was not believed to be associated with hepatotoxicity. A 23-year-old patient with idiopathic PAH experienced fulminant liver failure and biopsy confirmed drug-induced toxic hepatitis 13 months after initiation macitentan 10mg daily [31A]. The patient ultimately received a liver transplant. In addition, a recent case report described the presence of cotton-wool spots after using macitentan [32A]. A 76-year-old African American woman was incidentally found to have cotton wool spots while taking macitentan; however, they resolved spontaneously over the course of several weeks. The patient had no symptoms or adverse effects associated with the cotton wool spots.

PERIPHERAL VASODILATORS

Phosphodiesterase-3 Inhibitors

Milrinone

Inhaled milrinone in acute respiratory distress syndrome (ARDS) was compared to the effects of inhaled nitric oxide (20ppm) (iNO) and epoprostenol (10 mcg/mL for a total volume of 5mL) in a cross-over pilot study of 15 patients [33c]. Baseline characteristics included median age 57 (interquartile range (IQR=22)), APACHE-II on ICU admission 23 (IQR=7), and PaO2/FiO2 138 (IQR=68). After sequential randomization and administration of iNO or epoprostenol over 20min followed by a 30min washout period, patients subsequently received milrinone (1mg/mL for a total of 5mL) alone or in combination with iNO. Hemodynamic instability (defined as a reduction of systolic arterial blood pressure $\geq$10mmHg or lowering of the MAP $\geq$5mmHg), worsening oxygenation status (reduction in arterial oxygen saturation >10%), and renal failure were evaluated as adverse effects. The median increase in oxygenation measurements from baseline were 6mmHg (IQR=18.4), 8.8mmHg (IQR=16.3), 6mmHg (IQR=15.8), and 9.2mmHg (IQR=20.2), respectively, with epoprostenol, iNO, inhaled milrinone, and iNO in addition to milrinone. Only the combination of iNO and milrinone ($P=0.004$) and iNO alone ($P=0.036$) showed significant improvements from baseline oxygenation measurements. No significant adverse events or alterations in hemodynamic parameters related to study medications were observed. Although milrinone appeared safe in this pilot study it failed to improve oxygenation in ARDS [33c].

PERIPHERAL VASOCONSTRICTORS

Angiotensin II

Angiotensin II (AGII) is a naturally occurring hormone that is a byproduct from the classically described renin-angiotensin system (RAS) and is a result of renin converting angiotensinogen to angiotensin I, and cleavage of angiotensin I by angiotensin-converting enzyme to angiotensin II. The increase in blood pressure is a result of numerous mechanisms that act in unison: increasing intracellular calcium through signal transduction pathways of smooth muscle, resulting in contraction and vasoconstriction of peripheral vessels, increasing central production of arginine vasopressin, release of adrenocorticotropin hormone that increases production and release of cortisol, and may potentiate sympathetic effects via direct action on postganglionic sympathetic fibers [34R, 35c]. Given the proliferative avenues that angiotensin II increases blood pressure its use as a potent vasoconstrictive agents seems plausible.

The plasma half-life of IV AGII is approximately 30s, but its distribution in tissue may prolong the half-life to be as long as 15–30min [35c]. Although no specific studies to date have evaluated the distribution of AGII in high-output shock, the mean value for the apparent volume of distribution in normotensive subjects is 23L [36S, 37c]. Serum levels of AGII after a 3h infusion in those with septic or other distributive shock were similar from baseline, but interestingly the serum level of angiotensin I (the precursor to AGII) were reduced by 40% [36S]. The metabolism and excretion of AGII are not dependent upon renal or hepatic function, but a comprehensive overview of the various metabolic pathways can be found in Fig. 1 [38R].

A prospective, phase three, randomized control trial reported the safety and efficacy of Angiotensin II for the Treatment of High-Output Shock (ATHOS-3) this year after encouraging findings from their initial pilot study (ATHOS) that was published in 2014. The ATHOS pilot study set out to determine the appropriate dose of AGII in the treatment of high-output shock as the previously reported ranges of dosing were from 0.4 to 40ng/kg/min [35C]. AGII was initiated in those deemed to have high-output shock (cardiovascular sequential organ function assessment score of >4 and a cardiac index >2.4L/min/BSA1.73m^2). The study drug was administered over 6 h, initiated at 20ng/kg/min, and titrated by 10ng/kg/min hourly to a minimum rate of 5ng/kg/min and a maximum of 40ng/kg/min to maintain a mean arterial pressure (MAP) of 65mmHg. Norepinephrine doses were greatly reduced upon initiation of AGII (27.6±29.3 mcg/min vs 7.4±12.4 mcg/min at hour 1, P=0.06) and remained lower compared to the cohort

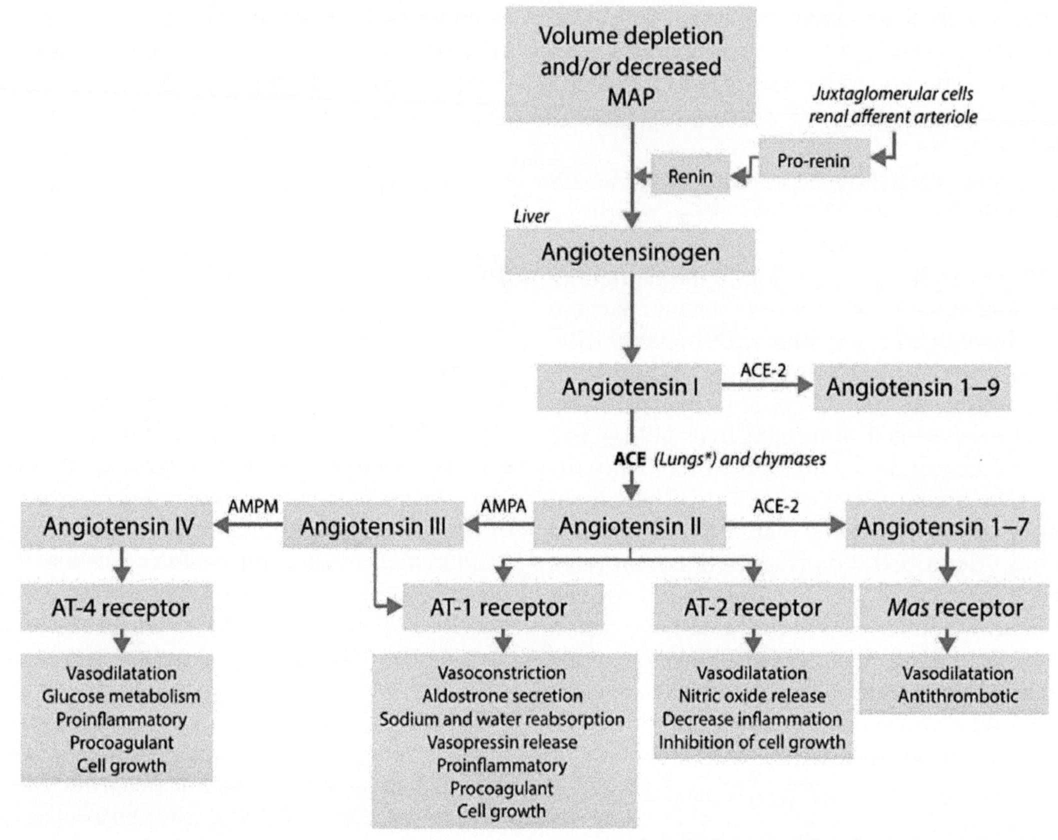

FIG. 1 Overview of the renin angiotensin pathway.

throughout the 6-h period. The most common adverse events experienced in the AGII group were alkalosis and hypertension which occurred in 20% of patients [35C]. The authors posited that the patients who experienced hypertension had been on an angiotensin-converting enzyme inhibitor prior to development of shock, but were unable to confirm this suspicion due to lack of data found in the chart-review [35C].

*Another consideration regarding the patients in the ATHOS pilot study who were extremely sensitive to AGII is pharmacogenomic polymorphisms of the angiotensin receptors [39R]. There are at least 14 angiotensin-1 receptor (AGT_1R) polymorphisms described relating to inherited predispositions in the development of essential hypertension or the effectiveness of angiotensin receptor blocking medications. The gene coding for the AGT_1R is located on human chromosome 3 (q22 band) and belongs to the G protein-coupled receptor family containing seven hydrophobic transmembrane segments that form α-helices in the lipid bilayer of the cell membrane [39R]. A particular polymorphism of interest is the silent A1166C single-nucleotide polymorphism (SNP) which has been associated with resistant hypertension in patients requiring two or more antihypertensive [40E]. Takahashi et al. identified seven polymorphisms in the 5′-flanking region of the AGT_1R gene, and found higher frequency of the $AT_1(-535)$*T allele in hypertensive Japanese patients [41R]. In patients that possess these polymorphisms it is possible they will be more sensitive to exogenous AGII and require a reduced starting dose to achieve the intended therapeutic effect.*

The ATHOS-3 trial evaluated the safety and efficacy of AGII in 321 severe vasodilatory shock (cardiac index >2.3 L/min/m^2 or as a central venous oxygen saturation of >70% and a central venous pressure >8 mmHg, with a MAP 55–70 mmHg) patients requiring high-dose vasopressors (>0.2 mcg/kg/min norepinephrine or equivalent) [42C]. The primary efficacy outcome was MAP response (defined as MAP increase ≥10 mmHg or MAP >75 mmHg without increase in background vasopressors) at 3 h. After 3 h and up to 48 h the study drug or placebo and other vasopressors were titrated to maintain a MAP between 65 and 75 mmHg. Significantly more patients in the AGII group compared to placebo met the primary end point of the MAP goal at hour 3 (69.9% vs 23.4%, $P<0.001$; odds ratio, 7.95; 95% confidence interval, 4.76–13.3) [42C]. Overall, serious adverse events of any grade with ≥5% frequency in AGII and the placebo group were 87.1% and 91.8%, respectively. The rates of adverse events of special interest were similar between the groups: intestinal ischemia (0.6% AGII group vs 1.9% with placebo), peripheral ischemia (4.3% vs 2.5%), ventricular tachycardia (3.1% vs 5.1%), and atrial fibrillation (13.5% vs 13.3%). The two adverse events that were statistically significant were infections and infestations (30.1% AGII group vs 19% with placebo, $P=0.029$), and delirium (5.5% vs 0.6%, $P=0.036$) [43C]. The package insert specifically draws *attention to adverse reactions occurring ≥4% of patients treated with AGII and >1.5% more often than in the placebo controlled patients: arterial and venous thromboembolic events (12.9% vs 5.1%), deep vein thrombosis (4.3% vs 0%), thrombocytopenia (9.8% vs 7%), tachycardia (8.6% vs 5.7%), fungal infection (6.1% vs 1.3%), delirium, acidosis (5.5% vs 0.6%), hyperglycemia (4.3% vs 2.5%), and peripheral ischemia [36S]. No differences were observed by death from any cause at 7 and 28 days [42C].*

While the addition of a third class of vasopressors to our armamentarium in those with vasodilatory shock is intriguing, clinicians should approach its use with caution. Particularly in patients who were excluded in ATHOS-3 (full inclusion and exclusion criteria can be found in the studies supplementary materials) [43C]. The safety of AGII in those with low cardiac output and its use as a single agent are questions that remained unanswered and may be potentially harmful. Lastly, given the significant increase in delirium observed with administration of AGII clinicians should be aware of the brain-RAS pathway. A neuroprotective pathway that enhances repair of damaged DNA, vasodilates via nitric oxide production, and results in neurite out-growth has been observed via activation of the angiotensin-2 receptor (AGT_2R) pathway [44R, 45R]. On the contrary activation of the AGT_1R may cause potentially neurologic injurious effects through vasoconstriction, inflammation, increased oxidative stress, blood–brain barrier disruption, and neurotoxicity [44R].

Clinicians should also be cognizant of a patient's medication history. Those on angiotensin converting enzyme inhibitors (e.g. benazepril, captopril, enalapril, fosinopril, lisinopril, perindopril, quinapril, ramipril, trandolapril) may have an increased response due to lack of the endogenous conversion of angiotensin I to AGII. Angiotensin-1 receptor blockers (e.g. azilsartan, candesartan, eprosartan, irbesartan, telmisartan, valsartan, losartan, olmesartan) may have a reduced response or theoretically worsen hypotension due to shunting AGII to bind to the AGT_2R which results in vasodilation and nitric oxide release [36S, 38R]. The half-life of the aforementioned medications should be considered prior to administration of AGII due to these potential drug–drug interactions.

Epinephrine

A cornerstone in the management of systemic anaphylaxis is the administration of intramuscular (IM) epinephrine to mitigate vasodilation and bronchoconstriction. A 21-year-old previously health male presented with an urticarial rash and difficulty in breathing after ingestion of prawns which was a known allergen to the patient [46c]. The patient received epinephrine 0.5mg IM, hydrocortisone 200mg IV, and chlorphenirami-ne10mg IV. Ten minutes after administration of epinephrine the patient developed palpitations and chest pain. The initial ECG revealed ST segment depressions in leads III, aVF and V1–V5. Subsequent ECG had persistent T wave inversions and troponin I 6h after the event was 0.69ng/mL. The authors concluded the patient likely developed a type II myocardial infarction (supply–demand mismatch) due to coronary vasospasm and was successfully treated with sublingual nitroglycerin. After anaphylaxis there are two possible reasons for a myocardial infarction to occur: (1) the allergen resulting in anaphylaxis results in an allergic myocardial infarction or (2) use of epinephrine and a supply–demand mismatch [46c].

Midodrine

Upon oral administration, midodrine undergoes rapid hepatic metabolism to the active metabolite desglymidodrine. As an α^1 agonist it is often used in the setting of orthostatic hypotension, prevention of intradialytic hypotension, refractory ascites, and hepatorenal syndrome [47S, 48R]. A randomized, open-label trial was performed to evaluate the efficacy and safety of pyridostigmine added to midodrine for a 3-month period to combat orthostatic hypotension [49C]. Pyridostigmine, an acetylcholinesterase inhibitor that permits increased nerve impulses across the neuromuscular junctions, may increase adrenergic tone only in the upright posture [50S]. The degree of orthostatic blood pressure drop at 3 months among the three treatment modalities (midodrine alone, pyridostigmine alone, midodrine + pyridostigmine) did not differ. Overall, midodrine was better at ameliorating symptoms compared to pyridostigmine. The combination demonstrated beneficial effects in controlling orthostatic blood pressure drops but failed to show better symptom improvement. Ten patients (11.5%) reported adverse events, and the proportion did not differ between treatment modalities. One patient (4.3%) in the midodrine-only group reported headache and aggravated dizziness. Six patients (25%) in the pyridostigmine-only group reported aggravated dizziness ($n=5$); headache ($n=2$); gastrointestinal symptoms, which included nausea and diarrhea ($n=2$); or limb tremors ($n=1$). Three patients in the midodrine + pyridostigmine group (11.5%) reported abdominal pain and nausea ($n=2$), dizziness ($n=1$), or visual disturbances ($n=1$) [49C].

A 20-year-old female ingested 350mg of midodrine as an intentional overdose and developed emesis, severe hypertension (blood pressure 210/100mmHg), and symptomatic bradycardia (heart rate 43–60 beats/min) [51A]. She received a 5mg glyceryl trinitrate patch for blood pressure control which resulted in an improved conscious state and a reduction in blood pressure to 124/81mmHg. Midodrine and desglymidodrine concentrations were obtained and the peak in blood pressure coincided with the serum concentrations. Given the short half-life of the parent drug and active metabolite, treatment of midodrine induced hypertension secondary to overdose can be treated with a vasodilator and supportive care.

Pseudoephedrine

Commonly prescribed for nasal congestion, pseudoephedrine is a sympathomimetic agent that is an agonist at α- and β-adrenergic receptors. A 63-year-old woman who was taking fexofenadine-pseudoephedrine daily for seasonal allergies developed atrial fibrillation (AF) and was referred to the emergency department. En route the patient received IV metoprolol and upon arrival the initial electrocardiogram (ECG) showed AF with non-specific inferolateral ST-T abnormalities. One hour after receiving additional IV and oral metoprolol she complained of chest pain, and a 12-lead ECG revealed AF, with slow ventricular response and inferior ST-segment elevation with reciprocal anterolateral ST depressions concerning for an acute inferior myocardial infarction. Coronary angiography revealed multivessel coronary vasospasm and intracoronary nitroglycerin was administered. Metoprolol is a selective β_1 receptor antagonist, and in this patient its use for rate control in AF may have potentiated the risk of pseudoephedrine-induced coronary vasospasm due to unopposed α-receptor stimulation and smooth muscle vasoconstriction [51A].

MISCELLANEOUS MEDICATIONS ASSOCIATED WITH CEREBROVASCULAR SEQUELAE

Acute Ischemic Stroke

Synthetic Cannabinoids (Spice, K2, etc.)

Several hundred synthetic agonists of cannabinoid(CB)-type receptors exist with varying degrees of affinity and potency for the CB-1 receptor [52R]. These psychoactive substances are created in clandestine laboratories, but were initially designed as potential pharmacologic tools to study the cannabinoid system with much higher potency than endogenous cannabinoids. Synthetic cannabinoids are chemicals with the potential to be 2–100 times more potent than Δ9-tetrahydrocannabinol (THC) in its ability to bind to cannabinoid receptors [53R]. While the pharmacologic fate of THC is well established through metabolism by CYP2C9 to a single biologic metabolite this may not be the case with synthetic cannabinoids. In fact the synthetics may produce several major metabolites that retain or produce even greater signaling activity compared to the parent compound [54R].

The fourth case of potentially spice-induced ischemic stroke was published describing a 25-year-old patient who presented after smoking 3g of a product called "Freeze" with left severe hemiparesis, left hypaesthesia, moderate dysarthria, and visual neglect [55A]. Utilizing gas chromatography the patient's urine sample was positive for the synthetic cannabinoid ADB-FUBINACA and MDMB-CHMICA, the former was also isolated in the product labeled "Freeze". The patient's urine drug screen was negative for other drugs of abuse, and besides alcohol and 10 cigarettes daily no other vascular risk factors were present. Angiographic imaging confirmed occlusion of the right middle cerebral artery, which is consistent with previous case descriptions of spice-induced ischemic stroke. The etiology of the stroke was likely cardioembolic due to a tachyarrhythmia [55A].

Thalidomide

Thalidomide may be used in the treatment of β-thalassemia major, which is a hereditary anemia due to defects in β-globin production, due to its ability to induce γ-globin chain expression and increase the proliferation of erythroid cells [56A]. A 7-year-old girl that had been receiving monthly packed cell blood transfusions was initiated on thalidomide 10 mg/kg/day 2 months prior in an effort to decrease frequency of transfusions. Upon presentation the child complained of a 2-week history of diffuse dull aching headache, generalized weakness of all four limbs, and the mother noticed an unsteady gait. An MRI revealed a hyperintense focus in the left cerebellum with restriction on diffusion weighted imaging, suggestive of an infarct. Subsequently, thalidomide was discontinued and the child was initiated on aspirin and iron chelators. Upon follow-up visits the child had complete recovery of the motor deficits [56A]. Increased thromboembolic complications are known to occur with thalidomide use in conjunction with anthracyclines and glucocorticoids, but the rate in children is largely unknown in the setting of β-thalassemia which carries a known complication of ischemic strokes [56A].

Reversible Cerebral Vasoconstriction Syndrome (RCVS) and Posterior Reversible Encephalopathy Syndrome (PRES)

Gemcitabine and Cisplatin

The pathogenesis of PRES is still unclear but is associated with malignant hypertension, renal dysfunction, and use of immunosuppressive agents. The vasogenic edema usually seen on T2-weighted MRIs is consistent with vasogenic edema and most commonly affects the posterior occipital and parietal lobes [57R]. It has been proposed that in patients receiving chemotherapy tumor destruction and subsequent dysfunctional cerebrovascular auto-regulation may result in cytotoxic edema resulting in PRES [58R].

Cherniawsky et al. describe a 58-year-old woman that was diagnosed with cholangiocarcinoma (T2bN0M0) who initially was treated surgically and was in remission for 18 months [59A]. Thereafter she was started on palliative cisplatin and gemcitabine. Cycle one consisted of gemcitabine alone, but the combination was used on cycles two through six of cisplatin 34 mg (20 mg/m^2) and gemcitabine 1700 mg (1000 mg/m^2) on days 1 and 8 of the cycle. After 24 h of cycle six day-1 therapy the patient had worsening somnolence over a 3 day period. A lumbar puncture was performed that was negative for viral or bacterial pathogens, but MRI revealed bilateral cortical and subcortical edema in the posterior regions of the occipital, parietal, and frontal lobes confirming the diagnosis of PRES. The patient was treated with magnesium for hypomagnesaemia and started on amlodipine 10 mg daily for blood pressure control. She was discharged 1 week after admission and had residual visual changes and loss of balance. However, she was able to ambulate and carry out all activities of daily living [59A].

Tacrolimus

Reversible cerebral vasoconstriction syndrome is characterized by hyperacute onset of severe headache, often described as a "thunderclap headache", and segmental vasoconstriction resembling vasculitis of cerebral arteries that resolves by 3 months [57R]. A 15-year-old girl, who had undergone a heart transplant 8 days prior, presented with generalized seizures without a headache [60A]. Her medication list included methylprednisolone 20 mg, tacrolimus 4 mg, mycophenolate mofetil 500 mg, diltiazem 100 mg, enalapril 2.5 mg, bisoprolol 1.25 mg, furosemide 20 mg, and spironolactone 25 mg. Of note her blood pressure upon presentation was 165/65 mmHg and the seizure promptly resolved after IV diazepam. Imaging revealed bilateral hyperintensity lesions in the medial parietal lobes, thin subarachnoid hemorrhage on the left convexity, and multi-segmental vasoconstrictions limited to the left posterior cerebral artery at the onset. The serum concentration of tacrolimus increased from 6.3 to 16.2 ng/mL (target range 10–15 ng/mL) on the day of the seizure and was discontinued. On MRI the multi-segmental vasoconstriction spread throughout the cerebral arteries bilaterally on day 3, diltiazem was continued for cerebral vasodilation, and the vasoconstrictions resolved on day 11 [60A].

References

[1] Kocasoy Orhan E, Baykan B. Medication Overuse Headache: the Reason of headache that common and preventable. Noro Psikiyatr Ars. 2013;50(Suppl 1):S47–51 [R].

[2] Piovesan EJ, et al. Referred pain after painful stimulation of the greater occipital nerve in humans: evidence of convergence of cervical afferences on trigeminal nuclei. Cephalalgia. 2001;21(2):107–9 [E].

[3] Bartsch T. Migraine and the neck: new insights from basic data. Curr Pain Headache Rep. 2005;9(3):191–6 [R].

[4] Karadas O, et al. Greater occipital nerve block in the treatment of triptan-overuse headache: a randomized comparative study. Acta Neurol Scand. 2017;135(4):426–33 [C].

[5] Novitskaya ES, et al. Triptans and third nerve paresis: a case series of three patients. Eye (Lond). 2017;31(3):503–5 [A].

[6] Lacchini R, Tanus-Santos JE. Pharmacogenetics of erectile dysfunction: navigating into uncharted waters. Pharmacogenomics. 2014;15(11):1519–38 [R].

[7] Cocci A, et al. Effectiveness and safety of Oro-dispersible sildenafil in a new film formulation for the treatment of erectile dysfunction: comparison between sildenafil 100-mg film-coated tablet and 75-mg Oro-dispersible film. J Sex Med. 2017;14(12):1606–11 [c].

[8] Scaglione F, et al. Phosphodiesterase type 5 inhibitors for the treatment of erectile dysfunction: pharmacology and clinical impact of the sildenafil citrate Orodispersible tablet formulation. Clin Ther. 2017;39(2):370–7 [R].

[9] Goren A, et al. Potential risk of developing herpes simplex encephalitis in patients treated with sildenafil following primary exposure to genital herpes. J Biol Regul Homeost Agents. 2017;31(3):679–82 [H].

[10] Abu El-Hamd M, Abdelhamed A. Comparison of the clinical efficacy and safety of the on-demand use of paroxetine, dapoxetine, sildenafil and combined dapoxetine with sildenafil in treatment of patients with premature ejaculation: a randomised placebo-controlled clinical trial. Andrologia. 2018;50(1):1–6 (e12829) [C].

[11] Karti DT, et al. Sildenafil-related cerebral venous sinus thrombosis and papilledema: a case report of a rare entity. Neurol Sci. 2017;38(9):1727–9 [c].

[12] Elkoundi A, et al. Sildenafil related cerebral venous thrombosis following spinal anesthesia. J Clin Anesth. 2017;42:47–8 [c].

[13] Graziano S, et al. Sildenafil-associated hepatoxicity: a review of the literature. Eur Rev Med Pharmacol Sci. 2017;21(1 Suppl):17–22 [R].

[14] Eltony SA, Abdelhameed SY. Effect of chronic administration of sildenafil citrate (Viagra) on the histology of the retina and optic nerve of adult male rat. Tissue Cell. 2017;49(2 Pt B):323–35 [E].

[15] Sajjad A, Weng CY. Vision loss in a patient with primary Pulmonary hypertension and long-term use of Sildenafil. Retin Cases Brief Rep. 2017;11(4):325–8 [A].

[16] Tu JH, et al. Long-term follow-up of Lymphatic malformations in children treated with Sildenafil. Pediatr Dermatol. 2017;34(5):559–65 [c].

[17] Greish K, et al. Sildenafil citrate improves the delivery and anticancer activity of doxorubicin formulations in a mouse model of breast cancer. J Drug Target. 2017;1–6 [E].

[18] Dunn L, et al. Sildenafil in Pregnancy: a systematic review of maternal tolerance and obstetric and Perinatal outcomes. Fetal Diagn Ther. 2017;41(2):81–8 [R].

[19] Bhattar R, et al. Safety and efficacy of silodosin and tadalafil in ease of negotiation of large ureteroscope in the management of ureteral stone: a prosective randomized trial. Turk J Urol. 2017;43(4):484–9 [c].

[20] Kim SW, et al. Efficacy and safety of a fixed-dose combination therapy of Tamsulosin and Tadalafil for patients with lower urinary tract symptoms and Erectile dysfunction: results of a randomized, double-blinded, active-controlled trial. J Sex Med. 2017;14(8):1018–27 [C].

[21] Takeda, M., et al., Safety and efficacy of the combination of once-daily tadalafil and alpha-1 blocker in Japanese men with lower urinary tract symptoms suggestive of benign prostatic hyperplasia: a randomized, placebo-controlled, cross-over study. Int J Urol, 2017. 24(7): p. 539–547 [C].

[22] Oelke M, et al. Efficacy and safety of tadalafil 5mg once daily in the treatment of lower urinary tract symptoms associated with benign prostatic hyperplasia in men aged >/=75 years: integrated analyses of pooled data from multinational, randomized, placebo-controlled clinical studies. BJU Int. 2017;119(5):793–803 [R].
[23] Numata K, et al. The development of Cerebral Venous Thrombosis after Tadalafil ingestion in a patient with antiphospholipid Syndrome. Intern Med. 2017;56(10):1235–7 [c].
[24] Kubo M, et al. Safety and dose-finding trial of tadalafil administered for fetal growth restriction: a phase-1 clinical study. J Obstet Gynaecol Res. 2017;43(7):1159–68 [E].
[25] Goldberg DJ, et al. Results of a phase I/II multi-center investigation of udenafil in adolescents after fontan palliation. Am Heart J. 2017;188:42–52 [c].
[26] Kang SG, Kim JJ. Udenafil: efficacy and tolerability in the management of erectile dysfunction. Ther Adv Urol. 2013;5(2):101–10 [R].
[27] Sabri MR, et al. Comparison of the therapeutic effects and side effects of tadalafil and sildenafil after surgery in young infants with pulmonary arterial hypertension due to systemic-to-pulmonary shunts. Cardiol Young. 2017;27(9):1686–93 [A].
[28] Darland LK, et al. Evaluating the safety of intermittent intravenous sildenafil in infants with pulmonary hypertension. Pediatr Pulmonol. 2017;52(2):232–7 [c].
[29] Rashid J, et al. Inhaled sildenafil as an alternative to oral sildenafil in the treatment of pulmonary arterial hypertension (PAH). J Control Release. 2017;250:96–106 [E].
[30] Hamaguchi Y, et al. Safety and tolerability of bosentan for digital ulcers in Japanese patients with systemic sclerosis: prospective, multicenter, open-label study. J Dermatol. 2017;44(1):13–7 [c].
[31] Tran TT, Brinker AD, Munoz M. Serious liver injury associated with macitentan: a case report. Pharmacotherapy. 2018;38:e22–4 [A].
[32] Khan, M.A., A.C. Ho, and M.J. Spirn, Cotton-wool spots after use of Macitentan for pulmonary arterial hypertension. Retin Cases Brief Rep, 2017. 11(1): p. 4–6 [A].
[33] Albert M, et al. Comparison of inhaled milrinone, nitric oxide and prostacyclin in acute respiratory distress syndrome. World J Crit Care Med. 2017;6(1):74–8 [c].
[34] Campbell DJ. Do intravenous and subcutaneous angiotensin II increase blood pressure by different mechanisms? Clin Exp Pharmacol Physiol. 2013;40(8):560–70 [R].
[35] Chawla LS, et al. Intravenous angiotensin II for the treatment of high-output shock (ATHOS trial): a pilot study. Crit Care. 2014;18(5):534 [c].
[36] Giapreza (R). Angiotensin II. package insert, San Diego, CA: La Jolla Pharmaceutical Company; 2017 [S].
[37] Wolf RL, et al. Metabolism of angiotensin II-I-131 in normotensive and hypertensive human subjects. Circulation. 1961;23:754–8 [c].
[38] Correa TD, Takala J, Jakob SM. Angiotensin II in septic shock. Crit Care. 2015;19:98 [R].
[39] Baudin B. Angiotensin II receptor polymorphisms in hypertension. Pharmacogenomic considerations. Pharmacogenomics. 2002;3(1):65–73 [R].
[40] Kainulainen K, et al. Evidence for involvement of the type 1 angiotensin II receptor locus in essential hypertension. Hypertension. 1999;33(3):844–9 [E].
[41] Takahashi N, et al. Association of a polymorphism at the 5′-region of the angiotensin II type 1 receptor with hypertension. Ann Hum Genet. 2000;64(Pt 3):197–205 [R].
[42] Khanna A, et al. Angiotensin II for the Treatment of Vasodilatory Shock. N Engl J Med. 2017;377(5):419–30 [C].
[43] Khanna A, English SW, Wang XS, et al. Angiotensin II for the treatment of vasodilatory shock. N Engl J Med. DOI: https://doi.org/10.1056/NEJMoa1704154. 2017 January 21, 2018]; Available from: https://emcrit.org/wp-content/uploads/2017/06/ATHOS-3_appendix.pdf [C].
[44] Farag E, et al. The renin angiotensin system and the brain: new developments. J Clin Neurosci. 2017;46:1–8 [R].
[45] Mogi M, Horiuchi M. Effect of angiotensin II type 2 receptor on stroke, cognitive impairment and neurodegenerative diseases. Geriatr Gerontol Int. 2013;13(1):13–8 [R].
[46] Jayamali WD, Herath H, Kulathunga A. Myocardial infarction during anaphylaxis in a young healthy male with normal coronary arteries—is epinephrine the culprit? BMC Cardiovasc Disord. 2017;17(1):237 [c].
[47] ProAmatine (R). Midodrine. package insert, Lexington, MA: Shire USA Inc; 2017 [S].
[48] Gutman LB, Wilson BJ. The role of Midodrine for hypotension outside of the intensive care unit. J Popul Ther Clin Pharmacol. 2017;24(3):e45–50 [R].
[49] Byun JI, et al. Efficacy of single or combined midodrine and pyridostigmine in orthostatic hypotension. Neurology. 2017;89(10):1078–86 [C].
[50] Mestinon (R). Pyridostigmine. package insert, Hayward, CA: Impax generics; 2015 [S].
[51] Wong LY, et al. Severe Hypertension and Bradycardia Secondary to Midodrine Overdose. J Med Toxicol. 2017;13(1):88–90 [A].
[52] Mills B, Yepes A, Nugent K. Synthetic cannabinoids. Am J Med Sci. 2015;350(1):59–62 [R].
[53] Huestis MA, Tyndale RF. Designer Drugs 2.0. Clin Pharmacol Ther. 2017;101(2):152–7 [R].
[54] Brents LK, Prather PL. The K2/spice phenomenon: emergence, identification, legislation and metabolic characterization of synthetic cannabinoids in herbal incense products. Drug Metab Rev. 2014;46(1):72–85 [R].
[55] Moeller S, et al. Ischemic stroke associated with the use of a synthetic cannabinoid (spice). Asian J Psychiatr. 2017; 25:127–30 [A].
[56] Gunaseelan S, Prakash A. Thalidomide-induced stroke in a child with Thalassemia major. J Pediatr Hematol Oncol. 2017;39(8): e519–20 [A].
[57] Miller TR, et al. Reversible cerebral vasoconstriction syndrome, part 1: epidemiology, pathogenesis, and clinical course. AJNR Am J Neuroradiol. 2015;36(8):1392–9 [R].
[58] Granata G, et al. Posterior reversible encephalopathy syndrome—insight into pathogenesis, clinical variants, and treatment approaches. Autoimmun Rev. 2015;14(9):830–6 [R].
[59] Cherniawsky H, et al. A case report of posterior reversible encephalopathy syndrome in a patient receiving gemcitabine and cisplatin. Medicine (Baltimore). 2017;96(8):e5850 [A].
[60] Kodama S, et al. Tacrolimus-Induced reversible Cerebral Vasoconstriction Syndrome with delayed multi-segmental Vasoconstriction. J Stroke Cerebrovasc Dis. 2017;26(5):e75–7 [A].

CHAPTER

20

Antihypertensive Drugs

Katie Traylor[1], *Holly Gurgle*, *Joshua Brockbank*

Department of Pharmacotherapy, University of Utah College of Pharmacy, Salt Lake City, UT, United States

[1]Corresponding author: katie.traylor@pharm.utah.edu

ANGIOTENSIN-CONVERTING ENZYME INHIBITORS [SEDA 39: 183–187]

Enalapril

Mouth and Teeth

In an observational study utilizing the French PharmacoVigilance Database, several antihypertensive medications were reportedly associated with gingival bleeding. Though rarely fatal, gingival bleeding is often considered a serious adverse drug reaction and should prompt discontinuation of the causative agent. Five reports of gingival bleeding were linked to enalapril following a causality assessment that evaluated chronology and presence of other potential causes. Though enalapril has not previously been associated with gingival bleeding, this study may be an early signal of a rare adverse drug reaction [1c].

Ramipril

Mouth and Teeth

In an observational study utilizing the French PharmacoVigilance Database, several antihypertensive medications were reportedly associated with gingival bleeding. Though rarely fatal, gingival bleeding is often considered a serious adverse drug reaction and should prompt discontinuation of the causative agent. Five reports of gingival bleeding were linked to ramipril following a causality assessment that evaluated chronology and presence of other potential causes. Though ramipril has not previously been associated with gingival bleeding, this study may be an early signal of a rare adverse drug reaction [1c].

ANGIOTENSIN RECEPTOR BLOCKERS/ ANGIOTENSIN II RECEPTOR ANTAGONISTS

Telmisartan

Cardiovascular

Angiotensin receptor blockers (ARBs) have been used for the treatment of Raynaud phenomenon. However, in a French national pharmacovigilance database study investigating reports of Raynaud phenomenon related to drug exposures, two instances of Raynaud phenomenon were reported in patients using telmisartan. In these two patients, authors found no concomitant use of medications known to induce Raynaud phenomenon and no other history of or risk factor for this complication. Due to the nature of adverse event reporting, they note the association may have resulted from notification bias. The case/noncase methodology use in this study is typically only useful for generating signals of potential associations, so further studies are warranted to investigate the associations noted [2c].

Gastrointestinal

While olmesartan has been associated with sprue-like enteropathy, limited evidence exists linking this adverse effect related to other ARBs. In a case report by Negro et al., a 52-year-old man who had been taking telmisartan 40mg daily for 3 years presented with signs symptoms of moderate sprue-like enteropathy. Symptoms resolved within 1 week of discontinuation of his telmisartan, and duodenal biopsies 3 months later showed progressive duodenal recovery. This case report suggests that olmesartan-associated sprue-like enteropathy may actually be a class effect of ARBs. Though this is only the third known report

Side Effects of Drugs Annual, Volume 40
ISSN: 0378-6080
https://doi.org/10.1016/bs.seda.2018.07.013

of telmisartan-related sprue-like enteropathy, authors recommend discontinuation of the medication in the setting of a sprue-like enteropathy due to the potentially life-threatening nature of this adverse effect [3A].

ANGIOTENSIN II RECEPTOR BLOCKER; ANGIOTENSIN RECEPTOR NEPRILYSIN INHIBITOR

Valsartan/Sacubitril

General Information

In two separate studies of valsartan/sacubitril use in patients with systolic hypertension, peripheral edema was identified as an adverse effect, in addition to other previously described adverse effects. Results from the study by Izzo et al. revealed three patients (2.1%) with peripheral edema in the sacubitril 400mg plus valsartan 320mg group, compared to two patients (3.4%) in the placebo group; one patient (0.7%) on valsartan 320mg monotherapy also experienced peripheral edema [4C]. In a second study by Williams et al., six patients (2.6%) on sacubitril 97mg and valsartan 103mg experienced peripheral edema, compared to two patients (0.9%) in the control group (olmesartan 20mg) [5C]. Neither study commented on whether these differences were statistically significant [4C, 5C].

Drug–Drug Interaction

An open-label study in 28 healthy Chinese males investigated the potential drug–drug interaction between sacubitril/valsartan 97/103mg and atorvastatin 80mg. While atorvastatin did not significantly alter levels of sacubitril/valsartan or active metabolites, sacubitril/valsartan did result in a twofold increase in the C_{max} of atorvastatin and its metabolites. The AUC of atorvastatin was only marginally increased (less than 1.3-fold), suggesting limited effect on atorvastatin absorption and clearance. One patient experienced an increase in alanine aminotransferase (ALT) levels during the study, though this was mild (ALT 90U/L). Overall, the combination was considered safe and well-tolerated over the 5 days of coadministration [6c]. A pharmacokinetic modeling study also described this mild drug–drug interaction between sacubitril/valsartan, as well as predicted interactions between sacubitril/valsartan and other statins. Interactions were deemed unlikely with rosuvastatin, fluvastatin, and lovastatin, and predicted interactions were less than 1.5-fold for pitavastatin and pravastatin. No interaction was found between sacubitril/valsartan and simvastatin [7E]. Many patients on sacubitril/valsartan are also likely indicated for statin use; based on the results of these studies, coadministration can likely be considered safe.

Nervous System

Based on the mechanism of sacubitril in inhibiting the enzyme neprilysin, theoretical concern existed regarding long-term effects on dementia formation though amyloid-β peptide accumulation. Cannon et al. described cognitive outcomes from several large-scale trials studying sacubitril/valsartan. Investigators identified 27 dementia-related adverse effects in the PARADIGM-HF trial, 12 of which occurred in the sacubitril/valsartan groups and 15 of which occurred in the comparator groups (enalapril) [hazard ratio 0.73, 95% confidence interval 0.33–1.59]. Similar rates of dementia-related adverse effects were also found in three other trials, with no evidence of increased concerns for dementia in patients taking sacubitril/valsartan vs enalapril. Further studies with longer durations and more sensitive methods to detect dementia are necessary to confirm this conclusion [8R].

BETA BLOCKERS

Atenolol

Mouth and Teeth

In an observational study utilizing the French PharmacoVigilance Database, several antihypertensive medications were reportedly associated with gingival bleeding. Though rarely fatal, gingival bleeding is often considered a serious adverse drug reaction and should prompt discontinuation of the causative agent. Four reports of gingival bleeding were linked to atenolol following a causality assessment that evaluated chronology and presence of other potential causes. Though atenolol has not previously been associated with gingival bleeding, this study may be an early signal of a rare adverse drug reaction [1c].

Carvedilol

Cardiovascular

Carvedilol has recently been used in the management of portal hypertension as prophylaxis of gastroesophageal varices. This beta blocker has been evaluated in a number of studies involving patients with cirrhosis, with doses ranging from 3.125 to 50mg daily. Maharaj et al. describe a 56-year-old male with cirrhosis (Child-Pugh score of 7) who suffered cardiogenic shock following administration of two doses of carvedilol 12.5mg (i.e., total dose of 25mg). The authors note that only two other reports of carvedilol toxicity have been published, and both were in the setting of overdose. While carvedilol is contraindicated in severe hepatic impairment, there are no suggested dose adjustments for mild to moderate hepatic impairment. Based on this case, authors suggest low starting doses of carvedilol (e.g., 3.125mg twice daily) in patients with hepatic impairment, followed by slow up-titration of the dose and close monitoring for signs of toxicity [9A].

Propranolol

Drug–Drug Interaction

A case report by Rouabhia et al. described a 77-year-old man treated with propranolol and sofosbuvir who suffered from ventricular extrasystoles. The manufacturer of sofosbuvir has warned against co-administration with amiodarone due to risk of bradycardia, stating that this risk may be increased in patients who are also taking beta blockers; however, they did not warn against co-administration of sofosbuvir and beta blockers in the absence of amiodarone use. The patient in this case report had never taken amiodarone and was not on any other medications reported to interact with sofosbuvir. Three hours after his first sofosbuvir dose, the patient complained of palpitations and electrocardiogram revealed numerous monomorphic ventricular extrasystoles accompanied by bradycardia. No further doses of sofosbuvir were given, and symptoms resolved after 24 hours. Authors of the case report suggest that patients taking sofosbuvir and propranolol and should undergo cardiac rhythm monitoring during the first several hours of co-administration to identify this potential drug–drug interaction [10A].

Nervous System

In a case report, propranolol was suspected as the cause of temperature instability in an infant with infantile hemangioma. The patient, a 25 1/7-week premature twin girl, was started on propranolol at 36 5/7 weeks corrected gestational age. After 3 days on propranolol, her average axillary temperature had declined from 36.7°C to 36.4°C, requiring the patient to be returned to a heated incubator. The oral propranolol dose was increased from 0.7mg every 8 hours to 1.4mg every 8 hours, leading to further temperature decline. Despite returning to the lower dose, the infant was unable to wean from the incubator until the propranolol was discontinued on day 19 of treatment. Though cold extremities have been noted in infants receiving propranolol for infantile hemangioma, this type of temperature instability has only been reported with use of topical timolol. Authors of this case report suggest that the high lipophilicity and nonselective nature of propranolol may contribute to this adverse effect and suggest further consideration for preferential use of β_1-selective, low lipophilicity beta blockers such as atenolol in this treatment setting [11A].

Drug Overdose

Hopkins et al. presented the first case of a propranolol drug bezoar in a case report of a 21-year-old female. The patient presented to the hospital following attempted suicide by medication overdose. Suspected ingested tablets included propranolol 40mg sustained-release, amlodipine 10mg, and olanzapine 10mg. Despite fluid resuscitation, intubation, and treatment with glucagon, adrenaline, insulin, and intravenous fat emulsion, the patient remained hypotensive. Endoscopy revealed a large pharmacobezoar, estimated at 200mL in total of tablets. Though rare, this type of bezoar may form due to delayed gastric absorption in the setting of persistent low cardiac output. Though endoscopy and gastric washout are not routinely recommended or performed in drug overdose, authors concluded that these interventions may be considered as a means to limit further drug absorption in patients who are refractory to standard treatments [12A].

CALCIUM CHANNEL BLOCKERS

Amlodipine

General Information

Chen et al. compared initiation of *S*-(−)-amlodipine 5mg (high dose) vs 2.5mg (low dose) among patients with mild–moderate hypertension. At 8 weeks, the incidence of adverse events was low in both arms with similar tolerability. The rate of adverse events was similar between high dose (20.0%, $n=70$) and low dose (17.7%, $n=62$) arms ($P=0.50$) [13c].

Nicardipine

Respiratory/Hematologic

Monaco et al. described findings from an Italian pharmacovigilance database which collects real-world reports of suspected adverse drug reactions. Among several cardiovascular agents reviewed within the database, a potential correlation was observed between nicardipine and acute pulmonary edema (off-label use as tocolytic during pregnancy) and between nicardipine and thrombocytopenia. The authors noted the benefit of pharmacovigilance databases which allow for continuous evaluation of the risk and benefit of medications, particularly in special populations such as pregnancy which are rarely studied in clinical trials [14c].

DRUGS THAT ACT ON THE SYMPATHETIC NERVOUS SYSTEM

Urapidil

Mouth and Teeth

In an observational study utilizing the French PharmacoVigilance Database, several antihypertensive medications were reportedly associated with gingival bleeding. Though rarely fatal, gingival bleeding is often considered a serious adverse drug reaction and should prompt discontinuation of the causative agent. Three reports of gingival bleeding were linked to the alpha-1 agonist urapidil following a causality assessment that evaluated chronology and presence of other potential causes. Though urapidil has not previously been associated with gingival bleeding, this study may be an early signal of a rare adverse drug reaction [1c].

FIXED-DOSE ANTIHYPERTENSIVE COMBINATION THERAPIES

Indapamide/Perindopril

General Information

An analysis of the ADVANCE and PROGRESS trial data was conducted by Atkins et al. to evaluate the side effects and tolerability of dual fixed-dose combination antihypertensive therapy. The analysis included more than 14000 patients with hypertension who were randomized to treatment with perindopril and indapamide vs placebo. Patients were stratified into five subgroups by baseline systolic blood pressure (mmHg): less than 120, 120–129, 130–139, 140–159, and 160mmHg or greater. Discontinuation rates due to hypotension or dizziness ranged from 1.3% to 3.6% depending on the degree of baseline blood pressure elevation. Overall side effects with combination antihypertensives, including renal adverse effects, were similar across all subgroups stratified by baseline systolic blood pressure [15R].

Triple-Combination Therapy

General Information

Hypertension practice guidelines emphasize that many patients will require the use of combination therapies to achieve adequate blood pressure control. As many as one-third of patients may require three or more antihypertensive agents. Düsing et al. reviewed four randomized control trials evaluating the efficacy and safety of triple vs dual combination antihypertensive therapy in a single pill. Combinations were made up of amlodipine, angiotensin receptor blockers, and hydrochlorothiazide. While more efficacious at blood pressure reduction, triple combination therapy was associated with similar rates of adverse effects as dual combination therapy. The most common adverse effects with triple combination therapy were dizziness, peripheral edema, and headache [16R].

Additional case study reports can be found in these reviews [17R, 18R].

References

[1] Bondon-Guitton E, Mourgues T, Rousseau V, et al. Gingival bleeding, a possible "serious" adverse drug reaction: an observational study in the French PharmacoVigilance database. J Clin Periodontol. 2017;44(9):898–904 [c].

[2] Bouquet É, Urbanski G, Lavigne C, et al. Unexpected drug-induced Raynaud phenomenon: analysis from the French national pharmacovigilance database. Therapie. 2017;72(5):547–54 [c].

[3] Negro A, De Marco L, Cesario V, et al. A case of moderate sprue-like enteropathy associated with telmisartan. J Clin Med Res. 2017;9(12):1022–5 [A].

[4] Izzo Jr. JL, Zappe DH, Jia Y, et al. Efficacy and safety of crystalline valsartan/sacubitril (LCZ696) compared with placebo and combinations of free valsartan and sacubitril in patients with systolic hypertension: the RATIO study. J Cardiovasc Pharmacol. 2017;69(6):374–81 [C].

[5] Williams B, Cockcroft JR, Kario K, et al. Effects of sacubitril/valsartan versus olmesartan on central hemodynamics in the elderly with systolic hypertension: the PARAMETER study. Hypertension. 2017;69(3):411–20 [C].

[6] Ayalasomayajula S, Pan W, Han Y, et al. Assessment of drug-drug interaction potential between atorvastatin and LCZ696, a novel angiotensin receptor neprilysin inhibitor, in healthy Chinese male subjects. Eur J Drug Metab Pharmacokinet. 2017;42(2):309–18 [c].

[7] Lin W, Ji T, Einolf H, et al. Evaluation of drug-drug interaction potential between sacubitril/valsartan (LCZ696) and statins using a physiologically based pharmacokinetic model. J Pharm Sci. 2017;106(5):1439–51 [E].

[8] Cannon JA, Shen L, Jhund PS, et al. Dementia-related adverse events in PARADIGM-HF and other trials in heart failure with reduced ejection fraction. Eur J Heart Fail. 2017;19(1):129–37 [R].

[9] Maharaj S, Seegobin K, Perez-Downes J, et al. Severe carvedilol toxicity without overdose—caution in cirrhosis. Clin Hypertens. 2017;23:25 [A].

[10] Rouabhia S, Baghazza S, Sadouki H, et al. Ventricular extrasystoles after first dose of sofosbuvir in a patient treated with propranolol but not with amiodarone: a case report. Rev Recent Clin Trials. 2017;12(3):159–61 [A].

[11] Burkey BW, Jacobs JA, Aziz H. Temperature instability in an infant treated with propranolol for infantile hemangioma. J Pediatr Pharmacol Ther. 2017;22(2):124–7 [A].

[12] Hopkins LE, Sunkersing J, Jacques A. Too many pills to swallow: a case of a mixed overdose. J Intensive Care Soc. 2017;18(3):247–50 [A].

[13] Chen Q, Huang QF, Kang YY, et al. Efficacy and tolerability of initial high vs low doses of S-(-)-amlodipine in hypertension. J Clin Hypertens (Greenwich). 2017;19(10):973–82 [c].

[14] Monaco L, Melis M, Biagi C, et al. Signal detection activity on EudraVigilance data: analysis of the procedure and findings from an Italian Regional Centre for Pharmacovigilance. Expert Opin Drug Saf. 2017;16(3):271–5 [c].

[15] Atkins ER, Hirakawa Y, Salam A, et al. Side effects and tolerability of combination blood pressure lowering according to blood pressure levels: an analysis of the PROGRESS and ADVANCE trials. J Hypertens. 2017;35(6):1318–25 [R].

[16] Düsing R, Waeber B, Destro M, et al. Triple-combination therapy in the treatment of hypertension: a review of the evidence. J Hum Hypertens. 2017;31(8):501–10 [R].

[17] Traylor K, Gurgle H, Turner K, Woods A, Brown K, Ray SD. Antihypertensive drugs. In: Ray SD, editor. Side effects of drugs: annual, A worldwide yearly survey of new data in adverse drug reactions, vol. 39. Elsevier; 2017. p. 183–7 [R].

[18] DelLLelis T, Keshshishyan S, Kovalevskaya V, et al. Antihypertensives. In: Ray SD, editor. Side effects of drugs: annual. A worldwide yearly survey of new data in adverse drug reactions, vol. 38. Elsevier; 2016. p. 179–84 [R].

CHAPTER

21

Diuretics

Michelle Friedman-Jakubovics[1], *Roman Fazylov*

Department of Pharmacy Practice, Touro College of Pharmacy, New York, NY, United States

[1]Corresponding author: michelle.friedman15@touro.edu

CARBONIC ANHYDRASE INHIBITORS [SEDA-37, 237; SEDA-38, 197; SEDA-39, 191]

Acetazolamide

Pulmonary Edema

Noncardiogenic pulmonary edema is a rare adverse effect that has been reported with acetazolamide use. Ono and colleagues reported a case of a 61-year-old man who developed pulmonary edema and severe hypoxia following intravenous acetazolamide administration. The patient was admitted to an intensive care unit (ICU) due to severe chest trauma from a motor vehicle accident. He remained on ventilator support even on day 17 of his ICU stay and 500mg of intravenous acetazolamide was administered to increase urinary output and correct metabolic alkalosis. An hour after acetazolamide administration, the patient developed hypertension, tachycardia, and hypoxemia, and a bilateral butterfly shadow was observed on chest radiography. Diuretics and vasodilators were administered and respiratory failure slowly resolved. However, 5 days later, 1 hour following a repeat dose of acetazolamide, the patient again developed respiratory failure, and acetazolamide-induced pulmonary edema was suspected. Acetazolamide was not administered again during the patient's hospital stay, and no further episodes of respiratory failure occurred. The mechanism of this adverse effect is unclear, but it has been hypothesized that pulmonary edema develops due to sulfonamide cross-sensitivity and its immune mediated mechanism. This case illustrates the need for clinicians to be cognizant of the risk for pulmonary edema when administering acetazolamide to critically ill patients [1A].

Brain Infarction

Acetazolamide-challenged brain single-photon emission computed tomography (SPECT) is a tool used to evaluate cerebral perfusion in patients with cerebrovascular diseases. Many adverse effects of acetazolamide-challenged brain SPECT have previously been reported; however, they have mostly been mild and transient in nature. Chong and colleagues reported a case of severe fatal brain infarction following acetazolamide-challenged brain SPECT in an 11-year-old girl with moyamoya disease, a cerebrovascular disorder caused by blocked arteries in the basal ganglia. According to Chong et al. this appears to be the first case of fatal severe brain infarction following acetazolamide-challenged brain SPECT in a patient with moyamoya disease. The patient developed seizures and then became unconscious with dilated, fixed pupils. Brain magnetic resonance imaging revealed acute infarction of the bilateral occipital lobes, thalami, brain stem and cerebellum resulting in brain death. Clinicians should be aware of this rare, but serious reaction that may occur as a complication of acetazolamide use [2A].

Choroidal Effusion

Hári-Kovács and colleagues, in an article (in Hungarian), reported two cases of choroidal effusion following acetazolamide use. Choroidal effusion is an abnormal accumulation of fluid in the suprachoroidal space. Numerous articles have previously been published describing choroidal effusion following sulfa drug administration, especially following acetazolamide use. In previous reports, choroidal effusion was associated with acute angle-closure glaucoma, transient myopia, or a combination of both as the leading symptoms. However,

Side Effects of Drugs Annual, Volume 40
ISSN: 0378-6080
https://doi.org/10.1016/bs.seda.2018.07.008

Hári-Kovács et al. reported two cases of acetazolamide-induced choroidal effusion that presented without myopic shift in refraction or acute elevation of intraocular pressure. To their knowledge, this was the first report describing this unusual presentation of drug-induced choroidal effusion. Authors concluded that these cases highlight the need for ophthalmic examination in patients taking a sulfa drug who develop acute visual deterioration [3A].

Safety of Acetazolamide for Preoperative Intraocular Pressure Reduction

Lorenz and colleagues conducted a randomized controlled trial to evaluate the safety and efficacy of topical dorzolamide plus timolol in comparison to oral acetazolamide plus topical dexamethasone when administered to patients with glaucoma undergoing trabeculectomy. Sixty-two patients were included in the study, 30 in the dorzolamide/timolol group and 32 in the acetazolamide/dexamethasone group. In terms of the primary efficacy endpoint of change in intraocular pressure at 3 months post-trabeculectomy, preoperative dorzolamide/timolol was non-inferior to oral acetazolamide/topical dexamethasone (adjusted mean change in intraocular pressure of −8.12mm Hg vs −8.30mm Hg; difference 0.18; 95% CI −1.91 to 2.26, $P=0.8662$). In terms of safety, 82.14% (23/28 patients) of the dorzolamide/timolol group experienced an adverse effect in comparison to 96.77% (30/31 patients) of the acetazolamide/dexamethasone group. All adverse effects observed were consistent with the known adverse effect profile of these drugs, with the exception of atrial fibrillation which developed in one patient receiving acetazolamide/dexamethasone. Authors concluded that preoperative topical dorzolamide/timolol is a safe and effective alternative to oral acetazolamide/topical dexamethasone [4c].

LOOP DIURETICS [SEDA-37, 237; SEDA-38, 185; SEDA-39, 189]

Respiratory-Related Morbidity and Mortality in Chronic Obstructive Pulmonary Disease

There are minimal and conflicting data regarding the respiratory effects of diuretic agents in patients with chronic obstructive pulmonary disease (COPD). It has been theorized that diuretics improve health outcomes in patients with COPD by reducing pulmonary hypertension, pulmonary edema, and cor pulmonale. However, it has also been proposed that loop diuretics and less so, other diuretic agents, may cause respiratory harm by increasing sodium bicarbonate and arterial pH which may result in hypercapnia, and by causing hypokalemia which has been linked to respiratory muscle weakness and acute respiratory failure. Vozoris and colleagues conducted a retrospective cohort study using health administrative data from Ontario, Canada, to assess COPD outcomes in patients treated with or without diuretic therapy. 99766 individuals aged 66 years and older with COPD were included, among whom, 51.7% received diuretic therapy, including, loop diuretics, potassium-sparing diuretics, thiazide diuretics, or carbonic anhydrase inhibitors. When results were analyzed for each diuretic subclass in comparison to nonusers of diuretics, the following findings were observed. Incident users of loop diuretics had increased rates of hospitalization for COPD or pneumonia (HR 1.36, 95% CI 1.16–1.60), increased rates of outpatient respiratory exacerbations (HR 1.11, 95% CI 1.02–1.21), increased emergency room visits for COPD or pneumonia (HR 1.62, 95% CI 1.38–1.90), increased rates of ICU admission for COPD or pneumonia (HR 1.67, 95% CI 1.08–2.58), and higher all-cause mortality (HR 1.31, 95% CI 1.13–1.51) when compared to the control group. In contrast, the only significant difference observed between new users of thiazide diuretics and the control group was a decreased rate of outpatient respiratory COPD exacerbations in patients receiving thiazide diuretics (HR 0.75, 95% CI 0.66–0.84). No significant differences were observed for the potassium-sparing diuretics and carbonic anhydrase inhibitors for any of the above outcomes when compared to the control group. While investigators adjusted their findings for 55 covariates, it is still possible that confounding by indication or influence of unmeasured clinical covariates played a role in the study's findings. This study identifies the potential for respiratory adverse effects with loop diuretic use; however, further studies are needed to elucidate whether the adverse respiratory outcomes observed in the loop diuretic group were caused by diuretic therapy or residual confounding [5MC].

Vertebral Fractures

A prospective cohort study by Paik and colleagues evaluated the risk of incident vertebral fracture in women receiving loop diuretic or thiazide diuretic therapy. 55780 women aged 55–82 years who had no prior history of fracture were included in the study. Researchers controlled for potentially confounding variables, such as, age, body mass index, and race, and evaluated the incidence of vertebral fracture among users and nonusers of loop and thiazide diuretics. During the study period of 2002–2012 there were 420 cases of vertebral fracture, and diuretic use was found to be independently associated with increased fracture risk. Risk was increased by 47% in thiazide users (RR 1.47, 95% CI 1.18–1.85) and 59% in users of loop diuretics (RR 1.59, 95% CI 1.12–2.25). Loop diuretics are thought to increase risk of vertebral fracture by increasing urinary calcium excretion resulting in decreased bone mineral density. The increased risk of vertebral fracture observed with thiazide diuretic use was unexpected as thiazide diuretics have previously been shown to be protective agents against hip and vertebral fractures. However, it is possible that thiazide diuretics may increase vertebral fracture risk by causing hyponatremia. Hyponatremia results in osteoclast activation which leads to increased bone resorption. Increased bone resorption without concurrent bone formation results in decreased bone quality and increased risk for fractures. Based on this study's findings, loop diuretics and thiazide diuretics should be used cautiously in patients at high risk for vertebral fracture [6C].

Furosemide

Risk of Fractures and Fall in the Elderly

Lai and colleagues conducted a retrospective case–control study utilizing the Taiwan National Health Insurance Program database to evaluate the association between furosemide and hip fracture in the elderly. 4523 patients 65 years of age and older who had experienced a hip fracture from 2000 to 2013 were compared to a control group of 4523 adults over age 65 who did not experience a hip fracture during that time period. Cases and controls were matched by sex, age, comorbidities, and index year and month of hip fracture diagnosis. Results were adjusted for potential covariates. The adjusted odds ratio for hip fracture was found to be elevated in current users of furosemide (OR 1.30, 95% CI, 1.14–1.48). Patients who were recent users or past users of furosemide did not have significantly increased odds of hip fracture when compared to nonusers [7MC]. These results indicate that extra caution is required when furosemide is used in the elderly. Additionally, since increased odds of hip fracture were only observed for current furosemide users, clinicians should regularly assess whether continued use of furosemide is indicated to ensure an appropriate duration of therapy.

Another study by Okada et al. evaluated whether decreasing doses of furosemide would result in decreased falls and fall-related fractures. Daily loop diuretic doses and other diuretic regimens for 50 special geriatric nursing home patients were reduced or replaced by spironolactone 12.5 mg orally once every other day beginning in July 2012. After the dose decreases, the number of falls decreased from 53 in 2011 to 29 and 27 in 2012 and 2014, respectively, and no fall-related proximal femoral fractures were observed within the following 3 years. As observed in this study, spironolactone may be an effective diuretic option in some geriatric patients and decreased use and doses of loop diuretics, such as, furosemide, may result in decreased falls and fall-related fractures in elderly patients [8c].

Increased Oxidative Stress in Acute Kidney Injury

Furosemide is sometimes used in patients with acute kidney injury (AKI) to increase urine output. However, it remains uncertain whether risks of therapy outweigh benefits in this patient population. Silbert and colleagues conducted a study in 30 critically ill patients with AKI treated with furosemide to assess the effect of furosemide on oxidative stress. Urine and plasma levels of F2-isoprostanes, markers of oxidative stress, were measured before and after furosemide use. Urine F2-isoprostane levels were found to be positively correlated with urine furosemide levels ($P = 0.001$). Furosemide-induced increase in urine F2-isoprostanes varied by severity of AKI ($P < 0.001$) and was greatest in those patients with the most severe AKI. However, no effects on plasma F2-isoprostane levels were observed with furosemide use. According to Silbert et al. this was the first study demonstrating increased renal oxidative stress with furosemide use in AKI. Follow-up randomized controlled trials are needed to further elucidate the significance of this adverse drug reaction with furosemide therapy and to determine whether benefits of furosemide use outweigh the risks in patients with AKI [9c].

Tolerance to Diuretic Therapy

Development of tolerance to furosemide therapy has been well described in adults. However, there are limited data regarding development of tolerance to furosemide in pediatric patients. Kim and colleagues conducted a retrospective chart review of pediatric patients who received furosemide for at least 3 consecutive days between June 1, 2013 and December 31, 2013 to evaluate whether pediatric patients developed tolerance to furosemide therapy. Daily net fluid balance, defined as, the difference in the patient's daily intake and urine output, normalized by weight and total daily dose of furosemide, was used as the marker of tolerance development in this study. Sixty-one patients aged 2 days to 20 years (median 3 years) were included in the study. The median daily dose of furosemide was 1.96 mg/kg/day. The mean net fluid balance increased significantly from a mean of 6.83 mL/kg/mg on day 1 of therapy to a mean of 26.66 mL/kg/mg on the last day of therapy ($P = 0.011$). No significant relationship was found between age and difference in net fluid balance from the first to the last day of furosemide therapy. Authors concluded that tolerance to furosemide therapy develops in pediatric patients as evidenced by increasing net fluid balance over time with furosemide therapy. To ensure continued efficacy of diuretic agents when used for an extended period of time, clinicians may consider increasing the dose of furosemide or adding a second diuretic agent which have both been shown to help decrease the net fluid balance per day [10c].

Acute Pancreatitis

A case of probable furosemide-induced acute pancreatitis was described in a 74-year-old male with a past medical history of coronary artery disease, sleep apnea, and gastroesophageal reflux disease. The patient was admitted to the hospital with complaints of epigastric pain described as 7-out-of-10 in intensity beginning the morning of admission. On admission, his blood glucose was 104 mg/dL, his triglyceride level was 80 mg/dL, and his renal and liver functions were within normal limits. His serum amylase (1022 U/L) and lipase levels (>600 U/L) were elevated, and no gallstones were visualized on ultrasound. The patient denied smoking and alcohol intake, and the only recent change in his medications was initiation of furosemide 6 weeks prior to admission for bilateral lower extremity swelling. On admission, furosemide was discontinued, intravenous fluids were started and oral intake was avoided. Within 24 hours, symptom improvement was observed. Due to the temporal relationship between furosemide initiation and the development of acute pancreatitis coupled with the lack of evidence of alternative causes of pancreatitis, such as gallstones, alcoholism, or hypertriglyceridemia, furosemide-induced acute pancreatitis was suspected. The Naranjo Adverse Drug Reaction Probability Scale score of 5 indicates furosemide was a probable cause of this patient's pancreatitis. Furosemide-induced pancreatitis has previously been described and is suggested to result from decreased pancreatic perfusion due to decreased extracellular fluid volume in response to diuresis. It has also been proposed that furosemide stimulates pancreatic exocrine stimulation, has a direct toxic effect on the pancreas, and can stimulate a hypersensitivity reaction affecting the pancreas. While it cannot be definitively determined that furosemide caused this patient's pancreatitis, furosemide should be considered as part of the differential diagnosis of a patient presenting with acute pancreatitis especially during the first few weeks of furosemide therapy [11A].

Severe Hypokalemia Resulting in Acute Transient Quadriparesis

Lybecker and colleagues reported a case of severe hypokalemia resulting in transient quadriparesis in a 48-year-old man treated with high doses of furosemide coupled with regular use of insulin. In the days preceding hospital admission, the patient experienced decreasing muscle tone until he was unable to walk. At the time of admission the patient had developed paralysis of both upper and lower extremities. Home medications included furosemide 180 mg daily, insulin NPH 18 units daily, insulin aspart 24 units daily, losartan 150 mg daily, and metoprolol 200 mg daily. The patient reported that he had not been in contact with his general practitioner for 3 years, and it was noted that the patient was not receiving potassium supplementation along with his furosemide therapy. On admission, the patient was found to have severe hypokalemia (<1.5 mmol/L; normal range 3.5–4.6 mmol/L), hypocalcemia (1.99 mmol/L; normal range 2.20–2.55 mmol/L), and hypomagnesemia (0.59 mmol/L; normal range 0.70–1.10 mmol/L). Electrocardiography (ECG) revealed sinus rhythm along with characteristic changes of hypokalemia, including ST segment depression, prolonged QTc interval of 581 ms, and U waves. Bartter and Gitelman syndrome and thyrotoxicosis, potential causes of hypokalemia, were ruled out based on laboratory findings. The patient was treated with oral and intravenous potassium replacement and his cardiac rhythm and serum potassium were closely monitored. The patient's potassium level normalized within 24 hours, and he slowly regained his strength. At discharge all laboratory findings were normal and the ECG revealed normal sinus rhythm with a normal QTc interval. Lack of regular follow-up by the patient with his general practitioner together with high dose furosemide use in combination with regular insulin therapy likely resulted in severe hypokalemia in this patient. Clinicians should be cognizant of the risk for severe hypokalemia with furosemide use and regular monitoring of potassium levels and potassium supplementation should be provided as clinically indicated [12A].

Hyperacute Leucopenia

A case of hyperacute leucopenia was described in a 72-year-old male treated with furosemide for an exacerbation of heart failure. On admission to the hospital the patient was started on furosemide 40 mg intravenously with two doses given 4 hours apart. Prior to starting furosemide, the patient's white blood cell count (WBC) was 9.8×10^9/L (reference range 4.1–9.3×10^9/L). Four hours after the second dose of furosemide, his WBC count had dropped to 2.4×10^9/L. This lab was repeated three times and all results were consistent with an acute drop in WBC count. Furosemide was discontinued and 13 hours after furosemide discontinuation the WBC count had increased to 7.1×10^9/L. For the following 3 days, furosemide was not administered and the WBC count remained within normal limits. On day 4, oral furosemide was initiated and there was a mild, but transient drop again observed in the patient's WBC count. While the patient was on furosemide prior to admission, his home dose was 20 mg once to twice daily, a significantly lower dose than he received in the emergency department. No other medications were added that might have contributed to this reaction. Additionally, this patient exhibited a similar drop in WBC count after administration of intravenous furosemide during four previous hospitalizations for worsening dyspnea due to heart failure. The Naranjo Adverse Drug Reaction Probability Scale score was 7, which indicates that this is a probable case of furosemide-induced leucopenia. It was hypothesized that hyperacute leucopenia was associated with pulse doses of intravenous furosemide, a phenomenon not previously reported in the literature. The mechanism of furosemide-induced leucopenia remains unclear, but it warrants further investigation. Clinicians should be aware of the potential for furosemide-induced hyperacute leucopenia and should consider discontinuation of furosemide in patients with an unexplained acute drop in WBC count during furosemide therapy [13A].

Hyperparathyroidism

Secondary hyperparathyroidism is a rare adverse effect that has been observed following long-term use of furosemide. Srivastava and colleagues reported a case of a 6-month-old, 28-week premature infant who developed severe secondary hyperparathyroidism and hypocalcemia while on chronic furosemide therapy for bronchopulmonary dysplasia. After furosemide was discontinued and calcium supplements were initiated, parathyroid hormone levels dropped from 553 to 238 pg/mL. Cinacalcet was added and the parathyroid hormone level decreased further and was maintained within normal limits. This case indicates that clinicians should be aware of the potential for furosemide-induced hyperparathyroidism in patients on chronic furosemide therapy and that a calcimimetic may be successful in the treatment of hyperparathyroidism when initial treatment fails [14A].

Torsemide

Effect on Warfarin Dosage Requirements

Drug interaction databases indicate that torsemide may potentiate the effects of warfarin; however, there are limited and conflicting data regarding this interaction. A similar interaction is not reported for furosemide or bumetanide. Lai and colleagues conducted a retrospective cohort study at the Veteran Affairs Healthcare System in San Diego, California, to evaluate the significance of this drug–drug interaction. Warfarin dosage requirements were evaluated in 18 patients changed from bumetanide to torsemide from March 2014 to July 2014. The mean weekly warfarin dose was similar before and after torsemide initiation (34 ± 15 mg vs 34 ± 13 mg, $P > 0.05$). Only two patients experienced INR elevation during the 3 months after torsemide was started, and they required warfarin dose reductions of 5.3%–18%. This study indicates that warfarin and torsemide do not have a significant interaction in most patients, and a preventative warfarin dosage decrease is not required when starting torsemide in a patient receiving warfarin. However, clinicians should be cognizant that the potential for interaction exists in some patients and increased monitoring is appropriate after torsemide initiation [15c].

THIAZIDE AND THIAZIDE-LIKE DIURETICS [SEDA-37, 239; SEDA-38, 199; SEDA-39, 192]

Hyperuricemia

Hyperuricemia continues to be a prevalent and significant adverse effect associated with the use of thiazide and thiazide-like diuretics. A number of trials have been conducted over the past year, and their results highlight the importance of hyperuricemia with the use of thiazide diuretics. Two randomized, double-blind, placebo-controlled clinical trials utilizing hydrochlorothiazide (HCTZ) and one randomized, double-blind, placebo-controlled clinical trial utilizing chlorthalidone either alone or in combination with other antihypertensives, primarily calcium channel blockers and angiotensin receptor antagonists, showed a dose-related relationship with the incidence of hyperuricemia and thiazide use. Doses of HCTZ $\geq$25 mg/day and chlorthalidone $\geq$12.5 mg/day were associated with hyperuricemia [16MC, 17MC, 18MC, 19M, 20M, 21M]. The clinical significance of thiazide diuretic-induced hyperuricemia is unclear at this time, because available data from prospective, randomized clinical trials goes up to 52 weeks, which is not a suitable amount of time to evaluate the health outcomes of hyperuricemia in a robust manner. An umbrella review published this year evaluating systematic reviews and meta-analyses of the effects of serum uric acid levels on health outcomes found mixed results when analyzing data from observational studies as compared to randomized controlled trials. They report that there were no convincing data to support an association between hyperuricemia and any health outcomes; however, results were highly suggestive of an association between hyperuricemia and nephrolithiasis and gout in randomized controlled trials and observational studies, respectively [22R]. At this point, the totality of evidence suggests that hyperuricemia may be harmful for patients with respect to the development of gout and nephrolithiasis; therefore, it is imperative that clinicians recognize this adverse effect with thiazide diuretics and monitor patients appropriately.

Hyponatremia

Hyponatremia is a common and potentially life threatening adverse effect related to thiazide diuretic use. Major risk factors related to the development of hyponatremia with thiazide use have been identified as increased age, low body mass index, and female sex [23c]. A recently published review has also identified high HDL levels as a risk factor for hyponatremia development with thiazide diuretic therapy, and it has shown a higher degree of association than all other risk factors combined [24R]. This review analyzed data from the participants of the SPRINT study to construct a predictive model of hyponatremia and validated this model with data from The National Health and Nutrition Examination Survey (NHANES) for a total of approximately 26000 patients [25MC, 26S]. When analyzing the association between patient baseline characteristics and the incidence of hyponatremia, HDL levels above 62mg/dL were associated with a 3.7-fold increase in hyponatremia development. This association was not only shown in the SPRINT study group, but also confirmed in the NHANES cohort as well. As far as we know, there are no other reports or data on the association of elevated HDL levels and the incidence of hyponatremia with the use of thiazide diuretics. This association is very peculiar, because elevated HDL levels in females have been previously associated with lower cardiovascular risk and anti-inflammatory and antithrombotic properties among others, yet none of these known effects of HDL can explain the association with hyponatremia [27R]. Regardless of the mechanism, these new data provide compelling evidence that HDL values should be taken into account when choosing antihypertensive therapy and assessing the risk for hyponatremia.

Further efforts to characterize risk factors for hyponatremia with thiazide use have been explored with respect to the phenotypic and genotypic expressions of hospitalized patients in a prospective cohort control study published earlier this year [28C]. Two cohorts of patients over two distinct time periods were evaluated with DNA analysis as well as clinical blood and urine tests to elucidate the physiological mechanisms and possible genetic causes of thiazide-induced hyponatremia (TIH). The first cohort of patients was designed to compare cases of TIH to normonatremic thiazide controls while the second cohort was designed to compare cases of TIH to normonatremic TIH patients now off of thiazides, normonatremic thiazide controls, and normonatremic nonthiazide controls. Their results showed that patients with TIH had an exaggerated increase in free water absorption and expansion of the intravascular compartment as evidenced by significantly lower 24-hour urine volume, osmolarity, and urea excretion. A genome wide association study found that gene SLCO2A1 that codes for a prostaglandin transporter and the SNP rs34550074 coding for variant p.A396T showed association with TIH in the first cohort at $P=0.0005$ (OR$=3.3$). When data were combined across both cohorts, the pooled effect estimate for the association between rs34550074 and severe TIH was OR$=2.13$ ($P=1.70\times10^{-4}$). SLCO2A1 p.A396T was twice as frequent in TIH cases compared with control and general populations, with approximately half of all TIH patients carrying this variant. While there was a statistical association between this genetic variant and TIH it is important to realize that since half of the affected patients in this study did not carry the affected gene and approximately a quarter of unaffected patients did carry the gene, it is clear that there are more determinant factors in the development of TIH.

Osteoporosis and Fracture Risk

Historically, thiazide diuretics have been hailed as the diuretic of choice to prevent osteoporosis and decrease fracture risk by preventing the urinary excretion of calcium. However, contemporary literature has begun to challenge this notion with compelling evidence to suggest that thiazides may actually increase the risk of osteoporosis and fractures. The rationale for this adverse effect is that thiazide-induced hyponatremia may result in bone damage as sodium is a crucial element found extensively in skeletal bone that may play an important role in bone and sodium homeostasis [29R]. A recent retrospective cohort study evaluated the effect of thiazide-induced hyponatremia on incidence of clinical vertebral fracture in approximately 240 female patients over age 70. Osteoporotic vertebral fractures were observed in 42.5% of patients with thiazide-induced hyponatremia and 8.8% of patients without thiazide-induced hyponatremia (OR 7.60; 95% CI 3.755–15.39) [30c]. However, this data should be taken with caution as the authors' findings were based on retrospective records in which they could not account for or possibly identify any confounding factors, and the mean age of patients in the TIH group was statistically higher than the patients in the normonatremia group, thus putting the patients in the TIH group at an increased baseline risk.

A population-based, propensity-matched cohort study published this year exploring the association between thiazide diuretics and risk of hip fracture after stroke in 7470 patients after 2 years found that patients on thiazide diuretics had a lower hazard ratio with respect to hip fractures as compared to controls (HR$=0.64$, 95% CI 0.46–0.89, $P=0.007$) [31c]. In relation to the previous study discussed, these results should be viewed as more robust with no statistical differences between patient groups and a high number of patients evaluated. Unfortunately, similar to the previous study, the data were based on the Taiwan's National Health Insurance Research Database and may be incomplete and/or not inclusive of other patient demographics which limits the external validity of these results. Further prospective randomized studies are necessary to further elucidate the true risk of osteoporosis and fracture with thiazide use.

Cancer Risk

Recent literature has begun to investigate the possible role of thiazide diuretics in increasing the risk of various cancers such as basal and squamous cell carcinomas. The risk of antihypertensive medications causing cancer is not a new trend; however, previous reports detailing the risks of various cancers have been inconclusive and have not made their way into tertiary literature sources. The majority of intriguing new literature with respect to the risk of cancer and thiazide diuretics has been found with the risk of lip cancer. Two recent matched case–control studies out of Denmark that used nationwide prescription, patient, and education registries both showed a significantly increased risk of squamous cell carcinoma in patients receiving thiazide diuretics. Both studies quantified the amount of HCTZ exposure based on available prescription filling data and found that patients with high cumulative doses of $\geq$50000 mg of HCTZ were found to have an odds ratio of developing squamous cell carcinoma ranging from 3.98 to 10.5 with the risk rising with increasing cumulative doses. The major proposed mechanism of this increased risk seems to be the photosensitivity effect of thiazide diuretics and prolonged exposure to these agents causing long-term irreparable damage to squamous cells. Unfortunately, both studies were done in Denmark and an overwhelming majority of patients were fair skinned, thus limiting the applicability of these results to various patient populations and possibly overestimating the risk when taking into account different patient populations. At the very least, these studies highlight that a risk of squamous cell carcinoma may be a significant risk after prolonged use and vigilant monitoring of such cancers should be implored by clinicians for patients at risk [32c, 33c].

Hydrochlorothiazide

Photosensitivity

Photosensitivity reactions with thiazide diuretics are well known and well documented. A recently published database analysis study sought to obtain new information on drug risk comparisons and on drug-induced photosensitivity (DIP) onset profiles and found intriguing results with respect to hydrochlorothiazide. The study analyzed DIP reports from the Japanese Adverse Drug Event Report (JADER) Database which is a spontaneous reporting system (SRS) that has been released by the Pharmaceuticals and Medical Devices Agency (PMDA), a regulatory agency in Japan. Their results showed statistically higher reporting odds ratios (RORs) for HCTZ-induced photosensitivity when combined with either ACEi or ARBs as compared to either agent alone. These data suggest that the risk of DIP with HCTZ may be increased with the addition of ACEi/ARBs despite the fact that when used alone these agents are not known to cause DIP. However, the authors could not comment on a plausible mechanism of this drug interaction. Unfortunately, the data at this stage should be considered hypothesis generating due to a number of limitations. Firstly, the authors report that HCTZ combination products are relatively new in Japan with the first report being generated from 2006. With the promotion of SRSs in the contemporary world, particularly with newly available medicines/formulations there exists a risk of notoriety bias. Additionally, the mechanism of HCTZ DIP has not yet been fully elucidated, thus there may be genetic or environmental factors which may increase the risk particularly for patients living in Japan as compared to other locations. Finally, the reports in the JADER database are not verified and thus drug causality could not be confirmed for any case. At this time, given the widespread use of combination products with HCTZ and ACEi/ARBs further associations should be explored about the risk of DIP with HCTZ when used alone and in combination with other anti-hypertensives [34c].

Lichen Planus

A recent case report describes a case of probable lichen planus due to administration of hydrochlorothiazide (HCTZ) 12.5 mg once daily. Lichen planus is a mucocutaneous inflammatory disease thought to be caused by activation of immune cells, particularly CD4 and CD8 cells, by exogenous compounds. This type of drug reaction to thiazides has been described in the past, although infrequently and often grouped with other drug-induced mucocutaneous reactions. The reports' most compelling point is the temporal relationship that is reported. The patient reported receiving HCTZ 1 year prior to presentation, but this was rapidly discontinued due to severe itching. It was later restarted due to uncontrolled hypertension, and the patient began to develop a rash after 3 days of HCTZ administration. Further evidence of an immune mediated mucocutaneous eruption was highlighted by the presence of eosinophilia and the absence of other lab abnormalities. Application of the Naranjo Adverse Drug Reaction Probability scale revealed a score of 8 which corresponds to a probable likelihood of HCTZ-induced lichen planus. This report highlights that thiazides possess a risk of causing iatrogenic mucocutaneous reactions and should be suspected in those patients where other causes have been excluded [35A].

ALDOSTERONE RECEPTOR ANTAGONISTS [SEDA-37, 240; SEDA-38, 201; SEDA-39, 193]

Spironolactone

Hyperkalemia-Induced Cardiac Arrest

A recently published case report portrays a sudden and unexpected cardiac arrest linked to acute hyperkalemia in a patient receiving spironolactone. The case describes a 74-year-old female with a history of atrial fibrillation, hypertension, pulmonary hypertension, stage 3 chronic kidney diseases, and diabetes presenting to the emergency room for an episode of symptomatic hypoglycemia successfully treated with 25g of intravenous dextrose. The patient was taking the following medications at home: aspirin, carvedilol, insulin, simvastatin, metolazone, potassium chloride, paroxetine, spironolactone, bumetanide, and NPH insulin (55units in the morning and 35units in the evening). On initial evaluation, the patient was found to have a potassium level 5.7mmol/L. At that time no changes were seen on electrocardiography and all other laboratory findings were consistent with baseline values. The patient was treated for asymptomatic hyperkalemia with cessation of potassium supplementation and intravenous hydration. Repeat point-of-care testing revealed hypoglycemia with blood glucose of 52mg/dL and a second infusion of 25g of dextrose was administered. Two and a half hours after infusion, correction of glycemic values prompted patient discharge. However, during transport to her vehicle, the patient was found to be unresponsive and pulseless. Return of spontaneous circulation (ROSC) was achieved minutes after cardiac arrest and repeat blood chemistry analysis revealed a serum potassium level of 6.6mmol/L. All other causes of cardiac arrest were ruled out with extensive workup and the identified cause was acute hyperkalemia resulting from shifting potassium homeostasis [36A]. This case illustrates an important risk of hyperkalemia when considering use of aldosterone antagonists, particularly in patients with insulin resistant diabetes. Shifts in potassium concentrations in vivo are largely controlled by endogenous insulin and aldosterone and have been shown to be affected by administration of exogenous glucose [37c, 38c]. Previous studies have shown that there are linear increases in both insulin and aldosterone levels in healthy individuals in response to hyperkalemia as a physiological mechanism to promote intracellular uptake of extracellular potassium. Intravenous infusion of dextrose has been shown to cause acute hyperkalemia due to a shift of intracellular potassium to the extracellular space. This particular patient had evidence of insulin resistance as well as hypoaldosteronism due to use of spironolactone and thus was unable to mount a physiologic response to the dextrose infusion. This case report highlights the importance of hyperkalemia as a potentially life-threatening adverse effect of aldosterone antagonists and clinicians should consider approaching these multifaceted types of cases with more caution.

Anemia

Anemias are among one of the most common conditions that are encountered for clinicians today and yet there are limited data evaluating the effects of drugs on key endogenous regulators of red blood cell production such as hepcidin. A recent genetic and animal confirmatory study sought to elucidate the effect of various drugs on the genetic expression of hepcidin from human hepatocarcinoma cells and donated human tissue from surgical procedures as well as to evaluate the in vivo effect in mice. The investigators found that spironolactone significantly decreased the expression of hepcidin mRNA in both hepatocarcinoma cells and healthy hepatic cells. After 2 weeks of administration of spironolactone at normal doses to mice, the investigators reported significantly decreased hepcidin mRNA levels, but blood count values did not differ from baseline [39E]. The risk causing or propagating anemia in patients treated with spironolactone is highly intriguing and an adverse effect that has not been identified in the past. The current level of data does not support causal relationship between the use of spironolactone and anemia, but it does offer a hypothesis generating thought that should provoke future randomized controlled trials to evaluate this risk in greater detail.

Intracranial Hypertension

Intracranial hypertension is a rare adverse effect that has not previously been associated with the use of aldosterone receptor antagonists. The following case report describes a 39-year-old female with a PMH of obesity, polycystic ovary syndrome (PCOS), hypothyroidism, hyperlipidemia and idiopathic intracranial hypertension that has not necessitated treatment for the past 13 years. Her medication list includes: levothyroxine 100mcg and atorvastatin 10mg. A chief complaint of androgenetic alopecia prompted therapy with spironolactone 50mg and topical solution of minoxidil. Two months later, the patient presented at the neurology clinic due to blurred vision, dizziness, pulsating tinnitus, and headache. These complaints were observed since the beginning of the treatment and were mild, but then progressively worsened up until presentation at the neurology clinic. The patient was diagnosed with recurrent idiopathic intracranial hypertension. Spironolactone was discontinued, minoxidil was continued, and acetazolamide was initiated. Six weeks post discontinuation of spironolactone all symptoms significantly improved and acetazolamide was stopped. The temporal relationship of the initiation of spironolactone and the presence of symptoms is supportive of an adverse drug event; however, the past medical history of idiopathic intracranial hypertension does cast doubt as to whether spironolactone was really responsible. Thus, this adverse drug event has been categorized as possible with a Naranjo score of 3. Nevertheless, clinicians should be aware of this possible adverse effect and should take caution in prescribing spironolactone to patients with a history of idiopathic intracranial hypertension [40A]. Additional case studies were reviewed by DeLLelis et al. [41R] and Traylor et al. [42R].

References

Carbonic Anhydrase Inhibitors

[1] Ono Y, Morifuso M, Ikeda S, et al. A case of non-cardiogenic pulmonary edema provoked by intravenous acetazolamide. Acute Med Surg. 2017;4:349–52 [A].
[2] Chong S, Park JD, Chae JH, et al. Extensive brain infarction involving deep structures during an acetazolamide-challenged single-photon emission computed tomography scan in a patient with moyamoya disease. Childs Nerv Syst. 2017;33(11):2029–33 [A].
[3] Hári-Kovács A, Soós J, Gyetvai T, et al. Case report on choroidal effusion after oral acetazolamide administration: an unusual manifestation of a well-known idiosyncratic effect? Orv Hetil. 2017;158(50):1998–2002 [A].
[4] Lorenz K, Wasielica-Poslednik J, Bell K, et al. Efficacy and safety of preoperative IOP reduction using a preservative-free fixed combination of dorzolamide/timolol eye drops versus oral acetazolamide and dexamethasone eye drops and assessment of the clinical outcome of trabeculectomy in glaucoma. PLoS One. 2017;12(2)e0171636 [c].

Loop Diuretics

[5] Vozoris NT, Wang X, Austin PC, et al. Incident diuretic drug use and adverse respiratory events among older adults with chronic obstructive pulmonary disease. Br J Clin Pharmacol. 2018;84(3):579–89 [MC].
[6] Paik JM, Rosen H, Gordon CM, et al. Diuretic use and risk of vertebral fracture in women. Am J Med. 2016;129(12):1299–306 [C].
[7] Lai SW, Cheng KC, Lin CL, et al. Furosemide use and acute risk of hip fracture in older people: a nationwide case-control study in Taiwan. Geriatr Gerontol Int. 2017;17:2552–8 [MC].
[8] Okada K, Okada M, Kamada N, et al. Reduction in diuretics and analysis of water and muscle volumes to prevent falls and fall-related fractures in older adults. Geriatr Gerontol Int. 2017;17:262–9 [c].
[9] Silbert BI, Ho KM, Lipman J, et al. Does furosemide increase oxidative stress in acute kidney injury? Antioxid Redox Signal. 2017;26(5):221–6 [c].
[10] Kim GJ, Capparelli E, Romanowski G, et al. Development of tolerance to chronic intermittent furosemide therapy in pediatric patients. J Pediatr Pharmacol Ther. 2017;22(6):394–8 [c].
[11] Ghatak R, Masso L, Kapadia D, et al. Medication as a cause of acute pancreatitis. Am J Case Rep. 2017;18:839–41 [A].
[12] Lybecker MB, Madsen HB, Bruun JM. Severe hypokalaemia as a cause of acute transient quadriparesis. BMJ Case Rep. 2017;1–3 [A].
[13] Ma BJ. Hyperacute leucopenia associated with furosemide. BMJ Case Rep. 2017;1–5 [A].
[14] Srivastava T, Jafri S, Truog WE, et al. Successful reversal of furosemide-induced secondary hyperparathyroidism with cinacalcet. Pediatrics. 2017;140(6):e20163789 [A].
[15] Lai S, Momper JD, Yam FK. Evaluation of the effect of torsemide on warfarin dosage requirements. Int J Clin Pharm. 2017;39(4):831–5 [c].

Thiazide and Thiazide-Like Diuretics

[16] Rakugi H, Shimizu K, Nishiyama Y, et al. A phase III, open-label, multicenter study to evaluate the safety and efficacy of long-term triple combination therapy with azilsartan, amlodipine, and hydrochlorothiazide in patients with essential hypertension. Blood Press. 2018;27(3):125–33 [MC].
[17] Hong SJ, Jeong HS, Han SH, et al. Comparison of fixed-dose combinations of amlodipine/losartan potassium/chlorthalidone and amlodipine/losartan potassium in patients with stage 2 hypertension inadequately controlled with amlodipine/losartan potassium: a randomized, double-blind, multicenter, phase III study. Clin Ther. 2017;39(10):2049–60 [MC].
[18] Higaki J, Komuro I, Shiki K, et al. Effect of hydrochlorothiazide in addition to telmisartan/amlodipine combination for treating hypertensive patients uncontrolled with telmisartan/amlodipine: a randomized, double-blind study. Hypertens Res. 2017;40(3):251–8 [MC].
[19] Bennett A, Chow CK, Chou M, et al. Efficacy and safety of quarter-dose blood pressure-lowering agents: a systematic review and meta-analysis of randomized controlled trials. Hypertension. 2017;70(1):85–93 [M].
[20] Sommerauer C, Kaushik N, Woodham A, et al. Thiazides in the management of hypertension in older adults—a systematic review. BMC Geriatr. 2017;17(Suppl 1):228 [M].
[21] Liang W, Ma H, Cao L, et al. Comparison of thiazide-like diuretics versus thiazide-type diuretics: a meta-analysis. J Cell Mol Med. 2017;21(11):2634–42 [M].
[22] Li X, Meng X, Timofeeva M, et al. Serum uric acid levels and multiple health outcomes: umbrella review of evidence from observational studies, randomised controlled trials, and Mendelian randomisation studies. BMJ. 2017;357:j2376 [R].
[23] Sharabi Y, Illan R, Kamari Y, et al. Diuretic induced hyponatraemia in elderly hypertensive women. J Hum Hypertens. 2002;16(9):631–5 [c].
[24] Israel A, Grossman E. Elevated high-density lipoprotein cholesterol is associated with hyponatremia in hypertensive patients. Am J Med. 2017;130(11):1324.e7–1324.e13 [R].
[25] SPRINT Research Group, Wright Jr. JT, Williamson JD, et al. A randomized trial of intensive versus standard blood-pressure control. N Engl J Med. 2015;373(22):2103–16 [MC].
[26] Centers for Disease Control and Prevention (CDC). National Health and Nutrition Examination Survey (NHANES). n.d. Available at: https://www.cdc.gov/nchs/nhanes/. Accessed March 3, 2018 [S].
[27] Alwaili K, Awan Z, Alshahrani A, et al. High-density lipoproteins and cardiovascular disease: 2010 update. Expert Rev Cardiovasc Ther. 2010;8(3):413–23 [R].
[28] Ware JS, Wain LV, Channavajjhala SK, et al. Phenotypic and pharmacogenetic evaluation of patients with thiazide-induced hyponatremia. J Clin Invest. 2017;127(9):3367–74 [C].
[29] Hannon MJ, Verbalis JG. Sodium homeostasis and bone. Curr Opin Nephrol Hypertens. 2014;23(4):370–6 [R].
[30] De vecchis R, Ariano C, Di biase G, et al. Thiazides and osteoporotic spinal fractures: a suspected linkage investigated by means of a two-center, case-control study. J Clin Med Res. 2017;9(11):943–9 [c].
[31] Lin SM, Yang SH, Cheng HY, et al. Thiazide diuretics and the risk of hip fracture after stroke: a population-based propensity-matched cohort study using Taiwan's National Health Insurance Research Database. BMJ Open. 2017;7(9)e016992 [c].
[32] Pedersen SA, Gaist D, Schmidt SAJ, et al. Hydrochlorothiazide use and risk of nonmelanoma skin cancer: a nationwide case-control study from Denmark. J Am Acad Dermatol. 2018;78(4):673–81 [c].
[33] Pottegård A, Hallas J, Olesen M, et al. Hydrochlorothiazide use is strongly associated with risk of lip cancer. J Intern Med. 2017;282(4):322–31 [c].
[34] Nakao S, Hatahira H, Sasaoka S, et al. Evaluation of drug-induced photosensitivity using the Japanese adverse drug event report (JADER) database. Biol Pharm Bull. 2017;40(12):2158–65 [c].

Aldosterone Receptor Antagonists

[35] Sin B, Miller M, Chew E. Hydrochlorothiazide induced lichen planus in the emergency department. J Pharm Pract. 2017;30(2):266–9 [A].

[36] Offman R, Paden A, Gwizdala A, et al. Hyperkalemia and cardiac arrest associated with glucose replacement in a patient on spironolactone. Am J Emerg Med. 2017;35(8):1214.e1–3 [A].
[37] Himathongkam T, Dluhy RG, Williams GH. Potassium-aldosterone-renin interrelationships. J Clin Endocrinol Metab. 1975;41(1):153–9 [c].
[38] Seldin D, Tarail R. Effect of hypertonic solutions on metabolism and secretion of electrolytes. J Clin Invest. 1949;159:160–74 [c].
[39] Mleczko-sanecka K, Da silva AR, Call D, et al. Imatinib and spironolactone suppress hepcidin expression. Haematologica. 2017;102(7):1173–84 [E].
[40] Albraidi H, Alzuman O, Alajlan A. Possible spironolactone induced intracranial hypertension in a patient with androgenetic alopecia: a case report. Open Dermatol J. 2018;12:1–4 [A].
[41] DeLLelis T, Keshshishyan S, Kovalevskaya V, et al. Antihypertensive. In: Ray SD, editor. Side effects of drugs annual: a worldwide yearly survey of new data in adverse drug reactions, vol. 38. Elsevier; 2016. p. 179–84 [R].
[42] Traylor K, Gurgle H, Turner K, et al. Antihypertensive drugs. In: Ray SD, editor. Side effects of drugs annual: a worldwide yearly survey of new data in adverse drug reactions, vol. 39. Elsevier; 2017. p. 183–7 [R].

CHAPTER

22

Metal Antagonists and Metals

Joshua P. Gray,[1], Natalia Amacher[†], Christina Ford[†], Sidhartha D. Ray[‡]*

*Department of Science, United States Coast Guard Academy, New London, CT, United States
[†]Department of Pharmaceutical Sciences, Manchester University College of Pharmacy, Fort Wayne, IN, United States
[‡]Department of Pharmaceutical and Biomedical Sciences, Touro University College of Pharmacy and School of Osteopathic Medicine, Manhattan, NY, United States
[1]Corresponding author: joshua.p.gray@uscga.edu

AMMONIUM TETRATHIOMOLYBDATE [SEDA-38, 206; SEDA-39, 198]

A review discussed the copper chelator ammonium tetrathiomolybdate which is currently in clinical trials for use in Wilson's disease and for cancer [1R].

Nervous System

Psychiatric disorders ($n=6$), gait disturbance ($n=1$), elevated liver enzymes (aminotransferases) ($n=2$), and decline in neurological functioning ($n=1$) were observed in a clinical trial of *bis*-choline tetrathiomolybdate (WTX101) for 28 patients with early stage Wilson's disease [2c].

Immunologic

Neutropenia was observed in 3.7% of 75 patients undergoing treatment with tetrathiomolybdate for treatment of breast cancer [3c].

CADMIUM [SEDA-39, 198]

Endocrine

Cd exposure increases the risk of diabetes in the Korean, Chinese, and U.S. populations most likely due to Cd-induced apoptosis of the pancreatic beta cell [4R].

Kidney

Cadmium (Cd) is a naturally occurring metal that poses risks to humans via contaminated foods (soil source), tobacco smoke, and polluted air. Women and children exhibit enhanced gastrointestinal and pulmonary absorption rates for Cd which increases their risks for toxicity. Once Cd is absorbed, it is deposited into the kidney. Due to a lack of transporters for excretion, it will accumulate in the renal cortex and cause tubular damage. It is known that Caucasian, Black, and Mexican-American women in the United States with a minimum blood Cd level of 0.4 μg/L had an association with an increased risk and prevalence of hypertension. When Cd is absorbed into the body, it forms a small complex with metallothionein (an inducible protein) that passes through the glomeruli and is reabsorbed by the proximal convoluted tubule. The complex degrades and releases free Cd ions leading to oxidative stress that damages tubular epithelial cells and other proteins. The Na/K-ATPase is oxidized and degraded which alters sodium transport activity. Low levels of 20-hydroxyeicosatetraenoic acid (20-HETE) are produced which lead to increased salt and water uptake into the systemic circulation. Subsequently, this raises blood pressure from chronic fluid and salt retention [4R].

COBALT

Cardiovascular

The use of metal-on-metal (MoM) joint implants has caused toxic serum levels of chromium (Cr) and cobalt

Side Effects of Drugs Annual, Volume 40
ISSN: 0378-6080
https://doi.org/10.1016/bs.seda.2018.08.013

(Co) in some patients. Localized Co accumulation in the body may induce inflammatory responses, tissue lesions, and metallosis while its widespread damage may involve the respiratory system, endocrine system, nervous system, and cardiovascular system. When Cr accumulates, its toxicity may cause immunosuppression and pro-inflammatory responses in its recipient.

In 2003, a case of a 58-year-old woman portrayed an example of Co toxicity [5A]. She received a MoM hip arthroplasty with an uneventful postoperative period. Unfortunately, the positioning of the acetabular component was subject to wear and tear. After 10 years, she was hospitalized with non-ischemic dilated cardiomyopathy and severe biventricular dysfunction. Laboratory values consisted of serum Co: 169 ppb and Cr: 31 ppb, which is above the acceptable ranges of <7 ppb following this procedure. In 2013, she underwent an emergent cardiac transplant and then 2 years later, replaced her MoM hip with a ceramic head alternative. During surgery, her MoM joint was found to be mispositioned leading to metallosis, osteolysis, and granuloma. Over time her condition improved as subsequent serum Cr and Co levels stabilized. To conclude, this case is evidence that Cr and Co toxicities can lead to cardiomyopathy. Although the exact mechanism of Co and Cr cardiotoxicity is unknown, the author states that it is the accumulation of Co in cardiac myocytes that leads to dilated cardiomyopathy. It is believed that histological findings usually result in undifferentiated fibrosis which complicates and prolongs the diagnosis of metal toxicity in patients. Therefore, an awareness of implant toxicity and early intervention could prevent devastating complications for future joint recipients.

A second case of Co toxicity occurred in a 46-year-old male who underwent a bilateral MoM hip arthroplasty [6A]. After 2 years, he experienced shortness of breath on exertion and an enlarged abdomen. He was diagnosed with idiopathic cardiomyopathy with a left ventricular ejection fraction (LVEF) of 20%. He progressively deteriorated over the following 6 years in which he was evaluated for cobaltemia. His serum Co level was 156 ppb and he had a negative cardiac tissue biopsy for cobalt. He received a left ventricular assist device to supplement his failing heart. Despite having a negative biopsy, cobaltemia was the suspected etiology of his cardiac dysfunction. Therefore, the patient underwent a left hip revision. His joint at the femoral head had severe metallosis and osteolysis at the acetabulum and his serum cobalt level was 114 ppb. The metal debris was removed, and bone graft was placed. Two weeks after surgery, the patient's Co level decreased to 27.2 ppb. Next, he underwent revision of his right hip with similar findings. Over time, upon removal of both MoM apparatuses, his hip pain improved, he recovered cardiac function (LVEF: 25%–35%), and his cobalt level decreased to 1.5 ppb. All in all, this case is another correlation of cobalt and its cardiotoxic effects.

DEFERASIROX [SEDA-35, 420; SEDA-36, 323; SEDA-37, 259; SEDA-38, 206; SEDA-39, 199]

Gastrointestinal

A new film-coated tablet formulation of deferasirox was compared with the original dispersible tablet for oral suspension formulation of deferasirox in a randomized, open-label, phase II ECLIPSE study [7c]. The film-coated tablet does not contain lactose or sodium lauryl sulfate which might reduce side effects. Also, the dose is 30% lower in the film-coated tablet form because of its greater bioavailability. Adverse effects were reported for similar proportions of patients, but severe events occurred less frequently in the film-coated tablets (19.5% vs 25.6%). Fewer gastrointestinal adverse effects were observed in a phase II clinical trial comparing the two treatments (NCT02125877) and the patients had higher compliance as measured by pill count (92.9% vs 85.3%). A review also discussed this finding [8R].

Gastrointestinal upset occurred in 1 of 12 patients in a prospective randomized trial comparing phlebotomy with deferasirox (10 mg/kg/day) for the treatment of iron overload in pediatric patients with thalassemia major following curative stem cell transplantation [9c].

Liver

Acute liver failure occurred in a 12-year-old male treated with deferasirox (three tablets of 250 mg/day) for homozygous sickle cell disease [10A]. The patient presented after being found unresponsive by his parents and covered in emesis. The previous day, the patient had presented to his primary care provider complaining of abdominal pain and 1 day of emesis. He had been diagnosed with viral gastroenteritis. Current treatments were three tablets of 250 mg deferasirox taken once daily for 6 years, folic acid 1-mg tablet daily, and penicillin VK 250 mg in 5 mL taken twice daily. Six months prior, the patient experienced intermittent abdominal pain. Laboratory evaluation demonstrated mild transaminitis and analysis for viruses was negative. The patient was hypoglycemic and was treated with several boluses of D50 to achieve blood glucose of 100 mg/dL, and was started on D10 intravenous fluids. Laboratory analysis showed elevated AST and ALT levels. Deferasirox was withheld, and substantial improvement in transaminase, bilirubin, prothrombin time, and ammonia levels occurred within 5 days. Hyperammonemia was treated with lactulose and rifamixin and decreased before discharge. The patient was switched from chronic transfusions to hydroxyurea.

Fulminant hepatitis followed by death occurred in a 43-year-old female 56 days following initiation of deferasirox treatment (20 mg/kg/day orally) [11A]. Transaminases levels increased fourfold and the drug was immediately discontinued. However, AST and ALT levels progressively elevated, and bilirubin levels were also increased. Hepatocellular drug-induced liver injury was classified as moderate (score 2) according to the Drug-Induced Liver Injury Network criteria. 13 days after admission, liver function deteriorated as demonstrated by a prolonged prothrombin time, minor bleeding, and the need of fresh-frozen plasma transfusions. The following day, the patient spiked a fever and was given antibiotics. On day 31, the patient experienced acute renal failure requiring dialysis and multi-organ failure which resulted in cardiac arrest and death.

Urinary Tract

Urinary beta2-microglobulin and serum cystatin-C levels were higher in patients taking deferasirox for transfusion-dependent thalassemia than in patients taking deferiprone in a study of 80 patients (50 with transfusion-dependent thalassemia and 30 controls) [12c]. Patients taking high-dose deferasirox had significantly greater urinary β2-microglobulin than those receiving 15–20 mg/kg or controls. The authors suggest that subclinical renal injury might be present.

Acute proximal tubular dysfunction (Fanconi syndrome) occurred in a 20-year-old man who was accidentally provided a one-time excess dose of 90 mg/kg of deferasirox instead of 20 mg/kg [13A]. The patient experienced nausea and vomiting immediately after ingestion. The patient then experienced signs of kidney dysfunction, including metabolic acidosis with glycosuria, hematuria, urine losses of phosphorous and potassium, and elevated β2-microglobulinuria. The patient was managed by replacement of fluid and electrolyte losses for 14 days. Urine β2-microglobulin levels dropped 100-fold from the initial incident and reached normal by 56 days.

Skin

Skin rash occurred in 1 of 12 patients in a prospective randomized trial comparing phlebotomy with deferasirox (10 mg/kg/day) for the treatment of iron overload in pediatric patients with thalassemia major following curative stem cell transplantation [9c].

Urticarial vasculitis occurred in a 77-year-old female under treatment for diabetes mellitus and coronary artery disease with metformin, furosemide, aldactazide, spironolactone, and hydrochlorothiazide for 3 years [14A]. The patient was also under treatment for myelodysplastic syndrome which was diagnosed 6 months prior with darbepoetin alfa (6 months), thalidomide (3 months), and deferasirox (3 days). Histological analysis of a punch biopsy of the skin lesions demonstrated erythrocyte extravasation, neutrophilic infiltration with eosinophils, and fibrinoid necrosis in the vessel walls of the dermis, indicative of vasculitic drug eruption. Both thalidomide and deferasirox therapies were stopped and treatment with 40 mg/day of methylprednisolone and antihistamines was started. Rash and edema ended within 5 days, and deferasirox was reintroduced after 10 days. The rash reappeared and the therapy was permanently halted.

Immunologic

Desensitization to deferasirox was successfully performed in a 4-year-old patient with beta-thalassemia major [15A]. One week after commencing deferasirox (125 mg dissolved in 100 mL of water), the patient developed widespread erythema multiforme major. She was rechallenged with the same dose 1 month later and developed mouth ulcers with associated swelling. A desensitization program was adopted from a previously published report of adults successfully desensitized to deferasirox.

Susceptibility Factor: Age

A phase II multicenter, single-arm study of deferasirox after hematopoietic stem cell transplantation in children with beta-thalassemia major investigated the safety and efficacy of this drug [16c]. Patients ($n=27$) receiving deferasirox were given 10 mg/kg/day, with increased dose to 20 mg/kg/day. 134 adverse effects were found in 25 patients during the study, with 29.6% experiencing grade 3 AEs. Of these, 10 AEs occurring in 4 patients were attributed to deferasirox, including ALT/AST increase ($n=4$) and urinary tract infection ($n=1$).

Fulminant liver failure occurred in a 3-year-old girl treated with deferasirox for chronic iron overload due to beta thalassemia [17A]. The girl received deferasirox treatment since 23 months of age. She presented to an emergency room with symptoms of rhinorrhea, cough, non-bloody diarrhea, and fever of 103.8°F. Her heart rate was 170 bpm, respiratory rate was 36/min, and blood pressure was 110/70 mmHg. Her ALT and AST were elevated (199 and 263 U/L, respectively). Deferasirox was discontinued on admission. She continued to develop metabolic acidosis, hypokalemia, and worsening hypophosphatemia demonstrating Fanconi syndrome in addition to acute liver failure. She followed with respiratory arrest and cerebellar tonsillar herniation, she died.

DEFERIPRONE [SEDA-35, 422; SEDA-36, 327; SEDA-37, 264; SEDA-38, 207]

Gastrointestinal

A multi-center, retro- and prospective, non-interventional cohort study in six Mediterranean countries investigating the long-term safety of deferiprone for treatment of beta-thalassemia major [18c]. Participants ($n=297$) had been diagnosed with beta-thalassemia major and transfusional iron overload, started deferiprone between age 1 month and 18 years, and had at least one dose of deferiprone. 172 adverse events in 104 patients were considered deferiprone-related. 25 non-splenectomized patients exhibited neutropenia (8.4% of total), and agranulocytosis occurred in two females in the first 12 months which ultimately resolved upon halting the treatment (0.7%). Gastrointestinal disorders had an incident between 0.3% and 3.4%. 31 patients had increased liver enzymes (elevated transaminases in 10.4%). Deferiprone was halted in 23.5% of patients due to adverse effects. More adverse effects subsided in children less than 10 years of age than in older children.

Commonest ADR was arthropathy, including arthralgia and swelling, primarily in the knee joints, which disabled 5 patients with severe symptoms. More than half of the affected nearly 50% of the patients experienced symptoms within the first year of treatment. European patients (3.8%) ($P<0.001$) were less frequently affected compared to North African patients (24.2%). Frequently encountered ADRs included mild rash and urticaria, moderate fatigue, minor weight increases and chromaturia. Conditions such as, hepatic fibrosis, audiological disorders, visual toxicities or renal impairments were not found and were not related to drug exposure. Either temporary DFP interruption or reduction in dose or even without intervention led to diminution of ADRs.

DEFEROXAMINE [SEDA-35, 423; SEDA-37, 265; SEDA-38, 207]

Ear–Nose–Throat

A meta-analysis found that hearing loss was associated in patients treated with deferoxamine for thalassemia [19M]. Sensorineural (10.6%; 95% CI: 5.7–18.8), conductive (14.6%; 95% CI: 10.5–20.6), and mixed hearing loss (9.1%; 95% CI: 5.6–14.6) were observed. However, there was no correlation ($P=0.30$) between hearing loss and average daily dose of deferoxamine.

Liver

An abscess of *Klebsiella pneumoniae* was observed following deferoxamine subcutaneous self-injection of a 47-year-old woman with glucose-6-phosphate deficiency and non-transfusion-dependent thalassemia [20A]. The patient reported to the hospital with abdominal pain to the lower quadrants and fever of 38°C. Two weeks prior she had been treated with oral amoxicillin/clavulanic because of a subcutaneous abscess in the mesogastric abdominal quadrant, the site of an auto-injection with deferoxamine. *K. pneumoniae* was isolated from blood cultures. The patient refused blood transfusion. The patient gradually improved and was discharged 4 weeks after admission with resolution of the leukocytosis. Antibiotics were continued for 11 weeks. Contrast MRI showed that the abscess had resolved.

DIMERCAPTOPROPANESULFONIC ACID

Genetic Factors

Hemolysis developed in two patients with glucose-6-phosphate dehydrogenase deficiency following treatment with sodium dimercaptosulphonate for treatment of Wilson's disease [21A]. The symptoms resolved upon withdrawal of treatment with DMPS.

EDETIC ACID (EDTA) [SEDA-35, 372; SEDA-37, 265]

Kidney

Oxalate nephropathy occurred in a 73-year-old man under treatment with edetic acid (for calcific atherosclerosis), intravenous vitamin C (to promote health), and lithium for the past 10 years [22A]. The patient previously had nonobstructive bilateral nephrolithiasis. Within the past 3 years, had creatinine concentrations between 0.9 and 1.3 mg/dL, corresponding to estimated glomerular filtration rates of 58–88 mL/min/1.73 m^2. One of the patient's daughters had end-stage renal disease due to diabetic nephropathy. The patient presented with an eGFR of 3.6 mg/dL and proteinuria (1.4 g/24 h). Serum creatinine concentrations progressively increased over the next few days necessitating hemodialysis. A kidney biopsy showed extensive intratubular calcium oxalate crystals with 70%–80% acute tubular injury,

10%–20% interstitial fibrosis, and 6 of 44 glomeruli globally sclerosed by light microscopy. The patient was advised to stop all supplements and he remained hemodialysis dependent. The patient was lost to follow-up.

HYDROXYUREA [SEDA-36, 330; SEDA-37, 266; SEDA-38, 208; SEDA-39, 200]

Hydroxyurea is used to treat sickle cell disease by increasing the expression of fetal hemoglobin. Several reviews cover the safety and efficacy [23R].

A meta-analysis of 17 studies showed that hydroxyurea was associated with more occurrences of acute chest syndrome and infections in the hydroxyurea and phlebotomy groups of two studies. In a primary prevention study, seven strokes occurred in the hydroxyurea and phlebotomy group compared with the transfusion and chelation group, causing early termination of that study [24M].

Respiratory

Interstitial pneumonitis occurred in a 69-year-old man with polycythemia vera 2 months after beginning treatment with hydroxyurea [25A]. Diffuse, bilateral, ground-glass opacities, bilateral pleural effusions, septal thickening, and subcentimeter pulmonary nodules were observed. Despite removal of the hydroxyurea, symptoms progressed to acute hypoxic respiratory failure. High dose prednisone caused a reversal of the symptoms, and resolution of the diffuse infiltrates and pulmonary nodules occurred.

Diffuse ground glass opacities, centrilobular low-density nodules, and minimal interstitial reticulation of the subpleural region occurred in an 83-year-old man patient treated for 4 months with hydroxyurea [26A]. Steroid therapy together with halting hydroxyurea treatment reversed the symptoms.

Nervous System

Excessive daytime sleepiness was associated with hydroxyurea treatment [27A].

Hematologic

Macrocytosis occurred in a 10-year-old male neutered Shetland Sheepdog being treated with hydroxyurea for a well-granulated mast cell tumor [28E].

Stroke occurred in a 2-year-old African American boy treated with hydroxyurea for sickle cell anemia [29A].

Skin

A review article discusses the dermatologic consequences of long-term hydroxyurea therapy [30r].

Hyperpigmentation, ichthyosis, plantar keratoderma, dermatomyositis-like eruptions, squamous cell carcinoma, and actinic keratoses occurred in a 67-year-old woman with history of polycythemia vera [31A].

Red and brown macules and traverse hyperpigmented bands occurred on the hands of an 85-year-old woman treated with hydroxyurea. Upon diagnosis of hydroxyurea-induced nail hyperpigmentation, she stopped treatment and the symptoms reversed [32A].

Medium-vessel vasculitis and panniculitis developed in an 80-year-old man treated with hydroxyurea (750 mg/d) for 5 months for polycythemia vera [33A]. Methotrexate successfully reversed the symptoms.

Hair Nails

Twenty-nail Traverse Melanonychia occurred in a 51-year-old female 3 months after starting treatment with hydroxyurea [34A].

Longitudinal melanonychia on multiple nails occurred in a 67-year-old Caucasian woman who had received treatment with hydroxyurea for 3 years for suspected onychomycosis [35A]. After discontinuing treatment for 6 weeks, outgrowth of the bands occurred.

Reproductive System

Oligospermia and azoospermia developed in 20% and 10% of sickle cell disease patients undergoing hydroxyurea therapy [36c]. Removal of hydroxyurea caused a reversal of symptoms.

Immunologic

Transient neutropenia (ANC <1000x106/L) occurred in 2.3% of laboratory monitoring in a study of 230 children treated with hydroxyurea for sickle cell anemia [37c].

Long-Term Effects: Tumorigenicity

Leukemia occurred in a 31-year-old woman who had received haploidentical transplantation of bone marrow cells following hydroxyurea-treated sickle cell disease for the treatment of sickle cell disease [38A].

Secondary malignancies occurred more often in patients with Philadelphia-negative myeloproliferative neoplasms who had been treated with hydroxyurea instead of interferon (OR of 4.01, 95% CI of 1.12–14.27, and P-value 0.023) [39c].

Susceptibility Factors: Age

An experimental study showed that central nervous system development is affected by hydroxyurea in the rat model [40E]. A review discusses the importance of halting hydroxyurea treatment during pregnancy [41R].

D-PENICILLAMINE [SEDA-35, 424; SEDA-36, 330; SEDA-37, 267; SEDA-38, 208; SEDA-39, 203]

Dermal

Elastosis perforans serpiginosa occurred in a 15-year-old Caucasian male treated with penicillamine (20 mg/kg/day) for 3 years [42A]. Cutaneous lesions appeared on the patient's neck and right arm. A biopsy showed granular cellular debris and fragmented elastic fibers surrounded by acanthotic epidermidis, configuring perforating channels. Topical steroid treatment did not resolve the lesions which progressed further. Three months following onset of symptoms, penicillamine was halted and replaced with zinc acetate. However, the cutaneous lesions persisted for 2 years following the halting of treatment with penicillamine.

Elastosis performans serpingosa, self-reported to have been present for 2 years, occurred in a 39-year-old man without any pruritus or pain [43A]. The patient had been on oral penicillamine for the treatment of hepatolenticular degeneration for 20 years. The rash was described as multiple reddish-brown keratinized papules forming irregular serpiginous plaques with central scaling and horns in the anterior and posterior of the cervical region. They also had central mild atrophy and depigmentations. A skin biopsy from the raised border of the plaque showed a transepidermal tunnel containing destroyed elastic fibers through the acanthotic epidermis. Abnormal elastic fibers were found under the reticular dermis. Collagen fibers and elastic fibers were found upon histopathological examination. Zinc was suggested in place of penicillamine.

POLYSTYRENE SULPHONATES AND RELATED DRUGS [SEDA-35, 427; SEDA-36, 333; SEDA-37, 268; SEDA-38, 209; SEDA-39, 203]

Gastrointestinal intolerance occurred in one patient of a retrospective study ($n=26$) investigating the use of low-dose sodium polystyrene sulfonate for long-term management of hyperkalemia in patients with chronic kidney disease [44c].

A review contrasted treatment of hyperkalemia with sodium polystyrene sulfonate, patriomer, and sodium zirconium cyclosilicate [45R]. The authors state that while all agents are associated with adverse GI effects, these occur less frequently with ZS9. Electrolyte abnormalities were associated with patiromer and SPS, whereas urinary tract infections, edema, and corrected QT-interval prolongations occurred with ZS9. A review argues in favor of the use of patriomer and sodium zirconium cyclosilicate in place of sodium polystyrene sulfonate, diuretics, and hemodialysis for the treatment of hyperkalemia [46R]. A review discusses the use of patiromer sorbitex calcium and sodium zirconium cyclosilicate for the treatment of hyperkalemia in patients with heart failure as alternatives to traditional treatments [47R].

A letter to the editor argues that sodium bicarbonate, sodium polystyrene sulfonate, and hemodialysis with low potassium dialysate are safe and useful if administered appropriately [48r].

Endocrine

Intact parathyroid hormone (iPTH) levels were increased following treatment with sodium polystyrene sulfonate for hyperkalemia [49c].

Gastrointestinal

Vancomycin-resistant *Enterococcus faecium* bacteremia occurred in an 80-year-old woman treated with sodium polystyrene sulfonate in sorbitol suspension for hyperkalemia [50A]. The diagnosis was given based on the presence of sodium polystyrene crystals in the colon biopsy.

Constipation occurred in 19 patients (8%, $n=247$) treated with calcium polystyrene sulfonate for hyperkalemia in patients with chronic kidney disease [51c]. Constipation was the most common side effect in patients on diuretic (7.6%) or off diuretic (14.4%) in a study of patients with chronic kidney disease ($n=243$) provided with 8.4 or 16.8 g patiromer daily [52c]. Diarrhea and constipation were the most common side effects in a study of 112 patients treated with patiromer, some of whom had chronic kidney disease (75.9%) or diabetes (82.1%) [53c]. Mild to moderate constipation was observed in 15% of patients of age 65 or older treated with patiromer (8.4 or 16.8 g/d) [54c].

Hematologic

Hypokalemia (31.6%), hypernatremia (26.3%), and hypocalcemia (21.1%) occurred in a retrospective cohort study of pediatric patients ($n=14$) treated with 19 courses of sodium polystyrene sulfonate for the treatment of chronic hyperkalemia secondary to chronic kidney disease [55c].

Physiological Factors

RDX7675 is polystyrene sulfonate with a calcium counter exchange ion in development for use in the treatment of hyperkalemia [56c]. Sodium was replaced with calcium to accommodate chronic kidney disease and/or heart failure. Palatability was greater for RDX7675 when compared with sodium polystyrene sulfonate.

TIOPRONIN [SED-15, 3430; SEDA-35, 373; SEDA-37, 268]

Tiopronin is used to treat cystinuria. It functions by reacting with urinary cystine, forming a disulfide bond which is subsequently excreted. Ethanol metabolism (0.8 g) was not affected by co-administration of tiopronin (500 mg) 1 hour afterwards in a randomized, double-blind, cross-over study of 13 healthy subjects [57c].

TITANIUM

Hair Nails

Yellow nail Syndrome is an uncommon disorder that presents as thickened yellow nails with respiratory problems (chronic cough and recurrent pneumonia) and lymphedema. Although rare, it normally affects those aged 40–60 years of age and is most likely attributed to titanium toxicity. Titanium can be found in dental implants, toothpastes, and can occur from occupational exposures.

A rare pediatric case of yellow nail syndrome presented in a 9-year-old girl. She had a 2-day fever, a cough occurring over 6 months, and non-improving pneumonia despite use of antibiotics a month prior. All 20 of her nails (fingers and toes) were thickened and yellowed for the previous year. Upon CT scan, she had maxillary sinusitis and right middle lobe bronchiectasis. Titanium was detected in her nail clippings by energy dispersive radiograph fluorescence with a level of 11.48 μg/g. This toxicity of titanium was caused by her habit of swallowing children's toothpaste. Her respiratory infection was effectively treated, and she was encouraged to stop swallowing toothpaste. Subsequently, new healthy nails grew, and she remained free from titanium detection 3 years later [58A].

VANADIUM

Vanadium, a transition metal, is proposed to be an essential trace element. Concentration determines whether its biological properties will elicit a therapeutic or toxic response. Vanadium is consumed in black pepper, mushrooms, spinach, and shellfish. Studies have suggested that obtaining 10 μg/day will prevent a biological deficiency. Vanadium deficiency results in adverse changes involving human growth, teeth and bone calcification, and the reproductive system. Prior animal studies have questioned vanadium safety, suggesting carcinogenicity and alveolar damage when inhaled by rats. Although classified as a group 2B carcinogen by the International Agency for Research on Cancer (IARC), there is no data supporting carcinogenicity in humans today.

Vanadium toxicity is known to cause adverse effects on tissues and organs. Those most sensitive include the kidneys and liver. It is thought that vanadium disintegrates cell membranes by inducing oxidative stress, lipid peroxidation, DNA degeneration, and protein denaturation. The valence and chemical form affect its potency in which the pentavalent forms are the most toxic. Animal studies have shown adverse effects on the stomach, liver, intestines, kidneys, and both male and female reproductive tracts. Most human cases of vanadium involve metal toxicity.

Vanadium was evaluated in a cohort study consisting of 7297 pregnant women in China [59c]. Metal urinary concentrations were measured using mass spectrometry, taking into account variable creatinine levels by creating a comparable ratio. The investigators found that high levels of vanadium increased risk for preterm delivery, early delivery, low birth weight, and small for gestational age infants. Exposure was attributed to the combustion of fossil fuels releasing vanadium into the atmosphere. Geographically, the concentration levels varied but the mean vanadium concentration of this study population was 1.73 μg/g creatinine.

Skin

A 62-year-old woman experienced more than 80 pruritic, nummular, eczematous plaques on her arms, legs, back, and buttocks for 3 weeks after wearing metal jewellery [60A]. 6-weeks prior, she had a vanadium plate (containing approximately 90% titanium, 6% aluminum, and 4% vanadium) placed in her foot. Upon evaluation of her skin lesions, a punch biopsy showed lymphoeosinophilic spongiosis, indicating a delayed hypersensitivity reaction. Topical steroid was applied to the lesions with minimal resolution. Next, a general allergen patch test was performed with negative results. In addition to the skin lesions, she continued to experience pain and swelling of her foot surgical site for 8 months. Subsequently, the hardware was removed due to implant failure. 3-weeks after removal, her skin lesions resolved, and another patch test specific for metals was performed, proving her systemic contact dermatitis was a result of the vanadium component of the hardware.

Death

A 24-year-old woman experienced a fatal level of vanadium, 6.22 mg/L (normal 0.07–1.1 mg/L) [61A]. She experienced abdominal pain, diarrhea, nausea, and vomiting. She had severe renal failure (GFR <21 mL/min), as vanadium is renally excreted. It was not explained how this woman encountered such high concentrations of vanadium. For the future, the authors encourage more research to assess the safety of vanadium in human subjects; especially since it is being considered as a potential anti-diabetic agent.

ZINC

A review contrasted the use of chelators (D-penicillamine and trientine) which work by increasing urinary copper excretion with zinc salts, and work by competitively inhibiting copper uptake in the gastrointestinal system [62r]. Another review discussed the lack of efficacy of Wilson's disease treatments in preventing neurological symptoms [63r]. A third review contrasts the usage of zinc vs chelation therapy in pediatric patients, suggesting that presymptomatic patients might be treated with zinc to avoid side effects [64R].

No side effects were observed after 3–24 months in 16 pediatric patients treated with zinc for Wilson's disease [65c].

Additional literature on these topics can be found in these reviews [66R,67R,68R,69R].

References

[1] Rupp C, Stremmel W, Weiss KH. Novel perspectives on Wilson disease treatment. Handb Clin Neurol 2017;142:225–30. https://doi.org/10.1016/B978-0-444-63625-6.00019-7. PubMed PMID: 28433106 [R].

[2] Weiss KH, Askari FK, Czlonkowska A, Ferenci P, Bronstein JM, Bega D, Ala A, Nicholl D, Flint S, Olsson L, Plitz T, Bjartmar C, Schilsky ML. Bis-choline tetrathiomolybdate in patients with Wilson's disease: an open-label, multicentre, phase 2 study. Lancet Gastroenterol Hepatol. 2017;2(12):869–76. https://doi.org/10.1016/S2468-1253(17)30293-5. PubMed PMID: 28988934 [c].

[3] Chan N, Willis A, Kornhauser N, Ward MM, Lee SB, Nackos E, Seo BR, Chuang E, Cigler T, Moore A, Donovan D, Vallee Cobham M, Fitzpatrick V, Schneider S, Wiener A, Guillaume-Abraham J, Aljom E, Zelkowitz R, Warren JD, Lane ME, Fischbach C, Mittal V, Vahdat L. Influencing the tumor microenvironment: a phase II study of copper depletion using Tetrathiomolybdate in patients with breast cancer at high risk for recurrence and in preclinical models of lung metastases. Clin Cancer Res 2017;23(3):666–76. https://doi.org/10.1158/1078-0432.CCR-16-1326. PubMed PMID: 27769988 [c].

[4] Satarug S, Vesey DA, Gobe GC. Kidney cadmium toxicity, diabetes and high blood pressure: the perfect storm. Tohoku J Exp Med 2017;241(1):65–87. https://doi.org/10.1620/tjem.241.65. PubMed PMID: 28132967 [R].

[5] Moniz S, Hodgkinson S, Yates P. Cardiac transplant due to metal toxicity associated with hip arthroplasty. Arthroplast Today 2017; 3(3):151–3. https://doi.org/10.1016/j.artd.2017.01.005. PubMed PMID: 28913397; PMCID: PMC5585818 [A].

[6] Charette RS, Neuwirth AL, Nelson CL. Arthroprosthetic cobaltism associated with cardiomyopathy. Arthroplast Today 2017; 3(4):225–8. https://doi.org/10.1016/j.artd.2016.11.005. PubMed PMID: 29204485; PMCID: PMC5712038 [A].

[7] Taher AT, Origa R, Perrotta S, Kourakli A, Ruffo GB, Kattamis A, Goh AS, Cortoos A, Huang V, Weill M, Merino Herranz R, Porter JB. New film-coated tablet formulation of deferasirox is well tolerated in patients with thalassemia or lower-risk MDS: results of the randomized, phase II ECLIPSE study, Am J Hematol 2017;92(5):420–8. https://doi.org/10.1002/ajh.24668. PubMed PMID: 28142202 [c].

[8] Shah NR. Advances in iron chelation therapy: transitioning to a new oral formulation. Drugs Context 2017;6:212502. https://doi.org/10.7573/dic.212502. PubMed PMID: 28706555; PMCID: PMC5499896 [R].

[9] Inati A, Kahale M, Sbeiti N, Cappellini MD, Taher AT, Koussa S, Nasr TA, Musallam KM, Abbas HA, Porter JB. One-year results from a prospective randomized trial comparing phlebotomy with deferasirox for the treatment of iron overload in pediatric patients with thalassemia major following curative stem cell transplantation. Pediatr Blood Cancer 2017;64(1):188–96. https://doi.org/10.1002/pbc.26213. PubMed PMID: 27576370 [c].

[10] Menaker N, Halligan K, Shur N, Paige J, Hickling M, Nepo A, Weintraub L. Acute liver failure during Deferasirox chelation: a toxicity worth considering. J Pediatr Hematol Oncol 2017;39(3):217–22. https://doi.org/10.1097/MPH.0000000000000786. PubMed PMID: 28221265 [A].

[11] Braga CCB, Benites BD, de Albuquerque DM, Alvarez MC, Seva-Pereira T, Duarte BKL, Costa FF, Gilli SCO, Saad STO. Deferasirox associated with liver failure and death in a sickle cell anemia patient homozygous for the -1774delG polymorphism in the Abcc2 gene. Clin Case Rep 2017;5(8):1218–21. https://doi.org/10.1002/ccr3.1040. PubMed PMID: 28781827; PMCID: PMC5538070 [A].

[12] Annayev A, Karakas Z, Karaman S, Yalciner A, Yilmaz A, Emre S. Glomerular and Tubular functions in children and adults with transfusion-dependent thalassemia. Turk J Haematol 2018;35(1):66–70. https://doi.org/10.4274/tjh.2017.0266. PubMed PMID: 28753129; PMCID: PMC5843777 [c].

[13] Shah L, Powell JL, Zaritsky JJ. A case of Fanconi syndrome due to a deferasirox overdose and a trial of plasmapheresis. J Clin Pharm Ther 2017;42(5):634–7. https://doi.org/10.1111/jcpt.12553. PubMed PMID: 28556939 [A].

[14] Polat AK, Belli AA, Karakus V, Dere Y. Deferasirox-induced urticarial vasculitis in a patient with myelodysplastic syndrome. An Bras Dermatol 2017;92(5 Suppl. 1):59–61. https://doi.org/10.1590/abd1806-4841.20176688. PubMed PMID: 29267448; PMCID: PMC5726679 [A].

[15] Davies GI, Davies D, Charles S, Barnes SL, Bowden D. Successful desensitization to deferasirox in a paediatric patient with beta-thalassaemia major. Pediatr Allergy Immunol 2017; 28(2):199–201. https://doi.org/10.1111/pai.12677. PubMed PMID: 27797415 [A].

[16] Yesilipek MA, Karasu G, Kaya Z, Kuskonmaz BB, Uygun V, Dag I, Ozudogru O, Ertem M. A Phase II, Multicenter, single-arm study to

evaluate the safety and efficacy of Deferasirox after hematopoietic stem cell transplantation in children with beta-thalassemia major. Biol Blood Marrow Transplant 2018;24(3):613–8. https://doi.org/10.1016/j.bbmt.2017.11.006. PubMed PMID: 29155313 [c].

[17] Ramaswami A, Rosen DJ, Chu J, Wistinghausen B, Arnon R. Fulminant liver failure in a child with beta-thalassemia on Deferasirox: a case report. J Pediatr Hematol Oncol 2017;39 (3):235–7. https://doi.org/10.1097/MPH.0000000000000654. PubMed PMID: 27479018 [A].

[18] Botzenhardt S, Felisi M, Bonifazi D, Del Vecchio GC, Putti MC, Kattamis A, Ceci A, Wong ICK, Neubert A, DEEP consortium (collaborative group). Long-term safety of deferiprone treatment in children from the Mediterranean region with beta-thalassemia major: the DEEP-3 multi-center observational safety study. Haematologica 2018;103(1):e1–e4. https://doi.org/10.3324/haematol.2017.176065. PubMed PMID: 29079595; PMCID: PMC5777195 [c].

[19] Badfar G, Mansouri A, Shohani M, Karimi H, Khalighi Z, Rahmati S, Delpisheh A, Veisani Y, Soleymani A, Azami M. Hearing loss in Iranian thalassemia major patients treated with deferoxamine: a systematic review and meta-analysis. Caspian J Intern Med 2017; 8(4):239–49. https://doi.org/10.22088/cjim.8.4.239. PubMed PMID: 29201313; PMCID: PMC5686301 [M].

[20] Furlan L, Graziadei G, Colombo G, Forzenigo LV, Solbiati M. K. Pneumoniae liver abscess following deferoxamine subcutaneous self-injection. Am J Hematol 2017;92(5):480–1. https://doi.org/10.1002/ajh.24675. PubMed PMID: 28188653 [A].

[21] Lai S, Huang YQ, Liu AQ, Wu HW. Haemolysis during sodium dimercaptosulphonate therapy for Wilson's disease in G6PD-deficient patients: first report of two cases. J Clin Pharm Ther 2017;42(6):783–5. https://doi.org/10.1111/jcpt.12576. PubMed PMID: 28635014 [A].

[22] Marques S, Santos S, Fremin K, Fogo AB. A case of oxalate nephropathy: when a single cause is not crystal clear. Am J Kidney Dis 2017;70(5):722–24. https://doi.org/10.1053/j.ajkd.2017.05.022. PubMed PMID: 28739328 [A].

[23] Cannas G, Poutrel S, Thomas X. Hydroxycarbamine: from an old drug used in malignant Hemopathies to a current standard in sickle cell disease. Mediterr J Hematol Infect Dis. 2017; 9(1):e2017015. https://doi.org/10.4084/MJHID.2017.015. PubMed PMID: 28293403; PMCID: PMC5333733 [R].

[24] Nevitt SJ, Jones AP, Howard J. Hydroxyurea (hydroxycarbamide) for sickle cell disease. Cochrane Database Syst Rev 2017;4: CD002202. https://doi.org/10.1002/14651858.CD002202.pub2. PubMed PMID: 28426137 [M].

[25] Kamal P, Imran M, Irum A, Latham H, Magadan J, 3rd. Hydroxyurea-induced interstitial pneumonitis: a rare clinical entity. Kans J Med 2017;10(2):47–9. PubMed PMID: 29472968; PMCID: PMC5733416 [A].

[26] Bargagli E, Palazzi M, Perri F, Torricelli E, Rosi E, Bindi A, Pistolesi M, Voltolini L. Fibrotic lung toxicity induced by Hydroxycarbamide. In Vivo 2017;31(6):1221–3. https://doi.org/10.21873/invivo.11194. PubMed PMID: 29102950; PMCID: PMC5756656 [A].

[27] Revol B, Joyeux-Faure M, Albahary MV, Gressin R, Mallaret M, Pepin JL, Launois SH. Severe excessive daytime sleepiness induced by hydroxyurea. Fundam Clin Pharmacol 2017;31(3):367–8. https://doi.org/10.1111/fcp.12260. PubMed PMID: 27998000 [A].

[28] Conrado FO, Weeden AL, Speas AL, Leissinger MK. Macrocytosis secondary to hydroxyurea therapy. Vet Clin Pathol 2017;46(3): 451–6. https://doi.org/10.1111/vcp.12511. PubMed PMID: 28582589 [E].

[29] Fridlyand D, Wilder C, Clay ELJ, Gilbert B, Pace BS. Stroke in a child with hemoglobin SC disease: a case report describing use of hydroxyurea after transfusion therapy. Pediatr Rep 2017;9(1):6984. https://doi.org/10.4081/pr.2017.6984. PubMed PMID: 28435652; PMCID: PMC5379224 [A].

[30] Mokni S, Fetoui Ghariani N, Aounallah A, Fathallah N, Boussofara L, Saidi W, Ben Salem C, Sriha B, Belajouza C, Denguezli M, Ghariani N, Nouira R. [Dermatologic complications of long-term hydroxyurea therapy]. Therapie 2017;72(3):391–4. https://doi.org/10.1016/j.therap.2016.05.009. PubMed PMID: 27912970 [r].

[31] Neill B, Ryser T, Neill J, Aires D, Rajpara A. A patient case highlighting the myriad of cutaneous adverse effects of prolonged use of hydroxyurea. Dermatol Online J 2017;23(11). Retrieved from https://escholarship.org/uc/item/8h64503t. PubMed PMID: 29447639 [A].

[32] Schoenfeld J, Tulbert BH, Cusack CA. Transverse melanonychia and palmar hyperpigmentation secondary to hydroxyurea therapy. Cutis. 2017;99(5):E2–E4. PubMed PMID: 28632807 [A].

[33] Mattessich S, Ferenczi K, Lu J. Successful treatment of hydroxyurea-associated panniculitis and vasculitis with low-dose methotrexate. JAAD Case Rep 2017;3(5):422–4. https://doi.org/10.1016/j.jdcr.2017.06.009. PubMed PMID: 28932785; PMCID: PMC5594232 [A].

[34] Osemwota O, Uhlemann J, Rubin A. Twenty-nail transverse Melanonychia induced by hydroxyurea: case report and review of the literature. J Drugs Dermatol 2017;16(8):814–5. PubMed PMID: 28809997 [A].

[35] Nguyen AL, Körver JE, Theunissen CCW. Longitudinal melanonychia on multiple nails induced by hydroxyurea. BMJ Case Rep 2017;2017. https://doi.org/10.1136/bcr-2016-218644. PubMed PMID: 28202485 [A].

[36] Sahoo LK, Kullu BK, Patel S, Patel NK, Rout P, Purohit P, Meher S. Study of seminal fluid parameters and fertility of male sickle cell disease patients and potential impact of hydroxyurea treatment. J Assoc Physicians India 2017;65(6):22–5. PubMed PMID: 28782309 [c].

[37] Estepp JH, Smeltzer MP, Kang G, Li C, Wang WC, Abrams C, Aygun B, Ware RE, Nottage K, Hankins JS. A clinically meaningful fetal hemoglobin threshold for children with sickle cell anemia during hydroxyurea therapy. Am J Hematol 2017;92(12):1333–9. https://doi.org/10.1002/ajh.24906. PubMed PMID: 28913922; PMCID: PMC5675769 [c].

[38] Janakiram M, Verma A, Wang Y, Budhathoki A, Suarez Londono J, Murakhovskaya I, Braunschweig I, Minniti CP. Accelerated leukemic transformation after haplo-identical transplantation for hydroxyurea-treated sickle cell disease. Leuk Lymphoma 2018; 59(1):241–4. https://doi.org/10.1080/10428194.2017.1324158. PubMed PMID: 28587497 [A].

[39] Khadartsev AA, Morgunova IN. Various parameters of the mechanics of respiration during expiratory stenosis of the trachea and bronchi. Ter Arkh. 1989;61(3):76–7. PubMed PMID: 2741124 [c].

[40] Marti J, Molina V, Santa-Cruz MC, Hervas JP. Developmental injury to the cerebellar cortex following hydroxyurea treatment in early postnatal life: an Immunohistochemical and Electron Microscopic Study. Neurotox Res 2017;31(2):187–203. https://doi.org/10.1007/s12640-016-9666-9. PubMed PMID: 27601242 [E].

[41] Ghafuri DL, Stimpson SJ, Day ME, James A, DeBaun MR, Sharma D. Fertility challenges for women with sickle cell disease. Expert Rev Hematol 2017;10(10):891–901. https://doi.org/10.1080/17474086.2017.1367279. PubMed PMID: 28891355 [R].

[42] Ranucci G, Di Dato F, Leone F, Vajro P, Spagnuolo MI, Iorio R. Penicillamine-induced elastosis perforans serpiginosa in Wilson disease: is useful switching to zinc? J Pediatr Gastroenterol Nutr 2017;64(3):e72-e3. https://doi.org/10.1097/MPG.0000000000000613. PubMed PMID: 25341025 [A].

[43] Yao XY, Wen GD, Zhou C, Liu BY, Du J, Chen Z, Zhang JZ. D-Penicillamine-induced elastosis perforans serpiginosa. Chin Med J 2017;130(16):2013–4. https://doi.org/10.4103/0366-6999.211899. PubMed PMID: 28776563; PMCID: PMC5555145 [A].

[44] Georgianos PI, Liampas I, Kyriakou A, Vaios V, Raptis V, Savvidis N, Sioulis A, Liakopoulos V, Balaskas EV, Zebekakis PE.

Evaluation of the tolerability and efficacy of sodium polystyrene sulfonate for long-term management of hyperkalemia in patients with chronic kidney disease. Int Urol Nephrol 2017;49(12):2217–21. https://doi.org/10.1007/s11255-017-1717-5. PubMed PMID: 29027620 [c].

[45] Beccari MV, Meaney CJ. Clinical utility of patiromer, sodium zirconium cyclosilicate, and sodium polystyrene sulfonate for the treatment of hyperkalemia: an evidence-based review. Core Evid 2017;12:11–24. https://doi.org/10.2147/CE.S129555. PubMed PMID: 28356904; PMCID: PMC5367739 [R].

[46] Pham AQ, Sexton J, Wimer D, Rana I, Nguyen T. Managing Hyperkalemia: Stepping into a new frontier. J Pharm Pract 2017; 30(5):557–61. https://doi.org/10.1177/0897190016665540. PubMed PMID: 27609505 [R].

[47] Sarwar CMS, Bhagat AA, Anker SD, Butler J. Role of hyperkalemia in heart failure and the therapeutic use of potassium binders. Handb Exp Pharmacol 2017;243:537–60. https://doi.org/10.1007/164_2017_25. PubMed PMID: 28382468 [R].

[48] Abuelo JG. Treatment of severe hyperkalemia: confronting 4 fallacies. Kidney Int Rep 2018;3(1):47–55. https://doi.org/10.1016/j.ekir.2017.10.001. PubMed PMID: 29340313; PMCID: PMC5762976 [r].

[49] Nakayama Y, Ueda K, Yamagishi SI, Sugiyama M, Yoshida C, Kurokawa Y, Nakamura N, Moriyama T, Kodama G, Minezaki T, Ito S, Nagata A, Taguchi K, Yano J, Kaida Y, Shibatomi K, Fukami K. Compared effects of calcium and sodium polystyrene sulfonate on mineral and bone metabolism and volume overload in pre-dialysis patients with hyperkalemia. Clin Exp Nephrol 2018;22 (1):35–44. https://doi.org/10.1007/s10157-017-1412-y. PubMed PMID: 28421299 [c].

[50] Cerrud-Rodriguez RC, Alcaraz-Alvarez D, Chiong BB, Ahmed A. Vancomycin-resistant *Enterococcus faecium* bacteraemia as a complication of Kayexalate (sodium polystyrene sulfonate, SPS) in sorbitol-induced ischaemic colitis. BMJ Case Rep 2017;2017. https://doi.org/10.1136/bcr-2017-221790. PubMed PMID: 29127125 [A].

[51] Yu MY, Yeo JH, Park JS, Lee CH, Kim GH. Long-term efficacy of oral calcium polystyrene sulfonate for hyperkalemia in CKD patients. PLoS One 2017;12(3):e0173542. https://doi.org/10.1371/journal.pone.0173542. PubMed PMID: 28328954; PMCID: PMC5362098 [c].

[52] Weir MR, Mayo MR, Garza D, Arthur SA, Berman L, Bushinsky D, Wilson DJ, Epstein M. Effectiveness of patiromer in the treatment of hyperkalemia in chronic kidney disease patients with hypertension on diuretics. J Hypertens 2017;35 Suppl. 1:S57–S63. https://doi.org/10.1097/HJH.0000000000001278. PubMed PMID: 28129247; PMCID: PMC5377986 [c].

[53] Pergola PE, Spiegel DM, Warren S, Yuan J, Weir MR. Patiromer lowers serum potassium when taken without food: comparison to dosing with food from an open-label, randomized, parallel group hyperkalemia study. Am J Nephrol 2017;46(4):323–32. https://doi.org/10.1159/000481270. PubMed PMID: 29017162; PMCID: PMC5804834 [c].

[54] Weir MR, Bushinsky DA, Benton WW, Woods SD, Mayo MR, Arthur SP, Pitt B, Bakris GL. Effect of Patiromer on hyperkalemia recurrence in older chronic kidney disease patients taking RAAS inhibitors. Am J Med 2018;131(5):555–64e3. https://doi.org/10.1016/j.amjmed.2017.11.011. PubMed PMID: 29180023 [c].

[55] Le Palma K, Pavlick ER, Copelovitch L. Pretreatment of enteral nutrition with sodium polystyrene sulfonate: effective, but beware the high prevalence of electrolyte derangements in clinical practice. Clin Kidney J 2018;11(2):166–71. https://doi.org/10.1093/ckj/sfx138. PubMed PMID: 29644055; PMCID: PMC5887418 [c].

[56] Zann V, McDermott J, Jacobs JW, Davidson JP, Lin F, Korner P, Blanks RC, Rosenbaum DP. Palatability and physical properties of potassium-binding resin RDX7675: comparison with sodium polystyrene sulfonate. Drug Des Devel Ther 2017;11:2663–73. https://doi.org/10.2147/DDDT.S143461. PubMed PMID: 28919716; PMCID: PMC5593397 [c].

[57] Nass F, Schneider B, Wilm S, Kardel B, Gabor E, Merges F, Kroll M. Influence of Tiopronin on the metabolism of alcohol in healthy subjects. Drug Res (Stuttg) 2017;67(4):204–10. https://doi.org/10.1055/s-0042-123826. PubMed PMID: 28142160 [c].

[58] Hsu TY, Lin CC, Lee MD, Chang BP, Tsai JD. Titanium dioxide in toothpaste causing yellow nail syndrome. Pediatrics 2017;139(1). https://doi.org/10.1542/peds.2016-0546. PubMed PMID: 27940507 [A].

[59] Hu J, Xia W, Pan X, Zheng T, Zhang B, Zhou A, Buka SL, Bassig BA, Liu W, Wu C, Peng Y, Li J, Zhang C, Liu H, Jiang M, Wang Y, Zhang J, Huang Z, Zheng D, Shi K, Qian Z, Li Y, Xu S. Association of adverse birth outcomes with prenatal exposure to vanadium: a population-based cohort study. Lancet Planet Health 2017;1(6): e230–e41. https://doi.org/10.1016/S2542-5196(17)30094-3. PubMed PMID: 29851608 [c].

[60] Engelhart S, Segal RJ. Allergic reaction to vanadium causes a diffuse eczematous eruption and titanium alloy orthopedic implant failure. Cutis 2017;99(4):245–9. PubMed PMID: 28492599 [A].

[61] Wilk A, Szypulska-Koziarska D, Wiszniewska B. The toxicity of vanadium on gastrointestinal, urinary and reproductive system, and its influence on fertility and fetuses malformations. Postepy Hig Med Dosw (Online) 2017;71(0):850–9. https://doi.org/10.5604/01.3001.0010.4783. PubMed PMID: 29039350 [A].

[62] Czlonkowska A, Litwin T. Wilson disease—currently used anticopper therapy. Handb Clin Neurol 2017;142:181–91. https://doi.org/10.1016/B978-0-444-63625-6.00015-X. PubMed PMID: 28433101 [r].

[63] Litwin T, Dusek P, Czlonkowska A. Symptomatic treatment of neurologic symptoms in Wilson disease. Handb Clin Neurol 2017;142:211–23. https://doi.org/10.1016/B978-0-444-63625-6.00018-5. PubMed PMID: 28433105 [r].

[64] Roberts EA, Socha P. Wilson disease in children. Handb Clin Neurol 2017;142:141–56. https://doi.org/10.1016/B978-0-444-63625-6.00012-4. PubMed PMID: 28433098 [R].

[65] Wiernicka A, Dadalski M, Janczyk W, Kaminska D, Naorniakowska M, Husing-Kabar A, Schmidt H, Socha P. Early onset of Wilson disease: diagnostic challenges. J Pediatr Gastroenterol Nutr 2017;65(5):555–60. https://doi.org/10.1097/MPG.0000000000001700. PubMed PMID: 28753182 [c].

[66] Gray JP, Ray SD. Metal antagonists. In: Ray SD, editor. Side Effects of Drugs Annual: A worldwide yearly survey of new data in adverse drug reactions. 37:Elsevier; 2015. p. 259–72 Chapter 23. [R].

[67] Roller LK, Baumgartner LJ. Metals. In: Ray SD, editor. Side Effects of Drugs Annual: A worldwide yearly survey of new data in adverse drug reactions. 38:Elsevier; 2016. p. 193–204 Chapter 20. [R].

[68] Gray JP, Ray SD. Metal Antagonists. In: Ray SD, editor. Side Effects of Drugs Annual: A worldwide yearly survey of new data in adverse drug reactions. 38:Elsevier; 2016. p. 205–10 Chapter 21. [R].

[69] Gray JP, Suhali-Amacher N, Ray SD. Metals and Metal Antagonists. In: Ray SD, editor. Side Effects of Drugs Annual: A worldwide yearly survey of new data in adverse drug reactions. 39:Elsevier; 2017. p. 197–208 Chapter 19. [R].

CHAPTER

23

Antiseptic Drugs and Disinfectants

Dirk W. Lachenmeier[1]

Chemisches und Veterinäruntersuchungsamt (CVUA) Karlsruhe, Karlsruhe, Germany

[1]Corresponding author: lachenmeier@web.de

ALL COMMONLY USED ANTISEPTICS AND DISINFECTANTS [*SEDA-37*, 273]

Respiratory

Longitudinal data from a cohort of 1695 young people (19–24 years) living in two major German cities were used to analyse the effects of domestic use of disinfectants on asthma. Compared with no use, high use of disinfectants was associated with a more than twofold increased risk of asthma (OR 2.79, 95% CI 1.14–6.93). The study size did not allow to examine specific effects of single agents. While the mechanism and causality remain questionable, the authors judged that the data support the hypothesis for an elevated risk of respiratory disease suggesting further studies with more detailed exposure assessment and mechanistic studies [1MC].

Immunologic

The comparative irritant and allergenic properties of commonly used antiseptics were reviewed with the observation that all compounds may induce cutaneous side effects. However, allergic contact dermatitis is uncommon but often misdiagnosed by practitioners, who confuse allergy and irritation. The irritant properties are mainly observed when antiseptics are misused, i.e. on an eczematous skin, under inadequate occlusion or at too high concentrations [2R].

Fertility

Women currently employed outside the home and trying to get pregnant ($n=1739$) in the Nurses' Health Study 3 cohort (2010–2014) were examined regarding occupational exposure to high-level disinfectants (HLD) of any type. Nurses exposed to HLD prior to baseline had a 26% (95% CI 8%–47%) longer median duration of pregnancy attempt compared to nurses who were never exposed. The use of protective equipment attenuated associations with fecundity impairments. The possibility that these associations are chance findings could not be ruled out; however, future research on mechanism was suggested [3C].

ALDEHYDES [*SED-15, 1439, 1513; SEDA-31, 409; SEDA-32, 437; SEDA-33, 479; SEDA-34, 377; SEDA-36, 339; SEDA-37, 273; SEDA-38, 211; SEDA-39, 209*]

Considering all disinfectants, aldehydes have a special status as they are able to pose occupational hazards even at very low concentrations in air (*SEDA-36, 339; SEDA-37, 273*).

Formaldehyde

Based on a review of studies published since 2013, Nielsen et al. [4R] have re-evaluated the World Health Organization (WHO) indoor air quality guideline for short- and long-term exposures to formaldehyde (FA) of 0.1 mg/m^3 (0.08 ppm) for all 30-min periods at lifelong exposure. The credibility of the guideline has not been challenged by the recent studies and the level is still judged as highly precautionary [4R].

Tumorigenicity

According to an assessment of the International Agency for Research on Cancer (IARC), formaldehyde was confirmed as carcinogenic to humans (Group 1). Formaldehyde causes cancer of the nasopharynx and leukemia [5S]. The IARC assessment and specifically the epidemiologic evidence on the association between

Side Effects of Drugs Annual, Volume 40
ISSN: 0378-6080
https://doi.org/10.1016/bs.seda.2018.06.005

formaldehyde exposure and risk of leukemia and other lymphohematopoietic malignancies have been previously discussed controversially [SEDA-36, 339; SEDA-37, 273]. An industry sponsored re-analysis [6R] of data on occupational exposure from China [7c] questions the original conclusion that formaldehyde may induce damage to hematopoietic cells that originate in the bone marrow. The authors suggest that the study needs to be replicated using a new study population, measured formaldehyde exposures, and validated laboratory tests [6R].

Immunologic

In a retrospective analysis of patch test data (1998–2014) of the North American Contact Dermatitis group, 132 patch-tested print machine operators were evaluated. Of these 4 patients (8%) tested positive for formaldehyde, which was among the most frequent allergens in this occupation [8c]. In a similar study in Denmark (2009–2013), positive patch tests for formaldehyde occurred in 137 out of 4485 (3.1%) patients diagnosed with non-occupational contact dermatitis and 44 out of 995 (4.4%) patients diagnosed with occupational contact dermatitis. Formaldehyde was the preservative most frequently registered in the Danish product register (894 products registered in 2014) [9C]. A multicenter study from 12 European countries (2009–2012) found 0.94% positive reactions to formaldehyde in 48676 patch tested patients [10MC]. A case of a 50-year-old man was presented who developed recurrent generalized urticaria after endodontic treatment using a *para*-formaldehyde-containing root canal sealant. Based on elevated formaldehyde-specific serum IgE, a type I hypersensitivity reaction was confirmed [11A]. Animal experiments in mice using inhalatory exposure (0.5 and 3.0 mg/m^3 for 6 h/day over 25 consecutive days) showed that formaldehyde induces Th2-type allergic responses in non-sensitized BALB/c and C57BL/6 mice [12E].

Nervous System

In a nested case–control study in Sweden, patients diagnosed with amyotrophic lateral sclerosis (ALS) ($n=5020$) were matched to five controls per case ($n=25100$). Among individuals younger than 65 years, an association with a higher risk of ALS was found for formaldehyde exposure (OR 1.20, 95% CI 1.00–1.65). Further studies are needed to confirm this finding and to clarify the potential mechanism [13C]. Animal experiments in mice exposed daily to formaldehyde (15.5 mg/kg day) for 1 week showed that early Alzheimer's disease-like changes (cognitive deficits, pathological alterations in the brain, accumulation of total β-amyloid plaques and hyper-phosphorylated tau in the cerebral cortex) were detected [14E]. The same group detected synergistic effects of co-exposure of formaldehyde with air particulate matter on the Alzheimer's disease-like changes in mice [15E].

Fetotoxicity

A case–control study conducted in 118 women in China with a diagnosed miscarriage at the first trimester and 191 healthy women who delivered at term was reported. The plasma levels of formaldehyde were significantly higher in women with miscarriage ($P<0.001$). Higher level of formaldehyde was significantly associated with risk of miscarriage (OR 8.06, 95% CI 4.96–13.09) [16c]. Animal experiments in mice confirmed that formaldehyde exposure during pregnancy may show significant toxicity in offspring. Formaldehyde induced oxidative stress and apoptosis of cardiomyocytes in both maternal mice and their offspring [17E].

Glutaraldehyde (Glutaral)

Cardiovascular

Allografts used for arch reconstruction may be treated with glutaraldehyde, but its use may cause tissue changes that predispose to recurrent obstruction. In a study of 206 infants who underwent Norwood procedure between 2000 and 2015, it was found that glutaraldehyde treatment (1% solution, 10 min) was associated with specific need for surgical reintervention (hazard ratio (HR) 4.05, 95% CI 1.19–13.77). The authors suggest that the advantages of decreased sensitization with glutaraldehyde treatment need to be balanced against the risk of aortic reobstruction [18c].

Immunologic

Glutaraldehyde was removed from the list of the American contact dermatitis society core allergen series because it was not felt to be a high-yield allergen [19S].

GUANIDINES

Chlorhexidine *[SED-15, 714; SEDA-31, 410; SEDA-32, 439; SEDA-33, 480; SEDA-34, 378; SEDA-36, 340; SEDA-37, 273; SEDA-38, 212; SEDA-39, 210]*

Drug Formulations

Chlorhexidine is used extensively in oral hygiene but can cause staining of the teeth and oral mucosa and adversely affect taste but rarely causes pain *[SEDA-30, 278; SEDA-31, 416; SEDA-34, 378; SEDA-36, 340; SEDA-37, 273; SEDA-38, 212; SEDA-39, 210]*. A Cochrane systematic review regarding chlorhexidine mouthrinses as an adjunctive treatment for gingival health identified a large increase in extrinsic tooth staining in participants using the mouthrinse at 4–6 weeks (8 trials, 415 participants analyzed, risk ratio 5.41, 95% CI 2.03–14.47). There was also a large increase in extrinsic tooth staining in participants using chlorhexidine mouthrinse at 7–12 weeks and 6 months. The other adverse effects most commonly reported were taste disturbance/alteration (reported in 11 studies), effects on the oral mucosa including soreness, irritation, mild desquamation and mucosal ulceration/erosions (reported in 13 studies) and a general burning sensation or a burning tongue or both (reported in 9 studies) [20M]. In a commentary regarding the Cochrane systematic review, the results were interpreted in a fashion that longer-term use of chlorhexidine needs to be carefully balanced with these recognized adverse effects [21r]. Teeth and tongue staining and taste disturbance were also reported as common side effects in a meta-analysis of clinical trials (no incidence data or statistical evaluations provided) [22R]. Some further trials also confirmed these effects. In a randomized clinical trial comparing the effects of oral rinse with 0.2% and 2% chlorhexidine on oropharyngeal colonization and ventilator associated pneumonia ($n = 114$), 2 patients developed tooth discoloration in the 2% chlorhexidine group (3.5%) and 1 patient in each group developed oral mucosa irritation (each 1.8%). The difference between the groups was not statistically significant ($P = 0.361$) [23c]. In another randomized clinical trial ($n = 59$) comparing 0.06%, 0.12% and 0.2% chlorhexidine for dental plaque prevention, statistical significant differences were observed with "loss of taste" and "numb feeling", the incidence for the effects was 65%–60% (0.2% group), 55%–40% (0.12% group) and 21%–26% (0.06% group) [24c]. Burning sensation in the mouth, bitter taste after rinsing and taste disturbance were reported as most frequent adverse events in a clinical trial ($n = 35$) comparing chlorhexidine 0.12% mouthrinse with and without alcohol. The frequency of adverse events was significantly higher ($P = 0.013$) when chlorhexidine was combined with alcohol [25c].

Skin

A review regarding the use of chlorhexidine for skin antisepsis in newborn infants suggested skin reactions as common side effect, including erythema and erosions [26r]. Five cases of chlorhexidine-related chemical burns occurring within 2 days of live in preterm infants confirmed the potential for adverse effects on the skin. The authors suggest to use the lowest effective chlorhexidine concentration, preferably $< 1\%$, for skin antisepsis of extremely low birth weight and very low birth weight infants in their first week of life, when their epidermal barrier is incomplete [27A]. Patients undergoing ventral hernia repair ($n = 3924$) were separated into two groups, either receiving preoperative chlorhexidine scrub or not. Unlike expectations, the preoperative chlorhexidine scrub group had higher rates of surgical site occurrence (OR 1.34, 95% CI 1.11–1.61) and surgical site infection (OR 1.46, 95% CI 1.03–2.07). The authors judged that the chlorhexidine scrub increases the risk of 30-day wound morbidity in patients undergoing ventral hernia repair and suggested that the use of preoperative chlorhexidine might not be desirable for all surgical populations [28C].

Immunologic

An observational study about 104 patients attending specialist allergy clinics in the UK following perioperative hypersensitivity reactions to chlorhexidine was described. Most of the cases were life-threatening. Specific IgE and skin prick tests were found as reasonable first-line tests for chlorhexidine allergy, but false negativity around 7% was detected. The authors concluded that awareness of the potential allergenicity of chlorhexidine should be part of the training of all health-care professionals [29c]. A case report about a 77-year-old female patient suffering from severe anaphylaxis due to polyhexanide was presented. Allergologic evaluation yielded both specific IgE antibodies to chlorhexidine and to polyhexanide. The authors assumed cross-reactivity between both compounds due to structural similarities and suggest using structurally different antiseptics in patients with either allergy [30A]. Two cases of systemic urticaria including anaphylaxis in children (3-year and 12-year-old boys) were described that occurred following topical chlorhexidine use (0.5% and 1%) prior to surgical procedures. Skin prick tests and sIgE results were positive in both cases [31A].

Sensory Systems

Evidence about the use of chlorhexidine antiseptic solution on the face and head was reviewed in reference to its corneal and middle ear toxicity. The authors summarize 11 sentinel cases from the 1980s describing severe chlorhexidine-related keratitis. The authors conclude that chlorhexidine antiseptic solutions pose a risk on the middle ear and have the potential to cause irreversible damage to the cornea with a minimal splash exposure. They recommend to either apply protective measures or substitute chlorhexidine with alternative antiseptics for use on the face and scalp [32R]. According to in vitro result, chlorhexidine cytotoxicity may be mediated by intracellular reactive oxygen species generation, mitochondrial membrane potential collapse, lysosomal membrane injury, lipid peroxidation, and depletion of glutathione [33E].

Mouth and Teeth

Using a population-based study in Taiwan, a cohort of 18231 head and neck cancer patients, including 941 osteoradionecrosis patients and 17290 matched controls was investigated. The use of chlorhexidine mouthrinse as prophylactic modality before radiotherapy was found to be highly correlated with osteoradionecrosis occurrence (hazard ratio 1.83–2.66). Chlorhexidine exposure increased the risk by 2.43-fold among oral cancer patients. The authors suggested cautious prescription of chlorhexidine mouthrinse in this group of patients [34C].

Polyhexamethylene Guanidine [SEDA-36, 341; SEDA-37, 273; *SEDA-38, 213; SEDA-39, 211*]

Polyhexamethylene guanidine (PHMG) has been used as an antiseptic, especially for the suppression of hospital infection in the Russian Federation and as a disinfectant for sterilization of household humidifiers in Korea [SEDA-36, 341; SEDA-37, 273; SEDA-38, 213; SEDA-39, 211].

Respiratory

Further evidence was gathered on the association of the disinfectant PHMG with lung disease (see SEDA-36, 341 and SEDA-37, 273 for description of cases). A nationwide study in Korean children ($n=1577$) detected that 75.6% had used a humidifier, and the rate of humidifier disinfectant was 31.1%. PHMG was found to have the highest usage rate (62.0%) [35MC]. In Korean preschool children ($n=214$) with humidifier disinfectant-associated lung injuries risk increased ≥twofold in a dose-dependent manner in the highest quartile of airborne concentrations (135–1443 μg/m^3) compared with that in the lowest quartile (<33 μg/m^3). PHMG was the disinfectant with the highest use (80% of patients). Significant differences due to type of disinfectant could not be proven in the study, due to the low prevalence of other disinfectants and coexposure to several types of disinfectants [36C]. Similar results were found in another study of 530 cases of humidifier disinfectant-associated lung injury in Korea; 67% of patients developed injury in less than a year of humidifier disinfectant use. Products containing PHMG were the most frequently used (55.7%) [37C]. Based on measurement of PHMG concentrations in mists generated from ultrasonic humidifiers in experimental settings, the mean airborne PHMG concentration in winter was estimated as 0.2 μg/m^3. Based on this value, and NOAEL levels from animal experiments, the margin of exposure (MOE) could range between 60 and 150. The authors commented that the MOE cannot fully explain the observed hazard and other factors could be involved in inducing lung injury [38E]. In vivo animal research in mice (28-day study, intratracheal instillation of PHMG at 1.2 mg/kg) detected inflammatory cell infiltration and fibrosis mainly in the terminal bronchioles and alveoli in the lungs. Continuous induction of inflammatory response by PHMG was judged to play an important role in the development of pulmonary fibrosis [39E].

BENZALKONIUM COMPOUNDS *[SED-15, 421; SEDA-32, 440; SEDA-33, 481; SEDA-34, 379; SEDA-36, 341; SEDA-37, 273; SEDA-38, 213; SEDA-39, 212]*

Sensory Systems

It is believed that eye drops containing benzalkonium chloride as preservative may contribute to ocular surface disease [see also SEDA-36, 341; SEDA-37, 273; *SEDA-38, 213; SEDA-39, 212*]. Using in vitro research on human corneal epithelial cells, benzalkonium chloride was found to inhibit mitochondrial ATP and oxygen consumption in a concentration-dependent manner. At its pharmaceutical concentrations (107–667 μM), it inhibited mitochondrial function >90%. The authors concluded that benzalkonium chloride should be avoided in patients with mitochondrial deficiency, including Leber hereditary optic neuropathy patients [40E].

Respiratory

Based on literature review, benzalkonium chloride used as preservative in continuous albuterol nebulizer solutions may produce bronchospasm that is dose dependent and cumulative. Benzalkonium chloride containing albuterol use during severe acute asthma exacerbations may antagonize the bronchodilator response to albuterol, prolong treatment, and increase the risk of albuterol-related systemic adverse effects. The authors recommend that only preservative-free albuterol products be used [41r].

TRICLOSAN *[SEDA-34, 379; SEDA-36, 342; SEDA-37, 276; SEDA-39, 212]*

Nervous System

Based on non-systematic literature review, the neurotoxic effects of triclosane were summarized. The limited number of investigations in this area point to adverse effects on the CNS, mainly through induction of apoptosis and oxidative stress. However, the authors conclude that further research is needed to better understand the potential adverse effects of triclosan on the brain [42R].

Immunologic

Triclosan was removed from the list of the American contact dermatitis society core allergen series because it was suggested as infrequent allergen, only previously included on the list because of its extensive use in the US [19S]. A case of a 26-year-old woman with a 7-month history of cough, wheeze and chest tightness was presented. The patient had used 5% liquid cleaner containing 0.05% triclosan for occupational cleaning purposes in a nursery. A non-IgE-mediated mechanism was assumed. Forced expiratory volume in 1 second significantly decreased following exposure to the cleaner, consistent with sensitization. The patient remained relatively symptom free when the cleaner was replaced in the workplace [43A]. Topical application in a murine model suggests the S100A8/A9-TLR4-pathway to play an early role in augmenting immunomodulatory responses with triclosan exposure [44E].

Metabolism

Longitudinal associations between childhood exposure to triclosan, and subsequent measures of adiposity among girls ($n = 1017$) enrolled in the Breast Cancer and the Environment Research Program between 2003 and 2007 were investigated. Exposure was based on urinary triclosan analyses at baseline. At least 3 anthropometric measurements were taken through 2015. A positive association between triclosan exposure and adiposity was observed, but only among girls that were already overweight at baseline. The authors hypothesized that thyroid function may be impaired in obesity and influence susceptibility to triclosan exposure [45MC].

HALOGENS

Sodium Hypochlorite *[SED-15, 3157; SEDA-28, 262; SEDA-34, 380; SEDA-36, 342; SEDA-37, 273; SEDA-38, 214; SEDA-39, 213]*

Teeth

Sodium hypochlorite is used to irrigate root canals in dentistry and can cause many adverse reactions *[SEDA-34, 380; SEDA-36, 342; SEDA-37, 273; SEDA-38, 214; SEDA-39, 213]*. A systematic review of 52 case reports published between 1974 and 2015 suggests sudden pain, profuse bleeding and almost immediate swelling as a triad of signs/symptoms pathognomonic of sodium hypochlorite extrusion. The authors also remarked that clinical cases in this area were reported rather unsystematically and some relevant information was missing in many cases. For this reason, the authors suggest a standardized template for future reporting of sodium hypochlorite accidents [46R].

IODOPHORS *[SED-15, 1896; SEDA-31, 411; SEDA-32, 440; SEDA-33, 485; SEDA-34, 380; SEDA-36, 342; SEDA-37, 273; SEDA-38, 215; SEDA-39, 213]*

Iodine

Endocrine

125 Preterm infants exposed to topical iodine (maternal or neonatal use) were compared to 48 unexposed infants. 3 Infants (exposed group) had transient hyperthyrotropinaemia. Mean thyrotropin levels were significantly higher on postnatal days 7, 14, and 28 in infants exposed to topical iodine prior to caesarean section. The authors conclude that exposure to iodine via caesarean section may cause thyroid dysfunction [47c].

Polyvinylpyrrolidone (Povidone) and Povidone-Iodine

Nervous System

In a study involving 63 patients, pleurodesis with instillation of povidone-iodine was performed for treatment of malignant pleural effusion; 6.3% of patients had mild pain and 20.6% showed moderate pain. After povidone-iodine instillation, 6 patients had air leak that was resolved after 1 day [48c].

Skin

In a retrospective assessment of 12 patients that were prescribed topical gel containing povidone-iodine (2%) in dimethylsulfoxide for treatment of molluscum contagiosum (twice daily application for 4-week intervals), only 2 patients demonstrated mild irritation in the form of dryness of the surrounding skin at application site [49c].

Additional case studies on this topic can be found in these reviews [50R, 51R].

References

[1] Weinmann T, Gerlich J, Heinrich S, et al. Association of household cleaning agents and disinfectants with asthma in young German adults. Occup Environ Med. 2017;74(9):684–90 [MC].

[2] Lachapelle JM. A comparison of the irritant and allergenic properties of antiseptics. Eur J Dermatol. 2014;24(1):3–9 [R].

[3] Gaskins AJ, Chavarro JE, Rich-Edwards JW, et al. Occupational use of high-level disinfectants and fecundity among nurses. Scand J Work Environ Health. 2017;43(2):171–80 [C].

[4] Nielsen GD, Larsen ST, Wolkoff P. Re-evaluation of the WHO (2010) formaldehyde indoor air quality guideline for cancer risk assessment. Arch Toxicol. 2017;91(1):35–61 [R].
[5] IARC Working Group on the Evaluation of Carcinogenic Risks to Humans. Formaldehyde. IARC Monogr Eval Carcinog Risks Hum. 2012;100F:401–35 [S].
[6] Mundt KA, Gallagher AE, Dell LD, et al. Does occupational exposure to formaldehyde cause hematotoxicity and leukemia-specific chromosome changes in cultured myeloid progenitor cells? Crit Rev Toxicol. 2017;47(7):592–602 [R].
[7] Zhang L, Tang X, Rothman N, et al. Occupational exposure to formaldehyde, hematotoxicity, and leukemia-specific chromosome changes in cultured myeloid progenitor cells. Cancer Epidemiol Biomarkers Prev. 2010;19(1):80–8 [c].
[8] Warshaw EM, Hagen SL, Belsito DV, et al. Occupational contact dermatitis in North American print machine operators referred for patch testing: retrospective analysis of cross-sectional data from the North American contact dermatitis group 1998 to 2014. Dermatitis. 2017;28(3):195–203 [c].
[9] Schwensen JF, Friis UF, Menné T, et al. Contact allergy to preservatives in patients with occupational contact dermatitis and exposure analysis of preservatives in registered chemical products for occupational use. Int Arch Occup Environ Health. 2017;90(4):319–33 [C].
[10] Giménez-Arnau AM, Deza G, Bauer A, et al. Contact allergy to preservatives: ESSCA* results with the baseline series, 2009–2012. J Eur Acad Dermatol Venereol. 2017;31(4):664–71 [MC].
[11] Jang JH, Park SH, Jang HJ, et al. A case of recurrent urticaria due to formaldehyde release from root-canal disinfectant. Yonsei Med J. 2017;58(1):252–4 [A].
[12] Li L, Hua L, He Y, et al. Differential effects of formaldehyde exposure on airway inflammation and bronchial hyperresponsiveness in BALB/c and C57BL/6 mice. PLoS One. 2017;12(6):e0179231 [E].
[13] Peters TL, Kamel F, Lundholm C, et al. Occupational exposures and the risk of amyotrophic lateral sclerosis. Occup Environ Med. 2017;74(2):87–92 [C].
[14] Liu X, Zhang Y, Wu R, et al. Acute formaldehyde exposure induced early Alzheimer-like changes in mouse brain. Toxicol Mech Meth. 2018;28(2):95–104 [E].
[15] Liu X, Zhang Y, Luo C, et al. At seeming safe concentrations, synergistic effects of PM2.5 and formaldehyde co-exposure induces Alzheimer-like changes in mouse brain. Oncotarget. 2017;8(58):98567–79 [E].
[16] Xu W, Zhang W, Zhang X, et al. Association between formaldehyde exposure and miscarriage in Chinese women. Medicine. 2017;96(26):1–4 [c].
[17] Wu D, Jiang Z, Gong B, et al. Vitamin E reversed apoptosis of cardiomyocytes induced by exposure to high dose formaldehyde during mice pregnancy. Int Heart J. 2017;58(5):769–77 [E].
[18] Martin BJ, Kaestner M, Peng M, et al. Glutaraldehyde treatment of allografts and aortic outcomes post-norwood: challenging surgical decision. Ann Thorac Surg. 2017;104(4):1395–401 [c].
[19] Schalock PC, Dunnick CA, Nedorost S, et al. American contact dermatitis society core allergen series: 2017 update. Dermatitis. 2017;28(2):141–3 [S].
[20] James P, Worthington HV, Parnell C, et al. Chlorhexidine mouthrinse as an adjunctive treatment for gingival health (Review). Cochrane Database Syst Rev. 2017;2017(3):CD008676 [M].
[21] Richards D. Chlorhexidine mouthwash plaque levels and gingival health. Evid Based Dent. 2017;18(2):37–8 [r].
[22] Cardona A, Balouch A, Abdul MM, et al. Efficacy of chlorhexidine for the prevention and treatment of oral mucositis in cancer patients: a systematic review with meta-analyses. J Oral Pathol Med. 2017;46(9):680–8 [R].
[23] Zand F, Zahed L, Mansouri P, et al. The effects of oral rinse with 0.2% and 2% chlorhexidine on oropharyngeal colonization and ventilator associated pneumonia in adults' intensive care units. J Crit Care. 2017;40:318–22 [c].
[24] Haydari M, Bardakci AG, Koldsland OC, et al. Comparing the effect of 0.06% -, 0.12% and 0.2% chlorhexidine on plaque, bleeding and side effects in an experimental gingivitis model: a parallel group, double masked randomized clinical trial. BMC Oral Health. 2017;17(1):1–8 [c].
[25] dos Santos GO, Milanesi FC, Greggianin BF, et al. Chlorhexidine with or without alcohol against biofilm formation: efficacy, adverse events and taste preference. Braz Oral Res. 2017;31:e32 [c].
[26] Ortegón L, Puentes-Herrera M, Corrales IF. Colonization and infection in the newborn infant: does chlorhexidine play a role in infection prevention? Arch Argent Pediatr. 2017;115(1): 65–70 [r].
[27] Neri I, Ravaioli GM, Faldella G, et al. Chlorhexidine-induced chemical burns in very low birth weight infants. J Pediatr. 2017;191:262–5 [A].
[28] Prabhu AS, Krpata DM, Phillips S, et al. Preoperative chlorhexidine gluconate use can increase risk for surgical site infections after ventral hernia repair. J Am Coll Surg. 2017;224(3):334–40 [C].
[29] Egner W, Helbert M, Sargur R, et al. Chlorhexidine allergy in four specialist allergy centres in the United Kingdom, 2009–13: clinical features and diagnostic tests. Clin Exp Immunol. 2017; 188(3):380–6 [c].
[30] Schunter JA, Stöcker B, Brehler R. A case of severe anaphylaxis to polyhexanide: cross-reactivity between biguanide antiseptics. Int Arch Allergy Immunol. 2017;173(4):233–6 [A].
[31] Lasa EM, González C, García-Lirio E, et al. Anaphylaxis caused by immediate hypersensitivity to topical chlorhexidine in children. Ann Allergy Asthma Immunol. 2017;118(1):118–9 [A].
[32] Steinsapir KD, Woodward JA. Chlorhexidine keratitis: safety of chlorhexidine as a facial antiseptic. Dermatol Surg. 2017;43(1): 1–6 [R].
[33] Salimi A, Alami B, Pourahmad J. Analysis of cytotoxic effects of chlorhexidine gluconate as antiseptic agent on human blood lymphocytes. J Biochem Mol Toxicol. 2017;31(8):1–8 [E].
[34] Chang CT, Liu SP, Muo CH, et al. Dental prophylaxis and osteoradionecrosis: a population-based study. J Dent Res. 2017;96(5):531–8 [C].
[35] Yoon J, Cho HJ, Lee E, et al. Rate of humidifier and humidifier disinfectant usage in Korean children: a nationwide epidemiologic study. Environ Res. 2017;155:60–3 [MC].
[36] Park DU, Ryu SH, Roh HS, et al. Association of high-level humidifier disinfectant exposure with lung injury in preschool children. Sci Total Environ. 2018;616–617:855–62 [C].
[37] Park DU, Ryu SH, Lim HK, et al. Types of household humidifier disinfectant and associated risk of lung injury (HDLI) in South Korea. Sci Total Environ. 2017;596–597:53–60 [C].
[38] Lee JH, Yu IJ. Human exposure to polyhexamethylene guanidine phosphate from humidifiers in residential settings: cause of serious lung disease. Toxicol Ind Health. 2017;33(11):835–42 [E].
[39] Kim MS, Kim SH, Jeon D, et al. Changes in expression of cytokines in polyhexamethylene guanidine-induced lung fibrosis in mice: comparison of bleomycin-induced lung fibrosis. Toxicology. 2018; 393:185–92 [E].
[40] Datta S, Baudouin C, Brignole-Baudouin F, et al. The eye drop preservative benzalkonium chloride potently induces mitochondrial dysfunction and preferentially affects LHON mutant cells. Invest Ophthalmol Vis Sci. 2017;58(4):2406–12 [E].
[41] Prabhakaran S, Abu-Hasan M, Hendeles L. Benzalkonium chloride: a bronchoconstricting preservative in continuous albuterol nebulizer solutions. Pharmacotherapy. 2017;37(5):607–10 [r].

[42] Ruszkiewicz JA, Li S, Rodriguez MB, et al. Is triclosan a neurotoxic agent? J Toxicol Env Health Part B Crit Rev. 2017;20(2):104–17 [R].
[43] Walters GI, Robertson AS, Moore VC, et al. Occupational asthma caused by sensitization to a cleaning product containing triclosan. Ann Allergy Asthma Immunol. 2017;118(3):370–1 [A].
[44] Marshall NB, Lukomska E, Nayak AP, et al. Topical application of the anti-microbial chemical triclosan induces immunomodulatory responses through the S100A8/A9-TLR4 pathway. J Immunotoxicol. 2017;14(1):50–9 [E].
[45] Deierlein AL, Wolff MS, Pajak A, et al. Phenol concentrations during childhood and subsequent measures of adiposity among young girls. Am J Epidemiol. 2017;186(5):581–92 [MC].
[46] Guivarc'h M, Ordioni U, Ahmed HMA, et al. Sodium hypochlorite accident: a systematic review. J Endod. 2017;43(1):16–24 [R].
[47] Williams FLR, Watson J, Day C, et al. Thyroid dysfunction in preterm neonates exposed to iodine. J Perinat Med. 2017;45(1):135–43 [c].
[48] Kahrom H, Aghajanzadeh M, Asgari M, et al. Efficacy and safety of povidone-iodine pleurodesis in malignant pleural effusions. Ind J Palliative Care. 2017;23(1):53 [c].
[49] Capriotti K, Stewart K, Pelletier J, et al. Molluscum contagiosum treated with dilute povidone-iodine: a series of cases. J Clin Aesthet Dermatol. 2017;10(3):41–5 [c].
[50] Lachenmeier DW. Chapter 22—Antiseptic drugs and disinfectants. In: Ray SD, editor. Side effects of drugs: Annual: A worldwide yearly survey of new data in adverse drug reactions. Elsevier, 2016;38:p. 211–216 [R].
[51] Lachenmeier DW. Chapter 20—Antiseptic drugs and disinfectants. In: Ray SD, editor. Side effects of drugs: Annual: A worldwide yearly survey of new data in adverse drug reactions. Elsevier, 2017;39:p. 209–215 [R].

CHAPTER

24

Beta-Lactams and Tetracyclines

Rebecca A. Buckler, Meghan T. Mitchell[†], Michelle M. Peahota[†,1], Jason C. Gallagher[‡]*

*Jefferson Health—Methodist Hospital Division, Philadelphia, PA, United States
[†]Thomas Jefferson University Hospital, Philadelphia, PA, United States
[‡]Temple University, Philadelphia, PA, United States
[1]Corresponding author: michelle.peahota@jefferson.edu

CARBAPENEMS

Ertapenem

Organs and Systems

NERVOUS SYSTEM

Seizure, a well-known adverse effect of the carbapenem class, has been described with ertapenem use [1r]. The association of ertapenem with other neurologic effects, such as hallucinations and delirium, is less commonly reported. Patel and colleagues published a case series describing four patients with spinal cord injury (SCI) who experienced mental status changes and hallucinations attributed to ertapenem. All four patients were male (age range 47–90 years), had SCI, and received ertapenem 1000mg intravenous (IV) every 24h, which was dosed appropriately for their renal function. The patients developed new delirium, visual hallucinations, or myoclonus shortly after the initiation of ertapenem (onset range: 1–6 days). Use of the Naranjo Probability Score indicated a probable relationship in three cases and possible in one. All four patients had chronic hypoalbuminemia, and since ertapenem is a highly protein-bound drug, low albumin levels result in elevated free drug concentrations. The authors underscore the importance or recognizing ertapenem-associated hallucinations in SCI patients [2A].

HEMATOLOGIC

A 68-year-old woman with a past medical history of diverticulosis, hypertension, diabetes, and peripheral artery disease presented with near syncope and abdominal pain. The patient was not taking any prescription medications for her medical conditions. On admission, her vital signs were normal, she was noted to have left lower quadrant tenderness, and a computed tomography scan of the abdomen and pelvis suggested diverticulitis with abscess. She was initiated on IV ciprofloxacin 400mg twice daily and metronidazole 500mg three times daily. Her baseline blood work showed a hemoglobin (Hb) of 11g/dL, white blood cell count (WBC) 20000/mm^3 and a platelet (PLT) count of 487000/mm^3. Due to a continued rise in her WBC during the next 48h, her antibiotics were changed to an undescribed dose of IV ertapenem on hospital day 4. During the next several days the woman clinically improved, but her platelets trended upwards, reaching a maximum value of 610000mm^3. Due to concerns of ertapenem-induced thrombocytosis, ertapenem was discontinued on hospital day 9 and her antibiotics were switched to oral ciprofloxacin and metronidazole. Twenty-four hours following ertapenem discontinuation her platelet count decreased to 465000mm^3 and her WBC was 10000mm^3. The mechanism of ertapenem-induced thrombocytosis, a rarely described adverse effect, is not well understood. This report adds to the small body of literature describing ertapenem-associated hematologic effects [3A].

Meropenem

Organs and Systems

HEMATOLOGIC

Drug-induced thrombocytopenia (DITP) is a serious drug reaction characterized by severe platelet depletion, possible bleeding, and presence of drug-dependent anti-platelet antibodies. Carbapenems have been described with the development of DITP; however, the published

Side Effects of Drugs Annual, Volume 40
ISSN: 0378-6080
https://doi.org/10.1016/bs.seda.2018.07.014

cases lack confirmatory testing, limiting the ability to draw a definitive conclusion. Huang and colleagues described what they believe to be the first case of confirmed meropenem-induced thrombocytopenia. A 59-year-old male was hospitalized for a shoulder and pulmonary infection and initiated on piperacillin–tazobactam 2.25 g three times daily. Due to concerns of ongoing infection, vancomycin 0.5 g twice daily was added on day 14 of hospitalization. On hospital day 26, he received wound incision and drainage. The drainage culture grew *Enterobacter cloacae*. Based on sensitivities, vancomycin and piperacillin/tazobactam were discontinued and he was initiated on meropenem 0.5 g three times daily and cefoperazone–sulbactam 2 g twice daily. The patient's infection appeared to be improving, however, on hospital day 34 his platelet count decreased to 25×10^9 L, a marked drop from his admission platelet count of 125×10^9 L. Both meropenem and cefoperazone–sulbactam were discontinued as DITP was suspected. Enzyme-linked immunosorbent assay, flow cytometry, and monoclonal antibody-specific immobilization of platelet antigen tests specifically detected the presence of meropenem-mediated antiplatelet antibodies, confirming meropenem-induced immune thrombocytopenia. The patient's platelet count began to increase on day 36 and returned to baseline on day 42 of hospitalization. The authors describe the first lab-confirmed case of meropenem-induced thrombocytopenia and hope to underscore the importance of identifying the causative agent in these situations so that future exposure can be avoided [4A].

ELECTROLYTE BALANCE

Anuhya and colleagues describe two cases of meropenem-induced hypokalemia. A 53-year-old female was prescribed meropenem 1 g three times daily for pneumonia and lower extremity cellulitis. Her home medications included pantoprazole, vitamin C, calcium and vitamin B-complex. Six days after meropenem initiation she complained of weakness but had no diarrhea or vomiting. A serum electrolyte panel revealed a potassium level of 2.3 mmol/L. Meropenem was thought to be the cause of hypokalemia; therefore, it was discontinued. Her potassium level returned to normal following meropenem discontinuation and oral potassium supplementation. After further investigation of her medical records, it was found that she had also experienced hypokalemia while she received meropenem during a previous hospital admission. The second case describes a 53-year-old female with right renal staghorn calculi with gross hydroureteronephrosis who received meropenem 1 g three times daily for a complicated urinary tract infection. Her concomitant medications included metformin, aspirin, clopidogrel, and atorvastatin. Her baseline potassium level was noted to be within normal limits. Two days after meropenem initiation, her potassium level decreased to 3 mmol/L. A nephrolithotomy was performed and she was discharged. Follow-up details were not available. The authors encourage awareness of meropenem-induced hypokalemia, an uncommon but potentially life-threatening adverse effect [5A].

Interactions

DRUG–DRUG INTERACTIONS

Concurrent use of carbapenem antibiotics and valproate derivatives may reduce serum valproate acid (VPA) concentrations to subtherapeutic levels. This known interaction, previously described in case reports and pharmacokinetic studies, is listed as a drug-interaction warning in the prescribing information of VPA and its derivatives. The mechanism of this interaction is not well understood. Sima and colleagues present two additional cases of this interaction, involving meropenem and subtherapeutic VPA levels to add to the body of literature. The authors highlight the importance for physicians to recognize this drug interaction. Concomitant use of a carbapenem and VPA (and its derivatives) necessitates frequent monitoring of serum VPA if the concurrent therapies are required. Ideally, alternative antibacterial or anticonvulsant therapy should be administered [6A].

PENICILLINS

Amoxicillin

Organs and Systems

RESPIRATORY

Amoxicillin is the most commonly prescribed antibiotic in the outpatient setting in Europe, responsible for about 40% of total outpatient antibiotic use. In a multicenter, randomized, placebo-controlled trial conducted at 16 primary care facilities in Europe, 2061 patients were randomized to receive amoxicillin or placebo for the diagnosis of lower respiratory tract infection (LRTI). In a secondary analysis to estimate adverse events the study found that at the end of 1 week of therapy there was a significantly higher proportion of adverse events (defined as diarrhea, nausea or rash) in patients prescribed amoxicillin when compared with those prescribed placebo (OR = 1.31, 95% CI 1.05–1.64). The number needed to harm (NNH) was 24. When evaluating each adverse effect individually, diarrhea was the only adverse effect with a statistically significantly higher rate in the amoxicillin group when compared to placebo (OR 1.43 95% CI 1.08–1.90, NNH = 29). Subgroup analysis demonstrated a significantly higher rate of rash in males prescribed amoxicillin [interaction term 3.72 95% CI 1.22–11.36; OR of amoxicillin in males 2.79 (95% CI 1.08–7.22)]. The authors acknowledge that the incidence of adverse events was not the primary objective of the trial, and that a larger sample size would be needed to achieve sufficient power for further subgroup analyses [7MC].

Ampicillin

Organs and Systems

NERVOUS SYSTEM

A 49-year-old female prescribed trimethoprim–sulfamethoxazole (160mg/800mg) twice daily for a urinary tract infection presented to the emergency department 5h following her first dose complaining of holocephalic headache, nausea and vomiting. Physical examination was positive only for terminal meningismus. The patient was slightly tachycardic (92bpm) with otherwise normal vital signs. A cerebral computed topography scan was unremarkable. Notable laboratory values included C-reactive protein of 5.9mg/L, WBC of 13000/μL and CSF analysis: 1671 WBC/μL (1599 granulocytes/μL, 72 lymphocytes/μL), protein 311mg/dL and glucose 61mg/dL. No pathogens were detected on CSF analysis. Empiric meningitis treatment was initiated with dexamethasone, ceftriaxone, ampicillin, and acyclovir. With administration of the fifth dose of ceftriaxone and ampicillin the patient began experiencing full body myoclonic twitches that were successfully managed with clonazepam. Ampicillin therapy was discontinued after listeria infection had been ruled out. To the author's knowledge, this is the first case report describing the combination of trimethoprim–sulfamethoxazole-induced aseptic meningitis with myoclonic twitches secondary to ampicillin administration [8A].

Nafcillin

Interactions

DRUG–DRUG INTERACTION

Tacrolimus, a calcineurin inhibitor, demonstrates many drug–drug interactions due to its metabolism via the CYP450 isoenzyme 3A4 system. An anti-staphylococcal penicillin, nafcillin, is a moderate CYP3A4 inducer. Much has been published regarding nafcillin and drug–drug interactions with other CYP3A4 substrates; however, Wungwattana and colleagues described the first case report demonstrating the interaction between nafcillin and tacrolimus. A 13-year-old male with a past medical history of cystic fibrosis, type 1 diabetes mellitus, pancreatic insufficiency and liver transplantation presented with a 2 week history of upper respiratory symptoms. He had failed 6 days of outpatient trimethoprim–sulfamethoxazole. The patient had a prior history of methicillin-sensitive *Staphylococcus aureus* infection and colonization. The patient's tacrolimus dose at home was 1.5mg by mouth twice daily with a goal trough of 3–5ng/mL. The patient's tacrolimus level was found to be 2.8ng/mL 7 days prior to admission and 4.2ng/mL the 1st day following admission. The patient was initiated on nafcillin (2g/dose, frequency not described) due to worsening respiratory status. The patient had received two doses prior to the first tacrolimus level being drawn. Following the administration of eight doses total over 3 days the level was noted to be 2.0ng/mL. Three days further (total of 16 doses) the level was found to be undetectable (<2.0ng/mL) without alteration in the tacrolimus dosing. Nafcillin was discontinued on discharge and 5 days post-discharge the patient's tacrolimus level rose to 3.0ng/mL. Two months following discharge, the level reached at 4.6ng/mL without any exposures to new CYP3A4 inducers/inhibitors. The authors concluded that co-administration of nafcillin and tacrolimus may result in sub-therapeutic tacrolimus levels [9A].

Penicillin

CARDIOVASCULAR

An 11-year-old girl with no past medical history presented with a 3 day history of malaise, fever and sore throat. She was diagnosed with streptococcal pharyngitis and given an intramuscular dose of penicillin (dose not indicated). One hour following administration she developed chills, acute malaise and shortness of breath. She went into cardiopulmonary arrest and expired. Twelve hours following death an autopsy was performed in which the cause of death was determined to be eosinophilic myocarditis. Drug-induced eosinophilic myocarditis is rare with penicillin-induced eosinophilic myocarditis being even less common. The onset of fulminant clinical deterioration in this case was comparatively very rapid as the onset is variable from hours to months. The authors conclude that this is one of the very few pediatric reports of penicillin-induced fatal hypersensitivity myocarditis [10A].

Piperacillin

Organs and Systems

CARDIOVASCULAR

Electrolyte disturbances are a well-described adverse event associated with the administration of penicillin-based antibiotics. Kumar and colleagues describe a case of hypokalemia secondary to piperacillin and imipenem that resulted in Torsades de Pointes (TdP). A 39-year-old female with no past medical history of syncopal episodes and no family history of sudden cardiac death or long QT syndrome was being managed for peritoneal abscess secondary to complications of an oophorectomy and ventral hernia repair. On admission the patient had a normal echocardiogram, normal QT interval and potassium of 5.2 mEq/L. The patient was treated with 6 days of imipenem–cilastatin (IC) and developed hypokalemia (3.1 mEq/L), metabolic alkalosis, and prolongation of QT interval (QTc 533 ms). Other electrolytes were within normal limits. The patient was switched to piperacillin–tazobactam and hypokalemia worsened despite adequate potassium repletion (80–120 mEq/day). No other medications with a known adverse effect of QT prolongation were administered in the 48 h prior to the episode of TdP. After 48 h of therapy with piperacillin–tazobactam, hypokalemia worsened to 2.9 mEq/L and QTc increased to 632 ms. The patient experienced three episodes of self-terminating TdP. The patient survived these events and was further medically managed and discharged to a long-term care facility. The authors state that there are a few prior case reports detailing piperacillin–tazobactam causing hypokalemia, but none to their knowledge of hypokalemia from IC. The authors caution that these antibiotics may result in hypokalemia unresponsive to repletion which has the potential to cause cardiac events [11A].

URINARY TRACT

There is mounting literature evaluating the potential nephrotoxicity of concomitant IV vancomycin with piperacillin–tazobactam when compared with the co-administration of IV vancomycin and alternative beta-lactams [12C, 13MC, 14c]. Navalkele and colleagues published a retrospective, matched, cohort study that evaluated the incidence of acute kidney injury (AKI) in 558 patients receiving either IV vancomycin and piperacillin–tazobactam or IV vancomycin and cefepime. Patients were included if they had received ≥48 h of dual therapy and had the two antibiotics initiated within 24 h of one another. Patients with baseline renal dysfunction (defined as having a serum creatinine >1.2 mg/dL or requiring renal replacement) were excluded. Patients were matched based on severity of sepsis and ICU status at antibiotic initiation, duration of combination therapy, number of concomitant nephrotoxic medications and total daily dose of vancomycin. Rates of AKI were significantly higher in the piperacillin–tazobactam group than in the cefepime group (81/279 [29%] vs 31/279 [11%]). Piperacillin–tazobactam administration was found to be an independent predictor for AKI in multivariate analysis (hazard ratio = 4.27; 95% confidence interval, 2.73–6.68). Additionally, median onset of AKI was shorter in the piperacillin group when compared with the cefepime group (3 vs 5 days, $P \leq 0.0001$). The authors caution that combination IV vancomycin and piperacillin–tazobactam therapy was associated with a fourfold increase in AKI, which occurred more rapidly when compared to IV vancomycin and cefepime [12C].

Mousavi and colleagues published a single-center, retrospective, matched, cohort study of 280 adults to address whether or not the mode of piperacillin–tazobactam administration (extended-infusion vs standard-infusion) impacted the nephrotoxicity of combination therapy with IV vancomycin. Patients were included if they had received at least 96 h of combination therapy and had no underlying renal dysfunction. Rates of AKI were defined by the risk, injury, failure, loss, and end-stage kidney disease (RIFLE) and the Acute Kidney Injury Network (AKIN) criteria. Acute kidney injury rates were similar between the extended-infusion and standard-infusion, respectively [RIFLE—17.9% vs 17.1% ($P = 1$); AKIN—32.9% vs 29.3% ($P = 0.596$)]. Conditional logistic regression analysis demonstrated no association between extended-infusion administration and AKI (OR 0.522, 95% confidence interval 0.043–6.295, $P = 0.609$). No significant difference was noted in regard to onset of nephrotoxicity between the groups. The authors conclude that the length of piperacillin–tazobactam infusion does not appear to influence the likelihood of developing AKI when co-administered with IV vancomycin. They recommend further prospective, randomized, controlled studies to confirm this finding [15C].

Evidence suggesting an increased risk of AKI associated with the co-administration of IV vancomycin and piperacillin–tazobactam has also been noted in the pediatric population. Downes and colleagues published a retrospective cohort evaluation of 1915 hospitalized children in one of six pediatric hospitals who received concomitant vancomycin and an antipseudomonal beta-lactam. Of those evaluated 1009 of the patients received piperacillin–tazobactam as their beta-lactam therapy. The median age was 5.6 (2.1–12.7) years. Overall, this study found that of the children receiving IV vancomycin with an antipseudomonal beta-lactam, 8.2% (157 of 1915 patients) experienced an antibiotic-associated AKI. Of those, the children receiving piperacillin–tazobactam experienced a higher than overall rate of AKI at 11.7% (117 of 1009 patients). Combination IV vancomycin and piperacillin–tazobactam therapy was associated with a higher risk of AKI for each hospital day when compared with vancomycin with an alternative antipseudomonal beta-lactam (adjusted odds ratio, 3.40; 95% CI, 2.26–5.14). The authors conclude that co-administration of IV vancomycin and piperacillin–tazobactam may increase the likelihood of pediatric antibiotic-associated AKI [13MC].

Similarly, a single-center, retrospective cohort study of 71 pediatric cystic fibrosis patients evaluating the difference in antibiotic-associated AKI between those treated with piperacillin–tazobactam or cefepime in combination with IV vancomycin and tobramycin found a significant increase in the rate of kidney injury in those treated with piperacillin–tazobactam. Acute kidney injury, as defined by modified pediatric risk, injury, failure, loss and end stage renal disease (pRIFLE criteria), occurred in 54.5% (18/33) of patients receiving piperacillin–tazobactam and 13.2% (5/38) of patients receiving cefepime ($P \leq 0.0001$). The authors conclude that cefepime may be the preferred antipseudomonal beta-lactam in pediatric cystic fibrosis patients receiving concomitant IV vancomycin and tobramycin [14c].

SKIN

Linear IgA bullous dermatosis (LABD) is an idiopathic or drug-induced autoimmune disease characterized by the linear deposition of IgA at the dermoepidermal junction with various clinical presentations. LABD can present similarly to Stevens–Johnson syndrome (SJS) and toxic epidermal necrolysis (TEN) overlap. It is most commonly associated with vancomycin administration. A 47-year-old man with past medical history significant only for obesity and hypertension was admitted for surgical debridement of skin necrosis of the foot secondary to a crush injury. Prior to admission the patient received 4 days of oral flucloxacillin with no improvement. On admission the patient was broadened to piperacillin–tazobactam. On the 3rd day of therapy the patient developed a large pruritic, erythematous rash on his trunk and extremities. Piperacillin–tazobactam was discontinued and the patient was initiated on vancomycin and metronidazole. On following day the patient became hemodynamically unstable and developed wide-spread bullae. The patient was treated with IVIG for 48h and subsequently transferred to a burn center for further care. There were no target lesions, nor ocular or mucosal manifestations; however, 15% of the patient's body surface area evidenced epidermal detachment. Histological and immunofluorescence findings were consistent with LABD in the recovery phase. The eruption improved rapidly while being managed with antimicrobial barrier dressings and corticosteroid ointments and the patient continued on systemic vancomycin and metronidazole. The authors acknowledge that flucloxacillin cannot entirely be excluded as the causative agent, however, given the timing of administration and the agent's short half-life it seems less likely. The authors conclude that this is the first report of LABD following exposure to piperacillin–tazobactam without co-administration of vancomycin [16A].

Susceptibility Factors

AGE

A single-center, retrospective cohort evaluation of 106 outpatient pediatric patients receiving parenteral antimicrobials for prolonged courses found an increased risk of adverse events with piperacillin–tazobactam when compared with other agents. All patients were receiving appropriate doses of piperacillin–tazobactam (240–300mg/kg/day piperacillin component divided every 6–8h for patients ≤40kg; 12–16g/d piperacillin component for patients >40kg). The median age was 10.5 years (range: 2 months–17.9 years). Thirty-six percent of patients were prescribed piperacillin–tazobactam (38/106). The other major agents prescribed were ceftriaxone ($n = 30/106$, 28%), cefazolin ($n = 11/106$, 10%) and vancomycin ($n = 9/106$, 8%). Thirty-nine percent (15/38) of the patients receiving piperacillin–tazobactam experienced at least one adverse event and the majority of those experiencing adverse events developed three or more concurrently ($n = 12/15$, 80%). The most common adverse events were fever (73%), transaminitis (67%), neutropenia (67%), rising CRP (67%) and ESR (60%) and abdominal pain (40%). The authors recommend considering drug-related adverse effects in pediatric patients who receive prolonged courses of piperacillin–tazobactam and present with these constellation of symptoms [17c].

CEPHALOSPORINS

Cefazolin

Organs and Systems

URINARY TRACT

Cefazolin is one of the first line therapy options for methicillin-susceptible *Staphylococcus aureus* (MSSA) bacteremia. A retrospective cohort study was conducted to compare the incidence of AKI in patients receiving cefazolin vs nafcillin for the treatment of these infections. Adult patients were included in the study if they had at least one blood culture positive for MSSA and had received cefazolin or nafcillin for at least 72h. Patients were excluded if they had pre-existing renal insufficiency (serum creatinine (SCr) > 2mg/dL) or were receiving any form of renal replacement therapy. AKI was defined as an increase of SCr from baseline ≥0.3mg/dL within 48h or any increase of 50% or greater as determined by blind assessment. Overall, AKI occurred in 13% (9 of 68) of patients receiving cefazolin compared to 32% (26 of 81) of patients receiving nafcillin. AKI with an increase of SCr of at least 50% from baseline occurred in seven patients receiving cefazolin (10.3%). The median duration of empiric vancomycin therapy prior to switching to cefazolin was 3 days. These results are consistent with previous literature suggesting first-generation cephalosporins such as cefazolin are less likely to cause nephrotoxicity than nafcillin. The authors state the incidence of AKI in this study was higher than in previous studies, which they attribute to differences in patient populations and severity of illness, though the use of a non-standard definition is also notable. They also note that kidney biopsies were not performed, so they were unable to classify the specific type of AKI that occurred in these patients [18c].

IMMUNOLOGIC

Hypersensitivity reactions to cephalosporins, including cefazolin, are well documented. Evidence suggests cases of selective hypersensitivity to cefazolin may be associated with an R1 side chain that differs from other beta-lactam antibiotics. Almeida and colleagues published two case reports of selective cefazolin hypersensitivity. A 49-year-old morbidly obese female developed acute generalized urticaria 3 h after undergoing a gastric banding procedure. Four years later she underwent sleeve gastrectomy, and shortly after induction with anesthesia she developed IV-grade anaphylaxis requiring treatment with epinephrine, hydrocortisone, and mechanical ventilation. The patient had no reported history of drug allergies and during both procedures she received rocuronium, propofol and cefazolin. During allergy workup, skin tests for all drugs involved and specific IgEs were negative for beta-lactam antibiotics. However, an intradermal test with cefazolin was positive at a concentration of 0.1 mg/mL. The patient was then given open drug challenges to penicillin, amoxicillin, cefuroxime and ceftriaxone, all of which were negative for immediate and delayed responses. A basophil activation test was positive for cefazolin.

A 36-year-old female, 38 weeks pregnant with no other significant medical history or medication allergies, underwent a cesarean section and developed facial rash, lip angioedema and hypotension immediately after receiving cefazolin, oxytocin and ephedrine. She was treated with fluids and hydrocortisone and demonstrated reversal of symptoms. Upon allergy work-up specific IgEs were negative for beta-lactams. Skin prick tests were all negative with the exception of a positive immediate intradermal test response for cefazolin. All open oral challenges for penicillin, amoxicillin, cefuroxime and ceftriaxone, as well as basophil activation test for cefazolin were negative. The patient subsequently tolerated several treatments with cefuroxime. The authors suggest the allergy work-up of both these patients confirms a selective hypersensitivity to cefazolin with tolerance to other beta-lactam antibiotics. This therefore suggests a lack of cross-reactivity among the beta-lactam class. The authors hypothesize this is likely due to the different R1 side chain in cefazolin compared to other beta-lactams, but admit these individual cases are insufficient to draw definitive conclusions about cross-sensitivity [19A].

Cefepime

Organs and Systems

NERVOUS SYSTEM

Neurotoxicity in patients receiving cefepime is thought to be related to inhibition of GABA-A receptors or inhibition of GABA release, but is not well characterized. Ayesha and colleagues conducted a systematic review of the literature in order to characterize the occurrence of neurotoxicity in patients receiving cefepime. Overall the authors included 198 patients receiving cefepime who experienced neurologic adverse events in their analysis. The average age of patients experiencing neurotoxicity was 67 years, and 51% of patients were women. Most patients (87%) had renal insufficiency, including 29% with end-stage renal disease (ESRD). The median cefepime dose was 4 g per 24 h, and this dosing exceeded the recommended maximum dose for renal function in 50% of cases. The mean onset of neurotoxicity was 5 days after cefepime was initiated. EEG was used for diagnosis in 81% of cases and the most common findings included triphasic waves consistent with toxic metabolic encephalopathy and epileptiform discharges. The majority of patients (80%) experienced delirium or encephalopathy, while 47% of patients demonstrated disorientation or agitation. Myoclonus occurred in 40% of cases, and non-convulsive status epilepticus occurred in 31% of cases. Eleven percent of patients experienced seizure activity and 9% experienced aphasia. Mortality during the hospital stay when neurotoxicity occurred was noted in 13% of patients, however, the authors note these deaths cannot be attributed to cefepime toxicity. The authors suggest that cefepime toxicity occurs more commonly in older adults and those with renal dysfunction, especially ESRD, and these patients should have their doses adjusted and be closely monitored [20M].

Renal insufficiency is a known risk factor for cefepime-induced encephalopathy, but the effectiveness of preventing this adverse effect by adjusting doses is unknown. Nakagawa and colleagues conducted a single-center retrospective observational study of 422 patients receiving dose-adjusted cefepime to assess the incidence of encephalopathy. The diagnostic criteria used for cefepime-induced encephalopathy included impaired consciousness, altered mental state or involuntary movements emerging several days after treatment cessation. Diagnostic EEG findings included rhythmic delta activity of triphasic-like waves in the absence of other more likely causes for neurologic symptoms. Six patients (1.4%) were determined to have cefepime-induced encephalopathy. These patients were being treated with cefepime for urinary infection (2 patients), pneumonia (2 patients) and febrile neutropenia (2 patients). Five of 67 patients (7.5%) with ESRD developed encephalopathy. The daily maintenance dose for these patients ranged from cefepime 0.5 to 1 g IV daily. Neurologic symptoms included mild to moderate impaired consciousness, confusion or stupor in all except one patient. Five patients had positive myoclonus. EEG abnormalities were found in all six patients, including triphasic-like waves and delta waves in four patients. Risk factors for encephalopathy in these patients included pre-existing central nervous system morbidity and low serum blood urea nitrogen (BUN). The authors observed that encephalopathy occurred in patients with ESRD regardless of their daily dose, as there was no significant association found between daily cefepime dose and risk of encephalopathy. The authors recommend monitoring patients with ESRD and possible dose reduction to prevent cefepime-induced encephalopathy [21c].

Ceftaroline

Organs and Systems

HEMATOLOGIC

Ceftaroline has been reported to cause neutropenia in patients receiving prolonged treatment courses. Turner and colleagues conducted a retrospective cohort study comparing the development of neutropenia for patients receiving prolonged courses of ceftaroline to patients receiving comparator antibiotics. Adult patients were included if they received 14 or more consecutive days of ceftaroline, cefazolin, daptomycin, linezolid, nafcillin or vancomycin. Fifty-three of the 753 patients included in the study received ceftaroline. The primary outcome was the development of neutropenia during antibiotic therapy. Neutropenia was defined as an absolute neutrophil count (ANC) $<1500\,cells/mm^3$. Secondary outcomes included the development of consecutive neutropenia, moderate neutropenia and discontinuation of the antibiotic due to neutropenia. Overall 36 patients (4.8%) in the study developed neutropenia. In all cases the Naranjo Adverse Drug Reaction Probability Scale indicated that the drug was a possible cause of neutropenia. Nine of 53 patients (17%) receiving ceftaroline developed neutropenia, and this occurred significantly more frequently than in the comparator group ($P<0.001$). Moderate neutropenia ($ANC<1000\,cells/mm^3$) occurred in four patients (7.6%) receiving ceftaroline. Consecutive neutropenia ($ANC<1500\,cells/mm^3$ for 2 or more days) occurred in seven patients (13.2%), and severe neutropenia ($ANC<500$ $cells/mm^3$) occurred in two patients (3.8%) receiving ceftaroline. Discontinuation of ceftaroline due to neutropenia occurred in two patients. These results suggest that prolonged ceftaroline therapy is associated with greater incidence of neutropenia compared to other antibiotics, and the authors suggest routinely monitoring ANC during therapy [22C].

Ceftazidime–Avibactam

General Adverse Drug Reactions

Temkin and colleagues presented a case series of 38 patients with infections caused by carbapenem-resistant *Enterobacteriaceae* or *Pseudomonas aeruginosa* treated with ceftazidime–avibactam. Thirty-six patients (94%) received ceftazidime/avibactam as salvage therapy after failed treatment with other antibiotics. The minimum length of treatment was 3 days. Twenty-four patients (63%) received ceftazidime–avibactam 2.5g every 8h IV, while 14 patients received renally adjusted doses. Six patients (16%) developed adverse effects, including increased blood alkaline phosphatase (2 patients), nausea and vomiting (1 patient), *Clostridium difficile*-associated diarrhea (1 patient), convulsions (1 patient) and disorientation which progressed to stupor (1 patient). Both patients who developed neurotoxicities were over the age of 70 and had normal renal function [23A].

Ceftolozane–Tazobactam

General Adverse Drug Reactions

Munita and colleagues conducted a multicenter retrospective study of patients with infections caused by carbapenem-resistant *P. aeruginosa* and treated with ceftolozane–tazobactam. Thirty-five patients were included in the study and the most common indication for treatment was pneumonia. All patients included were treated for a minimum of 72h with ceftolozane–tazobactam. Twenty-seven patients (77%) were prescribed ceftolozane–tazobactam as monotherapy and eight patients received concomitant dual antipseudomonal therapy. Of the patients with a creatinine clearance (CrCl) greater than 50mL/min, 11 patients received ceftolozane–tazobactam 1.5g IV every 8h and 9 patients received 3g IV every 8h. Four patients had a CrCl of 30–50mL/min and received doses ranging from 0.75 to 1.5g IV every 8h. Three patients were receiving intermittent hemodialysis and received 0.375g every 8h IV, while seven patients receiving continuous hemodialysis were administered doses ranging from 1.5 to 0.375g IV every 8h. There were two cases of adverse effects attributed to ceftolozane–tazobactam. One patient reported self-limited diarrhea, with a subsequent negative molecular assay for *C. difficile*. One patient was reported to have peripheral eosinophilia and eosinophiluria with possible interstitial nephritis. Renal injury never developed and the eosinophilia resolved after ceftolozane–tazobactam was discontinued after 14 days of therapy [24C].

Ceftriaxone

Organs and Systems

CARDIOVASCULAR

Ceftriaxone is not recommended to be administered to neonates ≤28 days old within 48h of receiving calcium-containing solutions due to the potential for an interaction that can result in crystalline deposition in the pulmonary or renal vasculature. A systematic review published in 2017 included three older studies that reported adverse cardiopulmonary events in neonates receiving ceftriaxone [25M]. A prospective case series reported respiratory events for 51 of 86 patients aged 11–59 days who received intramuscular (IM) ceftriaxone as outpatients. The authors state it is unclear which of these events could be attributed to ceftriaxone, as these events were not described in detail [26c]. Another study provided an assessment of eight cardiopulmonary events in neonates receiving concurrent ceftriaxone and calcium-containing products that were filed with the FDA Adverse Event Reporting System. Five patients were <3 weeks old, two patients were 4–8 weeks old and one patient was of unknown age. The ceftriaxone dosage varied and was not consistently reported among cases. Seven of eight patients with reported events died and had autopsy findings consistent with the presence of crystalline material or white precipitate in the lungs [27c]. In a prospective case series, three neonates ≤3 days old receiving ceftriaxone died of cardiopulmonary events including asphyxia and persistent pulmonary hypertension. In addition, 11 patients experienced thrombocytosis [28c]. The authors of the review acknowledge that these studies had significant methodological limitations, but further support that concurrent administration of IV ceftriaxone and calcium-containing solutions should be avoided in neonates due to the risk of cardiopulmonary adverse events.

NERVOUS SYSTEM

Delayed clearance of ceftriaxone has been reported in patients with ESRD, which may lead to toxicity. Inoue and colleagues report a case series of three patients with renal dysfunction who developed ceftriaxone-induced neurotoxicity. A 72-year-old male with a history of ESRD on hemodialysis, myocardial infarction, and ischemic stroke was admitted to the hospital with pneumonia. The patient received ceftriaxone 4 g IV daily for 7 days followed by 2 g IV daily for 3 days. On hospital day 8 the patient developed impaired consciousness with no response to haloperidol. The patient continued to deteriorate to a Glasgow Coma Scale (GCS) of 11 and experienced spasmodic activity in his legs. Neurological evaluation revealed positive Babinski reflexes and myoclonic movement in both legs. Laboratory studies and a head CT showed no abnormalities. EEG showed defuse slow-wave activity but no indication of epilepsy. The patient's symptoms gradually improved, with full recovery at day 14 (4 days after ceftriaxone discontinuation). A serum ceftriaxone trough concentration on day 8 revealed a high value of 472 mcg/mL, which the authors state supports a diagnosis of ceftriaxone-induced neurotoxicity.

A 75-year-old female with ESRD on hemodialysis, myelodysplastic syndrome and atrial fibrillation was administered ceftriaxone 2 g IV daily for 9 days for diverticulitis. On day 9 of therapy she developed agitation, hyperkinesia and confusion. Head CT and MRI showed no abnormalities and EEG showed diffuse slow-wave activity. Ceftriaxone was discontinued at this time. On day 13 her impaired consciousness gradually improved. Her serum ceftriaxone trough concentrations on day 4, 6, and 9 were 304, 331 and 422 mcg/mL, respectively.

A 68-year-old female with a history of type 2 diabetes and ESRD due to diabetic nephropathy was admitted to the hospital for pyogenic arthritis and given ceftriaxone 4 g IV daily for 7 days. The patient did not develop neurogenic symptoms during therapy, but the treating physician recommended a dose reduction to ceftriaxone 1 g IV daily on day 8. This dose was continued until day 23 when the patient was discharged. Serum ceftriaxone trough concentrations on days 2, 4, and 7 were 172, 178 and 188 mcg/mL, respectively. The authors suggest that this patient did not experience neurotoxicity because the serum ceftriaxone concentration was not as high as the first two patients reported above. Dose reduction of ceftriaxone should be considered in patients with ESRD [29A].

HEMATOLOGIC

Cephalosporins, including ceftriaxone, have been reported to rarely cause drug-induced hemolytic anemia. A 64-year-old female Jehovah's Witness was being treated with daily ceftriaxone IV infusions for group B streptococcal bacteremia secondary to endocarditis. Three weeks after being discharged from the hospital the patient was readmitted complaining of transient bilateral vision loss which subsequently improved within a few hours. The vision loss occurred immediately after her last two to three ceftriaxone infusions as an outpatient. She also reported increased fatigue and mild shortness of breath and denied chest pain, hemoptysis, change in stool, fevers, chills, abdominal pain or urinary symptoms. On admission the patient's blood pressure was 108/62 mmHg, heart rate 91 bpm, respiratory rate 18 bpm, temperature 37°C and oxygen saturation 96%. On physical exam the patient appeared pale and cardiac exam revealed a +3/6 systolic murmur. Laboratory measurements revealed a hemoglobin of 6 g/dL, hematocrit of 17%, haptoglobin of 101 mg/dL and lactate dehydrogenase of 348 U/L. A peripheral blood smear showed lymphocyte 45%, monocyte 48%, eosinophil 6%, basophil 1%, nucleated red blood cell 10%, anisocytosis, and polychromasia. The patient was diagnosed with symptomatic, likely hemolytic anemia. The infectious disease service was consulted and suspected drug-induced immune hemolytic anemia secondary to ceftriaxone. The patient was subsequently changed to vancomycin and ceftriaxone was discontinued. The patient refused blood products and was started on ferrous sulfate 325 mg daily, folic acid 1 mg daily, and IM vitamin B12 1000 mcg injections for 7 days. The patient was also initiated on a prednisone taper and epoetin alfa subcutaneously every other day. The patient's hemoglobin gradually improved over the next 20 days and the patient was discharged from the hospital to finish her antibiotic course with vancomycin. The patient's episodes of vision loss never returned and were attributed to ischemic optic neuropathy from severe anemia. The authors stressed the importance of close monitoring of patients receiving drugs known to cause hemolytic anemia. Standard treatment for hemolytic anemia includes blood product transfusion, but this case is unique in that the patient was able to be successfully treated in the absence of blood products [30A].

A 76-year-old female with a history of arterial hypertension and glaucoma was diagnosed with "chronic Lyme disease" to which she was receiving outpatient treatment with ceftriaxone 4g IV twice a week for 20 consecutive weeks. The patient experienced improvement of symptoms following the course of antibiotics. Several years later the patient again developed fatigue and diffuse arthralgia to which she received another 8-week course of ceftriaxone. During the 3rd week of this treatment course the patient developed throbbing pain to the lower back, general malaise, nausea and vomiting within 10min of ceftriaxone administration. These symptoms progressed over the next 2 days and the patient presented to the hospital for evaluation. On physical exam the patient appeared sick and pale. The patient had a temperature of 38.1°C, elevated blood pressure (171/78mmHg) and decreased oxygen saturation (86%). Laboratory studies showed normocytic anemia (Hgb 9.9g/dL), mild thrombocytopenia, leukocytosis, and elevated sedimentation rate and C-reactive protein. The patient also had an elevated SCr of 2.8mg/dL, increased LDH and indirect bilirubin. A direct Coombs test was positive, strongly suggesting a diagnosis of immune-mediated hemolytic anemia. Other causes of hemolytic anemia were ruled out (infection, malignancy, autoimmune), and the patient was diagnosed with drug-induced hemolytic anemia secondary to ceftriaxone. A direct antiglobulin test was positive in the presence of a weak positive test for IgG, a negative test for IgM and IgA, a weakly positive test for anti-C3c and a positive test for anti-C3d. The patient's ceftriaxone was discontinued and there was immediate clinical and biochemical improvement. The patient was discharged from the hospital 14 days after admission [31A].

Cefuroxime

Organs and Systems

NERVOUS SYSTEM

Neurotoxicity ranging from mild encephalopathy to coma is a rare adverse reaction from cephalosporin use and primarily reported to occur in older patients with renal dysfunction or previous neurological disease. A 61-year-old female with a history of COPD, diabetes, rheumatoid arthritis and contracted kidneys was admitted to the hospital with a 14-day history of fever, right flank pain and acute on chronic renal insufficiency. An abdominal CT scan showed right sided urolithiasis with urethral obstruction and pelvic dilatation requiring nephrostomy catheter placement. At this time the patient was given cefuroxime 1500mg IV three times daily and gentamicin 400mg IV one time for a suspected urinary tract infection. The patient improved clinically over the next 4 days. Cultures from the nephrostomy catheter grew *Proteus mirabilis.* Upon neurological exam on day 9 of admission, the patient showed a decline in GSC, horizontal nystagmus, myoclonus and tremors, and the patient was transferred to the ICU for management. Cephalosporin-induced neurotoxicity was suspected and cefuroxime was discontinued. The patient was treated with continuous veno-venous hemofiltration and cefuroxime levels rapidly declined over a few hours. The next day the patient had complete resolution of neurological symptoms. Caution should be taken in patients with renal insufficiency receiving cefuroxime [32A].

SENSORY SYSTEMS

Anterior and posterior inflammation has been reported in patients receiving high dose cefuroxime as an intracameral injection. Radua and colleagues performed complete ophthalmic examinations postoperatively on 19 patients who received intracameral cefuroxime during cataract surgery. Fourteen patients received cefuroxime 12.5mg/0.1mL and five received 10mg/0.1mL. Eight of these patients (42%) showed ocular side effects. One patient demonstrated noninfectious panuveitis and one patient showed serous macular detachment. Five patients showed disruption of the ellipsoid layer with drop in visual acuity, and one patient displayed color alteration. The authors suggest that protocols to avoid dilution errors of cefuroxime should be available when commercial preparations are not available to avoid high doses that could lead to toxicity [33A].

IMMUNOLOGIC

Antibiotics are common causes of perioperative anaphylaxis. Patients on concurrent antihypertensive medications can experience anaphylaxis that is refractory to conventional therapy. A 46-year-old female with a history of hypertension and hyperthyroidism was to undergo an open reduction and internal fixation of a fracture of the humerus. The patient was taking several antihypertensive medications including atenolol 25mg daily, losartan 50mg twice daily, prazosin 2.5mg daily, and nicardipine 20mg twice daily. The patient was also taking levothyroxine sodium 100 mcg daily. The patient had a reported penicillin allergy but endorsed previously tolerating amoxicillin and erythromycin without adverse reaction. The morning of the surgery she received all of her antihypertensive medications except losartan, and had an initial blood pressure of 171/75 mmHg and heart rate of 69 BPM. Within minutes of administration of 750 mg of cefuroxime for surgical prophylaxis, her heart rate dropped to 36 BPM and her blood pressure was unrecordable. At this time all anesthetics were stopped and the patient was switched to manual ventilation. She was unresponsive to an initial 0.6 mg of atropine, but her heart rate rose to 112 bpm after a second dose. The patient developed a diffuse maculopapular rash and angioedema of the eyelids, lips and face. The patient was then given IV fluids, hydrocortisone, epinephrine and promethazine. Her blood pressure became recordable at 81/30 mmHg. The patient was started on a dopamine infusion, but her blood pressure did not respond and she was subsequently started on norepinephrine. Her blood pressure increased to 92/43 mmHg and heart rate of 146 BPM. The patient was transferred to the critical care unit and the surgery was postponed. In the critical care unit the patient continued to need hemodynamic support for 24 h when her blood pressure and heart rate gradually improved. At this time all vasopressor medications were stopped and the patient was responsive and following commands. The patient had her antihypertensives restarted 3 days later. Two weeks later the patient returned for her operation and she was given ciprofloxacin for preoperative antibiotics. The authors state that although skin testing could have been administered prior to the procedure, there are no validated diagnostic tests for evaluating IgE-mediated reactions to antibiotics other than penicillin [34A].

TETRACYCLINES AND GLYCYCLINES

Doxycycline

Organs and Systems

SKIN

The overall incidence of cutaneous side effects from doxycycline administration is not well-defined and is even more poorly defined in the pediatric population. Bayhan and colleagues performed a retrospective analysis over 2 years evaluating 189 patients, aged 1 month to 18 years, who were diagnosed with brucellosis. Patients greater than 8 years of age ($n = 141$) were treated with rifampicin (15–20 mg/kg per day) and doxycycline (200 mg daily). Of the 141 patients evaluated, seven (5.0%) experienced cutaneous side effects. The average duration of treatment prior to onset of symptoms was 9.5 weeks. The majority of the patients suffered from nail hyperpigmentation and onycholysis (without painful lesions). Symptoms gradually improved over time while patients continued on doxycycline. Two of the seven patients discontinued treatment leading to complete resolution of symptoms. The first of these patients suffered from nail hyperpigmentation, onycholysis and tooth discoloration and the second had pronounced facial photosensitivity. The authors conclude that cutaneous side effects should be suspected in pediatric patients receiving doxycycline for extended durations [35c].

Photosensitivity from tetracyclines is well-described in the literature. There is, however, limited literature devoted to doxycycline-associated photosensitivity specifically. Given the frequency with which doxycycline is used in malaria prophylaxis for travelers who may be subjected to significant sun exposure, Goetze and colleagues published a systematic review summarizing all available reports on the clinical manifestations of and influencing factors for doxycycline-related phototoxicity. In total, 21 articles published between 1990 and 2015 in German and English were analyzed. There was inconsistency across the publications in the definition of "phototoxic" reaction. Clinical symptoms varied from sunburn-like manifestations with erythema and itching to large-area erythematous plaques. Photo-onycholysis was documented in several references and lichenoid reaction and actinic granuloma were reported in one case each, respectively. Rates of doxycycline-associated phototoxicity ranged from 6% to 42%. The triggering UV spectrum consisted mainly of UVA-1 (340–400 nm). The authors acknowledge that studies are required to prove the efficacy of sunscreen products, but suggest that all travelers taking doxycycline for malaria prophylaxis should be counseled to avoid sun exposure when possible [36R].

Susceptibility Factors

AGE

Gaillard and colleagues published a review of dental enamel-associated side effects of doxycycline in the pediatric population to re-evaluate its role in malaria treatment due to rising resistance to artemisinin-based combination therapy (ACT). All tetracycline compounds are contraindicated in patients less than 8 years of age because of the risk of tooth discoloration and dental enamel hypoplasia. The US Centers for Disease Control recommend the use of doxycycline in pediatric patient for the treatment of Q fever and rickettsial diseases. Doxycycline has a lower affinity for calcium chelation when compared with tetracycline and in animal studies tooth discoloration was associated with tetracycline more than chlortetracycline or oxytetracycline. The authors' review found no visible dental effects following short courses of doxycycline in children under 8 years of age in recent data. The authors conclude that doxycycline, in combination with quinine, should be recommended for the treatment of malaria in pediatric patients under 8 years of age when ACT fails or is unavailable [37r].

Minocycline

Organs and System

NERVOUS SYSTEM

Minocycline-induced drug fever is rare, and when described in prior case reports have exclusively occurred with concomitant skin eruptions. Gu and colleagues present a case report in which minocycline induced fever occurred without cutaneous manifestation. A 24-year-old woman was prescribed minocycline 100mg daily for the treatment of acne. The patient presented to the emergency department with a 6-day history of fever (39°C), dizziness, chest tightness and eye pain. Minocycline was discontinued and the patient was diagnosed with a cold and treated (treatment not described). The patient defervesced after 5 days. Eight days following the initial discontinuation, minocycline was restarted. Within a few hours of re-initiation the patient's symptoms returned. Other than fever, the patient's vitals were normal and there were no signs of infection on physical exam. Blood cultures were negative for bacteria and herpesvirus 6 IgG. Imaging of the chest and brain revealed no abnormalities. Laboratory findings were indicative of normal liver and kidney function, normal C-reactive protein and eosinophilia (WBC 7.5×10^9/L.; eosinophils, 1.2×10^9/L). Given the temporal relationship between symptom onset and medication administration the authors concluded that the patient was experiencing minocycline-induced fever without rash. Minocycline was discontinued and the patient was prescribed oral prednisolone 30mg per day. The patient's symptoms improved dramatically. The authors advise that physicians be vigilant in regard to unusual causes of drug-induced fever [38A].

SKIN

Minocycline-induced hyperpigmentation (MIH) of various tissues is a well-described adverse effect. Multiple case reports published in 2017 describe this adverse effect in the setting of unique circumstances. There are three distinguishable types of MIH: type I is present in an area of prior scarring and is typically blue-gray or black in color; type II occurs on the skin of the shins and forearms and is blue-gray; and type III is present in areas of high sun exposure and appears brown. Fernandez-Flores and colleagues detail a 50-year-old man who presented with blue-gray discoloration of the face, sclera and back not characteristic of any of these types. The patient had taken minocycline 100mg twice daily for 15 years as well as colloidal silver supplements for 2 years 16 years prior to presentation. The authors describe that argyria, the mucocutaneous discoloration that occurs with silver salt ingestion, may have contributed to the non-characteristic distribution and presentation of the hyperpigmentation. The histopathological findings were consistent with pigmentation due to minocycline only. This is the first case report, to the authors' knowledge, of hyperpigmentation in a patient with a history of both minocycline and silver ingestion [39A].

Haskes and colleagues presented a case of an 87-year-old man who presented with a 1-year history of worsening blue discoloration of his sclera bilaterally. The patient had been on minocycline 200mg daily for 13 years as suppressive therapy for a bacterial infection of his right knee replacement. No visual disturbance was noted and the decision was made that the patient would remain on minocycline therapy due the risk of recurrent join infection [40A].

Minocycline is used to treat severe papulopustular skin eruptions caused by epidermal growth factor receptor (EGFR) inhibitors. Bella and colleagues presented a case of an 87-year-old man who developed inflammatory follicular-based papules and pustules over 50% of his body within 2 months of therapy with erlotinib. The erlotinib dose was reduced the eruption was treated with minocycline 100mg twice daily for 8 months. At this time the patient developed blue-grey skin hyperpigmentation. Initially the discoloration was attributed to the erlotinib; however, a biopsy 30 months into therapy confirmed that the minocycline was the causative agent. The patient was changed from minocycline to doxycycline, and the hyperpigmentation resolved with laser therapy. The authors present this data to advise physicians that hyperpigmentation is not commonly associated with erlotinib, and that minocycline should be evaluated as the causative agent in these patients [41A].

Hyperpigmentation has been described in cardiovascular tissue of patients with long-term minocycline exposure (>30 years). A 56-year-old man with a past medical history significant for coronary artery atherosclerosis was undergoing mitral and aortic valve repair when a bluish discoloration of the aorta and coronary arteries was observed. The blue color appeared concerning for intramural thrombus and occult aortic dissection and the patient was placed on cardiopulmonary bypass. Tissue sent for frozen section microscopy revealed no structural damage but was noted to have intralaminar pigmentation without calcification of the vessel wall and pigment in the intimal layer of the aorta. No thrombus or dissection was detected. Discoloration of the sclera, nail beds, gums and lower extremity vasculature were further noted. Of the patient's home medications, only minocycline carried a known association with tissue discoloration. The patient had been on minocycline 100 mg three times daily for the prior 4 years. The authors present this data as the first known minocycline-induced hyperpigmentation of cardiovascular tissue advanced enough to appear similar to aortic and coronary artery dissection. This error may have resulted serious unnecessary morbidity and mortality for the patient [42A].

Tigecycline

Organs and Systems

HEMATOLOGIC

There are few cases in the literature associating tigecycline administration with coagulopathies and hypofibrinogenemia. A 47-year-old man was admitted to the intensive care unit (ICU) with severe acute cholangitis. He had no hereditary or past medical history of hepatic or hematologic disease. He was diagnosed with choledocoduodenal fistula and initiated on 1 g of imipenem–cilastatin IV every 8 h. On the 3rd day of hospitalization the patient clinically decompensated and was found to have a leukocytosis of 19.1×10^3/mL, increase in C-reactive protein (CRP) of 68 mg/L and a maximum temperature of 40°C. Blood cultures were positive for methicillin-resistant *Staphylococcus warneri* (MRSW) and bile cultures were positive for *Candida albicans*. The patient was initiated on continuous veno-venous hemofiltration (CVVH), caspofungin and high-dose tigecycline at a dose of 100 mg twice daily. On the 2nd day of tigecycline therapy the patient had clinically defervesced, however, was experiencing coagulopathies with an increased prothrombin time (24.3 from 13.6 s), international normalized ratio (2.09 from 1.18), activated partial thromboplastin time (94.5 from 35.1 s), thrombin time (53.0 from 18.2 s) and decreased fibrinogen (0.69 from 2.84 g/L). Tigecycline was discontinued and switched to vancomycin. Five days following discontinuation the patient's coagulopathies resolved and all values returned to baseline. The patient was further treated and eventually discharged. The authors acknowledge that the decrease in fibrinogen and coagulopathies may have been consequent to increased consumption or impaired synthesis due to critical illness. They cite the Naranjo Adverse Drug Reaction Probability Scale score of 5 as supporting an adverse drug event. The authors reviewed three prior similar case reports and found that female gender, renal insufficiency and high-dose tigecycline may be risk factors for these adverse events. Given the increase in multi-drug resistant organisms that may necessitate off-label high-dose tigecycline administration, the authors conclude that coagulation parameters should be closely monitored during administration in patients with these risk factors [43A].

General Adverse Drug Reactions

Patients who have received abdominal solid organ transplants (SOTs) are at an increased risk for polymicrobial and multi-drug resistant intra-abdominal infections (IAI) due to immunosuppression and prior antibiotic exposure. There is currently little literature evaluating tigecycline in the treatment of IAI in the SOT population. A retrospective cohort study of 81 patients evaluated the efficacy and safety of tigecycline vs comparator therapy (broad-spectrum beta-lactam with or without a Gram positive agent) for SOTs with IAI. Patients were matched 1:2 (tigecycline vs comparator) with 27 patients receiving tigecycline and 54 patients receiving comparator therapy. Adverse events occurred more frequently in the tigecycline group with at least one event occurring in 29.6% of patients (8 of 27) and in only 9.3% (5 of 54) of the comparator group ($P = 0.026$). Nausea was the most common adverse event in the tigecycline group occurring in 6 of 27 patients, requiring discontinuation in four of those patients. Other adverse events occurring less frequently included thrombocytopenia, hepatotoxicity and pancreatitis. Nausea and pancreatitis resolved following discontinuation of tigecycline. Neither thrombocytopenia nor hepatotoxicity resolved with discontinuation; however, the authors postulate that this may have been due to worsening hepatic function rather than tigecycline toxicity. Only five adverse events occurred in the comparator group including creatinine kinase elevations, nausea and myelosuppression. The authors conclude that careful risk benefit analyses should be undertaken prior to initiation of tigecycline as first-line therapy for IAIs in the abdominal SOT population [44c].

Additional case studies on the same topic can be found in another published review [45R].

References

[1] Aydm A, Aykan MB, Saglam K. Seizure induced by ertapenem in an elderly patient with dementia. Consult Pharm. 2017;32(10):561–2 [r].
[2] Patel UC, Fowler MA. Ertapenem-associated neurotoxicity in the spinal cord injury (SCI) population: a case series. J Spinal Cord Med. 2017;6:1–10. https://doi.org/10.1080/10790268.2017.1368960 [A].
[3] Docobo R, Bukhari S, Baloch ZQ. Ertapenem-induced thrombocytosis. Cureus. 2017;9(5):1–5. https://doi.org/10.7759/cureus.1263 [A].
[4] Huang R, Cai GQ, Zhang JH, et al. Meropenem-induced immune thrombocytopenia and the diagnostic process of laboratory testing. Transfusion. 2017;57:2715–9 [A].
[5] Anuhya TV, Acharya R, Madhystha S, et al. Meropenem induced hypokalemia. J Clin Diagn Res. 2017;11(8):5–6 [A].
[6] Sima M, Hartinger J, Rulisek J, et al. Meropenem-induced valproic acid elimination: a case report of clinically relevant drug interaction. Prague Med Rep. 2017;118(2):105–9 [A].
[7] Tandan M, Vellinga A, Bruyndonckx R, et al. Adverse effects of amoxicillin for acute lower respiratory tract infection in primary care: secondary and subgroup analysis of a randomised clinical trial. Antibiotics (Basel). 2017;6:36 [MC].

[8] Fisse AL, Straßburger-Krogias K, Gold R, et al. Recurrent trimethoprim-sulfamethoxazole induced aseptic meningitis with associated ampicillin-induced myoclonic twitches. Int J Clin Pharmacol Ther. 2017;55(7):627–9 [A].
[9] Wungwattana M, Savic M. Tacrolimus interaction with nafcillin resulting in significant decreases in tacrolimus concentrations: a case report. Transpl Infect Dis. 2017;19:12662 [A].
[10] Khelil M, Chkirbene Y, Milka M, et al. Penicillin-induced fulminant myocarditis: a case report and review of the literature. Am J Forensic Med Pathol. 2017;38(1):29–31 [A].
[11] Kumar V, Khosla S, Stancu M. Torsade de pointes induced by hypokalemia from imipenem and piperacillin. Case Rep Cardiol. 2017;2017:4565182 [A].
[12] Navalkele B, Pogue J, Karino S, et al. Risk of acute kidney injury in patients on concomitant vancomycin and piperacillin–tazobactam compared to those on vancomycin and cefepime. Clin Infect Dis. 2017;64(2):116–23 [C].
[13] Downes K, Cowden C, Laskin B, et al. Association of acute kidney injury with concomitant vancomycin and piperacillin/tazobactam treatment among hospitalized children. JAMA Pediatr. 2017;171(12)e173219 [MC].
[14] LeCleir L, Pettit R. Piperacillin-tazobactam versus cefepime incidence of acute kidney injury in combination with vancomycin and tobramycin in pediatric cystic fibrosis patients. Pediatr Pulmonol. 2017;52:1000–5 [c].
[15] Mousavi M, Zapolskaya T, Scipione M, et al. Comparison of rates of nephrotoxicity associated with vancomycin in combination with piperacillin-tazobactam administered as an extended versus standard infusion. Pharmacotherapy. 2017;37(3):379–85 [C].
[16] Adler NR, McLean CA, Aung AK, et al. Piperacillin–tazobactam-induced linear IgA bullous dermatosis presenting clinically as Stevens–Johnson syndrome/toxic epidermal necrolysis overlap. Clin Exp Dermatol. 2017;42:299–302 [A].
[17] Yusef D, Gonzalez B, Foster C, et al. Piperacillin–tazobactam-induced adverse drug events in pediatric patients on outpatient parenteral antimicrobial therapy. Pediatr Infect Dis J. 2017; 36:50–2 [c].
[18] Flynt LK, Kenney RM, Zervos MJ, et al. The safety and economic impact of cefazolin versus nafcillin for the treatment of methicillin-susceptible staphylococcus aureus bloodstream infections. Infect Dis Ther. 2017;6:225–31 [c].
[19] Almeida JP, Lopes A, Campos Melo A, et al. Selective hypersensitivity to cefazolin and contribution of the basophil activation test. Eur Ann Allergy Clin Immunol. 2017;49(2):84–7 [A].
[20] Appa AA, Jain R, Rakita RM, et al. Characterizing cefepime neurotoxicity: a systematic review. Open Forum Infect Dis. 2017;4(4)ofx170 [M].
[21] Nakagawa R, Sato K, Uesaka Y, et al. Cefepime-induced encephalopathy in end-stage renal disease patients. J Neurol Sci. 2017;376:123–8 [c].
[22] Turner RB, Wilson ED, Saedi-Kwon H, et al. Comparative analysis of neutropenia in patients receiving prolonged treatment with ceftaroline. J Antimicrob Chemother. 2018;73:772–8. https://doi.org/10.1093/jac/dkx452 [C].
[23] Temkin E, Torre-Cisneros J, Beovic B, et al. Ceftazidime-avibactam as salvage therapy for infections caused by carbapenem-resistant organisms. Antimicrob Agents Chemother. 2017;61(2):1–11 [A].
[24] Munita JM, Aitken SL, Miller WR, et al. Multicenter evaluation of ceftolozane-tazobactam for serious infections caused by carbapenem-resistance *Pseudomonas aeruginosa*. Clin Infect Dis. 2017;65(1):158–61 [C].
[25] Donnelly PC, Sutich RM, Easton R, et al. Ceftriaxone-associated biliary and cardiopulmonary adverse events in neonates: a systematic review of the literature. Paediatr Drugs. 2017; 19:21–34 [M].
[26] McCarthy CA, Powell KR, Jaskiewicz JA, et al. Outpatient management of selected infants younger than two months of age evaluated for possible sepsis. Pediatr Infect Dis J. 1990;9:385–9 [c].
[27] Bradley JS, Wassel RT, Lee L, et al. Intravenous ceftriaxone and calcium in the neonate: assessing the risk for cardiopulmonary adverse events. Pediatrics. 2009;123:e609–13 [c].
[28] Van Reempts PJ, Van Overmeire B, Mahieu LM, et al. Clinical experience with ceftriaxone treatment in the neonate. Chemotherapy. 1995;41:316–22 [c].
[29] Inoue Y, Doi Y, Aristo T, et al. Three cases of hemodialysis patients receiving high-dose ceftriaxone: serum concentrations and its neurotoxicity. Kidney Int Rep. 2017;2:984–7 [A].
[30] Tasks J. Ceftiraxone-induced hemolytic anemia in a Jehovah's witness. Am J Case Rep. 2017;18:431–5 [A].
[31] De Wilde M, Speeckaert M, Callens R, et al. Ceftriaxone-induced immune hemolytic anemia as a life-threatening complication of antibiotic treatment of 'chronic lyme disease'. Acta Clin Belg. 2017;72(2):133–7 [A].
[32] van Dam DGHA, Burgers DMT, Foudraine N, et al. Treatment of cefuroxime-induced neurotoxicity with continuous venovenous haemofiltration. Neth J Med. 2017;75(1):32–4 [A].
[33] Kamal-Salah R, Osoba O, Edward D. Ocular toxicity after inadvertent intracameral injection of high dose of cefuroxime during cataract surgery: a case series. Retin Cases Brief Rep. 2017:1–4 [A].
[34] Nag DS, Samaddar DP, Kant S, et al. Perianesthetic refractory anaphylactic shock with cefuroxime in a patient with history of penicillin allergy on multiple antihypertensive medications. Rev Bras Anestesiol. 2017;67(2):217–20 [A].
[35] Bayhan GI, Akbayram S, Yavuz GO, et al. Cutaneous side effects of doxycycline: a pediatric case series. Cutan Ocul Toxicol. 2017;36(2):140–4 [c].
[36] Goetze S, Hiernickel C, Elsner P, et al. Phototoxicity of doxycycline: a systematic review on clinical manifestations, frequency, cofactors, and prevention. Skin Pharmacol Physiol. 2017; 30:76–80 [R].
[37] Gaillard T, Briolant S, Madamet M, et al. The end of a dogma: the safety of doxycycline use in young children for malaria treatment. Malar J. 2017;16:148 [r].
[38] Gu W, Shi D, Mi N, et al. Physician, beware! Drug fever without skin rashes can be caused by minocycline. J Investig Allergol Clin Immunol. 2017;27(4):268–9 [A].
[39] Fernandez-Flores A, Nguyen T, Cassarino D, et al. Mucocutaneous hyperpigmentation in a patient with a history of both minocycline and silver ingestion. Am J Dermatopathol. 2017;39:916–9 [A].
[40] Haskes C, Shea M, Imondi D, et al. Minocycline-induced scleral and dermal hyperpigmentation. Optom Vis Sci. 2017;94(3):436–42 [A].
[41] Bella A, Romanb J, Gratrixb M, et al. Minocycline-induced hyperpigmentation in a patient treated with erlotinib for non-small cell lung adenocarcinoma. Case Rep Oncol. 2017;10:156–60 [A].
[42] Buckley T, Lee J, Gipson K, et al. Minocycline-induced hyperpigmentation mimicking aortic dissection. Ann Thorac Surg. 2017;103:e121–2 [A].
[43] Wu X, Zhao P, Dong L, et al. A case report of patient with severe acute cholangitis with tigecycline treatment causing coagulopathy and hypofibrinogenemia. Medicine. 2017;96(49)e9124 [A].
[44] Liebenstein T, Schulz L, Viesselmann C, et al. Effectiveness and safety of tigecycline compared with other broad-spectrum antimicrobials in abdominal solid organ transplant recipients with polymicrobial intraabdominal infections. Pharmacotherapy. 2017;37(2):151–8 [c].
[45] Buckler RA, Peahota MM, Gallagher JC. Chapter 21—beta-lactases and tetracyclines. In: Ray SD, editor. Side effects of drugs: annual: A worldwide yearly survey of new data in adverse drug reactions, vol. 39. Elsevier; 2017. p. 217–27 [R].

CHAPTER

25

Miscellaneous Antibacterial Drugs

Emily C. Tucker[1], *Matthew B. Roberts*, *Nicola A. Sweeney*, *David L. Gordon*

Department of Microbiology and Infectious Diseases, Flinders Medical Centre, Bedford Park, SA, Australia
[1]Corresponding author: emily.tucker@sa.gov.au

AMINOGLYCOSIDES [SEDA-35, 463; SEDA-36, 363; SEDA-37, 293; SEDA-38, 229; SEDA-39, 229]

Pediatric

Once vs thrice daily dosing of aminoglycosides (tobramycin) in adults and children with cystic fibrosis was examined in a Cochrane Systematic review involving 15 studies. Overall, there was no difference in ototoxicity (RR 0.56, 95% CI 0.04–7.96, $n=266$) between comparator arms. Nephrotoxicity was not altered in adults; however, in children, once-daily dosing appeared to have a lesser effect on percentage change in creatinine rise with mean difference being −8.25 (95% CI −15.32 to −1.08) compared to thrice-daily dosing. This finding suggests once daily dosing may reduce adverse effects in children, but requires more rigorous evaluation to confirm [1M].

Psychiatric

Neuropsychiatric effects of antibiotics, including aminoglycosides, were reviewed with minimal evidence of systemically administered aminoglycosides contributing to neuropsychiatric symptoms [2R].

Pulmonary

The use of aerosolized antibiotics, including aminoglycosides, was reviewed. Included studies were heterogeneous and therefore, direct comparisons cannot be drawn. Adverse event reported in this review included bronchospasm [3R].

The use of nebulised anti-infective agents in invasively and mechanically ventilated adults was reviewed in a meta-analysis involving 11 studies. There was no increase in nephrotoxicity or neurotoxicity related to the use of inhaled colistin or aminoglycoside. Four randomised controlled trials examining nebulised anti-infectives demonstrated a 9% increase (risk difference 0.09, 95% CI −0.01 to 0.18) in respiratory complications but with high heterogeneity. This area requires further investigation to guide clinicians on the expected adverse effects associated with inhaled antimicrobials [4R].

Sensory Systems

A review into the proposed mechanisms and potential pathways for development of aminoglycoside-related ototoxicity was published that includes future directions for aminoglycoside analogues to overcome this side effect [5R].

A Brazilian cohort of 599 patients with multi-resistant tuberculosis found 164 patients who reported auditory or vestibular complaints whilst receiving aminoglycosides. Amikacin and streptomycin were used in 145 and 19 patients, respectively. Aminoglycoside was ceased in 55 (33.5%) of patients due to these complaints; 70.7% of these complaints were hypoacusis. Only 12 cases had audiometric testing performed and hearing loss was confirmed in 11. This study confirms the known risk of ototoxicity with injectable aminoglycosides [6C].

The prevalence of aminoglycoside-related vestibular toxicity was reported to be 0%–60% in a review which included 27 studies. It included prevalence of abnormal test results in multiple types of objective tests for vestibular function with comment on the potential use of these. No conclusion could be reached on the exact prevalence with various aminoglycosides or the optimal objective vestibular test based on included evidence [7R].

Mechanisms of drug ototoxicity, including aminoglycosides, were outlined by Lanvers-Kaminsky and colleagues [8R].

Side Effects of Drugs Annual, Volume 40
ISSN: 0378-6080
https://doi.org/10.1016/bs.seda.2018.07.019

Amikacin [SEDA-35, 463; SEDA-36, 363; SEDA-37, 294; SEDA-38, 229; SEDA-39, 229]

Sensory Systems

Eighty patients with multidrug-resistant tuberculosis receiving 6.5mg/kg/day kanamycin or amikacin were included in a retrospective observational study examining the relationship between ototoxicity and therapeutic drug monitoring. Seventy patients had audiometric testing results available, 9 demonstrated hearing loss. Regression analysis showed daily dose of aminoglycoside correlated with decibel hearing loss at 8000Hz. Clinical outcomes were successful in 35 (67.3%) of patients. Hence, there was minimal hearing loss in a cohort with audiometric monitoring [9C].

The use of inhaled amikacin for difficult-to-treat non-tuberculous mycobacterial disease was examined in 26 patients in a single centre. Inhaled amikacin was associated with reversible tinnitus in one patient but no other adverse events were reported [10c].

A randomised controlled trial in 89 patients with treatment-refractory non-tuberculous pulmonary mycobacterial disease examined the addition of inhaled liposomal amikacin (590mg once daily) vs placebo to their usual multi-drug regime. Those receiving liposomal amikacin experience more dysphonia (43.2% vs 8.9%), bronchiectasis exacerbations (38.6% vs 20.0%) and cough (31.8% vs 13.3%). Audiovestibular outcomes were similar. One patient receiving amikacin experienced a reversible rise in creatinine [11c].

A retrospective analysis of 100 consecutive patients treated with either injectable amikacin or capreomycin for multidrug resistant tuberculosis at four centres demonstrated a fivefold greater risk of ototoxicity associated with amikacin (HR 5.2, 95% CI 1.2–22.6, $P=0.03$). However, amikacin was associated with less hypokalaemia than capreomycin (OR 0.26, 95% CI 0.11–0.72) [12c].

Paromomycin

Skin

The Cochrane collaborate published a systematic review that included analyses of topical paromomycin for treatment of Old World cutaneous leishmaniasis. Compared to placebo vehicles, topical paromomycin was associated with more localised skin reactions (RR 1.42, 95% CI 0.67–3.01; $n=713$); however, this result is without confidence. Use of twice daily topical paromomycin for 4 vs 2 weeks plus an additional 2 weeks of placebo vehicle was compared. There were no adverse events reported in either group ($n=233$) [13M].

Gentamicin [SEDA-35, 463; SEDA-36, 364; SEDA-37, 294; SEDA-38, 229; SEDA-39, 230]

Gastrointestinal

Intravesical instillation of gentamicin for 22 patients with neurogenic bladder was given for 6 months at the time of intermittent self-catheterisation. This was associated with the development of diarrhoea in one patient but direct causality was not proven [14c].

Sensory systems

A 21-year-old female experienced haemorrhagic occlusive retinal vasculitis probably secondary to inadvertent intravitreal injection of high dose gentamicin [15A].

A 65-year-old male experienced macular toxicity, postulated to result from subconjunctival injection of gentamicin which accessed the macula via an un-sutured 25 gauge sclerotomy. The report outlines the known risk of macular toxicity with gentamicin when applied in different settings, such as subconjunctival injection, where spread to the macula may occur [16A].

Urinary Tract

A systematic review including 26 studies examined the adverse events associated with a single dose of gentamicin (dose range 1mg/kg to 480mg). There was significant heterogeneity of the studies included, thus a meta-analysis could not be performed. A total of 15552 records were included with 2520 reports of acute kidney injury (AKI) [17R].

A nested case–control study among paediatric patients admitted to intensive care found use of gentamicin was significantly associated with development of renal failure when adjusting for other risk factors for AKI (adjusted OR 1.9, 95% CI 1.4–2.4) [18C].

Tobramycin [SEDA-35, 464; SEDA-36, 365; SEDA-37, 295; SEDA-38, 229; SEDA-39, 230]

Ear, Nose and Throat

Tobramycin inhalation powder was administered to 10 paediatric patients with cystic fibrosis and chronic *Burkholderia cepacia* complex infection. One patient experienced throat pain and difficulty swallowing on day 3 that was likely related to therapy, resulting in drug discontinuation [19c].

Pulmonary

A review of studies investigating the use of inhaled tobramycin in patients with cystic fibrosis reported the most common adverse events with tobramycin inhalers are cough and pulmonary exacerbations [20R].

FLUOROQUINOLONES [SEDA-36, 464; SEDA-36, 365; SEDA-37, 295; SEDA-38, 231; SEDA-39, 231]

Cardiovascular

A meta-analysis was conducted to assess the link between fluoroquinolones and serious cardiovascular arrhythmias. Seven of 16 studies included reported risk estimates for serious arrhythmias. The relative risk of serious arrhythmia associated with fluoroquinolone use was 2.29 (95% CI, 1.2–4.36, $P=0.1$). The relative risk was greatest with moxifloxacin (4.2, 95% CI 1.91–9.27, $P<0.001$) followed by levofloxacin (1.41, 95% CI 1.16–1.70, $P<0.001$). Ciprofloxacin was not significantly associated with an increased risk of serious arrhythmias (RR 1.73, 95% CI 0.89–3.37, $P=0.1$). When compared to no fluoroquinolone use, the use of fluoroquinolones was associated with 160 more cases of serious arrhythmias per 1 000 000 courses. Fluoroquinolone use was also found to be associated with an increased risk of cardiovascular death (RR 1.60, 95% CI: 1.17–2.20; $P=0.004$) but not all cause mortality, a finding which could not be explained from the results analysed [21M].

Gastrointestinal

Fluoroquinolone-induced collagen degradation has been proposed as one of the mechanisms behind some of the well-recognised adverse effects associated with this class of antibiotic. It is known that genetic disorders of collagen are associated with an increased incidence of gastrointestinal perforation and it has been postulated that fluoroquinolones may have a similar association. A Taiwanese nested case–control study utilized the national health insurance claims database to identify 17 510 cases of gastrointestinal perforation with each case matched to 100 controls. The study found that there was a non-negligible increased risk of gastrointestinal perforation in patients who had received recent fluoroquinolone therapy (RR 1.90, 95% CI 1.62–2.22) after adjusting by disease risk. No direct cause and effect relationship could be established and the authors acknowledged there were many potential confounders [22MC].

Musculoskeletal

Two cases of triceps tendon rupture were reported following fluoroquinolone use for urinary tract infections (UTIs). Both men sustained forceful elbow injuries approximately 3 months following fluoroquinolone use and neither had any other identifiable risk factor for tendon rupture. No histopathology was available in either case and the extent to which fluoroquinolones were implicated could not be proven. The authors noted that whilst there are increasing reports of tendon rupture associated with fluoroquinolone use, triceps tendon rupture is very rare [23A].

Reproductive System

A case–control study was conducted using data from the Quebec Pregnancy Cohort to assess the association between antibiotic use and spontaneous abortion. Of 182 369 pregnancies included, 8702 resulted in spontaneous abortion. The use of quinolones in early pregnancy was associated with an increased risk of spontaneous abortion (OR 2.72, 95% CI 2.27–3.27). A subgroup analysis to compare the rate of spontaneous abortion in women with UTIs treated with quinolones vs penicillin found higher rates in the quinolone group, adjusted OR 8.73 (95% CI 3.08–24.77) in 17 exposed cases [24MC].

Sensory System

A retrospective cohort study examined whether the use of quinolone ear drops in children following tympanostomy tube placement was associated with an increased risk of perforation requiring tympanoplasty. The use of any quinolone ear drop, ofloxacin and ciprofloxacin with or without the addition of hydrocortisone or dexamethasone, was compared to neomycin plus hydrocortisone in a cohort of 96 595 children. There was an increased incidence of perforation in the patients receiving quinolone ear drops (adjusted HR 1.61, 95% CI 1.15–2.26) compared to neomycin and this risk was further increased with the addition of steroids. There were limitations to the study, including insufficient capture of cases due to the extended follow-up required and the possibility of misclassification due to the nature of data collection through billing. The authors acknowledged that this reversible adverse effect is likely still favourable to the irreversible sensorineural hearing loss associated with the alternative, an aminoglycoside [25MC].

Levofloxacin [SEDA-35, 465; SEDA-36, 366; SEDA-37, 296; SEDA-38, 232; SEDA-39, 232]

Nervous System

A 78-year-old man with tuberculosis was commenced on levofloxacin after developing peripheral neuropathy with isoniazid. Four days after commencement he presented with brief and non-rhythmic involuntary movements. No metabolic cause was found and a neurologist assessed his symptoms as likely myoclonus induced by levofloxacin; symptoms resolved a few days after levofloxacin was ceased [26A].

A 56-year-old woman treated with levofloxacin for a UTI developed bilateral musculospiral paralysis 30 minutes after ingesting the drug. These symptoms resolved with discontinuation of the antibiotic. Fluoroquinolone use has previously been associated with the development of both reversible and permanent peripheral neuropathies [27A].

A 20-year-old male with well-controlled type 1 diabetes was prescribed a 10-day course of levofloxacin for treatment of suspected epididymitis. He developed diffuse myalgia and severe bilateral burning pain in his feet a few days into treatment requiring hospitalisation. This was initially diagnosed as complex regional pain syndrome (CRPS). Approximately 1 month into his admission, following review of the literature, the diagnosis was revised to fluoroquinolone-induced neuropathy; he did not have all the features required for a diagnosis of CRPS and the temporal relationship to his antibiotic therapy was considered pertinent. It is estimated that approximately 1% of patients taking fluoroquinolones develop peripheral neuropathy. A large number describe severe pain resulting in emergency presentation and the average duration of symptoms lasts 1–2 years [28A].

Pediatric

Fluoroquinolones have generally been avoided in the paediatric population due to the association with musculoskeletal adverse effects in animal studies. With the increasing number of resistant Gram-negative infections, there has been increased use of fluoroquinolones in the neonatal population. A case series evaluated the tolerability of levofloxacin in neonates. Six cases were included: 5 demonstrated clinical response, the sixth case had NICU support withdrawn and ultimately died due to a failure to respond to therapy. Overall levofloxacin was well tolerated in all cases; 2 neonates developed loose stools, with a third having a change in stool colour and consistency, but not diarrhoea, thought possibly related to levofloxacin. There was insufficient follow-up to comment on long-term musculoskeletal sequelae of treatment [29c].

Moxifloxacin [SEDA-35, 466; SEDA-36, 367; SEDA-37, 297; SEDA-38, 233; SEDA-39, 233]

Drug–Drug Interactions

A study was conducted in South Africa on 58 patients with tuberculosis using therapeutic drug monitoring to measure the effect of rifampicin co-administration on moxifloxacin concentration. The authors found, in keeping with previous studies, a reduced concentration of moxifloxacin when co-administered with rifampicin. The measured moxifloxacin concentrations were generally lower in this patient population than those previously reported, which may have been a result of genetic variations in drug metabolism. Additionally, many patients (72.4%) were also on anti-retroviral therapy (ART), with the majority treated with efavirenz-based ART. It was noted that there was a significant reduction in moxifloxacin concentrations in those on efavirenz, compared to those on no ART or on an alternative regimen. This is the first report of such an interaction [30c].

GLYCOPEPTIDES [SEDA-35, 466; SEDA-36, 368; SEDA-37, 298; SEDA-38, 234; SEDA-39, 234]

Dalbavancin [SEDA-37, 297; SEDA-38, 234; SEDA-39, 234]

Drug Studies

A review of the safety and efficacy of single dose dalbavancin for the treatment of acute bacterial skin and soft tissue infections was undertaken. Overall the tolerability profile was similar to 2 doses. The most common adverse effects were nausea, headache, vomiting, diarrhoea and dizziness which all occurred in less than 5% of patients. There was a low but increased risk of a reversible rise in alanine transaminase. Infusion reactions like flushing, urticaria, pruritus and rash associated with rapid infusion rates have been described. Animal studies have shown that dalbavancin may be associated with reproductive toxicity and reduced fertility, but no human studies have been conducted [31R].

A phase 1 open-labelled multi-centre study assessed the pharmacokinetics and safety of single dose dalbavancin in hospitalized paediatric patients aged 3 months to 11 years of age. Nineteen of 34 subjects experienced 36 treatment-emergent adverse events. Five of those were assessed as possibly or probably related to dalbavancin, including rash, urticaria, infusion site discomfort and a transient liver function test abnormality. Within the limitations of audiometric testing in this population no ototoxicity was detected [32c].

Teicoplanin [SEDA-35, 467; SEDA-36, 368; SEDA-37, 298; SEDA-38, 235; SEDA-39, 234]

Drug–Drug Interaction

A retrospective review of 49 patients compared the effect of teicoplanin vs vancomycin administration on the prothrombin time-international normalised ratio (PT-INR) in patients who were concurrently receiving warfarin therapy. PT-INR increased significantly more after teicoplanin administration than following vancomycin administration; 80.9% increase (52%–155.3%) and 30.6% (4.5%–44.5%), respectively. It was hypothesised that teicoplanin, which binds more strongly to albumin than vancomycin, could displace warfarin from its albumin binding site leading to an increase in free warfarin concentration. The authors recommended careful monitoring of PT-INR in patients on concurrent teicoplanin/warfarin therapy [33c].

Telavancin [SEDA-35, 467; SEDA-36, 369; SEDA-37, 298; SEDA-38, 234]

A review which included the tolerability profile of telavancin was undertaken. The most frequently observed adverse effects observed were dysgeusia, nausea, headache, foamy urine and QTc prolongation. Rapid infusion of telavancin had the potential to induce a red man syndrome-like reaction. Reversible nephrotoxicity was reported in ~5% of patients. Animal experiments have shown reduced foetal weight and an increased incidence of digit and limb malformation making it a Class C teratogenic agent. There is a paucity of human data. The authors recommended a pregnancy test for fertile women prior to starting therapy. Paediatric safety has yet to be established and the drug is currently not recommended for use in children [34R].

Urinary Tract

A review was undertaken to compare the efficacy and safety of telavancin and vancomycin in the treatment of MRSA infections. Seven studies were included in the analysis. Whilst telavancin appeared to be efficacious it had more serious adverse events reported, particularly nephrotoxicity (RR 2.13, 95% CI 1.72–2.64) when compared to vancomycin [35R].

Vancomycin [SEDA-35, 467; SEDA-36, 369; SEDA-37, 298; SEDA-38, 235; SEDA-39, 235]

Drug Dosage Regimens

The relationship between trough serum vancomycin concentration (SVC) and neutropenia was retrospectively evaluated in 1307 adult patients, without renal failure, treated with vancomycin at conventional doses. Patients with neutropenia ($n = 163$) showed lower peak and trough SVC than non-neutropenic subjects ($P < 0.0001$). The median estimated vancomycin half-life was significantly shorter in the neutropenic than the non-neutropenic group (7.4 hours (IQR 5.9–10.7 hours) vs 8.9 hours (IQR 6.9–12.0 hours), respectively ($P < 0.0001$). Neutropenia remained significantly associated with low SVC even after adjusting for other variables. The mechanism for suboptimal SVC is unknown. The authors recommended that vancomycin dosing should be closely monitored in neutropenic patients [36C].

A single-centre prospective cohort study investigated the relationship between the timing of antimicrobial prophylaxis with respect to surgical incision time and the rate of surgical site infections (SSI) in cardiac surgery patients. The antimicrobial prophylaxis protocol used cefazolin and vancomycin. The antibiotic prophylaxis timing protocol was violated in 305 of 741 (41.2%) patients, in all cases because skin incision occurred before the end of the vancomycin infusion. SSI occurred in 3% (13/346) of patients without violation of protocol and 15.4% (47/305) with protocol violation ($P < 0.001$). When patients were stratified into low- and high risk of SSI a significant difference in the incidence of SSI remained. There was a higher mortality in those that had a protocol violation; in the group considered at low risk of SSI without protocol violation the mortality was 3/236 (1.3%) vs 8/166 (4.8%) when the protocol was violated ($P = 0.03$) and in the high risk of SSI group without protocol violation the mortality was 9/200 (4.5%) vs 20/139 (14.4%) where the protocol was violated ($P < 0.0001$). The study provides evidence to support the need to complete vancomycin prior to skin incision [37C].

Hematologic

A literature review was undertaken to evaluate vancomycin-induced thrombocytopenia. Thrombocytopenia appeared to be duration dependent; the mean time to platelet nadir was 8 days. Time to thrombocytopenia was significantly shorter if the case was re-exposure. The nadir ranged from 2000 to 100000/mL. Vancomycin-specific antibodies were detected in 13/17 patients tested in published case reports. When vancomycin was ceased, the platelet count normalised within 5–6 days [38R].

Immunologic

A 3-year retrospective chart review was undertaken at 2 centres evaluating all cases of confirmed drug reaction with eosinophilia and systemic symptoms (DRESS). Of 32 cases identified, vancomycin was the most common cause, accounting for 12/32 cases (37.5%). The authors noted that there is currently a limited understanding of what predisposes patients to such hypersensitivity reactions and further research is required to identify which patient populations are most susceptible to vancomycin-induced DRESS [39c].

A case of DRESS secondary to vancomycin has been described in a 52-year-old man. He presented with symptoms that were initially thought to be sepsis and was treated with vancomycin and piperacillin–tazobactam. Three weeks prior to presentation he had been prescribed vancomycin for MRSA septic arthritis. On review of this presentation, he already had peripheral eosinophilia and deranged liver function tests, suggesting a delayed hypersensitivity reaction to the previously administered vancomycin and not sepsis. The re-exposure to vancomycin may have led to a more rapid deterioration. He responded to cessation of antibiotics and systemic steroids. The authors noted that a timely dermatology consultation and consideration of a hypersensitivity reaction when the patient initially presented could have led to an earlier diagnosis [40A].

Two episodes of vancomycin anaphylaxis were described in the same 59-year-old woman who received intravenous vancomycin as surgical prophylaxis. In the first instance anaphylaxis was erroneously attributed to midazolam due to the temporal relationship. Four weeks later she again received vancomycin as surgical prophylaxis and had a second anaphylactic episode. The authors highlighted the importance of vigilance when assigning a cause to anaphylactic reactions [41A].

A case of biopsy-proven Henoch-Schönlein purpura (HSP) was described in a 72-year-old man who presented with a generalised purpuric skin rash following administration of intravenous vancomycin for MRSA septicaemia secondary to an infected arteriovenous graft. The rash resolved 3 days after cessation of vancomycin without the need for immunosuppressants. Alternate causes were excluded. The authors noted it was only the third reported case of HSP secondary to vancomycin in the literature. Prompt discontinuation is the most effective approach and avoidance of further vancomycin use is recommended [42A].

Monitoring Therapy

A 64-year-old morbidly obese woman received 3 doses of vancomycin for surgical prophylaxis and ventilator associated pneumonia. She demonstrated increasing serum vancomycin concentrations for several days despite no further doses of vancomycin being administered. All vancomycin levels were drawn from a central venous catheter (CVC). The explanation for the increasing levels was uncertain but may have been related to drug leaching from the CVC or altered pharmacokinetics related to obesity. The authors recommended that clinicians should consider resampling from a peripheral site to rule out the possibility of drug leaching from the CVC if unexpected vancomycin drug levels occur in CVC-collected samples [43A].

Pediatric

A retrospective chart review was undertaken to evaluate both children with documented vancomycin reactions and those that had received linezolid as an alternative agent. Of the 78 vancomycin reactions, 72 (92%) were objectively consistent with red man syndrome (RMS), 5 could not be classified, and 1 was a non-RMS/non-IgE reaction. Of 60 children who received linezolid, 19 had previous reactions consistent with RMS, which should not have precluded further vancomycin administration and led to potentially unnecessary linezolid use [44c].

Sensory Systems

A retrospective case series was undertaken of 23 patients (36 eyes) with hemorrhagic occlusive retinal vasculitis (HORV). All eyes had received intraocular vancomycin via intracameral bolus (33/36), intravitreal injection (1/36) or through an irrigation bottle (2/36). Patients presented with HORV 1–21 days after surgery or intravitreal injection. Visual outcome was predominantly poor. The authors concluded that HORV was caused by a delayed hypersensitivity reaction to vancomycin and avoidance of additional intravitreal vancomycin is recommended if HORV is suspected [45c].

Urinary Tract

A review was undertaken to evaluate vancomycin-associated nephrotoxicity in adult patients. Based on available literature it was concluded that vancomycin used at currently recommended doses for non-critically ill patients was minimally nephrotoxic. Toxicity occurred more commonly in patients with multiple risk factors for acute kidney injury (AKI) who were more unwell. The authors noted that current consensus guidelines recommend clinicians aim for trough levels of >10 mg/L to prevent the development of resistance. It is unclear, from the literature, whether trough levels of 15–20 mg/L are more efficacious than 10–15 mg/L. Trough levels >15 mg/L were more clearly associated with nephrotoxicity than those <15 mg/L, but it was unclear if this was the cause or result of nephrotoxicity. The authors highlighted that further research is required. The authors noted that there is evidence that the combination of piperacillin–tazobactam and vancomycin is associated with a higher incidence of nephrotoxicity and the combination should be avoided or the duration minimised where possible [46R].

A further review of vancomycin-associated nephrotoxicity was undertaken. Nephrotoxicity has been linked to longer duration of hospitalisation, costs and risk of mortality. The authors argued that monitoring of vancomycin concentration, which consumes time and resources, could be saved by using alternative agents including daptomycin, linezolid or ceftaroline. No discussion was provided on the limitations of alternative therapies but it was noted that future research is required to evaluate the risk of nephrotoxicity with these agents as current data is limited [47R].

Several studies have been published that demonstrate an increased incidence of AKI when the combination of vancomycin/piperacillin–tazobactam was used compared to other beta-lactam/vancomycin combinations. The mechanism involved is unknown, but piperacillin–tazobactam interstitial nephritis augmented by vancomycin oxidative stress, or impaired clearance of vancomycin have been suggested [48R].

A single-centre retrospective matched cohort study of 4193 patients compared the incidence of AKI in patients who received vancomycin/piperacillin–tazobactam vs cefepime/vancomycin for at least 48 hours. Patients with severe chronic or structural kidney disease, dialysis, pregnancy or cystic fibrosis were excluded. After controlling for confounders vancomycin/piperacillin–tazobactam combination therapy was associated with 2.18 times the odds of AKI than the vancomycin/cefepime combination (95% CI 1.64–2.94 times). In this study vancomycin doses between 3 and 4 g daily were associated with an increased incidence of AKI compared to doses of 1–1.5 g/daily (adjusted OR, 0.61, 95% CI 1.11–2.32), and durations of vancomycin >7 days were associated with 1.47 times odds risk of AKI (95% CI 1.14–1.89 times) compared to less than 7 days [49MC].

A retrospective, matched-cohort study of 558 patients who received vancomycin/piperacillin–tazobactam or vancomycin/cefepime for ≥48 hours found AKI rates were significantly higher in the piperacillin–tazobactam group than the cefepime group (81/279 [29%] vs 31/279 [11%]). In multivariate analysis, combination therapy with vancomycin/piperacillin–tazobactam was an independent predictor for AKI (HR 4.27, 95% CI 2.73–6.68). Additionally, in the patients who developed AKI, the median onset was more rapid in the vancomycin/piperacillin–tazobactam group compared to the vancomycin/cefepime group (3 vs 5 days $P \leq 0.0001$) [50C].

A retrospective cohort study of 1915 children receiving combination therapy with vancomycin and an antipseudomonal beta-lactam antibiotic during their first week of hospitalization found 157/1915 children (8.2%) developed antibiotic-associated AKI. After adjustment for age, intensive care unit care, concomitant nephrotoxins and hospital, intravenous vancomycin plus piperacillin–tazobactam combination therapy was associated with higher odds of AKI each hospital day compared with vancomycin plus another antipseudomonal beta-lactam antibiotic combination (adjusted OR 3.40, 95% CI 2.26–5.14) [51C].

A prospective, open-label cohort study of 85 adult patients who received either piperacillin–tazobactam/vancomycin or cefepime or meropenem/vancomycin for >72 hours found the incidence of AKI was significantly higher with piperacillin–tazobactam/vancomycin (37.3%) compared with the cefepime or meropenem/vancomycin groups (7.7%, $P = 0.005$) [52c].

A multicentre cohort study of patients undergoing cardiac, orthopaedic joint replacement, vascular, colorectal and hysterectomy procedures assessed if combination of prophylaxis regimens with vancomycin plus a beta-lactam were associated with a lower incidence of surgical site infection (SSI) than single agent prophylaxis. The 7-day incidence of post-operative AKI was also measured. A significant benefit for reduced incidence of SSI was found for combination of prophylaxis in cardiac surgery only (adjusted RR 0.61, 95% CI 0.85–0.46). Across all surgical procedures, the risk of AKI was increased in the combination of prophylaxis group (2971/12508, 23.8%) vs vancomycin alone (1058/5089, 20.8%) vs a beta-lactam alone (7314/52504, 13.9%). The beta-lactams used were not specified. The authors noted that if the observation was causal then the number needed to harm from severe AKI was 167 [53MC].

A case series of 8 cases of presumed vancomycin-induced acute tubular necrosis (ATN) has been described, including three that required haemodialysis. The authors highlighted that vancomycin-associated kidney injury is modified by many variables and the optimal dosing to prevent kidney injury is uncertain [54c].

A retrospective review of intravenous vancomycin infusions >48 hours in critically ill patients with a target trough concentration between 20 and 30 mg/L was undertaken to evaluate the incidence and risk factors for AKI. Two levels of ≥20 mg/L were required for entry. AKI occurred in 31/107 courses (29%). Higher serum vancomycin levels were associated with AKI ($P < 0.01$). Factors independently associated with AKI were highest serum vancomycin concentration ≥40 mg/L (OR 3.75, 95% CI 1.40–10.37, $P < 0.01$), higher cumulative number of organ failures (OR 2.63, 95% CI 1.42–5.31, $P < 0.01$) and cirrhosis of the liver (OR 5.58, 95% CI 1.08–31.59, $P < 0.04$) [55C].

A 56-year-old woman with normal renal function developed unexplained AKI after administration of vancomycin, without hypovolemia or concomitant aminoglycosides. Renal biopsy showed obstructive tubular casts composed of non-crystal nanospheric vancomycin aggregates entangled with uromodulin. A new immunohistologic staining technique was developed to detect vancomycin in renal tissue and retrospectively confirmed vancomycin-associated casts in 8 additional patients who had ATN without hypovolemia. All patients had concomitant high vancomycin trough plasma levels. The authors subsequently demonstrated the toxic and obstructive nature of vancomycin-associated cast nephropathy in mice and concluded that the interaction of uromodulin with nanospheric vancomycin aggregates represented a new mode of tubular cast formation and an unsuspected mechanism of vancomycin-associated renal injury [56c].

A case of acute renal failure due to tubulointerstitial nephritis with a diffuse leukocytoclastic vasculitic rash following vancomycin infusion occurred in a 79-year-old man prescribed vancomycin for MRSA bacteraemia of unknown source. His skin lesions resolved with steroids and cessation of vancomycin but his renal function never recovered and he remained dialysis dependent. The authors noted that renal failure due to tubulointerstitial nephritis is a rare side effect of vancomycin treatment and ATN is more commonly seen [57A].

KETOLIDES [SEDA-35, 469; SEDA-36, 370; SEDA-37, 299; SEDA-38, 235; SEDA-39, 234]

Solithromycin [SEDA-35, 469; SEDA-36, 370; SEDA-37, 299; SEDA-38, 235; SEDA-39, 234]

Several reviews discuss the potential role of solithromycin for community-acquired pneumonia. Although efficacy is acceptable, concerns regarding rates of infusion-site reactions and, more significantly, confirmation of lack of hepatotoxicity are required [58R, 59R, 60R].

Cardiovascular

The effect of a single 800mg dose of intravenous solithromycin on QT interval was determined in 47 subjects. Heart rate increased by ~15bpm immediately after the infusion, but, unlike the moxifloxacin control, no QT prolongation or effect on PR and QRS interval was detected [61c]. This contrasts with most macrolides, although a small effect may have been masked by the heart rate increase [62R].

LINCOSAMIDES [SEDA-35, 469; SEDA-36, 371; SEDA-37, 299; SEDA-38, 236; SEDA-39, 236]

Clindamycin [SEDA-38, 235; SEDA-39, 236]

Gastrointestinal

A parallel, double-blinded randomised control trial in 420 patients was undertaken to compare the combination of flucloxacillin and oral clindamycin with flucloxacillin alone for the treatment of lower limb cellulitis. No significant changes were seen in improvement of cellulitis at day 5 when comparing the two arms, but there was a significant difference in the number of patients with diarrhoea at day 5 in the combination group (34/160) compared to the flucloxacillin alone group (16/176, 9%) (OR 2.7, 95% CI 1.41–5.07, $P=0.002$). No cases of *C. difficile* infection were reported [63C].

MACROLIDES [SEDA-35, 469; SEDA-36, 371; SEDA-37, 299; SEDA-38, 236; SEDA-39, 236]

Reproductive System

A Canadian study on antibiotic use in early pregnancy and risk of spontaneous abortion found the use of macrolides, excluding erythromycin, to be associated with an increased risk of spontaneous abortion. The risk of spontaneous abortion was increased by 65% with azithromycin and twofold with clarithromycin [24MC].

Sensory Systems

A literature review was performed to establish the prevalence of sensorineural hearing loss in patients treated with macrolide antibiotics. 44 studies were included, 3 prospective studies and the remainder retrospective. 78 cases of sensorineural hearing loss were identified. There was some evidence to suggest the effect may be dose dependent; however, in 5 of the reported cases sensorineural hearing loss occurred with standard oral dosing. In one study, 21% of patients receiving 4g of erythromycin experienced hearing impairment compared to none of those receiving 2g. Most cases resolved with macrolide cessation. The prevalence of this adverse effect was not estimable due to the nature of the review [64R].

Azithromycin [SEDA-35, 469; SEDA-36, 371; SEDA-37, 299; SEDA-38, 236; SEDA-39, 236]

Cardiovascular

A case–control study was conducted in Europe to assess the potential for azithromycin to cause ventricular arrhythmias. Seven databases of antibiotic users from 5 European countries were used. The study identified an increased incidence of ventricular arrhythmias among azithromycin users compared to those not on antibiotic therapy; however, the incidence was not higher when compared to patients on amoxicillin. The authors suggest this may be secondary to the underlying indication resulting in an increased risk of arrhythmia rather than azithromycin itself [65M].

Yang et al. conducted a study to assess the mechanism underlying the increased risk of cardiovascular and sudden cardiac death associated with azithromycin therapy. They used mice, cardiomyocytes and heterologously expressed human ion channels to evaluate the in vitro and in vivo effects of azithromycin on the cardiac conduction system. The study concluded that even in the absence of QT prolongation or structural cardiac abnormality, azithromycin can cause a rapid, polymorphic ventricular tachycardia. The authors reported this was in addition to the potential for QT prolongation, although they acknowledge that Torsades itself is rare with azithromycin therapy [66E].

Clarithromycin [SEDA-35, 470; SEDA-36, 372; SEDA-37, 299; SEDA-38, 237; SEDA-39, 236]

Cardiovascular

A retrospective cohort study compared the rate of cardiovascular (CV) events in patients with either upper respiratory (URTI) or lower respiratory (LRTI) infections treated with clarithromycin monotherapy compared with an alternative antibiotic monotherapy. The results of previous studies, which have suggested an increased rate of CV events associated with clarithromycin therapy, were not replicated in this study. This study reported a significant increase in CV events in those with LRTI, which appeared to be explained by increased age and comorbidity index in this group. Given clarithromycin is more commonly prescribed for LRTI than for URTI, the authors proposed that this may have confounded previous study results. Overall there was no evidence that clarithromycin was associated with more cardiac events than other commonly prescribed antibiotics [67M].

Drug–Drug

Co-administration of clarithromycin and calcium-channel blockers can result in significant hypotension. A 78-year-old male was admitted to intensive care for inotropic support following concurrent treatment with nifedipine, diltiazem and clarithromycin. Alternate causes of hypotension were excluded on clinical grounds. Clarithromycin inhibits CYP3A4, which metabolises calcium-channel blockers, thereby exaggerating the vasodilatory effect of the calcium-channel blocker [68A]. This type of drug interactions should be kept in mind while prescribing certain medications.

Psychiatric

A case of clarithromycin-induced psychosis, known as antibiomania, has been reported in a 57-year-old male treated with clarithromycin for *Helicobacter pylori* eradication. He had no past psychiatric or substance abuse history. His symptoms developed within 24 hours of antibiotic onset and resolved 48 hours after clarithromycin was discontinued. The authors outlined several other cases of clarithromycin-induced antibiomania and emphasized the importance of awareness of this rare adverse effect [69A].

Skin

A 43-year-old man developed "Baboon Syndrome" following treatment with clarithromycin for a respiratory tract infection. He had no known previous sensitization to clarithromycin. Baboon syndrome is an erythematous, well-demarcated rash that affects the groin and buttock region. The drug was discontinued and he received topical steroids and anti-histamines with resolution of his rash. Patch testing was later performed which confirmed an allergy to clarithromycin. This is the first described case of such a reaction secondary to clarithromycin [70A].

OXAZOLIDINONES [SEDA-36, 373; SEDA-37, 300; SEDA-38, 237; SEDA-39, 238]

Linezolid [SEDA-36, 373; SEDA-37, 300; SEDA-38, 237; SEDA-39, 238]

Age-Related

A 5-year observational retrospective study of 179 children (mean age 4 years, range 6 days–17 years) receiving >3 days of linezolid was conducted in Turkey. Adverse effects occurred in 20.1% of children, most commonly thrombocytopenia (14.5%). Elevated liver enzymes, leucopenia and anemia, renal impairment and skin reactions occurred in 0.6%–2.2%. No neuropathy was observed. Adverse effects were not related to treatment duration and occurred at a median of 7.5 days' treatment [71C].

Hematologic

A retrospective observational study of 549 Spanish subjects (non-ICU) who received >5 days of linezolid for various indications was conducted to develop a predictive score for haematological toxicity, which developed in ~30%. On multivariate analysis, independent variables associated with haematological toxicity (and Odds ratio) were basal platelet count $<90000/mm^3$ (OR 7.635), moderate or severe liver disease (OR 1.914), renal failure (OR 2.132) and cerebrovascular disease (OR 1.862). A weighted score of 8, 2, 2 and 2 points for each of these risk factors, respectively, allowed prediction of hematological toxicity from 6.2% (0–4 points) to 76.5% (>10 points) [72C].

Metabolic

A 77-year-old African-American man with type II diabetes experienced recurrent hypoglycemic episodes requiring two emergency visits and a hospitalisation 9 days after commencing linezolid for a possible urinary tract infection due to *Staphylococcus hominis*. Repeated administration of intravenous 50% dextrose was required to correct the blood glucose [73A]. There are a small number of previous reports of linezolid-associated hypoglycemia; this may be related to linezolid MAOI effects of promoting insulin production and insulin sensitivity. This report also defines the 'washout' time of hypoglycemia persistence to be 26–34 hours after linezolid cessation.

The pathogenesis of linezolid-induced lactic acidosis resulting from inhibition of mitochondrial protein synthesis, reduced aerobic energy production via anaerobic glycolysis and lactate production was reviewed. The authors recommend regular monitoring of lactate levels during prolonged treatment [74R]

Nervous System

Peripheral neuropathy developed in a 15-year-old boy after receiving 8 months of 450–600 mg linezolid daily for multidrug-resistant tuberculosis [75A].

Ophthalmic

A 34-year-old man developed visual loss (3/60 vision both eyes) and colour desaturation after taking linezolid 600 mg daily for 18 months for tuberculosis. He was diagnosed with optic neuropathy; of interest, on ceasing linezolid, there was rapid improvement in vision (6/6 both eyes and full recovery of colour vision) after 1 week. However, pattern electroretinography and visual evoked potentials remained abnormal after 3 weeks. No further follow-up was reported [76A].

Tedizolid [SEDA-37, 301; SEDA-38, 237]

Hematologic

Several cases have been reported in which patients who developed myelosuppression with linezolid were able to tolerate subsequent prolonged courses of tedizolid.

A 68-year-old female treated with co-trimoxazole and linezolid for cerebral nocardiosis had steadily declining white blood cell and platelet counts. Tedizolid was substituted for linezolid and she tolerated 6 months of therapy [77A].

An elderly male receiving linezolid 600mg bd for non-tuberculous mycobacterial pulmonary infection had a fall in hemoglobin from 13.5g/dL over 5 weeks. He was subsequently changed after a 26-day washout to tedizolid 200mg daily and tolerated this for 7 weeks before Hb levels fell from 11.2 to 9.6g/dL requiring cessation of tedizolid [78A].

Another 3 cases of patients who developed anaemia and/or thrombocytopenia on linezolid but tolerated 200mg tedizolid have been reported [79A].

These reports suggest that some patients with myelosuppression on linezolid may tolerate a switch to 200mg tedizolid daily. However, a review of the FDA adverse event reporting system observed similar increased risks of thrombocytopenia with both linezolid and tedizolid [80R].

POLYMYXINS [SEDA-35, 473; SEDA-36, 374; SEDA-37, 301; SEDA-38, 237; SEDA-39, 238]

Urinary Tract

A review comparing colistin to polymyxin B and renal outcomes summarised available literature at the time of publication. The authors conclude a lower rate of nephrotoxicity with polymyxin B may be advantageous in its use [81R].

A systematic review and meta-analysis of 5 studies comparing systemic colistin and polymyxin B were conducted. Colistin was associated with a higher unadjusted risk of nephrotoxicity (RR 1.55, 95% CI 1.36–1.78); however, this difference was not maintained when renal injury was stratified according to RIFLE criteria of risk, injury and failure, where risk was similar between colistin and polymyxin B. However, colistin was associated with more episodes of nephrotoxicity (HR 2.16, 95% CI 1.43–3.27) [82R].

Colistin and polymyxin B were evaluated in 414 patients with and without cystic fibrosis (CF). Rates of nephrotoxicity were similar between polymyxin B and colistin in patients both with CF (34.5% vs 29.8%, $P=0.77$) and without CF (42.9% vs 50.3%, $P=0.46$). Risk of acute kidney injury appeared to be lower in those with CF receiving polymyxin B (30.5% vs 48.5%, $P<0.001$); however, this rate was unadjusted [83C].

Colistin [SEDA-35, 473; SEDA-36, 374; SEDA-37, 301; SEDA-38, 237]

Nervous System

Intraventricular colistin for multidrug-resistant Gram-negative infections involving the central nervous system was not associated with any reports of adverse events in a case series of 7 patients. As a small case series, this result should be interpreted with caution [84c].

Pediatric

A retrospective observational study of 65 neonates receiving colistin for multiple drug-resistant Gram-negative infections identified adverse reactions as nephrotoxicity in 3, convulsions in 3 (2 without an underlying central nervous system disorder) and apnoea in one. No adverse events were clearly attributed to colistin. All episodes of nephrotoxicity were resolved [85c].

Pulmonary

An eosinophilic lung reaction probably secondary to inhaled colistin was reported in a patient with a history of bronchiectasis 8 days after commencing inhaled colistin [86A].

Urinary Tract

A retrospective analysis of 249 patients receiving colistin found the incidence of acute kidney injury (AKI) to be 12% at 48 hours and 29% at 7 days, with risk factors determined by multivariate regression to be colistin doses >5mg/kg, chronic liver disease and concomitant vancomycin. This study highlights the importance of early recognition of AKI and applies new definitions of nephrotoxicity based on KDIGO guidelines [87C].

Colistin therapy for extremely-drug-resistant *Pseudomonas aeruginosa* infections in 91 patients was associated with AKI in 13.2% of patients. Renal failure at the end of therapy was an independent predictor of mortality (OR 3.8, 95% CI 1.26–11.47) [88c].

Colistin therapy and renal outcomes were examined in a retrospective single-centre cohort study with 229 patients. 13.1% developed nephrotoxicity overall. Patients were more likely to develop nephrotoxicity if they were in intensive care (96% vs 79%, $P=0.02$), had abnormal baseline creatinine (36% vs 8%, $P=0.001$), or received vancomycin conjunction with colistin vs alone (42% vs 14.2%, $P=0.039$). Dosing of 6 vs 9 million units per day did not influence nephrotoxicity risk. However, numbers with nephrotoxicity compared to those without in the identified risk groups were small, so risk factor interpretation should be done with caution [89C].

A retrospective cohort study compared 29 patients with severe sepsis and AKI who received colistin, to 58 controls who did not receive colistin but had similar initial illnesses. Of those receiving colistin, 75% progressed to chronic kidney disease at 6 months after discharge, compared to 27% of those who did not receive colistin (OR 6.5, 95% CI 2.3–17, $P<0.001$). A dose of >5g was a predictor of progression to chronic kidney disease. This study suggests those who develop colistin-related AKI should be monitored long term for development of chronic kidney disease [90c].

Polymyxin B [SEDA-39, 239]

Skin

A cohort of 249 patients treated with intravenous polymyxin B was followed prospectively through therapy for skin hyperpigmentation. 8% of patients were noted to develop skin hyperpigmentation; 3 patients consented to skin biopsy and further evaluation. Areas of hyperpigmented skin in biopsied individuals demonstrated Langerhans cell hyperplasia and a higher number of dermal dendritic cells. The authors hypothesise an inflammatory process contributes to the development of hyperpigmentation [91c].

A case of hyperpigmentation was reported in a 65-year-old female receiving polymyxin B after 4 days of therapy [92A].

Urinary Tract

Combination of polymyxin B and vancomycin (group 1) was compared to polymyxin B alone (group 2) in a retrospective cross-sectional study in an intensive care unit to examine renal outcomes. Of 115 included patients 56 received polymyxin B alone. Mean duration of therapy was similar between the groups at 15 and 14 days, respectively. Cumulative polymyxin B dose was similar in both groups, but a higher number in group 1 had a cumulative dose >10 million IU (86.4% vs 71.4%, $P=0.06$). Group 1 had a higher incidence of AKI (62.7% vs 28.5%, $P=0.005$). However, multivariate analysis suggested this risk was independent of vancomycin use and may be due to another factor or limitations in sample size. Accumulated polymyxin B dose >10 million IU was an independent risk factor for development of AKI (OR 2.72, 95% CI 1.13–6.51, $P=0.024$). There was significant heterogeneity between groups with group 1 being younger (56 vs 63 years, $P=0.03$) and having less comorbid diseases such as stroke (15.2% vs 32.1%, $P=0.047$) and obstructive pulmonary disease (6.7% vs 19.6%, $P=0.05$), suggesting some selection bias. This study reinforces the risk of AKI with higher accumulated doses of polymyxin B, but cannot confirm concerns of vancomycin producing additive nephrotoxicity risk when co-administered with colistin [93c].

A retrospective cohort study examining high dose polymyxin B (>3mg/kg/day or total daily dose >250mg/day) involved 222 patients. Nephrotoxicity of any degree was found in 47% of the 115 patients who had renal data available. Total daily dose or dose per kilogram of body weight did not correlate with risk of renal failure. Two patients experienced infusion-related events [94C].

STREPTOGRAMINS

Nil studies

TRIMETHOPRIM AND TRIMETHOPRIM–SULFAMETHOXAZOLE [SEDA-35, 474; SEDA-36, 375; SEDA-37, 301; SEDA-38, 238; SEDA-39, 239]

Trimethoprim [SEDA-39, 239]

Nervous System

A case of presumed aseptic meningitis secondary to trimethoprim has been described in a 34-year-old woman who presented with headache, photophobia and neck stiffness 3 days after commencing trimethoprim for a presumed UTI. Four years earlier she had a similar presentation with aseptic meningitis and had also received prior treatment with trimethoprim for a UTI. In vitro lymphocyte proliferative and γ-interferon ELISpot assays were consistent with sensitisation to trimethoprim. The mechanism of drug-induced meningitis is unknown. The authors noted that trimethoprim/sulfamethoxazole (TMP/SMX) is the most commonly reported cause of antibiotic-associated aseptic meningitis followed by trimethoprim alone [95A].

Trimethoprim–Sulfamethoxazole [SEDA-35, 474; SEDA-36, 375; SEDA-37, 301; SEDA-38, 238; SEDA-39, 239]

Endocrine

A case of severe refractory hypoglycaemia presumed secondary to trimethoprim–sulfamethoxazole (TMP/SMX) was described in an 85-year-old diabetic man. He had no prior history of hypoglycaemic episodes. Glucose levels normalised 3 days after cessation of TMP/SMX. The authors highlighted that TMP/SMX contains the same sulphanilamide structural group as the sulfonylurea hypoglycaemia agents and may mimic their action on pancreatic islet cells by acting as a secretagogue, increasing endogenous insulin secretion. This was supported by the patient's unsuppressed endogenous serum levels of insulin and C-peptide [96A].

Immunologic

A case of drug-induced lupus erythematosus (DILE) secondary to TMP/SMX in a 62-year-old woman is described. The antibiotic was prescribed as pre-operative surgical prophylaxis. She presented with fever and malaise with bilateral, sterile, exudative pleural effusions and subsequently developed a large pericardial effusion causing cardiac tamponade. Serum C-reactive protein was 84 mg/L and antinuclear antibody was >1:1280 in a homogenous pattern. Antihistone antibody was 6.7 units (normal 0–0.9). The rest of the auto-immune antibody panel was negative. The patient improved within a week of hydrocortisone therapy. No further immunosuppressive therapy was required. The authors recommended consideration of DILE when confronted with sterile, exudative effusions and highlighted the importance of a thorough drug history [97A].

A case of drug-induced immune hemolytic anemia (DIIHA) and drug-induced immune thrombocytopenia (DIIT) secondary to TMP–SMX has been described in a 61-year-old woman post hematopoietic stem cell transplantation (HSCT). The TMP–SMX was commenced 24 days earlier for *Pneumocystis jirovecii* pneumonia prophylaxis. The patient's serum showed hemolysis of donor red blood cells in the presence of TMP–SMX and TMP–SMX-induced platelet antibodies. She improved with transfusions, hemodialysis, and immunosuppressive agents. The authors noted this is the first reported case of concurrent DIIHA and DIIT due to TMP–SMX-induced antibodies in a HSCT patient [98A].

Nervous System

A case of recurrent TMP/SMX-induced aseptic meningitis has been reported in a 44-year-old woman. The patient presented with low grade fever, headache and neck pain 5 hours after taking the first dose of TMP/SMX for a presumed UTI. The patient reported a similar episode of aseptic meningitis 10 years prior which also occurred after prescription of TMP/SMX for a suspected UTI. Whilst rare, the authors highlight that TMP/SMX is one of the most common drugs to cause aseptic meningitis and recommend consideration of this as a differential diagnosis in patients' who present with severe headache whilst taking this agent [99A].

OTHER ANTIMICROBIAL DRUGS

Daptomycin [SEDA-35, 474; SEDA-36, 375; SEDA-37, 302; SEDA-38, 238; SEDA-39, 240]

Respiratory

There have been increasing reports of eosinophilic pneumonia associated with daptomycin use. An analysis of the data submitted to the USFDA was performed to establish a case definition for eosinophilic pneumonia in daptomycin-treated patients, which was published in 2012 [18C]. However, a case series from Japan has suggested that these clinical criteria may be inadequate to identify all cases of daptomycin-induced eosinophilic pneumonia. Six cases of eosinophilic pneumonia were reported among 492 patients treated with daptomycin in 2 hospitals in Japan. All patients were male, >60 years, and presented with fever and peripheral eosinophilia, 5–52 days after daptomycin initiation. None had a history of allergy. A definite or probable case, as per the FDA criteria, requires the presence of dyspnoea and hypoxia. However, 4/6 Japanese cases had no hypoxia or other respiratory symptoms on presentation. 1/6 had no clinical findings on auscultation but all six had widespread changes on high-resolution computed tomography. A definite case as per the FDA criteria requires an eosinophil count >25% on bronchoalveloar lavage (BAL). Only 3/6 patients had BAL, and only one of these had an eosinophil count >25%. All cases responded to cessation of daptomycin. Given these cases differ from the case definition of drug-induced eosinophilic pneumonia, the authors have questioned whether this should be revised. To assist in possible case definitions, they reviewed 43 additional reported cases and their 6 cases. Fever was present in all cases, fine crackles in the majority (93.9%) and, notably, 22.4% did not develop hypoxia despite bilateral pulmonary infiltrates. A quarter of patients had normal peripheral eosinophil counts and just over half had ≥25% eosinophils in BAL [100c].

Additional case studies can be found in several other reviews [101R, 102R].

References

[1] Smyth AR, Bhatt J, Nevitt SJ. Once-daily versus multiple-daily dosing with intravenous aminoglycosides for cystic fibrosis. Cochrane Database Syst Rev. 2017;3:CD002009 [M].

[2] Zareifopoulos N, Panayiotakopoulos G. Neuropsychiatric effects of antimicrobial agents. Clin Drug Investig. 2017;37(5):423–37 [R].

[3] Wood GC, Swanson JM. An update on aerosolized antibiotics for treating hospital-acquired and ventilator-associated pneumonia in adults. Ann Pharmacother. 2017;51(12):1112–21 [R].

[4] Sole-Lleonart C, Rouby JJ, Blot S, et al. Nebulization of antiinfective agents in invasively mechanically ventilated adults: a systematic review and meta-analysis. Anesthesiology. 2017;126(5):890–908 [R].

[5] O'Sullivan ME, Perez A, Lin R, et al. Towards the prevention of aminoglycoside-related hearing loss. Front Cell Neurosci. 2017;11:325 [R].

[6] Vasconcelos KA, Frota S, Ruffino-Netto A, et al. The importance of audiometric monitoring in patients with multidrug-resistant tuberculosis. Rev Soc Bras Med Trop. 2017;50(5):646–51 [C].

[7] Van Hecke R, Van Rompaey V, Wuyts FL, et al. Systemic aminoglycosides-induced vestibulotoxicity in humans. Ear Hear. 2017;38(6):653–62 [R].

[8] Lanvers-Kaminsky C, Zehnhoff-Dinnesen AA, Parfitt R, et al. Drug-induced ototoxicity: mechanisms, pharmacogenetics, and protective strategies. Clin Pharmacol Ther. 2017;101(4):491–500 [R].
[9] van Altena R, Dijkstra JA, van der Meer ME, et al. Reduced chance of hearing loss associated with therapeutic drug monitoring of aminoglycosides in the treatment of multidrug-resistant tuberculosis. Antimicrob Agents Chemother. 2017;61(3) [C].
[10] Yagi K, Ishii M, Namkoong H, et al. The efficacy, safety, and feasibility of inhaled amikacin for the treatment of difficult-to-treat non-tuberculous mycobacterial lung diseases. BMC Infect Dis. 2017;17(1):558 [c].
[11] Olivier KN, Griffith DE, Eagle G, et al. Randomized trial of liposomal amikacin for inhalation in nontuberculous mycobacterial lung disease. Am J Respir Crit Care Med. 2017;195(6):814–23 [c].
[12] Arnold A, Cooke GS, Kon OM, et al. Adverse effects and choice between the injectable agents amikacin and capreomycin in multidrug-resistant tuberculosis. Antimicrob Agents Chemother. 2017;61(9) [c].
[13] Heras-Mosteiro J, Monge-Maillo B, Pinart M, et al. Interventions for Old World cutaneous leishmaniasis. Cochrane Database Syst Rev. 2017;12:1–282 [M].
[14] Cox L, He C, Bevins J, et al. Gentamicin bladder instillations decrease symptomatic urinary tract infections in neurogenic bladder patients on intermittent catheterization. Can Urol Assoc J. 2017;11(9):E350–4 [c].
[15] Querques L, Miserocchi E, Modorati G, et al. Hemorrhagic occlusive retinal vasculitis after inadvertent intraocular perforation with gentamycin injection. Eur J Ophthalmol. 2017;27(2):e50–3 [A].
[16] Heath Jeffery RC, Bowden FJ, Essex RW. Subconjunctival gentamicin-induced macular toxicity following sutureless 25-gauge vitrectomy. Clin Exp Ophthalmol. 2017;45(3):301–4 [A].
[17] Hayward RS, Harding J, Molloy R, et al. Adverse effects of a single dose of gentamicin in adults: a systematic review. Br J Clin Pharmacol. 2018;84(2):223–38 [R].
[18] Slater MB, Gruneir A, Rochon PA, et al. Identifying high-risk medications associated with acute kidney injury in critically Ill patients: a pharmacoepidemiologic evaluation. Paediatr Drugs. 2017;19(1):59–67 [C].
[19] Waters V, Yau Y, Beaudoin T, et al. Pilot trial of tobramycin inhalation powder in cystic fibrosis patients with chronic Burkholderia cepacia complex infection. J Cyst Fibros. 2017;16(4):492–5 [c].
[20] Hamed K, Debonnett L. Tobramycin inhalation powder for the treatment of pulmonary Pseudomonas aeruginosa infection in patients with cystic fibrosis: a review based on clinical evidence. Ther Adv Respir Dis. 2017;11(5):193–209 [R].
[21] Liu X, Ma J, Huang L, et al. Fluoroquinolones increase the risk of serious arrhythmias: a systematic review and meta-analysis. Medicine. 2017;96(44) [M].
[22] Hsu S-C, Chang S-S, Lee M-G, et al. Risk of gastrointestinal perforation in patients taking oral fluoroquinolone therapy: an analysis of nationally representative cohort. PLoS One. 2017;12(9): e0183813 [MC].
[23] Shybut TB, Puckett ER. Triceps ruptures after fluoroquinolone antibiotics: a report of 2 cases. Sports Health. 2017;9(5):474–6 [A].
[24] Muanda FT, Sheehy O, Bérard A. Use of antibiotics during pregnancy and risk of spontaneous abortion. Can Med Assoc J. 2017;189(17):E625–33 [MC].
[25] Alrwisan A, Antonelli PJ, Winterstein AG. Quinolone ear drops after tympanostomy tubes and the risk of eardrum perforation: a retrospective cohort study. Clin Infect Dis. 2017;64(8):1052–8 [MC].
[26] Kunder SK, Avinash A, Nayak V, et al. A rare instance of levofloxacin induced myoclonus. J Clin Diagn Res. 2017;11(7): FD01 [A].
[27] Pan L, Wang Z, Xu Y. Levofloxacin-induced transient musculospiral paralysis. Am J Emerg Med. 2017;35(2):375.e1–2 [A].
[28] Dukewich M, Danesh A, Onyima C, et al. Intractable acute pain related to fluoroquinolone-induced peripheral neuropathy. J Pain Palliat Care Pharmacother. 2017;31(2):144–7 [A].
[29] Newby BD, Timberlake KE, Lepp LM, et al. Levofloxacin use in the neonate: a case series. J Pediatr Pharmacol Ther. 2017;22(4):304–13 [c].
[30] Naidoo A, Chirehwa M, McIlleron H, et al. Effect of rifampicin and efavirenz on moxifloxacin concentrations when co-administered in patients with drug-susceptible TB. J Antimicrob Chemother. 2017;72(5):1441–9 [c].
[31] Garnock-Jones KP. Single-dose dalbavancin: a review in acute bacterial skin and skin structure infections. Drugs. 2017; 77(1): 75-83 [R].
[32] Gonzalez D, Bradley JS, Blumer J, et al. Dalbavancin pharmacokinetics and safety in children 3 months to 11 years of age. Pediatr Infect Dis J. 2017;36(7):645–53 [c].
[33] Nakano T, Nakamura T, Nakamura Y, et al. Effects of teicoplanin on the PT-INR controlled by warfarin in infection patients. Yakugaku Zasshi. 2017;137(7):909–16 [c].
[34] Das B, Sarkar C, Das D, et al. Telavancin: a novel semisynthetic lipoglycopeptide agent to counter the challenge of resistant Gram-positive pathogens. Ther Adv Infect Dis. 2017;4(2):49–73 [R].
[35] Liu Y, Wang J. A comparison of telavancin and vancomycin for treatment of methicillin-resistant Staphylococcus aureus infections: a meta-analysis. Int J Clin Pharmacol Ther. 2017;55(11):839–45 [R].
[36] Choi MH, Choe YH, Lee S-G, et al. Neutropenia is independently associated with sub-therapeutic serum concentration of vancomycin. Clin Chim Acta. 2017;465:106–11 [C].
[37] Cotogni P, Barbero C, Passera R, et al. Violation of prophylactic vancomycin administration timing is a potential risk factor for rate of surgical site infections in cardiac surgery patients: a prospective cohort study. BMC Cardiovasc Disord. 2017;17(1):73 [C].
[38] Mohammadi M, Jahangard-Rafsanjani Z, Sarayani A, et al. Vancomycin-induced thrombocytopenia: a narrative review. Drug Safety. 2017;40(1):49–59 [R].
[39] Lam BD, Miller MM, Sutton AV, et al. Vancomycin and DRESS: a retrospective chart review of 32 cases in Los Angeles, California. J Am Acad Dermatol. 2017;77(5):973–5 [c].
[40] Maxfield L, Schlick T, Macri A, et al. Vancomycin-associated drug reaction with eosinophilia and systemic symptoms (DRESS) syndrome: masquerading under the guise of sepsis. BMJ Case Rep. 2017;2017:1–4. bcr-2017-221898. [A].
[41] Evans T, Patel S. Two consecutive preoperative cardiac arrests involving vancomycin in a patient presenting for hip disarticulation: a case report. Anesth Analg. 2017;9(9):262–4 [A].
[42] Min Z, Garcia RR, Murillo M, et al. Vancomycin-associated Henoch-Schönlein purpura. J Infect Chemother. 2017;23(3): 180–184 [A].
[43] Kane SP, Hanes SD. Unexplained increases in serum vancomycin concentration in a morbidly obese patient. Intensive Crit Care Nurs. 2017;39:55–8 [A].
[44] Lin SK, Mulieri KM, Ishmael FT. Characterization of vancomycin reactions and linezolid utilization in the pediatric population. J Allergy Clin Immunol Pract. 2017;5(3):750–6 [c].
[45] Witkin AJ, Chang DF, Jumper JM, et al. Vancomycin-associated hemorrhagic occlusive retinal vasculitis: clinical characteristics of 36 eyes. Ophthalmology. 2017;124(5):583–95 [c].

[46] Filippone EJ, Kraft WK, Farber JL. The nephrotoxicity of vancomycin. Clin Pharmacol Ther. 2017;102(3):459 [R].
[47] Jeffres MN. The whole price of vancomycin: toxicities, troughs, and time. Drugs. 2017;77(11):1143–54 [R].
[48] Watkins RR, Deresinski S. Increasing evidence of the nephrotoxicity of piperacillin/tazobactam and vancomycin combination therapy—what is the clinician to do? Clin Infect Dis. 2017;65(12):2137–43 [R].
[49] Rutter WC, Cox JN, Martin CA, et al. Nephrotoxicity during vancomycin therapy in combination with piperacillin-tazobactam or cefepime. Antimicrob Agents Chemother. 2017;61(2):e02089-16 [MC].
[50] Navalkele B, Pogue JM, Karino S, et al. Risk of acute kidney injury in patients on concomitant vancomycin and piperacillin–tazobactam compared to those on vancomycin and cefepime. Clin Infect Dis. 2016;64(2):116–23 [C].
[51] Downes KJ, Cowden C, Laskin BL, et al. Association of acute kidney injury with concomitant vancomycin and piperacillin/tazobactam treatment among hospitalized children. JAMA Pediatr. 2017;171(12):e173219-e [C].
[52] Peyko V, Smalley S, Cohen H. Prospective comparison of acute kidney injury during treatment with the combination of piperacillin-tazobactam and vancomycin versus the combination of cefepime or meropenem and vancomycin. J Pharm Pract. 2017;30(2):209–13 [c].
[53] Branch-Elliman W, Ripollone JE, O'Brien WJ, et al. Risk of surgical site infection, acute kidney injury, and Clostridium difficile infection following antibiotic prophylaxis with vancomycin plus a beta-lactam versus either drug alone: a national propensity-score-adjusted retrospective cohort study. PLoS Med. 2017;14(7): e1002340 [MC].
[54] O'Donnell JN, Ghossein C, Rhodes NJ, et al. Eight unexpected cases of vancomycin associated acute kidney injury with contemporary dosing. J Infect Chemother. 2017;23(5):326–32 [c].
[55] Lacave G, Caille V, Bruneel F, et al. Incidence and risk factors of acute kidney injury associated with continuous intravenous high-dose vancomycin in critically ill patients: a retrospective cohort study. Medicine. 2017;96(7) [C].
[56] Luque Y, Louis K, Jouanneau C, et al. Vancomycin-associated cast nephropathy. J Am Soc Nephrol. 2017;28(6):1723–8 [c].
[57] Pingili CS, Okon EE. Vancomycin-induced leukocytoclastic vasculitis and acute renal failure due to tubulointerstitial nephritis. Am J Case Rep. 2017;18:1024 [A].
[58] Donald BJ, Surani S, Deol HS, et al. Spotlight on solithromycin in the treatment of community-acquired bacterial pneumonia: design, development, and potential place in therapy. Drug Des Devel Ther. 2017;11:3559 [R].
[59] Buege MJ, Brown JE, Aitken SL. Solithromycin: a novel ketolide antibiotic. Am J Health Syst Pharm. 2017;74:875–87. ajhp160934. [R].
[60] Metersky ML, Huang Y. Ketolide antibiotics: will they ever be used for community-acquired pneumonia? Ann Res Hosp. 2017;1(4):16 [R].
[61] Darpo B, Sager PT, Fernandes P, et al. Solithromycin, a novel macrolide, does not prolong cardiac repolarization: a randomized, three-way crossover study in healthy subjects. J Antimicrob Chemother. 2017;72(2):515–21. dkw428. [c].
[62] Mason JW. Antimicrobials and QT prolongation. J Antimicrob Chemother. 2017;72(5):1272–4 [R].
[63] Brindle R, Williams OM, Davies P, et al. Adjunctive clindamycin for cellulitis: a clinical trial comparing flucloxacillin with or without clindamycin for the treatment of limb cellulitis. BMJ Open. 2017;7(3):e013260 [C].
[64] Ikeda AK, Prince AA, Chen JX, et al. Macrolide-associated sensorineural hearing loss: a systematic review. Laryngoscope. 2018;128(1):228–36 [R].
[65] Trifirò G, de Ridder M, Sultana J, et al. Use of azithromycin and risk of ventricular arrhythmia. Can Med Assoc J. 2017;189(15): E560–8 [M].
[66] Yang Z, Prinsen JK, Bersell KR, et al. Azithromycin causes a novel proarrhythmic syndrome. Circ Arrhythm Electrophysiol. 2017;10(4):e003560 [E].
[67] Berni E, de Voogd H, Halcox JP, et al. Risk of cardiovascular events, arrhythmia and all-cause mortality associated with clarithromycin versus alternative antibiotics prescribed for respiratory tract infections: a retrospective cohort study. BMJ Open. 2017;7(1):e013398 [M].
[68] Takeuchi S, Kotani Y, Tsujimoto T. Hypotension induced by the concomitant use of a calcium-channel blocker and clarithromycin. BMJ Case Rep. 2017;2017: bcr2016218388. [A].
[69] Torres F. Acute psychotic episode secondary to Helicobacter pylori eradication treatment. Rev Esp Enferm Dig. 2017;109(2):168–9 [A].
[70] Moreira C, Cruz MJ, Cunha AP, et al. Symmetrical drug-related intertriginous and flexural exanthema induced by clarithromycin. An Bras Dermatol. 2017;92(4):587–8 [A].
[71] Bayram N, Düzgöl M, Kara A, et al. Linezolid-related adverse effects in clinical practice in children. Arch Argent Pediatr. 2017;115(5):470–5 [C].
[72] González-Del Castillo J, Candel F, Manzano-Lorenzo R, et al. Predictive score of haematological toxicity in patients treated with linezolid. Eur J Clin Microbiol Infect Dis. 2017;36(8):1511–7 [C].
[73] Johannesmeyer HJ, Bhakta S, Morales F. Linezolid-associated hypoglycemia. Drug Saf Case Rep. 2017;4(1):18 [A].
[74] Santini A, Ronchi D, Garbellini M, et al. Linezolid-induced lactic acidosis: the thin line between bacterial and mitochondrial ribosomes. Expert Opin Drug Saf. 2017;16(7):833–43 [R].
[75] Swaminathan A, Cros P, Seddon JA, et al. Peripheral neuropathy in a diabetic child treated with linezolid for multidrug-resistant tuberculosis: a case report and review of the literature. BMC Infect Dis. 2017;17(1):417 [A].
[76] Kreps EO, Brown L, Rennie IG. Clinical recovery in linezolid-induced optic nerve toxicity. Acta Ophthalmol. 2017;95(4):e341–2 [A].
[77] Matin A, Sharma S, Mathur P, et al. Myelosuppression-sparing treatment of central nervous system nocardiosis in a multiple myeloma patient utilizing a tedizolid-based regimen: a case report. Int J Antimicrob Agents. 2017;49(4):488–92 [A].
[78] Yuste JR, Bertó J, Del Pozo JL, et al. Prolonged use of tedizolid in a pulmonary non-tuberculous mycobacterial infection after linezolid-induced toxicity. J Antimicrob Chemother. 2016;72(2):625–8 [A].
[79] Khatchatourian L, Le Bourgeois A, Asseray N, et al. Correction of myelotoxicity after switch of linezolid to tedizolid for prolonged treatments. J Antimicrob Chemother. 2017;72(7):2135–6 [A].
[80] Lee EY, Caffrey AR. Thrombocytopenia with tedizolid and linezolid. Antimicrob Agents Chemother. 2018;62(1): e01453-17 [R].
[81] Zavascki AP, Nation RL. Nephrotoxicity of polymyxins: is there any difference between colistimethate and polymyxin B? Antimicrob Agents Chemother. 2017;61(3):e02319-16 [R].
[82] Vardakas KZ, Falagas ME. Colistin versus polymyxin B for the treatment of patients with multidrug-resistant Gram-negative infections: a systematic review and meta-analysis. Int J Antimicrob Agents. 2017;49(2):233–8 [R].
[83] Crass RL, Rutter WC, Burgess DR, et al. Nephrotoxicity in patients with or without cystic fibrosis treated with polymyxin

B compared to colistin. Antimicrob Agents Chemother. 2017;61(4):1–11. https://doi.org/10.1128/AAC.02329-16 [C].

[84] Gilbert B, Morrison C. Evaluation of intraventricular colistin utilization: a case series. J Crit Care. 2017;40:161–3 [c].

[85] Cagan E, Kiray Bas E, Asker HS. Use of colistin in a neonatal intensive care unit: a cohort study of 65 patients. Med Sci Monit. 2017;23:548–54 [c].

[86] Lepine PA, Dumas A, Boulet LP. Pulmonary eosinophilia from inhaled colistin. Chest. 2017;151(1):e1–3 [A].

[87] Shields RK, Anand R, Clarke LG, et al. Defining the incidence and risk factors of colistin-induced acute kidney injury by KDIGO criteria. PLoS One. 2017;12(3):e0173286 [C].

[88] Sorli L, Luque S, Segura C, et al. Impact of colistin plasma levels on the clinical outcome of patients with infections caused by extremely drug-resistant Pseudomonas aeruginosa. BMC Infect Dis. 2017;17(1):11 [c].

[89] Ghafur A, Gohel S, Devarajan V, et al. Colistin nephrotoxicity in adults: single centre large series from India. Indian J Crit Care Med. 2017;21(6):350–4 [C].

[90] Meraz-Munoz A, Gomez-Ruiz I, Correa-Rotter R, et al. Chronic kidney disease after acute kidney injury associated with intravenous colistin use in survivors of severe infections: a comparative cohort study. J Crit Care. 2017;44:244–8 [c].

[91] Mattos KPH, Cintra ML, Gouvea IR, et al. Skin hyperpigmentation following intravenous polymyxin B treatment associated with melanocyte activation and inflammatory process. J Clin Pharm Ther. 2017;42(5):573–8 [c].

[92] Lahiry S, Choudhury S, Mukherjee A, et al. Polymyxin B-induced diffuse cutaneous hyperpigmentation. J Clin Diagn Res. 2017;11: FD01–2 [A].

[93] Soares DS, Reis ADF, Silva Junior GBD, et al. Polymyxin-B and vancomycin-associated acute kidney injury in critically ill patients. Pathog Glob Health. 2017;111(3):137–42 [c].

[94] John JF, Falci DR, Rigatto MH, et al. Severe infusion-related adverse events and renal failure in patients receiving high-dose intravenous polymyxin B. Antimicrob Agents Chemother. 2018;62(1) [C].

[95] Lochlainn MN, Gooi HC, Ogese MO, et al. Trimethoprim-induced aseptic meningism. Br J Hosp Med. 2017;78(2):108 [A].

[96] Rossio R, Arcudi S, Peyvandi F, et al. Persistent and severe hypoglycemia associated with trimethoprim-sulfamethoxazole in a frail diabetic man on polypharmacy: a case report and literature review. Int J Clin Pharmacol Ther. 2018;56:86–9 [A].

[97] Jose A, Cramer A, Davar K, et al. A case of drug-induced lupus erythematosus secondary to trimethoprim/sulfamethoxazole presenting with pleural effusions and pericardial tamponade. Lupus. 2017;26(3):316–9 [A].

[98] Linnik YA, Tsui EW, Martin IW, et al. The first reported case of concurrent trimethoprim-sulfamethoxazole–induced immune hemolytic anemia and thrombocytopenia. Transfusion. 2017;57(12):2937–41 [A].

[99] Fisse AL, Strassburger-Krogias K, Gold R, et al. Recurrent trimethoprim-sulfamethoxazole-induced aseptic meningitis with associated ampicillin-induced myoclonic twitches. Int J Clin Pharmacol Ther. 2017;55(7):627–9 [A].

[100] Hirai J, Hagihara M, Haranaga S, et al. Eosinophilic pneumonia caused by daptomycin: six cases from two institutions and a review of the literature. J Infect Chemother. 2017;23(4):245–9 [c].

[101] Chaudhry S. Miscellaneous antibacterial drugs. In: Ray SD, editor. Side effects of drugs: annual: a worldwide yearly survey of new data in adverse drug reactions. 38:2016. p. 229–41 [chapter 24]. [R].

[102] Tucker E, Roberts MB, Gordon DL. Miscellaneous antibacterial drugs. In: Ray SD, editor. Side effects of drugs: annual: a worldwide yearly survey of new data in adverse drug reactions, vol. 39 2017. p. 229–43 [chapter 22]. [R].

CHAPTER

26

Antifungal Drugs

Dayna S. McManus,[1], Nika Bejou†*

*Department of Pharmacy Services, Yale-New Haven Hospital, New Haven, CT, United States

†University of Pittsburgh Medical Center Pinnacle, Harrisburg, PA, United States

[1]Corresponding author: dayna.mcmanus@ynhh.org

ANTIFUNGAL AZOLES [SEDA-35, 484; SEDA-36, 382; SEDA-37, 307; SEDA-38, 245]

Fluconazole

A retrospective cohort study of 91330 patients was conducted to review the association between non-vitamin K oral anticoagulants (NOACs) and interacting concurrent medications, including fluconazole. It concluded that concurrent use of fluconazole resulted in a significant increase in adjusted incidence rates per 1000 person-years of major bleeding vs NOACs alone (241.92 vs 102.7, respectively; 99% CI: 80.96–195.97) [1R].

A case of rhabdomyolysis triggered by an interaction between fluconazole and simvastatin was reported in a 76-year-old Caucasian women. The patient was taking fluconazole for 3 months for *Candida albicans* vertebral osteomyelitis, as well as simvastatin 20mg by mouth daily. She reported a 7-day history of weakness and pain in her arms and legs and was found to be in acute kidney injury with a creatinine of 1.42mg/dL (up from the patient's baseline of 0.8mg/dL) and a creatine kinase (CK) level of 8876U/L. Upon patient presentation, the fluconazole was discontinued and the patient continued on simvastatin. The patient received aggressive oral and intravenous hydration; after waiting ~32 hours, the half-life of fluconazole, the patient was safely continued on fluconazole for the remainder of therapy without incident [2A].

A review article evaluated the incidence and severity of hepatotoxicity associated with azole antifungals, including fluconazole. In a retrospective review, fluconazole was associated with greater risk of transaminitis vs amphotericin B (14% vs 10%; $P=0.43$). Further, a meta-analysis within the review article also estimated the pooled risk for mild elevation of liver enzymes to be 8.6% for empiric treatment with fluconazole (95% CI: 1–16.1) [3r].

A systematic review and meta-analysis of 37 studies involving 1259 patients evaluated the safety and efficacy of azole therapy for tegumentary leishmaniasis. The overall adverse event rate, including transaminitis, epigastric pain, nausea, vomiting, headaches, skin rash, and jaundice, was 11%. 7% of patients receiving fluconazole (CI95%: 3%–14%) experienced increased liver enzymes vs 12% with itraconazole (CI95%: 8%–19%) and 13% with ketoconazole (CI95%: 6%–29%); there was no significant difference amongst the azoles ($P=0.35$) [4M].

Itraconazole [SEDA-35, 485; SEDA-36, 38; SEDA-37, 307]

A case of severe fatigue was reported in a 72-year-old woman with allergic bronchopulmonary aspergillosis (ABPA) and asthma. The patient was being treated with itraconazole 200mg daily for 2 years, along with fluticasone and salmeterol inhalers once puff twice daily for 4 years. There were multiple unsuccessful attempts to discontinue the itraconazole over the course of the 2 years; however, the patient's disease recurred. The patient was admitted with a 2-week history of increasing tiredness and shortness of breath. During the workup, it was found that the patient's ACTH level was <10ng/mL and serum cortisol was 4nmol/L (normal 9 A.M. range: 138–635nmol/L). Other pituitary profile tests, including TSH, FSH, LH, and prolactin, were all within normal limits. It was determined that there was likely an interaction between the oral itraconazole and inhaled fluticasone, resulting in suppression of the hypothalamic–pituitary–adrenal (HPA) axis. Itraconazole is a potent inhibitor of CYP3A4, which is involved in the metabolism of corticosteroids, therefore decreasing their clearance. The patient's low ACTH and low cortisol levels were consistent with HPA axis suppression. The patient was given

Side Effects of Drugs Annual, Volume 40
ISSN: 0378-6080
https://doi.org/10.1016/bs.seda.2018.08.003

hydrocortisone, first via parenteral route then oral once clinically stable. Ultimately, the itraconazole was continued for the ABPA [5A].

Cardiotoxicity

A case series of cardiotoxicity was reported in two patients that developed acute systolic heart failure with itraconazole use. The first patient was a 31-year-old male admitted to the hospital for worsening dyspnea, orthopnea, edema, and weight gain for 2 months; he was on itraconazole for 6 months for cutaneous blastomycosis. On physical examination, patient had significant elevation of jugular venous pressure, with crackles on chest auscultation, and 2+ pitting edema of bilateral lower extremities. His ejection fraction (EF) was found to be 10%–15%. The second case was a 64-year-old male on itraconazole for biopsy-proven histoplasma cellulitis on his forearm, admitted with similar symptoms and found to have an EF of 40%–45%. No other etiology was identified in both patients despite an extensive workup that included cardiac catheterization. Itraconazole was discontinued in both the cases. The first patient required diuretics for symptom control; his cardiac function did not improve 3 months after cessation of itraconazole. Ultimately, the patient had an automatic implantable defibrillator (AICD) placed and the patient was referred for cardiac transplant evaluation. The second patient did improve and was able to be tapered off diuretics; a follow-up echocardiogram 6 months after itraconazole cessation revealed an EF of 55% [6A].

Further, another case of cardiac toxicity secondary to itraconazole use is reported in a 76-year-old male with hypertension and COPD who experienced heart failure after receiving the drug for aspergillosis. The patient's symptoms and ejection fraction resolved upon withdrawal of the itraconazole and initiation of diuretics [7A].

Hepatotoxicity

A review article evaluated the incidence and severity of hepatotoxicity associated with azole antifungals, including itraconazole. In a randomized clinical trial comparing the safety and efficacy of itraconazole to amphotericin B in neutropenic patients with proven or highly suspected fungal infection, there was found to be no difference in hepatotoxicity; AST elevations greater than 30% above pre-treatment value occurred in 40% and 55% of patients receiving itraconazole and amphotericin B, respectively. In another open, randomized, controlled, multicenter trial comparing itraconazole and amphotericin B as empiric antifungal therapy, fewer incidents of hepatotoxicity were reported with itraconazole compared to amphotericin (5% vs 54%, respectively; $P=0.001$). A large systematic review included in the article reported drug induced liver injury (DILI) in 31.6% of patients taking itraconazole vs 18.6% of those receiving amphotericin B; however, there was a wide variation in the definitions of DILI utilized in the studies [3r].

A case of liver dysfunction was reported in a 67-year-old woman who developed DILI after taking oral itraconazole for the treatment of kerion celsi. Initially, the patient had presented with a nodule with crusting, erosion, infiltration and hair loss on the temporal area. The patient was diagnosed with kerion celsi caused by *Trichophyton rubrum*, and confirmed by clinical, pathological, and mycological findings. Baseline labs demonstrated no pre-existing hepatic dysfunction. The patient was started on itraconazole 100mg by mouth daily; while the skin lesion improved, the patient developed severe hepatitis 1 month after admission (AST 232IU/L, ALT 465IU/L, T-bili 6.1mg/dL). The itraconazole was held and the patient was given oral ursodiol for the hepatotoxicity. Approximately 2 months after starting itraconazole, the patient's hepatobiliary enzymes returned to normal and the skin lesion had healed without requiring further treatment [8A].

Isavuconazole [SEDA-37, 307; SEDA-38, 246]

Isavuconazole is a relatively new triazole approved for use in invasive aspergillosis and Mucormycosis; it also has activity against *Candida* species. Often times, this drug is a particularly attractive option for the treatment of these fungal infections due to several characteristics. Given isavuconazole does not require a cyclodextran vehicle due to its high water solubility, it does not require therapeutic monitoring of levels. Isavuconazole has also demonstrated non-inferiority to voriconazole for treatment of Aspergillosis, with decreased rates of adverse effects [9H].

Isavuconazole is both a substrate and inhibitor of the CYP3A4 enzyme. As such, there is a propensity for multiple drug interactions. One such case was reported in a 23-year-old male with severe aplastic anemia who presented to the ED with petechial bleeding, fever, and chills and admitted with suspected acute leukemia. The patient was diagnosed with severe aplastic anemia on bone marrow biopsy. The patient was treated with antithymocyte antiglobulin (ATGAM) and cyclosporine. Three weeks after ATGAM treatment, the patient developed neutropenic fever; a CT scan revealed three subpleural nodules and the patient was started initially on IV voriconazole due to suspicion of invasive pulmonary aspergillosis. All diagnostic tests and fungal workup were negative. Despite utilizing a slow infusion rate of the voriconazole, the patient experienced severe visual disturbances and hallucinations, so he was switched to liposomal amphotericin B 3mg/kg. The patient developed a type 1 anaphylactic reaction with first infusion of the amphotericin. As such, he was switched to posaconazole, which had to be discontinued due to severe transaminitis. Therefore, the posaconazole was discontinued and the patient was started on oral isavuconazole 200mg twice daily for the first 2 days, then 200mg daily thereafter. Three days after initiating isavuconazole, the serum concentrations of cyclosporine doubled. The patient's cyclosporine dose was reduced and he was discharged 4 days later on isavuconazole 200mg daily [10A].

Further, a retrospective study of 55 solid organ transplant recipients was conducted to evaluate the drug interaction between tacrolimus and isavuconazole prophylaxis. Patients were required to have received tacrolimus and follow-up for at least 40 days after discontinuation of isavuconazole. Whole blood levels of tacrolimus were measured during and after isavuconazole prophylaxis. The study authors found that the median tacrolimus concentration/dose (C/D) ratio was highest on day 4, likely reflecting the effects of the isavuconazole loading dose, and stabilized by day 8. The median tacrolimus C/D was also higher amongst liver transplant recipients than other transplant types ($P = 0.002$). Finally, after discontinuation of isavuconazole, median tacrolimus C/D decreased gradually and stabilized after 28 days. Given the significant interpatient variability in the magnitude of the drug interaction, a uniform pre-emptive adjustment of tacrolimus levels is not recommended. Management of this CYP3A4 drug interaction should be guided by therapeutic drug level monitoring [11c].

Isavuconazole is also believed to have some inhibitory potential for other CYP enzymes, as well, including CYP1A2 and CYP2C9. Warfarin, an anticoagulant, is metabolized primarily by CYP2C9. A phase I trial evaluated the pharmacokinetic and pharmacodynamic interactions between isavuconazole and warfarin in healthy adults. Isavuconazole 372mg three times daily for 2 days was administered as a load, then 372mg daily thereafter was administered with and without single doses of warfarin 20mg. Coadministration of isavuconazole and warfarin resulted in increased mean area under plasma concentration-time curves by 11%, for the S-enantiomer, and 20%, for the R-enantiomer. There were, however, no treatment-emergent adverse effects and no subjects discontinued the study due to adverse effects. As such, the study authors found that isavuconazole has no clinically relevant effect on warfarin kinetics or dynamics [12c].

Isavuconazole is approved for treatment of invasive aspergillosis. A case report of a 56-year-old female with allergic bronchopulmonary aspergillosis (ABPA) was published. The patient's past medical history was significant for asthma; she presented with a flu-like illness, consisting of fever, malaise, and cough that worsened over several months. A serum IgE level was 813IU/mL, serum IgG antibodies to *Aspergillus fumigatus* were detected, and the sputum cultures grew *Aspergillus fumigatus*. Imaging was obtained and demonstrated hyperinflation of the lungs bilaterally. Pulmonary function tests (PFTs) revealed obstructive airway disease that did not improve after receiving bronchodilators; as such, the patient was diagnosed with APBA. Her symptoms did not resolve and she continued to experience dyspnea with cough. The following winter, the patient had recurrent bronchitis and pneumonia; a BAL revealed *Aspergillus* spp. The patient was started on voriconazole with significant clinical improvement. However, her serum ALT more than doubled and the voriconazole was discontinued. As such, the patient was started on isavuconazole 200mg every 8 hours for 2 days, then 200mg once daily for 10 weeks. Several weeks into her treatment course, the patient's sputum production and wheezing significantly improved such that she was able to hike 16miles 1 month later. Over the course of therapy, the patient did not experience any LFT abnormalities, but did note diarrhea, which resolved several weeks after discontinuing the isavuconazole [13A].

While the azoles are typically associated with QT interval prolongation, isavuconazole may decrease the QT interval [13A]. The cardiac effects of isavuconazole were studied clinically in a phase I trial with 161 patients; 40 patients were randomized to placebo, 41 patients to isavuconazole 200mg, and 40 patients to isavuconazole 600mg. Seven subjects, all in the isavuconazole 600mg arm, discontinued due to a treatment-related adverse effect. A decrease in baseline QT interval of >30ms was observed in 13 patients in the isavuconazole 600mg arm vs 7 patients in the isavuconazole 200mg arm. Palpitations were reported in 1 patient in the placebo arm (2.5%) vs 4 patients in the isavuconazole 600mg arm (10.3%); tachycardia was reported in 2 patients (5.1%) in the isavuconazole 600mg arm as well. There were no serious treatment-emergent adverse effects reported [14C].

Ketoconazole [SEDA-34, 430; SEDA-35, 486; SEDA-36, 383; SEDA-37, 307; SEDA-38, 246]

A phase I, open-label study was published evaluating the interaction between ketoconazole, a CYP3A4 inhibitor, and venetoclax. Twelve patients with non-Hodgkin lymphoma received venetoclax 50mg orally on day 1 and 8, as well as ketoconazole 400mg daily on day 5–11. It was found that ketoconazole increased the venetoclax mean maximum plasma concentration (C_{max}) by 2.3-fold (90% CI: 2.0–2.7). The area under the concentration-time curve also increased by 6.4-fold (90% CI: 4.5–9.2). As such, it was determined that coadministration of venetoclax and ketoconazole resulted in increased venetoclax exposure, likely due to CYP3A4 inhibition [15c].

Posaconazole [SEDA-33, 553; SEDA-34, 430; SEDA-35, 486; SEDA-36, 383; SEDA-37, 307; SEDA-38, 247]

There was a case of rhabdomyolysis reported in a patient receiving the extended-release tablet form of posaconazole for antifungal prophylaxis. The patient was a 55-year-old male with history of hypertension, benign prostatic hypertrophy (BPH), gastroesophageal reflux disease (GERD), and relapsed AML diagnosed a year prior to presentation. The patient was admitted with a 1-day history of fevers and chills in the setting of re-induction chemotherapy with FLAG-IDA 10 days prior to admission. He tolerated the chemotherapy well and was discharged on acyclovir, levofloxacin, pentamidine, and posaconazole for neutropenia prophylaxis. Initial workup revealed leukopenia (WBC 100/μL), anemia (Hgb 8.5g/dL), and thrombocytopenia (platelets 10000/μL). It was observed that the patient's urine was dark brown in color, and he began complaining of diffuse myalgias that had started since admission. A repeat UA revealed 2+ protein and 3+ blood, both of which were not seen on the UA obtained upon admission. A serum creatine kinase (CK) level was also obtained and was significantly elevated (73133units/L). Nephrology recommended IV fluids, mannitol, and bicarbonate for treatment of the rhabdomyolysis. The case received a score of 5 on the Naranjo adverse drug reaction probability scale, suggesting that the rhabdomyolysis was secondary to the extended-release posaconazole. Of note, the patient was not on any concomitant medications that interact with the CYP450 enzyme system [16A].

A case of posaconazole salvage therapy with concomitant sirolimus was reported in a liver transplant recipient. The patient was a 55-year-old female who developed rhinocerebral mucormycosis 3 months after orthotopic liver transplantation for cryptogenic cirrhosis. For maintenance immunosuppression, the patient was initially on tacrolimus, but was switched to sirolimus 3mg daily (goal 6–8ng/mL) 3 months post-transplant due to complaints of headaches and tingling sensations while on tacrolimus. Upon admission, the patient was complaining of brown, foul-smelling sinus discharge, as well as headache, nausea and vomiting. CT imaging revealed osseous destruction of the hard palate, and a tissue biopsy revealed fungal elements consistent with mucormycosis. The patient was started on liposomal amphotericin B and hydration; on day 14 of admission, the patient's serum creatinine (SCr) increased to 2.8mg/dL, and she was switched to posaconazole 200mg orally every 6 hours. Sirolimus levels were monitored on average every 4.6 days while the patient was on posaconazole; 13 sirolimus levels were drawn, in total, ranging from 4.3 to 11.8ng/mL. The patient did not experience any adverse effects associated with the sirolimus, such as anemia, pneumonitis, proteinuria, and mouth ulcerations. The patient completed 3 months of treatment for the mucormycosis; the posaconazole was discontinued due to elevated transaminases (AST 47U/L, ALT 90U/L, ALP 698U/L, total bilirubin 2.1mg/dL). At the time of publication, 3 years post-transplant, the patient remains free of recurrent mucormycosis. As such, this case outlines the concomitant use of posaconazole and sirolimus, a combination typically contraindicated given the potential for CYP3A4 inhibition [17A].

There was also a case of a possible drug interaction between posaconazole and cytarabine; often recommended for re-induction chemotherapy in acute myelogenous leukemia (AML) in patients less than 60 years of age. The patient was a 24-year-old female presenting with persistent bleeding after incision and drainage of right axillary abscess. The patient also presented with submandibular lymphadenopathy, tachycardia, and a systolic murmur. A peripheral smear was obtained and consistent with a diagnosis of AML. The patient was started on cytarabine and idarubicin; on day 5 of chemotherapy, she was also started on posaconazole delayed-release tablets 300mg by mouth daily as antifungal prophylaxis. Her induction chemotherapy was complicated by left eye blindness due to thrombocytopenia-induced pre-retinal hemorrhage, mucositis, and neutropenic fever. Repeat bone marrow biopsy at day 14 revealed induction failure with 30% blasts. Re-induction therapy with HiDAC was initiated; the patient continued on posaconazole throughout her induction chemotherapy and for the first 2 days of HiDAC, then held. After the cytarabine was completed, the posaconazole was resumed. The patient then began experiencing tingling and numbness in both hands and feet, followed by swelling, erythema, and severe pain, consistent with palmar-plantar erythrodysesthesia (PPE), also known as hand-foot syndrome. She was treated with aggressive pain management for the PPE. Another repeat bone marrow biopsy 14 days after HiDAC initiation revealed no evidence of residual blasts; however, the patient began experiencing grade 4 severe pancytopenia and remained transfusion-dependent throughout the rest of her hospitalization. Due to her severe neutropenia, she developed a variety of infections, including *Clostridium difficile* infection as well as bacteremias due to *Enterococcus faecalis* and *Enterobacter cloacae*. The patient unfortunately passed away due to these comorbidities 4 months after her initial diagnosis. It was believed that the severe adverse effects that the patient experienced may have been an interaction between the posaconazole, a P-gp inhibitor, and cytarabine, a P-gp substrate. While there is limited data, authors involved in the study concluded that the addition of posaconazole may have led to cytarabine accumulation and increased toxicities [18A].

A retrospective study was performed to evaluate the safety of oral posaconazole suspension in 45 hematopoietic stem cell transplant (HSCT) patients. The success rate at 12 weeks after posaconazole treatment was 57.8%. The posaconazole suspension did result in some mild–moderate adverse effects with a relatively short duration. Three subjects (6.7%) experienced nausea and vomiting likely related to the posaconazole. Two patients developed transaminitis (AST 67.2, ALT 150.9; AST 86.2, ALT 263.5) that resolved after dose adjustment of the posaconazole. Only one subject had to discontinue the posaconazole therapy due to intolerant nausea and vomiting [19c].

Finally, a multicenter, retrospective, observational study was done in adult cancer patients to assess whether delayed-release (DR) posaconazole levels correlate with increased liver enzymes and/or QTc prolongation. A total of 166 patients on DR posaconazole tablets were included from November 2013 to November 2014. The median posaconazole level was 1250 ng/mL (range: 110–4220 ng/mL); all levels were obtained between day 5 and 14 after starting posaconazole. There was a statistically significant increase in AST, ALT, alkaline phosphatase (ALK), and total bilirubin ($P<0.001$), as well as QTc ($P=0.05$) above baseline. Posaconazole levels themselves, however, were not associated with increased AST, ALT, ALK, or bilirubin. There was also no statistically significant correlation between posaconazole levels and QT prolongation. Therefore, this study adds to existing data that posaconazole is associated with significant increases in liver enzymes from baseline [20C].

Voriconazole [SEDA-35, 486; SEDA-36, 384; SEDA-37, 307; SEDA-38, 247]

Voriconazole is widely used in the treatment and prophylaxis of invasive fungal infections; however, its use can be limited by various toxicities, including hepatic, neurological, cardiac, and visual. A large meta-analysis of 4122 patients from 16 studies was performed to review the safety of voriconazole for prophylaxis and treatment of invasive fungal infections. The primary outcome of the meta-analysis was tolerability of voriconazole compared with the composite of other antifungals. The study authors found that voriconazole appeared to be slightly inferior to the composite (OR = 1.71; 95% CI = 1.21 = 2.40; $P=0.002$). The likelihood of neurotoxicity was twice higher for voriconazole compared to the composite group (OR = 1.99; 95% CI = 1.05–3.75; $P=0.03$); visual toxicities were also sixfold higher in the voriconazole group (OR = 6.50; 95% CI = 2.93–14.41; $P<0.00001$). Finally, hepatotoxicity was also more common in the voriconazole group (OR = 1.60; 95% CI = 1.17–2.19; $P=0.003$). The study authors found that the rate of adverse effects with voriconazole was higher than in the composite group, consistent with the other available literature [21M].

Another systematic review and meta-analysis with seven studies total: five studies comparing voriconazole to other antifungals as prophylaxis of invasive fungal infections (IFIs), and two studies for treatment. The objective of the review was to assess the efficacy and safety of voriconazole as compared with other antifungals for prophylaxis and treatment of IFIs in hematology–oncology patients. Voriconazole was found to be more effective than the comparator (RR = 1.17; 95% CI = 1.01–1.34). Risk of adverse events was relatively equal between both groups (RR = 1.06; 95% CI = 0.66–1.72), though there was significant heterogeneity ($P<0.01$). This review suggests that voriconazole was as safe as the comparator agents; however, the study had significant variability with regards to the sample sizes, age differences of the patient population, and differences between outcome measures in the various studies [22M].

Special Population: Pediatrics

Data evaluating the safety and efficacy of voriconazole in pediatric patients for the treatment of invasive Aspergillosis (IA) and invasive Candidiasis (IC) are relatively limited. A prospective, open-label, noncomparative study of voriconazole was conducted in pediatric patients (age 2–18 years of age) who received voriconazole for 6–12 weeks, dosed appropriately based upon age, weight, and indication. Fifty-three patients were treated with voriconazole: 31 patients with IA and 22 with IC. Of the 31 patients with IA, 14 had proven or probable IA; of the 22 patients with IC, 17 had confirmed disease. Transaminitis and hepatotoxicity were reported in 22.6% of IA patients and 22.7% of IC patients; visual toxicities were reported in 16.1% of IA patients and 27.3% of IC patients. All-cause mortality was 14.3% at week 6, but all deaths occurred in the AI group and none were attributed to the voriconazole itself. As such, this data suggests that there is a correlation between voriconazole use and hepatic events in pediatric patients, similar to that seen in adult patients [23c].

Squamous Cell Carcinoma

An international, multicenter, retrospective, cohort study was conducted in adult patients undergoing lung transplantation from 2005 to 2008. The objective was to evaluate the association between voriconazole and development of squamous cell carcinoma (SCC) in lung transplant recipients by adjusting for confounding variables, such as immunosuppression. Nine hundred lung transplant recipients were included in this study with a median follow-up time from transplantation of 3.51 years. A Cox regression analysis was utilized to assess the effect of voriconazole on the risk of developing biopsy-confirmed

SCC. It was found that exposure to voriconazole alone (adjusted HR 2.39; 95% CI 1.31–4.37) was associated with squamous cell carcinoma when compared to patients who were not exposed to voriconazole, after adjusting for confounding variables, such as immunosuppression. This case highlights the potential link between voriconazole use and SCC in lung transplant recipients; additional studies, including randomized control trials, are needed to investigate this potential association further [24C].

Cardiotoxicity

Voriconazole is typically a relatively safe drug in pediatric patients and may result in visual disturbances, hepatitis, or rash. Manifestations of cardiotoxicity, such as QT prolongation, torsades de pointes, and ventricular bigeminy, are rarely reported in the literature. A case of ventricular tachycardia associated with voriconazole was reported in an 11-year-old boy being treated for aplastic anemia. The patient was started on vancomycin and meropenem; the next day, he developed oral thrush and was started on amphotericin B. While the oral lesions did improve, the patient's blood urea nitrogen (BUN) level increased to 65mg/dL. The patient was then switched to voriconazole; on day 16 of his voriconazole course, he began complaining of transient visual disturbances. Several days later, he was also noted to be jaundiced; labs revealed a serum bilirubin of 3mg/dL, serum glutamic pyruvate transaminase 80IU/dL, and serum glutamic ornithine transaminase 76IU/dL. As such, the voriconazole therapy was held. A week later, the patient's liver enzymes normalized, but the oral lesions reappeared. The patient was then started on IV voriconazole; 2 days later, the patient began complaining of dizziness. His HR was 160bpm, RR 26bpm, and BP 75/50. Labs were notable for Hgb 6g/dL, WBC 1100/mm^3, and platelets 10000/mm^3. The patient's EKG also revealed monomorphic ventricular tachycardia. The team attempted cardioversion in light of the severe hypotension and ventricular tachycardia, but unfortunately the patient succumbed to his illness. This case highlights the potential for cardiac toxicity with voriconazole [25A].

A retrospective chart review was also conducted of all patients treating with voriconazole in a large tertiary care center from 2009 to 2015. The objective of the study was to determine the incidence of QT prolongation in patients who received voriconazole. Fifty-four patients were included total; mean QTc during voriconazole therapy (448.0 ± 5.9ms) was significantly longer compared to QTc off voriconazole (421.8 ± 42.2ms; $P=0.002$). Further, a QTc interval >30ms was found in 43% ($n=23$) of patients and a QTc interval >60ms was seen in 28% ($n=15$) of patients. A multivariate analysis also revealed that QT prolongation was correlated with having a baseline QTc interval >450ms ($P<0.01$) and low serum potassium levels ($P<0.01$). The study authors concluded that hematologic/oncologic patients who are treated with voriconazole are at increased risk for QT prolongation, especially if they have hypokalemia or a prolonged QT at baseline [26C].

Hepatotoxicity

A review article evaluated the incidence and severity of hepatotoxicity associated with azole antifungals, including voriconazole. In a randomized unblinded trial, voriconazole (4mg/kg IV BID with a 6mg/kg IV BID load on day 1; then 200mg BID by mouth) was compared to amphotericin B for treatment of definite or probable aspergillosis in a population consisting largely of patients with hematologic malignancies or those who had received allogeneic hematopoietic cell transplantation. In addition to having significantly better outcomes, voriconazole also had fewer severe drug-related adverse events ($P=0.008$). Seven patients receiving voriconazole presented with hepatotoxicity vs four patients receiving amphotericin B ($P=0.54$) [3r].

Voriconazole is frequently implicated in cases of hepatotoxicity and transaminitis; however, a case of successful voriconazole use was reported in a 21-year-old patient with subacute liver failure. The patient presented with liver failure secondary to a suspected viral infection; he was initially treated with ganciclovir and methylprednisolone, but his liver function did not improve (ALT 165U/L, AST 99U/L, T-bili 654mmol/L). His sputum smear was negative for bacteria, but did reveal fungal elements; sputum cultures obtained several days later grew *Aspergillus fumigatus*. Voriconazole was initiated at a dose of 400mg for 2 doses on day 1, then a maintenance dose of 100mg twice daily. The patient began experiencing various adverse effects, purportedly due to the voriconazole, such as tremors, muscle twitching and hallucinations. A voriconazole trough level was drawn and found to be supratherapeutic 8.1mg/mL. The dose was then adjusted to 100mg once daily and subsequent troughs drawn were within the 2.5–4.7mg/mL range; the patient reported no further side effects secondary to the voriconazole. This case adds to the evidence that low-dose voriconazole can be safe and effective for the treatment of Aspergillosis in a patient with liver failure [27A].

A retrospective study was published evaluating voriconazole, plus caspofungin, for invasive fungal infections (IFI) in 22 pediatric patients with acute leukemia in South Korea between 2009 and 2013. Of the 22 patients, 9 patients were diagnosed with probable IFI and 13 with possible IFI. Eleven patients had ALL and 11 had AML. None of the patients had undergone HSCT prior to the onset of IFI. Nodules (38.1%) and consolidation (33.3%) were the most common findings upon imaging. All patients had severe neutropenia ($ANC<500$). The patients were switched from antifungal monotherapy to voriconazole and caspofungin combination therapy.

The 100-day survival rate was 90.9%. The combination was well tolerated in most patients, with the exception of two patients who experienced elevated liver enzymes (grade 1). The voriconazole was continued for both patients, with a dose reduction, and the transaminitis resolved [28c].

Periostitis

Voriconazole has been associated with periostitis, due to the fluoride content of the drug. The fluoride accumulates, resulting in increased levels which usually subside once the voriconazole is stopped. Daily dosage of 400mg of voriconazole provides 65mg of fluoride, exceeding the recommended daily intake of 3–4mg. A systematic review was published to review and analyze voriconazole-induced periostitis. Twenty-five articles were included overall: 17 case reports, 6 case series, and 1 retrospective study. Patients in the case reports and series presented with myalgias and diffuse bone pain. Radiography showed multifocal periostitis, periosteal thickening, or periosteal reaction. Nineteen cases reported the dose of voriconazole with the majority of patients taking 400mg daily; three patients were taking 200, 600, and 800mg daily while on patient was on 8mg/kg IV daily. One of the studies found a significant difference was found in voriconazole dosing in patients who did and did not develop periostitis (780mg $\pm$ 43 vs 450mg $\pm$ 82; $P<0.001$). The bone pain improved 2 weeks to 4 months of stopping the antibiotic in some cases, but resolved much quicker in others (2–5 days) [29M].

There was also a case reported of voriconazole-induced periostitis in a 49-year-old male with cystic fibrosis given voriconazole for 3 months prophylactically following a lung transplant. He presented with painful, swollen hands and a technetium bone scan demonstrated findings of periostitis and exostosis. The serum concentration of voriconazole had been maintained within the target range of 1.5–4.5mg/L during the 3-month period. The patient's LFTs were also elevated upon presentation. The voriconazole was held and the patient experienced symptom relief 5 days later [30A].

ALLYLAMINES [SEDA-36, 381; SEDA-37, 307; SEDA-38, 243]

Terbinafine [SEDA-36, 381; SEDA-37, 307; SEDA-38, 243]

A case of a 34-year-old women who was prescribed both oral and topical terbinafine for treatment of a fungal infection developed a delayed reaction. After 3 days of therapy the patient developed itchy skin reaction on the gluteal area where the terbinafine had been applied. Oral and topical terbinafine were discontinued and the patient was given topic steroids. Three days after discontinuation the patients symptoms worsened with spreading of the lesions. After a 2-week course of oral prednisone her symptoms resolved. On follow-up with dermatology the patient was found to be negative for a skin prick test with terbinafine as well as a patch test. Three months after the reaction a lymphocyte transformation test (LTT) with terbinafine was performed to try to clear up the reaction, which was positive. Therefore, this demonstrates that the LTT advantages over patch and intradermal tests, including absolute safety and the assessment of a T-cell response to the drug, especially when performed 3–9 months after the onset of the reaction. This is the first reported case of confirmation of hypersensitivity to terbinafine through a positive LTT [31A].

A case of psoriatic erythroderma in an 84-year-old women thought to be caused by terbinafine has been reported. Initially, the patient noticed scaly erythematous plaques on the trunk and extremities which developed into a generalized scaly erythema. The patient was taking terbinafine for approximately 1 month prior to presentation. There was no history or family history of psoriasis that was reported. The patient had lab work as well as punch biopsy done and the findings were consistent with psoriatic erythroderma likely due to a medication. A lymphocyte stimulation test (LST) with terbinafine was done, which was positive. In addition a patch test with terbinafine after treatment was positive. Within 2 weeks of stopping the terbinafine and taking topical steroids along with oral methylprednisolone the patient's skin findings resolved [32A].

A case of baboon syndrome secondary to terbinafine was reported. A 60-year-old healthy male presented with a 1-week history of erythematous eruptions around the buttock, axillae and the groins. The patient was on terbinafine 125mg twice daily for 2 weeks about 10 days prior to presentation. He stated that the rash began about 3 days into treatment with the terbinafine. A patch test was conducted which was negative; however, no other cause of the syndrome were identified. He was treated with topical steroids and the area improved at 1 week follow-up [33A].

Lichen planus is a condition of the skin and mucous membranes which can drug-induced. There was a case of a 22-year old who presented with lichen planus after 2 weeks of terbinafine. The medication was stopped and 8 weeks later the lichen planus completed resolved. This is another example of a skin-related condition that can occur in relation to terbinafine. It is important to ensure patients are on the lookout for these dermatologic side effects [34A].

A study was performed that looked at genome-wide association (GWAS) to identify genetic risk factors for drug-induced liver injury (DILI) from drugs to see if there is a genetic association. A total of 862 patients with DILI and 10588 matched controls were included in the study. The study took place in multiple countries including the United States, Europe and South America. DNA samples were used to analyze HLA genes and single nucleotide polymorphisms. An association of DILI was found with rs114577328 (a proxy for A*33:01 a HLA class I allele; odds ratio [OR], 2.7; 95% confidence interval [CI], 1.9–3.8; $P=2.4\times10^{-8}$) and with rs72631567 on chromosome 2 (OR, 2.0; 95% CI, 1.6–2.5; $P=9.7\times10^{-9}$). The association with A*33:01 was mostly effected by for terbinafine-, fenofibrate-, and ticlopidine-related DILI. Therefore, caution should be applied in persons of European descent with HLA-A*33:01 as they are at a higher risk for DILI from terbinafine and possibly fenofibrate and ticlopidine. This study highlights the importance of pharmacogenetics. Given this information patients with this specific variant may be at higher risk for DILI from terbinafine and should not receive this medication for treatment [35C].

AMPHOTERICIN [SEDA-35, 483; SEDA-36, 382; SEDA-37, 307; SEDA-38, 244]

There are no new studies or reports of side effects related to amphotericin at this time.

PYRIMADINE ANALOGUES [SEDA-36, 383; SEDA-37, 307; SEDA-38, 251]

Flucytosine

There are no new studies or reports of side effects related to flucytosine at this time.

ECHINOCANDINS [SEDA-33, 556; SEDA-34, 434; SEDA-35, 489; SEDA-36, 388; SEDA-37, 307; SEDA-38, 251]

Micafungin [SEDA-35, 489; SEDA-36, 388; SEDA-37, 307; SEDA-38, 251]

There are no new studies or reports of side effects related to micafungin at this time.

Anidulafungin [SEDA-35, 489; SEDA-36, 388; SEDA-37, 307; SEDA-38, 252]

There are no new studies or reports of side effects related to anidulafungin at this time.

Caspofungin [SEDA-33, 556; SEDA-34, 434; SEDA-35, 490; SEDA-36, 389; SEDA-37, 307; SEDA-38, 252]

There are no new studies or reports of side effects related to caspofungin at this time.

Finally, a review article was published which looked at all the different available antifungal agents that treat invasive fungal infections. The review looked at safety toxicity as it relates to these agents and the risk for hepatotoxicity. Drug–drug interactions as well as mechanisms for the hepatotoxicity are reviewed. This review is a good guide for clinicians when treating invasive fungal infections to understand the risks associated with each different agent. Therapy can be tailored based on the infection that is being treated as well as the risk for hepatotoxicity [36R].

Additional case studies can be found in these SEDA reviews [37R, 38R, 39R].

References

[1] Chang SH, Chou IJ, Yeh YH, et al. Association between use of non-vitamin K oral anticoagulants with and without concurrent medications and risk of major bleeding in nonvalvular atrial fibrillation. JAMA. 2017;318(13):1250–9 [R].

[2] Charokopos A, Muhammad T, Surbhi S, et al. Weakness and pain in arms and legs dark urine history of vertebral osteomyelitis Dx? J Fam Pract. 2017;66(3):170–3 [A].

[3] Kyriakidis I, Tragiannidis A, Munchen S, et al. Clinical hepatotoxicity associated with antifungal agents. Expert Opin Drug Saf. 2017;16(2):149–65 [r].

[4] Galvao EL, Rabello A, Cota GF. Efficacy of azole therapy for tegumentary leishmaniasis: a systematic review and meta-analysis. PLoS One. 2017;12(10):e0186117 [M].

[5] Nalla P, Dacruz TA, Obuobie K. Tiredness in a patient treated with Itraconazole. BMJ. 2017;356:i6819 [A].

[6] Paul V, Rawal H. Cardiotoxicity with itraconazole. BMJ Case Rep. 2017;1–2. https://doi.org/10.1136/bcr-2017-219376 [A].

[7] Rodrigo-Troyano A, Mediavilla MM, Garin N, et al. Heart failure induced by itraconazole. Med Clin (Barc). 2017;148(2):69–70 [A].

[8] Ikeda E, Watanabe S, Sawada M, et al. A case of liver dysfunction requiring hospital admission after taking oral itraconazole for the treatment of kerion celsi. Med Mycol J. 2017;58(4):J105–11 [A].

[9] Murrell D, Bossaer JB, Carico R, et al. Isavuconazonium sulfate: a triazole prodrug for invasive fungal infections. Int J Pharm Pract. 2017;25(1):18–30 [H].

[10] Hoenigl M, Prattes J, Neumeister P, et al. Real-world challenges and unmet needs in the diagnosis and treatment of suspected invasive pulmonary aspergillosis in patients with haematological diseases: an illustrative case study. Mycoses. 2018;61(3):201–5 [A].

[11] Rivosecchi RM, Clancy CJ, Shields RK, et al. Effects of isavuconazole on the plasma concentrations of tacrolimus among solid-organ transplant patients. Antimicrob Agents Chemother. 2017;61(9). pii: e00970-17. [c].

[12] Desai A, Yamazaki T, Dietz AJ, et al. Pharmacokinetic and pharmacodynamic evaluation of the drug-drug interaction between isavuconazole and warfarin in healthy subjects. Clin Pharmacol Drug Dev. 2017;6(1):86–92 [c].

[13] Jacobs SE, Saez-Lacy D, Wynkoop W, et al. Successful treatment of allergic bronchopulmonary aspergillosis with isavuconazole: case

report and review of the literature. Open Forum Infect Dis. 2017;4(2):1–6. https://doi.org/10.1093/ofid/ofx040 [A].
[14] Keirns J, Desai A, Kowalski D, et al. QT interval shortening with isavuconazole: in vitro and in vivo effects on cardiac repolarization. Clin Pharmacol Ther. 2017;101(6):782–90 [C].
[15] Agarwal SK, Salem AH, Danilov AV, et al. Effect of ketoconazole, a strong CYP3A inhibitor, on the pharmacokinetics of venetoclax, a BCL-2 inhibitor, in patients with non-Hodgkin lymphoma. Br J Clin Pharmacol. 2017;83(4):846–54 [c].
[16] Mody MD, Ravindranathan D, Gill HS, et al. Rhabdomyolysis following initiation of posaconazole use for antifungal prophylaxis in a patient with relapsed acute myeloid leukemia: a case report. J Investig Med High Impact Case Rep. 2017;5(1):1–4. https://doi.org/10.1177/2324709617690747 [A].
[17] Deyo JC, Nicolsen N, Lachiewicz A, et al. Salvage treatment of mucormycosis post-liver transplant with posaconazole during sirolimus maintenance immunosuppression. J Pharm Pract. 2017;30(2):261–5 [A].
[18] Alzghari SK, Seago SE, Cable CT, et al. Severe palmar-plantar erythrodysesthesia and aplasia in an adult undergoing re-induction treatment with high-dose cytarabine for acute myelogenous leukemia: a possible drug interaction between posaconazole and cytarabine. J Oncol Pharm Pract. 2017;23(6):476–80 [A].
[19] Zhang S, He Y, Jiang E, et al. Efficacy and safety of posaconazole in hematopoietic stem cell transplantation patients with invasive fungal disease. Future Microbiol. 2017;12:1371–9 [c].
[20] Pettit NN, Miceli MH, Rivera CG, et al. Multicentre study of posaconazole delayed-release tablet serum level and association with hepatotoxicity and QTc prolongation. J Antimicrob Chemother. 2017;72(8):2355–8 [C].
[21] Xing Y, Chen L, Feng Y, et al. Meta-analysis of the safety of voriconazole in definitive, empirical, and prophylactic therapies for invasive fungal infections. BMC Infect Dis. 2017;17(1):798 [M].
[22] Rosanova MT, Bes D, Serrano Aguilar P, et al. Efficacy and safety of voriconazole in immunocompromised patients: systematic review and meta-analysis. Infect Dis (Lond). 2017;50:489–94. https://doi.org/10.1080/23744235.2017.1418531 [M].
[23] Martin JM, Macias-Parra M, Mudry P, et al. Safety, efficacy, and exposure-response of voriconazole in pediatric patients with invasive aspergillosis, invasive candidiasis or esophageal candidiasis. Pediatr Infect Dis J. 2017;36(1):e1–e13 [c].
[24] Hamandi B, Fegbeutel C, Silveira FP, et al. Voriconazole and squamous cell carcinoma after lung transplantation: a multicenter study. Am J Transplant. 2018;18(1):113–24 [C].
[25] Dewan P, Gomber S, Arora V. Ventricular tachycardia: a rare side effect of voriconazole. Indian J Pediatr. 2017;84(2):152–3 [A].
[26] Gueta I, Loebstein R, Markovits N, et al. Voriconazole-induced QT prolongation among hemato-oncologic patients: clinical characteristics and risk factors. Eur J Clin Pharmacol. 2017;73(9):1181–5 [C].
[27] Liu X, Su H, Tong J, et al. Significance of monitoring plasma concentration of voriconazole in a patient with liver failure: a case report. Medicine (Baltimore). 2017;96(42):e8039 [A].
[28] Lee K, Lim Y, Hah J, et al. Voriconazole plus caspofungin for treatment of invasive fungal infection in children with acute leukemia. Blood Res. 2017;52(3):167–73 [c].
[29] Adwan MH. Voriconazole-induced periostitis: a new rheumatic disorder. Clin Rheumatol. 2017;36(3):609–15 [M].
[30] Metayer B, Bode-Milin C, Ansquer C, et al. Painful and swollen hands 3 months after lungs graft: suracute voriconazole-induced periostitis and exostosis. Joint Bone Spine. 2017;84(1):97–8 [A].
[31] Gonzalez-Cavero L, Dominguez-Ortega J, et al. Delayed allergic reaction to terbinafine with a positive lymphocyte transformation test. J Investig Allergol Clin Immunol. 2017;27(2):136–7 [A].
[32] Oda T, Sawada Y, et al. Psoriatic erythroderma caused by terbinafine: a possible pathogenetic role for IL-23. J Investig Allergol Clin Immunol. 2017;27(1):63–4 [A].
[33] Janjua SA, Pastar Z, et al. Intertriginous eruption induced by terbinafine: a review of baboon syndrome. Int J Dermatol. 2017;56(1):100–3 [A].
[34] Zheng Y, Zhang J, et al. Terbinafine-induced lichenoid drug eruption. Cutan Ocul Toxicol. 2017;36(1):101–3 [A].
[35] Nicoletti P, Aithal GP, Bjornsson ES. Association of liver injury from specific drugs, or groups of drugs, with polymorphisms in HLA and other genes in a genome-wide association study. Gastroenterology. 2017;152(5):1078–89 [C].
[36] Kyriakidis I, Tragiannidis A, et al. Clinical hepatotoxicity associated with antifungal agents. Expert Opin Drug Saf. 2017;16(2):149–65 [R].

Further Reading

[37] McManus D. Antifungal drugs. In: Ray SD, editor. Side effects of drugs annual: a worldwide yearly survey of new data in adverse drug reactions, vol. 37. Elsevier; 2015. p. 307–19 [chapter 27]. [R].
[38] McManus D. Antifungal drugs. In: Ray SD, editor. Side effects of drugs annual: a worldwide yearly survey of new data in adverse drug reactions, vol. 38. Elsevier; 2016. p. 243–53 [chapter 25]. [R].
[39] McManus D. Antifungal drugs. In: Ray SD, editor. Side effects of drugs annual: a worldwide yearly survey of new data in adverse drug reactions, vol. 39. Elsevier; 2017. p. 245–58 [chapter 23]. [R].

CHAPTER

27

Antiprotozoal Drugs

Adrienne T. Black,[1], Dayna S. McManus†, Olawale D. Afolabi‡, Sidhartha D. Ray§*

*3E Services, Verisk 3E, Warrenton, VA, United States

†Department of Pharmacy Services, Yale-New Haven Hospital, New Haven, CT, United States

‡Milken institute School of Public Health, The George Washington University, Washington, DC, United States

§Department of Pharmaceutical and Biomedical Sciences, Touro University College of Pharmacy and School of Osteopathic Medicine, Manhattan, NY, United States

[1]Corresponding author: adrienne159@gmail.com

ANTI-MALARIAL DRUGS

Amodiaquine

Amodiaquine has a high efficacy against chloroquine-resistant *Plasmodium falciparum* malaria, although severe idiosyncratic agranulocytosis and hepatotoxicity are associated with administration of this drug. This toxicity is thought to occur through metabolism by various cytochrome P450 (CYP) enzymes to form a reactive metabolite. It is likely that individuals with greater than normal expression of CYP3A4, CYP2C8, CYP2C9 and CYP2D6 have a greater risk of hepatotoxicity when treated with amodiaquine for malaria due to the metabolism pathway of the drug [1H].

Artemether–Lumefantrine

A single-arm prospective evaluation study was conducted to determine the efficacy and safety of artemether–lumefantrine (AL) for treatment of uncomplicated falciparum malaria. A total of 92 patients (children and adults) with ongoing malaria received the standard regimen of AL (20/120 mg) given twice daily for 3 days with a 28-day follow-up. The most common adverse event was headache (8 cases), perioral ulcer (7 cases), anorexia (6 cases), diarrhea (3 cases), vomiting (2 cases), cough (6 cases), abdominal pain (5 cases), dizziness and nausea (4 cases), weakness or fatigue (4 cases) and sleep disorder (5 cases). No serious adverse events were reported [2c].

A series of case reports described two patients who developed delayed hemolytic anemia after receiving artemether–lumefantrine treatment for malaria. A 20-year-old man and a 27-year-old woman with severe falciparum malaria were both given daily artemether–lumefantrine (20/120 mg) orally for 3 days; the patients exhibited malarial symptoms with an increasing temperature throughout the treatment. Hemolytic anemia as determined by clinical symptoms and laboratory results was evident beginning 11 days after the artemether–lumefantrine treatment. A clear association between the artemether–lumefantrine treatment and the development of hemolytic anemia were not present, although previous case studies indicate that this drug-induced condition may occur [3A].

Oral artemether–lumefantrine medication is typically taken with high-fat food to increase the bioavailability of lumefantrine (by 16-fold) and artemether (by threefold). Nonadherence to the prescribed regimen such as taking the medication while fasting is known to result in treatment failure and the recurrence of malarial symptoms. Preclinical data indicate that the bioavailability of two new solid dispersion formulations of lumefantrine is increased fourfold over that of the standard tablets when taken under fasting conditions. A study of 49 males (18–44 years) received the new artemether–lumefantrine formulations (formulation 1: 480 mg lumefantrine; formulation 2: 960 mg lumefantrine). No serious adverse events were reported with either formulation. Mild to moderate (grade 1 or 2) adverse events were reported as postural dizziness, nasal congestion/hypersensitivity,

Side Effects of Drugs Annual, Volume 40
ISSN: 0378-6080
https://doi.org/10.1016/bs.seda.2018.08.004

headache, iridocyclitis, and catheter site pain and respiratory tract infection, and these effects occurred at a higher incidence with formulation 2 than with formulation 1 (37.5% and 25.0%, respectively) [4c].

Artesunate–Amodiaquine

A prospective, longitudinal, descriptive, non-comparative, non-interventional study assessed the efficacy and safety of an artesunate–amodiaquine (ASAQ; 25/67.5mg) combination treatment administered unsupervised to over 12000 patients with suspected uncomplicated falciparum malaria. ASAQ was given orally once per day for 3 days by the patient at home. A total of 2545 adverse events were reported but most were mild or moderate in nature and included asthenia (682 cases), vomiting (482 cases) and somnolence (174 cases). Serious adverse events (105 cases) included malaria recurrence (29 cases), anemia (28 cases), asthenia (17 cases), vomiting (13 cases) and pyrexia (9 cases), and transient extrapyramidal disorders (3 cases), although only 11 of these serious events were considered related to the ASAQ treatment [5MC].

Dihydroartemisinin–Piperaquine

Although dihydroartemisinin–piperaquine (DP) treatment is a well-tolerated, effective therapy for uncomplicated malarial infections, this drug combination is known to interfere with cardiac repolarization by prolonging QT intervals on electrocardiograms. It is thought that the risk of cardiotoxicity may be lessened by varying the DP treatment regimens. Accordingly, a review of 3 randomized clinical trials with 256 subjects compared the risk of cardiac repolarization due to the standard 3-day treatment regimen vs a recently introduced compressed 2-day regimen. Both groups received a total DP dose of 360/2880mg orally (3-day: 3 tablets/day; 2-day: 4.5 tablets/day). The analysis showed that the compressed 2-day regimen group had a twofold higher the risk of QT interval prolongation due to greater plasma concentrations of piperaquine as compared to the 3-day regimen group [6R].

A prospective, observational, open-label, non-comparative, multi-site study of 4563 patients assessed the safety of dihydroartemisinin–piperaquine (20/160 or 40/320mg with the specific dose based on patient weight) as a first-line combination treatment for acute uncomplicated falciparum malaria. The drug treatment was given over 3 days with the first dose administered in the clinic with the other doses self-administered at home. Follow-up was 28 days after the last dose. A total of 347 adverse events were reported and included infections (212), gastrointestinal disorders (47), itching/pruritus (7), dizziness (1), and skin lesions (1). No serious adverse events were reported. Prior administration of antimalarial medications was associated with a higher incidence of adverse events as compared to no prior medication (9.3% vs 6.1%, respectively) [7MC].

A case report described a 41-year-old man who received a single oral dose of dihydroartemisinin–piperaquine (2.1/16.8mg/kg) as treatment for uncomplicated falciparum malaria. He later presented to the hospital with involuntary body twitching and irregular movements that had begun 3 hours after receiving the drug treatment. Dihydroartemisinin–piperaquine was discontinued and the symptoms were resolved. A diagnosis of dihydroartemisinin–piperaquine-induced generalized choreoathetosis was established [8A].

PfSPZ Vaccine

It is known that repeated immunization with radiation-attenuated sporozoites is effective against malarial infections, although the standard methods of exposure to sporozoite-infected irradiated mosquitoes are unsuitable for mass vaccinations. The recently developed PfSPZ vaccine is a biological product of radiation attenuated cryopreserved sporozoites from a well-characterized isolate of *P. falciparum* that can be administered by IV injection and has been shown to be effective against malarial infection in the laboratory. As a next step, the safety and protective efficacy of the PfSPZ vaccine were assessed in a double-blind, randomized, placebo-controlled trial (NCT01988636). A total of 93 patients (18–35 years) received 5 IV immunizations of either 270000 PfSPZ ($n=46$) or placebo ($n=47$) at weeks 4, 8, 12, 16 and 24. Adverse events were similar between the vaccine and placebo groups and were mild (grade 1) in severity. The common adverse events in both groups were headache (vaccine: 3 (7%); placebo: 4 (9%)), fatigue (vaccine: 0; placebo: 1 (2%)), fever (vaccine: 0; placebo: 1 (2%)), and myalgia (vaccine and placebo: 1 (2%) each) [9c].

Primaquine

Primaquine is the drug of choice for treatment of *Plasmodium vivax* malarial infection but is known to cause hemolysis in patients with a glucose-6-phosphate-dehydrogenase (G6PD) enzyme deficiency. The G6PD gene is located on the X-chromosome, causing males to be either normal or deficient, while females are normal, heterozygous-deficient or homozygous-deficient. In addition, the G6PD genes in the red blood cells undergo a process known as lyonization or X-inactivation in which one of the X genes is inactivated. As a result, the G6PD deficiency phenotypes found in heterozygous females may vary widely. An evaluation of the current information showed that the risk and extent of hemolysis associated with primaquine treatment varied not only with the dose given but also with the level of G6PD activity and genetic variant of the individual being treated. As a result, the World Health Organization (WHO) as well as 7 malaria-endemic countries in the Asia-Pacific region recommends that the G6PD status be established prior to administration of primaquine to determine the appropriate dose in order to minimize the risk of hemolysis [10H].

Artesunate

A systematic review of 12 studies and 7 case reports was conducted to determine the efficacy and safety of artesunate treatment for imported severe malaria. A total of 624 patients who had contracted malaria in traveling to endemic areas were included in the studies. Of these, 23 deaths occurred but none were treatment related. The most common adverse event was the development of delayed hemolysis 7–30 days after artesunate therapy; this effect was present in 15% of the treated patients but no deaths or sequelae occurred. Additional adverse events included mild hepatitis, neurological, renal, cutaneous and cardiac manifestations but these effects were uncommon [11R].

Tafenoquine

Tafenoquine is a recently approved prophylactic anti-malarial drug whose efficacy and safety have been evaluated in 3 clinical trials (NCT02491606, NCT02488980 and NCT02488902) with a total of 1076 adults. The participants (18–55 years) received various combinations of tafenoquine, mefloquine or placebo in the following regimens: 200 or 400 mg tafenoquine for 3 days followed by 200 or 400 mg tafenoquine or placebo once weekly for 24 weeks, 250 mg mefloquine for 3 days followed by 250 mg mefloquine once weekly for 24 weeks, or placebo for 3 days followed by placebo once weekly for 24 weeks. The incidence of adverse events for the treatment groups was tafenoquine (resident of area, 67.6% and non-resident of area, 94.9%), mefloquine (80.6%), and placebo (64.1%). The groups receiving tafenoquine exhibited diarrhea, nausea, vomiting, gastroenteritis, nasopharyngeal tract infections, and back/neck pain at greater frequencies than those receiving the placebo. Although tafenoquine treatment increased the incidence of mild to moderate adverse events, the prophylactic treatment appeared to be effective against malarial infection [12R].

A number of clinical trials and studies have been conducted to compare the various artemether combination treatments for uncomplicated malarial infections. These results aid in identifying the risk of adverse events for each combination medication.

The efficacy and safety of adding supervised primaquine (PQ) administration to the standard artemether–lumefantrine (AL) or chloroquine (CQ) treatments for malaria were assessed in open-label randomized controlled trial with four arms (NCT01680406). Patients ($n=398$) with normal G6PD status and symptomatic *P. vivax* malarial infection received either CQ ($n=104$; weight-based dose given once daily for 3 days), CQ plus PQ ($n=102$; CQ: weight-based dose given once daily for 3 days, PQ: weight-based dose given once daily for 14 days), AL ($n=100$; weight-based dose given twice daily for 3 days), or AL plus PQ ($n=92$; AL: weight-based dose given twice daily for 3 days, PQ: weight-based dose given once daily for 14 days). No difference in the incidence of adverse effects occurred between the treatment regimens [13C].

A phase III, bicenter, open-label, randomized, three-arm trial was conducted in 571 children (age 12–60 months) with recurrent falciparum malaria infection who had received an initial oral treatment with artemether–lumefantrine (AL) (20/120 mg per tablet) in Uganda or artesunate-amodiaquine (ASAQ) (25/67.5 mg per tablet) in the Democratic Republic of Congo (NCT01374581 and PACTR201203000351114). The patients were divided into 3 oral treatment groups: AL [1–3 tablets/day for 3 days, ($n=240$)], ASAQ (1 tablet/day for 3 days; $n=233$) or quinine–clindamycin (quinine: 125 mg/tablet, 0.5–2 tablets/day; clindamycin: 10 mg/kg twice daily) for 7 days ($n=98$). Adverse events were mild across all treatment groups and no serious effects were reported. The most common adverse events in the AL group included anorexia (13%), asthenia (8%), coughing (7%), abnormal behavior (5%) and diarrhea (5%), while the most common adverse event in the ASAQ group was anorexia (6%). Anorexia (12%), abnormal behavior (6%), asthenia (6%) and pruritus (5%) were the reported adverse events in the quinine–clindamycin group [14C].

The safety and efficacy of three drug combinations, artemether–lumefantrine (AL; 20/120 mg per tablet), mefloquine–artesunate (MQAS; 100/220 mg per tablet) and dihydroartemisinin–piperaquine (DHAPQ; 40/320 mg per tablet) were assessed in pregnant women (second or third trimester) with falciparum malaria. A total of 900 patients received one of the three treatment regimens, 300 subjects per treatment group, given 3 tablets per day for 3 days and were followed for 63 days, at delivery and 1-year post-delivery. The adverse events were similar between the treatment and included dizziness, nausea, vomiting, headache and asthenia. These events occurred more often with MQAS (67.9%) than with AL (12.7%) or DHAPQ (23.3%) combinations but were not considered serious. A total of 7 serious adverse events occurred in the study with one case considered related to the MQAS treatment (severe vomiting) [15C]. A similar clinical trial compared dihydroartemisinin–piperaquine (DHAPQ) to artesunate–amodiaquine (ASAQ) for treatment of uncomplicated falciparum malaria during pregnancy. Pregnant women in their second and third trimester with asymptomatic malaria ($n = 417$) received either DHAPQ or ASAQ over 3 days with follow-up on days 1, 2, 3, 7, 14, 28 and 42, at delivery and at 6-week post-partum. Adverse effects included vomiting, dizziness, and general weakness and occurred more often with DHAPQ as compared to ASAQ. No effects on birth outcomes were reported in either study [16C].

A phase IIa, multicenter, open-label, safety-focused dose-ranging randomized study compared the safety and activity of ferroquine–artesunate to amodiaquine–artesunate for treatment of uncomplicated malaria in 72 men (18–65 years) (NCT00563914). The patients were randomized to one of four ferroquine groups: ferroquine (100, 200, 400, or 600 mg) plus artesunate (200 mg) or an amodiaquine group: amodiaquine (612 mg) plus artesunate (200 mg). Each oral treatment was given daily for 3 days. The most common adverse event was increases in alanine aminotransferase (ALT) levels with grade 1 toxicity present at 400 mg ferroquine and grade 2 toxicity with 200 and 600 mg ferroquine. Discontinuation of treatment occurred with 1 patient receiving the 100 mg ferroquine–artesunate treatment due to a prolonged QT interval [17c].

A systematic review and meta-analysis of 21 studies evaluated the treatment outcomes of anti-malarial drugs in use in Ethiopia with a total of 3123 participants. The occurrence of adverse events in response to anti-malarial treatment was reported in 7 studies that included 822 participants. The pooled results showed that the overall incidence of adverse events was 39.8% with artemether–lumefantrine and chloroquine treatments being 41.2% and 35.6%, respectively. The most commonly reported adverse events in response to these 2 drug combinations were gastrointestinal irritation and inflammation, headache, weakness or fatigue, anorexia, vomiting and oral ulcers [18MC].

A randomized, non-inferiority trial compared artemether–lumefantrine (AL), the standard treatment for uncomplicated falciparum malaria, to AQ-13, a recently developed 4-aminoquinoline product with modified side chains. Previous studies have shown that side chains of 2–3 carbons or 10–12 carbons are more effective against the malarial parasites than the 5-carbon side of existing anti-malarial drugs. Accordingly, AQ-13 was developed using a 3-carbon *n*-propyl side chain. Men ($n = 66$; ≥ 18 years) with uncomplicated malaria received either AL (80/480 mg) orally twice daily for 3 days ($n = 33$) or AQ-13 (638.50 mg) orally for 2 days with AQ-13 (319 mg) orally on day 3 ($n = 33$). All patients received a 5-week follow-up. No serious adverse events occurred in either group. A total of 453 adverse events of grade 1 or less was reported with 214 events in AL group and 239 in the AQ-13 group. These mild adverse events included fever, weakness, myalgias and arthralgias, headache, anorexia, nausea, vomiting, abdominal pain, diarrhea, cough, pruritus, tinnitus, influenza-like symptoms, pallor, and jaundice. Many of the reported adverse effects were those associated with the acute malarial disease state such as fever, weakness, myalgias and arthralgias, headache, anorexia, nausea, vomiting, and abdominal pain and most commonly occurred on days 1–3 [19c].

The efficacy, safety, tolerability and pharmacokinetics of artefenomel–piperaquine for treatment of uncomplicated falciparum malaria were assessed in a randomized, double-blind single-dose study (NCT020833380). A total of 437 patients (>6 months to ≤ 70 years) with uncomplicated malaria in Africa ($n = 355$) and Asia ($n = 82$) were included with approximately 85% of the study population being <5 years old. All patients received a single dose of artefenomel (800 mg) and one of 3 doses of piperaquine (640, $n = 143$; 960, $n = 148$; or 1440 mg, $n = 146$) and were followed for 42 or 63 days. A total of 364 adverse occurred in the study and generally at the same incidence between treatment groups; these adverse events included infections and infestations, QT prolongation, vomiting, gastrointestinal disorders, and fever. A total of 6 serious adverse events occurred in 4 patients and included severe anemia, febrile convulsions, elevated transaminase levels and neutropenia. Study discontinuation occurred in 1 case due to excessive vomiting [20C].

A retrospective, observational study assessed hepatic function in 57 hospitalized patients (≥ 18 years) with non-severe falciparum malaria received either artemether–lumefantrine (AL; $n = 19$) or quinine–doxycycline (QD; $n = 38$) treatments while hospitalized. During the treatment period, the AL group had a greater number of liver enzyme abnormalities as opposed to the QD group: 8/19 (42%) and 2/38 (5%), respectively. No other differences in toxicity or length of hospital stay were apparent between the groups. The liver enzymes returned to normal during out-patient follow-up after discharge [21c].

LEISHMANIASIS

Meglumine Antimoniate

A phase III, non-inferiority, randomized, single-blind, controlled trial of 72 patients (≥ 13 years) with cutaneous leishmaniosis received either megluime antimoniate (5 mg/kg/day) for 30 days or the standard 20 mg/kg/day treatment for 20 days (NCT01301924). Although the successful treatment rate was 77.8% and 94.4%, respectively, the 20 mg dose group experienced more serious adverse events at a greater rate overall as well as more serious adverse events per person. These effects resulted in a much greater discontinuation rate for the 20 mg group.

The major adverse events in the 20mg group were pancreatic toxicity, neutropenia, hypermylasemia and abnormal laboratory values with a relative risk of 2.08, a 108.3% increase over that of the 5mg group [22c].

A series of case reports described 3 patients with cutaneous leishmaniosis who had received intralesion megluime antimoniate treatment via subcutaneous injection. The specific dose administered was determined by observing the lesion during injection of the drug and continued until the area was swollen upon palpation. Previous experience has shown that effective doses are those with volumes greater than 5mL per injection. This injection process was administered 1–3 times at 3-week intervals. Adverse effects were reported in one patient and consisted of tinnitus, light local edema and transient hyperlipasemia. All lesions were successfully treated and the patients fully received within 8 months [23A].

Amphotericin B

An open-label, non-comparative randomized, parallel arm study compared the efficacy and safety of different therapy regimens of amphotericin B used as treatment for post kala-azar dermal leishmaniasis. A total of 50 hospitalized patients (5–65 years) received IV amphotericin B treatment at 0.5 mg/kg/day ($n=25$; cumulative dose=30mg/kg) or 1mg/kg/day ($n=25$; cumulative dose=60mg/kg) on alternate days. Each treatment was given for 20 days in 3 courses with 15 days between each course. Adverse effects including fever, rigor, nausea and vomiting occurred in both groups but were more frequent with the 1mg/kg dose. Similarly, acute nephrotoxicity with elevated serum creatinine levels occurred more often in the high dose group. All symptoms were resolved by time of discharge [24c]. A similar study was conducted in children and adolescents (<15 years) treated with IV amphoterin B (10mg/kg) for kala-azar visceral leishmaniasis. The most common adverse events were chills and rigor and were mild in nature; no serious adverse events were reported [25c].

A randomized, controlled, open-label, parallel group phase III clinical trial compared the safety and efficacy data for drug combinations of amphotericin B deoxycholate (AmB), miltefosine and paromomycin with amphotericin B monotherapy. Patients ($n=602$; 5–60 years) with visceral leishmaniosis were hospitalized and treated with one of four drug regimens: a single dose of IV AmB (5mg/kg) on day 1 ($n=159$), a single dose of IV AmB (5mg/kg) on day 1 and oral miltefosine (50–100mg/kg) on days 2–8 ($n=159$), a single dose of IV AmB (5mg/kg) on day 1 and IM paromomycin (11mg/kg/day) on days 2–11 ($n=142$), or IM paromomycin with oral miltefosine on days 1–10 ($n=142$). There were 12 serious adverse events in the study including 3 non-treatment related deaths. A total of 3 serious adverse events occurred in the miltefosine–paromomycin group and were nephropathy and ototoxicity 2 weeks after treatment, both in 1 person, and acute hepatitis in a second individual; all effects were resolved when the treatment was finished. A high grade fever, rash and swelling occurred in the AmB-miltefosine group that was resolved with AmB treatment alone. No other serious adverse events were related to any treatment. The remaining adverse events occurred with all treatment regimens and were generally mild or moderate in nature and included vomiting, anorexia and respiratory distress [26C].

A multicenter, randomized, open-label, controlled trial evaluated the efficacy and safety of various drug treatments for visceral leishmaniasis (NCT01310738). A total of 378 patients (6 months to 50 years) with visceral leishmaniasis received one of the following four IV drug regimens: amphotericin B deoxycholate (AmB; 1mg/kg/day) for 14 days ($n=45$), liposomal amphotericin B (LAMB; 3mg/kg/day) for 7 days ($n=109$), a combination of LAMB (10mg/kg single dose) plus meglumine antimoniate (MA; 20mg Sb+5/kg/day) for 10 days ($n=112$) or the standard treatment of MA (20mg/kg/day) for 20 days ($n=112$). Due to an unplanned interim safety analysis, the AmB treatment was suspended based on the Data Safety Monitoring Board (DSMB) recommendation, and this arm was dropped from the study, leaving 332 patients remaining to completion. Adverse effects occurred with all treatments and there were no differences between regimens. Adverse events occurred in 239 patients (71.2%) and of these, 66 cases were considered serious. Early withdrawal from the study due to adverse events occurred with MA (15 patients), LAMB plus MA (10 patients) and LAMB (1 patient) and included increased pancreatic and liver enzymes and cardiac toxicity. Serious adverse events also included 1 death to MA-related respiratory and hemodynamic distress. Overall, the LAMB monotherapy was found to have a lower incidence of adverse events than the other treatments [27C].

The efficacy and safety of amphotericin B deoxycholate and of *N*-methylglucamine antimoniate for treatment of visceral leishmaniasis were compared in a randomized, open-label, 2-arm, controlled trial with 101 pediatric patients (6 months to 12 years). The subjects received either *N*-methylglucamine antimoniate ($n=51$; 20mg/kg/day) for 20 days or amphotericin B deoxycholate ($n=50$; 1mg/kg/day) for 14 days. Adverse events occurred with all patients in both groups and no significant difference was found between the groups. Most adverse events were mild to moderate (gastrointestinal disturbances, electrolyte imbalances and myalgia). Serious adverse events occurred with *N*-methylglucamine antimoniate (3 cases) and amphotericin B deoxycholate (2 cases) and consisted of abnormal liver enzymes, anemia, and hypomagnesemia [28C].

CHAGAS DISEASE

Benznidazole

Although benznidazole is typically used as treatment for Chagas disease, adverse events occur in approximately half the patients prescribed this drug, leading to therapy discontinuation. The occurrence of adverse events was assessed in a study of 2075 patients who received benznidazole treatment for Chagas disease to help increase the understanding of factors involved in the drug-related reactions. Treatment was 5–7.5mg/kg/day 2–3 times per day for approximately 60 days. Benznidazole was permanently discontinued in 211 patients (10.2%) and was associated with mild and moderate adverse effects including skin disorders and occurred with a greater frequency in female patients [29C]. A retrospective follow-up study was also conducted to determine the safety of benznidazole by reviewing the medical records of 224 adults with chronic Chagas disease treated with the drug. The analysis showed that adverse events occurred in 205 cases (91.5%) and that 52 of these cases (23.2%) required stopping benznidazole treatment. The most common mild to moderate adverse events were rash, itching, epigastric pain, abdominal bloating and nausea. Serious adverse events included rash, urticarial reaction, angioedema, skin peeling, epigastric pain, bullous eruption, paraesthesia, drug reaction with eosinophilia and systemic symptoms (DRESS), sleep disturbance, depression and significantly elevated alanine aminotransferase (ALT) levels. All adverse events resolved with discontinuation of the drug. Risk factors for experiencing a serious adverse event were a benznidazole dose $\geq$6 mg/kg/day, female sex and presence of >3 adverse events [30C].

A case report described a 49-year-old woman received benznidazole (5mg/kg/day) for treatment of chronic Chagas disease. The patient developed an erythematous plaque 4 days after beginning treatment with abdominal pain, nausea and vomiting, malaise, and dysthermia 8 days later. Further examination showed a low-grade fever, extensive erythematous papules and elevated liver function values. Benznidazole treatment was discontinued and the symptoms were resolved with steroid administration over a 2-week period [31A].

OTHER PROTOZOAL INFECTIONS

Ornidazole

Ornidazole is a synthetic nitroimidazole used for treatment of infections caused by both anaerobic bacteria and protozoa. A case report described a 48-year-old woman who had received ornidazole as treatment for a genitourinary tract infection. A dermal fixed drug eruption occurred on the sole of the foot following ornidazole administration and resolved once the treatment ceased. A further investigation into the patient history indicated that ornidazole had been prescribed three times over the previous 6 months with the development of similar dermal reactions with each drug administration [32A].

Pyrimethamine

Pyrimethamine-based medications are used for the treatment of toxoplasmosis caused by *Toxoplasma gondii* infection. A systematic review of 31 studies with a total of 2975 patients evaluated the safety of pyrimethamine-based treatments for the three main Toxoplasma infections (toxoplasmic encephalitis, ocular toxoplasmosis, and congenital toxoplasmosis). Eleven of these studies with 1284 patients reported adverse events; the adverse events varied between the different infection types but were generally mild to moderate in nature and included altered liver function enzymes, fever, dizziness, headache, reversible aphasia, palpitations, stenocardia, emotional distress, arthralgia, facial swelling, and malaise. Hematological toxicity indicative of bone marrow suppression was a serious adverse event common to all treatments and conditions. Discontinuations occurred in each of these studies due to serious adverse events such as hematological toxicity (anemia, thrombocytopenia, leukopenia, agranulocytosis, hemoglobin depression) as well as skin rash, severe gastrointestinal distress, epistaxis and Steven–Johnson syndrome [33R].

Additional case studies can be found in these SEDA reviews [34R,35R,36R].

References

[1] Zhang Y, Vermeulen NPE, Commandeur JNM. Characterization of human cytochrome P450 mediated bioactivation of amodiaquine and its major metabolite N-desethylamodiaquine. Br J Clin Pharmacol. 2017;83:572–83 [H].

[2] Teklemariam M, Assefa A, Kassa M, et al. Therapeutic efficacy of artemether-lumefantrine against uncomplicated Plasmodium falciparum malaria in a high-transmission area in Northwest Ethiopia. PLoS One. 2017;12(4):e0176004 [c].

[3] Tsuchido Y, Nakamura-Uchiyama F, Toyoda K, et al. Development of delayed hemolytic anemia after treatment with oral artemether–lumefantrine in two patients with severe falciparum malaria. Am J Trop Med Hyg. 2017;96(5):1185–9 [A].

[4] Jain JP, Leong FJ, Chen L, et al. Bioavailability of lumefantrine is significantly enhanced with a novel formulation approach, an outcome from a randomized, open-label pharmacokinetic study in healthy volunteers. Antimicrob Agents Chemother. 2017;61(9), pii: e00868-17 [c].

[5] Assi SB, Aba YT, Yavo JC, et al. Safety of a fixed-dose combination of artesunate and amodiaquine for the treatment of uncomplicated Plasmodium falciparum malaria in real-life conditions of use in Côte d'Ivoire. Malar J. 2017;16:8 [MC].

[6] Vanachayangkul P, Lon C, Spring M, et al. Piperaquine population pharmacokinetics and cardiac safety in Cambodia. Antimicrob Agents Chemother. 2017;61(5), pii: e02000-16 [R].

[7] Oduro AR, Owusu-Agyei S, Gyapong M, et al. Post-licensure safety evaluation of dihydroartemisinin piperaquine in the three major ecological zones across Ghana. PLoS One. 2017;12(3): e0174503 [MC].

[8] Kadia BM, Morfaw C, Simo ACG. Choreoathetosis—an unusual adverse effect of dihydroartemisinin-piperaquine: a case report. J Med Case Reports. 2017;11(1):360 [A].

[9] Sissoko MS, Healy SA, Katile A, et al. Safety and efficacy of PfSPZ vaccine against Plasmodium falciparum via direct venous inoculation in healthy malaria-exposed adults in Mali: a randomised, double-blind phase 1 trial. Lancet Infect Dis. 2017;17(5):498–509 [c].

[10] Thriemer K, Ley B, Bobogare A, et al. Challenges for achieving safe and effective radical cure of *Plasmodium vivax*: a round table discussion of the APMEN Vivax Working Group. Malar J. 2017;16(1):141 [H].

[11] Roussel C, Caumes E, Thellier M, et al. Artesunate to treat severe malaria in travellers: review of efficacy and safety and practical implications. J Travel Med. 2017;24(2), taw093 [R].

[12] Novitt-Moreno A, Ransom J, Dow G, et al. Tafenoquine for malaria prophylaxis in adults: an integrated safety analysis. Travel Med Infect Dis. 2017;17:19–27 [R].

[13] Abreha T, Hwang J, Thriemer K, et al. Comparison of artemether-lumefantrine and chloroquine with and without primaquine for the treatment of Plasmodium vivax infection in Ethiopia: a randomized controlled trial. PLoS Med. 2017;14(5):e1002299 [C].

[14] Mavoko HM, Nabasumba C, da Luz RI, et al. Efficacy and safety of re-treatment with the same artemisinin-based combination treatment (ACT) compared with an alternative ACT and quinine plus clindamycin after failure of first-line recommended ACT (QUINACT): a bicentre, open-label, phase 3, randomised controlled trial. Lancet Glob Health. 2017;5(1):e60–8 [C].

[15] Nambozi M, Kabuya JB, Hachizovu S, et al. Artemisinin-based combination therapy in pregnant women in Zambia: efficacy, safety and risk of recurrent malaria. Malar J. 2017;16(1):199 [C].

[16] Osarfo J, Tagbor H, Cairns M, et al. Dihydroartemisinin-piperaquine versus artesunate-amodiaquine for treatment of malaria infection in pregnancy in Ghana: an open-label, randomised, non-inferiority trial. Trop Med Int Health. 2017;22(8):1043–52 [C].

[17] Supan C, Mombo-Ngoma G, Kombila M, et al. Phase 2a, open-label, 4-escalating-dose, randomized multicenter study evaluating the safety and activity of ferroquine (SSR97193) plus artesunate, versus amodiaquine plus artesunate, in African adult men with uncomplicated Plasmodium falciparum malaria. Am J Trop Med Hyg. 2017;97(2):514–25 [c].

[18] Gebreyohannes EA, Bhagavathula AS, Seid MA, et al. Anti-malarial treatment outcomes in Ethiopia: a systematic review and meta-analysis. Malar J. 2017;16(1):269 [MC].

[19] Koita OA, Sangaré L, Miller HD, et al. AQ-13, an investigational antimalarial, versus artemether plus lumefantrine for the treatment of uncomplicated Plasmodium falciparum malaria: a randomised, phase 2, non-inferiority clinical trial. Lancet Infect Dis. 2017;17(12):1266–75 [c].

[20] Macintyre F, Adoke Y, Tiono AB, et al. A randomised, double-blind clinical phase II trial of the efficacy, safety, tolerability and pharmacokinetics of a single dose combination treatment with artefenomel and piperaquine in adults and children with uncomplicated Plasmodium falciparum malaria. BMC Med. 2017;15(1):181 [C].

[21] Silva-Pinto A, Ruas R, Almeida F, et al. Artemether-lumefantrine and liver enzyme abnormalities in non-severe Plasmodium falciparum malaria in returned travellers: a retrospective comparative study with quinine-doxycycline in a Portuguese centre. Malar J. 2017;16(1):43 [c].

[22] Saheki MN, Lyra MR, Beoya-Pacheco SJ, et al. Low versus high dose of antimony for American cutaneous leishmaniasis: a randomized controlled blind non-inferiority trial in Rio de Janeiro, Brazil. PLoS One. 2017;12(5):e0178592 [c].

[23] Pimentel MIF, Vasconcellos ECFE, Ribeiro CO, et al. Intralesional treatment with meglumine antimoniate in three patients with New World cutaneous leishmaniasis and large periarticular lesions with comorbidities. Rev Soc Bras Med Trop. 2017;50(2):269–72 [A].

[24] Rabi Das VN, Siddiqui NA, Pal B, et al. To evaluate efficacy and safety of amphotericin B in two different doses in the treatment of post kala-azar dermal leishmaniasis (PKDL). PLoS One. 2017;12(3): e0174497 [c].

[25] Pandey K, Pal B, Siddiqui NA, et al. Efficacy and safety of liposomal amphotericin B for visceral leishmaniasis in children and adolescents at a tertiary care center in Bihar, India. Am J Trop Med Hyg. 2017;97(5):1498–502 [c].

[26] Rahman R, Goyal V, Haque R, et al. Safety and efficacy of short course combination regimens with AmBisome, miltefosine and paromomycin for the treatment of visceral leishmaniasis (VL) in Bangladesh. PLoS Negl Trop Dis. 2017;11(5):e0005635 [C].

[27] Romero GAS, Costa DL, Costa CHN, et al. Efficacy and safety of available treatments for visceral leishmaniasis in Brazil: a multicenter, randomized, open label trial. PLoS Negl Trop Dis. 2017;11(6):e0005706 [C].

[28] Borges MM, Pranchevicius MC, Noronha EF, et al. Efficacy and safety of amphotericin B deoxycholate versus N-methylglucamine antimoniate in pediatric visceral leishmaniasis: an open-label, randomized, and controlled pilot trial in Brazil. Rev Soc Bras Med Trop. 2017;50(1):67–74 [C].

[29] Sperandio da Silva GM, Mediano MFF, Hasslocher-Moreno AM, et al. Benznidazole treatment safety: the Médecins Sans Frontières experience in a large cohort of Bolivian patients with Chagas' disease. J Antimicrob Chemother. 2017;72(9):2596–601 [C].

[30] Olivera MJ, Cucunubá ZM, Valencia-Hernández CA, et al. Risk factors for treatment interruption and severe adverse effects to benznidazole in adult patients with Chagas disease. PLoS One. 2017;12(9):e0185033 [C].

[31] Coronel MV, Frutos LO, Muñoz EC, et al. Adverse systemic reaction to benznidazole. Rev Soc Bras Med Trop. 2017;50(1):145–7 [A].

[32] Emre S, Ahsen H, Aktas A. Ornidazole-induced fixed drug reaction on sole: case report and review of the literature. Cutan Ocul Toxicol. 2017;36(3):294–6 [A].

[33] Ben-Harari RR, Goodwin E, Casoy J. Adverse event profile of pyrimethamine-based therapy in toxoplasmosis: a systematic review. Drugs R D. 2017;17(4):523–44 [R].

[34] Thurston S, Hite GL, Petry AN, et al. Antiprotozoal drugs. In: Ray SD, editor. Side effects of drugs annual: a worldwide yearly survey of new data in adverse drug reactions. vol. 37. Elsevier; 2015. p. 321–7 [chapter 28]. [R].

[35] McManus D, Hall K, Ray SD. Antiprotozoal drugs. In: Ray SD, editor. Side effects of drugs annual: a worldwide yearly survey of new data in adverse drug reactions. vol. 38. Elsevier; 2016. p. 255–9 [chapter 26]. [R].

[36] McManus D, Ray SD. Antiprotozoal drugs. In: Ray SD, editor. Side effects of drugs annual: a worldwide yearly survey of new data in adverse drug reactions. vol. 39. Elsevier; 2017. p. 255–9 [chapter 26]. [R].

CHAPTER

28

Antiviral Drugs

Sreekumar Othumpangat[1], John D. Noti

Allergy and Clinical Immunology Branch, Health Effects Laboratory Division, National Institute for Occupational Safety and Health, Centers for Disease Control and Prevention, Morgantown, WV, United States

[1]Corresponding author: seo8@cdc.gov

Abbreviations

3TC lamivudine (dideoxythiacytidine)
ABC abacavir
ACV acyclovir
ADV adefovir dipivoxil
D4T stavudine (didehydrodideoxythymidine)
DCV daclatasvir
DTG dolutegravir
ETV entecavir
ETC emtricitabine
EFV efavirenz
EVG elvitegravir
FOS foscarnet
GCV ganciclovir
Peg-IFN peg-interferon-alfa-2a
IM imiquimod
LAM lamivudine
LDV ledipasvir
MVC maraviroc
RAL raltegravir
RBV ribavirin
RPV rilpivirine
SIM simeprevir
SOF sofosbuvir
TAF tenofovir alafenamide
TDF tenofovir disoproxil fumarate
VGV valganciclovir
ZDV zidovudine

DRUGS ACTIVE AGAINST CYTOMEGALOVIRUS

Cidofovir [SEDA-35, 503; SEDA-36, 401; SEDA-37, 329; SEDA-38, 261; SEDA-39, 269]

Human Cytomegalovirus (CMV) infection can complicate successful solid organ transplantation in patients. Seven male and two female transplant recipients for the kidney (two recipients), pancreas (two recipients), lung (four recipients), and small bowel (one recipient) received cidofovir for refractory CMV infections. Three recipients were CMV seronegative, but all nine received grafts from CMV-seropositive donors. Five patients were on antithymocyte globulin, four received daclizumab induction, and one suffered from immunodeficiency. Of these treated patients, seven experienced rejection. Eight patients had received prophylactic ganciclovir (GCV), and eight had been treated for CMV infection with one or more drugs (GCV—eight; CMV immunoglobulin—three; foscarnet—three). Four patients had mild nephrotoxicity, and three developed renal failure. Two patients died of uncontrolled infections and concurrent CMV disease, one with invasive aspergillosis and another with nocardiosis. The authors were of the opinion that use of GCV is effective in solid organ transplant patients and adverse events are of low grade (AEs) [1c].

Letermovir [SEDA-38, 261; SEDA-39, 269]

Letermovir acts on the CMV-terminase complex and inhibits viral replication and survival.

Observational Studies

In a Phase 3 double-blinded study in CMV-seropositive transplant recipients, letermovir or placebo was randomly administered orally or intravenously. A total of 565 patients underwent randomization and received letermovir or placebo beginning 9 days after transplantation. Letermovir was administered at a dose of 480 mg per day orally (or 240 mg per day in patients taking cyclosporine). The primary end point for the determination of drug efficiency was the lack of CMV DNA in patients. Among 495 patients with undetectable CMV DNA at the time of

Side Effects of Drugs Annual, Volume 40
ISSN: 0378-6080
https://doi.org/10.1016/bs.seda.2018.08.005

randomization, fewer patients in the letermovir group than in the placebo group had clinically significant CMV infection (122 of 325 patients vs 103 of 170). The frequency and severity of AEs were similar in the treated and placebo groups. Vomiting was reported in 18.5% of the patients who received letermovir and in 13.5% who received the placebo; edema was reported in 14.5% and 9.4%, and atrial fibrillation or flutter in 4.6% and 1.0%, respectively. The rates of myelotoxic and nephrotoxic events were similar in the letermovir group and the placebo group. The mortality at week 48 after transplantation was 20.9% in letermovir and 25.5% in placebo recipients. Letermovir prophylaxis resulted in a significantly lower risk of CMV infection than placebo. AEs with letermovir were mainly of low grade [2C].

Brincidofovir [SEDA-39, 270]

Brincidofovir (BCV), a lipid conjugated prodrug of cidofovir, is a long-acting, broad-spectrum antiviral that was evaluated for the prevention and treatment of CMV and adenovirus infections in healthy subjects in Phase 1 and in hematopoietic cell transplant (HCT) recipients in Phase 2/3 clinical trials. The most common AEs reported for BCV were mild gastrointestinal events and asymptomatic, transient elevations in serum transaminases.

The kinetics of viremia and toxicity following preemptive treatment with BCV in children and adolescents diagnosed with HCT-related adenoviremia were compared in a multicenter trial. The study included 333 subjects (from January 2015 to May 2016) undergoing allogeneic stem cell transplant from seven pediatric transplant centers. Patients with significant viremia (adenovirus levels $\geq$1000 copies per mL) on two consecutive occasions were treated with cidofovir (5 mg/kg) weekly for 2consecutive weeks, followed by 1 mg/kg, three times weekly. Patients with preexisting renal impairment, those that developed renal impairment on cidofovir, and those that did not respond to 2 weeks of cidofovir were treated with BCV, 2 mg/kg, twice weekly. Adenoviremia was reported in 47 (14.1%) patients and significant viremia requiring antiviral treatment was noted in 27 patients (8.1%). Thirteen patients (80%) treated with BCV cleared viremia compared to eight patients (35%) treated with cidofovir. Two patients treated with cidofovir died of disseminated adenoviral infection. BCV was well tolerated and only one patient required interruption of BCV therapy after 4 weeks, due to severe abdominal cramps and diarrhea. BCV-mediated diarrhea was reported as a frequent side effect in the Phase 3 double-blind placebo-controlled trial for preventing CMV infections. The authors suggested that it is important to distinguish BCV toxicity from gut graft-vs-host disease (GVHD) and virus-associated diarrhea in HCT transplant patients. Nine of 23 patients treated with cidofovir developed mild-to-moderate nephrotoxicity [3C].

Foscarnet [SEDA-35, 504; SEDA-36, 403; SEDA-37, 329; SEDA-38, 262; SEDA-39, 270]

Observational Study

A 4-month-old infant had severe combined immunodeficiency disease (SCID), with undetectable levels of immunoglobulins and severe lymphopenia (476 cells/mL). The patient also was diagnosed with CMV viremia. Further analysis showed CD19 (341 cells/mL) and CD56 (250 cells/mL) expressing cells with no cells expressing CD3, which is the basic diagnosis of T − B + NK + SCID. DNA sequence analysis revealed a G615T homozygous stop codon mutation in the IL-7 Ra gene. Treatment was initiated with ganciclovir (GCV), and within 3 weeks CMV copies increased (5.8×10^6). First course of foscarnet (FOS) was initiated (180 mg/kg/day). During this time, total serum calcium was 10.6 mg/dL. Concurrent medications given were: piperacillin–tazobactam, prophylactic trimethoprim/sulfamethoxazole, rifampicin, isoniazid, and fluconazole. After 2 weeks of treatment, the CMV copies doubled and FOS was discontinued. This regimen followed Cidofovir administration 5 mg/kg once weekly. There was no decrease in CMV copies and FOS (180 mg/kg/day) was reintroduced combined with GCV (10 mg/kg/day). A CMV point mutation was detected suggesting that the virus was drug resistant. At initiation of the combined treatment, serum creatinine and urea levels were within the normal range. However, after 30 days of FOS treatment the calcium level reached 15.8 mg/dL, although the patient did not have any symptoms of hypercalcemia. CMV levels were reduced in the patient over time and FOS was withdrawn from the treatment regimen. The patient's treatment was again supplemented with FOS at a lower dose along with GCV to reduce the CMV count. This patient underwent a bone marrow transplant and was administered the same dose of FOS and GCV, but the patient died due to respiratory failure. The authors suggested that FOS binds to the inorganic bone matrix creating hypercalcemia, and calcium levels should be monitored closely during the treatment with FOS [4C].

The clinical safety and efficacy of FOS prophylaxis and pre-emptive therapy for CMV infection in allogeneic hematopoietic stem cell transplantation (HSCT) have also been reported. Ninety-six patients undergoing HSCT were included in this study. FOS was given at 60 and 120 mg/kg/day in a prevention and a preemptive study, respectively. The side effects of FOS prophylaxis were mild without any hematologic toxicities [5c].

Ganciclovir and Valganciclovir [SEDA-35, 504; SEDA-36, 404; SEDA-37, 330; SEDA-38, 262; SEDA-39, 270]

Observational Study

A renal transplant patient showed delayed graft and renal impairment leading to resistance to valganciclovir (VGC) [6A]. In another case study, a 71-year-old CMV-seronegative Caucasian man having end-stage renal disease had a renal transplantation from a CMV-seropositive donor. He was on hemodialysis and received 2 doses of thymoglobulin (1.5mg/kg) and methylprednisolone (500mg) and tacrolimus, mycophenolate mofetil 1000mg (twice a day), and prednisone during the first month of post-transplantation. The patient was on sulfamethoxazole–trimethoprim and VGC 900 (mg each), daily, at the time of transplantation. Four months after transplantation, the recipient developed CMV viremia and acute thrombocytopenia. VGC was replaced with GCV 5mg/kg twice daily and mycophenolate mofetil was lowered to 500mg, twice daily. During the first 3 weeks of treatment, the patient's viral load progressively reduced, but after 27 days, the viral load doubled. GCV dosage was then increased to 10mg/kg, twice daily. Further analysis showed a UL97-resistant mutation, and the patient was administered with FOS. Seven weeks after FOS treatment, the viral load increased and resistant strains to FOS, cidofovir, and GCV were detected in the patient's serum. CMV immunoglobulin was administered three times per week to overcome the mutant strains and that resulted in complete viral clearance. No side effects were reported from these treatments [6A].

Combination Study

A 61-year-old man with a 6-year history of diarrhea was admitted for weight loss. He had chronic sinusitis and continuous cough, and his sputum cultures showed positive for *Haemophilus influenzae* and β-haemolytic *Streptococcus*. The patient was subsequently diagnosed with oesophageal candidiasis, CMV disease, and a bacterial chest infection. The patient's CMV infection was treated with intravenous (IV) FOS (60mg/kg, three times daily) for 3 weeks, followed by VGC (900mg, once daily). Three months after treatment, the patient became negative for blood CMV DNA, and VGC was removed from the treatment regimen. He subsequently relapsed with CMV viremia and diarrhea due to CMV colitis and was again retreated with VGC (900mg, twice daily) for the next 3 weeks, followed by long-term VGC prophylaxis (900mg, once daily). A VGC resistance mutation (UL97 mutation H520Q) was detected in CMV isolated from the patient's blood. Cidofovir (5mg/kg daily) was administered to reduce the CMV mutants. Oral Leflunomide (20mg, once daily) was added to his treatment regimen. This study demonstrated viral resistance to VGC and severe side effects from FOS, emphasizing the difficulties in long-term suppression of CMV in patients with persistent immunodeficiency [7A].

Valaciclovir

A 43-year-old diabetic patient reported having partial vision and was on oral steroids and anti-viral therapy. The right eye of the patient showed inferior and had temporal retinal thinning. The right eye also had pigmentation and periarterial whitish focal Kyrieleis' plaques. The patient's left eye had mild vitritis, optic disc pallor, arteriolar attenuation, and whitening of the retina. Serology for human immunodeficiency virus (HIV), herpes simplex virus (HSV), and CMV was all negative, but IgM positive for varicella zoster virus. Valaciclovir (1g, three times daily) was administered orally. No side effects of the drug in this patient were reported. The authors were of the opinion that a peripheral retinal examination must be done in cases with Kyrieleis' plaques to rule out retinitis and vasculitis [8A].

DRUGS ACTIVE AGAINST HERPES VIRUSES [SEDA-35, 507; SEDA-36, 407; SEDA-37, 332; SEDA-38, 263; SEDA-39, 271]

Acyclovir

Acyclovir (ACV) has been widely used to treat infections caused by herpes simplex virus (HSV) and varicella zoster virus (VZV). The common AEs of ACV included nausea, diarrhea, headache, dizziness and mental changes.

A randomized double-blind controlled trial compared the safety and efficacy of different strains of lactobacilli and ACV in female patients with recurrent genital HSV-2 infections. Patients were treated with multi-strain *Lactobacillus brevis* capsule every 12h and oral ACV (400mg, twice daily) for 6 months. Of the 53 patients treated, no differences were identified between ACV and probiotic for the primary and secondary efficacy end point, resolution of episode, lesion healing time, viral shedding, and percentage of pain. *L. brevis* was safe to treat patients compared to ACV. Some AEs were reported for ACV. The authors were of the opinion that multi-strain *L. brevis* could play an important role in suppression of recurrent genital HSV infection [9c].

A 67-year-old Chinese male who had VZV infection was on ACV (5mg/kg, IV for 8h) and developed severe thrombocytopenia within 10 days of starting ACV. ACV was stopped and the patient's condition improved. The

ACV-dependent platelet antibody test revealed that ACV was the causative agent for thrombocytopenia. This is the first case report of ACV-induced immune thrombocytopenia in VZV patients. The authors suggested that dentists should never overlook this rare AE of ACV, as rapid and appropriate treatment may prevent life-threatening complications [10A].

The effect of oral ACV administration in HSV-2-positive pregnant women on premature rupture of membranes (PROM) and risk of preterm delivery was reported in this study. A randomized, double-blind placebo-controlled trial among 200 HSV-2-positive pregnant women at 28 weeks of gestation was included in the study. Participants were assigned randomly to take ACV (400mg orally, twice daily) for 28–36 weeks. Both control and treated patients were administered ACV after 36 weeks of pregnancy until delivery. One hundred women were randomized and received ACV and 100 were assigned to the placebo arm. There was a reduction of incidence of PROM at 36 weeks but this was not statistically significant in the ACV and placebo arms. A significant reduction occurred in the incidence of preterm delivery (11.1% vs 23.5%) in the ACV and placebo arms, respectively. The authors concluded that oral ACV treatment for HSV-2-positive pregnant women from 28 to 36 weeks reduced the incidence of preterm delivery but did not reduce the incidence of pre-PROM [11C].

Famciclovir

General adverse reactions reported for famciclovir treatment were headache, nausea, and diarrhea.

Neurological

A cross-sectional observational study by means of a quantitative survey in Ireland explored the frequency of diagnosis, methods of treatment and cost of acute herpes zoster and post-herpetic neuralgia (PHN). Famciclovir and valaciclovir were preferred for anti-viral therapy, and pregabalin for the treatment of increasing pain. Mild opioids (32%) were the most common analgesic agents used for first-line acute herpes zoster pain, and pregabalin (37%) for second-line acute herpes zoster pain. According to the authors, acute herpes zoster and its complication resulting herpetic neuralgia is a challenge in aging population of patients [12C].

One hundred and forty-three fibromyalgia patients were enrolled in a 16-week, double-blinded, multicenter, placebo-controlled study. This study evaluated a famciclovir + celecoxib drug combination (IMC-1), against suspected herpes virus reactivation and infection, and for the treatment of fibromyalgia. Patients received either IMC-1 or placebo in a 1:1 ratio. A significant decrease in fibromyalgia-related pain was observed in patients on IMC-1 treatment compared to the placebo group. Gastrointestinal and nervous system AEs were reported less in the IMC-1 group compared to the placebo group [13C].

Valaciclovir

A randomized, double-blind, valaciclovir-controlled Phase 3 study was conducted to evaluate the efficacy and safety of amenamevir. Seven hundred and fifty-one herpes zoster patients were randomly assigned to receive either amenamevir (400 or 200mg) once daily or valaciclovir (1000mg) three times daily for 7 days. The primary efficacy end point was the proportion of cessation of new lesion formation. The day 4 cessation proportions for amenamevir 400 and 200mg and valaciclovir were 81.1% (197/243), 69.6% (172/247) and 75.1% (184/245), respectively. The study results revealed that amenamevir 400mg was not inferior to valaciclovir. To the number of days taken for the cessation of new lesion formation, complete crusting, healing, pain resolution and virus disappearance were evaluated as secondary end points. There were no significant differences in secondary end points in any of the treatment groups. Amenamevir 400 and 200mg and valaciclovir were all well tolerated. The proportions of patients who experienced drug-related AEs were 10.0% (25/249), 10.7% (27/252) and 12.0% (30/249) with amenamevir 400mg, 200mg, and valaciclovir, respectively. In conclusion, amenamevir 400mg appears to be effective and well tolerated for treatment of herpes zoster in immunocompetent patients [14C].

A meta-analysis to examine the effectiveness and/or safety of nucleoside antiviral drugs for recurrent herpes labialis was reported. This study included 16 publications reporting 25 randomized controlled trials (8453 patients). The parameters used to measure efficacy were time to healing of classic and all lesions, time to resolution of pain, and percentage of aborted lesions. Nucleoside antiviral drugs decreased the time to healing of all lesions and also reduced the time to resolution of pain and increased the percentage of aborted lesions. Valaciclovir more effectively reduced the time to healing of all lesions and the time to resolution of pain than aciclovir. Both nucleoside antiviral drugs increased the percentage of aborted lesions, but penciclovir and famciclovir failed to improve curing the lesions. The authors were of the opinion that the nucleoside antiviral drugs are safe and beneficial for the treatment of recurrent herpes labialis; valaciclovir is more effective than ACV, especially in reducing the time to healing lesions [15R].

An overview of the clinical impact of alpha and beta herpes viruses was published that highlights the mechanisms of action, pharmacokinetics, clinical indications, and AEs of antiviral drugs for the management of herpes simplex virus, VZV and CMV. The important AE, according to the authors, is the emergence of drug-resistant virus populations in organ transplant recipients [16R].

DRUGS ACTIVE AGAINST HEPATITIS VIRUSES

Adefovir [SEDA-35, 507; SEDA-36, 409; SEDA-37, 333; SEDA-38, 264; SEDA-39, 272]

A comparative study on the benefits of using tenofovir disoproxil fumarate (TDF) in chronic hepatitis B virus (HBV) patients has been reported. Forty-six chronic HBV patients were evaluated after LAM monotherapy vs the combination of lamivudine (LAM) plus adefovir dipivoxil (ADV). No significant differences between the two groups were noted in identifying AEs (53.8%, TDF vs 37.5%, LAM + ADV), and none of the AEs were serious [17c].

Multidrug-resistant HBV is still a continuing problem. Another study reported that the use of TDF + LAM is an effective therapy in LAM-resistant patients. In this study, 59 patients were included and were on TDF + LAM (300 mg/day) for 5 years. At the end of 5-year treatment, 75% (45/59) of the patients had achieved complete viral suppression. Throughout the study, only two patients had dose reduction due to AEs, such as increased glomerular filtration rate (eGFR) and non-Hodgkin lymphoma. This study concluded that long-term TDF treatment is safe and effective in patients with prior failure to LAM treatment and a suboptimal response to ADV therapy [18c].

Tenofovir alafenamide (TAF) is a novel prodrug of tenofovir that showed significant suppression of drug resistant-HBV in vitro and in patients [19E,20A].

TDF in chronic hepatitis B patients with gastric ulcers showed improvement on host immune response. Sixty patients were included in this study. The control group was treated with ADV, and the observation group was treated with TDF. The incidence of AEs was 10.00% in the observation group and 13.33% in the control group. There was no significant difference in the incidence of adverse drug reactions between the two groups [21c].

Another study was reported from the Netherlands, using peg-interferon-alfa-2a (peg-IFN) and nucleotide analogue combination therapy in patients with chronic HBV and low viral load. Participants were randomly assigned (1:1:1) to receive peg-IFN (180 μg/week) plus ADV (10 mg/day), peg-IFN (180 μg/week) plus TDF (245 mg/day), or no treatment for 48 weeks. The most frequent AEs (>30%) were fatigue, headache, fever, and myalgia, which were attributed to peg-IFN treatment. Two (4%) serious AEs were reported in the peg-IFN plus adefovir group, and one was admitted to a hospital for alcohol-related pancreatitis and another with severe depression [22c].

Adverse effects of oral antiviral therapy in HBV patients were reviewed [23R].

DIRECT-ACTING ANTIVIRAL PROTEASE INHIBITORS [DAA-PI] [SEDA-35, 508; SEDA-36, 409; SEDA-37, 335; SEDA-38, 267, 334; SEDA-39, 273]

Entecavir [SEDA-35, 512; SEDA-36, 411; SEDA-37, 335; SEDA-39, 273]

Observational Study

The efficacy and safety of the combined therapy of LAM and ADV, as well as entecavir (ETV) monotherapy in patients with hepatitis B-induced decompensated cirrhosis have been reported. One hundred and twenty-seven patients with decompensated cirrhosis were divided into four groups, and each group received different doses of regimens: initial combination of LAM and ADV, ADV add-on therapies with previous 12-week LAM, ADV add-on therapies with previous 24-week LAM, and ETV monotherapy. No serious AEs were observed in each group at the end of the 96-week treatment. The authors concluded that combination therapy and early ADV addition were preferred in the antiviral treatment of hepatitis B-induced decompensated cirrhosis patients [24C].

Ribavirin [SEDA-35, 512; SEDA-36, 412; SEDA-37, 335; SEDA-38, 267; SEDA-39, 273]

HIV/HCV co-infected patients treated with all-oral DAA regimens were included in this study. They were on a combination antiretroviral therapy (cART). Among these 323 patients, 60% were cirrhotic; 68% had previously received anti-HCV treatment, but the treatment was not effective. In this study the cART used was a protease inhibitor (PI)-based in 23%, non-nucleoside reverse transcriptase inhibitors (NNRTI)-based in 15%, and integrase inhibitor (II)-based in 38%, while 24% of patients received other regimens. SOF + DCV ± RBV was prescribed to 56.0% of patients, followed by SOF + ledipasvir (LDV) ± RBV in 20.4% of cases, SOF + RBV in 15.8%, and SOF + SMV ± RBV in 5.9%. Of the 164 patients with cirrhosis receiving at least two DAAs, 26 patients (16%) received a RBV-containing regimen for 12 weeks ($n = 9$) or 24 weeks ($n = 17$). Fifteen non-cirrhotic patients (14%) received a dual-DAA + RBV regimen for 12 weeks ($n = 11$) or 24 weeks. The most common AEs reported were fatigue and digestive disorders. AEs related to HCV therapy were reported in 94 patients (30%). Anemia occurred in 16 patients; of these, 11 patients received RBV. Eleven patients stopped their HCV therapy prematurely, two for intolerance and one for lack of virological response. HCV treatment modifications were reported for 36 patients, of whom 10 stopped RBV and one stopped DCV. Dose modifications were reported for 32 patients for the following drugs: DCV ($n=14$), RBV ($n=13$), SMV ($n=2$), asunaprevir ($n=1$), SOF ($n=1$) and LDV ($n=1$). The reasons for treatment modification were intolerance ($n=6$), under-dosing ($n=8$), and unknown reasons ($n=22$). The authors concluded that the new all-oral DAA regimens were well tolerated and had excellent viral clearance in HIV/HCV co-infected patients [25C].

Sofosbuvir [SEDA-37, 335; SEDA-38, 268; SEDA-39, 273]

Chronic hepatitis C is the leading problem in liver transplantation. A recent review analyzed DAAs that have advanced in treatment of HCV in terms of tolerability, and duration of therapy with significant increases in the rates of sustained virologic response (SVR) with low side effects [26c].

Another study reported certain AEs in chronic HCV patients after DAA therapy in Brazil. The most frequently reported AEs in these patients were fatigue (43%), headache (42%), neuropsychiatric symptoms (30%) and nausea (26%). The most frequent (38%) laboratory abnormality was the low levels of hemoglobin (<12mg/dL). Neuropsychiatric symptoms were the only AEs significantly different in treatment-experienced group compared to naïve patients [27c].

Mild AEs were reported in HCV patients treated orally with SOF based DAAs [28c].

A retrospective chart review was conducted in 204 HCV patients on SOF, and the role of psychological factors was evaluated. Depression or generalized anxiety had no role in viral clearance, but the use of cocaine influenced the SVR12 [29M].

Observational Study

A retrospective cohort study reported on 213 HCV patients. Seventy percent of patients received both SOF+RBV, whereas the remaining patients received triple therapy (SOF+RBV+IFN). The overall rate of SVR at 12 weeks post treatment was 72.9%. Most patients reported anemia and fatigue. The authors suggested that patients with HCV genotype 1 and 3 infection are better off with triple therapy compared to dual therapy [30C].

For the first time, a case was reported of an HCV infected patient treated for 12 weeks with the combination of SOF/ledipasvir plus RBV who developed a miliary tuberculosis (TB) infection. The authors are of the opinion that this case is relevant to increase the awareness of opportunistic infections such as TB during the treatment of HCV [31A].

One hundred patients with cirrhosis and infected with HCV genotypes 1 and 3 were included in a study that revealed the efficacy of SOF in combination with DCV and RBV. Patients were given 1 daily tablet of a combination pill (400mg SOF and 60mg DCV and weight-based RBV) for 12 weeks. One patient developed an increased creatinine level followed by severe diarrhea and gastroenteritis and was excluded from the study, one patient died due to unrelated reasons and four patients were lost to follow-up. Among the remaining 94 patients, 92 achieved SVR12 (98%). None of the patients reported any side effects [32C].

A case report showed two heart transplant recipients with HCV infection were treated with SOF+DCV and achieved SVR12 and neither patient showed any side effects [33A].

Twenty-four hundred cirrhotic patients with chronic HCV infection were treated with SOF and RBV for 24 weeks. The overall SVR12 rate was 71.2%. The most common AEs reported were fatigue, myalgia, headache, insomnia, and anemia. One hundred and thirty-five (5.63%) patients stopped treatment permanently due to the appearance of complications. The authors concluded that the use of SOF and RBV combination is safe and effective for treating HCV patients with liver cirrhosis [34C].

Combination Study

A multicenter, open-label, Phase 2 study evaluated the efficacy and safety of a fixed-dose combination of SOF-velpatasvir (400mg/100mg) plus weight-adjusted RBV administered for 24 weeks. Sixty-nine patients (HCV infected) did not achieve SVR by prior treatment with direct-acting antiviral regimens that included SOF plus the nonstructural protein 5A inhibitor, velpatasvir, with or without voxilaprevir. Most AEs were mild to moderate in severity. Other AEs reported were fatigue, nausea, headache, insomnia, and rash. One patient discontinued the study due to an AE (irritability) [35c].

SOF is not recommended for HCV patients with severe renal impairment. This study reported on 322 patients having renal dysfunction and infected with HCV. The patients received dual therapy of DCV and asunaprevir. Treatment discontinuation rates and AEs, including alanine aminotransferase elevation, anemia, and renal disorders, were not changed in dual therapy. According to the authors, the use of DCV and asunaprevir in combination therapy for HCV patients with renal dysfunction was safe and effective [36C].

Another study evaluated the use of ledipasvir/SOF without RBV for the treatment of recurrent HCV in post-liver transplant patients. There were no serious AEs and no discontinuation of treatment and 100% (60/60) SVR was achieved in 12 weeks of treatment.

This combination of ledipasvir/SOF was well tolerated without serious AEs or discontinuation [37C].

Sixty-three patients (median age 52 years; 80% males) with post-LT recurrent HCV were treated with SOF and RBV in a living donor liver transplant center in South Asia. Most (76.2%) were treatment experienced and predominantly HCV patients with either genotype 3 (77.7%) patients or genotype 1 (20.6%). AEs were noted in 34 patients; weakness and fatigue were the common side effects. Six patients showed a significant drop in hemoglobin (<8g/dL). The authors concluded that SOF+ RBV combination therapy for 24 weeks was safe and effective in treatment for post-LT recurrent HCV patients [38C].

Chinese patients with kidney transplant (KT) having HCV infection were safely treated with SOF and DCV without any noticeable side effects or AEs [39c].

The emergence of drug resistance-associated variants (RAVs) in a combination of SOF plus ledipasvir therapy has been reported. One hundred and seventy-six patients with chronic HCV genotype 1 infection were treated with SOF/ledipasvir for 12 weeks. Serum lipid-related markers were measured. SVR was achieved in 94.9% (167 out of 176) of patients. Serum low-density lipoprotein cholesterol and apolipoprotein B levels were significantly elevated at week 4 in SOF/ledipasvir-treated patients. These elevations were greater than in ombitasvir/paritaprevir/ritonavir-treated patients. The authors concluded that NS5A multi-RAVs are likely to develop in patients who fail to respond to SOF/ledipasvir therapy [40c].

Co-Infection

The risk of HCV infection is six times higher for HIV-positive patients than for the HIV-negative population. HIV infection seems to accelerate HCV-associated liver fibrosis. This study was conducted in 669 HIV/HCV co-infected patients, who were treated with DCV (60mg) plus SOF (400mg) once daily, for 24 weeks. Fifty-five patients experienced one or more serious AEs, and 26 experienced one or more AEs of grade 3 or 4. There were 10 deaths, mostly due to advanced liver disease; one was considered possibly related to HCV or HIV treatment, and two were due to multi organ failure plus septic shock plus intestinal obstruction, and hepatic carcinoma. The remaining seven deaths were not related to treatment regimen. There were seven discontinuations or AEs, of which three were subsequently fatal (hepatic carcinoma, decompensated cirrhosis/multi organ failure, respiratory distress) and four were nonfatal (lymphopenia, renal insufficiency, attempted suicide, and anxiety/ascites/hepatocellular carcinoma pneumonia/encephalopathy). The authors concluded that treatment involving DCV+ SOF+RBV could achieve high SVR12 and was well tolerated in HIV/HCV co-infected patients having liver disease [41c].

Another study also showed a similar outcome when treated with DCV and SOF in HCV and HIV co-infected patients [42c].

Twenty-two adult liver transplant (LT) recipients with HCV (16 mono-infected and 6 co-infected with HIV) received a 24-week course of SOF+DCV treatment. Viral suppression was very rapid with undetectable HCV-RNA in all patients by 12 weeks. All patients completed the 24-week treatment course without any significant side effects except for one case of severe bradycardia [43c].

Multiple studies reported the use of SOF and DCV in combination or monotherapy to improve the liver function in chronic HCV patients [44c,45c,46c].

Simeprevir [SEDA-38, 269; SEDA-39, 274]

The efficacy and safety of simeprevir (SIM)/SOF in patients with chronic HCV genotype 4 infection were reported. A multicenter observational study conducted in Egypt included 583 patients with HCV genotype 4 infection and were treated with SOF/SIM for 12 weeks. Side effects reported included rash in 21 patients, photosensitivity in 18 patients, pruritus in 44 patients and hyperbilirubinemia in 42 patients [47C].

Combination Study

A recurrent HCV study was reported in a cohort of 424 patients, treated for 24 weeks with SOF/RBV. In 55 patients, a treatment regimen with DCV or SIM was added. The outcome indicator SVR was 86.7% in patients treated with SOF/RBV and 98.3% (58/59) in patients who received a second antiviral (DCV/SIM). No significant AEs reported in all treatment groups and all patients tolerated the treatment [48c].

Twenty-three liver–kidney transplant recipients treated with DAAs were included in this study. Recipients had different HCV genotypes: genotype 1a in five cases, genotype 1b in nine cases, genotype 3 in five cases and genotype 4 in three cases. All the recipients received at least one NS5B inhibitor (SOF) in their antiviral regimen. Two patients received SOF+SIM, and other patients received SOF and ledipasvir or DCV or RBV or a combination of these two drugs. Mild to moderate AEs were reported in 20 recipients. The most common AE was anemia and was more common in RBV-treated patients. In recipients treated with RBV, the dose of RBV was reduced and then discontinued for 2 recipients. Serious AEs were reported in 9 recipients: severe infection in 3 recipients (CMV-induced colitis, pneumonia, septicemia related to urinary tract infection), and one case each of hematuria, basocellular carcinoma, stroke, acute leg ischemia, acute kidney failure, and anemia/leukopenia [49C].

Daclatasvir

Observational Study

A Phase 2, open-label, single-arm, multicenter study was conducted in 106 HCV patients on a dual therapy. Among the patients, 27% were aged >65 years, 39% had cirrhosis, 53% had an estimated GFR 30–89mL/min, 14% had diabetes, and 38% had arterial hypertension. The patients were administered with SIM (150mg)+ DCV (60mg) once daily for 12 or 24 weeks. Overall, 42/106 received 12 weeks of treatment and 64/106 received 24 weeks of treatment. Ninety-seven (92%) patients achieved SVR12 after the end of treatment. The reasons for failure were viral breakthrough ($n=7$) at weeks 4–16, early treatment discontinuation ($n=1$) and viral relapse ($n=1$). Seventy-four (70%) patients had more than one AE during treatment, including six (6%) patients with ≥1 serious AE. Three patients discontinued treatment due to AEs. SIM+DCV demonstrated strong antiviral activity and was well tolerated in patients with hepatitis C virus genotype 1b infection, but, viral breakthrough occurred in seven patients, making this treatment regimen unsatisfactory [50C].

Dual oral therapy with DCV and asunaprevir in real-life settings in Japan was well tolerated in HCV genotype 1b patients (651 patients) included in the study, with a similar safety profile and achieved similar SVR12. Of these, only 2.9% discontinued the therapy due to the elevation of alanine transaminase (ALT). Seven patients in clinical trial-met group and 20 in clinical trial-unmet group discontinued therapy because of AEs other than the ALT elevation [51C].

DRUGS ACTIVE AGAINST HUMAN IMMUNODEFICIENCY VIRUS: COMBINATIONS

Abacavir/Lamivudine/DTG

A retrospective clinical audit was conducted on HIV-patients who switched treatment to DTG/ABC/3TC fixed-dose combination therapy. Data from 443 HIV patients (97% male and 45% ≥50 years) were included in the study. Four hundred and forty-three patients participated in the study, and two patients discontinued from DTG/ABC/3TC after the study period. The most common reason for patients to switch therapy to DTG/ABC/3TC was simplification, toxicity/intolerance and patient preference. Fourteen patients (3.2%) discontinued DTG/ABC/3TC; none of these were due to virologic failure. Discontinuations were mainly related to AEs in 2.5% of the patients. Less than 1% of patients discontinued the treatment due to a psychiatric event [52C].

A randomized, open-label, Phase 3b study was conducted in adults with HIV-1 RNA (<50 copies/mL) and on antiretroviral therapy (ART) at the time of enrollment. Subjects were randomly assigned to switch to ABC/DTG/3TC once daily for 48 weeks (early-switch group) or continue current ART for 24 weeks and then switch to ABC/DTG/3TC (late-switch group). The primary end point was the proportion of subjects with HIV-1 RNA <50 copies/mL at the end of 24 weeks treatment. 553 subjects were enrolled in the study and 275 were randomly assigned to switch immediately to ABC/DTG/3TC (early switch group), whereas 278 continued on current ART (late switch group). At week 24, subjects who switched to ABC/DTG/3TC (85%) or remained on current ART (88%) showed viral suppression, indicating that ABC/DTG/3TC was not inferior in virological suppression. AEs were reported more frequently with ABC/DTG/3TC (66%) than with current ART (47%) by week 24, and in the late-switch group 60% of the subjects reported AEs [53C].

Elvitegravir/Cobicistat/FTC/Tenofovir

Observational Studies

Weight gain has been reported in several HIV patients who switched from EFV/TDF/FTC to DTG/ABC/3TC [54c].

An observational study reported on 542 HIV-1-infected adult patients who were on ART with DTG or EVG/COBI. The combination of ABC/3TC/DTG in a single pill was given to 195 patients, TDF/FTC/EVG/COBI for 151 patients, TAF/FTC/EVG/COBI for 116 patients and DTG combined with other anti-retrovirals (i.e. NRTIs and NNRTIs) to 80 patients. The global incidence of discontinuation was 8.5%, with a 6.6% discontinuation rate due to DTG side effects. The neuropsychiatric disorders were the main AEs leading to discontinuation of the treatment. Other AEs were related to neuropsychiatric disturbances (70.4%), gastrointestinal discomfort (22.2%), alterations of renal function (3.7%) and hematology toxicity (3.7%). Most patients experienced more than one neuropsychiatric toxicity, which included abnormal dreams, insomnia, headache, dizziness, nervousness, irascibility, anxiety, depressive symptoms and suicidal ideation. In the case of EVG/COBI, 63.1% treatment discontinuations were due to AEs: 50% gastrointestinal discomfort, 33.3% related to neuropsychiatric disorders, and 16.7% because of alterations in renal function. The rest of discontinuations in EVG/COBI were 21.1% due to pharmacological interactions and 15.8% due to virological failure. In most cases, all reported side effects disappeared quickly once these drugs were discontinued or withdrawn from the treatment regimen. The authors concluded that DTG and EVG/COBI showed high efficacy in treatment-naïve and pre-treated patients. DTG especially in concomitant use with ABC was identified as a predictor for discontinuations due to neuropsychiatric AEs [55C].

Another study evaluated the effectiveness, safety and costs of switching to a RPV/FTC/TDF regimen in 146 treatment-experienced HIV-1-infected patients. Of the 146 patients 25.3% discontinued RPV/FTC/TDF (mainly due to renal impairment). Throughout the 96 weeks, there were significant decreases in total cholesterol (TC) (14.0mg/dL), TC/HDL cholesterol ratio (0.4mg/dL) and triglycerides. Moreover, switching to RPV/FTC/TDF reduced the annual per-patient anti-retroviral cost. According to the authors, this treatment regimen was less favorable compared to other treatments available due to high rates of virological failure and AEs [56C].

DRUGS ACTIVE AGAINST HUMAN IMMUNODEFICIENCY VIRUS: NUCLEOSIDE ANALOGUE REVERSE TRANSCRIPTASE INHIBITORS (NRTI) [SEDA-35, 516; SEDA-36, 415; SEDA-37, 337; SEDA-38, 270; SEDA-39, 276]

Abacavir [SEDA-35, 516; SEDA-36, 415; SEDA-37, 337; SEDA-38, 270; SEDA-39, 276]

Observational Studies

A double-blinded, multicenter Phase 3 trial compared the initial treatment of HIV patients with bictegravir, emtricitabine, and TAF vs DTG, ABC, and LAM. Results from the study showed that the co-formulated bictegravir, emtricitabine, and TAF achieved virological suppression in 92% of previously untreated adults and were not inferior to co-formulated DTG, ABC, and LAM. Bictegravir, emtricitabine, and TAF were safe and well tolerated with better gastrointestinal tolerability than DTG, ABC, and LAM [57C].

Another study monitored the safety and evaluated the effectiveness of abacavir sulfate (300mg) in a Korean population. A total of 669 patients were enrolled in this study. Of these, 196 (29.3%) patients reported 315 AEs, and four patients reported seven serious AEs. Among the 97 adverse drug reactions that were reported from 75 patients, the most frequent included diarrhea (12 events), dyspepsia (10 events), and rash (9 events) [58C].

Lamivudine [SEDA-35, 517; SEDA-36, 416; SEDA-37, 338; SEDA-39, 276]

Observational Studies

The applicability of dual treatments based on integrase inhibitors is less studied. 94 individuals were included in a study that used the combination of lamivudine + DTG as an option when switching from standard cART in virologically suppressed patients. All patients were switched to a dual combination of dolutegravir (50mg once daily) plus lamivudine (300mg once daily). The lipid profile slightly changed after switching to the dual regimen. Neither virological failure, nor viral failure above 50 copies/mL was detected. Nineteen percent of the patients reported AEs [59C].

Zidovudine [SEDA-35, 517; SEDA-36, 417; SEDA-37, 338; SEDA-38, 272; SEDA-39, 276]

Also see Tenofovir.

A retrospective cohort analysis was conducted within the International Epidemiologic Database to evaluate AIDS (IeDEA) collaboration in West Africa. The analysis examined adult patients (age ≥16 years) living with HIV and initiating a first-line ART regimen between 2002 and 2014 that contained three or more drugs. This study showed that the risk of developing severe neutropenia was associated with ZDV-containing ART regimen [60C].

DRUGS ACTIVE AGAINST HUMAN IMMUNODEFICIENCY VIRUS: NUCLEOTIDE ANALOGUE REVERSE TRANSCRIPTASE INHIBITORS

Tenofovir [SEDA-35, 518; SEDA-36, 418; SEDA-37, 338; SEDA-38, 272; SEDA-39, 276]

Relationship between adverse prenatal outcomes and prenatal TDF use has been reported. The most frequent ART regimens used were TDF/3TC/EFV (39%) and AZT/3TC/NVP (34%); 49% of pregnancies had prenatal TDF exposure and 6% used a protease inhibitor. AEs reported included neonatal death (2%), preterm birth (8%), and pregnancy loss (12%). There were no differences between pregnancies with and without exposure to TDF in the loss of pregnancy. Preterm birth occurred less frequently among pregnancies of patients administered with TDF. The authors concluded that maternal TDF use did not adversely affect perinatal outcomes [61c].

A cohort study was conducted in a South African population that included 15156 patients' data. Patients included antiretroviral-naïve ≥16 years old who started tenofovir-containing anti-retroviral therapy between 2002 and 2013. Overall, 292 (1.9%) patients developed EGFR <30mL/min. Patients on tenofovir with baseline EGFR 90mL/min experienced small, but significant declines in EGFR over time [62MC].

Bone mineral density reductions at the hip and spine after TDF initiation have also been reported, but there was no difference in phosphaturia after TDF and ABC treatment [63C].

Case reports showed that high creatinine levels occurred in HIV patients suggestive of nephrotoxicity due to TDF treatment. A 54-year-old patient with known HIV infection and on combined anti-retroviral treatment consisting of TDF, 3TC and EFV was admitted to the hospital. The patient was oliguric with a urine output of 55 mL in 24 h, with no evidence of hematuria and body swelling. He had no previous illnesses apart from HIV infection. His serum creatinine was 1361 μmol/L, urea concentration 30 mmol/L and the electrolyte concentrations were sodium 126 mmol/L, potassium 5.8 mmol/L, chloride 94 mmol/L, and calcium 1.91 mmol/L. He started hemodialysis and the cART regimen changed to ZDV, renal-dosed 3TC and EFV. His creatinine level decreased and he continued on the same cART. His creatinine levels dropped to normal in a few weeks [64A].

In another case study, a 53-year-old male having HIV infection was admitted to the hospital with symptoms of abdominal pains, nausea, and vomiting (the vomitus contained food, but not blood stained). He also complained of dizziness, some cough and his urine output was reduced without change in color. He had no history of diarrhea or fever. His treatment for HIV was initiated with ZDV, 3TC and EFV and cotrimoxazole prophylaxis (960 mg daily). Later, he was switched to a second line of treatment with ritonavir-boosted lopinavir, 3TC and TDF. His creatinine was 559 μmol/L, urea 55 mmol/L, potassium 4.5 mmol/L and sodium of 114 mmol/L. TDF was discontinued and replaced by ABC, and 3TC was renal dosed. In 3 days, his creatinine level was 541 μmol/L, urea 35 mmol/L, potassium 2.5 mmol/L, chloride 86 mmol/L and sodium122 mmol/L. The change in treatment regimen brought down his creatinine levels and renal function to normal levels [64A].

DRUGS ACTIVE AGAINST HUMAN IMMUNODEFICIENCY VIRUS: NON-NUCLEOSIDE REVERSE TRANSCRIPTASE INHIBITORS (NNRTI) [SEDA-35, 519; SEDA-36, 420; SEDA-37, 339; SEDA-38, 273; SEDA-39, 277]

Efavirenz [SEDA-35, 519; SEDA-36, 420; SEDA-37, 339; SEDA-38, 273; SEDA-39, 277]

Observational Studies

A retrospective review reported in treatment-naive and treatment-experienced HIV-positive adult patients in a correctional facility. This study included 553 HIV patients. Patients were on either a single tablet regimen (STR) (efavirenz (EFV), rilpivirine, elvitegravir based) or multiple tablet regimen (MTR) (emtricitabine/tenofovir with ATV/ritonavir, darunavir/ritonavir, or RAL). No significant differences in virologic suppression were seen between the two groups (326 STR and 164 MTR patients). Similar proportions of patient-reported AEs, self-reported adherence, and discontinuation rates were found in both groups. Though there was no significant difference in the initial stages of treatment, at week 72, a significant difference in viral suppression was noted with MTR (97.5%) over STR (88.0%). Patients from both STR and MTR reported AEs (15.3% and 18.9%, respectively). The most common AEs found in STR patients were CNS symptoms (including difficulty concentrating, difficulty sleeping, or vivid/abnormal dreams; 7.7%), dizziness/lightheadedness (2.8%), psychiatric effects (depression/suicidal ideation; 1.5%), and gastrointestinal (GI)-related symptoms (nausea, vomiting, diarrhea; 1.5%). The AEs profile for patients on MTR included scleral icterus (9.1%), GI-related symptoms (8.5%), and rash (6.7%). Elevated total bilirubin was the most common lab-reported AE and occurred in 43.3% ($n=71$) of MTR patients. The overall rate of patients with at least one self- or laboratory-reported adverse reaction was significantly higher for MTR (53.6%) than STR (15.6%) [65C].

The drug–drug interactions that occurred in older HIV patients are reviewed in a study, which also provide the metabolic pathways of ART and highlight potential areas of concern for drug–drug interactions and suggest alternative approaches for treating HIV patients [66R,67R].

Among HIV-infected individuals, use of lopinavir/ritonavir compared with EFV was associated with lower cerebral vasoreactivity [68c].

Pediatric Patients

Data were collected from 51 children and adolescents aged ≤18 years who received EFV-based treatment from 1998 to 2014. Thirty patients (59%) subsequently stopped EFV—14 (29%) following virological failure, and 16 (30%) after reporting AEs. Most AEs reported for EFV were related to CNS (19.6%), including sleep disturbance, reduced concentration, headaches, mood change and psychosis. Four children developed gynaecomastia, two developed hypercholesterolemia, and one developed Stevens–Johnson syndrome. The authors concluded that pediatric patients may need an alternative treatment regimen and not EFV-based treatment [69c].

A prospective, Phase 1/2 open-label study was conducted of children with HIV infection (Cohort I) or HIV/TB co-infection (Cohort II). Participants were divided into two groups: 3 to less than 24 months (28 patients) and at least 24 to less than 36 months (19 patients) and were treated with an anti-retroviral regimen consisting of two nucleoside reverse transcriptase inhibitors and weight band-based EFV given as capsules opened into porridge, formula or expressed breast milk. An initial EFV [~1600 mg × (weight in kg/70)] was given to all participants. Fifteen of 47 participants (32%) experienced nonlife-threatening toxicities that were deemed at least related to anti-retroviral treatment regardless of grade. Six patients (13%) reported neurologic toxicities of lethargy, sleepiness or sleep disturbances. There were no deaths, hospitalizations or other serious AEs. Among participants with evaluable virologic data, 89% at week 4 and 82% at week 8 achieved virologic success [70c].

Nevirapine [SEDA-33, 593; SEDA-34, 460; SEDA-35, 521; SEDA-36, 421; SEDA-37, 339; SEDA-38, 274; SEDA-39, 278]

Efavirenz- or Nevirapine-based antiretroviral therapy in HIV-infected children from Africa has been reported. Four hundred and forty-five (53%) children received efavirenz (EFV) and 391 children (47%) received nevirapine. The initial non-nucleoside reverse transcriptase inhibitor (NNRTI) was permanently discontinued due to AEs in 7 of 445 (2%) children initiating EFV and 9 of 391 (2%) initiating nevirapine [71C].

A study reported on 322 African children treated with nevirapine, LAM, and either ABC, stavudine, or AZT for HIV. The viral suppression increased with increasing nevirapine concentration, and there was no clear concentration threshold predictive of this suppression. AEs considered to be nevirapine-related were hypersensitivity reactions, elevated liver enzymes and acute hepatitis [72C].

Rilpivirine [SEDA-35, 521; SEDA-36, 423; SEDA-37, 340; SEDA-38, 274; SEDA-39, 278]

A prospective Swiss HIV cohort study was conducted in 644 HIV patients. 48 (7.5%) were cART-naïve at initiation of the RPV/TDF/FTC co-formulation. Five hundred and ninety-eight patients were switched to RPV/TDF/FTC during the study period. Treatment simplification (266/596; 44.6%) and CNS toxicity (143/596; 24.0%) were the two main reasons for switching to the co-formulation. CNS toxicity was the prime cause of switching 126 patients on an EFV-based regimen. AEs reported were insomnia/sleep disturbances (26.9%; 53/197); abnormal dreams (18.8%; 37/197); depression (17.3%; 34/197); dizziness (15.2%; 30/197); fatigue/tiredness (13.7%; 27/197); and other reasons (8.1%; 16/197). Six months after the switch from EFV to RPV, 74.8% (92/123) of patients reported an improvement of CNS symptoms, 14.6% (18/123) reported a stable condition and 3.2% (4/123) described worsening CNS side effects. Viral suppression in cART naive patients (HIV-RNA <50 copies/mL) was achieved in 24 months. According to the authors, use of RPV is a favorable option with most patients experiencing CNS side effects from EFV treatment showing improvement after switching to RPV [73C].

Low side effects and long-term use of RPV have been discussed here [74c].

A Phase 3b, randomized, double-blind, non-inferiority study was conducted in HIV-1 adult patients in 119 hospital sites in 11 countries in North America (Canada and the USA) and Europe (Belgium, France, Germany, Italy, the Netherlands, Spain, Sweden, Switzerland, and the UK). Participants were either switched to a single-tablet regimen of 25 mg RPV, 200 mg emtricitabine, and 25 mg TAF or remained on a single-tablet regimen of 25 mg RPV, 200 mg emtricitabine, and 300 mg TDF. Six hundred and thirty participants received at least one dose of study drug. Of these 630 participants, 316 were randomized to switch to the TAF regimen. The remaining 314 participants remained on their previous TDF regimen. Viral suppression was maintained in 296 (94%) out of 316 participants in the TAF group and in 294 (94%) of 314 participants in the TDF group. Both treatments were well tolerated, with most AEs reported as mild or moderate. AEs leading to study drug discontinuation were uncommon. Participants in the TAF group had a lower incidence of drug-related AEs than those in the TDF group. One participant (<1%) in each treatment group had a study drug-related AE leading to discontinuation; fatigue (in the TAF group) and hypersensitivity (in the TDF group). Two individuals died in the study, one in each treatment group: cardiac arrest in the TAF group and carbon monoxide poisoning in the TDF group. Forty participants (13%) out of 315 in the TAF group had grade 3 or 4 laboratory abnormalities compared with 19 (6%) out of 314 in the TDF group. Other AEs reported in both groups were upper respiratory tract infection, diarrhea, nasopharyngitis, headache, bronchitis, and sinusitis [75C].

DRUGS ACTIVE AGAINST HUMAN IMMUNODEFICIENCY VIRUS: PROTEASE INHIBITORS [SEDA-35, 522; SEDA-36, 423; SEDA-37, 340; SEDA-38, 274; SEDA-39, 278]

Atazanavir

Observational Study

A large pool of HIV patients was screened to understand the effect of a protease inhibitor in HIV treatments. Seven hundred and five patients were screened for the study, and 499 were randomly assigned to receive study medication (250 in the DTG group vs 249 in the ATV group). Participants were randomly assigned (1:1) to receive either a fixed-dose combination single-tablet regimen of DTG (50 mg plus ABC 600 mg and lamivudine 300 mg) once a day or a three-tablet regimen (ATV 300 mg boosted with ritonavir 100 mg plus a fixed-dose combination of TDF 300 mg and emtricitabine 200 mg) once a day for 48 weeks. The overall rates of AEs were similar between treatment groups: 195 (79%) of 248 participants in the DTG group compared with 197 (80%) of 247 in the ATV group. Alanine aminotransferase concentrations increased to at least three times the upper limit of normal in 4 (2%) of 248 participants assigned to DTG. Changes in serum creatinine concentrations were greater at week 48 in participants who were on DTG than ATV patients. Creatinine phosphokinase was increased in seven participants in each group. Three grade 3 creatinine phosphokinase toxic effects occurred in the DTG group, and one grade 4 toxic effect occurred in the ATV group. There were no significant differences between the two treatment groups in total cholesterol or HDL or triglycerides concentration [76C].

RAL-treated patients were compared for cost and safety with DRV/r and ATV/r, each used in combination with FTC/TDF, in adults with HIV-1 infection [77C].

DRUGS ACTIVE AGAINST HUMAN IMMUNODEFICIENCY VIRUS: INHIBITORS OF HIV FUSION [SEDA-35, 525; SEDA-36, 428; SEDA-37, 341; SEDA-38, 275; SEDA-39, 278]

Enfuvirtide

Entry of HIV-1 virus into the target cell is initiated by binding of gp120, the surface subunit of HIV-1 envelope glycoprotein (Env), to the receptor CD4 and co-receptor CXCR4 or CCR5 on the target cell. This study demonstrated that the multivalent bispecific proteins 2Dm2m and 4Dm2m, which target both CD4bs and CoRbs in gp120, can effectively inactivate cell-free HIV-1 virions before attachment to the receptor CD4 and co-receptor, CCR5 or CXCR4, on the target cells. This study also provides information on the mechanism of fusion protein inhibitors functioning on HIV virus replication [78A].

Enfuvirtide (T20) is the only HIV viral fusion inhibitor used in combination therapy, but it has low antiviral activity and often develops resistant mutants. Recent studies showed that lipopeptide-based fusion inhibitors, such as LP-11 and LP-19, variants of LP-20, which target gp41, have greatly improved antiviral potency and in vivo stability [79A].

DRUGS ACTIVE AGAINST HUMAN IMMUNODEFICIENCY VIRUS: INTEGRASE INHIBITORS [SEDA-35, 525; SEDA-36, 428; SEDA-37, 342; SEDA-38, 275; SEDA-38, 276; SEDA-39, 278]

Dolutegravir (DTG)

Observational Studies

Psychiatric symptoms (PSs) are reported in HIV patients and are common in DTG-treated patients. These events are reported with low frequency and rarely necessitate DTG discontinuation. Drug withdrawal rates for PSs were higher for RAL than DTG [80c].

A retrospective analysis in a cohort of HIV-infected patients from a German outpatient clinic between 2007 and 2016 has been reported. The estimated rates of any AE and of neuropsychiatric AEs leading to discontinuation within 12 months were 7.6% and 5.6% for DTG, respectively [81C].

A retrospective case chart analysis of HIV-1-positive adults on DTG between July 2014 and September 2015 has been reported. One hundred and fifty-seven DTG patients on ART were included in the study, of these 106 (68%) were switched to DTG from another regimen, and 51 (32%) were ART-naïve. Overall, 56 reported side effects; 40 patients reported either difficulty with low mood, anxiety or sleep disturbance. Sixteen discontinued DTG, with 13 due to intolerable side effects [82c].

Raltegravir

Observational Study

Dual therapy with raltegravir (RAL) plus LAM in selected patients could be safe, well tolerated and a proven effective strategy to reduce the long-term side effects and costs of combination ART. In this study of 14 patients, 4 being treated with ABC, 1 with AZT and 9 with TDF were switched to raltegravir once daily. Patients showed significant improvement in creatinine level, lipid profile and ALT after switching [83c].

A prospective cohort study compared the AEs between RAL and DTG. Neuropsychiatric complaints were the most common toxic AEs reported and were more frequent in the DTG treatment group than the RAL treatment group [84C].

Some ARV drugs have been shown to be nephrotoxic and associated with worsening renal function in HIV patients. This review describes the novel antiviral agents, such as DTG, RAL, elvitegravir, cobicistat, TAF and ATV, that have shown some issues with renal function, creatinine handling and potential nephrotoxicity [85R].

DRUGS ACTIVE AGAINST HUMAN IMMUNODEFICIENCY VIRUS: CHEMOKINE RECEPTOR CCR5 ANTAGONISTS [SEDA-35, 528; SEDA-36, 430; SEDA-37, 343; SEDA-38, 276; SEDA-39, 279]

Maraviroc

Patients on 3-drug ART with stable HIV-1 RNA (<50 copies/mL) were randomized 1:1 to maraviroc (MVC) with darunavir/ritonavir per day (study arm) or continued on the current ART (continuation arm). One hundred and fifteen patients were included in the study (56 in the study arm, 59 in the continuation arm). Two participants in the study arm and 10 in the continuation arm discontinued therapy due to significant AEs. Femoral bone mineral density was significantly improved in the study arm. Switching to MVC with darunavir/ritonavir showed improved tolerability but was virologically inferior to 3-drug therapy [86C].

A retrospective multicenter study (27 centers in Spain) evaluated the efficacy and safety of MVC administered once daily in HIV-1 patients. Data were collected from the records of patients starting a regimen with MVC. Laboratory and clinical data were recorded every 3 months the first year and every 6 months thereafter. Among the 667 patients treated with MVC, 142 (21.3%) received MVC once daily: 108 (76.1%) at a dose of 150mg and 34 (23.9%) at a dose of 300mg. Patients had baseline HIV-RNA <50 copies/mL. MVC was administered for the following reasons: salvage therapy (36.6%), drug toxicity (31.2%), simplification (16.9%), and immune discordant response (7.1%). Twenty-five (17.6%) patients discontinued MVC for the following reasons: virologic failure (6), medical decision (5), and other reasons (14). Two patients showed grade 3 AEs (hypertransaminasemia, hypertriglyceridemia) without the need for MVC withdrawal, whereas MVC was discontinued in two patients due to gastrointestinal toxicity. The authors were of the opinion that the use of MVC once-daily combined with at least PI/r was virologically effective in pretreated patients [87C].

DRUGS ACTIVE AGAINST INFLUENZA VIRUSES: NEURAMINIDASE INHIBITORS [SEDA-35, 528; SEDA-36, 431; SEDA-37, 344; SEDA-38, 277; SEDA-39, 279]

The M2 inhibitors, amantadine and rimantadine, were historically effective for the prevention and treatment of influenza A, but all circulating strains are currently resistant to these drugs [88R].

Oseltamivir (Tamiflu)

An open-label randomized study evaluated the difference in viral dynamics and influenza symptoms in patients aged 4–12 years to intravenous peramivir, oral oseltamivir, inhaled zanamivir, or inhaled laninamivir. The time to virus clearance was significantly shorter with peramivir than with oseltamivir [89c].

A retrospective cohort study reported on adult (≥18 years) patients with suspected influenza viral infection (N=3743). Adults hospitalized for seasonal influenza were enrolled as a comparison group (n=312). This study evaluated whether RSV infection is associated with higher mortality than seasonal influenza and on oseltamivir treatment. The outcome of the study showed that oseltamivir had no significant effect on mortality of patients with influenza [90C].

A retrospective cohort analysis was conducted in 57 patients admitted to the intensive care unit (ICU) with confirmed influenza infection. Patients receiving high-dose of oseltamivir were compared to those receiving standard dosing. As compared to the standard doses of oseltamivir, a higher-dose of oseltamivir was not associated with improvement in any clinical outcomes [91C].

Combination Study

A double-blind, randomized, Phase 2 study was conducted in 50 sites in the USA, Thailand, Mexico, Argentina, and Australia, on a combination of oseltamivir, amantadine, and RBV vs oseltamivir monotherapy with matching placebo for the treatment of influenza. Participants who were diagnosed with influenza and were at increased risk of complications were randomly assigned (1:1) to receive either oseltamivir (75mg), amantadine (100mg), and RBV (600mg) combination therapy or oseltamivir monotherapy, twice daily for 5 days. Six hundred and thirty-three participants were randomly assigned to receive a combination of antiviral therapy (n=316) or monotherapy (n=317). The primary analysis included 394 participants, excluding 47 in the pilot phase, 172 without confirmed influenza, and 13 discontinued. Eighty participants in the combination group had detectable virus at day 3 compared with 97 participants in the monotherapy group. The most common AEs were gastrointestinal-related disorders, nausea (65 reported AEs in the combination group vs 63 reported AEs in the monotherapy group), diarrhea (56 vs 64), and vomiting (39 vs 23). Twenty-two serious AEs were also reported in 20 participants; 16 AEs in 14 participants in the combination group and 6 AEs in 6 participants in the monotherapy group. More than one participant showed asthma exacerbation, diarrhea, and pneumonia. There were four gastrointestinal serious AEs in the combination group and none in the monotherapy group. Only two of the serious AEs in the monotherapy group were related to study medication. One participant in the monotherapy group died from cardiovascular failure, and that was not drug related. Thirteen participants in the combination group and three participants in the monotherapy group were admitted to the hospital with other complications. This study showed that monotherapy is not inferior to the combination therapy in treating influenza patients [92C].

Peramivir

Peramivir is the first intravenously (IV) administered neuraminidase inhibitor for immediate delivery of an effective single-dose treatment in patients with influenza. A systematic meta-analysis compared the efficacy of IV peramivir with oral oseltamivir for treatment of patients with influenza. Seven trials (two randomized controlled trials and five non-randomized observational trials) involving 1676 patients were analyzed. The total numbers of peramivir- and oseltamivir-treated patients were 956 and 720, respectively. Overall, the time to alleviation of fever was lower in the peramivir-treated group compared with the oseltamivir-treated group. Mortality, length of hospital stay, change in virus titer, and the incidence of AEs were not significantly different between the two groups. IV peramivir therapy reduced the time to alleviation of fever compared with oral oseltamivir therapy in patients with influenza [93MC].

OTHER DRUGS

Imiquimod [SEDA-35, 530; SEDA-36, 431; SEDA-37, 344; SEDA-38, 277; SEDA-39, 280]

The most common AE reported for imiquimod (IM) is local skin irritation at the application site. Other AEs include headache, flu like symptoms and myalgia.

Dermatological Studies

Topical treatment with IM for penile cancer has been reported with very low AEs [94R].

A meta-analysis evaluated the efficacy and safety of photodynamic therapy (PDT), surgery excision (SE), cryotherapy (CT), IM, radiotherapy (RT), 5-fluorouracil (FU), and vehicle (VE) for non-melanoma skin cancer (NMSC) treatment. Data from 18 trials with 3706 patients showed IM were more likely to induce AEs than VE. The authors concluded that SE was the optimal regimen for NMSC with high efficacy considering CLR, CLC, and CRP [95MC].

Another study reported the use of IM to treat mycosis fungoides (MF) tumors. Two stage IIB MF patients, including one with large cell transformation, were treated with IM 5% cream after failing other therapies. One patient reported AE, such as application site irritation and flu-like symptoms [96c].

A recent review on the use of IM resulted in AEs as site reactions in patients with cutaneous molluscum contagiosum, caused by pox virus. Among the most common AEs of IM treatment in this patients were pain during application, erythema, and itching [97R].

Topical IM is sometimes used for lentigo maligna (LM) in situ melanoma instead of surgery, but frequency of cure is uncertain. A single-arm Phase 2 trial of 60 imiquimod applications over 12 weeks for LM was reported. Clinical evaluation showed that 13 of 28 patients had complete disappearance of the LM after IM treatment. Eleven of 29 patients (38%) had a severe local-site reaction over the study period; 10 (34%) had a moderate reaction and 8 (28%) had mild or no reaction [98c].

A Phase 2 prospective multicenter clinical trial assessed the safety and activity of TMX-101, a novel liquid formulation of IM. Enrolled patients in this study received six weekly intravesical administrations of 200 mg/50 mL TMX-101 0.4%. Overall, 75% of patients experienced treatment-related AEs, only one was >grade 2 (urinary tract infection). Two patients showed a negative cytology at 6 weeks of treatment. Significant increases in urinary cytokines, including IL-6 and IL-18, were also reported following IM treatment [99c].

Additional case studies on this topic can be found in these SEDA reviews [100R,101R,102R].

DISCLAIMER

The findings and conclusions in this review are those of the authors and do not necessarily represent the official position of the National Institute for Occupational Safety and Health, Centers for Disease Control and Prevention. Mention of any company or product does not constitute endorsement by the National Institute for Occupational Safety and Health, and Centers for Disease Control and Prevention.

References

[1] Bonatti H, Sifri CD, Larcher C, et al. Use of cidofovir for cytomegalovirus disease refractory to ganciclovir in solid organ recipients. Surg Infect. 2017;18(2):128–36 [c].

[2] Marty FM, Ljungman P, Chemaly RF, et al. Letermovir prophylaxis for cytomegalovirus in hematopoietic-cell transplantation. N Engl J Med. 2017;377(25):2433–44 [C].

[3] Hiwarkar P, Amrolia P, Sivaprakasam P, et al. Brincidofovir is highly efficacious in controlling adenoviremia in pediatric recipients of hematopoietic cell transplant. Blood. 2017;129(14):2033–7 [C].

[4] Rabinowicz S, Somech R, Yeshayahu Y. Foscarnet-related hypercalcemia during CMV treatment in an infant with SCID: a case report and review of literature. J Pediatr Hematol Oncol. 2017;39(3):e173–5 [C].

[5] Wu YM, Cao YB, Li XH, et al. Clinical study on foscarnet prophylaxis and pre-emptive therapy for cytomegalovirus infection in hematopoietic stem cell transplantation. J Leuk Lymphoma. 2017;26(6):331–5 [c].

[6] Echenique IA, Beltran D, Ramirez-Ruiz L, et al. Ganciclovir dosing strategies and development of cytomegalovirus resistance in a kidney transplant recipient: a case report. Transplant Proc. 2017;49(7):1560–4 [A].

[7] Bright PD, Gompels M, Donati M, et al. Successful oral treatment of ganciclovir resistant cytomegalovirus with maribavir in the

context of primary immunodeficiency: first case report and review. J Clin Virol. 2017;87(Suppl C):12–6 [A].
[8] Chawla R, Tripathy K, Sharma YR, et al. Periarterial plaques (Kyrieleis' arteriolitis) in a case of bilateral acute retinal necrosis. Semin Ophthalmol. 2017;32(2):251–2 [A].
[9] Mohseni AH, Taghinezhad-S S, Keyvani H, et al. Comparison of acyclovir and multistrain Lactobacillus brevis in women with recurrent genital herpes infections: a double-blind, randomized, controlled study. Probiotics Antimicrob Proteins. 2017;1–8 [in press], [c].
[10] Hong X, Wang X, Wang Z. A rare case report of acyclovir-induced immune thrombocytopenia with tongue hematomas as the first sign, and a literature review. BMC Pharmacol Toxicol. 2017;18(1) [in press], [A].
[11] Nakubulwa S, Kaye DK, Bwanga F, et al. Effect of suppressive acyclovir administered to HSV-2 positive mothers from week 28 to 36 weeks of pregnancy on adverse obstetric outcomes: a double-blind randomised placebo-controlled trial. Reprod Health. 2017;14:31 [C].
[12] Crosbie B, Lucey S, Tilson L, et al. Acute herpes zoster and post herpetic neuralgia in primary care: a study of diagnosis, treatment and cost. Eur J Clin Microbiol Infect Dis. 2017;1–5 [C].
[13] Pridgen WL, Duffy C, Gendreau JF, et al. A famciclovir + celecoxib combination treatment is safe and efficacious in the treatment of fibromyalgia. J Pain Res. 2017;10:451–60 [C].
[14] Kawashima M, Nemoto O, Honda M, et al. Amenamevir, a novel helicase–primase inhibitor, for treatment of herpes zoster: a randomized, double-blind, valaciclovir-controlled phase 3 study. J Dermatol. 2017;44(11):1219–27 [C].
[15] Chen F, Xu H, Liu J, et al. Efficacy and safety of nucleoside antiviral drugs for treatment of recurrent herpes labialis: a systematic review and meta-analysis. J Oral Pathol Med. 2017;46(8):561–8 [R].
[16] Abad CL, Razonable RR. Treatment of alpha and beta herpesvirus infections in solid organ transplant recipients. Expert Rev Anti Infect Ther. 2017;15(2):93–110 [R].
[17] Rodríguez M, Pascasio JM, Fraga E, et al. Tenofovir vs lamivudine plus adefovir in chronic hepatitis B: TENOSIMP-B study. World J Gastroenterol. 2017;23(41):7459–69 [c].
[18] Lim L, Thompson A, Patterson S, et al. Five-year efficacy and safety of tenofovir-based salvage therapy for patients with chronic hepatitis B who previously failed LAM/ADV therapy. Liver Int. 2017;37(6):827–35 [c].
[19] Liu Y, Miller MD, Kitrinos KM. Tenofovir alafenamide demonstrates broad cross-genotype activity against wild-type HBV clinical isolates and maintains susceptibility to drug-resistant HBV isolates in vitro. Antivir Res. 2017;139:25–31 [E].
[20] Fasano M, Maggi P, Leone A, et al. Long-term efficacy and safety of switching from lamivudine + adefovir to tenofovir disoproxil fumarate in virologically suppressed patients. Dig Liver Dis. 2017;49(5):530–4 [A].
[21] Yu YJ. Tenofovir dipivoxil for treatment of chronic hepatitis B patients with gastric ulcer: efficacy and impact on host immune response. World Chinese J Digestolo. 2017;25(12):1079–82 [c].
[22] de Niet A, Jansen L, Stelma F, et al. Peg-interferon plus nucleotide analogue treatment versus no treatment in patients with chronic hepatitis B with a low viral load: a randomised controlled, open-label trial. Lancet Gastroenterol Hepatol. 2017;2(8):576–84 [c].
[23] Kayaaslan B, Guner R. Adverse effects of oral antiviral therapy in chronic hepatitis B. World J Hepatol. 2017;9(5):227–41 [R].
[24] Zhang D, Zhao G, Li L, et al. Observation of combined/optimized therapy of lamivudine and adefovir dipivoxyl for hepatitis B-induced decompensated cirrhosis with baseline HBV DNA > 1,000 IU/mL. Acta Gastro-Enterol Belg. 2017;80(1):9–13 [C].
[25] Piroth L, Wittkop L, Lacombe K, et al. Efficacy and safety of direct-acting antiviral regimens in HIV/HCV-co-infected patients—French ANRS CO13 HEPAVIH cohort. J Hepatol. 2017;67(1):23–31 [C].
[26] Jhaveri M, Procaccini N, Kowdley KV. Update on hepatitis C treatment: systematic review of clinical trials. Minerva Gastroenterol Dietol. 2017;63(1):62–73 [c].
[27] Medeiros T, Salviato CM, do Rosário NF, et al. Adverse effects of direct acting antiviral-based regimens in chronic hepatitis C patients: a Brazilian experience. Int J Clin Pharm. 2017;39(6):1304–11 [c].
[28] Hlaing NKT, Mitrani RA, Aung ST, et al. Safety and efficacy of sofosbuvir-based direct-acting antiviral regimens for hepatitis C virus genotypes 1–4 and 6 in Myanmar: real-world experience. J Viral Hepat. 2017;24(11):927–35 [c].
[29] Janda M, Mergenhagen KA. The Effect of psychosocial factors on success rates of hepatitis C treatment. Psychosomatics. 2017;58(6):624–32 [M].
[30] Louie V, Latt NL, Gharibian D, et al. Real-world experiences with a direct-acting antiviral agent for patients with hepatitis C virus infection. Perm J. 2017;21 [C].
[31] Ballester-Ferré MP, Martínez F, Garcia-Gimeno N, et al. Miliary tuberculosis infection during hepatitis C treatment with sofosbuvir and ledipasvir plus ribavirin. World J Hepatol. 2017;9(3):161–6 [A].
[32] Merat S, Sharifi AH, Haj-Sheykholeslami A, et al. The efficacy of 12 weeks of sofosbuvir, daclatasvir, and ribavirin in treating hepatitis C patients with cirrhosis, genotypes 1 and 3. Hepat Mon. 2017;17(1):1–4 [C].
[33] Vitrone M, Andini R, Mattucci I, et al. Direct antiviral treatment of chronic hepatitis C in heart transplant recipients. Transpl Infect Dis. 2017; [in press], [A].
[34] Abd-Elsalam S, Sharaf-Eldin M, Soliman S, et al. Efficacy and safety of sofosbuvir plus ribavirin for treatment of cirrhotic patients with genotype 4 hepatitis C virus in real-life clinical practice. Arch Virol. 2017;1–6 [c].
[35] Gane EJ, Shiffman ML, Etzkorn K, et al. Sofosbuvir-velpatasvir with ribavirin for 24 weeks in hepatitis C virus patients previously treated with a direct-acting antiviral regimen. Hepatology. 2017;66(4):1083–9 [c].
[36] Suda G, Nagasaka A, Yamamoto Y, et al. Safety and efficacy of daclatasvir and asunaprevir in hepatitis C virus-infected patients with renal impairment. Hepatol Res. 2017;47(11):1127–36 [C].
[37] Shoreibah M, Orr J, Jones DA, et al. Ledipasvir/sofosbuvir without ribavirin is effective in the treatment of recurrent hepatitis C virus infection post-liver transplant. Hepatol Int. 2017;11(5):434–9 [C].
[38] Anand AC, Agarwal SK, Garg HK, et al. Sofosbuvir and ribavirin for 24 weeks is an effective treatment option for recurrent Hepatitis C infection after living donor liver transplantation. J Clin Exp Hepatol. 2017;7(3):165–71 [C].
[39] Xue Y, Zhang LX, Wang L, et al. Efficacy and safety of sofosbuvir and daclatasvir in treatment of kidney transplantation recipients with hepatitis C virus infection. World J Gastroenterol. 2017;23(32):5969–76 [c].
[40] Kan H, Imamura M, Kawakami Y, et al. Emergence of drug resistance-associated variants and changes in serum lipid profiles in sofosbuvir plus ledipasvir-treated chronic hepatitis C patients. J Med Virol. 2017;89(11): 1963–72 [c].
[41] Lacombe K, Fontaine H, Dhiver C, et al. Real-world efficacy of daclatasvir and sofosbuvir, with and without ribavirin, in HIV/HCV coinfected patients with advanced liver disease in a French early access cohort. J Acquir Immune Defic Syndr. 2017;75(1):97–107 [c].

[42] Rockstroh JK, Ingiliz P, Petersen J, et al. Daclatasvir plus sofosbuvir, with or without ribavirin, in real-world patients with HIV-HCV coinfection and advanced liver disease. Antivir Ther. 2017;22(3):225–36 [c].

[43] Castells L, Llaneras J, Campos-Varela I, et al. Sofosbuvir and daclatasvir in mono-and HIV-coinfected patients with recurrent hepatitis C After liver transplant. Ann Hepatol. 2017;16(1):86–93 [c].

[44] Mohamed MS, Hanafy AS, Bassiony MAA, et al. Sofosbuvir and daclatasvir plus ribavirin treatment improve liver function parameters and clinical outcomes in Egyptian chronic hepatitis C patients. Eur J Gastroenterol Hepatol. 2017;29(12):1368–72 [c].

[45] Wang C, Ji D, Chen J, et al. Hepatitis due to reactivation of hepatitis B virus in endemic areas among patients with hepatitis C treated with direct-acting antiviral agents. Clin Gastroenterol Hepatol. 2017;15(1):132–6 [c].

[46] Elfeki MA, Abou Mrad R, Esfeh JM, et al. Sofosbuvir/ledipasvir without ribavirin achieved high sustained virologic response for hepatitis C recurrence after liver transplantation: two-center experience. Transplantation. 2017;101(5):996–1000.

[47] El-Khayat HR, Fouad YM, Maher M, et al. Efficacy and safety of sofosbuvir plus simeprevir therapy in Egyptian patients with chronic hepatitis C: a real-world experience. Gut. 2017;66(11):2008–12 [C].

[48] Carrai P, Morelli C, Cordone G, et al. The Italian compassionate use of sofosbuvir observational cohort study for the treatment of recurrent hepatitis C: clinical and virological outcomes. Transpl Int. 2017;30(12):1253–65 [c].

[49] Dharancy S. Direct-acting antiviral agent-based regimen for HCV recurrence after combined liver-kidney transplantation: results from the ANRS CO23 CUPILT study. Am J Transplant. 2017;17(11):2869–78 [C].

[50] Hézode C, Almasio PL, Bourgeois S, et al. Simeprevir and daclatasvir for 12 or 24 weeks in treatment-naïve patients with hepatitis C virus genotype 1b and advanced liver disease. Liver Int. 2017;37(9):1304–13 [C].

[51] Sezaki H, Suzuki F, Hosaka T, et al. The efficacy and safety of dual oral therapy with daclatasvir and asunaprevir for genotype 1b in Japanese real-life settings. Liver Int. 2017;37(9):1325–33 [C].

[52] Ferrer PE, Bloch M, Roth N, et al. A retrospective clinical audit of general practices in Australia to determine the motivation for switch to dolutegravir/abacavir/lamivudine and clinical outcomes. Int J STD AIDS. 2018;29(3):300–5 [C].

[53] Trottier B, Lake JE, Logue K, et al. Dolutegravir/abacavir/lamivudine versus current ART in virally suppressed patients (STRIIVING): a 48-week, randomized, non-inferiority, open-label. Phase IIIb study Antiviral Therapy. 2017;22(4):295–305 [C].

[54] Norwood J, Turner M, Bofill C, et al. Brief report: weight gain in persons with HIV switched from efavirenz-based to integrase strand transfer inhibitor-based regimens. J Acquir Immune Defic Syndr. 2017;76(5):527–31 [c].

[55] Cid-Silva P, Llibre JM, Fernández-Bargiela N, et al. Clinical experience with the integrase inhibitors dolutegravir and elvitegravir in HIV-infected patients: efficacy, safety and tolerance. Basic Clin Pharmacol Toxicol. 2017;121(5):442–6 [C].

[56] Arrabal-Durán P, Rodríguez-González CG, Chamorro-de-Vega E, et al. Switching to a rilpivirine/emtricitabine/tenofovir single-tablet regimen in RNA-suppressed patients infected with human immunodeficiency virus 1: effectiveness, safety and costs at 96 weeks. Int J Clin Pract. 2017;71(8):12968 [C].

[57] Gallant J, Lazzarin A, Mills A, et al. Bictegravir, emtricitabine, and tenofovir alafenamide versus dolutegravir, abacavir, and lamivudine for initial treatment of HIV-1 infection (GS-US-380-1489): a double-blind, multicentre, phase 3, randomised controlled non-inferiority trial. Lancet. 2017;390(10107):2063–72 [C].

[58] Ann H, Kim KH, Choi HY, et al. Safety and efficacy of Ziagen (abacavir sulfate) in HIV-infected Korean patients. Infect Chemother. 2017;49(3):205–12 [C].

[59] Maggiolo F, Gulminetti R, Pagnucco L, et al. Lamivudine/dolutegravir dual therapy in HIV-infected, virologically suppressed patients. BMC Infect Dis. 2017;17(1) [C].

[60] Leroi C, Balestre E, Messou E, et al. Incidence of severe neutropenia in HIV-infected people starting antiretroviral therapy in West Africa. PLoS One. 2017;12(1):e0170753 [C].

[61] Pintye J, Baeten J, Celum C, et al. Maternal tenofovir disoproxil fumarate use during pregnancy is not associated with adverse perinatal outcomes among HIV-infected east african women: a prospective study. J Infect Dis. 2017;216(12):1561–8 [c].

[62] De Waal R, Cohen K, Fox MP, et al. Changes in estimated glomerular filtration rate over time in South African HIV-1-infected patients receiving tenofovir: a retrospective cohort study. J Int AIDS Soc. 2017;20(1):21317 [MC].

[63] Gupta SK, Yeh E, Kitch DW, et al. Bone mineral density reductions after tenofovir disoproxil fumarate initiation and changes in phosphaturia: a secondary analysis of ACTG A5224s. J Antimicrob Chemother. 2017;72(7):2042–8 [C].

[64] Some F, Koech M, Chesire E, et al. Reversal of tenofovir induced nephrotoxicity: case reports of two patients. Pan Afr Med J. 2017;27:126 [A].

[65] Merker A, Badowski M, Chiampas T, et al. Effectiveness of single- and multiple-tablet antiretroviral regimens in correctional setting for treatment-experienced HIV Patients. J Correct Health Care. 2018;24(1):52–61 [C].

[66] Chary A, Nguyen NN, Maiton K, et al. A review of drug-drug interactions in older HIV-infected patients. Expert Rev Clin Pharmacol. 2017;10(12):1329–52 [R].

[67] Chirch LM, Luthra P, Mirza F. Metabolic complications of HIV infection. J Clin Outcomes Manag. 2017;24(12) [in press], [R].

[68] Chow FC, Li Y, Hu Y, et al. Relationship between HIV infection, antiretroviral therapy, inflammatory markers, and cerebrovascular endothelial function among adults in urban China. J Acquir Immune Defic Syndr. 2017;74(3):339–46 [c].

[69] Wynberg E, Williams E, Tudor-Williams G, et al. Discontinuation of efavirenz in paediatric patients: why do children switch? Clin Drug Investig. 2018;38:231–8 [c].

[70] Moore CB, Capparelli EV, Samson P, et al. CYP2B6 genotype-directed dosing is required for optimal efavirenz exposure in children 3–36 months with HIV infection. AIDS. 2017;31(8):1129–36 [c].

[71] Kekitiinwa A, Szubert AJ, Spyer M, et al. Virologic response to first-line efavirenz- or nevirapine-based antiretroviral therapy in HIV-infected African children. Pediatr Infect Dis J. 2017;36(6):588–94 [C].

[72] Bienczak A, Denti P, Cook A, et al. Determinants of virological outcome and adverse events in African children treated with paediatric nevirapine fixed-dose-combination tablets. AIDS. 2017;31(7):905–15 [C].

[73] Sculier D, Gayet-Ageron A, Battegay M, et al. Rilpivirine use in the Swiss HIV cohort study: a prospective cohort study. BMC Infect Dis. 2017;17(1):476 [C].

[74] Viciana P. Rilpivirine: the key for long-term success. AIDS Rev. 2017;19(3):156–66 [c].

[75] Orkin C, DeJesus E, Ramgopal M, et al. Switching from tenofovir disoproxil fumarate to tenofovir alafenamide coformulated with rilpivirine and emtricitabine in virally suppressed adults with

HIV-1 infection: a randomised, double-blind, multicentre, phase 3b, non-inferiority study. Lancet HIV. 2017;4(5):e195–204 [C].

[76] Orrell C, Hagins DP, Belonosova E, et al. Fixed-dose combination dolutegravir, abacavir, and lamivudine versus ritonavir-boosted atazanavir plus tenofovir disoproxil fumarate and emtricitabine in previously untreated women with HIV-1 infection (ARIA): week 48 results from a randomised, open-label, non-inferiority, phase 3b study. Lancet HIV. 2017;4(12):e536–46 [C].

[77] Davis AE, Brogan AJ, Goodwin B, et al. Short-term cost and efficiency analysis of raltegravir versus atazanavir/ritonavir or darunavir/ritonavir for treatment-naive adults with HIV-1 infection in Spain. HIV Clin Trials. 2017;18(5–6): 214–22 [C].

[78] Qi Q, Wang Q, Chen W, et al. HIV-1 gp41-targeting fusion inhibitory peptides enhance the gp120-targeting protein-mediated inactivation of HIV-1 virions. Emerg Microbes Infect. 2017;6(6):e59 [A].

[79] Ding X, Zhang X, Chong H, et al. Enfuvirtide (T20)-based lipopeptide is a potent HIV-1 cell fusion inhibitor: implications for viral entry and inhibition. J Virol. 2017;91(18): e00831-17 [A].

[80] Fettiplace A, Stainsby C, Winston A, et al. Psychiatric symptoms in patients receiving dolutegravir. J Acquir Immune Defic Syndr. 2017;74(4):423–31 [c].

[81] Hoffmann C, Welz T, Sabranski M, et al. Higher rates of neuropsychiatric adverse events leading to dolutegravir discontinuation in women and older patients. HIV Med. 2017;18(1):56–63 [C].

[82] Todd SEJ, Rafferty P, Walker E, et al. Early clinical experience of dolutegravir in an HIV cohort in a larger teaching hospital. Int J STD AIDS. 2017;28(11):1074–81 [c].

[83] Cucchetto G, Lanzafame M, Nicole S, et al. Raltegravir plus lamivudine as 'maintenance therapy' in suppressed HIV-1-infected patients in real-life settings. J Antimicrob Chemother. 2017;72(7):2138–40 [c].

[84] Elzi L, Erb S, Furrer H, et al. Adverse events of raltegravir and dolutegravir. AIDS. 2017;31(13):1853–8 [C].

[85] Milburn J, Jones R, Levy JB. Renal effects of novel antiretroviral drugs. Nephrol Dial Transplant. 2017;32(3):434–9 [R].

[86] Rossetti B, Gagliardini R, Meini G, et al. Switch to maraviroc with darunavir/r, both QD, in patients with suppressed HIV-1 was well tolerated but virologically inferior to standard antiretroviral therapy: 48-week results of a randomized trial. PLoS One. 2017;12(11):e0187393 [C].

[87] Saumoy M, Llibre JM, Terrón A, et al. Short communication: maraviroc once-daily: experience in routine clinical practice. AIDS Res Hum Retrovir. 2017;33(1):29–32 [C].

[88] Ison MG. Antiviral treatments. Clin Chest Med. 2017;38(1):139–53 [R].

[89] Hirotsu N, Saisho Y, Hasegawa T, et al. Clinical and virologic effects of four neuraminidase inhibitors in influenza A virus-infected children (aged 4–12 years): an open-label, randomized study in Japan. Expert Rev Anti Infect Ther. 2018;16:173–82 [c].

[90] Kwon YS, Park SH, Kim MA, et al. Risk of mortality associated with respiratory syncytial virus and influenza infection in adults. BMC Infect Dis. 2017;17(1) [C].

[91] Noel Z, Bastin MLT, Montgomery A, et al. Comparison of high-dose versus standard dose oseltamivir in critically Ill patients with influenza. J Intensive Care Med. 2017;32(10):574–7 [MC].

[92] Beigel JH, Bao Y, Beeler J, et al. Oseltamivir, amantadine, and ribavirin combination antiviral therapy versus oseltamivir monotherapy for the treatment of influenza: a multicentre, double-blind, randomised phase 2 trial. Lancet Infect Dis. 2017;17(12):1255–65.

[93] Lee J, Park JH, Jwa H, et al. Comparison of efficacy of intravenous peramivir and oral oseltamivir for the treatment of influenza: systematic review and meta-analysis. Yonsei Med J. 2017;58(4):778–85 [MC].

[94] Manjunath A, Brenton T, Wylie S, et al. Topical therapy for non-invasive penile cancer (Tis)-updated results and toxicity. Translational Androl Urol. 2017;6(5):803–8 [R].

[95] Lv R, Sun Q. A network meta-analysis of non-melanoma skin cancer (NMSC) treatments: efficacy and safety assessment. J Cell Biochem. 2017;118(11):3686–95 [MC].

[96] Lewis DJ, Byekova YA, Emge DA, et al. Complete resolution of mycosis fungoides tumors with imiquimod 5% cream: a case series. J Dermatol Treat. 2017;28(6):567–9 [c].

[97] J.C. van der Wouden, R. van der Sande, E.J. Kruithof, et al., Interventions for cutaneous molluscum contagiosum, Cochrane Database Syst Rev 2017 (5): CD004767 [R].

[98] Marsden JR, Fox R, Boota NM, et al. Effect of topical imiquimod as primary treatment for lentigo maligna: the LIMIT-1 study. Br J Dermatol. 2017;176(5):1148–54 [c].

[99] Donin NM, Chamie K, Lenis AT, et al. A phase 2 study of TMX-101, intravesical imiquimod, for the treatment of carcinoma in situ bladder cancer. Urol Oncol. 2017;35(2):39e1–7 [c].

[100] Othumpanagat S, Noti JD, Ray SD. Antiviral drugs. In: Ray SD, editor. Side effects of drugs: annual: a worldwide yearly survey of new data in adverse drug reactions, vol. 37; Elsevier; 2015. p. 329–34 [chapter 29]. [R].

[101] Othumpanagat S, Ray SD, Noti JD. Antiviral drugs. In: Ray SD, editor. Side effects of drugs: annual: a worldwide yearly survey of new data in adverse drug reactions, vol. 38; Elsevier; 2016. p. 261–328 [chapter 27]. [R].

[102] Othumpanagat S, Ray SD, Noti JD. Antiviral drugs. In: Ray SD, editor. Side effects of drugs: annual: a worldwide yearly survey of new data in adverse drug reactions. vol. 39; Elsevier; 2017. p. 269–82 [chapter 25]. [R].

CHAPTER

29

Drugs in Tuberculosis and Leprosy

Meenakshi R. Ramanathan[*,†,1], *Crystal K. Howell*[*,‡], *James M. Sanders*[*,§]

*The University of North Texas Health Science Center System College of Pharmacy, Fort Worth, TX, United States

[†]Medical City Arlington, Arlington, TX, United States

[‡]Medical City Dallas, Dallas, TX, United States

[§]JPS Health Network, Fort Worth, TX, United States

[1]Corresponding author: meenakshi.ramanathan@unthsc.edu

Abbreviations

AE adverse effect
ALT alanine transaminase
AMG aminoglycoside
ART antiretroviral therapy
AST aspartate transaminase
ATT anti-tuberculosis treatment
BDQ bedaquiline
CFZ clofazamine
CKD chronic kidney disease
CS cycloserine
CYP cytochrome
DIC disseminated intravascular coagulation
DILI drug-induced liver injury
DM diabetes mellitus
DRESS drug reaction with eosinophilia and systemic symptoms
EMB ethambutol
FDA U.S. Food and Drug Administration
FDC fixed-dose combination
FQ fluoroquinolone
HAART highly active antiretroviral therapy
HCW healthcare workers
HIV human immunodeficiency virus
IM intramuscular
IV intravenous
INH isoniazid
IPT isoniazid preventive therapy
LFTs liver function tests
LNZ linezolid
LTBI latent tuberculosis infection
MAOI monoamine oxidase inhibitor
MDR-TB multidrug-resistant tuberculosis
ms milliseconds
NNRTI non-nucleoside reverse transcriptase inhibitor
OR odds ratios
PAS para-aminosalicylic acid
PE pulmonary embolism
PI protease inhibitor
PO by mouth
PZA pyrazinamide
QTc corrected Q-T interval
RBT rifabutin
RIF rifampin
RMP rifampicin
RPT rifapentine
SJS Stevens–Johnson syndrome
SNRI serotonin–norepinephrine reuptake inhibitor
SOT solid organ transplant
SSRI selective serotonin reuptake inhibitor
STR streptomycin
TB tuberculosis
TDM therapeutic drug monitoring
ULN upper limit of normal
US United States
XDR-TB extensively drug-resistant tuberculosis

AMINOGLYCOSIDES

Treatment of multidrug-resistant tuberculosis (MDR-TB) often necessitates the utilization of adjunctive parenteral therapy. The aminoglycosides (AMGs) (i.e., amikacin, kanamycin and streptomycin [STR]) are employed as second-line agents for the management of both MDR-TB and patients with intolerance to first-line agents [1S, 2S]. Reservation of these agents to second-line therapy results from the restriction to parenteral administration and the adverse effects (AEs) seen with these agents. Nephrotoxicity and ototoxicity constitute two of the more common and concerning adverse drug events seen with AMGs [1S, 3R].

- Sagwa and colleagues performed a case/non-case disproportionality analysis of individual case safety reports in the World Health Organization's database, VigiBase, between 1968 and 2014 for STR, amikacin, kanamycin, and capreomycin. There were 3194 cases

Side Effects of Drugs Annual, Volume 40
ISSN: 0378-6080
https://doi.org/10.1016/bs.seda.2018.06.014

(94%) of adverse reactions for STR, 40 cases (1%) for kanamycin, 40 cases (1%) for amikacin, and 117 cases (4%) for capreomycin. When defined by adverse reaction type, "other adverse reaction" was the most common at 2785 cases (83%) followed by vertigo (394, 12%), tinnitus (91, 3%), deafness (71, 2%), and unspecified ototoxicity (20, 1%). The reporting odds ratio (OR) for any ototoxicity, using STR as the reference, was 1.4 for kanamycin, 0.7 for amikacin, and 0.3 for capreomycin [4C].

Electrolyte Disturbances

Jan and colleagues reviewed the effect of amikacin on serum potassium and sodium levels in a cohort of 179 subjects receiving amikacin as second-line treatment for MDR-TB. The study found a significant link between utilization of amikacin and hypokalemia while showing no clear effect on sodium levels within the cohort. Average baseline potassium levels were 3.73 mmol/L and average levels were 3.54–3.58 mmol/L during the intensive and continuation phases of treatment [5C].

Arnold and colleagues compared amikacin and capreomycin AEs among a cohort of 100 patients receiving treatment for MDR-TB. The occurrence of hypokalemia was significantly less with amikacin relative to capreomycin (OR=0.28 [95% CI 0.11–0.72]). Of note, 44% of those with accessible potassium levels experienced hypokalemia while on one of the injectable agents. Capreomycin induced hypomagnesemia in 87% of patients tested with 4 requiring a switch in therapy and one early discontinuation [6C].

Ototoxicity

Arnold and colleagues performed a comparison of amikacin and capreomycin-induced ototoxicity and found a 14-fold higher chance of developing hearing impairment with amikacin. Overall, 55% of the cohort had ototoxicity demonstrated [6C]. A retrospective review of 164 cases of patients receiving MDR-TB treatment with AMGs therapy—majority received amikacin—in Brazil found 34% had confirmed cases of ototoxicity (vestibular and auditory), with hypoacusis reported most often, that resulted in treatment discontinuation. Of these cases, audiometric testing was only performed in 22% illustrating the potential need for audiometric testing to monitor and clarify the extent of damage [7C]. In a smaller series of 11 patients in France with MDR-TB receiving amikacin, 3 patients had documented hearing loss and 3 with possible hearing loss occurring with an average delay of 2.8 months [8c].

Pediatrics represents a special population that commonly has underrepresented adverse drug event reporting relative to adults. A review of 599 children in China receiving AMGs therapy for tuberculosis (TB) demonstrated 2 cases of auditory impairment both related to STR and both patients had full recovery following cessation of the drug [9C].

Therapeutic drug monitoring (TDM) of AMGs presents a possible means for preventing AEs [3R]. Van Altena and colleagues retrospectively studied patients treated with either amikacin or kanamycin for MDR-TB or extensively resistant tuberculosis (XDR-TB). Ototoxicity was seen in the amikacin and kanamycin groups in 9.1% and 22% of patients, respectively. Therapeutic targeted dosing resulted in lower doses, which correlated with changes in high frequency hearing loss suggesting a potential role for TDM [10c]. Renal replacement therapy represents another scenario that may require TDM. A case report of a 49-year-old man receiving amikacin while on continuous venovenous haemofiltration (CVVH) illustrates variable pharmacokinetics of this agent and the potential need for close monitoring to optimize therapy and prevent AEs [11A].

Nephrotoxicity

Arnold and colleagues documented 25% of the patients receiving either amikacin or capreomycin demonstrated nephrotoxicity as defined as an elevation in serum creatinine (SCr) ≥1.5 times baseline; of these cases, 90% returned to baseline. No discernable difference in the ability to induce nephrotoxicity was found between the amikacin and capreomycin [6C]. Similar results were seen in a comparison of amikacin and kanamycin with 23% and 35% of patients experiencing a SCr elevation ≥1.5 times baseline, respectively [10c]. In contrast, amikacin resulted in only one case out of 11 in patients on injectable treatment for MDR-TB or XDR-TB [8c]. In a large cohort of children only one experienced nephrotoxicity and fully recovered following discontinuation of amikacin [9C].

Drug–Drug Interactions

A pharmacokinetic model combining lopinavir, ritonavir and amikacin in human immunodeficiency virus (HIV)-infected children demonstrated no altered pharmacokinetic parameters of either lopinavir or ritonavir [12c]. This suggests concomitant administration of amikacin will not affect simultaneous antiretroviral therapy containing the lopinavir or ritonavir.

BEDAQUILINE

Bedaquiline (BDQ) remains a poorly studied drug outside of clinical trials for the treatment of (MDR-TB) [13R].

Potential serious AEs associated with BDQ include cardiac abnormalities (i.e., prolonged QTc [corrected Q-T interval]) and hepatotoxicity. These toxicities may be compounded with utilization of select concomitant TB medications [14R]. Guglielmetti and colleagues performed a retrospective review of 45 patients initiated on BDQ. The median treatment duration was approximately 1 year. Eleven percent of the patients reported Fridericia's formula QTc (QTcF) elevations greater than 500 ms with three patients discontinuing therapy. The median maximum QTcF increase was 36.2 ms. Prolonged therapy did not correlate with increased AE profiles [15c].

Off-label use of BDQ for MDR-TB in children was documented in 27 children from November 2014 to 2017. Of these children, 5 (19%) experienced a grade 3 or 4 AE related to BDQ. Four patients had QTcF elevation >60 ms from recorded baseline [16c].

Drug–Drug Interactions

The increase in MDR- and XDR-TB has resulted in the utilization of agents that may carry additive toxicity, for example, BDQ and delamanid [2S, 14R]. Migliori and colleagues reviewed the literature finding two cases of combination BDQ and delamanid reported and one having QT interval prolongation [17A]. Rifampin (RIF)-reduced BDQ exposure in healthy adults when compared to BDQ alone by almost 50% [18c].

CARBAPENEMS

Carbapenems are utilized within some MDR- and XDR-TB regimens. Overall this class of medications retains a favorable safety profile with hematologic, allergic, dermatologic, and epileptogenic AEs being reported [19R]. A pharmacokinetic model of 42 healthy volunteers to model ertapenem in MDR patients predicted exposure effectively and may serve as a model to optimize future dosing regimens [20c].

CLOFAZIMINE

Clofazimine (CFZ) demonstrates activity against multiple *Mycobacterium* species; therefore is utilized for the treatment of nontubercular mycobacterium (e.g., *Mycobacterium abscessus*), TB and leprosy [21R]. CFZ therapy often results in gastrointestinal AEs along with skin and fluid discoloration. In addition, it has the potential to prolong the QTc interval [14R, 22R]. The safety of CFZ was reviewed in a cohort of 112 adult and pediatric patients, including cystic fibrosis patients. Overall, 88% of the patients reported AEs with a median of 4 AEs. The top three reported events were skin discoloration (61%), nausea (33%), and fatigue (30%) [23C].

A comparison of CFZ containing MDR-TB regimens against pyrazinamide (PZA) containing regimens found that CFZ regimens resulted in more hyperpigmentation and less arthralgia. Overall, the CFZ containing regimens caused hyperpigmentation, arthralgia, and gastrointestinal intolerance; however, this was confounded by multiple other drugs within the regimen [24C].

A 6-year-old child treated for XDR-TB spondylodiscitis with moxifloxacin + CFZ + linezolid (LNZ) + isoniazid (INH) + amoxicillin-clavulanate + para-aminosalicylic acid + capreomycin reported no adverse drug events following 18 months of treatment [25A].

CYCLOSERINE

Cycloserine (CS) has limited utility in the treatment of TB based on a paucity of studies demonstrating efficacy that would outweigh its propensity to induce neuropsychiatric AEs [26R].

A 38-year-old male on CS for the treatment of TB presented with itchy lichenoid eruptions. His second-line treatment regimen consisted of ethambutol (EMB) and levofloxacin. Abnormal laboratory values included hypereosinophilia. Lab values and presenting symptoms resolved within approximately a month of stopping TB therapy. Patch and lymphocyte transformation test demonstrated positive responses to CS [27A].

A single-center pharmacokinetic study performed in Korea suggests that current recommended doses for CS may be suboptimal with 500–750 mg daily being more optimal in the population studied [28c]. This may aid future studies on dosing to optimize safety and efficacy.

DAPSONE

Dapsone constitutes a common treatment for leprosy and is often associated with hematologic and hypersensitivity reactions [29R]. A review of patients in Nepal on dapsone for leprosy demonstrated 0.82% rate of AEs over 4 years. Of the 18 patients reporting adverse drug events the following occurred in order of highest prevalence: jaundice, dermatitis, hemolytic anemia, fever, and headache. Most of the AEs were managed but notably four patients died [30c]. Dapsone safety in immune thrombocytopenia was measured by review of 42 patients treated with 100 mg daily. Thirty-one percent of the patients had treatment related AEs including skin rash, sulfone syndrome, methemoglobinemia, neuropathy, dyspnea, fatigue, and diarrhea. Withdrawal of dapsone therapy occurred in 22% of the patients [31c]. The overall prevalence of mental or neurological manifestations of dapsone are rare, 0.7%, as described in a systematic review of over 2000 patients [32C].

Hematologic Disorders

A 75-year-old female presented with sepsis thought to be from a urinary source and subsequently became cyanotic. Her methemoglobin level was 11% in conjunction with a dark red arterial blood sample. Methylene blue treatment resulted in improved symptoms and methemoglobin levels. In addition, the patient was found to have a pulmonary embolism (PE). History revealed that the patient was on dapsone for pneumocystis prophylaxis. Despite low methemoglobin levels the patient showed cyanosis that may have been related to concomitant PE [33A]. Nazir and colleagues reviewed patients with acute lymphoblastic leukemia and tolerability to dapsone vs trimethoprim/sulfamethoxazole. Patients on dapsone had higher absolute neutrophil counts and tolerated higher cytotoxic medications doses [34c]. The combination of dapsone and azathioprine resulted in significant anemia in the AZALEP trial; the trial studied azathioprine for the treatment of leprosy damage [35C].

Hypersensitivity

Maddala and colleagues report a case of dapsone hypersensitivity that included anatopic response with double sparing syndrome. In brief, a 26-year-old male with leprosy was treated with dapsone for 1 month that resulted in diagnosis of dapsone hypersensitivity. Of note, the patient's rash spared the leprosy patches and tinea versicolor lesions [36A]. Marchese and colleagues present a case of dapsone hypersensitivity with associated thyroiditis and fulminant type 1 diabetes mellitus (DM). On admission the patient presented with the following lab findings: thyroid-stimulating hormone = 0.01 Uiu/mL, free thyroxine = 3.94 ng/dL, thyroid-stimulating immunoglobulin = 21% and serum glucose = 450 mg/dL. Treatment included steroids, beta-blocker and insulin therapy. At follow-up thyroiditis had resolved and insulin requirements although decreased persisted [37A].

A review of literature on dapsone hypersensitivity among Chinese patients showed an overall prevalence of approximately 2% with the majority of patients receiving therapy for leprosy [38M]. Dapsone-related severe cutaneous reactions were significantly associated with the HLA-B*13:01 allele in Thai patients suggesting it may serve as a useful genetic marker [39c].

Hepatotoxicity

Devarbhavi and colleagues reviewed 128 cases of drug-induced acute liver failure within a drug-induced liver injury (DILI) registry. Dapsone was identified in 7 (5.5%) of the cases with a mortality rate of approximately 30% [40C]. A detailed review of the dapsone-induced hepatitis cases revealed a majority of the patients had a hypersensitivity reaction and/or jaundice. The majority had a delayed onset with a mean of 34 days [41c].

DELAMANID

Delamanid has been shown to be a viable option in the treatment of MDR-TB and XDR-TB. Delamanid is considered a nitroimadazole, which inhibits the biosynthesis of mycolic acid, a key component in the wax-rich mycobacterial cell wall. Generally well-tolerated, some patients may experience QTc interval prolongation while on the medication [17A]. A recent systematic review showed the cardiotoxicity risk of using BDQ and delamanid in combination due to their QT prolonging effects [17A].

ETHAMBUTOL

Ethambutol (EMB) is considered first-line treatment for active TB and *Mycobacterium avium* complex (MAC) infections. It is one of the four backbone agents in the treatment of active TB; however, it is generally the first of the agents taken off due to AE data involving dose-related optic neuropathy. EMB works by inhibiting the enzyme aribinosyl transferase III, which inhibits the production of arabinogalactan, a component of the cell of mycobacteria. The most common AEs of EMB are optic neuritis, generally manifesting as either decreased visual acuity or the inability to differentiate between red and green. This is considered both dose-dependent, duration-dependent, and generally reversible once EMB is discontinued. The use of vision exams are important in the monitoring of EMB toxicity [1S, 42S, 43S, 44R].

Visual Disturbances

A 75-year-old man previously treated with EMB for TB presented to the hospital with complaints of mistiness of vision. Patient had no family history of neuropathy. Upon further examination, the patient was found to have EMB-induced optic neuropathy (EON). Since the patient had already discontinued EMB, the patient received some neurotrophic agents and was asked to return in 1 month, where the results showed complete recovery of cecocentral scotoma, with good prognosis. Incidence with EON is thought to be 1% and is considered dose-related [45A].

Fixed Dose Combination Therapy

Silva and colleagues evaluated a four-drug fixed-dose combination (FDC) regimen in Rio de Janeiro, Brazil, primarily focusing on recurrence rates, cure, treatment abandonment, AEs, and death. This was a retrospective observational study comparing two groups: the RIF/INH/PZA group between January 2007 and October 2010 and the FDC regimen of RIF/INH/PZA/EMB between November 2010 and June 2013. A total of 208 patients were evaluated, but only 203 patients had AE data available. AEs occurred in 68.5% of patients ($n=139$) with 86 patients (69.9%) in the RIF/INH/PZA groups and 53 patients (66.3%) in the FDC group; however, no difference in AE occurrence was found ($P=0.35$). Twenty-four patients required a regimen switch due to AEs (18 patients in the RIF/INH/PZA group and 6 patients in the FDC group), with no statistical difference ($P=0.11$). Common AEs seen in both arms include acne/itching, arthralgia, anorexia/vomiting/abdominal pain/nausea, paresthesia, hepatotoxicity, exanthema. Neither of the groups reported optic neuritis [46C].

FLUOROQUINOLONES

The fluoroquinolones (FQs) are known as second-line agents in the treatment of active TB. Currently on the market, moxifloxacin, levofloxacin, and ofloxacin are the most common FQs used in the treatment of TB. They work by inhibiting DNA topoisomerase II and topoisomerase IV (also known as DNA gyrase), which are involved in DNA supercoiling. In recent years, more and more data are coming out on the increased risks vs the benefits of using the FQs in less severe infections, such as acute sinusitis, acute bronchitis, and uncomplicated urinary tract infections. Some AEs associated with the FQs include photosensitivity, neurotoxicity, tendon rupture, and QT interval prolongation. The FQs are also considered teratogenic, and therefore, should not be used in pregnant TB patients. Pediatric use should be limited due to the risk of cartilage damage in this population [1S, 2S, 42S, 43S, 44R, 47S, 48c].

Central Nervous System Disturbances

A case report on levofloxacin-induced myoclonus was reported by Kunder and colleagues. A 78-year-old male being treated for pulmonary TB was admitted to the hospital for difficulty walking and brief shock-like movements in all four limbs. He had a 40+ year history of smoking and drinking, but otherwise had no other comorbidities. The patient had initially been started on INH 300mg, rifampicin (RMP) 450mg, and EMB 800mg, but due to peripheral neuropathy from INH, the patient's INH was switched to levofloxacin 500mg by mouth (PO) daily. Four days after starting the new medication, the patient noted involuntary movements and on and off altered behavior (i.e., clapping, hitting himself, and altered speech). No gross abnormalities were found on workup and laboratory findings. Based on the neurologist's opinion, the patient was diagnosed with levofloxacin-induced myoclonus/chorea and levofloxacin was stopped. The patient was restarted on a fixed dose combination of RMP, INH, and EMB in addition to oral diazepam 2mg/day and oral pyridoxine 40mg/day. Three days after discontinuation of levofloxacin with the above mentioned interventions, the involuntary activities significantly reduced. The mechanism behind levofloxacin-induced myoclonus/chorea is thought to be the result of excitation of NMDA receptors and inhibition of GABA-A receptors or triggering of the myoclonic generator [49A].

ISONIAZID

Considered a backbone medication in both active TB and latent tuberculosis infection (LTBI), isoniazid (INH) was first approved by the US Food and Drug Administration (FDA) in 1953. It works by preventing the synthesis of mycolic acids in the cell wall. It is part of the standard four-drug TB regimen recommended for active TB patients upfront, and is considered one of the most active components of the regimen, in addition to RIF. It is also considered the drug of choice for LTBI, with INH 5mg/kg (maximum dose 300mg) PO daily for 9 months as the preferred regimen. However, INH is not without its AEs, which include hepatotoxicity (e.g., elevated transaminases), neuropathy (e.g., peripheral and optic), and gastrointestinal disturbances (e.g., diarrhea). It is oftentimes administered with pyridoxine, also known as vitamin B6, to reduce the risk of peripheral neuropathy [1S, 42S, 43S, 44R, 50S, 51S].

Active TB

- A retrospective observational study in Qatar compared a separate tablet (ST) vs a fixed-dose combination (FDC) of INH, RIF, PZA, and EMB in diabetics being treated with TB. A total of 148 patients were included: 90 patients in the FDC group and 58 patients in the ST group. Only visual AEs were noted to be statistically significant in the ST group (5.2% vs 0%, $P=0.03$). More musculoskeletal AEs were seen in the ST group (22% vs 11%, $P=0.06$), although not statistically significant. Overall, non-diabetics had higher rates of musculoskeletal AEs (25% vs 0%, $P=0.03$), while diabetics had higher rates of hepatotoxicity (18.2% vs 5.2%, $P=0.016$) and gastrointestinal AEs (54.5% vs 16.5%, $P<0.001$) compared to non-diabetic patients [52C].
- Sekaggya-Wiltshire and colleagues conducted a prospective observational study aimed at assessing the

correlation between anti-TB drug concentrations and drug-associated toxicities among TB/HIV-coinfected patients. Patient's drug concentrations were taken at weeks 2, 8, and 24 weeks after initiation of anti-tuberculosis treatment (ATT). Blood sampling was taken at 0 h (drug dosing), and at 1, 2, and 4 h after witnessed dosing of ATT and antiretroviral therapy (ART). Serum concentrations were analyzed via ultraviolet high-performance liquid chromatography. Those patients with no, mild, moderate, and severe hepatotoxicity were shown to have median maximum concentrations of 6.57 μg/mL (interquartile range (IQR) 4.83–9.41 μg/mL), 7.39 μg/mL (IQR 5.10–10.20 μg/mL), 7.00 μg/mL (IQR 6.05–10.95 μg/mL), and 3.86 μg/mL (IQR 2.81–14.24 μg/mL), respectively. The study noted that there was no difference among patients with RMP concentrations and hepatotoxicity or INH concentrations and peripheral neuropathy among TB/HIV co-infected patients [53C].

LTBI

- Arguello Perez and colleagues reviewed the management of LTBI in healthcare workers (HCW) at Memorial Sloan Kettering Cancer Center in New York City, New York over a 10-year period. During this time, 363 HCWs accepted treatment for LTBI, with 202 HCWs choosing INH, 106 HCWs choosing RIF, and 55 HCWs choosing rifapentine (RPT)/INH. In terms of completion rates, HCW on INH alone were less likely to complete their regimen (57%) compared to RIF (80%, $P<0.0001$) and RPT/INH (87%, $P<0.0001$). Discontinuation of therapy due to AEs was again higher in the INH group (29 HCWS, 35%) compared to the other two groups with RIF at 21% (11 HCWs, not statistically significant) and RPT/INH at 10% (4 HCWs, $P=0.0042$). No difference in discontinuation rates were seen between the RIF and RPT/INH groups. Increased aminotransferase levels were the most commonly cited INH AE reported (57 HCWs, 28%), followed by gastrointestinal symptoms (9 HCWs, 4%) [54C].
- Guirao-Arrabal and colleagues reviewed the efficacy and safety of a short course regimen with INH/RMP for LTBI in lung transplant candidates in Córdoba, Spain. Patients were given INH 300 mg PO daily and RMP 600 mg PO daily for 3 months in order to expedite the LTBI treatment prior to transplantation. A total of 398 lung transplant candidates were evaluated, with 92 patients being LTBI positive, but only 23 patients receiving treatment. None of the patients receiving LTBI treatment developed post-transplant TB, compared to 3/62 patients (4.8%) who did not receive LTBI treatment. Only two patients (8.7%) reported AEs, which later led to discontinuation: one patient with moderate epigastric pain and another patient with transaminase levels three times the upper limit of normal (ULN) along with biliary cholelithiasis. The authors concluded that the 3-month short course regimen was both safe and efficacious [55c].
- Zenner et al. performed a meta-analysis reviewing the various LTBI treatment regimens, primarily focusing on rates of hepatotoxicity and prevention of active TB. Because of the sparse data on hepatotoxicity outcomes, the meta-analysis focused on direct comparisons of odds ratios (OR). RMP (RMP)-only and RPT/INH regimens had lower rates of hepatotoxicity compared to INH-only regimens of 6, 9, or 12–72 months. RMP-INH regimens also had lower rates of hepatotoxicity when compared to INH-only regimens of 12–72 months. Higher hepatotoxicity rates were seen with PZA when compared with 6 months of INH or 12 weeks of RPT-INH. Unfortunately, data were not available for RFB-INH and INH-EMB regimens [56M].

Prophylaxis

In 2017, the primary AE data for INH were linked to its use as a TB prophylaxis agent in special populations, including pediatric patients, advanced HIV patients, and solid organ transplant (SOT) patients.

- A retrospective cohort study reviewed the incidence of TB pre- and post-implementation of an expanded INH prophylaxis program in its SOT patients. The study reviewed 1966 SOT patients between January 2003 and December 2012, of which 1391 patients were kidney transplant patients. All patients receiving a deceased donor kidney or living donor kidney with evidence of LTBI after January 2008 received INH 300 mg PO daily for 9 months as TB prophylaxis. Patients in the INH group were noted to have mild liver enzyme elevation, equivalent to $<2\times$ ULN without clinical symptoms [57C].
- Ayele and colleagues performed a prospective cross-sectional study reviewing isoniazid preventative therapy (IPT) for LTBI in patients with HIV/AIDS between March 10 and June 11, 2016. A total of 154 patients were given INH 300 mg PO daily for a minimum of 6 months, up to 36 months in areas of high TB prevalence and transmission. Forty-eight patients (31.2%) reported INH-related AEs, with the most common AEs noted to be abdominal pain, vomiting, skin rash, jaundice, and numbness. Three patients (2%) discontinued IPT due to AEs. It was noted that patients who experienced AEs were 36%

less likely to be adherent compared to those patients who did not experience AEs from INH. Major AEs leading to nonadherence included hepatitis and peripheral neuropathy [58C].

- Badje and colleagues performed a long-term follow-up trial of the TEMPRANO trial (Trial of Early Antiretrovirals and Isoniazid Preventive Therapy in Africa), a factorial 2×2 trial assessing the benefits of early ART and 6-month IPT) reviewing the risk of death of IPT in advanced HIV patients from west Africa. Four study groups were identified from TEMPRANO for review: (1) deferred ART, (2) deferred ART plus IPT, (3) early ART, (4) early ART plus IPT. Of the 12 IPT patients who discontinued IPT prematurely, six patients developed elevated aminotransferase concentrations (two grade 2 events, two grade 3 events, and two grade 4 events, none leading to death), four patients developed psychiatric side-effects (two grade 2 events and two grade 3 events), and two patients developed pruritus (two grade 2 events) [59C].
- Similarly, Hakim et al. performed a factorial open-label trial assessing enhanced prophylaxis vs standard prophylaxis with ART on 24-week mortality for advanced HIV patients in four African countries (Uganda, Zimbabwe, Malawi, and Kenya). Enhanced prophylaxis was defined as a co-formulation of trimethoprim–sulfamethoxazole plus INH-pyridoxine for at least 12 weeks, fluconazole for 12 weeks, azithromycin for 5 days, and albendazole single dose. Standard therapy was defined as sulfamethoxazole–trimethoprim alone. A total of 1805 patients (1733 adults and 72 children or adolescents) were randomized to either the enhanced prophylaxis group ($N=906$) or the standard prophylaxis group ($N=899$) and were followed for 48 weeks. Lower rates of serious AEs and grade 4 events were found in the enhanced prophylaxis group, although nonsignificant ($P=0.08$ and $P=0.09$, respectively). Fourteen patients in the enhanced prophylaxis group discontinued therapy due to toxicity involving the liver ($N=7$), skin ($N=4$), and blood ($N=3$) [60C].
- Maharaj and colleagues conducted a secondary analysis looking into the implementation of IPT in TB treatment experienced patients on ART. The original prospective cohort study entitled the Centre for the AIDS Programme of Research Africa's CAPRISA 005 TB Recurrence upon Treatment with HAART (TruTH) study occurred between October 2009 and October 2013 and assessed TB recurrence in patients started on ART. After 2011 and onwards, patients 18 years of age and older were eligible for IPT. A total of 212 patients were initiated on IPT, with 184 patients completing a 6-month course of INH. Of these patients, 167 patients completed IPT with no interruptions. Twenty-four patients (11.3%) permanently discontinued IPT, with seven patients (3.8%) discontinuing due to INH toxicity. Six patients (2.8%) experienced moderate to severe liver function abnormality, with only two of the patients requiring discontinuation. Median aspartate transaminase (AST) and alanine transaminase (ALT) levels were statistically significantly higher after IPT initiation ($P<0.001$), and also statistically significantly lower after IPT completion ($P<0.001$), with no difference noted between levels before and after IPT [61C].
- Okwara et al. conducted a prospective longitudinal cohort study by assessing the use of 6 months of IPT to at-risk children less than 5 years of age in household contact with an adult TB patient. Patients were monitored for new-onset TB, compliance, and AEs. A total of 414 patients received IPT, with 368 patients (88%) completing the 1-year follow-up. AEs were noted in 82 patients (22.2%): skin rash (15.2%), gastrointestinal AEs (i.e., nausea, anorexia, vomiting, 9.5%), and neurologic symptoms (i.e., irritability and/or weakness, paraesthesias, painful limbs, reduced play and altered sleeping patterns, 5.4%). Six patients (1.6%) were found to have yellow discoloration of the eyes or urine. Mean baseline ALT was 46.1 mmol/L and mean baseline AST was 29.6 mmol/L. After 1 month of IPT, mean baseline ALT increased to 90.9 mmol/L and mean baseline AST increased to 54.1 mmol/L (approximately a twofold increase). Severe hepatotoxicity, defined as a threefold increase, was found in three patients (0.08%), who were also on ART [62C].
- A Cochrane review on INH for preventing TB in HIV-infected children was published by Zunza and colleagues in 2017. The review compared TB preventive treatment vs placebo in HIV-positive children on acquiring active TB, death, and reported AEs. A total of three trials, including 991 patients, were included. AEs were split into clinical AEs and laboratory AEs. For HIV-positive children not on ART, one trial reported no clinical AEs in children not on ART; while also noting similar hematological and liver enzyme abnormalities. In HIV-positive children on ART, similar rates of peripheral neuropathy, hematologic abnormalities, and liver enzyme abnormalities were noted. The review concluded that both the clinical and laboratory AEs were similar between the INH prophylaxis and placebo groups [63M].

LINEZOLID

Linezolid (LNZ) is considered a second-line agent for the treatment of MDR-TB and XDR-TB. First of the oxazolidinones, it works as a protein synthesis inhibitor by binding to the 50S ribosomal subunit; thus, inhibiting the formation of the 70S initiation complex. LNZ is well-known for causing not only hematological AEs (e.g., bone marrow suppression in the forms of anemia, leukopenia, thrombocytopenia), but also neurological AEs (peripheral and optic neuropathy). Lactic acidosis may also be seen when taken over a longer period of time (months). Because it is a weak monoamine oxidase (MAOI) inhibitor, one should be aware of the risk of serotonin syndrome when taken with other serotonergic agents, such as the selective serotonin reuptake inhibitors (SSRIs) and the serotonin–norepinephrine reuptake inhibitors (SNRIs). In the literature, AEs associated with LNZ are thought to be due to use of higher dosages for a longer duration of time [1S, 2S, 43S, 48c, 64A].

- Singh and colleagues are currently in the process of performing a Cochrane Review on the safety (focusing on prevalence and severity of AEs) and efficacy of LNZ as a second-line agent for treatment of MDR-TB and XDR-TB. AE data will focus on all AEs, all serious AEs, AEs leading to discontinuation of anti-TB drugs or dose reduction, and AEs specifically attributed to LNZ (e.g., peripheral and optic neuropathy, anemia, thrombocytopenia, lactic acidosis, and serotonin syndrome) [65M].
- Swaminathan et al. report a case of a 15-year-old male from Tajikistan with type 1 DM, recently diagnosed with MDR-TB, who presented with peripheral neuropathy with the use of long-term linezolid (LNZ) treatment. His peripheral neuropathy consisted of paresthesia and mild intermittent pain in both feet. The patient had been diagnosed with type 1 DM 5 years prior and was being treated with insulin; however, his blood glucose monitoring was inconsistent and type 1 DM uncontrolled. His MDR-TB strain was found to be resistant to INH, RMP, STR, kanamycin, capreomycin, and amikacin; with susceptibilities to PZA, EMB, and ofloxacin. The patient was started on a nine-drug regimen including PZA, EMB, capreomycin, moxifloxacin, prothionamide, CS, LNZ, amoxicillin/clavulanate, and CFZ due to multi-drug resistance. The patient was originally started on LNZ 600 mg by mouth once daily, but then reduced to LNZ 450 mg by mouth once daily at the 6 month mark as the patient was showing good clinical and bacteriological response. It is unclear whether the source of the neuropathy was due to the LNZ or uncontrolled type 1 DM in this patient. The patient was also on two other medications which may have contributed to the peripheral neuropathy (CS and prothionamide), but is considered rare. After discontinuation of LNZ, the patient showed signs of clinical improvement with nerve conduction studies showing significant improvement in neuropathy. The patient was declared cured following 21 months of treatment. The authors emphasized the importance of more frequent follow-up in pediatric, diabetic patients who will be on long-term LNZ therapy for TB in terms of AE and blood glucose monitoring. This same patient also showed signs of transient thrombocytopenia at the 6 month mark, which resolved without a dose reduction [64A].
- Yi et al. evaluated the use of LNZ in 26 MDR-TB/XDR-TB patients in a single-center retrospective clinical analysis in Japan. LNZ dose was reduced or LNZ withdrawn altogether in 11 patients due to AEs, which included myelosuppression in 10 patients (6 cases of anemia, 7 cases of leukopenia, and 2 cases of thrombocytopenia) and severe anorexia in 2 patients. Neuropathy was not observed. A student's *t*-test showed that there was no difference in total LNZ dosage and incidence of AEs ($t=1.017$, $P=0.320$) [66c].

MACROLIDES

Macrolides are intracellular antibacterial agents that inhibit protein synthesis by binding to the 50S ribosomal subunit of susceptible bacteria. Clarithromycin is contraindicated in patients with cholestatic jaundice or hepatic dysfunction from prior clarithromycin use. Drug interactions are of significant concern with clarithromycin. Both clarithromycin and azithromycin have been associated with hepatotoxicity as well as QTc prolongation. According to international guidelines, these agents have an unclear role in the treatment of MDR-TB [42S]. In the 2016 update the guidelines no longer recommend macrolides for treatment of MDR-TB [2S].

- Van der Paardt and colleagues conducted a retrospective multi-site chart review to evaluate the safety and tolerability of clarithromycin in patients with MDR-TB. Thirty-nine patients were evaluated with a median ATT duration of 548 days (323–608 days). Overall, nine patients reported nausea, vomiting, and diarrhea. This led to discontinuation of clarithromycin in one patient though the correlation was not very strong with a Naranjo score of one. Another patient experienced a significant increase in their ALT prompting a dose reduction of the clarithromycin. A QTc elevation above 450 milliseconds (ms) was only noted in one patient who was also receiving a FQ [67r].

PYRAZINAMIDE

Pyrazinamide (PZA), a pyrazine analogue of nicotinamide, is considered a first-line agent in the treatment of TB. Despite original approval in 1971 by the FDA for treatment of TB, the exact mechanism of action of PZA remains unknown. Common AEs attributed to PZA include nausea, vomiting, hyperuricemia, arthralgia, anemia, and hepatotoxicity. There has been some evidence that the hepatotoxic effects could be attributed to drug-induced decreases in mitochondrial function, which are dose-related, and could occur throughout therapy [68E, 69S]. Thrombocytopenia, sideroblastic anemia, and increased serum iron concentrations can also occur. While rare, PZA has been associated with fever, porphyria, dysuria, and photosensitivity. PZA is contraindicated in individuals who have shown hypersensitivity reactions to it, individuals with severe liver damage, and individuals with acute gout [1S, 2S, 50S].

- Shamaei and colleagues performed a retrospective study from 2007 to 2010 to investigate the recurrence of drug-induced hepatitis when patients were on therapy with INH, RIF, EMB, and either ofloxacin or PZA. Nineteen patients in the ofloxacin group developed a recurrence of hepatitis whereas there were only four cases in the PZA group ($P=0.803$) [70c].
- A retrospective review of ATT-related serious AEs in children diagnosed with TB between 2008 and 2013 was performed by Li and colleagues. Twenty-one of 599 patients developed serious AEs. Hepatotoxicity was the most common in 11 of 599 patients (1.8%) and attributed to INH, RIF, and PZA. Rash with or without fever occurred in six patients and was considered to be due to the PZA [9C].
- Sheth and colleagues report a case of PZA-induced erythema multiforme in a patient with TB meningitis. A 65-year-old male presented to a dermatology physician complaining of itchy skin lesions on the dorsal aspect of both feet for the previous five days. He was found to have additional lesions spreading toward the legs and buttocks, and on both arms. The lesions were raised, erythematous, well-defined plaques with targetoid lesions bilaterally. On the buttocks and both arms the patient also had petechiae. The patient's medications for TB meningitis were discontinued and slowly reintroduced. When PZA was re-introduced, new lesions developed. The drug was removed to allow the lesions to heal. When PZA was re-introduced for a second time, the lesions were aggravated. The patient was then permanently taken off of PZA and was recovering at time of discharge [71c].
- A systematic review and meta-analysis by Marks and colleagues evaluated treatment of LTBI studies from 1994 to 2014. Pulled from 12 studies, the mean percentage of patients that experienced AEs leading to discontinuation of treatment was 19% (106/558, 95% CI 16%–22%). Rates of AEs were lower in the four studies including children at 5% (13/277). AEs were highest in PZA based regimens and led to the highest rate of discontinuation due to AEs (51%, 95% CI 44%–59%) [72M].

PARA-AMINOSALICYLIC ACID

Para-aminosalicylic acid (PAS) is not currently available in the United States (US), but is still used in some countries. Even though the drug has been in use since the 1940s, the mechanism of action is still unclear. According to product labeling, the mechanism of action of PAS is suspected to be the inhibition of folic acid synthesis by targeting dihydrofolate reductase and/or iron reduction through inhibition of mycobactin [73E]. Both *in vitro* and *in vivo*, PAS has demonstrated to be bacteriostatic against TB and is therefore listed in major guidelines only for combination therapy of MDR-TB [42S, 43S, 74S]. Historically, this drug has been associated with occurrences of drug-induced hepatitis and is contraindicated in renal disease. In addition, the drug is difficult to tolerate with gastrointestinal symptoms being the most common adverse reaction [74S].

POLYPEPTIDES (CAPREOMYCIN)

Capreomycin is an intravenous (IV) and intramuscular (IM) polypeptide antibiotic used as a second-line agent for MDR-TB. Its mechanism of action is not fully elucidated but does involve inhibition of protein synthesis at the ribosomal level. Although structurally different, capreomycin is often linked to AMGs due to similarities in proposed mechanisms of action and similar AEs including ototoxicity, nephrotoxicity, and electrolyte disturbances. Serial liver function tests (LFTs) are also recommended with capreomycin due to potential hepatotoxicity. In addition, capreomycin has been observed to cause leukocytosis and leukopenia. Neuromuscular blockade has occurred with rapid infusion of the IV product [75E, 76S].

- Arnold and colleagues performed a service evaluation cohort study at MDR-TB treatment centers in the UK in order to compare real world AEs experienced by patients taking either amikacin or capreomycin. Please see the above section on amikacin for details [6C].
- Sagwa and colleagues performed a case/non-case disproportionality analysis of individual case safety reports in the World Health Organization's database, VigiBase, between 1968 and 2014 for STR, amikacin, kanamycin, and capreomycin. Please see the above section on amikacin for details [4C].

- A 23-year-old female receiving capreomycin treatment for MDR-TB reported carpopedal spasms and tingling in her hands. Upon checking electrolytes, the patient was found to have hypocalcemia (calcium = 6.98 mg/dL), hyponatremia (sodium = 130 mEq/L), hypomagnesemia (magnesium = 0.5 mg/dL), hypokalemia (potassium = 1.8 mEq/L), hypercalciuria and hypochloremic metabolic alkalosis thought to be related to capreomycin. The patient's symptoms resolved and electrolytes remained within normal limits upon discontinuation of capreomycin and electrolyte supplementation [77A].

RIFAMYCINS

The rifamycins have continued to be a mainstay for multiple indications including active TB where it is most commonly used as a backbone agent, LTBI, and leprosy. There have been numerous attempts in the literature to compare fixed-dose combination regimens as well as shorter durations of treatment with rifamycins. Of the rifamycins, rifampicin (RMP) also known as rifampin (RIF) is the most common agent used in practice. Other agents in the rifamycin class used for TB include rifabutin (RBT) and rifapentine (RPT). As the rifamycins, especially RIF, are potent cytochrome (CYP) 3A4 inducers, one must take into account the multiple drug–drug interactions at play, especially in HIV co-infected patients. Many of the available antiretroviral therapies including protease inhibitors (PIs) and non-nucleoside reverse transcriptase inhibitors (NNRTIs) are dependent on this CYP P450 system for metabolism. Historically, rifamycins have been associated with hepatotoxicity including drug-induced hepatitis, flu-like symptoms, thrombocytopenia, leukopenia, anemia, and hyper-sensitivity reactions such as drug reaction with eosinophilia, and systemic symptoms (DRESS). The most notable AE for patients is typically the red staining of body fluids [78S].

- Chen and colleagues report a case of disseminated intravascular coagulation (DIC) associated with RMP. A 22-year-old Tibetan male who was otherwise healthy was diagnosed with TB and prescribed INH, RMP, EMB, and PZA. Within a week he had nasal hemorrhage and thrombocytopenia (platelet count of 0.4×10^9/L) prompting admission to the hospital. This was followed by hematochezia, hematuria, and purpura within 2 days. He was given fresh frozen plasma, platelets, and transferred hospitals where he appeared to have jaundiced sclera, pallor, purpura, tachycardia, and with diminished breath sounds in the left lower lung. His prothrombin time was 17.8 s, international normalized ratio of 1.52, fibrinogen of 1.13 g/L, D-dimer of 23.45 mg/L, fibrin degradation product 60.4 mg/L, and platelets of 2×10^9/L, indicating DIC thought to be due to his ATT. Other lab values, including LFTs were consistent with liver injury. All ATT was discontinued. EMB, moxifloxacin, and amikacin were started 5 days later. Eight days after admission his platelet counts started to recover. The patient was discharged on INH, EMB, levofloxacin, and STR 4 weeks after admission. Within 8 months of follow-up, there was no recurrence of DIC hemorrhage [79c].

LTBI

- Arguello Perez and colleagues evaluated INH vs RIF vs RPT + INH for LTBI of healthcare workers from 2005 to 2014 through a retrospective single-center review. Please see the above section in INH for details [54C].

Special Populations

- Simkins and colleagues retrospectively evaluated severe hypertension (≥180/110 mmHg) after RPT and INH initiation in 37 renal transplant candidates through chart review. All patients had stage 5 chronic kidney disease (CKD) with 31 of the 37 (91%) on dialysis. Severe hypertension occurred in eight of the patients (22%) with varying onsets (1–12 weeks) [80r].
- Simkins and colleagues also retrospectively evaluated patients treated for LTBI with either 12 weeks of RPT + INH or 9 months of INH. LFTs were only elevated in the INH monotherapy group (6 of 153 patients). Adverse reactions leading to discontinuation of treatment did not differ between the two groups [81c].
- Lung transplant candidates with LTBI were retrospectively reviewed by Guirao-Arrabal and colleagues when the candidate received 3 months of RMP and INH. Please see the above section on INH for details [55c].

TERIZIDONE

Terizidone is a bacteriostatic second-line agent for MDR-TB used internationally, but is not currently approved in the US. It is a combination of two molecules of CS. The mechanism of action includes competitively inhibiting L-alanine racemase and D-alanine ligase to inhibit cell wall synthesis. Known AEs are considered to be fewer than with CS, but still include seizures, dizziness, slurred speech, tremors, insomnia, confusion, depression, and suicidal tendency. For doses >1 g per day, there have been reports of hepatotoxicity, congestive heart failure, convulsions, and coma [82r]. Since terizidone is considered to be as efficacious as CS with fewer AEs, it can be used instead of CS when needed [42S, 82r].

THIOAMIDES

Thioamides including ethionamide and prothionamide are considered interchangeable second-line bacteriostatic agents for MDR-TB [43S]. The mechanism of action is not fully understood, but involves inhibiting peptide synthesis. Ethionamide is contraindicated in patients with severe hepatic impairment. Gastrointestinal AEs compose the majority of reported adverse reactions with a reported 50% of patients unable to tolerate 1 g of the drug. Psychotic disturbances have also been reported and rare AEs such as optic neuritis, diplopia, blurred vision, a pellagra-like syndrome, and peripheral neuritis have been reported historically. Ethionamide is also known to transiently increase LFTs and cause hepatitis. Product labeling also recommends to avoid taking with alcohol due to a possible psychotic reaction [83S]. Prothionamide has been historically associated with menstrual disturbances, and it is recommended that blood sugar levels should be measured in patients with DM [84r]. Due to gastrointestinal symptoms, ethionamide and prothionamide are usually avoided in combination with PAS [42S].

- Banerjee and colleagues report a case of a 22-year-old male with MDR-TB who developed gynecomastia on EMB, PZA, kanamycin, levofloxacin, ethionamide, CS, and pyridoxine. The patient had previously been treated for pulmonary TB twice. His sputum culture grew *Mycobacterium tuberculosis* resistant to INH and RMP. Based on his weight of 42 kg, he was put on EMB 800 mg, PZA 1250 mg, kanamycin 0.5 g IM, levofloxacin 750 mg, ethionamide 500 mg, CS 500 mg, and pyridoxine 100 mg all daily. After 2 months the patient complained of enlargement of the left breast. There was evidence of glandular tissue hyperplasia on ultrasonography. The patient denied other medication use and an endocrinologist was consulted. Ethionamide was a suspected cause for the gynecomastia so the patient care team replaced the ethionamide with PAS. After a month, the swelling and pain resolved. Ethionamide was then reintroduced and the same breast swelling and pain recurred within 3 weeks. The patient care team then changed the ethionamide to PAS and again the swelling and pain of the left breast subsided. The patient was diagnosed with ethionamide-induced gynecomastia and was later able to become smear negative for his TB [85c].

THIOACETAZONE

Thioacetazone is bacteriostatic against TB with a relatively unknown mechanism of action. There have been some reports that suggest one of the targets of thioacetazone is cyclopropane mycolic acid synthases [86E]. However, thioacetazone has fallen out of favor due to severe AEs including nausea, vomiting, diarrhea, anemia, thrombocytopenia, abdominal pain, agranulocytosis, ataxia, blurred vision, vertigo, seizures, deafness, and cerebral edema. Stevens–Johnson syndrome (SJS) and toxic epidermal necrolysis have also been historically reported to a significant enough degree that the drug has been discontinued and is contraindicated in HIV patients [87r]. According to international guidelines, this agent has an unclear role in the treatment of MDR-TB [42S].

Additional case studies can be found in these reviews [88R, 89R].

Acknowledgements

The authors are grateful to Brook Amen, the research and education librarian at UNTHSC, for her continued support throughout the project, especially in reviewing databases for background literature.

References

[1] American Thoracic Society. Centers for disease control and prevention, and infectious diseases society of America. Treatment of tuberculosis. MMWR Recomm Rep. 2003;52(RR-11):1–77 [S].

[2] WHO treatment guidelines for drug-resistant tuberculosis, 2016. update. WHO guidelines approved by the guidelines review committee. 2016:Geneva: World Health Organization; 2016 [S].

[3] Craig WA. Optimizing aminoglycoside use. Crit Care Clin. 2011;27(1):107–21 [R].

[4] Sagwa EL, Souverein PC, Ribeiro I, et al. Differences in VigiBase® reporting of aminoglycoside and capreomycin-suspected ototoxicity during tuberculosis treatment. Pharmacoepidemiol Drug Saf. 2017;26(1):1–8 [C].

[5] Jan F, Hassan M, Muhammad N, et al. Electrolytes imbalance caused by amikacin in patients receiving multi drug resistance-tuberculosis treatment at Hazara region Kpk, Pakistan. Tuberk Toraks. 2017;65(3):193–201 [C].

[6] Arnold A, Cooke GS, Kon OM, et al. Adverse effects and choice between the injectable agents amikacin and capreomycin in multidrug-resistant tuberculosis. Antimicrob Agents Chemother. 2017;61(9):e02586-16 [C].

[7] Vasconcelos KA, Frota SMMC, Ruffino-Netto A, et al. The importance of audiometric monitoring in patients with

multidrug-resistant tuberculosis. Rev Soc Bras Med Trop. 2017;50(5):646–51 [C].
[8] Quenard F, Fournier PE, Drancourt M, et al. Role of second-line injectable antituberculosis drugs in the treatment of MDR/XDR tuberculosis. Int J Antimicrob Agents. 2017;50(2):252–4 [c].
[9] Li Y, Zhu Y, Zhong Q, et al. Serious adverse reactions from anti-tuberculosis drugs among 500 children hospitalized for tuberculosis. Pediatr Infect Dis J. 2017;36(8):720–5 [C].
[10] van Altena R, Dijkstra JA, van der Meer ME, et al. Reduced chance of hearing loss associated with therapeutic drug monitoring of aminoglycosides in the treatment of multidrug-resistant tuberculosis. Antimicrob Agents Chemother. 2017;61(3):e01400-16 [c].
[11] Sin JH, Elshaboury RH, Hurtado RM, et al. Therapeutic drug monitoring antitubercular agents for disseminated *Mycobacterium tuberculosis* during intermittent haemodialysis and continuous venovenous haemofiltration. J Clin Pharm Ther. 2018;43:291–5 [A].
[12] van der Laan LE, Garcia-Prats AJ, Schaaf HS, et al. Pharmacokinetics and drug-drug interactions of lopinavir/ritonavir administered with first and second-line antituberculosis drugs in HIV-infected children treated for multidrug-resistant tuberculosis. Antimicrob Agents Chemother. 2018;62:e00420-17 [c].
[13] Yadav S, Rawal G, Baxi M. Bedaquiline: a novel antitubercular agent for the treatment of multidrug-resistant tuberculosis. J Clin Diagn Res. 2016;10(8):FM01–2 [R].
[14] Brigden G, Hewison C, Varaine F. New developments in the treatment of drug-resistant tuberculosis: clinical utility of bedaquline and delamanid. Infect Drug Resist. 2015;8:367–78 [R].
[15] Guglielmetti L, Jaspard M, Le Dû D, et al. Long-term outcome and safety of prolonged bedaquiline treatment for multidrug-resistant tuberculosis. Eur Respir J. 2017;49(3) (pii: 1601799) [c].
[16] Achar J, Hewison C, Cavalheiro AP, et al. Off-label use of bedaquiline in children and adolescents with multidrug-resistant tuberculosis. Emerg Infect Dis. 2017;23(10):1711–3 [c].
[17] Migliori GB, Pontali E, Sotgiu G, et al. Combined use of delamanid and bedaquiline to treat multidrug-resistant and extensively drug-resistant tuberculosis: a systematic review. Int J Mol Sci. 2017;18(2):341 [A].
[18] Healan AM, Griffiss JM, Proskin HM, et al. Impact of rifabutin or rifampin on bedaquiline safety, tolerability, and pharmacokinetics assessed in a randomized clinical trial with healthy adult volunteers. Antimicrob Agents Chemother. 2018;62(1):e00855-17 [c].
[19] Zhanel GG, Wiebe R, Dilay L, et al. Comparative review of the carbapenems. Drugs. 2007;67(7):1027–52 [R].
[20] van Rijn SP, Zuur MA, van Altena R, et al. Pharmacokinetic modeling and limited sampling strategies based on healthy volunteers for monitoring of ertapenem in patients with multidrug resistant tuberculosis. Antimicrob Agents Chemother. 2017;61(4):e01783-16 [c].
[21] Cholo MC, Steel HC, Fourie PB, et al. Clofazimine: current status and future prospects. J Antimicrob Chemother. 2012;67(2):290–8 [R].
[22] Ramachandran G, Swaminathan S. Safety and tolerability profile of second-line anti-tuberculosis medications. Drug Saf. 2015;38(3):253–69 [R].
[23] Martiniano SL, Wagner BD, Levin A, et al. Safety and effectiveness of clofazimine for primary and refractory nontuberculous mycobacterial infection. Chest. 2017;152(4):800–9 [C].
[24] Dalcolmo M, Gayoso R, Sotgiu G, et al. Effectiveness and safety of clofazimine in multidrug-resistant tuberculosis: a nationwide report from Brazil. Eur Respir J. 2017;49:1602445 [C].
[25] Shah SS, Goregaonkar AA, Goreagaonkar AB. Extensively drug-resistant tuberculosis of the lumbar spine in a six-year-old child: a case report. J Orthop Case Rep. 2017;7(2):40–3 [A].
[26] Tiberi S, Scardigli A, Centis R, et al. Classifying new anti-tuberculosis drugs: rationale and future perspectives. Int J Infect Dis. 2017;56:181–4 [R].
[27] Kim J, Park S, Jung CM, et al. A case of cycloserine-induced lichenoid drug eruption supported by the lymphocyte transformation test. Allergy Asthma Immunol Res. 2017;9(3):281–4 [A].
[28] Chang MJ, Jin B, Chae JW, et al. Population pharmacokinetics of moxifloxacin, cycloserine, p-aminosalicylic acid and kanamycin for the treatment of multi-drug-resistant tuberculosis. Int J Antimicrob Agents. 2017;49(60):677–87 [c].
[29] Kar HK, Gupta R. Treatment of leprosy. Clin Dermatol. 2015;33(1):55–65 [R].
[30] Guragain S, Upadhayay N, Bhattarai BM. Adverse reactions in leprosy patients who underwent dapsone multidrug therapy: a retrospective study. Clin Pharmacol. 2017;9:73–8 [c].
[31] Estève C, Samson M, Guilhem A, et al. Efficacy and safety of dapsone as second line therapy for adult immune thrombocytopenia: a retrospective study of 42 patients. PLoS One. 2017;12(10):e0187296 [c].
[32] Bitta MA, Kariuki SM, Mwita C, et al. Antimalarial drugs and the prevalence of mental and neurological manifestations: a systematic review and meta-analysis. Wellcome Open Res. 2017;2:13 [C].
[33] Amjad W, Sandhu R, Aung S. A case of cyanosis: dapsone induced methemoglobinemia and coexisting pulmonary embolism. Am J Ther. 2017;e493-e495 [A].
[34] Nazir HF, Elshinawy M, AlRawas A, et al. Efficacy and safety of dapsone versus trimethoprim/sulfamethoxazole for Pneumocystis jiroveci prophylaxis in children with acute lymphoblastic leukemia with a background of ethnic neutropenia. J Pediatr Hematol Oncol. 2017;39(3):203–8 [c].
[35] Lockwood DN, Darlong J, Govindharaj P, et al. AZALEP a randomized controlled trial of azathioprine to treat leprosy nerve damage and type 1 reactions in India: main findings. PLoS Negl Trop Dis. 2017;11(3):e0005348 [C].
[36] Maddala RR, Ghorpade A, Adulkar S, et al. Anatopic response: double sparing phenomenon in a patient with dapsone hypersensitivity syndrome. Indian J Dermatol Venereol Leprol. 2017;83(2):241–3 [A].
[37] Marchese M, Leinung M, Shawa H, et al. Drug-induced hypersensitivity reaction: a case of simultaneous thyroiditis and fulminant type 1 diabetes. Avicenna J Med. 2017;7(2):67–70 [A].
[38] Wang N, Parimi L, Liu H, et al. A review of dapsone hypersensitivity syndrome among Chinese patients with an emphasis on preventing adverse drug reactions with genetic testing. Am J Trop Med Hyg. 2017;96(5):1014–8 [M].
[39] Tempark T, Satapornpong P, Rerknimitr P, et al. Dapsone-induced severe cutaneous adverse drug reactions are strongly linked with HLA-B*13:01 allele in the Thai population. Pharmacogenet Genomics. 2017;27(12):429–37 [c].
[40] Devarbhavi H, Patil M, Reddy VV, et al. Drug-induced acute liver failure in children and adults: results of a single center study of 128 patients. Liver Int. 2018;38:1322–9 [C].
[41] Devarbhavi H, Raj S, Joseph T, et al. Features and treatment of dapsone-induced hepatitis, based on analysis of 44 cases and literature review. Clin Gastroenterol Hepatol. 2017;15(11):1805–7 [c].
[42] Treatment of Tuberculosis: Guidelines. WHO guidelines approved by the Guideline Review Committee. 4th ed. Geneva: World Health Organization; 2010 [S].
[43] Nahid P, Dorman SE, Alipanah N, et al. Official American thoracic society/centers for disease control and prevention/infectious diseases society of America clinical practice guidelines: treatment of drug-susceptible tuberculosis. Clin Infect Dis. 2016;63(7):e147–95 [S].

[44] Gallagher JC, MacDougall C. Antibiotics simplified. 3rd ed. Burlington, MA: Jones & Bartlett Learning; 2014 [R].
[45] Song W, Si S. The rare ethambutol-induced optic neuropathy: a case-report and literature review. Medicine (Baltimore). 2017;96(2): e5889 [A].
[46] Silva VDD, Mello FCQ, Figueiredo SCA. Estimated rates of recurrence, cure, and treatment abandonment in patients with pulmonary tuberculosis treated with a four-drug fixed-dose combination regimen at a tertiary health care facility in the city of Rio de Janeiro, Brazil. J Bras Pneumol. 2017;43(2):113–20 [C].
[47] U.S. Food and Drug Administration; 2016 [cited 2018 Jan 12]. Fluoroquinolone antibacterial drugs: drug safety communication—FDA advises restricting use for certain uncomplicated infections [Internet]. Available from: https://www.fda.gov/Drugs/DrugSafety/ucm500143.htm [S].
[48] Bartoletti M, Martelli G, Tedeschi S, et al. Liver transplantation is associated with good clinical outcome in patients with active tuberculosis and acute liver failure due to anti-tubercular treatment. Transpl Infect Dis. 2017;19(2):e12658 [c].
[49] Kunder SK, Avinash A, Nayak V, et al. A rare instance of levofloxacin induced myoclonus. J Clin Diagn Res. 2017;11(7): FD01–2 [A].
[50] U.S. Food and Drug Administration [Internet]. Silver Spring, MD: U.S. Food and Drug Administration; 12 Jan 2018 [12 Jan 2018]. Available from: http://www.fda.gov/ [S].
[51] Centers for Disease Control and Prevention (CDC). [Internet]. Atlanta, GA: Treatment regimens for latent TB infection (LTBI), 6 May 2016. [12 Jan 2018]; Available from: https://www.cdc.gov/tb/topic/treatment/ltbi.htm [S].
[52] Al-Shaer MH, Mansour H, Elewa H, et al. Treatment of outcomes of fixed-dose combination versus separate tablet regimens in pulmonary tuberculosis patients with or without diabetes in Qatar. BMC Infect Dis. 2017;17(1):118 [C].
[53] Sekaggya-Wiltshire C, von Braun A, Scherrer AU, et al. Anti-TB drug concentrations and drug-associated toxicities among TB/HIV-coinfected patients. J Antimicrob Chemother. 2017;72(4):1172–7 [C].
[54] Arguello Perez E, Seo SK, Schneider WJ, et al. Management of latent tuberculosis infection among healthcare workers: 10-year experience at a single center. Clin Infect Dis. 2017;65(12):2105–11 [C].
[55] Guirao-Arrabal E, Santos F, Redel J, et al. Efficacy and safety of short-term treatment with isoniazid and rifampicin for latent tuberculosis infection in lung transplant candidates, Clin Transplant. 2017;31(3): e12901. https://doi.org/10.1111/ctr.12901 [c].
[56] Zenner D, Beer N, Harris RJ, et al. Treatment of latent tuberculosis infection: an updated network meta-analysis. Ann Intern. 2017;167(4):248–55 [M].
[57] Al-Mukhaini SM, Al-Eid H, Alduraibi F, et al. Mycobacterium tuberculosis in solid organ transplantation: incidence before and after expanded isoniazid prophylaxis. Ann Saudi Med. 2017;37(2):138–43 [C].
[58] Ayele AA, Atnafie SA, Balcha DD, et al. Self-reported adherence and associated factors to isoniazid preventive therapy for latent tuberculosis among people living with HIV/AIDS at health centers in Gondar town, North West Ethiopia. Patient Prefer Adherence. 2017;11:743–9 [C].
[59] Badje A, Moh R, Gabillard D, et al. Effect of isoniazid preventive therapy on risk of death in west African, HIV-infected adults with high CD4 cell counts: long-term follow-up of the temprano ANRS 12136 trial. Lancet Glob Health. 2017;5(11): e1080–9 [C].
[60] Hakim J, Musiime V, Szubert AJ, et al. Enhanced prophylaxis plus antiretroviral therapy for advanced HIV infection in Africa. N Engl J Med. 2017;377(3):233–45 [C].
[61] Maharaj B, Gengiah TN, Yende-Zuma N, et al. Implementing isoniazid preventive therapy in a tuberculosis treatment-experienced cohort on ART. Int J Tuberc Lung Dis. 2017;21(5):537–43 [C].
[62] Okwara FN, Oyore JP, Were FN, et al. Correleates of isoniazid preventive therapy failure in child household contacts with infectious tuberculosis in high burden settings in Nairobi, Kenya—a cohort study. BMC Infect Dis. 2017;17(1):623 [C].
[63] Zunza M, Gray DM, Young T, et al. Isoniazid for preventing tuberculosis in HIV-infected children. Cochrane Database Syst Rev. 2017;8::CD006418 [M].
[64] Swaminathan A, du Cros P, Seddon JA, et al. Peripheral neuropathy in a diabetic child treated with linezolid for multidrug-resistant tuberculosis: a case report and review of the literature. BMC Infect Dis. 2017;17(1):417 [A].
[65] Singh B, Cocker D, Ryan H, et al. Linezolid for drug-resistant tuberculosis. Cochrane Database Syst Rev. 2017;11::CD012836. https://doi.org/10.1002/14651858.CD012836 [M].
[66] Yi L, Yoshiyama T, Okumura M, et al. Linezolid as a potentially effective drug for the treatment of multidrug-resistant tuberculosis in Japan. Jpn J Infect Dis. 2017;70(1):96–9 [c].
[67] Van der Paardt AL, Akkerman OW, Gualano G, et al. Safety and tolerability of clarithromycin in the treatment of multidrug-resistant tuberculosis, Eur Respir J. 2017;49 (pii: 1601612). https://doi.org/10.1183/13993003.01612-2016 [r].
[68] Elmorsy E, Attalla SM, Fikry E, et al. Adverse effects of anti-tuberculosis drugs on HepG2 cell bioenergetics. Hum Exp Toxicol. 2016;36(6):616–25 [E].
[69] Pyrazinamide tablets [package insert]. VersaPharm Incorporated; 2012 [S].
[70] Shamaei M, Mirsaeidi M, Baghaei P, et al. Recurrent drug-induced hepatitis in the tuberculosis-comparison of two drug regimens. Am J Ther. 2017;24:e144–9 [c].
[71] Sheth HJ, Shah AN, Malhotra SB, et al. An unusual case of pyrazinamide induced erythema multiforme in a patient of tuberculosis meningitis: a case report. Natl J Intergr Res Med. 2017;8:153–5 [c].
[72] Marks SM, Mase SR, Morris SB. Systemic review, meta-analysis, and cost-effectiveness of treatment of latent tuberculosis to reduce progression to multidrug-resistant tuberculosis. Clin Infect Dis. 2017;64:1670–7 [M].
[73] Zheng J, Rubin EJ, Bifani P, et al. Para-aminosalicylic acid is a prodrug targeting dihydrofolate reductase in mycobacterium tuberculosis. J Biol Chem. 2013;288:23447–56 [E].
[74] PASER-aminosalicylic acid granule, delayed release [package insert]. Princeton, NJ: Jacobus Pharmaceutical Company, Inc; 1996 [S].
[75] Capreomycin. Tuberculosis (Edinb). 2008;88:89–91 (PMID: 18486038) [E].
[76] Capastat sulfate [package insert]. Indianapolis, IN: Eli Lilly Company; 2008 [S].
[77] Sharma P, Sahay RN. Unusual complication of multidrug resistant tuberculosis. Case Rep Nephrol. 2017;2017:6835813 [A].
[78] Rifadin [package insert]. Bridgewater, NJ: Sanofi Aventis; 2017 [S].
[79] Chen G, He JQ. Rifampicin-induced disseminated intravascular coagulation in pulmonary tuberculosis treatment: a case report and literature review. Medicine (Baltimore). 2017;96: e6135 [c].
[80] Simkins J, Morris MI, Abbo LM, et al. Severe hypertension after initiation of rifapentine/isoniazid for latent tuberculosis in renal transplant candidates. Transpl Int. 2017;30:108–9 [r].
[81] Simkins J, Abbo LM, Camargo JF, et al. Twelve-week rifapentine plus isoniazid versus 9-month isoniazid for the treatment of latent tuberculosis in renal transplant candidates. Transplantation. 2017;101:1468–72 [c].

[82] Vora A. Terizidone. J Assoc Physicians India. 2010;58:267–8 [r].
[83] Trecator [package insert]. Philadelphia, PA: Wyeth Pharmaceutical, Inc; 2006 [S].
[84] Prothionamide. TB Online. http://www.tbonline.info/posts/2011/8/24/prothionamide/. Published 2011. Accessed December 2017 [r].
[85] Banerjee S, Rupam T, Salim Mallick M, et al. Ethionamide-induced gynecomastia: a rare case report. Egypt J Bronchol. 2017;11:70–3 [c].
[86] Alahari A, Trivelli X, Guérardel Y, et al. Thiacetazone, an antitubercular drug that inhibits cyclopropanation of cell wall mycolic acids in mycobacteria. PLoS One. 2007;2:e1343 [E].
[87] Isoniazid and Thiacetazone (Systemic). Drugs.com. https://www.drugs.com/mmx/isoniazid-and-thiacetazone.html. Published 1995. Accessed December 2017 [r].
[88] Smith M, Accinelli A, Tejada FR, et al. Chapter 28 - Drugs used in tuberculosis and leprosy. In: Ray SD, editor. Side Effects of Drugs-Annual: A worldwide yearly survey of new data in adverse drug reactions. Elsevier. 2016;38:283–93 [R].
[89] Ramanathan MR, Sanders JM. Chapter 26 - Drugs used in TB and leprosy. In: Ray SD, editor. Side Effects of Drugs- Annual: A worldwide yearly survey of new data in adverse drug reactions. Elsevier. 2017;39:283–93 [R].

CHAPTER

30

Antihelminthic Drugs

Igho J. Onakpoya[1]

Nuffield Department of Primary Care Health Sciences, Oxford, United Kingdom

[1]Corresponding author: igho.onakpoya@phc.ox.ac.uk

ALBENDAZOLE

Drug Combination Studies

In a randomized, single-blinded, non-inferiority trial, the safety of tribendimidine plus ivermectin was compared with tribendimidine alone, tribendimidine plus oxantel pamoate, and albendazole plus oxantel pamoate in the treatment of hookworm and concomitant soil-transmitted helminth infections in Tanzania and Côte d'Ivoire [1C]. In total 636 hookworm-positive participants aged 15–18 years were included. Most of the adverse events were mild in intensity. Moderate adverse events reported included vomiting ($n=1$; tribendimidine monotherapy); one headache and two vertigo ($n=3$; tribendimidine plus oxantel pamoate); and diarrhea and vomiting ($n=1$; albendazole plus oxantel pamoate). No serious adverse events were reported.

IVERMECTIN

Case Series

In a study of 1668 adverse reaction reports attributed to ivermectin identified from the WHO Vigibase (a computerized spontaneous reporting system), the occurrence of neurological adverse reactions attributed to ivermectin beyond its indication for onchocerciasis was examined [2A]. The most commonly reported adverse reactions were pruritus (25.3%), headache (13.9%), and dizziness (7.5%). A total of 426 neurological adverse reactions were reported, out of which 156 (36.6%) were classified as serious. Although the authors found that the likelihood of neurologic adverse events is rare, they suggested that consideration of individual risk factors as part of therapeutic decision-making could minimize the risk of harms. They also advocated further research into the potential for drug–drug interactions and polymorphisms in the mdr-1 gene.

Observational Study

The safety of ivermectin during mass drug administration (two oral doses of 200 μg/kg each) for treatment of Strongyloides seroprevalence has been examined in a survey of 1013 participants in a remote Australian Aboriginal community [3c]. No adverse events were reported by the surveyed participants based on self-reported questionnaire review 24–72 hours post-drug administration.

Comparative Studies

The pathophysiology of post-treatment adverse reactions after single-dose therapy with either diethylcarbamazine or ivermectin has been examined in randomized trial of 12 subjects with Loa loa infection [4c]. All 12 subjects experienced adverse events that were mild to moderate in intensity. Headache and pruritus were the most commonly reported adverse events observed. The adverse events peaked earlier in subjects who received diethylcarbamazine compared to those who received ivermectin (8 hours versus 24 hours); this coincided with a trend towards quicker and complete clearance of the microfilarial worms in the diethylcarbamazine group. However, the overall pattern of hematologic and immunologic changes suggested that the post-treatment reactions to both drugs shared a common pathophysiology.

Combination Studies

See Section 'Albendazole'.

Side Effects of Drugs Annual, Volume 40
ISSN: 0378-6080
https://doi.org/10.1016/bs.seda.2018.06.010

Nervous System

A case of encephalopathy attributed to ivermectin has been reported in Senegal [5A]

- A 35-year-old patient living in Mbirkilane (Kaffrine region of Senegal) with no relevant past medical or travel history, nor history of onchocerciasis was admitted in the neurology department of Fann National Teaching Hospital. He received two doses of ivermectin (150 mcg/kg body weight). The second ivermectin dose was taken by the patient in error because he did not notify the drug distributor about the previous dose of ivermectin. Two days after receiving the second dose, he presented with bilateral eye pains, blurred vision and moderate to severe headaches associated with vomiting, followed few hours later by generalized tonic–clonic seizures. He was first admitted and treated in the regional hospital without improvement. The onset of altered consciousness resulted in his transfer to the neurology department. Clinical examination revealed headaches, tonic–clonic seizures, obtundation (Glasgow coma score of 10/15), bilateral visual loss and unreactive mydriasis and bradycardia (51 beats per minute). Funduscopy revealed bilateral papilledema with retinal haemorrhages. The patient was urgently admitted to the intensive care unit. Neuroimaging showed multiple hypersignal lesions in T2 weighted and FLAIR (fluid-attenuated inversion recovery) brain MRI, located in the periventricular white matter and semi-oval centers. The visual evoked potentials showed bilateral signs of optic nerve axonal loss and demyelination. The results of cerebrospinal fluid analysis were: albumin 0.49 g/L (normal <0.5 g/L), glucose 0.69 g/L (normal $\geq$ half of concomitant fasting blood sugar that was 1.22 g/L), white blood cells <2 cells/mm^3 (normal <5 cells/mm^3). Laboratory tests for bacteria, viruses, fungi and parasites remained negative. Blood sample was collected the next morning for microfilariae testing, but anti-Acanthocheilonema vitae Immunoglobulin G results were negatives. The liver function test was impaired with SGOT at 396.8 IU/L (normal <40 IU/L), SGPT at 95.5 IU/L (normal <41 IU/L) and gamma-glutamyltransferase at 72 IU/L (normal: 12–64 IU/L). Blood titres for HBs antigen, anti-HCV antibodies and the alpha-fetoprotein were negative and abdominal ultrasound was normal. CRP was 14.74 mg/L and leukocytosis was 14660/mm^3 (normal: 4000–10000/mm^3) predominantly neutrophilic. A diagnosis of post-ivermectin encephalopathy was made and the patient was treated with corticosteroids: betamethasone 0.1 mg/kg/day then prednisone 1 mg/kg/day; calcium and potassium supplementation; omeprazole; analgesics; and hydration with normal saline. The evolution was marked by the return to consciousness 2 days after admission, the cessation of headaches, vomiting and seizures during the first week of hospitalization. However, the bilateral blindness persisted despite appropriate ophthalmologic treatment. The patient was subsequently discharged from the hospital and followed up in collaboration with ophthalmologists.

Immunologic

A case of ivermectin-induced Stevens–Johnson syndrome has been reported [6A]

- A 38-year-old adult Cameroonian male presented to our health facility with facial rash, painful oral sores, black eschars on lips and red tearing eyes 3 days following ingestion of ivermectin received during a nationwide anti-filarial campaign. He had no known chronic illness, no known allergies and was not on any medications prior to the campaign. Physical examination revealed discharging erythematous eyes, crusted and blister-like lesions with cracks on his lips and oral mucosa. His laboratory tests were unremarkable but for a positive Human Immunodeficiency Virus (HIV) test. A diagnosis of Ivermectin induced Stevens–Johnson syndrome in a newly diagnosed HIV patient was made. The patient was managed with supportive therapy and the evolution thereafter was favourable.

Toxic epidermal necrolysis secondary to ivermectin therapy has been reported [7A]

- A 45-year-old male presented with worsening rash involving his entire body 3 days after taking 15 mg of ivermectin which was prescribed in the emergency department for suspected body louse. On examination, he had oral blisters, conjunctivitis and a generalized erythematous maculopapular rash involving his entire body with denudation of the skin on his back. His HIV, hepatitis (A, B, C), urine toxicology, autoimmune screen and rapid plasma regain were all negative. Skin biopsy showed acute vacuolar interface dermatitis with many cytoids and epidermal necrosis consistent with TEN. He was managed with supportive care and made an unremarkable recovery.

LEVAMISOLE

Cases of adverse reactions attributed to levamisole-contaminated cocaine continue to accumulate in the published literature. In an institutional database review of a cases series of three patients with levamisole-induced necrosis syndrome (LINES)—a condition characterized by neutropenia, vasculitis, and subsequent progression to skin necrosis was described [8A]. The authors concluded that case presentations suggestive of LINES are expected to increase. The authors of another review showed that apart from agranulocytosis and vasculitis, levamisole-cocaine autoimmune syndromes affect other organs as well [9A]. They postulated that neutrophils appear to be major players in the development of cocaine/levamisole-associated autoimmune syndrome (CLAAS).

Nervous System

A case of levamisole-associated multifocal inflammatory leukoencephalopathy in a cocaine user has been confirmed [10A]

- *A 63-year-old woman with a history of asthma and hepatitis C infection presented with 3 days of progressive confusion, fevers and headache. Her family noted a change in her cognition over the preceding few days with disorientation, word finding problems, and difficulty carrying out simple tasks such as dressing. On initial assessment she was irritable with impaired memory recall, attention, naming, and motor planning but otherwise lacked focal neurologic deficits. A basic laboratory workup was unrevealing except for urine toxicology which was positive for cocaine. The patient denied any history of substance use; however, a friend reported that both she and the patient had been snorting cocaine together roughly once a week for the prior month. Non-contrast computerized tomography (CT) scan was negative. Over the next 48 hours, the patient became increasingly more obtunded and progressed to severe stupor, only responsive to noxious stimuli, with intact brainstem reflexes and spastic quadraparesis. A magnetic resonance image (MRI) of the brain was then obtained revealing innumerable FLAIR hyperintensities within the supratentorial subcortical white matter, many of which were ovoid shaped in a perivenular distribution. Most of these lesions demonstrated reduced diffusion, with corresponding low apparent diffusion coefficient (ADC) signal, as well as active enhancement with an incomplete ring pattern. An MRI of the spine was normal as was CT angiography. Her general examination revealed no new cutaneous lesions and her basic laboratory testing demonstrated a normal white blood cell count and differential. Cerebrospinal fluid (CSF) analysis demonstrated a lymphocytic pleocytosis with an elevated protein but no unique oligoclonal bands nor elevated IgG index. An infectious and autoimmune workup including aquaporin-4 antibody, Sjogren's antibodies, anti-nuclear antibody, anti-neutrophil cytoplasmic antibodies, antiphospholipid antibodies, HIV, RPR as well as HSV-1 PCR and VZV PCR in the CSF were negative. Overall her presentation was most consistent with a fulminant and aggressive demyelinating process and, given her history of cocaine exposure, levamisole-associated MIL was suspected. Liquid chromatography–tandem mass spectrometry analysis of her urine sample from admission returned positive for levamisole. Given her precipitous decline, she was initiated on high-dose intravenous methylprednisolone for 5 days followed by plasmapheresis without appreciable improvement. An MRI brain was obtained on hospital day 12 which revealed interval progression of disease with development of numerous new bilateral periventricular and subcortical FLAIR hyperintense lesions, many of which now appeared confluent, with continued reduced diffusion and enhancement. Due to the patient's lack of clinical or radiologic improvement despite aggressive immunosuppression, she was treated with intravenous cyclophosphamide (750 mg/m^2). She remained hospitalized for nearly 2 months and received a second dose of cyclophosphamide (1000 mg/m^2) 1 month after her first treatment. A percutaneous endoscopic gastrostomy tube was placed, and the patient was discharged to a nursing facility. Since discharge she has undergone multiple interval MRIs demonstrating resolution of enhancement and restricted diffusion, though with persistent FLAIR hyperintensities, marked cerebral volume loss, and atrophy of the bilateral caudate and corpus striatum. Ten months following her initial presentation, she has remained in a minimally conscious state with mutism, spasticity with hyperreflexia, and slight withdrawal to noxious stimuli in all extremities.*

Sensory System

A case of eye necrosis with secondary cicatricial ectropion attributed to levamisole-associated vasculitis has been reported [11A]

- *A 56-year-old female presented with eyelid necrosis secondary to systemic levamisole-induced vasculitis. Skin biopsy revealed necrotic epidermis with small-vessel thrombosis, fibrinoid reaction, and neutrophilic infiltration of vessel walls in the dermis with +pANCA. She was treated with plasmapheresis and steroids. Six months later, she developed severe, symptomatic cicatricial ectropion with marked anterior lamellar shortage and middle lamellar contracture. Scar release in the middle lamellar plane with lateral tarsal strip procedures was performed, with full-thickness skin grafts from the upper eyelids. She remained fully epithelialized postoperatively with improvement in symptoms, although she incomplete graft take due to her eyelid necrosis and compromised dermal blood supply.*

Skin

Painful purpura caused by suspected exposure to levamisole-adulterated cocaine has again been described [12A]

- *A 56-year-old man presented with a 2-day history of a painful rash. Examination revealed necrotic retiform purpura over the trunk and extremities. Laboratory testing was notable for leucopenia (3.2 cells/μL) and positive peripheral antineutrophil cytoplasmic antibody (p-ANCA) with antimyeloperoxidase specificity*

(3.5 U, normal < 0.4 U). Skin biopsy showed microvascular thrombosis and fibrinoid necrosis. On questioning, the patient endorsed regular cocaine use. The patient was diagnosed with levamisole-induced vasculitis. He was discharged after workup but was readmitted a week later with wound superinfection requiring antibiotic therapy, surgical debridement and skin grafting. He recovered well postoperatively with complete resolution of his wounds at 2-month follow-up.

Immunologic

A case of levamisole-induced leukocytoclastic vasculitis in a patient with negative serological markers has been reported [13A]

- *A 58-year-old African American man with a history of polysubstance abuse presented 4 days after last cocaine use, with sudden onset of painful pruritic rash and polyarthralgias. The rash initially started on both the upper limbs and then progressed to the lower limbs. He had a similar presentation in the past after cocaine use, and it had resolved spontaneously. There was no history of any known allergies, and the patient was not taking any medications at home. On examination, he was found to have normal vital signs, left fifth proximal interphalangeal joint tenderness and swelling, bilateral knee joint swelling, erythema and tenderness with impaired range of motion on both active and passive moments, and bilateral ankle tenderness on both active and passive range of motion, with no erythema or effusion. The skin exam was significant for centripetally distributed purpuric skin lesions around the arms, thighs, and legs bilaterally. Initially, patient was started on steroids 40 mg once daily, broad-spectrum antibiotics, and NSAIDS. Laboratory workup was positive for elevated CRP and leukopenia with a WBC of 2.3, but no neutropenia. The rest of the workup was negative, including, ANA, ESR, hepatic panel, coagulation profile, RF, HIV, HCV, CCP, anti-dsDNA, c-ANCA, p-ANCA and cryoglobulins. Blood cultures were negative and urine toxicology was positive for cocaine. Clinical improvement was noted on day 5 of hospital stay. By then, urine analysis by high-performance liquid chromatography was positive for levamisole. Ultimately, a final diagnosis was made by skin biopsy showing acute and chronic inflammation of the superficial dermis, consisting of neutrophils, eosinophils, and lymphocytes. The deeper dermis and subcutaneous fat showed mostly acute perivascular inflammation, acute vasculitis, and acute inflammation of the eccrine glands, with mixed eosinophilic infiltrate. Antibiotics were stopped. A clinical diagnosis of levamisole-induced vasculitis was made and patient was discharged home with a follow-up appointment with the Rheumatology Clinic, unfortunately patient did not keep the appointment.*

MEBENDAZOLE

Placebo-Controlled Trial

In a double-blinded, randomized study ($n=295$), the safety and efficacy of a novel chewable mebendazole (500 mg) for the treatment of *Ascaris lumbricoides* and *Trichuris trichiura* infestation were assessed in children aged 1–15 years and residing in Ethiopia and Rwanda [14c]. Adverse events commonly reported included abdominal pain and nasopharyngitis. The frequency of treatment emergent adverse events (TEAE) was not significantly different ***between groups: 9/144 versus 7/278 (RR 2.48, 95% CI 0.94–6.53, $P=0.07$). No deaths, serious TEAEs, discontinuations because of TEAEs were reported.

OXANTEL PAMOATE

See SEDA-38, 303; SEDA-39, 299

PRAZIQUANTEL

Comparative Study

See Section 'Tribendimidine'.

Placebo-Controlled Studies

The safety and efficacy of praziquantel (20, 40, or 60 mg/kg) in the treatment of schistosomiasis in 885 children aged 2–16 years have been examined in a randomised controlled, parallel-group, single-blind, dose-ranging, phase 2 trial [15C]. The most common adverse events reported 3 hours post-treatment in pre-school aged children (2–5 years; $n=660$) were diarrhea (9%) and stomach ache (8%) and in school-aged children (6–15 years; $n=225$) were diarrhea (28%), stomach ache (37%), and vomiting (15%). At 24 hours post-treatment, diarrhea (5%) and vomiting (3%) were the most common adverse events reported among pre-school aged children; stomach ache (16%), cough (11%) and diarrhea (10%) were the most common events reported among school-aged children. Cough, diarrhea and fever were the most common adverse events reported among pre-school aged children at 48 hours post treatment. Stomach ache, cough and diarrhea were the most common adverse events reported among school aged children at 48 and 72 hours post treatment. No serious adverse events were observed.

Meta-analysis

The safety and efficacy of praziquantel 40 mg/kg for treatment of schistosomiasis in preschool-aged and

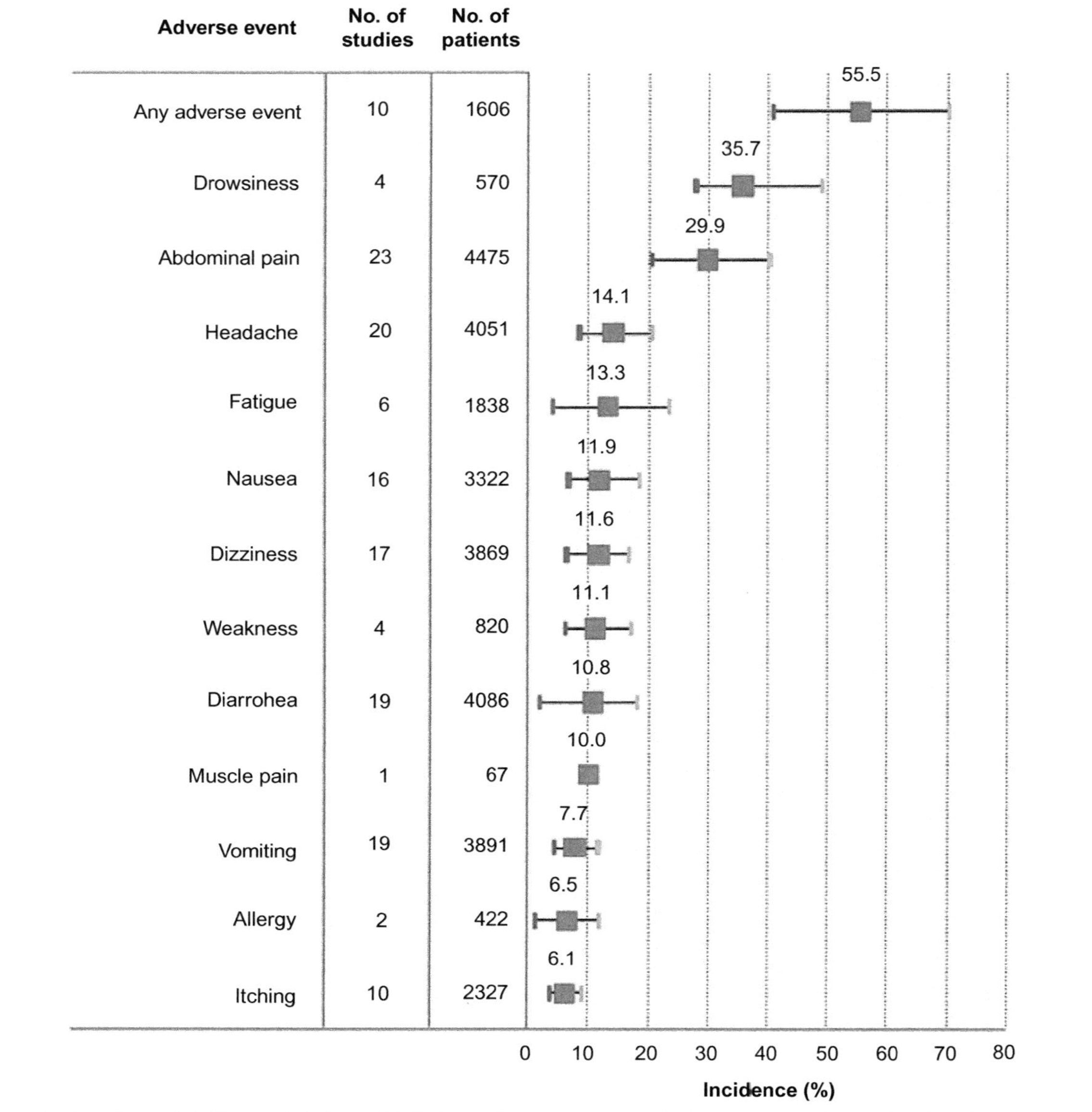

FIG. 1 Adverse event rates in preschool and school children administered PZQ 40 mg/kg for treatment of schistosomiasis. Reproduced with permission from Zwang J, Olliaro P. Efficacy and safety of praziquantel 40 mg/kg in preschool-aged and school-aged children: a meta-analysis. Parasit Vectors. 2017;10(1):47. https://doi.org/10.1186/s13071-016-1958-7 (web archive link).

school-aged children have been evaluated in a meta-analysis of 47 comparative and non-comparative studies ($n = 14340$) [16M]. Twenty-two included studies ($n = 6713$) reported adverse events data, out of which data from 10 studies ($n = 1066$) were pooled (see Fig. 1). The most common adverse events reported in the praziquantel group were drowsiness (36%), abdominal pain (30%), headache (14%), fatigue (13%), nausea (12%) and dizziness (12%).

TRIBENDIMIDINE

Comparative Study

The safety of tribendimidine versus praziquantel in the treatment of *Opisthorchis viverrini* infestation in children and adults has been evaluated in an open-label, randomised, non-inferiority, phase 2 trial including 607 participants [17C]. Participants received single oral doses of tribendimidine (200 mg in children and 400 mg in adolescents and adults) and two oral doses of praziquantel (50 and 25 mg/kg 6 hours apart). The adverse events observed were of mild to moderate intensity. The most common adverse events reported were headache, vertigo, nausea, and fatigue. The frequency of adverse events was significantly greater with praziquantel compared to tribendimidine: OR: 4.5, 95% CI: 3.2–6.3; $P < 0.0001$.

References

[1] Moser W, Coulibaly JT, Ali SM, et al. Efficacy and safety of tribendimidine, tribendimidine plus ivermectin, tribendimidine plus oxantel pamoate, and albendazole plus oxantel pamoate against hookworm and concomitant soil-transmitted helminth infections in Tanzania and Côte d'Ivoire: a randomised, controlled, single-blinded, non-inferiority trial. Lancet Infect Dis. 2017;17:1162–71. pii: S1473-3099(17)30487-5. [C].

[2] Chandler RE. Serious neurological adverse events after ivermectin—do they occur beyond the indication of onchocerciasis? Am J Trop Med Hyg. 2018;98:382–8. https://doi.org/10.4269/ajtmh.17-0042 [A].

[3] Kearns TM, Currie BJ, Cheng AC, et al. Strongyloides seroprevalence before and after an ivermectin mass drug administration in a remote Australian aboriginal community. PLoS Negl Trop Dis. 2017;11(5)e0005607 [c].

[4] Herrick JA, Legrand F, Gounoue R, et al. Posttreatment reactions after single-dose diethylcarbamazine or ivermectin in subjects with Loa loa infection. Clin Infect Dis. 2017;64(8):1017–25 [c].

[5] Massi DG, Mansare ML, Traoré M, et al. Post-ivermectin encephalopathy in senegal: a case report. Pan Afr Med J. 2017;27:202 [A].

[6] Aroke D, Tchouakam DN, Awungia AT, et al. Ivermectin induced Stevens-Johnsons syndrome: case report, BMC Res Notes. 2017;10(1):179 [A]. https://doi.org/10.1186/s13104-017-2500-5.

[7] Seegobin K, Bueno E, Maharaj S, et al. Toxic epidermal necrolysis after ivermectin. Am J Emerg Med. 2018;36:887–9. pii: S0735-6757(17)307453. [A].

[8] Fredericks C, Yon JR, Alex G, et al. Levamisole-induced necrosis syndrome: presentation and management. Wounds. 2017;29(3):71–6 [A].

[9] Cascio MJ, Jen KY. Cocaine/levamisole-associated autoimmune syndrome: a disease of neutrophil-mediated autoimmunity. Curr Opin Hematol. 2018;25(1):29–36 [A].

[10] Vitt JR, Brown EG, Chow DS, et al. Confirmed case of levamisole-associated multifocal inflammatory leukoencephalopathy in a cocaine user. J Neuroimmunol. 2017;305:128–30 [A].

[11] Ramesh S, Sobti D, Mancini R. Eyelid necrosis and secondary cicatrical ectropion secondary to levamisole-associated vasculitis. Ophthal Plast Reconstr. 2017;33(3S Suppl 1):S38–40 [A].

[12] Yek C, Li X, Mauskar M. Painful purpura associated with exposure to levamisole-adulterated cocaine. BMJ Case Rep. 2017;2017: pii: bcr2016219077. [A].

[13] Salehi M, Morgan MP, Gabriel A. Levamisole-induced leukocytoclastic vasculitis with negative serology in a cocaine user. Am J Case Rep. 2017;18:641–3 [A].

[14] Silber SA, Diro E, Workneh N, et al. Safety of a single-dose mebendazole 500 mg chewable, rapidly-disintegrating tablet for Ascaris lumbricoides and Trichuris trichiura infection treatment in pediatric patients: a double-blind, randomized, placebo-controlled, phase 3 study, Am J Trop Med Hyg. 2017;97:1851–6 [c]. https://doi.org/10.4269/ajtmh.17-0108.

[15] Coulibaly JT, Panic G, Silué KD, et al. Efficacy and safety of praziquantel in preschool-aged and school-aged children infected with Schistosoma mansoni: a randomised controlled, parallel-group, dose-ranging, phase 2 trial. Lancet Glob Health. 2017;5(7): e688–98 [C].

[16] Zwang J, Olliaro P. Efficacy and safety of praziquantel 40 mg/kg in preschool-aged and school-aged children: a meta-analysis. Parasit Vectors. 2017;10(1):47 [M].

[17] Sayasone S, Keiser J, Meister I, et al. Efficacy and safety of tribendimidine versus praziquantel against *Opisthorchis viverrini* in Laos: an open-label, randomised, non-inferiority, phase 2 trial. Lancet Infect Dis. 2018; pii: S1473–3099(17)30624–2. [C].

CHAPTER

31

Vaccines

Michael P. Veve*,†,[1], Vasilios Athans‡

*Department of Clinical Pharmacy and Translational Science, University of Tennessee Health Science Center College of Pharmacy, Knoxville, TN, United States

†Department of Pharmacy, University of Tennessee Medical Center, Knoxville, TN, United States

‡Department of Pharmacy, Cleveland Clinic, Cleveland, OH, United States

[1]Corresponding author: mveve1@uthsc.edu

Abbreviations

AVA anthrax vaccine adsorbed
BCG bacillus Calmette–Guérin
DTaP diptheria + tetanus toxoids + acellular pertussis vaccine
HAV hepatitis A vaccine
HBV hepatitis B vaccine
Hib haemophilus influenza type B
HPV human papillomavirus
HPV4 quadrivalent human papillomavirus vaccine
HZ/su herpes zoster subunit vaccine
HZV herpes zoster vaccine
IIV inactivated influenza vaccine
IPV inactivated poliovirus vaccine
JE Japanese encephalitis
JE-CV Japanese encephalitis chimeric vaccine
LAIV live-attenuated influenza vaccine
MenB *Neisseria meningitides* serogroup B
MenC *Neisseria meningitides* serogroup C
MMR measles + mumps + rubella
OPV oral poliovirus vaccine
PCV7 7-valent pneumococcal conjugate vaccine
PCV13 13-valent pneumococcal conjugate vaccine
PPSV23 23-valent pneumococcal polysaccharide vaccine
RV rotavirus
Tdap tetanus toxoid + diphtheria toxoid + acellular pertussis vaccine
VZV varicella zoster virus
YF yellow fever

VIRAL VACCINES

Dengue Vaccine [SEDA-39, 302–303]

General

A randomized controlled trial sought to determine the efficacy of CYD-TDV efficacy in flavivirus-naïve Australian adults. There were no safety concerns in patients who had vaccinal viremia or shedding into urine or saliva [1C]. Two patients experienced non-vaccine related severe adverse effects: one developed an exacerbation of anxiety and depression requiring hospitalization and discontinuation from the study, and the other patient developed pyrexia. A separate randomized trial assessed the efficacy of CYD-TDV in flavivirus-exposed or vaccinated patients, where patients who received CYD-TDV had greater incidence of rash (66%–6%, $P = < 0.001$) when compared to placebo for the first vaccine series [2C]. For patients who developed a rash, it was mild in 81% of cases and all cases self-resolved. Two severe adverse events occurred in one patient after vaccine administration, syncope (51 days post-vaccine) and transient ischemic attack (77 days post-vaccine); neither event was attributed to the vaccine. Four vaccinated patients developed short-lived neutropenia; two cases were considered mild, with the other 2 cases considered moderate.

AGE

Childhood vaccinations remain a crucial preventative measure against a number of infectious diseases. A systematic review of efficacy and safety of the 3-series CYD-TDV was performed in children and adolescents from 2 to 18 years old; 6 randomized controlled trials ($n = 2099$) were included [3M]. Severe, solicited injection-site and systemic adverse effects, and unsolicited adverse effects were similar to the test and control groups in all trials, but statistical comparisons were not included. From one of the studies included, 3 patients (0.3%) withdrew due to adverse effects thought to be secondary to vaccination (e.g., fever, rash, and worsening of cervical spondylosis).

Side Effects of Drugs Annual, Volume 40
ISSN: 0378-6080
https://doi.org/10.1016/bs.seda.2018.06.006

Interactions

DRUG–DRUG INTERACTIONS

Decreased vaccine immunogenicity has been demonstrated when multiple vaccines are co-administered [4M]. The immunogenicity and safety of CYD-TDV administered with the DTap–IPV//Hib vaccine were studied in healthy Mexican children 15 to 18 months old [5C]. Children were randomized to 2 groups and received either CYD-TDV sequentially ($n=295$) or co-administered with the DTap–IPV//Hib vaccine ($n=286$). The investigators found no observed impact on immunogenicity or safety from vaccine co-administration. Grade 3-solicated reactions were similar between sequential (18.8%) and co-administered (16.2%) vaccine groups; the most common reaction was appetite loss. Overall, 41 (5.7%) patients experienced at least one solicited adverse effect, but only 3 (febrile seizures) were considered related to study vaccine(s) by the sponsor.

Ebolavirus Vaccine [SEDA-39, 303–304]

Ebola virus disease (EVD) was first described during the 1970s in Central Africa, [6S, 7S], but since then there have been several EVD-related outbreaks throughout history. The most notable and significant EVD outbreak to date occurred in West Africa during February 2014 and has infected over 27000 people and caused 11000 deaths [8S]. Issues complicating the 2014 EVD outbreak were largely due to amplification and transmission of infections from hospitalized critically ill patients to other healthcare workers [9S, 10R]. Additional factors, such as unregulated or overpopulated boarders and lack of public healthcare infrastructure, are thought to be key variables associated with the 2014 EVD outbreak [9S]. Because the virus may be spread from human contact with infected subjects, EVD has had large impact on international travel and increased boarder control measures [9S, 10R]. The 2014 EVD outbreak spurred the development and implementation of investigational vaccinations as an effective preventative measure. To date, EVD vaccines have been efficacious with few side effects, but the widespread use of investigational vaccinations in such a short timespan may lead to undetected serious adverse effects.

Epidemiology

The first Central African EVD outbreak in 1976 was estimated to have about 300 cases and 200 deaths [6S, 7S, 11R]. Since then, there have been sporadic outbreaks mainly in Central/West Africa that were largely contained in rural villages. Nonetheless, mortality rates were extremely high (60%–90%) [6S, 7S, 9S]. The initial method of patient management was to quarantine infected patients, and many Central African medical institutions closed during the early outbreaks due to the large number of infected healthcare workers [6S, 7S].

The most recent EVD outbreak occurred in the West African country Guinea during late 2013, which was later confirmed by the World Health Organization (WHO) in March 2014 [12S] and subsequently spread to Sierra Leone and Liberia. The outbreak of 2014 has been the primary cause of over 27000 infections, primarily through human-to-human contact [8S]. The severity of this outbreak allowed the disease to spread to other countries outside of West Africa, most notably the United States, United Kingdom, Spain, and Germany [9S, 10R, 13S].

Transmission and Pathogenesis

The ebolavirus, a member of the Filoviridae, is an enveloped, single-stranded, negative sense RNA virus that consists of five different subspecies: Zaire, Bundibugyo, Sudan, Reston, and Tai Forest virus [10R]. The most common single species associated with infection is the zaire ebolavirus, which was identified as the causative organism of the 2014 EVD outbreak and is associated with the highest attributable mortality [14S]. While the specific natural reservoir for ebolavirus is unknown, experts hypothesize that frugivorus bats are mostly likely carrier and reason for viral spread. Infected bats then contaminate other large animals (e.g. cows, monkeys), or "bush meat" that are consumed by humans indignant to the area. EVD can be spread directly from infected animals, from eating undercooked contaminated meat, through human-to-human including contact with body fluids or close contact to a deceased human infected with EVD during ceremonial burial practices [10R, 11R, 15S]. Patients who have survived EVD cannot transmit the virus conventionally, but EVD may be transmitted from semen through sexual intercourse for up to 7 weeks [14S]. Unlike other viral illnesses, EVD cannot be spread through vectors like air, water or insects [12S, 14S].

The mechanism of pathogenesis EVD is not well understood. In vitro data support the hypothesis of EVD-included massive cytokine release and tissue necrosis of cellular progenitors (e.g. parenchymal and endothelial cells, macrophages), which has been reported in organ systems such as the liver, spleen, and kidneys [11R]. EVD also affects the human coagulation cascade and can lead to acute hemorrhages. Additional in vitro data show evidence of circulating glycoproteins that enable the virus to act with an immunological decoy and prevent a true immune response [11R]. The disease pathogenicity and mortality associated with EVD strongly support the need to develop therapeutic options and vaccinations to improve patient outcomes.

Clinical Aspects and Presentation

Patients with EVD may not present with signs of infection for up to 8–10 days after exposure, with an incubation period of around 2–21 days [10R, 14S]. Those with acute infection may have flu-like symptoms, are often febrile, and may have muscular aches. These non-descript clinical presentations could lead to unrecognized early-stage EVD. Most patients will progress to more severe gastrointestinal symptoms, such as profuse diarrhea with nausea and vomiting, all of which have been identified as independent risk factors for patient mortality [14S, 16S]. A cardinal symptom of EVD is hemorrhagic fever, defined loosely as patients presenting with bleeding from various orifices (e.g. mouth, anus, intravenous lines). The pathophysiology behind viral hemorrhagic fever is unknown but hypothesized to be due to EVD effect on platelets [14S]. Most patients presenting with viral hemorrhagic fever are near death and are seen in upwards of 50% of cases [17r]. While medical care has vastly improved over time, EVD mortality is reportedly around 25%–90% [10R, 17r]. This mortality likely due to the rapid progression of disease severity and lack of effective treatment. For patients who do survive long enough to constitute an immune response, many patients report debilitating long-term side effects such as severe pain, hearing loss, and ocular manifestations [18r].

Diagnosis, Management, and Outcomes

A diagnosis of EVD can also be supported through polymerase chain reaction testing or through other extreme laboratory values, most notably thrombocytopenia, severe renal dysfunction, and elevated liver enzymes [10R, 14S, 19R]. As no effective treatments are available, EVD management is largely based on supportive care to maintain patient homeostasis. This includes implementing renal replacement therapy for severe renal disease, mechanical ventilation, fluid resuscitation, and management of adverse effects with medications to curb diarrhea or vomiting [20R]. Additional treatment methods, such as convalescent plasma transfusions, have been attempted with little data to support routine use [19R]. The lack of EVD treatment has shifted focus towards widespread vaccination as an effective means to control outbreaks.

Vaccinations

EVD vaccine development began in the late 1970s, shortly after the discovery of the virus, but because EVD outbreaks have been rare and well-controlled, commercial vaccine manufacturers have demonstrated little urgency in advancing vaccines through clinical trials [21M]. The 2014 EVD outbreak led to a significant surge and fast-track development of Phase 1 clinical trials for EVD vaccines previously tested on animals. According to Clinicaltrial.gov, a clinical trial database registry, 46 clinical trials involving EVD vaccines were launched. The three main categories of EVD vaccine are based on (i) non-replicative vector-based EBV, (ii) replicative vector-based EBV, and (iii) other EBV vaccines. All vaccines have demonstrated effectiveness in vivo against the most common ebolavirus types, though the non-replicative vector-based EVD vaccines are thought to have favorable tolerability profiles without risks of viremia post-vaccination [21M]. One of the most promising vaccine candidates is a replicative vector-based vaccine, the recombinant vesicular stomatitis virus-based vaccine (rVSV-EBOV), which has demonstrated efficacy and safety in a number of clinical trials to date.

Public Health Implications and Prevention

Given the large number of patients affected by EVD and global threat of the virus, the WHO and Centers for Disease Control and Prevention (CDC) developed strategic plans to curtail the risks associated with EVD and prevent further spread to different countries [22S, 23S]. One of these methods included development and employment of investigational EVD vaccines for widespread prevention [23S].

While the use of effective experimental vaccines may help prevent the spread of deadly communicable diseases, the risk–benefit for adverse effects must be considered. Given the significant mortality seen from EVD in West Africa and transmission to other countries, the decision to utilize compassionate-use EVD vaccines at large scale was made. Initially, adults living in geographic rings of West Africa were randomized to receive EVD vaccines; after interim results were published showing the vaccine's efficacy in humans, most vaccines were opened to children at least 6 years old. Many clinical trials involving the EVD vaccine are required to utilize appropriate adverse effect screening and follow-up data to capture unanticipated adverse effects, but only now are these data being published after many trials were initiated during the 2014–2015 timeframe. It is therefore crucial for investigators to be cognizant of potential unexpected adverse effects when vaccines are used at such a large scale.

Conclusions

The threat of EVD highlighted the need to rapidly develop and distribute safe and effective vaccines to large populations in order to prevent further spread of the virus. While no vaccines were approved at the time of the EVD outbreak, safety precautions are required for many clinical trials. Patients can be observed for adverse effects immediately after vaccination, and long follow-up periods are used to assess for solicited adverse effects. While most trials have shown high efficacy in vaccinated patients, the concept of "herd immunity" and indirect protection of unvaccinated patients was also likely demonstrated. Regardless, the rapid distribution of compassionate use EVD vaccines with little long-term safety data should capture the attention of clinicians.

General

A Phase 2 trial of immunogenicity and safety of two EBV vaccines, the chimpanzee adenovirus 3 (ChAD3-EBO-Z) and recombinant vesicular stomatitis virus vaccine (rVSVΔG-ZEBOV-GP), was conducted in Liberia during 2015 [24C]. Uninfected adult patients were randomized 1:1:1 to receive either the ChAD3-EBO-Z ($n = 500$), the rVSVΔG-ZEBOV-GP ($n = 500$), or saline placebo ($n = 500$), and were followed for 12-months after receipt of study drug. The vaccines were similar in regards to efficacy at 12-month antibody response for the ChAd3-EBO-Z and rVSVΔG-ZEBOV-GP group compared to placebo (63.5% vs 79.5% vs 6.8%, $P<0.001$). Within 1-month post-injection, 20 participants had a serious adverse event: 1.2% in the ChAd3-EBO-Z group, 1.2% in the rVSVΔG-ZEBOV-GP group, and 1.6% in the placebo group ($P=0.68$). A total of 70% of the serious adverse events that occurred in the first 30-days post-injection were attributed to malaria. Over the 12-month follow-up time frame, serious adverse events occurred in 40 participants (8.0%) in the ChAd3-EBO-Z group, 47 (9.4%) in the rVSVΔG-ZEBOV-GP group, and in 59 (11.8%) in the placebo group. The majority of serious adverse events were attributed to malaria (71%). Malaria developed in fewer participants in the ChAd3-EBO-Z group than in the placebo group (5.2% vs 8.8%, $P=0.03$); a similar finding was seen in the rVSVΔG-ZEBOV-GP group (6.6%, vs 8.8%; $P=0.25$). At 12-months, 6 deaths had occurred in the 2 vaccine groups: 1 ChAd3-EBO-Z, 5 rVSVΔG-ZEBOV-GP. However, these deaths were not attributed to the study vaccines. During the week post-injection, injection-site reactions were reported in 28.5% of the participants in the ChAd3-EBO-Z group and 30.9% of those in the rVSVΔG-ZEBOV-GP group, as compared with 6.8% of those in the placebo group ($P<0.001$). The patients did not report high rates of arthralgia that have previously been reported.

Gsell and colleagues expanded on the safety of rVSV-ZEBOV during March 2016 of the EBV outbreak [25MC]. Using a four-ring vaccination approach, the vaccine was administered to eligible non-pregnant adults and children at least 6 years old who lived in the Guinée Forestière region of Guinea. Safety was assessed immediately and for 30-minutes post-vaccination. Patients ($n=1510$) were visited at home for additional safety checks at days 3, 14, and 21 post-vaccination. Overall, 17% of 6–17 year olds and 36% of adults reported at least one adverse event following vaccination, almost all of which occurred between 31 minutes and 3 days. The most common adverse effects reported in adults were: 15% headache, 13% muscle pain, 13% myalgia, 10% fatigue, and 7% arthralgia, which are consistent with previously published rVSV-ZEBOV trials. Children aged 6–17 years commonly reported: 12% headache, 4% muscle pain, and 3% myalgia. Arthralgia symptoms were reported by one vaccinee (<1%) aged 6–17 years, compared with 7% of adult vaccinees. Three serious adverse events not related to the study vaccine were reported that were: one case of stroke, one malaria case, and one case diagnosed with both salmonellosis and malaria.

Halperin and colleagues performed a Phase 3 study in patients who received low or high-dose rVSVΔG-ZEBOV-GP ($n=1194$) or placebo ($n=133$) [26MC]. The investigators determined the vaccine to be effective and adverse effects were similar to previous reports (81%–85% vaccinated patients reported ≥1 adverse event vs 44% placebo). In total, 1.7% cases of arthritis were reported in the vaccine group and none in the placebo group. Five pregnant patients were vaccinated with the study drug; 2 were loss to follow-up, 2 gave birth without complications, and 1 patient had a spontaneous abortion with limited data to assess associations with the study vaccine.

Other Phase 1 and safety data surrounding the rVSVΔG-ZEBOV-GP and ChAD3-EBO have been recently published, which are consistent with previously stated findings that the vaccine is safe and effective [27R, 28c, 29C, 30C, 31C, 32c, 33c].

Hepatitis A Virus Vaccine [SED-16, 255–293, 696–706; SEDA-38, 308; SEDA-39, 304]

Drug Administration

DRUG DOSAGE REGIMEN

A trial was conducted to assess the immunogenicity and safety of different HAV and hepatitis B virus vaccination strategies in Chinese children [34C]. Patients 18–24 months-old were randomized to receive 1 of 3 different HAV vaccine regimens: (i) 2 doses of inactivated HAV ($n=79$), (ii) 1 dose of inactivated HAV plus one dose of combined HAV and HBV vaccine ($n=85$), and (iii) 2 doses of combined HAV and HBV vaccine ($n=88$). The investigators found significantly increased seroprotection and geometic mean HAV and HBV concentrations 1 month after the second vaccine inoculation in patients who received a combined vaccine compared to those who did not ($P<0.001$). Adverse events were observed in 23.6% of patients: 15.3% Grade 1 events, 8.0% Grade 2 events, and 0.3% Grade 3 events. There were no statistical differences between adverse events between the three groups ($P=0.345$).

Organs and Systems

IMMUNOLOGIC

The immunogenicity and safety of the inactivated HAV was assessed in a randomized trial of Greek children with juvenile idiopathic arthritis (JIA) and compared to healthy controls [35C]. Children between 2 and 16 years old were included; JIA cases were required to be in clinical remission while receiving methotrexate therapy for at least 6 months. HAV immune response and seroconversion rates were similar between cases and controls. Reactogenicity was also similar between the JIA and control groups (10.5% vs 8.4%, $P=0.65$). Eighteen JIA flare episodes occurred over an 18-month period, but these flares were not attributed to the vaccine.

LIVER

A healthy, 22 year old woman presented with symptoms of malaise, diarrhea, and jaundice 1 month after receiving HAV and HBV vaccinations [36A]. Her laboratory values were suggestive of acute liver toxicity, and viral/toxoplasma serology was negative. The patient was diagnosed with hepatitis after vaccination and redeveloped jaundice with elevated liver enzymes 1-month after her initial visit. Serology for hepatitis and autoimmune workup were also not significant. A liver biopsy was performed that was consistent with autoimmune hepatitis (AIH) and the patient was subsequently treated with prednisone and azathioprine. The patient's successful response to treatment and other case-related factors were suggestive of probable AIH.

A similar case occurred with a previously healthy 28 year old woman, who presented with malaise, nausea, fever, and jaundice 10 days after HAV vaccination [36A]. Laboratory results were suggestive of liver injury and her viral serology was negative. An abdominal ultrasound was performed at time of presentation and showed no anomalies. A follow-up liver biopsy showed mild hepatitis with eosinophilia, which was considered toxic hepatitis. The patient responded to immunosuppressive therapy, but her ALT began to rise shortly after immunosuppressive therapy was discontinued. The patient was ultimately diagnosed with probably AIH secondary to the HAV vaccine.

Hepatitis B Virus Vaccine [SED-16, 255–293, 696–706; SEDA-37, 384; SEDA-38, 307–308; SEDA-39, 306–307]

General

Patients who struggle with addiction and inject illicit drugs are at high risk for blood-borne infectious diseases, including HBV. Standard-dose HBV vaccine non-response has been demonstrated in select populations, mainly those who are immunocompromised, and has caused investigators to study high-dose HBV vaccine in various scenarios. A randomized trial was conducted to assess the immunogenicity and safety of high-dose HBV among patients receiving methadone maintenance treatment for opioid dependence [37C]. Patients were randomized to receive 3-doses of either 20 or 60 mcg recombinant HBV vaccine. When compared to the low-dose HBV vaccine (n = 98), patients who received high-dose HBV vaccine (n = 97) had increased rates of seroconversion, response, and mean general concentrations of anti-HBs, but these data were not statistically significant ($P > 0.05$). No patients reported serious adverse reactions or became HBsAg positive during follow-up. Of all patients who received at least one dose of a vaccine, 7.7% reported solicited adverse reactions within 7 days post-vaccine, and 2.6% reported unsolicited adverse reactions within 28 days post-vaccine. The most common adverse effects were pain, erythema, induration, and fever within 7 or 28 days post-vaccine; none of the reported adverse reactions were significantly different between the two groups ($P > 0.05$).

Susceptibility Factors

PHYSIOLOGICAL FACTORS: RENAL DISEASE

A systematic review was performed to assess the safety of double-dose hepatitis B vaccination in patients with chronic kidney disease not on hemodialysis [38M]. Seven studies were included for quantitative analyses and found that double-dose vaccination with plasma-derived or recombinant HBV vaccine was associated with similar rates of seroconversion as standard vaccination for pre-hemodialysis and hemodialysis patients. Minor adverse events were reported in 9%–15% of injections for included studies, but sufficient description of adverse events precluded the investigators from performing meaningful secondary analyses.

PHYSIOLOGICAL FACTORS: HEPATIC DISEASE

A randomized trial was performed to assess the efficacy and safety of a electroporation-mediated dual-plasmid hepatitis B virus (HBV) DNA vaccine compared to placebo in patients with chronic HBV on lamivudine therapy [39C]. Efficacy was assessed by rate of undetectable HBV DNA or HBeAg seroconversion; safety was documented over the 72-week study period. After 9 days of HBsAg stimulation, patients in the vaccine group patients showed greater immunological response than placebo (86.4% 56.5%, $P = 0.027$). Vaccine injections were well tolerated, and minimal adverse effects were reported for DNA vaccine compared to placebo (12.6% vs 12.7%). Two (1.8%) patients in the DNA vaccine group and 1 (0.9%) patient in the placebo group reported severe, non-drug related adverse effects ($P > 0.05$).

Human Papillomavirus Vaccine (HPV) [SED-16, 255–293; SEDA-37, 384; SEDA-38, 308–309; SEDA-39, 306–307]

General

Costa and colleagues performed a meta-analysis on three clinical trials assessing the adverse effects of the HPV9 or -4 vaccines in women ($n = 27465$) [40M]. Pain (OR, 1.72; 95%CI, 1.62–1.82) and erythema (OR, 1.29; 95%CI, 1.21–1.36) occurred significantly more in the HPV9 group. However, there was no significant difference between the groups for the following adverse effects: headache (OR, 1.07; 95%CI, 0.99–1.15), dizziness (OR, 1.09; 95%CI, 0.93–1.27), and fatigue (OR, 1.09; 95%CI, 0.91–1.30), and the occurrence of serious events related to vaccination was similar among those vaccinated. The authors concluded there were no differences in safety profiles between the two vaccines. Chandler and colleagues performed an observational study exploring global adverse effect reporting patterns for the HPV vaccine [41C]. VigiBase, the WHO international database of suspected adverse drug reactions, was used as a data source. Overall, there were 39953 HPV vaccine reports in VigiBase, including serious adverse effects such as postural orthostatic tachycardia syndrome, complex regional pain syndrome, and chronic fatigue syndrome. The authors concluded causality was difficult to assess given the lack of sufficient adverse drug reaction details.

Second-Generation Effects

PREGNANCY

Two large retrospective studies of over 10500 periconceptional and/or pregnant women did not find associations of the HPV vaccine with adverse effects to mothers or infants [42C, 43C].

AUTOIMMUNE DISEASE

Autoimmune/inflammatory syndrome induced by adjuvants (ASIA) following vaccination has been described. Symptoms typically include manifestations of autoimmune disease and can range in severity [44R]. Jara and colleagues performed a systematic review of severe ASIA, which included major organ involvement, life-threatening conditions, intensive treatment, disability, hospitalization and outcome (survival and death) [45M]. Of the 305 severe ASIA cases identified, 72% of patients had received the HPV vaccine. However, no formal statistics were performed to determine causal associations with the vaccine. Conversely, two case reports were published of patients with a history of multiple keratinocyte carcinomas (KC) who experienced decreased rates of KC over 12 months after receiving the quadrivalent HPV vaccine [46A].

Influenza Vaccine [SED-16, 98–106; SEDA-37, 385; SEDA-38, 309; SEDA-39, 307–313]

General

Influenza, an RNA virus with A and B serotypes, is responsible for endemic and pandemic flu. Both influenza types are responsible for endemic flu, while type A, due to antigenic drift of its surface proteins hemagglutinin and neuraminidase, is responsible for flu pandemics [47R]. An estimated 140000–710000 U.S. patients are hospitalized each year due to influenza, with 12000 to 56000 influenza-related deaths [48S]. Widespread yearly influenza subsequently results in increased numbers of reported vaccine-related adverse events. Cai and colleagues developed a signal detection method for temporal variation of adverse effects with of the FLU3 influenza vaccine from the Vaccine Adverse Event Reporting System (VAERS) database between 1990 and 2013 [49C]. A statistical model was devised to detect vaccine safety signals by testing the heterogeneity of reporting rates of given vaccine-event combinations across reporting years using a random effects model. The authors concluded that evaluating temporal trends of reporting rates can suggest the potential impact of changes in vaccines on the occurrence of AEs and provide evidence for further investigations. Clothier and colleagues performed a cross-sectional, interrupted time series analyses to compare allergy-related adverse events following influenza vaccination in Australia [50C]. Adverse events (e.g., confirmed anaphylaxis, angioedema, urticarial, or generalized non-specific reactions) with the influenza vaccine were extracted from 2008 to 2015, and additional vaccine details were collected. Overall, 1010 events were reported over the timeframe, but none were associated with a specific influenza vaccine type. The investigators found an almost doubled allergy-related adverse event following influenza vaccination in 2015 (2.4 normalized per 100000 vaccine doses distributed) compared to the other years included in the study. Inconsistencies in Australian pharmacovigilance reporting and lack of casual inference were noted as study limitations.

Susceptibility Factors

AGE

Elderly patients and those with chronic comorbid conditions may be more likely to suffer from adverse effects from influenza virus, and recent data support that these populations may be more vulnerable to viral infection [51R]. This has led to formulation and use of higher-dose influenza vaccine in elderly patients. Kaka and colleagues surveyed adults older than 65 years who had received either high-dose ($n=547$) or standard-dose ($n=541$) influenza vaccines [52C]. Thirty-seven percent of high-dose recipients and 22% of standard-dose recipients reported a local or systemic side effect ($P<0.001$), most of which were mild to moderate. Only 7 of 547 (1.3%) high-dose recipients and 3 of 541 (0.6%) standard-dose recipients reported a severe side effect ($P=0.34$). The authors concluded that side effects were more common among patients who received high-dose influenza vaccines vs standard-dose. Wilkinson and colleagues performed a meta-analyses ($n=7$ trials) on the efficacy and safety of high-dose influenza vaccine in patients over 65 years old and found patients who received the high-dose vaccine were less likely to develop influenza than standard-dose (RR, 0.76; 95%CI, 0.65–0.90) without increases in adverse effects [53C].

IMMUNOLOGIC

Lieberman and Curtis describe a case of herpes simplex virus-2 reactivation and myelitis after influenza vaccination in a 62 year old woman [54A]. The patient's past medical history was significant for varicella, measles, and mumps infection prior to 10 years of age, in addition to HSV-2 manifesting at 24 years old. When she was 43 years old, she had ventriculoperitoneal shunts placed on her left side to relieve pressure from pesudotumor cerebri. After receiving the influenza vaccine at age 57, she developed strange sensational feelings of her left side and began vibrating. She developed 7 severe HSV-2 blisters on her back in addition to migraine headaches. The patient was later diagnosed with transverse myelitis.

Second-Generation Effects

PREGNANCY

Pregnant women are at risk for vaccine-preventable, disease-related complications and adverse pregnancy outcomes, including congenital anomalies, spontaneous abortion, preterm birth deliveries, and low birth weight [47R]. Pregnant women infected with influenza generally have worse outcomes when compared to the general population (e.g. hospitalizations, cardiopulmonary complications, and death) [47R]. Since 1997, the Advisory Committee on Immunization Practices (ACIP) recommended routine immunization of pregnant women with trivalent inactivated influenza vaccine (IIV3) after the first trimester; in 2004, this recommendation was expanded to all trimesters [55C]. It is therefore imperative for pregnant mothers and healthcare advocates to cognizant of potential adverse effects, or lack thereof, from the influenza vaccine in this setting.

Arriola and colleagues performed a retrospective cohort study to determine associations of the influenza vaccine and positive effects on birth outcomes in Nicaraguan mothers [56r]. Mothers were identified through medical record data and interviewed to determine self-reported influenza vaccine status and collect baseline demographics and birth outcomes. A total of 3268 mothers were interviewed and identified through medical record data (vaccinated $n=1789$, unvaccinated $n=1479$). After adjusting for confounding variables through propensity score matching and multivariable logistic regression, the investigators found protective associations with the influenza vaccination in the second or third trimester and birth outcomes (adjOR, 0.87; 95%CI, 0.75–0.99), and also for influenza vaccination in the second trimester and preterm birth (adjOR 0.80; 95%CI, 0.64–0.97).

Hviid and colleagues performed a retrospective cohort to identify associations with the pandemic influenza A (H1N1) vaccination in pregnant mothers and early childhood morbidity [57MC]. Live-born children birthed from mothers 18–44 years old who were pregnant during the Danish influenza A (H1N1) vaccination campaign were included. Health outcomes of children birthed from vaccinated mothers ($n=61359$) were compared to children birthed from unvaccinated mothers ($n=55048$). Health outcomes (e.g. morbidity) were assessed through the Danish National Patient Register, and the primary outcome included inpatient hospitalizations after birth. Propensity-score matching was used to even the distribution of potential confounding variables between groups. The investigators found that children of vaccinated mothers were not more likely to be hospitalized than unexposed children ($P>0.05$), regardless of the trimester the mother was vaccinated.

One large cross-sectional study surveyed the Vaccine Adverse Event Reporting System (VAERS) database for adverse events in pregnant mothers and their infants after seasonal influenza vaccines in pregnant mothers from 1990 to 2009 [55R]. The VAERS database was screened for reports of pregnant women (and their infants) in the USA who received the seasonal live-attenuated influenza vaccines (LIAV) ($n=127$) or IIV inactivated influenza vaccines ($n=544$) with, or without other vaccines, from 2010 to 2016. For each patient meeting these criteria, a primary diagnosis was established, and in the scenario of multiple diagnoses, the investigators assigned a primary diagnosis based on the best available evidence. Gestational age was calculated for all mothers. Serious events occurred in 61 (11.2%) reports after IIV and one (0.8%) report after LAIV. The most frequent pregnancy-specific condition reported was spontaneous abortion (11.4%); the median time from vaccination to occurrence of symptoms or signs associated with spontaneous abortion was 5.5 days (0–58 days). There were also 10 reports of stillbirth and 6 reports of preterm delivery. The most commonly reported non-pregnancy-specific condition was injection-site reactions in 10.1% of IIV reports. Immune system disorders, which included mostly non-anaphylaxis allergic reactions, were the second most commonly reported condition (7.2%), followed by respiratory events (6.9%). Twenty-two reports involved infant conditions and seven of these were major birth defects; vaccination occurred during the first trimester of pregnancy in five of these reports. The 61 serious reports after IIV included 21 pregnancy-specific conditions included: spontaneous abortion (7), stillbirth (5), preterm delivery (3), pre-eclampsia (3), preterm labor (1), abruptio placentae (1), fetal death (1). Thirty-six non-pregnancy-specific conditions were also reported: respiratory disorders (11), nervous system disorders (11), general administration disorders (5), immune system disorders (5), infections and infestations (2), neoplasia (1), and genitourinary system disorders (1). Four neonatal conditions were reported: one report each of laryngomalacia, cleft palate incomplete, multiple birth defects, and neonatal hypoxia. Among 47 reports with information on gestational age at time of vaccination, LAIV was administered during the first trimester in 51.1% of reports, second trimester in 31.9%, and third trimester 17.0% of reports. The authors highlighted a number of limitations to these aggregated data, including that the VAERS database is prone to biased reporting, has potential inconsistencies in the quality and completeness of reports, and data can be provided by individuals with little or no medical training may adversely affect the quality of the report.

A number of additional large cohorts were published looking at poor health outcomes of children for mothers vaccinated against influenza, none of which identified increased risks for pre-term or low infant birth weights [58MC] or autism spectrum disorders [59MC],

Organs and Systems

INFECTION RISK

Hibino and Kondo described two cases of interstitial pneumonia after influenza vaccination [60A]. Both cases were elderly Japanese women who developed symptoms of interstitial pneumonia 30–40 days after receiving the seasonal influenza vaccine. Both patients were successfully treated with corticosteroids.

NERVOUS SYSTEM

During the 2009 influenza A outbreak and subsequent widespread vaccination, case reports arose of excessive daytime sleepiness in children and young adults. Trogstad and colleagues performed a large retrospective review to assess the relation of narcolepsy and hypersomnia in Norwegian children and young adults [61MC]. Patients under 30 years old were included, and data were extracted from the Norwegian Institute of Public Health based on influenza laboratory diagnoses and clinical reports; these patients were subsequently matched to ICD-10 codes for narcolepsy and hypersomnia. The investigators found increased risk of narcolepsy after vaccination during the first 6 months of the pandemic (adjHR, 17.21; 95%CI, 6.28–47.14) and an increased overall risk of hypersomnia (adjHR, 1.54, 95%CI; 0.81–2.93). The mechanism behind this adverse event is unclear.

SENSORY SYSTEMS: EYES

Papke and colleagues describe a case of a 55 year old woman who developed optic neuropathy after influenza vaccination [62A]. The patient presented to a tertiary care center with complaints of a progressive gray spot in her peripheral vision, which began 2 days after receiving the inactivated influenza vaccine. She had also complained of sensory disturbances, including tinnitus, muscle spasms, bilateral knee pain associated with redness and swelling, a transient rash on her cheeks. Her medical history was most significant for lumbosacral neuritis, lumbar radiculopathy, and sciatica. The patient was diagnosed with right optic neuropathy secondary to a systemic reaction to the influenza vaccine and was treated with corticosteroids.

Japanese Encephalitis Vaccine [SED-16, 393–396; SEDA-37, 393; SEDA-38, 319; SEDA-39, 313]

General

A single-arm, open-label study was performed to assess human T-cell response to a live-attenuated JEV vaccine in adults living in areas endemic to JEV and dengue virus [63r]. The investigators sought to determine whether an immune-response was cross-reactive with dengue and other flaviviruses. Seventeen adult patients aged 18–50 years old were enrolled to the study and received the live-attenuated JEV vaccine; adverse effects were self-reported and kept via journal entries. The investigators found some degree of JEV immune responses can cross-react with dengue virus, but future research is needed to formulate a vaccine sufficient against both flaviviruses. Six adverse effects were reported in 3 of 17 patients; 2 Grade 1 and Grade 2 events occurred within the first 4 weeks post-vaccination, and 2 additional Grade 2 events occurred greater than 4 weeks post-vaccination. The most common adverse effects were febrile illnesses and dizziness, with and without headache. All adverse events resolved without intervention.

Drug Administration

DRUG FORMULATIONS

Infection with the Japanese encephalitis virus (JEV) is a common cause of viral encephalitis in children indignant to Southeast Asia and the Western Pacific. Concerns for adverse effects (mild-to-moderate systemic reactions) related to the inactivated mouse brain-derived vaccine have led to the pursuit of an alternate vaccine. A Phase 3 trial compared the safety of a new Vero-cell derived JEV vaccine to 2 control non-JEV vaccines [64C] in children. Patients from at least 2 months to 18 years were randomized to receive either the JEV vaccine or one of two non-JEV vaccines based on age strata (e.g. less than 1 year old, or greater than 1 year to 18 years old); the primary study endpoint was the rate of serious adverse effects up to 56 days after the first vaccination. Patients vaccinated against JEV received either 0.25 or 0.5 mL dose. In the under 1 year old cohort, at least one solicited or unsolicited adverse effect was reported in 84.0% of JEV vaccinated patients compared to 87.5% in the control group up to 56 days after first vaccination. In the 1-to-18 year old cohort, at least one solicited or unsolicited adverse effect was reported in 62.0% of JEV vaccinated patients compared to 59.6% in the control group up to 56 days after first vaccination. There were no differences between groups for serious and medically attended adverse effects up to 56 days after the first vaccination ($P > 0.05$), and none were contributed to the study vaccines. The most common medically attended adverse effect was upper respiratory tract infection. There were also no differences in serious adverse effects during the 7-month follow-up period (1.5%–2.5% of the cohort), none of which were deemed related to the study vaccines by the investigator and the DSMB. Non-serious adverse effects of special interest related to hypersensitivity/allergy and neurologic disorders were also similar between groups. Two mild events were found to be associated with the JEV vaccine: 1 rash and 1 papular rash. The authors concluded that the safety of the JEV vaccine was comparable to control vaccines, and that the most commonly observed adverse effects are typical for patients in the age strata studied.

Interactions

DRUG–DRUG INTERACTIONS

A Phase 4 trial was performed assessing the effectiveness of a live, attenuated JEV vaccine in Thai children as either a primary vaccination or a booster 1 year post-initial vaccination with inactivated JEV vaccine [65C]. Children 9 months to 5 years old were randomized to receive either primary JEV vaccination or JEV booster vaccination based on previous JEV immunization status; the groups were allocated to 70% primary vaccination and 30% booster vaccination. From the 10 000 patients included in the cohort, no Grade 3 immediate reactions were reported after either primary or booster vaccination. Serious adverse effects occurred in 3.0% of patients after primary vaccination and 1.9% after booster vaccination. Three (<1%) serious adverse effects reported in 2 patients were considered vaccine-related by the investigator. One patient developed urticaria 4 hours after vaccination that recovered 2 days later. The other patient experienced 2 events of urticaria (2 days after vaccination, lasting 7 days, and 16 days after vaccination, lasting 5 days). The 3 events were considered as moderate hypersensitivity and did not require hospitalization. There were no adverse effects of special interest related to the study vaccines as determined by the investigator, with febrile convulsion being the most common adverse effect.

Two similar randomized trials assessed the immunogenicity and safety of a booster JEV vaccine after initial vaccination with live-attenuated JEV vaccine [66C, 67c]. Both trials concluded that booster JEV vaccines were effective and are associated with very low rates of adverse effects, most commonly being solicited injection site reactions (up to 10% redness, swelling, or pain) and solicited systemic reactions (up to 12% fever, crying/irritability, drowsiness, loss of appetite, and rash).

Measles–Mumps–Rubella Vaccine [SED-15, 3555, 3566, 3567, 3569; SEDA-35, 575; SEDA-36, 473; SEDA-37, 391]

Organs and Systems

SKIN

Neven and colleagues described 6 cases of chronic skin granulomas in children with primary immunodeficiency (PID) post-vaccination for MMR [68A]. All patients presented similar cutaneous extensive and persistent cutaneous granuloma in the context of PID: ataxia-telangiectasia ($n=5$), hypomorphic RAG deficiency ($n=2$), activated phosphoinositide 3-kinase δ syndrome ($n=1$), and combined immunodeficiency of unknown cause ($n=1$). All patients had received 1 or 2 MMR vaccines during childhood, and granulomas developed at least 2–145 months post-vaccination. In all patients, skin biopsy results confirmed the diagnosis of non-sarcoidosis granuloma, and no infectious agent was identified by conventional methods.

SENSORY SYSTEMS

Kuniyoshi and colleagues describe a case of acute bilateral photoreceptor degeneration in a 13-month-old boy after MMR and *Haemophilus influenza* type B/Pneumonococcal conjugate vaccinations [69A]. The boy developed bilateral, acute loss of vision 31 days after *H. influenzae* type b and Pneumococcal conjugate vaccinations and 24 days after a MMR vaccine. He had developed a common cold 10 days prior to the vision loss. Ultrasonography showed an exudative retinal detachment 1 day after the onset of the visual reduction; however, his fundi appeared normal 4 days later. His eyes did not pursue objects and pupillary light reflexes were not present. No signs of anterior uveitis were noted. He was treated with corticosteroids, but his vision did not improve.

Poliovirus Vaccine [SED-16, 257, 847–853; SEDA-38, 320; SEDA-39, 315–316]

General

DRUG DOSAGE REGIMENS

Complications related to the transport and storage of the oral poliovirus vaccine (OPV), in addition to adverse events, have caused a shift in use of the inactivated poliovirus vaccine (IPV) in resource-limited countries, but IPV cost and supply constraints are challenging. A Phase 2, single-center study compared the efficacy and safety of three investigational inactivated poliovirus vaccines (IPV) with varying reduced doses to the standard IPV in Dominican children [70C]. Patients were at least 6 weeks of age and randomized to 1 of 3 reduced-dose IPV (either 1/3rd [$n=205$], 1/5th [$n=205$], or 1/10th [$n=205$] the standard IPV dose) or the standard dose IPV ($n=206$). The investigators determined that all 3 reduced-dose IPV were non-inferior to the standard dose IPV in regards to immunogenicity against poliovirus. Three serious adverse events occurred, all unrelated to the vaccinations: 2 events of bronchiolitis and 1 event of amoebic dysentery. Nineteen adverse events, including the 3 serious events, were reported: 3 mild injection site reactions ($\leq$25 mm) and 16 systemic adverse events (both less than 2% across groups). Similar dose-reduction studies have showed fractional and reduced dose IPV are effective when compared to standard-dosed IPV with no significant adverse effects [71C, 72C].

Organ and Systems

INFECTION RISK

Li and colleagues described a rare side effect possibly associated with OPV in a 3 month old male child [73A]. Eighteen days after the patient received a second dose of OPV, he developed cough and became febrile. Chest imaging was suggestive of pneumonia, and the patient's symptoms worsened while on antibiotics. No organisms were identified from an infectious workup, but a nasopharyngeal swab was performed that was positive for enterovirus RNA. A series of molecular serotyping was performed that identified Type 2 polio vaccine from the original enterovirus RNA. The case investigation further revealed the patient was hospitalized with respiratory distress 27 days after receiving the first OPV dose. The authors concluded it was unclear if the polio vaccine strain was a provoking factor or a contributing factor to pneumonia onset.

SKIN

Gianotti–Crosti syndrome (GCS) is a papular rash often associated with a number of viral infections and is sometimes reported as an adverse reaction from common vaccines. A cross-sectional analysis was performed to identify associations between GCS and OPV in children [74R]. The investigators found 116 cases of children under 5 years old that were diagnosed with GCS over an 18-month timeframe. Eleven (9.5%) and 105 (90.5%) children developed GCS 1 month before and 1 month after OPV administration, respectively (RR, 1.81; 95%CI, 1.40–2.35; $P<0.0001$). Three (2.6%) and 58 (50.0%) children developed GCS 1 week before and 1 week after OPV administration, respectively (RR, 1.90; 95%CI, 1.12–3.22; $P<0.0001$).

IMMUNOLOGIC

Immunocompromised patients who receive active formulations of the poliovirus vaccine, such as OPV, may be at risk for reversion to neurovirulence and transmissible characteristics of wild-type poliovirus strains. A 2 year old girl with common variable immunodeficiency developed sudden onset of left lower limb weakness for 1 month after receiving the first booster of the OPV [75A]. A collected stool sample was significant for vaccine-derived poliovirus type 2. The child was unable to close her left eye and displayed weakness in her lower limbs. Her weakness did not progress and she was effectively treated with intravenous immunoglobulin. The authors concluded the need to avoid OPV, especially in high-risk patients such as children with no prior immunodeficiency screening.

Rotavirus Vaccine [SED-16, 252–256; SEDA-36, 473; SEDA-37, 391; SEDA-39, 316]

Organs and Systems

GASTROINTESTINAL

Rotavirus is a common cause of diarrhea in children across the world. Widespread rotavirus vaccination efforts first lead to the discovery of intussusception as an adverse post-vaccination reaction in the late 1990s [76r]. Intussusception is a form of intestinal obstruction in which a segment of the bowel prolapses into a more distal segment. Prolonged intussusception puts patients at risk for intestinal perforation and other serious adverse effects, including death. Since the first rotavirus vaccine, RotaShield®, was developed, there have been numerous reports linking the vaccine to intussusception. While subsequent versions of the rotavirus vaccines, Rotateq® and Rotarix®, were thought not to carry that risk, additional data emerged but suggesting the risk [76r]. The mechanism of rotavirus-associated intussusception remains poorly understood.

Koch and colleagues performed a recent systematic review of the risk of intussusception after rotavirus vaccination [77M]. Of the 16 studies included in analyses, the authors found that the rotavirus vaccine was significantly associated with intussusception 1–7 days after the first vaccination (RR, 5.71; 95%CI, 4.5–7.25), with increased risk of intussusception after the second and third doses as well. The authors concluded that rotavirus vaccination should begin at 6–12 weeks old to avoid higher risk of intussusception seen in older children. Other studies have published conflicting data regarding post-vaccination risk. One retrospective cohort found no associations of rotavirus vaccine and intussusception in Korean children [78C], and quasi-experimental study found no difference in rates of intussusception in Canadian children after implementation of a rotavirus immunization program [79C].

Lopez and colleagues described a case report of an episode of acute gastroenteritis post-rotavirus vaccination in an 11 week old boy with short-bowel syndrome [80A]. The child had confirmed Hirschsprung's disease, where patients are born without parts of a function gastrointestinal tract, and had an ileostomy placed. The child received the Rotavix® rotavirus vaccine, and 3 days after developed increased stoma output of greater than 70mL/kg/day. On the 5th day following Rotarix® immunization, the patient developed bloody stomal output and became lethargic. An abdominal ultrasound was negative for intussusception. The investigators concluded the clinical and sonographic evidence supported a diagnosis of necrotizing enterocolitis, with an alternative diagnosis being Hirschprung's associated enterocolitis triggered by rotavirus vaccination.

IMMUNOLOGIC

Levin and colleagues performed a randomized trial to assess the safety and immunogenicity of the liver rotavirus vaccine on African infants exposed to the human immunodeficiency virus (HIV), with or without confirmed HIV infections [81C]. A total of 76 HIV positive and 126 HIV-status unknown patients were randomized receive rotavirus vaccine ($n=37$ HIV+, $n=62$ HIV-status unknown) or placebo ($n=39$ HIV+, $n=64$ HIV-status unknown). There were no significant differences found in adverse events, including serious adverse events ($P>0.05$).

Drug Administration

DRUG FORMULATION

Widespread rotavirus vaccination efforts in resource-limited countries can be complicated by storage requirements for the current formulation of the pentavalent rotavirus vaccine (RV), which should be kept refrigerated at 2–8°C. Martinón-Torres and colleagues assessed the efficacy and safety of a modified pentavalent rotavirus vaccine (MPRV) that is stable at room temperature [82C]. Children 6–12 weeks old were randomized to receive three doses of either the MPRV ($n=505$) or the current RV ($n=509$), and both vaccines were refrigerated to maintain an appropriate blinding protocol. The MPRV was determined to be non-inferior to the current RV in regards to immunogenicity, and no serious vaccine-related adverse effects occurred in the study. Adverse effects were reported by 86.5% of vaccinated subjects, with 50.8% of vaccinated subjects reporting a vaccine-related AE: diarrhea (24.4%), pyrexia (16.6%) and vomiting (14.1%). The majority of the adverse reported were of mild-to-moderate intensity and generally comparable between the 2 groups. Two subjects discontinued the study as a result of intussusception, which were considered serious. The 2 cases of intussusception reported were in the MPRV group and occurred more than 30 days post-dose 2 at ~5 months of age, the time of peak incidence for naturally occurring intussusception in the absence of rotavirus vaccination.

Varicella/Herpes Zoster Vaccine [SED-16, 260–365; SEDA-37, 391; SEDA-38, 320; SEDA-39, 316–319]

General

Harada and colleagues described a case of a 64 year old immunocompetent man who developed a widespread pruritic and vesicular rash 2 weeks after receiving the zoster vaccine [83A]. The patient's pertinent medical history included a Whipple procedure for a benign pancreatic cyst and diet-controlled diabetes mellitus. He first reported pimples on his chest 1-week post-vaccination, which then spread to his face, abdomen, arms, buttocks, and groin. The rash was associated with fatigue, decreased appetite, and fevers. Laboratory examination was significant for elevated liver enzymes, and the patient was found positive for VZV-specific PCR from serum, saliva, and skin swabs. The patient was treated with 7 days of intravenous acyclovir and recovered.

Kwak and colleagues performed sequencing analyses from samples of varicella zoster virus in 51 Korean children [84A]. The investigators found that two cases of VZV were caused by the vaccine-type virus in two patients without a history of varicella. Time to VZV presentation was 5 and 22 months post-vaccination in the 2 children.

Organs and Systems

SKIN

Cook described a case of a 70 year old woman who developed a cellulitic-like injection site reaction post-vaccination with the VZV vaccine [85A]. Her past medical history was significant for a splenectomy 8 years prior secondary to lymphoma. The patient received the vaccine in her left deltoid, and had red, tender swelling at the injection site later that evening. She sought medical attention the following day, was diagnosed with bacterial cellulitis, and was prescribed cephalexin. The reaction resolved spontaneously, with no effect from cephalexin, 3-weeks post-vaccination.

SENSORY SYSTEMS

Jastrzebeski and colleagues report a case of a 67 year old woman who presented with reactivated herpes zoster opthalmicus (HZO) after receiving the VZV vaccine [86A]. Patient history was significant for 5 years of prior recurrent unilateral herpes zoster keratitis. The patient continued to have mild recurrent bouts of VZV-related keratitis, iritis, and trabeculitis in the right eye, which had responded to medical management. The patient presented with another recurrence 2 weeks post-VZV vaccination, and 2 months later developed descemetocele. After 3 months, she presented with a corneal perforation managed by penetrating keratoplasty. Immunohistopathological examination was pertinent for positive VZV staining in most of the keratocytes surrounding the descemetocele and perforation. Grillo and Fraunfelder also describe a case series of 24 patients who developed kertatitis post-VZV vaccine [87A].

Yellow Fever Vaccine [SED-16, 537–540; SEDA-37, 392; SEDA-38, 321; SEDA-39, 318–319]

Organs and Systems

SKIN

A 31 year old woman presented to a travel medicine clinic to receive the yellow fever vaccine prior to departure throughout South Africa and Zimbabwe [88A]. Her past medical history was notable for chronic rhinitis, and her only active medications were loratadine and intranasal fluticasone. She received an intramuscular injection for typhoid fever and a subcutaneous injection for yellow fever and tolerated both. One day after vaccination, the patient reported a small blister overlying the yellow fever vaccine injection site. Multiple subsequent serosanguinous vesicles developed into a 4-cm bulla. The area was pruritic and erythematous with underlying induration. The patient reported no other symptoms, but was seen in an emergency department 5 days post-vaccination, where the lesions were unroofed. She was discharged home and received wound care with topical antibiotic ointment. The area eventually was noted to be hyper-pigmented, but the patient's symptoms resolved within 3 weeks. The investigators concluded the characteristics of the patient presentation were similar to previous case reports describing vesicular lesions at the yellow fever injection site.

NERVOUS SYSTEM

A 23 year old man presented with painful sensations and erythema to his right arm [89A]. The patient described electric shock-like pain on his skin and reduced visual acuity of his right eye. His past medical history was significant for right eye vision loss 4 weeks prior, which was untreated and considered idiopathic as diagnosed from an outside source. The patient had received the yellow fever vaccine 4 months earlier prior to a trip to Brazil. He had developed sub-febrile temperatures, nuchal headaches, hiccups, and vomiting episodes 2 weeks post-vaccination. An extensive neurological workup was performed that proved unremarkable, but symptoms were considered a non-specific reaction to the yellow fever vaccine. An additional cerebrospinal fluid analysis was performed that was significant for mild lymphomonocytic pleocytosis, elevated total protein with normal glucose, and positive oligoclonal bands. A MRI of the spine revealed a hyperintense lesion in the right dorsolateral spinal cord at C6/7. Re-evaluation of a MRI of the brain was suggestive of hyperintense lesions in the medullary brainstem bilaterally and involving the postrema area. AQP4-abs in the serum were elevated and a diagnosis of neuromyelitis optica was established. The patient was treated with high-dose corticosteroids and eventually rituximab, which prevented further relapses over 2 years. The investigators concluded this case to be the first suggested yellow fever vaccine to trigger neuromyelitis optica.

A 70 year old man presented to the emergency department with a 12-hour onset of expression aphasia, agraphia, dyscalculia, right–left disorientation and finger agnosia [90A]. The patient was admitted and treated for a suspected acute cerebrovascular accident, although the initial head CT was non-impressive for lesions. His past medical history was significant for type 2 diabetes mellitus, hypertension, hypercholesterolemia, and nicotine abuse. He suffered a myocardial infarction 2 years prior, and a vertebrobasilar stroke 1 year prior of which he recovered without limitations. While admitted, extensive cerebrovascular disease workup was performed. It was identified that the patient received the yellow fever vaccine 18 days prior to symptom onset in preparation for a trip to Angola. Laboratory workup, including cerebrospinal fluid analysis and neurotropic virus serology, was also normal. Yellow fever IgG was positive in serum and CSF, with negative IgM and virus PCR in CSF. An electroencephalograph was performed that was consistent with encephalopathy, and a diagnosis of yellow fever vaccine associated-neurological disease was made. The patient's condition began to spontaneously resolve, and the authors concluded the events were related to the vaccine.

A similar case was published for a 39 year old male, who was admitted to a neurological ward with 5 day complaints of a severe headache, malaise, and increasing fever [91A]. The patient had been vaccinated against yellow fever 2 weeks prior due to a planned trip to Panama. Other pertinent past medical included frequent sinusitis that last occurred 1 month prior to vaccination. Most of the patient's neurological workup showed no abnormalities, except for a CSF analyses suggestive of meningitis, and antibiotics were initiated. Further CSF workup was positive for yellow fever virus, and due to patient decline, immunoglobulin was started. Due to significant increases in the patient's liver enzymes, diagnostic workups were performed that suggested drug-induced damage of the liver. Yellow fever associated neurotropic disease with meningitis manifestation was diagnosed.

BACTERIAL VACCINES

Anthrax Vaccine [SED-16, 270, 527; SEDA-39, 319–320]

General

Anthrax is a serious infectious disease caused by exposure to *Bacillus anthracis*, a Gram-positive, non-motile spore forming bacteria. After entering a human or animal, *B. anthracis* spores are thought to germinate locally or be transported by phagocytic cells to the lymphatic system or regional lymph nodes, where they germinate (or both). Within hours of germination, *B. anthracis* produces toxins such as edema toxin (ET) and lethal toxin (LT), which have been associated with hypotension, myocardial depression, and progressive renal and hepatic dysfunction. Anthrax infection can be potentially fatal and may manifest in the skin (cutaneous), gastrointestinal tract, and/or lungs (inhalational anthrax) [92R]. Anthrax vaccination is targeted towards individuals with increased risk of exposure (e.g., military personnel, laboratory workers, certain individuals who handle high-risk animals or animal products) or those already exposed to *B. anthracis* (i.e., post-exposure prophylaxis). In June 2014, a laboratory incident involved potential aerosolization of *B. anthracis* spores at two Centers for Disease Control and Prevention (CDC) locations in Atlanta, Georgia. A survey of laboratory workers who underwent post-exposure prophylaxis with antimicrobials and immunization was conducted to assess adherence and tolerability. Of 28 individuals exposed to at least 1 dose of AVA, 24 (86%) reported an AE with the most common being injection site tenderness (75%), limited motion of arm (32%), fatigue (25%), headache (21%), and muscular pain (18%). Although most AEs were considered mild and commonly occurring, they were frequently cited by respondents as reasons for discontinuing post-exposure prophylaxis [93c].

Organs and Systems

IMMUNOLOGIC

The CDC recommends pre-exposure prophylaxis with the anthrax vaccine adsorbed (AVA) in U.S. military personnel due to their higher risk of exposure to weaponized *B. anthracis* spores compared to the civilian population. AVA was first licensed in 1970 in the U.S., and its first large-scale deployment was in 1991 during the Persian Gulf War. In 1998, the Anthrax Vaccine Immunization Program was established to protect U.S. active duty and reserve military members as well as emergency-essential individuals assigned to areas at risk for an anthrax attack. Eventually service member and public concerns relating to safety of AVA emerged, with particular attention to reports of arthralgia, one case of rheumatoid arthritis (RA), and two cases of systemic lupus erythematosus (SLE). In response to these early reports, a recent case–control study was completed to assess the relationship between RA or SLE and recent or remote exposure to AVA [94c]. Cases of RA or SLE were identified from ICD-9-CM codes in the Defense Medical Surveillance System (DMSS) database, following which cases were confirmed by medical record review and rheumatologist clinical adjudication. Among 77 RA cases identified, 13 (17%) had ever received AVA. Looking back 1095 days, there was no association between remote AVA exposure and RA (OR, 1.03; 95%CI, 0.48–2.19); however, there was a significant association if AVA was administered within 90 days of RA diagnosis (OR, 3.93; 95%CI, 1.08–14.27). Most cases (71%) who received AVA within 90 days before disease onset received more than the first 3 doses of the priming series, whereas only 2 (25%) of the controls received a fourth dose (or more) of the priming series. As the vaccine exposure interval increased from 90 days up to the maximum of 1095 days, the OR consistently trended towards 1. This indicates that exposure to AVA may accelerate RA development in certain individuals, but eventually those individuals would have developed RA as a result of other exposures and risk factors, irrespective of AVA administration. Among 39 SLE cases, 5 (13%) had ever received AVA, and there was no significant association between vaccine exposure and SLE diagnosis (OR, 0.91; 95%CI, 0.26–3.25) [94c]. This study suggests that AVA exposure may trigger early RA disease onset without modifying long-term RA risk. Specific autoimmune pathways may be activated by epitopes present in AVA; however, plausible biological mechanisms require further characterization.

Bacillus Calmette–Guérin Vaccines [SED-16, 267, 797–806]

General

The Bacillus Calmette–Guérin (BCG) vaccine contains a live, attenuated strain of *Mycobacterium bovis* administered to promote active immunity and prevent tuberculosis (TB). Alternatively, the BCG vaccine may be administered as an intravesical installation for local bladder cancer therapy [95S]. Although both uses have been clinically applicable for decades, the distribution of AEs between the preventative TB vaccine and therapeutic bladder cancer vaccine has not been differentiated until recently. A large data mining study and meta-analysis abstracted cases of BCG-associated AEs from the U.S. Vaccine Adverse Event Reporting System (VAERS) database and published cases available through a PubMed database query. A query of the VAERS database from 1990 to 2016 identified 397 AEs, of which 64 and 14 were significantly associated with BCG as a TB vaccine and cancer therapy, respectively. The most commonly associated TB vaccine AEs were lymphadenopathy (50%), cough (47%), and diarrhea (28%). The most frequently observed AE groups were 'immune system', 'respiratory system', and 'skin'. In contrast, AEs associated with the BCG cancer therapy were enriched in the urinary system (e.g. hematuria, dysuria, incontinence). A similar distribution of AE types was concluded following literature meta-analysis. Twenty-four cases of BCG-associated death were identified during the search period. Ten deaths occurred within 2 days of BCG administration, 21 deaths occurred in patients less than 1 year old, 12 deaths occurred following co-immunization with other vaccines, and most cases involved underlying immunodeficiency such as severe combined immunodeficiency (SCID) [96M].

Organs and Systems

RESPIRATORY

Previous systematic reviews and meta-analyses have yielded conflicting results with respect to BCG vaccination and an association with childhood asthma. Thus, the largest epidemiologic study to date was completed to assess potential linkage between administration of BCG vaccine and development of childhood asthma in Quebec, Canada. From 1974 to 1994, 35 612 children were administered BCG vaccine, mostly (92.3%) within the first year of life. From this birth cohort, 5870 (16.4%) eventually were identified as having developed asthma according to administrative records confirming asthma-related healthcare encounters. After adjusting for a number of previously identified confounding variables, no association between BCG vaccination and asthma was identified (OR, 0.95; 95%CI, 0.87–1.04) [97MC].

SKIN

Following a correctly administered intradermal BCG vaccination, a temporary wheal reaction is usually observed, which is often followed by a pustule and flat injection site scar. Brink and colleagues conducted a nested sub-study within a larger randomized multicenter trial to evaluate the association between post-BCG injection reaction, wheal size, and probability and size of scar development [98C]. A total of 4262 Danish infants were randomized to receive either BCG (SSI 1331 strain) vaccination or no intervention within 7 days of birth. Of the 492 BCG-vaccinated infants included in the sub-study, 427 (87%), 56 (11%), and 9 (2%) developed a wheal, bulge, or no reaction, respectively. In total, 442 (95%) infants developed a scar by the age of 13 months. Initial wheal, bulge, and non-reaction predicted eventual scar development at rates of 96%, 87%, and 56%, respectively. Notably, increased wheal size was associated with both probability and size of eventual scar development. As injection site scar development may be used as a surrogate for successful BCG vaccination, predictors of scar formation (i.e. formation and size of wheal) may be useful to confirm proper intradermal injection technique [98C]. Lack of immediate post-BCG injection site reaction may indicate suboptimal technique or underlying immunological abnormality.

In contrast to the previous study, Chen and colleagues investigated the potential utility of a novel BCG vaccine powder-laden, dissolvable microneedle array (MNA) for reducing the incidence of lesions following BCG vaccination. Following insertion of the MNA onto the skin of mice, the individual microneedle shafts melted away, allowing the BCG powder to dissolve and diffuse through interstitial fluid into the epidermis. In this small experimental study, there were no localized AEs following MNA vaccination in mice. Inoculation sites were indistinguishable from normal skin 1 day after vaccination, and BCG delivered in this powder-laden form was found to elicit an acceptable immune response [99E]. Delivery of BCG vaccine by MNA may be a painless, lesion-free, self-applied alternative to conventional intradermal BCG injection.

INFECTION RISK

A 3 year old boy presented approximately 2.5 years after neonatal BCG vaccination with a large erythematous plaque of the left shoulder where he had been immunized. The child was asymptomatic, and multiple skin-colored follicular and non-follicular papules were distributed on his face, anterior and posterior torso, and limbs. Axillary lymph nodes were enlarged and non-tender. Based on histopathologic and clinical examination, a reactive Mantoux test, and a positive DNA polymerase chain reaction for *M. bovis* from a torso papule, the boy was diagnosed with lupus vulgaris and co-existing lichen scrofulosorum (chronic cutaneous mycobacterial infection manifestations). Lupus vulgaris is a rare BCG AE (~5 per 1 million vaccinations) more commonly presenting in immunocompromised individuals and with systemic involvement. This case was unique in that the boy was deemed immunocompetent and constitutionally asymptomatic [100A].

A 2 month old girl with CHARGE syndrome (a rare genetic disorder which may include coloboma, heart defect, atresia choanae, retarded growth and development, genital hypoplasia, and ear anomalies/deafness) was administered BCG vaccine after a complicated perinatal admission. At 6 months of age, she was readmitted with fever and respiratory distress eventually requiring 100% oxygen. Laboratory analysis revealed SCID with low Immunoglobulin (Ig) G and M, as well as a low absolute lymphocyte count. Shortly thereafter, left axillary lymphadenitis was discovered and thoracic computerized tomography (CT) demonstrated consolidation and diffuse centrilobular nodules. Microbiologic examination revealed *M. bovis* from a tracheal aspirate and pus associated with the axillary lymphadenitis. Due to the risk of BCG dissemination, the patient received anti-tuberculous therapy and intravenous immune globulin (IVIG) but ultimately succumbed to BCG pneumonitis. This case underscores the need for thorough immunodeficiency screening and diagnosis prior to receipt of the live BCG vaccination, which can lead to disseminated infection in this vulnerable population [101A].

Susceptibility Factors

AGE AND ETHNICITY

As BCG vaccination for the prevention of TB typically occurs in newborns or children, they are a key demographic assessed for safety. A safety analysis conducted in Australia identified AEs following BCG immunization in children less than 7 years of age from 2009 to 2014. Using a national database of reported vaccine AEs, 110 were identified with BCG as a potential cause. Fifty-eight percent of reports were for males, and 68% were for children less than 1 year of age. The 110 AE reports encompassed 150 different reaction types, most commonly abscess (31%), injection site reaction (27%), and lymphadenopathy/lymphadenitis (17%). The rate of reported BCG-related AEs increased in Australia from 87 per 100000 doses in 2009 to 201 per 100000 doses in 2014. Although reasons for this trend are unclear, it may reflect national utilization of different BCG strains depending on changing vaccine product availability. Furthermore, the AE rate was found to increase with age at the time of BCG vaccination [102C].

In Brazil, intradermal BCG vaccination (Moreau Rio de Janiero strain) is administered to more than 99% of neonates at birth. A study conducted in Sao Paulo, Brazil identified 127 BCG-related AE in children less than 10 years of age from 2009 to 2011. Median vaccine age was 2 days and median time to first AE symptom was 2.5 months. The most common typical BCG AE were lymphadenitis/lymphadenopathy ($n=72$), abscess ($n=31$), and ulcer ($n=4$). Isoniazid was used to treat 85.7% of patients with typical AEs for a median treatment duration of 3.1 months. Most (90.6%) isoniazid-treated patients had complete AE resolution with a median time-to-resolution of 5.2 months. The most common atypical AEs were wart-like lesions ($n=7$) and BCG scar reactivation ($n=5$), and no cases of BCG dissemination were observed [103C].

Meningococcal Vaccines [SED-16, 269, 825–829; SEDA-37, 393; SEDA-38, 322; SEDA-39, 322]

General

Two quadrivalent meningococcal conjugate vaccines are licensed for use in the United States—MenACWY-D (Menactra®) and MenACWY-CRM (Menveo®). Menactra® received initial U.S. approval in 2005 for persons aged 11–55 years with subsequent approval in children as young as 9 months. As most safety data were generated pre-licensure, Hansen and colleagues sought to retrospectively describe the AE profile of Menactra® during the provision of routine care of patients at Kaiser Permanente Northern California from April 2005 to April 2006. This Phase 4, retrospective, observational study included all 11 to 55 year old patients who received Menactra® as part of routine clinical care during the study period ($n=31561$). Patients were followed for 6 months post-vaccination, and vaccine recipients were also matched to controls on the basis of age, sex, and receipt of a comparison vaccine. Overall, there were 21 outcomes with significantly elevated incidence rate ratios compared to controls. Although these included abdominal pain, vomiting, dyspnea, suicidal ideation, and febrile illness, none were found to be vaccine-related on manual chart review. Two SAEs (new-onset diabetes mellitus 4 weeks post-vaccination and new-onset juvenile rheumatoid arthritis 3 weeks post-vaccination) were considered possibly vaccine-related by study investigators. Overall, routine Menactra® vaccination in individuals aged 11–55 years was not associated with any unexpected safety signals [104MC].

Menveo® was approved in 2010 for use in persons aged 11–55 years with subsequent approval in children as young as 2 months. Pre-licensure studies of Menveo® found mostly mild, transient AEs consisting of injection site pain, redness, induration, irritability, headache, myalgia, and sleepiness. However, data regarding post-licensure AE incidence and categorization are limited. Therefore, Myers and colleagues extracted and characterized Menveo® AE reports from the CDC's VAERS database from January 2010 to December 2015. During the reporting period, 2614 AEs following Menveo® vaccination were retrieved, including 76 SAEs and 1 report of death. Concurrent vaccines were given in 1380 (53%) of reported cases. Most AE reports (74%) were in adolescents, and median age was 12 (range 0–89) years. The most frequently reported AEs were injection site pain, swelling, warmth, or redness, dizziness, headache, fever, or syncope. For SAEs, the most common system organ classifications of AEs included 'nervous system' (16 reports), 'immune system disorders' (10 reports), and 'cardiac disorders' (7 reports). Specifically, anaphylaxis (9 reports) and syncope (5 reports) were the most common SAEs. Furthermore, 14 reports of Menveo® administration during pregnancy (10/14 first trimester) were captured, resulting in no SAEs. Most vaccine recipients reported no AEs, with the exception of 1 vaccinee who experienced fever and myalgia. Menveo® is not contraindicated during pregnancy [105C].

Two serogroup B meningococcal vaccines, 4CMenB (Bexsero®) and rLP2086 (Trumenba®), were recently approved in the U.S. and Europe. Unlike their quadrivalent polysaccharide-based counterparts, these multicomponent vaccines contain protein antigens. Laboratory employees working with *N. meningitidis* exhibit a 65- to 184-fold higher risk of developing invasive meningococcal disease than the general population. As such, laboratory employees working with *N. meningitidis* are advised to receive quadrivalent conjugate meningococcal vaccine, and some countries have also updated their recommendations to include a serogroup B vaccine. In a small observational study conducted in France, 14 laboratory employees were offered Bexsero®, 12 volunteered to receive the vaccine, and 11 completed the two-dose series. In the 7 days following vaccination, ≥1 AE was self-reported in all recipients. Following the first dose, injection site reactions and sleepiness comprised 50% of AEs. Following the second dose, AEs were more common and included injection site pain (100%), swelling (88%), myalgia (50%), chills (50%), and fever (33%). Although AEs were common following Bexsero® administration in French laboratory employees, they were all mild and largely consistent with safety analyses contained in pre-licensure studies [106c].

Organs and Systems

SENSORY SYSTEMS

A 41 year old Caucasian man presented with sudden onset scotoma affecting the nasal potion of his right eye visual field, as well as mild ocular pain with right eye movement. Four days prior to symptom onset, the patient received concurrent live-attenuated yellow fever vaccine and meningococcal vaccine (formulation not specified). Clinical exam and multimodal imaging revealed a small retinal arteriolar occlusion, leading to a corresponding area of retinal greying and scotoma resembling a second physiologic blind spot. The patient's visual field improved with monitoring alone at 1-month follow-up. The event may have been alternatively explained by ischemic arteriole occlusion secondary to hypoperfusion (patient was bradycardic) or hypovolemia (patient was suffering from diarrhea). Although retinal vascular occlusion following vaccination is rare, this phenomenon has been previously reported following Hepatitis B vaccination. In this case, meningococcal vaccine was one of two vaccines the patient received shortly prior to retinal vascular occlusion, so it remains difficult to assign causality with a high degree of confidence. Ongoing surveillance for this rare, potentially vaccine-related AE should continue [107A].

IMMUNOLOGIC

Acute disseminated encephalomyelitis (ADEM) refers to a monophasic inflammatory demyelinating process affecting the central nervous system that tends to follow infection or vaccination. One hypothesis suggest that a temporary auto-immune response targets the myelin as a result of cross-reactive mimicry of viral or self-peptides. An alternative explanation suggests that the infectious agent or vaccine provokes non-specific clonal activation of autoreactive T cells, thus beginning the inflammatory cascade resulting in demyelination [108A]. A 55 year old woman with no significant past medical history presented with a tingling sensation in her left calf beginning a few hours after serogroup B meningococcal (Bexsero®) vaccination. The sensation progressed to her groin by 24 hours and to her left mammary region by day 5. A neurological exam revealed left tactile hypoesthesia with hypoalgesia at the C5 level. Cervicothoracic spinal magnetic resonance imaging demonstrated a spindle-shaped demyelinating lesion extending from C7 to T2. Cerebral images also displayed isolated bilateral frontal subcortical lesions and another precentral cortical lesion on the right side. All other studies, including examinations of the cerebrospinal fluid and serological tests, were unremarkable for alternative etiologies. Watchful waiting was employed and the functional impairment begin to resolve spontaneously. By 4 months the spinal lesion had disappeared and 6 months later the patient remained asymptomatic. Although rare, ADEM has been reported following vaccination with serogroup A and C meningococcal vaccine. This case represents the first instance of probably ADEM diagnosed following administration of serogroup B meningococcal vaccine [108A].

Urticarial vasculitis is a rare form of leukocytoclastic vasculitis characterized by erythematous wheal eruptions that clinically mimic urticaria, but these lesions persist beyond 24 hours and usually resolve with residual pigmentation. Although the cause of urticarial vasculitis is unknown, it has been reported in association with drugs, infections, and physical factors. Regarding vaccination, influenza vaccine has been most commonly linked to the AE of vasculitis, with a proposed mechanism of immune complex deposition in the blood vessel walls. A previously healthy 6 year old girl appeared to the emergency department with complaints of right ankle stiffness and asymptomatic limb lesions that had been present for 6 hours. On examination, multiple dusky, erythematous to violaceous, concentric, annular and arciform urticarial wheals were identified over her posterior lower limbs. She denied any systemic complaints including fever, ocular, or pulmonary involvement. Notable in the patient's medication history was receipt of serogroup B meningococcal vaccine (Bexsero®) 7 days before lesion onset. Expanded laboratory and infectious examinations did not reveal alternative etiologies. Cutaneous lesion biopsy was consistent with leukocytoclastic vasculitis. Urticarial lesions resolved over the next 2 weeks, leaving behind post-inflammatory hyperpigmentation. This represents the first case of possible urticarial vasculitis following serogroup B meningococcal vaccination [109A].

INFECTION RISK

Eculizumab is a humanized monoclonal antibody that acts as a terminal complement inhibitor used to treat paroxysmal nocturnal hemoglobinuria and atypical hemolytic uremic syndrome. An unwanted complication of complement inhibition is an increased risk of infection due to encapsulated bacteria, particularly *Neisseria meningitidis*. As such, patients receiving eculizumab should receive meningococcal vaccination at least 14 days prior to initiating therapy. Parikh and colleagues reported the first case of serogroup B meningococcal vaccine (Bexsero®) failure in a fully immunized 22 year old female receiving eculizumab, after she presented to the emergency department with fever, myalgia, lethargy, sore throat, and headache without photophobia or neck stiffness. There was no evidence of meningitis on lumbar puncture. The patient's blood culture, however, revealed a penicillin-resistant, vaccine-preventable strain of serogroup B meningococcus. She was treated with 7 days of intravenous ceftriaxone, followed by 10 days of oral ciprofloxacin with full resolution of infection. Six months prior, she was started on eculizumab (with penicillin chemoprophylaxis) for a diagnosis of atypical hemolytic uremia syndrome. At the time of diagnosis, she also received the MenACWY vaccine with two doses of serogroup B meningococcal vaccine (Bexsero®) given 1 month apart. This case highlights the difficulty of overcoming the complement inhibition of eculizumab, even with adequate meningococcal immunization and chemoprophylaxis. Current guidelines recommend testing antibody responses in patients receiving eculizumab before and 4–6 weeks after vaccination. Unfortunately, serogroup B titers are difficult to interpret in this context because the assay utilizes exogenous human complement, which is inactivated by eculizumab [110A].

Susceptibility Factors

AGE

Although Menveo® is approved for use in patients aged 2 months–55 years, there are limited data to describe AE occurrence during the course of routine clinical practice among specific age cohorts. Patients aged 2–10 years at Kaiser Permanente Southern California who received Menveo® from 2011 to 2014 were included in an observational cohort study. Among 327 vaccinees included in the analysis, most (72%) children received Menveo® between the ages of 7 and 10 years. A total of 45 'events of interest' were identified among the analysis population, but most were excluded based on temporality preceding vaccination. One confirmed asthma case occurring 237 days after vaccination was deemed incidental, though causality was difficult to establish. A total of 192 'serious medically attended events' were identified in the vaccine cohort, though most were excluded based on pre-existing conditions or non-plausible temporal association. The remaining 16 events were pneumonia, bronchitis, cough, febrile convulsion, and vomiting identified within 30 days of Menveo® vaccination. These findings are consistent with clinical trials of subjects 2–10 years of age, in which both pneumonia and febrile convulsion were reported within 30 days of vaccination [111C]. A similarly designed observational cohort study was conducted in 11 to 21 year old subjects receiving Menveo® as part of routine clinical practice at Kaiser Permanente Southern California from 2011 to 2013 ($n = 48899$). Approximately 72% of subjects received at least 1 concomitant vaccine with their index Menveo® administration. At total of 1127 'events of interest' were identified during the study period, of which 260 occurred after vaccination. Of these events, the majority were excluded upon further investigation by the case review committee. Bell's palsy, an 'event of interest' assessed by the case review committee, was found to have a statistically significant relative incidence (RI, 2.9; 95%CI, 1.1–7.5). Stratified analysis revealed that the risk was increased in subjects receiving concomitant vaccines (RI, 5.0; 95%CI, 1.4–17.8), but not in subjects receiving Menveo® alone (RI, 1.1; 95%CI, 0.2–5.5). Although the temporal association between Menveo® and development of Bell's palsy in subjects aged 11–21 years may be due to chance finding or concomitant vaccination, further investigation is needed to characterize this relationship [112MC].

Pertussis Vaccines (Including Diptheria–Tetanus–Acellular/Whole-Cell Pertussis Containing Vaccines) [SED-16, 257–258, 216, 269, 645–654, 764–767, 1011–1014; SEDA-37, 396; SEDA-38, 325; SEDA-39, 323–325]

Organs and Systems

SKIN

Extensive limb swelling after a booster dose of acellular pertussis (aP)-containing vaccine may cause concern and is commonly mistaken for cellulitis. In 2004 in the United Kingdom, whole-cell pertussis (wP) primary vaccination was replaced by a series of aP-containing vaccine at 2, 3, and 4 months of age. In children who have been primed with aP, redness and swelling can be increased following a booster dose of aP compared to those primed with wP vaccine. To quantify this phenomenon, Southern and colleagues assessed the frequency of extensive limb swelling (defined as swelling >100mm diameter) in 973 children in the United Kingdom who received a booster dose of aP in one of four vaccine presentations combined with inactivated polio (IPV): TdaP/IPV (Repevax®) and 3 presentations of DTaP/IPV, two of which also contained *H. influenzae* b (Hib) conjugate vaccine (Infanrix-IPV®, Infanrix-IPV+Hib®, Pediacel®). Local swelling reactions >50mm were recorded within 7 days in 2.2% of TdaP/IPV recipients and 6.6%–11.1% of DTaP/IPV recipients. Local redness reactions >50mm were recorded in 7.0% and 13.3%–17.7% of TdaP/IPV and DTaP/IPV recipients, respectively. There were no significant differences in local injection site pain or systemic symptoms across the four booster vaccine groups. A total of 13 (1.3%) children experienced extensive limb swelling, 3 after TdaP/IPV. These reactions resolved within a few days and were usually painless and without systemic involvement. Parents should be warned that a benign, transient limb swelling reaction may occur following administration of aP-containing booster vaccine in pre-school age children [113C]. It is not well understood why conversion from wP-to-aP vaccine has coincided with a higher rate of local AEs, including extensive limb swelling. A separate study comparing 4 year old children who developed pronounced local swelling or erythema ($n=30$) to controls who experienced no such AE ($n=30$) identified higher total IgE, vaccine antigen-specific IgG and IgG4, and levels of interleukin-13. This suggests that children with pronounced local reactions, including extensive limb swelling, develop more pronounced humoral and cellular immune responses to aP-containing vaccines. However, other biologic influences cannot be entirely ruled out [114c].

Second-Generation Effects

PREGNANCY

Due to the trend of increasing pertussis cases in the U.S., all pregnant patients are recommended to receive pertussis vaccination in the form of tetanus, diphtheria, and acellular pertussis (Tdap). Several recent studies and meta-analyses have demonstrated the safety of routine Tdap administration during pregnancy with respect to maternal and fetal outcomes [115C, 116M]. While large retrospective data sets and reports from the CDC's VAERS database have provided valuable insights, there are limited assessments focusing on prospective, patient-reported vaccine reactions. Perry and colleagues conducted a prospective observational study of solicited AEs within 1–7 days of Tdap administration in pregnant females from 2014 to 2016. Of 737 patients evaluated, 496 (67%; 95%CI, 64%–71%) reported at least 1 AE, and 187 (25%; 95%CI, 22%–29%) reported at least 2 AEs. The most common AEs reported were pain (65%) or swelling (15%) at the injection site, local redness (14%), myalgia (7%), and fever (3%). Two patients were admitted to the hospital following vaccination due to febrile syndromes, while 2 others presented to the emergency department with local reactions. However, it is unclear whether vaccine causality was suspected in these cases. While vaccination with Tdap during pregnancy was generally well-tolerated, 24 (3%; 95%CI, 2%–5%) of patients suggested they would not be willing to receive Tdap during a subsequent pregnancy because of the AEs they experienced [117C].

Pneumococcal Vaccines [SED-16, 836–840; SEDA-37, 395; SEDA-38, 327; SEDA-39, 325–326]

Organs and Systems

SKIN

A previously healthy 2.5 year old boy presented with a 2-day history involving a pruritic cutaneous eruption appearing 4 days after vaccination with 13-valent pneumococcal conjugate vaccine (PCV13). There were target-like papular lesions affecting extremities symmetrically with no involvement of the trunk or mucosae. The lesions contained three concentric zones: a central dark red zone, an intermediate pink zone, and an external erythematous one. Medication history, physical exam, and laboratory studies were all unremarkable with the exception of recent PCV13 administration. Although biopsy and histological analysis were not performed, the boy's clinical history in addition to morphological examination of skin lesions established a diagnosis of erythema multiforme. The cutaneous eruption resolved within 15 days of vaccination. Although erythema multiforme has been reported in association with other vaccines, this represents the first case correlated to pneumococcal vaccination [118A].

INFECTION RISK

In a series of similarly designed epidemiologic studies, Olarte and colleagues investigated the occurrence of invasive pneumococcal disease (IPD) in children at 8 pediatric hospitals in the United States during the pneumococcal conjugate vaccine era. Although infants ≤2 months of age are too young to receive pneumococcal vaccination, they may receive indirect protection from routine vaccination of older children. Following the introduction of PCV13 in 2010, IPD in infants aged 0–60 days decreased by approximately 30%, indicating a degree of indirect protection. Most neonatal IPD occurred during the first week of life (64%), and primarily during the first 48 hours. The most common clinical syndromes were bacteremia without a focus, meningitis, mastoiditis, and pneumonia. Despite reduction in neonatal IPD following introduction of PCV13, 60% of cases in the post-PCV13 era group were still caused by PCV13 serotypes, indicating only partial indirect protection [119c]. In a similar study of IPD occurring in pediatric solid organ transplant (SOT) and hematopoietic stem cell transplant (HSCT) recipients in the PCV13 era, 37% of cases were caused by PCV13 serotypes. Four cases of suspected vaccine breakthrough were identified, all in SOT recipients, with a time between transplant and IPI of 14–31.3 months. The change in serotype distribution (PCV13 vs non-PCV13 serotypes) was not statistically different before and after PCV13 vaccine introduction ($P=0.4$) [120c]. In a separate study, Olarte and colleagues evaluated the impact of routine PCV13 vaccination on the incidence of pneumococcal osteoarticular infection (OAI), including septic arthritis and osteomyelitis. During the overall study period (2000–2015), 97 (3.3%) of all IPD cases in patients ≤18 years old were categorized as OAI. Hospitalization for pneumococcal OAI caused by PCV13 serotypes decreased from 4.6 (95%CI, 3.4–6.2) to 0.9 (95%CI, 0.3–1.9) per 100000 admissions following PCV13 introduction (87% relative reduction; $P<0.0001$) [121c]. Lastly, Olarte and colleagues assessed the incidence of pediatric pneumococcal pneumonia (PP) following introduction of PCV13 in the United States. Overall, hospitalization for PP decreased from 53.6 to 23.3 per 100000 admissions post-PCV13, including a reduction in complicated PP (both $P<0.0001$). Additionally, hospitalization rates involving PCV13 serotype PP declined from 47.2 to 15.7 per 100000 admissions post-PCV13. PCV13 serotypes 19A, 3, 7F, 1, and 5 were still responsible for approximately 68% of pediatric PP cases from 2011 to 2014 [122C].

Susceptibility Factors

OTHER: ASPLENIA

A Phase 3, multicenter, open-label randomized controlled trial was conducted to assess immunogenicity and safety of the 10-valent pneumococcal non-typeable *Haemophilus influenzae* protein D conjugate vaccine (PHiD-CV) in 2–17 year old children with asplenia or splenic dysfunction. Participants were stratified according to the following age categories: 2–4 years, 5–10 years, and 11–17 years. Unprimed children received a 2-dose series of PHiD-CV (≥2 months apart), whereas primed children received a single dose. Of 45 participants included in the final analysis, all age strata experienced adequate immunogenicity for most or all serotypes included in PHiD-CV. No significant safety concerns were raised. The most common solicited local AE was pain at the injection site, whereas the most common solicited general AEs were loss of appetite (2–4 years), fatigue (5–17 years), and headache (5–17 years). The most common unsolicited AE was rhinitis after 4/54 (7.4%) doses and Grade 3 somnolence in one study participant. One non-fatal SAE (respiratory tract infection) was recorded in the 2–4 year unprimed age stratum, though it was considered by investigators to be unrelated to vaccination [123c].

OTHER: HEMATOLOGY AND ONCOLOGY

In a prospective, non-randomized, single-center cohort study, Hung and colleagues evaluated the immunogenicity and safety of PCV13 administered to Australian pediatric and adolescent oncology patients aged 1–18 years. Participants were grouped according to whether they were receiving active immunosuppressive therapy (Group 1; $n=46$) or completed immunosuppressive therapy in the prior 12 months (Group 2; $n=36$). The mean ages in Groups 1 and 2 were 8.8 and 7.1 years, respectively, with a male predominance (61% and 62%, respectively). Three (3.5%) SAEs were recorded during the study period. Two participants in Group 1 developed a fever >38.5°C within 48 hours of vaccination, resulting in admission for empiric intravenous antibiotics. Both participants were discharged 48-hours later after fever resolution and negative bacterial cultures. One additional participant in Group 1 developed non-fatal *Streptococcus pneumoniae* sepsis with a non-vaccine serotype 25 days after PCV13 vaccination. The most common solicited AEs were pain, redness, or swelling at the injection site, mostly mild and resolved after day 2. Local reactions and gastrointestinal symptoms (nausea, vomiting, and decrease appetite) were reported more frequently in

Group 1, whereas irritability was more commonly reported in Group 2. Furthermore, a lower proportion of Group 1 participants achieved protective antibody titers to PCV13 serotypes as compared to Group 2, which is biologically plausible [124c].

In an open-label, single-center vaccination study, 30 adult Japanese allogeneic HSCT recipients completed single dose immunization with the 23-valent pneumococcal polysaccharide (PPSV23) vaccine. Median age was 54 years (range 23–68), and median interval from HSCT to PPSV23 vaccination was 756 days (range 389–1903). Immune response 1 year post-vaccination was modest according to IgG titers for 7 serotypes (43%) and opsonophagocytic assay for 8 serotypes (55%). No SAEs assessed as PPSV23-related occurred during the study follow-up period. One participant developed non-fatal pneumococcal bacteremia at 998 days post-vaccination (1702 days post-HSCT), although serotype was unable to be assessed since she was hospitalized at an outside institution [125c].

OTHER: HUMAN IMMUNODEFICIENCY VIRUS

A Phase 3, open-label, single-center, partially randomized study was conducted to assess the immunogenicity and safety of PHiD-CV vaccination in South African children according to the following stratifications based on HIV status of mother and child: HIV-infected (HIV+; $n=83$), HIV-exposed-uninfected (HUE; $n=101$), and HIV-unexposed-uninfected (HUU; $n=100$). Children received PHiD-CV primary vaccination at age 6, 10, and 14 weeks (co-administered with routine childhood vaccines), followed by a booster dose at age 9–10 months. Immune response and safety were assessed for up to 14 months post-booster dose. Following primary vaccination, functional antibody response for vaccine serotypes were ≥72%, ≥81%, and ≥79% for HIV+, HEU, and HUU children, respectively. Following the booster dose, antibody response was ≥87% for all children. Injection site pain was the most common solicited AE in all groups, and irritability was the most common solicited general AE. Incidences of unsolicited AEs were similar between groups, with cough being the most frequent across all participants. Thirty-one (37%) HIV+, 25 (25%) HEU, and 20 (20%) HUU participants reported at least 1 SAE. One fatality (sudden infant death syndrome in HEU group) was considered possibly vaccine-related due temporality of the occurrence within 3 days following PHiD-CV. Two separate non-fatal SAEs were temporally associated with PHiD-CV, which included gastroenteritis in an HIV+ child and febrile seizure in an HUU child [126C].

OTHER: SICKLE CELL ANEMIA

The impact of sickle cell disease (SCD) on immunogenicity and safety of PHiD-CV was evaluated in a Phase 3, open-label, single-center, non-randomized study. Children with SCD (S) and without SCD (NS) were assigned to parallel groups according to age. Children 8–11 weeks of age received 3 primary doses and a booster dose of PHiD-CV coadministered with diphtheria–tetanus–whole cell pertussis–hepatitis B virus and *H. influenzae* type B vaccine (DTPw-HBV-Hib) and oral poliovirus vaccine, children 7–11 months of age received 2 primary doses and a booster dose of PHiD-CV, and children 12–23 months of age received 2 catch-up doses of PHiD-CV. Based on IgG titers and assays of functional opsonophagocytic activity, immune responses to PHiD-CV serotypes after age-appropriate vaccination were not influenced by SCD in children up to 2 years of age. Similarly, SCD did not appear to influence the incidence of solicited or unsolicited AEs in each of the parallel age cohorts. Pain was the most common solicited local AE (range 0%–24%), and fever was the most common solicited general AE (range 22%–80%). The most frequent unsolicited AEs post-primary vaccination were malaria, bronchitis, and gastroenteritis. Three deaths and 20 SAEs were reported during the study period; however, none of these occurrences met criteria for vaccine causality [127C].

Drug Administration

DRUG FORMULATIONS

Through serotype emergence or replacement phenomena, increasing rates of IPD due to non-vaccine serotypes is a public health challenge. As such, several investigatory vaccines target conserved surface proteins common to most or all pneumococcal strains, which may confer broader protection than existing polysaccharide antigen-based vaccines. PnuBioVax is one such candidate vaccine, which is produced from a genetically-modified *S. pneumoniae* TIGR4 strain in which the cholesterol-binding hemolytic cytotoxin pneumolysin (dPly) has been modified to remove toxicity whilst retaining immunogenicity. In a Phase 1, double-blinded, dose escalation trial, 36 subjects (18–40 years) were randomized to receive PnuBioVax, 28 days apart, at one of three dose levels (50, 200, or 500 mcg) or placebo. Following randomization to PnuBioVax ($n=27$) or placebo ($n=9$), there were 72 and 15 treatment-emergent AEs reported, respectively. A majority of AEs were classified as common vaccine-related AEs occurring in 13 (48%) PnuBioVax and 2 (22%) placebo recipients. The most common vaccine-related AE was headache, with a numerical trend suggesting a dose–response (22.0%, 33.3%, and 55.6% at the 50, 200, and 500 mcg doses, respectively). Mild-to-moderate local injection reactions of pain and tenderness were more commonly observed following PnuBioVax administration as compared to placebo (92.6% vs 44.4%), particularly with the 200 and 500 mcg doses. No SAEs were recorded during the study period [128c].

A Phase 2, multicenter, randomized controlled trial was conducted in infants from the Czech Republic, Germany, Poland, and Sweden to assess the safety and immunogenicity of two investigational protein-based pneumococcal vaccines. The investigational vaccines consisted of 10-valent pneumococcal conjugate vaccine co-formulated with dPly and either 10 or 30 mcg of pneumococcal histidine-triad protein D (PHiD-CV/dPly/PhtD-10 or PHiD-CV/dPly/PhtD-30). Infants were randomized (1:1:1:1) to receive a four-dose series of either investigational vaccine, PHiD-CV alone, or PCV13 alone. Both investigational vaccines were immunogenic for dPly and PhtD antigens, though the 30 mcg PhtD formulation was superior to the 10 mcg formulation. Incidence of AEs was similar across the 3 PHiD-CV groups, and most commonly consisted of injection site redness (33.6%–38.1%) and irritability (55.0%–56.6%). No fever >40.0°C was reported during primary vaccination, and local Grade 3 symptom incidence remained under 6% for all vaccination groups. The most frequent unsolicited AEs were conjunctivitis, bronchitis, nasopharyngitis, and rhinitis; however, incidence of unsolicited AEs thought to be vaccine-related was low at 0.2%–0.9%. One SAE in the PHiD-CV alone group (hypotonic-hyporesponsive episode on the day of first vaccination) was assessed as causally related to vaccination, and no fatalities occurred during the study [129C].

DRUG ADDITIVES

Vaccines administered from multi-dose vials generate several concerns, including contamination risk and potential for waste. PCV13 formulated with 2-phenoxyethanol, a preservative used in other pediatric vaccines, may allow expanded vaccine delivery in resource limited settings while meeting strict WHO requirements for antimicrobial suppression. A Phase 3, open-label, randomized controlled trial in Gambian infants assigned participants to receive PCV13 at ages 2, 3, and 4 months from either a multi-dose vial containing 2-phenoxyethanol ($n=245$) or a standard single dose syringe ($n=244$). Non-inferior immunogenicity of the multi-dose vial formulation was demonstrated for all serotypes according to IgG titers and opsonophagocytic assay. Local and systemic AEs following vaccination were mostly mild and comparable between groups. Local reactions occurred approximately 24-hours post-vaccination and resolved by 72 hours. No participant reported a severe local reaction. Fever rates were low (1.2%–3.6%) and comparable between groups, and no occurrences of fever >40.0°C were recorded. One sudden infant death occurred in the multi-dose vial group (day 20 after dose 3), but this SAE was considered unrelated to vaccination [130C].

DRUG DOSAGE REGIMENS

Goldblatt and colleagues investigated the immunogenicity and safety of a reduced priming PCV13 schedule (Group 1: doses at 3 and 12 months) as compared to the current schedule (Group 2: doses at 2, 4, and 12 months) in infants from the United Kingdom. In this multicenter, parallel group, randomized controlled trial, infants up to 13 weeks of age were assigned to receive either the Group 1 ($n=107$) or Group 2 ($n=106$) vaccine schedule along with other routine childhood immunizations. For 9 of 13 serotypes contained in PCV13, post-booster geometric mean IgG concentrations in Group 1 were equivalent or superior to immunogenicity observed in Group 2. Safety endpoints were limited; however, 26 SAEs were recorded in 21 (10%) participants across the study period: 18 and 8 SAEs in Groups 1 and 2, respectively. Of these, only 1 SAE (pyrexia and refusal to feed in a Group 2 participant) was considered potentially vaccine-related. Overall, a single dose PCV13 priming schedule followed by a booster dose was deemed protective and safe [131C]. A pair of Phase 3, open-label, multicenter trials were conducted to evaluate the immunogenicity and safety of a PCV13 booster dose in children depending on whether they were primed with PHiD-CV or PCV13. In the Czech Republic, children aged 12–15 months received a PCV13 booster dose following a 3-dose priming series with PHiD-CV ($n=53$) or PCV13 ($n=45$). In Slovakia, children aged 11–23 months received PCV13 following a 2-dose priming series with PHiD-CV ($n=39$) or PCV13 ($n=50$). Overall, robust IgG and opsonophagocytic assay responses were robust, regardless of initial priming schedule administered. Solicited local and systemic AEs were reported in 82.7% and 86.5% of participants in the 3-dose and 2-dose priming schedule groups, respectively. Approximately 40% of participants in both studies reported unsolicited AEs within 1 month of PCV13 booster, and these were most commonly mild respiratory symptoms. Four SAEs were reported in the 3-dose priming schedule group, while 3 SAEs were reported in the 2-dose priming schedule group. None were assessed as vaccine-related. This study suggests a degree of interchangeability between PHiD-CV and PCV13 in terms of safety and immunogenicity [132C].

Interactions

DRUG–DRUG INTERACTIONS

Belimumab, an intravenous immune globulin targeting B-lymphocyte stimulator protein, is an approved adjunctive therapy in patients with active, autoantibody-positive SLE. As belimumab inhibits the maturation and survival of B lymphocytes, it theoretically may affect vaccine response. In a Phase 4, open-label study conducted in the United States, participants diagnosed with SLE received PPSV23 either prior to commencing belimumab (pre-BEL cohort; $n=34$) or at week 24 of belimumab treatment (BEL cohort; $n=45$). Pneumococcal titers were assessed prior to first belimumab dose (week 4) in the pre-BEL cohort and at prior to the last dose of belimumab (week 28) in the BEL cohort. Most (91.1%) patients were female with a mean age of 39.6 years. At 4 weeks post-PPSV23 vaccination, a similar proportion of subjects in both cohorts achieved adequate IgG response to vaccine serotypes. Thirty-one (91.2%) and 39 (86.7%) of participants in the pre-BEL and BEL cohorts experienced ≥1 AE following PPSV23 vaccination, and the most common AEs were arthralgia and nausea. The proportion of AEs considered vaccine-related were 23.5% and 8.9%, respectively, for the pre-BEL and BEL cohorts. Four (11.8%) non-fatal SAEs were reported in the pre-BEL cohort, while 3 (6.7%) were reported in the BEL cohort. None were deemed potentially vaccine-related [133c].

Ixekizumab (IXE) is an interleukin-17A antagonist approved for the treatment of moderate-to-severe psoriasis in adults. In some cases, biologic therapies have been shown to modify the immune response to vaccines in patients with immune-mediated inflammatory diseases. Therefore, Gomez and colleagues assessed the safety and immunogenicity of tetanus (Boostrix®) and PPSV23 vaccines in the presence of absence of IXE. This Phase 1, multicenter, open-label, parallel group study randomized healthy adult participants to receive IXE ($n=41$) or control ($n=43$). Tetanus and PPSV23 vaccines were administered 2 weeks following the first 160-mg IXE dose, with a measurement of immune response 4 weeks post-vaccination. Fifty-one percent of participants were male with a mean age of 41.4 years. Non-inferior immune response was demonstrated for IXE-treated subjects based on difference in proportion of responders to tetanus (1.4%; 90% CI: −16.6 to 19.2) and PPSV23 (−0.8%; 90% CI: −12.9 to 11.0) vaccines. Fourteen (34.1%) and 6 (14.3%) of subjects in the IXE and control groups, respectively, reported a total of 43 AEs. All recorded AEs were mild in severity, and no SAEs or fatalities occurred during the study. Headache, injection site redness, and fatigue were the most common AEs reported in up to 5% of all subjects. In conclusion, IXE was not found to suppress humoral response to tetanus or PPSV23 vaccines and was well-tolerated in otherwise healthy subjects [134c].

Simultaneous administration of influenza and pneumococcal vaccines could be a viable strategy to enhance vaccine uptake in the elderly. Although comparable immunogenicity and safety of trivalent inactivated influenza vaccine (IIV3) co-administered with PPSV23 or PCV13 have been previously demonstrated, no such comparison exists for elderly patients specifically. A multicenter, randomized comparative trial was conducted during the 2012–2013 influenza season in South Korean adults aged ≥65 years. All eligible subjects received IIV3 given concomitantly with either PPSV23 ($n=110$) or PCV13 ($n=110$) in the contralateral deltoid. One month following concomitant vaccine administration, comparable immunogenicity for all three influenza strains in IIV3 was demonstrated in both groups. In contrast, pneumococcal serotype protection was not further characterized. The incidence and type of solicited AEs within 7 days of vaccination were statistically comparable between groups, and no unsolicited AEs were reporting during the study. The most common local AE was mild pain at the pneumococcal injection site (24.1%) or influenza injection site (1.8%). General AEs occurred with a frequency of 6.4% in the IIV3+PPSV23 group and 9.0% in the IIV3+PCV13 group ($P=0.572$). These were mostly mild occurrences of malaise, myalgia, or headache [135C]. Another multicenter South Korean study was conducted to assess potential interference and safety of PCV13 co-administered with MF59-adjuvanted trivalent inactivated influenza vaccine (MF59-aTIV) in adults age ≥60 years. Patients were randomized (1:1:1) to receive both vaccines simultaneously ($n=373$), PCV13 alone ($n=394$), or MF59-aTIV alone ($n=382$). Immunologic response and safety outcomes were evaluated at 1 month and 14 days post-vaccination, respectively. All study groups met non-inferior immunogenicity criteria for all 3 influenza subtypes and 13 pneumococcal serotypes, suggesting no immunological vaccine interaction. Pain at the injection site was more commonly reported in the concomitant vaccine group (55.2%) as compared to the PCV13 alone (41.4%) or MF59-aTIV alone (29.8%) group ($P<0.01$). PCV13 recipients were more likely to report redness than those receiving MF59-aTIV alone ($P=0.01$). The most frequently reported general AEs were headache (15.5%–18.0%), fatigue (21.2%–27.3%), chills (7.3%–11.8%), and myalgia (12.8%–20.1%). Myalgia was the only general AE found to be more common in PCV13-containing groups as compared to MF59-aTIV alone ($P=0.017$). No SAEs were recorded during the 14-day safety follow-up period [136C].

PARASITIC VACCINES

Malaria Vaccines [SED-16, 733–734; SEDA-39, 326–327]

General

Malaria poses a significant public health threat to humanity, with approximately 450 000 deaths occurring in 2016 according to data from the World Health Organization (WHO) [137S]. Although malaria case and mortality incidence have declined globally since 2010, the rate of decline has stalled and even reversed in some regions since 2014. There are five species of the genus *Plasmodium* which may cause disease in humans; however, *P. falciparum* accounts for greater than 90% of cases. The primary modes of malaria prevention include chemoprophylaxis in children and pregnant women residing in areas of moderate-to-high malaria transmission, as well as methods of vector control such as the use of insecticide-treated mosquito nets. Additionally, significant resources have been dedicated to the cultivation of a safe and effective malaria vaccine. More than 30 *P. falciparum* vaccine candidates are at either advanced preclinical or clinical stages of evaluation, many using recombinant protein antigens to target various stages of the parasitic life cycle. However, the RTS,S/AS01 vaccine [Mosquirix®] is the only candidate to have completed Phase 3 trials with a positive regulatory assessment [138S]. RTS,S/AS01 is a recombinant protein antigen vaccine targeting the pre-erythrocytic parasite stage. The vaccine consists of a recombinant protein virus-like particle (RTS,S), combined with the proprietary AS01 adjuvant system. The RTS,S immunogen contains a portion of *P. falciparum* circumsporozoite protein fused to the amino terminal end of the hepatitis B virus surface antigen (HBsAg) which is also used in hepatitis B vaccines. The vaccine itself is comprised of two separate preservative-free vials containing RTS,S lyophilized antigen powder and AS01 adjuvant suspension. The antigen powder is reconstituted with the adjuvant suspension, yielding 1 mL (2 doses) for intramuscular injection [139R]. Although RTS,S/AS01 is not yet licensed for routine use, the WHO has recommended an initial pilot in children aged 5–17 months of age, as a 4-dose schedule in regions with moderate-to-high malaria transmission [138S].

RTS,S/AS01 efficacy, immunogenicity, and safety were assessed in the Phase 3 Malaria-055 trial, which was conducted in 7 sub-Saharan African countries in 6537 infants 6–12 weeks of age and 8922 children 5–17 months of age. Participants in each age group were randomized (1:1:1) to receive 4 doses of RTS,S/AS01 vaccine at study months 0, 1, 2, and 20, or 3 doses of RTS,S/AS01 and 1 dose of control, or 4 doses of control vaccine. In patients aged 5–17 months, rabies vaccine was used for the first 3 control doses, whereas meningococcal serogroup C conjugate vaccine was used as the fourth control dose. Children were followed for an average of 48 months after first vaccination. Vaccine reactogenicity was evaluated in the first 200 participants enrolled at each study center who received at least one dose. Overall, incidence of solicited local and general symptoms was higher in the RTS,S/AS01 vaccine groups compared to control. Compared to rabies vaccine control, the most common local symptoms after the first 3 doses of RTS,S/AS01 were pain (12.4% vs 5.8%), swelling (9.6% vs 7.6%), and redness (3.1% vs 2.7%), while the most common general symptoms were fever (31.1% vs 13.4%), irritability (11.5% vs 5.3%), and loss of appetite (11.4% vs 7.4%). Grade 3 fever (axillary temperature >39.0°C) was reported in 2.5% of children after the first 3 doses and 5.3% of children after the fourth dose of RTS,S/AS01. No significant difference in the overall incidence of AEs was observed between RTS,S/AS01 and control vaccine with 86.1% vs 86.8% of participants, respectively, reporting at least one unsolicited AE within 30 days after vaccination. In addition to solicited symptoms, the most common unsolicited AEs potentially related to RTS,S/AS01 vaccination were injection site induration, vomiting, and diarrhea. The incidence of participants reporting at least one SAE was slightly lower in RTS,S/AS01 vaccine groups compared to controls. However, there was a non-statistically significant increase in the rate of generalized convulsive seizures within 7 days following the first 3 vaccine doses (1.04 cases per 1000 RTS,S/AS01 vaccine doses compared to 0.57 cases per 1000 rabies vaccine doses) (RR, 1.8; 95% CI, 0.6–4.9). Within 7 days following the fourth RTS,S/AS01 vaccine dose, the generalized convulsive seizure incidence was 2.5 per 1000 doses compared to 0.4 per 1000 doses of control vaccine. All febrile convulsions occurring within 7 days post-vaccination were considered to be potentially vaccine-related based on temporal association and biological plausibility [139R].

Additional safety signals warranting further evaluation include a higher incidence of post-vaccination meningitis and an association with rebound severe and/or cerebral malaria in the RTS,S/AS01 vaccination groups. Regarding meningitis, 16 cases were recorded in the 5–17 months age category following the initial 3 doses of RTS,S/AS01, as compared to only 1 case in the control group (RR, 8.0; 95%CI, 1.1–60.3). This effect was considered a chance finding due to the low number of cases (1) in the control group and lack of numerical meningitis imbalance found in the 6–12 weeks age category [139R]. In a follow-up systematic review of the RTS,S/AS01 data and surrounding literature, Gessner and colleagues contend that the meningitis safety signal observed in the RTS,S/AS01 trials was not due to malaria vaccination itself, but instead reflected a protective effect of rabies control vaccination. For this reason, the control group of the 5–17 months age category had an unusually low incidence of meningitis, whereas the rates observed with the 3- or 4-dose RTS,S/AS01 vaccination series approximated typical background meningitis rates. Further, the timing of meningitis with respect to malaria vaccination was highly variable (up to 1100 days post-dose) and did not follow any plausible cluster pattern [140M]. A separate analysis identified 22 (0.4%) cases of hospitalized malaria with low Blantyre Coma Score (a marker suggestive of cerebral malaria) following vaccination with RTS,S/AS01, as compared to 6 (0.2%) cases in the control group. Once again, this disparity in AE incidence was observed only in the 5–17 months age category and not the 6–12 weeks age category. The low biological plausibility of a pre-erythrocytic malaria vaccine impacting the pathogenesis of severe malaria at the blood stage, along with the variable time-to-onset observed, do not strongly support direct RTS,S/AS01 causality [139R]. Nonetheless, this safety signal will be further explored during the pilot vaccine implementation recommended by the WHO [138S].

Finally, a post-hoc analysis of all-cause mortality requested by the WHO identified gender-related disparities that will be further delineated in the pilot phase of RTS,S/AS01 implementation. For example, mortality was twofold higher in females receiving RTS,S/AS01 compared to control vaccine, whereas no such disparity was observed for males. Due to the limitations and potential for bias and confounding associated with post-hoc analyses, further characterization is required. No study fatalities were assessed as vaccine-related by study investigators.

Drug Administration

DRUG FORMULATIONS

In a small Phase 1b randomized controlled trial, Mensah and colleagues assessed the safety and immunogenicity of two experimental viral vector vaccines in 65 Gambian infants and neonates belonging to three age cohorts: 1–8 weeks, 8–16 weeks, and 16–24 weeks. The two experimental vaccines consist of chimpanzee adenovirus 63 (ChAd63) or modified vaccinia virus Ankara (MVA) used as viral vectors to boost the reactogenicity of a *P. falciparum* pre-erythrocytic antigen known as multiple epitope string thrombospondin-related adhesion protein (ME-TRAP). Participants received ChAd63 ME-TRAP intramuscularly via thigh injection at 1, 8, or 16 weeks, followed by MVA ME-TRAP 8 weeks later. All other routine vaccines were administered in a separate thigh according to the WHO Expanded Program on Immunization (EPI) schedule. Most AEs were mild, occurred within 24 hours of vaccination, and resolved within 48 hours of vaccination. Not unexpectedly, patients receiving malaria vaccination with concurrent EPI vaccines were found to have higher rates of fever, pain, and discoloration of the injection site. Overall, the co-administration of ChAds63 ME-TRAP and MVA ME-TRAP with EPI vaccines was considered safe and effective, but larger confirmatory studies are needed [141c].

The *P. falciparum* sporozoites (PfSPZ) vaccine is a metabolically active, non-replicating, whole malaria sporozoite vaccine that was found to be safe and protective against controlled human malaria infection in malaria-naïve individuals. Sissoko and colleagues evaluated the tolerability, safety, immunogenicity, and protective efficacy of PfSPZ in a Phase 1 randomized controlled trial conducted in malaria-experienced adults in Mali, West Africa. Adults aged 18–35 were randomized to receive 5 doses of PfSPZ or saline placebo at days 0, 28, 56, 84, and 140 during the dry season (period of low malaria transmission). Eighty eight participants (44 per group) were included in the safety analyses. PfSPZ vaccination was well-tolerated with no SAEs. All recorded PfSPZ AEs were Grade 1 (mild) and most commonly consisted of headache (7%) and myalgia (2%). The incidence of local and systemic AEs between the vaccine and placebo groups did not differ significantly, and preliminary efficacy analyses suggested significant protection against *P. falciparum* in African adults [142c].

CROSS REFERENCES

The following 2017 references discuss vaccine-related adverse events to some degree:

EBV

1. Dayer JA, Siegrist CA, Huttner A, et al. Volunteer feedback and perceptions after participation in a phase I, first-in-human Ebola vaccine trial: An anonymous survey. PLoS One. 2017 Mar 8;12(3): e0173148. [c]
2. Shukarev G, Callendret B, Luhn K, et al. A two-dose heterologous prime-boost vaccine regimen eliciting sustained immune responses to Ebola Zaire could support a preventive strategy for future outbreaks. Hum Vaccin Immunother. 2017 Feb;13(2):266-270. [M]

HBV

1. Zhu F, Deckx H, Roten R et al. Comparative Efficacy, Safety and Immunogenicity of Hepavax-Gene TF and Engerix-B Recombinant Hepatitis B Vaccines in Neonates in China. Pediatr Infect Dis J. 2017 Jan;36(1):94-101. [C]
2. Lee AW, Jordanov E, Boisnard F, et al. DTaP5-IPV-Hib-HepB, a hexavalent vaccine for infants and toddlers. Expert Rev Vaccines. 2017 Feb;16(2):85-92. [C]
3. Lalwani SK, Agarkhedkar S, Sundaram B et al. Immunogenicity and safety of 3-dose primary vaccination with combined DTPa-HBV-IPV/Hib in Indian infants. Hum Vaccin Immunother. 2017 Jan 2;13(1):120-127. [C]

4. Vesikari T, Becker T, Vertruyen AF, et al. A Phase III Randomized, Double-blind, Clinical Trial of an Investigational Hexavalent Vaccine Given at Two, Three, Four and Twelve Months. Pediatr Infect Dis J. 2017 Feb;36(2):209-215. [C]

HPV

1. Ozawa K, Hineno A, Kinoshita T, et al. Suspected Adverse Effects After Human Papillomavirus Vaccination: A Temporal Relationship Between Vaccine Administration and the Appearance of Symptoms in Japan. Drug Saf. 2017 Dec;40(12): 1219-1229. [c]
2. Carnovale C, Damavandi PT, Gentili M, et al. On the association between human papillomavirus vaccine and sleep disorders: Evaluation based on vaccine adverse events reporting systems. J Neurol Sci. 2017 Feb 15;373:179-181. [c]
3. Cramon C, Poulsen CL, Hartling UB, et al. Possible adverse effects of the quadrivalent human papillomavirus vaccine in the Region of Southern Denmark: a retrospective, descriptive cohort study. Dan Med J. 2017 Jul;64(7). pii: A5398. [c]
4. Andrews N, Stowe J, Miller E. No increased risk of Guillain-Barré syndrome after human papilloma virus vaccine: A self-controlled case-series study in England. Vaccine. 2017 Mar 23;35(13):1729-1732. [C]
5. Gallagher KE, Howard N, Kabakama S, et al. Lessons learnt from human papillomavirus (HPV) vaccination in 45 low- and middle-income countries. PLoS One. 2017; 12(6): e0177773. [M]
6. Setiawan D, Luttjeboer J, Pouwels KB, et al. Immunogenicity and safety of human papillomavirus (HPV) vaccination in Asian populations from six countries: a meta-analysis. Jpn J Clin Oncol. 2017 Mar 1;47(3):265-276. [M]
7. Signorelli C, Odone A, Ciorba V, et al. Human papillomavirus 9-valent vaccine for cancer prevention: a systematic review of the available evidence. Epidemiol Infect. 2017 Jul;145(10):1962-1982. [M].
8. Praditpornsilpa K, Kingwatanakul P, Deekajorndej T, et al. Immunogenicity and safety of quadrivalent human papillomavirus types 6/11/16/18 recombinant vaccine in chronic kidney disease stage IV, V and VD. Nephrol Dial Transplant. 2017 Jan 1;32 (1):132-136. [c]
9. Skufca J, Ollgren J, Ruokokoski E. Incidence rates of Guillain Barré (GBS), chronic fatigue/systemic exertion intolerance disease (CFS/SEID) and postural orthostatic tachycardia syndrome (POTS) prior to introduction of human papilloma virus (HPV) vaccination among adolescent girls in Finland, 2002-2012. Papillomavirus Res. 2017 Jun;3:91-96. [c]
10. Grimaldi-Bensouda L, Rossignol M, Koné-Paut I, et al. Risk of autoimmune diseases and human papilloma virus (HPV) vaccines: Six years of case-referent surveillance. J Autoimmun. 2017 May;79: 84-90. [A]

Influenza Virus

1. Airey J, Albano FR, Sawlwin DC, et al. Immunogenicity and safety of a quadrivalent inactivated influenza virus vaccine compared with a comparator quadrivalent inactivated influenza vaccine in a pediatric population: A phase 3, randomized noninferiority study. Vaccine. 2017 May 9;35(20):2745-2752. [MC]
2. Augustynowicz E, Lutyńska A, Piotrowska A, et al. The safety and effectiveness of vaccination against influenza and pertussis in pregnant women. Przegl Epidemiol. 2017;71(1):55-67. [M]
3. Baxter R, Eaton A, Hansen J, et al. Safety of quadrivalent live attenuated influenza vaccine in subjects aged 2-49 years. Vaccine. 2017 Mar 1;35(9):1254-1258. [MC]
4. Cordero E, Roca-Oporto C, Bulnes-Ramos A, et al. Two Doses of Inactivated Influenza Vaccine Improve Immune Response in Solid Organ Transplant Recipients: Results of TRANSGRIPE 1-2, a Randomized Controlled Clinical Trial. Clin Infect Dis. 2017 Apr 1;64(7):829-838. [C]
5. Demeulemeester M, Lavis N, Balthazar Y, et al. Rapid safety assessment of a seasonal intradermal trivalent influenza vaccine. Hum Vaccin Immunother. 2017 Apr 3;13(4):889-894. [c]
6. Dos Santos G, Seifert HA, Bauchau V, et al. Adjuvanted (AS03) A/H1N1 2009 Pandemic Influenza Vaccines and Solid Organ Transplant Rejection: Systematic Signal Evaluation and Lessons Learnt. Drug Saf. 2017 Aug;40(8):693-702. [M]
7. Duffy J, Lewis M, Harrington T, et al. Live attenuated influenza vaccine use and safety in children and adults with asthma. Ann Allergy Asthma Immunol. 2017 Apr;118(4):439-444. [MC]
8. Dunkle LM, Izikson R, Patriarca P, et al. Efficacy of Recombinant Influenza Vaccine in Adults 50 Years of Age or Older. N Engl J Med. 2017 Jun 22;376 (25):2427-2436. [C]
9. Dunkle LM, Izikson R, Patriarca PA, et al. Randomized Comparison of Immunogenicity and Safety of Quadrivalent Recombinant Versus Inactivated Influenza Vaccine in Healthy Adults 18-49 Years of Age. J Infect Dis. 2017 Dec 5;216 (10):1219-1226. [C]

10. Hung IFN, Yuen KY. Immunogenicity, safety and tolerability of intradermal influenza vaccines. Hum Vaccin Immunother. 2017 Jun 12:1-6. [M]
11. Treanor JT, Albano FR, Sawlwin DC, et al. Immunogenicity and safety of a quadrivalent inactivated influenza vaccine compared with two trivalent inactivated influenza vaccines containing alternate B strains in adults: A phase 3, randomized noninferiority study. Immunogenicity and safety of a quadrivalent inactivated influenza vaccine compared with two trivalent inactivated influenza vaccines containing alternate B strains in adults: A phase 3, randomized noninferiority study. Vaccine. 2017 Apr 4;35(15):1856-1864. [C]
12. Jing-Xia G, Yu-Liang Z, Jin-Feng L, et al. Safety and effectiveness assessment of 2011-2012 seasonal influenza vaccine produced in China: a randomized trial. Postgrad Med. 2017 Nov;129(8):907-914. [C]
13. Keam B, Kim MK, Choi Y, et al. Optimal timing of influenza vaccination during 3-week cytotoxic chemotherapy cycles. Cancer. 2017 Mar 1;123(5): 841-848. [c]
14. Leung DYM, Jepson B, Beck LA, et al. A clinical trial of intradermal and intramuscular seasonal influenza vaccination in patients with atopic dermatitis. J Allergy Clin Immunol. 2017 May;139(5):1575-1582. e8. [C]
15. Madan A, Ferguson M, Sheldon E, et al. Immunogenicity and safety of an AS03-adjuvanted H7N1 vaccine in healthy adults: A phase I/II, observer-blind, randomized, controlled trial. Vaccine. 2017 Mar 7;35(10):1431-1439. [c]
16. Meijer WJ, Wensing AMJ, Bos AA, et al. Influenza vaccination in healthcare workers; comparison of side effects and preferred route of administration of intradermal versus intramuscular administration. Vaccine. 2017 Mar 13;35(11):1517-1523. [C]
17. Menegay JL, Xu X, Sunil TS, et al. Live versus attenuated influenza vaccine uptake and post-vaccination influenza-like illness outcomes in HIV-infected US Air Force members. J Clin Virol. 2017 Oct;95:72-75. [c]
18. Millman AJ, Reynolds S, Duffy J, et al. Hospitalizations within 14days of vaccination among pediatric recipients of the live attenuated influenza vaccine, United States 2010-2012. Vaccine. 2017 Jan 23;35(4):529-535. [MC]
19. Norhayati MN, Ho JJ, Azman MY. Influenza vaccines for preventing acute otitis media in infants and children. Cochrane Database Syst Rev. 2017 Oct 17;10:CD010089. [M]
20. Pitisuttithum P, Boonnak K, Chamnanchanunt S, et al. Safety and immunogenicity of a live attenuated influenza H5 candidate vaccine strain A/17/turkey/Turkey/05/133 H5N2 and its priming effects for potential pre-pandemic use: a randomised, double-blind, placebo-controlled trial. Lancet Infect Dis. 2017 Aug;17(8):833-842. [c]
21. Ray GT, Lewis N, Goddard K, et al. Asthma exacerbations among asthmatic children receiving live attenuated versus inactivated influenza vaccines. Vaccine. 2017 May 9;35(20):2668-2675. [MC]
22. Sandhu SK, Hua W, MaCurdy TE, et al. Near real-time surveillance for Guillain-Barré syndrome after influenza vaccination among the Medicare population, 2010/11 to 2013/14. Vaccine. 2017 May 19;35(22):2986-2992. [MC]
23. Sarsenbayeva G, Volgin Y, Kassenov M, et al. Safety and immunogenicity of the novel seasonal preservative- and adjuvant-free influenza vaccine: Blind, randomized, and placebo-controlled trial. J Med Virol. 2018 Jan;90(1):41-49. [c]
24. Song JY, Cheong HJ, Hyun HJ, et al. Immunogenicity and safety of a 13-valent pneumococcal conjugate vaccine and an MF59-adjuvanted influenza vaccine after concomitant vaccination in ⩾60-year-old adults. Vaccine. 2017 Jan 5;35(2):313-320. [R]
25. Song JY, Choi MJ, Noh JY, et al. Randomized, double-blind, multi-center, phase III clinical trial to evaluate the immunogenicity and safety of MG1109 (egg-based pre-pandemic influenza A/H5N1 vaccine) in healthy adults. Hum Vaccin Immunother. 2017 May 4;13(5):1190-1197. [R]
26. Talaat KR, Halsey NA, Cox AB, et al. Rapid changes in serum cytokines and chemokines in response to inactivated influenza vaccination. Influenza Other Respir Viruses. 2017 Oct 9. [r]
27. Wijnans L, Dodd CN, Weibel D, et al. Bell's palsy and influenza(H1N1)pdm09 containing vaccines: A self-controlled case series. PLoS One. 2017 May 3;12(5):e0175539. [MR]
28. Zerbini CA, Ribeiro Dos Santos R, Jose Nunes M, et al. Immunogenicity and safety of Southern Hemisphere inactivated quadrivalent influenza vaccine: a Phase III, open-label study of adults in Brazil. Braz J Infect Dis. 2017 Jan - Feb;21(1):63-70. [r]

MMR

1. Timmermann CA, Budtz-Jørgensen E, Jensen TK, et al. Association between perfluoroalkyl substance exposure and asthma and allergic disease in children as modified by MMR vaccination. J Immunotoxicol. 2017 Dec;14(1):39-49. [C]

Poliovirus

1. López P, Arguedas Mohs A, Abdelnour Vásquez A, et al. A Randomized Controlled Study of a Fully Liquid DTaP-IPV-HB-PRP-T Hexavalent Vaccine for Primary and Booster Vaccinations of Healthy Infants

and Toddlers in Latin America. Pediatr Infect Dis J. 2017 Nov;36(11):e272-e28. [C]
2. Smith MJ, Jordanov E, Sheng X, et al. Safety and Immunogenicity of DTaP5-IPV Compared With DTaP5 Plus IPV as the Fifth Dose in Children 4-6 Years of Age. Pediatr Infect Dis J. 2017 Mar;36 (3):319-325. [C]
3. Li Y, Li RC, Ye Q, et al. Safety, immunogenicity and persistence of immune response to the combined diphtheria, tetanus, acellular pertussis, poliovirus and Haemophilus influenzae type b conjugate vaccine (DTPa-IPV/Hib) administered in Chinese infants. Hum Vaccin Immunother. 2017 Mar 4;13(3):588-598. [C]
4. Zhao D, Ma R, Zhou T, et al. Introduction of Inactivated Poliovirus Vaccine and Impact on Vaccine-Associated Paralytic Poliomyelitis - Beijing, China, 2014-2016. MMWR Morb Mortal Wkly Rep. 2017 Dec 15;66(49):1357-1361. [r]
5. O'Reilly KM, Lamoureux C, Molodecky NA, et al. An assessment of the geographical risks of wild and vaccine-derived poliomyelitis outbreaks in Africa and Asia. BMC Infect Dis. 2017 May 26;17(1):367. [M]

Rotavirus

1. Yen C, Shih SM, Tate JE, et al. Intussusception-related Hospitalizations Among Infants Before and After Private Market Licensure of Rotavirus Vaccines in Taiwan, 2001-2013. Pediatr Infect Dis J. 2017 Oct;36 (10):e252-e257. [C]
2. Kim A, Chang JY, Shin S, et al. Epidemiology and Factors Related to Clinical Severity of Acute Gastroenteritis in Hospitalized Children after the Introduction of Rotavirus Vaccination. J Korean Med Sci. 2017 Mar;32(3):465-474. [C]
3. Zaman K, Sack DA, Neuzil KM, et al. Effectiveness of a live oral human rotavirus vaccine after programmatic introduction in Bangladesh: A cluster-randomized trial. PLoS Med. 2017 Apr 18;14(4):e1002282. [C]
4. Vaarala O, Jokinen J, Lahdenkari M. Rotavirus Vaccination and the Risk of Celiac Disease or Type 1 Diabetes in Finnish Children at Early Life. Pediatr Infect Dis J. 2017 Jul;36(7):674-675. [C]
5. Gadroen K, Kemmeren JM, Bruijning-Verhagen PC, et al. Baseline incidence of intussusception in early childhood before rotavirus vaccine introduction, the Netherlands, January 2008 to December 2012. Euro Surveill. 2017 Jun 22;22(25). [C]
6. Lazarus RP, John J, Shanmugasundaram E, et al. The effect of probiotics and zinc supplementation on the immune response to oral rotavirus vaccine: A randomized, factorial design, placebo-controlled study among Indian infants. Vaccine. 2018 Jan 4;36 (2):273-279. [C]
7. Groome MJ, Koen A, Fix A, et al. Safety and immunogenicity of a parenteral P2-VP8-P[8] subunit rotavirus vaccine in toddlers and infants in South Africa: a randomised, double-blind, placebo-controlled trial. Lancet Infect Dis. 2017 Aug;17(8):843-853. [C]

VZV

1. Lotan I, Steiner I. Giant cell arteritis following varicella zoster vaccination. J Neurol Sci. 2017 Apr 15;375:158-159. [A]
2. Grupping K, Campora L, Douha M, et al. Immunogenicity and Safety of the HZ/su Adjuvanted Herpes Zoster Subunit Vaccine in Adults Previously Vaccinated With a Live Attenuated Herpes Zoster Vaccine. J Infect Dis. 2017 Dec 12;216(11):1343-1351. [C]
3. Vink P, Shiramoto M, Ogawa M, et al. Safety and immunogenicity of a Herpes Zoster subunit vaccine in Japanese population aged $\geq$50 years when administered subcutaneously vs. intramuscularly. Hum Vaccin Immunother. 2017 Mar 4;13(3):574-578. [C]
4. Su JR, Leroy Z, Lewis PW, et al. Safety of Second-Dose Single-Antigen Varicella Vaccine. Pediatrics. 2017 Mar;139(3). pii: e20162536. [C]
5. Parrino J, McNeil SA, Lawrence SJ, et al. Safety and immunogenicity of inactivated varicella-zoster virus vaccine in adults with hematologic malignancies receiving treatment with anti-CD20 monoclonal antibodies. Vaccine. 2017 Mar 27;35(14):1764-1769. [C]
6. Khan N, Shah Y, Trivedi C, et al. Safety of herpes zoster vaccination among inflammatory bowel disease patients being treated with anti-TNF medications. Aliment Pharmacol Ther. 2017 Oct;46 (7):668-672. [MC]
7. Schwarz TF, Aggarwal N, Moeckesch B, et al. Immunogenicity and Safety of an Adjuvanted Herpes Zoster Subunit Vaccine Coadministered With Seasonal Influenza Vaccine in Adults Aged 50 Years or Older. J Infect Dis. 2017 Dec 12;216(11):1352-1361. [C]
8. Godeaux O, Kovac M, Shu D, et al. Immunogenicity and safety of an adjuvanted herpes zoster subunit candidate vaccine in adults $\geq$ 50 years of age with a prior history of herpes zoster: A phase III, non-randomized, open-label clinical trial. Hum Vaccin Immunother. 2017 May 4;13(5):1051-1058. [c]

BCG

1. Loxton AG, Knaul JK, Grode L et al. Safety and immunogenicity of the recombinant *Mycobacterium bovis* BCG vaccine VPM1002 in HIV-unexposed newborn infants in South Africa. *Clin Vaccine Immunol*. 2017;24(2): pii: e00439-16. [c].

References

[1] Torresi J, Richmond PC, Heron LG, et al. Replication and excretion of the live attenuated tetravalent dengue vaccine CYD-TDV in a flavivirus-naive adult population: assessment of vaccine viremia and virus shedding. J Infect Dis. 2017;216(7):834–41 [C].
[2] Whitehead SS, Durbin AP, Pierce KK, et al. In a randomized trial, the live attenuated tetravalent dengue vaccine TV003 is well-tolerated and highly immunogenic in subjects with flavivirus exposure prior to vaccination. PLoS Negl Trop Dis. 2017;11(5): e0005584 [C].
[3] Agarwal R, Wahid MH, Yausep OE, et al. The immunogenicity and safety of CYD-tetravalent dengue vaccine (CYD-TDV) in children and adolescents: a systematic review. Acta Med Indones. 2017;49(1):24–33 [M].
[4] Skibinski DA, Baudner BC, Singh M, et al. Combination vaccines. J Glob Infect. 2011;3(1):63–72 [M].
[5] Melo FIR, Morales JJR, De Los Santos AHM, et al. Immunogenicity and safety of a booster injection of DTap-IPV// Hib (Pentaxim) administered concomitantly with tetravalent dengue vaccine in healthy toddlers 15-18 months of age in Mexico: a randomized trial. Pediatr Infect Dis J. 2017;36(6):602–8 [C].
[6] Report of an International Commission. Ebola haemorrhagic fever in Zaire, 1976. Bull World Health Organ. 1978;56:271–93 [S].
[7] Report of a WHO/International Study Team. Ebola haemorrhagic fever in Sudan, 1976. Bull World Health Organ. 1978;56:247–70 [S].
[8] Centers for Disease Control and Prevention. Outbreaks chronology: ebola virus disease, http://www.cdc.gov/vhf/ebola/outbreaks/history/chronology.html; 2018. Accessed 4 January 2018 [S].
[9] World Health Organization. Factors that contributed to undetected spread of the Ebola virus and impeded rapid containment, http://www.who.int/csr/disease/ebola/one-year-report/factors/en/; 2015. Accessed 1 February 2018 [S].
[10] Kadanali A, Karagoz G. An overview of Ebola virus disease. North Clin Istanb. 2015;2(1):81–6 [R].
[11] Peters CJ, LeDuc JW. An introduction to Ebola: the virus and the disease. J Infect Dis. 1999;179(Suppl 1):ix–xvi [R].
[12] World Health Organization. Ebola virus disease, http://www.who.int/csr/disease/ebola/en/; 2015. Accessed 1 February 2018 [S].
[13] BBC. Ebola: mapping the outbreak, http://www.bbc.com/news/world-africa-28755033; 2016. Accessed 1 February 2018 [S].
[14] Center for Disease Control and Prevention. Ebola virus disease, http://www.cdc.gov/vhf/ebola/index.html; 2015. Accessed February 1, 2018 [S].
[15] Center for Disease Control and Prevention. Life cycle of Ebola virus disease, http://www.cdc.gov/ncezid/dhcpp/vspb/images/carosel/ebola-ecology.jpg; 2018. Accessed 1 February 2018 [S].
[16] Li J, Duan HJ, Chen HY, et al. WHO: ebola virus disease, 2015. Int J Infect Dis. 2016;42:34–9 [S].
[17] Bah E, Lamah MC, Fletcher T, et al. Clinical presentation of patients with Ebola virus disease in Conakry, Guinea. NEJM. 2015;372:40–7 [r].
[18] Clark DV, Kibuuka H, Millard M, et al. Long-term sequelae after Ebola virus disease in Bundibugyo, Uganda: a retrospective cohort study. Lancet Infect Dis. 2015;15(8):905–12 [r].
[19] Kilgore PE, Grabenstein JD, Salim AM, et al. Treatment of ebola virus disease. Pharmacotherapy. 2015;35(1):43–53 [R].
[20] Fowler RA, Fletcher T, Fischer WA, et al. Caring for critically ill patients with ebola virus disease. Perspectives from West Africa. Am J Respir Crit Care Med. 2014;190(7):733–7 [R].
[21] Wang Y, Li J, Hu Y, et al. Ebola vaccines in clinical trial: the promising candidates. Hum Vaccin Immunother. 2017;13(1):153–68 [M].
[22] Center for Disease Control and Prevention. Ebola virus disease (EVD), http://www.cdc.gov/about/pdf/ebola/ebola-photobook-070915.pdf; 2015. Accessed 1 February 2018 [S].
[23] World Health Organization. WHO strategic plan response 2015: West Africa Ebola outbreak, http://www.who.int/csr/resources/publications/ebola/ebola-strategic-plan/en/; 2015. Accessed 1 February 2018 [S].
[24] Kennedy SB, Bolay F, Kieh M, et al. Phase 2 placebo-controlled trial of two vaccines to prevent Ebola in Liberia. N Engl J Med. 2017;377(15):1438–47 [C].
[25] Gsell PS, Camacho A, Kucharski AJ, et al. Ring vaccination with rVSV-ZEBOV under expanded access in response to an outbreak of Ebola virus disease in Guinea, 2016: an operational and vaccine safety report. Lancet Infect Dis. 2017;17(12):1276–84 [MC].
[26] Halperin SA, Arribas JR, Rupp R, et al. Six-month safety data of recombinant vesicular stomatitis virus-Zaire Ebola virus envelope glycoprotein vaccine in a phase 3 double-blind, placebo-controlled randomized study in healthy adults. J Infect Dis. 2017;215(12):1789–98 [MC].
[27] Lambe T, Bowyer G, Ewer KJ, et al. A review of phase I trials of Ebola virus vaccines: what can we learn from the race to develop novel vaccines? Philos Trans R Soc Lond B Biol Sci. 2017;372(1721):pii: 20160295 [R].
[28] Dolzhikova IV, Zubkova OV, Tukhvatulin AI, et al. Safety and immunogenicity of GamEvac-Combi, a heterologous VSV- and Ad5-vectored Ebola vaccine: an open phase I/II trial in healthy adults in Russia. Hum Vaccin Immunother. 2017;13(3):613–20 [c].
[29] Huttner A, Combescure C, Grillet S, et al. A dose-dependent plasma signature of the safety and immunogenicity of the rVSV-Ebola vaccine in Europe and Africa. Sci Transl Med. 2017;9(385) [C].
[30] Heppner Jr. DG, Kemp TL, Martin BK, et al. Safety and immunogenicity of the rVSV△G-ZEBOV-GP Ebola virus vaccine candidate in healthy adults: a phase 1b randomised, multicentre, double-blind, placebo-controlled, dose-response study. Lancet Infect Dis. 2017;17(8):854–66 [C].
[31] Agnandji ST, Fernandes JF, Bache EB, et al. Safety and immunogenicity of rVSVΔG-ZEBOV-GP Ebola vaccine in adults and children in Lambaréné, Gabon: a phase I randomised trial. PLoS Med. 2017;14(10):e1002402 [C].
[32] Regules JA, Beigel JH, Paolino KM, et al. A recombinant vesicular stomatitis virus Ebola vaccine. N Engl J Med. 2017;376(4): 330–41 [c].
[33] Ledgerwood JE, DeZure AD, Stanley DA, et al. Chimpanzee Adenovirus Vector Ebola Vaccine. N Engl J Med. 2017;376(10):928–38 [c].
[34] Li F, Hu Y, Zhou Y, et al. A randomized controlled trial to evaluate a potential hepatitis B booster vaccination strategy using combined hepatitis A and B vaccine. Pediatr Infect Dis J. 2017;36(5):e157–61 [C].
[35] Maritsi DN, Coffin SE, Argyri I, et al. Immunogenicity and safety of the inactivated hepatitis A vaccine in children with juvenile idiopathic arthritis on methotrexate treatment: a matched case-control study. Clin Exp Rheumatol. 2017;35(4):711–5 [C].
[36] van Gemeren MA, van Wijngaarden P, Doukas M, et al. Vaccine-related autoimmune hepatitis: the same disease as idiopathic autoimmune hepatitis? Two clinical reports and review. Scand J Gastroenterol. 2017;52(1):18–22 [A].
[37] Shi J, Feng Y, Gao L, et al. Immunogenicity and safety of a high-dose hepatitis B vaccine among patients receiving methadone maintenance treatment: a randomized, double-blinded, parallel-controlled trial. Vaccine. 2017;35(18):2443–8 [C].

[38] Mulley WR, Le ST, Ives KE. Primary seroresponses to double-dose compared with standard-dose hepatitis B vaccination in patients with chronic kidney disease: a systematic review and meta-analysis. Nephrol Dial Transplant. 2017;32(1):136–43 [M].
[39] Yang FQ, Rao GR, Wang GQ, et al. Phase IIb trial of in vivo electroporation mediated dual-plasmid hepatitis B virus DNA vaccine in chronic hepatitis B patients under lamivudine therapy. World J Gastroenterol. 2017;23(2):306–17 [C].
[40] Costa APF, Cobucci RNO, da Silva JM, et al. Safety of human papillomavirus 9-valent vaccine: a meta-analysis of randomized trials. J Immunol Res. 2017;2017:3736201 [M].
[41] Chandler RE, Juhlin K, Fransson J, et al. Current safety concerns with human papillomavirus vaccine: a cluster analysis of reports in VigiBase®. Drug Saf. 2017;40(1):81–90 [C].
[42] Lipkind HS, Vazquez-Benitez G, Nordin JD, et al. Maternal and infant outcomes after human papillomavirus vaccination in the periconceptional period or during pregnancy. Obstet Gynecol. 2017;130(3):599–608 [C].
[43] Scheller NM, Pasternak B, Mølgaard-Nielsen D, et al. Quadrivalent HPV vaccination and the risk of adverse pregnancy outcomes. N Engl J Med. 2017;376(13):1223–33 [C].
[44] Vera-Lastra O, Medina G, Cruz-Dominguez Mdel P. Autoimmune/inflammatory syndrome induced by adjuvants (Shoenfeld's syndrome): clinical and immunological spectrum. Expert Rev Clin Immunol. 2013;9(4):361–73 [R].
[45] Jara LJ, García-Collinot G, Medina G, et al. Severe manifestations of autoimmune syndrome induced by adjuvants (Shoenfeld's syndrome). Immunol Res. 2017;65(1):8–16 [M].
[46] Nichols AJ, Allen AH, Shareef S, et al. Association of human papillomavirus vaccine with the development of keratinocyte carcinomas. JAMA Dermatol. 2017;153(6):571–4 [A].
[47] Swamy GK, Heine RP. Vaccinations for pregnant women. Obstet Gynecol. 2015;125(1):212–26 [R].
[48] Centers for Disease Control and Prevention. Key facts about influenza vaccine. CDC; 2018. Available at:https://www.cdc.gov/flu/protect/keyfacts.htm. Accessed 1 February 2018 [S].
[49] Cai Y, Du J, Huang J, et al. A signal detection method for temporal variation of adverse effect with vaccine adverse event reporting system data. BMC Med Inform Decis Mak. 2017;17 (Suppl 2):76 [C].
[50] Clothier HJ, Crawford N, Russell MA, et al. Allergic adverse events following 2015 seasonal influenza vaccine, Victoria, Australia. Euro Surveill. 2017;22(20):pii: 30535 [C].
[51] Goodwin K, Viboud C, Simonsen L. Antibody response to influenza vaccination in the elderly: a quantitative review. Vaccine. 2006;24:1159–69 [R].
[52] Kaka AS, Filice GA, Myllenbeck S, et al. Comparison of side effects of the 2015-2016 high-dose, inactivated, trivalent influenza vaccine and standard dose, inactivated, trivalent influenza vaccine in adults ≥65 years. Open Forum Infect Dis. 2017;4(1) ofx001 [C].
[53] Wilkinson K, Wei Y, Szwajcer A, et al. Efficacy and safety of high-dose influenza vaccine in elderly adults: a systematic review and meta-analysis. Vaccine. 2017;35(21):2775–80 [C].
[54] Lieberman A, Curtis L. HSV2 reactivation and myelitis following influenza vaccination. Hum Vaccin Immunother. 2017; 13(3):572–3 [A].
[55] Moro P, Baumblatt J, Lewis P, et al. Surveillance of adverse events after seasonal influenza vaccination in pregnant women and their infants in the vaccine adverse event reporting system, July 2010-May 2016. Drug Saf. 2017;40(2):145–52 [C].
[56] Arriola CS, Vasconez N, Thompson MG, et al. Association of influenza vaccination during pregnancy with birth outcomes in Nicaragua. Vaccine. 2017;35(23):3056–63 [r].
[57] Hviid A, Svanström H, Mølgaard-Nielsen D, et al. Association between pandemic influenza A(H1N1) vaccination in pregnancy and early childhood morbidity in offspring. JAMA Pediatr. 2017;171(3):239–48 [MC].
[58] McHugh L, Andrews RM, Lambert SB, et al. Birth outcomes for Australian mother-infant pairs who received an influenza vaccine during pregnancy, 2012-2014: the FluMum study. Vaccine. 2017;35(10):1403–9 [MC].
[59] Zerbo O, Qian Y, Yoshida C, et al. Association between influenza infection and vaccination during pregnancy and risk of autism spectrum disorder. JAMA Pediatr. 2017;171(1):e163609 [MC].
[60] Hibino M, Kondo T. Interstitial pneumonia associated with the influenza vaccine: a report of two cases. Intern Med. 2017;56(2):197–201 [A].
[61] Trogstad L, Bakken IJ, Gunnes N, et al. Narcolepsy and hypersomnia in Norwegian children and young adults following the influenza A(H1N1) 2009 pandemic. Vaccine. 2017;35(15):1879–85 [MC].
[62] Papke D, McNussen PJ, Rasheed M, et al. A case of unilateral optic neuropathy following influenza vaccination. Semin Ophthalmol. 2017;32(4):517–23 [A].
[63] Turtle L, Tatullo F, Bali T, et al. Cellular immune responses to live attenuated Japanese encephalitis (JE) vaccine SA14-14-2 in adults in a JE/dengue co-endemic area. PLoS Negl Trop Dis. 2017;11(1): e0005263 [r].
[64] Dubischar KL, Kadlecek V, Sablan Jr. B, et al. Safety of the inactivated Japanese encephalitis virus vaccine IXIARO in children: an open-label, randomized, active-controlled, phase 3 study. Pediatr Infect Dis J. 2017;36(9):889–97 [C].
[65] Chotpitayasunondh T, Pruekprasert P, Puthanakit T, et al. Post-licensure, phase IV, safety study of a live attenuated Japanese encephalitis recombinant vaccine in children in Thailand. Vaccine. 2017;35(2):299–304 [C].
[66] Kosalaraksa P, Watanaveeradej V, Pancharoen C, et al. Long-term immunogenicity of a single dose of Japanese encephalitis chimeric virus vaccine in toddlers and booster response 5 years after primary immunization. Pediatr Infect Dis J. 2017;36(4):e108–13 [C].
[67] Sricharoenchai S, Lapphra K, Chuenkitmongkol S, et al. Immunogenicity [M1] of a live attenuated chimeric Japanese encephalitis vaccine as a booster dose after primary vaccination with live attenuated SA14-14-2 vaccine: a phase IV study in Thai children. Pediatr Infect Dis J. 2017;36(2):e45–7 [c].
[68] Neven B, Pérot P, Bruneau J, et al. Cutaneous and visceral chronic granulomatous disease triggered by a rubella virus vaccine strain in children with primary immunodeficiencies. Clin Infect Dis. 2017;64(1):83–6 [A].
[69] Kuniyoshi K, Hatsukawa Y, Kimura S, et al. Acute bilateral photoreceptor degeneration in an infant after vaccination against measles and rubella. JAMA Ophthalmol. 2017;135(5):478–82 [A].
[70] Rivera L, Pedersen RS, Peña L, et al. Immunogenicity and safety of three aluminium hydroxide adjuvanted vaccines with reduced doses of inactivated polio vaccine (IPV-Al) compared with standard IPV in young infants in the Dominican Republic: a phase 2, non-inferiority, observer-blinded, randomised, and controlled dose investigation trial. Lancet Infect Dis. 2017;17(7): 745–753 [C].
[71] Resik S, Tejeda A, Diaz M, et al. Boosting immune responses following fractional-dose inactivated poliovirus vaccine: a randomized, controlled trial. J Infect Dis. 2017;215(2): 175–182 [C].
[72] Lindgren LM, Tingskov PN, Justesen AH, et al. First-in-human safety and immunogenicity investigations of three adjuvanted reduced dose inactivated poliovirus vaccines (IPV-Al SSI) compared to full dose IPV vaccine SSI when given as a booster vaccination to adolescents with a history of IPV vaccination at 3, 5, 12months and 5years of age. Vaccine. 2017;35(4):596–604 [C].

[73] Li MZ, Zhang TG, Li AH, et al. A pneumonia case associated with type 2 polio vaccine strains. Chin Med J (Engl). 2017;130(1):111–2 [A].
[74] Zawar V, Chuh A. A case-control study on the association of pulse oral poliomyelitis vaccination and Gianotti-Crosti syndrome. Int J Dermatol. 2017;56(1):75–9 [R].
[75] Gomber S, Arora V, Dewan P. Vaccine associated paralytic poliomyelitis unmasking common variable immunodeficiency. Indian Pediatr. 2017;54(3):241–2 [A].
[76] Cale CM, Klein NJ. The link between rotavirus vaccination and intussusception: implications for vaccine strategies. Gut. 2002;50(1):11–2 [r].
[77] Koch J, Harder T, von Kries R, et al. Risk of intussusception after rotavirus vaccination. Dtsch Arztebl Int. 2017;114(15):255–62 [M].
[78] Kim KY, Kim DS. Relationship between pentavalent rotavirus vaccine and intussusception: a retrospective study at a single Center in Korea. Yonsei Med J. 2017;58(3):631–6 [C].
[79] Hawken S, Ducharme R, Rosella LC, et al. Assessing the risk of intussusception and rotavirus vaccine safety in Canada. Hum Vaccin Immunother. 2017;13(3):703–10 [C].
[80] Lopez RN, Krishnan U, Ooi CY. Enteritis with pneumatosis intestinalis following rotavirus immunisation in an infant with short bowel syndrome. BMJ Case Rep. 2017;2017:pii: bcr-2017-219482 [A].
[81] Levin MJ, Lindsey JC, Kaplan SS, et al. Safety and immunogenicity of a live attenuated pentavalent rotavirus vaccine in HIV-exposed infants with or without HIV infection in Africa. AIDS. 2017;31(1):49–59 [C].
[82] Martinón-Torres F, Greenberg D, Varman M, et al. Safety, tolerability and immunogenicity of pentavalent rotavirus vaccine manufactured by a modified process. Pediatr Infect Dis J. 2017;36(4):417–22 [C].
[83] Harada K, Heaton H, Chen J, et al. Zoster vaccine-associated primary varicella infection in an immunocompetent host. BMJ Case Rep. 2017;2017:pii: bcr-2017-221166 [A].
[84] Kwak BO, Lee HJ, Kang HM, et al. Genotype analysis of ORF 62 identifies varicella-zoster virus infections caused by a vaccine strain in children. Arch Virol. 2017;162(6):1725–30 [A].
[85] Cook IF. Herpes zoster vaccine (Zostavax®): cellulitic injection site reaction or bacterial cellulitis? Hum Vaccin Immunother. 2017;13(4):784–5 [A].
[86] Jastrzebski A, Brownstein S, Ziai S, et al. Reactivation of herpes zoster keratitis with corneal perforation after zoster vaccination. Cornea. 2017;36(6):740–2 [A].
[87] Grillo AP, Fraunfelder FW. Keratitis in association with herpes zoster and varicella vaccines. Drugs Today (Barc). 2017;53(7):393–7 [A].
[88] Wauters RH, Hernandez CL, Petersen MM. An atypical local vesicular reaction to the yellow fever vaccine. Vaccines (Basel). 2017;5(3). pii: E26 [A].
[89] Schöberl F, Csanadi E, Eren O, et al. NMOSD triggered by yellow fever vaccination—an unusual clinical presentation with segmental painful erythema. Mult Scler Relat Disord. 2017;11:43–4 [A].
[90] Beirão P, Pereira P, Nunes A, et al. Yellow fever vaccine-associated neurological disease, a suspicious case. BMJ Case Rep. 2017;2017: pii: bcr2016218706 [A].
[91] Florczak-Wyspiańska J, Nawotczyńska E, Kozubski W. Yellow fever vaccine-associated neurotropic disease (YEL-AND)—a case report. Neurol Neurochir Pol. 2017;51(1):101–5 [A].
[92] Hendricks KA, Wright ME, Shadomy SV, et al. Centers for disease control and prevention expert panel meetings on prevention and treatment of anthrax in adults. Emerg Infect Dis. 2014;20(2): e130687. https://doi.org/10.3201/eid2002.130687 [R].
[93] Nolen LD, Traxler RM, Kharod GA, et al. Postexposure prophylaxis after possible anthrax exposure: adherence and adverse events. Health Secur. 2016;14(6):419–23 [c].
[94] Bardenheier BH, Duffy J, Duderstadt SK, et al. Anthrax vaccine and the risk of rheumatoid arthritis and systemic lupus erythematosus in the U.S. military: a case-control study. Mil Med. 2016;181(10):1348–56 [c].
[95] Bacillus Calmette-Guérin Vaccine 2018 Lexi-Comp Online®, AHFS DI (Adult and Pediatric), Hudson, Ohio: Lexi-Comp, Inc. Accessed: January 3, 2018 [S].
[96] Xie J, Codd C, Mo K, et al. Differential adverse event profiles associated with BCG as a preventive tuberculosis vaccine or therapeutic bladder cancer vaccine identified by comparative ontology-based VAERS and literature meta-analysis. PLoS One. 2016;11(10):e0164792. https://doi.org/10.1371/journal.pone.0164792 [M].
[97] El-Zein M, Conus F, Benedetti A, et al. Association between Bacillus Calmette-Guérin vaccination and childhood asthma in the Quebec birth cohort on immunity and health. Am J Epidemiol. 2017;186(3):344–55 [MC].
[98] Birk NM, Nissen TN, Ladekarl M, et al. The association between Bacillus Calmette-Guérin vaccination (1331 SSI) skin reaction and subsequent scar development in infants. BMC Infect Dis. 2017;17(1):540 [C].
[99] Chen F, Yan Q, Yu Y, et al. BCG vaccine powder-laden and dissolvable microneedle arrays for lesion-free vaccination. J Control Release. 2017;255:36–44 [E].
[100] Angoori GR. Coexisting Bacillus Calmette-Guérin-induced lupus vulgaris involving the vaccination site and lichen scrofulosorum in an immunocompetent boy. Pediatr Dermatol. 2016;33(5): e274–5 [A].
[101] Kim HY, Kim YM, Park HJ. Disseminated BCG pneumonitis revealing severe combined immunodeficiency in CHARGE syndrome. Pediatr Pulmonol. 2017;52(2):E4–6 [A].
[102] Hendry AJ, Dey A, Beard FH, et al. Adverse events following immunization with bacille Calmette-Guérin vaccination: baseline data to inform monitoring in Australia following introduction of new unregistered BCG vaccine. Commun Dis Intell Q Rep. 2016;40(4):E470–4 [C].
[103] Moreira TN, Moraes-Pinto MI, Costa-Carvalho BT, et al. Clinical management of localized BCG adverse events in children. Rev Inst Med Trop Sao Paulo. 2016;58:84 [C].
[104] Hansen J, Zhang L, Klein NP, et al. Post-licensure safety surveillance study of routine use of quadrivalent meningococcal diphtheria toxoid conjugate vaccine. Vaccine. 2017;35(49 Pt B):6879–84 [MC].
[105] Myers TR, McNeil MM, Ng CS, et al. Adverse events following quadrivalent meningococcal CRM-conjugate vaccine (Menveo®) reported to the vaccine adverse event reporting system (VAERS), 2010-2015. Vaccine. 2017;35(14):1758–63 [C].
[106] Hong E, Terrade A, Taha MK. Immunogenicity and safety among laboratory workers vaccinated with Bexsero® vaccine. Hum Vaccin Immunother. 2017;13(3):645–8 [c].
[107] Moysidis SN, Koulisis N, Patel VR, et al. The second blind spot: small retinal vessel vasculopathy after vaccination against Neisseria meningitidis and yellow fever. Retin Cases Brief Rep. 2017;11(Suppl 1):S18–23 [A].
[108] Carrasco García de León SIRA, Manuel Flores Barragán J. Acute disseminated encephalomyelitis secondary to serogroup B meningococcal vaccine. J Neurol Sci. 2016;370:53–4 [A].
[109] Velasco-Tamariz V, Prieto-Barrios M, Tous-Romero F, et al. Urticarial vasculitis after meningococcal serogroup B vaccine in a 6-year-old girl. Pediatr Dermatol. 2018;35(1):e64–5 [A].
[110] Parikh SR, Lucidarme J, Bingham C, et al. Meningococcal B vaccine failure with a penicillin-resistant strain in a young adult on long-term eculizumab. Pediatrics. 2017;140(3):pii: e20162452. https://doi.org/10.1542/peds.2016-2452 [A].
[111] Tartof SY, Sy LS, Ackerson BK, et al. Safety of quadrivalent meningococcal conjugate vaccine in children 2-10 years. Pediatr Infect Dis J. 2017;36(11):1087–92 [C].

[112] Tseng HF, Sy LS, Ackerson BK, et al. Safety of quadrivalent meningococcal conjugate vaccine in 11- to 21-year-olds. Pediatrics. 2017;139(1):pii: e20162084 [MC].
[113] Southern J, Waight PA, Andrews N, et al. Extensive swelling of the limb and systemic symptoms after a fourth dose of acellular pertussis containing vaccines in England in children aged 3-6 years. Vaccine. 2017;35(4):619–25 [C].
[114] van der Lee S, Kemmeren JM, de Rond LGH, et al. Elevated immune response among children 4 years of age with pronounced local adverse events after the fifth diphtheria, tetanus, acellular pertussis vaccination. Pediatr Infect Dis J. 2017;36(9): e223–9 [c].
[115] Villareal Perez JZ, Ramirez Aranda JM, de la O Cavazos M, et al. Randomized clinical trial of the safety and immunogenicity of the Tdap vaccine in pregnant Mexican women. Hum Vaccin Immunother. 2017;13(1):128–35 [C].
[116] McMillan M, Clarke M, Parrella A, et al. Safety of tetanus, diphtheria, and pertussis vaccination during pregnancy: a systematic review. Obstet Gynecol. 2017;129(3):560–73 [M].
[117] Perry J, Towers CV, Weitz B, et al. Patient reaction to Tdap vaccination in pregnancy. Vaccine. 2017;35(23):3064–6 [C].
[118] Monastirli A, Pasmatzi E, Badavanis G, et al. Erythema multiforme following pneumococcal vaccination. Acta Dermatovenerol Alp Pannonica Adriat. 2017;26(1):25–6 [A].
[119] Olarte L, Barson WJ, Bradley JS, et al. Invasive pneumococcal disease in infants aged 0-60 days in the United States in the 13-valent pneumococcal conjugate vaccine era. J Pediatric Infect Dis Soc. 2017. https://doi.org/10.1093/jpids/pix034 [c].
[120] Olarte L, Lin PL, Barson WJ, et al. Invasive pneumococcal infections in children following transplantation in the pneumococcal conjugate vaccine era. Transpl Infect Dis. 2017;19(1). https://doi.org/10.1111/tid.12630 [c].
[121] Olarte L, Romero J, Barson W, et al. Osteoarticular infections caused by Streptococcus pneumoniae in children in the post-pneumococcal conjugate vaccine era. Pediatr Infect Dis J. 2017;36(12):1201–4 [c].
[122] Olarte L, Barson WJ, Barson RM, et al. Pneumococcal pneumonia requiring hospitalization in US children in the 13-valent pneumococcal conjugate vaccine era. Clin Infect Dis. 2017;64(12):1699–704 [C].
[123] Szenborn L, Osipova IV, Czajka H, et al. Immunogenicity, safety and reactogenicity of the pneumococcal non-typeable Haemophilus influenzae protein D conjugate vaccine (PHiD-CV) in 2-17-year-old children with asplenia or splenic dysfunction: a phase 3 study. Vaccine. 2017;35(40):5331–8 [c].
[124] Hung TY, Kotecha RS, Blyth CC, et al. Immunogenicity and safety of single-dose, 13-valent pneumococcal conjugate vaccine in pediatric and adolescent oncology patients. Cancer. 2017;123(21):4215–23 [c].
[125] Okinaka K, Akeda Y, Kurosawa S, et al. Pneumococcal polysaccharide vaccination in allogeneic hematopoietic stem cell transplantation recipients: a prospective single-center study. Microbes Infect. 2017;19(11):553–9 [c].
[126] Madhi SA, Koen A, Jose L, et al. Vaccination with 10-valent pneumococcal conjugate vaccine in infants according to HIV status. Medicine (Baltimore). 2017;96(2):e5881 [C].
[127] Sirima SB, Tiono A, Gansane Z, et al. Immunogenicity and safety of 10-valent pneumococcal non-typeable Haemophilus influenzae protein D conjugate vaccine (PHiD-CV) administered to children with sickle cell disease between 8 weeks and 2 years of age: a phase III, open, controlled study. Pediatr Infect Dis J. 2017;36(5): e136–50 [C].
[128] Entwisle C, Hill S, Pang Y, et al. Safety and immunogenicity of a novel multiple antigen pneumococcal vaccine in adults: a phase 1 randomised clinical trial. Vaccine. 2017;35(51):7181–6 [c].
[129] Prymula R, Szenborn L, Silfverdal SA, et al. Safety, reactogenicity and immunogenicity of two investigational pneumococcal protein-based vaccines: results from a randomized phase II study in infants. Vaccine. 2017;35(35 Pt B):4603–11 [C].
[130] Idoko OT, Mboizi RB, Okoye M, et al. Immunogenicity and safety of 13-valent pneumococcal conjugate vaccine (PCV13) formulated with 2-phenoxyethanol in multidose vials given with routine vaccination in healthy infants: an open-label randomized controlled trial. Vaccine. 2017;35(24):3256–63 [C].
[131] Goldblatt D, Southern J, Andrews NJ, et al. Pneumococcal conjugate vaccine 13 delivered as one primary and one booster dose (1 + 1) compared with two primary doses and a booster (2 + 1) in UK infants: a multicentre, parallel group, randomised controlled trial. Lancet Infect Dis. 2018;18(2):171–9 [C].
[132] Urbancikova I, Prymula R, Goldblatt D, et al. Immunogenicity and safety of a booster dose of the 13-valent pneumococcal conjugate vaccine in children primed with the 10-valent or 13-valent pneumococcal conjugate vaccine in the Czech Republic and Slovakia. Vaccine. 2017;35(38):5186–93 [C].
[133] Chatham W, Chadha A, Fettiplace J, et al. A randomized, open-label study to investigate the effect of belimumab on pneumococcal vaccination in patients with active, autoantibody-positive systemic lupus erythematosus. Lupus. 2017;26(14):1483–90 [c].
[134] Gomez EV, Bishop JL, Jackson K, et al. Response to tetanus and pneumococcal vaccination following administration of ixekizumab in healthy participants. BioDrugs. 2017;31(6): 545–54 [c].
[135] Seo YB, Choi WS, Lee J, et al. Comparison of immunogenicity and safety of an influenza vaccine administered concomitantly with a 13-valent pneumococcal conjugate vaccine or 23-valent polysaccharide pneumococcal vaccine in the elderly. Clin Exp Vaccine Res. 2017;6(1):38–44 [C].
[136] Song JY, Cheong HJ, Hyun HJ, et al. Immunogenicity and safety of a 13-valent pneumococcal conjugate vaccine and an MF59-adjuvanted influenza vaccine after concomitant vaccination in ≥60-year-old adults. Vaccine. 2017;35(2):313–20 [C].
[137] World Health Organization. World malaria report, 2017 [cited 2018 Jan 30]. Available from: http://apps.who.int/iris/bitstream/10665/259492/1/9789241565523-eng.pdf?ua=1 [S].
[138] World Health Organization. Malaria vaccine: WHO position paper, January 2016. Wkly Epidemiol Rec. 2016;91(4):33–52 [S].
[139] Vandoolaeghe P, Schuerman L. The RTS,S/AS01 malaria vaccine in children 5 to 17 months of age at first vaccination. Expert Rev Vaccines. 2016;15(12):1481–93 [R].
[140] Gessner BD, Knobel DL, Conan A, et al. Could the RTS,S/AS01 meningitis safety signal really be a protective effect of rabies vaccine? Vaccine. 2017;35(5):716–21 [M].
[141] Mensah VA, Roetynck S, Kanteh EK, et al. Safety and immunogenicity of malaria vectored vaccines given with routine expanded program on immunization vaccines in Gambian infants and neonates: a randomized controlled trial. Front Immunol. 2017;8:1551 [c].
[142] Sissoko MS, Healy SA, Katile A, et al. Safety and efficacy of PfSPZ vaccine against Plasmodium falciparum via direct venous inoculation in healthy malaria-exposed adults in Mali: a randomised, double-blind phase 1 trial. Lancet Infect Dis. 2017;17(5):498–509 [c].

CHAPTER

32

Blood, Blood Components, Plasma, and Plasma Products

Kristine Sobolewski,1, Sara Radparvar*,†, Cecilia Wong†,‡, Jackie Johnston†,§*

*RWJ Barnabas Health, Saint Barnabas Medical Center, Livingston, NJ, United States

†Ernest Mario School of Pharmacy, Rutgers University, Piscataway, NJ, United States

‡Saint Peter's University Hospital, New Brunswick, NJ, United States

§St. Joseph's University Medical Center, Paterson, NJ, United States

[1]Corresponding author: kristine.sobolewski@rwjbh.org

ALBUMIN [SEDA-15, 54; SEDA-37, 403; SEDA-38, 335; SEDA-39, 331]

To date, evidence has not demonstrated a benefit in mortality with albumin administration compared with crystalloid solution in large volume resuscitation. In fact, albumin may cause renal dysfunction and is associated with increased mortality in patient(s) with brain trauma due to elevated intracranial pressure [1r]. A retrospective study evaluated the impact of restricted albumin policy in 1401 cardiac surgery intensive care unit patients. After policy implementation, there was significant reduction in albumin use ($P<0.001$) and a trend towards reduced ventilator days. There were no differences in morbidity or mortality [2C]. A recent meta-analysis of 21 trials totaling 1277 patients with hepatocellular carcinoma free cirrhosis similarly demonstrated no mortality benefit with albumin (OR=0.78 (CI 95% 0.55–1.11); $P=0.17$) [3M]. Due to its unclear benefit over crystalloid solutions and greater cost, the appropriate use of albumin for many indications remains unclear.

Neonates

Albumin administration to preterm infants has been implicated in chronic lung disease, oxygen dependency, impairs gas exchange, pulmonary edema, and even cardiac decompensation. Over-administration has also been associated with risk of chronic lung disease and cardiac decompensation. The high sodium content of albumin may put neonates at risk of fluid retention and kidney injury, which has been demonstrated in adults [4r].

BLOOD TRANSFUSION [SEDA-15, 529; SEDA-37, 404; SEDA-38, 336; SEDA-39, 331]

Erythrocytes

Immunologic

A prospective observational center-based study assessed the risk of developing red blood cell (RBC) transfusion-associated necrotizing enterocolitis in preterm infants. Twenty infants who required RBC transfusion, specifically those with gestational age of less than 32 weeks or postnatal age more than 7 days, were evaluated for changes in serum cytokine levels and other inflammatory markers post transfusion. Following transfusion there was an increase in IL-1β, IL-8, IFN-γ, IL-17, MCP-1, IP-10 and ICAM-1, indicating a proinflammatory response and potential correlation for development of necrotizing enterocolitis [5c].

Infection Risk

A cohort review assessed 30-day readmission rates and peri- and post-operative complication rates for 160 spinal surgery patients who received transfusions intraoperatively. Intraoperative transfusion with packed red blood cells (PRBC) was found to be an independent predictor of 30-day readmission following hospital discharge (16.7% vs 5.0%, $P=0.01$). Intraoperative transfusion was also associated with increased hospital length of stay (8.88 vs 6.41 days, $P=0.02$) and rate of urinary tract infections (18.0% vs 5.0%, $P=0.0065$) [6C]. Retrospective analysis of a prospective observational cohort assessed

Side Effects of Drugs Annual, Volume 40
ISSN: 0378-6080
https://doi.org/10.1016/bs.seda.2018.06.011

the effect of one or more RBC transfusion in septic and septic shock patients within 24 hours. In the 6160 patients evaluated, no significant differences were found in 30-day mortality (hazard ratio [HR], 1.07; 95% CI, 0.88–1.30; $P=0.52$), despite an increase in nosocomial infections (HR, 2.77; 95% CI, 2.33–3.28; $P<0.01$) and severe hypoxemia (HR, 1.29; 95% CI, 1.14–1.47; $P<0.01$) [7C].

The risks associated with prolonged storage of RBCs were analyzed in a prospective, double-blind randomized trial. Of the 199 patients undergoing elective noncardiac surgery who received transfusions of RBCs ranging from 6 to 16 days old, occurrence of mortality and postoperative infection did not differ between groups (fresh blood 22% vs old blood 25%; relative risk [RR], 1.17; confidence interval [CI], 0.71–1.93). However, those who received old blood had greater frequency of wound infections (15% vs 5%; RR, 3.09; CI, 1.17–8.18), acute kidney injury (24% vs 6%; $P<0.001$) and longer length of stay (mean difference, 3.6 days; CI, 0.6–7.5) [8C].

Granulocytes

Granulocyte transfusions (GT), although seldom used, provide an additional therapy for neutropenic patients with severe infections refractory to antimicrobial medication. The American Society for Blood and Marrow Transplantation published a critical reappraisal review of granulocyte transfusions in 2017 revising key concepts emerging from past experiences with granulocyte transfusions [9R].

A recent review on granulocyte administration in patients with fungal infection summarized outcomes of 97 patients and found that 16% of patients reviewed experienced some form of adverse events including febrile and pulmonary reactions [10c]. A retrospective review at the National Institutes of Health Clinical Center evaluated patients with chronic granulomatous disease with severe and refractory infections who received GT for efficacy and safety. A total of 58 GT treated infectious episodes were evaluated, of which GTs were associated with 88% overall efficacy in infection outcomes stratified by cleared infection or partially cleared both denoting a form of improvement in the patients status. Of these, 12% experienced as adverse event involving fever and/or chills, transient dyspnea with one patient report of a transfusion related acute lung injury. None of the listed transfusions were associated with any lethal adverse events from the NIH review [11c]. A single-center, prospective study evaluated 40 patients with a diagnosis of acute myeloid leukemia or high risk myelodysplastic syndrome that received granulocyte transfusions during induction of chemotherapy or first salvage therapy. A median of 3 GT were administered per patient, of which all 40 patients reported more than one episode of neutropenic fever which at least one infectious episode per patient. Additionally, urticarial/pruritus ($n=1$) rash ($n=1$) and hypotension ($n=1$) were also identified and no patient experienced transfusion associated graft vs host disease [12c].

Platelets

Infection Risk

Auborn and colleagues reported results from a multicenter study evaluating 2250 patients who received platelets in the ICU to determine whether platelet transfusion in critically ill patients is associated with hospital acquired infection. After adjusting for patient confounders, platelet transfusions had an independent association with infection (adjusted OR 2.56 95% CI 1.98–3.31, $P<0.001$) and when determining the occurrence of bacteremia and bacteriuria, platelet transfusion was associated with increased infection outcome (adjusted OR bacteremia 3.30, 95% CI 2.30–4.74, $P<0.01$ and adjusted OR for bacteriuria 2.01, 95% CI 1.44–2.83, $P<0.01$) [13MC].

Cardiovascular

In a retrospective, observational study evaluating 169 patients receiving single early intraoperative platelet transfusion compared to matched reference group, those who received platelets experienced less blood loss but more often required vasoactive medication (OR 1.65; 95% CI, 1.03–2.11), prolonged mechanical ventilation (OR 1.47; 94% CI 1.03–2.11) and prolonged intensive care (OR 1.49; 95% CI, 104–2.12) [14C].

Gastrointestinal

Current expert opinion indicates platelet transfusion as a treatment in serious gastrointestinal bleed option even in patients with normal platelet counts. A single-center, retrospective cohort study of 204 patients found that platelet transfusion did not improve clinical outcomes and mortality was higher in patients who received transfusions after adjustment for potential confounders (adjusted OR 5.57, 95% CI 1.52–27.1) [15C].

Storage

Platelets are typically stored for 4–7 days depending on national guidelines and type of product. Although prolonged storage increases availability, the effect of storage time on outcomes after transfusion may influence efficacy and safety. A meta-analysis of 23 studies reported safety and efficacy outcomes which found most studies defined fresh platelets as <3 days and old as >4 days. The relative risk of transfusion reaction after administration of old platelets compared to fresh was 1.53 (CI, 104–2.25). However, the use of leucoreduction minimizes the risk of transfusion reactions for platelets stored for a prolonged period (RR 1.05; CI, 0.60–1.84). The bleeding risk was found to be greater in those receiving old platelets (RR 1.13; CI, 0.97–1.32) [16MC].

BLOOD SUBSTITUTES [SEDA-15, 84; SEDA-37, 406; SEDA-38, 339; SEDA-39, 333]

Hemoglobin-Based Oxygen Carriers (HBOC)

Dhar and colleagues administered PEGylated bovine carboxyhemoglobin to 12 patients with subarachnoid hemorrhage at risk of delayed cerebral ischemia. Doses of 160, 240, and 320 mg/kg were used, and no adverse events were noted during infusion. A transient rise in mean arterial pressure was noted which resolved within 24 hours [17c]. Gomez and colleagues published a case of a 46-year-old male with life threatening anemia who received 12 units of HBOC-201 over a period of 8 days. The patient experienced elevated arterial blood pressure after 3 units of HBOC-201, elevated methemoglobin concentration after onset of HBOC-201 administration (peaking at 22% after 10 units were administered) and pulse oximetry desaturation. All side effects were successfully managed with nicardipine, methylene blue and confirmation of oxygen saturation with arterial blood gas [18A].

PLASMA AND PLASMA PRODUCTS [SEDA-15, 84; SEDA-37, 407; 38, 340; SEDA-39, 333]

Alpha1-Antitrypsin

Earlier randomized controlled trials showed that purified alpha1 proteinase inhibitor (A1PI) slowed the progression of emphysema in patients with severe alpha1-antitrypsin deficiency. An open-label extension trial assessing treatment for up to 4 years with A1PI did not reveal any adverse events with prolonged treatment. A total of 56 patients experienced COPD exacerbations and 40 patients experienced nasopharyngitis; however, neither was thought to be related to drug administration [19c]. When comparing the safety of bi-weekly infusion of 120 mg/kg of A1PI vs weekly 60 mg/kg infusions, a higher infusion adjusted adverse event rate was noted in the 120 mg/kg and placebo group than the 60 mg/kg group (0.122 vs 0.1029) but the adverse events within the first 24 and 72 hours were comparable (0.0322 vs 0.0354 and 0.0579 vs 0.0568, respectively). COPD exacerbations and upper and lower respiratory tract infections were the most common adverse effects reported [20c].

C1 Esterase Inhibitor Concentrate

C1 esterase inhibitor concentrate (C1-INH) is the preferred treatment for hereditary angioedema (HAE) due to C1-inhibitor deficiency with several FDA approved products on the market. COMPACT was a phase 3 trial which assessed the efficacy of Haegarda compared to placebo at reducing the frequency of HAE attacks. A total of a 115 patients were enrolled, and no significant cardiac or hematological toxicities were reported in patients who received the active drug [21C]. A phase 2 trial assessing the safety and efficacy of recombinant human C1 inhibitor for hereditary angioedema prophylaxis did not find any concerning adverse events; headache and nasopharyngitis were the most common side effects reported [22c]. Thromboembolic events (TEE) have been reported when supratherapeutic doses of C1-INH were used for off-label indications in neonate and infants. Reviews of C1-INH use in pediatrics have not found any serious adverse event or TEE, and no patients developed treatment emergent antibodies to C1-INH [23R, 24M].

Cryoprecipitate

Adverse events associated with cryoprecipitate are similar to other blood products, including the risk of transmitting infectious diseases, transfusion related acute lung injury (TRALI) and transfusion associated circulatory overload (TACO), although this occurs at a lower frequency than seen with plasma. Cryoprecipitate can be used to manage disseminated intravascular coagulopathy (DIC). A retrospective study of 96 patients with acute myeloid leukemia and DIC sought to evaluate the rates of bleeding, transfusion reactions and volume overload when patients were treated with cryoprecipitate vs plasma. No significant differences were noted with any of the outcomes [25c].

Fresh Frozen Plasma

Usemann and colleagues conducted a retrospective case–control study of 20 extremely low birth weight (ELBW) infants with pulmonary hemorrhage and 40 matched controls. Their review found that fresh frozen plasma (FFP) transfusion may be an independent risk factor for pulmonary hemorrhage in ELBW infants caused by a developmental mismatch in hemostasis by transfusion of adult donor plasma [26c]. Friesenecker and colleagues report a 67-year-old male with hemorrhagic shock who received 15 units of FFP amongst other products for resuscitation. This led to further deterioration of hemostasis which was thought to be secondary to fibrinogen dilution. Administration of tranexemic acid, high dose fibrinogen concentrate, factor XIII, platelets and recombinant factor VIIa achieved hemostasis [27A].

PLASMA SUBSTITUTES [SEDA-37, 408; SEDA-38, 341; SEDA-39, 334]

Dextrans

A propensity-score matching study evaluated the effects of dextran-70 on outcomes in patients with severe sepsis or septic shock. This study investigated the effects on organ failure, incidence of bleeding and mortality by comparing individuals who received dextran-70 to those who received crystalloids and 5% or 20% albumin alone. Dextran- and non-dextran treated patients were propensity score matched to adjust for baseline differences. Of the 778 individuals matched, 490 were included in the analysis with 245 having received dextran-70. Mortality at 180 days was found to be lower in the dextran group (41.6% vs 50.2%, $P=0.046$), whereas mortality at earlier time points did not differ. An increase in renal dysfunction was identified demonstrated by higher acute kidney injury network (AKIN) scores in the dextran group compared to the control group ($P=0.06$). Additionally, there was no difference in the frequency of bleeding episodes between the dextran and control groups (18.0% vs 14.0%, $P=0.21$) [28M].

Etherified Starches

A prospective, randomized trial evaluated 6% hydroxyethyl starch (HES) to assess the effects on acute kidney injury in 120 pediatric cardiac surgery patients. This study found there was no difference in acute kidney injury for those that received HES when compared to control for both Pediatric Risk, Injury, Failure, Loss, End-stage renal disease (pRIFLE) and acute kidney injury network (AKIN) criteria (40.8% vs 30.0%; $P=0.150$ using pRIFLE; 19.6% vs 21.1%, $P=0.602$ using AKIN). Despite also finding no differences in mortality, mechanical ventilation duration or ICU length of stay, they did find differences in coagulation between the two groups. Those who received HES had prolonged clotting time as measured using rotational thromboelastometry (ROTEM), as well as decreased clot firmness after 10 minutes and maximal clot firmness [29C].

Hematologic

Effects on coagulation were also seen in an in vitro study that compared the adverse effects of balanced crystalloid, hydroxyethyl starch, and gelatin after 20% dilution of blood thromboelastometry. Blood samples obtained from healthy volunteers undergoing elective knee arthroscopy. This study found that blood exposed to HES had significantly impaired clot formation time and clot firmness ($P<0.05$) [30E]. Additionally, a single-center retrospective cohort study of adult patients admitted to the intensive care units for at least 24 hours which evaluated the impact of HES 70/0.5 on post-operative bleeding further supports these results. Of the 869 individuals analyzed, 653 met inclusion criteria. One hundred and nineteen matched pairs compared HES 70/0.5 to control. The primary outcome evaluated the drainage volume from surgical sites during the first 24 hours after ICU admission. This study found that drainage volume during the first 24 hours after ICU admission was significantly greater in the HES group than in the control group (400±479 vs 260±357 mL, $P<0.003$); therefore, the authors concluded the HES causes post-operative bleeding [31C].

Additional studies have demonstrated potential utilization for HES as fluid therapy in pediatric patients undergoing intracranial tumor resection. HES fluid therapy has also been evaluated for management of hypotension associated with spinal anesthesia in elective caesarean section with no differences in reported adverse reactions when compared to standard therapies [32c, 33C].

Urinary Tract

Hydroxyethyl starch (HES) has been associated with severe renal injury and mortality. A propensity-score matched cohort evaluated the effects of 6% HES 130/0.4 and utilization in those undergoing off-pump coronary arterial bypass grafting. The 413 patients who received intraoperative HES were matched with 249 who did not (NoHES). Post-operative bleeding did not differ (median [IQR]: 525 mL [350–760] in group HES vs 540 mL [400–670] in group NoHES, $P=0.203$), nor did the rate of postoperative bleeding related reoperation (2.9% [12/413] in group HES vs 1.2% [3/249] in group NoHES, OR 2.44 [95% CI: 0.64–9.34], $P=0.191$) between the two groups. Incidences of postoperative AKI defined by both KDIGO and RIFLE criteria were significantly higher in group HES than group NoHES (AKI by KDIGO: 10.7% [44/413] vs 3.6% [9/249], OR 3.43 [95% CI: 1.67–7.04], $P<0.001$; AKI by RIFLE: 5.8% [24/413] vs 2% [5/249], OR 3.32 [95% CI: 1.34–8.24], $P=0.01$) [34M]. A retrospective study in patients undergoing therapeutic plasma exchange also assessed the effects of HES on renal function when compared to albumin. In the 104 patients included serum creatinine values and blood urea nitrogen values decreased significantly after five cycles of therapeutic plasma exchange when compared to baseline [35C].

While HES utilization has been associated with renal dysfunction, a retrospective cohort study further investigated the dose-related risks. Adult patients undergoing elective or emergency cardiac surgery with or without cardiopulmonary bypass were divided into 2 groups whether they had received a cumulative HES dose of <30 mL/kg (Low HES) or >30 mL/kg (High HES) during the intra- and postoperative period. Of the1501 patients analyzed, 983 patients in the Low HES and 518 patients were in the High HES group. Development of acute kidney injury was less frequent in the Low HES group when compared to the High HES group (18.8% vs 23.0%, $P=0.06$). After case–control matching, acute kidney injury was still less frequent in the Low HES group compared to the High HES group (18.0% vs 23.0%, $P=0.19$). The results from this study demonstrated a decreased incidence of acute kidney injury in lower weight-adjusted dose of HES (Odds Ratio (95% CI) = 0.825 (0.727 ± 0.936); $P=0.003$) [36C].

Gelatin

A case report of a rare episode of perioperative anaphylaxis involving a 12-year-old girl detailed the management after the child experienced tachycardia, hypotension and ventilation difficulty after administration of Floseal®. Floseal® is a gelatin containing hemostatic agent and this case presents a rare cause of perioperative anaphylaxis and highlights the importance of thorough assessment of possible causes to minimize repeated exposures [37A].

It has been suggested that the use of hemostatic agents containing collagen/gelatin mixed with thrombin pose a risk of thromboembolic events. A retrospective, multicenter review found that in 53 out of 932 (5.6%) patients who underwent craniotomies for tumor removal experienced a perioperative symptomatic thromboembolic event. The use of gelatin matrix with or without thrombin did not result in a difference in thromboembolic event occurrence. Patients who received greater than 10 mL gelatin hemostatic agents (21/306) experienced a higher incident of thromboembolic events compared to those who received less than 10 mL (29/512) ($P<0.05$) [38C].

IMMUNOGLOBULINS [SEDA-15, 1719; SEDA-37, 409; SEDA-38, 342; SEDA-39, 335]

Intravenous Immunoglobulin

Intravenous immunoglobulin (IVIG) is fairly well-tolerated, although 20%–50% of patients experience mild adverse effects such as headache, nausea, fatigue, and mild fever. The incidence of these can be lowered with slower infusion rates. Rare, more serious adverse effects include thromboembolism, aseptic meningitis, acute renal failure, and anaphylaxis [39C].

Hematologic

IVIG administration is associated with several hematologic adverse effects including leukopenia, neutropenia, disseminated intravascular coagulation, thrombosis, and hemolysis [40A]. A recent multicenter, open-label, single-arm clinical trial in 57 patients with immune thrombocytopenia treated with IVIG which sought to determine the mechanism of IVIG-mediated hemolysis. Although the trial failed to observe any cases of clinically significant hemolysis, 18 patients, none with blood type O, met the post-hoc criteria for hemolytic laboratory reactions (hemoglobin decrease >1 g/dL). Laboratory analysis of all antiglobulin test-positive samples revealed no non-ABO blood group antibodies. Blood type A and B antigen density on red blood cells and IVIG dose (1 g/kg vs 2 g/kg) were significant risk factors for hemolysis. ($P=0.0006$ and $P=0.0049$, respectively) [41c]. A similar study evaluated 31 pediatric patients (median age 8.75 ± 5.23 years, range 2–18 years) receiving IVIG. All patients experienced a significant reduction in hemoglobin ($P<0.05$) and a rise in reticulocyte count ($P<0.05$) or direct bilirubin ($P<0.05$), but only six patients met criteria for IVIG-related hemolysis and only two required blood transfusions. There was no significant difference in hemolysis between blood type groups; of the six patients, four patients had blood type A, one had blood type AB and one had blood type O. The study found boys were more likely to be affected than girls ($P=0.013$) [42c]. Theoretically, a blood-type specific IVIG would carry a lower risk of hemolysis. Unfortunately, since IVIG is pooled from thousands of donors, blood-type specific IVIG is currently unfeasible. Blood type-specific mini-pool IVIG (MP-IVIG), however, is possible as each pool consists of only 20 plasma donations. A multicenter, open-label study comparing use of MP-IVIG with IVIG in 72 patients with immune thrombocytopenia demonstrated similar efficacy to IVIG; however, the impact on hemolysis risk could not be evaluated due to the small number of patients enrolled in each arm. Further studies are necessary to determine the efficacy and potential safety advantage of MP-IVIG [43c].

An 80-year-old female receiving IVIG 30 g/month hypogammaglobulinemia secondary to lymphoma experienced three documented episodes of thrombocytopenia over the course of 7 years. The lowest platelet nadir was $9 \times 109/L$; all three episodes were self-limiting, and none were associated with bleeding. Laboratory analysis through platelet immunofluorescence test and monoclonal antibody immobilization of platelet antigen demonstrated dose-dependent binding of IgG to the patient's platelets. The authors theorized that the patient's past medical history of lymphoproliferative disorders may have predisposed her to this interaction. There are other reports in the literature documenting a higher incidence of hematologic events due to IVIG in patient with autoimmune and inflammatory disorders [40A].

Neurologic

IVIG-induced aseptic meningitis is a rare adverse effect with reported incidence of 0.067%. A recent case report detailed a 47-year-old female who presented with headache, neck pain and rigidity while receiving IVIG for a myasthenic crisis. Cerebrospinal fluid was significant for an elevated white blood cell count with neutrophil predominance. Although the patient initially received antibiotics, they were discontinued after 24 hours due to high clinical suspicion of IVIG-induced aseptic meningitis. Forty-eight hours after discontinuation of IVIG and 24 hours after discontinuation of antibiotics, the patient's meningitis symptoms completely resolved [44A]. In another case, a 54-year-old woman who received IVIG for Guillain–Barré syndrome developed confusion and somnolence after 6 days of treatment. Her MRI revealed symmetric lesions in the splenium of the corpus callosum that follow-up imaging a month later demonstrated to be resolved. This was considered a result of cytotoxicity from IVIG-mediated increases in inflammatory cytokines or cerebral arterial vasospasm due an excitotoxic effect of IVIG in astrocytes [45A].

Renal

Renal dysfunction associated with IVIG administration is normally attributed to the sucrose component of certain IVIG formulations. A case of renal dysfunction was documented in an otherwise healthy 32-year-old male with common variable immunodeficiency (CVID) who received a sucrose-free IVIG product. The patient had received IVIG for over 6 years when after one monthly dose of IVIG (30 grams of 10% Gammagard®) laboratory values revealed a creatinine of 1.25 mg/dL. When the patient's creatinine returned to baseline, he received a 3-gram test dose of the same IVIG product again resulting in a rise in creatinine. Another sucrose-free product (Gammunex-C®) had a similar effect with full dose (30 grams) administration. As both products contained glycine as stabilizers, glycine was implicated as the cause of renal dysfunction [46A]. A retrospective cohort study of 20440 patients exposed to various IVIG products found 163 patients to have recorded-same day acute renal failure. Gamunex® was associated with a significantly lower risk of acute renal failure compared to Gammagard Liquid® (OR, 0.54; 95% CI, 0.32–0.93, and OR, 0.38; 95% CI, 0.19–0.78, respectively). Patients 45–64 years old (OR, 2.01; 95% CI, 1.26–3.21), those with a history of renal failure (OR, 2.28; 95% CI, 1.47–3.55), HIV/AIDS (OR, 4.17; 95% CI, 1.63–10.67) or fluid and electrolyte disorders (OR, 1.74; 95% CI, 1.15–2.63) had a significantly greater risk of acute renal failure. The majority (65%) of same-day acute renal failure events occurred after the first IVIG administration [47C].

Ophthalmic

A case report described a 44-year-old woman with Charcot–Marie–Tooth type I disease receiving IVIG 30 g/day for left brachial plexus neuritis. After the second dose of IVIG she developed bilateral acute uveitis. The patient experienced rapid improvement of her condition with cessation of IVIG and initiation of ophthalmic steroids. Two similar previously reported cases attributed the reaction to systemic vasculitis because of neutrophil cytoplasmic antibody (ANCA) contamination; however, the vasculitis screen was negative in this case [48A].

Autoimmune

Six patients developed cutaneous lupus erythematosus (cLE) during immunoglobulin treatment; five patients received IVIG and one received subcutaneous immunoglobulin. Two of the patients demonstrated recovery after discontinuation of immunoglobulin treatment and two others experienced symptom improvement after immunoglobulin withdrawal. In two of the patients cLE could not be completely contributed to immunoglobulin therapy as they were using other medications also associated with cLE. Drug-induced cLE may be a result of genetic predisposition and the use of photoactive medications leading to loss of immune regulation of sun-exposed skin. IVIG has previously been linked with photosensitivity with a rate of 0.1%–1% of patients. In these cases, three patients were switched to a different IVIG product resulting in symptom improvement but not remission [49A].

A post-authorization, prospective, open-label observational study of Flebogamma® DIF 5% and 10% was conducted in 66 adult and pediatric patients receiving a total of 265 infusions. A greater frequency of potentially treatment related adverse events was observed with administration of the higher concentration product (18.5% vs 2.2%, $P<0.0001$), although the type of reaction and severity was similar for both groups. The most frequently reported reactions included headache ($n=17$) and pyrexia ($n=6$) [50c]. A similar study investigating Gammaplex® 5% and Gammaplex® 10% in 33 adults and 15 children with primary immunodeficiency diseases found no difference in the rate of IVIG-related adverse effects while demonstrating bioequivalence based on mean IgG concentrations. The most common reactions reported by adult patients were headache (12.5% [Gammaplex® 10%] and 18.2% [Gammaplex® 5%]), migraine (6.3% and 6.1%, respectively), pyrexia (6.3% and 0%, respectively), fatigue (3.1% and 6.1%, respectively), and nausea (3.1% and 6.1%, respectively) [51c].

Subcutaneous Immunoglobulin

Subcutaneous immunoglobulin (SCIG) is an attractive alternative to IVIG due to its ease of administration. Due to variations in evaluation and documentation in studies, reported infusion site reaction rates range from 0.001 events/administration to 0.58 events/administration with discontinuation rates due to reactions ranging 0%–6.5% [52r, 53c, 54c, 55c, 56c, 57C]. A SCIG formulation with hyaluronidase, referred to as facilitated SCIG, appears to be similarly well-tolerated [58A].

In a multicenter, observation study of 23 patients receiving high-dose SCIG documented nervous system disorders were observed in 13.6% and cutaneous tissue disorders in 18.2% of patients. Two patients also experienced myocarditis and a cerebrovascular accident [59c]. Like IVIG, SCIG also carries a risk of anaphylaxis due to IgE or IgG components. A case series of four adults with common variable immunodeficiency receiving SCIG highlighted the risk of anaphylaxis with SCIG and the need for allergy screening and IgE testing as part of SCIG quality control [60A].

Anti-D Immunoglobulin

A retrospective analysis of 2518 patients receiving anti-D immunoglobulin (anti-D), IVIG, rituximab, romiplostim, or eltrombopag for immune thrombocytopenic purpura found a significantly lower rate of adverse events among patients receiving anti-D (13.8% vs IVIG: 21.1%, rituximab: 29.4%, romiplostim: 28.1%, eltrombopag: 22.4%). The most commonly documented reactions were nausea, vomiting, arthralgia and musculoskeletal pain [61C].

COAGULATION PROTEINS [SEDA-37, 411; SEDA-38, 344; SEDA-39, 336]

Factor I

Fibrinogen, also known as factor I, is the common pathway in the coagulation cascade. A randomized, placebo-controlled, double-blind clinical trial evaluated the effects of plasma-derived fibrinogen concentrate vs placebo for the treatment of intraoperative bleeding during high-risk cardiac surgery. Of the 120 individuals randomized, one-half received fibrinogen. A secondary analysis was performed to assess the safety and tolerability of fibrinogen administration. Fibrinogen administration was associated with more frequent in-hospital mortality (2 vs 0), stroke (4 vs 0), myocardial infarction (3 vs 1), renal insufficiency or failure (3 vs 2) and infections (3 vs 2) [62C].

Factor VIIa

Recombinant activated factor VIIa (rFVIIa) is utilized for treatment of hemophilia, factor VII deficiency, and Glanzmann's thrombasthenia. Additionally, rFVIIa can be utilized for refractory bleeding during or after cardiac surgery. A single-center, retrospective chart review evaluated the effects of rFVIIa on coagulation in 17 cardiac surgery patients following administration for management of uncontrolled bleeding during surgery with normal platelet and fibrinogen levels. With regards to efficacy, administration of rFVIIa decreased blood loss, red cell concentrate and fresh frozen plasma administration, and platelet transfusions during the 2 hours before and after rFVIIa administration. Additionally, there were marked improvements in coagulation parameters. Administration of rFVIIa was not associated with higher mortality. There were no deaths within the 24-h period following rFVIIa administration. Despite three deaths (17.6%), none were associated with thromboembolism [63c].

Factor VIII

Kovaltry® is an unmodified, full-length recombinant human FVIII product with improved manufacturing technologies that have resulted in a rFVIII product with a consistent purity. LEOPOLD II was a phase 2/3 study comparing prophylaxis vs on demand treatment in patients with hemophilia A with rFVIII. A recent subgroup analysis of Japanese patients revealed similar results in efficacy, safety, and pharmacokinetic properties compared to non-Japanese subjects. Despite neither group developing inhibitors following rFVIII administration, the Japanese subgroup did not experience any drug-related adverse events (0.0% vs 4.2%) [64c].

A recent case report investigated utilization of recombinant porcine FVIII (r-pFVIII) in a 5-year-old male with immune tolerance induction (ITI) refractory high-titer FVIII inhibitors who required cardiac surgery for repair of progressively symptomatic hypoplastic aortic arch. r-pFVIII received orphan drug designation for acquired hemophilia A (AHA) in adults and was licensed in 2014. Despite the presence of data for use in AHA, studies investigated use in congenital hemophilia A is limited. While more investigation is needed, this case report demonstrated r-pFVIII was able to provide effective periprocedural hemostasis with no adverse events [65A].

As FVIII prophylaxis is the optimal treatment for severe hemophilia A, investigation into longer-acting agents has gained recent attention. Adynovate® is an extended half-life form of Advate®, a full-length rFVIII agent. A phase 3, prospective, uncontrolled, multicenter, open-label study evaluated immunogenicity, pharmacokinetics, efficacy, safety and HRQoL of polyethylene glycol (peg)-ylated rFVIII in previously treated patients (PTP) with severe hemophilia A. PTPs <12 years without history of rFVIII inhibitors received twice-weekly infusions for ≥ 50 exposure days. There were 31 subjects in the two treatment groups, <6 years of age and 6 to <12 years of age. Efficacy for bleeding treatment was rated as 'excellent' or 'good' for 90% of subjects and <2 infusions were administered per bleeding event in the majority of cases{one infusion 82.9% (58/70); two infusions 8.6% (6/70); three infusions 8.6% (6/70)}. The results were comparable with those reported for other FVIII preparations in a variety of patient populations. This study also demonstrated that pegylation of rFVIII had no effect on product safety. FVIII inhibitors did not develop in any subject evaluated during the study. Nearly 66% of the subjects experienced one adverse effect. Four of the 156 adverse effects were serious and included febrile neutropenia, pancytopenia, acute gastritis and abdominal pain. Of the four serious adverse effects reports, none were fatal nor were they thought to be associated with pegylated rFVIII administration [66c].

Factor IX

A multicenter trial assessed the safety and efficacy of a long acting glycoPEGylated factor IX for patients with hemophilia B undergoing major surgery. All 13 patients received a single 80 unit/kg preoperative dose, and intraoperative hemostasis was rated as excellent or good in all cases. Two adverse events (pruritis, elevated serum ferritin) possibly related to glycoPEGylated factor IX were noted, although the elevated serum ferritin level remained within normal range. None of the patients developed factor IX neutralizing antibodies, and no deaths or thromboembolic events occurred [67c]. Kids B-LONG was a phase 3 trial assessing the safety and efficacy of recombinant factor IV Fc fusion protein (fFIXFc) in the treatment of hemophilia B in patients younger than 12 years old. 30 patients were given a median dose of 58.6 IU/kg per week and resulted in low bleeding rates. No patients developed factor IX neutralizing antibodies and the most common adverse events noted were falls and nasopharyngitis. One incident of head injury and one incident of ear infection were categorized as severe by the investigators [68c]. Interim results from B-YOND, the extension study assessing the long-term safety and efficacy of rFIXFc in pediatrics and adults, confirmed safety and no inhibitors were observed [69c]. A French post-marketing surveillance study of nonacog alfa (recombinant Factor IX) revealed one patient out of 58 patients enrolled developed FIX inhibitors during follow-up; however, the patient was asymptomatic [70c].

Prothrombin Complex Concentrate

A recent meta-analysis reviewed the safety and efficacy of 4 factor prothrombin complex concentrate (PCC) compared to FFP or no treatment in vitamin k antagonist associated bleeding. A total of 19 studies were included (18 cohort studies, 1 RCT). Thromboembolic complications were observed in 2.5% of PCC and 6.4% of FFP recipients [71M]. A prospective cohort study assessed the effectiveness and safety of PCC for the management of major bleeding events for patients on rivaroxaban or apixaban. A total of 84 patients received PCC at a median dose of 2000 IU (IQR 1500–2000 IU). Two patients developed ischemic strokes, 5 and 10 days after PCC administration [72c]. In a retrospective analysis of patients who received 4 factor PCC for warfarin reversal prior to orthotopic heart transplant, using 4F-PCC was associated with a shorter time to chest closure and less blood product utilization compared to plasma. No increases in acute kidney injury, thromboembolic complications or death were observed by the investigators [73c].

De Vlieger and colleagues reported a case of an 80-year-old female who underwent urgent neurosurgery for acute onset paraplegia due to spontaneous subdural spinal hematoma less than 5 hours after rivaroxaban ingestion. Fifty units/kg of PCC and 2 grams of tranexamic acid were administered pre-operatively, and good hemostasis was achieved. No additional doses of PCC were needed. One month after surgery, no thromboembolic complications were noted [74A].

von Willebrand Factor (VWF)/Factor VIII Concentrates

A pharmacovigilance report analyzing the safety of Humate-P® between 1982 and –2015 revealed 670 post-marketing case reports of adverse effects. From those case reports, 33 thromboembolic complications, 97 inhibitor formations and 110 hypersensitivity or allergic reactions were reported and deemed clinically relevant. Patients with thromboembolic complications had other risk factors for thromboembolic events, inhibitor formation occurred in patients using Humate-P for hemophilia A, and most patients with hypersensitivity reactions had von Willebrand disease [75R]. Other open-label studies involving VWF/factor VIII concentrates did not reveal any thrombotic events [76c, 77c]. One instance of mild injection site pruritus and once instance of mild eye edema were reported and both resolved within 4 days [76c].

ERYTHROPOIETIN AND DERIVATIVES [SEDA-37, 413; SEDA-38, 346; SEDA-39, 338]

Erythropoietin stimulating agents (ESA) serve as a treatment option for patients with anemia in chronic kidney disease or in oncologic disorders. In order to evaluate the benefit and risk profile of epoetin alpha biosimiliars vs ESA agents, a population based observation study evaluated a total of 13470 ESA users and found no difference with regard to risk of all-cause mortality, blood transfusion, major cardiovascular events, and blood dyscrasia in CKD patients (adjusted HR = 1.09; 91% CI 0.85–1.41) with comparable risk estimates observed in the oncology setting (adjusted HR 0.91, 0.79–1.06) [78MC].

The Dose Finding Trial of Pediatrics on Hemodialysis in Nephrology (DOLPHIN) specifically evaluated patients receiving continuous erythropoietin receptor activator-methoxypolyethylene glycol-epoetin beta (C.E.R.A) to determine a conversion factor for switching from other ESAs. They found a similar pattern of reported adverse events as observed in studies with adults. Of the 49 patients evaluated, 77% experienced one of more adverse event. Serious events reported include one patient who experienced an arteriovenous fistula thrombosis and another who experienced thrombosis in device. Other, more common adverse events reported include headache (14%), nasopharyngitis (14%), and hypertension (14%) in the core period [79c]. A Cochrane review of 34 studies including 3643 patients assessed the effectiveness and safety of ESAs when initiated in infants before 8 days after birth. This review found no significant difference in risk of stage >3 retinopathy of prematurity with early EPO treatment (RR 1.24; 95% CI 0.81–1.90) but identified significantly decreased rates of intraventricular hemorrhage (RR 0.60; 95% CI 0.43–0.85) and periventricular leukomalacia (RR 0.66; 95% CI, 0.48–0.92). Overall this study recommended awaiting the results of two ongoing large scale trails given the high heterogeneity of the analysis and conflicting results from previously published trials [80M].

THROMBOPOIETIN AND RECEPTOR AGONISTS [SEDA-15, 3409; SEDA-37, 414, SEDA-38, 347; SEDA-39, 339]

Given the associated mortality of thrombocytopenia in patients suffering from myelodysplastic syndrome, a single-blind, randomized controlled phase 2 superiority trial evaluated eltrombopag impact on patient outcomes. Greater platelet response was identified in patients in the eltrombopag group ($n=59$) compared to placebo ($n=31$) (OR 27; 95% CI 3.5–211.9; $P<0.0017$). Adverse events occurred in 46% of the eltrombopag treated patients compared to the 16% placebo treated patients ($P=0.0053$). Some of these grade 3–4 non-hematological adverse reactions included nausea and vomiting ($n=8$), lower respiratory tract infection ($n=6$), heart failure ($n=4$), and hypertransaminasaemia ($n=2$), with eight patients requiring permanent treatment discontinuation due to persistent drug related toxicity in the eltrombopag group [81c]. The National Heart, Lung and Blood Institute's investigator initiated, nonrandomized, historically controlled phase 1–2 study included 92 consecutive patients with acquired aplastic anemia to receive immunosuppressive therapy plus eltrombopag. Liver enzyme elevation resulted in a brief 2-week discontinuation in 7 patients receiving eltrombopag, and an additional two severe adverse events of cutaneous eruptions [82C]. A case report of a 77-year-old man, who underwent placement of a left atrial appendage occluder, detailed the developed of a large adherent thrombus in the atrial surface of the device 1 month following the initiation of a thrombopoietin receptor agonist. The patient required aortic valve replacement to mitigate any further thrombosis complications given his continued need for thrombopoietin receptor agonist therapy [83A].

℞ *Currently no treatment is approved for chemotherapy induced thrombocytopenia. A single-center, retrospective review of romiplostim use in cancer patients conducted identified 239 with cancer related usage of romiplostim [84c].*

A total of 15 venous thromboembolic events occurred with a calculated 11.6% VTE events per patient year which was determine to be at the same historical rate of thrombosis found in this patient population without romiplostim usage [84c].

TRANSMISSION OF INFECTIOUS AGENTS THROUGH BLOOD DONATION [SEDA-37, 414; SEDA-38, 347; SEDA-39, 340]

Major Review

A meta-analysis evaluating the impact of blood transfusion in patients undergoing total knee or hip arthroplasty found that patients who received allogeneic blood transfusion were at a significantly greater risk of post-operative surgical site infections (pooled OR 1.71; 95% CI: 1.23–2.40, $P=0.002$) [85M].

Virus

Zika virus transmission concerns continue to exist. A recent analysis using RNA quantification of 18 whole blood and 21 plasma samples found that Zika virus RNA persisted in whole blood for a median of 22 (range 14–100) days compared to 10 (range 7–37) days in plasma ($P=0.058$). The article offered an opinion to extend testing blood donations for Zika virus RNA in whole blood [86H]. Hepatitis E viral (HEV) infections are often self-limiting infections with severe complications primarily occurring in immunocompromised individuals. A case report of an immunocompetent trauma patient who underwent severe blunt splenic injury received massive transfusion and extending hospitalization. During this time abnormal liver function tests promoted histological examination of liver tissue and viral serologic tests identified HEV viremia reaching 1.8 × 105 copies/mL reported [87A].

STEM CELLS [SEDA-37, 415; SEDA-38, 348; SEDA-39, 340]

Hematopoietic Stem Cells

Hematopoietic stem cell transplantation (HSCT) continues to be studied for use in a growing number of indications. Studies have demonstrated that patients who are disease-free at 2 or 5 years post-HSCT have a 10-year survival rate over 80% [88R]. Unfortunately, many HSCT patients suffer from delayed effects, such as graft-vs-host disease (GVHD), cardiovascular disease, infection, and malignancy [88R, 89R]. Improved understanding of HLA matching, conditioning regimens, and recognition of late adverse effects has helped mitigate some adverse events; however, adverse reaction rates are still significant. A retrospective study of 1087 patients who underwent HSCT reported the cumulative incidence of non-malignant late effects at 5 years as 45% among autologous recipients and 79% among allogeneic recipients [88R].

Infusion Reactions

A retrospective, single-center evaluation of 1191 adult patients receiving a total of 1269 autologous hematopoietic stem cell transplants (HSCT) reported an adverse event rate of 37.8% during infusion. The most common reactions included facial flushing (39.4%), nausea/vomiting (38.1%), chest tightness (16.7%), cough (12.1%) and shortness of breath (8.3%). Patients receiving larger volumes of infusion per body weight were more likely to experience adverse events, perhaps due to greater exposure to dimethyl sulfoxide (DMSO) a cryoprotective agent associated with a high rate of adverse reactions (OR, 1.66; 95% CI, 1.29–2.15; $P<0.0001$). Women in this study also experienced a higher rate of adverse reactions (odds ratio [OR], 1.78; 95% confidence interval [CI], 1.40–2.26; $P<0.0001$). Infusions with greater amounts of mononuclear cells and granulocytes were independently associated with more adverse reactions (OR, 1.30; 95% CI, 1.01–1.67; $P=0.042$). Most patients in this study received HSCT for multiple myeloma (60%); patients with diagnoses other than multiple myeloma were more likely to experience adverse effects during infusions (OR, 1.44; 95% CI, 1.12–1.84; $P=0.004$), but this may be due to the lower volume of HSCT infusions for multiple myeloma [90c].

Dimethyl sulfoxide (DMSO) is a popular cryoprotectant used to store hematopoietic stem cells. Unfortunately, DMSO is notorious for adverse effects, most commonly malaise, nausea, vomiting, tremors, fever and tachycardia. One-hundred and forty-three patients were randomized to HSCT with products preserved in 10%, 7.5% or 5% DMSO. Significant differences in rates of reaction were seen between the 10% and 5% DMSO groups (41.7% vs 19.1%, respectively; $P=0.02$). There was no difference when comparing 10% DMSO and 7.5% DMSO, or 7.5% DMSO and 5% DMSO ($P=0.04$ and $P=0.12$, respectively). The most common reactions were nausea, vomiting, hypo- and hypertension. DMSO concentration had no impact on time to leukocyte recovery, neutrophil engraftment, hospital stay or number of platelet transfusions. The number of red blood cell transfusions was significantly higher in the 10% DMSO group when compared to the 7.5% group ($P=0.002$) and the 5% group ($P=0.02$). There was no statistical difference between 7.5% DMSO and 5% DMSO concentrations ($P=0.6$) [91C].

Cardiovascular

Transplantation-associated thrombotic microangiopathy (TA-TMA) is a severe complication of HCST characterized by renal insufficiency, neurological abnormalities and microangiopathic hemolytic anemia. The complete pathophysiology of TA-TMA is poorly understood; however, a recent laboratory analysis of blood samples from six children who received allogenic HSCT suggests that in addition to endothelial injury, HSCT may cause reductions in blood coagulation and fibrinolytic function [92c]. A prospective study of 654 consecutive patients undergoing HSCT, grades 2–4 acute GVHD (OR=5.55; 95% CI, 2.06–14.99; $P=0.001$) and cytomegalovirus viremia (OR=5.17; 95% CI, 1.10–24.28; $P=0.037$) were found to be independent risk factors for TA-TMA. Hypertension and a serum LDH level >500U/L were early signs of TA-TMA development ($P=0.004$ and $P=0.029$, respectively). The medial survival of patients with TA-TMA was 161 days post-HSCT. Three-year TA-TMA related mortality cumulative incidence was significantly higher for patients with either liver dysfunction or significant gastric bleeding, compared to patients without either condition ($83.3\% \pm 12.4\%$ vs $14.3\% \pm 9.8\%$, $P=0.0005$ and $83.3\% \pm 12.4\%$ vs $14.3\% \pm 9.7\%$, $P=0.0003$, respectively). These results suggest that patients with any of the mentioned risk factors should be monitored closely for TA-TMA [93C].

Renal

Acute kidney injury (AKI) is commonly observed after HSCT and is associated with chronic kidney disease and mortality. A retrospective analysis of 1057 pediatric patients found a cumulative incidence of AKI within the first 100 days to be $68.2\% \pm 1.4\%$. While there was no difference in mortality at year between patients without AKI and those with stage 1 or 2 AKI (66.1% vs 73.4% vs 63.9%, respectively), there was a significant increase in mortality among patients with stage 3 AKI with or without renal replacement therapy compared to those without AKI (66.1% vs 47.3% vs 7.5%, respectively; $P<0.001$). Age, year of transplantation, donor type, presence of sinusoidal obstruction syndrome and acute GVHD were all predictive of stage 1 through 3 AKI [94C].

Pediatric

A retrospective, international, multicenter study of 717 patients who underwent HSCT before the age of three and who were relapse free for at least 1 year discovered 30% of patients experienced at least one form of organ toxicity. The most commonly reported toxicities were hormone deficiency/growth disturbance (17%), cataracts (13%), hypothyroidism (11%), gonadal dysfunction/infertility requiring hormone replacement (4%), and stroke/seizure (3%). Unrelated bone marrow, stem cell and cord blood donor HSCT were associated with a greater risk of growth hormone deficiency/growth disturbance (HR, 2.9; 95% CI, 1.5–5.6; $P=0.0014$; HR, 2.2; 95% CI, 1.1–4.5; $P=0.023$, respectively). Patients between 2 to 3 years of age were more likely to suffer from cataracts than those treated at a younger age (HR, 2.0; 95% CI, 1.0–3.7; $P=0.044$). Females were more likely than males to experience gonadal dysfunction; however, the study was limited by the age of patients, as many were still too young to have gonadal dysfunction diagnosed (HR, 7.0; 95% CI, 2.0–24.0; $P=0.0019$). Total body irradiation was an independent risk factor for increased risk of hypothyroidism (HR, 5.3; 95% CI, 3.0–9.4; $P<0.001$), growth hormone deficiency/growth disturbance (HR, 3.5; 95% CI, 2.2–5.5; $P<0.001$), and cataracts (HR, 17.2; 95% CI, 7.4–39.8; $P<0.001$). The cumulative incidences of hypothyroidism, cataracts and growth hormone deficiency/growth disturbance increased over time, while the rate of stroke and seizure plateaued [95C].

Risk Factors

A retrospective evaluation of 44 pediatric patients who underwent allogeneic transplantation identified post-HSCT iron overload as a negative prognostic risk factor for sinusoidal obstruction syndrome (OR=17), osteoporosis (OR=6.8), pancreatic insufficiency (OR=17) and metabolic syndrome (OR=15.1). Patients with elevated iron levels also experienced greater mean time to engraftment of platelets (43.0 ± 35.3 days vs 22.1 ± 9.5 days, $P<0.05$) and neutrophils (23.1 ± 10.4 days vs 17.8 ± 4.6 days, $P<0.05$) [96c]. In adults, post-HSCT iron overload is similarly associated with higher rates of adverse effects, including infection, liver dysfunction, mucositis and hepatic sinusoidal obstruction syndrome. Several reports have also identified a link between iron overload and increased growth of *Aspergillosis* and other fungal infections [97R].

Mesenchymal Stem Cells

Mesenchymal stem cells (MSCs) are non-haematopoietic stem cell precursors present in various tissues. Typically derived from the bone marrow, adipose tissue and the umbilical cord, MSCs are easily cultured which makes them attractive for use in the treatment of cancer and several other indications [98R]. Typical adverse effects generally include mild headache and fever [99c, 100C, 101M]. Soreness and swelling at injection sites, or liposuction sites in the case of autologous adipose-derived stem cell transplantation, has also been reported [100C].

A meta-analysis of 10 studies comparing stem cell transplantation with rehabilitation treatment for 377 patients with spinal cord injury documented no instances of death, tumour or immune reaction. Documented adverse events included fever, headache, backache, numbness and abdominal distention, primarily due to spinal puncture [101M].

Cardiovascular

Thromboembolism is a recognized risk of HSTC; however, the incidence in MSC administration is less clear due to more limited clinical experience. Two double-blind, randomized controlled trials investigating intramyocardial administration of adipose-derived MSCs were terminated due to adverse events. Two out of the 17 treated patients suffered cardiac death, and another experienced a myocardial infarction. Two patients in the treatment arm and one in the placebo arm also suffered cerebrovascular events; however, all patients experienced complete or near complete recovery of symptoms. Similar risks have been documented with transendocardial administration. In a randomized, controlled trial one out of 21 patients suffered a myocardial infarction immediately after administration and died. A case series of 28 patients receiving adipose-derived MSCs through a similar administration method, resulted in the death of three patients at 1, 7 and 12 months post administration. Intravenous administration has also been associated with thromboembolism. A dose-escalating study of adipose-derived MSCs documented a case of lacunar infarction 8 days after administration. A randomized trial comparing HSCs, HSCs plus adipose-derived MSCs and placebo demonstrated no significant difference in the rate of cardiovascular deaths or all-cause mortality [102R]. It should be noted that the temporal relationship between these adverse events and MSCs administration is not completely clear and most of the studies did not have a control group.

A retrospective evaluation of a stem cell registry identifying 150 patients who received coronary artery bypass grafting with or without stem cell therapy identified no major differences in adverse events between groups.

In both groups, arrhythmias were most frequently observed 5 years after the procedure. Seventeen percent of patients from the stem cell group and 14% of patients from the control group required defibrillator implantation or resynchronization therapy ($P=0.799$). There was no difference in percentage of apoplexies during the 14-year follow-up period (8.8% stem cell group, 8.3% control group, $P=1.000$) [103c]. A clinical trial evaluating the safety in efficacy of umbilical-cord derived MSCs in 65 patients with heart failure found overall similar rates of adverse events compared to placebo. There were no documented new-onset arrhythmias diagnoses or observed with ECG Holter monitoring in the treatment group; however, the placebo group saw an increase in the among of premature ventricular complexes at 24-h follow-up [104c].

Oncologic

There are concerns regarding the risk of administering stem cells in the setting of previous malignancy as preclinical data has suggested that stem cell therapy may aggravate residual cancer cells. A study of 67 patients with previous breast cancer receiving autologous adipose-derived MSCs documented a single case of pelvic metastasis; however, the timing of this event was unclear. Local breast cancer recurrence was documented in a single patient out of 121 patients across two studies with 12-month follow-up periods. Studies evaluating previous prostate cancer patients with follow-up ranging from 3 to 6 months have not documented recurrence of malignancy; however, this period may be too short for evaluating oncologic safety. Longer term studies are necessary to confirm the safety of autologous stem cell transplantation in patients with history of malignancy [102R].

Intravitreal Administration

There are several reports concerning retinal detachment after intravitreal injections of adipose tissue-derived mesenchymal stem cells. Although stem cell therapy for age-related macular degeneration is still being investigated by clinical trials, "stem-cell clinics" around the United States are providing treatment without FDA-approval or oversight. It is postulated that administered stem cells undergo transformation into myofibroblasts resulting in vitreoretinopathy and retinal detachment. In case reports, patients presented with reduced visual acuity or even vision loss [105A, 106A].

A prospective, phase I, open-label study evaluated intravitreal injection of autologous MSCs in three patients. While two of the patients responded with visual improvement 2 weeks after the injection, the third patient suffered pre-retinal and vitreal fibrosis, resulting in retinal detachment and cataract formation [107c].

References

[1] El Gkotmi N, Kosemeri C, Filappatos TD, et al. Use of intravenous fluids/solutions: a narrative review. Curr Med Res Opin. 2017;33(3):459–71 [r].
[2] Rabin J, Meyenburg T, Lowery AV, et al. Restricted albumin utilization is safe and cost effective in a cardiac surgery intensive care unit. Ann Thorac Surg. 2017;104(1):42–8 [C].
[3] Kütting F, Schubert J, Franklin J, et al. Insufficient evidence of benefit regarding mortality due to albumin substitution in HCC-free cirrhotic patients undergoing large volume paracentesis. J Gastroenterol Hepatol. 2017;32(2):327–38 [M].
[4] Shalish W, Olivier F, Aly H, et al. Uses and misuses of albumin during resuscitation and in the neonatal intensive care unit. Semin Fetal Neonatal Med. 2017;22(5):328–35 [r].
[5] Dani C, Poggi C, Gozzini E, et al. Red blood cell transfusions can induce proinflammatory cytokines in preterm infants. Transfusion. 2017;57(5):1304–10 [c].
[6] Elsamadicy AA, Adogwa O, Vuong VD, et al. Association of intraoperative blood transfusions on postoperative complications, 30-day readmission rates, and 1-year patient-reported outcomes. Spine. 2017;42(8):610–5 [C].
[7] Dupuis C, Garrouste-orgeas M, Bailly S, et al. Effect of transfusion on mortality and other adverse events among critically ill septic patients: an observational study using a marginal structural Cox model. Crit Care Med. 2017;45(12):1972–80 [C].
[8] Spadaro S, Taccone FS, Fogagnolo A, et al. The effects of storage of red blood cells on the development of postoperative infections after noncardiac surgery. Transfusion. 2017;57(11):2727–37 [C].
[9] Valentini CG, Farina F, Pagano L, et al. Granulocyte transfusions: a critical reappraisal. Biol Blood Marrow Transplant. 2017;23:2034–41 [R].
[10] West KA, Gea-Banacloche J, Stroncek D, et al. Granulocyte transfusions in the management of invasive fungal infections. Br J Haematol. 2017;117:357–74 [c].
[11] Marciano BE, Allen ES, Cantilena CC, et al. Granulocyte transfusions in patients with chronic granulomatous disease and refractory infections: the NIH experience. J Allergy Clin Immunol. 2017;140:622–5 [c].
[12] Aung F, DiNardo C, Martinez F, et al. A phase II study of prophylactic non-irradiated granulocyte transfusions in AML patients receiving induction chemotherapy. J Blood Disord Transfus. 2017;8–376 [c].
[13] Auborn C, Flint AW, Bailey M, et al. Is platelet transfusion associated with hospital-acquired infections in critically ill patients? Crit Care. 2017;21:2 [MC].
[14] Hout FM, Hogervorst EK, Rosseel PM, et al. Does a platelet transfusion independently affect bleeding and adverse outcomes in cardiac surgery? Anesthesiology. 2017;126:441–9 [C].
[15] Zakko L, Rustagi T, Douglaas M, et al. No benefit from platelet transfusion for gastrointestinal bleeding in patients taking antiplatelet agents. Clin Gastroenterol Hepatol. 2017;15:46–52 [C].
[16] Kreuger AL, Caram-Deelder C, Jacobse J, et al. Effect of storage time of platelet products on clinical outcomes after transfusion: a systematic review and meta-analysis. Vox Sang. 2017;112:291–300 [MC].
[17] Dhar R, Misra H, Diringer MN. Sanguinate™ (PEGylated carboxyhemoglobin bovine) improves cerebral blood flow to vulnerable brain regions at risk of delayed cerebral ischemia after subarachnoid hemorrhage. Neurocrit Care. 2017;27(3):341–9 [c].
[18] Gomez MF, Aljure O, Ciancio G, et al. Hemoglobin-based oxygen carrier rescues double-transplant patient from life-threatening anemia. Am J Transplant. 2017;17:1941–7 [A].

[19] McElvaney NG, Burdon J, Holmes M, et al. Long term efficacy and safety of alpha1 proteinase inhibitor treatment for emphysema caused by severe alpha1 antitrypsin deficiency: an open-label extension trial (RAPID-OLE). Lancet Respir Med. 2017;5(1):51–60 [c].
[20] Greulich T, Chlumsky J, Wencker M, et al. Safety of bi-weekly intravenous therapy with alpha-1 antitrypsin. Eur Respir J. 2017;50(61):PA710 [c].
[21] Longhurst H, Cicardi M, Craig T, et al. Prevention of hereditary angioedema attacks with a subcutaneous C1 inhibitor. N Engl J Med. 2017;376:1131–40 [C].
[22] Riedl MA, Grivcheva-Panovska V, Moldovan D, et al. Recombinant human C1 esterase inhibitor for prophylaxis of hereditary angio-oedema: a phase 2, multicentre, randomized, double-blind, placebo-controlled crossover trial. Lancet. 2017;390(10102):1595–602 [c].
[23] Busse P, Baker J, Martinez-Saguer I, et al. Safety of Ci-inhibitor concentrate use for hereditary angioedema in pediatric patients. J Allergy Clin Immunol Pract. 2017;5(4):1142–5 [R].
[24] Baker JW, Reshef A, Moldovan D, et al. Recombinant human C1-esterase inhibitors to treat acute hereditary angioedema attacks in adolescents. J Allergy Clin Immunol Pract. 2017;5(4):1091–7 [M].
[25] Abu-Zeinah G, Al-Kawaaz M, Li VJ, et al. A single institutional experience comparing cryoprecipitate to plasma in patients with disseminated intravascular coagulopathy (DIC) from acute myeloid leukemia (AML). Blood. 2017;130(1):2409 [c].
[26] Usemann J, Garten L, Buhrer C, et al. Fresh frozen plasma transfusion—a risk factor for pulmonary hemorrhage in extremely low birth weight infants? J Perinat Med. 2017;45(5):627–33 [c].
[27] Friesenecker B, Fries D, Schmid S, et al. Deterioration of trauma induced coagulopathy during massive transfusion with fresh frozen plasma. Trauma Acute Care. 2017;2(4):1–5 [A].
[28] Bentzer P, Broman M, Kander T. Effect of dextran-70 on outcome in sever sepsis; a propensity-score matching study. Scand J Trauma Resusc Emerg Med. 2017;25(65):1–9 [M].
[29] Oh HW, Lee JH, Kim HC, et al. The effect of 6% hydroxyethyl starch (130/0.4) on acute kidney injury in paediatric cardiac surgery: a prospective, randomised trial. Anaesthesia. 2017;1–11 [C].
[30] Sevcikova S, Vymazal T, Durila M. Effect of balanced crystalloid, gelatin and hydroxyethyl starch on coagulation detected by rotational thromboelastometry in vitro. Clin Lab. 2017;63(10):1691–700 [E].
[31] Fukushima T, Uchino S, Fujii T, et al. Intraoperative hydroxyethyl starch 70/0.5 administration may increase postoperative bleeding: a retrospective cohort study. J Anesth. 2017;31(3):330–6 [C].
[32] Peng Y, Du J, Zhao X, et al. Effects of colloid pre-loading on thromboelastography during elective intracranial tumor surgery in pediatric patients: hydroxyethyl starch 130/0.4 versus 5% human albumin. BMC Anesthesiol. 2017;17(1):62 [c].
[33] Saghafinia M, Jalali A, Eskandari M, et al. The effects of hydroxyethyl starch 6% and crystalloid on volume preloading changes following spinal anesthesia. Adv Biomed Res. 2017;6:115 [C].
[34] Min JJ, Cho HS, Jeon S, et al. Effects of 6% hydroxyethyl starch 130/0.4 on postoperative blood loss and kidney injury in off-pump coronary arterial bypass grafting: a retrospective study. Medicine (Baltimore). 2017;96(18):e6801 [M].
[35] Radhakrishnan M, Batra A, Periyavan S, et al. Hydroxyethyl starch and kidney function: a retrospective study in patients undergoing therapeutic plasma exchange. J Clin Apher. 2017. https://doi.org/10.1002/jca.21598. [Epub ahead of print] [C].
[36] Momeni M, Nkoy ena L, Van dyck M, et al. The dose of hydroxyethyl starch 6% 130/0.4 for fluid therapy and the incidence of acute kidney injury after cardiac surgery: a retrospective matched study. PLoS One. 2017;12(10): e0186403 [C].
[37] Raveendran R, Khan D. Gelatin anaphylaxis during surgery: a rare cause of perioperative anaphylaxis. J Allergy Clin Immunol Pract. 2017;5(5):1466–7 [A].
[38] Gazzeri R, Galarza M, Conti C, et al. Incidence of thromboembolic events after use of gelatin-thrombin-based hemostatic matrix during intracranial tumor surgery. Neurosurg Rev. 2018;41:303–10 [C].
[39] Kaba S, Keskindemirci G, Aydogmus C, et al. Immediate adverse reactions to intravenous immunoglobulin in children: a single center experience. Eur Ann Allergy Clin Immunol. 2017;49(1):11–4 [C].
[40] Gurevich-Shapiro A, Bonstein L, Spectre G, et al. Intravenous immunoglobulin-induced acute thrombocytopenia. Transfusion. 2018;58:493–7 [A].
[41] Mielke O, Fontana S, Goranova-Marinova V, et al. Hemolysis related to intravenous immunoglobulins is dependent on the presence of anti-blood group A and B antibodies and individual susceptibility. Transfusion. 2017;57(11):2629–38 [c].
[42] Akman AO, Kara FK, Koksal T, et al. Association of hemolysis with high dose intravenous immunoglobulin therapy in pediatric patients: an open label prospective trial. Transfus Apher Sci. 2017;56(4):531–4 [c].
[43] Elafy M, Reda M, Elghamry I, et al. A randomized multicenter study: safety and efficacy of mini-pool intravenous immunoglobulin versus standard immunoglobulin in children aged 1-10 years with immune thrombocytopenia. Transfusion. 2017;57(12):3019–25 [c].
[44] Patel A, Potu KC, Strum T. A case of IVIG-induced aseptic chemical meningitis. S D Med. 2017;70(3):119–21 [A].
[45] Uygur Kucukseymen E, Yuksel B, Genc F, et al. Reversible splenial lesion syndrome after intravenous immunoglobulin treatment for Guillain-Barre syndrome. Clin Neuropharmacol. 2017;40(5):224–5 [A].
[46] Kim AS, Broide DH. Acute renal dysfunction caused by nonsucrose intravenous immunoglobulin in common variable immunodeficiency. Ann Allergy Asthma Immunol. 2017;118(2):231–3 [A].
[47] Ekezue BF, Sridhar G, Forshee RA, et al. Occurrence of acute renal failure on the same day as immune globulin product administrations during 2008 to 2014. Transfusion. 2017;57(12):2977–86 [C].
[48] Kocak ED, Wang BZ, Hall AJ. Bilateral uveitis following intravenous immunoglobulin administration. Am J Opthalmol Case Rep. 2017;6:74–6 [A].
[49] Adrichem ME, Starink MV, van Leeuwen EMM, et al. Drug-induced cutaneous lupus erythmatosis after immunoglobulin treatment in chronic inflammatory demyelinating polyneuropathy: a case series. J Peripher Nerv Syst. 2017;22(3):213–8 [A].
[50] Alsina L, Mohr A, Montañés M, et al. Surveillance study on the tolerability and safety of Flebogamma® DIF (10% and 5% intravenous immunoglobulin) in adult and pediatric patients. Pharmacol Red Prespect. 2017;5(5):e00345 [c].
[51] Wasserman RL, Melamed IR, Stein MR, et al. Evaluation of the safety, tolerability, and pharmacokinetics of Gammaplex® 10%

versus Gammaplex® in subjects with primary immunodeficiency. J Clin Immunol. 2017;37(3):301–10 [c].
[52] Ballow M, Wasserman RL, Jolles S, et al. Assessment of local adverse reactions to subcutaneous immunoglobulin (SCIG) in clinical trials. J Clin Immunol. 2017;37(6):517–8 [r].
[53] Vacca A, Marasco C, Malccio A, et al. Subcutaneous immunoglobulins in patients with multiple myeloma and secondary hypogammaglobulinemia. Clin Immunol. 2018;191:110–5 [c].
[54] Beecher G, Anderson D, Siggiqi ZA. Subcutaneous immunoglobulin in myasthenia gravis exacerbation: a prospective, open-label trial. Neurology. 2017;89(11):1135–41 [c].
[55] Borte M, Kriván G, Derfalvi B, et al. Efficacy, safety, tolerability and pharmacokinetics of a novel human immune globulin subcutaneous, 20%: a phase 2/3 study in Europe in patients with primary immunodeficiencies. Clin Exp Immunol. 2017;187(1):146–59 [c].
[56] Canessa C, Iacopelli J, Pecoraro A, et al. Shift from intravenous or 16% subcutaneous replacement therapy to 20% subcutaneous immunoglobulin in patients with primary antibody deficiencies. Int J Immunopathol Pharmacol. 2017;30(1):73–82 [c].
[57] van Schaik IN, Bril V, van Geloven N, et al. Subcutaneous immunoglobulin for maintenance treatment in chronic inflammatory demyelinating polyneuropathy (PATH): a randomised, double-blind, placebo-controlled, phase 3 trial. Lancet Neurol. 2018;17(1):35–46 [C].
[58] Pedini V, Savore I, Danieli MG. Facilitated subcutaneous immunoglobulin (fSCIg) in autoimmune cytopenias associated with common variable immunodeficiency. Isr Med Assoc J. 2017;19(7):420–3 [A].
[59] Hachulla E, Benveniste O, Hamidou M, et al. High dose subcutaneous immunoglobulin for idiopathic inflammatory myopathies and dysimmune peripheral chronic neuropathies treatment: observational study of quality of life and tolerance. Int J Neurosci. 2017;127(6):516–23 [c].
[60] Zdziarski P, Gamian A, Majda J, et al. Passive blood anaphylaxis: subcutaneous immunoglobulins are a cause of ongoing passive anaphylactic reaction. Allergy Asthma Clin Immunol. 2017;13:41 [A].
[61] Donga PZ, Bilir SP, Little G, et al. Comparative treatment-related adverse event cost burden in immune thrombocytopenic purpura. J Med Econ. 2017;20(11):1200–6 [C].
[62] Bilecen S, De groot JA, Kalkman CJ, et al. Effect of fibrinogen concentrate on intraoperative blood loss among patients with intraoperative bleeding during high-risk cardiac surgery: a randomized clinical trial. JAMA. 2017;317(7):738–47 [C].
[63] Tomita E, Takase H, Tajima K, et al. Change of coagulation after NovoSeven® use for bleeding during cardiac surgery. Asian Cardiovasc Thorac Ann. 2017;25(2):99–104 [c].
[64] Puppo V. Commentary on: an objective measure of splitting in parental alienation: the parental acceptance-rejection questionnaire, J Forensic Sci. 2018;63(1):342. https://doi.org/10.1111/1556-4029.13625 [c].
[65] Croteau SE, Abajas YL, Wolberg AS, et al. Recombinant porcine factor VIII for high-risk surgery in paediatric congenital haemophilia a with high-titre inhibitor. Haemophilia. 2017;23(2): e93–8 [A].
[66] Mullins ES, Stasyshyn O, Alvarez-román MT, et al. Extended half-life pegylated, full-length recombinant factor VIII for prophylaxis in children with severe haemophilia A. Haemophilia. 2017;23(2):238–46 [c].
[67] Escobar MA, Tehranchi R, Karim FA, et al. Low-factor consumption for major surgery in haemophilia B with long-acting recombinant glycoPEGylated factor IX. Haemophilia. 2017;23:67–76 [c].
[68] Fischer K, Kulkarni R, Nolan B, et al. Recombinant factor IX Fc fusion protein in children with haemophilia B (kids B-LONG): results from a multicenter, non-randomised phase 3 study. Lancet Haematol. 2017;4(2):e75–82 [c].
[69] Pasi JK, Fischer K, Ragni M, et al. Long-term safety and efficacy of extended-interval prophylaxis with recombinant factor IX Fc fusion protein (rFIXFc) in subjects with haemophilia B. Thromb Haemost. 2017;117(3):508–18 [c].
[70] Lambert T, Rothschild C, Volot F, et al. A national French noninterventional study to assess the long-term safety and efficacy of reformulated nonacog alfa. Transfusion. 2017;57:1066–71 [c].
[71] Brekelmans MPA, van Ginkel K, Daams JG, et al. Benefits and harms of 4-factor prothrombin complex concentrate for reversal of vitamin K antagonist associated bleeding: a systematic review and meta-analysis. J Thromb Thrombolysis. 2017;44:118–29 [M].
[72] Majeed A, Agren A, Holmstrom M, et al. Management of rivaroxaban or apixaban associated major bleeding with prothrombin complex concentrates: a cohort study. Blood. 2017;130(15):1706–12 [c].
[73] Sun GH, Patel V, Moreno-Duarte I, et al. Intraoperative administration of 4-factor prothrombin complex concentrate reduces blood requirements in cardiac transplantation. J Cardiothorac Vasc Anesth. 2017;17: S1053–0770. [c].
[74] De Vlieger J, Dietvorst S, Demaerel R, et al. Neurosurgery in a patient at peak levels of rivaroxaban: taking into account all factors. Res Pract Thromb Haemost. 2017;1:296–300 [A].
[75] Kouides P, Wawra-Hehenberger K, Sajan A, et al. Safety of a pasteurized plasma-derived factor VIII and von Willebrand factor concentrate: analysis of 33 years of pharmacovigilance data. Transfusion. 2017;57:2390–403 [R].
[76] Lissitchkov TJ, Buevich E, Kuliczkowski K, et al. Pharmacokinetics, efficacy, and safety of a plasma-derived VWF/FVIII concentrate (VONCENTO) for on-demand and prophylactic treatment in patients with von Willebrand disease (SWIFT-VWD study). Blood Coagul Fibrinolysis. 2017;28(2):152–62 [c].
[77] Srivastava A, Serban M, Werner S, et al. Efficacy and safety of a VWF/FVIII concentrate (wilate®) in inherited von Willebrand disease patients undergoing surgical procedures. Haemophilia. 2017;23:264–72 [c].
[78] Trotta F, Belleudi V, Fusco D, et al. Comparative effectiveness and safety of erythropoiesis-stimulating agents (biosimilars vs originators) in clinical practice: a population-based cohort study in Italy. BMJ. 2017;e011637:7 [MC].
[79] Fischbach M, Wühl E, Sylvie C, et al. Efficacy and long-term safety of C.E.R.A. Maintenance in pediatric Hemodialyis patients with Anemia of CKD. Clin J Am Soc Nephrol. 2018;13:81–90 [c].
[80] Amato L, Addis A, Saulle R, et al. Comparative efficacy and safety in ESA biosimilars vs. orginators in adult with chronic kidney disease: a systematic review and meta-analysis. J Nephrol. 2018;31(3):321–32 [M].
[81] Oliva EN, Alati C, Santini V, et al. Eltrombopag versus placebo for low-risk myelodysplastic syndromes with thrombocytopenia (EQoL-MDS): phase 1 results of a single-blind randomized, controlled, phase 2 superiority trial. Lancet Haematol. 2017;4: e127–36 [c].
[82] Townsley DM, Scheinberg P, Winkler T, et al. Eltrombopag added to standard immunosuppression for aplastic anemia. N Engl J Med. 2017;376:1540–50 [C].
[83] Cubero-Gallego H, Romaguera R, Teruel L, et al. Thrombosis of a left atrial appendage occluder after treatment with thrombopoietin receptor agonists. JACC Cardiovasc Interv. 2018;11(2):e15–6 [A].

[84] Li VJ, Miao Y, Wilkins C, et al. Efficacy and thrombotic adverse events of romiplostim use patients with thrombocytopenia related to underlying malignancies. Blood. 2017;130:2324 [c].
[85] Kim JL, Park JH, Han SB, et al. Allogeneic blood transfusion is a significant risk factor for surgical-site infection following total hip and knee arthroplasty: a meta-analysis. J Arthroplasty. 2017;32(1):320–32 [M].
[86] Mansuy JM, Mengelle C, Pasquier C, et al. Zika virus infection and prolonged viremia in whole-blood specimens. Emerg Infect Dis. 2017;23(5):863–5 [H].
[87] Loyrion E, Trouve-Buisson T, Pouzol P, et al. Hepatitis E virus infection after platelet transfusion in an immunocompetent trauma patient. Emerg Infect Dis. 2017;23(1):146–7 [A].
[88] Inamoto Y, Lee SJ. Late effects of blood and marrow transplantation. Haematologia. 2017;102(4):614–25 [R].
[89] Zeiser R, Blazar BR. Acute graft-versus-host disease—biologic process, prevention, and therapy. N Engl J Med. 2017;377(22):2167–79 [R].
[90] Otrock ZK, Sempek DS, Carey S, et al. Adverse events of cryopreserved hematopoietic stem cell infusions in adults: a single-center observational study. Transfusion. 2017;57(6):1522–6 [c].
[91] Mitrus I, Smagur A, Fidyk W, et al. Reduction of DMSO concentration in cryopreservation mixture from 10% to 7.5% and 5% has no impact on engraftment after autologous peripheral blood stem cell transplantation: results of a prospective, randomized study. Bone Marrow Transplant. 2018;53(3):274–80 [C].
[92] Ishiharaa T, Nogami K, Matsumoto T, et al. Potentially life-threatening coagulopathy associated with simultaneous reduction in coagulation and fibrinolytic function in pediatric acute leukemia after hematopoietic stem-cell transplantation. Int J Hematol. 2017;106(1):126–34 [c].
[93] Ye Y, Zheng W, Wang J, et al. Risk and prognostic factors of transplantation-associated thrombotic microangiopathy in allogenic haematopoietic stem cell transplantation: a nested case control study. Hemtol Oncol. 2017;35(4):821–7 [C].
[94] Koh KN, Sunkara A, Kang G, et al. Acute kidney injury in pediatric patient receiving allogenic hematopoietic cell transplantation: incidence, risk factors, and outcomes. Biol Blood Marrow Transplant. 2018;24(4):758–64 [C].
[95] Vrooman LM, Millard HR, Brazaukas R, et al. Survival and late effects after allogeneic hematopoietic cell transplantation for hematologic malignancy at less than three years of age. Biol Blood Marrow Transplant. 2017;23(8):1327–34 [C].
[96] Maximova N, Gregori M, Boz G, et al. MRI-based evaluation of multiorgan overload is a predictor of adverse outcome sin pediatric patients undergoing allogeneic hematopoietic stem cell transplantation. Oncotarget. 2017;8(45):79650–61 [c].
[97] Atilla E, Toprak SK, Demirer T. Current review of iron overload and related complication in hematopoietic stem cell transplantation. Turk J Haematol. 2017;34(1):1–9 [R].
[98] Mohr A, Zwacka R. The future of mesenchymal stem cell-based therapeutic approaches for cancer—from cells to ghosts. Cancer Lett. 2018;414:239–49 [R].
[99] Pang Y, Xiao HW, Zhang H, et al. Allogeneic bone marrow-derived mesenchymal stromal cells expanded in vitro for treatment of aplastic anemia: a multicenter phase II trial. Stem Cells Transl Med. 2017;6(7):1569–75 [c].
[100] Comella K, Parlo M, Daly R, et al. Safety analysis of autologous stem cell therapy in a variety of degenerative diseases and injuries using the stromal vascular fraction. J Clin Med Res. 2017;9(11):935–42 [C].
[101] Fan X, Wang JZ, Lin XM, et al. Stem cell transplantation for spinal cord injury: a meta-analysis of treatment effectiveness and safety. Neural Regen Res. 2017;12(5):815–25 [M].
[102] Toyserkani NM, Jørgensen MG, Tabatabaeifar S, et al. Concise review: a safety assessment of adipose-derived cell therapy in clinical trials: a systematic review of reported adverse events. Stem Cells Transl Med. 2017;6(9):1786–94 [R].
[103] Nesteruk J, Voronina N, Kundt G, et al. Stem cell registry programme for patients with ischemic cardiomyopathy undergoing coronary artery bypass grafting: what benefits does it derive? ESC Heart Fail. 2017;4(2):105–11 [c].
[104] Bartolucci J, Verdugo FJ, González PL, et al. Safety and efficacy of the intravenous infusion of umbilical cord mesenchymal stem cells in patients with heart failure: a phase 1/2 randomized controlled trial (RIMECARD trial [randomized clinical trial of intravenous infusion umbilical cord Mesenchymal stem cells on cardiomyopathy]). Circ Res. 2017;121(10):1192–204 [c].
[105] Saraf SS, Cunningham MA, Kuriyan AE, et al. Bilateral retinal detachments after intravitreal injection of adipose-derived 'stem cells' in a patient with exudative macular degeneration. Opthalmic Surg Lasers Imaging Retina. 2017;48(9):772–5 [A].
[106] Kuriyan AE, Albini TA, Townsend JH, et al. Vision loss after intravitreal injection of autologous "stem cells" for AMD. N Engl J Med. 2017;376(11):1047–53 [A].
[107] Satarian L, Nourinia R, Kanavi MR, et al. Intravitreal injection of bone marrow mesenchymal stem cells in patients with advanced retinitis pigmentosa; a safety study. J Ophthalmic Vis Res. 2017;12(1):58–64 [c].

Further Reading

[108] Opsha Y, Brophy A. Chapter 33 - Blood, Blood Components, Plasma, and Plasma Products. In: Ray SD, editor. Side Effects of Drugs- Annual: A worldwide yearly survey of new data in adverse drug reactions. 37; Elsevier; 2015. p. 403–18 [R].
[109] Brophy A, Opsha Y, Cardinale M. Chapter 31 - Blood, Blood Components, Plasma, and Plasma Products. In: Ray SD, editor. Side Effects of Drugs- Annual: A worldwide yearly survey of new data in adverse drug reactions. 38; Elsevier; 2016. p. 335–53 [R].
[110] Cardinale M, Owusu K, Malm T. Chapter 29 - Blood, Blood Components, Plasma, and Plasma Products. In: Ray SD, editor. Side Effects of Drugs: Annual: A worldwide yearly survey of new data in adverse drug reactions. 39; Elsevier; 2017. p. 331–43 [R].

CHAPTER

33

Vitamins, Amino Acids and Drugs and Formulations Used in Nutrition

Brittany Singleton, Shandrika Landry*, Sunil Sirohi†, Sara Al-Dahir*,[1]*

*Division of Clinical and Administrative Sciences, College of Pharmacy, Xavier University of Louisiana, New Orleans, LA, United States

†Laboratory of Endocrine and Neuropsychiatric Disorders, Division of Basic Pharmaceutical Sciences, College of Pharmacy, Xavier University of Louisiana, New Orleans, LA, United States

[1]Corresponding author: saaldah@xula.edu

VITAMIN A [SEDA-35, 607; SEDA-36, 503; SEDA-38, 355; SEDA-39, 345]

A randomized, double-blind, placebo-controlled trial focused on vitamin A and zinc supplementation among pregnant women to prevent placental malaria. Although neither supplement affected birthweight among the 2056 participants with birthweight data available, there was a non-significant tendency for higher placental weight among participants who received vitamin A compared with those who did not and among those who received zinc than those who did not. Vitamin A recipients showed an increased risk of severe anemia compared to those who did not. In addition, there was a statistically insignificant trend towards a lower risk of preterm birth among those who received vitamin A compared to those who did not. Neither vitamin A nor zinc supplementation was associated with fetal loss, perinatal death, infant mortality, low birthweight, smaller gestational age (SGA), or other outcomes measured. No interaction between vitamin A and zinc was observed (all $P > 0.05$). No safety concerns were identified for either intervention [1C].

Another study published their final analysis using retinoic acid and arsenic trioxide (ATRA-ATO) vs retinoic acid and chemotherapy (ATRA-CHT) in leukemia patients. A total of 95 serious adverse events (SAEs) were reported in 65 patients. Of these, 43 and 52 SAEs were reported in patients receiving ATRA-ATO and ATRA-CHT, respectively. Hematologic toxicity was significantly higher in patients in the ATRA-CHT arm compared with the ATRA-ATO arm, who experienced grade 3 or 4 neutropenia and thrombocytopenia lasting more than 15 days. Episodes of febrile neutropenia (including fever of unknown origin and documented infections counted together) were also significantly more frequent in the ATRA-CHT arm than in the ATRA-ATO arm. In the ATRA-ATO and ATRA-CHT arms, 30 and 75 episodes were documented during the induction course, respectively. Significant elevation of liver enzymes was found to be more frequent in the ATRA-ATO arm (44%) compared with the ATRA-CHT (3%) arm across all treatment cycles. Toxicity resolved in all patients with temporary discontinuation of ATO and/or ATRA or of low-dose CHT during maintenance. QTc prolongation (defined as QTc >450 milliseconds for men and >460 milliseconds for women with correction calculated using the Framingham formula) was observed in 15 patients (11%) in the ATRA-ATO arm and 1 patient in the ATRA-CHT arm throughout the treatment cycles. There were no cases of life-threatening cardiac arrhythmias; however, ATO was permanently discontinued in one of the 15 patients, and the patient went off the protocol. Neurotoxicity mainly consisting of reversible peripheral nerve neuropathy was observed in greater proportion of patients in the ATRA-ATO arm and occurred during consolidation. Finally, GI toxicity and cardiac function abnormalities were significantly more frequent in the ATRA-CHT arm. For patients receiving ATRA-ATO, adverse effects (AEs) mainly consisted of frequent increase of liver enzymes, QTc prolongation,

Side Effects of Drugs Annual, Volume 40
ISSN: 0378-6080
https://doi.org/10.1016/bs.seda.2018.08.008

and hyperleukocytosis. In almost all patients, this toxicity was reversible and manageable with temporary drug interruption and dose adjustments as per protocol recommendations, including the addition of hydroxyurea (the only cytotoxic agent allowed for the control of hyperleukocytosis) [2C].

Another study investigated the efficacy and safety of the administration of retinol palmitate (VApal) ophthalmic solution (500 IU/mL) for the treatment of patients with dry eye. Adverse events (AEs) occurred in three patients (9.1%) in the VApal group and two patients (6.1%) in the placebo group; however, the difference was insignificant. Furthermore, no significant difference was observed in adverse reactions (ADRs), which did not occur in the VApal group and occurred in one patient (3%) in the placebo group. AEs with an expression rate of ≥1% in the VApal group were urine protein detection, arthritis, and nasopharyngitis. No serious drug-related AEs occurred in any patient during the study. Slit lamp examination, visual acuity testing, intraocular pressure measurement, funduscopic examination, and clinical blood and urine analyses indicated no clinical problems [3c].

A multicenter, randomized, open-label, phase II study of adult patients (≥18 years) with primary immune thrombocytopenia was conducted in China. The investigators aimed to evaluate the efficacy and safety of all-*trans* retinoic acid (ATRA) plus danazol vs danazol in non-splenectomized patients with corticosteroid-resistant or relapsed primary immune thrombocytopenia. Only two grade 3 AEs were reported: one (2%) patient receiving ATRA plus danazol with dry skin, and one (2%) patient receiving danazol monotherapy with liver injury. There was no grade 4 or worse AE or treatment-related death in either group [4c].

A small study aimed to evaluate the effect of pharmacologic isotretinoin dose on serum uric acid (SUA) level. This was a cohort study in which 51 adult patients with severe acne vulgaris who were prescribed oral isotretinoin treatment (0.5 mg/kg). Dermatologic examination was performed and SUA levels were measured at study inclusion for each participant, and then repeated at the first and second months of therapy. SUA levels at the first month and the second month were significantly higher than baseline SUA levels (*P*: 0.001, 0.007, respectively). SUA levels at the second month were higher than SUA levels at the first month, but the difference did not reach statistical significance. Since hyperuricemia is associated with renal disease, hypertension, atherosclerosis and metabolic syndrome as well as gout, it is important for the dermatologist to be aware of this potential AE of isotretinoin, particularly in vulnerable population [5c].

A case series examined systemic retinoid use for acne and the risk of psychiatric disorders. Data were extracted from the French National Pharmaco Vigilance Database for systemic acne treatments, systemic retinoids and drugs used as comparators. Each report was subjected to double-blind analysis by two psychiatric experts. A disproportionality analysis calculated the number of psychiatric ADRs divided by the total number of notifications for each drug of interest. All 71 systemic acne treatment regimens reported severe psychiatric disorders involving isotretinoin; however, the highest proportion of mild/moderate psychiatric AEs was reported with isotretinoin (14.1%). This study hypothesizes that psychiatric disorders associated with isotretinoin are related to a class effect of retinoids, as a signal for concern for alitretinoin. The authors conclude that complementary studies are necessary to estimate the risk and further determine at-risk populations [6c].

A study explored possible factors that could trigger a relapse in patients with ulcerative colitis (UC). Patients were followed up for 1 year to determine the effect of clinical, dietary, and psychological factors on relapse. Ninety-seven patients (59 males, mean age 39 ± 11.9 years) were monitored for a mean duration of 9 ± 2.3 months. Eighteen (18.6%) relapsed with the median time to relapse being 3.5 months. A relapse rate of 18.6% was observed over a follow-up of 9 months in patients with UC in clinical, endoscopic, and histological remission. Independent predictors of relapse were history of NSAIDs use within 15 days of relapse and higher intake of vitamin A [7c].

A report compiled about 11 pediatric patients with low-risk acute promyelocytic leukemia (APL) treated with ATRA and ATO. All patients stayed in molecular remission. All suffered from hyperleukocytosis. Two patients experienced reversible severe side effects. One suffered from osteonecrosis at both femurs, seizures, and posterior reversible encephalopathy syndrome. The other patient had an abducens paresis [8c].

A 39-year-old female who used topical tretinoin presented with a history of psoriasis vulgaris in remission, atopy, and exaggerated reaction to insect bites. In addition, she reported having allergic reactions to products with nickel and intolerance to products containing tretinoin, manifested by erythema and scaling. She was treated at the dermatology clinic for melasma, using 4% hydroquinone at night and 16% vitamin C, combined with broad-spectrum photo-protection in the morning. Afterwards, a 5% tretinoin peeling in hydroalcoholic solution containing propylene glycol was performed and left for 6 hours. In <24 hours, the patient exhibited itching, accentuated swelling, and erythema on the entire face, with vesicles and blisters in the chin area. The patient was treated with 40 mg of prednisone/day for 5 days, 500 mg of azithromycin/day for 3 days, and 0.05% desonide cream twice a day for 10 days. The patient showed considerable progress and full recovery occurred after 7 days. The melasma did not worsen and there was no post-inflammatory hyperpigmentation [9A].

One case of eruptive facial milia was reported in the setting of isotretinoin treatment for recalcitrant acne [10A].

In a letter to the editor, two cases of lower eyelid retraction were reported, which were associated with topical retinoid use. Severe bilateral lower eyelid retraction was seen with long-term use of topical retinoids for cosmetic purposes (12 and 20 years of continuous use). The retraction occurred as a result of severe shortening of the anterior lamella. In both cases, patients complained of dry eye symptoms and were treated for many years by their primary ophthalmologist with aggressive lubrication and the use of punctual plugs. Patient reports presented in this report exhibited trace lagophthalmos, conjunctival injection, and inferior punctate epithelial erosions. Surgical management, including skin grafting, represented a challenge due to the patient's cosmetic expectations [11r].

The Scientific Committee on Consumer Safety (SCCS) released a final opinion on vitamin A (retinol, retinyl acetate and retinyl palmitate) in cosmetic products. In view of SCCS, the use of vitamin A in body lotions at the maximum concentration of 0.05% per se is safe. The SCCS considers that the use of vitamin A in rinse-off products at the maximum concentration of 0.3% per se is safe. The SCCS also states that the use of vitamin A in face/hand cream products at the maximum concentration of 0.3% per se is considered safe [12S].

VITAMINS OF THE B GROUP [SEDA-35, 607; SEDA-38, 355; SEDA-39, 346]

Cobalamins (Vitamin B12)

A prospective study aimed to investigate the associations between intakes of B-vitamins (dietary, supplemental, total) and breast cancer risk. Over 27 000 women aged ≥45 years were included, with a median follow-up time of 4.2 years. Dietary data were collected using repeated 24 h records. A specific questionnaire assessed dietary supplement use over a 12-month period. Supplemental and total pyridoxine intakes were inversely associated with breast cancer risk. Total thiamin intake was borderline inversely associated with breast cancer. Statistically significant interactions between alcohol consumption and B-vitamin (thiamin, riboflavin, niacin, pantothenic acid, pyridoxine, folate, and cobalamin) supplemental intake were observed, the latter being inversely associated with breast cancer risk in non-to-low alcohol drinkers but not in habitual drinkers. This large prospective study, including quantitative assessment of supplemental intake, suggests a potential protective effect of pyridoxine and thiamin on breast cancer risk in middle-aged women [13MC].

A study investigated how high serum folate and vitamin B12 (VB12) deficiency could collaboratively aggravate neuronal degeneration. In total, 146 older non-demented diabetic individuals with an average age of 75 ± 3.9 were recruited. The results showed significant gray matter atrophy of the right middle occipital gyrus and the opercular part of the inferior frontal gyrus in subjects with high folate status coupled with VB12 deficiency. Consistent with previous observational studies on cognitive function, this study lends support to the notion that high serum folate concentrations in older people with VB12 deficiency may be associated with increased neurodegeneration and adverse cognitive effects [14C].

A study aimed to determine the safety and efficacy of pregabalin and methylcobalamin combination (PG-B12) with diclofenac potassium (DP) usage in patients with chronic post-thoracotomy pain (CPTP). One hundred patients with CPTP after posterolateral/lateral thoracotomy were prospectively/randomly assigned and evaluated. 50 patients were given PG-B12 and another 50 patients were given DP treatment. Visual Analogue Scale (VAS) and the Leeds Assessment of Neuropathic Symptoms and Signs (LANSS) scorings were performed prior to the treatment (day 0) and on the 15th, 30th, 60th, and 90th days. AEs were documented. The number of patients with a VAS score < 5 at the latest follow-up (VAS90 < 5) was 44 (88%) and 18 (36%) in the PG-B12 and DP groups, respectively ($P < 0.05$). 44 patients (88%) in the PG-B12 group and 16 patients (32%) in the DP group had a LANSS score < 12 at the latest follow-up ($P < 0.05$). Minor AEs that did not mandate discontinuation of the treatment were observed in 14 patients (28%) in the PG-B12 group and 2 patients (4%) in the DP group. Adverse events in the PG-B12 group were nausea in 5 (10%), dizziness in 3 (6%), heartburn in 1 (2%), and sedation in 5 (10%) patients. The only AE observed in the DP group was heartburn in 2 patients (4%). The authors conclude that PB-B12 is safe and effective in the treatment of CPTP with minimal side effects and a high patient compliance [15C].

Folic Acid and Folinic Acid (Vitamin B9)

A multicenter, prospective, mother–child cohort study was performed to determine long-term effects of the maternal use of high dosages of folic acid supplements (FASs) that exceed the Tolerable Upper Intake Level (UL) (≥1000 μg/day) on child's neurocognitive outcomes. Study goal was to examine the association between the use of high dosages of FASs during pregnancy and child neuropsychological development at ages 4–5 years. During the periconception period, one-third of the women took FAS dosages ≥1000 μg/day. The use of FAS dosages ≥1000 μg/day in this period was negatively associated with several neuropsychological outcomes scores in children, such as global verbal, verbal memory, cognitive function of posterior cortex, and cognitive function of left posterior cortex. Overall the use of FAS dosages exceeding the UL (≥1000 μg/day) during the periconception period was associated with lower levels of cognitive development in children aged 4–5 years. The authors conclude that use of FAS dosages ≥1000 μg/day during pregnancy should be monitored and avoided as much as possible, unless medically prescribed [16MC].

A prospective study sought to investigate the potential adverse effect of high folate intake from both food and supplements during pregnancy on children's development of asthma. The authors conclude that pregnant women taking supplemental folic acid at or above the recommended dose, combined with a diet rich in folate, reach a total folate intake level associated with a slightly increased asthma risk in children [17MC].

A case–control study investigated the association between alcohol consumption, folate intake, and risk of pancreatic cancer. The results highlighted an inverse relationship between natural (food-based, but not from supplements) folate intake and the risk of pancreatic cancer. Moreover, there was a statistically significant increased pancreatic cancer risk among participants with low dietary folate intake compared with those who consume the highest amounts of dietary folate (≥267.66 mcg). This study advocates the protective effect of higher dietary folate intake against the risk of developing pancreatic cancer [18C].

A cohort study explored the association between duration of folic acid (FA) supplementation during pregnancy and risk of postpartum depression (PPD). Results showed that pregnant women who took FA supplements for >6 months had a lower prevalence of PPD, compared to those who took FA for ≤6 months. No AEs were reported. Thus, the authors proposed that prolonged FA supplementation during pregnancy may decrease the risk of PPD [19C].

A case of fatal folic acid toxicity was reported. Normally, acute or chronic ingestion of a large dose of folic acid manifests as neurological complications, which are usually reversible. This study reported an unusual outcome. A 23-year-old pregnant woman committed suicide by consuming several folic acid tablets and succumbed to death within 36 hours. Postmortem toxicological analysis detected folic acid in viscera. The authors remark that death following acute consumption of folic acid is rare and has been not reported in the literature, to the best of current knowledge [20A].

Pyridoxine (Vitamin B6)

This case reported pyridoxine (vitamin B6) toxicity in a 41-year-old woman. This patient experienced 2 years of progressive burning pain, numbness, tingling, and weakness in a stocking-glove distribution that pinpointed severe pyridoxine toxicity. Concurrent presence of large and small fiber nerve dysfunction was noted in the form of an abnormal electromyography. Nerve conduction study demonstrated a chronic sensory polyneuropathy. Autonomic testing demonstrated abnormal responses to quantitative sweat testing and cardiovagal function testing. This case highlights the need for consideration of assessment of small fiber nerve damage by obtaining autonomic testing in cases of pyridoxine toxicity [21A].

An in vitro study purports that supplementation with high concentrations of pyridoxine leads to decreased vitamin B6 function. Prior cases of sensory neuronal pain due to over supplementation of vitamin B6 have been reported in the literature. In this study, the neurotoxicity of the different forms of vitamin B6 was tested on SHSY5Y and CaCo-2 cells. Cells were exposed to pyridoxine, pyridoxamine, pyridoxal, pyridoxal-5-phosphate or pyridoxamine-5-phosphate for 24 hours, after which cell viability was measured using the MTT assay. The expression of Bax and caspase-8 was tested after the 24 hours exposure. The effect of the vitamers on two pyridoxal-5-phosphate-dependent enzymes was also tested. Pyridoxine induced SHSY5Y cell death in a concentration-dependent manner. The other vitamers did not affect cell viability. Pyridoxine significantly increased the expression of Bax and caspase-8 reflecting active apoptotic pathways. Moreover, both pyridoxal-5-phosphate dependent enzymes were inhibited by pyridoxine. This study concluded that the neuropathy observed after taking a relatively high dose of vitamin B6 supplements is due to pyridoxine. The inactive form pyridoxine competitively inhibits the active pyridoxal-5′-phosphate. Consequently, the authors report that the symptoms of vitamin B6 supplementation are similar to those of vitamin B6 deficiency [22E].

Riboflavin (Vitamin B2)

A study investigated the relationships between dietary components and serum alanine aminotransferase (ALT) activity in Taiwanese adolescents. Data were collected from 1941 adolescents aged 13–18 years who participated in the fourth National Nutrition and Health Survey in Taiwan. Increased dietary zinc and vitamin B2 intake was associated with higher serum ALT in adolescents [23MC].

A detailed report was published highlighting the importance of compounding riboflavin with the correct dextran solution when using it for collagen cross-linking

(CXL) procedures. Six eyes of 4 male patients with keratoconus aged from 20 to 38 years underwent CXL with substitution of 20% dextran (T-500) with 20% dextran sulfate in a compounded riboflavin 0.1% solution. Postoperatively, persistent corneal epithelial defects, stromal haze, and then scarring occurred. Corneal transplantation was performed for visual rehabilitation but was complicated by graft rejection followed by failure, dehiscence, cataract, post-laser ablation haze, and steroid-induced glaucoma. Thinning, vascularization, and scarring of the residual host tissue were noted. The authors warn that substitution of dextran (T-500) with dextran sulfate in riboflavin solutions during CXL may result in loss of vision from permanent corneal opacity [24A].

A systematic review of riboflavin as prophylaxis of migraine headaches was performed. A total of 11 clinical trials revealed a mixed effect of riboflavin in the prophylaxis of migraine headache. Five clinical trials show a consistent positive therapeutic effect in adults; four clinical trials show a mixed effect in pediatric and adolescent patients, and two clinical trials of combination therapy have not shown benefit. Adverse effects included gastric effects, vomiting, diarrhea, polyuria, tension headache, mild upper abdominal pain, mild facial erythema, and changes in urine color. Notably, orange colored urine was seen in a patient. ADRs with riboflavin were generally categorized as mild [25R].

VITAMIN C (ASCORBIC ACID) [SEDA-34, 531; SEDA-35, 609; SEDA-38; SEDA-39, 349]

A randomized, placebo-controlled trial was conducted to determine the efficacy of vitamin C on postoperative outcomes after posterior lumbar interbody fusion (PLIF). A total of 123 eligible patients were randomly assigned to either group A (62 patients with vitamin C) or group B (61 patients with placebo). Patient follow-up was continued for at least 1 year after surgery. The primary outcome measure was pain intensity in the lower back using a visual analogue scale. Pain intensity in the lower back was significantly improved in both groups compared with preoperative pain intensity, but no significant difference was observed between the 2 groups over the follow-up period. Postoperative pain intensity, the primary outcome measure, was not significantly different at 1 year after surgery between the 2 groups. However, vitamin C may be associated with improving functional status after PLIF surgery, especially during the first 3 postoperative months [26C].

A 12-month, single-blind, multicenter randomized control trial wished to investigate the combined effect of metformin with ascorbic acid vs acetyl salicylic acid on diabetes-related cardiovascular complications. The trial was aimed to determine the effect of vitamin C (ascorbic acid) and aspirin (acetylsalicylic acid) on metabolic markers (FBS, LDL, HDL, triglycerides, etc.) of type 2 diabetes mellitus. Results showed a significant reduction in FBS, HbA1c, diabetes related long-term complications, LDL-c, TG, Total-c, and Framingham score for 10-years risk of developing CVD with the ascorbic acid group compared to the control group. Investigators finally concluded that ascorbic acid with metformin is more effective against reducing risks for diabetes related long-term complications (including Albumin Creatinine Ratio); however, if the treatment goal is to reduce cardiovascular events or risks for CVD development, then aspirin with metformin is beneficial over ascorbic acid [27C].

A study investigated the analgesic effects of vitamin C in patients undergoing uvulopalatopharyngoplasty (UPPP) and tonsillectomy. This study included 40 patients that were evaluated in a randomized, double-blinded clinical trial. All patients underwent the same method of anesthesia and surgical procedure. During the first 30 min after the beginning of the surgery, group C (vitamin C) received infusion of 3 g vitamin C in 500 mL of Ringer and group P received 6 mL normal saline in 500 mL of Ringer. A significant difference in mean pain severity between the two groups was noted at recovery room, 6, 12 and 24 h after surgery (P-value = 0.001). Also, a significant difference in mean times was noted between the patients that requested an analgesic, time of first dose of analgesic and pethidine dose between the two groups (P-value < 0.05). Measurements of systolic blood pressure, diastolic blood pressure, mean arterial blood pressure and heart rate in different times between the two groups were insignificant (P-value > 0.05). Blood loss was similar in the two groups (P-value > 0.05). According to this study, administration of vitamin C, 3 g IV intraoperative, reduced postoperative pain without increased side effects in patients undergoing UPPP and tonsillectomy [28c].

A cohort study from New Zealand (adults ~50 years; $n = 404$) correlated vitamin C status with markers of metabolic and cognitive health. Participants with higher vitamin C levels (mean plasma concentration: 44.2 μmol/L) exhibited lower weight, BMI and waist circumference, and better measures of metabolic health, including HbA1c, insulin and triglycerides, all risk factors for type 2 diabetes. Lower levels of mild cognitive impairment were observed in those with the highest plasma vitamin C concentrations. Plasma vitamin C showed a stronger correlation with markers of metabolic health and cognitive impairment than dietary vitamin C. No adverse findings were reported [29c].

A case report documented acute oxalate nephropathy related to vitamin C intake within an intensive care unit (ICU). A 57-year-old female patient with septic shock related to pneumonia had a case complicated by acute respiratory distress syndrome and acute kidney injury and required renal replacement therapy for 75 days. Post a renal biopsy, severe acute oxalate nephropathy was discovered. The only cause identified was vitamin C intake received during hospitalization within the ICU (~30 g over 2.5 months). The authors suggest the use of high-dose vitamin C should be prescribed with caution in the critically ill population. [30A].

A meta-analysis was completed to assess the efficacy of vitamin C therapy in preventing complex regional pain syndrome type I (CRPS-I) after a wrist fracture. The authors remarked that vitamin C supplementation is safe in healthy individuals but may induce complications (fatigue and lethargy) when used in high dosages (28 g/day). None of the studies included in the literature reported complications with vitamin C supplementation at a daily dosage of 500 mg. The authors conclude that daily supplementation with 500 mg of vitamin C per day for 50 days decreases the 1-year risk of CRPS-I after wrist fracture [31M].

A meta-analysis evaluated the role of vitamin C for preventing atrial fibrillation in high-risk patients. It was hypothesized that atrial fibrillation (AF), a common arrhythmia contributing substantially to cardiac morbidity, is associated with oxidative stress, and that the antioxidant vitamin C might influence it. No AEs from vitamin C were reported in the trials. The results of the meta-analysis indicate that vitamin C may prevent post-operative atrial fibrillation in some countries outside of the USA, and it may also shorten the duration of hospital stay and ICU stay of cardiac surgery patients [32M].

A review article describing the role of vitamin C in the treatment of pain reported no side effects within the vitamin C treatment groups or any or side effects between the experimental and control groups [33R].

An in vitro study examined the synergistic enhancement of topotecan-induced cell death by ascorbic acid in human breast MCF-7 tumor cells. Ascorbic acid, which produces hydrogen peroxide in tumor cells, significantly increased topotecan cytotoxicity in MCF-7 tumor cells. The presence of ascorbic acid also increased both topoisomerase I-dependent topotecan-induced DNA cleavage complex formation and topotecan-induced DNA double-strand breaks, suggesting that ascorbic acid enhanced DNA damage induced by topotecan and that the enhanced DNA damage is responsible for the synergistic interactions of topotecan and ascorbic acid [34E].

VITAMIN D ANALOGUES [SEDA-34, 532; SEDA-35, 609; SEDA-38, 355; SEDA-39, 349]

A community-based study evaluated the association between serum 25-hydroxyvitamin D (25(OH)D) status and blood pressure (BP) and the influence of vitamin D supplementation on hypertension. Participants were provided vitamin D supplements to achieve a target serum 25(OH)D concentration of >100 nmol/L. Analysis showed that achieving 25(OH)D concentrations of ≥100 nmol/L in a hypertensive population was associated with a significant decrease in systolic and diastolic blood pressure and mean arterial pressure. To achieve and maintain a serum 25(OH)D concentration of 100 nmol/L, at least 4000 IU/day (100 μg/day) of vitamin D was required. No AEs were reported. The authors conclude that in addition to lifestyle modifications and prescription antihypertensive medications, vitamin D supplementation may offer a simple, safe and cost-effective method for reducing blood pressure in vitamin D insufficient and hypertensive individuals [35MC].

A clinical trial evaluated the effect of high-dose vs standard-dose vitamin D supplementation on the incidence of wintertime upper respiratory tract infections in young children. 349 participants were randomized to receive 2000 IU/day of vitamin D oral supplementation (high-dose group) and 354 participants were randomized to receive 400 IU/day (standard-dose group) for a minimum of 4 months. The difference in the number of laboratory-confirmed infections between groups was insignificant. No AEs were reported. These findings did not support the routine use of high-dose vitamin D supplementation in children for the prevention of viral upper respiratory tract infections [36C].

A randomized controlled trial examined the effects of vitamin D supplementation on bone turnover markers (BTMs) in hypertensive, middle-aged patients without osteoporosis with low 25(OH)D levels. No effect of high-dose vitamin D supplementation on markers of bone turnover—bALP, CTX, OC, and P1NP—was observed. In addition, the cross-sectional association between 25(OH)D and BTMs was insignificant. The number of AEs (i.e., hospitalization or hypercalcemia) did not significantly differ between vitamin D and placebo groups [37C].

A randomized, placebo-controlled trial compared the effects of daily supplementation with vitamin D3 4000 IU (100 μg), 2000 IU (50 μg) or placebo for 1 year on biochemical markers of vitamin D status. Supplementation with 4000 IU daily vitamin D3 compared with 2000 IU daily was associated with a significantly higher proportion of individuals achieving plasma levels of 25(OH)D > 90 nmol/L after 1 year of treatment. Mean plasma levels of PTH decreased significantly in both active vitamin D groups but were significantly lower in those allocated 4000 IU vitamin D daily compared to 2000 IU at both 6 and 12 months. The intake of vitamin D at these high doses was well tolerated and was not associated with any adverse clinical events at 1 year, including hypercalcemia or kidney stones. Vitamin D supplementation also had no detectable effects on cardiovascular risk factors or on measures of physical function after 1 year of treatment [38C].

A study examined the impact of vitamin D supplementation on quality of life and physical performance in knee osteoarthritis (OA) patients. From baseline to 6 months, there was a significant increase in mean serum 25 (OH)D level, while mean LDL cholesterol, protein carbonyl, and PTH all significantly decreased. Patient quality of life (SF-12) and pain (visual analog scale, VAS) scores both improved significantly from baseline to the 6-month time point. Knee OA patients demonstrated significant improvement in grip strength and physical performance measurements after vitamin D2 supplementation. During treatment, levels of serum calcium increased significantly, three OA patients developed hypercalcemia (Ca $>$ 10.5 mg/dL), and PTH decreased significantly ($P < 0.05$). The authors conclude that vitamin D2 supplementation for 6 months reduced oxidative protein damage, decreased pain (VAS), improved quality of life, and improved grip strength and physical performance in knee OA patients [39C].

A randomized placebo-controlled trial measured the effect of monthly, high-dose, long-term ($\geq$1 year) vitamin D supplementation on central blood pressure parameters. A total of 517 adults were recruited and randomized to receive, for 1.1 years, either vitamin D3 200 000 IU (initial dose) followed 1 month later by monthly 100 000-IU doses or placebo monthly. Results show that monthly, high-dose vitamin D supplementation increased serum 25(OH)D concentration by $>$50 nmol/L with respect to placebo but had little effect on BP parameters in the total sample. However, in the vitamin D-deficient sample, this supplementation did not significantly change brachial BP, but had clinically relevant, beneficial effects on central BP parameters. No AEs were reported [40C].

A multicenter, prospective, randomized, controlled clinical trial was conducted to determine the optimal maintenance regimen for topical treatment with calcipotriol monohydrate/betamethasone dipropionate gel in patients with psoriasis vulgaris, by comparing the efficacy of three 8-week maintenance regimens. Results showed that maintenance treatment with calcipotriol monohydrate/betamethasone dipropionate using a continuous daily regimen or an "as needed" daily regimen provided similar efficacy, whereas a twice-weekly regimen was significantly less efficacious than either of these regimens. The incidence of AEs did not differ significantly between the groups [41C].

A case–control study aimed to determine if vitamin D supplementation could improve pain management, quality of life (QoL) and decrease infections in palliative cancer patients. After 1 month, the vitamin D-treated group had a significantly lower fentanyl dose compared to the untreated group, with an increased difference at 3 months. The ESAS QoL-score improved in the vitamin D group within the first month. The vitamin D-treated group had significantly lower consumption of antibiotics after 3 months compared to the untreated group. Vitamin D was well tolerated by all patients and no AEs were reported [42c].

An 8-week, randomized, double-blind, placebo-controlled clinical trial evaluated the effect of vitamin D supplementation on psychiatric, cognitive and metabolic parameters in chronic clozapine-treated schizophrenia patients. Schizophrenia patients who had been maintained on clozapine treatment for at least 18 weeks and had low levels of vitamin D and total PANSS scores (to ascertain the presence of residual symptoms) were randomly allocated to either weekly oral drops of vitamin D (14 000 IU) or placebo and subsequently assessed at 2-week intervals for psychosis severity, mood, cognition and metabolic profile. There were no severe AEs noted during follow-up in both groups. The vitamin D group exhibited slightly more gastrointestinal complaints: two patients reported nausea, three reported diarrhea and one reported abdominal discomfort. Overall vitamin D supplementation was associated with a trend towards improved cognition, but did not affect psychosis, mood or metabolic status [43c].

A randomized clinical trial compared the efficacy of daily, weekly and monthly administration of vitamin D3. Patients were given either daily 1000 IU, weekly 7000 IU or monthly 30 000 IU vitamin D3. Dose responses for increases in serum vitamin 25(OH)D were statistically equivalent for each of the three groups. The treatment of subjects with selected doses restored 25(OH)D values to levels above 20 ng/mL in all groups. Treatment with distinct administration frequency of vitamin D3 did not exhibit any differences in safety parameters. The authors conclude that daily, weekly and monthly administration of daily equivalent of 1000 IU of vitamin D3 provide equal efficacy and safety profiles [44c].

A small study was performed to determine the effect of vitamin D supplementation on improving mood (depression and anxiety) and health status (mental and physical) in women with type 2 diabetes mellitus (T2DM). There was a significant decrease in depression and anxiety. An improvement in mental health status was also found. After controlling for covariates (race, season of enrollment, baseline vitamin D, baseline depression (PHQ-9), and body mass index), the decline in depression remained significant. There was a trend for a better response to supplementation for women who were not taking medications for mood. The medication was well tolerated. One participant was withdrawn due to significant itching. While the attribution of this event was possibly associated with vitamin D, no other events were determined to be possibly, probably, or definitely related to study therapy [45c].

A double-blind, randomized controlled trial was conducted in out-patient CKD clinics to assess the effect of vitamin D supplementation on vascular stiffness in CKD.

Change in pulse wave velocity (PWV) was measured after 6 months of treatment with a fixed dose of oral calcifediol (5000 IU 25-hydroxyvitamin D3), calcitriol (0.5 μg 1,25-dihydroxyvitamin D3), or placebo, thrice weekly. After 6 months, the PWV decreased in the calcifediol group, remained unchanged in the calcitriol group, and increased in the placebo group. Side effects were minor and rare [46c].

One trial found improvement in bone turnover markers (BTMs) in patients treated weekly with 35 000 IU of vitamin D for 1 month. Results showed improvement in the levels of parathyroid hormone (PTH), osteocalcin, and carboxy-terminal telopeptides of crosslinks of type 1 collagen (βCTX), which paralleled the increase in vitamin D levels [47c].

Two contrasting cases of hypervitaminosis D were reported. Patient 1 was a 75-year-old man who developed symptomatic hypercalcemia (peak serum calcium concentration of 15.3 mg/dL; reference range: 8.5–10.6 mg/dL), cardiac injury, and a high total serum vitamin D concentration of 243 ng/mL (30–80 ng/mL) as a result of daily consumption of prescribed 50 000 IU ergocalciferol (vitamin D2) and 500 mg calcium citrate for 1 year. Patient 2 was a 60-year-old woman who consumed 40 000 IU of cholecalciferol (vitamin D3) daily for >10 months with a peak total serum vitamin D concentration of 479 ng/mL (30–80 ng/mL) but did not present with symptoms related to vitamin D toxicity. These cases demonstrate that individual responses to supraphysiologic concentrations of vitamin D for extended periods of time vary widely, and that defining a toxic concentration of this vitamin is difficult [48A].

A case report described a female patient with Fabry disease who was treated with a high dose of paricalcitol as an antiproteinuric agent due to unsatisfactory double-RAAS blockage. This resulted in transient worsening of her cardiac and renal function. Despite the positive effects of paricalcitol as an antiproteinuric agent, this case highlights the possible serious AEs associated with the use of high doses of this drug [49A].

A meta-analysis examined the effect of vitamin D supplementation on the prevention of symptoms and structural progression of knee osteoarthritis (OA). The difference in the incidence of AEs between the vitamin D and placebo groups was insignificant. Vitamin D supplementation was effective in improving the WOMAC pain and function in patients with knee OA. However, it had no beneficial effect on the prevention of tibial cartilage loss. The authors conclude that there is currently a lack of evidence to support the use of vitamin D supplementation in preventing the progression of knee OA [50M].

A meta-analysis examined the link between vitamin D supplementation and the prevention of asthma exacerbations. The results show that vitamin D supplementation reduced the rate of asthma exacerbations requiring treatment with systemic corticosteroids overall. The authors did not find definitive evidence that the effects of this intervention differed across subgroups of patients. Results of the safety analysis demonstrate that vitamin D supplementation did not affect the risk of having at least one serious AE of any cause, and particularly that no participant experienced hypercalcemia or renal stones. In view of the low cost of this intervention and the major economic burden associated with asthma exacerbations, authors conclude that vitamin D supplementation represents a potentially cost-effective strategy to reduce this important cause of morbidity and mortality [51M].

Another meta-analysis examined the link between vitamin D supplementation and disease activity in patients with immune-mediated rheumatic diseases. Vitamin D supplementation is considered safe, with a low risk of hypercalcemia and urolithiasis and provides cardiovascular protection. Vitamin D supplementation is also known to prevent glucocorticoid-induced osteoporosis and to reduce fractures in elderly people with osteoporosis; however, vitamin D supplementation is not well established in immune-mediated rheumatic diseases such as systemic lupus erythematosus (SLE), RA, systemic sclerosis, vasculitis and Sjögren Syndrome. Results of the review show that vitamin D supplementation reduced anti-dsDNA positivity on SLE and could possibly reduce rheumatoid arthritis recurrence, although novel randomized clinical trials are needed to confirm and extend the benefits of this hormone in immune-mediated rheumatic diseases [52M].

A meta-analysis of several randomized controlled trials examined the role of vitamin D supplementation in preventing acute respiratory tract infections. Results show that vitamin D supplementation reduced the risk of acute respiratory tract infection among all participants. In subgroup analysis, protective effects were seen in those receiving daily or weekly vitamin D without additional bolus doses but not in those receiving one or more bolus doses. Among those receiving daily or weekly vitamin D, protective effects were stronger in those with baseline 25(OH)D levels <25 nmol/L than in those with baseline 25(OH)D levels ≥25 nmol/L. Vitamin D did not influence the proportion of participants experiencing at least one SAE, the risk of SAEs of any cause or death due to any cause. Instances of potential AEs to vitamin D were rare. Hypercalcemia was detected in 21/3850 (0.5%) and renal stones were diagnosed in 6/3841 (0.2%); both events were evenly represented between intervention and control arms. Stratification of this analysis by dosing frequency did not reveal any statistically significant increase in risk of AEs with either bolus dosing or daily or weekly supplementation. The authors concluded that vitamin D supplementation was safe, and it protected against acute respiratory tract infection overall with patients who were very vitamin D deficient, with those not receiving bolus doses experiencing the most benefit [53M].

A meta-analysis examined non-calcemic AEs and withdrawals in randomized controlled trials of long-term vitamin D2 or D3 supplementation. The results show that long-term vitamin D2 or D3 supplementation, compared with placebo, did not increase all AEs. Vitamin D also did not increase the risk of the most common non-calcemic AEs: gastrointestinal and dermatological symptoms. Vitamin D did not increase withdrawals from several studies; however, participants given vitamin D were more likely to report withdrawals than those given placebo in studies in which calcium was given in both arms. The authors suggest that vitamin D, by itself, does not increase the risk of non-calcemic AEs [54M].

A review evaluated the link between vitamin D deficiency and liver disease. Non-alcoholic fatty liver disease (NAFLD) is the most common chronic hepatic disease throughout the Western world and is recognized as the main cause of cryptogenic cirrhosis; however, the identification of an effective therapy for NAFLD is still a major challenge. Epidemiological studies point towards an association between hypovitaminosis D and the presence of non-alcoholic fatty liver disease (NAFLD) and non-alcoholic steatohepatitis (NASH). The authors remark that the action of vitamin D downstream metabolites, such as 25(OH) vitamin D and 125(OH)2 vitamin D, has been largely investigated, and evidence has shown an increased risk of hypercalcemia and hypercalciuria associated with these therapies [55R].

Another review examined the role of vitamin D supplementation in preventing or treating any clinical condition. Many studies found no reliable evidence of benefit or harm, or no difference in AEs. However, one study did reveal a higher risk of nephrolithiasis in a cholecalciferol-treated group. Based on moderate to high quality evidence, the review showed that there were benefits of vitamin D supplementation in pregnant women and asthma patients and no benefits for preventing fractures [56R].

The role of vitamin D in nociceptive and inflammatory pain was reviewed. Recent interventional studies have shown promising effects of vitamin D supplementation on cancer pain and muscular pain, but only in patients with insufficient levels of vitamin D when starting the intervention. The authors suggest that patients with deficient levels, defined as 25-hydroxyvitamin D (25-OH(D)) levels <30 nmol/L, are most likely to benefit from supplementation, while individuals with 25-OH(D) >50 nmol/L probably have little benefit from supplementation. Vitamin D supplementation was never associated with more AEs than in the placebo groups. The authors conclude that vitamin D may constitute a safe, simple and potentially beneficial way to reduce pain among patients with vitamin D deficiency, but that more randomized and placebo-controlled studies are needed before any firm conclusions can be drawn [57R].

VITAMIN E (TOCOPHEROL) [SEDA-35, 610; SEDA-36, 515; SEDA-38, 355; SEDA-39, 351]

In a 2017 systematic review of antioxidant use in kidney disease progression, no significant AEs were reported. A total of 15 articles encompassing 4345patients were included for analysis in the review. Antioxidants reviewed included vitamin E. The trials did not provide clear reporting of AEs yet simply stated that no difference existed between placebo and intervention (antioxidant) groups [58R].

A 2017 systematic review of 32 publications examined the impact of G6PD deficiency on the tolerability and efficacy of herbal and dietary supplements. Among the dietary supplements included in the review was vitamin E. Some of the severe adverse side effects included hemolysis requiring hospitalization and death in G6PD individuals. The authors indicated that these were with very low certainty based upon the evidence reported. Seven studies reported no AEs with vitamin E supplementation with doses up to 800 IU daily [59R].

In a Cochrane Systematic Review of vitamin E for Alzheimer's disease, there was no difference between placebo and vitamin E with regards to serious AEs over 4 months (RR 0.86, 95% CI 0.71–1.05) [60R].

A phase II study examined the impact of vitamin E at doses of 15 mg/kg/day for chronic hepatitis B in pediatric patients. No significant AEs were reported for the vitamin E group. No abnormalities were noted in hematologic or liver function blood assays. Overall, vitamin E was considered well tolerated by the intervention group [61c].

VITAMIN K ANALOGUES [SEDA-35, 610; SEDA-36, 515; SEDA-38, 355; SEDA-39, 352]

A 2017 study assessed the minimum dose of vitamin K1 necessary to treat vitamin K deficiency. Among the reviewed, serious AEs of vitamin K emphasized by the authors were the non-allergic anaphylactoid reaction induced by vitamin K1 injection. The authors note that the trigger of the anaphylactoid reaction is the solubilizing agent vehicle by which the medication is administered [62c].

A case report documents an erythematic, multiforme like reaction in a 6-year-old male. The child developed an itchy rash on his right face, neck and shoulders, to which his mother had applied Arnika gel, a vitamin K1 oxide. Within days after application, erythematous papules surfaced across the child's abdomen. Dermatologic and allergen testing revealed a severe contact dermatitis to the topical phytomenadione epoxide formulation. Two weeks of therapy with topical corticosteroids subsided the reaction. A similar reaction was noted in a 35-year-old female patient after a mesotherapy procedure [63A].

A 2017 review of the use of the injectable formulation of vitamin K delivered orally demonstrated that injectable vitamin K, delivered orally, is as tolerable as the oral formulation. The review did not note increased AEs or bleeding/clotting complications [64R].

AMINO ACIDS [SEDA-35, 610; SEDA-36, 515; SEDA-38, 355; SEDA-39, 352]

A large, randomized control trial examined the effects of parenteral feeding components among pediatric patients. A total of 1440 patients between the ages of newborn and 17 years old were randomized to early vs late feeding groups. Early initiation was defined as initiation of parenteral nutrition within 24 hours of admission to the pediatric ICU vs late initiation, which was initiation of parenteral nutrition 7 days post-ICU admission. In both groups, the introduction of amino acids was associated with harm as defined as increased risk for mechanical ventilation, acquiring infections, and increased length of stay in the PICU [65C].

Arginine

In a phase IIA trial, endothelium-independent relaxation was assessed in 22 patients with peripheral artery disease who received an L-arginine infusion. In patients receiving the L-arginine infusion, the therapy was well tolerated, and limb volumetric flow increased. Alterations in glucose homeostasis were noted, manifesting as both hypo- and hyperglycemia. No other AEs were reported, including anaphylaxis. The L-arginine infusion was well tolerated by all study subjects [66c].

A 2017 Cochrane Review of arginine supplementation for preventing necrotizing enterocolitis in preterm infants included a total of three eligible patients. Hypotension was noted in one study as compared to placebo, but the results did not reach significance (RR 1.03, 95% CI 0.41–2.59). Gastrointestinal AEs were noted in more than one trial, including nausea, vomiting and diarrhea [67R].

In a systematic review of the use of vitamins, mineral and amino acids for the treatment of cancer cachexia, the authors included in their analysis the amino acids arginine and glutamine. Protein mixtures including arginine were reported to have increased gastrointestinal AEs such as abdominal cramping, nausea and vomiting. The authors conclude that due to the tolerability profile, supplementation with arginine offers no additional benefits in the perioperative setting for these patients [68R].

PARENTERAL NUTRITION [SEDA-35, 610; SEDA-36, 515; SEDA-38, 355; SEDA-39, 353]

Electrolytes and Minerals

A 2016 case report of a 40-year-old woman initiated on long-term total parenteral nutrition (TPN) post radiotherapy for cervical cancer suggested that 2 months of TPN therapy resulted in copper deficiency related bone marrow suppression. According to Oo et al., copper deficiency results in cytoplasmic vacuole changes of the erythroid and myeloid precursor chains, resulting in hematologic abnormalities [69A].

A retrospective analysis of 147 ovarian cancer patients initiated on total parenteral nutrition demonstrated that TPN is associated with impaired clinical outcomes. All 147 cervical cancer patients in the study underwent tumor debulking and bowel resection. Patients on TPN had prolonged time to restoration of bowel function (5.77 days vs 4.7 days, $P<0.001$) and longer hospital stays (11.46 days vs 7.14 days, $P<0.001$). TPN patients also experienced a greater number of laboratory abnormalities, such as lower albumin levels (2.22 g/dL vs 2.97 g/dL) [70c].

A prospective, 7-day follow-up study of 90 patients in a highly specialized intestine failure unit on TPN found an increased risk of refeeding syndrome manifesting as electrolyte abnormalities. Increased risk was noticed among high-risk patients vs low-risk patients, defined by the 2006 NICE guidelines. Electrolyte abnormalities manifested as hypomagnesemia (27.5%) and hypophosphatemia (30%) while on TPN therapy. Individuals in the high-risk category receiving TPN required increased exogenous supplementation of electrolytes [71c].

Bretón et al. conducted an observational, retrospective cohort study of adult, non-critically ill hospitalized patients receiving a minimum of 7 days of TPN therapy. A total of 101 patients were included in their analysis. With an a priori definition of hypertriglyceridemia of plasma triglycerides >200 mg/dL, a total of 33% of the patients developed hypertriglyceridemia. Specific risk predictors of hypertriglyceridemia were increased body mass index (BMI) and a glucose infusion rate 3.1 g/kg/day [72c].

A 2017 review focused on parenteral nutrition-associated liver disease (PNALD). In the review, the mechanism of PNALD is defined as multifactorial, focusing on lipid emulsions serving as enhancers of proinflammatory markers. Definitive diagnosis is often complicated by multiple comorbidities in the typical TPN patient, such a postoperative hypoxic liver and infections. TPN is also associated with calculous and acalculous cholecystitis [73R].

In a retrospective, observational study conducted over 7 months, 222 hospitalized patients were followed after initiation of parenteral nutrition therapy. The authors hypothesize that serum sodium correction outside of the TPN is associated with an overestimation of true serum sodium level due to hypoproteinemia. Eighty-one percent of patients for whom the hyponatremia was corrected for low protein levels were diagnosed with hyponatremia vs 43% for whom no protein correction occurred ($P=0.001$). The authors conclude that the composition of the parenteral nutrition was not associated with hyponatremia among the patients [74c].

A historical cohort study of 195 pediatric patients (<19 years old) receiving TPN were assessed post initiation of TPN. Patients were monitored for liver dysfunction, described by elevated liver function tests (LFTs), hypertriglyceridemia and hyperbilirubinemia. Though the majority of patients maintained normal liver function while on TPN, bilirubin changes (68.5%) and hypertriglyceridemia (65.1%) were associated with TPN use and elevated liver function tests. The authors conclude that early monitoring of the hepatic profile during TPN initiation is necessary for pediatric patients [75c].

A 2017 case report describes a 21-year-old male with short bowel syndrome and long-term TPN dependency who was admitted to the emergency department for acute, severe epigastric pain. An abdominal ultrasound and endoscopic retrograde cholangiopancreatography (ERCP) revealed Mirizzi syndrome. Mirizzi syndrome is a hepatic duct obstruction commonly associated with total parenteral nutrition [76A].

A case of a 31-year-old female who developed endemic goiter while on TPN was documented. The patient was TPN dependent for 6 years due to mitochondrial disease-induced intestinal failure. After 6 years on TPN, the patient presented with hypothyroidism and endemic goiter due to an alteration of the TPN to a trace element formulation which did not contain iodine. The authors recommend 130mcg of iodine per day for patients that are total parenteral nutrition dependent [77A].

A comprehensive 2017 systematic review explored the correlation between TPN and refeeding syndrome. A total of 45 interventional and observational trials of a composite of 6608 patients were included for analysis. The incidence of refeeding syndrome ranged from 0% to 80%. Though the definition of refeeding syndrome varied greatly across the trials, with most focusing on electrolyte abnormalities, it was observed most often within the first 72 hours post parenteral nutrition initiation. The authors concluded that a definitive relationship between total parenteral nutrition and refeeding syndrome would require additional research [78M].

A comprehensive review documented the multiple complications associated with parenteral nutrition therapy. Among complications noted was persistent hyperglycemia. Though 50% of inpatients experienced episodes of hyperglycemia, the incidence was higher among individuals receiving parenteral nutrition, especially at an infusion rate >4 mg/kg/min [79R].

Contamination and Infections

In a retrospective analysis, 1184 patients undergoing pancreaticoduodenectomy (PD) were followed for nutritional intervention. A total of 19.6% of the patients undergoing PD received TPN perioperatively. Patients were monitored from initiation of TPN until completion of therapy. With a median duration of TPN of 9 days, hyperglycemia (35.3%) was noted as the most significant AE. Central line infections were rare, occurring at 2.6% total. The overall mortality rate was 3.4%, which was three times the institutional mortality rate at 0.8%. The institutional rate was not sub-stratified for surgical patients. The authors conclude that TPN remains a safe therapeutic option for patients in this surgical subset with delayed gastric emptying [80c].

A 2017 meta-review of patients in the United States discharged on total parenteral nutrition from 2001 to 2014 found a peak in TPN prescriptions in 2012, reaching 43350. The authors indicate that the rising trend toward outpatient TPN prescriptions should be reviewed with caution as infectious complications are common AEs associated with TPN [81R].

A 2017 retrospective review of 1367 hematopoietic stem cell transplant (HSCT) patients demonstrated that individuals on TPN ($n=197$) experienced increased infectious complications and mortality. Individuals on TPN vs those not on TPN had increased infections due to Gram-positive bacteria, Gram-negative bacteria and fungi (26% vs 15%, $P=0.0057$). All-cause mortality was significantly higher among patients receiving TPN (22% vs 7%, $P=0.0057$). Significant predictors of TPN-related infections included number of TPN-treatment days and number of days post-transplantation when TPN was initiated [82c].

In a 2017, retrospective, observational, cross-sectional study, 85 patients were followed post-initiation of TPN for 6 months. The primary outcome was defined as central venous catheter infections. Of individuals on TPN, 19% developed a catheter-related infection. Though catheter insertion was associated with multiple medication therapies, not just TPN, all patients were on concomitant TPN. The average time to development of infection post-catheter insertion was 78±64 days ($P=0.014$). Another significant risk factor was concomitant TPN use and surgery [83c].

A systematic literature review included 11 observational studies comprising a conglomerate of 2854 patients. The incidence of catheter-related bloodstream infection (CRBSI) was 0–6.6 per 1000 central venous access days (CVAD). No additional risk was noted for individuals on total parenteral nutrition vs those who were not [84M].

A 2017 review explains the relationship between infections, immune dysfunction and the gastrointestinal microbiome. Review of literature suggested how use of prolonged parenteral nutrition is associated with gut microbiome changes and subsequent gastrointestinal associated lymphoid tissue. This is exacerbated by impaired gut epithelium of the intestines and subsequent vulnerability to infections [85R].

In 2017 Wang et al. also reported on the relationship between gut microbiota, parenteral nutrition and the rate of infections. Fecal samples were collected from 18 infants with short bowel syndrome who were receiving parenteral nutrition. The authors report that the gut microbiota was significantly different among infants who had concomitant blood stream infections vs those who did not have blood stream infections. Pediatric patients with the infections were noted to have an overgrowth of Enterobacteriacae. Microbiome diversity was correlated with intestinal health and improved immune responses in individual receiving total parenteral nutrition [86c].

ENTERAL NUTRITION (NON-ORAL: GASTRIC AND JEJUNAL) [SEDA-35, 610; SEDA-36, 515; SEDA-38, 355; SEDA-39, 354]

NutritionDay ICU, a 7-year observational study across 46 countries, was a worldwide, survey-based study aimed at describing nutrition-based practices in intensive care units. The questionnaire focused on illness severity scores such as SAPS, SOFA and NEMS, and subsequent nutrition regimens prescribed to the most critically ill patients. Among individuals requiring nutrition support via enteral feeding, GI side effects such as constipation (9.8%) and diarrhea (5%) were most common. Elevated gastric residuals greater than 250mL and gastric reflux were also commonly associated side effects of enteral nutrition. These side effects correspond to commonly known side effects enteral feeding, both gastric and jejunal feeds [87MC].

According to a review by Abela in 2017 of post-surgical patients, gastrointestinal intolerance is indicated as an important AE when initiating and titrating enteral feeds. Vomiting was reported as more often present among early (21%) vs delayed (14%) initiation of enteral feedings, though the results did not reach statistical significance [88R].

El-Matary et al. completed a review of the use of enteral feedings in Crohn's disease patients in 2017. This review found no significant AEs associated with enteral nutrition in the 12 studies included, representing 1169 patients. Gastrointestinal AEs reported were diarrhea, vomiting, nausea and a sense of fullness. Patients also reported intolerance to the taste or smell of the enteral feeding formulas.[89R].

A prospective, randomized trial of critically ill patients compared the tolerability and efficacy of peptide based vs high protein based enteral formulas. 49 patients were included from medical, surgical and cardiothoracic intensive care units. There was no difference in AE rates between the two groups. Of note, the most often reported AEs were gastrointestinal, such as nausea, vomiting, high gastric residual, abdominal distention and others. The study was consistent with current literature regarding AEs among enteral feeding patients [90c].

Miyata et al. conducted a prospective, randomized study using omega-3 fatty acid enteral nutrition therapy to attenuate the AEs of neoadjuvant chemotherapy in patients with esophageal cancer. 61 patients were randomized to receive enriched omega-3 vs poor omega-3 formulas. Among patients receiving enriched formulas, there was a decrease in Grade ¾ chemotherapy associated diarrhea (16.1% vs 36.7%), though the results did not reach significance ($P=0.068$) [91c].

In a meta-analysis conducted by Tian et al., the authors reviewed 11 randomized control trials addressing the use of enteral energy supplementation among critically ill patients. The analysis focused on low-energy vs high-energy containing enteral formulas. Infection rates were similar between the two groups (RR 1.09, 95% CI 0.95–1.26, $P=0.23$) but gastrointestinal intolerance was observed at lower rates in the low-energy groups, with the results reaching significance (RR 0.79, 955 CI 0.65–0.97, $P<0.05$) [92M].

In a mixed prospective/retrospective study design, Kim and investigators analyzed the effects of an enteral feeding protocol on the improvement of enteral feeding in critically ill patients. Two hundred and seventy patients were included, with 134 representing pre-protocol and 136 representing post-protocol implementation. Only individuals receiving at least 24 hours of enteral therapy were included in the analysis. Implementation of the protocol was associated with an increased use of prokinetic agents, indicating gastric stasis. In addition, both enteral feeding groups had similar incidence of diarrhea, a common AE associated with enteral feedings [93c].

Hyperglycemia is identified as a common AE for individuals receiving enteral nutrition, reaching 30% among adults in long-term care facilities. Enteral feeding induced hyperglycemia is associated with additional metabolic derangements that are purportedly mediated by increased inflammatory markers such as cytokines. In a comprehensive review by Drincic et al., the authors recommend mitigating strategies including changing the enteral feeding formula and routine monitoring of blood glucose levels while on enteral feeding [94R].

Enteral feeding dependence is common among individuals with inflammatory bowel disease (IBD), such as Crohn's disease and ulcerative colitis. A 2017 systematic review by Gatti et al. described the role of enteral feeding dependence on alterations on gut microbiota. A total of 14 studies were included in the analysis representing case–control and prospective studies, including both adult and pediatric patients. All studies used laboratory-based analysis techniques for biological samples. Review results indicated that there was a significant decrease in intestinal microbiological diversity among enteral feeding patients, regardless of the bacterial analysis technique used. In studies using health controls, there was even a greater distance between enteral feeding formula patients than studies that used IBD controls. In turn, bacterial metabolism of short chain fatty acids was altered in patients that were enteral feeding dependent. The authors conclude that due to the variability in study results, a larger multicenter trial is necessary to compare the effects enteral feeding on this patient population [95R].

A retrospective cohort study of Auckland City Hospital database included 754 adult patients in the intensive care unit. Enteral tube feeding intolerance was documented for all patients. Thirty-three percent of enteral feeding patients experienced intolerance, resulting in an impediment to reaching their caloric goal. Enteral feeding intolerance was associated with poor clinical outcomes, though the underlying pathophysiology of the condition remains poorly described [96c].

In a retrospective, cohort study, a total of 107 patients admitted for acute respiratory failure were followed for metabolic complications. 107 of the subjects were initiated on enteral feeding. Serum albumin levels were lower in the enteral feeding group vs oral feeding group (2.7 mg/dL vs 3 mg/dL, $P=0.048$). In addition, 53% of the enteral feeding group developed airway complications, vs 32% in the control group ($P=0.03$). Finally, patients on enteral feeding experienced longer days on ventilation [97c].

In a retrospective chart review of 360 congenital heart disease patients receiving postoperative care, the investigators documented reasons for enteral feeding stops. A total of 32.2% of enteral feeding stops were associated with enteral feeding intolerance manifesting as gastrointestinal side effects, such as vomiting, gastrointestinal bleeding, diarrhea, constipation and elevated gastric residual volumes [98c].

Additional case studies can be found in these reviews [99R,100R].

CONFLICT OF INTEREST

Authors declare no conflict of interest.

References

[1] Darling AM, Mugusi FM, Etheredge AJ, et al. Vitamin A and zinc supplementation among pregnant women to prevent placental malaria: a randomized, double-blind, placebo-controlled trial in Tanzania. Am J Trop Med Hyg. 2017;96(4):826–34 [C].

[2] Platzbecker U, Avvisati G, Cicconi L, et al. Improved outcomes with retinoic acid and arsenic trioxide compared with retinoic acid and chemotherapy in non-high-risk acute promyelocytic leukemia: final results of the randomized Italian-German APL0406 trial. J Clin Oncol Off J Am Soc Clin Oncol. 2017;35(6):605–12 [C].

[3] Toshida H, Funaki T, Ono K, et al. Efficacy and safety of retinol palmitate ophthalmic solution in the treatment of dry eye: a Japanese phase II clinical trial. Drug Des Devel Ther. 2017;11:1871–9 [c].

[4] Feng F-E, Feng R, Wang M, et al. Oral all-trans retinoic acid plus danazol versus danazol as second-line treatment in adults with primary immune thrombocytopenia: a multicentre, randomised, open-label, phase 2 trial. Lancet Haematol. 2017;4(10):e487–96 [c].

[5] Solak B, Erdem T, Solak Y. Isotretinoin use for acne vulgaris is associated with increased serum uric acid levels. J Dermatol Treat. 2017;28(1):82–5 [c].

[6] Le Moigne M, Bulteau S, Grall-Bronnec M, et al. Psychiatric disorders, acne and systemic retinoids: comparison of risks. Expert Opin Drug Saf. 2017;16(9):989–95 [c].

[7] Dhingra R, Kedia S, Mouli VP, et al. Evaluating clinical, dietary, and psychological risk factors for relapse of ulcerative colitis in clinical, endoscopic, and histological remission. J Gastroenterol Hepatol. 2017;32(10):1698–705 [c].

[8] Creutzig U, Dworzak MN, Bochennek K, et al. First experience of the AML-Berlin-Frankfurt-Münster group in pediatric patients with standard-risk acute promyelocytic leukemia treated with arsenic trioxide and all-trans retinoid acid. Pediatr Blood Cancer. 2017;64(8) [c].

[9] Magalhães GM, Rodrigues DF, de Oliveira ER, et al. Tretinoin peeling: when a reaction is greater than expected. An Bras Dermatol. 2017;92(2):291–2 [A].

[10] Farmer W, Cheng K, Marathe K. Eruptive milia during isotretinoin therapy. Pediatr Dermatol. 2017;34(6):728–9 [A].

[11] Winkler KP, Black EH, Servat J. Lower eyelid retraction associated with topical retinol use. Ophthal Plast Reconstr Surg. 2017;33(6):483 [r].

[12] Scientific Committee of Consumer Safety—SCCS, Rousselle C. Opinion of the scientific committee on consumer safety SCCS—final version of the opinion on vitamin A (retinol, retinyl acetate and retinyl palmitate) in cosmetic products. Regul Toxicol Pharmacol RTP. 2017;84:102–4 [S].

[13] Egnell M, Fassier P, Lécuyer L, et al. B-vitamin intake from diet and supplements and breast cancer risk in middle-aged women: results from the prospective nutrinet-santé cohort. Nutrients. 2017;9(5) [MC].

[14] Deng Y, Wang D, Wang K, et al. High serum folate is associated with brain atrophy in older diabetic people with vitamin B12 deficiency. J Nutr Health Aging. 2017;21(9):1065–71 [C].

[15] Metin SK, Meydan B, Evman S, et al. The effect of pregabalin and methylcobalamin combination on the chronic postthoracotomy pain syndrome. Ann Thorac Surg. 2017;103(4):1109–13 [C].

[16] Valera-Gran D, Navarrete-Muñoz EM, Garcia de la Hera M, et al. Effect of maternal high dosages of folic acid supplements on neurocognitive development in children at 4–5 y of age: the prospective birth cohort Infancia y Medio Ambiente (INMA) study. Am J Clin Nutr. 2017;106(3):878–87 [MC].

[17] Parr CL, Magnus MC, Karlstad Ø, et al. Maternal folate intake during pregnancy and childhood asthma in a population-based cohort. Am J Respir Crit Care Med. 2017;195(2):221–8 [MC].

[18] Yellow W, Bamlet WR, Oberg AL, et al. Association between alcohol consumption, folate intake, and risk of pancreatic cancer: a case-control study. Nutrients. 2017;9(5) [C].

[19] Yan J, Liu Y, Cao L, et al. Association between duration of folic acid supplementation during pregnancy and risk of postpartum depression. Nutrients. 2017;9(11) [C].

[20] Devnath GP, Kumaran S, Rajiv R, et al. Fatal folic acid toxicity in humans. J Forensic Sci. 2017;62(6):1668–70 [A].

[21] Bacharach R, Lowden M, Ahmed A. Pyridoxine toxicity small fiber neuropathy with dysautonomia: a case report. J Clin Neuromuscul Dis. 2017;19(1):43–6 [A].

[22] Vrolijk MF, Opperhuizen A, Jansen EHJM, et al. The vitamin B6 paradox: supplementation with high concentrations of pyridoxine leads to decreased vitamin B6 function. Toxicol Vitro Int J Publ Assoc BIBRA. 2017;44:206–12 [E].

[23] Bai C-H, Chien Y-W, Huang T-C, et al. Increased dietary zinc and vitamin B-2 is associated with increased alanine aminotransferase in Taiwanese adolescents. Asia Pac J Clin Nutr. 2017;26(1): 78–84 [MC].

[24] Höllhumer R, Watson S, Beckingsale P. Persistent epithelial defects and corneal opacity after collagen cross-linking with substitution of dextran (T-500) with dextran sulfate in compounded topical riboflavin. Cornea. 2017;36(3):382–5 [A].

[25] Thompson DF, Saluja HS. Prophylaxis of migraine headaches with riboflavin: a systematic review. J Clin Pharm Ther. 2017;42(4):394–403 [R].

[26] Lee GW, Yang HS, Yeom JS, et al. The efficacy of vitamin c on postoperative outcomes after posterior lumbar interbody fusion: a randomized, placebo-controlled trial. Clin Orthop Surg. 2017;9(3):317–24 [C].

[27] Gillani SW, Sulaiman SAS, Abdul MIM, et al. Combined effect of metformin with ascorbic acid versus acetyl salicylic acid on diabetes-related cardiovascular complication; a 12-month single blind multicenter randomized control trial. Cardiovasc Diabetol. 2017;16(1):103 [C].

[28] Ayatollahi V, Dehghanpour Farashah S, Behdad S, et al. Effect of intravenous vitamin C on postoperative pain in uvulopalatopharyngoplasty with tonsillectomy. Clin Otolaryngol. 2017;42(1):139–43 [c].

[29] Pearson JF, Pullar JM, Wilson R, et al. Vitamin C status correlates with markers of metabolic and cognitive health in 50-year-olds: findings of the CHALICE cohort study. Nutrients. 2017;9(8):139–43 [c].

[30] Colliou E, Mari A, Delas A, et al. Oxalate nephropathy following vitamin C intake within intensive care unit. Clin Nephrol. 2017;88(12):354–8 [A].

[31] Aïm F, Klouche S, Frison A, et al. Efficacy of vitamin C in preventing complex regional pain syndrome after wrist fracture: a systematic review and meta-analysis. Orthop Traumatol Surg Res OTSR. 2017;103(3):465–70 [M].

[32] Hemilä H, Suonsyrjä T. Vitamin C for preventing atrial fibrillation in high risk patients: a systematic review and meta-analysis. BMC Cardiovasc Disord. 2017;17(1):49 [M].

[33] Carr AC, McCall C. The role of vitamin C in the treatment of pain: new insights. J Transl Med. 2017;15(1):77 [R].

[34] Sinha BK, van't Erve TJ, Kumar A, et al. Synergistic enhancement of topotecan-induced cell death by ascorbic acid in human breast MCF-7 tumor cells. Free Radic Biol Med. 2017;113:406–12 [E].

[35] Mirhosseini N, Vatanparast H, Kimball SM. The association between serum 25(OH)D status and blood pressure in participants of a community-based program taking vitamin D supplements. Nutrients. 2017;9(11): pii: E1244 [MC].

[36] Aglipay M, Birken CS, Parkin PC, et al. Effect of high-dose vs standard-dose wintertime vitamin D supplementation on viral upper respiratory tract infections in young healthy children. JAMA. 2017;318(3):245–54 [C].

[37] Schwetz V, Trummer C, Pandis M, et al. Effects of vitamin D supplementation on bone turnover markers: a randomized controlled trial. Nutrients. 2017;9(5): pii: E432 [C].

[38] Hin H, Tomson J, Newman C, et al. Optimum dose of vitamin D for disease prevention in older people: BEST-D trial of vitamin D in primary care. Osteoporos Int J Establ Result Coop Eur Found Osteoporos Natl Osteoporos Found USA. 2017;28(3):841–51 [C].

[39] Manoy P, Yuktanandana P, Tanavalee A, et al. Vitamin D supplementation improves quality of life and physical performance in osteoarthritis patients. Nutrients. 2017;9(8):799 [C].

[40] Sluyter JD, Camargo CA, Stewart AW, et al. Effect of monthly, high-dose, long-term vitamin d supplementation on central blood pressure parameters: a randomized controlled trial substudy. J Am Heart Assoc. 2017;6(10):e006802 [C].

[41] Lee J-H, Park C-J, Kim T-Y, et al. Optimal maintenance treatment with calcipotriol/betamethasone dipropionate gel in Korean patients with psoriasis vulgaris: a multicentre randomized, controlled clinical trial. J Eu Acad Dermatol Venereol JEADV. 2017;31(3):483–9 [C].

[42] Helde-Frankling M, Höijer J, Bergqvist J, et al. Vitamin D supplementation to palliative cancer patients shows positive effects on pain and infections-results from a matched case-control study. PLoS One. 2017;12(8):e0184208 [c].

[43] Krivoy A, Onn R, Vilner Y, et al. Vitamin D supplementation in chronic schizophrenia patients treated with clozapine: a randomized, double-blind, placebo-controlled clinical trial. EBioMedicine. 2017;26:138–45 [c].

[44] Takács I, Tóth BE, Szekeres L, et al. Randomized clinical trial to comparing efficacy of daily, weekly and monthly administration of vitamin D3. Endocrine. 2017;55(1):60–5 [c].

[45] Penckofer S, Byrn M, Adams W, et al. Vitamin D supplementation improves mood in women with type 2 diabetes. J Diabetes Res. 2017;2017:8232863 [c].

[46] Levin A, Tang M, Perry T, et al. Randomized controlled trial for the effect of vitamin d supplementation on vascular stiffness in CKD. Clin J Am Soc Nephrol CJASN. 2017;12(9):1447–60 [c].

[47] Sulimani RA, Mohammed AG, Alshehri SN, et al. A weekly 35,000 IU vitamin D supplementation improves bone turnover markers in vitamin D deficient Saudi adolescent females. Arch Osteoporos. 2017;12(1):85 [c].

[48] Kim S, Stephens LD, Fitzgerald RL. How much is too much? Two contrasting cases of excessive vitamin D supplementation. Clin Chim Acta. 2017;473:35–8 [A].

[49] Keber T, Tretjak M, Cokan Vujkovac A, et al. Paricalcitol as an antiproteinuric agent can result in the deterioration of renal and heart function in a patient with fabry disease. Am J Case Rep. 2017;18:644–8 [A].

[50] Gao X-R, Chen Y-S, Deng W. The effect of vitamin D supplementation on knee osteoarthritis: a meta-analysis of randomized controlled trials. Int J Surg Lond Engl. 2017;46: 14–20 [M].

[51] Jolliffe DA, Greenberg L, Hooper RL, et al. Vitamin D supplementation to prevent asthma exacerbations: a systematic review and meta-analysis of individual participant data. Lancet Respir Med. 2017;5(11):881–90 [M].
[52] Franco AS, Freitas TQ, Bernardo WM. Pereira RMR. Vitamin D supplementation and disease activity in patients with immune-mediated rheumatic diseases: a systematic review and meta-analysis. Medicine (Baltimore). 2017;96(23):e7024 [M].
[53] Martineau AR, Jolliffe DA, Hooper RL, et al. Vitamin D supplementation to prevent acute respiratory tract infections: systematic review and meta-analysis of individual participant data. BMJ. 2017;356:i6583 [M].
[54] Malihi Z, Wu Z, Mm Lawes C, et al. Noncalcemic adverse effects and withdrawals in randomized controlled trials of long-term vitamin D2 or D3 supplementation: a systematic review and meta-analysis. Nutr Rev. 2017;75(12):1007–34 [M].
[55] Barchetta I, Cimini FA, Cavallo MG. Vitamin D supplementation and non-alcoholic fatty liver disease: present and future. Nutrients. 2017;9(9)E1015 [R].
[56] Mateussi MV, de Latorraca COC, Daou JP, et al. What do cochrane systematic reviews say about interventions for vitamin D supplementation? Sao Paulo Med J Rev Paul Med. 2017;135(5): 497–507 [R].
[57] Helde-Frankling M, Björkhem-Bergman L. Vitamin D in pain management. Int J Mol Sci. 2017;18(10) [R].
[58] Bolignano D, Cernaro V, Gembillo G, et al. Antioxidant agents for delaying diabetic kidney disease progression: a systematic review and meta-analysis. PLoS One. 2017;12(6)e0178699 [R].
[59] Lee SWH, Lai NM, Chaiyakunapruk N, et al. Adverse effects of herbal or dietary supplements in G6PD deficiency: a systematic review. Br J Clin Pharmacol. 2017;83(1):172–9 [R].
[60] Farina N, Llewellyn D, MGEKN I, et al. Vitamin E for Alzheimer's dementia and mild cognitive impairment. Cochrane Database Syst Rev. 2017;27(1):CD002854 [R].
[61] Fiorino S, Loggi E, Verucchi G, et al. Vitamin E for the treatment of E-antigen-positive chronic hepatitis B in paediatric patients: results of a randomized phase 2 controlled study. Liver Int Off J Int Assoc Study Liver. 2017;37(1):54–61 [c].
[62] Mi Y-N, Ping N-N, Li B, et al. Finding the optimal dose of vitamin K1 to treat vitamin K deficiency and to avoid anaphylactoid reactions. Fundam Clin Pharmacol. 2017;31(5): 495–505 [c].
[63] Pastor-Nieto M-A, Gatica-Ortega M-E, Melgar-Molero V, et al. Erythema multiforme-like reaction resulting from vitamin K1 oxide (phytomenadione epoxide). Contact Dermatitis. 2017;77(5): 343–5 [A].
[64] Afanasjeva J. Administration of injectable vitamin K orally. Hosp Pharm. 2017;52(9):645–9 [R].
[65] Vanhorebeek I, Verbruggen S, Casaer MP, et al. Effect of early supplemental parenteral nutrition in the paediatric ICU: a preplanned observational study of post-randomisation treatments in the PEPaNIC trial. Lancet Respir Med. 2017;5(6):475–83 [C].
[66] Kashyap VS, Lakin RO, Campos P, et al. The largPAD trial: phase IIA evaluation of L-arginine infusion in patients with peripheral arterial disease. J Vasc Surg. 2017;66(1):187–94 [c].
[67] Shah PS, Shah VS, Kelly LE. Arginine supplementation for prevention of necrotising enterocolitis in preterm infants. Cochrane Database Syst Rev. 2017;4:CD004339 [R].
[68] Mochamat, Cuhls H, Marinova M, et al. A systematic review on the role of vitamins, minerals, proteins, and other supplements for the treatment of cachexia in cancer: a European palliative care research centre cachexia project. J Cachexia Sarcopenia Muscle. 2017;8(1):25–39 [R].
[69] Oo TH, Hu S. Copper deficiency-related bone marrow changes secondary to long-term total parenteral nutrition. Clin Case Rep. 2017;5(2):195–6 [A].
[70] Mendivil AA, Rettenmaier MA, Abaid LN, et al. The impact of total parenteral nutrition on postoperative recovery in patients treated for advanced stage ovarian cancer. Arch Gynecol Obstet. 2017;295(2):439–44 [c].
[71] Pantoja F, Patel P, Keane N, et al. PT10.2: re-feeding syndrome in adults receiving total parenteral nutrition: an audit in a highly specialized intestine failure unit. Clin Nutr. 2017;36: S50–1 [c].
[72] Ocón Bretón MJ, Ilundain Gonzalez AI, Altemir Trallero J, et al. Predictive factors of hypertriglyceridemia in inhospital patients during total parenteral nutrition. Nutr Hosp. 2017;34(3): 505–11 [c].
[73] Mitra A, Ahn J. Liver disease in patients on total parenteral nutrition. Clin Liver Dis. 2017;21(4):687–95 [R].
[74] Gómez-Hoyos E, Fernández-Peña S, Cuesta M, et al. Hyponatremia in patients receiving parenteral nutrition: the importance of correcting serum sodium for total proteins. The role of the composition of parenteral nutrition in the development of hyponatremia. Eur J Clin Nutr. 2018;72(3):446–51 [c].
[75] Golucci APBS, Morcillo AM, Hortencio TDR, et al. Hypercholesterolemia and hypertriglyceridemia as risk factors of liver dysfunction in children with inflammation receiving total parenteral nutrition. Clin Nutr ESPEN. 2018;23:148–55 [c].
[76] Sellers ZM, Thorson C, Co S, et al. Feeling the impact of long-term total parenteral nutrition. Dig Dis Sci. 2017;62(12):3317–20 [A].
[77] Pearson S, Donnellan C, Turner L, et al. Endemic goitre and hypothyroidism in an adult female patient dependent on total parenteral nutrition. Endocrinol Diabetes Metab Case Rep. 2017;2017:pii: 17-0030 [A].
[78] Friedli N, Stanga Z, Sobotka L, et al. Revisiting the refeeding syndrome: results of a systematic review. Nutrition. 2017;35:151–60 [M].
[79] Lappas BM, Patel D, Kumpf V, et al. Parenteral nutrition: indications, access, and complications. Gastroenterol Clin North Am. 2018;47(1):39–59 [R].
[80] Worsh CE, Tatarian T, Singh A, et al. Total parenteral nutrition in patients following pancreaticoduodenectomy: lessons from 1184 patients. J Surg Res. 2017;218:156–61 [c].
[81] John J, Seifi A. Total parenteral nutrition usage trends in the United States. J Crit Care. 2017;40:312–3 [R].
[82] Trifilio S, Rubin H, Fong JL, et al. Revisiting infectious complications in hematopoietic stem cell transplantation recipients who receive total parenteral nutrition. Open Forum Infect Dis. 2017;4(Suppl 1):S714–5 [c].
[83] Parra-Flores M, Souza-Gallardo LM, García-Correa GA, et al. Incidence of catheter-related infection incidence and risk factors in patients on total parenteral nutrition in a third level hospital. Cir Cir. 2017;85(2):104–8 [c].
[84] Gavin NC, Button E, Keogh S, et al. Does parenteral nutrition increase the risk of catheter-related bloodstream infection? A systematic literature review. JPEN J Parenter Enteral Nutr. 2017;41(6):918–28 [M].
[85] Pierre JF. Gastrointestinal immune and microbiome changes during parenteral nutrition. Am J Physiol Gastrointest Liver Physiol. 2017;312(3):G246–56 [R].
[86] Wang P, Wang Y, Lu L, et al. Alterations in intestinal microbiota relate to intestinal failure-associated liver disease and central line infections. J Pediatr Surg. 2017;52(8):1318–26 [c].
[87] Bendavid I, Singer P, Theilla M, et al. NutritionDay ICU: a 7 year worldwide prevalence study of nutrition practice in intensive care. Clin Nutr Edinb Scotl. 2017;36(4):1122–9 [MC].

[88] Abela G. The potential benefits and harms of early feeding post-surgery: a literature review. Int Wound J. 2017;14(5):870–3 [R].
[89] El-Matary W, Otley A, Critch J, et al. Enteral feeding therapy for maintaining remission in Crohn's disease: a systematic review. JPEN J Parenter Enteral Nutr. 2017;41(4):550–61 [R].
[90] Seres DS, Ippolito PR. Pilot study evaluating the efficacy, tolerance and safety of a peptide-based enteral formula versus a high protein enteral formula in multiple ICU settings (medical, surgical, cardiothoracic). Clin Nutr Edinb Scotl. 2017;36(3):706–9 [c].
[91] Miyata H, Yano M, Yasuda T, et al. Randomized study of the clinical effects of ω-3 fatty acid-containing enteral nutrition support during neoadjuvant chemotherapy on chemotherapy-related toxicity in patients with esophageal cancer. Nutrition. 2017;33:204–10 [c].
[92] Tian F, Gao X, Wu C, et al. Initial energy supplementation in critically ill patients receiving enteral nutrition: a systematic review and meta-analysis of randomized controlled trials. Asia Pac J Clin Nutr. 2017;26(1):11–9 [M].
[93] Kim S-H, Park C-M, Seo J-M, et al. The impact of implementation of an enteral feeding protocol on the improvement of enteral nutrition in critically ill adults. Asia Pac J Clin Nutr. 2017;26(1):27–35 [c].
[94] Drincic AT, Knezevich JT, Akkireddy P. Nutrition and hyperglycemia management in the inpatient setting (meals on demand, parenteral, or enteral nutrition). Curr Diab Rep. 2017;17(8):59 [R].
[95] Gatti S, Galeazzi T, Franceschini E, et al. Effects of the exclusive enteral nutrition on the microbiota profile of patients with Crohn's disease: a systematic review. Nutrients. 2017;9(8) [R].
[96] Wang K, McIlroy K, Plank LD, et al. Prevalence, outcomes, and management of enteral tube feeding intolerance: a retrospective cohort study in a tertiary center. JPEN J Parenter Enteral Nutr. 2017;41(6):959–67 [c].
[97] Kogo M, Nagata K, Morimoto T, et al. Enteral nutrition is a risk factor for airway complications in subjects undergoing noninvasive ventilation for acute respiratory failure. Respir Care. 2017;62(4):459–67 [c].
[98] Qi J, Li Z, Cun Y, et al. Causes of interruptions in postoperative enteral nutrition in children with congenital heart disease. Asia Pac J Clin Nutr. 2017;26(3):402–5 [c].
[99] Patel D, Holaway CS, Thomas S, et al. Vitamins, amino acids, and drugs and formulations used in nutrition. In: Ray SD, editor. Side effects of drugs annual: a worldwide yearly survey of new data in adverse drug reactions. 38. Elsevier; 2016. p. 355–64 [R]. [chapter 32].
[100] Al-Dahir S, Vithlani N, Smith A, et al. Vitamins, amino acids and drugs and formulations used in nutrition. In: Ray SD, editor. Side effects of drugs annual: a worldwide yearly survey of new data in adverse drug reactions. 39. Elsevier; 2017. p. 345–58 [R]. [chapter 30].

CHAPTER

34

Drugs That Affect Blood Coagulation, Fibrinolysis and Hemostasis

Kirk E. Evoy,†,‡,1, Jason Isch§, Hanna Raber¶*

*College of Pharmacy, University of Texas at Austin, Austin, TX, United States

†School of Medicine, University of Texas Health at San Antonio, San Antonio, TX, United States

‡University Health System, San Antonio, TX, United States

§College of Pharmacy, Natural & Health Sciences, Manchester University, Fort Wayne, IN, United States

¶College of Pharmacy, University of Utah, Salt Lake City, UT, United States

[1]Corresponding author: evoy@uthscsa.edu

COUMARIN ANTICOAGULANTS [SEDA-37, 419; SEDA-38, 365; SEDA-39, 359]

Warfarin [SEDA-37, 419; SEDA-38, 365; SEDA-39, 359]

Cardiovascular

A retrospective Taiwanese study assessed the impact of warfarin on cardiovascular outcomes in patients with atrial fibrillation (AF) on hemodialysis [1C]. In this matched cohort, patients on warfarin ($n=744$) were more likely than patients not receiving warfarin ($n=1767$) to develop congestive heart failure (CHF) (HR=1.82, $P<0.01$), peripheral artery disease (PAD) (HR=3.42, $P<0.01$), and aortic valve stenosis (HR=3.20, $P<0.05$), but not an increased risk of stroke (HR=0.91, 95% CI=0.59–1.39) or mortality (HR=1.04, 95%CI=0.88–1.23).

Pharmacogenomics

A sub-study of the Hokusai-venous thromboembolism study analyzed whether outcomes differed in patients with CYP2C9 and/or VKORC1 variants [2MC]. Based on the presence or absence of these genetic variants, patients were classified as normal, sensitive or highly-sensitive responders. Sensitive and highly-sensitive responders were more likely to require decreased warfarin doses ($P<0.001$), spend more time over-anticoagulated ($P<0.001$) and had increased bleeding risk (HR=1.38, $P=0.0035$ for sensitive and HR=1.79, $P=0.0252$ for highly-sensitive responders).

A multi-center randomized controlled trial (RCT) assessed outcomes associated with genetic-guided warfarin dosing in elderly patients following elective total knee (TKA) or hip arthroplasties (THA) [3MC]. 831 patients were treated with genotype-guided warfarin therapy while 819 received standard dosing. The primary outcome was composite major bleeding, international normalized ratio (INR) ≥4, VTE, or death. Patients receiving genotype-guided warfarin experienced less 30-day primary outcome events (87 (10.8%) vs 116 (14.7%); RR=0.73, 95%CI=0.56–0.95).

A single-center, prospective study of 460 Chinese patients assessed the impact of genetic polymorphisms on INR in the first week following heart valve replacement. Each patient received 2.5 mg warfarin daily without dose adjustment [4C]. The authors noted the effect of warfarin was correlated with VKORC1 rs7294 and CYP2C9*3 rs057910 beginning at day 3, with CYP4F2 rs2108622 from day 4, and ORM1 rs17650 from day 6 on, with 14.9% of the effect variation explained by these four genes. Notably, 42% of wild-type patients achieved an INR of 2.0–3.0 by day 7 with 2.5 mg dosing, while 90% of patients with VKORC1 rs7294 displayed subtherapeutic INRs below 1.63 at day 7.

Bleeding/Thromboembolism

A retrospective, multi-center cohort study of elderly Canadian patients with AF compared real-world hospital

Side Effects of Drugs Annual, *Volume 40*
ISSN: 0378-6080
https://doi.org/10.1016/bs.seda.2018.06.012

care associated with oral-anticoagulant-related hemorrhages with either warfarin or DOAC therapy, finding that patients on warfarin ($n=1542$) were more likely to receive reversal agents (72.9% vitamin K, 40.7% prothrombin complex concentrate (PCC), and 17.9% fresh frozen plasma) than previous RCTs have shown with warfarin (generally <40%, per authors) and more likely than patients receiving DOACs ($n=460$) in the present study (nonactivated PCC 12.3%, activated PCC 5.2%, and factor VIIa 2.2%) [5MC]. Despite higher reversal agent utilization, patients receiving warfarin experienced higher in-hospital mortality (15.2% vs 9.8%, ARR=0.66, 95%CI=0.49–0.89) than those receiving DOACs.

A single-center retrospective review of 9 young patients (median 14.4 years, range 7.1–22.8 years) with Kawasaki disease and ≥1 giant coronary artery aneurysm was conducted to assess the safety of guideline-based anticoagulation with warfarin plus aspirin [6c]. Time in therapeutic range (TTR) was 59%. During the 4-year study, three patients experienced a major bleed and two patients experienced an asymptomatic coronary thrombosis, but no myocardial infarctions (MI) or deaths were observed. This study is relevant because there is little documentation of outcomes associated with combined warfarin and aspirin therapy in children with giant aneurysms associated with Kawasaki disease in the medical literature, despite recommendation as the treatment of choice.

Analysis of the ORBIT-AF RCT sought to assess bleeding events associated with warfarin vs DOACs for AF in real-world practice [7MC]. 9967 patients from 244 sites were followed for a median of 360 days. Those treated with warfarin experienced 3.5 major bleeds/100 patient years vs 3.3 major bleeds/100 patient years with DOACs. Warfarin patients were more likely to receive blood product administration (76% vs 53%, $P=0.0004$) or correctional agents (35% vs 2.2%, $P<0.0001$) to treat a bleed than DOAC patients.

A Cochrane Review assessed five RCTs involving 12545 patients with AF and chronic kidney disease to compare warfarin to DOACs in this patient population and identified similar efficacy without increased risk of major bleeding [8M].

A case–control study was conducted within a single American health-system to assess the impact of obesity on warfarin-related bleeding risk in a real-world community setting. 265 cases with major bleeding while taking warfarin were compared with 305 controls. Obese patients had a significantly lower bleeding risk (OR=0.60, 95% CI=0.39–0.92) [9C]. An exploratory analysis of the interaction between obesity and genetic variants that alter warfarin bleeding risk (CYP2C9, VKORC1, CYP4F2) identified a significant interaction implying possible protective effects of obesity on risk of major bleeding for patients with wild-type CYP4F2*3 ($P=0.049$). This finding suggests that the protective effects of the CYP4F2*3 variant and obesity may be reduced when present together. The authors report this is the first study to assess association between body mass index and warfarin major bleeding risk.

A single-center, open-label, prospective RCT compared warfarin (INR goal 2.0–3.0) to aspirin (150mg daily) in 370 patients in their first 3 months following bioprosthetic aortic valve replacement, finding similar rates of thromboembolic events (11 (6.6%) vs 12 (7.5%), respectively, $P=0.83$) and 90-day mortality (8 (4.7%) vs 6 (3.7%), respectively, $P=0.79$), but more major bleeding associated with warfarin in multivariate analysis (OR=5.18, 95%CI=1.06–25.43) [10C]. Of note, patients in the warfarin group undergoing CABG ($n=63$) received concomitant aspirin 75mg daily while patients in the aspirin group with AF not resolved through cardioversion ($n=6$) received concomitant warfarin.

A meta-analysis of 11 RCTs including patients treated with vitamin K antagonists (VKAs) for AF ($n=25901$) or venous thromboembolism (VTE) ($n=15114$) assessed bleeding risk in high-risk patient populations [11M]. Increased bleeding risk (major bleeds in four studies, composite major and non-major bleeds in seven) with warfarin therapy was observed in patients aged >75 years (RR=1.62, $P<0.0001$), with low body weight (RR=1.20, $P=0.02$), and with impaired renal function (i.e., creatinine clearance (CrCl)=30–50mL/min; RR=1.59, $P<0.00001$). This risk was maintained when looking specifically at AF or VTE patients.

A single-center, retrospective, observational study of patients presenting to the emergency department (ED) with warfarin- ($n=342$) or DOAC-related (dabigatran ($n=33$), rivaroxaban ($n=32$), or apixaban ($n=30$)) bleed analyzed real-world treatment approach and outcomes, finding similar reversal agent utilization rates and similar mortality rates between patients treated with warfarin or DOAC [12C].

The first systematically conducted, prospective study of warfarin for treatment of bioprosthetic valve thrombosis of surgically implanted valves was conducted including 52 American patients at a single American hospital [13c]. In this single-arm study, warfarin (INR goal 2.0–3.0) produced positive response (≥50% reduction in prosthesis gradient) in 43 (83%) patients within 11 weeks (interquartile range (IQR)=6–22 weeks). Across a median follow-up of 86±24 weeks, two patients experienced bleeding events.

A small case series analyzed matched cohorts of patients admitted to the hospital following an intracranial hemorrhage (ICH) while concomitantly taking a DOAC ($n=9$; 8 on rivaroxaban and 1 on apixaban) or warfarin ($n=18$) to compare hematoma expansion rates, which has been shown to predict mortality following warfarin-associated ICH [14c]. Though underpowered, no significant difference was observed between groups, with 4/9 DOAC patients and 5/18 warfarin patients ($P=0.42$) experiencing hematoma expansion.

A retrospective study of four national Swedish patient quality registries ($n=3831$ patients) assessed the impact of warfarin treatment quality (measured through both TTR and INR variability) on outcomes in patients with mechanical heart valves [15MC]. Compared to the reference population of patients with high TTR (≥70%) and low INR variability above mean (≥0.40), patients with low TTR had significantly higher rates of bleeding (HR=1.50, 95%CI=1.01–2.38) and mortality (HR=2.44, 95%CI=1.58–3.78) but not thrombosis (HR=0.84, 95% CI=0.47–1.48), whereas patients with high INR variability had higher rates of bleeding (HR=1.41, 95%CI=1.07–1.84), thrombosis (HR=1.31, 95%CI=1.002–1.72), and mortality (HR=1.68, 95%CI=1.25–2.25). Patients with both low TTR and high INR variability had higher rates of bleeding (HR=2.50, 95%CI=1.99–3.15), thrombosis (HR=1.55, 95%CI=1.21–1.99) and mortality (HR=3.34, 95%CI=2.62–4.27). Additionally, patients with higher warfarin intensity (mean INR=2.8–3.2) displayed higher rates of bleeding (HR=1.29, 95%CI=1.06–1.58), death (HR=1.65, 95%CI=1.31–2.06), and total complications (HR=1.24, 95%CI=1.06–1.41) compared to patients with lower warfarin intensity (mean INR=2.2–2.7).

Utilizing data from the CALIBER study which linked population data from four national patient care registries in England, a large, retrospective study of 70206 patients with non-valvular AF measured ischemic stroke (IS) incidence and net clinical benefit (NCB) of warfarin at differing CHA_2DS_2-VASc scores [16MC]. The authors note this study overcame limitations of other similar studies by including both primary and secondary care data to more accurately identify stroke risk factors and by assessing NCB in addition to stroke risk. IS incidence rate (IR) was lower in patients using warfarin (1015 over 59006 patient-years, IR (95%CI)=1.7 (1.6–1.8)) vs patients not receiving anticoagulation (5990 over 157439 patient-years, IR (95%CI)=3.8 (3.7–3.9), $P=0.00$). IS rates were not significantly different in patients with CHA_2DS_2-VASc=0 (IR (95%CI)/100 patient-years=0.4 (0.2–0.8) vs 0.2 (0.1–0.3), $P=0.16$) but were for patients with CHA_2DS_2-VASc=1 (IR (95%CI)/100 patient-years=0.4 (0.3–0.7) vs 0.7 (0.6–0.8), $P=0.03$) and CHA_2DS_2-VASc=2 (IR (95%CI)/100 patient-years=0.8 (0.7–1.0) vs 1.4 (1.3–1.6), $P=0.00$). Factoring in both IS and hemorrhagic stroke risk, the overall NCB of warfarin was 1.9 (1.8–2.1) IS/100 patient-years. NCB (NCB (95%CI)/100 PY) was −0.3 (−0.8 to 0.1), 0.1 (−0.2 to 0.4), and 0.2 (−0.1 to 0.6)), respectively, for CHA_2DS_2-VASc=0, CHA_2DS_2-VASc=1 and CHA_2DS_2-VASc=2. The authors concluded that CHA_2DS_2-VASc accurately stratified stroke risk. In patients with CHA_2DS_2-VASc=1, in which there are differing opinions on whether or not to use warfarin, NCB was positive but not statistically significant.

A randomized, non-blinded, single-center, prospective Chinese study ($n=101$) compared low-intensity (INR goal=1.5–2.0) and standard-intensity (INR goal=2.0–2.5) uninterrupted warfarin therapy during radiofrequency catheter ablation for AF in elderly patients (age ≥70 years) [17C]. Patients receiving low-intensity warfarin experienced similar rates of thromboembolic complications (0 vs 1, $P=0.49$), major bleeding (0 vs 0), minor bleeding (1 vs 5, $P=0.1$), and asymptomatic cerebral emboli (6 vs 4, $P=0.82$) compared with those receiving standard-intensity warfarin, respectively.

A retrospective observational cohort study of the Danish National Patient Register was conducted to compare outcomes in patients resuming warfarin vs those not resuming anticoagulation following ICH [18MC]. 2415 patients were included, 1325 with an ICH secondary to hemorrhagic stroke and 1090 following a trauma. Within 1 year post-stroke, patients resuming warfarin experienced less IS or systemic embolisms (adjusted hazard ratio (AHR)=0.49, 95%CI=0.24–1.02) and increased recurrent ICH (AHR=1.31, 95% CI=0.24–1.02), though neither reached statistical significance. However, mortality was significantly reduced with warfarin resumption following both hemorrhagic stroke (AHR=0.51, 95%CI=0.37–0.71) and trauma (AHR=0.35, 95%CI=0.23–0.52).

A retrospective matched cohort study was conducted in the US and Canada to compare safety of DOACs to warfarin immediately following VTE in the community setting [19MC]. Of the 12489 DOAC ($n=11812$ rivaroxaban, $n=539$ dabigatran, and $n=104$ apixaban) and 47036 warfarin users compared, mortality (HR=0.99, 95% CI=0.84–1.16) and major bleed rates (HR=0.92, 95% CI=0.82–1.03) were similar after 90 days. It is notable that 94.6% of the DOAC population was rivaroxaban users. Thus these results are largely generalizable only to patients using rivaroxaban or warfarin.

A retrospective cohort study of the Swedish clinical registries compared outcomes of patients receiving warfarin ($n=36317$) vs DOACs ($n=12694$; 40.3% dabigatran, 31.2% rivaroxaban, 28.5% apixaban) [20MC]. Despite relatively high TTR in warfarin patients (70%), DOACs were associated with lower rates of major bleed (HR=0.78, 95%CI=0.67–0.92), intracranial bleed (HR=0.59, 95%CI=0.40–0.87), and hemorrhagic stroke (HR=0.49, 95%CI=0.28–0.86), though no difference was observed in IS (HR=1.04, 95%CI=0.75–1.43), MI (HR=0.95, 95%CI=0.72–1.24), GI bleeding (HR=1.14, 95%CI=0.88–1.46), or mortality (HR=0.94, 95%CI 0.82–1.07).

A meta-analysis of 31 studies assessed whether higher TTR was associated with major bleeding and thromboembolic events [21M]. Higher TTR was significantly associated with lower rates of major bleeding both in univariable model (TTR coefficient=(−)0.0029,

95%CI=(−)0.0048–(−)0.0009) and after adjusting for potential confounding variables (TTR coefficient=(−) 0.0019, 95%CI=(−)0.0036–(−)0.0002). However, higher TTR was associated with less thromboembolic events in the univariable model (TTR coefficient=(−)0.0012, 95%CI=(−)0.0020–(−)0.0004) but not in the multivariable model (TTR coefficient=(−)0.0002, 95%CI=(−) 0.0009–0.0006).

Drug Interactions

A large, retrospective observational cohort study of five American patient databases analyzed warfarin patients taking concomitant selective serotonin reuptake inhibitors (SSRIs) to determine if SSRIs that potently inhibit CYP2C9 (fluoxetine, fluvoxamine) negatively affect outcomes compared with other SSRIs [22MC]. Among the 49881 patients included (8000 taking potent CYP2C9 inhibitor SSRIs and 41881 taking other SSRIs), no significant difference was observed in bleeding (HR=1.14, 95%CI=0.94–1.38), ischemic or thromboembolic events (HR=1.03, 95%CI=0.87–1.21) or mortality (HR=0.90, 95%CI=0.72–1.14) for patients taking concomitant fluoxetine or fluvoxamine compared to other SSRIs.

A cohort cross-over study was conducted in Denmark to assess the impact on INR of patients receiving topical oral miconazole gel ($n=17$) or oral nystatin solution ($n=30$) while on warfarin [23c]. Of those prescribed miconazole gel, mean INR increased from 2.5 (95%CI=2.1–2.8) to 3.8 (95%CI=2.8–4.8) after exposure, whereas in patients receiving nystatin, mean INR was reduced from 2.7 (95%CI=2.3–3.1) to 2.5 (95% CI=2.2–2.9) after exposure. The authors concluded that nystatin solution may be a safer option than miconazole gel for oral candidiasis in patients receiving warfarin, though small sample size, potential confounders based on retrospective study-design, and measurement of a surrogate outcome (INR change) limit the ability to draw definitive conclusions.

Calciphylaxis

A small case series described warfarin-induced calciphylaxis, including three previously unreported cases from the authors' institution alongside 15 cases identified through literature review [24c]. Trends identified included predominantly female gender (15/18), mean 32 months of warfarin use prior to diagnosis, lesions predominantly found below the knees (12/18), high rate of survival to hospital discharge (15/18), and no reported elevated calcium-phosphate products (0/17). In all reports warfarin was either discontinued or there was no discussion of warfarin continuation or discontinuation.

Ophthalmic

A meta-analysis of 12 RCTs compared intraocular bleeding rates with warfarin ($n=44764$) vs DOACs (dabigatran, rivaroxaban, apixaban, and edoxaban; $n=57863$), identifying a significantly lower intraocular bleeding rate with DOACs (141 events with DOACS vs 133 events with warfarin; RR=0.78, 95%CI=0.61–0.99) [25M].

HEPARINS [SEDA-37, 419; SEDA-38, 367; SEDA-39, 360]

Enoxaparin [SEDA-37, 419; SEDA-38, 367; SEDA-39, 360]

Hematologic

A retrospective, single-center study of 151 patients undergoing pancreatic surgery evaluated enoxaparin efficacy and safety [26C]. 4000 IU/day of enoxaparin was used for pulmonary embolism (PE) prophylaxis with an average 5-day therapy duration (range of 0.5–17 days). The study reported no incidence of PE, the primary outcome. Five patients (3.3%) experienced major bleeds. No mortality was reported. The authors conclude use of enoxaparin for prevention of PE in patients undergoing pancreatic surgery is safe and effective but suggest further study to evaluate risk factors of minor bleeding with enoxaparin.

Unfractionated Heparin [SEDA-36, 530; SEDA-37, 419; SEDA-38, 367]

Respiratory

A prospective, randomized, single-center study of 40 patients with moderate-to-very-severe COPD admitted to a hospital's rehabilitation center and randomized to receive nebulized inhaled unfractionated heparin (UFH) (75000 or 150000IU twice daily) or placebo in addition to nebulized salbutamol (1mg) and beclomethasone diproprionate (400μg) for 21 days of treatment [27c]. Inhaled heparin at 150000IU significantly increased FEV_1 vs placebo after 7 days of treatment (+249±69mL, $P<0.05$). Inhaled heparin at both doses also increased trough FVC from baseline compared to placebo throughout the study ($P<0.01$). However, inhaled heparin resulted in higher hemoglobin concentrations in patients treated with 75000IU (13.77 vs 13.03g/dL, $P<0.05$), but this was the only blood coagulation parameter significantly different in either group. Neither of the inhaled heparin dose induced any serious adverse events, nor no significant differences were noted between the treatment arms and placebo in regards to cardiovascular, coagulation, or respiratory adverse effects ($P>0.05$). The authors suggest that this evidence further points to safe and effective use of UFH in moderate-to-very-severe COPD.

Drug Dosage Regimens

A meta-analysis of 8 studies ($n=2937$ patients) assessed the utility of UFH in transradial cardiac catheterization [28M]. Results favored using higher-dose UFH (5000 IU vs 2000–3000 IU), but authors were unable to offer a decisive conclusion given the observational nature of the study and significant heterogeneity in the results.

A retrospective cohort study of 197 patients requiring urgent anticoagulation for non-ST-segment elevation MI (NSTEMI), unstable angina (UA), or AF was assessed by time-to-event analysis for the achievement of therapeutic aPTT (60–80 seconds) with standard and aggressive UFH dosing protocols [29C]. The standard UFH dosing treatment arm ($n=71$) consisted of a 60 unit/kg bolus (maximum 4000 units) followed by an infusion of 12 units/kg/h (maximum 1000 units/h) while the aggressive treatment arm ($n=126$) received a 60 units/kg bolus (maximum 10000 units) followed by an infusion of 12 units/kg/h (maximum 2250 units/h). This resulted in significantly different bolus doses (aggressive = 59.6 vs standard = 53 units/kg, $P<0.0005$) and infusion rates (aggressive = 12 units/kg/h vs standard = 10.8 units/kg/h, $P<0.0005$) between groups. A significantly higher percentage of patients achieved therapeutic aPTT values within 6 hours in the aggressively-dosed vs standard-dosed group (23% vs 11%, $P=0.043$). Of note, there was also a significant improvement in achievement of aPTT within 6 hours for patients weighing between 80 and 99 kg in the aggressively-dosed vs standard-dosed group (48% vs 13%, $P=0.007$). No bleeding events were reported in either group. The authors concluded that aggressively-dosed UFH may result in quicker achievement of therapeutic anticoagulation without increased bleeding rates. A major limitation of this study is that a large percentage of patients still experienced considerable delay in achievement of therapeutic anticoagulation.

One Cochrane Review analyzed subcutaneous UFH for initial treatment of VTE [30R], and another Cochrane Review compared UFH vs low molecular weight heparins (LMWH) for avoiding heparin-induced thrombocytopenia during postoperative anticoagulation [31R].

DIRECT THROMBIN INHIBITORS [SEDA-37, 422; SEDA-38, 368; SEDA-39, 361]

Argatroban [SEDA-32, 632; SEDA-33, 717; SEDA-34, 544]

Hematologic

A multicenter RCT ($n=97$) assessed the safety of adding argatroban to recombinant tissue-type plasminogen activator (r-tPA) in patients treated for IS [32c]. Patients were randomized to standard-dose r-tPA alone ($n=29$) or in conjunction with argatroban (100 μg/kg bolus followed by either low-dose (1 μg/kg/min, $n=30$) or high-dose (3 μg/kg/min, $n=31$) infusions of argatroban for 48 hours). Similar asymptomatic ICH rates were observed between control (10%), low-dose (13%), and high-dose (7%) treatment arms. No major symptomatic bleeding events occurred. 90-day mortality rates were: r-tPA alone 17% ($n=5$, 95%CI = 6–36), low-dose argatroban 17% ($n=5$, 95%CI = 6–35), and high-dose argatroban 10% ($n=3$, 95%CI = 2–26). The authors of the study concluded that the evidence supports the safety of adjunctive argatroban for treating IS and further supports a need for larger efficacy trials to determine argatroban's place in therapy.

Bivalirudin [SEDA-35, 619; SEDA-38, 368; SEDA-39, 361]

Hematologic

A meta-analysis included 11 RCTs ($n=8428$ patients) evaluated bivalirudin vs heparin in patients with diabetes who underwent percutaneous coronary intervention (PCI) [33M]. The primary safety endpoint was major bleeding while secondary efficacy endpoints were net adverse clinical events (NACE), MI, and death. Bivalirudin decreased incidence of major bleeding compared to heparin (RR = 0.63, 95%CI = 0.52–0.75). Bivalirudin and heparin were not significantly different in regards to NACE (RR = 0.81, 95%CI = 0.61–1.07), death (RR = 0.75, 95%CI = 0.56–1.02) or MI (RR = 0.92, 95%CI = 0.67–1.26). A major limitation of this study was variability in heparin dosing between trials.

A retrospective analysis utilizing a Swedish national registry compared outcomes of 6006 patients receiving PCI for STEMI or NSTEMI via radial access while anticoagulated with bivalirudin or UFH [34MC]. Patients treated with bivalirudin ($n=3004$) were administered 0.75 mg/kg intravenous (IV) bolus followed by infusion of 1.75 mg/kg/h while patients treated with UFH ($n=3002$) received a total dose between 70 and 100 units/kg. The primary endpoint was composite death from any cause, MI, or major bleed at 180 days. No statistical difference was observed in the primary composite endpoint between the bivalirudin (12.3%) and heparin (12.8%) treatment groups (HR = 0.96, 95%CI = 0.83–1.10). Major bleeding rates were similar between groups (8.6% incidence in both, HR = 1.00, 95%CI = 0.84–1.19). Subgroup analyses of patients with either STEMI or NSTEMI identified no significant differences in outcomes. The open-label design of this study must be considered when evaluating practice impact.

A retrospective American patient registry analysis compared 67308 patients receiving PCI for STEMI via radial access anticoagulated with bivalirudin ($n=29600$)

or heparin ($n = 37708$) [35MC]. No significant difference in the composite endpoint (death, MI, stroke) was observed (OR = 0.95, 95%CI = 0.87–1.05). A significantly higher acute stent thrombosis rate was observed with bivalirudin (OR = 2.11, 95%CI = 1.73–2.57). Major bleeding rates were similar between groups (OR = 0.98, 95% CI = 0.91–1.05). This study's design and omission of treatment details (including dosing strategies and duration of therapy) limit the utility of these results.

A retrospective, propensity-matched cohort study of 3048 patients compared access site hematoma incidence in patients undergoing PCI while treated with bivalirudin or heparin [36MC]. The study favored bivalirudin compared to heparin with decreased access site hematoma incidence (2.4% vs 3.9%, $P = 0.018$), post-procedural hospitalization duration (1.0 vs 1.2 days, $P < 0.001$), and rates of discharge to a facility vs home (7.61% vs 9.73%, $P = 0.034$). In-hospital access site occlusion, distal embolization, and mortality rates were similar. Dosing strategies were not available. Further randomized comparisons are required to make definitive conclusions before incorporating results of this study in practice.

Gender Differences

A meta-analysis, including 5 RCTs and 4501 female patients, evaluated the efficacy and safety of bivalirudin alone vs heparin plus a glycoprotein IIb/IIIa inhibitors (GPI) during PCI [37M]. Doses varied between trials. Outcomes were measured after 30 days. All efficacy endpoints were statistically similar between bivalirudin and heparin + GPI; however, bivalirudin significantly reduced 30-day major bleeding compared to heparin + GPI (5.32% vs 9.20%, RR = 0.58, 95%CI = 0.47–0.72). The authors conclude that bivalirudin has equal efficacy compared to heparin + GPI while decreasing 30-day major bleeding incidence in females undergoing PCI. Another meta-analysis, published in 2015, reported similar efficacy and reduction of major bleeding in bivalirudin vs heparin + clopidogrel in women undergoing PCI [38M].

A BRAVO-3 trial sub-analysis examined sex-based differences in outcomes with bivalirudin or heparin for transcatheter aortic valve replacement (TAVR) [39R]. The cohort included 391 women (195 bivalirudin, 196 heparin) and 411 men (209 bivalirudin, 202 heparin). The primary endpoint of first incidence of major bleeding was similar between women and men (8.2% vs 7.8%, $P = 0.83$). All secondary endpoints of death, stroke or MI, major adverse cardiovascular events (MACE), or NACE were also similar between sexes. This was the first study to report similar rates of major bleeding and vascular complications between sexes during TAVR.

DIRECT THROMBIN INHIBITORS [SEDA-37, 422; SEDA-38, 368; SEDA-39, 361]

Dabigatran [SEDA-37, 422; SEDA-38, 369; SEDA-39, 362]

Gastrointestinal

A case report describing late-onset dabigatran-induced esophageal injury after being on dabigatran for >2 years was published [40A].

Another case report described an instance of intramural esophageal hematoma associated with dabigatran use [41A]. After the first dose, an 85-year-old female prescribed dabigatran for AF experienced hemoptysis and presented to the ED. Five days after presentation, esophageal mucosal tear and hematoma were identified via gastroscopy. With conservative treatment, she recovered without complications.

Psychiatric

Depression and anxiety status for 50 patients with nonvalvular AF switching from warfarin to dabigatran were compared [42c]. Depression and anxiety were assessed using the Beck Depression Scale (BDS) and Hamilton Anxiety Scale (HAS) at study initiation and 6 months later. Both BDS ($P < 0.001$) and HAS ($P < 0.001$) scores were significantly higher in patients while still taking warfarin compared with after they switched to dabigatran. The authors speculate that warfarin use is associated with higher rates of depression and anxiety vs dabigatran due to higher risk of complications and frequent follow-up.

Hematologic

A retrospective chart review examined clinical characteristics, interventions, and outcomes in patients with nonvalvular AF presenting with dabigatran-associated major bleeding [43c]. 118/191 (62%) had gastrointestinal hemorrhage, 36/191 (19%) had intracranial hemorrhage, 8/191 (4%) were nontraumatic cases and 28/191 (15%) were traumatic. 11/191 (6%) patients received purified coagulation factors and 12/191 (6%) were fatal.

Clinical outcomes associated with resumption of anticoagulation post-hemorrhage were assessed in a retrospective study [44C]. Patients were identified using Medicare claims data if they were diagnosed with AF, filled a prescription for dabigatran or warfarin between 2010 and 2012, and experienced a major bleed requiring hospitalization. Patients who filled a prescription for dabigatran or warfarin after the bleed, were followed to assess occurrence of stoke, recurrent bleeding event, or death. The incidence of recurrent major bleeding was higher for patients prescribed warfarin after the bleeding event than for those prescribed dabigatran (HR=2.31, 95%CI=1.19–4.76) or whose anticoagulation ceased (HR=1.556, 95%CI=1.10–2.22). Interestingly, incidence of major bleeding did not differ between patients restarting dabigatran and those discontinuing anticoagulation (HR=0.65, 95%CI=0.32–1.33).

A meta-analysis of 17 studies compared rivaroxaban with dabigatran and warfarin for stroke prevention in AF, finding major bleeding risk was significantly higher with rivaroxaban than dabigatran (HR=1.38, 95%CI=1.27–1.49) [45M].

A retrospective study of nonvalvular AF patients, who initiated oral anticoagulants between 2010 and 2014, was enrolled in MarketScan databases comparing bleeding risks with rivaroxaban and dabigatran [46C]. Rivaroxaban was associated with higher GI bleeding risk vs dabigatran [HR=1.28, 95%CI=1.06–1.54]. No statistical difference in intracranial bleeds was found.

A single-center, retrospective cohort study compared rates of severe bleeding requiring hospitalization in 1494 AF patients receiving dabigatran, warfarin, or antiplatelet therapy [47C]. An increased risk of GI bleeding was observed with dabigatran vs warfarin (13 (81%) vs 44 (48%) $P=0.02$). Alternatively, the dabigatran group had no instances of intracranial bleeding, whereas 14/782 (15%, $P=0.09$) of patients in the warfarin group experienced intracranial bleeding.

An open-label, multicenter RCT evaluated the safety of uninterrupted dabigatran in patients undergoing AF ablation [48MC]. Patients scheduled for catheter ablation were randomly assigned to receive either dabigatran or warfarin starting 4–8 weeks prior to the procedure and continued for 8 weeks after. Incidence of major bleeding events up to 8 weeks post-ablation was lower with dabigatran than warfarin (5 (1.6%) vs 22 (6.9%), $P<0.001$).

A multicenter RCT compared rates of major and clinically relevant nonmajor bleeding in patients receiving either triple therapy with warfarin plus aspirin plus clopidogrel OR ticagrelor vs dual therapy with dabigatran (110 or 150mg twice daily) plus clopidogrel OR ticagrelor post-PCI [49MC]. Patients outside the Unites States aged ≥80 years or ≥70 years in Japan were assigned to the 110mg dual-therapy group. Patients were followed for 14 months. 151/981 (15.4%) of patients in the dual-therapy group experienced a bleeding event vs 264/981 (26.9%) with triple-therapy group (HR=0.52, 95%CI=0.42–0.63).

A meta-analysis compared bleeding rates in AF patients treated with 110mg ($n=15848$) vs 150mg ($n=13416$) dabigatran [50M]. 110mg dabigatran was associated with significantly lower minor bleeding rates (OR=1.19, 95% CI=1.10–1.27). No difference was detected between doses for fatal or major bleeding.

Musculoskeletal

A Chinese propensity-matched, retrospective cohort study assessed osteoporotic fracture risk with dabigatran [51C]. Nonvalvular AF patients newly diagnosed between 2010 and 2014 and prescribed dabigatran ($n=3268$) or warfarin ($n=4884$) were included. Dabigatran was associated with significantly lower osteoporotic fracture risk vs warfarin [(−)0.68, 95%CI=(−)0.38-(−)0.86]. The authors hypothesize this difference is based on warfarin's mechanism of interaction interfering with a process leading to bone formation and that warfarin users are advised to limit dietary intake of vitamin K intake which may cause increased bone loss and fracture.

Drug–Drug Interactions

A case report detailing potential interaction between dabigatran and phenytoin resulting in left atrial thrombus was published [52A]. A 70-year-old presented to the hospital with generalized weakness and dyspnea and was diagnosed with left atrial thrombus. The patient had initiated dabigatran 1 year prior to admission and had started phenytoin 4 weeks prior. The authors propose that since phenytoin is a p-glycoprotein inducer it could affect the absorption and metabolism of dabigatran.

A population-based, case–control study of Canadian residents, older than 66 years of age, analyzed drug interactions between dabigatran and lovastatin or simvastatin [53c]. Of 2406 patients identified as taking dabigatran and diagnosed with major hemorrhage, 1117 received a statin within 60 days preceding the event. Use of simvastatin or lovastatin was associated with increased risk of major hemorrhage compared with other statins (OR=1.46; 95%CI=1.17–1.82). The proposed mechanism of interaction between dabigatran and lovastatin/simvastatin is that simvastatin and lovastatin inhibit intestinal p-glycoprotein, increasing systemic dabigatran exposure. The authors suggest that simvastatin and lovastatin should be avoided in favor of other statins in patients receiving dabigatran.

DIRECT FACTOR XA INHIBITORS [SEDA-37, 423; SEDA-38, 370; SEDA-39, 363]

Apixaban [SEDA-34, 546; SEDA-38, 370; SEDA-39, 363]

Hematologic

A single-center, retrospective matched cohort study compared apixaban ($n=73$) to warfarin ($n=73$) in patients with severe renal dysfunction (CrCl<25mL/min, serum creatinine >2.5mg/dL, or current dialysis) [54C]. Patients receiving apixaban experienced numerically, but not statistically significantly, lower major bleeding rates (9.6% vs 17.8%, $P=0.149$). No thromboembolisms were identified in either group.

A Danish real-world, observational cohort study compared reduced-dose apixaban (2.5mg twice daily; $n=4400$), rivaroxaban (15mg daily; $n=3476$), and dabigatran (110 mg twice daily, $n=8875$) to warfarin ($n=38893$) for AF over a mean 2.3-year follow-up [55MC]. While there are specific recommendations for dose reductions for renal dysfunction or older age for each drug, the cohort was not limited to patients with specific indications for dose reduction. At 1-year follow-up, apixaban trended towards higher rates of IS/systemic embolism (4.8%, HR=1.19, 95%CI=0.95–1.49) than warfarin (3.7%), while rivaroxaban (3.5%, HR=0.89, 95%CI=0.69–1.16) and dabigatran (3.3%, HR=0.89, 95%CI=0.77–1.03) showed slightly lower rates. However, none of these differences were statistically significant. Bleeding rates were similar between apixaban (5.1%, HR=0.96, 95%CI=0.73–1.27), rivaroxaban (5.6%, HR=1.06, 95%CI=0.87–1.29), and warfarin (5.1%) and slightly lower for dabigatran (4.1%, HR=0.80, 95%CI=0.70–0.92).

Subgroup analysis of the ARISTOTLE trial reported the incidence, location, and management of non-major bleeding associated with apixaban vs warfarin in AF patients [56R]. The most frequent sites of non-major bleeding were: hematuria (16.4%), epistaxis (14.8%), gastrointestinal (13.3%), hematoma (11.5%), and bruising (10.1%).

A retrospective, Danish registry study of 31522 AF patients compared efficacy and safety of direct oral anticoagulants (DOACs) for both standard and reduced doses of apixaban (5mg BID, $n=7203$; 2.5mg BID, $n=3861$) rivaroxaban (20mg daily, $n=6868$; 15mg daily, $n=2098$) and dabigatran (150mg BID, $n=7078$; and 110mg BID, $n=4414$), respectively [57MC]. Major bleeding at 1 year for the standard dosed DOACs were: apixaban (2.24%), rivaroxaban (2.78%), and dabigatran (1.84%). Rivaroxaban displayed a significant increase in major bleeding vs apixaban (AR 0.54; 95%CI, 0.05–0.99) and dabigatran (AR 0.93; 95%CI, 0.38%–1.45%). Reduced-dose DOACs displayed the following bleeding incidences: apixaban (3.71%), rivaroxaban (4.98%), and dabigatran (3.9%). Little evidence exists comparing DOACs, and further RCTs are needed to compare safety and efficacy between drugs in this class.

A prospective, single-center, open-label trial ($n=617$ patients) compared apixaban to VKA in AF patients after TAVR [58C]. 272 patients were found to be in AF after TAVR and were started on either apixaban ($n=141$) or a VKA ($n=131$) 48 hours post-TAVR and assessed at 30 days and 12 months. All patients received reduced dose apixaban (2.5mg twice daily). At 30 days, the early safety endpoint (composite of all-cause mortality, all stroke, life-threatening bleeding, acute kidney injury, coronary obstruction, major vascular complications, and valve dysfunction requiring intervention) was significantly less frequent in patients treated with apixaban compared to VKA (13.5% vs 30.5%; $P<0.01$). There was also a numerically but not statistically lower 30-day (2.1% vs 5.3%; $P=0.17$) and 12-month (1.2% vs 2.0%; $P=0.73$) stroke rate with apixaban vs VKA. This was the first trial to assess safety data in the TAVR patient population. The results of this trial along with results from the previously reported ARISTOTLE trial point to improved outcomes and safety with apixaban vs warfarin in elderly AF patients [59MC].

A prospective, randomized, registry study ($n=176$) within a Japanese hospital compared periprocedural use of apixaban and rivaroxaban in catheter ablation for AF [60C]. The incidence of asymptomatic cerebral microthromboembolism and hemopericardium in AF ablation was similar among the periprocedural use of apixaban and rivaroxaban compared to warfarin. However, existing data is conflicting regarding safety of periprocedural DOAC use during AF ablation.

Renal Disease

A retrospective, single-center cohort study of 160 patients with end-stage renal disease (ESRD) (defined as eGFR <15mL/min while on hemodialysis) received apixaban or warfarin for VTE treatment or prevention while hospitalized [61C]. Patients on apixaban ($n=40$) were dosed at either 2.5mg twice daily (57.5%) or 5mg twice daily (42.5%), while warfarin ($n=120$) was dosed to an INR goal of 2–3 (2.5–3.5 for patients with mechanical heart valves). 0% of apixaban patients vs 5.8% of warfarin patients experienced major bleeding ($P=0.34$). There were similar rates of clinically relevant non-major bleeding (5.8% vs 12.5%; $P=0.17$). The authors conclude that apixaban may be a cautious consideration in patients undergoing hemodialysis until further evidence is available. Currently, there are no RCTs assessing safety outcomes of apixaban in patients with ESRD. This study adds to the lacking data, but limitations in study design, including small sample size and retrospective design, must be considered.

Betrixaban

Cardiovascular

The APEX trial compared 35–42 days of betrixaban to 10±4 days of enoxaparin for thromboprophylaxis in hospitalized acutely ill patients [62MC]. The results were published in 2016, but several sub-studies were published in 2017. In one, full dose (80mg) betrixaban ($n=2986$) and reduced-dose (40mg, given to patients with severe renal insufficiency (i.e., CrCl=15–30mL/min) or those receiving concomitant strong P-glycoprotein inhibitors) betrixaban ($n=730$) were separately compared to enoxaparin (40mg, $n=2562$), unless patients had severe renal insufficiency in which case they received enoxaparin 20 mg ($n = 609$). Compared to enoxaparin, full-dose betrixaban produced lower primary outcome (asymptomatic proximal DVT, symptomatic or distal DVT, symptomatic nonfatal PE, or VTE-related death) rates (6.27% vs 8.39%, $P = 0.023$, RR = 0.26 (0.04–0.42)) without increasing major bleeding (RR = 0.75, $P = 0.51$). Low-dose betrixaban produced no difference in primary outcome events or major bleeding vs enoxaparin. Full dose betrixaban produced significantly higher plasma concentrations (19 ng/mL vs 11 ng/mL, $P < 0.001$) vs low-dose betrixaban.

A post-hoc analysis of the APEX trial focused specifically on safety and efficacy events that were either fatal or irreversible by comparing rates of a composite outcome of all efficacy (cardiopulmonary death, MI, PE, IS) and fatal or irreversible safety (fatal bleed or intracranial hemorrhage) events with betrixaban (80 or 40 mg for patients with severe renal insufficiency or concomitant P-glycoprotein inhibitor therapy) vs enoxaparin [63MC]. Researchers identified a significantly lower rate in patients receiving betrixaban (combined cohort receiving either dose) at 35–42 days (4.08% vs 2.90%, HR = 0.71, NNT = 86, $P = 0.006$) and 77 days (5.17% vs 3.64%, HR = 0.7, NNT = 65, $P = 0.002$).

A third APEX substudy utilized competing risks methods to identify whether similar results would be observed when factoring in the presence of non-VTE-related deaths that could alter or prevent a subject from experiencing the primary endpoint [64MC]. The proportion of non-VTE-related deaths was similar with betrixaban (3.6%) and enoxaparin (3.7%, $P = 0.85$). Utilizing the Fine and Gray method did not alter the results from the Cox model used in the original APEX analysis, identifying a sub-distribution hazard = 0.65, $P = 0.046$ in favor of betrixaban.

Another APEX substudy specifically compared stroke risk, identifying fewer all-cause (RR = 0.56, $P = 0.032$) and ischemic (RR = 0.53, $P = 0.026$) strokes with extended-duration betrixaban vs enoxaparin [65MC].

Special Review

Extended Duration Thromboprophylaxis in Acutely Ill Hospitalized Patients

Current guidelines call for short-term (6–14 days) use of LMWH, UFH, or fondaparinux in many patients hospitalized for acute illness due to increased risk of thromboembolic events [66R]. However, data suggests that increased thromboembolic risk extends beyond this traditional 1- to 2-week window and extended duration anticoagulation may provide further protection. Additionally, the availability of several DOAC agents, which boast faster onset of action than warfarin, offers the possibility of utilizing oral as opposed to injectable agents in this setting. In June 2017, betrixaban, a once-daily oral Factor Xa inhibitor, became the first, and currently only, agent approved for in-hospital and extended duration prophylaxis (35–42 days) for thrombotic risk associated with hospitalization in acutely ill patients. However, several studies have also assessed the efficacy and safety of low-dose, extended duration rivaroxaban, apixaban, and enoxaparin in this patient population. In 2017 two meta-analyses were published analyzing these trials.

A meta-analysis of four RCTs compared outcomes of extended-duration (24–47 days) vs short-duration (6–14 days) thromboprohylaxis in hospitalized medical patients [67M]. Among the 4 RCTs included, betrixaban (160 mg for one dose then 80 mg daily), rivaroxaban (10 mg daily), apixaban (2.5 mg twice daily), and enoxaparin (40 mg daily) were utilized in the extended-duration arm in one study each. Enoxaparin 40 mg daily was utilized in the short-duration arm of each study. In this meta-analysis, extended-duration was associated with less symptomatic DVTs (RR = 0.52, P = 0.001) and symptomatic non-fatal PEs (RR = 0.61, P = 0.04), but no statistically significant difference in VTE-related (RR = 0.69, P = 0.09) or all-cause mortality (RR = 1.00, P = 0.95) or fatal bleeding (RR = 2.01, P = 0.20). However, there was an increased risk of major bleeding with extended-duration (RR = 2.08, P < 0.0001). Notably though, the major bleeding risk was not increased when betrixaban was used (RR = 1.19, 95%CI = 0.67–2.12) but was much higher with each of the other three agents, enoxaparin (RR = 2.51, 95%CI = 1.21–5.22), apixaban (RR = 2.53, 95%CI = 0.98–6.50), and rivaroxaban (RR = 2.87, 95% CI = 1.60–5.16).

A similar study utilized data from the same four RCTs to analyze the NCB of extended duration thromboprophylaxis through bivariate analysis [68M]. In this study, extended-duration betrixaban displayed a favorable NCB based on superior efficacy and non-inferior safety profile vs enoxaparin. Apixaban, rivaroxaban, and extended-duration enoxaparin were all found superior in efficacy but inferior in safety vs standard-duration enoxaparin.

Currently available data suggests extended-duration betrixaban is a viable alternative to short-duration LMWH, UFH, or fondaparinux for thromboprophylaxis in patients hospitalized for acute illness, offering superior efficacy and non-inferior bleed-risk, and also appears to be associated with significantly less major bleeding risk when compared with extended-duration therapy utilizing other Factor Xa inhibitors.

Rivaroxaban [SEDA-37, 423; SEDA-38, 371; SEDA-39, 365]

Hematologic

A RCT of 27 395 patients <65 years old with stable atherosclerotic disease (documentation of atherosclerosis involving at least two vascular beds or two additional risk factors) evaluated rivaroxaban alone or in combination with aspirin for secondary cardiovascular prevention [69MC]. Patients received either rivaroxaban 2.5mg twice daily plus aspirin 100mg daily, or rivaroxaban 5mg twice daily alone, or aspirin 100mg once daily alone. The rivaroxaban-plus-aspirin group (4.1%) experienced fewer primary outcomes (composite cardiovascular death, stroke, or MI) vs aspirin alone (5.4%) (HR=0.76, $P<0.001$), and numerically lower mortality though this was not significant (3.4% vs 4.1%; HR=0.82, $P=0.01$ [threshold P-value for significance=0.0025]). However, major bleeding was more common with rivaroxaban-plus-aspirin (3.1%) vs aspirin-alone (1.9%) (HR=1.70, $P<0.001$). Intracranial or fatal bleeding did not differ between groups. 5mg rivaroxaban alone vs aspirin alone did not decrease primary outcome events, but did increase major bleeding. This trial was stopped for superiority in favor of the rivaroxaban-plus-aspirin group after 23 months of follow-up. The COMPASS trial authors conclude low dose rivaroxaban-plus-aspirin was superior to rivaroxaban alone or aspirin alone. Despite increased bleeding risk, this trial could have a significant impact on practice given its favorable efficacy. Of note rivaroxaban is currently not available in 2.5mg tablets.

Another RCT ($n=3396$) compared rivaroxaban to aspirin for extended VTE treatment [70MC]. After previously completing 6–12 months of anticoagulation, eligible patients were stratified to continue therapy with either rivaroxaban 10 ($n=1127$) or 20mg ($n=1107$) daily or aspirin 100mg daily ($n=1131$) for 12 additional months. The primary efficacy outcome (composite of symptomatic, recurrent fatal or nonfatal VTE and unexplained death) occurred significantly less in both rivaroxaban groups vs aspirin (1.2% rivaroxaban 10mg vs 4.4% aspirin; HR=0.26, $P<0.001$; 1.5% rivaroxaban 20mg vs 4.4% aspirin; HR=0.34, $P<0.001$). Major bleeding was not significantly different between groups (0.4% with rivaroxaban 10mg, 0.5% with rivaroxaban 20mg, and 0.3% with aspirin). Clinically relevant nonmajor bleeding was 2.7%, 2.0%, and 1.8%, respectively. The authors concluded that rivaroxaban at both treatment doses of 20mg daily or prophylaxis doses of 10mg daily were superior to aspirin without significantly increasing rates of bleeding. This trial also has the potential to impact practice in favor of extended rivaroxaban treatment following VTE.

A meta-analysis of 5 RCTs ($n=4895$) examined bleeding in patients taking 15 or 20mg rivaroxaban daily vs 110 or 150mg of dabigatran twice daily for AF [71M]. Outcomes were similar with rivaroxaban vs dabigatran, respectively, including bleeding (OR=1.28, $P=0.11$), GI bleeding (OR=0.98, $P=0.97$), or intracranial bleeding (OR=2.18, $P=0.29$). Rivaroxaban patients displayed numerically higher mortality (OR=1.42, $P=0.06$), but did not reach statistical significance.

An American retrospective, observational cohort study ($n=44793$) assessed stroke risk associated with nonvalvular AF (NVAF) in patients taking rivaroxaban by calculating CHA_aDS_s-VASc scores [72MC]. Rivaroxaban users had an overall major bleeding incidence rate (first major bleeding event) of 2.84/100 person-years (95%CI=2.69–3.00). Major bleeding rates among patients with a CHA_aDS_s-VASc of 0 were 0.30/100 patient-years (95%CI, 0.08–1.20) with the rate continually increasing to 5.40/100 patient-years (95%CI=5.00–5.84) in patients with CHA_aDS_s-VASc ≥ 5.

Another American retrospective, observational cohort study evaluated major bleeding incidence and characteristics in rivaroxaban users diagnosed with NVAF and diabetes mellitus (DM) [73MC]. Among 44793 rivaroxaban users, 12039 (26.9%) had DM. A significantly higher bleeding rate occurred in DM patients (3.68/100 person-years, 95%CI=3.37–4.03) vs those without DM (2.51/100 person-years, 95%CI=2.34–2.69). The study had several major limitations including lack of randomization and did not include the statistical analysis proving significance between cohorts.

A third retrospective, observational cohort study compared warfarin to apixaban ($n=2514$), dabigatran ($n=1962$), and rivaroxaban ($n=5208$) for NVAF with a prior history of stroke or transient ischemic attack (TIA) [74MC]. Compared to warfarin, apixaban and dabigatran had similar rates of the combined primary end point of IS or ICH while rivaroxaban reduced the combined endpoint (HR=0.45, 95%CI=0.29–0.72). All three DOACs had similar rates of major bleeding vs warfarin.

A retrospective, single-center analysis evaluated adverse drug events in patients receiving rivaroxaban after major orthopedic surgery vs warfarin or warfarin +enoxaparin [75C]. The results indicated a small, yet significant, increase in rivaroxaban-related bleeding vs warfarin and warfarin+enoxaparin. However, bleeding incidence was low regardless of treatment group and rivaroxaban caused no major bleeding.

A single-center, propensity-score matched analysis compared complications and reversal strategies of rivaroxaban to warfarin during trauma and acute surgery [76C].

Skin

A case report described a 68-year-old female with AF who developed a papulosquamous photodistributed pruritic eruption involving the forearms, trunk, neck and face 3 weeks after initiation of rivaroxaban (dose not specified) [77A]. Rivaroxaban was discontinued and monotherapy with clobetasol ointment was initiated with complete rash resolution at 4 months. This is the first reported reaction of drug-induced subacute lupus erythematosus. The authors note that rivaroxaban is currently being trialed for management of thrombotic antiphospholipid syndrome despite this patient population displaying higher risk of drug-induced lupus.

A small, single-center, open-label, retrospective case–control study compared rivaroxaban ($n = 44$) and warfarin ($n = 50$) following peripheral arterial procedures. No previous data has been published regarding the use of rivaroxaban for this procedure [78c]. There was no difference in need for re-intervention (20/44 (45.5%) vs 16/50 (32%), $P = 0.207$) or major bleeding (3/44 (6.8%) vs 8/50 (16%), $P = 0.209$) between rivaroxaban and warfarin, respectively. However, among patients aged ≤65 years, significantly fewer major bleeds were observed with rivaroxaban (0/27 (0%) vs 5/25 (20%), $P = 0.02$). More patients on warfarin received concomitant antiplatelet agents ($n = 47$ (94%) vs $n = 31$ (70%), $P = 0.016$).

THROMBOLYTIC DRUGS [SEDA-37, 424; SEDA-38, 371; SEDA-39, 366]

Alteplase [SEDA-37, 424; SEDA-38, 371; SEDA-39, 366]/Tenecteplase [SEDA-37, 424]

A Phase III, randomized, open-label, superiority trial was performed to investigate tenecteplase vs alteplase in patients with acute stroke who were eligible for IV thrombolysis within 4.5 hours of symptomatic onset [79MC]. Patients were given either IV tenecteplase ($n = 549$) 0.4 mg/kg (maximum 40 mg) or alteplase ($n = 551$) 0.9 mg/kg (maximum 90 mg). No significant differences in serious adverse effects were observed with tenecteplase (26%) vs alteplase (26%, $P = 0.74$). The groups were also found to be similar in efficacy and mortality. This trial had a very high percentage of patients with mild stroke. Further evidence is needed to ensure tenecteplase is non-inferior to alteplase in patients with a severe stroke.

GLYCOPROTEIN IIB–IIIA INHIBITORS [SEDA-37, 426; SEDA-38, 372; SEDA-39, 367]

Abciximab [SEDA-34, 548; SEDA-35, 622; SEDA-37, 426]

Hematologic

A review article discussed efficacy and adverse effects related to intra-arterial abciximab for treating thromboembolic complications associated with cerebral aneurysm coiling [80R].

Eptifibatide [SEDA-37, 426; SEDA-38, 372; SEDA-39, 367]

Hematologic

A retrospective, single-center study analyzed 30 patients with high risk or recent PCI while using eptifibatide for bridging therapy, finding no association with increased need for perioperative transfusion, bleeding rates, or increased length of stay [81c].

Tirofiban [SEDA-37, 426; SEDA-38, 372; SEDA-39, 367]

Hematologic

A retrospective study evaluated high-dose (25 μg/kg; $n = 140$) vs low-dose boluses (10 μg/kg; $n = 131$) of tirofiban followed by maintenance infusion (0.15 μg/kg/min) with concurrent clopidogrel 600 mg in patients with acute coronary syndrome (ACS) undergoing PCI [82c]. Major (2.1% vs 1.5%, $P = 1.000$) and minor bleeding (9.3% vs 8.4%, $P = 0.965$) rates were similar between high- and low-dose bolus groups, respectively. The authors concluded that high-dose tirofiban bolus improved outcomes but did not increase bleeding risk.

P2Y12 RECEPTOR ANTAGONISTS [SEDA-37, 427; SEDA-38, 373; SEDA-39, 367]

Cangrelor [SEDA-34, 427; SEDA-39, 367]

Hematologic

A meta-analysis of 15 RCTs ($n = 52025$) compared cangrelor ($n = 12475$), prasugrel ($n = 7455$), or ticagrelor ($n = 7192$) to clopidogrel ($n = 12475$) during PCI [83M]. All groups were statistically similar to clopidogrel in regards to efficacy (CV death, MI, MACE, and stent thrombosis) and major bleeding. However, ticagrelor had a significant increase in minor bleeding vs clopidogrel (OR = 1.59, 95%CI = 1.10–5.03). The study authors conclude that despite rapid platelet inhibition of cangrelor, the newer oral P2Y12 inhibitors all have comparable clinical outcomes.

A subgroup analysis of the CHAMPION series studies, 3 large RCTs, evaluated cangrelor vs clopidogrel in patients who did or did not receive a glycoprotein IIb/IIIa inhibitor during PCI [84MC]. Among 24902 patients, 3173 (12.7%) received a GPI, most commonly eptifibatide (69.4%). During the trials patients received either cangrelor (30 μg/kg bolus then 4 μg/kg/min infusion for ≥2 hours or the duration of PCI) or clopidogrel (300 or 600 mg), but administration and average dosing was variable across the trials. At 48 hours, rates of the composite primary endpoint (all-cause mortality, MI, ischemia-driven revascularization, or stent thrombosis) were lower with cangrelor vs clopidogrel in patients who did (4.9% vs 6.5%, OR=0.74, 95%CI=0.55–1.01) and did not receive GPIs (3.6% vs 4.4%, OR=0.82, 95%CI=0.72–0.94). Cangrelor did not increase severe or life-threatening bleeding (GUSTO-defined) incidence vs clopidogrel in patients who did (0.4% vs 0.5%, OR=0.71, 95%CI=0.25–1.99) or did not receive GPIs (0.2% vs 0.1%, OR=1.56, 95%CI=0.80–3.04). These findings point to superiority cangrelor efficacy and similar bleeding rates despite use of GPI administration for patients undergoing PCI.

Another subgroup analysis of the CHAMPION trials analyzed 1270 patients who had a cerebrovascular event >1 year prior to randomization [85MC]. The efficacy and bleeding profile of cangrelor in these patients was similar to the overall trial data.

P2Y12 RECEPTOR ANTAGONISTS [SEDA-37, 427; SEDA-38, 373; SEDA-39, 367]

Clopidogrel [SEDA-37, 427; SEDA-38, 373; SEDA-39, 367]

Hematologic

A RCT ($n=13885$) compared cardiovascular events in PAD patients treated with clopidogrel (75mg once daily) or ticagrelor (90mg twice daily) over a median of 30 months follow-up [86MC]. In each group, acute limb ischemia occurred in 1.7% of patients (HR=1.03, $P=0.85$) and major bleeding in 1.6% (HR=1.10, $P=0.49$).

Prasugrel [SEDA-37, 429; SEDA-38, 374; SEDA-39, 368]

Hematologic

A retrospective Japanese study evaluated the clinical significance of platelet reactivity (PR) in patients treated with prasugrel after MI undergoing PCI [87C]. Residual PR was assessed 2 hours after administration of a prasugrel loading dose and before the first maintenance dose. A total of 44/78 patients studied had a significantly higher PR 2 hours after loading dose administration. High PR was a significant predictor of MACE (cardiovascular death, non-fatal MI, stent thrombosis, stroke, and sustained ventricular tachycardia) during the 8-month follow-up (OR=4.911, $P=0.03$). Additionally, administration of prasugrel via nasogastric tube was a significant predictor of high PR (OR=43.1, $P=0.001$).

Ticagrelor [SEDA-37, 429; SEDA-38, 374; SEDA-39, 368]

Hematologic

A subgroup analysis compared the 3858 Asian patients enrolled in the SOCRATES multicenter RCT comparing ticagrelor with aspirin for major vascular events (stroke, myocardial infarction, or death) prevention in patients with acute IS or TIA [88MC]. In this Asian subgroup, 12 (0.6%) ticagrelor patients and 16 (0.8%) aspirin patients (HR=0.76, $P=0.47$) experienced major bleeding. A major or minor bleeding event was reported in 42 (2.4%) ticagrelor patients and 35 (1.9%) aspirin patients.

Another SOCRATES post-hoc analysis identified 31 major bleeds (0.5%) with ticagrelor vs 38 patients (0.6%) with aspirin (HR=0.83, 95%CI=0.52–1.34) [89MC]. The most common locations of major bleeds were intracranial and gastrointestinal with a total of nine fatal bleeds occurring in the ticagrelor group vs four with aspirin.

A systematic evaluation evaluated ticagrelor safety and efficacy for ACS in patients with diabetes mellitus [90M]. Among 22 studies ($n=35004$), no increased bleeding incidence was observed with ticagrelor vs prasugrel.

Systemic

A case report described the first reported incidence of ticagrelor-induced systemic inflammatory response syndrome [91A]. An 84-year-old male with a recent STEMI presented with dyspnea, fatigue, hypotension, tachycardia, and fever 15 days later. After 6 days of extensive work-up and antibiotic treatment, no infectious source was identified. After 7 days of persistent symptoms ticagrelor was replaced with clopidogrel, resulting in rapid improvement. No additional medication changes were made.

Drug–Drug Interactions

A case report detailed the first reported incidence of statin-and-ticagrelor-induced rhabdomyolysis in a patient with normal kidney function on low dose statin [92A]. A 72-year-old patient presented to the ED with muscle pain in both legs and elevated creatinine kinase. Five months prior to this presentation, the patient was hospitalized due to ACS and started on ticagrelor. Patient had previously been on simvastatin 20mg daily long-term. The patient was diagnosed with rhabdomyolysis, the simvastatin therapy was held, and the ticagrelor was changed to clopidogrel. Both simvastatin and ticagrelor are metabolized via CYP3A4 leading to a potential for increased levels of simvastatin when co-administered.

A 58-year-old Chinese male with a history of renal transplantation and recent PCI following ACS initiated on ticagrelor suffered from gum bleeding and bright red blood in the stool 8 days after PCI [93A]. The patient also experienced nausea, vomiting, hypotension, and tachycardia. The following day after initial symptom presentation the patient developed black stools approximately every 20 min and hemoglobin level dropped from 9.9 to 4.6 g/dL. Ticagrelor was held; however, cyclosporine was maintained for immunosuppression. Study authors suspect a drug interaction between ticagrelor and cyclosporine leading to an increased concentration of ticagrelor was responsible.

REVERSAL AGENTS [SEDA-38, 375; SEDA-39, 369]

Idarucizumab [SEDA-38, 375; SEDA-39, 369]

A prospective, open-label study was performed to assess 5 g of intravenous idarucizumab for the complete reversal of dabigatran anticoagulation in patients with uncontrolled bleeding ($n = 301$) or for patients requiring emergent procedures ($n = 202$) [94C]. After 90 days of observation, thrombotic events occurred in 6.3% of patients in the bleeding group and 7.4% of patients in the emergent procedure group, while mortality rates were 18.8% and 18.9%, respectively. No additional adverse effects were reported with idarucizumab. The trial authors concluded that idarucizumab can be used safely in situations where reversal of dabigatran are necessary.

Additional case studies can be found in these reviews [95R, 96R].

Acknowledgements

Jennifer Helmen, Saint Joseph Health System, Mishawaka, IN, USA.

References

[1] Lee K, Li S, Liu J, et al. Association of warfarin with congestive heart failure and peripheral artery occlusive disease in hemodialysis patients with atrial fibrillation. J Chin Med Assoc. 2017;80:277–82 [C].
[2] Vandell AG, Walker J, Brown KS, et al. Genetics and clinical response to warfarin and edoxaban in patients with venous thromboembolism. Heart. 2017;103:1800–5 [MC].
[3] Gage BF, Bass AR, Lin H, et al. Effect of genotype-guided warfarin dosing on clinical events and anticoagulation control among patients undergoing hip or knee arthroplasty: the GIFT randomized clinical trial. JAMA. 2017;318(12):1115–24 [MC].
[4] Liu J, Jiang HH, Wu DK, et al. Effect of gene polymorphisms on the warfarin treatment at initial stage. Pharmacogenomics J. 2017;17:47–52 [C].
[5] Xu Y, Schulman S, Dowlatshahi D, et al. Direct oral anticoagulant- or warfarin-related major bleeding: characteristics, reversal strategies, and outcomes from a multicenter observational study. Chest. 2017;152(1):81–91 [MC].
[6] Baker AL, Vanderpluym C, Gauvreau KA, et al. Safety and efficacy of warfarin therapy in Kawasaki disease. J Pediatr. 2017;189:61–5 [c].
[7] Steinberg BA, Simon DN, Thomas L, et al. Management of major bleeding in patients with atrial fibrillation treated with non-vitamin K antagonist oral anticoagulants compared with warfarin in clinical practice (from phase II of the outcomes registry for better informed treatment of atrial fibrillation [ORBIT-AF II]). Am J Cardiol. 2017;119:1590–5 [MC].
[8] Kimachi M, Furukawa TA, Kimachi K, et al. Direct oral anticoagulants versus warfarin for preventing stroke and systemic embolic events among atrial fibrillation patients with chronic kidney disease. Cochrane Database Syst Rev. 2017;11:1–69. Article Number CD011373 [M].
[9] Hart R, Veenstra DL, Boudreau DM, et al. Impact of body mass index and genetics on warfarin major bleeding outcomes in a community setting. Am J Med. 2017;130:222–8 [C].
[10] Rafiq S, Steinbruchel DA, Lilleor NB, et al. Antithrombotic therapy after bioprosthetic aortic valve implantation: warfarin versus aspirin, a randomized controlled trial. Thromb Res. 2017;150:104–10 [C].
[11] Di Minno MND, Ambrosino P, Dentali F. Safety of warfarin in "high-risk" populations: a meta-analysis of randomized and controlled trials. Thromb Res. 2017;150:1–7 [M].
[12] Singer AJ, Quinn A, Dasgupta N, et al. Management and outcomes of bleeding events in patients in the emergency department taking warfarin or a non-vitamin K antagonist oral anticoagulant. J Emerg Med. 2017;52(1):1–7 [C].
[13] Egbe AC, Connolly HM, Pellikka PA, et al. Outcomes of warfarin therapy for bioprosthetic valve thrombosis of surgically implanted valves. J Am Coll Cardiol Interv. 2017;10: 379–87 [c].
[14] Melmed KR, Lyden P, Gellada N, et al. Intracerebral hemorrhagic expansion occurs in patients using non-vitamin K antagonist oral anticoagulants comparable with patients using warfarin. J Stroke Cerebrovasc Dis. 2017;26(8):1874–82 [c].
[15] Grzymala-Lubanski B, Svensson PJ, Renlund H, et al. Warfarin treatment quality and prognosis in patients with mechanical heart valve prosthesis. Heart. 2017;103:198–203 [MC].
[16] Allan V, Banerjee A, Shah AD, et al. Net clinical benefit of warfarin in individuals with atrial fibrillation across stroke risk and across primary and secondary care. Heart. 2017;103:210–8 [MC].
[17] Xing Y, Xu B, Xu C, et al. Efficacy and safety of uninterrupted low-intensity warfarin for radiofrequency catheter ablation of atrial fibrillation in the elderly. Ann Pharmacother. 2017;51(9):735–42 [C].
[18] Nielsen PB, Larsen TB, Skjoth F, et al. Outcomes associated with resuming warfarin treatment after hemorrhagic stroke or traumatic intracranial hemorrhage in patients with atrial fibrillation. JAMA Intern Med. 2017;177(4):563–70 [MC].
[19] Jun M, Lix LM, Durand M, et al. Comparative safety of direct oral anticoagulants and warfarin in venous thromboembolism: multicenter, population based, observational study. BMJ. 2017;359: j4323 [MC].
[20] Sjogren V, Bystrom B, Renlund H, et al. Non-vitamin K oral anticoagulants are non-inferior for stroke prevention but cause fewer major bleedings than well-managed warfarin: a retrospective register study. PLoS One. 2017;12(7):e0181000 [MC].
[21] Vestergaard AS, Skjoth F, Larsen TB, et al. The importance of mean time in therapeutic range for complication rates in warfarin therapy of patients with atrial fibrillation: a systematic review and meta-regression analysis. PLoS One. 2017;12(11):e0188482 [M].

[22] Dong Y, Bykov K, Choudhry NK, et al. Clinical outcomes of concomitant use of warfarin and selective serotonin reuptake inhibitors. J Clin Psychopharmacol. 2017;37:200–9 [MC].
[23] Hellfritzsch M, Pottegard A, Pedersen AJT, et al. Topical antimycotics for oral candidiasis in warfarin users. Basic Clin Pharmacol Toxicol. 2017;120:368–72 [c].
[24] Yu WY, Bhutani T, Kornik R, et al. Warfarin-associated nonuremic calciphylaxis. JAMA Dermatol. 2017;153(3):309–14 [c].
[25] Sun MT, Wood MK, Chan W, et al. Risk of intraocular bleeding with novel oral anticoagulants compared with warfarin a systematic review and meta-analysis. JAMA Ophthalmol. 2017;135(8):864–70 [M].
[26] Imamura H, Adachi T, Kitasato A, et al. Safety and efficacy of postoperative pharmacologic thromboprophylaxis with enoxaparin after pancreatic surgery. Surg Today. 2017;47(8):994–1000 [C].
[27] Shute JK, Calzetta L, Cardaci V, et al. Inhaled nebulised unfractionated heparin improves lung function in moderate to very severe COPD: a pilot study. Pulm Pharmacol Ther. 2018;48:88–96 [c].
[28] Bossard M, Mehta SR, Welsh RC, et al. Utility of unfractionated heparin in transradial cardiac catheterization: a systematic review of meta-analysis. Can J Cardiol. 2017;33(10):1245–53 [M].
[29] Floroff CK, Palm NM, Steinberg DH, et al. Higher maximum doses and infusion rates compared with standard unfractionated heparin therapy are associated with adequate anticoagulation without increased bleeding in both obese and non-obese patients with cardiovascular indications. Pharmacotherapy. 2017;37(4):393–400 [C].
[30] Robertson L, Strachan J. Subcutaneous unfractionated heparin for the initial treatment of venous thromboembolism. Cochrane Database Syst Rev. 2017;2:CD006771 [R].
[31] Junqueira DR, Zorzela LM, Perini E. Unfractionated heparin versus low molecular weight heparins for avoiding heparin-induced thrombocytopenia in postoperative patients. Cochrane Database Syst Rev. 2017;4:CD007557 [R].
[32] Barreto AD, Ford GA, Shen L, et al. Randomized, multicenter trial of ARTSS-2 (Argatroban with recombinant tissue plasminogen activator for acute stroke). Stroke. 2017;48(6):1608–16 [c].
[33] Zhang J, Yang X. Efficacy and safety of bivalirudin versus heparin in patients with diabetes mellitus undergoing percutaneous coronary intervention: a meta-analysis of randomized controlled trials. Medicine (Baltimore). 2017;96(29):e7204 [M].
[34] Erlinge D, Omerovic E, Fröbert O, et al. Bivalirudin versus heparin monotherapy in myocardial infarction. N Engl J Med. 2017;377(12):1132–42 [MC].
[35] Jovin IS, Shah RM, Patel DB, et al. Outcomes in patients undergoing primary percutaneous coronary intervention for ST-segment elevation myocardial infarction via radial access anticoagulated with bivalirudin versus heparin: a report from the national cardiovascular data registry. JACC Cardiovasc Interv. 2017;10(11):1102–11 [MC].
[36] Ortiz D, Singh M, Jahangir A, et al. Bivalirudin versus unfractionated heparin during peripheral vascular interventions: a propensity-matched study. Catheter Cardiovasc Interv. 2017;89(3):408–13 [MC].
[37] Xu H, Wang B, Yang J, et al. Bivalirudin versus heparin plus glycoprotein IIb/IIIa inhibitors in women undergoing percutaneous coronary intervention: a meta-analysis of randomized controlled trials. PLoS One. 2017;12(1):e0169951 [M].
[38] Ng VG, Baumbach A, Grinfeld L, et al. Impact of bleeding and bivalirudin therapy on mortality risk in women undergoing percutaneous coronary intervention (from the REPLACE-2, ACUITY, and HORIZONS-AMI trials). Am J Cardiol. 2015;117:186–91 [M].
[39] Asgar A, Chandrasekhar J, Mikhail G, et al. Sex-based differences in outcomes with bivalirudin or unfractionated heparin for transcatheter aortic valve replacement: results from the BRAVO-3 randomized trial. Catheter Cardiovasc Interv. 2017;89(1):144–53 [R].
[40] Fujikawa K, Takasugi N, Goto T, et al. Very late-onset dabigatran-induced esophageal injury. Can J Cardiol. 2017;33(4):554.e15–6 [A].
[41] Trip J, Hamer P, Flint R. Intramural esophageal hematoma—a rare complication of dabigatran. N Z Med J. 2017;130(1456): 80–2 [A].
[42] Turker Y, Ekinozu I, Aytekin S, et al. Comparison of changes in anxiety and depression level between dabigatran and warfarin use in patients with atrial fibrillation. Clin Appl Thromb Hemost. 2017;23(2):164–7 [c].
[43] Milling TJ, Fromm C, Ganetsky M, et al. Management of major bleeding events in patients treated with dabigatran for nonvalvular atrial fibrillation: a retrospective, multicenter review. Ann Emerg Med. 2017;69(5):531–40 [c].
[44] Hernandez I, Zhang Y, Brooks MM, et al. Anticoagulation use and clinical outcomes after major bleeding on dabigatran or warfarin in atrial fibrillation. Stroke. 2017;48:159–66 [C].
[45] Bai Y, Deng H, Shantsila A, et al. Rivaroxaban versus dabigatran or warfarin in real-world studies of stroke prevention in atrial fibrillation. Stroke. 2017;48:970–6 [M].
[46] Norby FL, Bengtson LG, Lutsey PL, et al. Comparative effectiveness of rivaroxaban versus warfarin or dabigatran for the treatment of patients with non-valvular atrial fibrillation. BMC Cardiovasc Disord. 2017;17(1):238 [C].
[47] Riley TR, Gauthier-Lewis ML, Sanchez CK, et al. Evaluation of bleeding events requiring hospitalization in patients with atrial fibrillation receiving dabigatran, warfarin, or antiplatelet therapy. J Pharm Pract. 2017;30(2):214–8 [C].
[48] Calkins H, Willems S, Gerstenfeld EP, et al. Uninterrupted dabigatran versus warfarin for ablation in atrial fibrillation. N Engl J Med. 2017;376:1627–36 [MC].
[49] Cannon CP, Dl B, Oldgren J, et al. Dual antithrombotic therapy with dabigatran after PCI in atrial fibrillation. N Engl J Med. 2017;377(16):1513–24 [MC].
[50] Bundhun PK, Chaudhary N, Yuan J. Bleeding events associated with a low dose (110mg) versus a high dose (150mg) of dabigatran in patients treated for atrial fibrillation: a systematic review and meta-analysis. BMC Cardiovasc Disord. 2017;17(1):83 [M].
[51] Lau WC, Chan EW, Cheung C, et al. Association between dabigatran vs warfarin and risk of osteoporotic fractures among patients with nonvalvular atrial fibrillation. 2017;317(11):1151–8 [C].
[52] Hager N, Bolt J, Albers L, et al. Development of left atrial thrombus after coadministration of dabigatran etexilate and phenytoin. Can J Cardiol. 2017;554:e13–4 [A].
[53] Antoniou T, Macdonald EM, Hollands S, et al. Association between statin use and ischemic stroke or major hemorrhage in patients taking dabigatran for atrial fibrillation. CMAJ. 2017;189(1):e4–10 [c].
[54] Stanton BE, Barasch NS, Tellor KB. Comparison of the safety and effectiveness of apixaban versus warfarin in patients with severe renal impairment. Pharmacotherapy. 2017;37(4):412–9 [C].
[55] Nielsen PB, Skjoth F, Sogaard M, et al. Effectiveness and safety of reduced dose non-vitamin K antagonist oral anticoagulants and warfarin in patients with atrial fibrillation: propensity weighted nationwide cohort study. BMJ. 2017;356:j510 [MC].
[56] Bahit MC, Lopes RD, Wojdyla DM, et al. Non-major bleeding with apixaban versus warfarin in patients with atrial fibrillation. Heart. 2017;103(8):623–8 [R].
[57] Staerk L, Gerds TA, Lip GYH, et al. Standard and reduced doses of dabigatran, rivaroxaban and apixaban for stroke prevention in atrial fibrillation: a nationwide cohort study. J Intern Med. 2018;283(1):45–55 [MC].
[58] Seeger J, Gonska B, Rodewald C, et al. Apixaban in patients with atrial fibrillation after transfemoral aortic valve replacement. JACC Cardiovasc Interv. 2017;10(1):66–74 [C].
[59] Granger CB, Alexander JH, McMurray JJ, et al. Apixaban versus warfarin in patients with atrial fibrillation. N Engl J Med. 2017;365(11):981–92 [MC].

[60] Yoshimura A, Iriki Y, Ichiki H. Evaluation of safety and efficacy of periprocedural use of rivaroxaban and apixaban in catheter ablation for atrial fibrillation. J Cardiol. 2017;69(1):228–35 [C].
[61] Sarratt SC, Nesbit R, Moye R, et al. Safety outcomes of apixaban compared with warfarin in patients with end-stage renal disease. Ann Pharmacother. 2017;51(6):445–50 [C].
[62] Gibson CM, Halaby R, Korjian S, et al. The safety and efficacy of full-versus reduced dose betrixaban in the acute medically ill VTE (venous thromboembolism) prevention with extended-duration betrixaban (APEX) trial. Am Heart J. 2017;185:93–100 [MC].
[63] Gibson CM, Korjian S, Chi G, et al. Comparison of fatal or irreversible events with extended-duration betrixaban versus standard dose enoxaparin in acutely ill medical patients: an APEX trial substudy. J Am Heart Assoc. 2017;6:e006015 [MC].
[64] Arbetter DF, Jain P, Yee MK, et al. Competing risk analysis in a large cardiovascular clinical trial: an APEX substudy. Pharm Stat. 2017;16:445–50 [MC].
[65] Gibson CM, Chi G, Halaby R, et al. Stroke versus standard-dose enoxaparin among hospitalized medically ill patients. Circulation. 2017;135:648–55 [MC].
[66] Garland SG, DeRemer CE, Smith SM, et al. Betrixaban: a new oral factor Xa inhibitor for extended venous thromboembolism prophylaxis in high-risk hospitalized patients. Ann Pharmacother. 2018;52:554–61 [R].
[67] Liew AYL, Piran S, Eikelboom JW, et al. Extended-duration versus short-duration pharmacological thromboprophylaxis in acutely ill hospitalized medical patients: a systematic review and meta-analysis of randomized controlled trials. J Thromb Thrombolysis. 2017;43:291–301 [M].
[68] Chi G, Goldhaber SZ, Kittelson JM, et al. Effect of extended-duration thromboprophylaxis on venous thromboembolism and major bleeding among acutely ill hospitalized medical patients: a bivariate analysis. J Thromb Haemost. 2017;15:1913–22 [M].
[69] Eikelboom JW, Connolly SJ, Bosch J, et al. Rivaroxaban with or without aspirin in stable cardiovascular disease. N Engl J Med. 2017;377(14):1319–30 [MC].
[70] Weitz JI, Lensing AWA, Prins MH, et al. Rivaroxaban or aspirin for extended treatment of venous thromboembolism. N Engl J Med. 2017;376(13):1211–22 [MC].
[71] Bundhun PK, Soogund MZ, Teeluck AR, et al. Bleeding outcomes associated with rivaroxaban and dabigatran in patients treated for atrial fibrillation: a systematic review and meta-analysis. BMC Cardiovasc Disord. 2017;17(1):15 [M].
[72] Peacock WF, Tamayo S, Patel M, et al. CHA2DS2-VASc scores and major bleeding in patients with nonvalvular atrial fibrillation who are receiving rivaroxaban. Ann Emerg Med. 2017;69(5):541–50, e1 [MC].
[73] Peacock WF, Tamayo S, Sicignano N, et al. Comparison of the incidence of major bleeding with rivaroxaban use among nonvalvular atrial fibrillation patients with versus without diabetes mellitus. Am J Cardiol. 2017;119(5):753–9 [MC].
[74] Coleman CI, Peacock WF, Bunz TJ, et al. Effectiveness and safety of apixaban, dabigatran, and rivaroxaban versus warfarin in patients with nonvalvular atrial fibrillation and previous stroke or transient ischemic attack. Stroke. 2017;48(8):2142–9 [MC].
[75] Cieri NE, Kusmierski K, Lackie C, et al. Retrospective evaluation of postoperative adverse drug events in patients receiving rivaroxaban after major orthopedic surgery compared with standard therapy in a community hospital. Pharmacotherapy. 2017;37(2):170–6 [C].
[76] Myers SP, Dadashzadeh ER, Cheung J, et al. Management of anticoagulation with rivaroxaban in trauma and acute care surgery: complications and reversal strategies as compared to warfarin therapy. J Trauma Acute Care Surg. 2017;82(3):542–9 [C].
[77] McCarthy S, Foley CC, Dvorakova V, et al. A case of rivaroxaban-induced subacute lupus erythematosus. J Eur Acad Dermatol Venereol. 2017;31(1):e58–9 [A].
[78] Talukdar A, Wang SK, Czosnowski L, et al. Safety and efficacy of rivaroxaban compared with warfarin in patients undergoing peripheral arterial procedures. J Vasc Surg. 2017;66:1143–8 [c].
[79] Logallo N, Novotny V, Assmus J, et al. Tenecteplase versus alteplase for management of acute ischaemic stroke (NOR-TEST): a phase 3, randomised, open-label, blinded endpoint trial. Lancet Neurol. 2017;16(10):781–8 [MC].
[80] Martínez-Pérez R, Lownie SP, Pelz D. Intra-arterial use of abciximab in thromboembolic complications associated with cerebral aneurysm coiling: the London Ontario experience. World Neurosurg. 2017;100:342–50 [R].
[81] Waldron NH, Dallas T, Erhunmwunsee L, et al. Bleeding risk associated with eptifibatide (Integrilin) bridging in thoracic surgery patients. J Thromb Thrombolysis. 2017;43(2):194–202 [c].
[82] Ulus T, Şenol U, Tahmazov S, et al. High-dose bolus tirofiban versus low-dose bolus in patients with acute coronary syndrome undergoing percutaneous coronary intervention. Turk Kardiyol Dern Ars. 2017;45(2):126–33 [c].
[83] Westman PC, Lipinski MJ, Torguson R, et al. A comparison of cangrelor, prasugrel, ticagrelor, and clopidogrel in patients undergoing percutaneous coronary intervention: a network meta-analysis. Cardiovasc Revasc Med. 2017;18(2):79–85 [M].
[84] Vaduganathan M, Harrington RA, Stone GW, et al. Cangrelor with and without glycoprotein IIb/IIIa inhibitors in patients undergoing percutaneous coronary intervention. J Am Coll Cardiol. 2017;69(2):176–85 [MC].
[85] Sawlani NN, Harrington RA, Stone GW, et al. Impact of cerebrovascular events older than one year on ischemic and bleeding outcomes with cangrelor in percutaneous coronary intervention. Circ Cardiovasc Interv. 2017;10(1), pii: e004380 [MC].
[86] Hiatt WR, Fowkes FG, Heizer G, et al. Ticagrelor versus clopidogrel in symptomatic peripheral artery disease. N Engl J Med. 2017;36(1):32–40 [MC].
[87] Sato T, Namba Y, Kashihara Y, et al. Clinical significance of platelet reactivity during prasugrel therapy in patients with acute myocardial infarction. J Cardiol. 2017;70:35–40 [C].
[88] Wang Y, Minematsu K, Wong K, et al. Ticagrelor in acute stroke or transient ischemic attack in Asian patients. Stroke. 2017;48(1):167–73 [MC].
[89] Easton JD, Aunes M, Albers GW, et al. Risk for major bleeding in patients receiving ticagrelor compared with aspirin after transient ischemic attack or acute ischemic stroke in the SOCRATES study. Circulation. 2017;136:907–16 [MC].
[90] Tan Q, Jiang X, Huang S, et al. The clinical efficacy and safety evaluation of ticagrelor for acute coronary syndrome in general ACS patients and diabetic patients: a systematic review and meta-analysis. PLoS One. 2017;12(5):e0177872 [M].
[91] Krisai P, Haschke M, Buser PT, et al. Ticagrelor induced systemic inflammatory response syndrome. BMC Cardiovasc Disord. 2017;17(1):14 [A].
[92] Mrotzek SM, Rassaf T, Totzeck M. Ticagrelor leads to statin-induced rhabdomyolysis: a case-report. Am J Case Rep. 2017;18:1238–41 [A].
[93] Zhang C, Shen L, Cui M, et al. Ticagrelor-induced life-threatening bleeding via the cyclosporine-mediated drug interaction. Medicine. 2017;96(37):e8065 [A].
[94] Pollack Jr. CV, Reilly PA, van Ryn J, et al. Idarucizumab for dabigatran reversal—full cohort analysis. N Engl J Med. 2017;377(5):431–41 [C].
[95] Isch J, Nguyen D, Ali AN. Chapter 33 - Drugs that affect blood coagulation, fibrinolysis and hemostasis. In: Ray SD, editor. Side Effects of Drugs- Annual: A worldwide yearly survey of new data in adverse drug reactions. Elsevier. 2016;38:365–77 [R].
[96] Raber H, Isch J, Evoy K. Chapter 31 - Drugs that affect blood coagulation, fibrinolysis and hemostasis. In: Ray SD, editor. Side Effects of Drugs- Annual: A worldwide yearly survey of new data in adverse drug reactions. Elsevier. 2017;39:359–72 [R].

CHAPTER

35

Gastrointestinal Drugs

Kirby Welston, Dianne May*[†,1]

*Charlie Norwood VA Medical Center, Augusta, Georgia

[†]University of Georgia College of Pharmacy on Augusta University Campus, Augusta, Georgia

[1]Corresponding author: dimay@augusta.edu

ACID-IMPACTING AGENTS

Proton-Pump Inhibitors (PPIs) *[SED-16, 1040–1045; SEDA-37, 435; SEDA-38, 379–385, SEDA-39, 374–379]*

℞ *Does the co-administration of clopidogrel with a PPI decrease efficacy/increase adverse events due to significant drug interactions?*

While not a new topic, concomitant use of PPIs and clopidogrel continues to be a clinical controversy. Results from randomized controlled trials and observational studies are conflicting. Just in 2016, there were at least five publications on the co-administration of PPIs with clopidogrel and its potential sequelae.

A systematic review and meta-analysis of 11 studies representing 84729 patients from 2012 to 2016 was performed to determine if newer data revealed an association between increased cardiovascular adverse outcomes after percutaneous coronary interventions (PCI) in those receiving clopidogrel and a PPI [1M]. Patients who had received a PCI with subsequent clopidogrel therapy were divided into those who also received a concomitant PPI ($n = 29235$) and those who did not receive a PPI ($n = 55494$). Short-term mortality and target vessel revascularization favored those not on a PPI ($OR = 1.55$; $CI_{95\%}$: 1.43–1.68; $P < 0.00001$ and OR 1.26; $CI_{95\%}$: 1.06–1.49; $P < 0.009$, respectively). Other significant findings favoring the non-PPI group were seen with long-term major adverse cardiac events (MACEs), myocardial infarction, stent thrombosis, and target vessel revascularization. Long-term mortality was not statistically significant between groups.

Another systematic review and meta-analysis were performed assessing patient-centered outcomes in patients on a PPI combined with medications from 10 other drug classes [2M]. A total of 49 studies were included. Using 4 databases, reports of morbidity and mortality associated with PPIs used concomitantly with other medications were collected. Morbidity and mortality were not significantly different when PPIs were used concomitantly with antiplatelets (clopidogrel, prasugrel, ticagrelor) in the randomized controlled trials. However, observational studies showed conflicting results with an association in adverse cardiac events, myocardial infarction, and/or stroke when clopidogrel was co-administered with PPIs. Acute coronary syndrome was increased in diabetic patients with drug-eluting stents for coronary artery disease when a PPIs was used concomitantly with clopidogrel.

Three other studies looking at the interaction between concomitant use of PPIs and clopidogrel are described below.

Cardiovascular

A longitudinal observational study assessed the risk of cardiovascular disease and nephropathy in patients with type 2 diabetes who were receiving renin–angiotensin blocking therapy with ($n = 174$) or without ($n = 686$) a PPI from the Fremantle Diabetes Study Phase II [3C]. The time between assessments was approximately 2 years. Also evaluated were those who started a PPI ($n = 109$) and those who discontinued scheduled PPI therapy ($n = 67$) during the longitudinal observational study. The primary cardiovascular outcome was 5-year CVD risk. Patients starting a PPI showed an increase in 5-year CVD risk from 13% to 17.5% ($P < 0.001$) over the 2-year follow-up period.

Gastrointestinal

A review article evaluated PPI use and the risk of developing biopsy-confirmed microscopic colitis [4M]. A total of 19 publications were identified representing a total of 32 cases of microscopic colitis, 27 of which were

Side Effects of Drugs Annual, Volume 40
ISSN: 0378-6080
https://doi.org/10.1016/bs.seda.2018.07.009

with lansoprazole. Microscopic colitis occurred within 1 month in 15 cases. However, onset of symptoms was seen out as far as 6 years. The average age at time of diagnosis was 69.9 years ± 10.1 years and 53% were female. The risk of microscopic colitis based off of specific PPI used or the dose used was not assessed. While the review stated there could be a possible association between PPIs and microscopic colitis, larger studies are needed to determine the magnitude of risk and if it is dependent on the specific PPI used or the dose of PPI used.

Urinary Tract

A retrospective study assessed the change from baseline of blood urea and serum creatinine in 175 Indian patients who received at least 7 consecutive days of PPI therapy [5C]. All patients were at least 18 years old and were seen as inpatients or outpatients within an Internal Medicine practice in India. The incidence of acute kidney injury was determined from laboratory values obtained from the patient's medical record. Patients were excluded if their serum creatinine or blood urea was abnormally elevated prior to the study, if they were on concomitant nephrotoxic medications, or if they had a history of dialysis or renal transplant. Acute kidney injury was seen in 19/175 (10.8%) of patients. Of those with acute kidney injury, pantoprazole was the most prescribed proton pump inhibitor (84.2%). Male patients over 50 years old were most associated with the development of acute kidney injury. The mechanism of injury is not clearly defined. Proposed mechanisms include hypomagnesemia, oxidative stress, or an immunologic effect.

Genetic Factors

A total of 55 healthy Japanese subjects who had been part of a previously conducted randomized, double-blind, placebo-controlled trial were recruited to determine if polymorphisms of the CYP2C19 gene contributed to the celecoxib-induced small bowel injury seen in the first study [6c]. Subjects received celecoxib 200 mg twice daily with or without rabeprazole 20 mg daily for 2 weeks. Genotyping for CYP2C19 was performed and patients were labeled as poor metabolizers or extensive metabolizers. The difference in celecoxib-induced small bowel injury was compared between poor metabolizers and extensive metabolizers in each group. In patients receiving rabeprazole, small bowel injuries were seen more often in poor metabolizers compared with extensive metabolizers (85.7% vs 31.6%; [$P=0.26$], respectively).

Long-Term Effects

A review article described the most current literature regarding long-term risks associated with PPI use [7R]. Included in the review was a discussion of vitamin B12 deficiency, bone fractures, dementia, drug interactions, kidney disease, *Clostridium difficile* infections, pneumonia, gastric polyps and cancer associated with PPI use.

Drug–Drug Interaction, Alisertib

A phase I, open-label study evaluated the effects of multiple doses of esomeprazole on the pharmacokinetics of alisertib, an aurora a kinase inhibitor, in 18 patients with advanced solid tumors or lymphoma [8c]. The median age was 61 years old (age range: 32–80 years). Patients received one dose of alisertib 50 mg with or without esomeprazole 40 mg daily. Alisertib blood levels were measured for up to 72 h after the dose and area under the curve and maximum concentration (C_{max}) were determined. Patients on esomeprazole had a 28% increase in exposure to alisertib compared to those who were not on esomeprazole. Half-life was similar regardless of esomeprazole use. Alisertib solubility was increased under less acidic conditions seen with esomeprazole use.

Drug–Drug Interaction, Clopidogrel

An open-label, crossover study evaluated the antiplatelet effects of clopidogrel in 41 healthy Japanese subjects who received a low dose PPI [9c]. Patients received clopidogrel 75 mg daily for 7 days with or without low dose PPIs (10–15 mg). The PPI was given concomitantly at 8 a.m. or separated from the clopidogrel and administered at 8 p.m. The inhibition of platelet aggregation (IPA) was measured 4 h after the last dose of clopidogrel on day 7. Patients were crossed over to receive clopidogrel with a different PPI after a 14 day washout period. A second study was performed with rabeprazole 10 mg given at 12 noon and clopidogrel the following day at 8 a.m. (20 h after rabeprazole) for 7 days. IPA was significantly decreased when PPIs were given concomitantly (IPA = 43.2%–47.2%) compared to clopidogrel alone (IPA = 56%); (all P values <0.001). The IPA was also decreased when the PPI was separated from clopidogrel by 12 h (administered at 8 p.m.). Interestingly, when rabeprazole was administered 20 h prior to clopidogrel (at 12 noon), the IPA was similar to that seen with clopidogrel alone (51.6%); $P=0.114$. Separating PPIs from clopidogrel by 20 h may decrease the impact of the drug interaction between rabeprazole and clopidogrel.

A review article summarized data on use of PPIs with clopidogrel in East Asian patients [10R]. In addition to the drug interaction due to CYP2C19 inhibition, the review also addressed the impact of CYP2C19 generic polymorphisms in this patient population. Guidance was provided on the most appropriate PPI for coadministration with clopidogrel in East Asian patients. Overall, pantoprazole was identified to interact the least with clopidogrel.

Drug–Drug Interaction, Clopidogrel, Prasugrel

A comparative, crossover study evaluated the effects of vonoprazan, a potassium competitive acid blocker, and esomeprazole on antiplatelet function of clopidogrel and prasugrel using a P2Y12 assay in 31 healthy Japanese subjects [11c]. Fourteen subjects were found to be CYP2C19 homo-extensive, 9 were hetero-extensive, and 8 were poor metabolizers. Esomeprazole is a CYP2C19 inhibitor while vonoprazan inhibits both CYP2C19 and CYP3A4. Clopidogrel is a prodrug that must be converted to its active form by CYP2C19 and CYP3A4. Prasugrel is also a prodrug that relies on CYP3A4 primarily to convert to its active form. Results from this crossover study showed that regardless of genotype groups, vonoprazan decreased the median inhibition of platelet aggregation (IPA) of both clopidogrel and prasugrel more than esomeprazole due to its CYP2C19 and CYP3A4 inhibitory effects. The IPA for clopidogrel alone, clopidogrel plus esomeprazole, and clopidogrel plus vonoprazan was 34%, 33%, and 13%, respectively. For prasugrel, the median IPA for prasugrel alone, prasugrel plus esomeprazole, and prasugrel plus vonoprazan was 59%, 47%, and 37%, respectively. Esomeprazole did not decrease inhibition of platelet aggregation with prasugrel regardless of genotype.

Drug–Drug Interaction, Clozapine

A review article discussed the potential for concomitant PPI use to worsen hematological adverse effects, such as neutropenia and agranulocytosis, seen with clozapine [12R]. A direct toxic or immune-allergic effect may be involved but does not appear to be dose-related. Other proposed mechanisms for hematologic toxicity are related to the accumulation of the clozapine metabolites, nitrenium ion and *N*-desmethylclozapine, by induction of common metabolic pathways involving CYP1A2 or CYP3A4, or competitive inhibition of CYP2C19. This potential drug interaction has been most commonly reported with omeprazole or lansoprazole.

Drug–Drug Interaction, Methotrexate

A retrospective study evaluated the effect of PPIs on plasma methotrexate elimination in 43 hospitalized patients who had received high-dose methotrexate therapy [13c]. The study included a total of 73 cycles of high-dose methotrexate. Organic anion transporter 3 (OAT3) plays a significant role in the uptake of methotrexate into the renal proximal tubule cells. Therefore, inhibitors of organic anion transport 3 (OAT3) may delay the elimination of methotrexate leading to increased hematologic toxicities. An in vitro study was also performed to investigate the involvement of OAT3 in the drug interaction between PPIs or famotidine and methotrexate using human (h) OAT3 expressing HEK293 cells. Methotrexate concentrations were significantly higher in the patient receiving PPIs at both 48 and 72h ($P=0.0002$) compared with those receiving famotidine. The in vitro study demonstrated PPIs inhibit hOAT3-mediated uptake of methotrexate in a concentration-dependent manner with lansoprazole being the most likely to affect methotrexate levels followed by esomeprazole, rabeprazole, and omeprazole. Famotidine did not demonstrate an inhibitory effect on hOAT3-mediated methotrexate uptake.

Drug–Drug Interaction, Mycophenolic Acid

A retrospective study was performed in 125 renal transplant patients in Istanbul to evaluate the effects of lansoprazole or rabeprazole on the pharmacokinetic profile of mycophenolic acid (MPA) derivatives [14C]. In addition, the effects of CYP2C19 polymorphisms on this drug interaction were assessed at 1-year post-renal transplant. There were 71 men and 51 women aged 43.34 ± 12.13 years included in the study. Patients received mycophenolate mofetil or mycophenolate sodium along with tacrolimus and steroids. In addition, patients received either lansoprazole or rabeprazole or no PPI. The mean MPA concentrations were higher in those receiving mycophenolate sodium vs mycophenolate mofetil. Patients receiving lansoprazole had a lower mean MPA than those who received rabeprazole or no PPI ($P<0.05$). The mean concentration of mycophenolate mofetil was significantly lower in those who received lansoprazole compared to those who received rabeprazole or no PPI when the CYP2C19*2/*2 genotype was also present ($P<0.05$). The CYP2C19*2/*2 genotype appears to be an extensive metabolizer for mycophenolate sodium and mycophenolate mofetil. Patients who received mycophenolate mofetil with lansoprazole had a higher incidence of gastrointestinal side effects compared to those who received mycophenolate sodium with lansoprazole, regardless of genotype ($P<0.05$).

Drug–Drug Interaction, Neratinib

An open-label, two-period, fixed-sequence study was conducted in 15 healthy subjects to evaluate the effect of lansoprazole on the absorption and safety of neratinib, a pan-HER tyrosine kinase inhibitor [15c]. Most patients were Hispanic or Latino (73%) with a median age of 35 years old (age range: 28–54 years). During the first period, all subjects received one dose of neratinib 240mg followed by a 14-day washout period. Pharmacokinetic sampling was performed for up to 72h after dose. During period 2, subjects received 7 days of lansoprazole 30mg PO once daily. On day 5, subjects received a second dose of neratinib 240mg. Pharmacokinetic sampling again was performed for 72h after neratinib dose. The absorption of neratinib was significantly reduced in subjects who were also receiving lansoprazole vs neratinib alone. The mean maximum concentration (C_{max}) of neratinib was decreased approximately 70% with the addition of lansoprazole. The median time to maximum concentration (T_{max}) was also delayed by 1.5h and the area under the curve was decreased by up to 70%.

Drug–Drug Interaction, Momelotinib

Momelotinib is a selective, small-molecule JAK1/2 inhibitor used for myeloproliferative neoplasms [16c]. An open-label, phase I, fixed-sequence crossover study was performed in healthy subjects to determine the bioavailability of the tablet vs capsule formulation and to assess the effect of food and omeprazole on the pharmacokinetics of momelotinib after a single-dose. The C_{max} for momelotinib was moderately increased when taken with food (38% increase for low-fat meal; 28% increase for high-fat meal). Area under the curve (AUC) was also increased when taken with food (16% with low-fate meal and 28% for high-fat meal). Omeprazole decreased C_{max} by 36% and AUC by 33%.

Drug–Drug Interaction, Palbociclib

Palbociclib is a cyclin-dependent kinase inhibitor. Because it is considered a weak base with a highly pH-dependent solubility, there is a potential concern that co-administration with acid suppressing therapies may affect absorption. A phase I, open-label, fixed-sequence, 2-period crossover study was conducted in 26 healthy volunteers to evaluate the impact of multiple doses of rabeprazole on the absorption of palbociclib. Males and females aged 18–55 years old were included [17c]. Group A patients fasted for 10h overnight and then 4h following palbociclib 125mg given orally. Group B patients received rabeprazole 40mg (2 × 20mg tablets) once daily 30min before a meal for approximately 6 days. The 7th day of rabeprazole was administered after a 10-h overnight fast and 4h prior palbociclib 125mg given orally. Fasting continued for 4h after receiving palbociclib. There was a 10-day washout period between each sequence. In study 1, multiple doses of rabeprazole reduced palbociclib's AUC and C_{max} by 62% and 80%, respectively. A second phase I, open-label, fixed-sequence, 3-period study was performed in 27 healthy subjects to evaluate the impact of rabeprazole, famotidine, or an antacid had on the pharmacokinetics of palbociclib. In study 2 under fed conditions, co-administration with rabeprazole reduced the maximum plasma concentration of palbociclib by 41%, however, the AUC was only minimally affected. These results suggest that food limits the impact of increased gastric pH seen with acid suppression therapies and reduces the variability seen with palbociclib exposure.

Drug–Drug Interaction, Venlafaxine

A retrospective study, using a large database, analyzed venlafaxine and *O*-desmethylvenlafaxine concentrations in over 4000 patients to determine the extent of any drug–drug interactions with omeprazole or pantoprazole [18c]. Patients were divided into three groups: (1) No PPI group (n=906); (2) omeprazole group (n=40); (3) pantoprazole group (n=40). Doses of venlafaxine were similar between all groups. Patients receiving a PPI showed significant increases in the plasma concentrations for venlafaxine, *O*-desmethylvenlafaxine and the active moiety. Venlafaxine concentration increased by 27%, *O*-desmethylvenlafaxine increased by 36%–55%, and the active moiety increased by 29%–36% (all *P* values were statistically significant). Increased venlafaxine concentrations are most likely due to CYP2C19 inhibition leading to an increased metabolism via CYP2D6 to *O*-desmethylvenlafaxine.

ANTICONSTIPATION AND PROKINETIC

Domperidone *[SED-16, 1067–1068; SEDA-37, 445; SEDA-38, 385; SEDA-39, 379–380]*

Cardiovascular

A retrospective chart review was performed to identify all patients who received domperidone [19c]. Co-administration with other medications known to prolong the QTc interval was collected. Of the 155 patients on domperidone, 108 (69.7%) were also on another medication known to prolong the QTc interval. Baseline EKG showed that 15.3% of patients had QTc interval elevations present before domperidone was initiated. The 40 patients (25.8%) who had a follow-up EKG, the incidence of QTc interval prolongation increased to 32.5%. All of these patients were co-administered medications that are known to prolong the QTc interval. A review of the Federal Adverse Event Reporting System (FAERS) database identified 221 cases of nonfatal cardiac events in patients on domperidone. Of these, 163 patients (73%) were receiving medications known to increase the QTc interval. The risk of domperidone-induced QTc interval elevations was more common when co-administered with other medications known to cause QTc prolongation.

Linaclotide

Gastrointestinal

Linaclotide, for irritable bowel syndrome with predominant constipation (IBS-C), is most commonly associated with diarrhea (20% compared to 3% in placebo-treated patients) in pooled IBS-C placebo controlled trials [20R]. Nearly half of reported diarrhea cases occurred during the first week of treatment. Almost one-third of cases resolved within 7 days with continued treatment. Diarrhea and other adverse events resulted in 29% of patients having their dose reduced. Diarrhea can be minimized by taking linaclotide 30min before a meal and avoiding co-administration with other laxatives or food [21R]. A phase 2, randomized, double-blind placebo-controlled study showed that diarrhea was higher in every linaclotide group compared to placebo [22C]. Overall, adverse events were reported in 191 of 559 patients (34.1%). Adverse events were reported in 41 patients receiving placebo, 39 patients receiving 0.0625mg/day, 56 patients receiving 0.125mg/day, 46 patients receiving 0.25mg/day and 50 patients receiving 0.5mg/day. Adverse events most commonly reported included gastrointestinal related (diarrhea and abdominal pain), nasopharyngitis, and increased triglycerides. Severe adverse events included hepatic dysfunction ($n=2$) and diverticulitis ($n=1$).

Lubiprostone

Gastrointestinal, Nausea

Lubiprostone is approved for chronic idiopathic constipation (CIC), opioid-induced constipation (OIC) in patients with noncancer pain (24mcg twice daily) and irritable bowel syndrome with constipation (IBS-C) (8mcg twice daily). Seven studies evaluating CIC included 239 patients on treatment, four evaluating OIC included 663 patients on treatment and four evaluating IBS-C included 832 patients receiving treatment were included [23R]. These studies were a combination of double-blind, randomized, open-label and active treatment studies ranging from 3 to 48 weeks duration. Incidence of nausea was higher in randomized-controlled trials with lubiprostone vs placebo: (CIC 29.3% vs 6.3%, OIC 13.4% vs 6.4%; IBS-C, 10.9% vs 6.4%). Mild, moderate and severe nausea was reported more frequently with lubiprostone in the CIC trials; however, there was no difference in severe nausea in the OIC trials. Moderate nausea was reported more frequently than placebo in the IBS-C trials only. The majority of nausea events (64.3%) occurred during the first 5 days of treatment. Rates of nausea were significantly higher in women in the CIC trials ($P=0.007$) and OIC trials ($P=0.050$). A significant different was seen in age (<65 years vs ≥65 years) in the CIC trials ($P=0.028$) only. Rates of nausea were higher in patients with a BMI <25 in IBS-C trials ($P=0.0140$ for nausea, $P=0.0160$ for severe nausea). More patients discontinued treatment due to nausea (8.7%) in the CIC trials compared to 2.1% and 1.3% in the OIC and IBS-C trials. In randomized controlled trials alone, significantly more patients discontinued treatment in the CIC and OIC trials ($P<0.001$).

Metoclopramide [SED-16, 976–980; SEDA-37, 446; SEDA-38, 385; SEDA-39, 380]

Nervous System

A retrospective review of the Federal Adverse Event Reporting System (FAERS) from 2004 to 2015 reported on harmful effects of medications used for dyspepsia and gastroparesis [24MC]. A total of 1245 adverse effects were reported for gastroparesis treatments and 21243 adverse effects were reports for dyspepsia treatments. PPIs and metoclopramide were the most commonly reported adverse effects. Metoclopramide adverse effects peaked around 2011 going from approximately 20 reports every year to over 2700 then back down to baseline by 2015. The most common metoclopramide adverse effects were associated with tardive dyskinesia and other neurological problems.

Case Report

A 40-year-old female with bipolar disorder receiving mirtazapine, lamotrigine, aripiprazole, and lorazepam presented to the clinic with complaints of nausea which was subsequently treated with intramuscular metoclopramide [25A]. Within 24h, she presented to the Emergency Department with mydriasis, abnormal sweating, lethargy, myoclonus, and muscle rigidity. Labs were normal except for an elevated creatine phosphokinase level of 362 IU/L. The patient was diagnosed with serotonin syndrome possibly caused by the administration of metoclopramide, an agent with weak 5-HT3 receptor blocking effects. The patient began to improve 3 days after treatment consisting of increased lorazepam dose, decreased mirtazapine dose, and cyproheptadine, a histamine-1 receptor antagonist with serotonin-1A and serotonin-2A antagonistic properties.

A 16-year-old female from Cameroon received metoclopramide 10mg IV three times daily for nausea and vomiting associated with the intravenous quinine she was receiving for malaria [26A]. Minutes after the second dose she developed neck pain and stiffness which is in line with the 36h window when this type of reaction can occur. In addition, she complained of a protruding tongue with an inability to keep it in her mouth. She had difficulty swallowing. Metoclopramide-induced dystonia was diagnosed and she received chlorpheniramine 8mg orally. Within 4h, her dystonic symptoms resolved. At 2-week follow-up she was doing well. The dopaminergic antagonistic properties of metoclopramide are thought to cause an imbalance between the dopaminergic and cholinergic pathways leading to release of acetylcholine over dopamine.

Methylnaltrexone [SED-16, 953; SEDA-39, 380]

A study of OIC patients ($n=134$) receiving methylnaltrexone showed the most common adverse effects included abdominal pain, nausea, and urinary tract infections [27R]. Adverse effects occurred more often over the

4 week trial with methylnaltrexone (43.3%) vs placebo (32.8%). In a phase 2 trial of OIC patients ($n=288$), abdominal pain (27.9% vs 9.8% placebo), flatulence (13.3% vs 5.7%) and nausea (10.9% vs 4.9%) were reported more frequently with methylnaltrexone 0.15 and 0.30 mg/kg vs placebo [28R].

In a phase 3, multicenter, open-label trial, 1034 adults were given methylnaltrexone 12 mg once daily for 48 weeks for chronic noncancer pain [29MC]. Adverse effects occurred in 817 patients (79%). Most commonly reported adverse effects were gastrointestinal related including abdominal pain (24%), diarrhea (16.4%), and nausea (15.1%). Psychiatric disorders (anxiety, depression, insomnia) were reported in 10.7% of patients. Adverse events were predominately mild or moderate. Discontinuation rate due to adverse event was 15.2% and were related most commonly to abdominal pain (4.7%), nausea (2.5%) and diarrhea (2.3%). Serious adverse events were reported by 10.1% (104) of patients including pneumonia, back pain, myocardial infarction, abdominal pain, hypoesthesia, asthma, chronic obstructive pulmonary disease, diarrhea and non-cardiac chest pain, however, only four were possibly drug-related.

Cardiology

Of 1034 patients with chronic noncancer pain receiving methylnaltrexone 12 mg once daily for 48 weeks, 9 experienced a cardiac adverse event [AE] (AE rate=1.5 per 100 patient-years) [29MC]. Cardiac events were classified as: acute myocardial infarction (4 patients, AE rate=0.7), cardiac arrest (1 patient, AE rate=0.2), cardiac failure congestive (1 patient, AE rate=0.2), coronary artery disease (2 patients, AE rate=0.3), and ischemic coronary artery disorders (3 patients, AE rate=0.5). Three of the events were adjudicated as major adverse coronary events and were deemed not related to methylnaltrexone. Abnormal EKG measurements were found in 80.5% (832 patients) at least at one point during the study. Nearly 70% of patients had an abnormal EKG at baseline and at 1 h following administration of methylnaltrexone but were not considered clinically meaningful. The QT prolongation returned to baseline after discontinuation of methylnaltrexone.

Naloxegol *[SEDA-39, 380–381]*

Patients taking 30 to 1000 morphine equivalents daily for noncancer pain who developed OIC were randomized to receive naloxegol 25 mg daily vs usual care (laxatives) in a 52-week open-label randomized, parallel-group phase 3 study. Safety analysis included 804 patients of which 81.8% of patients receiving naloxegol experienced an adverse effect compared to 72.2% receiving usual care. Adverse events more frequently observed with naloxegol included abdominal pain and flatulence (17.8% vs 3.3%), diarrhea (12.9% vs 5.9%), nausea (9.4% vs 4.1%), headache (9% vs 4.8%), and upper abdominal pain (5.1% vs 1.1%). Rates of overall adverse events in a pooled analysis from two randomized trials of naloxegol 25 mg daily, 12.5 mg daily vs placebo ($n=720$) were higher in the naloxegol 25 mg group (63.1%) vs similar rates of almost 50% in the comparative groups [27R, 28R]. Discontinuation rates were higher with naloxegol (10.5%) vs usual care (1.8%) predominately due to diarrhea, abdominal pain, and vomiting. Serious GI related adverse events were not more common in the naloxegol group compared to usual care [30R].

Naldemedine

COMPOSE III, IV, and V are phase 3 trials reporting adverse events of naldemedine for opioid-induced constipation (OIC). COMPOSE III evaluated naldemedine 0.2 mg daily in non-cancer pain for 52 weeks ($n=621$) compared to standard of care ($n=619$). Most common adverse events reported more with naldemedine were abdominal pain (11% vs 5%), diarrhea (7% vs 3%), nausea (6% vs 5%), vomiting (3% vs 2%) and gastroenteritis (3% vs 1%). Adverse events were similar at 12 and 52 weeks [31R]. COMPOSE IV was a 2-week randomized, double-blind, placebo-controlled trial comparing naldemedine 0.2 mg daily ($n=97$) to placebo ($n=96$) in patients diagnosed with OIC and cancer pain. COMPOSE V was a 12-week extension study in which 62 patients were receiving naldemedine and 69 were receiving placebo. The majority of patients were male (approximately 60%) and approximately 64 years old. Treatment emergent adverse events were significantly higher with naldemedine in COMPOSE IV (44.3% vs 26.0%, $P=0.01$). During the 12-week extension, 80.2% of patients had an adverse event in the naldemedine group. Gastrointestinal adverse effects were reported most frequently with diarrhea predominating with nearly 20% of patients reporting the event in both COMPOSE IV and V. Diarrhea was the most common cause of study discontinuation and occurred in 5.2% of patients. Overall discontinuation was observed in 9.3% of patients in COMPOSE IV and 9.2% of patients in COMPOSE V. Diarrhea ($n=2$), vomiting ($n=1$) and abnormal liver function test ($n=1$) were reported as serious but nonfatal adverse events with naldemedine [32C].

Plecanatide

Plecanatide, for CIC, works to create an ionic gradient to allow for fluid secretion into the bowel and facilitate bowel movements. The safety of plecanatide was evaluated in 1394 patients. Patients were assigned to plecanatide

3mg ($n=474$), 6mg (457) or placebo ($n=458$) [33C]. Most adverse events were mild to moderate and were experienced by 35.4% of patients receiving plecanatide 3mg, 33% receiving plecanatide 6mg and 32.8% receiving placebo. The most common adverse events with an incidence of >2% were diarrhea (28% plecanatide 3mg, 26% plecanatide 6mg, 6% placebo), nasopharyngitis (4% plecanatide 3mg, 11% plecanatide 6mg, 8% placebo), and sinusitis (10% plecanatide 3mg, 3% plecanatide 6mg, 3% placebo). Discontinuation due to an adverse event was seen in 5.1% (3mg), 5.3% (6mg) and 1.3% (placebo). Diarrhea was the most common reason for discontinuation.

A study evaluated the safety of plecanatide 3mg ($n=443$), 6mg ($n=449$) and placebo ($n=445$) [34C]. Overall, 372 patients (26.5%) experienced one or more adverse events. Rates of adverse events were similar among all groups. Diarrhea was the most frequently reported with 4.5% (6mg), 3.2% (3mg) vs 1.3% (placebo). Discontinuation was more similar amongst groups compared to the other phase III trial with 3.8% and 3.2% of patients discontinuing in the plecanatide 6 and 3mg group, respectively, and 3% in the placebo group. Overall, mild to moderate diarrhea is the most frequent adverse event reported with plecanatide and it appears to be dose-dependent.

Polyethylene Glycol

Electrolyte Balance

Case Report

A 68-year-old female presented to the Emergency Department experiencing a generalized tonic–clonic seizure following 2 days of nausea, vomiting and diarrhea secondary to colonoscopy prep with polyethylene glycol [35A]. Initial evaluation revealed hyponatremia (serum sodium 106mmol/L) and hypokalemia (serum potassium 3.1mmol/L). The patient was treated in the intensive care unit with hypertonic (3%) sodium chloride and later recovered after sodium levels normalized over 5 days. Medical history included hypertension, for which, patient was treated with hydrochlorothiazide. Hyponatremia likely resulted due to excessive fluid loss from colonoscopy preparation combined with sodium excretion caused by hydrochlorothiazide.

Prucalopride

Twenty-nine patients with mild to moderately severe enteric dysmotility symptoms in systemic sclerosis (SSc) were evaluated in a crossover study to receive prucalopride 2mg daily or placebo for 1 month then crossed over after a 2-week washout period to the opposite treatment [36c]. Seven patients (17.5%) reported adverse effects that were not tolerable and included headache ($n=3$), abdominal pain ($n=2$), dizziness ($n=1$), and sensation of feeling sick ($n=1$). Four of seven cases were observed on the 1st day of treatment and considered severe enough to discontinue treatment.

Sodium Picosulfate and Magnesium Citrate *[SED-16, 729–732; SEDA-39, 381]*

Case Report

A 69-year-old male preparing for colonoscopy presented with epigastric pain for 1 day. He had ingested sodium picosulfate/magnesium citrate powder, however, had forgotten to dissolve it in water [37A]. After 10min of realizing his mistake, he drank approximately 2L of water, however, experienced discomfort after 6h. Esophagogastroduodenoscopy (EGD) performed 12h after ingestion revealed multiple longitudinal ulcers with hematin on the entire gastric body and antrum. Evaluation of lesions showed infiltration of lymphoid cells and neutrophils were present with fresh hemorrhage in the lamina propria. No mucosal abnormality was seen in the esophagus or duodenum. Lansoprazole 30mg daily for 6 weeks was used and symptoms completely resolved. Repeat EGD showed ulcer scar in healing state.

ANTIDIARRHEAL AND ANTISPASMODIC AGENTS

Loperamide *[SED-16, 668; SEDA-37, 450; SEDA-38, 386–387; SEDA-39, 382]*

Cardiology

Post-marketing reports using the Food and Drug Administration Adverse Event Reporting System (FAERS) database from December 28, 1976 through December 14, 2015 were reviewed to identify serious cardiac adverse events associated with loperamide which included: sudden cardiac death, cardiac arrest, torsade de points, arrhythmia, QT-interval prolongation, ventricular fibrillation, ventricular tachycardia, ventricular arrhythmia or syncope [38c]. Forty-eight cases were found. The most frequently reported cardiac cases were syncope ($n=24$), cardiac arrest ($n=13$), QT-interval prolongation ($n=13$), ventricular tachycardia ($n=10$), and torsade de pointes ($n=7$). The most commonly reported reason for use was drug abuse ($n=22$) while diarrhea treatment was lower ($n=17$). Four of 22 drug abuse cases described use of drugs known to be CYP3A4, CYP2C8, or P-gp inhibitors. The cardiac events resolved after discontinuation of 8 of the 22 drug abuse cases with 5 having a positive re-challenge. Drug abuse resulted in 14 hospitalizations and 7 deaths. Median daily dose was 250mg (range 70–1600mg). Of the cases reported for diarrhea, 5 were misusing loperamide at higher than recommended doses (median dose of 80mg, range 16–100mg) and 11 were taking at recommended doses.

In a similar study, 19 cases of cardiac dysrhythmias associated with loperamide use and abuse was reported [39A]. Loperamide doses ranged from 70 to 792mg/day with use from several days to years. Loperamide levels ranged from 22 to 140ng/mL (therapeutic level=0.24–3.1ng/mL). Syncope was the most common presentation (63%). QTc prolongation and ventricular tachycardia were also common. Thirteen patients were successfully treated while two died after aggressive treatment. Two patients died at home and two left hospital against medical advice. In follow-up, two patients continued to abuse loperamide and eventually died.

Simethicone

Drug–Drug Interaction, Levothyroxine

Although evidence is weak, the concomitant use of simethicone and levothyroxine should be avoided [40R]. A pediatric patient receiving both agents continued to have elevated TSH levels despite appropriate dosing. After discontinuing simethicone, levothyroxine dose had to be reduced to achieve a normal TSH level. The result of this drug interaction is likely related to decreased absorption of levothyroxine through the intestinal mucosa.

Teduglutide *[SEDA-39, 381]*

Pediatric patients age 1–17 years requiring parenteral nutrition (PN) and who showed minimal or no advance in enteral nutrition (EN) with intestinal failure associated with short bowel syndrome (SBS-IF) were enrolled in a 12-week open-label study with three teduglutide cohorts (0.0125mg/kg/day [n=8], 0.025mg/kg/day [n=14], 0.05mg/kg/day [n=15]) or standard of care [41c]. All patients reported at least one adverse event. The most common adverse event reported by all patients receiving teduglutide was vomiting and was more common with higher doses of 0.05mg/kg/day (47%) and 0.025mg/kg/day (36%). Gastrointestinal related adverse events (diarrhea, abdominal distention, flatulence, hematochezia, fecal volume increase) occurred in 67% (0.05mg/kg/day), 71% (0.025mg/kg/day), 50% (0.0125mg/kg/day) compared to only 20% in standard of care. There were no reports of intestinal obstruction, fluid overload, gallbladder, biliary or pancreatic disease in this pediatric group and no adverse events lead to treatment discontinuation. Neutralizing antibodies were not identified in any patient.

Case Report

A 71-year-old male with short bowel syndrome (SBS) and parenteral nutrition (PN) dependence with no history of colon polyps was started on teduglutide 4.2mg/day [42A]. Endoscopy at 10 months post-therapy initiation was negative but at 26 weeks another endoscopy performed for evaluation of gastrointestinal pain revealed multiple duodenal polyps (2 large polyps >1 cm in the duodenal bulb and 10 small polyps <5mm in the first portion of the duodenum). Polyps were hyperplastic in nature without features of advanced neoplasia. After 15 months, surveillance endoscopy showed a 30%–50% increase in size of polyps, which were removed. Pathology showed hamartomatous polyps without dysplasia. Teduglutide has the potential to cause intestinal neoplasia as it induces mucosal growth in the small and large intestine in patients with SBS.

ANTIEMETIC AGENTS

Aprepitant and Fosaprepitant *[SED-16, 657; SEDA-36, 544; SEDA-37, 450; SEDA-39, 382]*

Nervous System

Case Report

A 25-year-old male with recurrent B-cell Philadelphia-negative acute lymphoblastic lymphoma developed ifosfamide-induced encephalopathy after receiving aprepitant [43c]. Symptoms of encephalopathy developed 24h after receiving ifosfamide as a 1-h infusion; patient complained of headache and vomiting. A seizure and disorientation was also noted. All labs, EKG, and brain MRI were normal. The patient started to improve within 24h after drug discontinuation and treatment with complete recovery seen within 48–72h. Treatment consisted of hydration and methylene blue 50mg administered intravenously over 5min given six times daily until complete resolution of symptoms seen. The patient was also treated with ondansetron twice daily over 3 days. Aprepitant inhibits CYP3A4 which may increase ifosfamide metabolites leading to accumulation and risk for encephalopathy.

Drug–Drug Interaction, Bosutinib

A phase I, open-label, randomized, crossover study evaluated the pharmacokinetic and safety profile of bosutinib, an oral dual Src and Abl tyrosine kinase inhibitor, after a single dose of aprepitant in 19 healthy subjects (age range: 18–55 years old) [43c]. There were two treatment crossover sequences separated by a 14-day washout. Sequence 1 included bosutinib alone

(100 mg × 5) × 1 dose or bosutinib (100 mg × 5) × 1 dose in combination with aprepitant 125 mg × 1 dose. Sequence 2 was reversed from the treatment obtained in the first sequence. Bosutinib is primarily metabolized by CYP3A4 and aprepitant is considered a moderate CYP3A4 inhibitor. The AUC and C_{max} were increased by 99% and 53%, respectively. Treatment-emergent adverse events were similar between the groups and generally well-tolerated.

Ondansetron *[SED-16, 343–347; SEDA-37, 453, SEDA-38, 388; SEDA-39, 383]*

Cardiology

A recent editorial reiterated risk factors associated with ondansetron-induced torsades de pointes including female sex, hypomagnesemia, hypocalcemia, bradycardia, diuretics, hypothermia, and history of heart disease [44r]. Notably, these risk factors should be considered in patients receiving ondansetron due to increased risk of QTc prolongation leading to arrhythmias.

ANTIINFLAMMATORY AGENTS

Aminosalicylates *[SED-16, 242–254; SEDA-37, 454; SEDA-38, 389; SEDA-39, 383]*

Hematologic

Case Report, Fetal Anemia

A 43-year-old woman taking mesalazine 4 g daily had a hydropic fetus at 31 weeks with massive ascites and cardiomegaly [45A]. Fetal hemoglobin at cordocentesis was 51 g/L and platelets 94 × 10^9/L. During four visits over a 3-week course, 250 mL of packed erythrocytes were administered. Mesalazine was discontinued. Hemoglobin and platelets slowly rose and at completion were 123 g/L and 149 × 10^9/L, respectively, and the ascites was almost undetectable. At 37 weeks she delivered a healthy baby with hemoglobin 154 g/dL. Maternal and fetal concentrations of mesalazine decreased over time. At 12.5 h after last dose, maternal and fetal concentrations were similar at 3.89 and 3.48 ng/mL, respectively. After 162 h the concentrations had decreased more in the fetus to 0.22 ng/mL and to 1.08 ng/mL in the mother. By 282 h, mesalazine level was undetectable in both. Risk vs benefit should be considered during pregnancy.

Urinary Tract

Case Report, Kidney Failure

Severe kidney failure with massive crystalluria was reported in a 72-year-old male on sulfasalazine 2000 mg three times daily for Crohn's disease for approximately 10 years. He had no history of kidney disease and was admitted secondary to confusion and elevated serum creatinine (SCr) of 9.7 mg/dL [46A]. Three months prior his SCr was 1 mg/dL and had never been higher than 1.1 mg/dL in the previous 8 years. Urine microscopy showed amorphous crystals (1+) and uric acid crystals (4+) and elevated fractional excretion of sodium of 5.5%. Patient was treated with intravenous 0.9% sodium chloride. Renal ultrasound showed mild right hydronephrosis and possible small nonobstructing renal calculi. Patient voided numerous small stones measuring 0.5–2 mm and SCr decreased to 6.5 mg/dL. Patient was discharged with a SCr of 3.1 mg/dL and sulfasalazine was discontinued. Six months later, SCr was 0.95 mg/dL. Stones were analyzed and were composed of acetylated sulfapyridine compounds which are metabolites of the sulfasalazine derivative sulfapyridine. Kidney failure was attributed to precipitation and crystallization of the sulfasalazine metabolites.

Thiopurines (Azathioprine, 6-Mercaptopurine) *[SED-16, 759–781; SEDA-37, 455; SEDA-38, 389–390; SEDA-39, 383–384]*

Hematologic/Infection Risk

LEUKOPENIA AND INFECTION

Risk factors for nonthiopurine *S*-methyltransferase (TPMT) variants were assessed in 695 patients with the majority receiving azathioprine (64.5%) [mean dose 2.19 ± 0.20 mg/kg] while 35.5% received 6-mercaptopurine (6-MP) [mean dose 1.20 ± 0.15 mg/kg] [47c]. To compare the two doses, a conversion was created and a converted dose of 1.05 ± 0.09 mg/kg was used for statistical analysis. Median onset of leukopenia was 56 days (range: 29–112 days) with 6.5% of patients developing within the first 5 months of treatment. Moderate or grade 2 leukopenia (WBC count between ≥2 × 10^9/L and <3 × 10^9/L) developed in 41 (5.9%) patients and severe or grade 3 leukopenia (WBC count <2 × 10^9/L) developed in four (0.6%) patients. Treatment was discontinued (24.4%), dose reduced (44.5%) or no intervention was required (26.7%) as leukopenia self-resolved. Metabolite levels were not increased in patients with leukopenia, however, large amount of patients had dose reductions and steady-state concentrations were not routinely available. No correlation was found between baseline leukopenia and thiopurine induced leukopenia (HR = 0.78; $CI_{95\%}$ = 0.70–0.88). Smokers less often had leukopenia in a univariate analysis (HR = 0.37; $CI_{95\%}$ = 0.13–1.03). Infections were identified in 9.4% of patients within the 5-month period with 8.5% being classified as grade 2 or higher. Thirty-two patients experienced a grade 2 infection (oral intervention), 25 patients a grade 3 infection (intravenous treatment or operational intervention) and 2 patients a grade 4 infection (life-threatening consequences). Leukopenia-associated infection was associated with 45 patients and 5 patients (11.1%) had severe leukopenia. Infection rate without leukopenia was 8.3%, $P=0.41$. Associated factors for infection occurrence was age (HR=2.07; $CI_{95\%}$=1.18–3.63, $P=0.01$) with increased age having higher occurrence and concomitant use of biologic drugs (HR=2.15; $CI_{95\%}$=1.14–4.07; $P=0.02$).

A case series of 24 patients with 6-MP leukopenia evaluated association between drug level concentrations and myelotoxicity [48c]. Therapeutic levels of 6-thioguanine nucleotides (6-TGN) and 6-methylmercaptopurine (6-MMP) are 230–450 pmol/8×10^8 red blood cells (RBC) and <5700 pmol/8×10^8 RBC. Dosages ranged from 50 to 200 mg per day and 0.8 to 2.3 mg/kg per day. Nine patients were concomitantly receiving mesalazine. Onset of leukopenia occurred after a median of 11 weeks after duration of therapy (range 4–1000 weeks) with a mean WBC count of $2.7\pm0.9\times10^9$/L. Hemoglobin decreased to under lower reference limit in 75% of patients (6.9×10^9/L, range 3.2–8.4). Thrombocytopenia occurred in 17% of patients. Median 6-MMP concentration was 14500 pmol/8×10^8 RBC. 6-TGN concentrations were in therapeutic range in nine patients and lower than therapeutic range in other 15 patients. Dose reduction in patients typically resulted in therapeutic concentrations and resolution of leukopenia and other blood count abnormalities in a median time of 4 weeks in 83% of patients. Therapy switch to agents that did not undergo complex metabolism also resulted in recovery of blood counts.

Reproductive System

PRECONCEPTION USE

A study of 735 children in the nationwide Danish registry born to fathers who received a prescription for azathioprine (AZA)/mercaptopurine (6-MP) within 3 months prior to conception were evaluated for adverse long-term outcomes of malignancies, schizophrenia, psychosis, attention deficit hyperactivity disorders (ADHD), and autism spectrum disorders (ASD) per the Danish National Patient registry [49c]. Children born from 1997 up until age 19 or May 2015 who had a discharge diagnosis in the Danish National Patient registry of were compared to over one million children who had the same diagnosis whose father did not receive AZA/6-MP within 3 months prior to conception. There were no major differences found between the exposed and unexposed groups, with one incidence of leukemia (0.14%), ASD/schizophrenia (0.14%) and three with ADHD (0.41%).

ANTICHOLINERGIC AGENTS

Scopolamine

Drug Withdrawal

Case Report

A 15-year-old male with quadriplegic cerebral palsy was being treated for sialorrhea with botulinum-toxin A in addition to usual management of spasticity with botulinum-toxin A and ethanol using the SEMLC procedure (single event, multilevel chemoneurolysis) [50A]. He was on scopolamine patch for more than 4 years and had suffered anticholinergic adverse events including constipation and chest congestion. Botulinum-toxin A to the salivary glands was going to replace the scopolamine patch. Upon follow-up, he developed 4 days of severe nausea, decreased appetite and emesis with any oral intake and was admitted with severe hydration. All labs were normal. The patient was given scopolamine patch taper instructions, however, it was discovered this was not carried out and the patient's symptoms were attributed to scopolamine withdrawal.

MISCELLANEOUS AGENTS

Misoprostol *[SED-16, 1063–1068; SEDA-38, 391; SEDA-39, 384–385]*

Reproductive System

A post-hoc analysis of a double-blind randomized controlled trial evaluated intrapartum adverse events which resulted in misoprostol (MVI) or dinoprostone (DVI) vaginal insert retrieval [51C]. There were 678 women who received the MVI and 680 who received the DVI. Insert was removed due to intrapartum adverse event in 11.4% and 4% of women who received the MVI and DVI, respectively. Uterine tachysystole with fetal heart rate (FHR) involvement was the most common reason for removal in both groups with incidence of 5.3% and 1.2% in the MVI and DVI groups, respectively. Category II/III FHR pattern adverse event was also common with incidence of 3.2% and 1.9% in the MVI and DVI groups, respectively. The only significant predictor for adverse event was choice of agent with higher frequency of events with MVI ($P<0.001$). Baseline characteristics were similar and none were identified to be a predictor. No neonatal deaths were reported.

Body Temperature

FEVER

A retrospective study evaluated the characteristics of fever in second trimester termination of pregnancy with misoprostol from January 2008 to October 2012 [52c]. Patients were excluded if fever or antibiotic therapy occurred prior to misoprostol or if no temperatures were in medical record. Misoprostol 600 mcg was administered vaginally followed by 400 mcg every 3 h, maximum five doses per day. Population included 403 women and 168 (42%) developed a fever. Rectal temperature between 38°C and 39°C which developed within 10 h after start of induction was seen in 71.5% of patients. Intra-uterine infection was examined by assessment of placenta. In patients who had a fever, 76% (121) placentas were examined and 3% (5 placentas) showed a stage II (intermediate, acute chorioamnionitis) or III (advanced, necrotizing chorioamnionitis) reaction, indicative of intrauterine infection. In patients without fever, 2% (4 placentas) showed stage II or III reaction. Mean misoprostol dose was higher in patients with fever than without (1946 ± 1101 mcg vs 1254 ± 611 mcg, respectively, $P < 0.001$). Fever had a sensitivity of 55%, specificity of 58% and a positive predictive value for infection of 4%. This study shows that fever is a common adverse effect of misoprostol but is not a good predictor of intra-uterine infection.

Case Report

A 21-year-old female, primigravida at 39 weeks, with normal vitals and labs on admission spontaneously delivered the baby and misoprostol 600 mcg was used to manage third stage labor [53A]. After 30 min, patient was administered parenteral antipyretic and antiemetic due to severe headache, vomiting, high-grade fever (107°F), rigors and irrelevant talking. After another 30 min, patient developed three episodes of convulsions requiring intravenous lorazepam and more paracetamol for fever. Patient had another convulsion 15 min later and was managed in the intensive care unit with phenytoin 600 mg followed by 100 mg three times daily. Neurological deficits were not noted. Patient was discharged on phenytoin and remained on anticonvulsant at follow-up. Misoprostol was thought to lower seizure threshold. Timely management of an uncommon adverse effect of hyperpyrexia and convulsion prevented further complications.

Octreotide *[SEDA-38, 387]*

Twenty-nine patients from birth to 18 years who received octreotide in post-cardiac surgery chylothorax were evaluated for safety [54c]. The most common adverse event was hyperglycemia (45%) with glucose range 185–450 mg/dL. Tachycardia was also common. Baseline heart rate averaged 122 beats per minute (bpm) on day 1 and increased to 130 bpm on day 9 with 3 patients having a heart rate over 150 bpm. Abdominal distention was observed in one patient and elevated liver function tests were seen in two patients; however, one patient had severe sepsis with multi-organ failure. No adverse events required permanent discontinuation of octreotide.

Transjugular intrahepatic portosystemic shunts (TIPs) or long-acting octreotide were evaluated in patients to prevent variceal bleeding in 24 patients with portal hypertension [55c]. Liver function tests at baseline, 1 week and last follow-up were similar among groups with no statistical differences. Commonly reported adverse events related to long-acting octreotide were diarrhea ($n = 3$), flatulence ($n = 2$), dizziness ($n = 2$), cramps ($n = 2$), and hypoglycemia ($n = 1$) but were all noted as mild. No patients receiving long-acting octreotide developed gallstones, hypertension, biliary colic or hepatic encephalopathy.

Hematologic

Case Report

An 87-year-old patient was hospitalized with constipation and straining for 3 days. Labs showed WBC $4.1 \times 10^3/mm^3$ with 75.4% neutrophils and absolute neutrophil count (ANC) $3.1 \times 10^3/mm^3$ [56A]. Patient was given phosphate enema and magnesium hydroxide on day 1 without results then senna 8.6 mg/docusate 50 mg, polyethylene glycol 17 g and bisacodyl suppositories as needed. Computed tomography revealed distal colonic obstruction secondary to a mass that was later diagnosed as colon cancer. Octreotide 100 mcg every 8 h was initiated on day 4. After three doses, the patients WBC decreased to $1.6 \times 10^3/mm^3$ with neutrophils 62% and ANC $0.99 \times 10^3/mm^3$ and subsequent fever. WBC counts continued to stay down despite antibiotic administration so octreotide was discontinued on day 7. WBC counts increased on day 8 and by day 9 were $6.4 \times 10^3/mm^3$ with neutrophils 63.9% and ANC $4.1 \times 10^3/mm^3$. Infection source not identified. Most probable cause for neutropenia was octreotide.

Orlistat *[SED-16, 392–394; SEDA-38, 391; SEDA-39, 385]*

Gastrointestinal symptoms are the most commonly reported adverse events with orlistat and include diarrhea, flatulence, bloating, abdominal pain and dyspepsia [57R]. Due to its mechanism of action, fat-soluble vitamins deficiencies have been reported and daily multivitamins are often recommended while on therapy.

Drug–Drug Interaction, Levothyroxine

Two weeks after a patient, who was on a stable dose of levothyroxine 250 mcg/day, started orlistat she began to experience symptoms of hypothyroidism [40R].

Levothyroxine was increased to 300 mcg/day and orlistat was discontinued. Thyroid levels returned to normal. It remains slightly controversial given the adjustment of levothyroxine if the orlistat was fully implicated in this drug–drug interaction. It is possible that orlistat use with levothyroxine could require increased levothyroxine doses. Avoid concomitant use or separate by at least 2 h.

BIOLOGICS

See chapter 36 for further information.

STEROIDS

See chapters 38, 39 and 40 for further information.

Additional literature on this topic can be found in a SEDA-39 review [58R].

References

[1] Bundhun PK, Teeluck AR, Bhurtu A, et al. Is the concomitant use of clopidogrel and proton pump inhibitors still associated with increased adverse cardiovascular outcomes following coronary angioplasty?: a systematic review and meta-analysis of recently published studies (2012–2016). BMC Cardiovasc Disord. 2017;17:3. https://doi.org/10.1186/s12872-016-0453-6 [M].

[2] Shamliyan TA, Middleton M, Borst C. Patient-centered outcomes with concomitant use of proton pump inhibitors and other drugs. Clin Ther. 2017;39(2):404–27 [M].

[3] Davis TME, Drinkwater J, Davis W. Proton pump inhibitors, nephropathy, and cardiovascular disease in type 2 diabetes: the Fremantle Diabetes Study. J Clin Endocrinol Metab. 2017;102(8):2985–93. https://doi.org/10.1210/jc.2017-00354 [C].

[4] Law EH, Badowski M, Hung YT, et al. Association between proton pump inhibitors and microscopic colitis: implications for practice and future research. Ann Pharmacother. 2017;51(3):253–63. https://doi.org/10.1177/1060028016673859 [M].

[5] Avinash A, Patil N, Kunder SK, et al. A retrospective study to assess the effect of proton pump inhibitors on renal profile in a South Indian Hospital. J Clin Diagn Res. 2017;11(4):FC09–12. https://doi.org/10.7860/JCDR/2017/26097.9752 [C].

[6] Nuki Y, Umeno J, Washio E, et al. The influence of CYP3C19 polymorphisms on exacerbating effect of rabeprazole in celecoxib-induced small bowel injury. Aliment Pharmacol Ther. 2017;46331–6. https://doi.org/10.1111/apt.14134 [c].

[7] Eusebi LH, Rabitti S, Artesiani ML, et al. Proton pump inhibitors: risks of long-term use. J Gastroenterol Hepatol. 2017;32:1295–302. https://doi.org/10.1111/jgh.13737 [R].

[8] Zhou X, Pant S, Nemunaitis J, et al. Effects of rifampin, itraconazole, and esomeprazole on the pharmacokinetics of alisertib, an investigational aurora a kinase inhibitor in patients with advanced malignancies. Investig New Drugs. 2018;36(2):248–58. https://doi.org/10.1007/210637-017-0499-z [c].

[9] Furuta T, Sugimoto M, Kodaira C, et al. Influence of low-dose proton inhibitors administered concomitantly or separately on the anti-platelet function of clopidogrel. J Thromb Thrombolysis. 2017;43:333–42. https://doi.org/10.1007/s11239-016-1460-2 [c].

[10] Zou D, Goh KL. East Asian perspective on the interaction between proton pump inhibitors and clopidogrel. J Gastroenterol Hepatol. 2017;32:1152–9. https://doi.org/10.1111/jgh.13712 [R].

[11] Kagami T, Yamade M, Suzuki T, et al. Comparative study of effects of vonoprazan and esomeprazole on antiplatelet function of clopidogrel or prasugrel in relation to CYP2C19 genotype. Clin Pharmcol Ther. 2018;103(5):906–13. https://doi.org/10.1002/cpt.863 [c].

[12] Wicinski M, Ewclewicz MM, Mietkiewicz M, et al. Potential mechanisms of hematological adverse drug reactions in patients receiving clozapine in combination with proton pump inhibitors. J Psychiatr Pract. 2007;23(2):114–20. https://doi.org/10.1097/PRA.0000000000000223 [R].

[13] Narumi K, Sato Y, Kobayashi M, et al. Effects of proton pump inhibitors and famotidine on elimination of plasma methotrexate: evaluation of drug-drug interactions mediated by organic anion transporter 3. Biopharm Drug Dispos. 2017;38:501–8. https://doi.org/10.1002/bdd.2091 [c].

[14] Ciftci HS, Karadeniz MS, Tefik T, et al. Influence of proton pump inhibitors on mycophenolic acid pharmacokinetics in patients with real transplantation and the relationship with cytochrome 2C19 gene polymorphism. Transplant Proc. 2017;49:490–6. https://doi.org/10.1016/j.transproceed.2017.01.029 [C].

[15] Keyvanjah K, DiPrimeo D, Li A, et al. Pharmacokinetics of neratinib during coadministration with lansoprazole in healthy subjects. Br J Clin Pharmacol. 2017;83:554–61 [c].

[16] Xin Y, Shao L, Maltzman J, et al. The relative bioavailability, food effect, and drug interaction with omeprazole of momelotinib tablet formulation in health subjects. Clin Pharmacol Drug Dev. 2018;7(3):277–86. https://doi.org/10.1002/cpdd.397 [c].

[17] Sun W, Klamerus KJ, Yuhas LM. Impact of acid-reducing agents on the pharmacokinetics of palbociclib, a weak base with pH-dependent solubility, with different food intake conditions. Clin Pharmacol Drug Dev. 2017;6(6):614–26 [c].

[18] Kuzin M, Schoretsanitis G, Haen E, et al. Effects of the proton pump inhibitors omeprazole and pantoprazole on the cytochrome P450-mediated metabolism of venlafaxine. Clin Pharmacokinet. 2018;57(6):729–37. https://doi.org/10.1007//s40262-017-0591-8 [c].

[19] Ehrenpreis ED, Roginsky G, Alexoff A, et al. Domperidone is commonly prescribed with QT-interacting drugs: review of a community-based practice and a postmarketing adverse drug event reporting database. J Clin Gastroenterol. 2017;51(1):56–62 [c].

[20] Chandar AK. Diagnosis and treatment of irritable bowel syndrome with predominant constipation in the primary-care setting: a focus on linaclotide. Int J Gen Med. 2017;10:385–93 [R].

[21] Rey E, Mearin F, Alcedo J, et al. Optimizing the use of linaclotide in patients with constipation-predominant irritable bowel syndrome: an expert consensus report. Adv Ther. 2017;34(3):587–98 [R].

[22] Fukudo S, Nakajima A, Fujiyama Y, et al. Determining an optimal dose of linaclotide for use in Japanese patient with irritable bowel syndrome and constipation: a phase II randomized, double-blind, placebo-controlled study. Neurogastroenterol Motil. 2018;30: e13275 [C].

[23] Cryer B, Drossman D, Chey W, et al. Analysis of nausea in clinical studies of lubiprostone for the treatment of constipation disorders. Dig Dis Sci. 2017;62:3568–78 [R].

[24] Bielefeldt K. From harmful treatment to secondary gain: adverse event reporting in dyspepsia and gastroparesis. Dig Dis Sci. 2017;62:2999–3013. https://doi.org/10.1007/s10620-017-4633-8 [MC].

[25] Harada T, Hirosawa T, Morinaga K, et al. Metoclopramide-induced serotonin syndrome. Intern Med. 2017;56:737–9. https://doi.org/10.2169/internalmedicine.56.7727 [A].
[26] Tianyi FL, Agbor VN, Njim T. Metoclopramide induced acute dystonic reaction: a case report. BMC Res Notes. 2017;10:32. https://doi.org/10.1186/s13104-016-2342-6 [A].
[27] Pergolizzi Jr. JV, Raffa RB, Pappagallo M, et al. Peripherally acting μ-opioid receptor antagonists as treatment options for constipation in noncancer pain patients on chronic opioid therapy. Patient Prefer Adherence. 2017;11:107–19 [R].
[28] Streicher JM, Bilsky EJ. Peripherally acting μ-opioid receptor antagonists for the treatment of opioid-related side effects: mechanism of action and clinical implications. J Pharm Pract. 2017; [epub ahead of print]. [R].
[29] Webster LR, Michna E, Khan A, et al. Long-term safety and efficacy of subcutaneous methylnaltrexone in patients with opioid-induced constipation and chronic noncancer pain: a phase 3, open-label trial. Pain Med. 2017;18:1496–504 [MC].
[30] Bowers B, Crannage AJ. The evolving role of long-term pharmacotherapy for opioid-induced constipation in patients being treated for noncancer pain. J Pharm Pract. 2017; [epud ahead of print]. [R].
[31] Markham A. Naldemedine: first global approval. Drugs. 2017;77:923–7 [R].
[32] Katakami N, Harada T, Murata T, et al. Randomized phase III and extension studies of naldemedine in patients with opioid-induced constipation and cancer pain. J Clin Oncol. 2017;35(34):3859–70 [C].
[33] Miner PB, Koltun WD, Wiener GJ, et al. A randomized phase III clinical trial of plecanatide, a uroguanylin analog, in patients with chronic idiopathic constipation. Am J Gastroenterol. 2017;112:613–21 [C].
[34] DiMicco M, Barrow L, Hickey B, et al. Randomized clinical trial: efficacy and safety of plecanatide in the treatment of chronic idiopathic constipation. Ther Adv Gastroenterol. 2017;10(11):837–51 [C].
[35] Samad N, Fraser I. Severe symptomatic hyponatremia associated with the use of polyethylene glycol-based bowel preparation. Endocrinol Diabetes Metab Case Rep. 2017;1–5:pii: 16–0119. [A].
[36] Vigone B, Caronni M, Severino A, et al. Preliminary safety and efficacy profile of prucalopride in the treatment of systemic sclerosis (SSc)-related intestinal involvement: results from the open label cross-over PROGRASS study. Arthritis Res Ther. 2017;19(145):1–8 [c].
[37] Ze EY, Choi CH, Kim JW. Acute gastric injury caused by undissolved sodium picosulfate/magnesium citrate powder. Clin Endosc. 2017;50:87–90 [A].
[38] Swank KA, Wu E, McAninch J, et al. Adverse event detection using the FDA post-marketing drug safety surveillance system: cardiotoxicity associated with loperamide abuse and misuse. J Am Pharm Assoc. 2017;57(2S):S63–7 [c].
[39] Riaz IB, Khan MS, Kamal MU, et al. Cardiac dysrhythmias associated with substitutive use of loperamide: a systematic review. Am J Ther. 2017; [epub ahead of print]. [A].
[40] Skelin M, Lucijanic T, Amidzic Klaric D, et al. Factors affecting gastrointestinal absorption of levothyroxine: a review. Clin Ther. 2017;39(2):378–403 [R].
[41] Carter BA, Cohran VC, Cole CR, et al. Outcomes from a 12-week, open-label, multicenter clinical trial of teduglutide in pediatric short bowel syndrome. J Pediatr. 2017;181:102–11 [c].
[42] Ukleja A, Alkhairi B, Bejarano P, et al. De novo development of hamartomatous duodenal polyps in a patient with short bowel syndrome during teduglutide therapy: a case report. J Parenter Enteral Nutr. 2018;42(3):658–60 [A].
[43] Hsyu PH, Pignataro DS, Matschke K. Effect of aprepitant, a moderate CYP3A4 inhibitor, on bosutinib exposure in healthy subjects. Eur J Clin Pharmacol. 2017;73:49–56 [c].
[44] Cerit Z. Arrhythmic side of ondansetron alongside antiemetic effect. Pediatr Emerg Care. 2017;33(7):e9 [r].
[45] Ek S, Rosenborg S. Mesalazine as a cause of fetal anemia and hydrops fetalis. Medicine. 2017;96(50)e9277 [A].
[46] Durando M, Tiu H, Kim JS. Sulfasalazine-induced crystalluria causing severe acute kidney injury. Am J Kidney Dis. 2017;70(6):869–73 [A].
[47] Broekman MMTJ, Coenen MJH, Wanten GJ, et al. Risk factors for thiopurine-induced myelosuppression and infections in inflammatory bowel disease patients with a normal TPMT genotype. Aliment Pharmacol Ther. 2017;46:953–63 [c].
[48] Meijer B, Kreinje JE, van Moorsel SAW, et al. 6-Methylmercaptopurine-induced leukocytopenia during thiopurine therapy in inflammatory bowel disease patients. J Gastroenterol Hepatol. 2017;32(6):1183–90 [c].
[49] Friedman S, Larsen MD, Magnussen B, et al. Paternal use of azathioprine/6-mercaptopurine or methotrexate within 3 months before conception and long-term health outcomes in the offspring-a nationwide cohort study. Reprod Toxicol. 2017;73:196–200 [c].
[50] Chowdhury NA, Sewatsky ML, Kim H. Transdermal scopolamine withdrawal syndrome case report in the pediatric cerebral palsy population. Am J Phys Med Rehabil. 2017;96:e151–4 [A].
[51] Rugarn O, Tipping D, Powers B, et al. Induction of labour with retrievable prostaglandin vaginal inserts: outcomes following retrieval due to an intrapartum adverse event. BJOG. 2017;124(5):796–803 [C].
[52] Nijman TA, Voogdt K, Teunissen PW, et al. Association between infection and fever in terminations of pregnancy using misoprostol: a retrospective cohort study. BMC Pregnancy Childbirth. 2017;17:7 [c].
[53] Sharma N, Das R, Ahanthem SS, et al. Misoprostol induced convulsion: a rare side effect of misoprostol. J Clin Diagn Res. 2017;11(2):QD01–2 [A].
[54] Aljazairi AS, Bhuiyan TA, Alwadai AH, et al. Octreotide use in post-cardiac surgery chylothorax: a 12-year perspective. Asian Cardiovasc Thorac Ann. 2017;25(1):6–12 [c].
[55] Cui PJ, Yao J, Zhu Y, et al. Effects of a long-acting formulation of octreotide on patients with portal hypertension. Gastroenterol Res Pract. 2017;2017:1–6 [c].
[56] Tse SS, Kish T. Octreotide-associated neutropenia. Pharmacotherapy. 2017;37(6):e32–7 [A].
[57] Narayanaswami V, Dwoskin LP. Obesity: current and potential pharmacotherapeutics and targets. Pharmacol Ther. 2017;170:116–47 [R].
[58] Welston K, Dianne May D. Chapter-32 Gastrointestinal drugs. In: Ray SD, editor. Side effects of drugs: annual: a worldwide yearly survey of new data in adverse drug reactions, vol. 39; p. 373–87 [R].

Further Reading

[59] Kataria PS, Kendre PP, Patel AA. Ifosfamide-induced encephalopathy precipitated by aprepitant: a rarely manifested side effect of drug interaction. J Pharmacol Pharmacother. 2017;8(1):38–40 [A].

CHAPTER

36

Drugs That Act on the Immune System: Cytokines and Monoclonal Antibodies

Kelley Ratermann[*,†,1], *Jessica Cox*[*,†], *Lydia Benitez*[*,†], *Frank Davis*[*]

*University of Kentucky HealthCare, Lexington, KY, United States

†University of Kentucky College of Pharmacy, Lexington, KY, United States

[1]Corresponding author: klrater200@uky.edu

CYTOKINES

Avatrombopag

Immunologic

Avatrombopag is a new, second-generation thrombopoietin receptor agonist. This study assessed three drug–drug interactions of avatrombopag with dual or selective CYP2C9/3A inhibitors and inducers to assess for safety data. A total of 48 healthy subjects received single 20mg doses of avatrombopag alone or with one of 3 CYP2C9/3A inhibitors or inducers: fluconazole 400mg once daily for 16 days, itraconazole 200mg twice daily on Day 1 and 200mg once daily on Days 2–16, or rifampicin 600mg once daily for 16 days. Itraconazole had a mild increase on both avatrombopag pharmacokinetics/dynamics compared to fluconazole, which led to clinically significant elevation in platelet counts. Conversely, administration of concomitant rifampicin caused a slight decrease in AUC but has no impact on maximum platelet count. Coadministration with interacting drugs was found to be generally safe and well-tolerated [1c].

Colony-Stimulating Factors [SEDA-35, 659; SEDA-36, 563; SEDA-37, 461]

Infectious

Although autologous stem cell transplantation can achieve excellent responses in patients with polyneuropathy, organomegaly, endocrinopathy, monoclonal protein, skin changes (POEMS) syndrome, the optimal regimen for peripheral blood stem cell (PBSC) collection is still controversial. This retrospective review investigated the safety and efficacy of 41 PBSC collecting procedures in 37 patients with POEMS syndrome. Mobilization was performed using cyclophosphamide+granulocyte colony-stimulating factor (G-CSF) (CG, $n=14$) or G-CSF alone (G, $n=27$). Of the included patients, 12 (85.7%) patients in the CG group and all (100%) patients in the G group received induction chemotherapy prior to collection of PBSCs. The proportions of good mobilizers were comparable between the two groups; however, two (14.3%) patients in the CG group developed severe capillary leak symptoms (CLS) during the PBSC mobilization period, whereas no patient in the G group experienced severe adverse events. Authors conclude that this rare occurrence of CLS may be due to the combination of induction strategies; however, the use of G-CSF warrants further investigation in this setting as these results contradict those previously reported by Li et al. in 2013 [2c].

INTERLEUKINS

CT-P13

Neurologic

CT-P13 is a biosimilar of Remicade®, an agent approved for use in inflammatory bowel disease. Controlled clinical trials previously demonstrated the efficacy and safety of CT-P13 in rheumatic diseases, but not in inflammatory bowel disease. This prospective, single-center, observational study assessed both the effectiveness and safety of CT-P13 in 80 Crohn's disease patients and 40 ulcerative colitis patients. Patients were either treatment-naive or switched to anti-TNF treatment from the reference infliximab (Remicade®) to the study drug.

Side Effects of Drugs Annual, Volume 40
ISSN: 0378-6080
https://doi.org/10.1016/bs.seda.2018.07.010

Adverse events occurred in 7.5% of patients (9/120) including one skin reaction, one abdominal pain, two headaches and two paresthesias during infusion treatment, one Sweet's Syndrome, and two polyarthralgia. Overall, only seven patients discontinued treatment because of adverse events. Authors concluded that CT-P13 was efficacious and well tolerated in patients with inflammatory bowel disease [3C].

Infusion Reactions

This study assessed the frequency and characteristics of infusion reactions during CT-P13 therapy across 14 European centers. A total of 384 patients were included of which 28 (9.6%) developed infusion reaction during treatment, with 64.3% having been previously exposed to anti-TNF therapy. Nearly 40% of patients developing infusion reaction continued CT-P13 therapy with subsequent use of premedication. Therapy was stopped in 17 patients who developed severe infusion reaction and switched to adalimumab in 12 patients. Notably, previous anti-TNF-alpha exposure (30% vs 3.1%, $P<0.001$, OR 6.3) and anti-drug antibody positivity (32.6% vs 4.1%, $P<0.001$, OR 19) during induction therapy were predictive factors for infusion reactions [4C].

GP2015

Infectious

GP2015 is a newly proposed etanercept (ETN) biosimilar and the objective of the EGALITY trial was to prove equivalent efficacy and comparable safety between these products in patients with moderate-to-severe chronic plaque-type psoriasis. Over 530 patients were randomized 1:1 to self-administer GP2015 or ETN twice weekly subcutaneously and then were re-randomized based on improvement in symptoms or not. All efficacy endpoints showed equivalence and the incidence of treatment-emergent adverse events up to week 52 was comparable between continued GP2015 (59.8%) and ETN (57.3%). The most commonly reported adverse effect was nasopharyngitis. Additionally, when treatments were switched the safety profiles remained comparable. Antidrug antibodies, all non-neutralizing, were limited to five patients on ETN during treatment period 1, and one patient in the switched ETN group. Therefore, authors conclude that study results provide the strong clinical confirmation of biosimilarity between these two products [5C].

Infliximab [SEDA-37, 464; SEDA-38, 397; SEDA-39, 474]

Immunologic

Immunosuppressive therapy for inflammatory bowel disease (IBD) in pediatric patients has been thought to increase the risk of malignancy and lymphoproliferative disorders, including hemophagocytic lymphohistiocytosis (HLH). A prospective study comparing unadjusted incidence rates of malignancy and HLH in pediatric patients with IBD exposed to infliximab with patients not exposed to biologics. Data from 5766 participants were analyzed, with 15 patients developing a malignancy and 5 patients developing HLH. Thirteen of the 15 patients who developed a malignancy and all 5 of the patients who developed HLH had been exposed to thiopurines while 10 patients with malignancy had also been exposed to a biologic agent. Unadjusted incidence rates showed no increased risk of malignancy (0.46/1000 patient-years) or HLH (0.0/1000 patient-years) in patients exposed to infliximab as the only biologic vs those unexposed to biologics (malignancy: 1.12/1000 patient-years; HLH: 0.56/1000 patient-years). Additionally, standardized incidence ratios did not demonstrate an increased risk of malignancy among patients exposed to infliximab (SIR, 1.69; 95% CI, 0.46–4.32) when compared to patients not exposed to a biologic agent (SIR, 2.17; 95% CI, 0.59–5.56), even when stratified by thiopurine exposure. Therefore, the authors concluded infliximab exposure was not associated with increased risk of malignancy or HLH in pediatric patients with IBD [6MC].

Infectious

The first case of disseminated nocardiosis associated with infliximab treatment in a patient with ulcerative colitis was reported. A 57-year-old African American female with a past medical history of ulcerative colitis and arthritis was treated with infliximab and prednisone. She presented to the emergency department with acute chest pain, shortness of air, and a 2-week history of productive cough. Examination showed hypoxia and tachypnea with decreased and coarse bilateral breath sounds and erythematous masses on her trunk and groin. The patient was initially treated for a presumed community-acquired pneumonia until blood cultures grew *Nocardia farcinica* and treatment was switched to trimethoprim–sulfamethoxazole [7A].

MONOCLONAL ANTIBODIES

Adalimumab [SEDA-37, 462; SEDA-38, 397]

Neurologic

A case of mania induced by adalimumab treatment for ankylosing spondylitis was reported in an adult with a prior history of dysthymia. The 25-year-old male was started on adalimumab and manic symptoms subsequently developed over 7–8 months. His condition did not improve with outpatient management and he was admitted to an inpatient psychiatric unit.

Adalimumab was discontinued, substituted with non-steroidal anti-inflammatory agents, and antipsychotic therapy with valproate and aripiprazole was initiated. Mood symptoms resolved within days and the patient was discharged home. Upon follow-up, the patient was stable and remained compliant to his psychotropic medications. He was ultimately started on a different immunomodulatory agent for his ankylosing spondylitis [8A].

Genetics

Differences in responses to adalimumab therapy relating to the male or female sex have been previously described within rheumatologic patient populations. A clinical cohort of consecutive patients with Crohn's disease starting adalimumab in a single tertiary center was formed between March 2006 and February 2011. There were 107 female and 81 male patients followed for a median of 6.0 years (range 0.3–9.2). Drug survival was higher in male than female patients (48.1% vs 30.8%, $P=0.016$) and side effects were reported more often by female patients (81.3% vs 64.2%, $P=0.008$). Female patients also discontinued adalimumab more often due to side effects (35.4% vs 18.4%, $P=0.017$). The authors concluded that female sex is negatively associated with adalimumab drug survival [9C].

Atezolizumab [SEDA-39, 473]

Endocrine

It has been previously established that inhibitors of programmed cell death-1 and programmed cell death ligand 1 (PD-1/PD-L1) can induce significant immune-related adverse events (irAEs) including endocrinopathies such as hypophysitis, thyroid dysfunction, and rarely autoimmune diabetes. A case of PD-L1 inhibitor-induced autoimmune diabetes was reported in a previously euglycemic 57-year-old male with urothelial cancer. The patient presented to clinic with dehydration after his fifth cycle of atezolizumab treatment. Blood tests demonstrated rapid onset hyperglycemia, ketosis, and a low C-peptide level (0.65 ng/mL) confirming the diagnosis of type 1 diabetes. He responded well to insulin therapy and was discharged with stable blood glucose levels. Due to the growing use of PD-1/PD-L1 inhibitors in cancer treatment, the authors called for increased awareness of this rare yet treatable irAE [10A].

Avelumab

Musculoskeletal

In a multi-center phase Ib expansion cohort, urothelial carcinoma patients having progressed after platinum-based chemotherapy and unselected for PD-L1 expression received avelumab 10 mg/kg intravenously every 2 weeks. Safety and tolerability were primarily evaluated in addition to response rate and survival. Forty-four patients were treated and followed for a median of 16.5 months. The most frequent treatment-related adverse events (any grade) were fatigue/asthenia (31.8%), infusion-related reaction (20.5%), and nausea (11.4%). Grades 3–4 treatment-related adverse events occurred in three patients (6.8%) and included asthenia, AST elevation, creatine phosphokinase elevation, and decreased appetite [11c].

Bevacizumab [SEDA-37, 465; SEDA-38, 398; SEDA-39, 465]

Ocular

Acute angle closure is a unique complication of intravitreal injection, which is a procedure commonly performed to treat retinal vein occlusion. A case of acute angle closure occurring immediately after administration of intravitreal bevacizumab to treat branch retinal vein occlusion was reported. A 65-year-old woman was referred to a retina clinic for the treatment of macular edema secondary to branch retinal vein occlusion. A gonioscopy exam demonstrated narrow angle of over 180 degrees in both eyes. The patient received two doses of bevacizumab, each injected into her right eye intravitreally, with adequate response. The macular edema recurred 5 months later and a third injection was performed. One day later, the patient went to the emergency department complaining of persistent ocular pain in her right eye. The right pupil had dilated to 6 mm diameter and was fixed. Slit lamp exam revealed diffuse corneal edema in her right eye with an intraocular pressure of 56 mmHg. The patient was given intravenous mannitol with disappearance of corneal edema and decreased intraocular pressure. Although acute angle closure is a rare complication of intravitreal bevacizumab injection, authors noted it can occur patients with risk factors such as hyperopic eye or narrow angle [12A].

Brodalumab

Infectious

A meta-analysis evaluated the safety and efficacy of brodalumab for moderate-to-severe psoriasis through review of six clinical trials ($n=4118$ patients) comparing brodalumab to placebo. The rate of overall adverse events was slightly higher in the brodalumab group (RR = 1.13, 95% CI 1.06–1.22); however, no individual adverse events were significantly higher in the treatment group upon individual trial review. Authors concluded that larger studies with longer follow-up periods are required to confirm the maintenance of these results [13M].

Cetuximab [SEDA-37, 466; SEDA-39, 469]

Hematologic

A 72-year-old male presented with pretibial edema, acne-like skin rash, and nephrotic syndrome after receiving his sixth dose of cetuximab for advanced squamous cell carcinoma of the head and neck. Renal biopsy revealed features of thrombotic microangiopathy, with the expansion of the subendothelial zone, reduplication of the glomerular basement, and swelling of the endothelial cells. His pretibial edema disappeared and proteinuria decreased following discontinuation of cetuximab. Authors concluded that monitoring following cetuximab treatment should include parameters for development of nephrotic syndrome [14A].

Denosumab

Metabolic

Limited data are available concerning the use of denosumab in hemodialysis; however, it is known this drug does not require renal dose adjustment. A retrospective chart review of 12 osteoporotic hemodialysis patients showed that those who received a single 60-mg subcutaneous dose of denosumab every 6 months had gradual improvement of bone metabolism over a 24-month observation period. Few cases of hypocalcemia were seen but were more significant after the first and second injection. Careful monitoring of serum calcium and timely therapy adjustment allowed appropriate management of serum calcium levels [15c].

Musculoskeletal

A retrospective study of denosumab in solid tumor patients with bone metastasis analyzed denosumab both use and its adverse effects. 104 patients were reviewed, of whom 86 (82.7%) received concomitant outpatient cancer treatment and 39 (38%) had previously received zoledronate. The median number of denosumab doses was 7.5 (range 1–29) with the main reasons for treatment discontinuation being disease progression (20.2%) and adverse effects (25%). Baseline albumin-corrected calcium levels were available for 48 patients (46.2%) and 70 (67.3%) were receiving calcium/vitamin D supplements. Baseline and follow-up monitoring of calcium levels was poor. Hypocalcemia was recorded in 38.5% of patients and osteonecrosis of the jaw in 12.5%, a higher incidence than previously reported in the literature. The authors noted that poor adherence to published recommendations on calcium supplementation and guidelines on calcium monitoring probably affected outcomes [16C].

Dinutuximab [SEDA-38, 399]

Neurologic

Three patients were described with clinical symptoms of transverse myelitis shortly following initiation of dinutuximab therapy. Diagnosis was confirmed via magnetic resonance imaging, shortly thereafter. Dinutuximab treatment ceased in all patients and urgent supportive care treatment ensued with rapid improvement in symptoms and functional recovery [17A].

Fenretinide

Immunologic

Fenretinide is a synthetic retinoid that induces apoptotic cell death in B-cell non-Hodgkin lymphoma (NHL) and acts in synergy with rituximab in preclinical models. A phase I/II study of fenretinide in combination with rituximab for treatment of both indolent B-cell NHL and mantle cell lymphoma. The phase I design de-escalated from fenretinide at 900 mg/m^2 PO BID for days 1–5 of a 7-day cycle and the phase II portion added weekly rituximab IV 375 mg/m^2 on weeks 5–9 then every 3 months. Treatment was continued until progression or intolerance. A total of 32 patients were treated (7 in phase I and 25 in phase II). No dose-limiting toxicities were observed. The most frequent grade 3 or higher treatment-related adverse events included rash ($n=3$) and neutropenia ($n=3$) [18c].

Golimumab [SEDA-37, 464]

Ocular

The first documented case of retrobulbar optic neuropathy associated with golimumab was reported. A 48-year-old man was admitted with a 3-week history of progressive visual loss in his left eye. Notably, this occurred soon following his second infusion of golimumab for ankylosing spondylitis. Enhancement of both optic nerves seen on magnetic resonance imaging and visual evoked potentials was consistent with demyelinating bilateral optic neuropathy. The patient's condition was not improved with cessation of therapy or corticosteroids. Demyelinating complications related to TNF alpha inhibitors have been previously described, though less information is available with golimumab given its more recent development [19A].

Ipilimumab [SEDA-38, 400; SEDA-39, 472]

Cardiac

Ipilimumab has been associated with serious late immune-mediated side effects, including acute pericarditis with tamponade at 12 weeks after the last dose and polyarthralgia rheumatica; however, reports of recurrent monoarthritis were lacking. Unique cases of a patient with late-onset autoimmune pleuropericarditis leading to cardiac tamponade at 24 weeks post-ipilimumab and recurrent late immune knee arthritis at 8 and 32 weeks, respectively, were reviewed. Authors theorize that this late-onset toxicity might also be expected with the PD-1 inhibitors currently in clinical use [20A].

Gastrointestinal

A retrospective review of all metastatic melanoma patients who were treated with ipilimumab between March 2011 and May 2014 was performed. A total of 114 patients received a standard regimen of intravenous ipilimumab 3mg/kg every 3 weeks for four doses or until therapy was stopped due to toxicity or disease progression. Sixteen patients (14%) developed grade 2 or worse diarrhea which equated to an increase of four or more daily bowel movements. All 16 patients received high-dose corticosteroids (1–2mg/kg prednisone daily or equivalent), of which 9 (56%) had ongoing diarrhea despite high-dose steroids. Steroid-refractory patients received one dose of intravenous infliximab 5mg/kg and all but one had prompt resolution of diarrhea. Fourteen patients underwent colonoscopy or sigmoidoscopy with endoscopic findings ranging from mild erythema to colonic ulcers. Seven (88%) of eight patients with ulcers developed steroid-refractory symptoms requiring infliximab. With a median follow-up of 264 days, no major adverse events associated with prednisone or infliximab were reported. Authors concluded the presence of colonic ulcers on endoscopy was associated with a steroid-refractory course [21C].

Immunologic

Clinical trials for immune therapies commonly exclude patients with preexisting autoimmune diseases limiting clinical experience in this patient population. A retrospective review was conducted of metastatic melanoma patients with preexisting autoimmune disorders or previous ipilimumab-related immune-related adverse events who were receiving treatment with programmed cell death 1 (PD-1) inhibitors. Evaluation of response, flare of preexisting autoimmunity and development of new, not preexisting immune-related adverse events were performed. Forty-one patients had either preexisting autoimmunity ($n=19$, group A) or ipilimumab-triggered adverse event ($n=22$, group B). At initiation of PD-1 inhibitor therapy, six patients in group A and two patients in group B required immunosuppressive therapy. In group A, a flare of preexisting autoimmune disorders was seen in 42% of patients and new immune-related adverse event in 16%. Conversely in group B, 4.5% of patients had a flare of ipilimumab-triggered adverse events and 23% had a new immune-related adverse event. All flares of preexisting autoimmune disorders or events were managed by immunosuppressive and/or symptomatic therapy and did not require termination of PD-1 inhibitor therapy. Tumor responses (32% in group A and 45% in group B) were unrelated to occurrence of autoimmunity [22c].

Immunologic

Currently no validated biomarkers that predict ipilimumab toxicity or efficacy exist though it is known that body composition is an important predictor of toxicity and outcome of other drugs. A study of 84 patients with metastatic melanoma who were treated with ipilimumab between 2009 and 2015 was performed. Body composition was assessed by computed tomography at baseline and after four cycles of ipilimumab. At baseline, 24% of patients were sarcopenic and 33% had low muscle attenuation. Both sarcopenia and low muscle attenuation were significantly associated with high-grade adverse events (OR=5.34, 95% CI: 1.15–24.88, $P=0.033$; OR=5.23, 95% CI: 1.41–19.30, $P=0.013$, respectively), and low muscle attenuation was associated with high-grade immune-related adverse events (OR=3.57, 95% CI: 1.09–11.77, $P=0.036$). Longitudinal analysis of 59 patients showed significant reductions in skeletal muscle area, total body fat-free mass, fat mass (all $P<0.001$), and muscle attenuation ($P=0.030$). Mean reduction of skeletal muscle area was 3.3% over 100 days. A loss of 7.5% skeletal muscle area over 100 days was shown to be a significant predictor of overall survival (HR 2.1, $P=0.046$) Authors concluded that patients with sarcopenia and low muscle attenuation are more likely to experience severe treatment-related toxicity to ipilimumab [23c].

Namilumab

Immunologic

Namilumab (AMG203) is an immunoglobulin G1 monoclonal antibody that binds with high affinity to the GM-CSF ligand. A phase Ib, randomized, double-blind study (PRIORA) was conducted to assess the safety and tolerability of namilumab in active, mild-to-moderate rheumatoid arthritis. Patients received three subcutaneous injections of namilumab 150mg or placebo (cohort 1) or 300mg or placebo (cohort 2) on days 1, 15, and 29 and were followed for 12 weeks. The incidence of treatment-emergent adverse events was similar across three groups (namilumab 150mg: 63%; namilumab 300mg: 57%; placebo: 56%). Treatment-emergent adverse events in greater than or equal to 10% of patients were nasopharyngitis (17%) and exacerbation/worsening of rheumatoid arthritis (13%). No anti-namilumab antibodies were detected and overall subcutaneous namilumab was generally well tolerated [24c].

Natalizumab [SEDA-37, 467]

Immunologic

Natalizumab is a monoclonal antibody highly effective in the treatment of relapsing remitting multiple sclerosis. Recent reports have outlined concerns regarding the risk of progressive multifocal leukoencephalopathy (PML), a brain infection caused by John Cunningham virus, particularly after 24 doses of natalizumab and in patients that have previously received immunosuppressive therapy. Previously, no strategy has been discussed to avoid PML risk and in parallel reduce clinical and radiological rebound activity. A review of clinical trials and case reports was performed that evaluated natalizumab continuation and discontinuation followed by either full therapeutic suspension or therapy switch to second-line alternatives. Discontinuation of all therapy for multiple sclerosis after natalizumab increases occurrence of multiple sclerosis activity. However, there are no established guidelines regarding natalizumab treatment after 24 doses and choice should be based on professional experience and patients' clinical features and preferences [25M].

Infectious

In some patients, discontinuation of natalizumab is necessary due to the risk of progressive multifocal leukoencephalopathy though severe clinical and radiological worsening known as a rebound effect have been described after drug cessation. A retrospective study of patients switched from natalizumab to Fingolimod after presenting with a rebound of symptoms following natalizumab discontinuation was performed. Four patients were positive for JC virus. The mean disease duration was 9.5 years (SD: 4.12) with a mean time of 3.1 years on natalizumab. Neurological deterioration began at a mean time of 3.5 months (SD: 2.08) with multifocal involvement including motor disturbances (75%), cognitive impairment (50%), and seizures (25%). All patients received 5 days of IV methylprednisolone, and all patients started Fingolimod within 3–4 months. After the rebound, three patients continued treatment with Fingolimod and one patient resumed natalizumab [26A].

Nivolumab [SEDA-38, 401; SEDA-39, 473]

Gastrointestinal

A 76-year-old Japanese woman with advanced NSCLC was initially treated with nivolumab. After the second dose, she developed a severe rash with mucous involvement. A diagnosis of Stevens–Johnson syndrome was identified and the patient was started on oral prednisolone 1 mg/kg (50 mg). The rash completely resolved after initiation of steroids, but the severe pruritis was unable to be managed with emollients, antihistamines, and steroids. The patient's medical team eventually chose to administer aprepitant, an oral neurokinin-1 receptor antagonist, for her refractory pruritus and symptoms improved within 5 days. Severe refractory pruritus can arise from immune checkpoint inhibitors and aprepitant may be a useful treatment for this toxicity [27A].

Immunologic

An analysis of nivolumab immunogenicity and its impact on pharmacokinetics, safety, and efficacy in patients with solid tumors enrolled in six clinical studies was performed. The incidence and prevalence of antidrug antibodies were determined by validated electrochemiluminescence assays in samples collected during and up to 100 days after last nivolumab treatment. Among 1086 nivolumab-treated patients included in the analysis, 12.7% were positive for antidrug antibodies. The presence of these antibodies was not associated with increased hypersensitivity, infusion reactions, or loss of efficacy and had minimal impact on nivolumab clearance. The authors concluded that the reported immunogenicity of nivolumab is not clinically meaningful [28R].

Immunologic

Few data are available regarding anti-PD-1 safety in patients who develop ipilimumab-related severe adverse events that are grade 3 or worse. A single institution-based observational study included patients treated with an anti-PD-1 (nivolumab or pembrolizumab) and previously treated with ipilimumab for stage III or IV unresectable melanoma. Patients enrolled were classified according to the occurrence of ipilimumab-related adverse events (IPI-related AEs): group A (no previous IPI-related AEs); group B (mild-to-moderate IPI-related AEs); group C (severe to life-threatening IPI-related AEs) and evaluated for safety. Groups A, B, and C included, respectively, 16, 13, and 10 patients and the incidence of severe anti-PD-1-related adverse events (grades 3–4) was 12%, 23%, and 10% in each corresponding group. The number of patients was too small for a meaningful statistical comparison and authors did not observe any difference in anti-PD-1 toxicity onset incidence [29c].

Immune Mediated Complications Post Allogeneic Stem Cell Transplant

Patients with relapsed/refractory Hodgkin lymphoma who have failed high-dose chemotherapy followed by autologous stem-cell transplant have a dismal prognosis [30r]. In the extended follow-up of the CheckMate 205 phase II trial in patients with relapsed/refractory Hodgkin lymphoma, treatment with nivolumab resulted in 69% (95% CI: 63–75) objective response rate (53% partial responses and 16% complete responses) [31C]. For patients who respond to nivolumab, allogeneic stem-cell transplantation (allo-HCT) allows for the potential of a long-term cure thus candidates for this intervention may be taken to allo-HCT post-nivolumab therapy. However, given nivolumab's propensity to cause immunologic adverse effects, its safety in the pre- and post allogeneic transplant setting have been investigated.

An international retrospective review of patients with lymphoma (79% with Hodgkin lymphoma) treated with PD-1i at some point prior to allo-HCT was conducted between 2003 and 2016 [32R]. The median time from last dose of PD-1i to transplant was 62 days (range 7–260). The cumulative incidences of grades 2–4, grades 3–4, and grade 4 acute graft vs host disease (GVHD) were 44%, 23%, and 13% respectively. The median time from allo-HCT to acute GVHD onset was 27 days. The 1-year incidence of chronic GVDH was 41%. Furthermore, there were three deaths attributed to acute GVHD within 14 days from allo-HCT. In addition, seven patients (18%) developed a prolonged period of fevers 1–7 weeks post allo-HCT, which included elevated transaminases in three patients and a rash in four patients. Receipt of a bone marrow graft source resulted in lower rates of acute GVHD compared to peripheral stem-cell source (0% vs 32%, P=0.036).

The effects of nivolumab therapy in patients with Hodgkin lymphoma who relapse after allo-HCT have also been retrospectively evaluated [33c]. In a retrospective cohort of 20 patients, six patients (30%) developed nivolumab induced GVHD within 1 week of the initial infusion. In three of these patients, the GVHD was refractory to high-dose steroids. All patients who developed nivolumab induced GVHD had a prior history of acute GVHD. Furthermore, time from allo-HCT to nivolumab infusion was shorter in those who developed GVHD (median 8.5 months) vs those who did not (median 28.5) (P=0.0082).

These reports have resulted in an FDA issuing a 'Warning and Precaution' statement for complications of allogeneic HSCT after nivolumab, including severe or hyperacute graft-vs-host disease, other immune-mediated ARs, and transplant-related mortality [34S]. As a result, the risks and benefits of proceeding to allo-HCT after receipt of nivolumab or other PD-1 checkpoint inhibitors should be carefully considered and discussed with potential candidates.

Neurologic

Nivolumab is an immune checkpoint inhibitor currently undergoing further study for the treatment of pediatric glioblastoma. A case of a 10-year-old girl with glioblastoma treated with nivolumab under compassionate-use guidelines was presented. After the first dose of nivolumab the patient developed hemiparesis, cerebral edema, and significant midline shift due to severe tumor necrosis. She was treated with intravenous dexamethasone and discharged on a steroid taper. The patient was re-challenged with a second dose of nivolumab and her condition rapidly deteriorated again, demonstrating hemiplegia, seizures, and eventually unresponsiveness with a fixed and dilated left pupil. Computed tomography of her brain revealed malignant cerebral edema requiring emergency decompressive hemicraniectomy. Repeat imaging demonstrated increased size of the lesion, reflecting immune-mediated inflammation and tumor necrosis. The patient remained densely hemiplegic, but became progressively more interactive and was ultimately extubated. Nivolumab was resumed several weeks later, but again her condition deteriorated with headache, vomiting, swelling at the craniectomy site, and limited right-sided facial movement following the sixth dose. MRI demonstrated severe midline shift and uncal herniation despite her craniectomy. Ultimately, the patient's condition declined dramatically and she died several days later. This represents the first case of malignant cerebral edema requiring operative intervention following nivolumab treatment for glioblastoma in a pediatric patient [35A].

Pulmonary

Despite published literature establishing that nivolumab is known to cause immune-related interstitial lung diseases, the detailed characteristics of said disease are not fully understood. A 68-year-old man treated with nivolumab for unresectable sinonasal melanoma. He achieved a complete response soon after the initiation of the therapy and this response was maintained for 30 weeks until the patient experienced dyspnea of subacute onset. CT images revealed patchy infiltrates, ground-glass opacifications, and bronchoalveolar lavage fluid, which contained elevated percentages of lymphocytes (53%) and neutrophils (30%). A transbronchial lung biopsy revealed intra-alveolar fibrin balls without hyaline membranes, consistent with the pattern of acute fibrinous and organizing pneumonia. This is the first report of nivolumab-induced acute fibrinous and organizing pneumonia and adds to the body of literature regarding nivolumab induced interstitial lung disease [36A].

Pembrolizumab [SEDA-38, 402; SEDA-39, 473]

Dermatologic

Three patients receiving pembrolizumab developed eruptive keratoacanthomas. This sudden onset of multiple lesions on sun-exposed extremities occurred after a median of 13 months (range 4–18 months) of pembrolizumab therapy. Biopsy of the lesion showed a lichenoid infiltrate in the underlying dermis, predominantly composed of CD3+ T cells, scattered CD20+ B cells, and relatively few PD-1+ T cells. This is an immunophenotypic pattern observed in other cases of anti-PD-1-induced lichenoid dermatitis. All patients reviewed were treated with topical clobetasol and intralesional triamcinolone, while only two patients underwent open superficial cryosurgery. Keratoacanthomas resolved and no new lesions occurred during close follow-up.

Pembrolizumab treatment was continued without disruption in each case and it should be noted that each patient had a complete response documented [37A].

Metabolic

Pembrolizumab is an approved first-line therapy for unresectable metastatic melanoma that is often continued beyond complete response despite the fact that immune-related toxicities gradually increase with continuation of therapy. A case highlighting the occurrence of serious induced immune-related adverse events attributed to pembrolizumab was reviewed. A patient with metastatic melanoma obtained a complete response after 6.5 months of therapy. The patient remains disease-free 5.5 months after discontinuation of pembrolizumab although mild treatment-related toxicity persisted. The authors concluded that cessation of pembrolizumab should be strongly considered in patients who achieve a complete response due to the ongoing risk of toxicity with prolonged administration [38A].

Neurologic

Two cases of ocular myasthenia gravis were reported to have occurred after treatment with pembrolizumab, an anti-programmed death 1 monoclonal antibody for advanced melanoma in responding patients. One case was an 81-year-old man and the second an 86-year-old woman, both with BRAF-negative metastatic melanoma receiving pembrolizumab with symptoms appearing 7 and 11 weeks after the initiation of therapy, respectively. Neither patient had other evidence of neurological cause for presentation and symptoms improved rapidly with administration of steroids. Both patients showed good oncological response to treatment and one patient successfully continued treatment without further complications [39A].

Ocular

A 54-year-old female was diagnosed with a large choroidal malignant melanoma at which time her affected eye was enucleated. Nearly a year and a half later, she was diagnosed with metastatic uveal melanoma to the liver. Her disease progressed through treatment with ipilimumab and therapy was changed to pembrolizumab. After four infusions, the patient presented to clinic with floaters and blurred vision. Work-up revealed a nongranulomatous panuveitis characterized by perivascular retinal pigment epithelium changes, retinal venous sheathing, vitreous cellular reaction and haze, and optic disk edema. A sustained release steroid implant was administered and the uveitis regressed. A relapse of symptoms occurred but quickly subsided with repeat injection. Authors concluded that patients should be educated and monitored for signs of uveitis with pembrolizumab [40A].

Pinatuzumab

Hematologic

Pinatuzumab vedotin (PV) is an anti-CD22 antibody conjugated with the potent anti-microtubule agent monomethyl auristatin conjugated via a protease-cleavable linker. A phase I study evaluated safety, tolerability, and recommended dosing scheme for further studies alone and in combination for relapsed or refractory non-Hodgkin lymphoma. Seventy-five patients received single-agent PV. Dose-limiting toxicities of grade 4 neutropenia lasting greater than 7 days in one of three patients and less than 7 days in two of three patients treated at 3.2mg/kg occurred. No dose limiting toxicities occurred at 2.4mg/kg, which was the recommended phase II dose. The most common grade 3 adverse event was neutropenia. Administration of concomitant rituximab did not affect safety, tolerability, or pharmacokinetics of PV [41c].

Rituximab [SEDA-37, 467; SEDA-38, 403]

Hematologic

The most common adverse effects seen with rituximab are infusion reactions due to cytokine release. The drug is otherwise well tolerated though there has been an increased incidence of cytopenias as rituximab use has increased. A 39-year-old patient with precursor B-cell acute lymphoblastic leukemia was started on rituximab infusion. The patient developed a cytokine-release syndrome with hemodynamic instability followed by rapid-onset cytopenias and disseminated intravascular coagulation abnormalities that were characterized by coagulopathy with fibrinolysis and mucocutaneous bleeding. The patient was treated with supportive care. Prior to this case, only four cases of coagulopathy occurring with rituximab had been reported [42A].

Immunologic

A randomized, open-label phase III study spanning 30 countries aimed to show the pharmacokinetic non-inferiority of subcutaneous rituximab to intravenous rituximab in follicular lymphoma. Study participants received rituximab 375mg/m^2 intravenously or 1400mg subcutaneously in addition to plus six to eight cycles of chemotherapy (cyclophosphamide, doxorubicin, vincristine, and prednisone [CHOP] or cyclophosphamide, vincristine, and prednisone [CVP]). Therapy was administered every 3 weeks during induction followed by rituximab maintenance every 8 weeks. A total of 410 patients were randomly assigned equally to intravenous and subcutaneous groups. Frequency of all grade and high-grade adverse reactions (ADRs) was similar in both groups. The most common high-grade adverse event was neutropenia, occurring in 21% of the intravenous group subjects and 26% of the subcutaneous group. Of note, administration-related reactions occurred in 35% of the intravenous group and 48% patients in the subcutaneous group (typically low-grade local injection-site reactions). There were no new safety concerns noted between subcutaneous and intravenous administration of rituximab and no compromised anti-lymphoma activity was noted [43C].

Infectious

Rituximab has been an increasingly used immunotherapeutic agent for women of reproductive age for treatment of autoimmune diseases, leukemias, and lymphomas. Rituximab can cross the placenta but literature regarding it effect on the developing fetus is limited, though there is little to suggest that exposure places an infant at increased risk of immunosuppression and subsequent infection. A case of in utero rituximab exposure is reported to be associated with two severe septic episodes with *Enterococcus faecalis* in a premature infant of 29 weeks' gestational age. The patient was noted to have a depressed B-lymphocyte subset of 10% and undetectable immunoglobulins at 37 weeks' postmenstrual age. Notably, both incidences coincided with transition from donor human milk to formula. The patient was treated with intravenous immunoglobulin, antibiotics, and donor human milk. Authors postulated there may be a connection between the placental transfer of rituximab, prematurity, and the low levels of protective maternal antibodies and the increased susceptibility of this patient to sepsis by *E. faecalis*, a resident of normal gut flora [44A].

Tocilizumab [SEDA-37, 468]

Hepatic

Phases III and IV clinical trial data were pooled to determine liver enzyme elevations and hepatic adverse events during long-term tocilizumab treatment for rheumatoid arthritis. Patients received intravenous tocilizumab (4, 8, or 10 mg/kg) with or without disease-modifying antirheumatic drugs [DMARDs]. More than 16 200 patient-years of tocilizumab exposure were evaluated for a total of 4171 patients. Alanine aminotransferase (ALT) and aspartate aminotransferase (AST) elevations greater than the upper limit of normal (ULN) occurred in 70.6% and 59.4% of patients, respectively, and most elevations occurred during the first year of treatment. ALT/AST elevations were >1–$3\times$ ULN in 59%/55% of patients, >3–$5\times$ ULN in 8.9%/3.3% of patients, and $>5\times$ ULN in 2.9%/0.9% of patients. Elevations $>3\times$ ULN returned to normal in 80% of patients with a median time of 5.6 weeks to normalization. Overall, seven serious hepatic adverse events occurred in the pooled studies. Approximately 2.5% of patients withdrew from tocilizumab treatment following liver enzyme elevations [45R].

Ustekinumab [SEDA-38, 403]

Skin

A retrospective, multicenter, observational study included patients who received ustekinumab for a minimum of 3 months from 2009 to 2015 for treatment of psoriasis or psoriatic arthritis. Of the 58 patients included, there was low discontinuation due to side effects or rheumatological lack of efficacy. Discontinuation of ustekinumab was correlated with more number of obese patients, and higher number of previous biological therapies, and less with presence of plaque psoriasis [46c].

Vedolizumab

Infectious

Safety data from six trials of vedolizumab were integrated and reported. Adverse events were evaluated in patients who received one or more doses of vedolizumab or placebo and were reported as exposure-adjusted incidence rates. Among the 2830 patients with vedolizumab exposure, no increased risk of any infection was seen, although serious Clostridial infections, sepsis, and tuberculosis were reported infrequently ($\leq 0.6\%$ of patients). Independent risk factors for serious infection in ulcerative colitis were prior failure of a tumor necrosis factor alpha antagonist and narcotic analgesic use. Risk factors for serious infection in Crohn's disease were younger age, corticosteroid, or narcotic analgesic use. It was also noted that 18 vedolizumab-exposed patients ($<1\%$) were diagnosed with a malignancy [47M].

YS110

Dermatologic

YS110 is a humanized IgG1 monoclonal antibody with high affinity for the CD26 antigen. A Phase I study of YS110 in patients with CD26-expressing tumors was performed to assess the safety, tolerance, and preliminary efficacy. Thirty-three patients received a median of three (range 1–30) YS110 infusions across six dose levels (0.1–6 mg kg). While the maximal tolerated dose was not reached, a dose-limiting toxicity of infusion hypersensitivity reactions led to the inclusion of systemic premedication. Low-grade asthenia (30.3%), hypersensitivity (27.3%), nausea (15.2%), flushing (15.2%), chills (12.1%), and pyrexia (12.1%) were reported as adverse reactions [48c].

Additional case studies pertaining to this topic can be found in reviews authored by Lindfelt [49R] and Reeves et al. [50R].

References

[1] Nomoto M, Zamora CA, Schuck E, et al. Pharmacokinetic/pharmacodynamic drug-drug interactions of avatrombopag when coadministered with dual or selective CYP2C9 and CYP3A interacting drugs. Br J Clin Pharmacol. 2018;84(5):952–60 [c].

[2] Muto T, Ohwada C, Takaishi K, et al. Safety and efficacy of granulocyte colony-stimulating factor monotherapy for peripheral blood stem cell collection in POEMS syndrome. Biol Blood Marrow Transplant. 2017;23(2):361–3 [c].

[3] Arguelles-Arias F, Guerra Veloz MF, Perea Amarillo R, et al. Effectiveness and safety of CT-P13 (biosimilar infliximab) in patients with inflammatory bowel disease in real life at 6 months. Dig Dis Sci. 2017;62(5):1305–12 [C].

[4] Balint A, Rutka M, Vegh Z, et al. Frequency and characteristics of infusion reactions during biosimilar infliximab treatment in inflammatory bowel diseases: results from central European nationwide cohort. Expert Opin Drug Saf. 2017;16(8):885–90 [C].

[5] CEM G, Thaci D, Gerdes S, et al. The EGALITY study: a confirmatory, randomized, double-blind study comparing the efficacy, safety and immunogenicity of GP2015, a proposed etanercept biosimilar, vs. the originator product in patients with moderate-to-severe chronic plaque-type psoriasis. Br J Dermatol. 2017;176(4):928–38 [C].

[6] Hyams JS, Dubinsky MC, Baldassano RN, et al. Infliximab is not associated with increased risk of malignancy or hemophagocytic lymphohistiocytosis in pediatric patients with inflammatory bowel disease. Gastroenterology. 2017;152(8) 1901-14.e3. [MC].

[7] Garner O, Ramirez-Berlioz A, Iardino A, et al. Disseminated nocardiosis associated with treatment with infliximab in a patient with ulcerative colitis. Am J Case Rep. 2017;18:1365–9 [A].

[8] Ghossoub E, Habli M, Uthman I, et al. Mania induced by adalimumab in a patient with ankylosing spondylitis. Int J Psychiatry Med. 2016;51(6):486–93 [A].

[9] Lie MR, Kreijne JE, van der Woude CJ. Sex is associated with adalimumab side effects and drug survival in patients with Crohn's disease. Inflamm Bowel Dis. 2017;23(1):75–81 [C].

[10] Hickmott L, De La Pena H, Turner H, et al. Anti-PD-L1 atezolizumab-induced autoimmune diabetes: a case report and review of the literature. Target Oncol. 2017;12(2):235–41 [A].

[11] Apolo AB, Infante JR, Balmanoukian A, et al. Avelumab, an anti-programmed death-ligand 1 antibody, in patients with refractory metastatic urothelial carcinoma: results from a multicenter, phase Ib study. J Clin Oncol. 2017;35(19):2117–24 [c].

[12] Jeong S, Sagong M, Chang W. Acute angle closure attack after an intravitreal bevacizumab injection for branch retinal vein occlusion: a case report. BMC Ophthalmol. 2017;17(1):25 [A].

[13] Attia A, Abushouk AI, Ahmed H, et al. Safety and efficacy of brodalumab for moderate-to-severe plaque psoriasis: a systematic review and meta-analysis. Clin Drug Investig. 2017;37(5):439–51 [M].

[14] Koizumi M, Takahashi M, Murata M, et al. Thrombotic microangiopathy associated with cetuximab, an epidermal growth factor receptor inhibitor. Clin Nephrol. 2017;87(1):51–4 [A].

[15] Festuccia F, Jafari MT, Moioli A, et al. Safety and efficacy of denosumab in osteoporotic hemodialysed patients. J Nephrol. 2017;30(2):271–9 [c].

[16] Manzaneque A, Chaguaceda C, Mensa M, et al. Use and safety of denosumab in cancer patients. Int J Clin Pharmacol. 2017;39(3):522–6 [C].

[17] Ding YY, Panzer J, Maris JM, et al. Transverse myelitis as an unexpected complication following treatment with dinutuximab in pediatric patients with high-risk neuroblastoma: a case series. Pediatr Blood Cancer. 2018;65(1):e26732 [A].

[18] Cowan AJ, Stevenson PA, Gooley TA, et al. Results of a phase I-II study of fenretinide and rituximab for patients with indolent B-cell lymphoma and mantle cell lymphoma. Br J Haematol. 2017;176(4):583–90 [c].

[19] de Frutos-Lezaun M, Bidaguren A, de la Riva P, et al. Bilateral retrobulbar optic neuropathy associated with golimumab. Clin Neuropharmacol. 2017;40(3):149–51 [A].

[20] Dasanu CA, Jen T, Skulski R. Late-onset pericardial tamponade, bilateral pleural effusions and recurrent immune monoarthritis induced by ipilimumab use for metastatic melanoma. J Oncol Pharm Pract. 2017;23(3):231–4 [A].

[21] Jain A, Lipson EJ, Sharfman WH, et al. Colonic ulcerations may predict steroid-refractory course in patients with ipilimumab-mediated enterocolitis. World J Gastroenterol. 2017;23(11):2023–8 [C].

[22] Gutzmer R, Koop A, Meier F, et al. Programmed cell death protein-1 (PD-1) inhibitor therapy in patients with advanced melanoma and preexisting autoimmunity or ipilimumab-triggered autoimmunity. Eur J Cancer. 2017;75:24–32 [c].

[23] Daly LE, Power DG, O'Reilly A, et al. The impact of body composition parameters on ipilimumab toxicity and survival in patients with metastatic melanoma. Br J Cancer. 2017;116(3):310–7 [c].

[24] Huizinga TW, Batalov A, Stoilov R, et al. Phase 1b randomized, double-blind study of namilumab, an anti-granulocyte macrophage colony-stimulating factor monoclonal antibody, in mild-to-moderate rheumatoid arthritis. Arthritis Res Ther. 2017;19(1):53 [c].

[25] Clerico M, Artusi CA, Liberto AD, et al. Natalizumab in multiple sclerosis: long-term management. Int J Mol Sci. 2017;18(5):940 [M].

[26] Gonzalez-Suarez I, Rodriguez de Antonio L, Orviz A, et al. Catastrophic outcome of patients with a rebound after natalizumab treatment discontinuation. Brain Behav. 2017;7(4)e00671 [A].

[27] Ito J, Fujimoto D, Nakamura A, et al. Aprepitant for refractory nivolumab-induced pruritus. Lung Cancer. 2017;109:58–61 [A].

[28] Agrawal S, Statkevich P, Bajaj G, et al. Evaluation of immunogenicity of nivolumab monotherapy and its clinical relevance in patients with metastatic solid tumors. J Clin Pharmacol. 2017;57(3):394–400 [R].

[29] Amode R, Baroudjian B, Kowal A, et al. Anti-programmed cell death protein 1 tolerance and efficacy after ipilimumab immunotherapy: observational study of 39 patients. Melanoma Res. 2017;27(2):110–5 [c].

[30] Arai S, Fanale M, DeVos S, et al. Defining a Hodgkin lymphoma population for novel therapeutics after relapse from autologous hematopoietic cell transplant. Leuk Lymphoma. 2013;54(11):2531–3 [r].

[31] Armand P, Engert A, Younes A, et al. Nivolumab for relapsed/refractory classic Hodgkin lymphoma after failure of autologous hematopoietic cell transplantation: extended follow-up of the multicohort single-arm phase II CheckMate 205 trial. J Clin Oncol. 2018;36(14):1428–39 [C].

[32] Merryman RW, Armand P, Wright KT, et al. Checkpoint blockade in Hodgkin and non-Hodgkin lymphoma. Blood Adv. 2017;1(26):2643–54 [R].

[33] Herbaux C, Gauthier J, Brice P, et al. Efficacy and tolerability of nivolumab after allogeneic transplantation for relapsed Hodgkin lymphoma. Blood. 2017;129(18):2471–8 [c].

[34] Opdivo. Bristol-Myers Squibb Company. Princeton, NJ. 2017 [S].

[35] Zhu X, McDowell MM, Newman WC, et al. Severe cerebral edema following nivolumab treatment for pediatric glioblastoma: case report. J Neurosurg Pediatr. 2017;19(2):249–53 [A].

[36] Ishiwata T, Ebata T, Iwasawa S, et al. Nivolumab-induced acute fibrinous and organizing pneumonia (AFOP). Intern Med. 2017;56(17):2311–5 [A].

[37] Freites-Martinez A, Kwong BY, Rieger KE, et al. Eruptive keratoacanthomas associated with pembrolizumab therapy. JAMA Dermatol. 2017;153(7):694–7 [A].

[38] Hsieh AH, Faithfull S, Brown MP. Risk of cumulative toxicity after complete melanoma response with pembrolizumab. BMJ Case Rep. 2017;2017 [A].
[39] Nguyen BH, Kuo J, Budiman A, et al. Two cases of clinical myasthenia gravis associated with pembrolizumab use in responding melanoma patients. Melanoma Res. 2017;27(2):152–4 [A].
[40] Aaberg MT, Aaberg Jr TM. Pembrolizumab administration associated with posterior uveitis. Retin Cases Brief Rep. 2017;11(4):348–51 [A].
[41] Advani RH, Lebovic D, Chen A, et al. Phase I study of the anti-CD22 antibody-drug conjugate pinatuzumab vedotin with/without rituximab in patients with relapsed/refractory B-cell non-Hodgkin lymphoma. Clin Cancer Res. 2017;23(5):1167–76 [c].
[42] Rafei H, Nassereddine S, Garcia IF. Disseminated intravascular coagulation-like reaction following rituximab infusion. BMJ Case Rep. 2017;2017 [A].
[43] Davies A, Merli F, Mihaljevic B, et al. Efficacy and safety of subcutaneous rituximab versus intravenous rituximab for first-line treatment of follicular lymphoma (SABRINA): a randomised, open-label, phase 3 trial. Lancet Haematol. 2017;4(6):e272–82 [C].
[44] Hay S, Burchett S, Odejide O, et al. Septic episodes in a premature infant after in utero exposure to rituximab. Pediatrics. 2017;140(3):e20162819 [A].
[45] Genovese MC, Kremer JM, van Vollenhoven RF, et al. Transaminase levels and hepatic events during tocilizumab treatment: pooled analysis of long-term clinical trial safety data in rheumatoid arthritis. Arthritis Rheumatol. 2017;69(9):1751–61 [R].
[46] Almirall M, Rodriguez J, Mateo L, et al. Treatment with ustekinumab in a Spanish cohort of patients with psoriasis and psoriatic arthritis in daily clinical practice. Clin Rheumatol. 2017;36(2):439–43 [c].
[47] Colombel JF, Sands BE, Rutgeerts P, et al. The safety of vedolizumab for ulcerative colitis and Crohn's disease. Gut. 2017;66(5):839–51 [M].
[48] Angevin E, Isambert N, Trillet-Lenoir V, et al. First-in-human phase 1 of YS110, a monoclonal antibody directed against CD26 in advanced CD26-expressing cancers. Br J Cancer. 2017;116(9):1126–34 [c].
[49] Lindfelt T. Drugs that act on the immune system: cytokines and monoclonal antibodies. In: Ray SD, editor. Side effects of drugs: annual. A worldwide yearly survey of new data in adverse drug reactions, vol. 38; 2016. p. 395–405. chapter 35. [R].
[50] Reeves D. Cytostatic agents: monoclonal antibodies utilized in the treatment of solid malignancies. In: Ray SD, editor. Side effects of drugs: annual. A worldwide yearly survey of new data in adverse drug reactions, vol. 39; 2016. p. 465–82. chapter 40. [R].

CHAPTER

37

Drugs That Act on the Immune System: Immunosuppressive and Immunostimulatory Drugs

Marley L. Watson[*,1], Bridgette K. Schroader[†], Heather D. Nelkin[‡]*

*Emory Healthcare, Winship Cancer Institute, Atlanta, GA, United States

[†]University of Kentucky Markey Cancer Center, Lexington, KY, United States

[‡]Novant Health Oncology Specialists, Winston-Salem, NC, United States

[1]Corresponding author: marley.watson@emoryhealthcare.org

IMMUNOSUPPRESSIVE DRUGS

Belatacept [SEDA-34, 609; SEDA-37, 471; SEDA-38, 407; SEDA-39, 389]

Belatacept prevents the activation of T-lymphocytes by binding to antigen presenting cells leading to immunosuppression. Belatacept has been shown to be a reasonable alternative to calcineurin inhibitors (CNI) in the renal transplant patient population based on comparable efficacy, similar adverse effects (AEs), lower risk of metabolic and cardiovascular side effects, and an improvement in renal function as well as quality of life [1R, 2R, 3R, 4MC, 5MC].

Comparative Studies

A prospective, phase II, randomized controlled trial evaluated the safety and efficacy outcomes 3 years after kidney transplant patients were either randomized to switch to belatacept-based immunosuppression ($n=84$) or continue CNI-based immunosuppression therapy ($n=89$) [4MC]. The frequency of AEs was similar; however, more patients receiving belatacept developed any-grade viral or fungal infections including influenza, herpes, cytomegalovirus viremia, onychomycosis and tinea versicolor. Kidney function was evaluated based on estimated glomerular filtration rate (eGFR) which significantly favored the belatacept-based immunosuppression arm ($P=0.01$).

A retrospective cohort study using national registry data compared allograft survival in patients receiving belatacept-based immunosuppression ($n=657$) vs tacrolimus-based immunosuppression ($n=3210$) [6MC]. The most common cause of death in both arms was cardiovascular disease and infection which were similar between the two cohorts. PTLD was also found to be comparable between the two arms.

Immunologic

A 28-year-old African American female received a mismatched living unrelated kidney transplant for end-stage renal disease secondary to lupus nephritis [7A]. After receiving antibody mediated rejection treatment, she developed de novo atypical hemolytic uremic syndrome (aHUS). The patient discontinued tacrolimus and received both eculizumab and belatacept for anti-CD80 costimulatory blockade. Five months later, the patient restarted tacrolimus, prednisone and mycophenolate mofetil. The patient maintained allograft function and stable serum creatinine for approximately 2.5 years post-transplant. This case report suggests the potential combination of eculizumab and belatacept to reverse acute aHUS exacerbation with the ultimate goal of salvaging the allograft.

Cyclosporine (Ciclosporin) [SEDA-34, 609; SEDA-37, 473; SEDA-38, 409; SEDA-39, 390]

Cyclosporine is a calcineurin inhibitor utilized in a myriad of disease states with the ability to cause

Side Effects of Drugs Annual, Volume 40
ISSN: 0378-6080
https://doi.org/10.1016/bs.seda.2018.07.006

dose-dependent and dose-independent AEs. Common cyclosporine-related toxicities include nephrotoxicity, hypertension, hypertriglyceridemia, lymphoma, infections, seizures, paresthesia, tremors, hypertrichosis, skin cancers and anaphylaxis. Dose-dependent toxicities may be reduced by monitoring blood levels [8R, 9R, 10R].

Systematic Reviews

A systematic literature search of safety, efficacy and pharmacokinetic data was utilized to determine the therapeutic index of cyclosporine, tacrolimus, and sirolimus [11M]. Common AEs for cyclosporine included hyperlipidemia, hypertension, cardiac events, gingival hyperplasia, and increased serum creatinine.

A meta-analysis evaluated immunosuppression treatment effectiveness and tolerance for idiopathic membranous nephropathy ($n=137$) [12M]. Infection was observed with all immunosuppressive agents while those receiving cyclosporine were at risk for hypertension (9%) and relapse after remission of proteinuria (21%). Tacrolimus and cyclosporine were the best tolerated immunosuppressive agents.

Comparative Studies

The long-term efficacy and safety of cyclosporine in steroid-refractory acute severe ulcerative colitis patients were evaluated utilizing registry data [13MC]. Patients received either cyclosporine ($n=377$), infliximab ($n=131$) or sequential rescue therapy ($n=63$). Cyclosporine and sequential rescue therapy had higher colectomy rates (24.1% and 32.7%, respectively) compared to infliximab (14.5%; $P=0.01$). Serious AEs were lower in the cyclosporine arm (15.4%) compared to the infliximab patients (26.5%) and sequential rescue therapy arm (33.4%; $P<0.001$). Infection was the most common AE in patients receiving cyclosporine (48.4%).

Observational Studies

The idea of minimizing CNI exposure leading to recovery of kidney graft function by changing therapy to everolimus was studied in 56 kidney transplant patients (tacrolimus, $n=34$; cyclosporine, $n=22$) [14c]. AEs included increases in total cholesterol for both tacrolimus and cyclosporine, but only a significant rise in triglyceride levels for cyclosporine patients.

An observational study of hemophagocytic lymphohistiocytosis treatment assessed the addition of cyclosporine to the initial treatment, etoposide and dexamethasone [15C]. AEs thought to be directly correlated to cyclosporine use included neurotoxicity, hypertension, and posterior reversible encephalopathy syndrome (PRES). Corticosteroids can also cause PRES, and therefore the authors concluded no evidence of an increase of cyclosporine-induced PRES.

Tumorigenicity

A cohort study reviewed renal transplant patients from the United States Renal Data System (USRDS) database to determine the number of patients diagnosed with melanoma after transplantation [16C]. Risk factors significant for developing melanoma appeared to be older age of recipient and donor, male sex of recipient and donor, white race, less than 4 HLA mismatches, living donors, sirolimus, tacrolimus, and cyclosporine therapy. The estimated median time to melanoma was 1.45 years (95% CI, 1.31–1.70 years).

Cardiovascular

This large multicenter series of liver transplant recipients analyzed the prevalence and evolution of cardiovascular morbidity and mortality post-transplant ($n=1819$) [17MC]. Patients treated with cyclosporine were found to have a higher risk of developing metabolic syndrome ($P<0.001$), arterial hypertension ($P<0.001$) and de novo dyslipidemia ($P<0.001$) compared to patients who received tacrolimus. Cyclosporine use was also correlated to a reduced risk of de novo diabetes and better glycemic control compared with tacrolimus.

Infection Risk

This retrospective study evaluated the incidence of tuberculosis (TB) and the timeline of development of TB post renal transplant in patients receiving either tacrolimus ($n=1082$) or cyclosporine ($n=582$) [18C]. Duration of dialysis, positive tuberculin skin test, use of induction, mycophenolate use, cytomegalovirus infection, and new onset diabetes after renal transplant were significantly less in the tacrolimus arm. The incidence of TB was also significantly less in the tacrolimus arm compared to the cyclosporine group (6.1% vs 19.9%, $P<0.001$).

Reproductive System

A multicenter, observational study evaluated differences in AEs based on gender and menopausal status in females with plaque psoriasis receiving cyclosporine [19C]. Postmenopausal women had significantly less grades 1–2 AEs and higher grades 3–4 AEs compared to fertile women.

Drug–Drug Interactions

A 58-year-old Chinese male on cyclosporine 200 mg daily 5 years after kidney transplant was initiated on ticagrelor for the treatment of acute coronary syndrome [20A]. Eight days later, the patient developed bleeding of his gums, bright red blood in his stool, and nausea and vomiting. His hemoglobin dropped from 9.9 to 4.6 g/dL. Ticagrelor was discontinued and cyclosporine was continued. He then had required agkistrodon snake venom hemocoagulase in addition to packed red blood cells and plasma. Ticagrelor was changed to clopidogrel with no additional issues. Ticagrelor is a substrate and weak inhibitor of P-glycoprotein (P-gp) while cyclosporine is a potent inhibitor of P-gp. The concomitant use of these agents likely led to increased exposure of ticagrelor.

A cross-sectional study was performed to evaluate the effect cyclosporine and everolimus have on the pharmacokinetic parameters of mycophenolic acid (MPA) [21c]. Low-dose cyclosporine (<180 mg/day) led to a reduction in MPA AUC exposure below the therapeutic window which could lead to ineffective treatment. High-dose cyclosporine (doses >240 mg/day) correlated with a greater concentration of MPA which could lead to increased toxicity.

Cyclosporine inhibits CYP3A4, organic anion-transporting polypeptides, and P-gp which are metabolic pathways utilized by many statins [22c]. A pharmacist-led intervention was established to convert simvastatin, pitavastatin, lovastatin and atorvastatin to rosuvastatin, pravastatin or fluvastatin with no significant changes in clinical efficacy.

When given concomitantly with cyclosporine, letermovir is administered at half the recommended dose. In pharmacokinetic studies, cyclosporine increased the concentration of letermovir [23C]. At the same time, letermovir was found to also increase cyclosporine concentrations leading to an increased need for monitoring cyclosporine levels. The mechanism by which cyclosporine increases letermovir concentration is thought to be organic anion-transporting polypeptide (OATP1B1/1B3) while letermovir increases cyclosporine concentrations by inhibition of CYP3A4.

Everolimus [SEDA-36, 592; SEDA-37, 474; SEDA-38, 410; SEDA-39, 391]

Everolimus is a mammalian target of rapamycin (mTOR) inhibitor that has the ability to block cell proliferation and angiogenesis. Common toxicities include stomatitis, mouth ulceration, infections, fatigue, skin toxicity, headache, non-infectious pneumonitis, hypercholesterolemia, hypertriglyceridemia, anemia, nausea, peripheral edema, diarrhea, and nasopharyngitis. Drug interactions are common with everolimus as it is a substrate of P-gp, CYP3A4, CYP3A5, and CYP2C8 [24R, 25R, 26R, 27c].

Systematic Review

A meta-analysis evaluated the discontinuation rate of everolimus ($n=3366$) compared to placebo ($n=2265$) and found the incidence of discontinuing everolimus for AEs was 12.3% [28M]. The rate of discontinuation varied greatly depending on tumor-type. The most common AEs leading to discontinuation were fatigue, pneumonitis, stomatitis, diarrhea, neuropathy, and hemorrhage.

Comparative Studies

In a phase III multicenter, randomized controlled trial, postmenopausal women with human epidermal growth factor receptor 2 positive (HER2+) advanced breast cancer were randomly assigned to receive everolimus ($n=198$) or placebo ($n=105$), plus trastuzumab and paclitaxel [29C]. The most common toxicities observed with everolimus in the Asian subset were stomatitis (62.2%), diarrhea (48%), rash (43.4%), alopecia (41.3%) and neutropenia (42.3%) while the most common AEs in the non-Asian subset included stomatitis (69.6%), diarrhea (62.7%), peripheral neuropathy (40.2%), alopecia (50.7%) and cough (41.3%).

A final overall survival analysis was performed for a phase II study comparing first-line everolimus followed by sunitinib ($n=238$) vs first-line sunitinib followed by everolimus ($n=233$) in metastatic renal cell carcinoma (RCC) [30C]. The most commonly reported AEs with first-line everolimus were stomatitis (53%), fatigue (47%), and diarrhea (40%). Second-line everolimus led to more fatigue (35%), stomatitis (32%), and anemia (32%).

Observational Studies

Long-term effects were evaluated in patients with renal angiomyolipoma correlated with tuberous sclerosis complex or sporadic lymphangioleiomyomatosis who received prolonged use of everolimus [31C]. The most common AEs associated with everolimus included stomatitis (42%), hypercholesterolemia (30.4%), acne (25.9%), aphthous stomatitis, and nasopharyngitis (each 21.4%).

A prospective, observational study assessed AEs associated with everolimus in metastatic RCC after receiving anti-vascular endothelial growth factor therapy ($n=274$) [32C]. Treatment-related stomatitis (40%) and noninfectious lung disease (15%) were reported. Mouthwash was the primary treatment option for stomatitis and corticosteroids for noninfectious lung disease.

Relapsed or refractory indolent lymphoma patients were treated with everolimus 10 mg daily in 55 patients [33c]. The most commonly reported AEs were hematologic including grades 3–4 anemia (15%), neutropenia (22%), and thrombocytopenia (33%).

Cardiovascular

A 50-year-old male post simultaneous liver and kidney transplantation was admitted with large pleural effusion, pericardial effusion, and ascites [34A]. After everolimus was discontinued, the patient's pleural effusion and cardiomegaly started to improve. At 1 month, there was no evidence of pericardial effusion and only a small right-sided pleural effusion. The patient continued tacrolimus, prednisolone, and low-dose mycophenolate without recurrence of pericardial effusion, pleural effusion, or peripheral edema.

Mouth and Teeth

A 62-year-old Moroccan male with metastatic clear-cell renal carcinoma developed disease progression to retina on first-line therapy [35A]. Second-line treatment with everolimus 10mg orally once daily was initiated. At 4 months, the patient had improved symptoms, complete recovery of visual function, and a significant decrease in size of retinal metastases. He developed grade 2 mucositis which improved after supportive treatment.

A multicenter, phase-2 study analyzed the use of 10 milliliters of alcohol free dexamethasone 0.5 milligrams per 5 milliliters oral solution in metastatic hormone-receptor positive, HER-2 negative breast cancer being treated with everolimus and exemestane [36MC]. Patients receiving oral dexamethasone had an incidence of stomatitis of 2% while a historical control group had an incidence of 33%. The most common grade 3 or 4 AE reported was hyperglycemia (8%) and dyspnea (3%). This study concluded the prophylactic use of dexamethasone oral solution significantly reduced the frequency and severity of stomatitis in this patient population.

Cardiovascular

In a prospective trial, 120 kidney transplant patients were evaluated before transplant and 1 year after transplant [37C]. All patients were initiated on tacrolimus, MPA, and prednisolone. Between 3 and 6 months, MPA was switched to everolimus in group one ($n=58$) while group two patients continued with MPA ($n=62$). When compared to group two, group one showed more improvement in all cardiac parameters: ejection fraction ($P=0.02$), left ventricular diastolic diameter ($P=0.03$), posterior wall thickness ($P=0.04$), and left ventricular hypertrophy ($P=0.04$). A multivariate analysis determined everolimus to be an independent factor in improving cardiovascular function.

Respiratory

In an observational study, 85 metastatic RCC patients receiving everolimus were evaluated to determine if the presence of pneumonitis while taking everolimus was associated with improved clinical outcomes [38C]. A combined multivariate analysis ($n=233$) showed longer overall survival ($P=0.03$) in patients with CT-verified pneumonitis.

Endocrine

A 46-year-old female with metastatic breast cancer receiving everolimus and anastrozole was on treatment for 6 weeks when she developed hyperglycemia and hypertriglyceridemia [39A]. She presented to the emergency department with severe epigastric abdominal pain and nausea. She was diagnosed with diabetic ketoacidosis and treated with intravenous insulin. Her serum lipase was also elevated leading to a diagnosis of acute pancreatitis which was thought to be everolimus-exacerbated.

Fertility

A pooled analysis assessed the incidence of fertility events reported in patients receiving everolimus [40R]. Of the 112 patients evaluated, 38.4% reported at least one menstrual irregularity. Amenorrhea (24.1%) and irregular menstruation (17%) were most common.

Skin

Wound healing AEs were evaluated in kidney transplant patients on tacrolimus and prednisone with either everolimus or mycophenolate sodium [41C]. Those receiving everolimus were found to have a higher incidence of both clinical and subclinical wound healing AEs.

Drug–Drug Interactions

The pharmacokinetics of octreotide when given concomitantly with everolimus in advanced neuroendocrine tumors resulted in an increase in the minimum concentration of plasma octreotide [42MC]. This had no significance on clinical efficacy yet could lead to increased risk of pulmonary and metabolic events.

Fingolimod [SEDA-34, 616; SEDA-35, 703; SEDA-37, 471; SEDA-38, 407; SEDA-39, 392]

Fingolimod is an oral sphingosine-1-phosphate receptor modulator that blocks lymphocytes from leaving lymph nodes leading to a reduction in lymphocytes in the central nervous system (CNS) decreasing central inflammation. Common toxicities include infection, hypertension, leukopenia, macular edema, decreased pulmonary function tests and cardiotoxicity [43R, 44R].

Systematic Review

A systematic review was conducted to assess efficacy, safety and tolerability of fingolimod for the treatment of relapsing-remitting multiple sclerosis (MS) [45M]. Lymphopenia is often viewed as a toxicity of fingolimod although it is an expected event based on the mechanism of action. In this analysis, no correlation was found between lymphocyte count and infection. Viral and bacterial infections reported were mild herpes virus infections, serious varicella-zoster virus, herpes simplex virus encephalitis and cryptococcal infections. Due to the effect of fingolimod on atrial myocytes and vascular smooth muscle cells, cardiac effects are increased including vasoconstriction leading to elevated blood pressure and bradycardia requiring cardiovascular monitoring for the first 6 hours post administration. Patients with a history of diabetes mellitus or uveitis have a higher risk of developing macular edema.

Observational Studies

A retrospective study in patients with relapsing-remitting MS ($n=249$) evaluated the effectiveness and safety of fingolimod [46C]. Side effects were reported in 48% of patients receiving fingolimod with the most frequent including lymphopenia (21.7%), gamma-glutamyltransferase elevation (6.8%), hypertransaminasemia (3.6%), cardiotoxicity (3.6%) and infections (3.6%).

A Danish study observed 496 MS patients for cardiac and pulmonary toxicities while receiving fingolimod [47C]. In a subset of patients ($n=204$), 3.9% required prolonged cardiac monitoring for bradycardia and/or second degree AV block type I and 5.4% developed respiratory toxicities.

A multicenter, observational study was conducted to evaluate the real-world outcomes including safety and efficacy [48MC]. Of the 240 patients included, 35% reported an AE. Reasons for discontinuation included macular edema ($n=3$), lymphopenia ($n=3$), LFT elevation ($n=2$), hepatopathy ($n=2$), pain ($n=2$) and gynecological infection ($n=4$).

Safety and tolerability with fingolimod use in relapsing-remitting MS were studied in an open-label, multicenter trial ($n=906$) [49MC]. AEs were reported in 35.4% of patients with the most common AEs being headache (4.1%), influenza (2.1%), lymphopenia (1.8%), asthenia (1.8%) and pyrexia (1.8%). Three patients developed macular edema.

Tumorigenicity

A 61-year-old male with relapsing-remitting MS receiving fingolimod developed a painful skin lesion on his arm which was confirmed by biopsy to be Merkel cell carcinoma (MCC) [50A]. This is the second case report of MCC associated with fingolimod use.

A 39-year-old female presented with a 1 month history of a red and irritated right eye. She had been on fingolimod for the past 2 years [51A]. On examination, she was diagnosed with conjunctival tumors found to be ocular adnexal MALT lymphoma. Fingolimod was discontinued and rituximab was utilized to treat both MS and lymphoma.

Cardiovascular

An open-label, multicenter study analyzed 3951 patients for the incidence of bradycardia and second/third degree AV blocks during fingolimod treatment initiation [52MC]. Thirty-one patients (0.8%) had bradycardia and 62 patients (1.6%) developed transient second degree AV blocks. This large study showed that while present, cardiac abnormalities associated with the initiation of fingolimod are infrequent and short-lived.

Sensory Systems

Outside of clinical trials other ocular toxicities besides macular edema have been reported including retinal hemorrhage and retinal vein occlusion [53R]. The most common symptom in patients with fingolimod-associated macular edema (FAME) is painless, blurred vision. In one study, 84% of patients had resolution of FAME after discontinuing fingolimod although some patients required treatment with topical prednisolone and ketorolac.

Drug Withdrawal

A case series including four patients experiencing severe rebound syndrome were observed among 28 patients who discontinued fingolimod [54c]. These patients experienced significant worsening and increased inflammation in the CNS with multiple contrast-enhancing lesions.

Glatiramer [SEDA-34, 617; SEDA-36, 593; SEDA-37, 471; SEDA-38, 407; SEDA-39, 393]

Systematic Review

Glatiramer is an immunomodulatory agent that is thought to hinder antigen-presenting function of specific immune cells blocking pathogenic T-cell function. Another proposed mechanism is the ability to trigger T-lymphocyte suppressor cells targeted for a myelin antigen and may lead to neuroprotective repair [55M].

Comparison Studies

An open-label extension of a placebo-controlled study evaluated the effects of early start and delayed start of glatiramer acetate 40 mg/mL three times weekly [56MC]. The most common toxicities were influenza (6%), nasopharyngitis (14.5%), upper respiratory tract infection (7.9%), urinary tract infection (7.7%), back pain (7.1%), headache (11.1%), and injection-site reactions (39.8%). Immediate post-injection reactions (IPIRs) were observed in 10.4% of patients with dyspnea (4.1%) and vasodilation (3.5%) being the most common.

Observational Studies

A database analysis of MS patients exposed to glatiramer ($n=4076$) was assessed for long-term safety and tolerability [57C]. The most common AEs included injection-site reactions (ISR) in 49% of patients with erythema at site of injection (29.2%) being the most common type of ISR. IPIRs were reported in 24% of patients with dyspnea (12.1%) being the most common. Other toxicities observed were rash (15%), headache (14.1%), infection (12.2%), dyspnea (12.1%), vasodilation (11.3%) and pain (9.9%).

Immunologic

While anaphylaxis has only been reported in less than 0.1% of glatiramer treated patients, IPIRs are commonly reported (10%–15%) [58c]. IPIRs typically appear immediately post injection, resolve within minutes, and present as flushing, dyspnea, chest tightness, palpitations, and anxiety. A placebo-controlled challenge test utilizing skin testing was conducted in 18 patients who had glatiramer discontinued due to a reaction. After the challenge, all patients were able to continue treatment. No IgE-mediated immediate hypersensitivity reactions were reported.

Leflunomide *[SED-15, 2015; SEDA-36, 594; SEDA-37, 471; SEDA-38, 407; SEDA-39, 393]*

Systematic Review

Six trials in a meta-analysis ($n=1984$) including four trials comparing leflunomide to methotrexate in the treatment of rheumatoid arthritis (RA) reported safety outcomes [59M]. Known effects of elevated liver enzymes, gastrointestinal complaints and non-severe infections were reported. In two trials, patients discontinued leflunomide due to diarrhea; otherwise, rates of drug withdrawal were similar.

Comparative Studies

Patients with IgA nephropathy were assigned to either leflunomide plus prednisone or prednisone. AEs in the combination group occurred in 12 of 40 patients and included liver, lung and gastrointestinal side effects [60c].

Patients with RA were assigned to rituximab or leflunomide. Of 19 patients on leflunomide, one reported diarrhea and two reported cardiac events, including one death [61c].

Observational Studies

A retrospective review of the Canadian database RAPPORT of patients on leflunomide revealed 29% (73/249) experienced AEs, most commonly being hair loss, nausea, stomach pain and diarrhea [62C].

A retrospective, single-center, cohort study was reported including 18 patients with dermatomyositis activity on adjuvant leflunomide [63c]. One patient reported lower limb paresthesia and one patient reported persistent diarrhea.

A case series of 12 lung transplant patients reported the use of leflunomide as treatment and secondary prophylaxis for cytomegalovirus infections [64c]. While 9 patients did not report any leflunomide induced AEs, one patient reported anxiety, one patient reported pancytopenia and one patient reported anemia, diarrhea and increased creatinine. Authors note patients should be informed of potential side effects.

Respiratory

A 67-year-old woman developed a cavitary pulmonary infection with *Mycobacterium avium intercellulare* after 2.5 years of treatment with leflunomide 20mg for her RA [65r]. The infection recurred after reintroduction of leflunomide 10mg daily. The authors note other pulmonary *Mycobacterium tuberculosis* cases with leflunomide were almost always in combination with other drugs, so it is difficult to determine if this is sporadic coincidence or causation.

Hematologic

A 60-year-old male presented to a hospital in China with mouth ulcers, thrombocytopenia, neutropenia, anemia, and a positive fecal occult blood test after 28 days of leflunomide 20mg daily plus methotrexate 10mg weekly as treatment for RA. He recovered over the next 28-day period [66A].

Mouth and Teeth

A 59-year-old woman on prednisolone 7.5mg daily plus leflunomide 20mg daily for the past 6 years for RA and inflammatory polyarthritis presented with a non-healing ulcer [67A]. Authors concluded the traumatic ulcer was unable to heal due to leflunomide.

Liver

An in vitro study was completed using leflunomide in HepG2, HepaRG and primary human hepatocytes [68E]. The authors found potential pathways for leflunomide toxicity including induction of endoplasmic reticulum stress and alteration of the MAPK pathway branches—JNK and ERK1/2. The authors note the role of CYPs in patients may contribute to individual susceptibility to AEs.

Nails

A 12-year-old girl on leflunomide for juvenile idiopathic arthritis presented a year into treatment with lichenoid drug eruption on her dorsal feet and ankles [69A]. Uniquely, it further spread to her nails 2 months after discontinuation of drug.

Pregnancy

A population based cohort study from the Quebec Pregnancy Cohort discussed the risk of leflunomide during pregnancy [70MC].

Mycophenolic Acid *[SED-15, 2402; SEDA-35, 704; SEDA-36, 594; SEDA-37, 478; SEDA-38, 413; SEDA-39, 395]*

Systematic Reviews

A meta-analysis of trials utilizing mycophenolate mofetil in patients with IgA nephropathy did not report any novel AEs [71M].

Comparative Studies

Mycophenolate mofetil was evaluated for use in neuromyelitis optica and neuromyelitis optica spectrum disorder, moderately-severe Henoch-Schonlein Purpura Nephritis, idiopathic pulmonary fibrosis and moderate-to-severe Graves' orbitopathy [72C, 73c, 74c, 75C, 76c]. No novel AEs were reported.

Respiratory

A 50-year-old male on mycophenolate mofetil as part of his kidney transplant immunosuppression regimen presented with fatal pulmonary fibrosis [77A]. Authors note to use caution in patients with underlying pulmonary disease.

Fluid Balance

A 45-year-old female developed intractable ascites after her kidney-pancreas transplant and exposure to mycophenolate sodium [78A]. Mycophenolate should be considered in the differential for unexplained ascites.

Hematologic

A chart review of kidney transplant patients at seven centers was completed and 68 were found to have mycophenolate-related leukopenia [79C]. Of these, 38 were included in a matched pair gene association study showing the UGT2B7-900A > G allele was associated with an increased risk of leukopenia in patients treated with non-depleting antibodies and two SNPs in the IMPDH1 gene was associated with delayed time to leukopenia.

Gastrointestinal

A 54-year-old male presented with deep ulcers in the ileum with non-specific pathologic findings [80A]. While diarrhea is well described, this is one of three case reports describing pathologic findings of mycophenolate mofetil induced ulcers.

A 47-year-old man was given mycophenolate mofetil 1000mg twice daily as part of his immunosuppression regimen for small intestine and kidney transplant [81A]. He developed mycophenolate-induced enterocolitis mistaken for acute cellular rejection. Authors suspected mycophenolate toxicity due to changes in the native intestine and the patient improved with drug discontinuation.

Skin

A 51-year-old woman developed therapy-resistant verrucae while on mycophenolate mofetil and tacrolimus post kidney transplant that resolved upon dose reduction [82A]. It is unclear if the verrucae cleared due to reduced mycophenolate dose or overall immunosuppression.

A 45-year-old woman developed a generalized fixed drug eruption while on mycophenolic acid 720mg twice daily as treatment for undifferentiated systemic rheumatic disease [83A]. This is an uncommon side effect for immunosuppressants.

Tumorigenicity

A 56-year-old female on mycophenolate mofetil for systemic lupus erythematosus (SLE) developed primary CNS lymphoma [84A]. SLE patients on immunosuppression have a higher risk of lymphoma but it is unclear if the drug caused the lymphoma as this is only the fifth reported case.

Pregnancy

A retrospective cohort study included 382 cases of kidney transplant patients who became pregnant while on mycophenolic acid [85C]. Authors conclude current FDA guidelines are appropriate.

Teratogenicity

A population based retrospective cohort assessed birth outcomes in 350 children fathered by 230 kidney transplant males on mycophenolate and found no increased risk of adverse birth outcomes [86C].

A review noted the unique presentation of mycophenolate mofetil embryopathy and the importance of a craniofacial clinical examination [87R].

The US National Transplantation Pregnancy Registry wrote a letter to the editor expressing their concern regarding exposure to mycophenolate during the first trimester and disagreement with increased risk of graft loss if stopped less than 6 weeks preconception [88r].

Sirolimus *[SEDA-15, 3148; SEDA-35, 705; SEDA-36, 596; SEDA-37, 479; SEDA-38, 415; SEDA-39, 396]*

Comparative Studies

Osteosarcoma patients ($n=35$) were enrolled on a multicenter, single-arm phase II trial of gemcitabine plus sirolimus and had predicted side effects of fatigue ($n=21$), anemia ($n=16$), neutropenia ($n=15$), thrombocytopenia ($n=12$) and mucositis ($n=11$) [89c].

Patients with tuberous sclerosis complex were randomized to topical sirolimus gel vs placebo in a double-blind, dose-escalation, phase II trial [90c]. No serious AEs occurred. Another study investigated topical sirolimus vs Vaseline in this population [91c]. Of the 12 children enrolled, 3 patients had redness and irritation.

Cardiovascular

A woman in her 70s post renal transplant with cutaneous squamous cell carcinoma presented with a venous stasis ulcer [92A]. Authors note sirolimus decreases the risk of squamous cell carcinoma but may impair wound healing and exacerbate ulcers due to venous stasis from antiangiogenesis.

Serum lipid and PCSK9 levels were measured in 51 heart transplant patients switched from CNIs to sirolimus [93c]. The authors noted an elevation in PCSK9 levels from 316 to 343 ng/mL, which did not correlate to an increase in lipid levels.

Respiratory

A 21-year-old Kuwaiti female renal transplant recipient on sirolimus plus mycophenolate mofetil and steroids developed combined PRES and lymphocytic pneumonitis that resolved after cessation of sirolimus [94A].

Metabolism

An in vitro study of sirolimus in pancreatic beta cells showed sirolimus causes depletion of intracellular Ca^{2+} stores as well as alters mitochondrial fitness, leading to decreased insulin release from beta cells [95E].

Fluid Balance

A 32-year-old female kidney transplant patient presented with lymphedema in her transplanted kidney and ipsilateral pleural effusion after switching to sirolimus 4 months prior [96A]. Authors postulate sirolimus impaired healing of lymphatic channels damaged during surgery and induced anti-angiogenesis.

Skin

A 62-year-old man on sirolimus plus prednisone and tacrolimus for immunosuppression following a heart transplant developed persistent pruritus after a year of therapy [97A]. This is the second case of pruritus in transplant recipients.

Kidney transplant patients receiving no immunosuppressive therapy, CNIs, or sirolimus provided blood and skin samples [98c]. Sirolimus patients had increased numbers of multiple T-cell subsets in sun exposed skin. Authors note this may help explain why patients on sirolimus develop fewer cutaneous squamous cell carcinomas.

Drug–Drug Interactions

A phase I study included 21 healthy subjects who received isavuconazole and sirolimus [99c]. The sirolimus area under the concentration-time curve increased by 84%, and the maximum concentration increased by 65%. Patients reported feeling hot ($n=5$) and somnolent ($n=4$).

Interference With Diagnostic Tests

Everolimus metabolite cross-reactivity was reported in the Siemens sirolimus immunoassay across the whole blood therapeutic ranges, resulting in falsely elevated sirolimus concentrations. Authors recommend testing with LC-MS/MS in patients receiving both drugs [100r].

Tacrolimus *[SEDA-15, 3279; SEDA-35, 705; SEDA-36, 596; SEDA-37, 480; SEDA-38, 415; SEDA-39, 396]*

Systematic Reviews

A meta-analysis of 13 trials of patients receiving tacrolimus for myasthenia gravis found previously reported side effects; the most common were increase in hemoglobin A1c and neutrophil counts [101M]. A second review reported known side effects in renal transplant patients receiving tacrolimus [102R].

Comparative Studies

Comparative studies of tacrolimus eye drops were reported for vernal keratoconjunctivitis (2) and corneal endothelial rejection (1) [103c, 104c, 105c]. Reported side effects were mild and included burning or stinging.

Patients with idiopathic membranous nephropathy received tacrolimus or cyclophosphamide and steroids [106c]. No novel side effects were reported.

Observational Studies

A retrospective, observational review followed 34 patients with moderate to severe ulcerative colitis treated with tacrolimus [107c]. Known AEs including neurotoxicity, increased creatinine, infections, hyperglycemia, alopecia and hypertension were reported.

Nervous System

Predictive factors for developing neurological complications in 90 pediatric living donor liver transplant patients were reviewed [108c]. Multivariate analysis showed total bilirubin in the first week after transplant was an independent predictor for neurologic complications.

A 15-year-old female on methylprednisolone 20 mg, tacrolimus 4 mg and mycophenolate mofetil 500 mg per day after a heart transplant developed seizures (authors note supratherapeutic tacrolimus level recorded as

16.2 ng/L) [109A]. Her neurologic status improved with discontinuation of tacrolimus but she developed multi-segmental vasoconstriction on day 3 that resolved with diltiazem on day 11.

Sensory Systems

A review of topical tacrolimus in anterior segment inflammatory disorders noted patients with higher body surface area per weight and with skin abnormalities may absorb enough tacrolimus ointment to cause systemic immunosuppression [110R].

Metabolism

ADVANCE was a phase 4, multicenter, prospectively randomized, open-label trial investigating posttransplant diabetes mellitus rates in 1166 renal transplant patients [111MC]. They concluded a prolonged-release tacrolimus based regimen with a short period of posttransplant steroids could be effective while minimizing the risk of post-transplant diabetes.

Liver transplant patients ($n = 719$) were randomized to everolimus plus reduced tacrolimus, tacrolimus or tacrolimus elimination to evaluate weight gain [112C]. Weight gain at 12 and 24 months was highest in the tacrolimus alone arm, although posttransplant metabolic syndrome rates were comparable.

A chart review of 155 Hispanic and 124 Caucasian kidney transplant recipients on tacrolimus based immunosuppression found no difference in incidence of post-transplant diabetes between Hispanic and Caucasian patients (14.2% (22) and 10.5% (13), respectively) [113C].

Hematologic

A 14-month-old infant post heart transplant presented 2 months after switching to tacrolimus-based immunosuppression with concurrent autoimmune hemolytic anemia and acquired pure red cell aplasia [114A]. Authors hypothesize tacrolimus induced immune dysregulation resulting in destruction of erythrocytes and erythroid precursors by a common lineage-specific antibody.

Gastrointestinal

A 74-year-old heart transplant male patient was receiving tacrolimus and mycophenolate mofetil when he presented with diffuse progressive ulcers and cutaneous pathergy [115A]. This is the first reported incidence.

Urinary Tract

A retrospective review of 186 lung transplant patients in the Netherlands reported increased whole blood tacrolimus levels as a risk factor for acute kidney injury, although authors note multiple factors may influence the tacrolimus levels as well as development of acute kidney injury [116C].

The genetic polymorphism FOXP3 rs3761548 AA and AC phenotypes increased the risk for tacrolimus induced acute nephrotoxicity 10-fold in a review of 114 Chinese renal transplant patients on tacrolimus based immunosuppression followed for 2 years [117C].

Skin

Vitiligo patients treated with topical tacrolimus reported only mild AEs, including three patient reports of irritation [118c].

A 58-year-old man received 0.1% tacrolimus ointment for post-irradiation morphea and developed acute radiation dermatitis [119r].

Infection Risk

A 17-year-old received 0.1% tacrolimus ointment for plaque morphea and developed a dermatophyte injection at the site of application [120A].

Genetic Factors

Brazilian kidney transplant patients on tacrolimus-based immunosuppression with the *PPC3CA* c.249G > A variant did not have different clinical outcomes [121C].

The *ABCC2* c.3972C > T polymorphism increased dose adjusted trough blood concentrations of tacrolimus in Brazilian kidney transplant patients receiving tacrolimus, mycophenolate and prednisone for 90 days after transplant [122C]. Patients with *CYP2J2* c.-76G > T single nucleotide polymorphisms had higher rates of nausea and vomiting.

CYP3A5 genotype affected the adverse event rate of 25 patients being treated with tacrolimus for autoimmune disease [123c]. Only patients who did not express the *1 allele reported adverse events. Patients with ulcerative colitis treated with tacrolimus were more likely to have nephrotoxicity and a higher incidence of AEs if they expressed *CYP3A5* A6986G [124c].

Drug Overdose

Red blood cells were infused into a 60-year-old woman post liver transplant with inadvertent tacrolimus overdose to aid in erythrocyte bound tacrolimus clearance [125A].

Drug–Drug Interactions

A 13-year-old liver transplant and cystic fibrosis (CF) male patient on tacrolimus 1.5 mg twice daily was started on nafcillin [126A]. After 16 doses of nafcillin 2000 mg, his tacrolimus concentration was undetectable. This was reproducible on a second admission. Of note, CF patients frequently have altered pharmacokinetics.

Temsirolimus [SEDA-36, 592; SEDA-37, 471; SEDA-38, 407; SEDA-39, 389]

Systematic Reviews

Bleeding risk of temsirolimus was described in a review of clinical trials in patients with metastatic RCC [127M]. Serious bleeding adverse events occur in about 2% of patients and included hematuria, epistaxis, rectal hemorrhage and non-site-specific hemorrhage.

A review of controlled trials described stomatitis associated with mTOR inhibitors as a unique oral inflammatory condition that is distinguishable from chemotherapy and radiation induced mucosal toxicity [128M]. The lesions resemble aphthous stomatitis or herpetic lesions and are typically found on non-keratinized mucosa.

Comparative Studies

An international, open-label, randomized, phase III trial compared ibrutinib with temsirolimus in patients with relapsed or refractory mantle cell lymphoma [129C, 130C]. The temsirolimus group ($n = 141$) reported higher rates of grade 3 or 4 thrombocytopenia (42%), anemia (20%) neutropenia (17%) and fatigue (7%). Twenty-six percent of patients in the temsirolimus arm discontinued treatment due to AEs.

Observational Studies

A multicenter, single-arm, phase II trial explored the safety of temsirolimus in 54 bladder cancer patients [131c]. The most common side effects, of any grade, reported included gastrointestinal and constitutional symptoms such as fatigue, fever, sweating, weight loss and insomnia (62.3%). Grade 3 or 4 toxicities included hematological toxicity (18.9%), gastrointestinal toxicity (11.3%) and constitutional symptoms (18.9%). Overall, a total of 50 patients experienced at least a grade 1 or 2 toxicity (94%) and 28 patients experienced a grade 3 or 4 adverse event (52.8%).

Another phase II, single institution study assessed the toxicity of temsirolimus 25 mg plus low-dose weekly carboplatin AUC 1.5 and paclitaxel 80 mg/m^2 in patients with recurrent and/or metastatic head and neck squamous cell carcinoma [132c]. Two-thirds of patients experienced a toxicity of any grade and the most commonly reported AEs included hyperglycemia (94%), anemia (94%), fatigue (89%), leukopenia (81%), hypercholesterolemia (78%), thrombocytopenia (75%) and hypoalbuminemia (75%). The most common grade 3 toxicities were lymphopenia (61%), leukopenia (31%), dysphagia (25%), neutropenia (22%) and anemia (19%). Neutropenic fever occurred in eight (3%) patients. Grade 4 toxicities occurred in one patient and included leukopenia, anemia, pneumonia, and hypercalcemia.

Safety of temsirolimus plus sorafenib was evaluated in a phase II trial in the treatment of thyroid cancer [133c]. Thirty-six patients were evaluated for toxicities. Overall, 23 patients (64%) experienced a minimum of one grade 3 or greater toxicity while on treatment. The most common grades 3 and 4 toxicities associated with both medications were hyperglycemia (19%), fatigue (14%), anemia (11%) and oral mucositis (8%). One of two sudden deaths that occurred while on study was categorized as possibly drug-related.

A phase 1 trial explored the safety of temsirolimus in combination with cetuximab in 35 patients with advanced solid tumors [134c]. Three patients were treated with cetuximab 150 mg/temsirolimus 15 mg and cetuximab 200 mg/temsirolimus 15 mg. No dose limiting toxicities were observed. One of seven patients treated with cetuximab 200 mg/temsirolimus 20 mg experienced a grade 3 reaction of thrombocytopenia. One of six patients treated with cetuximab 250 mg/temsirolimus 20 mg experienced a grade 3 mucositis. One of seven patients treated with cetuximab 250 mg/temsirolimus 25 mg experienced a grade 3 rash. In the expansion cohort of patients who received cetuximab 250 mg/temsirolimus 25 mg, 3 out of 10 patients experienced a grade 3 dose limiting toxicity of mucositis.

A retrospective study assessed the time to onset of diffuse interstitial lung disease (DILD) in patients who received small molecule molecularly-targeted drugs [135c]. Data was pulled from the Japanese Adverse Drug Event Report (JADER) from April 2004 to March 2017. The median onset for DILD for temsirolimus was reported to range from 1 to 2 months.

Respiratory

An 86-year-old male developed delayed exacerbation of interstitial lung disease (ILD) after temsirolimus was discontinued for treatment of recurrent RCC [136A]. The patient had a pre-existing diagnosis of ILD before initiation. Temsirolimus was discontinued after hospitalization due to dyspnea and symptoms resolved. Forty days after discontinuation the patient returned to hospital due to rapidly progressive dyspnea was treated with corticosteroids but the symptoms did not resolve, and the patient expired.

Mouth and Teeth

Incidence of mTOR inhibitor associated stomatitis at any grade ranges from 14% to 40% in patients treated with temsirolimus. Incidence of grade 3 or higher stomatitis is less than 7% [137c].

Thiopurines [SEDA-34, 633; SEDA-35, 709; SEDA-36, 598; SEDA-37, 483; SEDA-38, 417; SEDA-39, 399]

Systematic Reviews

A retrospective study looked at 105 patients, who received thiopurine treatment between 2005 and 2009 ($n=60$, cohort 1) and thiopurine plus or minus allopurinol between 2010 and 2014 ($n=45$, cohort 2). Overall, patients remained on therapy longer in cohort 2 (10.8 months vs 34.1 months, $P<0.05$) due to decrease in hepatotoxicity with addition of allopurinol [138M]. In cohort 1, 29 patients (48%) experienced one or more AEs compared with 14 (31%) of the patients in cohort 2 ($P=0.07$). The only difference in side effects experienced was hepatotoxicity occurring in 10 patients (17%) in cohort 1 and no patients in cohort 2. The common toxicities leading to discontinuation of treatment in both groups included pancreatitis (7% cohort 1, 4% cohort 2), myelotoxicity (1% and 4%) and subjective toxicities (27% and 29%) such as flu-like illness, gastrointestinal complaints, alopecia, arthralgia, headache, dizziness or tingling of arms/ft.

Comparative Studies

A multicenter, randomized, double-blind, placebo-controlled trial looked at MP compared to placebo to prevent or delay recurrence of Crohn's disease following surgical resection in 240 patients [139MC]. AEs occurred in 121 out of 128 patients who received MP. The majority were mild (n = 14, 11.6%) to moderate (n = 65, 53.7%). Severe AEs occurred in 41 patients (33.9%). The side effects reported in the MP group include gastrointestinal symptoms, rash, infections, arthralgia and subjective symptoms. Two cases of pancreatitis (0.1%, 1 in the MP group) and four malignancies (0.2%, 3 in the MP group) were reported. The malignancies included basal cell carcinoma, breast cancer, and two cases of lentigo maligna.

Observational Studies

A subset analysis of the first 272 patients enrolled in TOPIC, a randomized, multi-center trial in irritable bowel disease patients starting thiopurine therapy drew blood samples 1 week after thiopurine therapy was initiated, for metabolites 6-thioguanine nucleotide (6-TGN) and 6-methylmercaptopurine ribonucleotide (6MMPR) [140c]. Thirty-two patients with leukopenia and 162 without were analyzed. Leukopenia rates were higher in patients treated with mercaptopurine (MP), compared with azathioprine and in combination with anti-TNF therapy. Further analysis revealed that elevations in concentrations of both 6-TGN and 6-MMPR 1 week after initiation in 80% of patients resulted in the highest risk for leukopenia. Validation is needed before implemented into clinical practice.

Liver

A review of hepatotoxicity due to immunosuppressant drugs discussed the risk from thiopurines to be between 0% and 17%, depending on definition used [141M]. Fifty percent of thiopurine dose-independent liver toxicity (DILI) is seen within 3 months of starting azathioprine and mercaptopurine and is linked to non-IgE mediated hypersensitivity and idiosyncratic cholestatic reaction. Symptoms include hepatomegaly, fever, lymphadenopathy or rash, atypical lymphocytosis and eosinophilia which can be observed along with elevations in liver function tests.

Data from the Drug-Induced Liver Injury Network Prospective Study was analyzed for patients with hepatotoxicity due to thiopurines [142M]. Twenty-two patients were identified and about 75% had liver injury that was characterized as self-limiting, cholestatic hepatitis that started within 3 months of initiation or dose increase.

Skin

A 50-year-old male with sarcoidosis and antisynthetase syndrome developed multiple new moles on his upper and lower extremities and trunk 3 weeks after initiation of azathioprine 100 mg daily. Pathology confirmed lesions were eruptive melanocytic nevi. Patient continued treatment with azathioprine and had more frequent skin checks [143A].

A case series reported two patients with azathioprine hypersensitivity syndrome that presented as febrile neutrophilic dermatosis 2 weeks after initiating azathioprine [144A]. Both patients presented with leukocytosis, elevated neutrophils and increased C-reactive protein. Patients had red, swollen lesions located on thighs, buttocks and lower back. Histopathology revealed a dense infiltrate of neutrophils in the hair follicles. Resolution occurred after increasing steroids and discontinuing azathioprine.

Another case series presents three patients with azathioprine-induced Sweet's Syndrome [145A]. Patients presented with nausea, joint pain, fever, fatigue and headaches along with a painful rash. The rash was pustular with red, swollen plaques and nodules located on face, hands, trunk and lower extremities. Sweet's syndrome was confirmed by histopathology.

Teratogenicity

A review including 28 studies, 14 were observational studies in humans, showed thiopurines in doses 2–20 times that of the human dose affected germ cell lines in various rodents [146M]. In human studies of 83 men, a relationship between thioguanines and impaired testicular function and sperm quality was not evident. In pooled data from all studies, 53 out of 973 offspring had congenital anomalies (5.4%), possibly as a result of paternal thioguanine exposure. The risk of congenital anomalies was not significantly increased when compared with offspring without paternal thiopurine exposure (4.7%).

Hematologic

A retrospective study of 40 011 patients showed azathioprine increases risk for development of a myeloid neoplasm [147MC]. They found 86 patients with treatment induced cancer. Myelodysplastic syndrome (MDS) occurred in 55 patients (64%), de novo acute myeloid leukemia (AML) in 21 patients (24%) and AML with history of MDS in 10 patients (12%).

Genetic Factors

A retrospective single-center study of 149 Korean dermatologic patients, who received azathioprine and had thiopurine *S*-methyltransferase (TPMT) testing, was included [148M]. The frequency of TPMT mutation was 4% (6 patients). Results suggest TPMT polymorphisms in Korean patients are similar to those previously reported in Asians with the most common mutant allele being TPMT*3C.

A retrospective review showed NUDT15 p.R139C to be strongly associated with development of leukopenia and severe alopecia in Korean patients [149M].

Toxicities of mercaptopurine were described in a cohort of 1574 patients with irritable bowel disease [150MC]. Bone marrow suppression was the most common cause of treatment discontinuation (14 of 47 withdrew) and was more common in patients with intermediate TPMT activity.

IMMUNOENHANCING DRUGS

Levamisole [SED-15, 2028; SEDA-34, 638; SEDA-35, 710; SEDA-36, 460; SEDA-37, 484; SEDA-38, 420; SEDA-39, 399]

Observational Studies

A randomized, international, multicenter, double-blind trial was performed to assess the efficacy of levamisole vs placebo in the treatment of 100 patients with steroid-sensitive idiopathic nephrotic syndrome [151C]. More patients in the levamisole (58%, 29/50 patients) group had at least one adverse event. The most clinically significant toxicities were similar between groups and included fever, nasopharyngitis, cough, and mild neutropenia (ANC 1000–1500 cells/μL). Ten cases of moderate neutropenia (ANC 500–1000 cells/μL) occurred in four patients receiving levamisole. Three hospitalizations occurred in the levamisole group due to fever, pulmonary infection, and abdominal pain.

A review of AEs of levamisole in cocaine abusers list neutropenia/agranulocytosis (>200 case reports), leukoencephalopathy (>5 case reports) and vasculitis (>100 case reports) as the most commonly reported toxicities. Other complications include pulmonary hypertension, nephrotic failure, hyponatremia, acute coronary syndrome, Wegener's granulomatosis and pyoderma gangrenosum [152R].

Hematologic

A significant association with use of levamisole and development of agranulocytosis is seen in patients who carry the HLA B27 genotype. The authors state that it is uncertain if levamisole or metabolites are directly responsible for HLA-dependent agranulocytosis [152R].

A case series describes two cases of agranulocytosis caused by levamisole [153A]. A 43-year-old male with known cocaine use admitted to the intensive care unit with fever, tachycardia and severe right lower abdominal pain. Laboratory results showed leukopenia and neutropenia. Histopathology revealed bowel infarct along with necrotizing enterocolitis. Urinalysis revealed cocaine and levamisole. The authors concluded that levamisole explained the agranulocytosis. A 49-year-old female presented to transplant clinic for evaluation for an allogeneic stem cell transplant secondary to chronic neutropenia for the last 5 years. Patient reported frequent skin and upper respiratory tract infections. She had been treated with granulocyte-colony stimulating factor, steroids, cyclosporine, anti-thymocyte globulin, and alemtuzumab over the course of 5 years with no response. She had also tested positive for HLA-B27 genotype and c-ANCA. She admitted to cocaine use and diagnosis of levamisole-induced agranulocytosis was concluded.

Nervous System

A 63-year-old woman presented with 3-day history of fevers, headaches and progressive confusion [154A]. Her urine was positive for cocaine and levamisole. Patient progressed to stupor and spastic quadraparesis during hospital stay. "MRI of brain revealed innumerable FLAIR hyper intensities with restricted diffusion and incomplete ring-enhancement". Infections and autoimmune workup were negative. This is the first case to confirm levamisole was present in a patient with leukoencephalopathy associated with use of cocaine.

Cardiovascular

A 10-year-old female with purpuric rash and nodules on upper and lower extremities receiving levamisole 2.5 mg/kg every other day for the last 2.5 years had a renal biopsy that showed pauci immune focal necrotizing and crescentic glomerulonephritis [155r].

R

Special Review: Vasculopathy/Vasculitis Associated With Levamisole-Adulterated Cocaine

The use of levamisole-adulterated cocaine has been associated with vasculopathy often involving the skin. Other organs such as the kidneys, lung, liver and brain can also be affected. Levamisole is used as an adulterant due to its ability to enhance psychotropic properties by potentiating dopaminergic effects of cocaine, its availability due to use in veterinary medicine, and the inability to detect it as an adulterant due to similar physical properties to cocaine. Cocaine-levamisole-induced vasculopathy syndrome is a diagnosis of exclusion based on clinical findings with 91% of cases presenting with cutaneous plaques and/or blisters. The plaques are purplish and retiform with bright red, swollen edges and necrotic center. Painful hemorrhagic blisters can appear suddenly anywhere on the body but are often seen on ears, lower extremities, nose and cheek bones. The lesion can progress to necrotic ulcers and increase risk of infection. The skin manifestations may also be accompanied by general symptoms such as fever, arthralgia and fatigue. Hematologic manifestations present in 60% of patients with cutaneous disease and 90%–95% are anti-neutrophil cytoplasmic antibodies (ANCA)-positive. Other hematological findings may include agranulocytosis, leukopenia, and neutropenia [156R].

Multiple case reports and case series have demonstrated the occurrence of vasculitis-associated toxicities following exposure to levamisole-adulterated cocaine [155r, 157A, 158A, 159A, 160A, 161A].

A 37-year-old female presenting with skin lesions on upper and lower extremities and ears had a positive drug screen for cocaine. On labs, ANCA was positive but biopsy did not show vasculitis. The differential included levamisole-contaminated cocaine induced pyoderma gangrenosum vs *levamisole vasculitis. Based on serology, a diagnosis of vasculitis was concluded [157A].*

A 50-year-old man presented with symptoms of bilateral pulmonary emboli. Patient's history of cocaine abuse along with history of skin ulcers, venous thrombotic events, and febrile illnesses led to the diagnosis of levamisole-induced pseudo vasculitis [158A].

A 32-year-old male presented with painful ulcerated and necrotic lesions on lower extremities. The patient was afebrile with no systemic manifestations. Drug test was negative. A livedoid vasculitis was suspected, and patient was started on aspirin. Six months later, he presented with identical lesions and ANCA and anticardiolipin IgM antibodies were detected. He tested positive for cocaine at this visit. He was diagnosed with levamisole vasculopathy [159A].

A 58-year-old male presented with painful pruritic rash and polyarthralgias four days after cocaine use. Rash was on upper and lower extremities. Workup revealed elevated C-reactive protein and leukopenia along with negative ANA, c-ANCA, p-ANCA, HIV, HCV, CCP, anti-dsDNA, ESR, hepatic panel, and coagulation profile. Skin biopsy was used to make the final diagnosis of levamisole-induced vasculitis [160A].

A case series reports vasculitis in three cocaine positive female patients leading to levamisole-induced necrosis syndrome (LINES) [161A]. All three had lesions that developed into full thickness necrotic wounds reported on face, torso and extremities. One patient had skin biopsies positive for perinuclear-ANCA and ANA and pathology that showed perivascular lymphocytic infiltrate with thrombi in small vasculatures.

References

[1] Wong TC, Lo CM, Fung JY. Emerging drugs for prevention of T-cell mediated rejection in liver and kidney transplantation. Expert Opin Emerg Drugs. 2017;22(2):123–36 [R].

[2] Kälble F, Schaier M, Schäfer S, et al. An update on chemical pharmacotherapy options for the prevention of kidney transplant rejection with a focus on costimulation blockade. Expert Opin Pharmacother. 2017;18(8):799–807 [R].

[3] Kumar D, LeCorchick S, Gupta G. Belatacept as an alternative to calcineurin inhibitors in patients with solid organ transplants. Front Med (Lausanne). 2017;4:60 [R].

[4] Grinyó JM, Del Carmen Rial M, Alberu J, et al. Safety and efficacy outcomes 3 years after switching to belatacept from a calcineurin inhibitor in kidney transplant recipients: results from a phase 2 randomized trial. Am J Kidney Dis. 2017;69(5):587–94 [MC].

[5] Florman S, Becker T, Bresnahan B, et al. Efficacy and safety outcomes of extended criteria donor kidneys by subtype: subgroup analysis of BENEFIT-EXT at 7 years after transplant. Am J Transplant. 2017;17(1):180–90 [MC].

[6] Cohen JB, Eddinger KC, Forde KA, et al. Belatacept compared with tacrolimus for kidney transplantation: a propensity score matched cohort study. Transplantation. 2017;101(10): 2582–9 [MC].

[7] Dedhia P, Govil A, Mogilishetty G, et al. Eculizumab and belatacept for de novo atypical hemolytic uremic syndrome associated with CFHR3-CFHR1 deletion in a kidney transplant recipient: a case report. Transplant Proc. 2017;49(1):188–92 [A].

[8] Kinoshita Y, Saeki H. A review of the active treatments for toxic epidermal necrolysis. J Nippon Med Sch. 2017;84(3):110–7 [R].

[9] Kim WB, Jerome D, Yeung J. Diagnosis and management of psoriasis. Can Fam Physician. 2017;63(4):278–85 [R].

[10] Biancone L, Annese V, Ardizzone S, et al. Safety of treatments for inflammatory bowel disease: clinical practice guidelines of the Italian group for the study of inflammatory bowel disease (IG-IBD). Dig Liver Dis. 2017;49(4):338–58 [R].

[11] Ericson JE, Zimmerman KO, Gonzalez D, et al. A systematic literature review approach to estimate the therapeutic index of selected immunosuppressant drugs after renal transplantation. Ther Drug Monit. 2017;39(1):13–20 [M].

[12] Ren S, Wang Y, Xian L, et al. Comparative effectiveness and tolerance of immunosuppressive treatments for idiopathic membranous nephropathy: a network meta-analysis. PLoS One. 2017;12(9):e0184398 [M].

[13] Ordás I, Domènech E, Mañosa M, et al. Long-term efficacy and safety of cyclosporine in a cohort of steroid-refractory acute severe ulcerative colitis patients from the ENEIDA registry (1989-2013): a nationwide multicenter study. Am J Gastroenterol. 2017;112(11):1709–18 [MC].

[14] Nojima M, Yamada Y, Higuchi Y, et al. Immunosuppression modification by everolimus with minimization of calcineurin inhibitors recovers kidney graft function even in patients with very late conversion and also with poor graft function. Transplant Proc. 2017;49(1):41–4 [c].
[15] Bergsten E, Horne A, Aricó M, et al. Confirmed efficacy of etoposide and dexamethasone in HLH treatment: long-term results of the cooperative HLH-2004 study. Blood. 2017;130(25):2728–38 [C].
[16] Ascha M, Ascha MS, Tanenbaum J, et al. Risk factors for melanoma in renal transplant recipients. JAMA Dermatol. 2017;153(11):1130–6 [C].
[17] D'Avola D, Cuervas-Mons V, Martí J, et al. Cardiovascular morbidity and mortality after liver transplantation: the protective role of mycophenolate mofetil. Liver Transpl. 2017;23(4):498–509 [MC].
[18] Agarwal SK, Bhowmik D, Mahajan S, et al. Impact of type of calcineurin inhibitor on post-transplant tuberculosis: single-center study from India. Transpl Infect Dis. 2017;19(1) [C].
[19] Colombo D, Banfi G, Cassano N, et al. The gender attention observational study: gender and hormonal status differences in the incidence of adverse events during cyclosporine treatment in psoriatic patients. Adv Ther. 2017;34(6):1349–63 [C].
[20] Zhang C, Shen L, Cui M, et al. Ticagrelor-induced life-threatening bleeding via the cyclosporine-mediated drug interaction. Medicine (Baltimore). 2017;96(37):e8065 [A].
[21] Noreikaitė A, Saint-Marcoux F, Marquet P, et al. Influence of cyclosporine and everolimus on the main mycophenolate mofetil pharmacokinetic parameters: cross-sectional study. Medicine (Baltimore). 2017;96(13):e6469 [c].
[22] Lamprecht Jr. DG, Todd BA, Denham AM, et al. Clinical pharmacist patient-safety initiative to reduce against-label prescribing of statins with cyclosporine. Ann Pharmacother. 2017;51(2):140–5 [c].
[23] Marty FM, Ljungman P, Chemaly RF, et al. Letermovir prophylaxis for cytomegalovirus in hematopoietic-cell transplantation. N Engl J Med. 2017;377(25):2433–44 [C].
[24] González-Larriba JL, Maroto P, Durán I, et al. The role of mTOR inhibition as second-line therapy in metastatic renal carcinoma: clinical evidence and current challenges. Expert Rev Anticancer Ther. 2017;17(3):217–26 [R].
[25] Davies M, Saxena A, Kingswood JC. Management of everolimus-associated adverse events in patients with tuberous sclerosis complex: a practical guide. Orphanet J Rare Dis. 2017;12:35 [R].
[26] Gajate P, Martínez-Sáez O, Alonso-Gordoa T, et al. Emerging use of everolimus in the treatment of neuroendocrine tumors. Cancer Manag Res. 2017;9(R):215–24.
[27] Pascual T, Apellániz-Ruiz M, Pernaut C, et al. Polymorphisms associated with everolimus pharmacokinetics, toxicity and survival in metastatic breast cancer. PLoS One. 2017;12(7): e0180192 [c].
[28] Rogers SC, Garcia CA, Wu S. Discontinuation of everolimus due to related and unrelated adverse events in cancer patients: a meta-analysis. Cancer Invest. 2017;35(8):552–61 [M].
[29] Toi M, Shao Z, Hurvitz S, et al. Efficacy and safety of everolimus in combination with trastuzumab and paclitaxel in Asian patients with HER2+ advanced breast cancer in BOLERO-1. Breast Cancer Res. 2017;19(1):47 [C].
[30] Knox JJ, Barrios CH, Kim TM, et al. Final overall survival analysis for the phase II RECORD-3 study of first-line everolimus followed by sunitinib versus first-line sunitinib followed by everolimus in metastatic RCC. Ann Oncol. 2017;28(6):1339–45 [C].
[31] Bissler JJ, Kingswood JC, Radzikowska E, et al. Everolimus long-term use in patients with tuberous sclerosis complex: four-year update of the EXIST-2 study. PLoS One. 2017;12(8): e0180939 [C].
[32] Joly F, Eymard JC, Albiges L, et al. A prospective observational study on the evaluation of everolimus-related adverse events in metastatic renal cell carcinoma after first-line anti-vascular endothelial growth factor therapy: the AFINITE study in France. Support Care Cancer. 2017;25(7):2055–62 [C].
[33] Bennani NN, LaPlant BR, Ansell SM, et al. Efficacy of the oral mTORC1 inhibitor everolimus in relapsed or refractory indolent lymphoma. Am J Hematol. 2017;92(5):448–53 [c].
[34] Kim K, Jeong DW, Lee YH, et al. Everolimus-induced systemic serositis after simultaneous liver and kidney transplantation: a case report. Transplant Proc. 2017;49(1):181–4 [A].
[35] Essadi I, Lalya I, Kriet M, et al. Successful management of retinal metastasis from renal cancer with everolimus in a monophthalmic patient: a case report. J Med Case Reports. 2017;11(1):340 [A].
[36] Rugo HS, Seneviratne L, Beck JT, et al. Prevention of everolimus-related stomatitis in women with hormone receptor-positive, HER2-negative metastatic breast cancer using dexamethasone mouthwash (SWISH): a single-arm, phase 2 trial. Lancet Oncol. 2017;18(5):654–62 [MC].
[37] Cakir U, Alis G, Erturk T, et al. Role of everolimus on cardiac functions in kidney transplant recipients. Transplant Proc. 2017;49(3):497–500 [C].
[38] Penttilä P, Donskov F, Rautiola J, et al. Everolimus-induced pneumonitis associates with favourable outcome in patients with metastatic renal cell carcinoma. Eur J Cancer. 2017;81:9–16 [C].
[39] Acharya GK, Hita AG, Yeung SJ, et al. Diabetic ketoacidosis and acute pancreatitis: serious adverse effects of everolimus. Ann Emerg Med. 2017;69(5):666–7 [A].
[40] Sparagana S, Franz DN, Krueger DA, et al. Pooled analysis of menstrual irregularities from three major clinical studies evaluating everolimus for the treatment of tuberous sclerosis complex. PLoS One. 2017;12(10):e0186235 [R].
[41] Ueno P, Felipe C, Ferreira A, et al. Wound healing complications in kidney transplant recipients receiving everolimus. Transplantation. 2017;101(4):844–50 [C].
[42] Pavel ME, Becerra C, Grosch K, et al. Effect of everolimus on the pharmacokinetics of octreotide long-acting repeatable in patients with advanced neuroendocrine tumors: an analysis of the randomized phase III RADIANT-2 trial. Clin Pharmacol Ther. 2017;101(4):462–8 [MC].
[43] Fernández Ó. Is there a change of paradigm towards more effective treatment early in the course of apparent high-risk MS? Mult Scler Relat Disord. 2017;17:75–83 [R].
[44] Dyckman AJ. Modulators of sphingosine-1-phosphate pathway biology: recent advances of sphingosine-1-phosphate receptor 1 (S1P1) agonists and future perspectives. J Med Chem. 2017;60(13):5267–89 [R].
[45] Thomas K, Proschmann U, Ziemssen T. Fingolimod hydrochloride for the treatment of relapsing remitting multiple sclerosis. Expert Opin Pharmacother. 2017;18(15):1649–60 [M].
[46] Izquierdo G, Damas F, Páramo MD, et al. The real-world effectiveness and safety of fingolimod in relapsing-remitting multiple sclerosis patients: an observational study. PLoS One. 2017;12(4):e0176174 [C].
[47] Voldsgaard A, Koch-Henriksen N, Magyari M, et al. Early safety and efficacy of fingolimod treatment in Denmark. Acta Neurol Scand. 2017;135(1):129–33 [C].
[48] Tichá V, Kodým R, Počíková Z, et al. Real-world outcomes in fingolimod-treated patients with multiple sclerosis in the Czech Republic: results from the 12-month GOLEMS study. Clin Drug Investig. 2017;37(2):175–86 [MC].
[49] Laroni A, Brogi D, Brescia Morra V, et al. Safety and tolerability of fingolimod in patients with relapsing-remitting multiple sclerosis: results of an open-label clinical trial in Italy. Neurol Sci. 2017;38(1):53–9 [MC].

[50] Mahajan KR, Ko JS, Tetzlaff MT, et al. Merkel cell carcinoma with fingolimod treatment for multiple sclerosis: a case report. Mult Scler Relat Disord. 2017;17:12–4 [A].
[51] Christopher KL, Elner VM, Demirci H. Conjunctival lymphoma in a patient on fingolimod for relapsing-remitting multiple sclerosis. Ophthal Plast Reconstr Surg. 2017;33(3):e73–5 [A].
[52] Limmroth V, Ziemssen T, Lang M, et al. Electrocardiographic assessments and cardiac events after fingolimod first dose—a comprehensive monitoring study. BMC Neurol. 2017;17(1):11 [MC].
[53] Mandal P, Gupta A, Fusi-Rubiano W, et al. Fingolimod: therapeutic mechanisms and ocular adverse effects. Eye (Lond). 2017;31(2):232–40 [R].
[54] Gündüz T, Kürtüncü M, Eraksoy M. Severe rebound after withdrawal of fingolimod treatment in patients with multiple sclerosis. Mult Scler Relat Disord. 2017;11:1–3 [c].
[55] Filippini G, Del Giovane C, Clerico M, et al. Treatment with disease-modifying drugs for people with a first clinical attack suggestive of multiple sclerosis. Cochrane Database Syst Rev. 2017;4:CD012200 [M].
[56] Khan O, Rieckmann P, Boyko A, et al. Efficacy and safety of a three-times-weekly dosing regimen of glatiramer acetate in relapsing-remitting multiple sclerosis patients: 3-year results of the glatiramer acetate low-frequency administration open-label extension study. Mult Scler. 2017;23(6):818–29 [MC].
[57] Ziemssen T, Ashtamker N, Rubinchick S, et al. Long-term safety and tolerability of glatiramer acetate 20 mg/ml in the treatment of relapsing forms of multiple sclerosis. Expert Opin Drug Saf. 2017;16(2):247–55 [C].
[58] Amsler E, Autegarden JE, Gaouar H, et al. Management of immediate hypersensitivity reaction to glatiramer acetate. Eur J Dermatol. 2017;27(1):92–5 [c].
[59] Alfaro-Lara R, Espinosa-Ortega HF, Arce-Salinas CA, et al. Systematic review and meta-analysis of the efficacy and safety of leflunomide and methotrexate in the treatment of RA. Reumatol Clin. 2017; pii: S1699-258X(17)30196-1. [M].
[60] Min L, Wang Q, Cao L, et al. Comparison of combined leflunomide and low-dose corticosteroid therapy with full-dose corticosteroid monotherapy for progressive IgA nephropathy. Oncotarget. 2017;8(29):48375–84 [c].
[61] Wijesinghe H, Galappatthy P, de Silva R, et al. Leflunomide is equally efficacious and safe compared to low dose rituximab in refractory RA given in combination with methotrexate: results from a randomized double blind controlled trial. BMC Musculoskelet Disord. 2017;310:18 [c].
[62] Shultz M, Keeling SO, Katz SJ, et al. Clinical effectiveness and safety of leflunomide in inflammatory arthritis: a report from the RAPPORT database with supporting patient survey. Clin Rheumatol. 2017;36(7):1471–8 [C].
[63] de Souza RC, de Souza FHC, Miossi R, et al. Efficacy and safety of leflunomide as an adjuvant drug in refractory dermatomyositis with primarily cutaneous activity. Clin Exp Rheumatol. 2017;35:1011–3 [c].
[64] Silva JT, Perez-Gonzalez V, Lopez-Medrano F, et al. Experience with leflunomide as treatment and as secondary prophylaxis for cytomegalovirus infection in lung transplant recipients: a case series and review of the literature. Clin Transplant. 2017;32(2)e13176 [c].
[65] Winetsky DE, Myers J, Schulz S, et al. A case of pulmonary *Mycobacterium intracellulare* infection in the setting of leflunomide treatment. J Clin Rheumatol. 2017;23(4):231–2 [r].
[66] Qu C, Lu Y, Leu W. Severe bone marrow suppression accompanying pulmonary infection and hemorrhage of the digestive tract associated with leflunomide and low-dose methotrexate combination therapy. J Pharmacol Pharmacother. 2017;8(1):35–7 [A].
[67] Kalogirous EM, Katsoulas N, Tosios KI, et al. Non-healing tongue ulcer in a RA patient medicated with leflunomide. An adverse drug event? J Clin Exp Dent. 2017;9(2):e325–8 [A].
[68] Ren Z, Chen S, Qing T, et al. Endoplasmic reticulum stress and MAPK signaling pathway activation underlie leflunomide-induced toxicity in HepG2 cells. Toxicology. 2017;392:11–21 [E].
[69] May C, Fleckman P, Brankdlin-Bennett HA, et al. Lichenoid drug eruption with prominent nail changes due to leflunomide in a 12-year-old child. Pediatr Dermatol. 2017;34(4):e225–6 [A].
[70] Berard A, Zhaoi JP, Shui I, et al. Leflunomide use during pregnancy and the risk of adverse pregnancy outcomes. Ann Rheum Dis. 2018;77:500–9. pii: annrheumdis-2017-212078. [MC].
[71] Du B, Jia Y, Zhou W, et al. Efficacy and safety of mycophenolate mofetil in patients with IgA nephropathy: an update meta-analysis. BMC Nephrol. 2017;18:245 [M].
[72] Chen H, Qiu W, Zhang Q, et al. Comparisons of the efficacy and tolerability of mycophenolate mofetil and azathioprine as treatments for neuromyelitis optica and neuromyelitis optica spectrum disorder. Eur J Neurol. 2017;24:219–26 [C].
[73] Nambiar AM, Anzueto AR, Peters JI. Effectiveness and safety of mycophenolate mofetil in idiopathic pulmonary fibrosis. PLoS One. 2017;12(4):e0176312 [c].
[74] Lu Z, Song J, Mao J, et al. Evaluation of mycophenolate mofetil and low-dose steroid combined therapy in moderately severe Henoch-Schonlein purpura nephritis. Clin Res. 2017;23:2333–9 [c].
[75] Ye X, Bo X, Hu X, et al. Efficacy and safety of mycophenolate mofetil in patients with active moderate-to-severe graves' orbitopathy. Clin Endocrinol (Oxf). 2017;86:247–55 [C].
[76] Boulos D, Ngian GS, Rajadurai A, et al. Long-term efficacy and tolerability of mycophenolate mofetil therapy in diffuse scleroderma skin disease. Int J Rheum Dis. 2017;20:481–8 [c].
[77] Takahashi K, Go P, Stone CH, et al. Mycophenolate mofetil and pulmonary fibrosis after kidney transplantation: a case report. Am J Case Rep. 2017;18:399–404 [A].
[78] Weber NT, Sigaroudi A, Ritter A, et al. Intractable ascites associated with mycophenolate in a simultaneous kidney-pancreas transplant: a case report. BMC Nephrol. 2017;18:360 [A].
[79] Varnell CD, Fukuda T, Kirby CL, et al. Mycophenolate mofetil-related leukopenia in children and young adults following kidney transplantation: influence of genes and drugs. Pediatr Transplant. 2017;21:e13033 [C].
[80] Sonoda A, Wada K, Mizukami K, et al. Deep ulcers in the ileum associated with mycophenolate mofetil. Intern Med. 2017;56:2883–6 [A].
[81] Apostolov R, Asadi K, Lokan J, et al. Mycophenolate mofetil toxicity mimicking acute cellular rejection in a small intestinal transplant. World J Transplant. 2017;7(1):98–102 [A].
[82] Ash MM, Jolly PS. A case report of the resolution of multiple recalcitrant verrucae in a renal transplant recipient after a mycophenolate mofetil dose reduction. Transplant Proc. 2017;49:213–5 [A].
[83] Georgesen C, Lieber S, Lee H. A generalized fixed drug eruption associated with mycophenolate. JAAD Case Rep. 2017;3:98–9 [A].
[84] Balci MA, Pamuk GE, Unlu E, et al. Development of primary central nervous system lymphoma in a systemic lupus erythematous patient after treatment with mycophenolate mofetil and review of the literature. Lupus. 2017;26:1224–7 [A].
[85] King RW, Baca MJ, Armenti VT, et al. Pregnancy outcomes related to mycophenolate exposure in female kidney transplant recipients. Am J Transplant. 2017;17:151–60 [C].
[86] Midtvedt K, Bergan S, Reisaeter AV, et al. Exposure to mycophenolate and fatherhood. Transplantation. 2017;101(7): e214–7 [C].
[87] Perez-Aytes A, Marin-Reina P, Boso V, et al. Mycophenolate mofetil embryopathy: a newly recognized teratogenic syndrome. Eur J Med Genet. 2017;60:16–21 [R].

[88] Moritz MJ, Constantinescu S, Coscia LA, et al. Mycophenolate and pregnancy: teratology principles and national transplantation pregnancy registry experience. Am J Transplant. 2017;17:581–2 [r].
[89] Martin-Broto J, Redondo A, Valverde C, et al. Gemcitabine plus sirolimus for relapsed and progressing osteosarcoma patients after standard chemotherapy: a multicenter, single-arm, phase II trial of Spanish group for research on sarcoma (GEIS). Ann Oncol. 2017;28:2994–9 [c].
[90] Wataya-Kaneda M, Nakamura A, Tanaka M, et al. Efficacy and safety of topical sirolimus therapy for facial angiofibromas in the tuberous sclerosis complex. JAMA Dermatol. 2017;153(1):39–48 [c].
[91] Cinar SL, Kartal D, Bayram AK, et al. Topical sirolimus for the treatment of angiofibromas in tuberous sclerosis. J Dermatol Venereol Leprol. 2017;83:27–32 [c].
[92] Totonchy MB, Colegio OS, Christensen SR. Sirolimus-associated rapid progression of leg ulcers in a renal transplant recipient. JAMA Dermatol. 2017;153(1):105–6 [A].
[93] Simha V, Qin S, Shah P, et al. Sirolimus therapy is associated with elevation in circulating PCSK9 levels in cardiac transplant patients. J Cardiovasc Trans Res. 2017;10(1):9–15 [c].
[94] Gheith O, Cerna M, Halim MA, et al. Sirolimus-induced combined posterior reversible encephalopathy syndrome and lymphocytic pneumonitis in a renal transplant recipient: case report and review of the literature. Exp Clin Transplant. 2017;(Suppl 1):170–4 [A].
[95] Lombardi A, Gambardella J, Du XL, et al. Sirolimus induces depletion of intracellular calcium stores and mitochondrial dysfunction in pancreatic beta cells. Sci Rep. 2017;7:15823 [E].
[96] Rashi-Farokhi F, Afshar H. Lymphedema of the transplanted kidney and abdominal wall with ipsilateral pleural effusion following kidney biopsy in a patient treated with sirolimus: a case report and review of the literature. Am J Case Rep. 2017;18:1370–6 [A].
[97] Cheng JY, Cohen PR. Sirolimus-associated pruritus: case report and review. Cureus. 2017;9(6):e1398 [A].
[98] Burke TM, Sambira Nahum LC, Isbel NM, et al. Sirolimus increases T-cell abundance in the sun exposed skin of kidney transplant recipients. Transplant Direct. 2017;3:e171 [c].
[99] Groll AH, Desai A, Han D, et al. Pharmacokinetic assessment of drug-drug interactions of isavuconazole with the immunosuppressants cyclosporine, mycophenolic acid, prednisolone, sirolimus and tacrolimus in health adults. Clin Pharmacol Drug Dev. 2017;6(1):76–85 [c].
[100] Bowen R, Rieta R, Joshi R, et al. Falsely high sirolimus concentrations due to everolimus cross-reactivity in the Siemens sirolimus immunoassay: corrective actions implemented. Clin Chim Acta. 2017;S0009-8981(17):30421–7 [r].
[101] Zhang Z, Yang C, Zhang L, et al. Efficacy and safety of tacrolimus in myasthenia gravis: a systematic review and meta-analysis. Ann Indian Acad Neurol. 2017;20(4):341–7 [M].
[102] Shrestha BM. Two decades of tacrolimus in renal transplant: basic science and clinical evidences. Exp Clin Transplant. 2017;1:1–9 [R].
[103] Zanjani H, Aminifard MN, Ghafourian A, et al. Comparative evaluation of tacrolimus versus interferon alpha-2b eye drops in the treatment of vernal keratoconjunctivitis: a randomized, double-masked study. Cornea. 2017;36(6):675–8 [c].
[104] Ghaffari R, Ghassemi H, Zarei-Ghanavati M, et al. Tacrolimus eye drops as adjunct therapy in severe corneal endothelial rejection refractory to corticosteroids. Cornea. 2017;36:1195–9 [c].
[105] Muller EG, dos Santos MS, Freitas D, et al. Tacrolimus eye drops as monotherapy for vernal keratoconjunctivitis: a randomized controlled trial. Arq Bras Oftalmol. 2017;80(3):154–8 [c].
[106] Liang Q, Li H, Xie X, et al. The efficacy and safety of tacrolimus monotherapy in adult-onset nephrotic syndrome caused by idiopathic membranous nephropathy. Ren Fail. 2017;39(1):512–8 [c].
[107] Olmedo-Martin RV, Amo-Trillo V, Gonzalez-Grande R, et al. Medium to long-term efficacy and safety of oral tacrolimus in moderate to severe steroid refractory ulcerative colitis. Rev Esp Enferm Dig. 2017;109(8):559–65 [c].
[108] Sato K, Kobayashi Y, Nakamura A, et al. Early post-transplant hyperbilirubinemia is a possible predictive factor for developing neurological complications in pediatric living donor liver transplant patients receiving tacrolimus. Pediatr Transplant. 2017;21:e12843 [c].
[109] Kodama S, Mano T, Masuzawa A, et al. Tacrolimus-induced reversible cerebral vasoconstriction syndrome with delayed multi-segmental vasoconstriction. J Stroke Cerebrovasc Dis. 2017;26(5):e75–7 [A].
[110] Shoughby SS. Topical tacrolimus in anterior segment inflammatory disorders. Eye Vis. 2017;4:7 [R].
[111] Mourad G, Glyda M, Albano L, et al. Incidence of posttransplantation diabetes mellitus in de novo kidney transplant recipients receiving prolonged-release tacrolimus-based immunosuppression with 2 different corticosteroid minimization strategies: ADVANCE, a randomized controlled trial. Transplantation. 2017;101(8):1924–34 [MC].
[112] Charlton M, Rinella M, Patel D, et al. Everolimus is associated with less weight gain than tacrolimus 2 years after liver transplantation: results of a randomized multicenter study. Transplantation. 2017;101(12):2873–82 [C].
[113] Baron PW, Infante S, Peters R, et al. Post-transplant diabetes mellitus after kidney transplant in Hispanics and Caucasians treated with tacrolimus-based immunosuppression. Ann Transplant. 2017;22:309–14 [C].
[114] Abongwa C, Abusin G, El-Sheikh A. Successful treatment of tacrolimus-related pure red cell aplasia and autoimmune hemolytic anemia with rituximab in a pediatric cardiac transplant patient. Pediatr Blood Cancer. 2017;64:e26674 [A].
[115] Pan D, Shaye O, Kobashigawa JA. Tacrolimus-associated diffuse gastrointestinal ulcerations and pathergy: a case report. Transplant Proc. 2017;49(1):216–7 [A].
[116] Sikma MA, Hunault CC, van de Graaf EA, et al. High tacrolimus blood concentrations early after lung transplantation and the risk of kidney injury. Eur J Clin Pharmacol. 2017;73:573–80 [C].
[117] Wu Z, Xu Q, Qiu X, et al. FOXP3 rs3761548 polymorphism is associated with tacrolimus-induced acute nephrotoxicity in renal transplant patients. Eur J Clin Pharmacol. 2017;73: 39–47 [C].
[118] Rokni GR, Golpour M, Gorji AH, et al. Effectiveness and safety of topical tacrolimus in treatment of vitiligo. J Adv Pharm Technol Res. 2017;8(1):29–33 [c].
[119] Chu CH, Cheng YP, Liang CW, et al. Radiation recall dermatitis induced by topical tacrolimus for post-irradiation morphea. JEADV. 2017;31:e80–1 [r].
[120] Verma KK. Tacrolimus induced dermatophyte infection overlying a plaque morphea. Dermatol Ther. 2017;30:e12395 [A].
[121] Salgado PC, Genvigir FDV, Felipe CR, et al. Association of the *PPP3CA* c.249G > A variant with clinical outcomes of tacrolimus-based therapy in kidney transplant recipients. Pharmgenomics Pers Med. 2017;10:101–6 [C].
[122] Genvigir FDV, Nishikawa AM, Felipe CR, et al. Influence of *ABCC2, CYP2C8* and *CYP2J2* polymorphisms on tacrolimus and mycophenolate sodium—based treatment in Brazilian kidney transplant recipients. Pharmacotherapy. 2017;37(5):535–45 [C].
[123] Muraki Y, Mizuno S, Nakatani K, et al. Monitoring of peripheral blood cluster of differentiation 4^+ adenosine triphosphate activity and CYP3A5 genotype to determine the pharmacokinetics, clinical effects and complications of tacrolimus in patients with autoimmune diseases. Exp Ther Med. 2017;15:532–8 [c].

[124] Asada A, Bamba S, Morita Y, et al. The effect of *CYP3A5* genetic polymorphisms on adverse events in patients with ulcerative colitis treated with tacrolimus. Dig Liver Dis. 2017;49(1):24–8 [c].
[125] Spallanzani V, Bindi L, Bianco I, et al. Red blood cell exchange as an approach for treating a case of severe tacrolimus overexposure. Transfus Apher Sci. 2017;56(2):238–40 [A].
[126] Wungwattana M, Savic M. Tacrolimus interaction with nafcillin resulting in significant decreases in tacrolimus concentrations: a case report. Transpl Infect Dis. 2017;19:e12662 [A].
[127] Crist M, Hansen E, Chablani L, et al. Examining the bleeding incidences associated with targeted therapies in metastatic renal cell carcinoma. Crit Rev Oncol Hematol. 2017;120:151–62 [M].
[128] Chambers MS, Rugo H, Litton J, et al. Stomatitis associated with mammalian target of rapamycin inhibition. A review of pathogenesis, prevention, treatment and clinical implications for oral practice in metastatic breast cancer. J Am Dent Assoc. 2018; pii: S0002–8177(17)31019-X [M].
[129] Dreyling M, Jurczak W, Jerkeman M, et al. Ibrutinib versus temsirolimus in patients with relapsed or refractory mantle-cell lymphoma: an international, randomised, open-label, phase 3 study. Lancet. 2016;387:770–8 [C].
[130] Rule S, Jurczak W, Jerkeman M, et al. Ibrutinib vs temsirolimus: three-year follow-up of patients with previously treated mantle cell lymphoma from the phase 3, international, randomized, open-label RAY study. Hematol Oncol. 2017;101:143–4, Abstract 134 [C].
[131] Pulido M, Roubaud G, Cazeau A, et al. Safety and efficacy of temsirolimus as second line treatment for patients with recurrent bladder cancer. BMC Cancer. 2018;18(1):194–202 [c].
[132] Dunn LA, Fury MG, Xio H, et al. A phase II study of temsirolimus added to low-dose weekly carboplatin and paclitaxel for patients with recurrent and/or metastatic (R/M) head and neck squamous cell carcinoma (HNSCC). Ann Oncol. 2017;28(10):2533–8 [c].
[133] Sherman E, Dunn L, Ho AL, et al. Phase 2 study evaluating the combination of sorafenib and temsirolimus in the treatment of radioactive iodine-refractory thyroid cancer. Cancer. 2017;123(21):4114–21 [c].
[134] Hollebecque A, Baheleda R, Faivre L, et al. Phase 1 study of temsirolimus in combination with cetuximab in patients with advanced solid tumours. Eur J Cancer. 2017;81:81–9 [c].
[135] Komada F. Analysis of time-to-onset of interstitial lung disease after the administration of small molecule molecularly-targeted drugs. Yagkugaku Zasshi. 2018;138(2):229–35 [c].
[136] Matsuki R, Okuda K, Mitani A, et al. A case of delayed exacerbation of interstitial lung disease after discontinuation of temsirolimus. Respir Med Case Rep. 2017;22:158–63 [A].
[137] Vigarios E, Epstein J, Sibaud V. Oral mucosal changes induced by anticancer targeted therapies and immune checkpoint inhibitors. Support Care Cancer. 2017;25(5):1713–39 [c].
[138] Meijer B, Seinen M, Egmond R, et al. Optimizing thiopurine therapy in inflammatory bowel disease among 2 real-life intercept cohorts effects of allopurinol. Inflamm Bowel Dis. 2017;23:2011–7 [M].
[139] Satsangi J, Kennedy N, Mowat C, et al. A randomized, double-blind, parallel-group trial to assess mercaptopurine versus placebo to prevent or delay recurrence of Crohn's disease following surgical resection (TOPPIC). Efficacy Mech Eval. 2017;4(4):5–59 [MC].
[140] Wong D, Coenen M, Vermeulen S, et al. Early assessment of thiopurine metabolites identifies patients at risk of thiopurine-induced leukopenia in inflammatory bowel disease. J Crohn's Colitis. 2017;11(2):175–84 [c].
[141] Tran-Minh M, Sousa P, Maillet M, et al. Hepatic complications induced by immunosupressants and biologics in inflammatory bowel disease. World J Hepatol. 2017;9(15):613–26 [M].
[142] Bjornsson E, Gu J, Kleiner D, et al. Azathioprine and 6-mercaptopurine induced liver injury: clinical features and outcomes. J Clin Gastroenterol. 2017;51(1):63–9 [M].
[143] Seinweg SA, Halvorson CR, Kao GF, et al. Eruptive melanocytic nevi during azathioprine therapy for antisynthetase syndrome. Cutis. 2017;99:268–70 [A].
[144] Aleissa M, Nicol P, Godeau M, et al. Azathioprine hypersensitivity syndrome: two cases of febrile neutrophilic dermatosis induced by azathioprine. Case Rep Dermatol. 2017;9:6–11 [A].
[145] McNally A, Ibbetson J, Sidhu S. Azathioprine-induced Sweet's syndrome: a case series and review of literature. Australas J Dermatol. 2017;58(1):55–7 [A].
[146] Simsek M, Lambalk CM, Wilschut JA, et al. The associations of thiopurines with male fertility and paternally exposed offspring: a systematic review and meta-analysis. Hum Reprod Update. 2017;28:1–15 [M].
[147] Ertz-Archambault N, Kosiorek H, Taylor G, et al. Association of therapy for autoimmune disease with myelodysplastic syndromes and acute myeloid leukemia. JAMA Oncol. 2017;3(7):936–43 [MC].
[148] Lee M, Seo J, Bang D, et al. Thiopurine S-methyltransferase polymorphisms in Korean dermatologic patients. Ann Dermatol. 2017;29(5):529–35 [M].
[149] Kim S, Shin J, Park J, et al. NUDT15 p.R139C variant is common and strongly associated with azathioprine-induced early leukopenia and severe alopecia in Korean patients with various neurological diseases. J Neurol Sci. 2017;378:64–8 [M].
[150] Bermerjo F, Algaba A, Lopez-Duran S, et al. Mercaptopurine and inflammatory bowel disease: the other thiopurine. Rev Esp Enferm Dig. 2017;109(1):10–6 [MC].
[151] Gruppen M, Bouts A, Jansen-van der Weide M, et al. A randomized clinical trial indicates that levamisole increases the time to relapse in children with steroid-sensitive idiopathic nephrotic syndrome. Kidney Int. 2018;93(2):518–9 [C].
[152] Brunt TM, van den Berg J, Pennings E, et al. Adverse effects of levamisole in cocaine users: a review and risk assessment. Arch Toxicol. 2017;91(6):2303–13 [R].
[153] Srivastava R, Rizwan M, Jamil MO, et al. Agranulocytosis—sequelae of chronic cocaine use: case series and literature review. Cureus. 2017;9(5):e1221 [A].
[154] Vitt JR, Brown EG, Chow DS, et al. Confirmed case of levamisole-associated multifocal inflammatory leukoencephalopathy in a cocaine user. J Neuroimmunol. 2017;305(4):128–30 [A].
[155] Ranawake R, Gamage M, Lokuge K, et al. Levamisole induced pauci immune focal necrotizing and crescentic glomerulonephritis. Indian J Pediatr. 2018;85(7):599 [r].
[156] Marquez J, Aguirre L, Munoz C, et al. Cocaine-levamisole-induced vasculitis/vasculopathy syndrome. Curr Rheumatol Rep. 2017;19(6):36 [R].
[157] Ghias A, Brine G. Vanishing vasculitis: a case of acute necrotic skin findings without pathologic features of vasculitis from adulterated cocaine. J Community Hosp Intern Med Perspect. 2017;7(5):321–4 [A].
[158] Fan T, Macareag J, Haddad T, et al. A case report on suspected levamisole-induced pseudovasculitis. WMJ. 2017;116(1):37–9 [A].
[159] Fernandez Armenteros JM, Veà Jódar A, Matas Nadal C, et al. Severe and recurrent levamisole-induced cutaneous vasculopathy. J Cutan Pathol. 2018;45(4):309–11 [A].
[160] Salehi M, Morgan M, Gabriel A. Levamisole-induced leukocytoclastic vasculitis with negative serology in a cocaine user. Am J Case Rep. 2017;18:641–3 [A].
[161] Fredericks C, Yon J, Alex G, et al. Levamisole-induced necrosis syndrome: presentation and management. Wounds. 2017;29(3):71–6 [A].

CHAPTER

38

Corticotrophins, Corticosteroids, and Prostaglandins

Melissa L. Thompson Bastin[*,†,1], *Brittany D. Bissell*[*,†]

*Medical Intensive Care Unit/Pulmonary, Lexington, KY, United States

[†]Department of Pharmacy Practice and Science, University of Kentucky HealthCare, University of Kentucky College of Pharmacy, Lexington, KY, United States

[1]Corresponding author: mlthompson@uky.edu

SYSTEMIC GLUCOCORTICOIDS

Two systematic reviews were published during this time. The first meta-analysis reviewed elective non-cardiac surgery patients being treated for post-operative nausea and vomiting with glucocorticoids with low (dexamethasone <8mg), medium (dexamethasone 8–16mg) and high (>16mg) dosing regimens. There was no difference in wound infection outcomes (odds ratio, 0.8; 95% CI, 0.6–1.2). The glucocorticoid use increased peak post-operative glucose concentrations by 20.0mg/dL; CI, 11.4–28.6; $P<0.001$, and decreased C-reactive protein concentrations postoperatively by a mean weighted difference of 22.1mg/L; CI, −31.7 to −12.5; $P<0.001$). There were no other differences in side effects, nor was there a dose-side effect relationship in the 3 dosing arms [1M]. A systematic review of 8 clinical trials evaluating glucocorticoid use in treating systemic lupus erythematosus described adverse event rate in patients receiving oral steroids. The overall adverse event rate for hyperglycemia or diabetes was 9/100 patients/year. The infection rate was 25/100 patients/year, and avascular hip necrosis was 12/100 patients/year. The high doses used in these studies (up to and above 30mg/day prednisone) are higher than recommended doses for this condition (typically <7.5mg/day). The authors conclude that steroid sparing therapies and strategies to decrease steroid doses are needed in this patient population [2M]. A Cochrane review of 10 studies identified risk factors for adrenal insufficiency in childhood acute lymphoblastic leukemia. 298 patient observations were evaluated; however, due to heterogeneity in the data, the individual outcomes were not pooled. The authors found no differences in the rates of adrenal insufficiency between prednisone and dexamethasone. One study found a faster rate of recovery from adrenal insufficiency in children receiving prednisone over dexamethasone. Several studies described the effects of concomitant fluconazole use prolonging the duration of adrenal insufficiency, especially at doses higher than 10mg/kg/day [3M].

Cardiovascular

A retrospective study in 176 Kawasaki pediatric patients evaluated the cardiovascular side effects of prednisolone. The prevalence of bradycardia was higher in the prednisolone group over the IVIG group (79.1% vs 7.1%; $P<0.001$), with the mean time of bradycardia development of 63 hours. The average decrease in heart rate in the prednisolone group was 15.1 beat/min (95% CI 10.2–20.0; $P<0.001$) [4c]. A case of severe hemorrhagic pheochromocytoma was attributed to dexamethasone use in a 68-year-old healthy patient. The patient had undergone an elective orbital repair, following a fall, and was discharged on dexamethasone 4mg three times daily. Four days later, he developed acute epigastric pain and headache. Work-up revealed a suspected adrenal tumor, which had ruptured. The patient's hospital stay was complicated by multi-organ failure, for which he eventually recovered and underwent adrenalectomy. This case adds to the previous 15 cases reported prior in the literature, suggesting glucocorticoids can have a

Side Effects of Drugs Annual, Volume 40
ISSN: 0378-6080
https://doi.org/10.1016/bs.seda.2018.08.006

role in propagating an episode of pheochromocytoma. The suggested mechanism involves augmented catecholamine release, and in this case, peripheral vasodilatation reducing blood flow to the adrenal tumor, resulting in rupture [5A].

Central Nervous System

A case of a 41-year-old woman being treated for leukocytoclastic vasculitis with prednisone developed a 6-week history of headache and intermittent vision loss. She was ultimately diagnosed with idiopathic intracranial hypertension, which was attributed to her underlying autoimmune vasculitis, and possibly prednisone use. Treatment included reduction in prednisone dose, weight loss, acetazolamide and topiramate [6A].

A case study done with the French Pharmacovigilance database investigated risk factors for development of Progressive multifocal leukoencephalopathy (PML). 101 cases of PML were identified, with glucocorticoids identified as a risk factor in 19 cases. The time to development of PML ranged between 2 and –11 months of chronic steroid therapy. The odds ratio of developing PML with dexamethasone exposure was 17.9 (5.7, 56.6) <0.001; with prednisolone exposure 4.1 (1.0, 16.6) 0.050; and with prednisone 22.2 (12.6, 39.1) <0.001. Most of the patients studied had other causes of immune suppression (transplant, autoimmune). PML monitoring should be included in counseling and risk benefit ratio considerations in this population [7C].

Ninety-six healthy volunteers participated in a stress-learning experiment to determine the role of cortisol on stress and learning. Participants received hydrocortisone, placebo or yohimbine prior to completing a memory test. The hydrocortisone arm performed worse on the schema-based learning over the two other groups, suggesting learning is impaired from cortisol-based activation of stress. This research is important in discovering the effects of stress on learning [8c]. A case of persistent hiccups was described in a 72 male following injection of dexamethasone (0.8 mg) into the lumbar space for management of spinal stenosis. The hiccups appeared at 4–7 episodes per minute, lasting for 36 hours on two occasions of the intra lumbar injection of dexamethasone. While the mechanism is not well understood, it is thought that central nervous system activation and neuro-excitatory pathways can contribute. Hiccups is an important side effect of this procedure, which has mostly been described in younger patients <65 years or age [9A].

A retrospective review of 162 patients with reversible cerebral vasoconstriction syndrome identified glucocorticoid use as a significant risk factor for worsening symptoms. Glucocorticoid use was associated with a 10.2-fold risk of clinical worsening when controlling for sex, infarction size and intra-arterial vasodilator use [OR (95% CI) 10.2 (3.30–31.6)], $P = 0.001$. Glucocorticoid use was associated with a 4.23-fold increased risk of a poor outcome (modified rankin score of >3), when controlled for the same variables, [OR (95% CI) 4.23 (1.20–14.9)], $P = 0.025$. The mechanism is postulated to potentiate the vasoconstrictor effects of endogenous vasoconstrictors norepinephrine, angiotensin II, endothelin. The authors conclude glucocorticoids do not reduce poor outcomes in this patient population, and in fact may worsen outcomes. Glucocorticoids should be avoided in this population [10C]. A case of paralysis from glucocorticoid use was described in a 39-year-old male. The patient received an ultrasound-guided trigger point injection of methylprednisolone and bupivacaine in the left iliopsoas tendon. Within 12 hours following the injection, his serum potassium dropped to 1.7 mmol/L, and he experienced severe paresis of the upper extremities, along with abnormal EKG findings. After aggressive potassium replacement, the symptoms resolved. Mineralocorticoid effects and activation of the beta-2 adrenergic receptors through sodium–potassium pump upregulation is thought to explain this phenomenon. Electrolyte monitoring should be performed following this procedure [11A].

Endocrine

A prospective observational cohort study of 39 pediatric patients (<16 years old) receiving chronic glucocorticoid therapy for rheumatic diseases reports a 54.8% rate of adrenal suppression. Median age of the participant was 12.9 years, and median duration of glucocorticoid therapy was 39.6 weeks. The median (IQR) glucocorticoid dose (in prednisone equivalents) was 8.9 (6.7–12.7) mg/m^2. The only predictor of adrenal suppression was a higher glucocorticoid dose, ROC 0.72, $P=0.04$. Most children had resolution of the adrenal suppression within 1 year after discontinuation of the glucocorticoid. Patients and providers should be aware of the prolonged effect on adrenal function [12c]. A national incidence survey study of adrenal suppression in Canadian children identified inhaled and parenteral routes of administration of glucocorticoids were identified as risk factors. The annual incidence reported was 0.35/100 000 children aged 0–18 years (95% CI 0.26–0.47). Growth failure was the most commonly reported symptom in 35% of cases. The study did not differentiate between routes of administration on risk of development of adrenal suppression. The authors recommend close monitoring for adrenal suppression in children receiving inhaled and parenteral glucocorticoids [13C].

Metabolic

A large retrospective review of 1628 Danish patient records diagnosed with Giant Cell Arteritis and 342 patients with Granulomatosis with Polyangiitis were reviewed for the incidence and risk factors of development of diabetes mellitus. Glucocorticoid use is common in this patient population, as 99% of patients received prescriptions filled, with the median cumulative prednisolone/prednisone dose was $\geq$5.6 g during the first year and $\leq$2.5 per year afterwards. Risk of diabetes was 7- and 10-fold higher than the general population in the Giant Cell Arteritis and Granulomatosis with Polyangiitis groups, respectively. Regression modeling found for every 10 mg increase in the daily prednisone/prednisolone in the first year, the risk of developing diabetes increased by 30%, OR 1.3 (95% CI 1.01–1.8). Diabetes is a common adverse effect found in the population, and screening programs should be in place for early diagnosis [14C]. A survey of the general population of the Netherlands determined risk of metabolic syndrome with systemic and local corticosteroid use. 140 879 participants completed the survey, of which 10.9% were currently using a corticosteroid product. Corticosteroid use in women was associated with metabolic syndrome (obesity, dyslipidemia, hypertension) adjusted OR 1.24 (CI 1.17–1.32], $P = 0.001$. For systemic only use of corticosteroids in both men and women, the odds of having metabolic syndrome was even higher at OR, 1.68 (95% CI, 1.34–2.10), $P < 0.001$. Corticosteroids use in women, in this cohort of patients, was significantly associated with metabolic syndrome criteria [15C]. A retrospective review of 184 children with standard and medium risk acute lymphoblastic leukemia (ALL) evaluated effects of glucocorticoids on body mass index. Glucocorticoids are commonly prescribed as part of treatment regimens for childhood ALL, making this an important side effect to measure in this pediatric patient population. The proportion of overweight/obese children was 21.3% at diagnosis, rising to 29% at the end of treatment. This number remained unchanged at 29.1% and 27.7% at 5 and 7 years post treatment, respectively. Controlling for diagnosis, in patients exposed to dexamethasone, the increase in mean body mass index *z*-score was (1.08, $P < 0.001$) at the end of induction therapy, which remains throughout therapy intensification (0.81, 95% CI: 0.64–0.98, $P < 0.001$), maintenance and up to 7 years after diagnosis (0.76, 95% CI: 0.47–1.04, $P = 0.002$). The authors conclude glucocorticoids contribute to the complex pathophysiology of childhood obesity during the treatment of childhood ALL [16C].

Skin

A cross-sectional cohort survey study sought to quantify the effect of glucocorticoids adverse reactions in patients being treated for rheumatoid arthritis. Two centers provided data via a survey, totaling 381 patients. Most patients 88 (41%) received an average prednisone dose of 5–7.5 mg/day, followed by 66 (31%) on <5 mg/day, and then 16 (7.5%) on >7.5 mg/day. As the dose increased, the proportion of patients reporting side effects of Cushing syndrome-like body habitus, bruising, skin atrophy and diminished wound healing increased ($P < 0.0001$). The proportion of patients taking the highest doses reported Cushing syndrome-like body habitus at 38%, bruising 36%, skin atrophy 41%, and diminished wound healing at 29%. This study highlights the important cutaneous side effects of glucocorticoids across the dosing spectrum [17C]. A case report summarized the effects of ultrasound guided triamcinolone acetonide injection for tendonitis on the development of linear spread of atrophy of surrounding tissues and skin. The patient was 64 years old and has received 2 injections to her left extensor pollicis brevis and abductor pollicis longus tendon sheaths for tendonitis. The symptoms of purpura, erythematous telangiectasia, and lipoatrophy developed 1 year after the injections. The atrophy spread into a 10 cm space on her left forearm was novel and not previously reported in the literature. Clinicians should be aware of this side effect and complications of spreading beyond the original injection site [18C].

Musculoskeletal

A case study reviewed the effects of glucocorticoid therapy on musculoskeletal side effects in a Japanese child with nephrotic syndrome. A 12-year-old boy was admitted to the hospital and started on prednisone 60 mg/day (~2 mg/kg/day) for 4 weeks, followed by a 40 mg/day every other day as part of a treatment regimen for nephrotic syndrome. He had 2 relapses in the condition, requiring the 60 mg/day prednisone therapy × 14 days, then requiring methylprednisolone pulsed therapy at 500 mg/day × 3 days, and 1000 mg/day × 3 days. He was also started on additional immune suppression agents and subsequently was weaned from the steroids. The cumulative steroid dose he received was 13 792.5 mg (496 mg/kg). The patient began to complain of back pain on hospital day 186 and was ultimately diagnosed with compression fractures extending from the seventh vertebra to the first lumbar. He was ultimately diagnosed with osteoporosis through bone mineral density tests. It is imperative to monitor the bone effects of glucocorticoids in pediatric patients [19A]. A case series review of Indian children with Crohn's disease evaluated risk factors for bone mineral density reduction. 30 children included in the study, of which 87% reported to be taking corticosteroids (dose/duration not given). In a bivariate analysis, corticosteroid use was associated with reduced bone mineral density tests 76.9% vs 25.0%,

$P=0.069$. Other risk factors were not controlled for in this study, notably many of the children were underweight and malnourished [20A]. A prospective observational prevalence study in Chinese rheumatoid arthritis patients described the impact of glucocorticoid use on osteoporosis. 798 rheumatoid patients were compared to 158 controls matched on age, sex, body mass index. The mean (SD) daily dose of glucocorticoids was 10mg (5mg) for a median duration of 20 months. Osteoporosis was significantly higher in the rheumatoid group, than controls (36.8% vs 13.9%, $P<0.0001$). The prevalence of osteoporosis in those taking corticosteroids in the rheumatoid group was 41.6% vs no steroids in the rheumatoid group 29.4%. Treatment duration was a stronger predictor of osteoporosis development than daily dose [21C]. A multicenter retrospective analysis of previously published studies on glucocorticoids in pediatric patients for various conditions sought to establish a correlation between novel genetic markers and bone mineral density tests in the children from these studies. A total of 520 children from an acute lymphoblastic leukemia trial and an asthma study were included. The researchers have no variants that met their endpoint for genomic-wide significance; however, they did find a SNP (rs6461639) that was significantly associated with bone mineral density tests in the leukemia patients, after adjusting for concurrent chemotherapeutic agents. Future studies should investigate the pharmacogenomics interactions between prednisone dose and bone mineral density scores in pediatric patients with long-term steroid use [22R].

A secondary analysis of a previously published study evaluated the relationship among duration of corticosteroid use, dose, activity and changes in body mass index. The analysis included 197 patients who were apart of the Rituximab in ANCA-Associated Vasculitis (RAVE) trial. Newly diagnosed patient has a lower mean BMI, than those with relapsed disease in the original study ($28.0\pm5.7\,kg/m^2$ vs $29.6\pm6.8\,kg/m^2$), respectively. Body mass index increased the most during the first 6 months of the trial ($+1.1\pm2.2\,kg/m^2$, $P<0.0001$), and this was statistically associated with glucocorticoid use, Rituxan use and disease activity improvement. Changes in body mass index in this population are associated with steroid exposure, as well as improvement in disease [23C]. A case report of a 12-year-old girl who received triamcinolone injection to her right knee for Osgood–Schlatter syndrome. Two months after the injection, the patient reported skin tenderness and purple striated markings behind her knee. This is one of a few cases reported in the literature, and the first reported side effect from triamcinolone. This benign side effect needs to be a part of the counseling process for patients receiving this procedure [24A].

Other

An online cross-sectional survey study of United Kingdom adults taking glucocorticoids assessed patient specific outcomes, including most concerning side effects. 604 survey respondents noted that weight gain, cardiovascular disease, diabetes, eye disease and infections were of the most importance on a scale of 1–10. Side effects of acne were not as important to the responders. The most worrisome side effect was weight gain median score (IQR) at 9 (6–10), then insomnia and moon face which scored equally at 8 (5–10). This survey study focusing on patient specific outcomes highlights the perspectives of our patients and suggests taking patient preference into account when prescribing glucocorticoids [25C]. A retrospective analysis of a pediatric acute lymphoblastic leukemia measured associations with N363S and Bcl*I* polymorphisms and side effect reporting in the 49 patients of the trial. Dyspepsia was associated more with the CG than in the CC (wild-type) genotype (36.4% vs 5.3%, $P=0.018$) groups. Depression was reported more frequently in the G allele (CG+GG) than the CC genotype (39.3% vs 10.5%, $P=0.031$). The Bcl*I* polymorphism is associated with a higher proportion of reported side effects in this patient population and should be considered when selecting treatment regimens [26c]. A retrospective study of insurance-based claims in patients with pemphigus receiving oral steroids for treatment evaluated the effects of dose on adverse events. 644 patients were included, and the mean daily dose of oral steroids was 29mg/day. The rate of adverse events was 0.46 events/patient year. For every 1g total exposure of oral steroids, the rate of general adverse events increased (hazard ratio [HR], 1.01; $P=0.03$), the risk of cataract increased (HR, 1.02; $P<0.001$) and the risk of fracture increased (HR, 1.01; $P=0.03$). Additional analysis revealed that developing these adverse events was associated with higher overall medical costs, in hospital costs and pharmacy costs. This study highlights the overall economic burden of the adverse events of corticosteroids on this patient population, and the need for steroid sparing treatment options [27C].

Infection

Infectious complications of corticosteroids have been demonstrated in several routes of administration for various conditions. A 59-year-old man receiving systemic corticosteroids in additional to immunosuppressive medications for liver transplantation developed a *Nocardia* liver abscess 6 weeks after liver transplant. Fever resolved with steroid taper and appropriate antibiotic treatment [28A]. A 4-month-old boy receiving betamethasone dipropionate topically followed by oral prednisolone and intramuscular betamethasone for miliaria rubra-like lesions worsened the dermatitis later classified as scabies. Treatment included steroid discontinuation and initiation of appropriate antibiotics [29A]. In a retrospective review of 21 cases of confirmed *Pneumocystis jiroveci* pneumonia (PJP), an average dose of 36mg of prednisone was received. The authors confirmed that those receiving greater than 20mg daily of prednisone equivalents should receive prophylaxis. No patients receiving less than that dose who developed PJP died. Eighty-six-patients were receiving greater than 20mg when PJP developed [30c]. A 14-year old receiving 8 weeks of dexamethasone for *Mycobacterium tuberculosis* meningeal infection developed secondary *Cryptococcus neoformans* without any other risk factors other than steroid use [31A]. The largest trial in 2017 looked at oral methylprednisolone 0.6–0.8mg/kg/day to a maximum 48mg daily vs placebo for patients with IgA nephropathy. This trial stopped early after 262 patients were randomized for potential harm. Patients receiving steroids had a significantly higher incidence of severe adverse events, driven primarily by a significant increase in severe infections, including respiratory, *Pneumocystis* pneumonia, and others [32C]. In regards to topical administration, one case reported a 24-year-old man with Tinea who developed a worsening rash after receiving a clobetasol-containing cream secondary to acute symptom resolution and disease exacerbation. The rash resolved with steroid discontinuation and appropriate antifungal coverage [33A]. A 52-year-old female presenting with bilateral corneal infiltration and hypoon was found to have *Morganella morganii* keratitis secondary corticosteroid-induced rosacea-like dermatitis preceding infection onset. The patient was unaware of the steroid utilized but was used for 5 months prior. Steroid discontinuation and antibiotic eye drops allowed for infection resolution [34A].

Death

A retrospective analysis of 392 patients with non-functioning pituitary adenoma receiving daily glucocorticoids examined overall outcomes from treatment. Patients receiving greater than 20mg of hydrocortisone equivalent per day for adrenal insufficiency were found to have a higher mortality when compared to placebo (HR 1.88, CI 1.06–3.33). A difference in mortality was not seen for doses lower than this [35c]. A cross-sectional study of 357 patients diagnosed with systemic lupus erythematosus determined the impact of chronic high-dose glucocorticoids. Glucocorticoids were associated with a worse disease outcome (HR 1.02, CI 1.006–1.035). Patients who developed cataracts or osteoporosis were more likely to have received glucocorticoids ($P < 0.001$, $P = 0.041$, respectively). The dose of glucocorticoid received to classify a patient for high-dose was not defined [36c].

ADRENOCORTICOTROPIN

Adrenocorticotropin (ACTH) has been studied for the treatment of infantile spasms (West Syndrome).

Drug Dosage Regimens

A retrospective review sought to evaluate the effectiveness and safety of three alternative dose and duration regimens of ACTH for the treatment of infantile spasms. Of 36 patients receiving 1U/kg/day, two required drug discontinuation secondary to liver function and second-degree atrioventricular block, respectively. Of those 59 patients receiving 1–1.9U/kg/day, four required discontinuation. One developed a lung infection with hypertension, two developed dysphoria, and another had treatment-related hypertension. In the group of 118 patients receiving 2–4U/kg/day, ACTH was discontinued in five patients for drug-related concerns. Two patients developed lung infections with dysphoria or poor spirit, one with an unspecified allergy, 1 for bradycardia, 1 for electrolyte disturbances and abdominal distention. No statistical difference was found between dosing groups in regards to incidence of side effects (47% vs 52.7% vs 61.3%). The most common side effect in all dosing regimens was pneumonia (20.6%–23.6%) followed by hypertension (13.5%–23.6%). Other side effects with a greater than 5% incidence included electrolyte disturbances, bronchitis, and diarrhea, all which appeared dose-dependent. Vomiting and thrush were also reported in both the lowest and highest dose ranges [37c].

Immunologic

A 9-month-old female treated with ATH 100U/m^2 for 2 weeks followed by a 4-week taper required hospital admission 1 month after initiation of treatment for *Legionella pneumophila* pneumonia. The patient developed a mild parapneumonic effusion requiring drainage. While all cultures, including those from the drain, remained negative, patient tested positive for a *Legionella* urine antigen and chest radiography demonstrated a right upper lobe infiltrate with an elevated C-reactive protein. Patient required mechanical ventilation for 1 week after a second decline in respiratory status while hospitalized secondary to aspiration. Patient received 2 weeks total of levofloxacin and was able to return to baseline and discharge home [38A]. A 14-month-old male treated with ATH 100 U/m^2 for 2 weeks followed by a 6-week taper required hospitalization 2 weeks into therapy for hypertension. Within the third week, patient developed an elevated white blood cell count and fever with a chest radiograph demonstrating right upper lobe consolidation. Despite 3 days of treatment, patient remained febrile. Blood and viral respiratory polymerase chain reaction panel were negative, but a *Legionella* urine antigen was positive. The patient received 2 weeks of levofloxacin and was able to be discharged home. Both cases were suspected secondary to the potential immunologic effects of ACTH, primarily decreased lymphocyte counts, in combination with home nebulizers containing tap water [38A].

ALPROSTADIL

Recent studies of alprostadil have evaluated its efficacy and safety for the treatment of intermittent claudication and peripheral arterial occlusive disease. In a randomized controlled trial of 541 patients with intermittent claudication, 35% of patients receiving alprostadil had treatment emergent adverse events compared to 33.7% in the alternative treatment group receiving pentoxifylline [39C]. In another randomized controlled trial of 840 patients with Fontaine Stage IV peripheral arterial occlusive disease, a significantly higher rate of treatment emerged adverse events during treatment were demonstrated with alprostadil compared to placebo. Specifically, there was no difference in overall mortality, stroke, or myocardial infarction [40C].

EPOPROSTENOL

Both intravenous and inhaled epoprostenol have been evaluated, in patients with acute respiratory distress syndrome and pulmonary hypertension. In a 61-year-old female receiving long-term IV epoprostenol, focal dilatation of her 8-year-old tunneled catheter in which the infusion was running resulted in a focal area of "ballooning" near her clamp, likely secondary to weakened catheter walls. When this area was compressed, the patient experienced sudden onset lightheadedness and flushing likely secondary, presumably related to a small epoprostenol bolus received [41A]. Inhaled epoprostenol (10 μg/mL) was prospectively compared to nitric oxide and milrinone in the management of acute respiratory distress syndrome within 15 patients. No significant difference was seen in hemodynamic parameters and no significant adverse events were seen [42c].

ILOPROST

A variety of literature have surrounded iloprost use, both inhaled and intravenously. Iloprost utilization has ranged from the treatment of acute respiratory distress syndrome, pulmonary hypertension, and systemic sclerosis.

Cardiovascular

In a preterm 0.9 kg female, inhaled iloprost for acute respiratory distress syndrome, administration was associated with a decrease in mean arterial pressure requiring vasopressor agents and subsequent drug discontinuation [43A]. However, in a retrospective evaluation of patients with idiopathic or connective tissue disease associated-pulmonary arterial hypertension receiving intravenous iloprost, treatment was not associated with a change in hemodynamic parameters [44c]. Iloprost inhalation in 67 patients with pulmonary hypertension secondary to chronic obstructive pulmonary disease (COPD) was well tolerated with only five reports of cough and no worsening dyspnea [45C].

Respiratory

In a 3.2 kg neonate male receiving inhaled iloprost for pulmonary hypertension, administration was associated with severe hypoxia secondary to drug connection to the endotracheal tube. No significant difference was seen in hemodynamic parameters and no significant adverse events were seen [43A]. Long-term iloprost infusions of greater than 2 years for systemic sclerosis related digital ulcers or Reynaud's Phenomenon were studied in a cohort of 50 patients reported. No severe adverse events were reported, but hypotension, flushing, and headache occurred [46c].

MISOPROSTOL

Misoprostol complications during its use for medical induction vs dilation and evacuation for second-trimester uterine evacuations were studied in a cohort of 465 women. Misoprostol 400 mcg vaginally every 4–6 hours was associated with a significantly higher risk of complication with the most common type including infection (11.6% of patients) followed by retained products of conception (9.9%) [47c].

PROSTAGLANDINS AND ANALOGUES

Prostaglandins as a general class have continued to be studied for the management of pulmonary hypertension with associated secondary diseases.

Cardiovascular

In a retrospective review of 27 patients with pulmonary arterial hypertension and peripheral pulmonary stenosis, the impact of prostacyclin therapy on the incidence of adverse events within 30 days of lung transplant, including death, intrathoracic hematoma and bleeding, cardiac congestion or shock, cerebral infarction, and pulmonary embolism, was evaluated. Patients who received long-term preoperative prostacyclin (≥770 days) had a significantly higher incidence of the composite endpoint compared to those receiving short-term prostacyclin (≤668 days). Intrathoracic hematoma and bleeding was the most common side effect in both groups, 38.8% and 50% in long- and short-term respectively. This is suspected secondary to neovascularization, which has been previously observed with long-term prostacyclin therapy as well as its inhibition of platelets [48c].

Nervous System

Within case series of 11 female patients receiving prostacyclin therapy for pulmonary arterial hypertension with concomitant leg pain, 45% were receiving treprostinil and 55% received epoprostenol. While epoprostenol was administered intravenously (IV) only, including both Veletri and Flolan brands, patients receiving treprostinil varied in route of administration, with subcutaneous being most common ($n=2$) followed by IV, oral, and inhaled ($n=1$ each). All patients reported constant neuropathic pain, most commonly described as burning (91%) or prickling/tingling (74%) and all developed such pain within 2 months of therapy initiation. Prostacyclin-induced neuropathy is suspected to a combined prostacyclin toxicity and metabolic derangement preceding initiation. Furthermore, endogenous prostacyclins are responsible for potentiation of nociceptive pain responses. Other adverse events included dysautonomia (45%), early satiety (27%), sudomotor changes (18%), and orthostatism (27%) [49c].

SELEXIPAG

Major Review

A major review focused on the use of selexipag for the treatment of pulmonary arterial hypertension. In healthy subjects, selexipag doses up to 1600 mcg twice daily were associated with mild adverse events including headache, nausea, and vomiting. Doses above this resulted in moderate adverse events in 50% of patients. In a randomized controlled trial, selexipag was associated with an increase in heart rate; however, no effect was seen on conduction intervals. Hypotension requiring discontinuation occurred in 4 of 91 patients. When the previous data was reviewed, no clear correlation was seen between discontinuation of selexipag and the dose received. Headache, myalgia, and nausea were the most common reason for discontinuation and adverse events are more common when administered in a fasting state. Specifically within the pulmonary arterial hypertension population, selexipag was most commonly discontinued for headache, diarrhea, nausea, or jaw pain in a phase 2 study. Other adverse events in phase 2 trials include hyperthyroidism (1%), anemia (8.3%), hypotension (5%), and syncope (6.4%). Adverse events are more common during titration intervals of the medication [50R].

Additional case studies in the same or related topics can be found in other reviews [51R,52R].

References

[1] Toner AJ, Ganeshanathan V, Chan MT, et al. Safety of perioperative glucocorticoids in elective noncardiac surgery: a systematic review and meta-analysis. Anesthesiology. 2017;126(2):234–48 [M].

[2] Sciascia S, Mompean E, Radin M, et al. Rate of adverse effects of medium- to high-dose glucocorticoid therapy in systemic lupus erythematosus: a systematic review of randomized control trials. Clin Drug Investig. 2017;37(6):519–24 [M].

[3] Rensen N, Gemke RJBJ, van Dalen EC, et al. Hypothalamic-pituitary-adrenal (HPA) axis suppression after treatment with glucocorticoid therapy for childhood acute lymphoblastic leukaemia. Cochrane Database Syst Rev. 2017;11:Cd008727 [M].

[4] Nagakura A, Morikawa Y, Sakakibara H, et al. Bradycardia associated with prednisolone in children with severe Kawasaki disease. J Pediatr. 2017;185:106–111.e101 [c].

[5] Yeoh CJ, Ng SY, Goh BKP. Pheochromocytoma multisystem crisis triggered by glucocorticoid administration and aggravated by citrate dialysis. A A Case Rep. 2017;8(3):58–60 [A].

[6] Cestari DM, Cunnane ME, Rizzo JF, et al. Case 2–2018. A 41-year-old woman with vision disturbances and headache. N Engl J Med. 2018;378(3):282–9 [A].

[7] Colin O, Favrelier S, Quillet A, et al. Drug-induced progressive multifocal leukoencephalopathy: a case/noncase study in the French pharmacovigilance database. Fundam Clin Pharmacol. 2017;31(2):237–44 [C].

[8] Kluen LM, Nixon P, Agorastos A, et al. Impact of stress and glucocorticoids on schema-based learning. Neuropsychopharmacology. 2017;42(6):1254–61 [c].

[9] Odonkor CA, Smith B, Rivera K, et al. Persistent singultus associated with lumbar epidural steroid injections in a Septuagenarian: a case report and review. Am J Phys Med Rehabil. 2017;96(1):e1–4 [A].

[10] Singhal AB, Topcuoglu MA. Glucocorticoid-associated worsening in reversible cerebral vasoconstriction syndrome. Neurology. 2017;88(3):228–36 [C].

[11] Soriano PK, Bhattarai M, Vogler CN, et al. A case of trigger-point injection-induced hypokalemic paralysis. Am J Case Rep. 2017;18:454–7 [A].

[12] Ahmet A, Brienza V, Tran A, et al. Frequency and duration of adrenal suppression following glucocorticoid therapy in children with rheumatic diseases. Arthritis Care Res. 2017;69(8):1224–30 [c].

[13] Goldbloom EB, Mokashi A, Cummings EA, et al. Symptomatic adrenal suppression among children in Canada. Arch Dis Child. 2017;102(4):338–9 [C].

[14] Faurschou M, Ahlstrom MG, Lindhardsen J, et al. Risk of diabetes mellitus among patients diagnosed with giant cell arteritis or Granulomatosis with Polyangiitis: comparison with the general population. J Rheumatol. 2017;44(1):78–83 [C].
[15] Savas M, Muka T, Wester VL, et al. Associations between systemic and local corticosteroid use with metabolic syndrome and body mass index. J Clin Endocrinol Metab. 2017;102(10):3765–74 [C].
[16] Touyz LM, Cohen J, Neville KA, et al. Changes in body mass index in long-term survivors of childhood acute lymphoblastic leukemia treated without cranial radiation and with reduced glucocorticoid therapy. Pediatr Blood Cancer. 2017;64:e26344 [C].
[17] Amann J, Wessels A-M, Breitenfeldt F, et al. Quantifying cutaneous adverse effects of systemic glucocorticoids in patients with rheumatoid arthritis: a cross-sectional cohort study. Clin Exp Rheumatol. 2017;35(3):471–6 [C].
[18] Willardson HB, Buck S, Neiner J. Linear atrophy and vascular fragility following ultrasoundguided triamcinolone injection for DeQuervain tendonitis. Dermatol Online J. 2017;23(1) [C].
[19] Ashida A, Fujii Y, Matsumura H, et al. A case of multiple vertebral compression fractures due to glucocorticoid-induced osteoporosis in a pediatric patient with nephrotic syndrome. Int J Clin Pharmacol Ther. 2017;55(3):264–9 [A].
[20] Khan SS, Patil SS. Bone density in pediatric Crohn's disease: a cross-sectional observation from South India. Indian J Gastroenterol. 2017;36(3):184–8 [A].
[21] Ma CC, Xu SQ, Gong X, et al. Prevalence and risk factors associated with glucocorticoid-induced osteoporosis in Chinese patients with rheumatoid arthritis. Arch Osteoporos. 2017;12(1):33 [C].
[22] Park HW, Tse S, Yang W, et al. A genetic factor associated with low final bone mineral density in children after a long-term glucocorticoids treatment. Pharmacogenomics J. 2017;17(2):180–5 [R].
[23] Wallace ZS, Milostavsky EM, Cascino M, et al. effect of disease activity, glucocorticoid exposure, and rituximab on body composition during induction treatment of antineutrophil cytoplasmic antibody-associated vasculitis. Arthritis Care Res. 2017;69(7):1004–10 [C].
[24] Wise K, Warren D, Diaz L. Unilateral striae distensae of the knee after a steroid injection for the treatment of Osgood-Schlatter disease. Dermatol Online J. 2017;23(3) [A].
[25] Costello R, Patel R, Humphreys J, et al. Patient perceptions of glucocorticoid side effects: a cross-sectional survey of users in an online health community. BMJ Open. 2017;7(4):e014603 [C].
[26] Kaymak Cihan M, Karabulut HG, Kutlay NY, et al. Association between N363s and BCLI polymorphisms of the glucocorticoid receptor gene (NR3C1) and glucocorticoid side effects during childhood acute lymphoblastic leukemia treatment. Turk J Haematol. 2017;34(2):151–8 [c].
[27] Wormser D, Chen DM, Brunetta PG, et al. Cumulative oral corticosteroid use increases risk of glucocorticoid-related adverse events in patients with newly diagnosed pemphigus. J Am Acad Dermatol. 2017;77(2):379–81 [C].
[28] Hanchanale P, Jain M, Varghese J, et al. Nocardia liver abscess post liver transplantation—a rare presentation. Transpl Infect Dis. 2017;19(2) [A].
[29] Lima F, et al. Crusted scabies due to indiscriminate use of glucocorticoid therapy in infant. An Bras Dermatol. 2017;92(3):383–5 [A].
[30] Mecoli CA, Saylor D, Gelber AC, et al. Pneumocystis jiroveci pneumonia in rheumatic disease: a 20-year single-centre experience. Clin Exp Rheumatol. 2017;35(4):671–3 [c].
[31] Nidhi A, Meena A, Sreekumar A, et al. Corticosteroid-induced cryptococcal meningitis in patient without HIV. BMJ Case Rep. 2017. https://doi.org/10.1136/bcr-2016-216496 [A].
[32] Lv J, Zhang H, Wong MG, et al. Effect of oral methylprednisolone on clinical outcomes in patients with IgA nephropathy: the TESTING randomized clinical trial. JAMA. 2017;318(5):432–42 [C].
[33] Verma S. Steroid modified tinea. BMJ. 2017;356:j973 [A].
[34] Zhang B, Pan F, Zhu K. Bilateral Morganella Morganii keratitis in a patient with facial topical corticosteroid-induced rosacea-like dermatitis: a case report. BMC Ophthalmol. 2017;17(1):106 [A].
[35] Hammarstrand C, Ragnarsson O, Hallen T, et al. Higher glucocorticoid replacement doses are associated with increased mortality in patients with pituitary adenoma. Eur J Endocrinol. 2017;177(3):251–6 [c].
[36] Tarr T, Papp G, Nagy N, et al. Chronic high-dose glucocorticoid therapy triggers the development of chronic organ damage and worsens disease outcome in systemic lupus erythematosus. Clin Rheumatol. 2017;36(2):327–33 [c].
[37] Yin J, Lu Q, Yin F, et al. Effectiveness and safety of different once-daily doses of adrenocorticotropic hormone for infantile spasms. Paediatr Drugs. 2017;19(4):357–65 [c].
[38] Shachor-Meyouhas Y, Ravid S, Hanna S, et al. Legionella pneumophila pneumonia in two infants treated with adrenocorticotropic hormone. J Pediatr. 2017;186:186–188.e181. [A].
[39] Schellong SM, von Bilderline P, Gruss JF, et al. Intravenous alprostadil treatment compared to oral pentoxifylline treatment in outpatients with intermittent claudication—results of a randomised clinical trial. Vasa. 2017;46(5):403–5 [C].
[40] Lawall H, Pokrovsky A, Chcinsky P, et al. Efficacy and safety of alprostadil in patients with peripheral arterial occlusive disease fontaine stage IV: results of a placebo controlled randomised multicentre trial (ESPECIAL). Eur J Vasc Endovasc Surg. 2017;53(4):559–66 [C].
[41] LeVarge BL, Law AC, Murphy B. Occult catheter rupture causing episodic symptoms in a patient treated with epoprostenol. Pulm Circ. 2017;8(1):1–3 [A].
[42] Albert M, Crsilli D, Williamson DR, et al. Comparison of inhaled milrinone, nitric oxide and prostacyclin in acute respiratory distress syndrome. World J Crit Care Med. 2017;6(1):74–8 [c].
[43] Unal S, et al. Iloprost instillation in two neonates with pulmonary hypertension. J Coll Physicians Surg Pak. 2017;27(4):257–9 [A].
[44] Ramjug S, et al. Long-term outcomes of domiciliary intravenous iloprost in idiopathic and connective tissue disease-associated pulmonary arterial hypertension. Respirology. 2017;22(2):372–7 [c].
[45] Wang L, Jin Y-Z, Zhao Q-H, et al. Hemodynamic and gas exchange effects of inhaled iloprost in patients with COPD and pulmonary hypertension. Int J Chron Obstruct Pulmon Dis. 2017;12:3353–60 [C].
[46] Colaci M, Lumetti F, Giuggioli D, et al. Long-term treatment of scleroderma-related digital ulcers with iloprost: a cohort study. Clin Exp Rheumatol. 2017;35(106):S179–83 [c].
[47] Sonalkar S, Ogden SN, Tran LK, et al. Comparison of complications associated with induction by misoprostol versus dilation and evacuation for second-trimester abortion. Int J Gynaecol Obstet. 2017;138(3):272–5 [c].
[48] Akagi S, Oto S, Kobayashi M, et al. High frequency of acute adverse cardiovascular events after lung transplantation in patients with pulmonary arterial hypertension receiving preoperative long-term intravenous postacyclin. Int Heart J. 2017;58(4):557–61 [c].
[49] Pagani-Estevez GL, Swetz KM, McGoon MD, et al. Characterization of prostacyclin-associated leg pain in patients with pulmonary arterial hypertension. Ann Am Thorac Soc. 2017;14(2):206–12 [c].
[50] Bruderer S, Hurst N, Remenova T, et al. Clinical pharmacology, efficacy, and safety of selexipag for the treatment of pulmonary arterial hypertension. Expert Opin Drug Saf. 2017;16(6):743–51 [R].
[51] Brophy A, Ray SD. Glucocorticoids. In: Ray SD, editor. Side effects of drugs annual: a worldwide yearly survey of new data in adverse drug reactions, vol. 38. Elsevier; 2017. p. 425–31 [chapter 34]. [R].
[52] Kaplan JB, Brophy A. Corticotrophins, corticosteroids, and prostaglandins. In: Ray SD, editor. Side effects of drugs annual: a worldwide yearly survey of new data in adverse drug reactions, vol. 39. Elsevier; 2017. p. 407–15 [chapter 34]. [R].

CHAPTER

39

Sex Hormones and Related Compounds, Including Hormonal Contraceptives

Caitlin M. Gibson[1], *Amulya Tatachar*

System College of Pharmacy, University of North Texas Health Science Center, Fort Worth, TX, United States

[1]Corresponding author: caitlin.gibson@unthsc.edu

ESTROGENS AND DERIVATIVES [SEDA-35, 731; SEDA-36, 615; SEDA-37, 499–500; SEDA-38, 433–439; SEDA-39, 417–418]

Diethylstilbestrol (DES) [SEDA-35, 731; SEDA-36, 615; SEDA-37, 499–500; SEDA-38, 433–439, SEDA-39, 417–418]

DES is an estrogen derivative that was historically prescribed to reduce the risk of miscarriage and other complications during pregnancy as well as for breast and endocrine cancers. It was withdrawn from the US market for its link to clear cell carcinoma and received international attention for a variety of adverse effects including cancers, genital malformations, and psychiatric disorders in patients and their descendants. A recent study of 69 individuals from 30 Caucasian families in which at least one sibling was exposed to DES in utero was conducted. Analysis of participant DNA revealed no differences in methylation at sites of interest. However, higher level DES-exposed individuals had schizophrenia diagnoses than individuals who were not exposed to DES, and patients with schizophrenia had more methylation in the ZPF57 region. ZPF57 is a transcriptional regulator involved in maintenance of imprinting and DNA methylation. The authors conclude that more studies are needed to determine the potential role of ZPF57 in psychotic disorders and schizophrenia [1c].

Hormone Replacement Therapy (HRT) [SEDA-35, 732; SEDA-36, 616; SEDA-317, 500–501; SEDA-38, 434–435]

The 2017 Hormone Therapy Replacement Statement of The North American Menopause Society (NAMS) updated the 2012 Hormone Therapy Position Statement of NAMS. NAMS establishes hormone therapy as the most effective treatment for vasomotor symptoms and genitourinary syndrome of menopause and has been shown to prevent bone loss and fracture. The risks depend on a variety of factors including route of administration, timing of initiation, dose, formulation, duration of use, type, and whether progestin should be used [2S]. For clarification, hormone replacement therapy (HRT) is hormone replacement and it can be a combination of estrogen and progestin therapy or just estrogen alone.

Transgender Population

DEEP VENOUS THROMBOEMBOLISM

Venous thromboembolism (VTE) is a well-recognized potential adverse event associated with estrogen use. However, few cases describe the development of VTE in male-to-female transgender patients taking estrogen therapy. A recent case described the diagnosis of deep vein thrombosis (DVT) in a 44-year-old transgender woman who was status post orchiectomy and taking intramuscular estradiol 25 mg injections weekly plus progesterone 200 mg daily. After spending 8 h on a flight and 8 h in a car, the woman presented to the hospital with a painful swollen leg and was diagnosed with a DVT. She was treated with an oral anticoagulant. Usual care would also include discontinuation of hormone therapy, however, the patient stated that discontinuation of her gender-affirming hormones would cause distress regarding her physical and physiologic identity. Ultimately, the patient was switched from estradiol injections to a transdermal estradiol patch in lieu of discontinuing hormone therapy. In light of increasing attention to transgender healthcare, a discussion of estrogen-induced VTE in this unique patient population is warranted [3A].

Side Effects of Drugs Annual, Volume 40
ISSN: 0378-6080
https://doi.org/10.1016/bs.seda.2018.06.007

PROSTATE CANCER

In another case report, use of estrogen in a male-to-female transgender patient may have contributed to the development of prostate cancer. A 56-year-old woman who had received unknown estrogen injections as androgen suppression therapy for approximately 20 years was diagnosed with prostate cancer after an abnormal digital rectal exam, elevated prostate specific androgen levels, and a positive prostate biopsy. She permanently discontinued estrogen therapy and 2 months later underwent a prostatectomy. After 1 year, the patient's testosterone levels remained low likely due to long-term estrogen treatment permanently shutting down testosterone production. There are multiple hypothesized mechanisms for development of prostate cancer in transgender patients receiving estrogen therapy including estrogen-sensitive prostate lesions, androgen receptor variants, and preexisting but unidentified prostate lesions prior to initiation of estrogen therapy. The authors caution that screening for prostate cancer is still critical among male-to-female transgender patients as the incidence of prostate cancer will likely increase in this population as they age [4A].

Reactivation of Endometriosis

A recent systematic review describes 16 historical cases and case series of estrogen-containing HRT leading to reactivation and/or malignant transformation of endometriosis. In the review, 17 patients experienced endometriosis recurrence which often presented with pain and abnormal bleeding. Twenty-five patients experienced malignant transformation of endometriosis, 13 of which had a history of endometriosis in more than one site and 19 of which occurred in women taking unopposed estrogen for HRT. Although the most recent case report in this review is from 2009, the authors argue that the number of cases in the literature should precipitate a renewed look at the risk vs benefit profile of HRT in perimenopausal women with a history of endometriosis. Women require comprehensive counseling regarding the possibility of disease recurrence and should seek medical attention if they experience symptoms consistent with endometriosis so they can make informed decisions about their healthcare [5M].

Hepatic Adenomas

A recent case series of one adolescent and two young women receiving norethindrone acetate (NET-A) therapy for endometriosis suggest that progestin administration may result in hepatic adenomas in women with endometriosis. All three women were prescribed a maximum dose of NET-A 15mg per day between ages of 15–19 years old. After a period of 12–30 months, the women experienced abdominal pain and/or nausea and vomiting, and were ultimately diagnosed with hepatic adenoma. Of note, all three women had previous exposure to estrogens (oral contraceptives of conjugated estrogens). Two women with limited hepatic disease continued progestin therapy after diagnosis and repeat imaging revealed stability of lesions. The third woman with extensive disease discontinued estrogen use and the hepatic lesions subsequently reduced in size. The mechanism is likely peripheral conversion of NET-A to ethinyl estradiol, which has previously been linked to development of hepatic adenomas. The author concludes NET-A in a dose higher than 10mg per day may be associated with an increased risk for hepatic adenomas [6A].

Another case described a 43-year-old woman who had recently been prescribed an unknown oral contraceptive pill. She presented with abdominal pain, fatigue, malaise, night sweats, and arthralgias. Her past medical history was consistent with obesity and gastric stapling. Upon work-up she was found to have elevated erythrocyte sedimentation rate and C-reactive protein, elevated alkaline phosphatase, low serum albumin, and multiple hepatic adenomas were discovered on imaging. She was diagnosed with likely adenoma-driven systemic inflammation and her oral contraceptives were discontinued. After 4 years of discontinuing her oral contraceptive pill, her hepatic lesions had decreased in size and her inflammatory markers had normalized [7A].

Glycemic Profile Deterioration

A case of deterioration of the glycemic profile of a 54-year-old Japanese woman with type 1 diabetes mellitus taking HRT (estrogen plus medroxyprogesterone) for vasomotor symptoms related to menopause was recently reported. She presented to the hospital with altered mental status and was diagnosed with diabetic ketoacidosis and treated with insulin and fluids until resolution of symptoms and laboratory parameters. Twenty-one days after presentation, the patient took medroxyprogesterone 2.5mg by mouth in addition to percutaneous estradiol gel. The patient's previously controlled blood glucose level quickly rose to 398mg/dL (22.1mmol/L), a significant increase from baseline ($P<0.001$). HRT was discontinued and the patient's glycemic profile. There are few reports of HRT impact on glycemic control in type 1 diabetics, but the authors postulate that the glycemic fluctuations were due to medroxyprogesterone-induced worsening of insulin sensitivity, especially during the luteal phase [8A].

Hormonal Contraceptives [SEDA-35, 722; SEDA-36, 618, 622–623; SEDA-37, 501–504, 506; SEDA-38, 436–439]

Mental Health

The potential consequences of hormonal contraceptives on mental health have been gaining renewed attention.

A recent report describes the case of a 19-year-old female with a history of bipolar affective disorder (in remission for 3 years) who was prescribed medroxyprogesterone acetate (MPA) for secondary amenorrhea which likely developed as a consequence of risperidone therapy. After starting MPA 10 mg orally daily, she developed manic episodes such as insomnia, irritable mood, anger, talkativeness, expression of increased libido, and increased goal-directed activity. She was admitted to the hospital and treated with antipsychotics and mood stabilizers, and her mania resolved after 3 weeks. While the patient's manic episodes may have been due to poor adherence to her psychiatric medications, the authors postulate that the commencement of MPA is most likely the cause because she had no other recent changes in her medications. The mechanism of MPA-induced mania is unknown and previous literature has reported conflicting effects of progestin administration on psychiatric symptoms [9A].

Chronic Urticaria

A case series of seven women aged 26–46 years who presented to a single medical institution with mostly daily symptoms of urticarial, dermatographism, and angioedema, and identified the use of intrauterine devices (IUDs) as the likely cause. IUDs had been implanted 1 day to 2 years prior to symptoms. Six of the women had levonorgestrel IUDs, and skin tests for progesterone sensitivity were positive in all women who were able to discontinue antihistamines for testing purposes. All women experienced some symptom relief after removal of the IUD. Potential mechanisms include prior sensitization to progesterones via systemic use. However, because one woman did experience a reaction to a copper IUD, the mechanism may not be exclusively related to progestin exposure [10A].

Menangiomas

The potential relationship between meningioma incidence and hormone use has been previously described, but a recent case series of two patients further supports the relationship between hyperemia and tumor growth. Both cases involve use of cyproterone acetate (CPA), steroid with antiandrogenic, antigonadotropic, and progestin-like activity used for prostate cancer, paraphilia, and acne. CPA is available in Europe, Canada, and Mexico but is not available in the United States due to concerns of hepatotoxicity. The first case of meningioma linked to CPA use was of a 26-year-old woman with a history of an intramedullary glial tumor status post-surgery and chemotherapy who presented to the hospital after a seizure. Imaging revealed two menangiomas. The woman had been receiving CPA 50 mg daily for acne which was discontinued during hospital admission. One year later, the two tumors had shrunk by about 80% each, vessels within the tumor were no longer visible on imaging, and the patient's neurologic status remained normal. The second patient was a 43-year-old female diagnosed with multiple meningiomas after workup for seizures and short-term memory deficits. This patient was also receiving CPA 25 mg daily for 20 years for acne, which was discontinued. The patient was a candidate for surgery, but a more conservative approach was elected. At 1 year all tumors had decreased in size on imaging. The authors conclude that these cases add to the evidence supporting a conservative approach to meningioma treatment in which the first treatment step is discontinuing progestin-containing medications with close follow-up, even in tumors amenable to surgical intervention [11A].

Aromatase Inhibitors: Anastrozole [SEDA-35, 735; SEDA-36, 619; SEDA-37, 504; SEDA-39, 423]

Autoimmune Hepatitis

Anastrozole is an aromatase inhibitor frequently used in the setting of estrogen receptor-positive breast cancer in postmenopausal women. Anastrozole prevents the conversion of androgens to estrogens, thereby suppressing plasma estrogen levels. A case of a 71-year-old female with breast cancer who developed autoimmune hepatitis (AIH) was recently described. The patient had no prior history of liver disease. Routine blood work revealed elevated liver function tests (LFTs), and she was found to have hepatic steatosis but no liver metastases on imaging. Her LFTs continued to rise over the course of 5 months. Subsequent workup showed elevated antinuclear antibodies (ANAs) and antibodies to Sjögren's-syndrome-related antigen A. Liver biopsy revealed grade 2 chronic hepatitis that was classified as AIH. The patient discontinued anastrozole therapy and was treated with prednisone and azathioprine. Her LFTs improved and have remained normal for 2 years. The authors report they were able to find two additional cases of anastrozole-induced AIH in the literature. One proposed mechanism is modulation of the immune system by anastrozole. Clinicians should be aware of this rare but potentially serious complication of anastrozole use [12A].

Gonadotropins, Clomiphene Citrate, and hCG [SEDA-36, 660; SEDA-37, 539; SEDA-38, 463; SEDA-39, 447]

Malformations

A population-based retrospective cohort study evaluated the association of major malformations overall or with specific fetal anomalies when pregnant women in southern Israel were exposed to clomiphene citrate (CC). No association was detected between CC exposure and rates of major malformations in 1872 pregnant women (aOR 1.08, 95% CI 0.88–1.32). Exposure was not associated with anencephaly (aOR 2.27, 95% CI 0.44–11.71) or esophageal atresia (aOR 3.681, 95% CI 0.65–20.76). This large retrospective cohort study demonstrated lack of major malformations with CC use [13c].

Ovarian Hyperstimulation Syndrome

A retrospective, cohort study assessed the efficacy of two different gonadotropin-releasing hormone (GnRH) agonists, triptorelin and leuprolide, in final oocyte maturation in patients with increased risk of ovarian hyperstimulation syndrome (OHSS). One mild case of OHSS occurred in both the leuprolide and triptorelin treatment groups in which both patients complained of lower abdominal pain, mild nausea, enlarged ovaries, and vomiting. Both patients were <25 years of age, had elevated estradiol levels >4000 pg/mL, and >25 oocytes collected. Both medications were comparable in terms of clinical pregnancy and OHSS rates as compared to placebo [14c].

Another retrospective cohort study at an academic medical center evaluated pregnancy outcomes and OHSS using a sliding scale hCG protocol in 10427 fresh in vitro fertilization—intracytoplasmic sperm injections. The incidence of moderate to severe OHSS was 0.13% ($n=14$) and severe OHSS was 0.03% ($n=4$) of cycles. This was one of the few studies that determined the lowest threshold dose of hCG to maintain high pregnancy rates while decreasing risk of OHSS [15c].

Endometrial Cancer

This review evaluated the association between ovary-stimulating drugs for treatment of infertility and risk for endometrial cancer. Five studies assessed subfertile women exposed to CC found a positive association (RR 1.32, 95% CI 1.01–1.71; 88618 participants, very low quality evidence) at high dosage (RR 1.69, 95% CI 1.07–2.68, 2 studies, 12073 participants) and high number of cycles (RR 1.69, 95% CI 1.16–2.47; 3 studies, 13757 participants) with endometrial cancer. Although studies were low quality evidence, CC was associated with increased risk of endometrial cancer with high doses >2000 mg and high number of cycles >7. Other studies included in the review examining exposure to other gonadotropins did not provide robust evidence for an increased risk of endometrial cancer [16R].

Palinopsia

A case report of a 22-year-old man noted parotitis and orchitis after starting CC 100 mg daily for preservation of sperm activity. Palinopsia persisted more than 1 year after discontinuation of CC. Physical and laboratory examination were unremarkable with exception of patient's complaints of blurred images of moving objects. Palinopsia associated with CC may be dose-related but the mechanism is still unknown. The authors suspected the side effect may be related to alterations of serotonin in the brain, which was treated with tianeptine, a serotonin reuptake inhibitor. Despite treatment, the patient's clinical condition did not improve [17A].

A 30-year-old Asian women complained of persistent shadow-like visual afterimages following treatment with CC 100 mg per day for 5 consecutive days for anovulation. The symptoms started 4 days after CC initiation and occurred 5–7 times in a month lasting for 5–10 min. Patient denied pain, redness, photophobia, migraines, or decreased vision. She conceived after two cycles of CC. Symptom severity and frequency decreased but persisted for more than 1 year after discontinuation of CC. Physical examination was unremarkable. Illusory palinopsias are caused by diffuse neuronal pathology (i.e. global alterations in neurotransmitters). Estrogen can directly or indirectly stimulate the visual cortex, triggering visual hallucinations. CC has agonistic properties when endogenous estrogen levels are low and may contribute to ocular side effects [18A].

ANABOLICS, ANDROGENS, AND RELATED COMPOUNDS [SEDA-36, 627; SEDA-37, 507–508; SEDA-38, 439–441; SEDA-39, 423]

Testosterone [SEDA-36, 631; SEDA-37, 509; SEDA-38, 439–440; SEDA-39, 423–424]

Anthropometric and Biochemical Profiles

A prospective study assessed the associations between testosterone treatment in patients with late-onset hypogonadism and wide range of characteristics that included hormonal, anthropometric, and biochemical features. Eighty-eight patients received 1000 mg of intramuscular testosterone undecanoate for 1 year. Testosterone treatment increased serum testosterone levels and significantly decreased body mass index (BMI), white count, total cholesterol, triglycerides, and hemoglobin A1c from initial baseline levels in all patients. Further, testosterone injections significantly increased hemoglobin and hematocrit at the end of the 12-month follow-up period. No significant increases in liver function enzymes were observed [19c].

Central Serous Retinopathy

Providers have increasingly prescribed off-label testosterone to postmenopausal women for improvement of sexual function. A retrospective case series from a referral retina practice describes three postmenopausal women who presented with prolonged visual changes diagnosed as central serous retinopathy (CSR) after receiving exogenous testosterone and progesterone therapy for symptoms of menopause and libido loss. CSR is characterized by serous detachment of the neurosensory retina causing metamorphopsia and visual loss. The women were treated with testosterone for 36–87 months before presentation of visual symptoms. The average age at presentation was 54.7 years and all three patients had unilateral findings from optical coherence tomography. Discontinuation of androgen therapy, foveal sparing oscillatory photodynamic therapy (OPDT), and intravitreal bevacizumab resulted in complete resolution of CSR in all three women [20A].

Anabolic Steroids [SEDA-35, 738; SEDA-36, 627; SEDA-37, 507–508; SEDA-38, 440–441; SEDA-39, 424]

Focal Nodular Hyperplasia

Focal nodular hyperplasia (FNH) is the second most common benign tumor of the liver but clinically relevant cases of FNH are rare in US studies. A 30-year-old man taking dietary and herbal supplements including anabolic steroids, carnitine, and L-arginine, with no significant PMH, including liver disease or risk factors for viral hepatitis, presented with diffuse abdominal pain and weight loss. Examination revealed an enlarged liver with a focal lesion but no biliary obstruction. Magnetic resonance imaging (MRI) determined that the lesion was about 14 × 9cm in diameter, hypointense on diffusion weighted images. Liver biopsy was not done. FNH was not initially diagnosed but suspected due to increase in size of mass during 2 years of continuous monitoring. Surgery was recommended and FNH was diagnosed histologically. Anabolic-androgenic steroids (AASs), carnitine, L-arginine are precursors of nitric oxide (NO), an endogenous mediator of vasodilation. Long-term use of AASs may cause changes of liver tissue via NO synthesis. It is recommended to regularly monitor liver function in patients taking AASs [21A].

Renal Infarction

A 43-year-old male with PMH of obsessive compulsive disorder and prior appendectomy presented with 2 days of left flank pain. The patient reported taking both testosterone and the injectable AAS trenbolone acetate intermittently over 5 years with last use 2 weeks prior to admission. A contrast CT indicated new, wedge-shaped hypodensity in the superolateral pole of the left kidney and dilatation of the left renal artery. Serum creatinine increased from baseline (<1.2mg/dL) to 1.7mg/dL. The patient was diagnosed with left renal parenchymal infarction and acute kidney injury. A transthoracic echocardiogram and hypercoagulable workup were unremarkable. Patient was ultimately bridged with enoxaparin and placed on warfarin therapy. This is one of few studies reporting renal infarct secondary to AAS use. Given the delayed diagnosis, this case highlights the importance of asking all patients, especially young, athletic men, about supplements as well as AAS use. Direct oral anticoagulants have not been established as treatment for renal infarcts [22A].

Androgen Deprivation Therapy (ADT) and Anti-Androgens [SEDA-35, 740; SEDA-36, 628; SEDA-37, 508–509]

Abiraterone is FDA approved for treatment of metastatic castration-resistant prostate cancer (MCRPC) in adult men who are asymptomatic or mildly symptomatic after failure of androgen deprivation therapy in whom chemotherapy is not clinically indicated. Three cases are described in this case series study after 1 year of implementation of a MCRPC protocol for abiraterone candidates. An 85-year-old patient with prostate adenocarcinoma and PMH of hypertension, history of ischemic heart disease (ejection fraction 44%) and aortic valve stenosis was given 1000mg abiraterone with prednisone. Forty-eight hours later, the patient complained of two episodes of crushing chest pain both relieved by nitroglycerin. His blood pressure was 190/108mmHg and he required admission to emergency services. The second case reports a 92-year-old male with PMH of controlled hypertension and renal disease. After 6 days of abiraterone, the patient reported pruritic erythematous lesions on the trunk that persisted despite discontinuing abiraterone and administering an antihistamine. The patient was admitted to the hospital and treated with topical corticosteroids (methylprednisone) leading to almost complete disappearance of lesions 1 month later. The third patient was a 56-year-old with no significant PMH, including normal liver function. After 4 weeks of abiraterone, transaminase levels substantially increased (GOT 243IU/L, GPT 568IU/L, GGT 24IU/L). Abiraterone treatment was subsequently discontinued. Transaminase levels peaked at 2 weeks and declined at 4 weeks after therapy cessation. The authors concluded the importance of close monitoring of cardiac function, renal function, electrolytes, and liver function for at least 4 weeks after starting treatment [23A].

Optic Neuropathy

A 21-year-old female presented with 4-day history of right brow pain exacerbated by eye movement and 3-day history of blurring of the right temporal field of vision. The patient was taking desogestrel 75mg and cyproterone acetate 50mg, an ADT, for 2 months for hormone imbalance. The patient displayed signs of hirsutism and acne and was overweight. Laboratory workup was unremarkable and Goldmann perimetry showed right enlarged blind spot with temporal visual loss. Discontinuation of antiandrogen therapy led to resolution in eye pain and improved visual acuity within 3 weeks. There was a temporal correlation between discontinuation of antiandrogen therapy and improvement in visual symptoms. This was one of few cases demonstrating ocular side effects of cyproterone. A possible hypothesis of isolated optic neuropathy is lower systemic levels of testosterone or estrogen [24A].

Bilateral Femoral Neck Fractures

ADT in prostate cancer is associated with decreased bone mineral density (BMD) and increased risk of fracture. An 83-year-old man with PMH significant for lumbar spondyloarthrosis and prostate adenocarcinoma being actively treated by ADT was admitted with bilateral femoral neck fracture. The patient denied history of falls or injury. Bilateral cemented hemiarthroplasty was performed with favorable results. Patient was able to stand and walk some steps with walker-aid and was transferred to a rehabilitation hospital. While limited evidence exists, calcium and vitamin D supplementation is recommended to improve BMD for men on ADT [25A].

References

[1] Rivollier F, Chaumette B, Bendjemaa N, et al. Methylomic changes in individuals with psychosis, prenatally exposed to endocrine disrupting compounds: lessons from diethylstilbestrol. PLoS One. 2017;12(4):e0174783 [c].

[2] The NAMS 2017 Hormone Therapy Position Statement Advisory Panel. The 2017 hormone therapy position statement of The North American Menopause Society. Menopause. 2017;24(7):728–53. New York, N.Y. [S].

[3] Chan W, Drummond A, Kelly M. Deep vein thrombosis in a transgender woman. CMAJ. 2017;189(13):E502–4 [A].

[4] Sharif A, Malhotra NR, Acosta AM, et al. The development of prostate adenocarcinoma in a transgender male to female patient: could estrogen therapy have played a role? Prostate. 2017;77(8):824–8 [A].

[5] Gemmell LC, Webster KE, Kirtley S, et al. The management of menopause in women with a history of endometriosis: a systematic review. Hum Reprod Update. 2017;23(4):481–500 [M].

[6] Brady PC, Missmer SA, Laufer MR. Hepatic adenomas in adolescents and young women with endometriosis treated with norethindrone acetate. J Pediatr Adolsc Gynecol. 2017;30(3):422–4 [A].

[7] Sinclair M, Schelleman A, Sandhu D, et al. Regression of hepatocellular adenomas and systemic inflammatory syndrome after cessation of estrogen therapy. Hepatology. 2017;66(3):989–91 [A].

[8] Tanaka S, Hishiki M, Ogasawara J, et al. The deterioration of the glycemic profile during hormone replacement therapy in a patient with fulminant type 1 diabetes. Intern Med. 2017;56(5):531–4 [A].

[9] Aydin M, Ilhan C, Akyildiz A, et al. Medroxyprogesterone acetate-induced mania in a patient with bipolar affective disorder. Düsünen Adam. 2017;30:66–9 [A].

[10] Wunschel J, Poole JA. Intrauterine uterine contraception and chronic urticaria: a case series. Ann Allergy Asthma Immunol. 2017;118(3):378–80 [A].

[11] Kalamarides M, Peyre M. Dramatic shrinkage with reduced vascularization of large meningiomas after cessation of progestin treatment. World Neurosurg. 2017;101:814.e7–14.e10 [A].

[12] Klapko O, Ghoulam E, Jakate S, et al. Anastrozole-induced autoimmune hepatitis: a rare complication of breast cancer therapy. Anticancer Res. 2017;37(8):4173–6 [A].

[13] Weller A, Daniel S, Koren G, et al. The fetal safety of clomiphene citrate: a population-based retrospective cohort study. BJOG. 2017;124(11):1664–70 [c].

[14] Sukur YE, Ozmen B, Ozdemir ED, et al. Final oocyte maturation with two different GnRH agonists in antagonist co-treated cycles at risk of ovarian hyperstimulation syndrome. Reprod Biomed Online. 2017;34(1):5–10 [c].

[15] Gunnala V, Melnick A, Irani M, et al. Sliding scale HCG trigger yields equivalent pregnancy outcomes and reduces ovarian hyperstimulation syndrome: analysis of 10,427 IVF-ICSI cycles. PLoS One. 2017;12(4):e0176019 [c].

[16] Skalkidou A, Sergentanis TN, Gialamas SP, et al. Risk of endometrial cancer in women treated with ovary-stimulating drugs for subfertility. Cochrane Database Syst Rev. 2017;3: Cd010931 [R].

[17] Choi SY, Jeong SH, Kim JS. Clomiphene citrate associated with palinopsia. J Neuro-Oncol. 2017;37(2):220–1 [r].

[18] Venkatesh R, Gujral GS, Gurav P, et al. Clomiphene citrate-induced visual hallucinations: a case report. J Med Case Rep. 2017;11(1):60 [A].

[19] Canguven O, Talib RA, El Ansari W, et al. Testosterone therapy has positive effects on anthropometric measures, metabolic syndrome components (obesity, lipid profile, diabetes mellitus control), blood indices, liver enzymes, and prostate health indicators in elderly hypogonadal men. Andrologia. 2017;49(10) [c].

[20] Conway MD, Noble JA, Peyman GA. Central serous chorioretinopathy in postmenopausal women receiving exogenous testosterone. Retin Cases Brief Rep. 2017;11(2):95–9 [A].

[21] Romano A, Grassia M, Esposito G, et al. An unusual case of left hepatectomy for focal nodular hyperplasia (FNH) linked to the use of anabolic androgenic steroids (AASs). Int J Surg Case Rep. 2017;30:169–71 [A].

[22] Colburn S, Childers WK, Chacon A, et al. The cost of seeking an edge: recurrent renal infarction in setting of recreational use of anabolic steroids. Ann Med Surg. 2017;14:25–8 [A].

[23] Ramudo-Cela L, Balea-Filgueiras J, Vizoso-Hermida JR, et al. Study of cases of abiraterone discontinuation due to toxicity in pre-chemotherapy after 1 year's experience. J Oncol Pharm Pract. 2017;23(8):615–9 [A].

[24] Ni Mhealoid A, Cunniffe G. Optic neuritis secondary to antiandrogen therapy. Ir J Med Sci. 2017;186(3):565–70 [A].

[25] Marttini Abarca J, Carrillo Garcia P, Mora-Fernandez J, et al. Atraumatic bilateral femoral neck fractures in an octogenarian male associated with androgen deprivation therapy: a case report. Eur Geriatr Med. 2016;8:77–8 [A].

CHAPTER

40

Miscellaneous Hormones

Amulya Tatachar[*,1], *Sidhartha D. Ray*[†], *Caitlin M. Gibson*[*]

*System College of Pharmacy, University of North Texas Health Science Center, Fort Worth, TX, United States

[†]Department of Pharmaceutical and Biomedical Sciences, Touro University College of Pharmacy and School of Osteopathic Medicine, Manhattan, NY, United States

[1]Corresponding author: amulya.tatachar@unthsc.edu

SOMATOTROPIN (HUMAN GROWTH HORMONE, hGH) [SEDA-36, 661; SEDA-37, 542; SEDA-38, 463; SEDA-39, 448]

Limited systemic evidence exists on evaluating adverse effects related to growth hormone abuse in humans. Overuse/abuse may be associated with unknown adverse effects given that these medications are often combined with other agents and administered at higher doses than recommended. The first documented report of recombinant human growth hormone (rhGH)-abuse-induced diabetes was reported in 2007 in a 36-year-old male who presented to the emergency department with acute renal failure after 3 years of rhGH use and 15 years of steroid abuse. Other side effects are similar to those observed in acromegaly (i.e., hypertension, carpal tunnel syndrome, diabetes, neuropathy), frail elderly (i.e., edema, gynecomastia), and healthy adults [1R].

A multicentre, randomized, controlled, open-label phase III trial ($n = 210$ children) evaluated the safety profile of biosimilar rhGH for 84 months of treatment. Commonly reported adverse effects with a frequency of at least 0.05 events per patient-year included increased hemoglobin A1c, headache, and injection-site hematoma. One serious side effect of worsening pre-existing scoliosis requiring several hospitalizations occurred. A multicenter, non-comparative phase III study ($n = 51$) showed adverse effects occurred in >5% of patients and included eosinophilia (12%), injection site reactions (10%), hypothyroidism (8%), and scoliosis (6%). The patients treated with Omnitrope® (PATRO) Children Study, a long-term, post-marketing surveillance program, found 38.4% ($n = 5007$) experienced mild to moderate intensity adverse effects, with headache being the most common (75 patients, 1.5%). The main objective of the PATRO Children is to assess long-term safety of biosimilar rhGH. Another ongoing, prospective, open-label, non-comparative, multicenter, phase IV study assessing safety and efficacy of Omnitrope® in children with short stature born with small for gestational age (SGA) is currently enrolling patients. Currently, 249 patients have completed 2 years of treatment and no child has developed diabetes mellitus during the first 2 years of trial. Twenty-four drug-related adverse effects have occurred in 17 patients (6.1%), including hypothyroidism, headache, impaired fasting plasma glucose, hematoma, upper respiratory tract infections. PATRO Adults is an ongoing, observational, multicenter, open, longitudinal study assessing safety and efficacy of rhGH in adults treated in routine practice. As of March 2016, 1043 adult patients have been enrolled and 2025 adverse drug events (ADEs) have been reported in 597 patients (54%). Main ADEs reported are related to nervous system disorders, musculoskeletal/connective tissue disorders, administration site conditions, and increased insulin-like growth factor-1 (IGF) levels. Of the 159 who discontinued treatment, 39 discontinued due to ADEs [2R].

GeNeSIS is a prospective, multinational, open-label, pediatrician surveillance program designed to examine long-term safety and efficacy of GH (Humatrope®) administered for treatment of short stature as a result of SHOX deficiency. Five hundred and fourteen patients with SHOX deficiency were followed for 2.8 ± 2.0 years. Fourteen side effects were reported in 12 patients and included appendicitis, chronic renal failure, adenoidectomy, death, hypertension, ligament injury, limb operation, lower limb fracture, neoplasm recurrence, osteomyelitis, post-streptococcal glomerulonephritis, and spondylolisthesis. The death occurred in 18 months after discontinuing GH and the neoplasm recurrence

Side Effects of Drugs Annual, Volume 40
ISSN: 0378-6080
https://doi.org/10.1016/bs.seda.2018.06.008

occurred about 1 year and 11 months after GH initiation. The adverse effects for GH-treated patients did not indicate any new concerns [3r].

OXYTOCIN AND ANALOGUES [SEDA-36, 665; SEDA-37, 546; SEDA-38, 465; SEDA-39, 450]

Autism Spectrum Disorder

Due to oxytocin's role in influencing social behavior, it has recently been studied as a potential treatment option in Autism Spectrum Disorder (ASD). A review was recently published to examine the safety of intranasal oxytocin in this unique patient population. Six studies met inclusion criteria for the review; all were crossover trials and all were published since 2015. Intranasal oxytocin doses varied from 16 to 48 international units per day. The majority of the adverse effects reported in the trial were well-known side effects such as nasal discomfort, fatigue, and irritability. One study reported three cases of hyperactivity and aggression; two occurred during week 1 of oxytocin administration and one occurred during week 1 of placebo administration. All three cases resolved after oxytocin discontinuation. Another study reported two patients who experienced seizures. One patient was likely non-adherent to previously prescribed anti-epileptic therapy, and one experienced seizures both during the oxytocin and control phases of the study. The authors concluded that oxytocin use is likely safe in patients with ASD and that the severe adverse events reported in trials were unlikely to be linked to oxytocin [4M].

Breastfeeding Deficits

Endogenous oxytocin is critical for regulating neuroendocrine pathways responsible for onset of lactation. However, manipulation of oxytocin concentrations with administration of exogenous oxytocin may alter lactation capacity, and exogenous oxytocin can pass into breast milk and potentially impact neonates as well. A review of 26 studies examining the impact of exogenous oxytocin on breastfeeding outcomes was recently conducted. There were significant heterogeneity between trials, and not all trials were conducted to answer the clinical question of oxytocin's impact on breastfeeding. Some studies suggested delayed initiation of breastfeeding, reduced duration of breastfeeding, and/or negatively impacted infant behaviors or feeding-related reflexes. On the contrary, several studies also showed mixed results or no impact of oxytocin on these endpoints. The authors concluded that due to the high variability of studies included in this review, healthcare providers should counsel patients that there is no proven effect of synthetic oxytocin on breastfeeding outcomes or neonate feeding behaviors. Screening for oxytocin exposure in the assessment for suboptimal breastfeeding may lend to early intervention. The authors concluded that more research is needed in this area [5R].

Atrioventricular Block

Ergonovine is an ergot-derived oxytocic agent frequently used on obstetrics. A recent case of its use in a 38-year-old pregnant female with no medical history aside from infertility suggests ergonovine may induce atrioventricular (AV) block. While the patient was receiving care for in vitro fertilization, she complained of swollen right adnexa and was diagnosed with a hydrosalpinx. She underwent laparoscopic surgery and later suffered from vicinal bleeding which was treated with oral ergonovine 0.2 mg three times daily. After the fourth dose, the patient complained of chest tightness and weakness, and an electrocardiogram (EKG) revealed complete AV block. According to the Naranjo adverse drug reaction causality score, ergonovine was the probable cause of the adverse reaction. The agent was subsequently discontinued and normal sinus rhythm was restored the next day. Although ischemic heart disease was ruled out in this patient, one proposed mechanism of this ADR is coronary vasospasm-induced changes in cardiac electrophysiology. Other proposed mechanisms include modulation of calcium channels or alpha receptors in the myometrium. While the authors state that this is likely the first case of complete AV block, bradycardia has been previously reported with methylergonovine use. The authors call for increased attention to this potentially fatal adverse effect of a commonly used obstetric drug [6A].

SOMATOSTATIN (GROWTH HORMONE RELEASE-INHIBITING HORMONE) AND ANALOGUES [SEDA-35, 794; SEDA-36, 666; SEDA-37, 549; SEDA-38, 466]

Octreotide

Neutropenia

Octreotide is a somatostatin analog frequently prescribed for the treatment of variceal bleeds, acromegaly, and severe diarrhea. A recent case described the course of an 87-year-old man who received three doses of octreotide 100 μg subcutaneously every 8 h for the treatment of a malignant bowel obstruction. His white blood cell (WBC) count decreased from 4100 cells/mm^3 (75% neutrophils) on admission to 1600 cells/mm^3 (62% neutrophils) the day after octreotide administration.

The neutropenia was confirmed with a repeat WBC of 1300 cells/mm^3 (60% neutrophils). The patient also developed a fever but it was deemed unlikely that he was experiencing infection. Octreotide was discontinued and the patient's WBCs increased to 4900 cells/mm^3 the next day. There was a probable relationship between octreotide and neutropenia per the Naranjo Adverse Drug Reaction Probability Scale. The authors state that this case is the first known report of neutropenia associated with octreotide use. The mechanism is unknown, but the authors postulate that the most likely explanation is an immune-related response due to the prompt decrease in WBC count and antibody production in response to octreotide. This adverse effect has been previously reported [7A].

Necrotizing Enterocolitis

Soon after birth, a newborn received Octreotide 20 μg/kg/day for hypoglycemia and diagnosed with congenital hyperinsulinism. A month later, the newborn failed to respond to medical therapy indicated for hypoglycemic attacks and therefore, Octreotide 30 μg/kg/day was administered again. At 4 months old, the patient was readmitted with abdominal distention, emesis, and bloody diarrhea. He was diagnosed with acute hemorrhagic gastroenteritis and necrotizing enterocolitis. He also tested positive for Rota virus. He was treated with anti-infectives and subsequently discharged. Some aspects of the patient's development were delayed including speech and hearing. The authors postulate that octreotide may have contributed to the patient's necrotizing enterocolitis, possibly due to a dose-dependent decrease in splanchnic blood flow [8A].

Lanreotide

Hepatic Cyst Infection

An interim analysis of an ongoing clinical trial of lanreotide in autosomal dominant polycystic kidney disease (ADPKD) suggested that there is an increased risk of hepatic cyst infection with lanreotide use. Among 309 patients enrolled in the trial, there were eight cases of hepatic cyst infection among seven patients (2%) compared to no patients in the standard care (lanreotide-free) group. The patients ranged from 46 to 57 years of age. There were no differences in baseline characteristics between the groups except that more patients in the lanreotide group had a history of hepatic cyst infection (29% vs 0.7%). The mechanism is unclear but may involve bacterial translocation from the gut to the liver, a process that may be promoted by the reduced blood flow in the gut which occurs as a consequence of somatostatin analog use. Another proposed mechanism is somatostatin-analog induced reduction in gallbladder contractility which may promote bacterial translocation through the biliary tract. This adverse effect is likely a class effect as it has previously been described in the literature with octreotide. Regardless, most cases of hepatic cyst infection require hospitalization and close follow-up, so the authors urge health care practitioners to consider the benefit to risk ratio when prescribing this medication [9c].

Glycemic Profile Alterations

Alterations of glycemic profile has been well described with somatostatin and somatostatin analogs. The package inserts for these agents cautions about the risk of hypoglycemia, especially in patients receiving concurrent insulin therapy. A recent case of an 80-year-old female with a history of diabetes mellitus detailed the course of glycemic alterations after initiating lanreotide, a somatostatin analog. The patient was receiving multiple insulin injections daily to maintain euglycemia when she was diagnosed with acromegaly and received a dose of lanreotide 90 mg intramuscularly. The next day the patient's blood glucose dropped significantly. The patient was readmitted to the hospital for a repeat bout of hypoglycemia 4 days after the lanreotide injection. Labs revealed low urine and serum C-peptide. The patient received a dextrose infusion and her blood glucose normalized on day 5. This article adds to the evidence that lanreotide contributes to dysregulation of glucose control in 30%–40% of patients [10A]. This phenomenon was also reflected in a recent trial of in acromegaly in which hemoglobin A1c was significantly decreased from baseline in patients receiving lanreotide [11c].

Pasireotide

Interestingly, the safety and tolerability results of pasireotide during the acromegaly, open-label, multicenter, safety monitoring program (ACCESS) for treating patients who have a need to receive medical care report that 45.5% of patients experienced hyperglycemia-related adverse events. All 43 patients with evaluable glycemic parameters experienced fasting blood glucose levels >99 mg/dL (5.5 mmol/L) after pasireotide administration compared to 18 patients (41.9%) at baseline. Four patients ultimately discontinued participation in the trial due to hyperglycemia. The authors note that the low discontinuation rate suggests that the hyperglycemia-related adverse effects are likely manageable with standard anti-diabetic therapy. A previous study suggests that the mechanism is related to a somatostatin-analog induced reduction in insulin secretion. Of note, six patients (13.6%) experienced hypoglycemia during trial enrollment [12c].

DESMOPRESSIN (N-AMINO-8-D-ARGININE VASOPRESSIN, DDVAP AND ANALOGUES) [SEDA-35, 798; SEDA-36, 669; SEDA-37, 552; SEDA-38, 467, SEDA-39, 453]

Arginine Vasopressin

Transient Diabetes Insipidus

While two case reports on transient diabetes insipidus (DI) after arginine vasopressin (AVP) discontinuation have been published, this is the first case series of this phenomenon and the first report to include neurosurgical patients. The series describes six patients with critical neurological illness including subarachnoid, intraparenchymal, and intraventricular hemorrhages, glioblastoma, epidural hematoma, and traumatic brain injury. In all cases, AVP was used for hemodynamic support, usually as a second- or third-line agent. Half of the patients experienced hyponatremia from syndrome of inappropriate antidiuretic hormone (SIADH) and/or cerebral salt wasting prior to development of DI, often requiring treatment with demeclocycline and/or hypertonic saline. In all cases, discontinuation of AVP infusion resulted in a sudden increase in serum sodium levels up to 180 mEq/L and sudden increase in urine output of up to 1700 mL/h. Many cases were treated by restarting AVP infusion and/or desmopressin doses of 2 or 5 μg. To prevent and treat DI after AVP discontinuation, the authors recommend prompt attention to sudden increases in serum sodium levels, which might normally result in suspicion of laboratory error. Additionally, hypertonic saline and drugs which can induce DI such as demeclocycline should be held for 12–24 h before discontinuing AVP. Close monitoring of urine output and serum sodium levels are critical, and any resulting DI should be treated with desmopressin 2 μg twice daily or, in severe cases, resumption of AVP infusion. The mechanism of DI in these patients is not well understood but the authors argue that despite response to desmopressin therapy, these cases may represent nephrogenic DI caused by downregulation of vasopressin 2 receptors in response to prolonged supraphysiologic AVP dosing [13A].

Desmopressin

Reset Osmostat

Reset osmostat is a rare condition which hyponatremia occurs due to preservation of osmoregulation of vasopressin but antidiuretic hormone (ADH) osmolality threshold then is lowered. A 72-year-old male physician developed hyponatremia secondary to desmopressin acetate superimposed on chronic hyponatremia caused by reset osmostat. Patient's past medical history consisted of self-prescribed oral desmopressin acetate (DDAVP) 100 μg three times a day despite lack of urinary measurements or water deprivation testing. Patient also reported taking propafenone, warfarin for deep vein thrombosis secondary to factor V Leiden deficiency, atenolol, pyridoxine, and salt tablets. The desmopressin dose was adjusted to achieve a goal of ~2 L of urine output per day. Patient's serum sodium level prior to admission was normal at ~140 mEq/L. Patient denied recent alcohol use or exposure to glycine, sorbitol, or mannitol. Cardiovascular and pulmonary examination were mostly unremarkable with exception of irregularly irregular rhythm without murmurs or rubs. At admission, sodium was 115 mEq/L, urine osmolality 759 mOsm/kg, and urine Na 63 mEq/L. All other labs were within normal limits. At patient presentation, DDAVP and salt tablets were discontinued, dextrose 5% in water was given, and at discharge, patient was recommended to allow thirst to guide his water intake. One-year post discharge, patient's sodium was 131 mEq/L. This unusual case presents a euvolemic man with profound hyponatremia with a diagnosis of reset osmostat. The cause of reset osmostat is unknown given he was overall healthy at presentation but may be a potential sequela of long-term DDAVP exposure [14A]. See the previous edition of this textbook for more information on cases of hyponatremia with desmopressin use[15R].

Central Pontine Myelinolysis (CPM)

CPM consists of central nervous system demyelination after rapid correction of hyponatremia. This case study describes CPM in a pregnant patient, a condition relatively unknown among obstetricians and gynecologists. A 39-year-old pregnant woman in week 25 presented to the hospital for pre-term labor. The patient was given atosiban, an oxytocin antagonist, to cease uterine contractions and prevent pre-term labor and betamethasone acetate for fetal lung maturation. After first cycle of atosiban started on day of admission, sodium level showed 129 mEq/L. The second cycle (started on day 6 of admission), patient showed a sodium level of 115 mEq/L on day 7 and 113 mEq/L on day 14. Atosiban was given for three cycles, uterine contractions ceased, and patient was discharged. At 1-week follow-up, patient showed slowed speech, decreased lower limb strength and immediately referred to Neurology. Neurological exam showed abnormalities. Sodium levels were taken and was 137 mEq/L demonstrating rapid correction of sodium. Patient underwent emergent caesarean section at week 31 and gave birth to two males. A drug alert published by Spanish Agency of Medicines and Health Products (AEMPS) suspended atosiban preparation due to hyponatremia-related cases in pregnancy. Investigations confirmed atosiban being contaminated by desmopressin and the lots recalled were located at the hospital the patient was admitted to. In this case, desmopressin-contaminated atosiban shows a certain probability (score = 8) in the Karch–Lasagna algorithm for a causality relationship between hyponatremia and drug [16A].

VASOPRESSIN RECEPTOR ANTAGONISTS [SEDA-35, 797; SEDA-36, 668; SEDA-37, 552; SEDA-38, 466]

Tolvaptan

Tolvaptan is a vasopressin type 2 receptor antagonist prescribed to correct hyponatremia or for the treatment of autosomal dominant polycystic kidney disease. The agent carries a Boxed Warning in the United States because rapid correction of hyponatremia may lead to central pontine demyelination. A case of amiodarone-induced syndrome of inappropriate antidiuretic hormone secretion (SIADH) which was masked by concomitant use of tolvaptan was recently described. The patient was admitted to the hospital for new heart failure which was treated with continuous milrinone and dobutamine. His serum sodium at admission was 139 mEq/L and he was initiated on furosemide at an unknown dose and tolvaptan 7.5 mg daily. The patient suffered atrial tachycardia which could not be controlled with beta blockers due to symptomatic hypotension. Amiodarone was initiated as an intravenous infusion which was converted to oral therapy after normal sinus rhythm was achieved. The patient remained inotrope-dependent and required a left ventricular assist device (LVAD) implantation 9 months after hospital admission. Tolvaptan was discontinued after the surgery and the patient was continued on amiodarone 100 mg daily. One month after surgery his sodium had decreased from 143 to 116 mEq/L. The patient's antidiuretic hormone (ADH) level was elevated at 3.6 pg/mL, so he was diagnosed with SIADH and treated with water restriction. Three months later his sodium improved to 131 mEq/L and he was discharged from the hospital. His serum sodium remained low, and at 10 months post-surgery amiodarone was thought to be the culprit in drug-induced SIADH, so his dose was reduced to 50 mg daily. Three months later he was admitted to the hospital with lethargy and malaise and his sodium was found to be 108 mEq/L. Amiodarone was discontinued and he was treated with fludrocortisone and eventually discharged. Unfortunately, 2 days after discharge he became delirious and his LVAD was damaged and he required urgent pump exchange. One month later his serum sodium improved to 126 mEq/L and his ADH level improved to 0.6 pg/mL. This case cautions about the potential for masked drug-induced SIADH in the setting of tolvaptan use, and sodium levels should be closely monitored after discontinuation of vasopressin antagonists [17A].

CALCIMIMETICS

Entelcalcetide

The USFDA approved entelcalcetide for treatment of secondary hyperparathyroidism (HPT) in adult patients with chronic kidney disease (CKD) on hemodialysis in February 2017. Approval was based on two 26-week, randomized, double-blind, placebo controlled studies. An aggregate of 1023 patients with moderate-to-severe HPT on hemodialysis were randomized to receive intravenous entelcalcetide ($n = 503$) or placebo ($n = 513$) three times a week at the end of dialysis sessions in addition to standard of care. Entelcalcetide significantly reduced baseline PTH levels compared to placebo at the end of 6 months. FDA noted risks associated with entelcalcetide were generally consistent with risks expected for the calcimimetic class of drugs. Gastrointestinal (GI) adverse reactions (i.e., nausea, vomiting), hypocalcemia, hypophosphatemia, and oversuppression of parathyroid hormone (adynamic bone suppression) were adverse reactions with this drug class and entelcalcetide. Cases of worsening heart failure and fatal upper GI bleeds have occurred. In clinical studies, two patients treated with entelcalcetide in 1253 patient-years of exposure had upper GI bleeding at the time of death. The exact cause of GI bleeding in these patients are unknown and given the small number of cases, the investigators were unable to determine whether cases were related to entelcalcetide. Patients with risk factors for upper GI bleeding such as gastritis, esophagitis, ulcers, or severe vomiting may be at increased risk for GI bleeding with entelcalcetide [18C].

A randomized, double-blind, double-dummy active clinical trial comparing IV entelcalcetide vs oral placebo and oral cincalcet vs IV placebo in 683 patients receiving hemodialysis with PTH levels above 500 pg/mL demonstrated non-inferiority of entelcalcetide as compared to cinacalcet in reducing PTH concentrations over 26 weeks and no differences in major adverse effects. Both medications decreased blood calcium (68.9% vs 59.8%) [19C].

Cinacalcet

A review of cinacalcet after kidney transplantation summarizing 24 articles consisting of two randomized controlled trials and 22 observational studies was recently published. A total of 713 patients were treated for a wide range of time (from 2 weeks up to 53 months) after renal transplantation. Overall, 37 (6.4%; $n = 578$) patients treated with cinacalcet discontinued treatment due to persistent hypercalcemia or side effects such as GI problems, paraesthesias, and hypocalcemia. Doses of cinacalcet ranged from 30 to 180 mg daily [20R].

A case report described a 58-year Thai woman with severe hyperparathyroidism who developed allograft dysfunction on day 2 post-renal transplantation. Allograft biopsy demonstrated calcium phosphate deposition in distal tubules. The patient was treated with cinacalcet and aluminum hydroxide which successfully decreased serum intact PTH level and calcium level and improved graft function. Cinacalcet was given for 2 days prior to transplantation, restarted on day 2 after transplantation, and then withheld for 24 h due to high urine calcium/creatinine (Ca/Cr) until the ratio decreased to 0.08 consistently for 1 week. During acute kidney injury, the elevated urine Ca/Cr ratio was found to be associated with furosemide and cinacalcet therapy. The authors could not determine if the increase in urine Ca/Cr ratio was due to high urine calcium or low urine creatinine. Overall, the patient's kidney function improved within 1 week of starting cinacalcet treatment [21A].

References

[1] Anderson LJ, Tamayose JM, Garcia JM. Use of growth hormone, IGF-I, and insulin for anabolic purpose: pharmacological basis, methods of detection, and adverse effects. Mol Cell Endocrinol. 2017;464:65–74. https://doi.org/10.1016/j.mce.2017.06.010 [R].

[2] Borras Perez MV, Kristrom B, Romer T, et al. Ten years of clinical experience with biosimilar human growth hormone: a review of safety data. Drug Des Devel Ther. 2017;11:1497–503 [R].

[3] Benabbad I, Rosilio M, Child CJ, et al. Safety outcomes and near-adult height gain of growth hormone-treated children with SHOX deficiency: data from an observational study and a clinical trial. Horm Res Paediatr. 2017;87(1):42–50 [r].

[4] Cai Q, Feng L, Yap KZ. Systematic review and meta-analysis of reported adverse events of long-term intranasal oxytocin treatment for autism spectrum disorder. Psychiatry Clin Neurosci. 2017;72(3):140–51. https://doi.org/10.1111/pcn.12627 [M].

[5] Erickson EN, Emeis CL. Breastfeeding outcomes after oxytocin use during childbirth: an integrative review. J Midwifery Womens Health. 2017;62(4):397–417 [R].

[6] Wang HT, Liu WH, Chen YL. Transient sick sinus syndrome with complete atrioventricular block associated with ergonovine intake: a case report. Medicine (Baltimore). 2017;96(44):e8559 [A].

[7] Tse SS, Kish T. Octreotide-associated neutropenia. Pharmacotherapy. 2017;37(6):e32–7 [A].

[8] Alsaedi AA, Bakkar AA, Kamal NM, et al. Late presentation of necrotizing enterocolitis associated with rotavirus infection in a term infant with hyperinsulinism on octreotide therapy: a case report. Medicine (Baltimore). 2017;96(40):e7949 [A].

[9] Lantinga MA, D'Agnolo HM, Casteleijn NF, et al. Hepatic cyst infection during use of the somatostatin analog lanreotide in autosomal dominant polycystic kidney disease: an interim analysis of the randomized open-label multicenter DIPAK-1 study. Drug Saf. 2017;40(2):153–67 [c].

[10] Tanaka S, Haketa A, Yamamuro S, et al. Marked alteration of glycemic profile surrounding lanreotide administration in acromegaly: a case report. J Diabetes Investig. 2018;9(1):223–5 [A].

[11] Caron PJ, Petersenn S, Houchard A, et al. Glucose and lipid levels with lanreotide autogel 120 mg in treatment-naive patients with acromegaly: data from the PRIMARYS study. Clin Endocrinol (Oxf). 2017;86(4):541–51 [c].

[12] Fleseriu M, Rusch E, Geer EB. Safety and tolerability of pasireotide long-acting release in acromegaly-results from the acromegaly, open-label, multicenter, safety monitoring program for treating patients who have a need to receive medical therapy (ACCESS) study. Endocrine. 2017;55(1):247–55 [c].

[13] Bohl MA, Forseth J, Nakaji P. Transient diabetes insipidus after discontinuation of vasopressin in neurological intensive care unit patients: case series and literature review. World Neurosurg. 2017;97:479–88 [A].

[14] Andreoli DC, Whittier WL. Reset osmostat: the result of chronic desmopressin abuse? Am J Kidney Dis. 2017;69(6):853–7 [A].

[15] McCafferty R, Fawzy R. Miscellaneous hormones. In: Side Effects of Drugs Annual. Elsevier: 2017;39:p. 447–55 [R].

[16] Sanchez-Ferrer ML, Prieto-Sanchez MT, Orozco-Fernandez R, et al. Central pontine myelinolysis during pregnancy: Pathogenesis, diagnosis and management. J Obstet Gynaecol. 2017;37(3):273–9 [A].

[17] Nakamura M, Sunagawa O, Kugai T, et al. Amiodarone-induced hyponatremia masked by tolvaptan in a patient with an implantable left ventricular assist device. Int Heart J. 2017;58(6):1004–7 [A].

[18] Block GA, Bushinsky DA, Cunningham J, et al. Effect of etelcalcetide vs placebo on serum parathyroid hormone in patients receiving hemodialysis with secondary hyperparathyroidism: two randomized clinical trials. JAMA. 2017;317(2):146–55 [C].

[19] Block GA, Bushinsky DA, Cheng S, et al. Effect of etelcalcetide vs cinacalcet on serum parathyroid hormone in patients receiving hemodialysis with secondary hyperparathyroidism: a randomized clinical trial. JAMA. 2017;317(2):156–64 [C].

[20] Dulfer RR, Franssen GJH, Hesselink DA, et al. Systematic review of surgical and medical treatment for tertiary hyperparathyroidism. Br J Surg. 2017;104(7):804–13 [R].

[21] Cheunsuchon B, Sritippayawan S. Successful treatment of early allograft dysfunction with cinacalcet in a patient with nephrocalcinosis caused by severe hyperparathyroidism: a case report. BMC Res Notes. 2017;10(1):153 [A].

CHAPTER

41

Thyroid Hormones, Iodine and Iodides, and Antithyroid Drugs

Vishakha Bhave, Anuj Patel*, Rahul Deshmukh†, Ajay Singh‡, Vicky Mody*,[1]*

*Department of Pharmaceutical Sciences, PCOM School of Pharmacy, Suwanee, GA, United States

†Department of Pharmaceutical Sciences, Rosalind Franklin University of Medicine and Science, College of Pharmacy, North Chicago, IL, United States

‡Department of Pharmaceutical Sciences, South University School of Pharmacy, Savannah, GA, United States

[1]Corresponding author: vickymo@pcom.edu

THYROID HORMONES [SEDA-34, 679; SEDA-35, 747; SEDA-36, 635; SEDA-37, 513; SEDA-39, 427]

Thyroid hormones such as eprotirome, levothyroxine (T4), and triiodothyronine (liothyronine, T3) are used in the treatment of hypothyroidism. Patients with overt or subclinical are presented with elevated TSH levels. Clinically, overt hypothyroidism is further defined by low levels of free triiodothyronine (fT3) and free levothyroxine (fT4). Patients presenting with overt or subclinical hypothyroidism are at increased risk of developing various metabolic disorder which can lead to cardiovascular disease hence care has to be sought as soon as discovered.

Eprotirome

Eprotirome, a thyroid mimitic hormone, has been shown to lower serum low-density lipoprotein (LDL) cholesterol concentrations in patients with dyslipidemia. Mechanistically it acts on the hepatic β receptor. Abnormal lipid levels put patients at risk for cardiovascular disease, which sometimes becomes difficult to reverse with stand-alone statin therapy [1R]. The use of thyroid hormone mimetics in these patients can help lower levels of LDL cholesterol. Various adverse events have been reported during the use of eprotirome. All of them have been listed below.

Cardiovascular Disorders

A low-density lipoprotein (LDL) has been correlated with the increased cardiovascular disease (CVD) risk in various populations, such as postmenopausal women. Anagnostis et al. published a systematic review to evaluate the effect of various drugs including eprotirome on patient's cardiovascular function. Eprotirome is a liver selective thyroid hormone receptor agonist on LDL receptors [2R]. Authors found that eprotirome causes upregulation of LDL receptors at a maximum dose of 200 mcg/d [2R].

Cartilage Defect in Dogs

It was found by Lindemann and coworkers that a 12–15-month eprotirome treatments in dog lead to cartilage defects in them. This was unexpected as there is no evidence of defect in cartilage occurring in humans [3M,3a].

Liver Disorders

Aspartate aminotransferase (AST) and alanine aminotransferase (ALT) are enzymes that indicate liver damage or disease at high concentrations in the blood. Sjouke et al. found that AST and ALT levels increased past the upper limit of normal (ULN) for eprotirome dosages of 50 and 100 μg [4C].

Side Effects of Drugs Annual, Volume 40
ISSN: 0378-6080
https://doi.org/10.1016/bs.seda.2018.08.011

Sex Hormones

Ladenson et al. discovered that those men that were given eprotirome vs placebo experienced an increase in serum follicle stimulating hormone (FSH), luteinizing hormone (LH), and total testosterone concentration. However, there were no adverse effects recorded in sexual dysfunction or libido by the men on the eprotirome treatment [5C].

Levothyroxine

Levothyroxine is one of the most important medications which are used to treat hypothyroidism. However, there is no recourse from this agent and it becomes a life-long companion of patient. Hence, it is also associated with various adverse effects from quality of life, to neurological developmental issues on fetus to its effects on bone. These effects have to be carefully reviewed and monitored over the period of time so as soon as patient becomes symptomatic needful action is taken.

Acute Mania

Yu et al. presented a case report about a 34-year-old Filipino male who presented to ED with Hashimoto's thyroiditis (or disease). The patient arrived with an enlarged thyroid gland with elevated thyroid peroxidase antibodies and elevated thyroglobulin antibodies, hallmark of Hashimoto's thyroiditis. A subsequent ECG also discovered a high-grade atrioventricular block. Based on his presentation, a full replacement dose of 100 μg levothyroxine was administered to target the hypothyroidism. Although the patient had no prior history of psychiatric issues, the patient developed acute mania symptoms 24 hours after levothyroxine administration, which included agitation, irritability and restlessness. Also seen was an increase in his speed and volume of speech, distractedness, and insomnia. The rapid conclusion of the patient's mania (prolonged in Hashimoto's) and its abrupt presentation post levothyroxine administration lead authors to believe that the acute mania was a result of levothyroxine-induced change in thyroid function [6A].

Bone Loss

Kim et al. examined the effect of levothyroxine (LT4) on bone losses in 93 patients after initiating levothyroxine therapy [7c]. Authors found that there was a mean bone loss in the lumbar spine, femoral neck, and total hip. Additionally, the bone loss was more prominent in postmenopausal women. It was found that postmenopausal women who received supplementation of calcium/vitamin D showed less bone loss as compared to others who did not. Thus, the authors concluded that levothyroxine therapy can accelerate bone loss predominantly in postmenopausal women who are not taking calcium/vitamin D supplement. These losses can be reduced by adding calcium/vitamin D to the therapy.

Nyandege and co-workers suggested that the concomitant use of bisphosphonates and other medications can stimulate bone metabolism. For example, acid-suppressive therapy, levothyroxine, thiazolidinediones (TZDs), and selective serotonin reuptake inhibitors (SSRIs) can increase the risk of fracture [8R]. Authors found that the concomitant use of acid suppressive agents with bisphosphonate can precipitate the risk of fracture. However, they suggested that TZDs, SSRIs, and levothyroxine can have similar implications based on their pharmacological action. Hence, precaution should to be taken while using bisphosphonates along with these other medications.

Cardiovascular Disorders

Bakiner et al. conducted a prospective, controlled, single-blind study to determine plasma Fetuin A levels in hypothyroid patients ($n = 39$) before and after levothyroxine treatment [9C]. Authors found that there was no correlation between Fetuin A levels and cardiovascular risk factors; however, the mean HDL cholesterol levels decreased in those patients from 49.3 (23–83.9) to 44 (28.3–69.0) [9C]. Patients that suffer from thyroid cançer are sometimes treated with a radioactive iodine therapy. However, this requires the removal of whatever thyroid hormone treatment they are on. An et al. documented the severity of the adverse effects suffered by patients who went through levothyroxine withdrawals after its removal. Cardiometabolic parameters were recorded prior to removal of the levothyroxine treatment and were subsequently recorded during the withdrawal phase. The authors discovered the removal of levothyroxine caused a severe aggravation of the patient's LDL-C, triglyceride, free fatty acid and apolipoprotein B levels. The authors concluded that removal of a levothyroxine therapy plan may cause an increased risk of cardiovascular disease [10c].

Drug–Drug Interaction

Levothyroxine can interact with various other drugs. In an observational study carried out by Irving and coworkers, authors evaluated the effect of drugs co-administered with thyroxine [11C]. In this study, authors evaluated 10999 patients (mean age 58 years, 82% female) who were prescribed levothyroxine on at least three occasions within a 6-month period, prior to the start of a study. They found that both iron and calcium supplements, proton pump inhibitors, and oestrogen were responsible for the increase in serum TSH concentration (7.5%, 4.4%, 5.6% and 4.3%), respectively. However, there was a decrease in the TSH concentration (0.17 mU/L) for those patients on statins. Hence, authors concluded that there is a significant interaction between levothyroxine and iron, calcium, proton pump inhibitors, statins and oestrogens. Co-administration of these drugs with levothyroxine may reduce the effectiveness of levothyroxine therapy, and hence the TSH concentrations in those patients should be carefully monitored.

The amount of levothyroxine absorbed can also be affected by the co-administration of other drugs such as ciprofloxacin or rifampin [12A]. In a randomized, double-blind, placebo-controlled study on 8 healthy volunteers who received either 1000 μg of levothyroxine and placebo, or 1000 μg of levothyroxine and 750 mg ciprofloxacin, or 1000 μg of levothyroxine and 600 mg rifampin authors found that the co-administration of ciprofloxacin significantly decreased the area T4 levels by ~39% ($P=0.035$), whereas rifampin co-administration significantly increased T4 by 25% ($P=0.003$). Patients that were switching from ritonavir-boosted protease inhibitors to dolutegravir-based HAART (highly active antiretroviral therapy) while taking levothyroxine was discovered to have developed clinical and biological hyperthyroidism according to Berger et al. It was suggested the side effects arose from the increased elimination of levothyroxine, necessitating a higher dose [13A].

Drug Overdose

Toxic effects of high doses of levothyroxine on cardiac cells were reported by Stuijver et al. [14A]. Authors reported a case of 23-year-old woman who attempted suicide by ingesting 25 mg of levothyroxine. Patient experienced hypercoagulation and a hypofibrinolytic effect, reflecting an increased risk of venous thrombosis [14A]. Similarly, a 61-year-old patient who accidentally ingested 1000 times excess of levothyroxine rather than the actual dose of 50 mg exhibited altered mental consciousness, acute respiratory failure, and atrial fibrillation [15A]. The toxic effects of levothyroxine were documented by Flores-González et al. in the case of a 14-year-old boy who presented to the ED with lethargy and walking instability after an attempted suicide via ingestion of 75 tablets of levothyroxine 75 μg. Patient also suffered from auricular flutter and J-wave syndrome. Symptoms stabilized with the resolution of the abnormal thyroid function induced by the levothyroxine overdose [16A].

Multiple Subungual Pyogenic Granulomas

A case study was reported for a 54-year-old woman on levothyroxine for 3 months presented with rapidly growing lesions in her nail beds [17A]. During examination it was found that red nodules had invaded patients nail plates. A wide excision was performed but the symptoms recurred as noted during the 3-month follow-up period. During this time levothyroxine treatment was stopped and the symptoms did not recur [17A].

Neurodevelopmental Issues

Korevaar et al., conducted a population based prospective cohort study on pregnant women in Netherlands [18M]. They followed women requiring thyroid therapies during pregnancy until their children were 6 years old. They found that there was an association between high maternal free thyroxin and IQ level of the child [18M]. In fact, they showed a reduction in child's average IQ of 1.4–3.8 as compared to the control. Authors concluded that the thyroid hormone is involved in brain development and levothyroxine therapy during pregnancy might put unborn child at the risk of neurodevelopment issues.

Osteoporosis

Laine et al. reviewed a case of a 38-year-old woman who presented to an endocrine outpatient clinic due to stress fractures and other severe symptoms due to osteoporosis. The patient had been on 300 μg of levothyroxine since her teens after a thyroidectomy when she was 12. A frequent runner, she presented to the clinic due to pain in both her thighs which were due to bilateral femoral stress fractures. She has a past history of fractures, such as a low-impact forearm fracture and 8 separate rib fractures. A blood analysis showed serum free T4 level of 37 ng/L (normal range 10–22 ng/L). A reduction of her levothyroxine to 100 μg a day led to her femoral pain subsiding in a 6-month follow-up. Interestingly, a 20-year follow-up showed the patient had not suffered from fractures or bone pains since the reduction of the levothyroxine dose [19A].

Quality of Life

Quality of life (QOL) can be effected by primary hypothyroidism. The association between QOL and various parameters in hypothyroid patients who were taking levothyroxine was studied by Kelderman-Bolk et al. [20R]. The authors evaluated the QOL for 90 patients (20 males and 70 females) who were treated for primary hypothyroidism. The evaluation was done using Short-Form 36, Hospital Anxiety & Depression Scale, and MFI20. Authors found an inverse relationship between QOL and BMI. Similarly, an inverse relationship was found between QOL and hypothyroid patients who were on thyroxine treatment. Authors concluded that the weight gain might be responsible for the lower QOL of the patients who were on levothyroxine therapy.

Rheumatoid Arthritis (RA)

Patients with autoimmune disorders taking levothyroxine over a period of time have shown to develop increased the risk of developing rheumatoid arthritis (RA) [21R]. In this study, authors compared 1998 patients using levothyroxine and 2252 controls to assess the incident of RA cases. They found that patients on levothyroxine were at twofold higher risk for RA [21R].

Small Intestinal Bacterial Overgrowth

Levothyroxine was found to contribute to small intestinal bacterial overgrowth (SIBO) in a study by Brechmann et al. SIBO, or an increase in bacteria in the upper GI tract, was found to proliferate due to various triggers. A retrospective study was done, reviewing patients who were given a lactulose or a glucose hydrogen breath test. Using a clinical algorithm, each patient was reviewed to deduce the cause of the SIBO via clinical examination, lab tests, endoscopy and transabdominal ultrasound. From a study summary of 1809 patients, individuals on levothyroxine therapy showed a significantly higher prevalence for SIBO (17.1% vs 6.5%). The authors concluded that levothyroxine, impaired intestinal clearance and immunosuppression were the strongest contributors, with levothyroxine being the highest of them all [22C].

Vitamin D Deficiency

The effect on levothyroxine due to vitamin D deficiency was evaluated by comparing the levels of 25-hydroxyvitamin D and parathyroid hormone (PTH) on 59 non-lactating women [23C]. Patients were divided into four groups, A, B, C, and D. Group A and B, both consisted of 14 hypothyroid and euthyroid females with post-partum thyroiditis, respectively. Group C included 16 female with non-autoimmune hypothyroidism, whereas group D was a control which included 15 healthy euthyroid females. The patients in both groups A and C were treated with L-thyroxine for 6 months. After 6 months, it was found that the serum levels of 25-hydroxyvitamin D were lower in group A than in group B, as well as in group C in comparison with group D. Hence, the authors concluded that there is an association of vitamin D status with post-partum thyroiditis and levothyroxine treatment.

Octreotide

The use of preoperative administration of octreotide in cohort of patients with TSH secreting pituitary adenomas (TSHoma) was evaluated by Fukuhara et al. [24c]. Authors discovered that of 81 patients who underwent surgery for TSHoma at Toranomon Hospital between January 2001 and May 2013, 44 received preoperative short-term octreotide. Among these 44 patients 19 received octreotide as a subcutaneous injection, and 24 patients received octreotide as a long-acting release (LAR) injection, and one of them was excluded due to side effects. It was found that the use of short-term preoperative octreotide administration was highly effective in suppressing TSHoma shrinkage. Some common side effects such as mild diarrhea (5 patients), constipation (1 patient), nausea and elevated bilirubin (1 patient) were observed. Hence, the authors concluded that preoperative octreotide are effective in suppressing TSHoma and should be recommended to patients to avoid problems with hyperthyroidism.

IODINE AND IODIDES [SEDA-35, 752; SEDA-36; SEDA-37, 514; SEDA-39, 429]

Dietary Iodine

Dietary iodine is essential for thyroid hormone production, and regulation of many important biochemical processes in the body. Iodine deficiency can have multiple adverse effects on growth and development, and it also can critically jeopardize children's mental health and often their very survival. Iodine deficiency disorders (IDD) in pregnant women can result in stillbirth, spontaneous abortion, and congenital abnormalities such as cretinism, a grave, irreversible form of mental retardation. The World Health Organization (WHO) recommends that iodine deficiency in patients should be corrected through intake of iodized salt. Significant progress has been made to eliminate iodine deficiencies worldwide since WHO's initiation in 1990 [25R]. Moreover, iodine deficiency and its side effects routinely reported worldwide [26R,27C,28M,29C], particularly from the underdeveloped countries of Africa and Asia [30R].

A case study of a 27-year-old gravid 1 at 27 weeks 6 days with a history of hypothyroidism had an ultrasound that demonstrated a $3.9 \times 3.2 \times 3.3$-cm well-circumscribed anterior neck mass, an extended fetal head, and polyhydramnios. Further characterization by MRI showed a fetal goiter. Patient's history revealed she was on a nutritional iodine supplement for treating her hypothyroidism which might be an underlying cause of the fetal goiter. Patient was ingesting 62.5 times the recommended amount of daily iodine in pregnancy. The excessive iodine consumption caused the fetal hypothyroidism and goiter formation. After the iodine supplement was discontinued, the fetal goiter decreased in size. At delivery, the airway was not compromised. The infant was found to have reversible hypothyroidism and bilateral hearing loss postnatally [31A].

In a data analysis of large-scale cross-country (46 countries) studies (89 national surveys between 1994 and 2012), a general association between the household unavailability of iodized salt and child growth was determined [32R]. Data consisted of 390328 children for the stunting analysis, 397080 children for the underweight analysis, 384163 children for the wasting analysis, and 187744 children for the low-birth-weight analysis. Lack of use of iodized salt was associated with 3% higher odds of being stunted (95% CI of ORs: 1.00, 1.06; $P=0.04$), 5% higher odds of being underweight (95% CI: 1.02, 1.09; $P<0.01$), and 9% higher odds of low birth weight (95% CI: 1.02, 1.17; $P=0.01$).

Iodine-Containing Solutions

Excessive iodine intake due to a long-term topical exposure like iodine solution dressings or by intravenous administration of iodine-containing agents can lead to hyperthyroidism [33R]. Excessive iodine intake can also lead to hypothyroidism.

A unique case of a 3-day-old neonate born with a giant omphalocele, who was being treated with topical povidone-iodine dressings to promote escharification, was presented with a suppressed thyroid stimulating hormone of 0.59 μIU/mL, elevated free thyroxine of 5.63 ng/dL, and frank cardiovascular manifestations of thyrotoxicosis. After replacement of the topical iodine dressings with iodine-free silver sulfadiazine, the thyroid status gradually improved with complete resolution of hyperthyroidism by 17th DOL [34A].

Radioactive Iodine

Radioactive iodine which is used to treat Graves' disease can also lead to multiple adverse effects.

Hoarseness Post Radioiodine Therapy

Ahmad et al. reviewed the case of a 29-year-old lady with Graves' disease. After a multitude of treatments, she received 15 mCi of radioactive iodine. The day after administration, the patient began complaining of severe neck pain along with hoarseness of voice. Treatment with NSAIDs proved helpful only for the pain and soreness but the hoarseness remained. Two laryngoscopies came up negative. While the symptom subsided over a 6-month span, this trend of vocal issues post radioiodine therapy, though rare, seems to be a trend according to the authors [35A].

Salivary Gland Dysfunction

Moreddu et al. reached out to patients who had received adjuvant radioiodine therapy from January 2011 to December 2012. The questionnaires sent out looked to investigate radioiodine-induced side effects. 36% of the patients who responded back claimed they suffered from xerostomia [36c].

Weight Gain

Radioiodine therapy used to treat patients with Graves' disease with prior obesity saw significant weight gain in the first year post radioiodine treatment. Chen et al. labeled patients within the category 'overweight', who saw a doubling of their weight, while patients that were labeled 'obese' saw their weight more than triple [37R].

ANTITHYROID DRUGS [SEDA-35, 754; SEDA-36, 638; SEDA-37, 518; SEDA-39, 430]

Thionamides, a class of antithyroid drugs (ATDs), are compounds that are known to inhibit thyroid hormone synthesis. Iodine is incorporated into thyroglobulin for the production of thyroid hormone, which is achieved after the oxidation of iodide by peroxide. Thioamides inhibit organification of iodine to tyrosine residues in thyroglobulin and the subsequent coupling of iodotyrosines [38R]. The commonly available thionamides are propylthiouracil (PTU) and methimazole (MMI). Compared to propylthiouracil, MMI exhibits longer half-life. This pharmacokinetic advantage allows for once daily dosing and better patient compliance. The safety profile of MMI is also significantly better compared to PTU, with less hepatotoxicity observed with MMI. Carbimazole (CBZ), an another agent available as an ATD, is a prodrug of methimazole and is currently not available in the Unites States.

Common Side Effects

Some of the common side effects associated with PTU and MMI include pruritus, rash, urticaria, arthralgias, arthritis, fever, abnormal taste sensation, nausea, and vomiting. These adverse effects were observed in 13% of patients ($n=389$) taking thionamide drugs in one study [39C].

Agranulocytosis

Agranulocytosis, although not common, is a serious complication of thionamide therapy. A prevalence of as high as 0.5% has been observed within the first 2 months of treatment with thionamide drugs [3M,40R].

A 6-year-old girl was reported to develop agranulocytosis after about 18 months on MMI therapy. In most instances, this condition develops within few months of therapy. A late onset though not common is still observed in about 4% of children [41A]. A study aimed to estimate the incidence, mortality, and risk of the drugs associated with agranulocytosis in Chinese patients was carried out using clinical data analysis and monitoring system, between January 2004 and December 2013. The study concluded that carbimazole had the highest risk of agranulocytosis (adjusted OR 416.7, 95% confidence interval 51.5–3372.9) with an incidence of 9.2 (95% confidence interval 6.9–12.1) per 10000 users and 3.6 (95% confidence interval 2.7–4.8) per 10000 user-years [42c].

A study aimed to identify genetic variants associated with antithyroid drug-induced agranulocytosis in a white European population using Genome-wide association studies concluded that drug-induced agranulocytosis was associated with HLA-B*27:05 and with other SNPs on chromosome 6. The study further concluded that to avoid one case of agranulocytosis, based on the possible risk reduction roughly 238 patients would need to be genotyped for all the three SNPs identified [43C]. A Japanese study by Kubato et al. indicated that if patients show allergic cutaneous reactions to initial MMI treatment for Graves' disease, there is a strong likelihood that the patient will be able to tolerate lower doses of MMI after the initial symptoms associated with Grave's Disease have subsided.

ANCA-Positive Vasculitis

Propylthiouracil (PTU)-associated vasculitis is normally associated with tetrad of fever, sore throat, arthralgia, and skin lesions but may also involve multiple systems. Recently, it was reported that a perinuclear anti-neutrophil cytoplasmic antibody-associated vasculitis developed during treatment with PTU for Grave's disease [44R]. A patient suffering from hyperthyroidism exhibited propylthiouracyl-induced vasculitis with renal involvement. The vasculitis reversed completely after cessation of PTU therapy. This was followed by treatment with thiamazole that caused agranulocytosis with fever. After transient lithium carbonate therapy a successful thyroidectomy was performed. This case highlights some of the challenges and complications encountered in the management of hyperthyroidism [45A].

A 60-year-old woman was admitted because of hyperthyroidism and leukopenia. A terminal diagnosis of PTU-induced AAV was made. After PTU withdrawal and use of steroid, the patient recovered well and then accepted RAI therapy [46A]. Another fatal case of 41-year-old woman was attributed to extreme PTU associated vasculitis, and skin necrosis that eventually led to death due to septic shock and multisystem organ failure. The patient tested positive for p-ANCA, MPO-ANCA and proteinase 3-ANCA. The skin lesions caused infection requiring debridement and leg amputation but also led to septic shock [47A]. In another case study, a 27-year-old woman presenting with refractory hypoxaemic respiratory failure, haemoptysis, and thyrotoxicosis was attributed by the authors as a rare manifestation of propylthiouracil therapy resulting from the development of c-ANCA [48A]. A 34-year-old female was being treated for autoimmune hyperthyroidism; six weeks later she developed purpuric plaques with central necrosis on the gluteal areas [49A]. Laboratory results showed the presence of cryoglobulin, cryofibrinogen, and c-ANCA. PTU is considered to be the most common inducer of ANCA-associated microscopic polyangiitis [50R]. When PTU was stopped and replacing it with MMI, the skin lesions improved within a week, but the cryoglobulins, cryofibrinogens, c-ANCA and anti-SSA remained positive even after 5 months.

Hepatotoxicity

Drug-induced liver injury (DILI) is a major problem for pharmaceutical industry and drug development. Although MMI has been associated with liver disease, it is typically due to cholestatic dysfunction, not hepatocellular inflammation [51R]. Due to the idiosyncratic nature of the injury, the understanding of the mechanism of these toxicities is limited. It appears that reactive metabolite formation and immune-mediated toxicity may play a role in antithyroids liver toxicity, especially those caused by MMI, though other plausible mechanisms may include reactive metabolites formation, oxidative stress induction, intracellular targets dysfunction [52H]. A recent report presents two women with cholestatic jaundice due to methimazole treatment. Before initiating therapy, both women had normal liver function and complete blood counts. DILI was inferred based on the relationship between methimazole therapy initiation and cholestasis onset. After methimazole was discontinued, the symptoms gradually resolved. DILI is a rare condition that is more likely to occur in females and has dose dependent relationship [53A]. Cholestatic jaundice was found in a case report of an Asian man with no prior hepatic history but with multiple comorbidities, including hyperthyroidism, gastritis, and epilepsy. The authors suggest that Asian men may be more susceptible to hepatotoxicity, because of oral methimazole use [54A]. A similar case was seen in a 16-year-old Chinese girl who used PTU for 5 months. She was diagnosed with acute liver failure due to PTU exposure [55A].

Neonatal Malformation

Anti-thyroid exposure to pregnant women to treat Graves' diseases has now been indicated to trigger neonatal malformations and congenital anomalies. Meta-analysis performed by Chen et al. on 12 independent studies involving 8028 women participants showed women who were exposed to MMI/CMZ anti-thyroid drugs were at a higher risk of neonatal congenital anomalies (OR 1.88; 95% CI 1.33–2.65; $P=0.0004$). Exposure to PTU did not increase such risk when compared to no antithyroid drug exposure (OR 0.81; 95% CI 0.58–1.15; $P=0.24$) [56M].

Pancreatitis

Pancreatitis has very rarely been reported in association with MMI treatment [57A]. A population-based case–control study analyzing the database of the Taiwan

National Health Insurance Program involving 5764 individuals aged 20–84 years with a first attack of acute pancreatitis from 1998 to 2011 was evaluated to estimate the relative risk of acute pancreatitis associated with the use of MMI. The study did not detect a substantial association between the use of MMI and risk of acute pancreatitis [58C]. Recently, a sixth case of MMI-induced pancreatitis (according to the authors), in a patient with toxic multinodular goiter, was reported. The authors suggest exploring the possibility of pancreatitis in subjects treated with MMI in the presence of suggestive symptoms. Discontinuation of the drug is recommended if the diagnosis is confirmed by elevated pancreatic enzymes [59A].

Respiratory Disorders

A case report documented by Gaspar-da-Costa et al. records a case where a 75-year-old man was diagnosed with unilateral pleural effusion. The chest pain arose 6 days post methimazole usage. After ruling out other causes for the chest pain, the authors concluded that the diagnosis can be attributed to hypersensitivity reaction to MMI. The authors also note that previous literature review also revealed 5 additional cases where patients reported of pleural effusion after beginning an anti-thyroid drug treatment [60A].

Miscellaneous Side Effects

For the first time a case of an erythema annulare mimicking a figurate inflammatory dermatosis of infancy has been reported in a 11-year-old female with GD on MMI treatment. Fifteen days after MMI administration, generalized itching erythematous rash developed all over the body. Erythema disappeared slowly after MMI withdrawal. The clinicians suspect the rash to be due to methimazole use [41A]. A 76-year-old female with history of hyperthyroidism also exhibited hypotension and low blood glucose which was attributed to use of MMI. Though this condition is rare, prior use of MMI in patients exhibiting hypoglycemia should be evaluated. Once discontinuing her methimazole and treating with dextrose and octreotide, the hypoglycemic episodes resolved [61A]. A case study was published, for the first time, on methimazole-induced chronic cutaneous lupus erythematosus (CCLE). A 30-year-old female with autoimmune thyroid disease developed an erythematous patch on her nasal pyramid. The patch appeared after 1 month on methimazole therapy and worsened on sun exposure. After discontinuing MMI therapy the symptoms gradually resolved [62A].

Additional case studies can be found in these reviews [63R,64R,65R].

References

[1] France M, Schofield J, Kwok S, et al. Treatment of homozygous familial hypercholesterolemia. Clin Lipidol. 2014;9(1):101–18. https://doi.org/10.2217/clp.13.79 [R].

[2] Anagnostis P, Karras S, Lambrinoudaki I, et al. Lipoprotein(a) in postmenopausal women: assessment of cardiovascular risk and therapeutic options. Int J Clin Pract. 2016;70(12):967–77. https://doi.org/10.1111/ijcp.12903 [R].

[3] Andersen SL, Olsen J, Wu CS, et al. Birth defects after early pregnancy use of antithyroid drugs: a Danish nationwide study. J Clin Endocrinol Metab. 2013;98(11):4373–81. Epub 2013/10/24, https://doi.org/10.1210/jc.2013-2831. PubMed PMID: 24151287. [M].; (a) Lindemann JL, Webb P. Sobetirome: the past, present and questions about the future. Expert Opin Ther Targets. 2016;20(2):145–9

[4] Sjouke B, Elbers LPB, van Zaane B, et al. Effects of supra-physiological levothyroxine dosages on liver parameters, lipids and lipoproteins in healthy volunteers: a randomized controlled crossover study. Sci Rep. 2017;7(1):14174. https://doi.org/10.1038/s41598-017-14526-2. PubMed PMID: 29074892; PMCID: PMC5658438. [C].

[5] Ladenson PW, Kristensen JD, Ridgway EC, et al. Use of the thyroid hormone analogue eprotirome in statin-treated dyslipidemia. N Engl J Med. 2010;362(10):906–16. https://doi.org/10.1056/NEJMoa0905633. PubMed PMID: 20220185. [C].

[6] Yu MG, Flores KM, Isip-Tan IT. Acute mania after levothyroxine replacement for hypothyroid-induced heart block. BMJ Case Rep. 2017;2017:1–3. https://doi.org/10.1136/bcr-2016-218819 [A].

[7] Kim MK, Yun KJ, Kim MH, et al. The effects of thyrotropin-suppressing therapy on bone metabolism in patients with well-differentiated thyroid carcinoma. Bone. 2015;71:101–5. Epub 2014/12/03, https://doi.org/10.1016/j.bone.2014.10.009. PubMed PMID: 25445448. [c].

[8] Nyandege AN, Slattum PW, Harpe SE. Risk of fracture and the concomitant use of bisphosphonates with osteoporosis-inducing medications. Ann Pharmacother. 2015;49(4):437–47. Epub 2015/02/11, https://doi.org/10.1177/1060028015569594. PubMed PMID: 25667198. [R].

[9] Bakiner O, Bozkirli E, Ertugrul D, et al. Plasma fetuin—a levels are reduced in patients with hypothyroidism. Eur J Endocrinol/Eur Feder Endocrine Soc. 2014;170(3):411–8. Epub 2013/12/25, https://doi.org/10.1530/eje-13-0831. PubMed PMID: 24366942. [C].

[10] An JH, Song KH, Kim DL, et al. Effects of thyroid hormone withdrawal on metabolic and cardiovascular parameters during radioactive iodine therapy in differentiated thyroid cancer. J Int Med Res. 2017;45(1):38–50. https://doi.org/10.1177/0300060516664242. PubMed PMID: 27856930; PMCID: PMC5536594. [c].

[11] Irving SA, Vadiveloo T, Leese GP. Drugs that interact with levothyroxine: an observational study from the thyroid epidemiology, audit and research study (TEARS). Clin Endocrinol (Oxf). 2015;82(1):136–41. Epub 2014/07/22, https://doi.org/10.1111/cen.12559. PubMed PMID: 25040647. [C].

[12] Goldberg AS, Tirona RG, Asher LJ, et al. Ciprofloxacin and rifampin have opposite effects on levothyroxine absorption. Thyroid. 2013;23(11):1374–8. Epub 2013/05/08, https://doi.org/10.1089/thy.2013.0014. PubMed PMID: 23647409. [A].

[13] Berger J-L, N'Guyen Y, Lebrun D, et al. Early neuropsychological adverse events after switching from PI/r to dolutegravir could be related to hyperthyroidism in patients under levothyroxine. Antivir Ther. 2017;22(3):271–2. https://doi.org/10.3851/IMP3107 [A].

[14] Stuijver DJ, van Zaane B, Squizzato A, et al. The effects of an extremely high dose of levothyroxine on coagulation and

fibrinolysis. J Thromb Haemost. 2010;8(6):1427–8. https://doi.org/10.1111/j.1538-7836.2010.03854.x. PubMed PMID: 20345725. [A].

[15] Kreisner E, Lutzky M, Gross JL. Charcoal hemoperfusion in the treatment of levothyroxine intoxication. Thyroid. 2010;20(2):209–12. https://doi.org/10.1089/thy.2009.0054. PubMed PMID: 20151829. [A].

[16] Flores-González JC, Grujic B, Lechuga-Sancho AM. Thyroid hormone intoxication as a not yet described cause of J-wave syndrome in a pediatric patient. Endocrine. 2017;55(3):989–91. https://doi.org/10.1007/s12020-017-1228-2 [A].

[17] Keles MK, Yosma E, Aydogdu IO, et al. Multiple subungual pyogenic granulomas following levothyroxine treatment. J Craniofac Surg. 2015;26(6):e476–7. Epub 2015/09/12, https://doi.org/10.1097/scs.0000000000001922. PubMed PMID: 26355986. [A].

[18] Korevaar TIM, Muetzel R, Medici M, et al. Association of maternal thyroid function during early pregnancy with offspring IQ and brain morphology in childhood: a population-based prospective cohort study. Lancet Diabetes Endocrinol. 2016;4(1):35–43. https://doi.org/10.1016/S2213-8587(15)00327-7 [M].

[19] Laine CM, Landin-Wilhelmsen K. Case report: fast reversal of severe osteoporosis after correction of excessive levothyroxine treatment and long-term follow-up. Osteoporos Int. 2017;28(7):2247–50. https://doi.org/10.1007/s00198-017-3981-8 [A].

[20] Kelderman-Bolk N, Visser TJ, Tijssen JP, et al. Quality of life in patients with primary hypothyroidism related to BMI, Eur J Endocrinol/Eur Feder Endocrine Soc. 2015;173(4):507–15. Epub 2015/07/15, https://doi.org/10.1530/eje-15-0395. PubMed PMID: 26169304. [R].

[21] Bengtsson C, Padyukov L, Källberg H, et al. Thyroxin substitution and the risk of developing rheumatoid arthritis; results from the Swedish population-based EIRA study. Ann Rheum Dis. 2014;73(6):1096–100. https://doi.org/10.1136/annrheumdis-2013-203354 [R].

[22] Brechmann T, Sperlbaum A, Schmiegel W. Levothyroxine therapy and impaired clearance are the strongest contributors to small intestinal bacterial overgrowth: results of a retrospective cohort study. World J Gastroenterol. 2017;23(5):842–52. https://doi.org/10.3748/wjg.v23.i5.842. PubMed PMID: 28223728; PMCID: PMC5296200. [C].

[23] Krysiak R, Kowalska B, Okopien B. Serum 25-hydroxyvitamin D and parathyroid hormone levels in non-lactating women with post-partum thyroiditis: the effect of L-thyroxine treatment. Basic Clin Pharmacol Toxicol. 2015;116(6):503–7. Epub 2014/11/15, https://doi.org/10.1111/bcpt.12349. PubMed PMID: 25395280. [C].

[24] Fukuhara N, Horiguchi K, Nishioka H, et al. Short-term preoperative octreotide treatment for TSH-secreting pituitary adenoma. Endocr J. 2015;62(1):21–7. Epub 2014/10/03, https://doi.org/10.1507/endocrj.EJ14-0118. PubMed PMID: 25273395. [c].

[25] Pearce EN, Andersson M, Zimmermann MB. Global iodine nutrition: where do we stand in 2013?, Thyroid. 2013;23(5):523–8. Epub 2013/03/12, https://doi.org/10.1089/thy.2013.0128. PubMed PMID: 23472655. [R].

[26] Laillou A, Sophonneary P, Kuong K, et al. Low urinary iodine concentration among mothers and children in cambodia. Nutrients. 2016;8(4):172. Epub 2016/04/09, https://doi.org/10.3390/nu8040172. PubMed PMID: 27058551; PMCID: Pmc4848647. [R].

[27] Mizehoun-Adissoda C, Desport JC, Houinato D, et al. Evaluation of iodine intake and status using inductively coupled plasma mass spectrometry in urban and rural areas in Benin, West Africa. Nutrition. 2016;32(5):560–5. Epub 2016/01/23, https://doi.org/10.1016/j.nut.2015.11.007. PubMed PMID: 26796150. [C].

[28] Yang J, Zhu L, Li X, et al. Iodine status of vulnerable populations in henan province of China 2013–2014 after the implementation of the new iodized salt standard. Biol Trace Elem Res. 2016;173(1):7–13. Epub 2016/01/19, https://doi.org/10.1007/s12011-016-0619-1. PubMed PMID: 26779621. [M].

[29] Edmonds JC, McLean RM, Williams SM, et al. Urinary iodine concentration of New Zealand adults improves with mandatory fortification of bread with iodised salt but not to predicted levels. Eur J Nutr. 2016;55(3):1201–12. Epub 2015/05/29, https://doi.org/10.1007/s00394-015-0933-y. PubMed PMID: 26018655. [C].

[30] Gernand AD, Schulze KJ, Stewart CP, et al. Micronutrient deficiencies in pregnancy worldwide: health effects and prevention. Nat Rev Endocrinol. 2016;12(5):274–89. https://doi.org/10.1038/nrendo.2016.37. PubMed PMID: PMC4927329. [R].

[31] Overcash RT, Marc-Aurele KL, Hull AD, et al. Maternal iodine exposure: a case of fetal goiter and neonatal hearing loss. Pediatrics. 2016;137(4):e1–5 [A].

[32] Krämer M, Kupkà R, Subramanian S, et al. Association between household unavailability of iodized salt and child growth: evidence from 89 demographic and health surveys. Am J Clin Nutr. 2016;104(4):1093–100. https://doi.org/10.3945/ajcn.115.124719 [R].

[33] Deshmukh R, Singh AN, Martinez M, et al. Thyroid hormones, iodine and iodides, and antithyroid drugs. In: Sidhartha DR, editor. Side effects of drugs annual. Elsevier; 2016. p. 443–52 [chapter 39]. [R].

[34] Malhotra S, Kumta S, Bhutada A, et al. Topical iodine-induced thyrotoxicosis in a newborn with a giant omphalocele. AJP Rep. 2016;06(02). https://doi.org/10.1055/s-0036-1584879. e243-5. [A].

[35] Ahmad T, Ulhaq I, Islam N. Unexplained hoarseness of voice after radioactive iodine therapy; a rare complication. Open Access J Endocrinol. 2017;1(2):000109 [A].

[36] Moreddu E, Baumstarck-Barrau K, Gabriel S, et al. Incidence of salivary side effects after radioiodine treatment using a new specifically-designed questionnaire. Br J Oral Maxillofac Surg. 2017;55(6):609–12. https://doi.org/10.1016/j.bjoms.2017.03.019 [c].

[37] Chen M, Lash M, Nebesio T, et al. Change in BMI after radioactive iodine ablation for graves disease. Int J Pediatr Endocrinol. 2017;2017(5):1–5. https://doi.org/10.1186/s13633-017-0044-z. PubMed PMID: 28588625; PMCID: PMC5455212. [R].

[38] Cooper DS. Antithyroid drugs. N Engl J Med. 2005;352(9):905–17. https://doi.org/10.1056/NEJMra042972. PubMed PMID: 15745981.[R].

[39] Werner MC, Romaldini JH, Bromberg N, et al. Adverse effects related to thionamide drugs and their dose regimen. Am J Med Sci. 1989;297(4):216–9. PubMed PMID: 2523194. [C].

[40] Watanabe N, Narimatsu H, Noh JY, et al. Antithyroid drug-induced hematopoietic damage: a retrospective cohort study of agranulocytosis and pancytopenia involving 50,385 patients with Graves' disease. J Clin Endocrinol Metab. 2012;97(1):E49–53. https://doi.org/10.1210/jc.2011-2221. PubMed PMID: 22049174. [R].

[41] Arrigo T, Cutroneo PM, Vaccaro M, et al. Lateralized exanthem mimicking figurate inflammatory dermatosis of infancy after methimazole therapy. Int J Immunopathol Pharmacol. 2016;29(4):707–11. https://doi.org/10.1177/0394632016652412. PubMed PMID: 27272160. [A].

[42] Sing CW, Wong IC, Cheung BM, et al. Incidence and risk estimate of drug-induced agranulocytosis in Hong Kong Chinese. A population-based case-control study. Pharmacoepidemiol Drug Saf. 2017;26:248–55. https://doi.org/10.1002/pds.4156. PubMed PMID: 28083886. [c].

[43] Hallberg P, Eriksson N, Ibañez L, et al. Genetic variants associated with antithyroid drug-induced agranulocytosis: a genome-wide association study in a European population, Lancet Diabetes Endocrinol. 2016;4(6):507–16. https://doi.org/10.1016/S2213-8587(16)00113-3 [C].

[44] Criado PR, Grizzo Peres Martins AC, Gaviolli CF, et al. Propylthiouracil-induced vasculitis with antineutrophil cytoplasmic antibody. Int J Low Extrem Wounds. 2015;14(2):187–91. https://doi.org/10.1177/1534734614549418. PubMed PMID: 25256279. [R].
[45] Sohar G, Kovacs M, Gyorkos A, et al. Rare side effects in management of hyperthyroidism. Case report. Orv Hetil. 2016;157(22):869–72. https://doi.org/10.1556/650.2016.30465. PubMed PMID: 27211356. [A].
[46] Yi X-Y, Wang Y, Li Q-F, et al. Possibly propylthiouracil-induced antineutrophilic cytoplasmic antibody-associated vasculitis manifested as blood coagulation disorders: a case report. Medicine. 2016;95(41):e5068. https://doi.org/10.1097/MD.0000000000005068. PubMed PMID: PMC5072949. [A].
[47] Wall AE, Weaver SM, Litt JS, et al. Propylthiouracil-associated leukocytoclastic necrotizing cutaneous vasculitis: a case report and review of the literature. J Burn Care Res. 2017;38:e678–85. https://doi.org/10.1097/bcr.0000000000000464 9000;Publish Ahead of Print. PubMed PMID: 01253092-900000000-98491 [A].
[48] Ortiz-Diaz EO. A 27-year-old woman presenting with refractory hypoxaemic respiratory failure, haemoptysis and thyrotoxicosis: a rare manifestation of propylthiouracil therapy. BMJ Case Rep. 2014;2014:1–3. https://doi.org/10.1136/bcr-2014-204915. Epub 2014/08/26. PubMed PMID: 25150234. [A].
[49] Akkurt ZM, Ucmak D, Acar G, et al. Cryoglobulin and antineutrophil cytoplasmic antibody positive cutaneous vasculitis due to propylthiouracil. Indian J Dermatol Venereol Leprol. 2014;80(3):262–4. Epub 2014/05/16, https://doi.org/10.4103/0378-6323.132261. PubMed PMID: 24823411. [A].
[50] Bonaci-Nikolic B, Nikolic MM, Andrejevic S, et al. Antineutrophil cytoplasmic antibody (ANCA)-associated autoimmune diseases induced by antithyroid drugs: comparison with idiopathic ANCA vasculitides. Arthritis Res Ther. 2005;7(5):R1072–81. https://doi.org/10.1186/ar1789. PubMed PMID: 16207324; PMCID: 1257438. [R].
[51] Arab DM, Malatjalian DA, Rittmaster RS. Severe cholestatic jaundice in uncomplicated hyperthyroidism treated with methimazole. J Clin Endocrinol Metab. 1995;80(4):1083–5. https://doi.org/10.1210/jcem.80.4.7714072. PubMed PMID: 7714072. [R].
[52] Heidari R, Niknahad H, Jamshidzadeh A, et al. An overview on the proposed mechanisms of antithyroid drugs-induced liver injury. Adv Pharm Bull. 2015;5(1):1–11. https://doi.org/10.5681/apb.2015.001. PubMed PMID: 25789213; PMCID: PMC4352210. [H].
[53] Zou H, Jin L, Wang LR, et al. Methimazole-induced cholestatic hepatitis: two cases report and literature review. Oncotarget. 2016;7(4):5088–91. https://doi.org/10.18632/oncotarget.6144. PubMed PMID: 26498145; PMCID: PMC4826268. [A].
[54] Ji H, Yue F, Song J, et al. A rare case of methimazole-induced cholestatic jaundice in an elderly man of Asian ethnicity with hyperthyroidism: a case report. Medicine. 2017;96(49):e9093. https://doi.org/10.1097/MD.0000000000009093. PubMed PMID: 29245333; PMCID: PMC5728948. [A].
[55] Wu DB, Chen EQ, Bai L, et al. Propylthiouracil-induced liver failure and artificial liver support systems: a case report and review of the literature. Ther Clin Risk Manag. 2017;13:65–8. https://doi.org/10.2147/TCRM.S122611. PubMed PMID: 28138249; PMCID: PMC5238756. [A].
[56] Song R, Lin H, Chen Y, et al. Effects of methimazole and propylthiouracil exposure during pregnancy on the risk of neonatal congenital malformations: a meta-analysis. PLoS One. 2017;12(7): e0180108. https://doi.org/10.1371/journal.pone.0180108. PubMed PMID: 28671971; PMCID: PMC5495385. [M].
[57] Yang M, Qu H, Deng HC. Acute pancreatitis induced by methimazole in a patient with Graves' disease. Thyroid. 2012;22(1):94–6. https://doi.org/10.1089/thy.2011.0210. PubMed PMID: 22136208. [A].
[58] Lai S-W, Lin C-L, Liao K-F. Use of methimazole and risk of acute pancreatitis: a case–control study in Taiwan. Indian J Pharm. 2016;48(2):192–5 [C].
[59] Agito K, Manni A. Acute pancreatitis induced by methimazole in a patient with subclinical hyperthyroidism. J Investig Med High Impact Case Rep. 2015;3(2). https://doi.org/10.1177/2324709615592229. 2324709615592229. PubMed PMID: 26425645; PMCID: PMC4557366. [A].
[60] Gaspar-da-Costa P, Duarte Silva F, Henriques J, et al. Methimazole associated eosinophilic pleural effusion: a case report. BMC Pharmacol Toxicol. 2017;18(1):16. https://doi.org/10.1186/s40360-017-0121-1. PubMed PMID: 28320470; PMCID: PMC5360045. [A].
[61] Jain N, Savani M, Agarwal M, et al. Methimazole-induced insulin autoimmune syndrome. Ther Adv Endocrinol Metab. 2016;7(4):178–81. https://doi.org/10.1177/2042018816658396 PubMed PMID: 27540463; PMCID: PMC4973408. [A].
[62] Venturi M, Ferreli C, Pinna AL, et al. Methimazole-induced chronic cutaneous lupus erythematosus. J Eur Acad Dermatol Venereol. 2017;31(2). https://doi.org/10.1111/jdv.13857. e116-7. PubMed PMID: 27519167. [A].
[63] Mody V, Singh AN, Desmukh R, et al. Thyroid hormones, iodine and iodides, and antithyroid drugs. In: Ray SD, editor. Side effects of drugs annual: a worldwide yearly survey of new data in adverse drug reactions, vol. 37. Elsevier; 2015. p. 513–9 [chapter 40]. [R].
[64] Mody V, Singh AN, Desmukh R, et al. Thyroid hormones, iodine and iodides, and antithyroid drugs. In: Ray SD, editor. Side effects of drugs annual: a worldwide yearly survey of new data in adverse drug reactions, vol. 38. Elsevier; 2016. p. 443–52 [chapter 39]. [R].
[65] Ethredge H, Najafabadi KI, Deshmukh R, et al. Thyroid hormones, iodine and iodides, and antithyroid drugs. In: Ray SD, editor. Side effects of drugs annual: a worldwide yearly survey of new data in adverse drug reactions, vol. 39. Elsevier; 2017. p. 427–33 [chapter 36]. [R].

CHAPTER

42

Insulin and Other Hypoglycemic Drugs

*Laura A. Schalliol**, *Jasmine M. Pittman*†,1, *Sidhartha D. Ray*‡

*South College School of Pharmacy, Knoxville, TN, United States
†Parkwest Medical Center, Knoxville, TN, United States
‡Department of Pharmaceutical and Biomedical Sciences, Touro University College of Pharmacy and School of Osteopathic Medicine, Manhattan, NY, United States
[1]Corresponding author: jmckee@covhlth.com

ALPHA-GLUCOSIDASE INHIBITORS (AGIs) [SEDA-31, 691; SEDA-32, 772; SEDA-33, 893; SEDA-36, 647; SEDA-37, 523; SEDA-39, 435]

Tumorigenicity

In an observational, longitudinal cohort study of a Taiwanese population investigating the effects of a nationwide pay-for-performance diabetes program on cancer incidence and mortality, a total of 42373 participants were included. α-glucosidase inhibitor (AGI) use was associated with less incidence of cancer [adjusted subdistribution hazard ratio (aSHR) 0.62, 95% confidence interval (CI) 0.55–0.69, $P<0.001$], breast cancer (aSHR 0.60, 95% CI 0.40–0.91, $P=0.02$), colorectal cancer (CRC) (aSHR 0.60, 95% CI 0.44–0.81, $P=0.001$), liver cancer (aSHR 0.47, 95% CI 0.36–0.61, $P<0.001$), lung cancer (aSHR 0.60, 95% CI 0.42–0.86, $P=0.005$), and prostate cancer (aSHR 0.38, 95% CI 0.20–0.73, $P=0.004$). No difference was reported in oral cancer ($P=0.08$), cervical cancer ($P=0.24$), stomach cancer ($P=0.30$), and bladder cancer rates ($P=0.07$) [1MC].

Acarbose

Liver

In a randomized multicenter trial conducted in 237 Chinese centers, 5535 patients with uncontrolled type 2 diabetes (T2DM) were treated with metformin and sitagliptin therapy for 20 weeks. Those who did not achieve glycated hemoglobin A1c (HbA1c) goals were randomized into a triple therapy group with glimepiride, gliclazide, repaglinide, or acarbose for an additional 24 weeks. Abnormal hepatic function was reported in a patient in the acarbose triple therapy group [2MC].

BIGUANIDES [SEDA-34, 687; SEDA-36, 647; SEDA-37, 523–526; SEDA-38, 459–461; SEDA-39, 435–436]

Metformin

Cardiovascular

A meta-analysis investigating the association between metformin use and cardiovascular disease (CVD) included 53 observational and experimental studies found a decreased risk of CVD in metformin users compared to non-users (hazard ratio (HR) 0.76, 95% CI 0.66–0.87, $P<0.0001$). This finding was primarily supported by comparing metformin users to insulin users (HR 0.78, 95% CI 0.73–0.83, $P<0.00001$), and there was no difference when metformin users were compared to sulfonylurea (SU) users. Additionally, a significant decrease in stroke risk was associated with metformin use when compared to non-users (HR 0.70, 95% CI 0.53–0.93, $P=0.01$). There was a significantly lower incidence of both macrovascular morbidity and mortality when metformin users were compared to insulin users (HR 0.34, 95% CI 0.21–0.56, $P=0.001$). Of note, there was no significant difference in myocardial infarction (MI) risk when comparing metformin users to non-users (HR 0.63, 95% CI 0.28–1.42, $P=0.27$). It is important to point out that many of these studies had significant differences between the metformin and control groups at baseline, were not controlled for confounding variables, or the follow-up was too short to detect the development of CVD [3M].

In a meta-analysis that included data from 17 observational studies investigating heart failure (HF) readmissions, metformin users were less likely to be readmitted for HF when compared to non-users (HR 0.87, 95% CI 0.78–0.97, $P=0.009$). There was no difference in cardiovascular (CV)

Side Effects of Drugs Annual, Volume 40
ISSN: 0378-6080
https://doi.org/10.1016/bs.seda.2018.06.009

mortality when comparing metformin users to non-users in those with HF [4M].

When metformin added to lifestyle intervention (metformin + lifestyle) was compared to lifestyle intervention alone in 54 individuals with prediabetes, the metformin + lifestyle group saw a significant difference in carotid intima media thickness (mm) over the course of the study (baseline 0.63 ± 0.1, value 0.56 ± 0.08, $P=0.009$), which was not present in the lifestyle intervention group (baseline 0.59 ± 0.08, posttreatment 0.5 ± 0.09, $P=0.06$) [5c].

In a multicenter randomized controlled trial (RCT) investigating the effects of metformin on preventing or delaying diabetes that included 643 participants, there was a significant difference in coronary artery calcium severity between male metformin-users when compared to male non-users (mean of 40.2 vs 63.7, $P<0.05$) and to those undergoing lifestyle modifications (mean of 40.2 vs 70.1, $P<0.05$) [6MC].

In a RCT investigating the cardiometabolic effects of sitagliptin and metformin use in a non-diabetic Middle Eastern population with hypertension and "borderline" lipid profiles, metformin use was associated with decreased systolic blood pressure (SBP) ($P \leq 0.05$), diastolic blood pressure (DBP) ($P \leq 0.05$), total cholesterol (TC) ($P \leq 0.05$), triglycerides (TG) ($P \leq 0.01$), and low density lipoprotein (LDL-C) ($P \leq 0.01$) from baseline. Use of metformin was also associated with a higher high density lipoprotein (HDL—C) ($P \leq 0.01$) [7c].

A 19-year-old female who overdosed on an "unquantifiable" amount of metformin experienced hypoglycemia, coma, metabolic acidosis, and a cardiac arrest which was caused by a prolonged QT interval. She was resuscitated, but eventually died 42h after the overdose [8c].

Respiratory

In a retrospective cohort study investigating metformin's proposed anti-inflammatory effects on asthma-related outcomes, 1332 Taiwanese participants were included and followed over 11 years. At baseline, metformin non-uses had a longer duration of asthma (43 vs 39.7 months, $P<0.01$). Metformin use was significantly associated with a lower incidence of asthma-related hospitalizations (0.9% vs 3.3%, $P<0.01$) and exacerbations (2.3% vs 5.0%, $P<0.02$). There was no difference in inhaled corticosteroid use between the two groups, but a higher percentage of metformin users had a short-acting β_2-agonist (30.2% vs 24.1%, $P=0.02$), methylxanthine (42.8% vs 32.8%, $P<0.01$), or systemic steroid (24.1% vs 18.6%, $P=0.02$) when compared to non-users. These differences in asthma medications could have influenced the results seen in this study [9MC].

Nervous System

A meta-analysis investigating the association between metformin use and the development of diseases of aging in a diabetic population included 53 observational and experimental studies. This analysis found metformin users were less likely to develop cognitive impairment in comparison to non-users (odds ratio (OR) 0.49, 95% CI 0.25–0.95, $P<0.05$). When stratified by treatment duration, less cognitive impairment was seen in those who took metformin for greater than 6 years (OR 0.27, 95% CI 0.12–0.60, $P<0.05$), but not in those who had taken metformin for a shorter duration [3M].

Sensory Systems

A meta-analysis that included 53 observational and experimental studies that aimed to investigate the effects of metformin use of conditions associated with aging found less risk of open angle glaucoma associated with metformin use in diabetic participants when compared to those not on metformin (HR 0.75, 95% CI 0.59–0.95, $P<0.05$) [3M].

In a post hoc analysis of the Singapore Epidemiology of Eye Diseases study that included 8063 participants, metformin use was significantly associated with increased intraocular pressure (0.70 mmHg higher) when compared to those not on metformin (95% CI 0.50–0.90 mmHg, $P<0.001$) [10MC].

Death

In a meta-analysis investigating the effects of metformin use on conditions related to aging that included 53 observational and experimental studies, it was found that diabetic patients on metformin experienced significantly less risk of mortality when compared to the general population (HR 0.93, 95% CI 0.88–0.99, $P=0.03$). Those with diabetes on metformin had lower mortality in comparison to those with diabetes not on metformin (HR 0.72, 95% CI 0.65–0.80, $P<0.00001$), on SUs (HR 0.80, 95% CI 0.66–0.97, $P=0.02$), or on insulin (HR 0.68, 95% CI 0.63–0.75, $P<0.00001$) [3M].

In a meta-analysis including information from 17 observational studies that investigated all-cause mortality in diabetic patients with moderate–severe chronic kidney disease, metformin users had less risk of mortality compared to non-users (HR 0.78, 95% CI 0.63–0.96, $P<0.001$). When comparing all-cause mortality in diabetic patients with HF, metformin users had less risk of mortality compared to non-users (HR 0.78, 95% CI 0.71–0.87, $P=0.003$) [4M].

Tumorigenicity

A major topic throughout the literature this year was the relationship between metformin use and cancer incidence and outcomes. Many articles found that metformin use, overall, was associated with less incidence or better cancer-related outcomes [1MC, 3M, 11MC, 12C, 13C, 14C, 15C]. Metformin use was associated with less incidence of larynx, pharynx, upper gastrointestinal (GI), liver, and cervical cancers [1MC, 3M, 16MC, 17C, 18MC].

Metformin use was associated with no significant difference in acute myeloid leukemia (AML), thyroid, esophageal, oral, stomach, bladder, urothelial, and endometrial cancer rates [1MC, 3M, 15C, 17C, 19M, 20M]. Evidence is mixed regarding the relationship between metformin and head and neck, lower GI, prostate, breast, lung, pancreatic, and renal cancers [1MC, 3M, 16MC, 17C, 19M, 25M, 21M, 22M, 23C, 24c, 28MC, 29MC, 30MC, 31MC, 32M, 33M, 34M, 35MC, 26C, 27MC]. A selection of this year's literature investigating metformin's association with cancer is summarized below.

In a meta-analysis including information from 53 observational and experimental studies, diabetic participants on metformin had a lower cancer incidence than the general population (rate ratio 0.94, 95% CI 0.92–0.97, $P=0.0003$). When compared to non-users, metformin use was associated with a lower incidence of hepatocellular cancer (OR 0.79, 95% CI 0.75–0.85, $P<0.0001$), breast cancer (HR 0.71, 95% CI 0.54–0.92, $P=0.01$), and lung cancer (HR 0.80, 95% CI 0.65–0.98, $P=0.03$). Initially, there was no significant association between metformin use and CRC development, but after sensitivity analysis, a decreased risk of CRC development was associated with metformin use in comparison to those not on metformin (HR 0.92, 95% CI 0.85–0.99, $P=0.02$). Some additional findings from this meta-analysis include finding no significant association between metformin use and renal cell carcinoma, thyroid, head and neck, esophageal, pancreatic, bladder, or prostate cancers when differentiating between different types of cancer, even though metformin use was associated with less incidence of cancer [3M].

After reviewing the China National Knowledge Infrastructure, Embase, PubMed, and the VIP Library of Chinese Journal, 20 articles were included in a meta-analysis. One of the included articles was a RCT. Seven articles were case-control studies and 12 of the articles were cohort studies. This analysis found metformin use to be significantly associated with decreased CRC incidence when compared to those not on metformin (OR 0.73, 95% CI 0.58–0.90, $P<0.01$). Metformin users were also less likely to develop advanced adenoma than non-users (OR 0.52, 95% CI 0.38–0.72, $P<0.01$) [22M].

In an effort to assess the effects of metformin use on pancreatic cancer survival, a meta-analysis included eight retrospective cohort studies and two RCTs to assess a total population of 2986 participants. Four of these studies took place in Europe, three in the United States, and three took place in Korea. Follow-up in these studies ranged from 9.26 to 28.1 months, although follow-up was not reported in 3 of the 10 studies. Metformin use was significantly associated with improvement in overall survival (OS) (HR 0.78, 95% CI 0.66–0.92, $P=0.017$) in cohort studies. When conducting the analysis using randomized controlled trials, no association was found between metformin use and pancreatic cancer OS or progress-free survival (PFS) [34M].

A meta-analysis investigating the use of metformin as adjuvant therapy in the treatment of solid tumors included information from 23 publications and 4 conference abstracts. One of the studies was a prospective cohort study, with the rest of the included studies being retrospective cohort studies. A total of 24 178 participants were included in this study. Metformin use was associated with improvements in CRC recurrence (HR 0.63, 95% CI 0.47–0.85, $P=0.015$) and OS (HR 0.69, 95% CI 0.58–0.83, $P=0.000$) when compared with metformin non-users. Although there was initially a difference in CRC cancer-specific survival (CSS) associated with metformin use, that difference did not remain through various models explored in the study. The study also found that metformin use was associated with improvements in OS (HR 0.82, 95% CI 0.73–0.93, $P=0.000$) and CSS rates (HR 0.58, 95% CI 0.37–0.93, $P=0.017$) in prostate cancer. No significant association with metformin was found in the recurrence, OS, or CSS in breast or urothelial cancer. It is worth noting that all studies that met inclusion criteria were included; none were excluded for quality [19M].

A meta-analysis that included a total of eight retrospective studies published from 2012 to 2016 investigated the influence of metformin therapy on CRC outcomes in patients with T2DM. A total of 11 052 participants were included in this study, with one of the studies taking place in Korea, three in the United States, and four in Europe. This study found metformin use was significantly associated with improved OS (HR 0.82, 95% CI 0.77–0.87, $P=0.073$) when compared to those not on metformin [25M].

In a meta-analysis that included 17 observational studies, 10 of which were multicenter trials, the association between metformin use and CRC outcomes was investigated. Three of these studies took place in Europe, six in Asia, and eight took place in North America. More than 250 000 participants were included in the pooled study results. Metformin use was significantly associated with an increased OS rate (HR 0.69, 95% CI 0.61–0.77, $P<0.01$). Metformin use was not significantly associated with CSS or disease-free survival [21M].

In a meta-analysis that included 12 cohort studies and a total of 124 533 participants that investigated the effects of metformin use on lung cancer survival, these participants were followed for a range of 17 months to 19 years. This study found that there was no association between metformin use and death from any cause. Of note, metformin use was associated with improved OS in lung cancer in Chinese studies (HR 0.47, 95% CI 0.32–0.70, $P<0.01$), but not in US studies. When comparing the effects of metformin on different subtypes of lung cancer, metformin use was associated with improved lung cancer mortality in those with small-cell lung cancer (SCLC) (HR 0.52, 95% CI 0.29–0.91, $P=0.02$) but not for non-small-cell lung cancer (NSCLC). Analyzing only studies that investigated metformin being started after diagnosis of lung cancer, there was an improvement in OS in those patients that were on metformin (HR 0.79, 95% CI 0.72–0.87, $P<0.01$). Metformin use was associated with improved PFS in those with lung cancer (0.62, 95% CI 0.39–0.96, $P=0.03$) [32M].

Metformin use was associated with a favorable increase in OS when compared to non-users (HR 0.77, 95% CI 0.66–0.90, $P<0.001$) in a meta-analysis that investigated the effects of metformin use on lung cancer survival in those with diabetes. This study included a total of nine retrospective cohort studies and one case-control study, which provided a total of 4052 participants. Follow-up in these studies ranged from 10.8 to 68 months, although follow-up was not reported in 4 of the 10 studies. Additionally, metformin use is associated with a significant increase in PFS rate when compared to non-use (HR 0.53, 95% CI 0.41–0.68, $P<0.001$). When specifically looking at NSCLC, metformin use was significantly associated with improved OS (HR 0.77, 95% CI 0.71–0.84, $P=0.002$) and improved PFS (HR 0.53, 95% CI 0.39–0.71, $P<0.001$) in those with diabetes. These improvements were also seen with SCLC, with improvements in OS (HR 0.52, 95% CI 0.29–0.91, $P=0.022$) and PFS (HR 0.54, 95% CI 0.34–0.84, $P=0.007$) associated with metformin [33M].

In a case series reporting the effects of metformin as adjuvant therapy for dysplastic lesions in nondiabetic patients, metformin 500mg PO BID was associated with improvement in all three patients. A 71-year-old female with oral cavity squamous cell carcinoma (SCC) was treated with partial maxillectomy. Over 36 months, recurrence required six more surgeries. She was started on metformin therapy, and after 33 months, her recurrent lesions regressed. A 67-year-old male with glottic SCC was treated with radiation and partial laryngectomy. Over the next 4 years, he continued to experience recurrent cricotracheal leukoplakia which was treated through ablation and eventually partial internal cricoid lamina resection. He was started on metformin therapy, and after 20 months only required one ablation. In the 14 months following that ablation, he has not experienced any more recurrence. A 71-year-old male with SCC of the base of his tongue was treated with brachytherapy and neck dissection. He had experienced recurrent multifocal glottic leukoplakia that required laser treatment. He was then started on metformin, and after 3 months of follow-up, no recurrent lesions were identified [36A].

A 24-year-old female with adrenocortical carcinoma was treated with chemotherapy and a right hepatectomy and right adrenalectomy. Within months, the cancer had metastasized to the left lower lobe of the lung and the left side of the liver; the recurrence was resistant to standard chemotherapy. At the age of 26 years, the patient underwent a partial left hepatectomy and was started on melatonin 20mg PO HS, metformin 500mg PO BID, and a statin. Her condition seemed to improve. After 2 years, the spot on her lung was removed; testing of the spot showed growth inhibition. After 8 years of follow-up, the patient was still cancer free and on melatonin and metformin to help prevent recurrent cancer [37A].

A 52-year-old male was diagnosed with a liver hemangioma in 2006; on ultrasound, the mass was on the left lobe of the liver and measured 20 × 25mm. A CT showed fatty liver and a high-density mass in the lesion; a biopsy was not conducted due to the patient's bleeding risk. Each year, imaging showed no change in the tumor until 2011. In February 2012, the tumor increased to 35 × 30mm over a span of 4 months. Metformin 750mg/day was started for glycemic control in May 2012, and it was noted that the size of the tumor decreased. In April 2015, metformin was increased to 1250mg/day to further improve T2DM control, and in October 2015, another ultrasound revealed the tumor was undetectable [38A].

Pregnancy

A meta-analysis investigating the outcomes of metformin use on fetal and maternal outcomes in obese women without diabetes included information from two RCTs. The purpose of metformin use in this population was to reduce maternal and neonatal weight gain during pregnancy. No significant difference was found in birth weight between those on metformin and those not on metformin [mean difference (MD) −0.09, 95% CI −0.23 to 0.06, $P=0.25$]. Those in the metformin group experienced favorable maternal gestational weight gain compared to those not on metformin, (MD −1.35, 95% CI −2.07 to −0.62, $P=0.0003$). Incidence of adverse events associated with maternity did not significantly differ between metformin and placebo, including c-section, postpartum hemorrhage, preeclampsia, pregnancy-induced hypertension, and spontaneous early preterm birth. There was also no difference found in reported fetal adverse events, including congenital anomalies, fetal death, and neonatal death [39M].

DIPEPTIDYL PEPTIDASE-4 INHIBITORS (DPP4Is) [SEDA-33, 894; SEDA-34, 688; SEDA-36, 648; SEDA-37, 526–528; SEDA-38, 454–457; SEDA-39, 436–439]

Musculoskeletal

In a study that investigated the association between fracture risk and DPP4Is in a German population, those on DPP4Is were less likely to develop a bone fracture in comparison to those not on DPP4Is (6.4% vs 8.3%, HR 0.67, 95% CI 0.54–0.84, $P<0.001$) [40MC].

In a retrospective cohort study investigating the association between T2DM and hypoglycemia resulting in hip and forearm fractures, information from the National Health Insure Research Database in Taiwan was used to identify 7761 participants. Patients on DPP4Is were more likely to develop a hip fracture than those not on DPP4Is (HR 5.82, 95% CI 2.16–15.69). When adjusted for potential confounders, this association remained (HR 6.14, 95% CI 2.24–16.81) [41MC].

Saxagliptin

AUTACOIDS

A case of angioedema was reported in an 83-year-old woman who presented with acute tongue swelling. The patient reported taking ramipril and saxagliptin. Clinicians assumed the patient was experiencing ACE inhibitor-induced angioedema since she had been taking ramipril regularly. Ramipril was discontinued, and the patient did not respond to treatment with glucocorticoids and antihistamines. She was subsequently treated with C1-esterase-inhibitor (human); however, her swelling progressed and she required intubation. The next day swelling persisted and treatment with icatibant, a selective and specific bradykinin 2 receptor antagonist, was required. Angioedema began to subside in 8 h and she was extubated 24 h later. Authors presumed the unusual therapy-resistant angioedema was due to inhibition of two bradykinin degradation enzymes by ramipril and saxagliptin [42A].

Sitagliptin

CARDIOVASCULAR

In a study investigating the effects of sitagliptin use in a non-diabetic Middle Eastern population, sitagliptin use was associated with decreased SBP 15.8% ($P \leq 0.01$), DBP 12.2% ($P \leq 0.01$), TC 20.2% ($P \leq 0.05$), TG 13.8% ($P \leq 0.05$), and LDL-C 23.7% ($P \leq 0.05$); use of sitagliptin is associated with a higher HDL-C 11.2% ($P \leq 0.05$) from baseline [7c].

GLUCAGON-LIKE PEPTIDE-1 (GLP-1) RECEPTOR AGONISTS [SEDA-33, 896; SEDA-34, 690; SEDA-36, 650; SEDA-37, 528–530; SEDA-38, 457–458; SEDA-39, 439]

Cardiovascular

A review investigated 24-h heart rate monitoring in patients with T2DM treated with either short-acting GLP-1s or long-acting GLP-1s ($N = 1112$; active-treatment arms). The review also included two independent head-to-head trials of lixisenatide and liraglutide ($N = 202$; active-treatment arms). Short-acting GLP-1s were associated with a transient 24-h heart rate increase of 1–3 beats per minute (bpm), while long-acting GLP-1s were associated with more pronounced increases in 24-h heart rate. Highest heart rate increases were seen with liraglutide and albiglutide (6–10 bpm) compared with dulaglutide and exenatide LAR (3–4 bpm). In the two head-to-head trials, lixisenatide was associated with a small increase in heart rate, while liraglutide was associated with a larger increase which persisted over 24 h. The cause for increased heart rate is unknown but could be related to effects at the sinus node or sympathetic nervous system stimulation [43R].

INSULINS [SEDA-34, 685; SEDA-36, 645–647; SEDA-37, 521–523; SEDA-38, 453–454; SEDA-39, 439–441]

Cardiovascular

In a retrospective cohort study of 5238 participants with T2DM in England and Wales that investigated the effect of DPP4Is on CV outcomes in comparison to insulin, insulin was associated with an increased risk of a composite of non-fatal acute MI, all-cause death, or non-fatal stroke (adjusted HR 2.6, 95% CI 1.9–3.4). Insulin use was also associated with an increased risk of CV events (adjusted HR 2.0, 95% CI 1.5–2.8). Notably, there was no difference between insulin and DPP4I use in the rate of CV death (adjusted HR 2.6, 95% CI 0.8–8.9) [44MC].

Musculoskeletal

Of the 7761 Taiwanese participants included in a retrospective cohort analysis, those participants on insulin were more likely to develop a hip fracture than those not on insulin (HR 1.59, 95% CI 1.25–2.02) in a study that investigated the association between hypoglycemia and hip fracture rates in patient with T2DM. When adjusted for potential confounders, this association did not stand (HR 1.12, 95% CI 0.84–1.33), but when this analysis was not adjusted for hypoglycemia, the increased risk associated with insulin use stood (HR 1.89, 95% CI 1.25–2.84). Of note, when insulin was combined with SUs, there was still an increased risk of developing hip fractures (crude HR 2.16, 95% CI 1.46–3.20; adjusted HR 2.03, 95% CI 1.33–3.10). When insulin was combined with meglitinides, there was also an increased risk associated with the combination in comparison to those on no hypoglycemia agents (crude HR 2.92, 95% CI 1.41–6.06; adjusted HR 2.54, 95% CI 1.19–5.42) [41MC].

Death

The effects of multiple anti-diabetic agents on the survival rate of diabetic patients on dialysis were investigated in a retrospective cohort analysis of Taiwanese patients that used information from 912 participants. Those on only insulin had significantly less risk of death when compared to those not on treatment for diabetes (HR 0.70, 95% CI 0.54–0.91, $P<0.01$). This reduced risk remained when those on both insulin and oral antihyperglycemic agents were compared to those not on treatment for diabetes (HR 0.38, 95% CI 0.29–0.49, $P<0.01$) [45MC].

In a retrospective cohort analysis of over 5000 participants with T2DM from England and Wales that investigated the effect of DPP4Is on CV outcomes in comparison to insulin, insulin use was associated with an increased risk of all-cause death (adjusted HR 3.7, 95% CI 2.7–5.2) [44MC].

Tumorigenicity

In a prospective cohort study that included 612846 Swedish participants and followed them for a mean duration of 5 ± 3 years, insulin or SU use for more than 1 year was significantly associated with a decreased risk of prostate cancer (HR 0.73, 95% CI 0.55–0.98) when compared to those who had T2DM for a longer duration than 1 year [29MC].

A retrospective cohort study investigating the effects of baseline T2DM therapy use and cancer outcomes that included information from 924 cases found that those with AML that were on insulin at baseline had worse OS than those not on insulin (HR 2.03, 95% CI 1.01–4.06, $P=0.040$). When adjusted for potential confounders, this association did not stand [15C].

In a multicenter case-control study investigating risk factors associated with cancer incidence in 432 participants with T2DM, insulin use was not significantly associated with cancer incidence for men (OR 1.39, 95% CI 0.79–2.46), but it was for women (OR 1.85, 95% CI 1.10–3.10, $P=0.027$). When adjusting for other potential confounders, this association did not hold [13C].

Another multicenter retrospective case-control study investigating risk factors associated with cancer in patients with T2DM included information from 203 cases and 203 controls. This study found that insulin use was significantly associated with cancer incidence (OR 1.781, 95% CI 1.202–2.641, $P=0.005$). This association was present, no matter the insulin dose, even when adjusted for BMI, T2DM duration, and metabolic control. Of note, this association with an increased cancer incidence was not present in individuals on insulin for 10 or more years (OR 1.377, 95% CI 0.705–2.690, $P=0.442$) [14C].

MEGLITINIDES

Cardiovascular

In a retrospective cohort study using information from the National Health Insurance claims database in Taiwan, a total of 327154 participants were included. Those participants on meglitinides had the highest risk of hospitalization for heart failure when compared to those on SUs or acarbose (adjusted HR 1.53, 95% CI 1.24–1.88) [46MC].

Death

The effects of multiple anti-diabetic agents on the risk of death in diabetic patients on dialysis were investigated in a retrospective cohort analysis that included 912 Taiwanese participants that were followed for 5 years. There was no difference in risk of death when comparing those on meglitinides to those not on any treatment, but there was a significant decrease in death when insulin was added to meglitinides in comparison to those not on treatment for diabetes (HR 0.62, 95% CI 0.45–0.86, $P<0.01$) [45MC].

SODIUM-GLUCOSE COTRANSPORTER 2 (SGLT2) INHIBITORS [SEDA-33, 898; SEDA-34, 695; SEDA-36, 652; SEDA-37, 530–531; SEDA-38, 458–459, SEDA-39, 441–442]

Canagliflozin

Acid–Base Balance

Diabetic ketoacidosis (DKA) was reported in three patients with a history of T2DM who had recently started a SGLT2 inhibitor. The first report was a 76-year-old man who started canagliflozin 3 weeks prior. He presented with nausea, vomiting, mild dehydration, and elevated troponins. He was treated for DKA and acute coronary syndrome. The second report of DKA was a 75-year-old man who started dapagliflozin 4 weeks prior. The third patient took canagliflozin for 1 week and presented to the hospital with DKA after an acute ingestion of alcohol. In all three cases, DKA resolved within 24h after administration of insulin and fluids [47A].

A 27-year-old Asian woman with T2DM who was treated with canagliflozin for 3 months developed euglycemic DKA and persistent diuresis. The patient's DKA was treated with intravenous fluids and insulin and canagliflozin was discontinued. On day 3, the patient was switched to multiple daily injections of insulin. The patient's nocturnal urination had not resolved; nonetheless she was discharged on the 8th day of treatment [48A].

Urinary Tract

In a study that investigated the effects of canagliflozin on the progression of kidney function decline in diabetic patients when compared to glimepiride, the estimated glomerular filtration rate for those on canagliflozin 100mg declined by 0.5mL/min per $1.73m^2$ when compared to a decline of 3.3mL/min per $1.73m^2$ for those on glimepiride ($P < 0.001$). Those on canagliflozin 300 mg experienced a decline of 0.9 mL/min per 1.73 m^2, which was also statistically significant when compared with glimepiride ($P = 0.002$). Those on canagliflozin 300 mg experienced a statistically significant decrease in urine albumin to creatinine ratio by 11.2% when compared to those on glimepiride (95% CI 3.6–18.3, $P < 0.01$); there was no significant difference between canagliflozin 100 mg and glimepiride [49C].

Musculoskeletal

The CANVAS Program included 10142 patients with T2DM and high CV risks. Patients were randomized to receive canagliflozin or placebo and were followed for a mean of 188.2 weeks. Patients in the canagliflozin group had an increased risk of lower extremity amputation (6.3 vs 3.4 participants per 1000 patient-years, HR 1.97, 95% CI 1.41–2.75). Risk of amputation was highest in patients with a history of amputation or peripheral vascular disease [50C].

Empagliflozin

Acid–Base Balance

A case of rapid-onset severe ketoacidosis was reported 2 days after empagliflozin was added to metformin, sitagliptin, and gliclazide in a patient with uncontrolled T2DM with unrecognized acromegaly. Intravenous fluid and insulin replacement resulted in rapid metabolic recovery. Clinicians noticed the patient's appearance was suggestive of acromegaly. Diagnosis was confirmed upon the discovery of elevated insulin-like growth factor 1 and growth hormone concentrations. Magnetic resonance imaging revealed a macroadenoma. The patient underwent a transsphenoidal resection of the growth-hormone secreting macroadenoma which restored normal growth hormone and insulin-like growth factor 1 concentrations and the patient's diabetes was well controlled thereafter. Authors hypothesized that SGLT2 inhibitors may possibly precipitate ketoacidosis in patients with active acromegaly and diabetes mellitus due to the additive effects of both conditions on free fatty acid and ketone-body metabolism [51A].

SULFONYLUREAS (SUs) [SEDA-34, 695; SEDA-36, 652; SEDA-37, 531–532; SEDA-38, 461; SEDA-39, 442–443]

Cardiovascular

In a meta-analysis including information from 26 observational studies and 82 RCTs that compared the effects of SUs on CV outcomes to other antidiabetic agents, there was an increased risk of CV mortality associated with SUs (HR 1.46, 95% CI 1.21–1.77). The increased risk of CV death was statistically significant when compared to TZDs (HR 3.05, 95% CI 1.79–5.54), SGLT-2 inhibitors (42.6, 95% CI 1.71–359.1), insulins (HR 1.30, 95% CI 1.02–1.66), GLP-1 agonists (HF 45.4, 95% CI 2.07–362.8), and DPP4Is (HR 4.42, 95% CI 1.92–13.0). The risk of an acute MI was greater with SUs than with other antidiabetic agents. The increased risk was statistically significant when SUs were compared to SGLT-2 inhibitors (HR 41.8, 95% CI 1.64–360.4) and DPP4Is (HR 2.54, 95% CI 1.14–6.57). Additionally, there was a significantly increased risk of stroke with SUs vs TZDs (HR 1.75, 95% CI 1.20–2.69), insulins (HR 1.46, 95% CI 1.01–2.14), GLP-1 inhibitors (HR 45.4, 95% CI 1.99–362.7), and DPP4Is (HR 9.40, 95% CI 3.27–41.9) [52M].

Sensory Systems

A post hoc analysis of the Singapore Epidemiology of Eye Diseases study included information from 8063 participants and found SU use was significantly associated with an increased intraocular pressure (0.83 mmHg higher) when compared to those not on SUs (95% CI 0.61–1.05 mmHg, $P < 0.001$) [10MC].

Musculoskeletal

In a retrospective cohort study investigating the association between hypoglycemia and hip fracture rates in Taiwanese patients with T2DM, information from 7761 participants were included. SU use was significantly associated with the development of a hip fracture when compared to those not on SUs (HR 1.35, 95% CI 1.03–1.76). When adjusted for potential confounders, this association remained (HR 1.44, 95% CI 1.07–1.92). Additionally, when SUs were combined with insulin, the increased risk remained (crude HR 2.16, 95% CI 1.46–3.20; adjusted HR 2.03, 95% CI 1.33–3.10) [41MC].

Death

A meta-analysis that included information from 82 RCTs and 26 observational studies found there was an increased risk of all-cause mortality when comparing SUs to other antidiabetic agents (HR 1.26, 95% CI 1.10–1.44). The comparisons between SUs and TZDs (HR 1.54, 95% CI 1.14–2.10), insulins (HR 1.21, 95% CI 1.01–1.45), DPP4Is (HR 2.03, 95% CI 1.22–3.58), and biguanides (HR 1.37, 95% CI 1.03–1.84) were statistically significant [52M].

The effects of multiple anti-diabetic agents on the risk of death in diabetic patients on dialysis were investigated in a retrospective cohort analysis that included information from 912 Taiwanese participants that were followed for 5 years. Those on SUs had significantly less risk of death when compared to those not on treatment for diabetes (HR 0.69, 95% CI 0.52–0.92, $P=0.01$). This risk further decreased when SUs combined with insulin were compared to those not on treatment (HR 0.65, 95% CI 0.47–0.89, $P<0.01$) [45MC].

Tumorigenicity

In a cohort study investigating the associations between various antihyperglycemic therapies and cancer rates including information from 1847051 Austrian participants, men on SUs had a higher incidence of cancer in comparison to other diabetic patients (OR 1.10, 95% CI 1.03–1.19, $P<0.05$). Women on SUs also had a higher incidence of cancer when compared to other diabetic patients (OR 1.19, 95% CI 1.08–1.30, $P<0.05$) [11MC].

An observational, longitudinal cohort study that included 42373 Taiwanese participants found SU use was significantly associated with an increased risk of developing colorectal cancer (aSHR 1.81, 95% CI 1.12–2.93, $P=0.02$). There was no significant difference in the development of overall cancer ($P=0.53$), breast cancer ($P=0.13$), oral cancer ($P=0.72$), liver cancer ($P=0.54$), lung cancer ($P=0.12$), cervical cancer ($P=0.60$), prostate cancer ($P=0.43$), stomach cancer ($P=0.90$), and bladder cancer ($P=0.69$) [1MC].

In a prospective cohort study investigating the association between T2DM and prostate cancer risk in a Swedish population that included 612846 participants, insulin or SU use for more than 1 year was associated with a significantly decreased risk of prostate cancer (HR 0.73, 95% CI 0.55–0.98) when compared to those who had T2DM for a longer duration than 1 year [29MC].

A prospective cohort study investigating the effects of antidiabetic medications on pancreatic adenocarcinoma survival rates in those who underwent pancreaticoduodenectomy that included 414 participants found that, for those with DM, improved survival was seen in those on SUs when compared to those not on SUs (27.5 months vs 14.6 months, $P<0.05$). Those with or without DM treated with SUs experienced improved survival when compared to those not on SUs (27.5 months vs 16.4 months, $P<0.05$) [53C].

THIAZOLIDINEDIONES (TZDs) [SEDA-33, 899; SEDA-34, 696; SEDA-36, 653; SEDA-37, 532–534; SEDA-39, 443–444]

Musculoskeletal

A retrospective study that investigated the association between hypoglycemia and hip fracture rates in patients with T2DM included information from 7761 Taiwanese participants. Participants on TZDs were less likely to develop a hip fracture than those not on TZDs (HR 0.44, 95% CI 0.32–0.61). When adjusted for potential confounders, this association remains (HR 0.48, 95% CI 0.34–0.66) [41MC].

Death

The effects of multiple anti-diabetic agents on the risk of death in diabetic patients on dialysis were investigated in a retrospective cohort analysis that included 912 Taiwanese participants. These participants were followed for 5 years. Those on TZDs had significantly less risk of death when compared to those not on treatment for diabetes (HR 0.48, 95% CI 0.29–0.79, $P<0.01$). Interestingly, this decreased risk was not present in those on a combination of TZDs and insulin in comparison to those not on treatment (HR 0.72, 95% CI 0.48–1.09, $P=0.12$) [45MC].

Tumorigenicity

In a cohort study investigating the association between antihyperglycemic therapies and cancer rates in an Austrian population, 1847051 participants were included for this assessment. Men on TZDs had less incidence of cancer in comparison to other diabetic patients (OR 0.86, 95% CI 0.76–0.96, $P<0.05$). Women on TZDs also had less risk of cancer in comparison to other diabetic patients (OR 0.83, 95% CI 0.71–0.96, $P<0.05$) [11MC].

An observational, longitudinal cohort study of 42373 participants found that TZD use was associated with less incidence of overall cancer (aSHR 0.78, 95% CI 0.70–0.86, $P<0.001$), colorectal cancer (aSHR 0.58, 95% CI 0.43–0.79, $P=0.001$), and liver cancer (aSHR 0.76, 95% CI 0.60–0.96, $P=0.02$). There was no difference in the incidence of breast cancer ($P=0.18$), oral cancer ($P=0.50$), lung cancer ($P=0.31$), cervical cancer ($P=0.31$), prostate cancer ($P=0.15$), stomach cancer ($P=0.92$), and bladder cancer ($P=0.30$) [1MC].

Pioglitazone

CARDIOVASCULAR

In a prospective investigational study evaluating the effects of pioglitazone use on CV function, 24 participants were included. Those with T2DM were naïve to TZD use. The group without T2DM differed from the T2DM group at baseline in mean HbA1c (5.5 ± 0.4 vs $6.7\pm1.3\%$, $P<0.005$), fasting plasma glucose (FPG) (93 ± 6 vs 149 ± 48 mg/dL, $P<0.005$), HDL-C (55.7 ± 9.8 vs 38.8 ± 11.9 mg/dL, $P<0.005$), TG (128 ± 94 vs 265 ± 155 mg/dL, $P<0.005$), resting heart rate (63.3 ± 6.8 vs 78.1 ± 10.5 bpm, $P<0.01$), stroke volume/BSA (42.7 ± 5.0 vs 37.7 ± 7.3 mL/m^2, $P<0.05$), transmitral E/A flow ratio (1.48 ± 0.37 vs 1.04 ± 0.28, $P<0.01$), end diastolic volume (EDV)/BSA (69.1 ± 10.3 vs 61.9 ± 9.1 mL/m^2,

$P < 0.05$), and peak left ventricular filling rate (PLVFR)/BSA (196 ± 33 vs 171 ± 52 mL/s·m^2, $P < 0.05$). When compared to baseline, after 24 weeks of pioglitazone treatment in the group with T2DM, significant decreases were found in HbA1c (5.6 ± 0.8 vs 6.7 ± 1.3%, $P < 0.01$), FPG (112 ± 23 vs 149 ± 48 mg/dL, $P < 0.05$), and TG (153 ± 74 vs 265 ± 155 mg/dL, $P < 0.01$). Significant increases were found in HDL-C (41.5 ± 9.7 vs 38.8 ± 11.9 mg/dL, $P < 0.005$), ejection fraction (65.6 ± 6.9 vs 60.7 ± 6.3%, $P < 0.05$), stroke volume/BSA (41.7 ± 8.5 vs 37.7 ± 7.3 mL/m^2, $P < 0.05$), transmitral E/A flow ratio (1.25 ± 0.38 vs 1.04 ± 0.28, $P < 0.01$), and PLVFR/BSA (212 ± 54 vs 171 ± 52 mL/s·m^2, $P < 0.01$) [54c].

Additional case studies can be found in these reviews [55R, 56R].

References

[1] Hsieh HM, He JS, Shin SJ, et al. A diabetes pay-for-performance program and risks of cancer incidence and death in patients with type 2 diabetes in Taiwan. Prev Chronic Dis. 2017;14:E88 [MC].

[2] Xu W, Mu Y, Zhao J, et al. Efficacy and safety of metformin and sitagliptin based triple antihyperglycemic therapy (STRATEGY): a multicenter, randomized, controlled, non-inferiority clinical trial. Sci China Life Sci. 2017;60(3):225–38 [MC].

[3] Campbell JM, Bellman SM, Stephenson MD, et al. Metformin reduces all-cause mortality and diseases of ageing independent of its effect on diabetes control: a systemic review and meta-analysis. Ageing Res Rev. 2017;40:31–44 [M].

[4] Crowley MJ, Diamantidis CJ, McDuffie JR, et al. Clinical outcomes of metformin use in populations with chronic kidney disease, congestive heart failure, or chronic liver disease: a systematic review. Ann Intern Med. 2017;166(3):191–200 [M].

[5] Arslan MS, Tutal E, Sahin M, et al. Effect of lifestyle interventions with or without metformin therapy on serum levels of osteoprotegerin and receptor activator of nuclear factor kappa B ligand in patients with prediabetes. Endocrine. 2017;55(2):410–5 [c].

[6] Goldberg RB, Aroda VR, Bluemke DA, et al. Effect of long-term metformin and lifestyle in the diabetes prevention program and its outcome study on coronary artery calcium. Circulation. 2017;136(1):52–64 [MC].

[7] Hussain M, Atif MA, Ghafoor MB. Beneficial effects of sitagliptin and metformin in non-diabetic hypertensive and dyslipidemic patients. Pak J Pharm Sci. 2016;29(6 Suppl):2385–9 [c].

[8] Ebrahim I, Blockman M. Metabolic acidosis in a patient with metformin overdose. S Afr Med J. 2017;107(2):110–1 [c].

[9] Li CY, Erickson SR, Wu CH. Metformin use and asthma outcomes among patients with concurrent asthma and diabetes. Respirology. 2016;21(7):1210–8 [MC].

[10] Ho H, Shi Y, Chua J, et al. Association of systemic medication use with intraocular pressure in a multiethnic Asian population: the Singapore epidemiology of eye diseases study. JAMA Ophthalmola. 2017;135(3):196–202 [MC].

[11] Kautzky-Willer A, Thurner S, Klimek P. Use of statins offsets insulin-related cancer risk. J Intern Med. 2017;281(2):206–16 [MC].

[12] Kusturica J, Kulo Ćesić A, Gušić E, et al. Metformin use associated with lower risk of cancer in patients with diabetes mellitus type 2. Med Glas (Zenica). 2017;14(2):176–81 [C].

[13] Dąbrowski M, Szymańska-Garbacz E, Miszczyszyn Z, et al. Differences in risk factors of malignancy between men and women with type 2 diabetes: a retrospective case-control study. Oncotarget. 2017;8(40):66940–50 [C].

[14] Dąbrowski M, Szymańska-Garbacz E, Miszczyszyn Z, et al. Risk factors for cancer development in type 2 diabetes: a retrospective case-control study. BMC Cancer. 2016;16(1):785 [C].

[15] Ceacareanu AC, Nimako GK, Wintrob ZAP. Missing the benefit of metformin in acute myeloid leukemia: a problem of contrast? J Res Pharm Pract. 2017;6(3):145–50 [C].

[16] Figueiredo RA, Weiderpass E, Tajara EH, et al. Diabetes mellitus, metformin and head and neck cancer. Oral Oncol. 2016;61:47–54 [MC].

[17] Nimako GK, Wintrob ZA, Sulik DA, et al. Synergistic benefit of statin and metformin in gastrointestinal malignancies. J Pharm Pract. 2017;30(2):185–914 [C].

[18] Kasmari AJ, Welch A, Liu G, et al. Independent of cirrhosis, hepatocellular carcinoma risk is increased with diabetes and metabolic syndrome. Am J Med. 2017;130(6) 746:e1–746.e7 [MC].

[19] Coyle C, Cafferty FH, Vale C, et al. Metformin as an adjuvant treatment for cancer: A systematic review and meta-analysis. Ann Oncol. 2016;27(12):2184–95 [M].

[20] Guo J, Xu K, An M, et al. Metformin and endometrial cancer survival: a quantitative synthesis of observational studies. Oncotarget. 2017;8(39):66169–77 [M].

[21] Du L, Wang M, Kang Y, et al. Prognostic role of metformin intake in diabetic patients with colorectal cancer: an updated qualitative evidence of cohort studies. Oncotarget. 2017;8(16):26448–59 [M].

[22] Hou YC, Hu Q, Huang J, et al. Metformin therapy and the risk of colorectal adenoma in patients with type 2 diabetes: a meta-analysis. Oncotarget. 2017;8(5):8843–53 [M].

[23] Han MS, Lee HJ, Park SJ, et al. The effect of metformin on the recurrence of colorectal adenoma in diabetic patients with previous colorectal adenoma. Int J Colorectal Dis. 2017;32(8):1223–6 [C].

[24] Miranda VC, Braghiroli MI, Faria LD, et al. Phase 2 trial of metformin combined with 5-fluorouracil in patients with refractory metastatic colorectal cancer. Clin Colorectal Cancer. 2016;15(4):321–8 [c].

[25] Tian S, Lei HB, Liu YL, et al. The association between metformin use and colorectal cancer survival among patients with diabetes mellitus: an updated meta-analysis. Chronic Dis Trans Med. 2017;3(3):169–75 [M].

[26] Frouws MA, Sibinga Mulder BG, Bastiaannet E, et al. No association between metformin use and survival in patients with pancreatic cancer: an observational cohort study. Medicine (Baltimore). 2017;96(10):e6229 [C].

[27] Ki YJ, Kim HJ, Kim MS, et al. Association between metformin use and survival in nonmetastatic renal cancer treated with a curative resection: a nationwide population study. Cancer Res Treat. 2017;49(1):29–36 [MC].

[28] Joentausta RM, Kujala PM, Visakorpi T, et al. Tumor features and survival after radical prostatectomy among antidiabetic drug users. Prostate Cancer Prostatic Dis. 2016;19(4):367–73 [MC].

[29] Häggström C, Van Hemelrijck M, Zethelius B, et al. Prospective study of type 2 diabetes mellitus, anti-diabetic drugs and risk of prostate cancer. Int J Cancer. 2017;140(3):611–7 [MC].

[30] Sonnenblick A, Agbor-Tarh D, Bradbury I, et al. Impact of diabetes, insulin, and metformin use on the outcome of patients with human epidermal growth factor receptor 2-positive primary breast cancer: analysis from the ALTTO phase III randomized trial. J Clin Oncol. 2017;35(13):1421–9 [MC].

[31] Lega IC, Fung K, Austin PC, et al. Metformin and breast cancer stage at diagnosis: a population-based study. Curr Oncol. 2017;24(2):e85–91 [MC].

[32] Zhong S, Wu Y, Yan X, et al. Metformin use and survival of lung cancer patients: meta-analysis findings. Indian J Cancer. 2017;54(1):63–7 [M].
[33] Cao X, Wen ZS, Wang XD, et al. The clinical effect of metformin on the survival of lung cancer patients with diabetes: a comprehensive systematic review and meta-analysis of retrospective studies. J Cancer. 2017;8(13):2532–41 [M].
[34] Dong YW, Shi YQ, He LW, et al. Effects of metformin on survival outcomes of pancreatic cancer: a meta-analysis. Oncotarget. 2017;8(33):55478–88 [M].
[35] Jang WI, Kim MS, Kang SH, et al. Association between metformin use and mortality in patients with type 2 diabetes mellitus and localized resectable pancreatic cancer: a nationwide population-based study in Korea. Oncotarget. 2017;8(6):9587–96 [MC].
[36] Lerner MZ, Mor N, Paek H, et al. Metformin prevents the progression of dysplastic mucosa of the head and neck to carcinoma in nondiabetic patients. Ann Otol Rhinol Largyngol. 2017;126(4):340–3 [A].
[37] Brown RE, Buryanek J, McGuire MF. Metformin and melatonin in adrenocortical carcinoma: morphoproteiomics and biomedical analytics provide proof of concept in a case study. Ann Clin Lab Sci. 2017;47(4):457–65 [A].
[38] Ono M, Sawada K, Okumura T. A case of liver hemangioma with markedly reduced tumor size after metformin treatment: a case report. Clin J Gastroenterol. 2017;10(1):63–7 [A].
[39] Elmaraezy A, Abushouk A, Emara A, et al. Effect of metformin on maternal and neonatal outcomes in pregnant obese non-diabetic women: a meta-analysis. Int J Reprod Biomed (Yazd). 2017;15(8):461–70 [M].
[40] Dombrowski S, Kostev K, Jacob L. Use of dipeptidyl peptidase-4 inhibitors and risk of bone fracture in patients with type 2 diabetes in Germany—a retrospective analysis of real-world data. Osteoporos Int. 2017;28(8):2421–8 [MC].
[41] Hung YC, Lin CC, Chen HJ, et al. Severe hypoglycemia and hip fracture in patients with type 2 diabetes: a nationwide population-based cohort study. Osteoporos Int. 2017;28(7):2053–60 [MC].
[42] Hahn J, Trainotti S, Hoffmann T, et al. Drug-induced inhibition of angiotensin converting enzyme and dipeptidyl peptidase 4 results in nearly therapy resistant bradykinin induced angioedema: a case report. Am J Case Rep. 2017;18:576–9 [A].
[43] Lorenz M, Lawson F, Owens D, et al. Differential effects of glucagon-like peptide-1 receptor agonists on heart rate. Cardiovasc Diabetol. 2017;16(1):6 [R].
[44] Jil M, Rajnikant M, Richard D, et al. The effects of dual-therapy intensification with insulin or dipeptidylpeptidase-4 inhibitor on cardiovascular events and all-cause mortality in patients with type 2 diabetes: a retrospective cohort study. Diab Vasc Dis Res. 2017;14(4):295–303 [MC].
[45] Hsiao PJ, Wu KL, Chiu SH, et al. Impact of the use of anti-diabetic drugs on survival of diabetic dialysis patients: a 5-year retrospective cohort study in Taiwan. Clin Exp Nephrol. 2017;21(4):694–704 [MC].
[46] Lee YC, Chang CH, Dong YH, et al. Comparing the risks of hospitalized heart failure associated with glinide, sulfonylurea, and acarbose use in type 2 diabetes: a nationwide study. Int J Cardiol. 2017;228:1007–14 [MC].
[47] Ahmed M, McKenna M, Crowley R. Diabetic ketoacidosis in patients with type 2 diabetes recently commenced on SGLT-2 inhibitors: an ongoing concern. Endocr Pract. 2017;23(4):506–8 [A].
[48] Adachi J, Inaba Y, Maki C. Euglycemic diabetic ketoacidosis with persistent diuresis treated with canagliflozin. Intern Med. 2017;56(2):187–90 [A].
[49] Heerspink HJ, Desai M, Jardine M, et al. Canagliflozin slows progression of renal function decline independently of glycemic effects. J Am Soc Nephrol. 2017;28(1):368–75 [C].
[50] Neal B, Perkovic V, Mahaffey K, et al. Canagliflozin and cardiovascular and renal events in type 2 diabetes. N Engl J Med. 2017;377(7):644–57 [C].
[51] Quarella M, Walser D, Brandle M, et al. Rapid onset of diabetic ketoacidosis after SGLT2 inhibition in a patient with unrecognized acromegaly. J Clin Endocrinol Metab. 2017;102(5):1451–3 [A].
[52] Bain S, Druyts E, Balijepalli C, et al. Cardiovascular events and all-cause mortality associated with sulphonylureas compared with other antihyperglyaemic drugs: a Bayesian meta-analysis of survival data. Diabetes Obes Metab. 2017;19(3):329–35 [M].
[53] Toomey P, Teta A, Patel K, et al. Sulfonylureas (not metformin) improve survival of patients with diabetes and resectable pancreatic adenocarcinoma. Int J Surg Oncol (N Y). 2017;2(3):e15 [C].
[54] Clarke GD, Solis-Herrera C, Molina-Wilkins M, et al. Pioglitazone improves left ventricular diastolic function in subjects with diabetes. Diabetes Care. 2017;40(11):1530–6 [c].
[55] Trovinger SN, Hrometz SL, Keshishyan S, Ray SD. Chapter 40 - Insulin Other Hypoglycemic Drugs. In: Ray SD, editor. Side Effects of Drugs- Annual: A worldwide yearly survey of new data in adverse drug reactions, vol. 38. Elsevier; 2016. p. 453–62 [R].
[56] Pittman J, Schalliol LA, Ray SD. Chapter-37: Insulin and other hypoglycemic drugs. In: Ray SD, editor. Side Effects of Drugs: Annual: A worldwide yearly survey of new data in adverse drug reactions, vol. 39. Elsevier; 2017. p. 435–46 [R].

CHAPTER

43

Use of Antidiabetic, Antihypertensive, and Psychotropic Drugs in Pregnancy

*Dana R. Fasanella**,1,2, *Sarah E. Luttrell**,2, *Frederick R. Tejada*†,2, *Harris Ngokobi*‡, *Manyo Ayuk-Tabe*‡

*Department of Pharmacy Practice and Administration, University of Maryland Eastern Shore School of Pharmacy, Princess Anne, MD, United States

†Department of Pharmaceutical Sciences, University of Maryland Eastern Shore School of Pharmacy, Princess Anne, MD, United States

‡University of Maryland Eastern Shore School of Pharmacy, Princess Anne, MD, United States

[1]Corresponding author: drfasanella@umes.edu

ANTIDIABETIC DRUGS

Pregnancy in women with preexisting diabetes is characterized by lower glucose levels and lower insulin requirements which changes quickly as insulin resistance increases exponentially during the second and early third trimesters then levels off toward the end of the third trimester. Treatment needs to be adjusted appropriately to avoid hyperglycemia for women with gestational diabetes mellitus (GDM) or preexisting diabetes [1S]. Although some small randomized controlled trials have shown efficacy and short-term safety of metformin and glyburide for the treatment of GDM, insulin is still the first-line recommended treatment of GDM. Moreover, the current insulin preparations have not been shown to cross the placenta [2c, 3M, 4c]. Currently, insulin analogs used in clinical practice are mature two-chain peptides which possess different pharmacokinetic profiles [5H].

Metformin is a biguanide that work by inhibiting glucose production in the liver, decreases the absorption of glucose in the intestine, and stimulates glucose uptake in the peripheral tissues. Glyburide is a sulfonylurea that binds to the beta cells of the pancreas and increases insulin secretion and insulin sensitivity of the peripheral tissues. Both metformin and glyburide have been recommended for the treatment of GDM, but there are some safety concerns that are regarding the use of these oral agents. While insulin does not cross the placenta, both metformin and glyburide cross the placenta which presents long-term safety concerns. The American College of Obstetricians and Gynecologists (ACOG) states that metformin crosses the placenta with levels that can be as high as the maternal concentrations [6S]. Also, according to the American Diabetes Association (ADA), about 70% of maternal levels of glyburide cross the placenta. It is still questionable if these drugs used in the treatment of diabetes in pregnancy are as safe and effective as insulin [1S].

Pregnant women with GDM have high risk of preeclampsia, maternal weight gain, pregnancy induced hypertension, and respiratory distress syndrome (RDS). Some studies have also reported cases of depression, and incidence of cesarean section. Also, children born to mothers with GDM are at higher risk of macrosomia, large for gestational age (LGA), admission to the neonatal intensive care unit (NICU), perinatal mortality (fetal and neonatal death), hypoglycemia, jaundice, weight gain and increased risk of developing type 2 diabetes [7M]. Feig et al. claimed that poor glycemic control in mothers with diabetes increases the risk of respiratory distress syndrome, neonatal hypoglycemia and NICU admissions [8c].

Allergic Reactions

The risk of allergic reactions to insulin administration during pregnancy is uncommon with only 20 reported cases worldwide [9A]. Recently, three cases of patients who presented with localized reactions at subcutaneous

[2] Authors contributed equally to the work.

Side Effects of Drugs Annual, Volume 40
ISSN: 0378-6080
https://doi.org/10.1016/bs.seda.2018.07.005

injection sites within 1 week of treatment with insulin detemir for GDM were reported [9A]. In the first case, a 32-year-old pregnant woman of Chinese descent, who was given 20 units of detemir at 34 weeks gestation, noted pruritic, painful, erythematous reactions developing at the injection site the previous night. Similarly, a 26-year-old pregnant woman of Indian descent with GDM who was given detemir at 28 weeks gestation noticed generalized pruritus as well as tender, painful, erythematous nodules that appeared approximately 8 h after each insulin injection. For both cases, the allergic reaction resolved over a period of 4–5 days. The third case was a 39-year-old pregnant woman of Chinese descent who also noted the development of pruritic tender erythematous nodules 24 h after injecting detemir in her thighs. There were no further reactions at injection sites on changing to insulin glargine. However, it is possible that the allergic reactions may be due to non-insulin additives in the detemir preparations and to reactions to natural rubber latex in the insulin vial membrane [9A].

Shoulder Dystocia

A retrospective study of 19236 births that occurred between April 1, 2011 and July 25, 2013 in five hospitals located in different states (Wisconsin, Florida, Maryland, Michigan, and Alabama) reported that only births to the women treated with insulin were at an increased risk of shoulder dystocia (odds ratio = 2.10, 9 confidence interval [1.01, 4.37]) as opposed to women treated with other glycemic agents or through diet [10C]. Shoulder dystocia is a perinatal complication which occurs when a fetus's shoulders fail to exit the birth canal after the head [10C].

Neonatal Outcomes

Macrosomia

In a pairwise meta-analysis of 32 randomized controlled trials (RCTs), 25 studies involving 3412 patients with GDM showed that the incidence of macrosomia for metformin compared to insulin was significantly lower (OR, 0.729; 95% CI, 0.545–0.974) [7M]. Macrosomia was also significantly lower for metformin compared with glyburide (OR, 0.5411; 95% CI, 0.2385–0.9855) based on the network meta-analysis. These findings indicate that metformin has the lowest risk of macrosomia [7M]. In a prospective randomized controlled study comparing metformin to glyburide, the women with GDM in the glyburide group (glyburide 2.5–20 mg/day 30 min before a meal and/or at 10:00 P.M.) showed higher incidence of macrosomia than the women in the metformin (850–2550 mg/day right after meals and/or at 10:00 P.M.) group [11c]. A possible explanation is that glyburide readily crosses the placenta [11c]. A meta-analysis of pregnant women with polycystic ovary syndrome that included 13 studies involving 1606 pregnant women, only 1 study provided macrosomia data. The 1 study reported the rate of macrosomia was 0 in the metformin group compared to 12.5% in the placebo group [12M]. Although the abovementioned meta-analyses have indicated that glyburide was associated with a higher incidence of macrosomia, a meta-analysis that included 10 randomized control trials involving 1194 women with GDM indicated that the risk of macrosomia [RR, 1.69; 95% CI, 0.57–5.08; $P = 0.35$] did not differ between the glyburide group and insulin group [13M]. However, when the newest and largest trial was excluded from the analysis due to the use of a lower dose (0.2 U/kg), the findings indicated macrosomia occurred significantly more often with the glyburide group than insulin group [13M].

Large for Gestational Age (LGA)

According to a meta-analysis from 15 randomized control trials that involved 1813 patients with GDM, the incidence of LGA for metformin was significantly lower than insulin (OR, 0.647; 95% CI, 0.438–0.956) and glyburide (OR, 0.431; 95% CI, 0.229–0.814) [7M]. Song et al. reported that the incidence of LGA for glyburide actually increased, but there was no significant difference in the incidence of LGA between glyburide and insulin [RR, 2.54; 95%CI, 0.98–6.57; $P = 0.05$] [13M].

Hypoglycemia

Neonatal hypoglycemia occurs when in the first 24 h post-delivery the blood glucose is less than 40 mg/dL or in the second day of life the blood glucose is less than 50 mg/dL [11c]. Comparing insulin and the oral antidiabetics, a pairwise meta-analysis that included 26 studies involving 3360 patients with GDM indicated that metformin had lower incidence of neonatal hypoglycemia than insulin (OR, 0.636; 95% CI, 0.486–0.832), and insulin was lower than glyburide (OR, 0.647; 95% CI, 0.423–0.991) [7M]. Similarly, based on network meta-analysis, metformin was significantly lower compared to insulin (OR, 0.6331; 95% CI, 0.3987–0.9331), and glyburide (OR, 0.3898; 95% CI, 0.1989–0.6558), but insulin was significantly lower than glyburide (OR, 0.6236; 95% CI, 0.3464–0.9992) [7M]. Moreover, a prospective randomized controlled study involving 53 patients being treated with glyburide and 51 with metformin reported that glyburide was associated with higher incidence of neonatal hypoglycemia [11c].

Preterm Birth

Based on a recent review, the largest study of metformin use in GDM (MiG) trial involving 751 women with GDM reported metformin with a higher incidence of preterm birth (12.1% vs 7.6%; $P=0.04$) compared to insulin [14R]. The women on the metformin group were started at 500mg once or twice daily and titrated to a maximum of 2500mg as necessary plus insulin when required [14R]. However, a network meta-analysis that included 32 randomized controlled trials showed that glyburide, compared to metformin and insulin, has the highest incidence of preterm birth [7M].

Birth Weight

For birth weight, a meta-analysis of 32 studies with 4060 GDM patients indicated that glyburide is ranked worst with higher mean birth weight compared to metformin and insulin [7M]. The pairwise meta-analysis showed that the mean birth weight for metformin was significantly lower than that for insulin (SMD, 0.111; 95% CI, 0.194–0.028) and glyburide (SMD, 0.235; 95% CI, 0.399–0.071). Also, the mean birth weight for insulin was significantly lower when compared to that of glyburide (SMD, 0.180; 95% CI, 0.327–0.033). Based on network meta-analysis, it was also observed that metformin had significantly lower birth weight compared to glyburide (SMD, 0.2591; 95% CI, 0.4383–0.08446) [7M]. Contrary to most meta-analyses, a recent meta-analysis of 10 randomized control trials involving 1194 participants reported no significant difference in birth weight between the GDM patients in the glyburide group and the insulin group [mean difference (MD), 79; 95% CI, −64 to 221.99; $P=0.28$] [13M]. The study further concluded that glyburide is safe and effective for use in GDM [13M]. A retrospective study of 820 GDM pregnancies treated between January 2011 and September 2014 in a university and non-university hospital compared the pregnancy outcomes between two treatment groups: diet-only and additional insulin therapy [15c]. The results have shown that neonates born in the insulin-group had a lower birth weight compared with the diet-group (3364 vs 3467g, $P=0.005$) [15c].

Other Neonatal Outcomes

In a pairwise and network meta-analyses of 13 studies which included 2008 GDM patients, the findings showed no significant difference between the metformin, insulin, and the glyburide groups for the incidence of RDS [7M]. Similarly, the pair-wise and network meta-analysis showed no significant difference between the metformin, insulin, and the glyburide groups for the incidence of hyperbilirubinemia [7M]. Moreover, a systematic review that looked at the risk of hyperbilirubinemia in infants whose mothers had been treated with metformin or glibenclamide reported that there was no significant difference between the groups (RR 0.68, 95% CI 0.37–1.25; two studies, $n=205$ infants) [16R].

Preeclampsia

A meta-analysis of 10 randomized control trials involving 1194 participants showed no differences in the risk of preeclampsia [RR, 0.98; 95% CI, 0.56–1.74], a primary indicator of maternal outcome [13M]. The meta-analysis included women with GDM who were not controlled with lifestyle modifications and thus required drug treatment. The treatment schedule in the control was insulin and the interventional group was glyburide [13M]. This result was consistent with the findings of a previous meta-analysis of 11 studies that involved 1754 GDM patients [7M].

ANTIHYPERTENSIVE DRUGS

According to The American College of Obstetricians and Gynecologists (ACOG), hypertensive disorders in pregnancy include preeclampsia, eclampsia, chronic hypertension (existing prior to pregnancy), chronic hypertension with superimposed preeclampsia, and gestational hypertension (elevated blood pressure after 20 weeks of gestation without evidence of preeclampsia) [17S]. Globally, a major cause of both maternal and perinatal morbidity and mortality is hypertensive disorders which complicate about 10% of pregnancies [17S]. Treatment of chronic hypertension in pregnancy is complicated by concerns that excessive lowering of blood pressure can negatively affect the fetus and a lack of good quality evidence for specific treatment thresholds [17S]. ACOG recommends the use of antihypertensive drug for pregnant women with systolic blood pressure greater or equal to 160mmHg and a diastolic blood pressure greater or equal to 105mmHg [17S]. Recommendations for pharmacologic therapy identify labetalol, nifedipine, and methydopa as preferred above all other options [17S]. Renin–antgiotensin system blockers and mineralocorticoid receptor antagonists are explicitly recommended against [17S]. The ACOG 2017 Committee Opinion on Emergent Therapy for Acute Onset, Severe Hypertension During Pregnancy and the Postpartum period recommends administration of intravenous labetalol, intravenous hydralazine, or oral immediate release nifedipine for prevention of maternal stroke within 30–60min of diagnosis of severe hypertension [18S].

Two studies have reported on analysis of data from The National Birth Defects Prevention Study (NBDPS) for risk of cardiac malformation in infants exposed in utero to antihypertensive drugs. NBDPS is a population-based, case–control study in the United States. The 2009 publication in Hypertension by Caton et al. analyzed data from pregnancies with estimated due dates from 1997 to 2003 [19C]. There were 5021 cases and 4796 controls [19C]. In 2017,

Fisher et al. published a follow-up analysis of NBDPS data for pregnancies with estimated due dates from 2004 to 2011 [20C]. The analysis included 10625 cases and 11137 controls [20C]. As in the 2009 publication sample [19C], cases were more likely to be older, overweight or obese, and report cigarette smoking [20C]. Early pregnancy antihypertensive use and late pregnancy antihypertensive use and untreated hypertension were associated with statistically significant increased risk of coarctation of the aorta (CoA), pulmonary valve stenosis (PVS), perimembranous ventricular septal defects (VSD-PM), and secundum atrial defects (ASD2) [20C]. Analysis of association by class was performed for beta-blockers, renin–antgiotensin system blockers, centrally-acting antiadrenergics, diuretics and calcium channel blockers [20C].

Any Antihypertensive Drug

Coarctation of the Aorta (CoA)

No significant risk increase was shown for CoA in the analysis of exposure during the first trimester to specific classes or specific drugs in the 2009 analysis of NBDP data [19C]. Exposure to any antihypertensive drug in the first trimester did show an increase in CoA OR 3.0; 95% CI, 1.3–6.6) [19C]. The 2017 analysis of NBDP data showed an increase in CoA for both early and late exposure to any antihypertensive (OR 2.5; 95% CI, 1.52–4.11 and OR 2.31; 95% CI, 1.26–4.24) [20C].

Pulmonary Valve Stenosis (PVS)

Both first trimester and late in pregnancy exposure to any antihypertensive were associated with an increased risk of PVS (OR 2.6; 95% CI, 1.3–5.4 and OR 2.4; 95% CI, 1.1–5.4) [19C]. The 2017 analysis of NBDP data showed an increase in PVS for both early and late exposure to any antihypertensive (OR 2.19; 95% CI, 1.44–3.34 and OR 1.93; 95% CI, 1.10–3.37) [20C].

Ebstein Malformation

First trimester exposure to any antihypertensive showed increased risk of Ebstein malformation (OR 11.4; 95% CI, 2.8–34.1) [19C]. The 2017 analysis of NBDPS data also showed an increased risk of Ebstein malformation with early pregnancy exposure to any antihypertensive (OR 3.89; 95% CI, 1.51–10.06) [20C].

Septal Defects: Perimembranous Ventricular Septal Defects (VSD-PM) and Secundum Atrial Defects (ASD2)

Risk of ASD2 was increased for exposure to any antihypertensive in both the first trimester and late in pregnancy initiation (OR 2.4; 95% CI, 1.3–4.4 and OR 2.4; 95% CI, 1.3–4.4.) [19C]. Risk of VSD-PM was only increased with late initiation of antihypertensives (OR 2.3; 95% CI, 1.2–4.6) [19C]. The 2017 analysis of NBDPS data also shows and increased risk of ASD2 with exposure to any antihypertensive in early and late pregnancy (OR 1.94; 95% CI, 1.36–2.79 and OR 2.61; 95% CI, 1.75–3.89) [20C]. There was also an increased risk of VSD-PM associated with exposure to any antihypertensive in both early and late pregnancy (OR 1.90; 95% CI, 1.09–3.31 and OR 1.85; 95% CI, 1.02–3.37) [20C].

Beta-Blockers

The 2009 analysis of NBDPS showed an increase risk of PVS with first trimester use of beta-blockers (OR 5.0; 95% CI, 1.8–13.8) [19C]. Analysis of exposure to atenolol and labetalol specifically did not show a statically significant increase [19C]. In the 2017 analysis of NBPDPS data beta-blockers were associated with an increased risk of CoA (OR 2.61; 95% CI, 1.25–5.45), PVS (aOR, 3.03; 95% CI, 1.68–5.46), VSD-PM (aOR, 4.13; 95% CI, 1.82–9.37) and ASD2 aOR, (2.35; 95% CI, 1.37–4.04) [20C]. Labetalol was used by 43.8% of cases and 42.5% of controls exposed to beta blocker in early pregnancy and the defects reported were the same as for the class [20C].

Centrally-Acting Adrenergic Agents

Cardiovascular Effects

The 2009 analysis of NBDPS data showed an increase risk of Ebstein malformation with use of any centrally-acting antiadrenergic (OR 16.9; 95% CI, 3.0–62.1) and methyldopa specifically (OR 12.7; 95% CI, 1.4–58.3) with first trimester exposure [19C].

A case report published in 2014 describes hypertensive crisis and acute cardiac failure in a 2-week-old infant [21A]. No immediate underlying cause was identified [21A]. The mother had begun treatment for gestational hypertension with methyldopa 500mg twice daily at 27 weeks gestation [21A]. At 30 weeks gestation the methyldopa dose was increased to 500mg three times daily which controlled maternal blood pressure until cesarean delivery at 37 weeks gestation [21A]. Upon delivery the infant was small for gestational age and experiencing respiratory distress requiring admission to the neonatal intensive care [21A]. She was then discharged after a week [21A]. At 2 weeks of age she presented to the emergency department and was diagnosed with hypertensive crisis and acute cardiac failure which required admission to the pediatric intensive care unit (PICU) [21A]. The total duration of hospitalization was 2 weeks with 1 week spent in the PICU [21A]. The final diagnosis was withdrawal effects due to in utero exposure to methyldopa [21A].

Hepatitis

Another case report published in 2014 describes a 34-year-old woman who presented with severe jaundice and hepatitis at 8 weeks after delivering her fourth child [22A]. No previous history of hypertensive disorders in pregnancy, renal disease, or chronic hypertension [22A]. During this fourth pregnancy she was diagnosed with preeclampsia at 24 weeks gestation [22A]. Methyldopa 500mg three times daily and hydralazine 50mg three times daily were used to treat her hypertension [22A]. Delivery was induced at 38 weeks due to severe preeclampsia [22A]. Three weeks after delivery her nephrologist discontinued hydralazine and reduced the dose of methyldopa to 250mg three times daily [22A]. Liver function tests during pregnancy were within normal limits [22A]. She presented for her 8 weeks postpartum check-up with signs of jaundiced and lethargic and methyldopa was discontinued [22A]. The patient had no personal or family history of liver disease, no IV drug use or alcoholism [22A]. After excluding other causes, she was diagnosed with methyldopa induced hepatitis [22A]. Two months later liver function tests had decreased to near normal levels but residual hepatic fibrosis remained [22A].

Preterm Birth

This prospective observational cohort study was established to assess the degree of major birth defects and spontaneous abortions in pregnant women taking methyldopa for chronic hypertension [23C]. The German Embryotox pharmacovigilance institute evaluated the outcomes due to exposure to methyldopa in the first trimester of pregnancy [23C]. The study examined patients from January 1, 2000 to December 31, 2014 [23C]. Of those patients in the German Embryotox pharmacovigilance institute who were exposed to antihypertensive drugs, 261 of them were pregnant women who were exposed to methyldopa during their first trimester and had complete follow-ups done throughout their pregnancies to assess pregnancy outcome [23C]. The control group consisted of 526 random pregnant women without chronic hypertension [23C]. Most baseline characteristics between both groups were similar with the exception of maternal BMI which was elevated in the methyldopa cohort (mean BMI, 28mg/m^2 vs 23mg/m^2) [23C]. This case cohort also included fewer women with an academic education and less smokers (8% vs 17%) [23C]; the medium dose was 500mg daily and about 54% of women were on methyldopa prior to conception [23C]. About 127 women in their first trimester were on single therapy methyldopa [23C]. Results showed a higher cumulative incidence of spontaneous abortions in the methyldopa cohort compared to the control; however, this was not statistically significant (17% vs 13%) [23C]. Elective terminations occurred less frequently in the methyldopa group [23C]. Females in this group also had a higher likelihood of developing gestational diabetes mellitus, abruption placentae, and Cesarean section. In the exposed/methyldopa cohort, the risk of preterm birth was significantly higher than the control (27% vs 10%; adjusted odds ratio, 4.11; 95% CI, 2.4–7.1) [23C]. The results of this study should be interpreted with caution due to the small sample size [23C]. Furthermore, the lack of comparison to women with untreated hypertension makes it difficult to determine if the adverse effects seen are due to methyldopa treatment or the disease itself [23C].

Renin–Angiotensin System Blockers

The 2009 analysis of NBDPS data showed an increase risk of Ebstein malformation with use of any renin–angiotensin system blocker (OR 26.4; 95% CI, 2.3–306) with first trimester exposure [19C]. The 2017 NBDPS analysis showed renin–angiotensin system blockers were associated with an increased risk of PVS (aOR, 3.74; 95% CI, 1.39–10.17), VSD-PM (aOR, 6.58; 95% CI, 1.55–27.97), and ASD2 (aOR, 3.25; 95% CI, 1.29–8.20, respectively) [20C].

Diuretics

The 2009 analysis of NBDPS data showed an increase risk of septal defects with use of any diuretic in the first trimester (OR 13.2; 95% CI, 1.1–692) [19C]. The 2017 NBDPS analysis showed diuretics were associated with an increased risk of ASD2 (OR 3.22; 95% CI, 1.30–7.99) [20C].

Calcium Channel Blockers

A 27-year-old woman was diagnosed with gingival hyperplasia after taking nifedipine for 9 weeks [24A]. This diagnosis was made during a hospital visit for preeclampsia [24A]. Upon presentation, nifedipine was discontinued and the patient was started on methyldopa for her hypertension [24A]. After about 48h, the gingival hyperplasia seemed to be resolving [24A]. She successfully gave birth 2 weeks later with a complete resolution of her gingival hyperplasia [24A]. Likelihood that the gingival hyperplasia was caused by nifedipine was confirmed with a Naranjo score [24A]. Neither the 2009 nor the 2011 NBDPS found any specific association with calcium channel blocker association and congenital heart defects [19C, 20C].

PSYCHOTROPIC DRUGS

Antidepressants

Based on US studies, it is estimated that nearly 10% of pregnant women fill antidepressant prescriptions during gestation [25R, 26R]. The use of any antidepressant increases the risk of hypertension or preeclampsia up to 1.5-fold [27MC, 28C]. The risk is directly correlated to duration of exposure, with a higher relative risk when prescribed for 4 or more months vs 2 months or less (1.47 and 1.05, respectively) [29MC]. The association with dose is unclear.

All antidepressants are associated with neonatal adaptation syndrome (NAS) shortly after birth [30MC, 31R, 32R, 33M, 34R]. These symptoms are usually mild, last several days, and may include respiratory complications, irritability, tremor, increased crying, hypo- or hypertonia and hyperreflexia [34R]. Of the serotonin reuptake inhibitors (SRIs), NAS has been documented in up to 30% of infants [34R, 35R], and is most commonly seen with paroxetine or venlafaxine [31R].

Bupropion

The lack of serotonergic effect of bupropion is attractive, as alterations in serotonin are potentially linked to fetal development abnormalities [36C, 37R]. In addition, the benefit of bupropion as a smoking cessation agent adds to its appeal [36C].

General Malformations and Risks

General malformation risks, as well as miscarriage or fetal demise, appear to be low with bupropion. Birth defect rates are approximately 3.6%, which is similar to the general population [36C]. The Bupropion Pregnancy Registry assessed pregnant women maintained on bupropion between 1997 and 2008, and found no increase in spontaneous miscarriage [38S]. A small prospective study, though, found an increase in miscarriage in bupropion-exposed mothers compared to controls (14.7% vs 4.5%, $P=0.009$) [39C]. However, this is within the general population miscarriage rate (14%–22%) [40R], however, several additional studies found no significant difference, or a lower risk, for premature birth and lower gestational weight at birth compared to controls [39C, 41MC].

Cardiovascular

Cardiac malformations are frequently associated with bupropion, but the data are inconclusive [31R]. A case–control study assessing first-trimester exposure of bupropion found an adjusted odds ratio of 1.6 (95% CI 1.0–2.8) for ventricular septal defect with any bupropion use. In women exposed only to bupropion and no other antidepressants, the risk was slightly higher (aOR 2.5, 1.3–5.0). No increased risk of left sided defects was found [42MC].

Mirtazapine

Mirtazapine has the benefit of replacing benzodiazepines and sedative hypnotics for the treatment of insomnia and other sleep disorders that may occur during pregnancy. This is to avoid the concerns with benzodiazepine-like drugs such as dependency, neonatal withdrawal and 'floppy baby syndrome' [33M]. In addition, its antiemetic and antinausea properties makes it attractive for women with depression and significant morning sickness. While only case reports are available regarding the use for pregnancy-related nausea and vomiting, it appears to be effective [43c].

General Malformations

Use of mirtazapine and serotonin and norepinephrine reuptake inhibitors (SNRIs) had a higher risk of preterm birth in over 700 women, with an odds ratio of 1.60 (95% CI 1.19–2.15). No increased risk for low birth weight, intrauterine death or infant death was identified [44M]. A prospective study in 104 women found a statistically significant increase in preterm birth with mirtazapine vs known non-teratogenic drugs ($P=0.04$). However, there was no difference compared to other antidepressants ($P=0.61$). There was also no difference in fetal demise, gestational age at birth or birth weight [33M, 45C]. Overall, three studies ($N=266$) reported a higher risk of preterm birth after mirtazapine exposure [44M, 45C, 46c]. However, multiple confounders were uncontrolled such as concomitant drugs, gestational or maternal age, smoking or alcohol use, and severity of depression.

Over 300 cases have been reported with no increased risk of major malformations compared to antidepressants or known non-teratogenic drugs [33M, 44M, 45C, 47c, 48C]. However, there may be risks that have been undetected due to the limited number of cases.

Serotonin and Norepinephrine Reuptake Inhibitors (SNRIs)

Congenital Malformations

A systematic review found relative risks for major congenital malformations with first-trimester use of venlafaxine of 1.12 (95% CI 0.92–1.35), and 0.80 (0.46–1.29) for duloxetine. This is approximately a major malformation rate of 3.36% and 2.40%, respectively, and is comparable to the background rate [49M]. Additional studies have not found an increase in major malformations with duloxetine ($P=0.99$) [50C, 51C]. A possible increase in spontaneous abortion has been reported, but with several confounding factors such as depression itself and concurrent drugs [52R].

Maternal Risks

As SNRIs increase norepinephrine availability, gestational hypertension and preeclampsia are expected risks. The estimated aRR for SNRIs is 0.75 for gestational hypertension, and between 1.49 and 1.95 for preeclampsia

[53M]. There may be a higher risk for preeclampsia with venlafaxine over duloxetine (aRR 1.57 vs 0.89, respectively) [31R, 54MC]. Blood pressure should be monitored regularly. Data remain limited, especially with the newer SNRIs such as desvenlafaxine or levomilnacipran, and additional studies are needed.

Selective Serotonin Reuptake Inhibitors (SSRIs)

Infants exposed to any SRIs such as SSRIs or SNRIs have higher rates of preterm birth, lower birth weight, and respiratory problems than the general population [30MC, 35R].

Major Malformations

Rates for major malformations with SSRIs are within the background rate of approximately 3% [35R]. However, fluoxetine is associated with higher rates (OR 1.14, CI 1.01–1.30) [55R]. For this reason, and its prolonged half-life, fluoxetine is recommended to use with caution.

CARDIOVASCULAR

All SSRIs are associated with a modest increase in congenital cardiac abnormalities, but paroxetine has the highest risk (OR 1.44, CI 1.12–1.86) [31R, 55R]. Paroxetine is a pregnancy category D and is not recommended during pregnancy.

Respiratory

Persistent pulmonary hypertension of the neonate (PPHN) is associated with SSRI use [56R]. Most respiratory issues are mild, requiring only temporary supportive care. Regardless, there is a statistically significant increase in PPHN with SSRIs, with the highest risk during the second-trimester (OR up to 4.28) [56R, 57MC].

Neurocognitive

A large cohort study reported an adjusted hazard ratio for autism spectrum disorder (ASD) of 1.59 (95% CI 1.17–3.17) with exposure to any serotonergic antidepressant. However, significance was lost after controlling for confounding factors (HR 1.61, CI 0.997–2.59) [58MC]. With SSRIs specifically, a meta-analysis identified an odds ratio of 1.45 (95% CI 1.15–1.82). Significance was lost after comparing to women with untreated psychiatric disorders (OR 0.96, CI 0.57–1.63). The authors concluded that there may be an increased risk with antidepressant exposure, but maternal diagnoses may be a confounder [59M]. Despite significant overlap in included studies, a second systematic review and meta-analysis found an association between any prenatal SSRI exposure and ASD (OR 1.82, CI 1.59–2.10, $P=0.00$) [60M]. The true correlation of ASD with antidepressants and maternal mental health is unclear, but should not prevent treatment [61MC].

Maternal Risks

Postpartum hemorrhage is a known risk with SSRIs [31R], with the highest risk during late pregnancy (RR 1.53, 95% CI 1.25–1.86) [34R]. Gestational hypertension is also associated with SSRIs, especially sertraline, paroxetine and fluvoxamine [28C, 30MC]. If left uncontrolled, this may progress into preeclampsia. The adjusted relative risk (aRR) for preeclampsia with SSRIs is between 1.05 and 3.16 [53M]. Regardless of the antidepressant, there is a higher risk of hypertension and preeclampsia when prescribed after 12 weeks of gestation [53M].

Dose Adjustment

It is often recommended to taper the dose potentially to discontinuation during the third trimester, and resume after birth [34R]. However, due to pharmacokinetic changes during pregnancy, SSRIs may require dose increases to maintain adequate concentrations [55R]. If a dose is reduced, close monitoring for psychiatric decompensation is required.

Tricyclic Antidepressants (TCAs)

General Malformations and Serious Risks

While TCAs were originally associated with cleft palate, diaphragmatic hernia and limb abnormalities, recent studies show the risk is much lower than believed [37R]. The large European study found women on TCAs had similar rates of spontaneous abortion and late fetal deaths to the general population, and 97% of infants had no congenital defects [37R]. The American Psychiatric Association (APA) and ACOG have stated they believe there is no association with TCAs and teratogenicity [37R].

However, the data are mixed. A Quebec study assessing first-trimester exposure of antidepressants in nearly 400 women found that infants exposed to amitriptyline had higher rates of ear, eye and neck malformations (aOR 2.45, 95% CI 1.05–5.72), and digestive system abnormalities (aOR 2.55, 95% CI 1.40–4.66) compared to other agents [62MC]. Other studies have estimated the risk for congenital defects to be similar to SSRIs (RR 0.9, 95% CI 0.6–1.2 vs 0.9, CI 0.7–1.2) [37R].

Neurocognitive

There may be a correlation between TCA use and ASD. A Swedish cohort study found that, after controlling for maternal depression, prenatal TCA exposure was statistically significantly associated with ASD (aOR 2.69, 1.04–6.96) [63MC]. However, a more recent Canadian cohort study did not find any statistical correlation (aOR 1.03, CI 0.23–4.61) [64R].

Maternal Risks

TCAs also have a risk for hypertension and preeclampsia. The adjusted relative risk for TCAs is often reported as 1.10 for hypertension, and between 0.35 and 3.23 for preeclampsia [53M]. However, one study showed a decreased risk when TCAs were used in comparison to other antidepressants [29MC]. Blood pressure should be regularly monitored.

Antiepileptics and Mood Stabilizers

Carbamazepine

MAJOR MALFORMATIONS

Carbamazepine is associated with multiple malformations including neural tube defects, microcephaly, skeletal abnormalities and reduced growth rate [37R, 64R]. Major anomalies were identified 6.7% of mothers prescribed carbamazepine vs 2.5% of controls or untreated epilepsy [65M]. This is approximately twice the general population rate and was confirmed in a 2016 cohort study [66C]. Spina bifida (OR 3.6, 95% CI 1.1–4.5) and cleft palate (OR 2.4, CI 1.1–4.5) are also reported risks with carbamazepine [67M]. All risks appear to be dose related, with higher rates with daily doses greater than 1000mg [68MC].

Divalproex Sodium

MAJOR MALFORMATIONS

When taken during the first trimester, there is a 20-fold risk for neural tube defects, cardiovascular abnormalities, developmental delay, and endocrine disorders compared to the general population [37R, 69R]. Highest risks are with daily doses greater than 1100mg compared to other AEDs (OR 7.3, $P < 0.0001$) [70C].

An ongoing case–control study assessed first-trimester AED exposure and found a higher risk of congenital malformations with divalproex, including neural tube defects (aOR 9.8, 95% CI 3.4–27.5), oral cleft (aOR 4.4, 1.6–12.2), heart defects (aOR 2.0, 0.78–5.3) and hypospadias (aOR 2.4, 0.62–9.0) [71MC]. Other defects found statistically more frequently than other AEDs included skeletal defects and genitourinary malformations [68MC].

Children exposed to divalproex during the first 3 months of gestation may experience fetal valproate syndrome (FVS). Some common features include facial deformities including connecting epicanthal folds, flat nasal bridge, small nose, anteverted nostrils, shallow philtrum, and small mouth with downturned angles, thin upper vermilion border, as well as delayed neurological development, microcephaly and other skeletal or muscular abnormalities [72c]. Sufficient folic acid should be given while taking divalproex to prevent neural tube defects [72c].

NEUROCOGNITIVE

Developmental delay and ASD are seen with divalproex. Developmental delay is associated with lower daily doses than other malformations (greater than 800mg) [73C, 74C, 75C]. ASD has been identified up to 10 times more frequently in those exposed to divalproex than the general population [37R].

LAMOTRIGINE

Most studies of in utero exposure have not found increased malformations rates with lamotrigine [37R, 64R]. In addition, there are no cases of developmental delay with lamotrigine use up to 6 years of age [76C, 77R].

However, lamotrigine doses must be increased during pregnancy due to estrogen level changes. In epilepsy, doses must be increased by an average of 250% [69R, 77R, 78c]. While doses are lower in mood disorders, patients should be closely monitored for destabilization of symptoms during the third trimester, and doses may need to be increased.

Lithium

GENERAL RISKS

Lithium freely crosses the placenta, so fetal and maternal blood levels are equivalent [37R]. Despite this, significant fetal toxicity is relatively limited. Reported symptoms include lower Apgar score at birth, hypotonicity, lethargy, goiter, increased birth weight, fetal heart failure and diabetes insipidus [37R, 64R].

The reported risks with lithium appear to be dose related. A small study of 20 infants compared high (>0.64mEq/L) vs low exposure (<0.64mEq/L), and found infants exposed to high concentrations had significantly lower Apgar scores, more days in the hospital, lower birth weight and were more likely to be born prematurely (all $P < 0.05$). In addition, over twice as many high concentration infants had cardiac complications [79c].

EPSTEIN'S ANOMALY

A 1976 study found a 10-fold increased rate of Epstein's anomaly than the general population [80C]. More recent studies indicate a lower risk than originally believed, but higher than the general population [37R, 74C, 81M]. A fetal echocardiogram may be indicated at 16–20 weeks [37R, 69R].

Antipsychotics

MAJOR MALFORMATIONS

Umbilical cord to maternal plasma ratios range between 25% and 70% for first-generation antipsychotics (FGAs) and second-generation antipsychotics (SGAs) [37R]. Of these, olanzapine has the highest ratio, but has little data demonstrating major congenital malformations [37R, 82M, 83M].

Some of the FGAs, namely, phenothiazines such as perphenazine, have no identifiable teratogenicity. A prospective cohort study of over 1300 pregnancies taking FGAs showed no increase in congenital abnormalities or death [37R, 84MC]. A large national registry did not find an increased risk in congenital defects with FGA or SGA use [85MC, 86C]. Another study of over 400 women found a nonsignificant increase in malformations in women prescribed SGAs vs those who stopped before pregnancy, or controls [87MC]. However, meta-analyses have found a more than twofold increased odds in congenital and cardiac malformations with antipsychotic use [88M, 89M]. Risperidone may have a slightly higher comparative risk for major malformations (RR 1.26, CI 1.02–1.56), especially cardiac (1.26, CI 0.88–1.81) [85MC].

STILLBIRTH OR ABORTION

Stillbirth or spontaneous abortion are reported with SGAs, but the data are conflicting [82M, 89M]. The lowest risk agents are quetiapine and olanzapine. However, these have the highest risk for gestational diabetes which is associated with other pregnancy complications [90R, 91C].

NEUROCOGNITIVE

There may be a risk of neurodevelopmental delay or behavioral disorders with in utero antipsychotic exposure, but data are limited [87MC]. Infants exposed to SGAs showed delayed cognitive, motor, social and emotional functioning development, and adaptive behavior at 2 months old vs unexposed infants ($P<0.01$ for all) [92c]. By 6 or 12 months, there were no significant differences [92c].

Clozapine may be the highest risk agent for developmental delay. Infants exposed to clozapine had additional signs of developmental delay at 2 and 6 months vs olanzapine, quetiapine or risperidone. Despite no significant difference at 12 months, clozapine should be limited to treatment-refractory patients [93c].

WITHDRAWAL

NAS is a risk after antipsychotic exposure. A prospective cohort study found SGA withdrawal syndromes occurred in 15% of cases, most frequently with quetiapine and olanzapine [94C]. However, most women were taking multiple psychotropic agents, so causality is difficult to assess [94C]. The most frequent symptoms include agitation, tremors, abnormal muscle tone, sleeping or feeding difficulties and respiratory complications [95R]. Matched cohort analyses have determined the risk for serious symptoms is small [96MC].

DOSE ADJUSTMENT

Serum concentrations are significantly lower during the third trimester for quetiapine (−76%, CI −83%, −66%; $P<0.001$) and aripiprazole (−52%, CI −62%, −39%; $P<0.001$), but not for olanzapine ($P=0.40$) [97C]. This may be related to a decrease in plasma proteins [97C]. In addition, several P450 enzymes are induced during pregnancy such as CYP3A4. This may affect antipsychotics, but no definitive studies have been completed [82M, 97C].

Readers are also referred to the review published on the related topic [98R].

References

[1] Cefalu WT. American diabetes association: standards of medical care in diabetes. J Clin Appl Res Educ. 2017;40:1–132 [S].

[2] Rowan JA, Hague WM, Gao W, et al. Metformin versus insulin for the treatment of gestational diabetes. N Engl J Med. 2008;358(19C):2003–15 [c].

[3] Gui J, Liu Q, Feng L. Metformin vs insulin in the management of gestational diabetes: a meta-analysis. PLoS One. 2013;8(5): e64585 [M].

[4] Langer O, Conway DL, Berkus MD, et al. A comparison of glyburide and insulin in women with gestational diabetes mellitus. N Engl J Med. 2000;343(16):1134–8 [c].

[5] Hossain MA, Bathgate RAD. Challenges in the design of insulin and relaxin/insulin-like peptide mimetics. Bioorganic Med Caahem. 2018;26(10):2827–41 [H].

[6] Silver R. Practice bulletin no. 181: prevention of Rh D alloimmunizatione. Obstet Gynecol. 2017;130(2):57–70 [S].

[7] Liang H, Ma S, Xiao Y, et al. Comparative efficacy and safety of oral antidiabetic drugs and insulin in treating gestational diabetes mellitus. Medicine (Baltimore). 2017;96(38):e7939 [M].

[8] Feig DS, Murphy K, Asztalos E, et al. Metformin in women with type 2 diabetes in pregnancy (MiTy): a multi-center randomized controlled trial. BMC Pregnancy Childbirth. 2016;16(1):173 [c].

[9] Morton A, Laurie J. Allergic reactions to insulin detemir in women with gestational diabetes mellitus. Aust Fam Physician. 2017;4(7):485–6 [A].

[10] Santos P, Hefele JG, Ritter G, et al. Population-based risk factors for shoulder dystocia. JOGNN-J Obstet Gynecol Neonatal Nurs. 2018;47(1):32–42 [C].

[11] Nachum Z, Zafran N, Salim R, et al. Glyburide versus metformin and their combination for the treatment of gestational diabetes mellitus: arandomized controlled study. Diabetes Care. 2017;40(3):332–7 [c].

[12] Zeng X-L, Zhang Y-F, Tian Q, et al. Effects of metformin on pregnancy outcomes in women with polycystic ovary syndrome: a meta-analysis. Medicine (Baltimore). 2016;95(36)e4526 [M].

[13] Song R, Chen L, Chen Y, et al. Comparison of glyburide and insulin in the management of gestational diabetes: a meta-analysis. PLoS One. 2017;12(8):1–18 [M].

[14] Mukerji G, Feig DS. Pharmacological management of gestational diabetes mellitus. Drugs. 2017;77(16):1723–32 [R].

[15] Koning SH, Hoogenberg K, Scheuneman KA, et al. Neonatal and obstetric outcomes in diet- and insulin-treated women with gestational diabetes mellitus: a retrospective study. BMC Endocr Disord. 2016;16(1):52 [c].

[16] Brown J, Martis R, Hughes B, et al. Oral anti-diabetic pharmacological therapies for the treatment of women

with gestational diabetes. Cochrane Database Syst Rev. 2017;1(1):1–111 [R].

[17] Roberts JM, Druzin M, August PA, et al. ACOG guidelines: Hypertension in pregnancy. American College of Obstetricians and Gynecologists; 2013. p. 1–100 [S].

[18] Practice C on O. Emergent therapy for acute-onset, severe hypertension during pregnancy and the postpartum period. Committee opinion No. 692. Obstet Gynecol. 2017;129:1–4 [S].

[19] Caton AR, Bell EM, Druschel CM, et al. Antihypertensive medication use during pregnancy and the risk of cardiovascular malformations. Hypertension. 2009;54(1):63–70 [C].

[20] Fisher SC, Van Zutphen AR, Werler MM, et al. Maternal antihypertensive medication use and congenital heart defects: updated results from the national birth defects prevention study. Hypertens (Dallas, Tex 1979). 2017;69(5):798–805 [C].

[21] Su JA, Tang W, Rivero N, et al. Prenatal exposure to methyldopa leading to hypertensive crisis and cardiac failure in a neonate. Pediatrics. 2014;133(5):e1392–5. [Internet]. Available from: http://pediatrics.aappublications.org/cgi/doi/10.1542/peds.2013-1438 [A].

[22] Kashkooli S, Baraty B, Kalantar J. α-Methyldopa-induced hepatitis during the postpartum period. BMJ Case Rep. 2014;2014: (feb27 1): bcr2014203712-bcr2014203712 [A].

[23] Hoeltzenbein M, Beck E, Fietz A-K, et al. Pregnancy outcome after first trimester use of methyldopa—a prospective cohort study. Hypertension. 2017;70:201–8 [C].

[24] Brochet MS, Harry M, Morin F. Nifedipine induced gingival hyperplasia in pregnancy: a case report. Curr Drug Saf. 2017;12:3–6. [Internet]. Available from:http://www.ncbi.nlm.nih.gov/pubmed/27113951 [A].

[25] Ray S, Stowe ZN. The use of antidepressant medication in pregnancy. Best Pract Res Clin Obstet Gynaecol. 2014;28(2014):71–83 [R].

[26] Angelotta C, Wisner KL. Treating depression during pregnancy: are we asking the right questions? Birth Defects Res. 2017;109(2017):879–87 [R].

[27] Reis M, Kllén B. Delivery outcome after maternal use of antidepressant drugs in pregnancy: an update using Swedish data. Psychol Med. 2010;40(10):1723–33 [MC].

[28] De Vera MA, Bérard A. Antidepressant use during pregnancy and the risk of pregnancy-induced hypertension. Br J Clin Pharmacol. 2012;74(2):362–9 [C].

[29] Avalos LA, Chen H, Li D-K. Antidepressant medication use, depression, and the risk of preeclampsia. CNS Spectr. 2015;20(1):39–47. [Internet]. Available from: https://www.cambridge.org/core/product/identifier/S1092852915000024/type/journal_article [MC].

[30] Yonkers KA, Gilstad-Hayden K, Forray A, et al. Association of panic disorder, generalized anxiety disorder, and benzodiazepine treatment during pregnancy with risk of adverse birth outcomes. JAMA psychiat. 2017;74(11):1145–52 [MC].

[31] Carvalho AF, Sharma MS, Brunoni AR, et al. The safety, tolerability and risks associated with the use of newer generation antidepressant drugs: a critical review of the literature. Psychother Psychosom. 2016;85:270–88 [R].

[32] Gentile S. On categorizing gestational, birth, and neonatal complications following late pregnancy exposure to antidepressants: the prenatal antidepressant exposure syndrome. CNS Spectr. 2010;15:167–85 [R].

[33] Smit M, Dolman KM, Honig A. Mirtazapine in pregnancy and lactation—a systematic review. Eur Neuropsychopharmacol. 2016;26(2017):126–35 [M].

[34] Latendresse G, Elmore C, Deneris A. Selective serotonin reuptake inhibitors as first-line antidepressant therapy for perinatal depression. J Midwifery Women's Heal. 2017;62(3):317–28 [R].

[35] Ornoy A, Koren G. Selective serotonin reuptake inhibitors during pregnancy: do we have now more definite answers related to prenatal exposure? Birth Defects Res. 2017;109(2017):898–908 [R].

[36] Hendrick V, Suri R, Gitlin MJ, et al. Bupropion use during pregnancy: a systematic review. Prim Care Companion CNS Disord. 2017;19(5). 17r02160, [Internet]. Available from: http://www.psychiatrist.com/PCC/article/Pages/2017/v19n05/17r02160.aspx [C].

[37] Ornoy A, Weinstein-Fudim L, Ergaz Z. Antidepressants, antipsychotics, and mood stabilizers in pregnancy: what do we know and how should we treat pregnant women with depression. Birth Defects Res. 2017;109(2017):933–56 [R].

[38] GlaxoSmithKline. The bupropion pregnancy registry. Final report [Internet], Wilmington NC: Kendle International Inc; 2008. Available from: http://pregnancyregistry.gsk.com/documents/bup_report_final_2008.pdf [S].

[39] Chun-Fai-Chan B, Koren G, Fayez I, et al. Pregnancy outcome of women exposed to bupropion during pregnancy: a prospective comparative study. Am J Obstet Gynecol. 2005;192(3):932–6 [C].

[40] Kline J, Stein Z, Susser M. Conception to birth: Epidemiology of prenatal development. vol. 14. Oxford University Press; 1989 [R].

[41] Bérard A, Zhao JP, Sheehy O. Success of smoking cessation interventions during pregnancy. Am J Obstet Gynecol. 2016;215(5) 611.e1–611.e8. [MC].

[42] Louik C, Kerr S, Mitchell AA. First-trimester exposure to bupropion and risk of cardiac malformations. Pharmacoepidemiol Drug Saf. 2014;23(10):1066–75 [MC].

[43] Omay O, Einarson A. Is mirtazapine an effective treatment for nausea and vomiting of pregnancy? A case series. J Clin Psychopharmacol. 2017;37(2):260–1 [c].

[44] Lennestål R, Källén B. Delivery outcome in relation to maternal use of some recently introduced antidepressants. J Clin Psychopharmacol. 2007;27(6):607–13 [M].

[45] Djulus J, Koren G, Einarson TR, et al. Exposure to mirtazapine during pregnancy: a prospective, comparative study of birth outcomes. J Clin Psychiatry. 2006;67(8):1280–4. [Internet]. Available from: http://www.ncbi.nlm.nih.gov/pubmed/16965209 [C].

[46] Manakova E, Hubickova L. Antidepressant drug exposure during pregnancy. a small prospective study. Neuro Endocrinol Lett. 2011;32(Supp 1):53–6. [Internet]. Available from: http://www.ncbi.nlm.nih.gov/pubmed/22167208 [c].

[47] Yaris F, Kadioglu M, Kesim M, et al. Newer antidepressants in pregnancy: prospective outcome of a case series. Reprod Toxicol. 2004;19(2):235–8 [c].

[48] Einarson A, Choi J, Einarson TR, et al. Incidence of major malformations in infants following antidepressant exposure in pregnancy: results of a large prospective cohort study. Can J Psychiatry. 2009;54(4):242–6. [Internet]. Available from: http://www.ncbi.nlm.nih.gov/pubmed/19321030 [C].

[49] Lassen D, Ennis ZN, Damkier P. First-trimester pregnancy exposure to venlafaxine or duloxetine and risk of major congenital malformations: a systematic review. Basic Clin Pharmacol Toxicol. 2016;118:32–6 [M].

[50] Einarson A, Smart K, Vial T, et al. Rates of major malformations in infants following exposure to duloxetine during pregnancy: a preliminary report. J Clin Psychiatry. 2012;73:1471 [C].

[51] Hoog SL, Cheng Y, Elpers J, et al. Duloxetine and pregnancy outcomes: safety surveillance findings. Int J Med Sci. 2013;10(4):413–9 [C].

[52] Andrade C. The safety of duloxetine during pregnancy and lactation. J Clin Psychiatry. 2014;75(12):e1423–7 [R].

[53] Uguz F. Is there any association between use of antidepressants and preeclampsia or gestational hypertension? A systematic review of current studies conclusions: although some studies have suggested a moderately in. J Clin Psychopharmacol. 2017;37:72–7 [M].

[54] Palmsten K, Huybrechts KF, Michels KB, et al. Antidepressant use and risk for preeclampsia. Epidemiology. 2013;24(5):682–91 [MC].
[55] Womersley K, Ripullone K, Agius M. What are the risks associated with different selective serotonin re-uptake inhibitors (SSRIs) to treat depression and anxiety in pregnancy? An evaluation of current evidence. Psychiatr Danub. 2017;29(Supp 3):629–44 [R].
[56] Huybrechts KF, Bateman BT, Palmsten K, et al. Antidepressant use late in pregnancy and risk of persistent pulmonary hypertension of the newborn, JAMA. 2015;313(21A):2142. [Internet]. Available from: http://jama.jamanetwork.com/article.aspx?doi=10.1001/jama.2015.5605 [R].
[57] Bérard A, Sheehy O, Zhao J-P, et al. SSRI and SNRI use during pregnancy and the risk of persistent pulmonary hypertension of the newborn. Br J Clin Pharmacol. 2017;83(2017):1126–33 [MC].
[58] Brown HK, Ray JG, Wilton AS, et al. Association between serotonergic antidepressant use during pregnancy and autism spectrum disorder in children. JAMA—J Am Med Assoc. 2017;317(15):1544–52 [MC].
[59] Kobayashi T, Matsuyama T, Takeuchi M, et al. Autism spectrum disorder and prenatal exposure to selective serotonin reuptake inhibitors: a systematic review and meta-analysis. Reprod Toxicol. 2016;65(2016):170–8 [M].
[60] Andalib S, Emamhadi MR, Yousefzadeh-Chabok S, et al. Maternal SSRI exposure increases the risk of autistic offspring: a meta-analysis and systematic review. Eur Psychiatry. 2017;45(2017):161–6 [M].
[61] Viktorin A, Uher R, Reichenberg A, et al. Autism risk following antidepressant medication during pregnancy. Psychol Med. 2017;47:2787–96 [MC].
[62] Bérard A, Zhao J-P, Sheehy O. Antidepressant use during pregnancy and the risk of major congenital malformations in a cohort of depressed pregnant women: an updated analysis of the Quebec pregnancy cohort. BMJ Open. 2017;7(1)e013372 [MC].
[63] Rai D, Lee BK, Dalman C, et al. Parental depression, maternal antidepressant use during pregnancy, and risk of autism spectrum disorders: population based case-control study. BMJ. 2013;346: f2059 [Internet]. [MC].
[64] Wichman CL. Managing your own mood lability: use of mood stabilizers and antipsychotics in pregnancy. Curr Psychiatry Rep. 2016;18(1):1–5 [R].
[65] Matalon S, Schechtman S, Goldzweig G, et al. The teratogenic effect of carbamazepine: a meta-analysis of 1255 exposures. Reprod Toxicol. 2002;16(1):9–17 [M].
[66] Vajda FJE, O'Brien TJ, Graham J, et al. Is carbamazepine a human teratogen? J Clin Neurosci. 2016;23:34–7 [C].
[67] Gilboa SM, Broussard CS, Devine OJ, et al. Influencing clinical practice regarding the use of antiepileptic medications during pregnancy: modeling the potential impact on the prevalences of spina bifida and cleft palate in the United States. Am J Med Genet Part C Semin Med Genet. 2011;157(3):234–46 [M].
[68] Campbell E, Kennedy F, Russell A, et al. Malformation risks of antiepileptic drug monotherapies in pregnancy: updated results from the UK and Ireland epilepsy and pregnancy registers. J Neurol Neurosurg Psychiatry. 2014;85(9):1029–34 [MC].
[69] Hogan CS, Freeman MP. Adverse effects in the pharmacologic management of bipolar disorder during pregnancy. Psychiatr Clin North Am. 2016;39(3):465–75 [R].
[70] Vajda FJE, Hitchcock A, Graham J, et al. Foetal malformations and seizure control: 52 months data of the Australian Pregnancy Registry. Eur J Neurol. 2006;13(6):645–54 [C].
[71] Werler MM, Ahrens KA, Bosco JLF, et al. Use of antiepileptic medications in pregnancy in relation to risks of birth defects. Ann Epidemiol. 2011;21(11):842–50 [MC].
[72] Mutlu-Albayrak H, Bulut C, Çaksen H. Fetal valproate syndrome. Pediatr Neonatol. 2017;58(2):158–64 [c].
[73] Meador KJ, Baker GA, Browning N, et al. Cognitive function at 3 years of age after fetal exposure to antiepileptic drugs, N Engl J Med. 2009;360(16):1597–605. [Internet]. Available from: http://www.scopus.com/inward/record.url?eid=2-s2.0-64749112939&partnerID=tZOtx3y1 [C].
[74] Baker GA, Bromley RL, Briggs M, et al. IQ at 6 years after in utero exposure to antiepileptic drugs: a controlled cohort study. Neurology. 2015;84(4):382–90 [C].
[75] Bromley RL, Mawer G, Love J, et al. Early cognitive development in children born to women with epilepsy: a prospective report. Epilepsia. 2010;51(10):2058–65 [C].
[76] Meador KJ, Baker GA, Browning N, et al. Fetal antiepileptic drug exposure and cognitive outcomes at age 6 years (NEAD study): a prospective observational study. Lancet Neurol. 2013;12(3):244–52 [C].
[77] Khan SJ, Fersh ME, Ernst C, et al. Bipolar disorder in pregnancy and postpartum: principles of management. Curr Psychiatry Rep. 2016;18(13):1–11 [R].
[78] Clark CT, Klein AM, Perel JM, et al. Lamotrigine dosing for pregnant patients with bipolar disorder, Am J Psychiatry. 2013;170(11):1240–7. [Internet]. Available from: http://www.pubmedcentral.nih.gov/articlerender.fcgi?artid=4154145&tool=pmcentrez&rendertype=abstract [c].
[79] Newport DJ, Viguera AC, Beach AJ, et al. Lithium placental passage and obstetrical outcome: implications for clinical management during late pregnancy. Am J Psychiatry. 2005;162(11):2162–70 [c].
[80] Weinstein MR. The international register of lithium babies. Ther Innov Regul Sci. 1976;10(2):94–100 [C].
[81] McKnight RF, Adida M, Budge K, et al. Lithium toxicity profile: a systematic review and meta-analysis. Lancet. 2012;379(9817):721–8 [M].
[82] Tosato S, Albert U, Tomassi S, et al. A systematized review of atypical antipsychotics in pregnant women: balancing between risks of untreated illness and risks of drug-related adverse effects. J Clin Psychiatry. 2017;78(6):e477–89 [M].
[83] Brunner E, Falk DM, Jones M, et al. Olanzapine in pregnancy and breastfeeding: a review of data from global safety surveillance. BMC Pharmacol Toxicol. 2013;14(1): 38. [Internet]. Available from: http://www.pubmedcentral.nih.gov/articlerender.fcgi?artid=3750520&tool=pmcentrez&rendertype=abstract [M].
[84] Slone D, Siskind V, Heinonen OP, et al. Antenatal exposure to the phenothiazines in relation to congenital malformations, perinatal mortality rate, birth weight, and intelligence quotient score. Am J Obstet Gynecol. 1977;128(5):486–8 [MC].
[85] Huybrechts KF, Hernández-Díaz S, Patorno E, et al. Antipsychotic use in pregnancy and the risk for congenital malformations. JAMA Psychiatry. 2016;73(9):938–46 [MC].
[86] Cohen LS, Viguera AC, McInerney KA, et al. Reproductive safety of second-generation antipsychotics: current data from the Massachusetts general hospital national pregnancy registry for atypical antipsychotics. Am J Psychiatry. 2016;173(3): 263–70 [C].
[87] Petersen I, McCrea RL, Sammon CJ, et al. Risks and benefits of psychotropic medication in pregnancy: cohort studies based on UK electronic primary care health records. Health Technol Assess. 2016;20(23C):1–176 [MC].
[88] Coughlin CG, Blackwell KA, Bartley C, et al. Obstetric and neonatal outcomes after antipsychotic medication exposure in pregnancy. Obstet Gynecol. 2015;125(5):1224–35 [M].
[89] Terrana N, Koren G, Pivovarov J, et al. Pregnancy outcomes following in utero exposure to second-generation antipsychotics: a systematic review and meta-analysis. J Clin Psychopharmacol. 2015;35:559–65 [M].
[90] Gentile S. Clinical utilization of atypical antipsychotics in pregnancy and lactation. Ann Pharmacother. 2004;38:1265–71 [R].
[91] Boden R, Lundgren M, Brandt L, et al. Risks of adverse pregnancy and birth outcomes in women treated or not treated with mood stabilisers for bipolar disorder: population based cohort study.

BMJ. 2012;345:e7085. [Internet]. Available from: http://www.bmj.com/cgi/doi/10.1136/bmj.e7085 [C].
[92] Peng M, Gao K, Ding Y, et al. Effects of prenatal exposure to atypical antipsychotics on postnatal development and growth of infants: a case-controlled, prospective study. Psychopharmacology (Berl). 2013;228(4):577–84 [c].
[93] Shao P, Ou J, Peng M, et al. Effects of clozapine and other atypical antipsychotics on infants development who were exposed to as fetus: a post-hoc analysis. PLoS One. 2015;10(4): e0123373 [c].
[94] Kulkarni J, Worsley R, Gilbert H, et al. A prospective cohort study of antipsychotic medications in pregnancy: the first 147 pregnancies and 100 one year old babies. PLoS One. 2014;9(5) [C].
[95] Whitworth AB. Psychopharmacological treatment of schizophrenia during pregnancy and lactation. Curr Opin Psychiatry. 2017;30(3):184–90 [R].
[96] Vigod SN, Gomes T, Wilton AS, et al. Antipsychotic drug use in pregnancy: high dimensional, propensity matched, population based cohort study. BMJ. 2015;350:h2298. [Internet]. Available from: http://www.bmj.com/cgi/doi/10.1136/bmj.h2298 [MC].
[97] Westin AA, Brekke M, Molden E, et al. Treatment with antipsychotics in pregnancy: changes in drug disposition. Clin Pharmacol Ther. 2018;103(3):477–84 [C].
[98] Pittman J, Schalliol LA, Ray SD. Insulin and other hypoglycemic drugs. In: Ray SD, editor. Side effects of drugs: annual: a worldwide yearly survey of new data in adverse drug reactions, 39; 2017. Elsevier; 2017. p. 435–46 [R].

Further Reading

[99] Boukhris T, Sheehy O, Mottron L, et al. Antidepressant use during pregnancy and the risk of autism spectrum disorder in children. JAMA Pediatr. 2015;170(2):1. [Internet]. Available from: http://archpedi.jamanetwork.com/article.aspx?doi=10.1001/jamapediatrics.2015.3356%5Cnhttp://www.ncbi.nlm.nih.gov/pubmed/26660917 [MC].

CHAPTER

44

Drugs That Affect Lipid Metabolism

Rebecca Tran,[1], Kerry Anne Rambaran†*

*Department of Pharmacy Practice, Keck Graduate Institute School of Pharmacy, Claremont, CA, United States

†Department of Pharmacy Practice, Keck Graduate Institute School of Pharmacy and Health Sciences, Claremont, CA, United States

[1]Corresponding author: rtran@kgi.edu

BILE ACID SEQUESTRANTS [SED-15, 1902; SEDA-36, 676; SEDA-37, 559]

Cholestyramine, Colesevelam and Colestipol

No relevant publications from the review period were identified.

CHOLESTEROL ABSORPTION INHIBITOR [SEDA-39, 457; SEDA-35, 810; SEDA-36, 677]

Ezetimibe

Liver

A case of severe hepatotoxicity was reported with ezetimibe in combination with atorvastatin in an overweight woman in her 70s with a past medical history of hypothyroidism, hypertension, coronary artery disease (CAD), and diabetes mellitus (DM) type 2. The patient had previously experienced elevated serum alanine aminotransferase (ALT) while on atorvastatin 20 and 40mg daily but levels normalized within 2 months of discontinuation. The patient was then started on pravastatin 40mg daily without adverse effects. After switching to the combination of atorvastatin 20mg and ezetimibe 10mg, the patient developed diffuse pain, lower extremity urticaria and rash, muscle weakness, fatigue, chest discomfort, and nausea. Aspartate aminotransferase (AST) and ALT ranged between 1224–2142U/L and 2003–2715U/L, respectively. Bilirubin ranged between 4.3 and 4.7mg/dL and pancreatic amylase and lipase were slightly elevated. Tests for causes such as autoimmune and viral hepatitis were negative and the liver ultrasound was unremarkable. The authors reported that the final diagnosis was drug-induced liver injury without liver failure. After normalization of transaminases, the patient was restarted on pravastatin 40mg daily without further adverse effects [1A].

A multicenter, open-label, prospective study randomized 109 patients with DM type 2 in Japan to either ezetimibe 10mg plus a baseline statin (atorvastatin 10mg or pitavastatin 1mg) or a double-dose of a baseline statin. In the ezetimibe plus statin group ($n=53$), there was one report each of the following: a common cold, hepatocellular carcinoma, eczema, adenocarcinoma of unknown origin, and ileus by abdominal abscess. There were statistically significant increases in transaminases in the ezetimibe plus statin group. AST increased from 21.3 ± 7.8 to 24.1 ± 11.3IU/L and ALT increased from 23.5 ± 12.4 to 26.9 ± 18.0IU/L. However, authors concluded overall that the two treatment arms were similar in terms of tolerability and safety from a clinical significance perspective [2C].

Musculoskeletal, Nervous System

In a retrospective, pharmacoepidemiological analysis, the use of ezetimibe in Australia from 2006 to 2015 was evaluated. The analysis was limited to ezetimibe prescriptions, alone and in combination products with simvastatin, atorvastatin, and rosuvastatin, within the administrative database of the Pharmaceutical Benefits Scheme (PBS) in Australia. There were 575 adverse event case reports in the Therapeutic Goods Administration (TGA) Database of Adverse Event Notifications between 2004 and 2015 for all ezetimibe products, with 456 cases (79%) for ezetimibe only. Within these ezetimibe-only cases, musculoskeletal disorders were commonly reported, such as myalgia (42%), arthralgia (10%), and muscle spasms (9%). Gastrointestinal symptoms were also often reported, such as nausea (19%), diarrhea (11%), and abdominal pain (10%). Within the 119 case

Side Effects of Drugs Annual, Volume 40
ISSN: 0378-6080
https://doi.org/10.1016/bs.seda.2018.07.016

reports with ezetimibe plus statin combination products, the most common adverse events were musculoskeletal and connective tissue disorders such as rhabdomyolysis, myalgia, and myopathy (76%), then nervous disorders such as seizures and amnesia (24%), and general disorders such as pain and asthenia (24%). Inferential statistics were not performed [3R].

Nervous System

In a post-hoc analysis of the IMPROVE-IT trial, a randomized, placebo-controlled trial comparing the combination of ezetimibe 10mg plus simvastatin 40mg to simvastatin 40mg alone, patients were stratified by the presence of DM at enrollment. In patients without DM, rates of elevated transaminases were similar between the simvastatin alone group vs simvastatin/ezetimibe group (2.3% vs 2.3%, $P=0.91$). In patients with DM, the rates of elevated transaminases were also similar (2.3% in simvastatin alone group vs 2.9% in simvastatin/ezetimibe group, $P=0.25$). Rates of cancer were similar in patients with DM and without DM. Patients with DM had higher rates of gall bladder-related adverse events (4.3% in simvastatin alone group vs 3.9% in simvastatin/ezetimibe group, $P=0.52$) and muscle-related adverse events (0.8% in simvastatin alone group vs 0.7% in simvastatin/ezetimibe group, $P=0.62$), although differences were not statistically significant. Rates of hemorrhagic stroke were highest among patients with DM in the simvastatin/ezetimibe group (0.9% vs 0.4% in simvastatin alone group, $P=0.023$; P-value for subgroup interaction$=0.092$). Rates of hemorrhagic stroke in patients without DM were 0.5% and 0.5% in the simvastatin/ezetimibe and simvastatin alone groups, respectively ($P=0.81$). The authors concluded that the benefits of treatment with simvastatin and ezetimibe outweighed the possible increased risk of hemorrhagic stroke with the combination therapy [4MC].

In another post-hoc analysis of the IMPROVE-IT trial, patients were stratified by sex. There was a higher rate of permanent study drug discontinuation in women (14.5%) compared to men (12.7%). Within the safety outcomes, rates of adverse events tended to be higher in women although differences were not statistically significant. In women, rates of increased transaminases were 2.3% ($n=50$) in the simvastatin alone group and 2.4% ($n=52$) in the simvastatin/ezetimibe group ($P=0.92$). In men, rates of increased transaminases were 2.1% ($n=143$) vs 2.3% ($n=153$) in the simvastatin alone and simvastatin/ezetimibe group ($P=0.56$), respectively. With respect to gallbladder-related adverse effects, rates were higher in women [3.4% ($n=72$) with simvastatin alone vs 2.9% ($n=63$) with simvastatin/ezetimibe, $P=0.38$] compared to men [2.7% ($n=180$) with simvastatin alone vs 2.3% ($n=151$) with simvastatin/ezetimibe, $P=0.13$]. Similar trends were seen for muscle-related symptoms. Rates were higher in women [1.0% ($n=21$) in simvastatin alone vs 0.6% ($n=12$) in simvastatin/ezetimibe, $P=0.12$] compared to men [0.5% ($n=34$) in simvastatin alone vs 0.5% ($n=35$) in simvastatin/ezetimibe, $P=0.90$]. Only cancer was higher in men compared to women [6.7% ($n=453$) in simvastatin alone vs 6.8% ($n=454$) in simvastatin/ezetimibe for men; 4.8% ($n=102$) in simvastatin alone vs 5.1% ($n=111$) in simvastatin/ezetimibe for women] [5MC].

NICOTINIC ACID DERIVATIVE [SEDA-39, 457; SEDA-15, 2512; SEDA-36, 679]

Niacin

Liver

Authors reported a 22-year-old female athlete who ingested a single dose of native niacin 20000mg for athletic performance enhancement and presented with abdominal pain. The patient developed acute liver failure within 3 days, with International Normalized Ratio (INR) >8.9, aspartate aminotransferase (AST) of 14985U/L, alanine aminotransferase (ALT) of 12594U/L, and alkaline phosphatase of 165U/L. The patient had no evidence of viral or autoimmune causes of hepatotoxicity. Liver function tests normalized quickly after a liver transplantation. The authors stated that a causality assessment by the Roussel Uclaf Causality Assessment Method (RUCAM) suggested that niacin was a "probable" cause [6A].

Gastrointestinal

An observational, cohort analysis was performed using data from the Mini-Sentinel program of the US Food and Drug Administration (FDA) to compare the rates of major gastrointestinal bleeding and intracranial hemorrhage between patients on extended-release niacin and fenofibrate. The Mini-Sentinel pilot is now the Sentinel surveillance system, a national database that monitors the safety of medical products and is controlled by participating Data Partners, who contribute and update administrative claims and clinical information. Using data from January 1, 2007 to August 31, 2013, the primary analysis compared individuals initiating single-drug or combination extended-release niacin products with fenofibrate initiators. In the matched analyses using a Propensity Score Matching (PSM) tool, there were no differences between the groups in terms of intracranial hemorrhage (HR 1.21; 95% CI 0.66–2.22) and gastrointestinal bleed (HR 0.98; 95% CI 0.82–1.18). The authors concluded that they did not see an association between niacin use and intracranial hemorrhage or gastrointestinal bleeding as was observed in the 2007 Heart Protection Study 2—Treatment of HDL to Reduce the Incidence of Vascular Events (HPS2-THRIVE) trial with niacin–laropiprant [7S].

FIBRIC ACID DERAVITIVES [SEDA-39, 459; SEDA-15, 1358; SEDA-35, 812]

Fenofibrate

Liver

The authors presented seven cases of liver injury attributed to fenofibrate therapy. The patients were enrolled in the Drug-Induced Liver Injury Network (DILIN), a prospective, multicenter, observational cohort study ($n=1723$) funded by the National Institutes of Health (NIH). All patients were evaluated for viral, autoimmune, and biliary causes of liver injury. Causality between the drug and liver injury was evaluated using a standardized scoring system and the RUCAM. All patients were prescribed doses between 48 and 160mg daily and duration of treatment ranged from 5 to 56 weeks. All patients had symptoms at presentation, with the most common being jaundice ($n=6$), fatigue ($n=3$), nausea ($n=3$), itching ($n=3$), and abdominal pain ($n=2$). The median initial serum ALT was 533U/L (range 78–2297), AST 245U/L (63–917), and alkaline phosphatase 218U/L (79–518). In the six patients with jaundice, peak bilirubin levels ranged between 4.7 and 37.3mg/dL. Injury in five of the patients were cholestatic or mixed in nature and the other two patients had hepatocellular-type injury. Two of the patients were considered to have severe liver injury with persistent elevated liver enzymes beyond 6 months, with one undergoing liver transplantation at 8 months and another dying of renal failure at 26 months. In terms of liver histopathology from four patients, one case had chronic hepatitis, two showing cholestasis, and one with chronic cholestasis and prominent zone three bile accumulation. Three patients showed "zone 3 cholestasis associated with duct injury that was characterized by reactive epithelial changes." Three of the seven fenofibrate patients were found to be carriers of the HLA type HLA-A*33:01, which has been associated with DILI. Fenofibrate was assigned causality scores of highly likely in five of the seven patients and probable in the other two [8MC].

Skin

A case series presented four patients with fenofibrate-induced photosensitization. The first patient was a 33-year-old woman with a history of hepatitis B and hyperlipidemia who had "erythematous papulovescicles with itching and burning sensation on the face, neck, bilateral forearms, dorsa of arms, and bilateral shins for 1 week." The patient was on fenofibrate 200mg daily and silymarin (active ingredient of milk thistle) 150mg daily for 3 weeks. The skin biopsy results showed spongiosis, focal basal vacuolarization, a superficial perivascular lymphocytic infiltrate, and dermal edema. The second patient was a 46-year-old man who presented with multiple pruritic papules and plaques, ranging in color from red to purple on the face, neck, bilateral forearms, and hands for 1 month. The patient was on fenofibrate 200mg daily for 3 weeks and the skin biopsy showed necrotic keratinocytes, basal vacuolarization, and a superficial perivascular and lichenoid lymphohistiocytic infiltrate. The third patient was a 28-year-old man with pruritic, scaly erythematous patches on the face, bilateral forearms, and dorsa of the hands after taking fenofibrate for 3 weeks. The skin biopsy results included basal vacuolarization, superficial perivascular lymphocytes, and few eosinophils infiltrate. The fourth patient was a 53-year-old man with pruritic erythematous papules over the neck, bilateral forearms, and knees for 2 weeks after taking fenofibrate for 4 weeks. The skin biopsy results included basal vacuolarization, focally dermal edema, superficial perivascular lymphocytes, histiocytes and eosinophils infiltrate, and pigmentary incontinence. All four patients had either Fitzpatrick skin phototype V (patients 1 and 2) or phototype IV (patients 3 and 4). All four patients' lesions resolved after discontinuing fenofibrate and treatment with systemic steroids, topical steroids, and/or systemic antihistamines [9c].

Decrease in HDL-Cholesterol

A case series described four patients who experienced sudden decreases in plasma HDL-c concentration after initiation of fenofibrate or ezetimibe in addition to existing fenofibrate. The first patient was a 55-year-old Caucasian man with no history of major disease or cardiovascular (CV) risk factors. In 2009, his HDL-c was 51mg/dL when he was switched from rosuvastatin 10mg to fenofibrate 200mg due to muscle intolerance. In 2012, his HDL-c dropped to 27mg/dL and then 13mg/dL while on fenofibrate. The drug was discontinued and HDL-c returned to baseline, 52mg/dL. The second patient was a 52-year-old Caucasian man with heterozygous familial hypercholesterolemia (HeFH) on atorvastatin 10mg/day, fenofibrate 200mg/day, acebutolol 200mg/day, nicorandil 10mg/day, aspirin 100mg/day, and LDL apheresis twice a month. Ezetimibe 10mg was started and HDL-c decreased from 31 to 5mg/dL within a month. Discontinuing ezetimibe had no effect on HDL-c so fenofibrate was also discontinued. The third patient was also a 52-year-old Caucasian man with HeFH. The patient's medications included micronized fenofibrate 160mg/day, atorvastatin 80mg/day, acebutolol 200mg/day, amlodipine 10mg/day, and aspirin 100mg/day. The patient was stable on this combination until ezetimibe 10mg/day was added. HDL-c decreased from 38mg/dL at baseline to 26 and 24mg/dL within 3 and 4 months, respectively. After ezetimibe was discontinued, HDL-c increased to 30mg/dL. HDL-c increased further to 48mg/dL when fenofibrate was also discontinued and switched to nicotinic acid. Ezetimibe was reintroduced without further adverse effects on HDL-c. The fourth patient was a 54-year-old overweight man with CAD and mixed dyslipidemia. His medications included nadolol 80mg/day, quinapril 20mg/day, aspirin 100mg/day, and fenofibrate 200mg/day. After ezetimibe 10mg/day was initiated, HDL-c dropped from 29 to 14mg/dL and triglycerides increased from 252 to 556mg/dL. After ezetimibe was discontinued, triglycerides returned to baseline within 2 months and HDL-c returned to near baseline within 5 months. Authors discuss that the prevalence of this "paradoxical" phenomenon of HDL-c reduction with fibrates alone or with glitazones ranges in literature from 0.02% to 9.6%. The mechanism of this paradoxical effect is unknown but the authors consider the possibility of predisposition to polymorphisms in genes regulating HDL metabolism. The authors also state that the mechanism of interaction between ezetimibe and fibrates leading to HDL-c decrease is also unknown but suggest that patients should be carefully monitored for this effect [10c].

HMG-COA REDUCTASE INHIBITORS [SEDA-39, 459; SEDA-15, 1632; SEDA-35, 812]

Musculoskeletal

A case report described a 74-year-old man who developed statin-induced rhabdomyolysis 1 week after initiation of atorvastatin. The patient had a history of DM type 2 and hypertension and his medications included amlodipine 10mg and metformin 500mg twice daily. The patient was started on aspirin 150mg daily, ticagrelor 90mg twice daily, and atorvastatin 40mg daily after experiencing an acute coronary syndrome event and coronary angioplasty with stent placement eluting was performed. One week later, the patient developed severe muscle tenderness with rapidly progressive weakness in the bilateral lower extremities along with acute kidney injury. Peak laboratory values were CK of 46488U/L, creatinine 3.4mg/dL, AST 1125IU/L, and ALT 314IU/L. The patient was diagnosed with statin-induced rhabdomyolysis and atorvastatin was changed to rosuvastatin. Laboratory values and symptoms resolved and the patient returned to asymptomatic baseline [11A].

This is a case report of statin-associated autoimmune myopathy in a 70-year-old man who was admitted with a large subdural hematoma after falling. Prior to the fall, the patient had bilateral pain and weakness in the shoulder and hip girdle region. His medications included atorvastatin 40mg daily and aspirin 81mg daily without any prior adverse effects. Creatine kinase (CK) was initially 12300IU/L and it remained elevated despite discontinuation of atorvastatin. The patient developed respiratory failure, requiring mechanical ventilation and sedation. A muscle biopsy showed a pauci-immune necrotizing myopathy and serologic testing detected autoantibodies to HMG-CoA reductase consistent with an autoimmune process and established the diagnosis of statin-associated autoimmune myopathy (SAM). The patient was subsequently treated with high-dose corticosteroids and rituximab. After some mild improvement, patient developed respiratory distress requiring a tracheostomy and ventilator support and ultimately died due to ventilator-associated pneumonia [12A].

A case report describes an 82-year-old man who was admitted with a 2-week history of asthenia, muscle pain and tenderness, and decreased muscle strength 3 weeks after the patient was started on a beta blocker, calcium channel blocker, ranolazine 375mg twice daily, atorvastatin 80mg daily, aspirin, ticagrelor, and ursodesossicolic acid. His past medical history included CAD, hypertension, prediabetes, dyslipidemia, and chronic kidney disease (CKD). At admission, serum creatinine was 2.67mg/dL, CK >3145U/L, lactate dehydrogenase was 262U/L. CK continued to rise 6h later to 5533U/L and the patient was diagnosed with rhabdomyolysis. During the next 24h, atorvastatin and ranolazine were discontinued but laboratory values continue to rise with CK peaking at 124095U/L several months later. Other laboratory tests included LDH 2105U/L, AST 1521U/L, ALT 343U/L, and myoglobin 4000ng/mL. The patient was treated with methylprednisolone 1mg/kg/day. Muscle biopsy results of the gluteus reveled lytic necrosis of the myocytes at different stages and significant macrophage infiltration with limited presence of CD4+ cells. The patient was transferred to another unit after resolution of the acute phase [13A].

This is a case report of a 72-year-old man with sudden onset of weakness in both legs which caused inability to walk, impaired memory and confusion, and abnormal behavior for 3 days. Past medical history included hypertension and dyslipidemia and medications included losartan, aspirin, hydrochlorothiazide, diltiazem, and atorvastatin. The patient had decreased muscle power and diminished tendon reflexes in all four extremities. Muscle tenderness was present at admission, peaked at hospital day 3, and improved and subsided by day 8. At admission, CK was 3169U/L, ALT 77U/L, AST 229U/L, and alkaline phosphatase 207U/L. Serum creatinine remained within normal limits. The patient was diagnosed with atorvastatin-induced rhabdomyolysis and atorvastatin was discontinued. On the 3rd day of hospitalization, the patient developed one episode of convulsions that lasted for less than 5min. On day 4, ceftriaxone and dexamethasone were initiated for 7-day empiric treatment of persistent confusion, neurophilia, and elevated CRP levels. The patient also experienced an episode of myoclonus of masseter muscles and generalized muscle tone increase on day 4. Nerve conduction velocity studies showed predominantly sensory axonal polyneuropathy and marginally low motor conduction velocity. The patient's muscle symptoms gradually decreased and mental status improved and was discharged on day 8. One week later, the patient was able to walk and CK was 213 U/L. The patient was suspected to have hypothyroidism based upon CK levels, hyponatremia, hypokalemia, and slow relaxing ankle jerk. Thyroid stimulating hormone (TSH) was found to be 86.1 mIU/L and antithyroid peroxidase (TPO) antibody level was >1000 IU/mL. CK levels decreased to normal levels once levothyroxine was initiated and titrated upwards. Retrospectively, the authors considered the patient's confusion, seizure, and myoclonus to be symptoms of Hashimoto's encephalopathy, secondary to undiagnosed Hashimoto's disease. The authors concluded that the patient's untreated hypothyroid status along with the drug interaction between atorvastatin and diltiazem were likely causes of rhabdomyolysis [14A].

This is a case report of a 63-year-old Guyanese woman with complaints of generalized weakness, malaise, dark urine, and muscle and skin pain throughout her body. Past medical history included hypertension, hyperlipidemia, heart failure, atrial fibrillation, and peripheral vascular disease. Her medications included atorvastatin 40 mg daily, carvedilol 25 mg twice daily, digoxin 0.125 mg daily, furosemide 40 mg daily, spironolactone 25 mg daily, rivaroxaban 15 mg daily, and sacubitril/valsartan 24/26 mg twice daily. The patient was on these medications for at least 1 year with the exception of sacubitril/valsartan, which was started 22 days prior. At admission, CK was 58 349 U/L. After discontinuation of sacubitril/valsartan, both CK elevation and symptoms resolved 23 days after initial presentation. The patient was able to tolerate rosuvastatin 5 mg afterwards. The authors indicate that a possible interaction between sacubitril/valsartan and atorvastatin may have been the cause of the patient's adverse event. The authors refer to possible single nucleotide polymorphisms within the *SLCO1B1* and/or *SLCO1B3* genes as a potential cause for this adverse effect. However, the authors were unable to arrange for *SLCOB* testing for this patient [15A].

This is a case report of a 72-year-old man with CAD, hypertension, and chronic obstructive pulmonary disease (COPD) who presented to an emergency department with muscle pain in both legs and cramping in the calves. His medications included acetylsalicylic acid 100 mg daily, ticagrelor 90 mg twice daily, ramipril 2.5 mg twice daily, pantoprazole 20 mg daily, simvastatin 20 mg daily, tiotropium inhalation twice daily, and beclomethasone/formoterol inhalation twice daily. The patient had elevated CK (4117 U/L) and myoglobin (426 ug/L) upon presentation but renal function tests remained within normal limits. Simvastatin was discontinued and ticagrelor was changed to clopidogrel. CK levels and symptoms resolved quickly and the patient remained symptom-free 1 week later. The authors concluded that the patient's rhabdomyolysis was secondary to concomitant use of low-dose simvastatin and ticagrelor as ticagrelor is a weak inhibitor of CYP 3A and is metabolized through CYP 3A4/5 [16A].

The authors report a case of a 58-year-old man with elevated CK, likely secondary to concomitant ticagrelor and atorvastatin. The patient had a past medical history of CAD, cerebrovascular accident, and CKD. The patient's long-term medications included aspirin 81 mg daily, atorvastatin 40 mg daily, and metoprolol tartrate 50 mg twice daily. Three months after starting ticagrelor 90 mg twice daily, the patient was hospitalized with recurrent chest pain and was found to be in acute diastolic heart failure. CK was 2360 IU/L and serum creatinine ranged from 2.01 to 2.62 mg/dL. Liver function tests were within normal limits and the patient did not complain of any muscle pain or weakness. Atorvastatin was temporarily discontinued for 3 days and resumed when CK level was reduced. Ticagrelor was then discontinued and switched to clopidogrel 75 mg daily. Two weeks later, CK had normalized to baseline at 57 IU/L and the patient was continued on the same regimen. The author suggests that a probable interaction occurred between ticagrelor and atorvastatin, thus leading to elevated creatine kinase [17A].

This is a case report of a 74-year-old Maltese woman who experienced an unwitnessed collapse following several days of generalized weakness. She had a history of ST elevation myocardial infarction (STEMI), hypertension, depression, osteoarthritis with a total hip replacement, and osteoporosis. The patient's medications included amlodipine/atorvastatin 5/80 mg, low-dose aspirin, and ticagrelor 90 mg twice daily. In the emergency department, the patient presented with hypotension and minimal urine output and was initially diagnosed with septic shock and acute kidney injury. The patient appeared confused and had elevations in ALT (746 U/L), AST (1142 U/L), ALP (260 U/L), total bilirubin (53 micromol/L), creatinine (480 micromol/L), pH (9.34), bicarbonate (11 mmol/L) and INR (1.8). A diagnosis of acute cholecystitis was made based upon computed tomography and ultrasound of the bladder which revealed a calculus thickened gall bladder with pericholecystic fluid. Continuous venovenous hemodialysis and filtration (CVVHDF) was started due to worsening metabolic acidosis and acute anuric renal failure. The patient subsequently developed worsening muscle pain and progressive weakness in the upper and lower limbs. CK levels peaked at almost 100 000 U/L with progressive elevations in liver function test results. ALT peak level was 1605 U/L and AST peak level was 2591 U/L.

On day 7, magnetic resonance imaging (MRI) of the musculoskeletal system and biopsy were performed. The patient also developed disseminated intravascular coagulation leading to epistaxis, upper gastrointestinal bleeding, and subcutaneous hemorrhage. A presumed diagnosis of autoimmune myositis was made. The patient became more fatigued, drowsy, and hypoxic, thus requiring endotracheal intubation and ventilation. The muscle biopsy revealed extensive myonecrosis, consistent with toxin induced necrotizing myositis. The renal biopsy confirmed acute tubular necrosis with myoglobin-related cast formation. The patient gradually improved and was eventually discharged. The authors state that rhabdomyolysis occurred in this patient secondary to an interaction between the concomitant use of atorvastatin, ticagrelor, and amlodipine [18A].

This is a case of a 60- to 65-year-old woman who presented with a 7-day history of progressive generalized weakness. The patient was on atorvastatin 40mg daily for years when she was diagnosed with estrogen receptor positive breast cancer, which metastasized to the liver, spleen, and bone 1 year later. After two cycles of palbociclib, the patient developed decreased stamina, pain in the proximal bilateral lower extremities which progressed to the upper extremities. Upon hospitalization, CK level was 14572. An antibody panel for myositis and myopathy was negative. A muscle biopsy showed scattered necrotic fibers without a lymphocytic or plasma cell inflammatory infiltrate, indicating acute necrotizing rhabdomyolysis. The patient deteriorated with increasing weakness and pain and died on hospital day 8. The authors speculate that palbociclib, a weak CYP 3A4 inhibitor, may have caused a CYP-mediated drug interaction with atorvastatin [19A].

Skin

This is a case report of a 69-year-old man with early onset dermatomyositis associated with simvastatin 20mg daily. The patient presented with a 1-week history of erythroderma, beginning 3 days after starting simvastatin. His past medical history included hypertension, DM type 2, and COPD due to smoking. Two months after statin discontinuation and initiation of systemic corticosteroids, the patient was admitted for muscle weakness and ulcer formation. Patient had heliotrope rash, Gottron's papules and periungual telangiectasis. An ulcer biopsy showed basal layer vasculopathic degeneration of keratinocytes with eosinophilic infiltrate and leukocyoclastic vasculitis. The patient was diagnosed with statin-induced dermatomyositis. After 2 additional weeks of high-dose steroid therapy, the patient developed dysphagia which progressed to aspiration pneumonia and sepsis, and died several days later [20A].

The authors present a case of an 82-year-old man with scaly, confluent, well-circumscribed erythematous plaques on the upper arms. The patient's past medical history includes gout, hypercholesterolemia, and multiple non-melanoma skin cancers. Current medications included allopurinol 300mg, hydrochlorothiazide 25mg, aspirin 81mg, pravastatin 20mg, and colchicine 0.6mg. The patient had been on pravastatin for many years but the dose was increased from 10 to 20mg several months prior to the onset of the rash. A punch biopsy of the forearms showed compact orthokeratosis and significant spongiosis with perivascular lympthocytic infiltrate, consistent with eczema. Despite treatment with triamcinolone ointment 0.0.25%, lesions had spread onto his anterior thighs and lower back 2 weeks later. The patient was then treated with triamcinolone acetonide 0.1% intramuscularly. Symptoms did not resolve 1 month later and the patient was started on methotrexate 10mg weekly for 8 weeks. Pravastatin therapy was then discontinued and the patient was switched to cholestyramine for 8 weeks, at which time the cutaneous lesions resolved [21A].

This is a case report of an obese 50-year-old woman who developed diffuse erythematous scaly plaques and exfoliation on her palm and soles 10 days after starting atorvastatin 40mg daily. She had a history of intermittent psoriasis vulgaris for 3 years that was stable for 10 months on cyclosporine prior to the episode. Other past medical history included hypertension and DM and other medications included telmisartan, metformin, glicazide, and pioglitazone. The atorvastatin was discontinued but the patient continued on cyclosporine and recovered back to baseline in 3 weeks. The patient had taken atorvastatin in the past without concomitant cyclosporine without any adverse effect. The authors hypothesized that the reaction was likely due to a drug interaction between atorvastatin and cyclosporine as both drugs are metabolized by cytochrome P450 3A4 [22A].

Endocrine

A population-based, case–control study was performed using the database from the Taiwan National Health Insurance (NHI) Program to evaluate a possible association between fluvastatin use and acute pancreatitis. Patients between 20 and 84 years old were identified from the database using the ICD-9 code for acute pancreatitis between 1998 and 2011 ($n=3501$). Individuals without acute pancreatitis ($n=8373$) were selected randomly from the database for the control group and matched for sex, age, and index year. After adjusting for multiple confounders, the multivariable analysis did not show elevated risk of acute pancreatitis in current users of fluvastatin (OR 1.17, 95% CI 0.69–1.97) or late users of fluvastatin (OR 1.82; 95% CI 0.41–8.19). The authors concluded that there was no association between fluvastatin use and acute pancreatitis [23MC].

PROPROTEIN CONVERTASE SUBTILISIN/ KEXIN TYPE 9 (PCSK9) INHIBITORS [SEDA-39, 462; SEDA-38, 474]

A meta-analysis of 11 randomized controlled trials (9 smaller early-phase and 2 larger outcome trials) assessed the safety of PCSK9 inhibitors, alirocumab and evolocumab. Overall, PCSK9 inhibitors did not significantly increase the risk for serious adverse events (OR 1.00; 95% CI 0.88–1.15; $P=0.96$), musculoskeletal adverse events (OR 1.01; 95% CI 0.87–1.13; $P=0.90$), neurocognitive events (OR 1.29; 95% CI 0.64–2.59; $P=0.47$), and stroke (OR=1.44; 95% CI 0.57–3.65; $P=0.45$). Neurocognitive events included delirium, confusion, cognitive and attention disorders and disturbances, dementia and amnestic conditions, disturbances in thinking and perception, and mental impairment disorders. In a sub-analysis of neurocognitive events in the two larger outcome studies, there was an increased risk of neurocognitive adverse events (OR 2.81; 95% CI 1.32–5.99; $P=0.007$). The authors suggested close monitoring in ongoing outcome studies and post-marketing surveillance [24M].

Alirocumab

The ODYSSEY COMBO II trial evaluated the safety and efficacy of alirocumab compared to ezetimibe. The study included 720 patients with high cardiovascular (CV) risk on maximally tolerated statin dose who were not at pre-specified LDL-c levels. Results from a prespecified analysis of the first 52 weeks were previously published. This reports the final results from the full 2 years. Adverse events of special interest specified in the protocols included local injection-site reactions, general allergic events, neurological events, hepatic disorders, and ophthalmological events. Among these, patients on alirocumab experienced higher rates of injection-site reactions [13 patients (2.7%) vs 3 patients (1.2%)], allergic reactions [38 patients (7.9%) vs 17 patients (7.1%)], and ophthalmological disorders [9 patients (1.9%) vs 4 patients (1.7%)]. Cataract conditions were reported in 10 patients (2.1%) of the alirocumab group and 6 patients (2.5%) of the ezetimibe group. Elevated transaminases were also higher in the alirocumab group. When comparing rates of adverse events between the 2 study years, rates were generally similar or lower during the 2nd year, including local injection-site reactions [12 patients (2.5%) 1st year vs 1 patient (0.2%) 2nd year]. Persistent anti-drug antibodies (ADA) were reported in six patients (1.3%) in the alirocumab group compared to one patient (0.4%) of the ezetimibe group but these ADA did not have a clinical impact on the pharmacokinetics or safety of alirocumab. There were seven patients (1.5%) in the alirocumab group who had neutralizing antibodies detected by immunoassay but these were transient and at isolated time points. Inferential statistics were not performed [25C].

In a sub-group analysis of the ODYSSEY COMBO II study, the effects of alirocumab were compared between patients with and without diabetes mellitus (DM). Among 225 patients with DM and 495 patients without DM, rates of adverse effects were similar. Rates of DM or diabetic complications were higher in alirocumab patients with DM compared to those without diabetes (11.5% vs 5.4%). Rates of adjudicated cardiovascular events were also higher in alirocumab patients with DM compared to those without DM (8.8% vs 5.4%). Inferential statistics were not performed [26C].

In a phase 3b, double-blind, placebo-controlled, multicenter trial, patients with insulin-treated DM type 1 or 2 and established atherosclerotic cardiovascular disease (ASCVD) and/or at least 1 additional CV risk factor were randomized in a 2:1 fashion to either alirocumab 75 mg or placebo administered subcutaneously every 2 weeks. The safety analysis included 344 patients on alirocumab and 170 patients on placebo. Treatment Emergent Adverse Effects (TEAEs) were reported in 64.5% in patients in the alirocumab group and 64.1% in patients in the placebo group. Rates of treatment discontinuation due to a TEAE were 4.9% in the alirocumab group and 2.4% in the placebo group. The most common TEAEs in the alirocumab group were nasopharyngitis (4.9%), urinary tract infection (4.4%), diarrhea (4.4%), and myalgia (4.4%). Other effects which occurred in more than 2% of the alirocumab group were: arthralgia, hypertension, dizziness, headache, influenza, and bronchitis [27C].

A post-hoc analysis was performed from pooled data of 4974 patients in 10 phase 3 ODYSSEY trials. TEAEs that occurred in at least 5% of the alirocumab population included nasopharyngitis (12.6% in placebo-controlled trials and 6% in ezetimibe-controlled trials), injection site reaction (7.2% and 2.9%), upper respiratory tract infection (7.0% and 7.2%), influenza (6.3% and 4.3%), urinary tract infection (5.5% and 2.4%), back pain (5.3% and 3.8%), diarrhea (5.3% and 3.5%), headache (5.1% and 5.0%), arthralgia (5.1% and 4.9%), myalgia (4.8% and 7.2%), and accidental overdose (1.3% and 6.3%) [28MC].

The safety of alirocumab was evaluated in patients with at least 2 consecutive LDL-c levels <25 mg/dL from pooled data from 14 phase 2 and 3 studies in the ODYSSEY program. Among 3340 participants in the alirocumab groups, 839 patients (25%) achieved 2 consecutive LDL-c levels <25 mg/dL and 314 (9.4%) achieved levels <15 mg/dL. Rates of TEAEs were 72.7% and 71.7% in patients on alirocumab with LDL-c <25 and <15 mg/dL, respectively, compared to 77.8% in the overall alirocumab group. Rates of TEAEs leading to permanent discontinuation were lower in the LDL <25 mg/dL (4.2%)

and <15 mg/dL (5.4) groups compared to the overall alirocumab group (6.9%). There was a higher incidence of cataracts in patients with LDL-c <25 mg/dL compared to those with LDL-c >25 mg/dL (2 per 100 patient-years vs 0.08 per 100 patient-years). A propensity score analysis was performed on patients from phase 3 trials to control for baseline factors. In this analysis, patients with LDL-c <25 mg/dL still had higher rates of cataracts (2.6% vs 0.8%, HR 3.40; 95% CI 1.58–7.35). The authors hypothesized that "acute LDL-c lowering accelerates underlying aging or metabolic syndrome or diabetes-related changes, contributing to cataracts" [29MC].

A meta-analysis from 14 double-blind, phase 2 and 3 trials of alirocumab specifically studied the incidence of neurocognitive TEAEs as mandated by the FDA. Neurocognitive events were self-reported by participants and were recorded by the investigators. The study did not use screening tools or formal neurocognitive testing. Rates of any neurocognitive TEAE were generally low, 0.9% in alirocumab group vs 0.7% in placebo group in placebo-controlled studies (HR 1.24; 95% CI 0.57–2.68) and 1.2% in alirocumab group vs 1.3% in ezetimibe group in ezetimibe-controlled studies (HR 0.81; 95% CI 0.32–2.08). Four patients in the alirocumab group (0.1%) had events that were considered serious. Only one patient in the pooled analysis discontinued alirocumab due to a neurocognitive TEAE (mild memory impairment). The most common individual events in the alirocumab group were confusion (0.2% in placebo-controlled studies and ezetimibe-controlled studies), amnesia (0.2% and 0.1%, respectively), and memory impairment (0.2% and 0.5%, respectively). All others, including aphasia, dementia, disorientation, transient global amnesia, and frontotemporal dementia, occurred in <0.01% of the study group. When study participants were grouped by age (<65, 65 to <75, ≥75 years), there were no differences in risk of neurocognitive TEAEs. In a subgroup analysis of alirocumab patients with LDL-c <25 mg/dL, the incidence rate of TEAEs was low (0.6% vs 1.0% in patients with LDL-c ≥25; HR 0.39; 95% CI 0.14–1.10). The authors concluded that the overall rates of neurocognitive TEAEs were low and comparable to rates in placebo and ezetimibe groups [30M].

A double-blind, placebo-controlled, parallel-group, multicenter study of patients with high CV risk was conducted in South Korea and Taiwan. Patients were randomized to receive either alirocumab 75 mg q2 weeks ($n=97$) or placebo ($n=102$). Patients were also on maximally tolerated statin therapy, which was defined as atorvastatin 40 to 80 mg daily, rosuvastatin 20 mg daily, or simvastatin 40 mg daily. The most common TEAEs in the alirocumab group were nasopharyngitis (6.2% vs 3.9% with placebo) and dizziness (6.2% vs 2.9% with placebo). Rates of serious TEAEs were higher in the alirocumab group compared to the placebo group (17.5% vs 9.8%). Among TEAEs of special interest, the rate of newly-developed diabetes was higher in the alirocumab group compared to placebo (6.2% vs 3.1%, $P=0.6801$). There were five patients (5.2%) in the alirocumab group who developed anti-drug antibodies (ADA) and one of these patients had a positive neutralizing status at a single follow-up visit. The authors stated that the positive ADA and neutralizing responses did not "appear to impact LDL-c efficacy" [31C].

A case report described a 62-year-old African American woman with cardiac disease, dyslipidemia, stage 4 CKD, and hypertension who developed acute tubular injury and necrosis (ATN) after starting treatment with alirocumab 75 mg subcutaneously q2 weeks. Her concurrent medications included aspirin, metoprolol succinate, doxazosin, clopidogrel, and ramipril. The patient's serum creatinine increased from 2.3 to 5.0 mg/dL a few weeks after initiation of alirocumab. Evaluations for causes of acute kidney injury did not reveal significant findings. A kidney biopsy was performed and "light microscopy showed ATN with secondary focal segmental glomerulosclerosis. Electron microscopy confirmed tubular injury and necrosis." Renal function improved and returned back to baseline after discontinuation of the drug. The authors concluded that alirocumab was the cause based upon the clinical presentation and findings [32A].

Evolocumab

The FOURIER trial was a double-blind, placebo-controlled trial of patients with ASCVD and LDL-c levels of 70 mg/dL or more on statin therapy. Patients were randomized to receive evolocumab 140 mg every 2 weeks or 420 mg monthly ($n=13784$) or subcutaneous placebo ($n=13780$). Rates of any adverse event, serious adverse event, cataract, rhabdomyolysis, muscle-related event, and neurocognitive event were not different between groups. The evolocumab group experienced more injection site reactions compared to the placebo group (2.1% vs 1.6%, $P<0.001$). In the evolocumab group, 43 patients (0.3%) developed new binding antibodies but none of the patients developed neutralizing antibodies [33MC].

The EBBINGHAUS study involved a subgroup of patients ($n=1974$) from the FOURIER trial. The study prospectively evaluated patients' cognition utilizing the Cambridge Neuropsychological Test Automated Battery (CANTAB). There was no statistically significant difference between groups (evolocumab vs placebo) as it pertained to the spatial working memory strategy index of cognitive function, working memory or episodic memory. As the dose of evolocumab increased, the psychomotor speed changed, albeit not statistically significant. The authors report cognitive adverse events in 11 (1.9%) and 8 (1.3%) patients in the evolocumab and placebo groups, respectively. Overall, there was no association between evolocumab and cognitive adverse events, when compared to placebo [34MC].

This study reported the results of a multiyear extension of the OSLER-1 trial, which was an open-label, randomized, and controlled study. The study randomized participants in a 2:1 fashion to evolocumab 420 mg subcutaneously monthly plus standard of care (SOC) or SOC alone. Rates of serious adverse events decreased as time increased from 6.9% in the 1st year, 6.8% in the 2nd year, 7.8% in the 3rd year, 4.8% in the 4th year, and 3.5% beyond 4 years. This was similar for muscle-related adverse events, neurocognitive-related adverse events, and potential injection-site reactions. Rates of muscle-related adverse events were 8.1%, 5.9%, 4.3%, 3.1%, and 1.3% for the 1st year, 2nd year, 3rd year, 4th year, and >4 years, respectively. Rates of neurocognitive-related adverse events were 0.6%, 0.5%, 0.6%, 0.2%, and 0% for the 1st year, 2nd year, 3rd year, 4th year, and >4 years, respectively. CV event rates also remained low during the 2nd, 3rd, and 4th years of evolocumab treatment (1.2%, 1.2%, and 0.9%, respectively) [35MC].

The GLAGOV trial was a multicenter, double-blind, placebo-controlled, randomized clinical trial that randomized patients with coronary stenosis on coronary angiography on a stable statin in a 1:1 fashion to either evolocumab 420 mg ($n=484$) or placebo ($n=484$) via monthly subcutaneous injection. Patients in the evolocumab group had higher rates of injection site reaction (0.4% vs 0%), myalgia (7% vs 5.8%), and neurocognitive events (1.4% vs 1.2%). There was one patient who developed anti-evolocumab binding antibodies and no cases of anti-evolocumab neutralizing antibodies [36C].

In a pooled safety analysis of 12 randomized trials comparing evolocumab with either placebo or ezetimibe, adverse event rates were similar. Musculoskeletal adverse events occurred in 14.7% ($n=581$) of evolocumab patients and 13.7% ($n=284$) of control patients. Neurocognitive adverse effects occurred in 0.1% ($n=5$) of evolocumab patients and 0.3% ($n=6$) of control patients. No neutralizing antibodies to evolocumab were detected. Inferential statistics were not performed [37MC].

Additional case studies on lipid lowering agents can be found in these reviews [38R, 39R].

References

[1] Bergland Ellingsen S, Nordmo E, Lappegard KT. Recurrence and severe worsening of hepatotoxicity after reintroduction of atorvastatin in combination with ezetimibe. Clin Med Insights Case Rep. 2017;10:1–4 [A].

[2] Sakamoto K, Kawamura M, Watanabe T, et al. Effect of ezetimibe add-on therapy over 52 weeks extension analysis of prospective randomized trial (RESEARCH study) in type 2 diabetes subjects. Lipids Health Dis. 2017;16:122 [C].

[3] Hollingworth SA, Ostino R, David MC, et al. Ezetimibe: use, costs, and adverse events in Australia. Cardiovasc Ther. 2017;35:40–6 [R].

[4] Giugliano RP, Cannon CP, Blazing MA, et al. Benefit of adding ezetimibe to statin therapy on cardiovascular outcomes and safety in patients with and without diabetes: results from IMPROVE-IT (improved reduction of outcomes: vytorin efficacy international trial). Circulation. 2018;137:1571–15827 [MC].

[5] Kato ET, Cannon CP, Blazing MA, et al. Efficacy and safety of adding ezetimibe to statin therapy among women and men: insight from IMPROVE-IT (improved reduction of outcomes: vytorin efficacy international trial). J Am Heart Assoc. 2017;6: e006901 [MC].

[6] Schaffellner S, Stadlbauer V, Sereinigg M, et al. Niacin-associated acute hepatotoxicity leading to emergency liver transplantation. Am J Gastroenterol. 2017;112:1345–6 [A].

[7] Gagne JJ, Houstoun M, Reichman ME, et al. Safety of niacin in the US Food and Drug Administration's mini-sentinel system. Pharmacoepidemiol Drug Saf. 2018;27:30–7 [S].

[8] Ahmad J, Odin JA, Hayashi PH, et al. Identification and characterization of fenofibrate-induced liver injury. Dig Dis Sci. 2017;62:3596–604 [MC].

[9] Tsai KC, Yang JH, Hung SJ. Fenofibrate-induced photosensitivity—a case series and literature review. Photodermatol Photoimmunol Photomed. 2017;33:213–9 [c].

[10] Nobecourt E, Cariou B, Lambert G, et al. Severe decrease in high-density lipoprotein cholesterol with the combination of fibrates and ezetimibe: a case series. J Clin Lipidol. 2017;11:289–93 [c].

[11] Kareem H, Sahu D, Rao MS, et al. Statin induced rhabdomyolysis with non oliguric renal failure: a rare presentation. J Clin Diagn Res. 2017;11:FD01–02 [A].

[12] Sweidan AJ, Leung A, Kaiser CJ, et al. A case of statin-associated autoimmune myopathy. Clin Med Insights Case Rep. 2017;10: 1179547616688231 [A].

[13] Canzonieri E, De Candia C, Tarascio S, et al. A severe myopathy case in aged patient treated with high statin dosage. Toxicol Rep. 2017;4:438–40 [A].

[14] Ehelepola NDB, Sathkumara SMBY, Bandara HMPAGS, et al. Atorvastatin-diltiazem combination induced rhabdomyolysis leading to diagnosis of hypothyroidism. Case Rep Med. 2017;2017:8383251 [A].

[15] Faber ES, Gavini M, Ramirez R, et al. Rhabdomyolysis after coadministration of atorvastatin and sacubitril/valsartan (Entresto™) in a 63-year-old woman. Drug Saf Case Rep. 2016;3:14 [A].

[16] Mrotzek SM, Rassaf T, Totzeck M. Ticagrelor leads to a statin-induced rhabdomyolysis: a case report. Am J Case Rep. 2017;18:1238–41 [A].

[17] Beavers JC. Elevated creatine kinase due to potential drug interaction with ticagrelor and atorvastatin. J Pharm Pract. 2017;897190017740282 (Jan 1) [A].

[18] Banakh I, Haji K, Kung R, et al. Severe rhabdomyolysis due to presumed drug interactions between atorvastatin with amlodipine and ticagrelor. Case Rep Crit Care. 2017;2017:3801819 [A].

[19] Nelson KL, Stenehjem D, Driscoll M, et al. Fatal statin-induced rhabdomyolysis by possible interaction with palbociclib. Front Oncol. 2017;7:150 [A].

[20] Chemello RML, Benvegnu AM, Dallazem LND, et al. Aggressive and fatal statin-induced dermatomyositis: a case report. Oxf Med Case Reports. 2017;12:242–5 [A].

[21] Salna MP, Singer HM, Dana AN. Pravastatin-induced eczematous eruption mimicking psoriasis. Case Rep Dermatol Med. 2017;2017:3418204 [A].

[22] Ardeshna KP, Someshwar S, Rohatgi S, et al. A case of psoriasis vulgaris aggravated with atorvastatin, aided by concomitant cyclosporine. Indian J Dermatol. 2017;62:537–8 [A].

[23] Liao KF, Huang PT, Lin CC, et al. Fluvastatin use and risk of acute pancreatitis: a population-based case-control study in Taiwan. Biomedicine (Taipei). 2017;7:17 [MC].

[24] Khan AR, Bavishi C, Riaz H, et al. Increased risk of adverse neurocognitive outcomes with proprotein convertase subtilisin-kexin type 9 inhibitors. Circ Cardiovasc Qual Outcomes. 2017;10: e003153 [M].

[25] El Shahawy M, Cannon CP, Blom DJ, et al. Efficacy and safety of alirocumab versus ezetimibe over 2 years (from ODYSSEY COMBO II). Am J Cardiol. 2017;120:931–9 [C].

[26] Leiter LA, Zamorano JL, Bujas-Bobanovic M, et al. Lipid-lowering efficacy and safety of alirocumab in patients with or without diabetes: a sub-group analysis of ODYSSEY COMBO II. Diabetes Obes Metab. 2017;19:989–96 [C].

[27] Leiter LA, Cariou B, Muller-Wieland D, et al. Efficacy and safety of alirocumab in insulin-treated individuals with type 1 or type 2 diabetes and high cardiovascular risk: the ODYSSEY DM-INSULIN randomized trial. Diabetes Obes Metab. 2017;19:1781–92 [C].

[28] Ray KK, Ginsberg HN, Davidson MH, et al. Reductions in atherogenic lipids and major cardiovascular events: a pooled analysis of 10 ODYSSEY trials comparing alirocumab with control. Circulation. 2016;134:1931–43 [MC].

[29] Robinson JG, Rosenson RS, Farnier M, et al. Safety of very low low-density lipoprotein cholesterol levels with alirocumab. J Am Coll Cardiol. 2017;69:471–82 [MC].

[30] Harvey PD, Sabbagh MN, Harrison JE, et al. No evidence of neurocognitive adverse events associated with alirocumab treatment in 3340 patients from 14 randomized phase 2 and 3 controlled trials: a meta-analysis of individual patient data. Eur Heart J. 2018;39(5):374–81 [M].

[31] Koh KK, Nam CW, Chao TH, et al. A randomized trial evaluating the efficacy and safety of alirocumab in South Korea and Taiwain (ODYSSEY KT). J Clin Lipidol. 2017;12(1):162–72,e6 [C].

[32] Jhaveri KD, Barta VS, Pullman J. Praluent (alirocumab)-induced renal injury. J Pharm Pract. 2017;30:7–8 [A].

[33] Sabatine MS, Giugliano RP, Keech AC, et al. Evolocumab and clinical outcomes in patients with cardiovascular disease. N Engl J Med. 2017;376:1713–22 [MC].

[34] Giugliano RP, Mach F, Zavitz K, et al. Cognitive function in a randomized trial of Evolocumab. NEJM. 2017;377:633–43 [MC].

[35] Koren MJ, Sabatine MS, Giugliano RP, et al. Long-term low-density lipoprotein cholesterol-lowering efficacy, persistence, and safety of evolocumab in treatment of hypercholesterolemia: results up to 4 years from the open-label OSLER-1 extension study. JAMA Cardiol. 2017;2:598–607 [MC].

[36] Nicholls SJ, Puri R, Anderson T, et al. Effect of evolocumab onprogression of coronary disease in statin-treated patients: theGLAGOV randomized clinical trial. JAMA. 2016;316: 2373–84 [C].

[37] Toth PP, Descamps O, Genest J, et al. Pooled safety analysis of evolocumab in over 6000 patients from double-blind and open-label extension studies. Circulation. 2017;135:1819–31 [MC].

[38] Ali AN, Kim JJ, Pisano ME, et al. Chapter 39—drugs that affect lipid metabolism. In: Ray SD, editor. Side effects of drugs annual: A worldwide yearly survey of new data in adverse drug reactions, 39. Elsevier; 2017. p. 457–63 [R].

[39] Beckett RD, Wilhite AL, Robinson N, et al. Chapter 42—drugs that affect lipid metabolism. In: Ray SD, editor. Side effects of drugs annual: A worldwide yearly survey of new data in adverse drug reactions, vol. 38. Elsevier; 2016. p. 469–77 [R].

CHAPTER

45

Cytostatic Agents

Sipan Keshishyan, Henry L. Nguyen†, Olawale D. Afolabi‡, Carmen Moore§, Adrienne T. Black¶, Sidhartha D. Ray‖,1*

*Department of Pharmaceutical Sciences, Manchester University College of Pharmacy, Natural and Health Sciences, Fort Wayne, IN, United States

†Department of Dermatology, Mayo Clinic, Rochester, MN, United States

‡Milken Institute School of Public Health, The George Washington University, Washington, DC, United States

§Department of Internal Medicine, Guy's and St. Thomas' Hospital, London, United Kingdom

¶3E Services, Verisk 3E, Warrenton, VA, United States

‖Department of Pharmaceutical and Biomedical Sciences, Touro University College of Pharmacy and School of Osteopathic Medicine, Manhattan, NY, United States

[1]Corresponding author: micrornagenomics@gmail.com

FLUOROPYRIMIDINES

5-Fluorouracil

The fluoropyrimidine chemotherapeutics (5-fluorouracil (5-FU) and its prodrug, capecitabine) are commonly used for treatment of multiple malignancies including breast cancer and multiple gastrointestinal tumors. Both medications are known to have cardiotoxicity (EKG changes, angina-like chest pain, coronary vasospasms, myocardial infarction, sudden cardiac death, etc.), gastrointestinal effects (mucositis, nausea, vomiting, etc.) and hand-foot syndrome. This cardiotoxicity was investigated in a case series of 11 patients with suspected fluoropyrimidine-induced coronary vasospasm receiving IV 5-FU or oral capecitabine treatment for colorectal, esophageal or breast carcinomas. In all cases, there was a latency period between treatment initiation and onset of symptoms; this result showed that there was no difference between bolus and infusion administration. Although the patients experienced some cardiotoxic effects with the initial treatments, no cardiac events occurred during re-challenge with either drug. The treatment regimens were completed in all cases along with the use of calcium blockers and oral nitrate therapy [1c].

A review of 44 randomized clinical trials with 23,510 adult patients compared the efficacy and safety of oral and IV fluoropyrimidine drugs for treatment of colorectal cancer. Adverse events (grade $\geq$3 neutropenia or granulocytopenia; stomatitis or any other grade $\geq$3 event) occurred less frequently with oral administration of fluoropyrimidine than with IV infusions. However, grade $\geq$3 hand-foot syndrome was more frequent with oral administration. No differences in administration method impacted the incidence of diarrhea (grade $\geq$3), febrile neutropenia, vomiting, nausea, mucositis and hyperbilirubinemia [2R].

A case report of a 76-year-old woman with facial actinic keratosis received a skin biopsy that showed verrucous squamous hyperplasia. She then received a 5-fluorouracil (5%) cream for application to the face twice per day for 2 weeks. However, the patient did not follow the prescribing instructions and applied the cream to the actinic keratosis lesions seven times per day. After 7 days of this application, she presented with painful blisters on the forehead and cheeks and around the eyes. An infection that was diagnosed as likely to be impetigo was present on the right cheek. 5-Fluoruracil was discontinued and the skin was healing and the blisters were gone following a regimen of topical petroleum jelly antibiotics after 1 week [3A].

A 68-year-old woman with squamous cell carcinoma was treated with topical 5-fluoruracil twice daily. An erythematous and scaly patch area on her neck developed 1 week after starting treatment and, 12 days later, had expanded across her neck and chest with edematous, crusting plaques and erythema. She then received

Side Effects of Drugs Annual, Volume 40
ISSN: 0378-6080
https://doi.org/10.1016/bs.seda.2018.08.012

levofloxacin, vancomycin, fluconazole, and valacyclovir for suspected bacterial, fungal, and viral infections as well as prednisone for the 5-fluoruracil drug reaction. A subsequent chest biopsy showed subcorneal and spongiform pustules with a perivascular and interstitial infiltrate. These findings were indicative of a pustular drug eruption most likely due to 5-fluoruracil. 5-Fluoruracil and the antibiotics were discontinued and the rash improved with the steroid [4A].

5-Fluorouracil (5-FU) is also used in combination with irinotecan, oxaliplatin and leucovorin for treatment of metastatic pancreatic cancer. Previous studies have shown that this drug combination was associated with greater incidence of serious adverse events (neutropenia, diarrhea and sensory neuropathy). A multicenter prospective phase II study of the 5-FU combination therapy (5-FU, 2400 mg/m^2; oxaliplatin, 85 mg/m^2; irinotecan, 150 mg/m^2; leucovorin, 200 mg/m^2) was administered every 2 weeks in 31 patients (49–72 years) with metastatic pancreatic cancer. Adverse events (grade $\geq$3) occurred in 27 of the 31 patients and included neutropenia (83.9%), febrile neutropenia (16.1%), thrombocytopenia (6.5%), peripheral sensory neuropathy (9.7%), diarrhea (6.5%), anorexia (6.5%), and vomiting (3.2%). Four patients discontinued the treatment due to adverse events (vomiting, peripheral sensory neuropathy and neutropenia) [5c].

5-Fluorouracil, in combination with irinotecan and oxaliplatin, comprise the first-line treatment for metastatic colorectal cancer although this drug combination has been associated with a number of adverse events including diarrhea, stomatitis, and myelosuppression. A recent recommendation to include bevacizumab in the therapy regimen has increased the potential for the occurrence of grade 3–4 neutropenia, diarrhea and stomatitis adverse events. Due to this increased potential, the association of genetic polymorphisms in the DPYD and UGT genes was conducted to evaluate the relationship of these genetic variants with the observed treatment-related adverse events in patients receiving the newly recommended drug combination. The genetic status for the *DPYD* and *UGT1A1* alleles and the incidence of treatment-related adverse events (grade $\geq$3) were determined for 443 patients who received the drug combination. An increased risk of severe drug-related adverse events was found in patients with the *DPYD* c.1905+1G/A and c.2846A/T and the *UGT1A1**1/*28 and *UGT1A1**28/*28 genotypes [6C].

Capecitabine

A dose escalation study in 12 patients with metastatic colorectal cancer evaluated the safety of different doses of a capecitabine–oxaliplatin–irinotecan combination therapy. The patients were randomized to 3 treatment groups: all groups received oxaliplatin (100 mg/m^2) and cetuximab (400 mg/m^2) on day 1 of the cycle and (250 mg/m^2/week) given every 3 weeks followed by capecitabine (1700 mg/m^2/day) on days 2–15 of each cycle. The dose of irinotecan varied between the groups (100, 120, and 150 mg/m^2) on day 1 of each cycle. All patients received 7 cycles of this treatment combination. The most common adverse events (grade $\geq$3) for all treatment groups were neutropenia (50%), diarrhea (17%), and febrile neutropenia (8%) and 1 case of grade 4 neutropenia was reported in the 120 mg/m^2 irinotecan group [7c].

A phase II study assessed the efficacy and safety of a capecitabine–oxaliplatin combination for treatment of locally advanced and metastatic pancreatic ductal adenocarcinoma. A total of 40 patients self-administered oral capecitabine (1000 mg/m^2) twice daily on days 1 through 14 of a 21-day cycle and IV oxaliplatin (130 mg/m^2) was given in the hospital on day 1 of the cycle. Evaluations were made at the completion of 2 cycles. Serious adverse events (grade 3–4) were fatigue (19%), nausea (17%), diarrhea (14%) anemia (8%), hand-foot reaction (3%), neuropathy (3%), vomiting (3%), anorexia (3%). Less severe adverse events (grade 1–2) included all these adverse events with the addition of thrombocytopenia (19%). Discontinuations due to adverse events occurred in 17% of the patients and dose reductions due to drug toxicity occurred in 25% of the subjects [8c].

A case report described a 73-year-old woman with stage IV (HER2 negative) breast cancer with metastases to the bone who had a photosensitive dermal reaction to capecitabine. Her medication history included anastrozole, zolendronic acid, exemestane, orteronel, and fulvestrant as treatment for her breast cancer over the past 12 years. A year ago, fulvestrant was discontinued and she was started on capecitabine (2500 mg) twice daily for 14 days of a 21-day cycle. Capecitabine was well tolerated for 3 cycles but she developed a mild rash on the face, arms and chest that worsened after 2 more treatment cycles and capecitabine was then discontinued. A shave biopsy revealed a lichenoid tissue reaction with eosinophils, indicating a drug-induced reaction and her clinical symptoms indicate a photosensitive reaction. The symptoms were resolved with topical triamcinolone (0.1%) and hydrocortisone (2.5%) ointments and sun avoidance. At a 2-week follow-up, capecitabine was restarted without further rash eruption [9A].

Genotyping of 274 patients with advanced colon cancer was conducted to determine if genetic variations were associated with capecitabine-induced toxicity. The analysis showed that the rs744591 and rs745666 microRNA polymorphisms were significantly related to the occurrence of adverse events following capecitabine treatment. It was suggested that identification of these genetic variations (*DPYD* status of affected patients) would be useful in determining the appropriate use of capecitabine for treatment of advanced colon cancer [10C].

MISCELLANEOUS AGENTS

Gemcitabine

Although gemcitabine is a commonly used chemotherapeutic for the treatment of solid organ cancers such as those in the lung, breast, ovary, and pancreas, use of this drug has been associated with the development of myelosuppressive conditions. The efficacy and safety of gemcitabine combined with nab-paclitaxel for treatment of metastatic pancreatic cancer was assessed in a single-center cohort study in a Korean population. A total of 81 patients with metastatic pancreatic cancer received gemcitabine and nab-paclitaxel (1000 and 125 mg/m^2, respectively) as the first-line chemotherapy with a median follow-up period of 10.7 months. Serious adverse events (grade$\geq$3) were neurotoxicity (18.5%), neutropenia (46.9%), febrile neutropenia (16.0%) and gastrointestinal disorders (19.8%). These adverse events caused dose reductions in 60.5% of patients. Although this combination therapy was effective, close monitoring of the patients was required due to the occurrence of treatment related adverse events [11c].

A randomized controlled phase III clinical trial compared the efficacy and safety of a gemcitabine–docetaxel combination to doxorubicin alone as the as first-line treatment for advanced or metastatic soft-tissue sarcoma (European Clinical Trials No. 2009-014907-29 and the International Standard Randomised Controlled Trial registry: ISRCTN07742377). Patients ($n=257$) with advanced or metastatic soft-tissue sarcoma received one of the following treatments: IV doxorubicin (75 mg/m^2 on day 1 every 3 weeks; $n=129$) or IV gemcitabine–docetaxel (gemcitabine: 675 mg/m^2 on days 1 and 8 and docetaxel: 75 mg/m^2 on day 8 every 3 weeks; $n=128$). Adverse events (grade 3–4) occurred with both doxorubicin and gemcitabine–docetaxel: neutropenia (32 cases (25%) and 25 cases (20%), respectively), febrile neutropenia (26 cases (20%) and 15 cases (12%)), fatigue (8 cases (6%) and 17 cases (14%)), oral mucositis (18 cases (14%) and 2 cases (2%)), and pain (10 cases (8%) and 13 cases (10%)). The most common serious adverse events were febrile neutropenia (27 cases (17%) in the doxorubicin group and 15 cases (12%) in the gemcitabine–docetaxel group), fever (18 cases (12%) and 19 cases (15%)), and neutropenia (22 cases (14%) and 10 cases (8%)). In addition, the pharmacogenomics portion of the study showed that the *SLC22A16* rs723685 minor allele was associated with a reduced incidence of grade 3–4 adverse events with doxorubicin treatment but not with gemcitabine–docetaxel therapy [12C].

A report describes 2 patients who were evaluated for atypical hemolytic uremic syndrome (aHUS) following gemcitabine use. The first patient was a 54-year-old woman with pancreatic adenocarcinoma, post-surgical resection, who had received gemcitabine and nab-paclitaxel and presented to the hospital with ongoing headache, elevated blood pressure and abnormal laboratory results. The symptoms were not resolved with an initial treatment of plasma exchange with IV nicardipine and antihypertensive medications. Nicardipine was stopped. Forty-eight hours later the symptoms resolved on administration of IV eculizumab (900 mg). The patient was discharged 2 days later and maintained on IV eculizumab (900 mg) once weekly for 4 weeks and then with IV eculizumab (1200 mg) once weekly for 2 weeks. The second patient a 61-year-old woman with multiple myeloma was treated with lenalidomide, bortezomib, and dexamethasone and, upon disease progression, cyclophosphamide, bortezomib, and dexamethasone. After a hematopoietic stem cell infusion and radiotherapy, the disease was in remission for several years. Upon relapse, she received bortezomib and dexamethasone and, upon worsening of disease status, carfilzomib and dexamethasone when she presented to the hospital with chest discomfort and shortness of air on day 5, cycle 9 of this treatment. Further examination showed an elevated blood pressure and decreased blood counts with rapidly declining renal function that required hemodialysis. Eculizumab (900 mg) was given IV weekly and, 2 weeks later, her symptoms and clinical status were significantly improved and she was discharged. As an out-patient, she received IV eculizumab (900 mg) weekly for 4 weeks, followed by IV eculizumab (1200 mg) every 2 weeks starting on the fifth week as a maintenance dose [13A].

A second case series described 4 patients with gemcitabine-related acute lipodermatosclerosis (ALDS) skin eruptions. The first, a 73-year-old man, was treated with gemcitabine and paclitaxel for metastatic pancreatic adenocarcinoma and presented with lower leg edema, venous hypertension and thrombocytopenia eruption 5 days after the first dose. ALDS-like eruption secondary to gemcitabine treatment was diagnosed and the symptoms resolved with twice daily topical clobetasol 0.05% ointment, leg elevation and compression therapy. The second, a 79-year-old woman, with pancreatic adenocarcinoma with liver metastases was initially treated with nivolumab, gemcitabine and abraxane. Nivolumab was discontinued following a severe infusion hypersensitivity reaction and edema, tenderness and erythema of the bilateral lower extremities that occurred 2 weeks after the second dose of gemcitabine and abraxane. The patient was admitted with an initial diagnosis of cellulitis which was later changed to a gemcitabine-associated ALDS-like eruption diagnosis. The symptoms resolved with compression therapy, high-potency topical steroids twice daily, discontinuation of antibiotics, and resumption of gemcitabine. The third patient, a 76-year-old man, with pancreatic adenocarcinoma received chemotherapy with gemcitabine. He developed swelling and pain of the lower legs after 5 doses of gemcitabine and a physical examination found tender, erythematous plaques with edema on both lower legs. Gemcitabine-induced ALDS-like eruption was diagnosed and treated with clobetasol 0.05% ointment twice daily for 2 weeks and compression therapy and the symptoms resolved in 2 weeks. At a 2-month follow-up, it was found that the skin eruption had recurred 2 days after gemcitabine infusion along when the patient had discontinued leg compression and topical steroids. Although the compression and steroid were reinstated, the patient had leukopenia, thrombocytopenia, and bilateral pleural effusions with the deterioration of pulmonary function, gemcitabine treatment was discontinued. The fourth, a 73-year-old man, with recurrent invasive transitional cell carcinoma of the bladder presented with painful lower legs with plaques and edema 4 days after the second dose of gemcitabine. Gemcitabine-induced ALDS-like eruption was diagnosed and was treated with twice-daily topical clobetasol 0.05% ointment and leg elevation with no compression therapy [14A].

A report described the case of a 51-year-old male with cholangiocarcinoma who presented with bilateral lower extremity swelling, rash and erythema that had been progressing over 2 weeks. At the time of the reactions, the patient was receiving an infusion of gemcitabine treatment weekly. An initial diagnosis of cellulitis was treated with vancomycin but was then changed to gemcitabine-induced pseudocellulitis [15A].

A case report of a 62-year-old woman with metastatic pancreatic adenocarcinoma presented with skin rash and swelling in both lower extremities with a moderate-to-severe sharp and shooting pain. She had received diphenhydramine 2 days earlier and a second dose of IV gemcitabine (1200 mg) was given 7 days prior to the antihistamine; no side effects occurred with either administration. At the time of presentation, IV ampicillin/sulbactam and IV vancomycin were given but her condition was not improved 48 hours later. A skin biopsy showed neutrophilic inflammation without migration into adjacent tissue and a diagnosis of gemcitabine-related pseudocellulitis was made [16A].

A 56-year-old woman with stage IIB pancreatic cancer presented with anasarca, hypoalbuminemia and hypotension following treatment gemcitabine for stage IIB pancreatic cancer. Gemcitabine (1000 mg/m^2) was given on days 1, 8 and 15 of a 28-day cycle. Ten days after day 1 of cycle 3, she presented with complaints of worsening shortness of breath and anasarca. Bilateral pleural effusions and hypotension with elevated vascular endothelial growth factor levels were identified upon further evaluation. A diagnosis of gemcitabine-induced systemic capillary leak syndrome was determined and the symptoms resolved with administration of corticosteroids [17A].

A case report described a 67-year-old woman with follicular lymphoma who experienced cardiomyopathy (congestive heart failure) with difficulty breathing following cycle 2 of a combination gemcitabine–rituximab–oxaliplatin (R-GemOx) treatment. Her previous medications included several chemotherapy combinations: rituximab–cyclophosphamide–doxorubicin–vincristine (R-CHOP); rituximab–ifosfamide–carboplatin–etoposide (R-ICE); brentuximab–rituximab; rituximab–dexamethasone–cytarabine–platinol (R-DHAP). Left ventricular systolic dysfunction with global hypokinesis was present 1-month after starting R-DHAP treatment which was then discontinued, and the symptoms improved. Gemcitabine treatment was subsequently started and congestive heart failure was present after the second cycle of gemcitabine. Gemcitabine was then discontinued and the cardiac symptoms resolved 2 months later [18A].

Doxorubicin

An international, open-label, randomized, phase III, multicenter trial compared the efficacy and safety of a doxorubicin–evofosfamide combination to doxorubicin monotherapy for treatment of locally advanced, unresectable or metastatic soft-tissue sarcoma (NCT01440088). A total of 640 patients (>15 years) received one of 2 treatments: doxorubicin (75 mg/m^2) via either bolus injection given over 5–20 minutes or IV infusion for 6–96 hours on day 1 of a 21-day cycle for up to 6 cycles or doxorubicin (the same dosing protocol) plus IV evofosfamide (300 mg/m^2) for 30–60 minutes on days 1 and 8 of a 21-day cycle for up to 6 cycles. Adverse events occurred in both treatment groups. The most common adverse events (grade $\geq$3) were anemia (48% in the doxorubicin–evofosfamide group and 21% in the doxorubicin monotherapy group), neutropenia (15% and 30%), febrile neutropenia (18% and 11%), and leukopenia (7% and 6%). Additional adverse events occurred more frequently with the combination treatment than with the monotherapy (thrombocytopenia (grade 3–4): 14% vs 1%, respectively, and stomatitis (grade 3–4): 8% vs 2%, respectively). A total of 5 patients (2%) died from drug-related adverse events in the combination treatment group (sepsis (2), septic shock (1), congestive cardiac failure (1) and an unknown cause (1)). In comparison, 1 patient died from a treatment related adverse event in the doxorubicin group (lactic acidosis (1)) [19C].

A case report described a 45-year-old woman with single metastatic breast cancer and metastasis to the pelvic bone who was treated surgically by bone osteotomy combined with a local doxorubicin (50 mg) injection. No recurrence of the metastatic lesions occurred in a 6-month observation period and no tumor cells were found in the pelvic bone during a surgical re-evaluation [20A].

Bevacizumab

A prospective, multicenter, (open-label, two-arm, parallel group, active control, randomized comparative clinical study was conducted to determine the pharmacokinetics, efficacy, and safety of a bevacizumab biosimilar (BevaciRel) compared to the standard bevacizumab combination treatment (CTRI/2013/05/003699). A total of 119 patients with metastatic colorectal cancer received either IV bevacizumab (5 mg/kg) every 2 weeks along with an IV regimen of 5-fluorouracil (400 mg/m^2), leucovorin (400 mg/m^2) and irinotecan (180 mg/m^2), ($n=83$) or the biosimilar drug ($n=33$). Adverse events ($n=715$) occurred at a higher rate with the biosimilar treatment as compared to the bevacizumab combination (70 vs 27 events, respectively). A total of 50 serious adverse events occurred with the biosimilar (27) and bevacizumab combination (23) and included gastrointestinal disorders, blood and lymphatic system disorders and general disorders and administration site conditions, infections and infestations. One patient in the biosimilar group discontinued treatment due to the adverse events [21C].

Oxaliplatin

A study was conducted in 103 patients with stage III and high-risk stage II colorectal cancer to determine the relationship between oxaliplatin treatment and increased splenic volume. The patients received either 5-fluorouracil, leucovorin, and oxaliplatin (mFOLFOX6; $n=37$) or oral 5-fluorouracil and leucovorin ($n=66$) treatment post-surgery. Splenic volumes were determined prior to surgery, immediately after treatment and 1 year after treatment. The volumes were increased after mFOLFOX6 as compared to the initial measurement but returned to baseline values after 1 year. No changes were found in the fluorouracil and leucovorin group at any time [22C].

SMALL MOLECULE KINASE INHIBITORS

Cancer is the second leading cause of death globally according to the World Health Organization and was responsible for 8.8 million deaths in 2015. The proliferation of cancerous cells arises from both genetic and epigenetic factors including: physical carcinogens (ultraviolet and ionizing radiation), chemical carcinogens (tobacco smoke, asbestos), and biological carcinogens (bacterial and viral infections) [23s]. Although traditional cancer therapies have been effective in cancer treatment, they do come with vast variety of adverse effects which at times limits their usage. One of the emerging fields in cancer treatment is the small molecule inhibitors. These compounds have many benefits over the traditional cytostatic agents, for example they are often administered orally. That said, these molecules do still have wide variety of adverse effects. This section will focus on the most commonly used group, small molecule kinase inhibitors [24M].

Small molecule kinase inhibitors are categorized by their mechanism of action of enzyme inhibition into Type I, II or III molecules. Type I kinase inhibitors interact directly with the ATP binding pocket. Type II kinase inhibitors bind to the ATP binding pocket as well as a hydrophobic pocket near to the ATP binding site. Type III kinase inhibitors bind to a hydrophobic pocket away from the ATP binding site, creating a conformation change which stops the kinase activity [25M].

Everolimus

Everolimus is a mammalian target of rapamycin inhibitor (mTOR inhibitor) and has been used for the management of several advanced cancers such as advanced pancreatic neuroendocrine tumors and renal angiomyolipoma. Everolimus frequently causes drug induced lung injury (DILI). According to a single-center, retrospective study of 45 patients, who had received Everolimus at a hospital setting, 15 patients (33%) developed DILI. Furthermore, 6 patients (40%) were asymptomatic and were diagnosed with DILI on routine follow-up chest CT findings. The median time from the start of everolimus administration to the onset of DILI was 64 days (4–277 days). When everolimus was discontinued 11 out of 15 patients (73%) saw an improvement of the DILI and 4 patients (27%) saw an improvement only with usage of systemic steroid [26c].

Acute kidney injury secondary to everolimus therapy occurs more frequently in patients with underlying kidney disease. It is a rare side effect in patient with normal kidney functions. A series of case reports described 2 women (54 and 56 years old) with breast carcinomas who received everolimus (10 mg) as part of their treatment. In both cases, a rise in serum creatinine levels occurred several months later, at which time, everolimus treatment was stopped. Symptoms including breathlessness, lethargy and anorexia occurred 1 week later and both patients were hospitalized. Their renal functions continued to worsen and hemodialysis was started in both cases. Although their symptoms and renal function improved the next few weeks, both patients remained dialysis dependent and no challenge with everolimus was attempted [27A].

In a small phase II study, 8 patients with metastatic renal cell cancer received oral everolimus (10 mg) once daily plus oral prednisone (5 mg) twice daily (NCT02479490). Previous clinical trial data showed that noninfectious pneumonitis was reported in up to 51% of patients and grade 3 noninfectious pneumonitis in 25% of patients treated with everolimus. In this study, 3 patients (38%) developed grade 2–3 noninfectious pneumonitis and discontinued study treatment due to the drug-related adverse event. Management of noninfectious pneumonitis consists of dose reduction or drug discontinuation and treatment with corticosteroids [28A].

Patients exposed to higher doses of everolimus (>10mg/day) are at a greater risk of being immunocompromised, allowing reactivation of latent tuberculosis (TB) infection. A 64-year-old man was admitted to hospital with fever, cough, nasal drip, and sputum. A chest X-ray showed pneumonia with lesions. He was previously diagnosed with stage I kidney cancer that later metastasized to the lung and pancreas. An initial treatment with sunitinib was replaced with everolimus (10mg) daily 14 months later. The patient was hospitalized 75 days after initiation of everolimus with fever, cough, nasal drip and sputum. Upon further testing, the patient was diagnosed with active TB, and appropriate anti-TB therapy of isoniazid, rifabutin, ethambutol, and pyrazinamide were started. Four weeks later, the anti TB therapy was changed to rifabutin/ethambutol/pyrazinamide/moxifloxacin for 8 weeks, and changed again to ethambutol/pyrazinamide/moxifloxacin/cycloserin and maintained for 9 months. By the fourth month of TB medication, lung metastatic progression was observed. Everolimus was replaced with pazopanib, which was taken for 21 months [29c].

SUNITINIB

Sunitinib is an oral multi tyrosine kinase inhibitor targeting platelet derived growth factor receptors, vascular endothelial growth factor receptors, FMS-like tyrosine kinase-3 (FLT3), colony-stimulating factor type 1, and glial cell-line-derived neurotrophic factor receptor. Due to its anti-tumor and anti-angiogenic activity, sunitinib is widely used in several types of cancer. Sunitinib has a similar side effect profile as other medications in this therapeutic class, and adverse events commonly include hypertension, diarrhea, nausea, asthenia, fatigue, vomiting, hand-foot syndrome, and hematologic toxicity [30C].

A previous case study described a 68-year-old female with unresectable pancreatic neuroendocrine tumor with hepatic metastases and persistent diarrhea. She was treated with sunitinib as a third-line chemotherapy but was previously treated with long-acting octreotide analogue and everolimus. After sunitinib initiation, the patient showed a partial response to the treatment but developed a watery diarrhea. No definite cause of the persistent diarrhea was evident and it persisted for over a month even with loperamide administration. The dose of the sunitinib was reduced to 25mg/day and loperamide was continued. The diarrhea symptoms improved for a short duration; however, the patient was re-hospitalized with severe diarrhea. Sunitinib was discontinued at this time. A diagnosis of pneumatosis cystoides intestinalis was made based on typical findings from a CT scan. Subsequent follow-up CT scans and an abdominal X-ray showed improvement and, 2 weeks later, she completely recovered from pneumatosis cystoides intestinalis. The diarrhea also improved a few days after the discontinuation of sunitinib and commencement of conservative management such as hydration and loperamide [31A].

A leading cause of renal failure is due to the development of acute interstitial nephritis, an inflammation of the renal tubules, and is likely drug-induced in 60%–70% of cases. A 69-year-old male with stage IV kidney clear cell carcinoma with metastases to the spine and brain presented with hematuria, oliguria, fever, fatigue, and decreased appetite for 4 days. Sunitinib treatment (50mg) daily had been started 2 weeks earlier. Sunitinib was discontinued but the hematuria persisted. A renal biopsy showed widespread interstitial inflammation and edema that was indicative of drug-induced acute intestinal nephritis. Oral steroids, hemodialysis, and blood transfusions were administered. Hematuria was still present while the renal function began to recover. Unfortunately, the patient succumbed to hospital-acquired pneumonia 2 months later. [32A].

A retrospective study determined the association of sunitinib and the development of hypothyroidism. A total of 27 patients with stage IV metastatic renal cell carcinoma received sunitinib on a standard schedule: initial dose 50mg/day, 4 weeks on, 2 weeks off. None of the patients had a history of hypothyroidism and thyroid-stimulating hormone (TSH) serum concentrations were measured at the baseline and every 12 weeks afterward. Hypothyroidism was detected in 12 patients (44%). Interestingly, sunitinib-induced hypothyroidism was associated with a longer progression-free survival compared to those patients with normal thyroid function (28.3 months vs 9.8 months, respectively) [33c].

Sunitinib-induced myelosuppression is typically mild and transient. With resolution at the end of each cycle when no drug is taken. However, a case of severe grade 3 neutropenia and thrombocytopenia was reported after 2 weeks of a 6-week cycle. The cytochrome P450 (CYP) 3A4 status was determined and it was found that the patient was heterozygous for CYP 3A4*22, which may partially explain the timing and severity of the myelosuppression. Using this pharmacogenetic information, a re-challenge of sunitinib with a reduced dose was subsequently administered and was well tolerated without additional effects [34A].

IMATINIB

A 63-year-old man with hypertension, hyperlipidemia, and chronic myeloid leukemia presented with painless grey–blue hyperpigmentation of the hard palate. His leukemia was treated with imatinib (400 mg) daily and had received the drug for approximately 9 years. He had also never received hydroxyurea, minocycline, or any anti-malarial agent. A diagnosis of imatinib-induced oral pigmentation was made based on the clinical examination, laboratory results and the medical history. A further examination at the 6-month follow-up did not show any swollen lymph nodes or other morphological changes. The hyperpigmentation was still present and no symptoms were reported [35A].

A 71-year-old woman with chronic myeloid leukemia (CML), Sjogren's syndrome and hypothyroidism presented with jaundice and generalized weakness. She had been receiving imatinib for 6 months as treatment for CML but this medication was suspended to avoid possible hepatotoxicity. All immune levels and infection assessments were negative. A biopsy indicated drug-induced liver damage. Prednisone (20 mg) was started and, within 2 days, her liver enzyme levels decreased and continued to do so until discharge. As an out-patient, she continued to receive prednisone (20 mg) for 1 month which was subsequently reduced to 5 mg before being discontinued. However, 2 weeks later, she was readmitted for progressive weakness, fatigue, and abdominal pain. An elevated white blood cell count was found, prednisone was stopped to avoid any gastrointestinal bleeding and IV proton-pump inhibitor treatment was started. The white blood cell count continued to rise and a petechial rash over the lower legs developed. A diagnosis of cellulitis was made and treated with antibiotics. A worsening kidney function was attributed to nonsteroidal anti-inflammatory drug-induced acute tubular necrosis. The patient expired 24 days after the second admission [36A].

A 62-year-old man with CML was receiving imatinib (400 mg) daily. No other comorbidities or medications were present. Five years later, he was diagnosed with acute pancreatitis and imatinib treatment was suspended for almost a month. Imatinib treatment was re-initiated at a lower dose and gradually increased to 400 mg/day. A year later, the patient experienced visual alterations and deficit with gravative acute and transient ocular pain in his right eye which were evaluated by an ophthalmologist. The examination showed optic disk edema and an inferior and superior arcuate scotoma that was not due to any thrombotic events. Imatinib treatment was suspended and within 1 week, the patient had visual improvement and another ophthalmological evaluation showed optic disk swelling. The ocular symptoms were almost completely resolved 2 months later, and a second-line treatment with nilotinib was initiated [37A].

A 48-year-old man with chronic Philadelphia chromosome positive CML and previous treatment for pulmonary tuberculosis 25 years earlier was receiving imatinib (400 mg) daily. However, he developed a dry cough and dyspnea on exertion and presented to the clinic with progressive nonproductive cough 10 months later. All reactants and immunoglobulin levels were within normal limits and lung function demonstrated a slight impairment of gas exchange although normal airways were present. A diagnosis of nonspecific interstitial pneumonitis was made based on the histopathological analysis of a transbronchial lung biopsy that showed thickened alveolar septa with infiltration of chronic inflammatory cells and interstitial fibrosis. These results indicate imatinib-induced interstitial pneumonitis. Imatinib was discontinued and the patient received daily administration of nilotinib (200 mg) and predisone (30 mg initially that was gradually reduced to 10 mg over 2 months). His condition gradually improved in the following months [38A].

DASATINIB

Dasatinib is a second-line BCR–ABL tyrosine kinase inhibitor for treatment of Imatinib-resistant or Imatinib-intolerant CML. A 48-year-old male presented with progressive shortness of breath for 1 month. He had received imatinib treatment for CML for 3 years, but due to poor treatment response and development of resistance, the treatment was changed to dasatinib (100 mg/day) 2 years ago. A chest X-ray showed bilateral pleural effusion, on the right side more than left, with cardiomegaly, and echocardiography showed massive pericardial effusion all around the heart with collapse of right atrium and right ventricle during diastole. The inferior vena cava also was dilated. Pericardiocentesis removed approximately 1200 mL of pericardial fluid over 2 days Dasatinib treatment was discontinued and imatinib restarted. A follow-up echocardiography 2 weeks later did not reveal any significant increase in pericardial fluid collection, indicating that the symptoms were likely due to dasatinib-induced cardiac tamponade and pleural effusion [39A].

A 79-year old woman presented to the dermatology clinic with facial redness apparent since the onset of dasatinib administration for the treatment of melanoma. These symptoms were accompanied by facial swelling, erythema of the cheeks, forehead and chin. Symptoms began 1–2 days after starting dasatinib. An examination showed erythema with slight scaling on the forehead, malar cheeks, upper lip, and chin area, and this erythema and scaling indicated a seborrheic dermatitis-like eruption. Topical desonide ointment was applied twice daily but within a few weeks, the patient experienced dehydration and failure to thrive which required hospitalization for rehydration and correction of electrolyte imbalances. Dasatinib was stopped; facial rash resolved [40A]. Additional case studies on this topic can be found in these reviews [41R, 42R, 43R].

References

[1] Clasen SC, Ky B, O'Quinn R, et al. Fluoropyrimidine-induced cardiac toxicity: challenging the current paradigm. J Gastrointest Oncol. 2017;8(6):970–9 [c].
[2] Chionh F, Lau D, Yeung Y, et al. Oral versus intravenous fluoropyrimidines for colorectal cancer. Cochrane Database Syst Rev. 2017;7:CD008398 [R].
[3] Chughtai K, Gupta R, Upadhaya S, et al. Topical 5-fluorouracil associated skin reaction. Oxf Med Case Reports. 2017;2017(8) omx043 [A].
[4] Nemer KM, Solus JF, Schaffer A, et al. Generalized pustular drug eruption caused by topical 5-fluorouracil. Dermatol Ther. 2017;30(5)e12520 [A].
[5] Yoshida K, Iwashita T, Uemura S, et al. A multicenter prospective phase II study of first-line modified FOLFIRINOX for unresectable advanced pancreatic cancer. Oncotarget. 2017;8(67):111346–55 [c].
[6] Cremolini C, Del Re M, Antoniotti C, et al. DPYD and UGT1A1 genotyping to predict adverse events during first-line FOLFIRI or FOLFOXIRI plus bevacizumab in metastatic colorectal cancer. Oncotarget. 2017;9(8):7859–66 [C].
[7] Sato Y, Hirakawa M, Ohnuma H, et al. A triplet combination with capecitabine/oxaliplatin/irinotecan (XELOXIRI) plus cetuximab as first-line therapy for patients with metastatic colorectal cancer: a dose escalation study. Cancer Chemother Pharmacol. 2017;80(6):1133–9 [c].
[8] Bullock A, Stuart K, Jacobus S, et al. Capecitabine and oxaliplatin as first and second line treatment for locally advanced and metastatic pancreatic ductal adenocarcinoma. J Gastrointest Oncol. 2017;8(6):945–52 [c].
[9] Shah RA, Bennett DD, Burkard ME. Photosensitive lichenoid skin reaction to capecitabine. BMC Cancer. 2017;17(1):866 [A].
[10] Mao Y, Zou C, Meng F, et al. The SNPs in pre-miRNA are related to the response of capecitabine-based therapy in advanced colon cancer patients. Oncotarget. 2017;9(6):6793–9 [C].
[11] Cho IR, Kang H, Jo JH, et al. Efficacy and treatment-related adverse events of gemcitabine plus nab-paclitaxel for treatment of metastatic pancreatic cancer "in a Korean" population: a single-center cohort study. Semin Oncol. 2017;44(6):420–7 [c].
[12] Seddon B, Strauss SJ, Whelan J, et al. Gemcitabine and docetaxel versus doxorubicin as first-line treatment in previously untreated advanced unresectable or metastatic soft-tissue sarcomas (GeDDiS): a randomised controlled phase 3 trial. Lancet Oncol. 2017;18(10):1397–410 [C].
[13] Gosain R, Gill A, Fuqua J, et al. Gemcitabine and carfilzomib induced thrombotic microangiopathy: eculizumab as a life-saving treatment. Clin Case Rep. 2017;5(12):1926–30 [A].
[14] Mittal BA, Jonathan SL. Gemcitabine-associated acute lipodermatosclerosis like eruption: an underrecognized phenomenon. JAAD Case Rep. 2017;3(3):190–5 [A].
[15] Gill D, Schrader J, Kelly M, et al. Gemcitabine associated pseudocellulitis: a missed diagnosis. J Oncol Pharm Pract. 2017;1078155217719584 [A].
[16] Strouse C, Epperla N. A rash diagnosis: gemcitabine-associated pseudocellulitis. J Oncol Pharm Pract. 2017;23(2):157–60 [A].
[17] Bajwa R, Starr J, Daily K. Gemcitabine-induced chronic systemic capillary leak syndrome. BMJ Case Rep 2017; 2017. pii: bcr-2017-221068 [A].
[18] Mohebali D, Matos J, Chang JD. Gemcitabine induced cardiomyopathy: a case of multiple hit cardiotoxicity. ESC Heart Fail. 2017;4(1):71–4 [A].
[19] Tap WD, Papai Z, Van Tine BA, et al. Doxorubicin plus evofosfamide versus doxorubicin alone in locally advanced, unresectable or metastatic soft-tissue sarcoma (TH CR-406/SARC021): an international, multicentre, open-label, randomised phase 3 trial. Lancet Oncol. 2017;8(8):1089–103 [C].
[20] Bohatyrewicz A, Karaczun M, Kotrych D, et al. Solitary breast cancer metastasis to pelvic bone treated with a unique method of surgery combined with local doxorubicin administration. Contemp Oncol (Pozn). 2017;21(4):306–10 [A].
[21] Apsangikar PD, Chaudhry SR, Naik MM, et al. Comparative pharmacokinetics, efficacy, and safety of bevacizumab biosimilar to reference bevacizumab in patients with metastatic colorectal cancer. Indian J Cancer. 2017;54:535–8 [C].
[22] Iwai T, Yamada T, Koizumi M, et al. Oxaliplatin-induced increase in splenic volume; irreversible change after adjuvant FOLFOX. J Surg Oncol. 2017;116(7):947–53 [C].
[23] World Health Organization. Cancer. World Health Organization; 2018, www.who.int/news-room/fact-sheets/detail/cancer [s].
[24] Lavanya V, Mohamed A, Neesar A, et al. Small molecule inhibitors as emerging cancer therapeutics. Integr Cancer Sci Therap. 2014;1(3):39–46 [M].
[25] Chahrour O, Cairns D, Omran Z. Small molecule kinase inhibitors as anti-cancer therapeutics. Mini Rev Med Chem. 2012;12(5):399–411 [M].
[26] Abe M, Tsushima K, Ikari J, et al. Evaluation of the clinical characteristics of everolimus-induced lung injury and determination of associated risk factors. Respir Med. 2018;134:6–11 [c].
[27] Chandra A, Rao N, Malhotra K, et al. Everolimus-associated acute kidney injury in patients with metastatic breast cancer. Indian J Nephrol. 2017;27(5):406–9 [A].
[28] Lolli C, Galla V, Schepisi G, et al. A phase II study of everolimus plus oral prednisone in patients with metastatic renal cell cancer. Oncologist. 2017;22:784. e74 [A].
[29] Jeon SY, Yhim HY, Lee NR, et al. Everolimus-induced activation of latent *Mycobacterium tuberculosis* infection in a patient with metastatic renal cell carcinoma. Korean J Intern Med. 2017;32:365–8 [c].
[30] Escudier B, Porta C, Bono P, et al. Randomized, controlled, double blind, cross-over trial assessing treatment preference for pazopanib versus sunitinib in patients with metastatic renal cell carcinoma: PISCES study. J Clin Oncol. 2014;32(14):1412–8 [C].
[31] Lee YS, Han JJ, Kim SY, et al. Pneumatosis cystoides intestinalis associated with sunitinib and a literature review. BMC Cancer. 2017;17:732 [A].
[32] Azar I, Esfandiarifard S, Sinai P, et al. Sunitinib-induced acute interstitial nephritis in a thrombocytopenic renal cell cancer patient. Case Rep Oncol Med. 2017;2017:6328204 [A].
[33] Buda-Nowak A, Kucharz J, Dumnick P, et al. Sunitinib-induced hypothyroidism predicts progression-free survival in metastatic renal cell carcinoma patients. Med Oncol. 2017;34(4):68 [c].
[34] Patel ND, Chakrabory K, Messmer G, et al. Severe sunitinib-induced myelosuppression in a patient with a CYP 3A4 polymorphism. J Oncol Pharm Pract. 2017;1. 1078155217724863 [A].
[35] Bombeccari G, Garagiola U, Pallotti F, et al. Hyperpigmentation of the hard palate mucosa in a patient with chronic myeloid leukaemia taking imatinib. Maxillofac Plast Reconstr Surg. 2017;39(1):37 [A].

[36] Bhatty O, Selim M, Kassim T, et al. A case of imatinib-induced hepatitis. Cureus. 2017;9(6)e1302 [A].
[37] Napolitano M, Santoro M, Mancuso S, et al. Late onset of unilateral optic disk edema secondary to treatment with imatinib mesylate. Clin Case Rep. 2017;5(10):1573–5 [A].
[38] Luo ZB, Xu N, Huang XP, et al. Imatinib-induced interstitial pneumonitis successfully switched to nilotinib in a patient with prior history of Mycobacterium tuberculosis infection. Turk J Haematol. 2017;34:356–81 [A].
[39] Wattal S, Rao MS, Chandra N, et al. Dasatinib induced cardiac tamponade—a rare association. J Clin Diag Res. 2017;11(2): FD03–4 [A].
[40] Riahi RR, Cohen PR. Dasatinib-induced seborrheic dermatitis-like eruption. J Clin Aesthet Dermatol. 2017;10(7):23–7 [A].
[41] Keshishyan S, Sehdev V, Reeves D, et al. Cytostatic agents. In: Ray SD, editor. Side effects of drugs annual: a worldwide yearly survey of new data in adverse drug reactions, vol. 37. Elsevier; 2015. p. 567–81 [chapter 44]. [R].
[42] Reeves D. Cytostatic agents—tyrosine kinase inhibitors utilized in the treatment of solid malignancies. In: Ray SD, editor. Side effects of drugs annual: a worldwide yearly survey of new data in adverse drug reactions, vol. 38. Elsevier; 2016. p. 479–91 [chapter 43]. [R].
[43] Reeves D. Cytostatic agents: monoclonal antibodies utilBized in the treatment of solid malignancies. In: Ray SD, editor. Side effects of drugs annual: a worldwide yearly survey of new data in adverse drug reactions, vol. 39. Elsevier; 2017. p. 465–82 [chapter 40]. [R].

CHAPTER

46

Radiological Contrast Agents and Radiopharmaceuticals

Manoranjan S. D'Souza, Arindam Basu Sarkar†,1*

*Department of Pharmaceutical and Biomedical Sciences, Raabe College of Pharmacy, Ohio Northern University, Ada, OH, United States

†Department of Pharmaceutical Sciences, College of Pharmacy, University of Findlay, Findlay, OH, United States

[1]Corresponding author: basusarkar@findlay.edu

INTRODUCTION

Contrast agents are the compounds or elements that absorb energy in a differential manner than the body fluid, tissue or bone when introduced in the body cavity or localized tissue. They produce an image when a specific form of energy is passed through the body and captured on the other side. Different types of contrast agents are available and typically used as image enhancing agents for diagnostic purposes (see Fig. 1). The contrast agents may be radioactive (like iodine) or non-radioactive (like gadolinium). Common imaging techniques include different energy forms. The X-ray-based techniques include computed tomography (CT) and radiography and generally use iodine or barium based contrast agents. Nuclear magnetic resonance imaging (NMRI, commonly called MRI) is a technique that utilizes the radio frequency that is emitted by certain molecules present in the physiological system (like hydrogen) when subjected to strong magnetic field. Most MRI applications do not need administration of any contrast agents. However, iron oxide and salts of the rare-earth element gadolinium are used for special purposes to improve the contrast. The ultrasound imaging uses low-frequency sound energy and the image is created as the sound wave reflecting back from the tissue/organ is captured. Ultrasound imaging technique has started using suspension of gas bubbles as the contrast agent only recently. Radiopharmaceuticals are the compounds that produce ionizing radiation and are used for diagnosis and treatment. Radiopharmaceuticals are commonly used in cancer treatment. The contrast agents and radiopharmaceuticals do not have any intended pharmacology, but these agents have toxicity or side effects which could be chronic, acute or even fatal. Overall, the incidence of side effects associated with these contrast agents is low and generally self-limited.

This chapter discusses side effects associated with iodinated contrast agents (ICAs), gadolinium-based contrast agents (GBCAs), agents used for echocardiography/ultrasonography, iron oxide-based contrast agents and radiopharmaceuticals that have been reported in the literature. We reviewed case reports, randomized clinical trials, meta-analysis and review articles over the last 1–2 years. Further, the chapter will discuss strategies to prevent and minimize some of the serious side effects associated with radiological contrast agents.

IODINE-BASED CONTRAST AGENTS [SEDA-33, 963; SEDA-34, 749; SEDA-35, 863; SEDA-36, 695; SEDA-37, 583; SEDA-38, 493]

Iodinated contrast agents (ICAs) are commonly used in contrast-enhanced imaging to enhance the ability to see blood vessels and organs during X-ray or CT-scan imaging. ICAs are administered orally, rectally or intravenously and are classified based on osmolarity as high and low osmolar agents (see Table 1). The low osmolar agents are further classified as either ionic or non-ionic type. Currently FDA-approved ICAs and their side effects discussed in this chapter are shown in Figs. 1 and 2. The American College of Radiology classifies ICA-induced side effects based on timing; as acute or delayed [1S]. Acute reactions occur almost immediately

Side Effects of Drugs Annual, Volume 40
ISSN: 0378-6080
https://doi.org/10.1016/bs.seda.2018.07.007

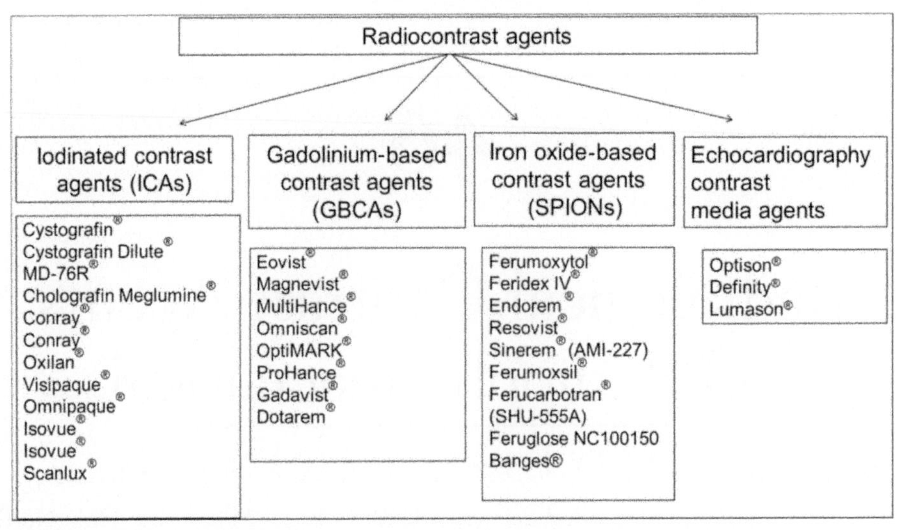

FIG. 1 Types of radiocontrast agents available for use in the United States and all over the world.

after administration of ICA. In contrast, delayed side effects may develop from 30–60min to up to 1 week following exposure to the contrast agent; with the majority occurring between 3h and 2 days. ICA-induced side effects can also be classified based on severity as mild (signs and symptoms are self-limited without evidence of progression), moderate (signs and symptoms are more pronounced and require some kind of medical management) and severe (signs and symptoms are often life-threatening and can result in permanent morbidity or death if not managed appropriately).

The overall incidence of ICA-induced side effects is lower with non-ionic ICAs compared to ionic ICAs. The risk factors for development of ICA-induced adverse events include history of asthma, hypertension, diabetes, cardiac and cerebrovascular disease. However, a retrospective study conducted in China evaluated approximately 120822 cases in patients who underwent enhanced CT examination over a 2 year period and found no increase in incidence of adverse effects in the presence of underlying disease states [2C]. The study reported that the incidence of adverse reaction was 0.4% and 0.44% in patients with and without an underlying disease, respectively ($P=0.378$). The risk of adverse effect was also dependent on history of the patient (higher risk with previous h/o of adverse effect with ionic ICA), the dose of ICA used (higher risk with dose >100mL) and speed of ICA administration (higher risk when ICA is administered rapidly >5mL/s). A limitation of these findings was that this was a single-site retrospective study conducted in a single country. Another single-institution retrospective study reported additional risk factors for occurrence of acute side effects and included approximately 137473 patients undergoing ICA-enhanced CT using the non-ionic low-osmolar ICAs iopromide and iopamidol [3C]. These additional risk factors include female patients, patients undergoing CT for emergency medical conditions, patients aged 50–60 years and patients underwent coronary CT angiography. The most common mild side effects were cutaneous (50.52%; rash, urticaria, flushing and itching) in nature. In contrast, 62.91% of the moderate and 48.28% of the severe side effects involved cardiovascular symptoms (e.g., transient chest pain and stuffiness, tachycardia, hypotension and cardiac arrest). Patients with moderate and severe side effects were treated with oral and/or intravenous hydration, antihistamines, steroids, and epinephrine as required. The majority of the patients (96.5%) with side effects recovered without sequelae. Fluid hydration is suggested as a possible pretreatment that can help to reduce the incidence of acute side effects. However, a prospective randomized controlled study conducted in approximately 5959 patients determined that fluid hydration had no effect on the incidence (4.3%; 254/5959) of acute side effects (occurring within the first 30min) after ICA administration during abdominal and pelvic CT in outpatients [4C]. One of the limitations of the study was that it was restricted to one institution in China and the status of hydration of patients prior to ICA administration was not known.

ICA-Induced Allergic Reactions

Case reports have suggested several allergic types of side effects after administration of non-ionic low osmolar ICAs, including urticaria, anaphylactic shock, malaise and enlargement of salivary gland [5A] (see Fig. 2). Non-ionic ICA-induced allergic reactions are possibly mediated by activation of mast cells and basophils by either IgE-dependent, complement-dependent and/or direct membrane effects [6E]. Premedication in

TABLE 1 United States Food and Drug Administration (USFDA) Approved Iodinated Contrast Agents (ICAs)

Brand Name(s)	Generic Name	Chemical Nature	Indication
Cystografin® (30%) Cystografin Dilute® (18%)	Diatrizoate meglumine	Ionic High-osmolar	Identification of tumors, infections, stones and other abnormalities of the urinary bladder
MD-76R®	Diatrizoate meglumine and Diatrizoate sodium	Ionic High-osmolar	Used in excretion urography, aortography, pediatric angiocardiography, peripheral arteriography, selective renal arteriography, visceral arteriography, coronary arteriography with or without left ventriculography, contrast enhancement of computed tomographic brain imaging and for intravenous digital subtraction angiography
Cholografin Meglumine®	Iodipamide meglumine	Ionic High-osmolar	Indicated for intravenous cholangiography and cholecystography
Conray® 30, 43	Iothalamate meglumine	Ionic High-osmolar	Indicated for excretory urography, cerebral angiography, peripheral arteriography, venography, arthrography, direct cholangiography, endoscopic retrograde cholangiopancreatography, contrast enhancement of CT brain images, cranial computerized angiotomography, intravenous digital subtraction angiography, arterial digital subtraction angiography, enhancement of CT scans for evaluation of lesions in the liver, pancreas, kidneys, abdominal aorta, mediastinum, abdominal cavity and retroperitoneal space
Oxilan®-300, 350	Ioxilan	Ionic High-osmolar	Indicated for cerebral arteriography, coronary angiography, left ventriculography and peripheral arteriography
Visipaque 270, 320®	Iodixanol	Non-ionic Iso-osmolar	Indicated for intra-arterial digital subtraction angiography, angiocardiography, peripheral and visceral arteriography, cerebral arteriography, CT imaging head, neck and urography, CT angiography, peripheral venography
Omnipaque® 140, 180, 240, 300, 350	Iohexol	Non-ionic Low-osmolar	Intrathecal administration in adults including myelography (lumbar, thoracic, cervical, total columnar) and in contrast enhancement for CT (myelography, cisternography, ventriculography)
Isovue®-200, 250, 300, 370 Isovue®-M 200, 300 Scanlux®-300, 370	Iopamidol	Non-ionic Low-osmolar	Indicated for angiography throughout the cardiovascular system, including cerebral and peripheral arteriography, coronary arteriography and ventriculography, pediatric angiocardiography, selective visceral arteriography and aortography, peripheral venography (phlebography), and adult and pediatric intravenous excretory urography and intravenous adult and pediatric contrast enhancement of CT head and body imaging
Ultravist® 150, 240, 300, 370	Iopromide	Non-ionic Low-osmolar	Indicated for cerebral arteriography, coronary arteriography, left ventriculography, visceral angiography and aortography, peripheral venography, excretory urography; CT of the head and body (intrathoracic, intra-abdominal and retroperitoneal regions) for the evaluation of neoplastic and non-neoplastic lesions
Optiray® 240, 300, 320, 350	Ioversol	Non-ionic Low-osmolar	Indicated for cerebral arteriography, peripheral arteriography, visceral and renal arteriography, aortography, coronary arteriography and left ventriculography, angiocardiography, CT imaging of head and body, venography, intravenous excretory urography, intravenous digital subtraction angiography
Hexabrix® 320	Ioxaglate meglumine and Ioxaglate sodium	Ionic Low-osmolar	Indicated for cerebral angiography, peripheral arteriography, selective coronary arteriography with or without left ventriculography, pediatric angiocardiography, intravenous digital subtraction angiography and intravenous contrast enhancement of CT of the brain and body. It is also indicated in phlebography, arthrography, excretory urography and hysterosalpingography

the form of antihistamines and steroids can help to reduce incidence of ICA-induced side effects. The effects of premedication based on past history of severity of side effects to ICAs were assessed in approximately 850 patients [7C]. In this study, patients with a history of mild side-effects to ICAs were administered an antihistamine (chlorpheniramine, 4 mg). Patients with a history of moderate side effects to ICA were administered a steroid (methylprednisolone, 40 mg). Finally, patients with a history of severe side effects to ICA were administered several doses steroids (methylprednisolone, 40 mg). Despite severity-tailored premedication, patients with a history of mild side effects who only received antihistamine as pretreatment showed breakthrough reactions after ICA exposure (approximately 89%), but did not need treatment.

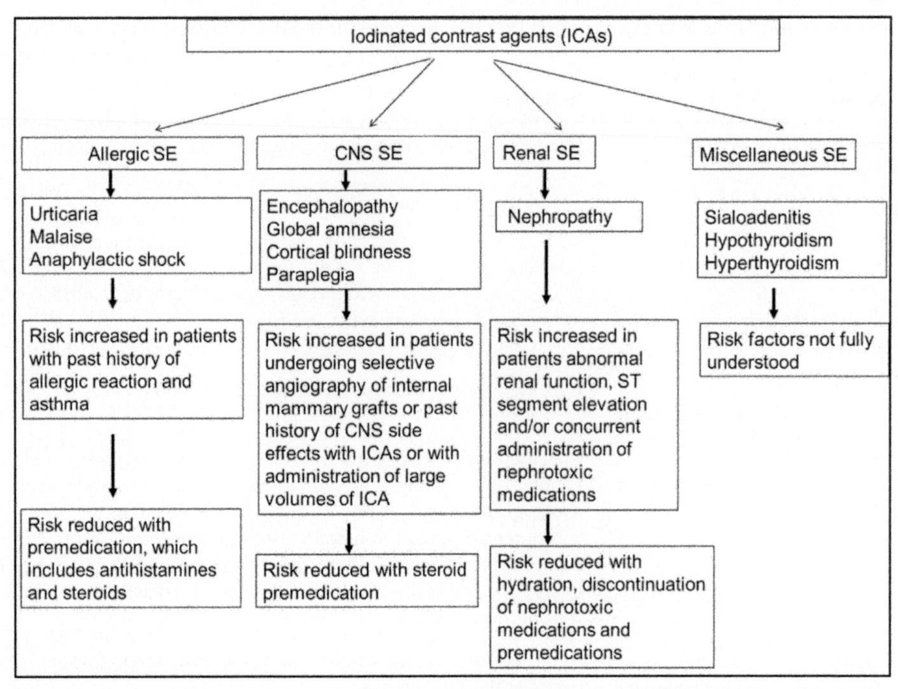

FIG. 2 Side effects (SE) associated with iodinated contrast agents (ICAs) and risk factors that increase occurrence of ICA-induced SEs.

Importantly, the study found that addition of a corticosteroid did not decrease the occurrence of breakthrough reactions in patients with h/o mild adverse reaction. The study also reported that the rate of breakthrough reactions was lower in patients with a history of severe side effects, who received more than one dose of steroid. Changing to another ICA also helped in reducing breakthrough reactions in patients with history of ICA-induced side effects. Overall, the study suggested that severity-based premedication protocol might be a more efficacious way to reduce the number of ICA-induced side effects. In summary, the incidence of ICA-induced allergic reactions is low and can be further reduced and/or effectively managed using a severity-based premedication protocol.

ICA-Induced CNS Effects

ICA-induced encephalopathy is an acute and reversible adverse reaction sometimes observed after intra-arterial administration of ICA. A case of repeated ICA-induced encephalopathy was reported in a 65-year-old patient undergoing coronary angiography with low-osmolar non-ionic monomeric ICA (Ultravist® 370 (Iopromide; 110mL), Bayer Healthcare, Pittsburg, PA, USA) [8A]. Symptoms, which included global aphasia, disorientation, drowsiness, confusion, weakness of the legs, and mild respiratory distress, appeared immediately after administration of the contrast agent and disappeared after 24–48h with supportive medication. This was the second such episode for this patient with a history of similar reaction in the past. The adverse effect occurred despite premedication treatment with intravenous hydrocortisone and promethazine. Upon recovery the patient had no recollection of the angiography procedure. Disruption of blood–brain barrier and/or direct neurotoxicity was considered as possible causes for ICA-induced encephalopathy. Risk factors for ICA-induced encephalopathy include hypertension, diabetes mellitus, renal impairment, the administration of large volumes of ICA or selective angiography of internal mammary grafts, or history of previous similar adverse reaction to ICA.

Transient global amnesia was reported after administration of Iomeprol (Iomeron® 300, Braco Imaging SpA, Milan, Italy) in a 54-year-old woman with an 8 day old left upper limb weakness [9A]. A brain CT, performed in the patient immediately on admission, revealed a subacute ischemic stroke in the right middle cerebral artery (MCA), posterior cerebral artery (PCA) watershed territory, two hypodense lesions in the right cerebral artery territory suggestive for subacute ischemic stroke and a right thalamic lacunar infarct. Stenosis (approximately 90%) of the right internal carotid artery and right anterior cerebral artery was also confirmed by doppler ultrasound. Four days after admission the patient underwent diagnostic digital subtraction cerebral angiography via right femoral approach with Iomeprol (100mL), which lasted for about 20min. A few minutes after the procedure the patient showed signs of anterograde amnesia and was unable to remember anything from the time of her admission in the hospital. Neurological examination revealed that the patient lacked orientation in time and place, ability to acquire and retain new memories, but had intact memory of personal data. Over a period of 24h the patient's amnesia gradually improved. A brain MRI performed 22h after the angiography procedure suggested hippocampal ischemia. A repeat neurological examination 7 days after the angiography, suggested a memory gap of a day when angiography was performed without other significant memory problems. An MRI performed 2 months after the procedure showed no lingering imaging changes in the hippocampal area, but memory of the angiography procedure remained deficient. The authors believed that this Iomperol-induced transient amnesia could be due to transient hippocampal ischemia. However, they did not rule out the possibility of direct neurotoxic effect of the contrast media on the hippocampal CA-1 neurons.

Transient cortical blindness was reported in a 60-year-old patient with a previous history of coronary bypass surgery using a left internal mammary artery graft, who underwent coronary angiography with a ICA for repeated angina [10A]. During the angiography procedure, the patient initially complained of blurring vision, which deteriorated to complete blindness within minutes. The angiography was immediately abandoned and a CT scan performed subsequently showed symmetrical subarachnoid and gyral hyperdensity in both occipital lobes. Over the course of 3 days the patient's vision slowly improved and returned to normal by the 4th day. The patient had a history of exposure to ICA, but had not reported any side effects to the same previously. The onset and course of ICA-induced cortical blindness reported in this case-report is consistent with other previously reported studies [11A, 12A]. ICA-induced cortical blindness can be avoided by limiting exposure to ICA, especially in patients with a history of ICA-induced cortical blindness, and/or pretreatment with steroids.

ICA-induced paraplegia was reported in a 76-year-old man with spinal dural arteriovenous fistula [13A]. The patient underwent a spinal digital subtraction angiography (DSA) with 100mL of Iomperol (Imeron® 300). One hour after the DSA procedure, the patient developed paraplegia of the legs. The paraplegia resolved completely within 24h after administration of dexamethasone. The patient underwent another DSA with the same compound prior to surgery and re-experienced paraplegia of the legs despite dexamethasone pretreatment, which lasted for another 72h. Additionally, the patient experienced hypesthesia below the L-1 levels as well as an atonic bladder sphincter muscle. In addition to dexamethasone, the patient also received clemastine (4mg) and prednisolone (250mg). The patient's surgery for the arteriovenous fistula was successful and the patient did not have any postoperative complications. ICA-induced paraplegia is a rare transient complication and the authors speculated that the paraplegia could be the result of ICA-induced neurotoxicity. Further, the authors suggested that patients with spinal cord vascular abnormalities must be selected carefully for ICA-induced CT. Use of such technique must be avoided especially if the patient has a history of neurological deficits like paraplegia following administration of ICAs as the long-term consequences of this transient paraplegia are not fully understood. In summary, risk factors for ICA-induced CNS side effects have not been fully identified. Although, ICA-induced CNS side effects are rare, they are serious and the occurrence of these side effects can be minimized by careful history taking and avoiding ICAs in patients with previous history of ICA-induced CNS side effects.

ICA-Induced Salivary SE

A case was reported of a 70-year-old woman, who complained of painful swelling in her upper neck region. The patient had difficulty swallowing approximately 24h after a CT scan of the thorax and abdomen using an ICA [14A]. Her past history included a diagnosis of follicular carcinoma and myelodysplastic syndrome approximately 15 years prior to the CT scan and was currently under remission. Physical examination of the patient showed hyperaemia of the upper neck and bilateral diffuse submandibular enlargement. The patient's symptoms resolved within a few days. The patient was diagnosed with ICA-induced sialadenitis (also called iodide mumps). The risk of developing ICA-induced sialadenitis is increased in patients with impaired renal function due to impaired iodine elimination, which causes swelling of the mucosal duct of the salivary gland resulting in impaired salivary excretion. ICA-induced sialadenitis has also been reported previously in patients with normal renal function [15A]. An additional case of a 67-year-old patient with dysphagia, neck swelling and hoarseness several hours after coronary angiography was recently reported [16A]. Patient's past history included cardiac transplantation due to idiopathic dilated cardiomyopathy complicated by chronic renal disease. The history of this patient included a similar reaction to ICA despite receiving hydrocortisone and diphenhydramine as premedication. Physical examination of this patient suggested enlargement and tenderness of submandibular glands. CT scan suggested symmetric enlarged submandibular glands with no indication of stones or fat stranding surrounding the glands. A diagnosis of ICA-induced sialadenitis was made based on the above described physical and imaging findings. The patient's symptoms resolved over a period of 5 days without treatment. In summary, ICA-induced sialadenitis (also called iodide mumps) is a rare, self-limited adverse effect that cannot be prevented with premedication.

ICA-Induced Endocrine Effects

A case-controlled single institution study suggested that children who underwent procedures with ICAs were at high risk of thyroid dysfunction [17C]. The study reviewed records of 1640 of children under 18 years of age over a 14 year (2001–2015) period. The risk of developing hypothyroidism was higher in children exposed to ICAs (odds ratio 2.60; 95% CI 1.43–4.72; $P<0.01$). The median interval between exposure and onset of hypothyroidism was approximately 10.8 months (interquartile range 6.6–17.9 months). Based on this finding, the authors suggested monitoring of thyroid function in children for the first year after exposure to ICAs.

Interestingly, a study reported a low risk of developing hyperthyroidism after exposure to ICAs, even in areas where iodine is deficient. A prospective study of 102 adults exposed to ICAs, only 2% of patients developed subclinical hyperthyroidism [18c]. Another case report suggested development of ICA-induced hyperthyroidism in a euthyroid patient, who had undergone hysterosalpingography for infertility [19A]. The patient returned 1 week after the exposure with palpitation, hand tremor, fatigue and excessive sweating. After laboratory testing, the patient was diagnosed with ICA-induced hyperthyroidism. She recovered spontaneously over a period of 1 month without any anti-thyroid treatment. The authors suggested that hyperthyroidism was due to excessive iodine absorption. Risk factors for developing hyperthyroidism after ICA exposure during hysterosalpingography include older patients, and patients who are iodine-deficient or have diffuse nodular goiter or Graves' disease. In summary, abnormalities in thyroid function can occur in at risk populations and practitioners must warn at risk patients to report symptoms associated with thyroid dysfunction.

ICA-Induced Nephropathy

ICA-induced nephropathy is defined as an abrupt deterioration in renal function [acute elevation (>0.5mg/dL or >25% of baseline) of serum creatinine (SCr)] within 72h after the injection of the ICA, in the absence of another etiology. Several previously published articles have reviewed ICA-induced nephropathy [1S, 20R], SEDA-36, 695; SEDA-37, 583; SEDA-38, 493. The overall of incidence of ICA-induced nephropathy is low in patients with normal kidney function. Multiple patient- and contrast media-related factors have been identified that predispose patients to ICA-induced nephropathy [21R, 22R] (see Table 2). The pathophysiology of ICA-induced nephropathy is not fully understood. It is hypothesized that vasoconstriction and direct tubular injury are two broad mechanisms that play a role in ICA-induced nephropathy [21R, 22R]. The ICA-induced vasoconstriction occurs due to release of factors from endothelial cells such as endothelin-A and/or activation of adenosine-A1 receptors. The direct tubular injury possibly occurs due to free non-complexed iodine production, which can be directly injurious to renal tubular cells. Another reason for direct tubular injury is possibly due to production of oxygen free radicals. The challenge in management of ICA-induced nephropathy is early and timely detection. Currently, ICA-induced nephropathy is defined by an elevation of serum creatinine as described above. However, several new biomarkers have been identified, which can help detect tubular injury within a few hours. These biomarkers include Cystanin C, neutrophil gelatinase-associated lipocalin, fatty-acid binding protein, insulin-like growth factor binding protein-7 and tissue inhibitor of metalloproteinase-2 [22R]. These markers may help in both detection of subclinical ICA-induced nephropathy and early detection of severe ICA-induced nephropathy, which will eventually help in early intervention to prevent long-term deterioration of renal function.

TABLE 2 Patient-, Procedure- and Contrast-Related Factors That Predispose to Increased Incidence of ICA-Induced Nephropathy

Patient-Related Factors	Contrast- and Procedure-Related Factors
Dehydration	Low-osmolar non-ionic ICAs
Advanced age	High dose of ICAs used for the procedure
History of malignancy	Intra-arterial administration of ICA during procedure
Pre-existing renal disease, e.g., Diabetic nephropathy	Repeated exposure to ICA (especially within 24–48h)
ST-segment elevation myocardial infarction (STEMI)	Percutaneous coronary intervention and coronary angiography especially when it is done under emergency conditions
Concomitant administration of nephrotoxic drugs, e.g., metformin, non-steroidal anti-inflammatory drugs (NSAIDs), Angiotensin converting enzyme (ACE) inhibitors, angiotensin II antagonists, and chemotherapeutic drugs	

Strategies to Reduce ICA-Induced Nephropathy

Several pharmacological and procedural strategies have been tried to minimize the occurrence of ICA-induced nephropathy. These include avoiding contrast-enhanced CT if possible in patients with compromised renal function (estimated glomerular filtration rate <30 mL/min/1.73 m^2). In patients with normal renal function, discontinue metformin and other nephrotoxic drugs such as angiotensin-converting enzyme inhibitors and angiotensin-2 receptor antagonists at least 24–48 h prior to contrast media administration [22R, 23R]. Another approach to reduce the incidence of ICA-induced nephropathy is ensuring adequate hydration either orally or via intravenous administration to facilitate rapid excretion of the contrast media [24M]. Other strategies that may minimize ICA-induced nephropathy include using the lowest dose of contrast media possible for the procedure and using in-line devices to minimize excessive contrast injection [25R]. Finally, a recent meta-analysis suggested using non-ionic iso-osmolar ICA compared to low-osmolar ICAs to reduce incidence of ICA-induced nephropathy [26M]. The meta-analysis compared different low-osmolar ICAs (iohexol, iopamidol, iopromide, ioxaglate) and also compared low-osmolar ICAs to iso-osmolar ICAs [26M]. The meta-analysis consisted of five trials (approximately 826 patients) comparing different low-osmolar ICAs and 25 trials (approximately 5053 patients) comparing iso-osmolar ICAs to low-osmolar ICAs. The studies comparing different low-osmolar ICAs found no significant statistical difference between different low-osmolar ICAs. In contrast, the iso-osmolar ICA iodixanol, when compared to low-osmolar ICAs, was associated with statistically significant lower incidence of ICA-induced nephropathy (RR = 0.80; 95% CI, 0.65–0.99). Over the long-term, however, there was no difference in clinical consequences such as need for dialysis, mortality or cardiac events between iso-osmolar ICA iodixanol compared to low-osmolar ICAs.

Another approach to reduce ICA-induced nephropathy is to use the RenalGuard® therapy [27C]. This therapy utilizes saline and furosemide that is delivered via the RenalGuard® system. The RenalGuard® system is a fluid balance system that measures and coordinates volume of intravenous fluid infusion with urine output. Basically, this therapy maintains a high urinary filtration rate (>300 mL/h) during and 4 h after the procedure. A study analyzed the incidence of ICA-induced nephropathy in 400 patients with compromised renal function [glomerular filtration rate ≤30 mL/min per 1.73 m^2 and/or a high predicted risk, i.e., Mehran score ≥11], who underwent percutaneous coronary intervention [28C]. The study concluded that RenalGuard® therapy was safe and effective in reaching a high urinary filtration rate in patients with compromised renal function and a mean intra-procedural urinary filtration rate of >450 mL/h was optimal for prevention of ICA-induced nephropathy in high risk patients.

In addition to these procedural strategies described above, several pharmacological strategies in the form premedication have been tried to reduce incidence of ICA-induced nephropathy. A meta-analysis assessed the relative efficacy of 12 different pharmacologic strategies for prevention of ICA-induced nephropathy [29M]. The different pharmacologic therapies included the following drugs in combination with hydration: *N*-acetylcysteine, theophylline, fenoldopam, iloprost, alprostadil, prostaglandin E1, statins, statins plus *N*-acetylcysteine, sodium bicarbonate, sodium bicarbonate plus *N*-acetylcysteine, ascorbic acid (vitamin C), tocopherol (vitamin E), α-lipoic acid, atrial natriuretic peptide, B-type natriuretic peptide, and carperitide. The trial network included 150 trials with 31 631 participants and 4182 ICA-induced nephropathy events assessing the 12 different pharmacologic strategies described above. Overall, the study concluded that combining statins and hydration with or without *N*-acetylcysteine might be the preferred treatment strategy to prevent ICA-induced nephropathy in patients undergoing diagnostic and/or interventional procedures requiring contrast media. Similarly, another meta-analysis assessed several premedications (*N*-acetylcysteine, sodium bicarbonate, statins and vitamin C) used to reduce ICA-induced nephropathy [24M]. Amongst these premedications, the study found that *N*-acetylcysteine and intravenous saline hydration with or without a statin was the most effective combination of premedications in reducing ICA-induced nephropathy.

In addition, a meta-analysis of 15 randomized clinical trials also reported decreased incidence of ICA-induced nephropathy in patients undergoing percutaneous coronary intervention or coronary angiography after premedication with moderate to high dose (20–40 mg/day) of rosuvastatin, but not with low dose (<10 mg/day) rosuvastatin [30M]. Similarly, another meta-analysis of 19 randomized controlled clinical trials also concluded that premedication with statins reduced the incidence of ICA-induced nephropathy in patients undergoing angiography [31M]. Furthermore, a meta-analysis of 49 randomized clinical trials that evaluated the efficacy of antioxidants like *N*-acetylcysteine in the prevention of ICA-induced nephropathy concluded that pretreatment with *N*-acetylcysteine compared to vitamin E and vitamin C reduced incidence of ICA-induced nephropathy in patients undergoing angiography [32M]. However, the study concluded that *N*-acetylcysteine did not reduce mortality associated with angiography.

Alprostadil, a prostaglandin, has also been evaluated by several studies for its effects on preventing ICA-induced nephropathy after percutaneous coronary intervention in diabetic patients. A meta-analysis, which included eight studies involving 969 patients (487 patients treated with alprostadil and 482 controls) [33M], reported that the incidence of ICA-induced nephropathy in the alprostadil treated group was significantly lower than that in the control group [OR = 0.28, 95% CI (0.18–0.42)].

All clinical parameters of kidney function such as BUN, serum creatinine, β2-microglobin and CysC were better in alprostadil treated group compared to controls. Overall, the findings of the meta-analysis suggested that premedication with alprostadil may be useful in reducing the incidence of ICA-induced nephropathy after percutaneous coronary angiography in diabetic patients. Due to the comparatively small number of patients included in the studies, further work is required to make firm conclusions on the use of alprostadil to prevent ICA-induced nephropathy.

Brain natriuretic peptide, a peptide released by ventricular myocytes in response to myocardial stretch, has been shown to increase glomerular filtration rate and thus improve renal hemodynamics and tubular function. In addition, brain natriuretic peptide has diuretic and natriuretic effects, which facilitates excretion of ICAs. Therefore, it has been hypothesized that pretreatment with brain natriuretic peptide may help in reducing the incidence of ICA-induced nephropathy in patients undergoing cardiovascular procedures. Consistent with this hypothesis, a meta-analysis of seven randomized clinical trials involving 1441 patients reported that pretreatment with brain natriuretic peptide treatment was associated with lower incidence of ICA-induced nephropathy in patients undergoing percutaneous coronary intervention (RR=0.38, 95% CI 0.27–0.54, $P<0.001$) [34M]. This study also reported a lower incidence of major cardiac events in patients undergoing percutaneous coronary intervention (RR=0.47, 95% CI 0.24–0.95, $P=0.034$) after pretreatment with brain natriuretic peptide. Another clinical trial also reported that premedication with brain natriuretic peptide and hydration compared to hydration alone reduced the incidence of ICA-induced nephropathy in patients with chronic kidney disease undergoing coronary angiography [35R].

Oxygen and theophylline have also been used for prevention of ICA-induced nephropathy. A recent study suggests decreased incidence of ICA-induced nephropathy after administration of nasal oxygen (2–3 L/min) and hydration in patients undergoing elective angiography [36c]. In this randomized controlled trial, patients in the treatment group [176 patients in the treatment group and 176 controls; total 348 patients] were administered oxygen 10 min prior to the procedure until the end of the procedure. A total of 105/348 patients developed ICA-induced nephropathy (30.2%; 95% CI, 25.4%–35%). However, a significant number of these 105 patients, i.e., 41.5% (73) received only hydration, while only 18.6% (32) in the treatment group developed ICA-induced nephropathy. This was statistically significant ($P<0.01$). Overall, the study suggested that supplementation with nasal oxygen in addition to standard hydration could be an effective strategy in reducing ICA-induced nephropathy. The study had several limitations such as small sample size, the moderate effect size and the fact that the study was conducted at a single site in a single country. Nevertheless, supplemental oxygen therapy prior to angiography needs to be further investigated, especially in patients at high risk of developing ICA-induced nephropathy. Activation of adenosine-A1 receptors plays a role in ICA-induced vasoconstriction, which has been implicated in ICA-induced nephropathy. Theophylline is an adenosine antagonist and small randomized single-site controlled study reported decreased incidence of ICA-induced nephropathy after premedication with theophylline, sodium bicarbonate and hydration [37c]. However, further work is required to determine the protective effect of theophylline in ICA-induced nephropathy.

A major shortcoming of most studies that have been conducted to assess interventions to prevent ICA-induced nephropathy is that they have mainly focused on changes in creatinine levels. Larger clinical trials involving large number of patients with adequate power, which document persistent clinically significant outcome over a period of time such as decline in renal function, need for dialysis and death are required [38r]. In summary, ICA-induced nephropathy is a serious side effect and many risk factors that predispose individuals to ICA-induced nephropathy have been identified. Future work must focus on identifying effective biomarkers that will help in early diagnosis of ICA-induced nephropathy and premedications that will help reduce long-term deterioration in renal function.

GADOLINIUM-BASED CONTRAST AGENTS [SEDA-33, 968; SEDA-34, 754; SEDA-35, 866; SEDA-36, 701; SEDA-37, 588; SEDA-38, 497]

Gadolinium-based contrast agents (GBCA) play an important role in detecting a wide range of pathologic processes that would otherwise not be detected with unenhanced MRI or other imaging modalities. All GBCAs share a common structure and consist of a central gadolinium heavy metal ion that is kept in stable, safe and soluble configuration by a tightly bound peripheral organic ligand. Based on their biochemical structure GBCAs are classified as linear or macrocyclic, which are further subdivided according to their charge as ionic or non-ionic [1S]. Macrocyclic chelates are more stable than linear chelates and ionic linear GBCAs are more stable than nonionic linear GBCAs. Current FDA-approved GBCAs are shown in Table 3. Toxicity associated with GBCAs ranges from acute to chronic serious side effects such as nephrogenic systemic fibrosis and gadolinium deposition in different organs (mainly in the brain, bone and skin) [also see Fig. 3].

TABLE 3 United States Food and Drug Administration (USFDA) Approved Gadolinium-Based Contrast Agents (GBCAs)

Brand Name	Generic Name	Chemical Nature	Indication
Eovist® (0.25 mmoL/mL)	Gadoxetate disodium	Ionic Linear	MRI of liver to detect and characterize lesions in patients with known or suspected focal liver disease
Magnevist® (0.5 mmoL/mL)	Gadopentetate dimeglumine	Ionic Linear	Visualize disrupted blood–brain barrier or lesions with abnormal vascularity in the brain, spine head, neck and other parts of the body
MultiHance® (529 mg/mL)	Gadobenate dimeglumine	Ionic Linear	Visualize disrupted blood–brain barrier or lesions with abnormal vascularity in the brain, spine and associated tissue. Additionally, used in patients with suspected renal or aorto-ilio-femoral occlusive disease
Omniscan® (287 mg/mL)	Gadodiamide	Ionic Linear	Visualize disrupted blood–brain barrier or lesions with abnormal vascularity in the brain, spine and associated tissues. Additionally, used to visualize lesions with abnormal vascularity within the abdominal, thoracic, pelvic and retroperitoneal space
OptiMARK® (330.9 mg/mL)	Gadoversetamide	Non-ionic Linear	Visualize disrupted blood–brain barrier or lesions with abnormal vascularity in the brain, spine and associated tissues. Additionally, used to visualize lesions with abnormal vascularity within the liver
ProHance® (273.9 mg/mL)	Gadoteridol	Non-ionic Macrocyclic	Visualize disrupted blood–brain barrier or lesions with abnormal vascularity in the brain, spine, head and neck
Gadavist® (1 mmoL/mL)	Gadobutrol	Non-ionic Macrocyclic	Visualize disrupted blood–brain barrier or lesions with abnormal vascularity in the brain, spine, head and neck. Assess presence and extent of breast malignant tissues. To evaluate known or suspected supra-aortic or renal artery disease in adults or children (including term neonates)
Dotarem® (0.5 mmoL/mL)	Gadoterate meglumine	Ionic Macrocyclic	Visualize disrupted blood–brain barrier or lesions with abnormal vascularity in the brain, spine, head and neck

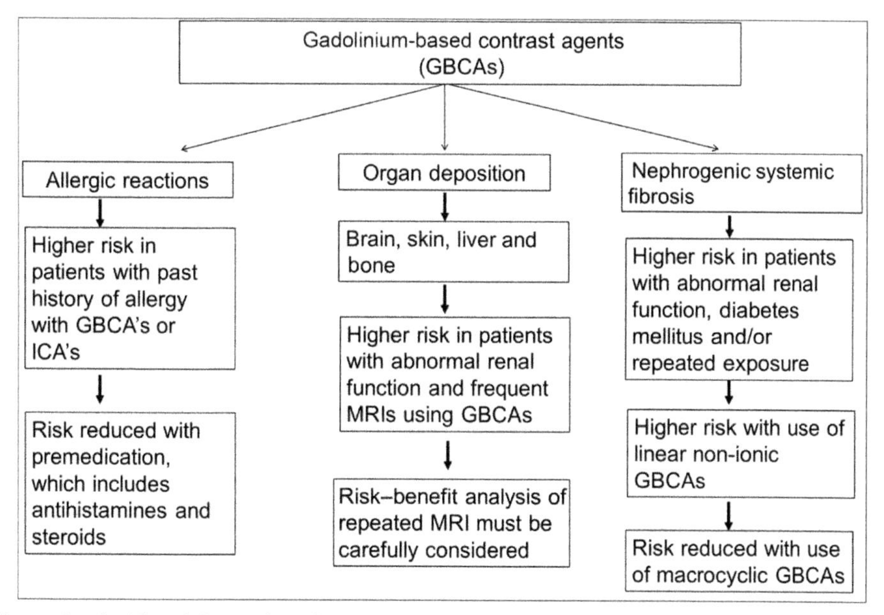

FIG. 3 Side effects (SE) associated with gadolinium-based contrast agents (GBCAs) and risk factors that increase occurrence of GBCA-induced SEs.

The incidence of acute side effects after administration of GBCA is low and ranges from 0.07% to 2.4% [1S, 39R, 40R]. Acute side effects of GBCAs can be categorized based on mechanism or severity. Based on mechanism, acute side effects include non-allergic and allergic type. A self-reported survey conducted in approximately 50 patients assessed the type and incidence of non-allergic acute side effects in patients who received a GBCA [41c]. All patients reported occurrence of side effects and most patients (66%) reported side effects immediately after GBCA exposure. However, some patients reported (33%) side effects up to 6 weeks after

GBCA exposure. The most common acute side effects reported included joint pain, headache, vision, and hearing changes. Other acute side effects reported included flu-like symptoms, skin changes, chest symptoms, gastrointestinal symptoms (nausea, vomiting, and diarrhea) and generalized whole body symptoms.

GBCA-induced allergic-type reactions include immediate hypersensitivity reactions and occur within seconds or minutes after injection and involve immune-mediated cellular response to whole or part of GBCA. The incidence of GBCA-induced anaphylaxis is very small (0.004%–0.01%). A severe anaphylactic life-threatening condition was reported after administration of a GBCA in a 48-year-old woman, who came in for a magnetic resonance cholangiopancreatography [42A]. The patient reported abdominal pain, acute onset diffuse maculopapular rash and hypotension within minutes of GBCA administration, suggesting an anaphylactic reaction. The patient responded to treatment with epinephrine, methylprednisolone, famotidine and diphenhydramine. The patient history was significant for an allergic reaction of unknown severity to a contrast agent during an intravenous pyelogram procedure approximately 15 years before without significant sequelae. Management of GBCA-induced acute side effects is similar to that of other contrast media and includes administration of adrenaline, corticosteroids, antihistamines, antacids, oxygen and hydration via intravenous fluid administration [39R].

Based on severity, the American College of Radiology classifies GBCA-induced side effects as mild, moderate or severe [1S]. Mild reactions are self-limited events, showing no significant progression and generally do not need medical treatment. In contrast, moderate reactions generally require medical treatment and/or admission to the emergency clinic. Finally, severe side effects are life-threatening and need emergency treatment. A recent retrospective analysis assessed acute side effects in 10608 Caucasian patients, who received a GBCA for non-CNS indications [43C]. The study reported 0.3% acute side effects in these patients, which occurred predominantly in women and in patients with a history of seasonal allergic rhinitis in the absence of any premedication. Amongst the total GBCA-induced side effects (0.3%), 75% of the side effects were mild, 12.5% were of moderate severity and 12.5% of the side effects were of severe nature. Among the six different GBCAs that were included in the study, the occurrence and severity of acute side effects were highest with MultiHance® (Gd-BOPTA), which is a linear ionic GBCA. Overall, the study concluded that the occurrence of acute side effects with GBCAs is low. Another retrospective study evaluated acute side effects associated with three GBCA agents: MultiHance® (Bracco Diagnostics, Inc., Princeton, NJ), Magnevist® (Bayer healthcare Pharmaceuticals, Wayne, NJ) and Gadavist® (Bayer healthcare Pharmaceuticals, Wayne, NJ) in 2393 pediatric patients [44C]. Of the 2393 patients, who received one of the above mentioned GBCA agents, approximately 40 patients reported acute side effects. A majority of these patients (30 patients) reported nausea and vomiting, the incidence of which was higher in non-sedated patients. Amongst the three agents, the incidence of nausea and vomiting was highest with MultiHance®. Overall, there was no significant difference in incidence of allergic or other acute side effects between the three GBCAs.

The incidence of GBCA-induced side effects especially acute side effects was higher in female patients compared to male patients and higher in patients with history of asthma, atopy, multiple allergies, multiple GBCA exposures, anaphylactic reactions to medications other than contrast agents, and patients who had experienced an allergic-reaction previously to GBCAs or ICAs [45M]. Interestingly, age did not determine either incidence or severity of acute GBCA-induced side effects. The incidence of GBCA-induced side effects can be reduced with use of premedication in patients with high risk of developing GBCA-induced side effects and/or history of GBCA-associated side effects [46R]. Premedication treatment can include antihistamines (H1 blockers), corticosteroids and/or acid reducers (H2 blockers). In summary, GBCAs are safe with low incidence of side effects, which can be mitigated by using appropriate premedication in patients at higher risk of developing GBCA-induced side effects.

Gadolinium Deposition: Brain, Skin, Bone and Liver

Deposition of GBCA in different organ systems in patients with renal dysfunction is well-known. Several reports suggest, however, deposition of GBCA in the brain, skin and bone in patients with normal renal function [40R, 47R, 48R, 49R, 50R]. In the brain, deposition of gadolinium after multiple exposure to GBCAs can occur in several areas such as the dentate nucleus, globus pallidus and pulvinar of the thalamus [51R]. The substantia nigra, which directs voluntary movement via the globus pallidus, is a region affected in Parkinson's disease. Therefore, a population-based study using a Canadian database assessed the effect of GBCA exposure on occurrence of Parkinson's disease [52r]. The study included 246557 patients (median age, 73 years; women 54.9%), who underwent at least one MRI for an indication other than those involving the brain and spinal cord. Approximately 40% of these patients had at least one GBCA exposure. The most common indication (non-brain and non-spine) for the GBCA-enhanced MRI was associated with abdominal abnormalities. Approximately 1.17% of these patients developed Parkinson's disease-like symptoms. This was not different (1.16%) from individuals, who underwent non-GBCA MRI. Overall, the study concluded that exposure to GBCAs does not increase the risk of developing Parkinson's disease. However, the study had some limitations such as lack of information on the specific type of GBCA used, lack of inclusion of younger patients, small number of patients who received more than four GBCA exposures and data from a single country. A case-study recently reported progressive increase in T1 signal intensity in the dentate nucleus and globus pallidus on unenhanced T1-weighted MR images in the brain of a pediatric patient after multiple GBCA exposures [53A]. The patient was a 13-year-old girl, who underwent a cervical spine MRI for arm-related complaints and was diagnosed to have a clival chordoma. The patient received gadopentetate dimeglumine and underwent six post-contrast brain MRI from 13 to 18 years of age. Hyperintensity in the dentate nucleus and globus pallidus was initially observed during unenhanced MRI after the 4th dose and was clearly observed after the 6th dose. This case report suggests that further research is required and a risk–benefit analysis of multiple exposure to GBCAs in pediatric patients must be conducted.

Similar to the brain, a case study reported deposition of gadolinium in the skin of a patient with normal renal function [54R]. This patient had undergone nearly 61 contrasted brain MRI scans over a period of 11 years. Skin biopsies from the forearm and lower extremity showed deposition of gadolinium within deep layers of the skin. The patient had no history of skin disease or nephrogenic systemic fibrosis but did have significant joint contractures of unknown etiology. The case study thus suggests that exposure to high cumulative doses of GBCAs can lead to gadolinium deposition in the skin even in patients with normal renal function. Skin deposition after multiple GBCA exposures has been previously reported in patients with nephrogenic systemic fibrosis [55c].

In addition to the brain and skin, gadolinium also deposits in the bone [40R, 56R]. A postmortem study reported deposition in both the brain and bone in nine decedents with normal renal function, who had a history of a single exposure to a macrocyclic GBCA and MRI during their lifetime. In fact, gadolinium levels deposited in the bone correlated with levels in the brain but were nearly 23 times higher than in the brain. Importantly, the study suggested that a single exposure to macrocyclic GBCA agent in patients with normal renal function could result in deposition of gadolinium in both the bone and brain and that bone deposition could serve as a surrogate for brain deposition. In summary, GBCAs can deposit in different organs irrespective of renal function status. The risk of GBCA deposition is higher with repeated exposure and children may be at higher risk of long-term effects of GBCA deposition.

GBCA-Induced Nephrogenic Systemic Fibrosis

GBCA-induced nephrogenic systemic fibrosis is an idiopathic adverse reaction of the skin characterized by thickening and hardening of the skin of the extremities and sometimes the trunk and was first reported in 1997 [57R, 58R]. Risk factors for development of GBCA-induced nephrogenic systemic fibrosis include repeated GBCA exposure, chronic renal disease and diabetes mellitus. The prevalence of GBCA-induced nephrogenic systemic fibrosis is higher after two or more exposures (approximately 36%) compared to a single exposure (12%), suggesting a cumulative effect. The incidence of GBCA-induced nephrogenic systemic fibrosis is higher with linear non-ionic agents like gadodiamide (Omniscan®, Gd-DTPA-BMA) and has dramatically decreased with increased use of macrocyclic GBCAs [57R, 58R].

A recently published retrospective study reported no incidence of nephrogenic systemic fibrosis in patients with chronic kidney disease when the GBCA was gadobenate dimeglumine [59C]. Nearly 3819 patients were included in the study and patients were followed for 501 days. Patients included in the study had severe chronic kidney disease and were undergoing either peritoneal or hemodialysis. Some of the limitations of the study included the fact that patients were enrolled from only one center, long-term monitoring was not carried out and a majority of the patients had multisystem disorders requiring a GBCA-enhanced MRI to make clinical decisions. A recently published case-study discussed a case of GBCA-induced nephrogenic systemic fibrosis in a 67-year-old male patient approximately 1 month after exposure to GBCA and MRI to rule out osteomyelitis of the foot [60A]. The patient presented with a painful nodular hardening of the skin on upper and lower limbs, gluteal region and the back and showed signs of acute kidney injury (serum creatinine >4.2 mg/dL). Additionally, the patient developed oropharyngeal muscle spasm, which required placement of a feeding tube. The lesion did not respond to hemodialysis and extracorporeal photopheresis.

Improvement in kidney function has been shown to improve GBCA-induced nephrogenic systemic fibrosis. A small retrospective study assessed the effects of renal transplantation and/or return of normal renal function in patients with acute kidney injury in patients diagnosed with GBCA-induced nephrogenic systemic fibrosis over a period of 15 years [61c]. The study reported that renal transplantation in patients with end-stage renal disease improved outcomes in patients with GBCA-induced nephrogenic systemic fibrosis. Similarly, return of renal function in patients with acute kidney injury also improved the symptoms of GBCA-induced nephrogenic systemic fibrosis. Importantly, no improvement in GBCA-induced nephrogenic systemic fibrosis was observed in patients, who did not show improvement in renal function. Overall, the study concluded that response to GBCA-induced nephrogenic systemic fibrosis correlated with return of renal function. In summary, the risk of GBCA-induced nephrogenic systemic fibrosis is high in patients with renal disease and outcomes in patients suffering from the disease improve after improvement in renal function. Thus in conclusion clinicians must be alert to the occurrence of GBCA-induced nephrogenic systemic fibrosis especially in patients with multiple exposures to GBCAs. Further, clinicians must explain to high-risk patients possible manifestations of the disease so that early intervention to prevent progression of the disease is possible. Finally, gadobenate dimeglumine may be a safer GBCA to use to avoid nephrogenic systemic fibrosis especially in high-risk patients.

In Utero

Gadolinium is to be used with caution during pregnancy and should be used only in conditions when the potential benefits exceed the potential risks. However, a recent prospective study assessed the potential side effects of *in utero* gadolinium exposure in high-risk premature infants [62C]. A total of 104 infants with gestational age between 24 and 33 weeks and who had at least one exposure to gadolinium were assessed for adverse outcomes in the study. The study reported that *in utero* gadolinium exposure was not associated with adverse neonatal outcomes. However, longitudinal studies are required to assess long-term neurotoxic and carcinogenic effects in these children before concluding that *in utero* use of gadolinium is safe.

Future Strategies to Prevent GBCA-Induced Side Effects

GBCA-induced side effects could be avoided by using zwitterion coated exceedingly small superparamagnetic iron oxide nanoparticles, which show high T1 contrast. An experimental study in rats and mice reported that zwitterion coated small superparamagnetic iron oxide nanoparticles were effective as contrast agents for both magnetic resonance angiography (MRA) and magnetic resonance imaging [63E]. No clinical studies have so far been conducted using this contrast agent in humans.

IRON OXIDE-BASED CONTRAST AGENTS [SEDA-33, 970; SEDA-34, 757; SEDA-37, 590; SEDA-38, 499]

Superparamagnetic iron oxide nanoparticles (SPIONs) have been approved by the FDA for treatment of anemia in adult patients with chronic renal disease [64R]. Ferumoxytol, due to its properties, can be used as a contrast agent for MRI as an alternative to GBCAs in patients with compromised renal function including renal failure [65R]. SPIONs could also be used for noninvasive diagnosis of chronic liver diseases (non-alcoholic fatty liver disease, nonalcoholic steatohepatitis, cirrhosis and liver tumors), magnetic resonance angiography (abdominal aortic aneurysms, assessment of renal artery stenosis and evaluation of endoleaks after stent graft repair of aneurysms), lymph node imaging, bone marrow imaging, and atherosclerotic plaque imaging [65R].

A recent retrospective study showed no difference in the occurrence of side effects between ferumoxytol and other intravenous iron products such as sodium ferric gluconate, iron sucrose or iron dextrose in patients being treated for iron deficiency anemia [66c]. The common side effects reported with ferumoxytol in this study were hypersensitivity and hypotension. Patients with chronic kidney disease were not included in the analysis. Another single-center study consisting of 217 patients reported no significant increase in side effects of ferumoxytol compared to gadofosveset trisodium amongst patients across a wide range of age, renal functions and indications [67c]. The most common side effect seen with ferumoxytol in this study was severe nausea.

A study evaluated the diagnostic role of ferumoxytol enhanced MR venography in children and young adults [$n=20$; age range = 3–21 years; mean weight 31 kg] with chronic kidney disease at high risk of developing GBCA-induced nephrogenic systemic fibrosis [68c]. Amongst the patients included were children with hemodialysis-dependent chronic kidney disease ($n=16$), children with functional renal transplants ($n=2$), children with acute kidney injury ($n=1$) and children with normal renal function ($n=1$). None of the patients had a documented allergy to intravenous iron replacement products and did not report any side effects. The study also reported that radiologists were able to diagnose with confidence and the diagnoses were confirmed with follow-up imaging or invasive procedures. The authors concluded that ferumoxytol could be used as an alternative for GBCAs for contrast enhanced MR venography in children with chronic kidney disease.

One challenge with use of SPIONs in patients with chronic liver disease is possibly that of iron overload. This could possibly be avoided or minimized by better design of SPIONs, which includes designing an ultrasmall iron oxide core (<8 nanometers), using coating materials that bind tightly to the iron oxide core, using zwitterions as part of the surface of the iron oxide core, minimizing serum protein binding via covalent modification of the coating polymer and finally modification of ultrasmall SPIONs for renal clearance of MRI [63E]. In conclusion, SPIONs can be used as an alternative to currently approved contrast agents for certain indications described above.

ECHOCARDIOGRAPHY/ ULTRASONOGRAPHIC CONTRAST MEDIA SIDE EFFECTS

Echocardiography is the major tool for cardiac imaging. Contrast agents in echocardiography are used in technically difficult studies in intensive care settings and in improving the accuracy of stress echocardiography. Additionally, contrast-enhanced echocardiography improves the ability to detect left ventricular thrombi and improves the accuracy of estimating left ventricular volume and ejection fraction. Currently three ultrasound/echocardiographic agents have been approved by the FDA and include: (1) Optison® (perfusion protein-type A microspheres; GE healthcare, UK), (2) Definity® (perflutren lipid microspheres, Lantheus Medical imaging, MA) and (3) Lumason® (sulfur hexafluoride lipid-type A microspheres, Bracco Diagnostics Inc., NJ). These agents should be used with caution in patients with a history of respiratory failure, severe emphysema or pulmonary emboli. The agents should be used in pregnant women only if absolutely necessary as the safety studies during pregnancy and their effects on the fetus are lacking. The incidence of allergic reactions with echocardiographic agents is low (approximately 1:10000) [69R]. Overall, these agents are generally safe and a very small number of side effects have been reported with these agents.

RADIOPHARMACEUTICALS [SEDA-15, 3017; SEDA-33, 973; SEDA-34, 758; SEDA-35, 869; SEDA-36, 703]

Radiopharmaceuticals are the compounds that contain radioisotopes emitting ionizing radiation and are used in treatment or modification of disease. Radiopharmaceuticals are used routinely in oncology either in the form of ionic complexes or as a part of the targeted peptide.

RADIUM ($^{223}R_a$)

Radium ($^{223}R_a$) is an alpha particle emitting radiopharmaceutical that forms complexes with bone mineral hydroxyapatite in areas of bone metastasis producing DNA destruction and cell death [70E]. The relative short path length (<100 micron) of alpha particles ensure minimum toxic effect to the surrounding tissue including bone marrow, making it one of the safest radiopharmaceuticals [71E]. Recently, a case of acute myelogenous leukemia (AML) following use of $^{223}R_a$ therapy in a patient suffering from prostate cancer was reported [72A]. A 69-year-old man with a metastatic castration resistant prostate cancer (mCRPC) and coronary artery disease was transferred to hospital following a 3-day history of abdominal pain and diarrhea. A complete blood count revealed pancytopenia and with a 40%–50% blast count which indicates AML. On closer scrutiny, the medical history revealed numerous therapies, including radiotherapy involving $^{223}R_a$. The AML could have possibly resulted from $^{223}R_a$ therapy ultimately causing accumulated bone marrow insult. The physicians (Varkaris et al.) could not establish the etiologic role of $^{223}R_a$ treatment in the AML with certainty because of numerous other simultaneous treatments. This recent report establishes a possible positive correlation between bone targeting radiopharmaceuticals and the development of AML. The clinicians need to be aware of this possibility and check for the signs of AML while using $^{223}R_a$ therapy.

YITTRIUM (^{90}Y)

Yittrium is a trivalent transitional metal used in medicine as a heavy isotope (^{90}Y) for radioembolization of different type of malignancies including renal carcinoma [73c], infiltrative hepatocellular carcinoma [74c], intrahepatic cholangiocarcinoma [75c], in bulk neuroendocrine tumors [76c], and other liver metastasis from breast [77c] and colorectal cancer [78c]. In most cases evidence from the small single site retrospective studies described above, the treatment was found to be safe with acceptable side effects. In one large retrospective study carried out recently at the School of Medicine, University of Alabama, Robertson and his colleagues found severe toxicity events (grade 3 or greater in toxicity scale as per Common Terminology Criteria for Adverse Events V4.0or CTAE) when ^{90}Y microspheres were used. In this retrospective study involving radioembolism with ^{90}Y resin microspheres in cases of unresectable primary and secondary hepatic malignancies, 21.5% of 79 treatments involving 58 patients were included in the study [79c]. Severe alkaline phosphatase (17.7%), albumin (12.7%) and bilirubin (10.1%) toxicities were the most common types. The group of physicians further identified positive correlation between certain conditions with the occurrence of severe toxicity to liver following radioembolism with Yittrium. Hepatic toxicity was correlated with decreases pre-treatment albumin (OR=26.2) and increase pretreatment international normalization ratio (INR, OR=17.7). Severe albumin toxicity was associated with increase pre-treatment levels of aspartate aminotransferase (AST, OR=7.4) and decreased pretreatment hemoglobin (OR=12.5). Increased pretreatment model for end-stage liver disease (MELD) score (OR=5.4) was correlated with severe total bilirubin toxicity. Colorectal adenocarcinoma histology was related to severe alkaline phosphatase toxicity (OR=5.4). This is a significant clinical finding because as much as 21.5% of the cases with Yittrium radioembolism show severe toxicity and the pathological conditions triggering the severe toxicity have been identified. The clinicians should check and carefully consider the pretreatment levels of albumin, INR, AST, hemoglobin, MELD and colorectal histology before selecting the patient for ^{90}Y microsphere therapy.

EMERGING PRODUCT ($^{99m}Tc\text{-}Al_2O_3$ NANOCOLLOID)

A new Russian radiopharmaceutical product namely $^{99m}Tc\text{-}Al_2O_3$ is available in nanocolloid form. The product contains nanocolloidal technitium adsorbed in alumina and coated with organic material. The $^{99m}Tc\text{-}Al_2O_3$ particles lose their organic coating during the passage along lymphatic vessels and get firmly fixed to sentinel lymph nodes (SLN) with no significant redistribution in the body. The relative short biological half-life ($t_{1/2}=6.02\,h$) coupled with restricted distribution of the product in the body helps to address some of the safety concerns associated with the radiopharmaceuticals. Yet, the product supplies sufficient penetrating power for radiometric measurement. The radioactive output of the product is 160–200 MBq per milliliter with the human dose of 0.007 mL/Kg (0.5 mL per person) [80E].

The acute, chronic and cumulative toxicity of $^{99m}Tc\text{-}Al_2O_3$ has been investigated in rats ($n=80$), mice ($n=180$) and Chinchilla rabbits ($n=12$). The results indicate that the product belongs to moderate-toxicity group as per toxicity classification. $^{99m}Tc\text{-}Al_2O_3$ shows no cumulative properties. No incidence of death occurred in the animals even after intraperitoneal dose of 50 and 25 mg/kg to rats and mice, respectively, and subcutaneous injection of 50 mL/Kg to both mice and rats. Microscopic and macroscopic investigation of the internal organs did not reveal any pathological changes. At the preclinical level of experimentation $^{99m}Tc\text{-}Al_2O_3$ seems to be a safe and effective radiological contrast agent to be used for lymph node visualization. Practitioners need to be aware of the existence and safety of $^{99m}Tc\text{-}Al_2O_3$, a novel radiological contrast agent, which is likely to be in human use in the foreseeable future.

References

[1] ACR. Manual on contrast media, version 10.3. American College of Radiology; 2017 [S].

[2] Li X, Liu H, Zhao L, et al. Clinical observation of adverse drug reactions to non-ionic iodinated contrast media in population with underlying diseases and risk factors. Br J Radiol. 2017;90:20160729 [C].

[3] Zhang B, Dong Y, Liang L, et al. The incidence, classification, and management of acute adverse reactions to the low-osmolar iodinated contrast media isovue and ultravist in contrast-enhanced computed tomography scanning. Medicine (Baltimore). 2016;95:e3170 [C].

[4] Motosugi U, Ichikawa T, Sano K, et al. Acute adverse reactions to nonionic iodinated contrast media for CT: prospective randomized evaluation of the effects of dehydration, oral rehydration, and patient risk factors. AJR Am J Roentgenol. 2016;207:931–8 [C].

[5] Sessa M, Rossi C, Rafaniello C, et al. Campania preventability assessment committee: a focus on the preventability of the contrast media adverse drug reactions. Expert Opin Drug Saf. 2016;15:51–9 [A].

[6] Zhai L, Guo X, Zhang H, et al. Non-ionic iodinated contrast media related immediate reactions: a mechanism study of 27 patients. Leg Med (Tokyo). 2017;24:56–62 [E].

[7] Lee SY, Yang MS, Choi YH, et al. Stratified premedication strategy for the prevention of contrast media hypersensitivity in high-risk patients. Ann Allergy Asthma Immunol. 2017;118:339–44 [C].

[8] Spina R, Simon N, Markus R, et al. Contrast-induced encephalopathy following cardiac catheterization. Catheter Cardiovasc Interv. 2017;90:257–68 [A].

[9] Tiu C, Terecoasa EO, Grecu N, et al. Transient global amnesia after cerebral angiography with Iomeprol: a case report. Medicine (Baltimore). 2016;95:e3590 [A].

[10] Balasingam S, Azman RR, Nazri M. Contrast media induced transient cortical blindness. QJM. 2016;109:121–2 [A].

[11] Studdard WE, Davis DO, Young SW. Cortical blindness after cerebral angiography. Case report. J Neurosurg. 1981;54:240–4 [A].

[12] Yazici M, Ozhan H, Kinay O, et al. Transient cortical blindness after cardiac catheterization with iobitridol. Tex Heart Inst J. 2007;34:373–5 [A].

[13] Mielke D, Kallenberg K, Hartmann M, et al. Paraplegia after contrast media application: a transient or devastating rare complication? Case report. J Neurosurg Spine. 2016;24:806–9 [A].

[14] Menegatti M, Pirisi M, Bellan M. An unusual adverse reaction to iodine-based contrast agent. Eur J Intern Med. 2016;32:e5–6 [A].

[15] Zhang G, Li Y, Zhang R, et al. Acute submandibular swelling complicating arteriography with iodide contrast: a case report and literature review. Medicine (Baltimore). 2015;94:e1380 [A].

[16] Afshar M, Alhussein M. Iodide-associated sialadenitis. N Engl J Med. 2017;376:868 [A].

[17] Barr ML, Chiu HK, Li N, et al. Thyroid dysfunction in children exposed to iodinated contrast media. J Clin Endocrinol Metab. 2016;101:2366–70 [C].

[18] Jarvis C, Simcox K, Tamatea JA, et al. A low incidence of iodine-induced hyperthyroidism following administration of iodinated contrast in an iodine-deficient region. Clin Endocrinol (Oxf). 2016;84:558–63 [c].

[19] Ma G, Mao R, Zhai H. Hyperthyroidism secondary to hysterosalpingography: an extremely rare complication: a case report. Medicine (Baltimore). 2016;95:e5588 [A].

[20] Chalikias G, Drosos I, Tziakas DN. Contrast-induced acute kidney injury: an update. Cardiovasc Drugs Ther. 2016;30:215–28 [R].

[21] Tao SM, Wichmann JL, Schoepf UJ, et al. Contrast-induced nephropathy in CT: incidence, risk factors and strategies for prevention. Eur Radiol. 2016;26:3310–8 [R].

[22] McCullough PA, Choi JP, Feghali GA, et al. Contrast-induced acute kidney injury. J Am Coll Cardiol. 2016;68:1465–73 [R].

[23] Homma K. Contrast-induced acute kidney injury. Keio J Med. 2016;65:67–73 [R].

[24] Subramaniam RM, Suarez-Cuervo C, Wilson RF, et al. Effectiveness of prevention strategies for contrast-induced nephropathy: a systematic review and meta-analysis. Ann Intern Med. 2016;164:406–16 [M].

[25] Rear R, Bell RM, Hausenloy DJ. Contrast-induced nephropathy following angiography and cardiac interventions. Heart. 2016;102:638–48 [R].

[26] Eng J, Wilson RF, Subramaniam RM, et al. Comparative effect of contrast media type on the incidence of contrast-induced

nephropathy: a systematic review and meta-analysis. Ann Intern Med. 2016;164:417–24 [M].

[27] Briguori C, Visconti G, Focaccio A, et al. Renal insufficiency after contrast media administration trial II (REMEDIAL II): renalguard system in high-risk patients for contrast-induced acute kidney injury. Circulation. 2011;124:1260–9 [C].

[28] Briguori C, Visconti G, Donahue M, et al. RenalGuard system in high-risk patients for contrast-induced acute kidney injury. Am Heart J. 2016;173:67–76 [C].

[29] Su X, Xie X, Liu L, et al. Comparative effectiveness of 12 treatment strategies for preventing contrast-induced acute kidney injury: a systematic review and Bayesian network meta-analysis. Am J Kidney Dis. 2017;69:69–77 [M].

[30] Liang M, Yang S, Fu N. Efficacy of short-term moderate or high-dose rosuvastatin in preventing contrast-induced nephropathy: a meta-analysis of 15 randomized controlled trials. Medicine (Baltimore). 2017;96:e7384 [M].

[31] Thompson K, Razi R, Lee MS, et al. Statin use prior to angiography for the prevention of contrast-induced acute kidney injury: a meta-analysis of 19 randomised trials. EuroIntervention. 2016;12:366–74 [M].

[32] Ali-Hasan-Al-Saegh S, Mirhosseini SJ, Ghodratipour Z, et al. Strategies preventing contrast-induced nephropathy after coronary angiography: a comprehensive meta-analysis and systematic review of 125 randomized controlled trials. Angiology. 2017;68:389–413 [M].

[33] Ye Z, Lu H, Guo W, et al. The effect of alprostadil on preventing contrast-induced nephropathy for percutaneous coronary intervention in diabetic patients: a systematic review and meta-analysis. Medicine (Baltimore). 2016;95:e5306 [M].

[34] Wei XB, Jiang L, Liu XR, et al. Brain natriuretic peptide for prevention of contrast-induced nephropathy: a meta-analysis of randomized controlled trials. Eur J Clin Pharmacol. 2016;72:1311–8 [M].

[35] Liu J, Xie Y, He F, et al. Recombinant brain natriuretic peptide for the prevention of contrast-induced nephropathy in patients with chronic kidney disease undergoing nonemergent percutaneous coronary intervention or coronary angiography: a randomized controlled trial. Biomed Res Int. 2016;2016:5985327 [R].

[36] Minoo F, Lessan-Pezeshki M, Firouzi A, et al. Prevention of contrast-induced nephropathy with oxygen supplementation: a randomized controlled trial. Iran J Kidney Dis. 2016;10:291–8 [c].

[37] Huber W, Huber T, Baum S, et al. Sodium bicarbonate prevents contrast-induced nephropathy in addition to theophylline: a randomized controlled trial. Medicine (Baltimore). 2016;95:e3720 [c].

[38] Weisbord SD, Palevsky PM. Prevention of contrast-associated acute kidney injury: what should we do? Am J Kidney Dis. 2016;68:518–21 [r].

[39] Thomsen HS. How to manage (treat) immediate-type adverse reactions to GBCA. Top Magn Reson Imaging. 2016;25:269–74 [R].

[40] Ramalho J, Ramalho M, Jay M, et al. Gadolinium toxicity and treatment. Magn Reson Imaging. 2016;34:1394–8 [R].

[41] Burke LM, Ramalho M, AlObaidy M, et al. Self-reported gadolinium toxicity: a survey of patients with chronic symptoms. Magn Reson Imaging. 2016;34:1078–80 [c].

[42] Pourmand A, Guida K, Abdallah A, et al. Gadolinium-based contrast agent anaphylaxis, a unique presentation of acute abdominal pain. Am J Emerg Med. 2016;34:1737:e1731-1732. [A].

[43] Granata V, Cascella M, Fusco R, et al. Immediate adverse reactions to gadolinium-based MR contrast media: a retrospective analysis on 10,608 examinations. Biomed Res Int. 2016;2016:3918292 [C].

[44] Neeley C, Moritz M, Brown JJ, et al. Acute side effects of three commonly used gadolinium contrast agents in the paediatric population. Br J Radiol. 2016;89:20160027 [C].

[45] Jung JW, Kang HR, Kim MH, et al. Immediate hypersensitivity reaction to gadolinium-based MR contrast media. Radiology. 2012;264:414–22 [M].

[46] Murphy KJ, Brunberg JA, Cohan RH. Adverse reactions to gadolinium contrast media: a review of 36 cases. AJR Am J Roentgenol. 1996;167:847–9 [R].

[47] Semelka RC, Ramalho J, Vakharia A, et al. Gadolinium deposition disease: initial description of a disease that has been around for a while. Magn Reson Imaging. 2016;34:1383–90 [R].

[48] Semelka RC, Ramalho M, AlObaidy M, et al. Gadolinium in humans: a family of disorders. Am J Roentgenol. 2016;207:229–33 [R].

[49] Ramalho J, Semelka RC, Ramalho M, et al. Gadolinium-based contrast agent accumulation and toxicity: an update. AJNR Am J Neuroradiol. 2016;37:1192–8 [R].

[50] Tedeschi E, Caranci F, Giordano F, et al. Gadolinium retention in the body: what we know and what we can do. Radiol Med. 2017;122:589–600 [R].

[51] Kanda T, Nakai Y, Oba H, et al. Gadolinium deposition in the brain. Magn Reson Imaging. 2016;34:1346–50 [R].

[52] Welk B, McArthur E, Morrow SA, et al. Association between gadolinium contrast exposure and the risk of parkinsonism. JAMA. 2016;316:96–8 [r].

[53] Roberts DR, Holden KR. Progressive increase of T1 signal intensity in the dentate nucleus and globus pallidus on unenhanced T1-weighted MR images in the pediatric brain exposed to multiple doses of gadolinium contrast. Brain Dev. 2016;38:331–6 [A].

[54] Roberts DR, Chatterjee AR, Yazdani M, et al. Pediatric patients demonstrate progressive T1-weighted hyperintensity in the dentate nucleus following multiple doses of gadolinium-based contrast agent. AJNR Am J Neuroradiol. 2016;37:2340–7 [R].

[55] Christensen KN, Lee CU, Hanley MM, et al. Quantification of gadolinium in fresh skin and serum samples from patients with nephrogenic systemic fibrosis. J Am Acad Dermatol. 2011;64:91–6 [c].

[56] Gibby WA, Gibby KA, Gibby WA. Comparison of Gd DTPA-BMA (Omniscan) versus Gd HP-DO3A (ProHance) retention in human bone tissue by inductively coupled plasma atomic emission spectroscopy. Invest Radiol. 2004;39:138–42 [R].

[57] Thomsen HS. Nephrogenic systemic fibrosis: a serious adverse reaction to gadolinium—1997-2006-2016. Part 2. Acta Radiol. 2016;57:643–8 [R].

[58] Thomsen HS. Nephrogenic systemic fibrosis: a serious adverse reaction to gadolinium—1997-2006-2016. Part 1. Acta Radiol. 2016;57:515–20 [R].

[59] Martin DR, Kalb B, Mittal A, et al. No incidence of nephrogenic systemic fibrosis after gadobenate dimeglumine administration in patients undergoing dialysis or those with severe chronic kidney disease. Radiology. 2018;286:113–9 [C].

[60] Koratala A, Bhatti V. Nephrogenic systemic fibrosis. Clin Case Rep. 2017;5:1184–5 [A].

[61] Wilson J, Gleghorn K, Seigel Q, et al. Nephrogenic systemic fibrosis: a 15-year retrospective study at a single tertiary care center. J Am Acad Dermatol. 2017;77:235–40 [c].

[62] Amin R, Darrah T, Wang H, et al. Editor's highlight: in utero exposure to gadolinium and adverse neonatal outcomes in premature infants. Toxicol Sci. 2017;156:520–6 [C].

[63] Wei H, Bruns OT, Kaul MG, et al. Exceedingly small iron oxide nanoparticles as positive MRI contrast agents. Proc Natl Acad Sci U S A. 2017;114:2325–30 [E].

[64] Finn JP, Nguyen KL, Han F, et al. Cardiovascular MRI with ferumoxytol. Clin Radiol. 2016;71:796–806 [R].

[65] Daldrup-Link HE. Ten things you might not know about iron oxide nanoparticles. Radiology. 2017;284:616–29 [R].

[66] Wetmore JB, Weinhandl ED, Zhou J, et al. Relative incidence of acute adverse events with ferumoxytol compared to other intravenous iron compounds: a matched cohort study. PLoS One. 2017;12:e0171098 [c].
[67] Nguyen KL, Yoshida T, Han F, et al. MRI with ferumoxytol: a single center experience of safety across the age spectrum. J Magn Reson Imaging. 2017;45:804–12 [c].
[68] Luhar A, Khan S, Finn JP, et al. Contrast-enhanced magnetic resonance venography in pediatric patients with chronic kidney disease: initial experience with ferumoxytol. Pediatr Radiol. 2016;46:1332–40 [c].
[69] Muskula PR, Main ML. Safety with echocardiographic contrast agents. Circ Cardiovasc Imaging. 2017;10 [R].
[70] Abou DS, Ulmert D, Doucet M, et al. Whole-body and microenvironmental localization of radium-223 in naive and mouse models of prostate cancer metastasis. J Natl Cancer Inst. 2016;108 [E].
[71] Henriksen G, Fisher DR, Roeske JC, et al. Targeting of osseous sites with alpha-emitting 223Ra: comparison with the beta-emitter 89Sr in mice. J Nucl Med. 2003;44:252–9 [E].
[72] Varkaris A, Gunturu K, Kewalramani T, et al. Acute myeloid leukemia after Radium-223 therapy: case report. Clin Genitourin Cancer. 2017;15:e723–6 [A].
[73] Kis B, Shah J, Choi J, et al. Transarterial yttrium-90 radioembolization treatment of patients with liver-dominant metastatic renal cell carcinoma. J Vasc Interv Radiol. 2017;28:254–9 [c].
[74] McDevitt JL, Alian A, Kapoor B, et al. Single-center comparison of overall survival and toxicities in patients with infiltrative hepatocellular carcinoma treated with Yttrium-90 radioembolization or drug-eluting embolic transarterial chemoembolization. J Vasc Interv Radiol. 2017;28:1371–7 [c].
[75] Jia Z, Paz-Fumagalli R, Frey G, et al. Resin-based yttrium-90 microspheres for unresectable and failed first-line chemotherapy intrahepatic cholangiocarcinoma: preliminary results. J Cancer Res Clin Oncol. 2017;143:481–9 [c].
[76] Kong G, Callahan J, Hofman MS, et al. High clinical and morphologic response using (90)Y-DOTA-octreotate sequenced with (177)Lu-DOTA-octreotate induction peptide receptor chemoradionuclide therapy (PRCRT) for bulky neuroendocrine tumours. Eur J Nucl Med Mol Imaging. 2017;44:476–89 [c].
[77] Pieper CC, Meyer C, Wilhelm KE, et al. Yttrium-90 radioembolization of advanced, unresectable breast cancer liver metastases—a single-center experience. J Vasc Interv Radiol. 2016;27:1305–15 [c].
[78] Kalva SP, Rana RS, Liu R, et al. Yttrium-90 radioembolization as salvage therapy for liver metastases from colorectal cancer. Am J Clin Oncol. 2017;40:288–93 [c].
[79] Roberson Ii JD, McDonald AM, Baden CJ, et al. Factors associated with increased incidence of severe toxicities following yttrium-90 resin microspheres in the treatment of hepatic malignancies. World J Gastroenterol. 2016;22:3006–14 [c].
[80] Varlamova NV, Churin AA, Fomina TI, et al. Studying the general toxicity and cumulative properties of a radiopharmaceutical nanocolloid, $^{99m}Tc-Al_2O_3$. Bull Exp Biol Med. 2016;161(3):371–3 [E].

Further Reading

[81] Hasegawa M, Gomi T. Chapter 45 - Radiological contrast agents and radiopharmaceuticals. In: Ray SD, editor. Side Effects of Drugs: Annual: A worldwide yearly survey of new data in adverse drug reactions. Elsevier. 2017;37:583–93 [R].
[82] Hasegawa M, Gomi T. Chapter 44 - Radiological contrast agents and radiopharmaceuticals. In: Ray SD, editor. Side Effects of Drugs: Annual: A worldwide yearly survey of new data in adverse drug reactions. Elsevier. 2016;38:493–501 [R].
[83] Hasegawa M, Gomi T. Chapter 41 - Radiological contrast agents. In: Ray SD, editor. Side Effects of Drugs: Annual: A worldwide yearly survey of new data in adverse drug reactions. Elsevier. 2017;39:483–9 [R].

CHAPTER

47

Drugs Used in Ocular Treatment

Lisa V. Stottlemyer,[1], Victoria L. Dzurinko*†*

*Wilmington VA Medical Center, Wilmington, DE, United States

†The Pennsylvania College of Optometry, Salus University, Elkins Park, PA, United States

[1]Corresponding author: lisa.stottlemyer@va.gov

INTRODUCTION

Many different classes of drugs are utilized for the treatment of ocular disease and all pose risk of side effects, some severe. Ocular medications can be administered topically, orally or injected into the eye, and each route allows for potential of local and systemic side effects. For example, the most common side effects of any topically applied eye drops are burning upon installation and conjunctival hyperemia, which may be related to either the therapeutic agent or preservatives [1M], while medication injected into the vitreous cavity often leaves the patient reporting floaters or a spot in their vision as they can appreciate the shadow of the depot of drug [2A].

By far, topical application is the most common route of administration of medications used for the treatment of ocular disease and some conditions, such as glaucoma, require frequent, chronic dosing. Often the preservatives therein contribute ocular surface disease [1M]. Symptoms such as redness, stinging, burning and foreign body sensation are reported to be present in up to 60% of patients using topical anti-glaucoma medications [3A].

There are several classic side effects for each class of medications that are so prevalent, we almost expect them. For example, beta blockers classically cause bradycardia even when topically applied, and steroids are well known to cause increased IOP and posterior sub-capsular cataracts. Systemic side effects of topically applied ocular medications have been attributed to the "first pass" effect that up to 80% of topically applied medication drains through the nasolacrimal system and is absorbed into the bloodstream through the nasal mucosa. By this route, organ systems are exposed to concentration of drug prior to the drugs first pass through the liver where many are metabolized [4A, 5A]. Infants and children may be more at risk for serious systemic side effects due to the fact that there is often no weight adjusted dose available and struggling with an uncooperative child may lead to a greater number of drops instilled [5A]. This chapter is meant to serve as a review of both the common and lesser occurring side effects of ocular medications reported in the literature.

Medications Used to Treat Glaucoma

Topical therapy is the mainstay of glaucoma management. Life threatening adverse events including central nervous system depression and cardiogenic shock are described in two infants treated with topical antiglaucoma medications (brimonidine and timolol) [5A]. These cases highlight the fact that systemic absorption and lack of weight adjusted dosages likely subject infants to greater risk of systemic side effects.

Benzalkonium chloride (BAK) is the most commonly use preservative for ocular medications and is associated with conjunctival inflammation and corneal compromise [6C]. BAK interferes with the lipid component of the tear film which results in a more rapid tear breakup time (TBUT) [7c] and therefore frequent dosing of BAK containing glaucoma medications are well-known to increase the symptoms of ocular surface disease (OSD).

Topical medication related ocular surface disease (OSD) symptoms often lead to reduced compliance with prescribed regimens [8c] and yet, often there is a disconnect between the clinical severity of OSD and the symptoms reported by patients. Romero-Diaz de Leon

Side Effects of Drugs Annual, Volume 40
ISSN: 0378-6080
https://doi.org/10.1016/bs.seda.2018.07.003

et al. explained this lack of correlation is related to reduced corneal and conjunctival sensitivity with chronic topical glaucoma regimens. A case controlled study ($n=182$ eyes) reported a mean corneal sensitivity of 58.98±2.25mm in the control group vs 52.97±6.41mm in patients using topical medication. Mean conjunctival sensitivity was 18.80±5.40mm in the control group and 11.76±5.45mm in the treated group. Eyes on regimens containing timolol showed greater reduction of sensitivity values when compared to other topical glaucoma agents [9C].

PROSTAGLANDIN ($PGF_{2\alpha}$) ANALOGUES

Latanoprost, Bimatoprost, Travoprost, Tafluprost

GENERAL

Prostaglandin analogue (PGA) side effects are local and involve conjunctival hyperemia, increased lash growth, periocular skin pigmentation and increased iris pigmentation, iris cysts, cystoid macular edema and prostaglandin-associated periorbitopathy (PAG) consisting of deepening of the upper eyelid sulci, sunken eye appearance and periorbital fat atrophy [10A, 11A]. Hyperemia is the most commonly noted side effect of PGAs. A systematic review and meta-analysis of 72 randomized controlled trials reporting data on safety and efficacy of glaucoma treatments was performed by Li et al., who reported the highest risk of hyperemia occurred in patients treated with PGAs when compared to patients treated with beta blockers and dual therapy regimens that did not contain PGAs [12M]. No systemic adverse events have been attributed to prostaglandin therapy [6C].

Abedi et al. report a case of prostaglandin-associated periorbitopathy in a 73-year-old woman treated with 0.004% Travatan both eyes for 7 years that resulted in an audible clicking in both eyes with every blink [10A]. Fong et al. report a similar case of audible clicking sounds on blinking in a 72-year-old woman treated with bimatoprost 0.3% in both eyes for 3 years followed by combination therapy travoprost 0.004%/timolol 0.5% for 2 years [11A]. Both women in these reports exhibited enophthalmos with anterior displacement of the lateral canthus leaving a gap for air to collect between the lateral canthi and the globes [10A, 11A].

Prostaglandin analogues have also been reported to cause anterior uveitis and CME presumably because the prostaglandin activity causes a pro-inflammatory state [13M]. A prospective study ($n=121$ eyes) evaluated anterior chamber flare and central macular thickness (CMT) in patients treated with topical glaucoma therapy and reported a statistically significant increase in flare in the bimatoprost and latanoprost treated groups while no statistically significant flare or increase in CMT values occurred in the timolol maleate and control groups. The authors note that while flare and increased CMT occurred on prostaglandin therapy, these changes were not associated with uveitis or clinically evident macular edema, and therefore were determined not to be clinically significant [13M]. While reports of CME associated with topical PGA use have been reported in patients with other accepted risk factors for CME such as complicated cataract surgery, diabetes epiretinal membrane etc., Makri et al. report a case of a 65-year-old Caucasian female who developed CME 7 months after initiation of preservative-free latanoprost 0.005% in an eye with a history of uncomplicated phacoemulsification surgery 19 months prior. The patient's vision was reduced to 20/30, OCT showed cystic spaces in the macula and fluorescein angiogram confirms petaloid leakage from the parafoveal capillaries which promptly resolved upon discontinuation of latanoprost (replaced with brinzolamide 1.0% BID) and initiation of Nevanac (1mg/mL) TID; vision returned to 20/20 1 month later [14A].

A commonly known side effect of PGAs is increased periocular skin pigmentation; however, Lin et al. report a case of paradoxical deep pigmentation of periocular skin lightening that occurred in a 55-year-old African-American man treated with 0.005% latanoprost that was initiated in both eyes at different times and occurred within 1 year of PGA treatment [15A]. Depigmentation improved upon discontinuation, but did not fully resolve.

Mohite et al. report a case of a 62-year-old Caucasian male with a history of bilateral laser peripheral iridotomies (LPIs) who developed bilateral iris pigment epithelial and ciliary body cysts 41 months after initiation of treatment with latanoprost. The patient was referred to an ocular oncologist who diagnosed benign secondary iris and choroidal cysts related to topical prostaglandin therapy. Latanoprost was discontinued and replaced with brinzolamide. The cysts slowly resolved over a 9 month period [16A].

Reversible reductions in corneal hysteresis (CH), corneal resistance factor (CRF) and central corneal thickness (CCT) are reported with chronic use of topical PGAs [17c]. Schrems et al. performed a retrospective single center study ($n=130$) aimed at evaluating the rates of change per year of CCT after glaucoma drug administration and reported a statistically significant decrease in CCT for eyes treated with prostaglandin monotherapy (−3.1μm/year) [17c]. A retrospective study ($n=46$) of patients with normal tension glaucoma reported a significant reduction of CCT during treatment with latanoprost

(544.4 ± 35.8 μm vs 531.4 ± 32.5 μm) which fully reversed over 2 years after discontinuation of the drug [18c].

Topical PGAs are also implicated in the development of meibomian gland dysfunction (MGD), a condition that leads to tear film instability and ocular surface disease. Mocan et al. performed a prospective cross-sectional study in which 70 eyes treated with topical glaucoma medications for greater than 12 months were separated into PGA and non-PGA subgroups and showed increased MGD in PGA subgroup when compared to non-PGA subgroup (92% vs 58%) [19c]. The authors suggest that the mechanism is likely related to the keratinization of the meibomian glands in response to exposure to PGA, much like the process involved in hypertrichosis and hyperpigmentation of periocular skin [19c].

Beta-Blockers

While, and localized side effects of beta blockers are in keeping with the other classes of glaucoma medications (predominantly burning upon installation of drops and conjunctival hyperemia), the prevalence of systemic effects is a frequent limitation of these agents [1M]. Beta blockers effect on heart rate and respiration is widely reported and includes bradycardia, blood pressure decreases, worsening of asthma attacks and chronic obstructive pulmonary disease (COPD). Morales et al. performed a population based, case controlled study matching moderate and severe asthma exacerbations in patients on topical beta blockers with asthma ($n = 4865$) and followed with meta-analysis of clinical trials that evaluated changes in lung function associated with beta blocker treatment [20M]. This study concluded that despite known risk and availability of safer treatment options, 14% of people with asthma and ocular hypertension were prescribed a non-selective beta blocker eyedrop which led to a 4.8-fold increase in relative incidence of moderate asthma exacerbations [20M].

Timolol

Nanda et al. report a case series of four patients (ranging in age from 66 to 93 years) that experienced visual hallucination with topical ocular timolol 0.5% use. The described visual phenomenon was similar in all four cases and involved colorful geometric patterns and/or change in color vision that resolved upon discontinuation of timolol and returned with re-trial of the drug [21A].

CDP2D6, a polymorphically expressed enzyme, is responsible for 90% of timolol metabolism in the liver therefore patients lacking functional CDP2D6 enzyme or those concurrently taking CDP2D6 inhibitors for comorbidities will experience greater timolol serum concentration and are at greater risk for serious cardiac adverse events (i.e. bradycardia, syncope, conduction defects and falls) [22R].

Carbonic Anhydrase Inhibitors (CAI)

A single dose of acetazolamide is frequently prescribed by cataract surgeons and to prevent an acute rise of IOP postoperatively. A case report by Grigera and Grigera details the onset of bilateral ciliary effusion causing myopic shift and appositional angle closure that occurred in a 49-year-old female within hours of dosing 500 mg of acetazolamide and promptly resolved with discontinuation of the drug along with cycloplegic and topical steroid therapy [23A].

Sympathetic alpha2-Receptor Antagonists

Local ocular side effects of sympathomimetic agents include conjunctival hyperemia, irritation, pupil dilation and allergic conjunctivitis [1M].

RHO KINASE (ROCK) INHIBITORS

Rho kinase inhibitors are a new class of topical anti-glaucoma therapy presently in development. ROCKET-1 and ROCKET-2 (phase 3, double masked, randomized, multicenter, parallel group studies) that evaluated the safety and efficacy of netarsudil 0.02%, and reported most common side effects were conjunctival hyperemia (which occurred in 50%–59% of patients), conjunctival hemorrhage and cornea verticillata (whorl deposits). In ROCKET-2, conjunctival hemorrhage occurred in 15% (37/251) of netarsudil 0.02% QD and 17% (43/253) netarsudil BID patients; these hemorrhages were deemed to be treatment related 43% and 47% of the time, respectively. In ROCKET-2, cornea verticillata were present in 9% (22/251) and 15% (37/253) of netarsudil QD vs BID treated patients, respectively. In ROCKET-1, cornea verticillata were reported in 5.4% (11/203) of netarsudil QD treated group; the onset of corneal deposits ranged from 6 to 13 weeks, had no effect on visual acuity and all resolved upon cessation of the drug [24C].

Mydriatics/Cycloplegics

Mydriatics and cycloplegics cause pupillary dilation and congestion of the anterior chamber angle yielding increased risk for acute angle closure in at risk patients. Additional side effects include potential central nervous system toxicity.

Cyclopentolate

Oner, Bulut and Oter studied ($n=74$) the effects of cyclopentolate 1% and tropicamide 1%, dosed three times with 10-min intervals, and reported statistically significant increase from baseline in subfoveal choroidal thickness in the cyclopentolate group only; no choroidal thickening was observed in the tropicamide treated group [25C].

Atropine

Gizzi et al. report a case of a 45-year-old white man, with history of long-term use of atropine 1% gel q 5 days since early childhood, who presented with pigment dispersion syndrome and elevated IOP in both eyes. It is postulated that because atropine binds to melanin in the iris, chronic exposure with long-term use may lead to damage to the iris pigment epithelium (IPE) making it more fragile, while at the same time hyperplasia of the iris dilator may increase iris thickness making contact with lens zonules more likely [26A].

Cyclomidril (0.2% Cyclopentolate and 1% Phenylephrine Combination)

Cardiopulmonary arrest was reported in a pre-term infant after topical administration of cyclomidril to effect dilation for a ROP screening [4A]. Three sets of eye drops were administered at 15-min intervals. Fifteen minutes after the third set, the baby became unresponsive. The baby was resuscitated but experienced a second event 3h later. Follow-up ROP screenings with 1.0% tropicamide and 2.5% phenylephrine (instead of cyclomydril) were uneventful [4A].

Phenylephrine

Lee and Nguyen describe a case of catecholamine induced inflammation of the myocardium in a 53-year-old Vietnamese woman who developed chest pain immediately after an intraocular injection of 10% phenylephrine for the treatment of anterior uveitis. She was hypotensive, with acute pulmonary edema and left ventricular systolic dysfunction with apical akinesis noted on transthoracic echocardiogram. Left ventriculogram findings confirmed diagnosis of Takotsubo cardiomyopathy despite the lack of acute psychological stressors typically associated with this condition. Her ocular treatment continued with dexamethasone only eye drops and oral ACE inhibitor and beta blocker treatment was initiated to address her systolic dysfunction. Her ventricular function returned to normal when reevaluated 3 weeks later [27A].

Anti-Inflammatory Agents

Corticosteroids

PREDNISOLONE, BETAMETHASONE, DEXAMETHASONE, TRIAMCINOLONE, FLUOCINOLONE ACETONIDE

It is widely known that topical corticosteroids frequently cause elevated intraocular pressure, decreased corneal healing and posterior sub-capsular cataracts.

Cushing's syndrome is a severe adverse reaction associated mainly with long-term, high dose systemic glucocorticoid treatment; association with topical treatment is rare [28A]. Two separate case reports detail development of iatrogenic Cushing's disease in young patients treated with topical glucocorticoids. Fukuhara et al. describe a case of a 9-year-old girl treated for bilateral iridocyclitis with betamethasone sodium phosphate 0.1% eye drops up to six times per day for 6 months duration that presented to her pediatrician with purple skin striae, truncal obesity, buffalo hump and moon face. Laboratory testing confirmed low endogenous cortisol and ACTH levels. Oral methotrexate treatment was initiated to allow tapering of the steroid eyedrops and her clinical signs and symptoms resolved over the next 6 months [28A]. Similarly, Rainsbury et al. report a case of a 15-year-old HIV-infected male treated for severe sight threatening vernal keratoconjunctivitis (VKC) with topical dexamethasone 0.1% eyedrops BID who, in response to a large corneal epithelial defect, was prescribed increased to hourly dosing of the dexamethasone drops and 2 weeks later developed moon-shaped face, change in mood and reduced serum cortisol. A rapid taper of the topical steroid as well as change in his anti-retroviral therapy (ART) yielded complete resolution of Cushing's associated facial features and improved serum cortisol levels within 12 weeks [29A].

Intravitreal injections and implants offer high potency and sustained dose of corticosteroid and major concerns with use are elevated intraocular pressure and cataract formation, among other effects. According to the FAME randomized controlled trial, the most common reported adverse event of fluocinolone acetonide (FAc) 190μg intravitreal implant was cataract surgery occurring 75%–85% in the treated group vs 23% of the sham group after 24 months of follow-up [30C].

Local intraocular immune suppression is suggested as causative for reactivation of viral eye disease in a case of, previously quiescent, HSV keratitis that developed in a 90-year-old man 3 weeks after receiving a dexamethasone implant (DEX) for the treatment of macular edema occurring with branch retinal vein occlusion [31A].

Central serous chorioretinopathy (CSCR) has been reported with topical and systemic steroid use. Eris et al.

now attribute a case of bilateral CSCR in a 46-year-old male to bilateral treatment with 0.7 mg dexamethasone intravitreal implants used for the treatment of diabetic macular edema as onset of serous macular detachment coincided with peak time of dexamethasone release and resolved slowly over the following months [32A].

A 48-year-old reported a large, linear "floater" in his central vision shortly after the FAc injection due to unintended positioning of the implant in the middle of the vitreous cavity along the visual axis. After weeks of observation without change, Nd:YAG vitreolysis severed the vitreous attachment allowing the implant to fall away from visual axis [33A].

A single center, retrospective, consecutive, interventional case series of patients undergoing intravitreal triamcinolone acetonide (IVTA) ($n=81$) reported a 12.3% incidence of presumed sterile endophthalmitis [34c].

Anti-Vascular Endothelial Growth Factor (VEGF) Medications

Pegaptanib, Bevacizumab, Aflibercept, Ranibizumab, Conbercept

GENERAL

Anti-VEGF medications are injected into the vitreous cavity through the pars plana and are used for the treatment of many conditions including age-related macular degeneration (AMD), central retinal vein occlusion (CRVO), diabetic macular edema and proliferative diabetic retinopathy, and retinopathy of prematurity (ROP). Reported adverse events following anti-VEGF injections include vitreous hemorrhage, retinal tear, retinal detachment, macular hole formation, endophthalmitis, intraocular inflammation, increased intraocular pressure, and retinal vascular occlusions [35R] as well as systemic events such as blood pressure elevation, myocardial infarction, stroke, and death.

Intraocular air bubbles have been identified as a transient side effect of intravitreal injection of anti-VEGF. A retrospective review of the medical records of 148 infants undergoing intravitreal ranibizumab, aflibercept, or bevacizumab showed an incidence of 3.37%. A temporary increase in intraocular pressure was also seen, but no sterile or infectious endophthalmitis occurred. Air bubbles spontaneously resolved within 72 h without intervention [36c].

Ocular hypertension is an emerging complication from intravitreal anti-VEGF treatments. Immediate rise in IOP can be linked to the volume of medication injected into the eye and is typically of limited duration and asymptomatic. Studies to date have shown little conclusive evidence of the benefits of prophylactic treatment with anti-hypotensive medications. Ocular hypertension can become chronic in some cases. The exact mechanism for this is unknown, but theories include trabecular meshwork obstruction by particles within the medication or syringe, a toxic or inflammatory response from the trabecular meshwork, or some other effect on aqueous outflow. The number of treatments, an increase in the frequency of injections, as well as a pre-existing diagnosis of glaucoma is risk factor for chronic elevated IOP associated with anti-VEGF treatment. The risk of vision loss from untreated macular pathology should be weighed against the risk of vision loss from glaucoma when deciding whether to discontinue anti-VEGF therapy in the presence of ocular hypertension [37R].

A retrospective study at the Tays Eye Centre analyzed the rate of endophthalmitis in 1349 eyes in 1117 patients undergoing 11 562 treatments with intravitreal anti-VEGF medications. The incidence was 0.086%, concurring with previous large studies. Vision returned to at least baseline in 7 out of 10 of the patients in this study [38R].

The Pan-American Collaborative Retina Study Group (PACORES) conducted a retrospective multicenter, interventional case series ($n=81$) and reported an incidence of 5.2% of eyes with proliferative diabetic retinopathy treated with bevacizumab developed a tractional retinal detachment, thought to result from contraction of fibrovascular tissue following a brisk halt to proliferating neovascular vessels, that required vitrectomy [39c].

A systematic review of all intravitreal anti-VEGF injections performed at a tertiary care clinic in Singapore evaluated both the ocular and systemic adverse events associated with this increasingly utilized procedure. A total of 14 001 intravitreal injections were performed on 2225 patients from January 1, 2007 to December 31, 2014. The distribution of ant-VEGF drugs used was as follows: 9992 bevacizumab (71.4%), 3306 ranibizumab (23.6%) and 703 aflibercept (5.0%) injections. Ocular complications were low with only one (0.007%) patient experiencing endophthalmitis, three (0.021%) developing traumatic cataracts, and one (0.007%) suffering a retinal detachment [40C].

Results from the Bevacizumab-Ranibizumab International Trials Group revealed no large disparity in the occurrence of serious adverse events for patients being treated with either drug during larger-scale randomized trials. When evaluating individual patient data involving 5 trials and 3052 patients, the adjusted relative risk of utilizing bevacizumab vs ranibizumab was 1.06 for developing ≥ 1 serious adverse event. The relative risk of secondary outcomes, including death, arteriothrombotic events, events associated with systemic anti-VEGF therapy, and events not associated with anti-VEGF therapy were 0.99, 0.89, 1.10, and 1.11, respectively [41MC].

Bevacizumab

A prospective study out of the Shinshu University School of Medicine in Japan looked at the suppression of plasma VEGF at 1 and 4 weeks post intravitreal injection of bevacizumab, aflibercept and ranibizumab ($n=42$). Patients undergoing treatment with intravitreal bevacizumab or aflibercept continued to demonstrate significant reductions from baseline in measured plasma VEGF at weeks 1 and 4 post injection; but no similar plasma reduction was seen in patients receiving ranibizumab. Bevacizumab, a full-length monoclonal antibody, has a long systemic half-life of approximately 20 days when administered intravenously. Aflibercept is a soluble fusion protein with a plasma half-life of 5–6 days. Both medications have an Fc antibody region which allows for endothelial binding to the neonatal Fc receptor and subsequent recycling remaining in circulation. As ranibizumab lacks this region, its systemic half-life is much shorter, at only 2h [42c].

Large scale studies of this emerging ROP treatment are needed to evaluate systemic toxicity given this developmentally fragile patient base; two trials are presently underway, one of which the Pediatric Eye Disease Investigative Group (PEDIG) is assessing some safety outcomes with bevacizumab [43R].

In a study by Huang et al., 14 patients with ROP were treated with intravitreal aflibercept or bevacizumab (5 and 9 patients, respectively). All that underwent treatment showed complete resolution of abnormal vascular proliferation due to ROP. Serum VEGF was measured utilizing enzyme-linked immunosorbent assays at 2, 4, 8 and 12 weeks post treatment. While all patients demonstrated reduced serum VEGF levels post injection, patients treated with bevacizumab showed more pronounced systemic suppression of VEGF after 3 months of treatment as compared to those managed with aflibercept [44c].

Kana et al. published a retrospective analysis of 23 patients in a South African hospital who underwent intravitreal bevacizumab as a first line treatment for retinopathy of prematurity. Birthweight ranged from 810 to 1480g. Data were available for 22 patients (43 eyes). With a mean follow-up period of 9 months, 41 eyes (95.3%) demonstrated complete regression or stabilization of ROP, while 2 eyes demonstrated progression. In this series, no adverse events were reported [45c].

In a series of infants ($n=16$) diagnosed with stage 3+ ROP at <27 weeks gestational age, the BEAT-ROP multicenter trial evaluated medical and neurodevelopmental outcomes after treatment including weight, length, head circumference, occurrence of cerebral palsy, and Bayley scores at 2 year follow-up. At 18–28 months corrected age, the growth percentiles and Bayley developmental scores were all reduced relative to age matched normal, healthy infants. There were no statistically significant differences in the measured developmental outcomes. However, infants receiving bevacizumab ($n=7$) required a median of 42 fewer days in the hospital vs those treated with laser ($n=9$) [46c].

Ranibizumab

The TREND study group recently completed a 12 month, phase 3b multicenter, randomized interventional study to determine the efficacy and safety of ranibizumab 0.5mg in a treat and extend (T&E) schedule as opposed to treating monthly. A total of 650 (T&E=323, monthly=327) patients were part of the study, with a safety set of 649 (T&E=323, monthly=326) patients ultimately available for analysis. Ocular adverse events (AEs) occurred in 36.4% of the study patients and non-ocular events were reported by 47.9% of the participants. Ocular AEs included elevated IOP (T&E=8.4%, monthly=8.6%), subconjunctival hemorrhage (T&E=4.3%, monthly=5.8%), and reduced VA (T&E=4.6%, monthly=3.7%), while non-ocular AEs included nasopharyngitis (T&E=5.6%, monthly=8.0%), hypertension (T&E=7.1%, monthly=4.0%), influenza (T&E=2.8%, monthly=3.7%), and bronchitis (T&E=2.5%, monthly=3.7%). Incidence of ocular serious adverse events (SAEs) was the same between the two treatment regimens (1.2% each), while non-ocular SAEs occurred 11.4% of patients overall, but differed between the two treatment groups (T&E=11.1%, monthly=11.7%). Only 1 patient (0.3%) from the monthly treatment group suffered from a serious ocular SAE (endophthalmitis). Five patients in the T&E group suffered non-ocular SAEs including lack of drug efficacy ($n=2$), and 1 patient each reported blindness, headache, recurrent neovascular AMD, retinal hemorrhage, and subretinal fluid. For those undergoing the monthly treatment 2 patients reported drug ineffectiveness and one suffered a macular hole. Three deaths occurred in the T&E groups and 4 in the monthly group; however, these were considered unrelated to the treatment per the study investigators. Overall, these findings are in line with what has been reported in the literature previously [47MC].

A review by Michael Steward involved several clinical trials including RISE and RIDE, RESTORE, and DRCR.net Protocols I and T for efficacy of ranibizumab and occurrence of adverse events. Ranibizumab was the first anti-VEGF approved for the treatment of patients suffering from DME in isolation and DME and proliferative retinopathy presenting simultaneously. The safety profile of ranibizumab was considered "excellent" with reported incidences of ocular and systemic AEs being between 1% and 2% overall. Occurrence of endophthalmitis has been decreasing as physicians have become more adept at performing intravitreal injections. Given the increased risk of cardiac events in the diabetic patient population, adverse events involving emboli need to be evaluated with this framework in mind. When comparing the incidence of hypertensive and arterial thromboembolic events in the RESOLVE trial, no significant difference was seen between the ranibizumab and sham patient sets. While the Antiplatelet Trialists' Collaboration showed rates differing between those treated with sham/laser (7.2%), 0.3-mg ranibizumab (0.8%), and 0.5-mg ranibizumab (10.4%) at 3 years, and the Protocol T trial had rate of 12% in patients treated with ranibizumab. However, the READ-2, Protocol I, and RESTORE studies found no significant differences between the ranibizumab treated patients and those receiving laser. The cumulative incidence of serious systemic adverse events for patients treated with ranibizumab is in line with reports from studies involving other drugs in the anti-VEGF class [48R].

Aflibercept

Integrated results from the COPERNICUS and GALLILEO studies were recently published. The assimilation of data from these two studies allowed the authors to better evaluate the safety profile of aflibercept given the increased number of studied subjects included. Patient data were available out to week 52 of treatment. The total number of patients included in the study was 360. A total of 34 patients (9.4%) experienced ocular SAEs in the two studies. Vitreous hemorrhage (7.8%), macular edema (1.7%) and glaucoma (1.4%) were the most prevalent ocular SAEs. Only 1 patient (0.3%) was diagnosed with endophthalmitis in this series. Of note, many of these adverse events were attributed to the underlying disease state rather than the treatment itself [49C].

A case report of a maculopapular-type drug induced skin eruptions after treatment with aflibercept was recently published. A 60-year-old Japanese male patient had received 11 injections of 0.5 mg ranibizumab and 23 injections of 2 mg aflibercept for polypoidal choroidal vasculopathy. Ten hours after his 24th intravitreal aflibercept injection, the patient returned to the hospital with erythema and itching involving his trunk and extremities, as well as redness of the pharynx. Laboratory testing was significant for increased leukocyte cell count (8300/mL; lymphocytes, 23.6%; eosinophils, 4.2%) and an elevated C-reactive protein (0.04 mg/dL). The lesions were biopsied as well confirming the presence of upper dermal lymphocytes and telangiectasias. The rash was attributed to the aflibercept administered earlier that day. The condition was recalcitrant to oral antihistamine and topical steroid treatments, and ultimately required 30 mg of oral prednisolone per day for a week to resolve [50A].

Conbercept

Conbercept is a new anti-VEGF medication approved by China's Food and Drug Administration for use in the treatment of neovascular AMD. Conbercept has been shown to bind to and deactivate VEGF-A and all its isoforms, making its affinity superior to that of ranibizumab. Given its newness to the market, long-term studies and agreement regarding this drug's clinical safety and efficacy, effective dosing regimen, best practices in clinical applications have yet to be ascertained [51H].

The FALCON study, a Phase II, nonrandomized, noncontrolled, 9-month trial, was conducted to evaluate both the efficacy and safety of intravitreal conbercept, when used to treat macular edema associated with either branch or central retinal vein occlusions. Patient who underwent at least 1 injection of conbercept were included in the safety date. Ocular adverse events (BRVO = 90%, CRVO = 86.67%) were like those seen in studies involving other intravitreal anti-VEGF medications: subconjunctival hemorrhage, vitreous floaters, transient increased IOP, and reduced visual sensitivity. One patient was forced to withdraw after experiencing a retinal tear with concurrent focal retinal detachment at month 2. A second patient developed retinal neovascularization at month 9 after undergoing monthly injections for the first 3 months of the study. At month 9, two patients in the BRVO and three patients in the CRVO groups experienced a total of seven SAEs, with the incidence of SAEs being 8.33% across all 60 study participants. These events did not involve the eye and were considered unconnected to conbercept or its delivery technique [52c].

A 10-month prospective, single-center, single-dose, randomized controlled clinical trial on patients with proliferative diabetic retinopathy was performed at the First People's Hospital of Yunnan Province, Kunming, Yunnan, China, from March 1, 2014, to January 7, 2015. A total of 107 eyes were randomized to receiving intravitreal conbercept prior to undergoing pars plana vitrectomy (54 eyes) or undergoing pars plana vitrectomy alone (53 eyes). Transient increase in IOP was seen in patients undergoing intravitreal conbercept. The IOP increase was mild (23.55 ± 4.62 mmHg) from baseline (16.74 ± 3.21 mmHg) and resolved without intervention after 1–2 h. Follow-up was limited to 1 month, but there were no reports of the development of endophthalmitis, iris neovascularization, or tractional retinal detachment. Further, no patients experienced cardiovascular or embolic SAEs during the short follow-up period [53c].

Antimicrobials

General

Antibiotics exert their effects via five main mechanisms: inhibition of cell wall synthesis (beta-lactams, vancomycin, bacitracin), inhibition of protein synthesis (aminoglycosides, tetracyclines, macrolides), alteration of cell membranes (polymyxin, bacitracin), inhibition of nucleic acid synthesis (quinolones), and anti-metabolite activity (sulfonamides, trimethoprim) [54H]. These drugs can cause local ocular side effects ranging from transient corneal deposition to permanent optic neuropathy. Systemically, topical antibiotics can exert their affects beyond the ocular surface through absorption via the nasolacrimal system. Oral antibiotics are also important is treating ocular conditions from blepharitis to toxoplasmosis.

Sulfonamides

Renal calculi can be traced to drug treatment in 1%–2% of cases. Poorly soluble drugs that are excreted in the urine such as sulfadiazine for the treatment of toxoplasmosis are among those that can become crystallized in the urine. Diagnostically, metabolites of the drug will be found on calculi analysis using X-ray diffraction or infrared spectroscopy. Drugs can also promote renal calculi formation by altering the pH of the renal environment. It is important to take a thorough medication history including supplements in order to mitigate the risk of kidney complications with these medications [55H].

Traditional therapy for ocular toxoplasmosis has been a combination treatment of pyrimethamine, sulfadiazine and folinic acid along with systemic steroids. Helfenstein and his group performed a retrospective analysis to evaluate this "classic" treatment regimen for complications. The records of 49 patients seen in the University Hospital Zurich Department of Ophthalmology between December 2011 and December 2015 were reviewed for adverse events while undergoing classic treatment. A total of 65 active ocular toxoplasmosis cases were found, and 54 (83.0%) of the episodes were treated with the classic regimen. A total of 37 patients underwent classic treatment. Out of the 37 patients, 9 (24.3%) had to discontinue treatment due to an AE. Among those AEs were elevated creatinine (5.4%) or liver enzymes (5.4%), vomiting (5.4%), rash (5.4%) and facial swelling (2.7%). The AE required a change in treatment regimen in 5 patients and complete discontinuation of treatment in 4 others. There were no reports of SAEs (anaphylaxis or pancytopenia) in this cohort. Severe side effects, such as potentially life-threatening allergic reactions or pancytopenia. Given the number of AEs seen with this treatment regimen, lab monitoring and coordination with other specialists was recommended for patients undergoing "classic" toxoplasmosis treatment [56c].

Substituting trimethoprim–sulfamethoxazole for sulfadiazine and pyrimethamine is an alternative therapy for ocular toxoplasmosis. However, trimethoprim–sulfamethoxazole is not without its concerning side effects. As an inhibitor of folic acid metabolism, bone marrow suppression can occur and care must be taken to establish the patient's baseline blood levels to quickly identify the development of pancytopenia. Baseline complete blood count, liver enzymes, and serum creatinine levels should be performed and repeated within the next 5–7 days after initiating treatment. Hematology should be alerted, and therapy discontinued if bone marrow suppression is identified. Careful patient education regarding the importance of reporting the development of symptoms associated with hypersensitivity reactions including fever and rash within the first 1–2 weeks of treatment will allow prompt intervention. Internal organ toxicity may occur, but be asymptomatic, and anaphylaxis is unusual. Treatment with sulfamethoxazole, in rare cases, may incite toxic epidermal necrolysis and Stevens–Johnson syndrome. Sulfamethoxazole treatment is the precipitating factor of the highest percentage of Stevens–Johnson cases secondary to antibiotic therapy around the globe. Finally, this treatment regimen is contraindicated in patient under the age of 2 months, pregnant females, patients with known hepatic or renal insufficiency, and anyone with an established sensitivity to trimethoprim and sulfonamides [57c].

AMINOGLYCOSIDES

A 21-year-old woman suffered a hemorrhagic occlusive retinal vasculitis (HORV) after undergoing a subconjunctival gentamicin injection for a recalcitrant infectious keratitis. The patient noted severe eye pain and a sudden loss of vision in her left eye. Vision was reduced to counting fingers and retinal evaluation revealed diffuse hemorrhaging including intraretinal, perivascular, and white centered hemorrhaging along with retinal edema. Occlusive vasculitis with macular ischemia was confirmed by both intravenous fluorescein angiography and optical coherence tomography. The diagnosis of HORV was made and attributed to a high dose of intraocular gentamycin [58A].

FLUOROQUINOLONES

Ciprofloxacin, Gatifloxacin, Moxifloxacin, Levofloxacin, Ofloxacin, Tosufloxacin

General Fluoroquinolones (FQ) are a class of antibiotics which have enjoyed extensive utilization due to having a broad spectrum of activity, favorable bioavailability, and general tolerability by patients [59H]. Fluoroquinolones exert their antibacterial effects by inhibiting bacterial topoisomerase and gyrase, thus preventing the unwinding and duplication of bacterial DNA necessary for replication. Multiple systemic side effects have been associated with FQ including tendon rupture, central and peripheral nervous system disruption, cardiovascular events, hepatotoxicity, nephrotoxicity, and even insulin resistance leading to type 2 diabetes. The FDA updated warnings regarding FQ use, including a black box warning for both the oral and injectable forms.

Fluoroquinolone antibiotics have been linked to the development in white, crystalline deposits in up to

18% of patients. The deposits represent a precipitation of the drug in the cornea. Corneas without intact epithelial barriers are most susceptible to deposition. The deposits tend to resolve over weeks to months without intervention, but vision can be permanently compromised. Delayed epithelial healing may necessitate surgical debridement in some cases [60R].

A post penetrating keratoplasty patient suffered corneal stromal deposition with the use of topical tosufloxacin. A 60-year-old female was experiencing delayed epithelialization of her new graft post op day 4. Treatment included topical 1.5% levofloxacin, 0.5% cefmenoxime, 0.1% betamethasone, 0.1% hyaluronate sodium, and 3% acyclovir. Due to suspicion of the levofloxacin causing the delayed healing, the patient was switched to 0.3% tosufloxacin. Deposits at the site of the epithelial defect began on day 6. Tosufloxacin was immediately discontinued, and the deposits slowly dissipated over the next 5 months with complete resolution [61A].

Some patients have experienced cardiovascular side effects during treatment. Various studies of FQ cardiac AEs involve some effect on normal heart rhythm including prolonged QT intervals, torsades de pointes, ventricular tachycardia or fibrillation, and even sudden cardiac death. The underlying mechanism of these FQ-induced cardiac events has yet to be elucidated. However, proposed mechanisms include FQ interference with cardiac rapid delayed rectifier potassium channels, exacerbation of other risk factors including electrolyte imbalances, hypothyroidism, simultaneous use of anti-arrhythmic medications, or the presence of pro-inflammatory cytokines during the infectious process [62M].

Xiao et al. conducted a review of several databases (Cochrane Collaboration, PubMed, and China National Knowledge Infrastructure) through August of 2017 and identified those studies which discussed relative risk (RR) with confidence intervals (CI) to quantify the association between FQ use and cardiac adverse events. A total of 16 studies were included, with 7 involving serious arrhythmias, 3 cardiovascular death, and 11 all-cause death. The total number of patients evaluated was 6139004, with 54.6% being female [62M].

Overall, the RR from FQ use were as follows: 2.29 (95% CI: 1.20–4.36, $P=0.01$) for serious arrhythmias; 1.60 (95% CI: 1.17–2.20, $P=0.004$) for CV death; and 1.02 (95% CI: 0.76–1.37, $P=0.92$) for all-cause death. However, this was not equal between the different FQ drugs. Gatifloxacin has the highest RR for serious arrhythmias at 6.27, moxifloxacin was next with a RR of 4.20, with ciprofloxacin (1.73) and levofloxacin (1.41) showing lower overall RR. Of note, ciprofloxacin has been associated with fewer serious arrhythmias due to its lower propensity to affect the rapidly acting delayed rectifier current (I_{kr}). Gatifloxacin, moxifloxacin, and levofloxacin have a comparably higher propensity for inhibition of I_{kr}, potentially leading to elongation of the QT interval resulting in a higher risk for a serious arrhythmia [62M].

A retrospective study was conducted at the Western New York Veteran's Affairs Heath System. Charts were reviewed for veterans treated with a FQ (ciprofloxacin, moxifloxacin, and levofloxacin) for a minimum of 48h in the hospital setting ($n=631$). The ages of the study population ranged from 22 to 95 years (mean=71.5 years). Overall, a rate of FQ-associated delirium/psychosis in this setting was determined to be 3.7% ($n=23$). Ciprofloxacin-treated patients ($n=14/391$) had a rate of 3.6%, moxifloxacin ($n=9/200$) 4.5%, and no cases were seen in those treated with levofloxacin ($n=0/40$). Age (1.8 increase in risk with every 10-year increase in age) and concurrent use of antipsychotic agents (odds ratio: 5.4 with 95% CI: 1.4–1.67) were identified as the most significant risk factors for developing fluoroquinolone-induced psychosis in this study [63C].

Caution should be taken in patients diagnosed with myasthenia gravis when prescribing FQ treatment. A retrospective study identifying adverse events in myasthenia patients treated with FQ was conducted. The authors were able to identify 37 cases via the FDA Adverse Event Reporting system as well as through an internet search. Patients were treated with the following FQ: levofloxacin ($n=11$), moxifloxacin ($n=6$), ciprofloxacin ($n=10$), ofloxacin ($n=3$), gatifloxacin ($n=2$), norfloxacin ($n=2$) and trovafloxacin ($n=1$), pefloxacin ($n=1$) and prulifloxacin ($n=1$). Symptoms developed quickly, with the median onset being 1-day post drug administration. Respiratory symptoms included dyspnea ($n=19$; 51%) and myasthenic crisis necessitating intervention with ventilation ($n=11$; 30%). Muscle affects ranged from generalized weakness ($n=20$; 54%) to trouble swallowing ($n=9$; 24%) and ocular symptoms of diplopia ($n=6$; 16%) and ptosis ($n=6$; 16%). Two deaths were also reported (5%) [64C].

Bilateral acute iris transillumination (BAIT) is a uveitic condition that involves a sudden onset of uveitis with iris pigment dispersion, transillumination defects, and a dilated pupil. Patients diagnosed with this condition often relate a recent upper respiratory tract infection or treatment with a systemic antibiotic. The clinical course of BAIT has been described as mild and self-limiting with resolution over a few weeks, to refractory leading to glaucomatous optic neuropathy or other incurable ocular complications. A case report involved a 55-year-old female who presented for evaluation of bilateral pain, photophobia, and increased intraocular pressure unsuccessfully treated by other providers over a 4-month period. She had suffered from pneumonia that was treated with oral moxifloxacin a few weeks prior to the onset of her symptoms. Initially, she was diagnosed with severe iritis and pigment dispersion syndrome (PDS).

Topical therapy did not successfully relieve her symptoms. The patient had developed cataracts, posterior synechia, as well as measurable retinal nerve fiber layer loss in her left eye. Treatment included laser peripheral iridotomy in the left eye, cataract extractions and ab internal trabeculotomy with Trabectome in each eye, and was managed topically with an ocular hypotensive drop and steroid. Systemic fluoroquinolones (ciprofloxacin, ofloxacin, gatifloxacin, levofloxacin, norfloxacin, moxifloxacin) have all been implicated in the development of uveitis post treatment, with moxifloxacin being the most commonly cited. Patients, particularly middle-aged females, presenting with photophobia, a relatively *mild* bilateral uveitis, iris transillumination defects, and dilated, mildly responsive pupils must be questioned regarding recent upper respiratory tract infections and/or recent fluoroquinolone treatment [65A].

Iris pigment is produced by an enzyme known as melanogenic tyrosinase (MTY). While pigment cells can be viewed in vivo with a slit lamp, MTY cannot be quantified. Fluoroquinolones have been associated with dermal melanocyte toxicity. While release of iris pigment into the anterior chamber of the eye, as in BAIT, has been associated with moxifloxacin treatment, a second condition, bilateral acute depigmentation of the iris (BADI), is similar in appearance but has not been linked to systemic FQ treatment. Mahanty et al. conducted a study to determine what effect, if any topical FQ had on melanocyte toxicity. Eighty-two patients scheduled to undergo cataract surgery were randomized to receive moxifloxacin ($N=27$, preservative-free) or ciprofloxacin ($n=29$, with preservative) or Tobramycin ($n=26$, with preservative, as a control). MTY activity in the aqueous humor was higher in those patients receiving ciprofloxacin vs those receiving moxifloxacin ($P<0.0001$) or tobramycin ($P<0.0001$). The reason for the discrepancy between ciprofloxacin and moxifloxacin is that while moxifloxacin is toxic to melanocytes due to the presence of MTY in the aqueous, its higher concentration inhibits the activity of MTY. No patient developed BAIT of BADI in this study, suggesting that these conditions may involve multiple mechanisms [66c].

VANCOMYCIN

The American Society of Cataract and Refractive Surgery (ASCRS) and the American Society of Retina Specialists (ASRS) create a joint task force study HORV and guide treatment to improve outcomes for patients suffering this condition. An online registry was created, and an online literature search was performed to identify the clinical characteristics of HORV. HORV typically is a delayed onset condition. Post-operative day 1 findings are normal. Patients then proceed to develop painless vision loss, anterior and posterior chamber inflammation, retinal hemorrhaging, and ischemia. Twenty-three patients and 36 eyes were diagnosed with HORV in this series. All identified patients had received vancomycin during surgery either via intracameral bolus (33/36), intravitreal injection (1/36), or irrigation bottle (2/36). Onset of symptoms varied from 1 to 21 days post op. Visual acuity was severely affected, with 61% of patients showing acuities of 20/200 or less. Twenty-two percent of the patients has no light perception (NLP) after onset of HORV. Three eyes were administered corticosteroid via intravitreal injection with two eyes recovering vision to 20/40 and 20/70, and the third eye measured at hand motion. Seven eyes were administered a second dose of intravitreal vancomycin, with five of the eyes becoming NLP at follow-up. Neovascular glaucoma occurred in 56% ($n=20$) of the eyes in this study. The joint task force concluded that HORV is a delayed hypersensitivity response to vancomycin. Early intervention with corticosteroids may be of some benefit. Neovascular glaucoma, which occurred in over half the eyes in this series, can be treated with anti-VEGF drugs and pan-retinal photocoagulation. Finally, doctors should avoid administering any supplementary vancomycin if patient presentation is highly suspicious for HORV [67c].

Special Review on Drug Delivery *The eye is replete with anatomic and physical barriers (i.e. nasolacrimal drainage, blinking, and multilayered cornea with hydrophilic and lipophilic components) that serve as a complex defense system. The viscosity, concentration, chemical structure of a compound and frequency of application all impact the bioavailability of a topically applied ophthalmic medication and yet still, the bioavailability of most topically instilled ophthalmic medications is generally reported to be less than 5% [68E]. Viscous formulations of eye drops are less affected by tear dilution which allows for longer ocular contact time; therefore, much attention is focused on gel systems and mucoadhesive concepts to maximize absorption [69E]. Nanoemulsions are colloidal carriers with nanoscopic droplets, usually oil droplets in aqueous medium that make them highly soluble and yield great potential to bridge the defense mechanisms of the eye and allow utilization of lower drug concentrations while offering higher therapeutic efficacy [68E]. The formulation of fully dissolving microneedles, comprised of besifloxacin that penetrate 200 μm into the cornea and dissolve within 5 min offers increased corneal penetrance and antibacterial activity in early investigational studies [70E].*

Conditions that require frequent topical dosing constitute a significant treatment burden for patients, which often lead to poor patient adherence to prescribed regimens. Novel controlled release delivery systems, such as a travoprost punctum plug (OTX–TP) and a bimatoprost-impregnated ring that fits into the ocular fornix are being studied as a means to offers sustained IOP control without requiring patients to instill eye drops at all. Another option being explored entails the use of brimonidine-loaded microspheres deposited into the supraciliary space, adjacent to the ciliary body, via a microneedle is showing promise in early animal studies [71E].

The presence of the blood ocular barrier, comprised of blood-aqueous and blood-retinal barriers, poses another challenge to drug delivery, accounting for the reason that systemic and IV medications are infrequently employed for the treatment of eye disease. High molecular weight drugs must be given intravitreally to offer the most efficient drug delivery to the posterior segment. High clearance rates, unfortunately, require frequent injections which expose patients repeatedly to potential adverse effects. Drug delivery systems that reduce the frequency of injection are presently being explored. A capsule drug ring (CDR) has been developed which researchers suggest may extend the duration of action of anti-VEGF agents to 90 days [72R]. Transscleral iontophoresis is being investigated as a way to drive drug through the scleral barrier utilizing a low electric current [73E]. Finally, the discovery of extracellular vesicles (EVs) offers a potential mechanism to easily bridge the protective barriers of the eye and may be more reliable than investigation cell-based therapies [74R].

DISCLOSURE

The authors have no proprietary or commercial interest in any materials discussed in this article.

References

[1] Inoue K. Managing adverse effects of glaucoma medications. Clin Ophthalmol. 2014;8:903–13 [M].

[2] Charalampidou S, Nolan J, Ormonde GO, et al. Visual perceptions induced by intravitreous injections of therapeutic agents. Eye. 2011;25(4):494–501 [A].

[3] Shah A, Modi Y, Wellik S, et al. Brimonidine allergy presenting as vernal-like keratoconjunctivitis. J Glaucoma. 2015;24(1):89–91 [A].

[4] Lee J, et al. Cardiopulmonary arrest following administration of cyclomydril eyedrops for outpatient retinopathy of prematurity screening. J AAPOS. 2014;18(2):183–4 [A].

[5] Kiryazov K, Stefova M, Iotova V. Can ophthalmic drops cause central nervous system depression and cardiogenic shock in infants? Pediatr Emerg Care. 2013;29(11):1207–9 [A].

[6] Peace J, et al. Polyquaternium-1-preserved travoprost 0.003% or benzalkonium chloride-preserved travoprost 0.004% for glaucoma and ocular hypertension. Am J Ophthalmol. 2015;60(2):266–74 [C].

[7] Walimbe T, et al. Effect of benzalkonium chloride free latanoprost ophthalmic solution on ocular surface in patients with glaucoma. Clin Ophthalmol. 2016;10:821–7 [c].

[8] Saini M, et al. Ocular surface evaluation in eyes with chronic glaucoma on long term topical antiglaucoma therapy. Int J Ophthalmol. 2017;10(6):931–8 [c].

[9] Romero-Díaz de León L, et al. Conjunctival and corneal sensitivity in patients under topical antiglaucoma treatment. Int Ophthalmol. 2016;36(3):299–303 [C].

[10] Abedi F, Chappell A, Craig JE. Audible clicking on blinking: an adverse effect of topical prostaglandin analogue medication. Clin Experiment Ophthalmol. 2017;45(3):304–6 [A].

[11] Fong CS, et al. Audible blink in prostaglandin-associated periorbitopathy. Clin Experiment Ophthalmol. 2016;44(7): 630–1 [A].

[12] Li F, Huang W, Zhang X. Efficacy and safety of different regimens for primary open-angle glaucoma or ocular hypertension: a systematic review and network meta-analysis. Acta Ophthalmol. 2018;96:e277–84 [M].

[13] Selen F, Tekeli O, Yanık Ö. Assessment of the anterior chamber flare and macular thickness in patients treated with topical antiglaucomatous drugs. J Ocul Pharmacol Ther. 2017;33(3):170–5 [M].

[14] Makri OE, et al. Cystoid macular edema associated with preservative-free latanoprost after uncomplicated cataract surgery: case report and review of the literature. BMC Res Notes. 2017;10(1):127 [A].

[15] Lin M, Schmutz M, Mosaed S. Latanoprost-induced skin depigmentation. J Glaucoma. 2017;26(11):e246–8 [A].

[16] Mohite AA, et al. Latanoprost induced iris pigment epithelial and ciliary body cyst formation in hypermetropic eyes, Case Rep Ophthalmol Med. 2017;2017:9362163. 4 pages, https://doi.org/10.1155/2017/9362163 [A].

[17] Meda R, et al. The impact of chronic use of prostaglandin analogues on the biomechanical properties of the cornea in patients with primary open-angle glaucoma. Br J Ophthalmol. 2017;101(2):120–5 [c].

[18] Yoo R, Choi YA, Cho BJ. Change in central corneal thickness after the discontinuation of Latanoprost in normal tension Glaucoma-change in central corneal thickness after stop of Latanoprost. J Ocul Pharmacol Ther. 2017;33: 57–61 [c].

[19] Mocan MC, Uzunosmanoglu E, et al. The association of chronic topical prostaglandin analog use with meibomian gland dysfunction. J Glaucoma. 2016;25(9):770–4 [c].

[20] Morales DR, et al. Respiratory effect of beta-blocker eye drops in asthma: population-based study and meta-analysis of clinical trials. Br J Clin Pharmacol. 2016;82(3):814–22 [M].

[21] Nanda T, et al. Ophthalmic timolol hallucinations: a case series and review of the literature. J Glaucoma. 2017;26(9): e214–6 [A].

[22] Mäenpää J, Pelkonen O. Cardiac safety of ophthalmic timolol. Expert Opin Drug Saf. 2016;15(11):1549–61 [R].

[23] Grigera JD, Grigera ED. Ultrasound biomicroscopy in acetazolamide-induced myopic shift with appositional angle closure. Arq Bras Oftalmol. 2017;80(5):327–9 [A].

[24] Serle JB, et al. Two phase 3 clinical trials comparing the safety and efficacy of netarsudil to timolol in patients with elevated intraocular pressure: Rho kinase elevated IOP treatment trial 1 and 2 (ROCKET-1 and ROCKET-2). Am J Ophthalmol. 2018;186:116–27 [C].

[25] Öner V, Bulut A, Öter K. The effect of topical anti-muscarinic agents on subfoveal choroidal thickness in healthy adults. Eye (Lond). 2016;30(7):925–8 [c].

[26] Gizzi C, Mohamed-noriega J, Murdoch I. A case of bilateral pigment dispersion syndrome following many years of uninterrupted treatment with atropine 1% for bilateral congenital cataracts. J Glaucoma. 2017;26:e225–8 [A].

[27] Lee A, Nguyen P. Takotsubo cardiomyopathy due to systemic absorption of intraocular phenylephrine. Heart Lung Circ. 2016;25(12):e159–61 [A].

[28] Fukuhara D, et al. Iatrogenic Cushing's syndrome due to topical ocular glucocorticoid treatment. Pediatrics. 2017;139(2): e1–5 [A].
[29] Rainsbury PG, et al. Ritonavir and topical ocular corticosteroid induced Cushing's syndrome in an adolescent with HIV-1 infection. Pediatr Infect Dis J. 2017;36(5):502–3 [A].
[30] Holden SE, Currie CJ, Owens DR. Evaluation of the clinical effectiveness in routine practice of fluocinolone acetonide 190 µg intravitreal implant in people with diabetic macular edema. Curr Med Res Opin. 2017;33(sup2):5–17 [C].
[31] Jusufbegovic D, Schaal S. Quiescent herpes simplex keratitis reactivation after intravitreal injection of dexamethasone implant. Retin Cases Brief Rep. 2017;11(4):296–7 [A].
[32] Eris E, et al. A new side effect of intravitreal dexamethasone implant (Ozurdex®), Case Rep Ophthalmol Med. 2017;2017:6369085. 3 pages, https://doi.org/10.1155/2017/6369085 [A].
[33] Moisseiev E, Morse LS. Fluocinolone acetonide intravitreal implant in the visual axis. JAMA Ophthalmol. 2016;134(9): 1067–8 [A].
[34] Fong AH, Chan CK. Presumed sterile Endophthalmitis Afer intravitreal triamcinolone (Kenalog)-more common and less benign than we thought? Asia Pac J Ophthalmol (Phila). 2017;6(1):45–9 [c].
[35] Pozarowska D, Pozarowski P. The era of anti-vascular endothelial growth factor (VEGF) drugs in ophthalmology, VEGF and anti-VEGF therapy. Cent Eur J Immunol. 2016;41(3):311–6 [R].
[36] Sukgen EA, et al. Occurrence of intraocular air bubbles during intravitreal injections for retinopathy of prematurity. Int Ophthalmol. 2016;37(1):215–9 [c].
[37] Bracha P, et al. The acute and chronic effects of intravitreal anti-vascular endothelial growth factor injections on intraocular pressure: a review. Survey of ophthalmology. Surn Ophthalmol. 2018;63:281–95 [R].
[38] Kataja M, et al. Outcome of anti-vascular endothelial growth factor therapy for neovascular age-related macular degeneration in real-life setting. Br J Ophthalmol. 2018;102:959–65 [C].
[39] Arevalo J, et al. Intravitreal bevacizumab for proliferative diabetic retinopathy: results from the Pan-American Collaborative Retina Study Group (PACORES) at 24 months of follow-up. Retina. 2017;37(2):334–43 [c].
[40] Xu Y, Tan CS. Safety and complications of intravitreal injections performed in an Asian population in Singapore. Int Ophthalmol. 2016;37(2):325–32 [C].
[41] Maguire MG, et al. Serious adverse events with bevacizumab or ranibizumab for age-related macular degeneration: meta-analysis of individual patient data. Ophthalmol Retina. 2017;1(5):375–81 [MC].
[42] Hirano T, et al. Changes in plasma vascular endothelial growth factor level after intravitreal injection of bevacizumab, Aflibercept, or Ranibizumab for diabetic macular edema. Retina. 2017;1–8 [c].
[43] Hartnett M. Role of cytokines and treatment algorithms and retinopathy of prematurity. Curr Opin Ophthalmol. 2017;28:282–8 [R].
[44] Huang C-Y, et al. Changes in systemic vascular endothelial growth factor levels after intravitreal injection of aflibercept in infants with retinopathy of prematurity. Graefes Arch Clin Exp Ophthalmol. 2018;256:479–87 [c].
[45] Kana H, et al. The efficacy of intravitreal antivascular endothelial growth factor as primary treatment of retinopathy of prematurity: experience from a tertiary hospital. S Afr Med J. 2017;107(3):215 [c].
[46] Kennedy KA, Mintz-Hittner HA. Medical and developmental outcomes of bevacizumab versus laser for retinopathy of prematurity. J AAPOS. 2018;22(1):61-65.e1 [c].
[47] Silva R, et al. Treat-and-extend versus monthly regimen in neovascular age-related macular degeneration. Ophthalmology. 2018;125(1):57–65 [MC].
[48] Stewart MW. A review of Ranibizumab for the treatment of diabetic retinopathy. Ophthalmol Therapy. 2017;6(1):33–47 [R].
[49] Pielen A, et al. Integrated results from the COPERNICUS and GALILEO studies. Clin Ophthalmol. 2017;11:1533–40 [C].
[50] Nagai N, et al. Maculopapular rash after intravitreal injection of an antivascular endothelial growth factor, aflibercept, for treating age-related macular degeneration. Medicine. 2017;96(21):e6965 [A].
[51] Cui C, Hong L. Clinical observations on the use of new anti-VEGF drug, conbercept, in age-related macular degeneration therapy: a meta-analysis. Clin Interv Aging. 2017;13:51–62 [H].
[52] Sun Z, et al. Efficacy and safety of intravitreal Conbercept injections in macular edema secondary to retinal vein occlusion. Retina. 2017;37(9):1723–30 [c].
[53] Yang X, et al. A randomized controlled trial of Conbercept pretreatment before vitrectomy in proliferative diabetic retinopathy. J Ophthalmol. 2016;2016:1–8 [c].
[54] Kapoor G, et al. Action and resistance mechanisms of antibiotics: a guide for clinicians. J Anaesthesiol Clin Pharmacol. 2017;33(3):300 [H].
[55] Daudon M, et al. Drug-induced kidney stones and crystalline nephropathy: pathophysiology, prevention and treatment. Drugs. 2017;78(2):163–201. https://doi.org/10.1007/s40265-017-0853-7 [H].
[56] Helfenstein M, et al. Ocular toxoplasmosis: therapy-related adverse drug reactions and their management. Klin Monbl Augenheilkd. 2017;234(04):556–60 [c].
[57] Ozgonul C, Besirli CG. Recent developments in the diagnosis and treatment of ocular toxoplasmosis. Ophthalmic Res. 2016;57(1):1–12 [c].
[58] Tang R, Tse R. Acute renal failure after topical fortified gentamicin and vancomycin eyedrops. J Ocul Pharmacol Ther. 2011;27(4):411–3 [A].
[59] Raizman MB, et al. Drug-induced corneal epithelial changes. Surv Ophthalmol. 2017;62(3):286–301 [r].
[60] Elia M, et al. Corneal crystalline deposits associated with topically applied Gatifloxacin. Cornea. 2014;33(6):638–9 [A].
[61] Liu X, et al. Fluoroquinolones increase the risk of serious arrhythmias. Medicine. 2017;96(44):e8273 [M].
[62] Malladi A, Liew E, Ng X, et al. Ciprofloxacin eyedrops-induced subtherapeutic serum phenytoin levels resulting in breakthrough seizures. Singapore Med J. 2014;55(7):e114–5 [A].
[63] Zeitler K, Jariwala R, Montero J. Adverse events related to antibiotic use in patients with myasthenia gravis. Open Forum Infectious Diseases. 2017;4(Suppl_1):S342 [c].
[64] Den Beste KA, Okeke C. Trabeculotomy ab interno with Trabectome as surgical management for systemic fluoroquinolone-induced pigmentary glaucoma. Medicine. 2017;96(43):e7936 [A].
[65] Mahanty S, et al. Aqueous humor tyrosinase activity is indicative of iris melanocyte toxicity. Exp Eye Res. 2017;162:79–85 [c].
[66] Myers W. Faculty of 1000 evaluation for Vancomycin-associated hemorrhagic occlusive retinal vasculitis: clinical characteristics of 36 eyes. Ophthalmology. 2017;124(5):583–95. F1000—Post-Publication peer review of the biomedical literature. [c].
[67] Mahboobian MM, Seyfoddin A, Rupenthal ID, et al. Formulation development and evaluation of the therapeutic efficacy of Brinzolamide containing Nanoemulsions. Iran J Pharm Res. 2017;16(3):847–57 [E].

[68] Tofighia P, Soltani S, Montazam SH, et al. Formulation of Tolmetin Ocuserts as carriers for ocular drug delivery system. Iran J Pharm Res. 2017;16(2):432–41 [E].
[69] Bhatnagar S, Saju A, Cheerla KD, et al. Corneal delivery of besifloxacin using rapidly dissolving polymeric microneedles. Drug Deliv Transl Res. 2018;8:473–83 [E].
[70] Chiang B, Kim YC, Doty AC, et al. Sustained reduction of intraocular pressure by supraciliary delivery of brimonidine-loaded poly (lactic acid) microspheres for the treatment of glaucoma. J Control Release. 2016;228:48–57 [E].
[71] Agrahari V, Agrahari V, Mandal A, et al. How are we improving the delivery to back of the eye? Advances and challenges of novel therapeutic approaches. Expert Opin Drug Deliv. 2017;14:1145–62 [R].
[72] Li SK, Hao J. Transscleral passive and iontophoretic transport: theory and analysis. Expert Opin Drug Deliv. 2018;15:283–99 [E].
[73] van der Merwe Y, Steketee MB. Extracellular vesicles: biomarkers, therapeutics, and vehicles in the visual system. Curr Ophthalmol Rep. 2017;5(4):276–82 [R].
[74] Querques L, et al. Hemorrhagic occlusive retinal vasculitis after inadvertent intraocular perforation with gentamycin injection. Eur J Ophthalmol. 2017;27(2):e50–3 [A].

Further Reading

[75] Stottlemyer LV and Polnariev A. Drugs used in ocular treatment. In: Ray S, editor. Side effects of drugs Annual. Amsterdam: Elsevier. 2017;39:p. 491–501.

CHAPTER

48

Safety of Complementary and Alternative Medicine (CAM) Treatments and Practices

Renee A. Bellanger,[1], Christina M. Seeger†, Helen E. Smith‡*

*Department of Pharmacy Practice, University of the Incarnate Word, Feik School of Pharmacy, San Antonio, TX, United States

†Pharmacy Librarian, University of the Incarnate Word, Feik School of Pharmacy, San Antonio, TX, United States

‡Department of Pharmaceutical Science, University of the Incarnate Word, Feik School of Pharmacy, San Antonio, TX, United States

[1]Corresponding author: bellangc@uiwtx.edu

INTRODUCTION

Consumer interest in complementary and alternative medicine (CAM) continues to escalate in the United States and globally [1R, 2r]. The global market for herbal supplements was projected to reach US$107 billion in 2017, and to exceed US$140 billion by 2024, [3R]. This is due in large part to the "presumptive belief in some therapeutic efficacy" and "absence of serious adverse effects" of botanicals and natural products "as evidenced by a long history of use in traditional medicine" [4R]. It is clear "natural" does not always equate to "safe" [1R]. Other factors fueling growth in the herbal and dietary supplement market in the U.S. and elsewhere include a more widespread acceptance of functional foods, an increase in consumer confidence to use herbs in preventive health and alternative medicine regimens, supplier innovations, and the highly anticipated release of Current Good Manufacturing Practices (CGMP) for dietary supplements by the FDA [5A].

Research on herbal and dietary supplements remains of poor quality and quantity, although there are now more than 50000 entries in the Dietary Supplement Label Database (https://ods.od.nih.gov/Research/Dietary_Supplement_Label_Database.aspx). In the U.S., dietary supplements are regulated under the Dietary Supplement Health and Education Act of 1994, which requires such products to be labeled with the term dietary supplement (or a similar term). This regulation also prohibits the marketing of adulterated or misbranded products. Although dietary supplements do not require FDA approval for safety and efficacy, in contrast to conventional medications, the agency monitors mandatory reporting of serious adverse effects and potential misbranding related to dietary supplements [2r]. The Office of Dietary Supplements was mandated in 2001 "to review current scientific literature on the efficacy and safety of dietary supplements" [6S].

A series of invited review articles addressed the efficacy, safety and U.S. regulations of herb and dietary supplements with recommendations for improvements and included published case reports of dietary supplement toxicity in the specific areas of liver, kidney, heart and cancer [1R]. The author contends that "less than 1 percent of Americans experience adverse events related to dietary supplements, and the majority was classified as minor". Dietary supplements with the highest potential for adverse events are those for sexual enhancement, weight loss and sports performance/body building, and these are often actually "tainted products marketed as dietary supplements" [7S] due to adulteration or contamination [1R]. Despite that the majority of dietary supplements appear inherently safe, case reports have shown a variety of adverse effects, some potentially life-threatening, although it is important to note that case studies do not verify causation or association [1R].

METHODS

A literature search of PubMed, CINAHL, Science Direct, Cochrane library, Web of Science, Google Scholar

Side Effects of Drugs Annual, Volume 40
ISSN: 0378-6080
https://doi.org/10.1016/bs.seda.2018.06.013

and MedWatch for English-language articles of meta-analyses, case studies and case reports from January 1, 2017 through January 1, 2018 was conducted. The following terms were used in the literature search: CAM therapies, Chinese herbal drugs, chiropractic manipulation, dietary supplements, homeopathy, mind–body therapies, musculoskeletal manipulations, phytotherapy, traditional Chinese medicine, yoga, and adverse effects (psychiatric, cardiac, endocrine, gastrointestinal, pulmonary, neurotoxic, nephrotoxicity, hepatotoxicity), contaminants, and side effects. Specific products searched for in the literature included: *Aconitum carmichaelii*, aloe, *Angelica sinensis* (dong quai), *Camellia sinensis*, *Citrus aurantium* (Bitter Orange), fish oils, *Ginkgo biloba*, mushroom poisoning, *Pausinystalia* (yohimbe), *Piper methysticum* (kava), probiotics, soybean, and *Zingiber officinale* (ginger). Pertinent articles found using these search terms are presented in this document categorized by body system, following an overview of cases showcasing the concern.

General Reviews

Several reviews giving an overview of the frequency and effects of adverse events in herbal and dietary supplements were published in 2017. Timbo and others from the FDA report on data gathered on dietary supplement associated adverse events during the time period of 2004 through 2013 from the Center for Food Safety and Applied Nutrition (CFSAN) adverse event reporting system database (CAERS). A total of 15430 adverse events were analyzed. The largest number of reports were from patients greater than 65 years of age (24.8%), while persons under 20 years of age were less frequently involved (4.1%). Life-threatening conditions were cited in 1218 (7.9%) events, serious outcomes including hospitalizations by 3927 (25.4%) and preventive surgical or medical intervention after exposure was noted in 789 reports (5.1%). Adverse event reports of death were at 2.2% overall. Herbal/botanical products were reported in 544 (3.5%) of total events, although some other product categories such as weight loss, cardiovascular health products, bodybuilding, immunity and infection may have herbal components in their composition and combined are almost 34% of the reported adverse events [8S].

Knapik and colleagues surveyed active duty U.S. military personnel regarding the use and any subsequent adverse events from dietary supplements. A random sample of 9672 service members from the Navy and Marine Corps was sent a questionnaire, by postal service or e-mail, requesting information on demographic characteristics and dietary supplements used more than once weekly. Of the 1708 completed questionnaires, 1683 were analyzed. Many service members (73%) used dietary supplements more than once weekly and 31% used more than 5 different supplements daily. Adverse events from dietary or nutritional supplement use were reported by 22.1%. The most commonly reported supplements causing adverse events were combination products (28.8%), purported prohormone products (9.4%), herbals (8.9%) and multivitamin/mineral products (8.4%). The adverse effects reported were palpitations, gastrointestinal symptoms, muscle cramping, weakness or pain, CNS symptoms, including seizures from various products [9C].

Three reports from cross-sectional prospective studies on the risks of dietary and herbal supplements in hospitalized patients were published by Levy and colleagues. In each of these studies questionnaires were administered by trained health personnel interviewing the patients. Drug interactions with dietary and herbal supplements in hospitalized patients were studied using the questionnaire data and a review of medical records. In this report, 927 hospitalized patients agreed to answer the questionnaire and 458 (49%) reported dietary supplement use in the previous year. Of the hospitalized patients that had used dietary supplements, 47% were at risk for at least one potential drug–supplement interaction. Risks for higher rates of potential interactions were increased with age, male gender, and comorbidities of endocrine, neurologic or gastrointestinal disease. The authors found that only 12% of the dietary supplements used by the patients were found in medical record documentation [10C, 11C].

Another report from Levy and colleagues details a subgroup of the 526 patients hospitalized for surgery from the larger group studied above. An analysis of the patients' use of dietary supplements and other medications before surgery to assess significant drug–nutrient interactions, including bleeding risk after surgery is described. Of these surgical patients, 230 reported using dietary and herbal supplements within the past year. On analysis of the data from the questionnaire and the patient's records after surgery, potential dietary supplement interactions with anesthesia occurred in 38 patients (16.5%). The authors felt that sage, chamomile and green tea have the most potential to interact with general anesthetic agents, but no adverse events were reported in the patient's files. The other potential adverse event in the surgical population was increased bleeding risk. Twenty four of the surgical patients were given anticoagulants or antithrombotic agents peri-operatively, and concomitant dietary supplements with the potential to increase bleeding risk included fish oil, green tea, rosemary, magnesium, flaxseed or ginger. Among these surgical patients, three had reported interoperative hemorrhage, all were taking fish oil supplements prior to surgery. The adverse events were determined to be probably associated with the consumption of the fish oil and the medication [12C].

The American College of Clinical Pharmacy (ACCP) published a white paper on natural products. A review is made for the clinician of the regulation of natural products by entities of the U.S. including governmental agencies and independent quality control organizations gives the reader an overview of what can be expected of the products on the U.S. market including safety and efficacy evaluations, which are not FDA mandated [2r].

DIETARY AND HERBAL SUPPLEMENTS

Adverse Effects: Central Nervous System

There is some potential for herbal and other dietary supplements to have adverse effects on the central nervous system. Only one article discussing two cases of a central nervous system (CNS) adverse effect of any herbal product was published in the English language literature within the last year. Kang et al. [13A] reported on two cases of acute cerebral infarction after the consumption of *Aconitum carmichaelii* (Chuanwu, aconite), a Chinese herb used to treat arthritis and neuralgia. A 68-year-old man with a history of chronic obstructive pulmonary disease ingested a tea made with aconite to treat joint pain. He presented to the emergency room with impaired consciousness, dyspnea, vomiting, and palpitations. He had low partial carbon dioxide pressure, respiratory alkalosis, and ventricular tachycardia. The patient was diagnosed with a cerebral infarct. The second reported case of aconite-associated cerebral infarction involved a 61-year-old woman. She had no significant medical history when she presented to the emergency room with chest pain and impaired consciousness. She had consumed water boiled with aconitine 30min before admission. This patient was also assessed and determined to have suffered a cerebral infarct associated with aconitine consumption. Aconitine consumption has been reported to cause many poisonings, but most cases are of patients experiencing adverse cardiovascular effects due to its ability to block open-state sodium channels. These cases of aconitine poisoning resulting in cerebral infarct were felt by the authors to be unusual.

Adverse Effects: Hematologic Systems

Omega-3 fatty acids are polyunsaturated and recommended for a variety of health benefits including anti-inflammatory effects, lowering risk of coronary artery disease and decreasing serum triglycerides. Omega-3 fatty acids, including fish oil supplements, have additional properties such as, reduction of platelet aggregation and increased bleeding risk, especially when used concomitantly with antiplatelet or anticoagulant medications. A case of an 83-year-old man with a subdural hematoma after a traumatic head injury is described. The patient was taking multiple medications for coronary heart disease, atrial fibrillation, dementia, and hypertension, including warfarin and omega-3 fatty acid fish oil supplements. The patient was given vitamin K and prothrombin to decrease his elevated INR, which was 2.8 on admission. His INR did not decline as expected although he was given additional interventions of fresh frozen plasma and prothrombin. His hematoma progressed with cerebral edema. Multiple attempts at lowering the INR with available treatments were attempted with the lowest INR at 1.3 after 22h. Due to the risk of bleeding, the patient did not meet criteria for any surgical procedure to reduce the hematoma and the patient succumbed. The authors suggest that the drug–nutrient interaction between the fish oil supplement and warfarin may have contributed to the inability to reverse the elevated INR, especially at high daily doses of fish oil [14A].

Adverse Effects: Cardiac

A weight loss dietary supplement containing green coffee extract 200mg, caffeine 200mg, yohimbe 20mg and 215mg of a proprietary blend including Garcinia cambogia, L-theanine, autumn fruit, ashwagandha and mint was thought to be the cause of a young woman developing dilated cardiomyopathy. The woman had chest pain and elevated laboratory values of beta-natriuretic peptide and troponin. Her heart rate was elevated to 99 beats per minute with a normal blood pressure. She had reduced movement of her left ventricular wall and her ejection fraction was 30%–35% on an echocardiogram and 20%–35% on subsequent cardiac catheterization. She had no personal or family history of cardiac disease or other contributing factors. She had started the dietary supplement 1 month prior to admission to enhance weight loss. After discontinuance of the supplement and start of appropriate cardiac therapeutic regimen, the patient recovered function of her heart [15A].

Adverse Effects: Vascular

A weight-loss dietary supplement containing green tea extract, L-carnitine and conjugated linoleic acid is considered the cause of a 50-year-old woman exhibiting acute onset reversible cerebral vasoconstriction syndrome (RCVS). The woman was admitted to the hospital via the emergency department where she presented with severe headache that was resistant to treatment. She had a medical history that included migraine and hypertension. She was diagnosed when she exhibited multiple arterial constrictions but no vascular malformations based on MRI and digital angiography. The patient recovered with discontinuance of the dietary supplement and on revised medication management of her hypertension. The authors correlated the RCVS in middle aged women with supplements containing green tea extract or L-carnitine in this and another case from the literature [16A].

Adverse Effects: Muscle

A drug–dietary supplement interaction was illustrated in a case of acute compartment syndrome in a 28-year-old man. The patient was concurrently using multiple medications for bipolar disorder and post-traumatic stress syndrome along with many different dietary supplements for energy and to induce weight loss. The patient presented with signs of rhabdomyolitis which progressed despite aggressive fluid therapy. Symptoms and pain increased despite appropriate medical therapy, and the patient was taken to surgery for a fasciotomy on two separate limbs. The patient was cautioned about using the appropriate amount and mix of dietary supplements while on prescription medications, including sertraline, which may have potentiated the risk of rhabdomyolysis when taken with the serotonin precursors found in the concurrently administered dietary supplements [17A].

Adverse Effects: Renal

The potential nephrotoxic effects of herbals and herbal medications are of some concern. Only a few publications on the potential nephrotoxicity of these compounds were published in the English language literature within the last year. One case of an herb–drug interaction resulting in nephrotoxicity and one compilation of cases was published.

A case of a kidney injury was a reported due to a drug–food interaction between the herb turmeric and the immunosuppressant tacrolimus [18A]. In this case, a 56-year-old male liver transplantation recipient suffered from nephrotoxicity when he ingested large amounts of turmeric while on his normal regimen of tacrolimus. He had been at a stable dose of tacrolimus with no indications of renal damage until he presented to the hospital with increasing edema of his extremities. A serum creatinine of 4.2mg/dL along with hypokalemia was identified on admittance. He stated he had been compliant with his drug regimen. However, the patient did admit to ingesting about 15 spoonsful of turmeric added to his food each day for 10 days prior to this admission. His tacrolimus blood level on admission was 29.9ng/mL while his last level was 9.7ng/mL at a 12-h trough. A review of the literature by his caregivers found that a drug interaction caused by inhibition of tacrolimus metabolism by turmeric was documented in rats. No other possible drug–drug or drug–herb interactions were found to explain the elevated tacrolimus levels felt to be associated with the renal dysfunction in this patient.

An in-depth review of kidney toxicity associated with herbal and dietary supplements was published by Brown [1R]. This article is one in a series of articles presenting reviews of various toxicities herbals and dietary supplements are known to cause. The review articles are accompanied by online tabular compilations of published cases. The on-line tables were created to provide a current list of published human cases of herbal-induced and dietary supplement-induced kidney injury that can be updated as more cases are reported. The original tables were compiled by identifying all human cases of herbal and dietary supplement-induced kidney toxicity cases published in PubMed from 1966 through June 2016. There were several exclusion criteria for the cases, resulting in only cases of a single herb or dietary supplements being included. For example, cases involving self-harm and illegal substances were not included.

Adverse Effects: Hepatic

Many dietary and herbal supplements have been well documented as causing hepatotoxicity. In the past year, there have been several additions to the medical literature regarding the safety of these products. Three case reports, a number of reviews, several observational studies, and one hypothesis-driven study of herbal and dietary supplement-associated liver injury have been published in the English literature within the last year.

One case report of acute liver failure caused by the consumption of an Yogi Detox Tea was reported by Kesavarapu et al. [19A]. Yogi Detox Tea is an herbal tea marketed as a hepatoprotectant. A 60-year-old female presented to a hospital with weakness and lethargy that had been increasing over a 2-week period. She had no other symptoms and her medical status at the time was unremarkable with the exception of a past history of hypertension being treated with hydrochlorothiazide and current obesity. She had no other relevant medical or social factors. She stated she had been drinking Yogi Detox Tea three times a day for 2 weeks before her symptoms appeared. Her hepatic function tests on admission were all significantly elevated and continued to increase over the next 2 weeks while hospitalized. A liver biopsy indicated necrosis with portal, periportal and panlobular inflammation with lymphocyte, neutrophils, and plasma cell infiltration but few eosinophils. The patient did not improve and passed on day 17 of hospitalization. Her liver failure was determined to be the result of Yogi Detox Tea consumption.

A case of Black Cohosh causing hepatotoxicity presenting as autoimmune hepatitis was also published in the past year [20A]. A 69-year-old female developed black cohosh-induced hepatotoxicity with autoimmune hepatitis. Black cohosh is often used by women to treat postmenopausal symptoms. This patient was admitted to the hospital after suffering significant abdominal pain in the right upper quadrant for 2 weeks and dark urine with clay-like stools for 3 days. She had been taking amlodipine, metoprolol and omeprazole for a number of years. One week before her hospital admission she began ingesting black cohosh at a dose of 150mg per day prepared as the root-standardized extract. A liver biopsy revealed autoimmune hepatitis. Her use of black cohosh was discontinued. The patient recovered after 3 months of treatment with prednisone and 6 months of treatment with azathioprine. Using the CIOMS causality assessment, the patient was found to have a CIOMS score of 10, indicating her use of black cohosh was highly likely to have been the cause of her hepatotoxicity. Only two cases of black cohosh-associated autoimmune hepatitis had been reported up to the time the publication of this case.

Efferth et al. [21A] published a case of a patient suffering hepatotoxicity due to a drug–herbal interaction. In this article, a 65-year-old female with a diagnosis of glioblastoma multiform was being treated with temozolomide and radiotherapy. She was also taking the anti-seizure medications levetiracetam, lorazepam and clobazam. Several months after initiating chemotherapy, she started taking the Chinese herbal medicines *Artemisia annua* (ART) and *Coptis-Kush* decoct. She ingested 200mg/day of ART for a little over a month. She ingested 20 doses of the Coptis-Kush decoct (0.5g *Rhizoma Coptidis chenensis*; 3.0g *Herba Siegesbeckiae orientalis*; 6.0g *Herba Artemisiae scoparieae*; and 2g *Radix Dictamni dascycarpi*) for 2 ½ weeks. The patient's liver enzymes became elevated after initiating the ART and Coptic-Kush decoct. She also suffered weight loss, heartburn, thoracic pain, nausea, weakness, fatigue and depression. After stopping both herbals, the patient's liver enzymes returned to normal and her other symptoms resolved. As the patient was taking several medications or herbal medicines that have potential hepatotoxicity, the causality of the herbal medicines in this case is difficult to assess. However, the resolution of the signs and symptoms of hepatotoxicity after the discontinuation of the herbal medications makes it plausible that these products were causative in this patient's hepatotoxicity.

Several reviews of the impact of the use of herbal dietary supplements and traditional Chinese herbal medicines were published in the last year. One of the publications addresses findings regarding the potential hepatotoxicity of one product while other reviews are more general and discuss many products. Pantano [22R] published a review of the possible effects of greater celandine (*Chelidonium majus*) (CM) on the liver. These authors conducted an extensive systematic search of the literature to find reports of CM-induced hepatotoxicity. This herb is used for a variety of reasons, although the evidence of its effectiveness when used is questionable. It is used as a hepatoprotectant. However, cases of hepatotoxicity associated with its use have been reported and the use of the herb was considered to be probable or highly probable as the cause of the observed hepatotoxicity.

Lopez-Gil [23R] reviewed the potential hepatotoxicity of herbal products commonly used in Latin America. Several herbal products are discussed and grouped by the active compound in the plant thought to be associated with their potential hepatotoxicity. Liu [24R] also published a review of potentially hepatotoxic herbal products, focusing on those plants used in Chinese Herbal Medicines. The authors list many herbal Chinese medicines known to cause herb-induced liver injury. This review also presents information on the diagnosis and treatment of herb-induced liver injury in China and methods of evaluating herbals for hepatotoxicity.

Brown published an in-depth review of liver toxicity associated with herbal and dietary supplements similar to the review published regarding the nephrotoxicity associated with herbal and dietary supplements [25R]. This article was the second in the series of articles presenting reviews of the various types of toxicities herbals and dietary supplements are known to cause, discussed above. This article was accompanied also by an online tabular compilation of cases published in PubMed between 1966 and June 2016 that can be updated as more cases are published. The exclusion criteria for cases of herbal and dietary supplement-induced kidney injury was used for herbal and dietary supplement-induced hepatotoxicity.

Several observational studies evaluating various aspects of herbals and dietary supplements and their potential to cause hepatotoxicity were published in the last year as well. A study of a series of reported cases evaluated the association between the use of herbal/dietary supplements (HDS) and cases of acute liver failure in Hawaiian patients. Liver transplant cases in which patients with acute hepatic necrosis had used HDS were identified in the Scientific Registry of Transplant Recipients between 2003 and 2015. Many cases did not have the reason for the need for transplant detailed. However, 625 out of the 2408 adult transplant cases were due to drug-induced liver injury. Of those 625, 21 were specifically associated with herbal/dietary supplement use. HDS use was the fourth largest category of the different classes of medications thought to cause drug-induced liver injury [26c].

DeKlotz et al. [27c] published a retrospective observational study evaluating the risk of hepatotoxicity in patients taking isotretinoin to treat acne while on protein or herbal dietary supplements. Eight patients being treated for acne vulgaris at an outpatient dermatology clinic were identified who were ingesting supplements while on isotretinoin and had abnormal liver function tests. The abnormal liver function tests were identified during routine patient screenings. A variety of supplements were being consumed by the patients in a variety of combinations. The supplements identified were protein supplements, muscle-building supplements, energy drinks, creatine, and green tea supplements. One patients with elevated liver function tests was also taking minocycline. Two identified as being sick. All patients were instructed to stop taking the supplements that they were consuming. All but one did so. Liver function tests returned to normal at the time of follow-up after for all the patients that discontinued using the supplements. The authors of this study concluded that the ingestion of dietary supplements in conjunction with isotretinoin was associated elevated liver enzymes, which was reversible on discontinuation of the supplements.

Larger observation studies were published in the last year as well. Ettel et al. [28C] reported on the results of an observational study evaluating the frequency and pathological characteristics of hepatotoxicity caused by the use of a variety of hepatotoxic drugs and herbal products. Cases of hepatitis from a medical center between 2012 and 2016 that had been confirmed by liver biopsy were evaluated for drug-induced liver injury (DILI). Of 604 biopsy-confirmed hepatitis cases, 11.6% were confirmed to be drug-induced. Supplements and herbal products were associated with the bulk of the DILI cases (31.4%), followed by antimicrobials, chemotherapeutics, antilipidemics, and immunomodulatory medications. The authors found that most DILI was hepatitis, bile duct injury, or mixed.

Cho also conducted a large observational study published in 2017, but the focus of this work was to investigate the incidence of herb-induced liver injury (HILI) [29C]. These authors evaluated this incidence across South Korea by conducting a prospective study between April 2013 and January 2016, capturing 1001 inpatients from 10 hospitals being treated with herbal medications. RUCAM scores between 4 and 7 were used to confirm HILI. Six female patients were identified with HILI, resulting in an incidence of 0.60% (95% confidence interval 0.12–1.08) for all patients and an incidence of 0.95% (95% confidence interval 0.19–1.68) for female patients.

One study of reported cases by Bonkovsky et al. [30C] evaluated the impact of drug, herbal or dietary supplement use in causing bile duct loss with liver injury in a major observation study. Cases of liver injury entered in the Drug Induced Liver Injury Network (DILIN) database over a 10-year period were evaluated for bile duct loss identified from histological data. 363 cases with histological data available were in the DILIN database; 26 of these had bile duct loss. The frequency, causes, clinical features of the bile duct loss and outcomes of these cases were queried. The likelihood of the medications, herbals, or dietary substances in causing bile duct loss in these cases was assessed. Two cases were ranked with the known cause of the bile duct loss assessed as definite; in 14 cases, the suspected causative agent was considered to have been highly likely the causative agent; in 10 cases, the suspected agent was considered to have been a probable cause of bile duct loss. Compounds found to be most commonly associated with bile duct loss included amoxicillin/clavulanate, herbal dietary supplements of various kinds, azithromycin, and fluoroquinolones. These agents, however, were also commonly associated with drug-induced liver injury without bile duct loss. The ingestion of multiple agents complicated the assessment of causality. Cases that exhibited bile duct loss were likely to develop chronic liver injury.

Several brief commentaries on the potential hepatotoxicity of herbals and dietary supplements have been published in the past year. Hume and de Boer and Sherker [31r, 32r] published reviews of the potential of various herbal and dietary supplements in causing hepatotoxicity. The de Boer review included information on the epidemiology of the problem, regulation of herbals and dietary supplements and information on clinical presentation and diagnosis. Both these review articles presented the types of liver injury seen with the use of specific herbal products. Asher et al. [33r] discussed the potential for commonly used herbal dietary supplements to have drug interactions with prescription medications. The authors summarized the potential for 15 herbal products to have drug interactions based on a survey of the literature and clinical experience. Goldenseal (*Hydrastis canadensis*) and St. John's wort (*Hypericum perforatum*) were listed as herbal dietary substances with a high risk of having drug interactions. The authors state that of the 15 herbal products they reviewed, these were the only two with data that consistently supported this ranking of a high likelihood for having drug interactions and that very few had data that indicated clearly there was no risk of drug interaction. Most herbal dietary supplements are likely to have the potential for drug interactions, making it very important that clinicians ask their patients about their use of herbal dietary supplements to evaluate each situation.

A review of the proceedings from the May 2015 symposium Liver Injury from Herbal and Dietary Supplements sponsored by the American Association for the Study of Liver Disease and the National Institutes of Health was published by Navarro and colleagues [34r]. The symposium included discussions of 130 cases of herbal and dietary supplement-induced liver injury cases enrolled in the DILIN Prospective Study. Forty-five of the liver injuries were due to consumption of products sold as body-building agents containing anabolic steroids. Liver injury resulting from the use of these products usually presents as jaundice that can be severe and last a long time, but is rarely fatal. The remaining DILIN cases were attributed to products with multiple ingredients. Green tea extract consumption accounted for a large number of these case. HDL-induced liver injury as a result of green tea extract consumption manifests as acute hepatitis which is generally self-limiting but has been fatal in some instances. A product highlighted in the symposium was the weight loss product OxyElitePro. This product has been associated with cases of severe hepatitis that may have been caused by the addition of aegeline to the product.

A study was published evaluating the network pharmacology assay method for identifying the HILI potential of traditional Chinese herbal medications. Hong et al. [35H] investigated the likelihood of Xiao-Chai-Hu-Tang (XCHT) and *Radix Polygoni Multiflori* (Heshouwu), both Chinese herbal medications thought to be hepatoprotective, for their potential to cause liver injury. XHCT is a composite of multiple herbal ingredients while Heshouwu is a single herbal product. By using a network pharmacology approach, two components of XCHT and one component of Heshouwu were found to have multiple liver toxicity targets. The results from this study support concerns that these two Chinese herbal medication products marketed as hepatoprotectants may have a significant risk of actually causing hepatotoxicity.

Adverse Effects: Dermatology

Yellowing of the palms and soles for 6 months without history or symptoms of liver dysfunction or disease in a 29-year-old woman was described. She was previously healthy, did not consume alcohol, and had no family history of liver disease. She began taking a nutritional supplement containing beta carotene for a period of about 6 months prior to the symptom manifestation. She had no other dietary changes during this period, in particular, no increased intake of carrots or other yellow- or orange-colored fruits or vegetables. Physical examination revealed yellow-orange pigmentation of her palms and soles. There was no scleral icterus. All laboratory test results were normal, including liver function, hemoglobin A_{1c}, bilirubin levels, renal function, and metabolic tests. It was concluded that the patient was suffering from carotenemia, which should be clinically differentiated from jaundice. The patient was advised to stop taking the nutritional supplements containing carotene and to decrease carotene-rich foods. With these changes, her skin discoloration had faded at a 10-month follow-up visit [36A].

Adverse Effects: Gastrointestinal

The use of *Saccharomyces boulardii* probiotic (a subtype of *Saccharomyces cerevisiae*) leading to sepsis in an 8-year-old child was reported from Turkey. The child was in the surgical intensive care unit (ICU) and had a tracheostomy and central venous (CVC) line in place. He was given a probiotic containing multiple organisms including *S. boulardii* for reduction of diarrhea. The product was opened and diluted in water at the bedside for administration via gastrostomy. He had recurring fever on broad-spectrum antibiotics for an infected decubitus ulcer. Fungal infection was determined on CVC blood culture. A 14-day course of liposomal amphotericin B and removal of the CVC device successfully treated the infection. The authors did a literature review and found 15 additional pediatric cases of *S. cerevisiae* fungemia secondary to probiotic use in patients from newborn to 16 years of age. An IV catheter was present in 14 of 15 patients. One patient in this group succumbed to the fungemia. A caution is given by the authors for clinicians to weigh the risks of probiotic use for pediatric patients with immunocompromise, long-term hospital stay, especially in the ICU setting, CVC devices, and broad spectrum antibiotic use [37A].

Seven cases of *Saccharomyces* fungemia in ICU patients, neonatal and adult, in India were associated with probiotic use. All but one of these confirmed cases had received probiotic products prior to diagnosis of fungemia, the one patient that had not received probiotics was in close proximity to patients who had. *S. cerevisiae* was present in blood isolates from the patients as well as in the probiotic products used by the patients. The *S. cerevisiae* were confirmed as the same organism using D1/D2 rDNA genotyping and antifungal susceptibility testing. The authors of this case series recommend avoiding use of probiotics for patients who are critically ill; especially those with CVC access [38A].

A case of a middle-aged man receiving broad coverage antibiotics for recurrent UTI and community acquired pneumonia, who was intubated, had a CVC placed and admitted to an ICU, with subsequent *S. boulardii* fungemia was reported. He had received a probiotic containing *S. boulardii* from admission via nasogastric tube. The probiotic was opened and mixed in water at bedside. The organism causing the fungemia was typed as being the same as the organism from the probiotic product used by D1/D2 rDNA genotyping and proteomic analysis. The institution changed policy on patient selection and method of probiotic administration after investigating this event and based on the conclusions of a utilization review showing no benefit in patients from probiotic use in the ICU [39A].

Appel da Silva and colleagues present a case report of *S. cerevisiae* var. *boulardii* in an oncology patient with a CVC in Brazil. The patient was prescribed several broad spectrum antibiotic courses for urinary tract infections which subsequently caused diarrhea. Due to this and a fever, the patient received a course of antiparasitic drug and a multiple organism based probiotic product at this point during therapy. Continued fever and a positive *Clostridium difficile* stool antigen screen prompted the discontinuation of the antiparasitic and the addition of metronidazole to the probiotic. Fever continued, so a central line culture was taken and revealed the growth of *S. cerevisiae* var. *boulardii*. Antibiotic therapy was altered, mycafungin therapy was started and probiotic treatment was discontinued. The central line was not removed. After this event, the authors revised their institution's protocol to decrease utilization of probiotic products in critically ill patients with central lines to minimize the risk of fungemia [40A].

Two adult patient cases from Turkey described *S. boulardii* systemic infections. The first was an 88-year-old man with urosepsis, subsequent septic shock and intubation. He received *S. boulardii* containing probiotic therapy to alleviate diarrhea caused by several days of carbapenem therapy. The patient developed a fever after probiotic institution and two subsequent blood cultures were *S. boulardii* positive. Probiotics were discontinued and fluconazole instituted with good results. The other case was a 38-year-old woman with pneumonia who was admitted to the ICU and intubated. She received probiotic therapy for diarrhea secondary to antibiotic therapy. Blood cultures revealed *S. boulardii*. This patient received fluconazole but later developed septic shock and died. Mycologic studies of the organisms from both cases were confirmed by genotyping to be the same as the organisms from the administered probiotic product. These authors also suggest caution against the use of probiotic products for patients in the ICU or who are immunocompromised [41A].

The use of probiotics in the ICU should be limited to patients that may benefit. The preparation of probiotics for administration via enteral device should be in an area away from patients, especially those with central venous catheters or immunosuppression. Patients with conditions that warrant CVC devices, who have or are currently receiving broad spectrum antibiotics and who are immunocompromised should not receive probiotic products due to increased risk of developing fungemia or sepsis from species included in the products [37A, 38A, 40A, 41A].

A case of pyogenic liver abscess and sepsis secondary to the use of a lactobacillus containing probiotic is described by Sherid and colleagues. The patient had a medical history of diabetes, hypertension and end stage renal disease. She had recently completed a course of antibiotics for *Clostridium difficile* colitis after complicated gall bladder surgery. She presented with non-specific symptoms—fever, malaise, nausea and vomiting. The liver abscess was found on CT scan of the abdomen after the discovery of hepatomegaly on physical examination. Cultures from the abscess included *Lactobacilli* species and gram-negative organisms which were not speciated, nor was the probiotic product. The abscess resolved on discontinuation of probiotic and the use of additional antibiotics [42A].

CONCERNS WITH HERBAL PRODUCT CONTAMINATION

Lead contamination occurs in some herbal supplements. The cause of observed adverse symptoms associated with lead may be misdiagnosed by the belief that they are due to the herbal content of the product. A case of a 73-year-old man in India who presented with abdominal pain, gastrointestinal symptoms, general body aches and pallor who was found to have lead toxicity was published by Chambial and colleagues. The patient was taking a compounded herbal powder for 8 months to treat diabetes. His initial laboratory assessments were as follows: blood lead concentration was 118.5 mcg/dL, red blood cell count 3.79×10^{12}/L, hemoglobin 9.8, and hematocrit 28.2. The herbal product was assayed for heavy metals and contained lead 638 Gm/kg (WHO permissible limit in ppm for herbal formulations is 10 mg/kg). The patient was advised to cease intake of the preparation and given chelation therapy (D-penicillamine). His blood lead concentration declined over time to acceptable levels and his anemia resolved [43A].

Another case of lead poisoning in a 48-year-old man who was taking Traditional Chinese Medicine (TCM) remedies is described by Tsai and others. He presented with abdominal pain and anemia. A misdiagnosis of porphyria was given due to his symptoms and ensuing treatment did not improve his outcome. A more in-depth medical history revealed that the patient's abdominal pain began when he started taking six TCM herbal pills daily. Further blood and laboratory analysis revealed a blood lead level of 62.8 mcg/dL (normal range <40 mcg/dL), a urine lead level of 823 mcg/dL (normal range <23 mcg/dL) and anemia with basophilic stippling. The TCM product was analyzed and found to have a lead content >90 ppm (greater than 90 times maximum allowed in food additives by Taiwanese standards). The patient was treated with the chelation therapy calcium disodium edetate. Clinical symptoms subsequently improved and a repeat blood lead level decreased to 31 mcg/dL. The patient had no relapse of symptoms [44A].

MIND–BODY THERAPIES

Manipulative and Invasive Therapies

A comprehensive review of 140 randomized controlled trials (RCTs) of spinal manipulative therapies was conducted to determine the extent and completeness of adverse event reporting in the literature from these manipulations. Spinal manipulative therapy involves high-velocity, low amplitude techniques at the spinal joint and may include manual or mechanical efforts. In this review, search strategies were made difficult due to lack of terms in EMBASE or MEDLINE specifically related to manipulative therapies. Most search terms for adverse effects or events are biased toward pharmaceutical agents. Therefore, of the reviewed RCTs, only 3 were found to have indexing terms assigned that related to adverse events in both EMBASE and MEDLINE, and only 12 or 13, respectively. In 91 RCTs, no adverse event related indexing terms were assigned. The authors conclude that it is difficult for practitioners to gage the safety of various practices due to lack of standardized search terms and advocate for changes in reporting and indexing [45R].

A retrospective case review to determine the rate of cervical artery dissection in chiropractic manipulation of patients from a single institution in the US over a 4-year period was conducted. To confirm any artery dissections or strokes, a board-certified radiologist reviewed all pertinent imaging. Of the 141 patients from the institution diagnosed with vascular artery or cervical artery dissection, 15 had chiropractic manipulation within days of presentation and of these, 12 were shown to be temporally related to the adverse event. These 12 patients had confirmed acute strokes by radiographic evidence. One patient died and nine had permanent sequelae from the stroke. The authors state that this data is similar to an earlier report of similar findings of stroke after chiropractic manipulation and suggest a large case control study to assess the risk more accurately [46A].

A healthy human subject of a clinical trial experienced a fainting episode during a study to establish electroencephalogram (EEG) and peripheral nerve system data during acupuncture. The researchers noted a burst of EEG signal increase at needle insertion and just prior to the fainting episode. The researchers noted a burst of EEG signal increase at needle insertion and just prior to the fainting episode. Although fainting occurs quite frequently during acupuncture treatment, this is the first case with data available regarding EEG or brain wave changes during an episode [47A].

A case is described of a man who presented to the emergency department with extensive sudden onset back pain. The pain extended to the shoulder and left arm, with a loss of strength and paranesthesias of his left sided upper and lower extremities. The patient was treated for left lumbosciatic pain caused by a disc herniation in the lumbar region using acupuncture the day prior. The emergency room physicians associated the large spinal hematoma from C2 to T12 found on MRI to the acupuncture procedure. The patient recovered with intravenous steroids and rest without long-term complications. The authors state that patients that have had spinal acupuncture be monitored for slow bleeding into the spinal tract and hospitalized for symptoms, especially if anticoagulated [48A].

Adverse Effects: Homeopathy

Homeopathy is a complex alternative medical system that uses personalized consultations, lifestyle modifications and oral or sublingual homeopathic medications. Homeopathic medicines are produced from natural botanical, animal, or metallic substances that are thought to stimulate healing by causing symptoms of disease. Serial dilutions are most often made of the substances with water and alcohol until the final product has very little of the original substance. Agitation, *potentisation*, between dilution steps is an important part of the process [49R]. The mechanism of action of these remedies has not been elucidated. Homeopathic practitioners use the term "aggravations" to describe what might be classed as "adverse effects or adverse events". A homeopathic aggravation is part of the treatment plan where the symptoms of disease may be exacerbated for a limited time during therapy but followed by improvement [50M, 51S]. Adverse effects may also be found in the literature of homeopathic pathogenic trials, where healthy volunteers are given a relatively high concentration of a substance to cause specific symptoms, so that then the most appropriate dilution of that substance can be prepared to decrease the symptoms in ill patients. Often, homeopathic practitioners prepare these diluted products individually for a specific patient, not on a large or commercial scale [52R].

Adverse effects are reported on homeopathic preparations and homeopathic OTC products in the literature. From clinical trials, examples of adverse effects are headache, GI complaints, skin rash and other mild symptoms to pollen products, arnica and individualized preparations. From case reports, mixed plant tinctures, pollen, and low toxic concentrations of various substances such as Belladonna, Ipecac, Phosphorus, Borax, or Sulfur caused a variety of mild and transient symptoms or aggravations of symptoms [52R]. More serious adverse symptoms were ascribed to commercial large manufacturer products sold on the OTC market as homeopathic or to non-diluted tinctures of toxic materials, either plant or metal based [50M, 52R]. A meta-analysis of randomized controlled trials using homeopathic treatments indicates that these were not more likely to be associated with adverse events than placebo, conventional medicine, herbal medicine or usual care [49M]. A reanalysis of this data published in a letter suggests that homeopathic remedies are more likely to be associated with adverse events than placebo or control but less than conventional medicine [53r].

Mass manufactured homeopathic teething products, tablets and gels are currently under intense FDA scrutiny due to an increase in MedWatch reports. Reported adverse events over the past 6 years include many CNS effects—seizure, tremor, lethargy, agitation and irritability, as well as respiratory difficulty, fever, constipation and vomiting. Out of more than 400 documented adverse effects, death as an outcome of the use of these products was reported 10 times. Belladonna extracts in higher than labeled concentrations were found in some commercially available products and may contribute to the adverse findings. The manufacturers of these products were asked to perform a recall [54r].

CONCLUSIONS

Growth in the global acceptance and use of complementary and alternative medicine (CAM) continues due to factors of belief and ease. Belief that the CAM therapy is historically used, less harmful and more beneficial than conventional medicine therapies. Ease of self-care options and use of medically active ingredients through foodstuffs rather than reliance on a healthcare practitioner can motivate some users [5A]. The FDA is increasing vigilance of adverse events associated with dietary supplements and publishing the results of the CAERS network and other databases [8S]. Homeopathy is not overlooked as a source of potential harm and some homeopathic products have been recalled by the FDA [50M, 54r].

References

[1] Brown AC. An overview of herb and dietary supplement efficacy, safety and government regulations in the United States with suggested improvements. Part 1 of 5 series. Food Chem Toxicol. 2017;107:449–71 [R].

[2] Gabay M, Smith JA, Chavez ML, et al. White paper on natural products. Pharmacotherapy. 2017;37(1):e1–e15 [r].

[3] Global Industry Analysts, Inc. Herbal supplements and remedies market trends. [Internet]. Oct [cited Feb 21, 2018]. Available from, http://www.strategyr.com/MarketResearch/ViewInfoGraphNew.asp?code=MCP-1081; 2017 [R].

[4] Schiff Jr PL, Srinivasan VS, Giancaspro GI, et al. The development of USP botanical dietary supplement monographs, 1995–2005. J Nat Prod. 2006;69(3):464–72 [R].

[5] Global Industry Analysts, Inc. Global herbal supplements and remedies market to reach US$107 billion by 2017, according to new report by global industry analysts, Inc. [Internet]. Feb 21 [cited Feb 21, 2018]. Available from, http://www.prweb.com/releases/herbal_supplements/herbal_remedies/prweb9260421.htm; 2018 [A].

[6] Camire ME, Kantor MA. Dietary supplements: nutritional and legal considerations. Food Technol. 1999;53(7):87 [S].

[7] U.S. Food and Drug Administration Tainted products marketed as dietary Supplements_CDER [Internet]. 2017 Dec 14 [cited Feb 21, 2018]. Available from: https://www.accessdata.fda.gov/scripts/sda/sdnavigation.cfm?sd=tainted_supplements_cder [S].

[8] Timbo BB, Chirtel SJ, Ihrie J, et al. Dietary supplement adverse event report data from the FDA Center for Food Safety and Applied Nutrition Adverse Event Reporting System (CAERS), 2004–2013. Ann Pharmacother. 2018;52(5):431–8 [S].

[9] Knapik JJ, Trone DW, Austin KG, et al. Prevalence, adverse events, and factors associated with dietary supplement and nutritional supplement use by US navy and marine corps personnel. J Acad Nutr Diet. 2016;116(9):1423–42 [C].

[10] Levy I, Attias S, Ben-Arye E, et al. Use and safety of dietary and herbal supplements among hospitalized patients: what have we learned and what can be learned?—A narrative review. World J Surg. 2017;41:927–34 [C].

[11] Levy I, Attias S, Ben-Arye E, et al. Adverse events associated with interactions with dietary and herbal supplements among inpatients. Br J Clin Pharmacol. 2017;83(4):836–45 [C].

[12] Levy I, Attias S, Ben-Arye E, et al. Perioperative risks of dietary and herbal supplements. World J Surg 2016: 1–8 [C].

[13] Kang HG, Lee SJ, Cheong JS. Acute cerebral infarction following aconitine ingestion. Neurol Asia. 2017;22(1):65–8 [A].

[14] Gross BW, Gillio M, Rinehart CD, et al. Omega-3 fatty acid supplementation and warfarin: a lethal combination in traumatic brain injury. J Trauma Nurs. 2017;24(1):15–8 [A].

[15] Murtaza G, Adhikari S, Siddiqui I, et al. Cardiomyopathy related to a weight loss supplement: a case report and review of literature. J Investig Med High Impact Case Rep. 2017;5(2). 2324709617711462. [A].

[16] Costa I, Mendonca MD, Cruz E, et al. Herbal supplements association with reversible cerebral vasoconstriction syndrome: a case report. J Stroke Cerebrovasc Dis. 2017;26(3): 673–6 [A].

[17] Patel YA, Marzella N. Dietary supplement-drug interaction-induced serotonin syndrome progressing to acute compartment syndrome. Am J Case Rep. 2017;18:926–30 [A].

[18] Nayeri A, Wu S, Adams E, Tanner C, Meshman J, Saini I, Reid W. In: Acute calcineurin inhibitor nephrotoxicity secondary to turmeric intake: a case report. Transplant Proc; Elsevier; 2017. p. 198–200 [A].

[19] Kesavarapu K, Kang M, Shin JJ, et al. Yogi detox tea: a potential cause of acute liver failure. Case Rep Gastrointest Med. 2017;2017: 3540756 [A].

[20] Franco DL, Kale S, Lam-Himlin DM, et al. Black cohosh hepatotoxicity with autoimmune hepatitis presentation. Case Rep Gastroenterol. 2017;11(1):23–8 [A].

[21] Efferth T, Schottler U, Krishna S, et al. Hepatotoxicity by combination treatment of temozolomide, artesunate and chinese herbs in a glioblastoma multiforme patient: case report review of the literature. Arch Toxicol. 2017;91(4):1833–46 [A].

[22] Pantano F, Mannocchi G, Marinelli E, et al. Hepatotoxicity induced by greater celandine (chelidonium majus L.): a review of the literature. Eur Rev Med Pharmacol Sci. 2017;21(1 Suppl): 46–52 [R].

[23] Lopez-Gil S, Nuno-Lambarri N, Chavez-Tapia N, et al. Liver toxicity mechanisms of herbs commonly used in latin america. Drug Metab Rev. 2017;49(3):338–56 [R].

[24] Liu C, Fan H, Li Y, et al. Research advances on hepatotoxicity of herbal medicines in China. Biomed Res Int. 2016;2016 [R].

[25] Brown AC. Liver toxicity related to herbs and dietary supplements: online table of case reports. Part 2 of 5 series. Food Chem Toxicol. 2017;107:472–501 [R].

[26] Wong LL, Lacar L, Roytman M, et al. Urgent liver transplantation for dietary supplements: an under-recognized problem. Transplant Proc. 2017;49(2):322–5 [c].

[27] DeKlotz CMC, Roby KD, Friedlander SF. Dietary supplements, isotretinoin, and liver toxicity in adolescents: a retrospective case series. Pediatrics. 2017;140(4):2940 [c].
[28] Ettel M, Gonzalez GA, Gera S, et al. Frequency and pathological characteristics of drug-induced liver injury in a tertiary medical center. Hum Pathol. 2017;68:92–8 [C].
[29] Cho J, Oh D, Hong S, et al. A nationwide study of the incidence rate of herb-induced liver injury in Korea. Arch Toxicol. 2017;91(12):4009–15 [C].
[30] Bonkovsky HL, Kleiner DE, Gu J, et al. Clinical presentations and outcomes of bile duct loss caused by drugs and herbal and dietary supplements. Hepatology. 2017;65(4):1267–77 [C].
[31] Dietary supplements may be overlooked as cause of liver injury. Pharm Today. 2017;23(12):21 [r].
[32] de Boer YS, Sherker AH. Herbal and dietary supplement-induced liver injury. Clin Liver Dis. 2017;21(1):135–49 [r].
[33] Asher GN, Corbett AH, Hawke RL. Common herbal dietary supplement-drug interactions. Am Fam Physician. 2017;96(2):101–7 [r].
[34] Navarro VJ, Khan I, Bjornsson E, et al. Liver injury from herbal and dietary supplements. Hepatology. 2017;65(1):363–73 [r].
[35] Hong M, Li S, Tan HY, et al. A network-based pharmacology study of the herb-induced liver injury potential of traditional hepatoprotective chinese herbal medicines. Molecules. 2017;22(4):632 [H].
[36] Wang ZS, Liu XK, Li J. Woman with yellow palms and soles. JAMA. 2017;317(15):1574–5 [A].
[37] Atıcı S, Soysal A, Cerit KK, et al. Catheter-related *Saccharomyces cerevisiae* fungemia following *Saccharomyces boulardii* probiotic treatment: in a child in intensive care unit and review of the literature. Med Mycol Case Rep. 2017;15:33–5 [A].
[38] Roy U, Jessani LG, Rudramurthy SM, et al. Seven cases of saccharomyces fungaemia related to use of probiotics. Mycoses. 2017;60(6):375–80 [A].
[39] Martin IW, Tonner R, Trivedi J, et al. *Saccharomyces boulardii* probiotic-associated fungemia: Questioning the safety of this preventive probiotic's use. Diagn Microbiol Infect Dis. 2017;87(3):286–8 [A].
[40] Appel-da-Silva MC, Narvaez GA, Perez LR, et al. *Saccharomyces cerevisiae* var. *boulardii* fungemia following probiotic treatment. Med Mycol Case Rep. 2017;18:15–7 [A].
[41] Kara İ, Yıldırım F, Özgen Ö, et al. *Saccharomyces cerevisiae* fungemia after probiotic treatment in an intensive care unit patient. J Mycol Med. 2018;28(1):218–21 [A].
[42] Sherid M, Samo S, Sulaiman S, et al. Liver abscess and bacteremia caused by lactobacillus: role of probiotics? Case report and review of the literature. BMC Gastroenterol. 2016;16(1). y. [A].
[43] Chambial S, Bhardwaj P, Mahdi AA, et al. Lead poisoning due to herbal medications. Indian J Clin Biochem. 2017;32(2):246–7 [A].
[44] Tsai M, Huang S, Cheng S. Lead poisoning can be easily misdiagnosed as acute porphyria and nonspecific abdominal pain. Case Rep Emerg Med. 2017;9050713 [A].
[45] Gorrell LM, Engel RM, Lystad RP, et al. Assignment of adverse event indexing terms in randomized clinical trials involving spinal manipulative therapy: an audit of records in MEDLINE and EMBASE databases. BMC Med Res Methodol. 2017;17(1). x. [R].
[46] Kennell KA, Daghfal MM, Patel SG, et al. Cervical artery dissection related to chiropractic manipulation: one institution's experience. J Fam Pract. 2017;66(9):556–62 [A].
[47] Kwon OS, Choi KH, Kim J, et al. Case report: fainting during acupuncture stimulation at acupuncture point LI4. BMC Complement Altern Med. 2017;17(1):9 [A].
[48] Domenicucci M, Marruzzo D, Pesce A, et al. Acute spinal epidural hematoma after acupuncture: personal case and literature review. World Neurosurg. 2017;102: 695.e14. [A].
[49] Kayne SB. Introduction to homeopathy. In: FASTtrack: complementary and alternative medicine. Chicago, IL: Pharmaceutical Press; 2008. p. 29–42 [R].
[50] Stub T, Musial F, Kristoffersen AA, et al. Adverse effects of homeopathy, what do we know? A systematic review and meta-analysis of randomized controlled trials. Complement Ther Med. 2016;26:146–63 [M].
[51] Food and drug administration plans crackdown on risky homeopathic remedies. [Internet]. December 18, [cited December 18, 2017]. Available fromhttps://www.npr.org/sections/health-shots/2017/12/18/571666553/food-and-drug-administration-plans-crackdown-on-risky-homeopathic-remedies; 2017 [S].
[52] Dantas F. Do homeopathic medicines cause drug-dependent adverse effects or aggravations? Revista De Homeopatia. 2017;80(3/4):142–50 [R].
[53] Mathie RT, Roberts R, Rutten AL. Adverse effects of homeopathy: we clearly need more details. Complement Ther Med. 2016;29:235 [r].
[54] Abbasi J. Amid reports of infant deaths, FTC cracks down on homeopathy while FDA investigates. JAMA. 2017;317(8):793–5 [r].

Further Reading

[55] Bellanger RA, Seeger CM, Smith HE. Safety of complementary and alternative medicine (CAM) treatments and practices. In: Ray SD, editor. Side Effects of Drugs Annual. Elsevier; 2017. p. 503–12 [R].

Reviewers

Reviewer 1st Name	Surname	Affiliation
Asima N.	Ali, PharmD, CPP	Clinical Assistant Professor – Pharmacy Practice, Campbell University College of Pharmacy and Health Sciences, Buies Creek, NC, USA; Wake Forest Baptist Health, Outpatient Internal Medicine Clinic, Winston-Salem, NC, USA.
Robert D.	Beckett, PharmD, BCPS	Associate Professor of Pharmacy Practice, Director of the Drug Information Center, Manchester University College of Pharmacy, Natural and Health Sciences, Fort Wayne, Indiana, USA.
Renee A.	Bellanger,Pharm.D., BCNSP	Professor, Department of Pharmacy Practice, University of the Incarnate Word, Feik School of Pharmacy.
Adrienne T.	Black, PhD, DABT	3E Services/3E Company, Verisk 3E, Warrenton., VA, USA 20187.
Alison	Brophy, Pharm D, BCPS, BCCCP	Assistant Director of Pharmacy for Clinical Services, Saint Barnabas Medical Center, 94 Old Short Hills Road, Livingston, NJ 07039.
Sheeva	Chopra, PharmD, BCACP	Assistant Professor, College of Pharmacy at UT Tyler, Parkland Health & Hospital System, 5200 Harry Hines Blvd., Dallas, Texas 75235.
Benjamin	Dagraedt, PharmD, MBA, BCPS	Chief Resident, Assistant Clinical Instructor, Texas Tech University Health Science Center, School of Pharmacy – Room 1B220, 3601 4th Street STOP 8162 Lubbock, Texas 79430-8162.
Alex	Ebied, PharmD, BCCCP	Clinical Assistant Professor, University of Florida College of Pharmacy, Gainesville, FL.
Kathleen	Eddy, PharmD	Baylor Scott & White All Saints Medical Center - Fort Worth, 1400 8th Ave., Fort Worth, TX 76104.
Dana R.	Fasanella, PharmD, CDE, BCACP	Assistant Professor, Department of Pharmacy Practice and Administration, School of Pharmacy and Health Professions, University of Maryland Eastern Shore, 212 Somerset Hall, Princess Anne, MD 21853.
Michelle	Friedman, Pharm.D., BCPS, BCGP	Assistant Professor of Pharmacy Practice, PGY-2 Internal Medicine Pharmacy Residency Program Director, Kingsbrook Jewish Medical Center, Touro College of Pharmacy, 2090 Adam Clayton Powell Blvd., Room 515, New York, NY 10027.
Christa M.	George, Pharm.D., BCPS, CDE, BCACP	Associate Professor, Clinical Pharmacy & Family Medicine, University of Tennessee Health Science Center, 881 Madison Avenue, Suite 215, Memphis, Tennessee 38163.
Cait	Gibson, PharmD, BCPS	Assistant Professor, Department of Pharmacotherapy, UNT System College of Pharmacy, 3500 Camp Bowie Blvd – RES 411D, Fort Worth, TX 76107.
Joshua P.	Gray, PhD	Professor of Chemistry & Biochemistry, Department of Science, United States Coast Guard Academy, New London, CT, USA.
Cecilia	Hui, PharmD	Baylor Scott & White All Saints Medical Center - Fort Worth, 1400 8th Ave., Fort Worth, TX 76104.
Holly	Lowe, PharmD, BCPS	Assistant Professor of Pharmacy Practice, South College School of Pharmacy, 400 Goody's Lane, Knoxville, TN 37922.

Reviewer 1st Name	Surname	Affiliation
Yekaterina (Kate)	Opsha, Pharm D., BCPS-AQ Cardiology	Clinical Assistant Professor, Ernest Mario School of Pharmacy, Rutgers, The State University of New Jersey, Clinical Specialist-Cardiovascular Medicine, Saint Barnabas Medical Center, Livingston, NJ 07039.
Sidhartha D.	Ray, PhD, FACN	Department of Pharmaceutical and Biomedical Sciences, Touro University College of Pharmacy and School of Osteopathic Medicine, Manhattan, NY, United States.
Connie F.	Rust, BPharm, MSSW, PhD	Assistant Professor of Pharmacy Practice, South College School of Pharmacy; 400 Goody's Lane, Knoxville, TN 37922.
Samit K.	Shah, RPh, MBA, PhD	Associate Dean of Academic Affairs, Professor of Biopharmaceutical Sciences, Sidney J. Weinberg Professor, Keck Graduate Institute College of Pharmacy, Claremont, CA.
Keaton S.	Smetana, PharmD, BCCCP	Specialty Practice Pharmacist – Neurocritical Care, The Ohio State University Wexner Medical Center, 410 W. 10th Ave, Doan Hall Room 368, Columbus, OH, 43210.
Sidney J.	Stohs, RPh, PhD, FAAT, FACN, ATS	Dean Emeritus, Creighton University College of Pharmacy & Health Professions, Omaha, NE, USA.
Amulya	Tatachar, PharmD, BCACP	Assistant Professor of Pharmacotherapy, UNT System College of Pharmacy, Office: 817-735-0490, University of North Texas Health Science Center, 3500 Camp Bowie Blvd., RES-411B, Fort Worth, TX 76107.
Katherine	Teare, PharmD, BCCCP	Clinical Pharmacist - Emergency Medicine, Saint Thomas Rutherford Hospital, 1700 Medical Center Parkway, Murfreesboro, TN 37129.

Index of Drugs

For drug–drug interactions see the separate index. In pages 631–634
Note: Page numbers followed by "*f*" indicate figures and "*t*" indicate tables.

B

Index of Drug-Drug Interactions

V

W

X

Z

Index of Adverse Effects and Adverse Reactions

Note: Page numbers followed by "t" indicate tables.

B

E